PRINCIPLES OF
POLYMERIZATION

PRINCIPLES OF POLYMERIZATION

Third Edition

GEORGE ODIAN
The College of Staten Island
The City University of New York
Staten Island, New York

A Wiley-Interscience Publication

JOHN WILEY & SONS, INC.

New York / Chichester / Brisbane / Toronto / Singapore

In recognition of the importance of preserving what has been
written, it is a policy of John Wiley & Sons, Inc., to have books
of enduring value published in the United States printed on
acid-free paper, and we exert our best efforts to that end.

Library of Congress Cataloging in Publication Data:
Odian, George G., 1933–
    Principles of polymerization / George Odian.—3rd ed.
       p.   cm.
    "A Wiley-Interscience publication."
    Includes index.
    ISBN 0-471-61020-8
    1. Polymerization.   I. Title.
QD281.P603 1991
541.3'93—dc20   90-24785
                           CIP

Printed in the United States of America

10 9 8 7 6 5 4 3 2

Printed and bound by Courier Companies, Inc.

PREFACE

This book describes the physical and organic chemistry of the reactions by which polymer molecules are synthesized. The sequence I have followed is to introduce the reader to the characteristics that distinguish polymers from their much smaller sized homologs (Chap. 1) and then proceed to a detailed consideration of the three types of polymerization reactions—step, chain, and ring-opening polymerizations (Chaps. 2–5, 7). Polymerization reactions are characterized according to their kinetic and thermodynamic features, their scope and utility for the synthesis of different types of polymer structures, and the process conditions that can be used to carry them out. Polymer chemistry has advanced to the point where it is often possible to tailor-make a variety of different types of polymers with specified molecular weights and structures. Emphasis is placed throughout the text on understanding the reaction parameters that are important in controlling polymerization rate, polymer molecular weight, and structural features such as branching and crosslinking. It has been my intention to give the reader an appreciation of the versatility that is inherent in polymerization processes and that is available to the synthetic polymer chemist.

The versatility of polymerization resides not only in the different types of reactants that can be polymerized but also in the variations allowed by copolymerization and stereospecific polymerization. Chain copolymerization is the most important kind of copolymerization and is considered separately in Chap. 6. Other copolymerizations are discussed in the appropriate chapters. Chapter 8 describes the stereochemistry of polymerization with emphasis on the synthesis of polymers with stereoregular structures by the appropriate choice of polymerization conditions. In the last chapter (Chap. 9), the reactions of polymers that are useful for modifying or synthesizing new polymer structures and the use of polymeric reagents, substrates, and catalysts are discussed. The literature has been covered through late 1990.

This book is intended for chemists with no background in polymers as well as the experienced polymer chemist. The text can serve as a self-educating introduction to polymers for the former. Each topic is presented with minimal assumptions regarding the reader's background, except for undergraduate organic and physical chemistry. The book is also intended to serve as a classroom text. With the appropriate selection of materials, the text can be used at either the undergraduate or graduate level. Each chapter contains a selection of problems. A solutions manual for the problems is available directly from the author.

I would like to take this opportunity to thank my colleagues Richard Brotzman, Howard Haubenstock, Fred Naider, Peter H. Plesch, Albert Rossi, and Arthur Woodward who graciously gave their time to read and comment on various portions of the text. Their helpful suggestions for improvements and corrections are gratefully acknowledged. Helpful discussions with Ira Blei and Nan-loh Yang are also acknowledged. I am grateful to The College of Staten Island for a sabbatical leave in the Fall 1988 semester, during which much of the planning and literature search for this book was accomplished.

GEORGE ODIAN

Staten Island, New York
June 1991

CONTENTS

3   RADICAL CHAIN POLYMERIZATION   198

PRINCIPLES OF
POLYMERIZATION

CHAPTER 1

INTRODUCTION

Polymers are macromolecules built up by the linking together of large numbers of much smaller molecules. The small molecules that combine with each other to form polymer molecules are termed *monomers*, and the reactions by which they combine are termed *polymerizations*. There may be hundreds, thousands, tens of thousands, or more monomer molecules linked together in a polymer molecule. When one speaks of polymers, one is concerned with materials whose molecular weights may reach into the millions.

1-1 TYPES OF POLYMERS AND POLYMERIZATIONS

There has been and still is considerable confusion concerning the classification of polymers. This is especially the case for the beginning student who must appreciate that there is no single generally accepted and unambiguous classification system for polymers. During the development of polymer science, two types of classifications of polymers have come into use. One classification divides polymers into *condensation* and *addition* polymers, and the other divides them into *step* and *chain* polymers. Confusion and error arise because the two classifications are usually used interchangeably without careful thought. The terms *condensation* and *step* are usually used synonymously as are the terms *addition* and *chain*. Although these terms can often be used synonymously as noted, this is not always the case because the two classifications arise from two different bases of classification. The condensation-addition classification is primarily applicable to the composition or structure of polymers. The step–chain classification is based on the mechanism of the polymerization reactions.

1-1a Polymer Composition and Structure

Polymers were originally classified by Carothers [1929] into condensation and addition polymers on the basis of the compositional difference between the polymer and the monomer(s) from which it was synthesized. Condensation polymers were those polymers that were formed from polyfunctional monomers by the various condensation reactions of organic chemistry with the elimination of some small molecule such as water. An example of such a condensation polymer is the polyamides formed from diamines and diacids with the elimination of water according to

$$n\text{H}_2\text{N—R—NH}_2 + n\text{HO}_2\text{C—R}'\text{—CO}_2\text{H} \rightarrow$$
$$\text{H——(—NH—R—NHCO—R}'\text{—CO——)}_n\text{—OH} + (2n - 1)\text{H}_2\text{O} \qquad (1\text{-}1)$$

where R and R' are aliphatic or aromatic groupings. The unit in parentheses in the polyamide formula repeats itself many times in the polymer chain and is termed the *repeating unit*. The composition of the repeating unit differs from that of the two monomers by the elements of water. The polyamide synthesized from hexamethylene diamine, R = $(\text{CH}_2)_6$, and adipic acid, R' = $(\text{CH}_2)_4$, is the extensively used fiber and plastic known commonly as nylon-6,6 or poly(hexamethylene adipamide). Other examples of condensation polymers are the polyesters formed from diacids and diols with the elmination of water

$$n\text{HO—R—OH} + n\text{HO}_2\text{C—R}'\text{—CO}_2\text{H} \rightarrow$$
$$\text{H——(—O—R—OCO—R}'\text{—CO——)}_n\text{—OH} + (2n - 1)\text{H}_2\text{O} \qquad (1\text{-}2)$$

and the polycarbonates from the reaction of an aromatic dihydroxy reactant and phosgene with the elimination of hydrogen chloride

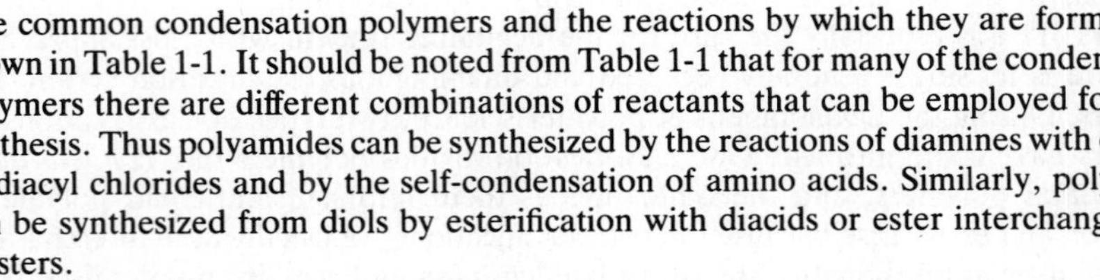

$$(1\text{-}3)$$

The common condensation polymers and the reactions by which they are formed are shown in Table 1-1. It should be noted from Table 1-1 that for many of the condensation polymers there are different combinations of reactants that can be employed for their synthesis. Thus polyamides can be synthesized by the reactions of diamines with diacids or diacyl chlorides and by the self-condensation of amino acids. Similarly, polyesters can be synthesized from diols by esterification with diacids or ester interchange with diesters.

Some naturally occurring polymers such as cellulose, starch, wool, and silk are classified as condensation polymers, since one can postulate their synthesis from certain hypothetical reactants by the elimination of water. Thus cellulose can be thought of as the polyether formed by the dehydration of glucose. Carothers included such polymers by defining condensation polymers as those in which the formula of the repeating unit lacks certain atoms that are present in the monomer(s) from which it is formed or to which it may be degraded. In this sense cellulose is considered a condensation polymer, since its hydrolysis yields glucose, which contains the repeating unit of

TABLE 1-1 Typical Condensation Polymers

Type	Characteristic Linkage	Polymerization Reaction
Polyamide	—NH—CO—	H_2N—R—NH_2 + HO_2C—R'—CO_2H → H$(\!-$NH—R—NHCO—R'—CO$-\!)_n$OH + H_2O H_2N—R—NH_2 + ClCO—R'—COCl → H$(\!-$NH—R—NHCO—R'—CO$-\!)_n$Cl + HCl H_2N—R—CO_2H → H$(\!-$NH—R—CO$-\!)_n$OH + H_2O
Protein, wool, silk	—NH—CO—	Naturally occurring polypeptide polymers; degradable to mixtures of different amino acids. H$(\!-$NH—R—CONH—R'—CO$-\!)_n$OH + H_2O
Polyester	—CO—O—	HO—R—OH + HO_2C—R'—CO_2H → H$(\!-$O—R—OCO—R'—CO$-\!)_n$OH + H_2O HO—R—OH + R"O_2C—R'—CO_2R" → H$(\!-$O—R—OCO—R'—CO$-\!)_n$OH + R"OH HO—R—CO_2H → H$(\!-$O—R—CO$-\!)_n$OH + H_2O
Polyurethane	—O—CO—NH—	HO—R—OH + OCN—R'—NCO → $(\!-$O—R—OCO—NH—R'—NH—CO$-\!)_n$
Polysiloxane	—Si—O—	Cl—SiR_2—Cl $\xrightarrow[-HCl]{H_2O}$ HO—SiR_2—OH → H$(\!-$O—$SiR_2$$-\!)_n$OH + H_2O
Phenol-formaldehyde	—Ar—CH₂—	
Urea-formaldehyde	—NH—CH₂—	H_2N—CO—NH_2 + CH_2O → $(\!-$HN—CO—NH—$CH_2$$-\!)_n$ + H_2O
Melamine-formaldehyde	—NH—CH₂—	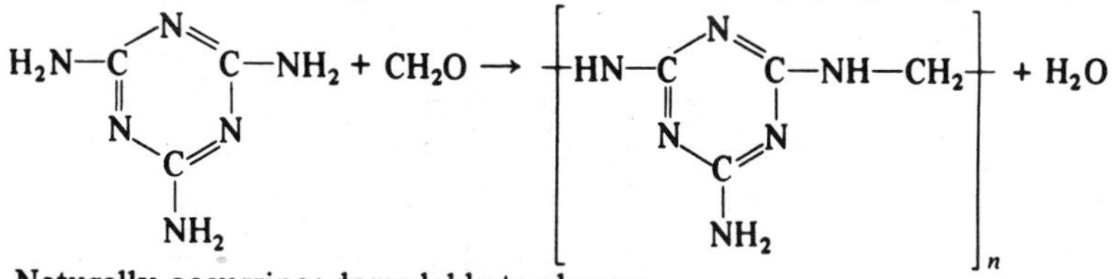
Cellulose	—O—C—	Naturally occurring; degradable to glucose $(\!-C_6H_{12}O_4\!-\!)_n$ + H_2O → $C_6H_{12}O_6$
Polysulfide	—S_m—	Cl—R—Cl + Na_2S_m → $(\!-S_m$—R$-\!)_n$ + NaCl
Polyacetal	—O—CH—O— R	R—CHO + HO—R'—OH → $(\!-$O—R'—OCHR$-\!)_n$ + H_2O

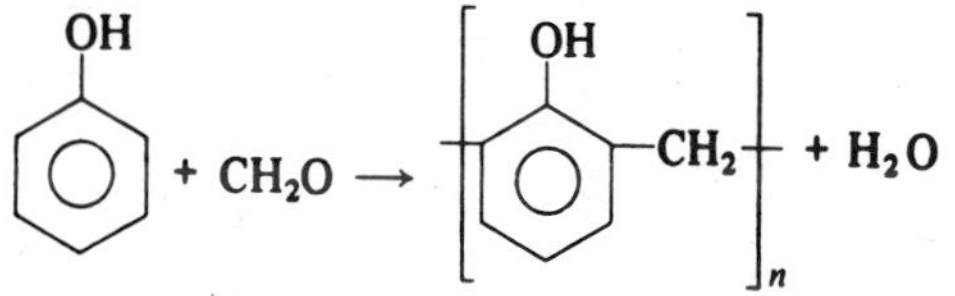

cellulose plus the elements of water

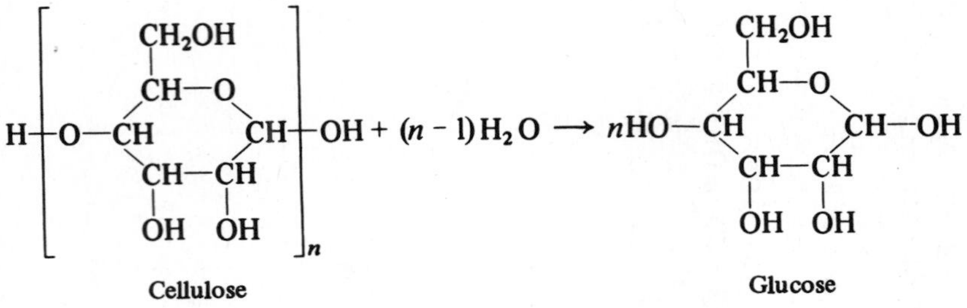

$$
\begin{bmatrix} & \text{CH}_2\text{OH} \\ & | \\ & \text{CH—O} \\ \text{H—O—CH} & \qquad \text{CH—OH} + (n-1)\text{H}_2\text{O} \longrightarrow n\text{HO—CH} \qquad \text{CH—OH} \\ & \text{CH—CH} \\ & | \quad | \\ & \text{OH} \quad \text{OH} \end{bmatrix}_n
$$

Cellulose   Glucose

$$(1\text{-}4)$$

Addition polymers were classified by Carothers as those formed from monomers without the loss of a small molecule. Unlike condensation polymers, the repeating unit of an addition polymer has the same composition as the monomer. The major addition polymers are those formed by polymerization of monomers containing the carbon–carbon double bond. Such monomers will be referred to as *vinyl monomers* throughout this text. (The term *vinyl*, strictly speaking, refers to a CH_2=CH—group attached to some substituent. Our use of the term *vinyl monomer* is broader—it applies to all monomers containing a carbon–carbon double bond, including monomers such as methyl methacrylate, vinylidene chloride, and 2-butene as well as vinyl chloride and styrene. The term *substituted ethylenes* will also be used interchangeably with the term vinyl monomers.) Vinyl monomers can be made to react with themselves to form polymers by conversion of their double bonds into saturated linkages, for example,

$$n\text{CH}_2{=}\text{CHY} \rightarrow \,{-}\!\!{\left(\text{CH}_2{-}\text{CHY}\right)}_{\overline{n}} \tag{1-5}$$

where Y can be any substituent group such as hydrogen, alkyl, aryl, nitrile, ester, acid, ketone, ether, and halogen. Table 1-2 shows many of the common addition polymers and the monomers from which they are produced.

The development of polymer science with the study of new polymerization processes and polymers showed that the original classification by Carothers was not entirely adequate and left much to be desired. Thus, for example, consider the polyurethanes, which are formed by the reaction of diols with diisocyanates without the elimination of any small molecule:

$$n\text{HO—R—OH} + n\text{OCN—R}'\text{—NCO} \rightarrow$$

$$\text{HO}{-}\!\!{\left(\text{R—OCONH—R}'\text{—NHCO—O}\right)}_{\overline{(n-1)}} \text{R—OCONH—R}'\text{—NCO} \tag{1-6}$$

Using Carothers's original classification, one would classify the polyurethanes as addition polymers, since the polymer has the same net composition as the monomer. However, the polyurethanes are structurally much more similar to the condensation polymers than to the addition polymers.

To avoid the obviously incorrect classification of polyurethanes as well as of some other polymers as addition polymers, polymers have also been classified from a consideration of the chemical structure of the groups present in the polymer chains. Condensation polymers have been defined as those polymers whose repeating units

TABLE 1-2 Typical Addition Polymers

Polymer	Monomer	Repeating Unit
Polyethylene	$CH_2{=}CH_2$	$-CH_2-CH_2-$
Polyisobutylene	$CH_2{=}\underset{\underset{CH_3}{\mid}}{\overset{\overset{CH_3}{\mid}}{C}}$	$-CH_2-\underset{\underset{CH_3}{\mid}}{\overset{\overset{CH_3}{\mid}}{C}}-$
Polyacrylonitrile	$CH_2{=}CH-CN$	$-CH_2-\underset{\underset{CN}{\mid}}{CH}-$
Poly(vinyl chloride)	$CH_2{=}CH-Cl$	$-CH_2-\underset{\underset{Cl}{\mid}}{CH}-$
Polystyrene	$CH_2{=}CH-\phi$	$-CH_2-\underset{\underset{\phi}{\mid}}{CH}-$
Poly(methyl methacrylate)	$CH_2{=}\underset{\underset{CO_2CH_3}{\mid}}{\overset{\overset{CH_3}{\mid}}{C}}$	$-CH_2-\underset{\underset{CO_2CH_3}{\mid}}{\overset{\overset{CH_3}{\mid}}{C}}-$
Poly(vinyl acetate)	$CH_2{=}CH-OCOCH_3$	$-CH_2-\underset{\underset{OCOCH_3}{\mid}}{CH}-$
Poly(vinylidene chloride)	$CH_2{=}\underset{\underset{Cl}{\mid}}{\overset{\overset{Cl}{\mid}}{C}}$	$-CH_2-\underset{\underset{Cl}{\mid}}{\overset{\overset{Cl}{\mid}}{C}}-$
Polytetrafluoroethylene	$\underset{\underset{F}{\mid}}{\overset{\overset{F}{\mid}}{C}}{=}\underset{\underset{F}{\mid}}{\overset{\overset{F}{\mid}}{C}}$	$-\underset{\underset{F}{\mid}}{\overset{\overset{F}{\mid}}{C}}-\underset{\underset{F}{\mid}}{\overset{\overset{F}{\mid}}{C}}-$
Polyisoprene (Natural rubber)	$CH_2{=}\underset{\underset{CH_3}{\mid}}{C}-CH{=}CH_2$	$-CH_2\diagdown\underset{CH_3\diagup}{}C{=}CH\diagup^{CH_2-}$

are joined together by functional units of one kind or another such as the ester, amide, urethane, sulfide, and ether linkages. Thus the structure of condensation polymers has been defined by

$$-R-Z-R-Z-R-Z-R-Z-R-Z-$$

I

where R is an aliphatic or aromatic grouping and Z is a functional unit such as $-OCO-$, $-NHCO-$, $-S-$, $-OCONH-$, $-O-$, $-OCOO-$, and $-SO_2-$. Addition polymers, on the other hand, do not contain such functional groups as part of the polymer chain. Such groups may, however, be present in addition polymers as pendant substituents hanging off the polymer chain. According to this classification, the polyurethanes are readily and more correctly classified as condensation polymers.

It should not be taken for granted that all polymers that are defined as condensation polymers by Carothers's classification will also be so defined by a consideration of the polymer chain structure. Some condensation polymers do not contain functional groups in the polymer chain. An example is the phenol-formaldehyde polymers produced by the reaction of phenol (or substituted phenols) with formaldehyde

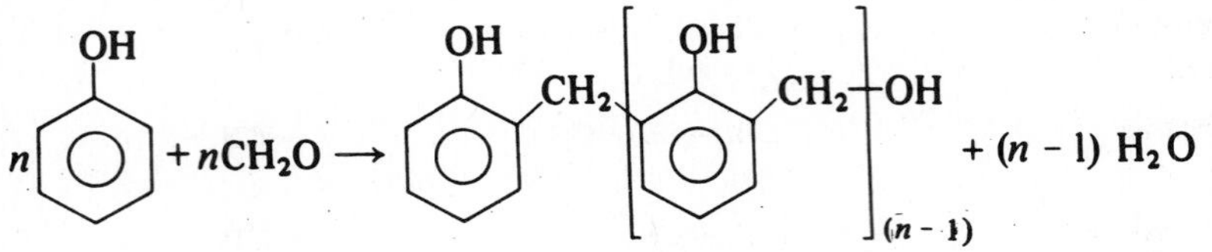

$$n \text{ (phenol-OH)} + n\text{CH}_2\text{O} \longrightarrow \text{ } + (n-1)\text{ H}_2\text{O} \quad (1\text{-}7)$$

These polymers do not contain a functional group within the polymer chain but are classified as condensation polymers, since water is split out during the polymerization process. Another example is poly(p-xylylene), which is produced by the oxidative coupling (dehydrogenation) of p-xylene:

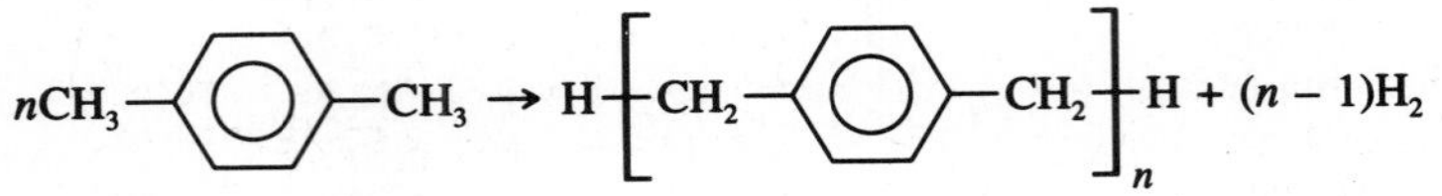

$$n\text{CH}_3 \text{—} \text{⬡} \text{—} \text{CH}_3 \longrightarrow \text{H} \text{—} [\text{CH}_2 \text{—} \text{⬡} \text{—} \text{CH}_2]_n \text{—} \text{H} + (n-1)\text{H}_2 \quad (1\text{-}8)$$

In summary, a polymer is classified as a condensation polymer if its synthesis involves the elimination of small molecules, or it contains functional groups as part of the polymer chain, or its repeating unit lacks certain atoms that are present in the (hypothetical) monomer to which it can be degraded. If a polymer does not fulfill any of these requirements, it is classified as an addition polymer.

1-1b Polymerization Mechanism

In addition to the structural and compositional differences between polymers, Flory [1953] stressed the very significant difference in the mechanism by which polymer molecules are built up. Polymerizations are classified into step and chain polymerizations based on the polymerization mechanism. (The terms *step-reaction polymerization* and *chain-reaction polymerization* are also used instead of step polymerization and chain polymerization, respectively.) Step polymers are, then, those produced by step polymerization and chain polymers, those produced by chain polymerization. The characteristics of the two polymerizations are considerably different. The two reactions differ basically in terms of the time-scale of various reaction events. More specifically, step and chain polymerizations differ in the length of time required for the complete growth of full-sized polymer molecules.

Step polymerizations proceed by the stepwise reaction between the functional groups of reactants as in reactions such as those described by Eqs. 1-1 through 1-3 and Eqs. 1-6 through 1-8. The size of the polymer molecules increases at a relatively slow rate in such polymerizations. One proceeds slowly from monomer to dimer, trimer, tetramer, pentamer, and so on:

Monomer + monomer → dimer
Dimer + monomer → trimer

Dimer + dimer → tetramer
Trimer + monomer → tetramer
Trimer + dimer → pentamer
Trimer + trimer → hexamer
Tetramer + monomer → pentamer
Tetramer + dimer → hexamer
Tetramer + trimer → heptamer
Tetramer + tetramer → octamer
etc.

until eventually large polymer molecules containing large numbers of monomer molecules have been formed. Any two molecular species can react with each other throughout the course of the polymerization. The situation is quite different in chain polymerizations where full-sized polymer molecules are produced almost immediately after the start of the reaction.

Chain polymerizations require an initiator from which is produced an initiator species R* with a reactive center. The reactive center may be either a free radical, cation, or anion. Polymerization occurs by the propagation of the reactive center by the successive additions of large numbers of monomer molecules in a chain reaction happening, in a matter of a second or so at most, and usually in much shorter times. Monomer can react only with the propagating reactive center, not with monomer. By far the most common example of chain polymerization is that of vinyl monomers. The process can be depicted as

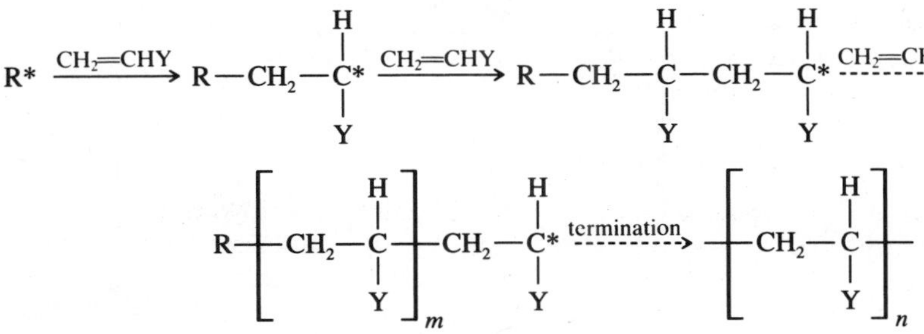

$$
R^* \xrightarrow{CH_2=CHY} R-CH_2-\underset{\underset{Y}{|}}{\overset{\overset{H}{|}}{C}}{}^* \xrightarrow{CH_2=CHY} R-CH_2-\underset{\underset{Y}{|}}{\overset{\overset{H}{|}}{C}}-CH_2-\underset{\underset{Y}{|}}{\overset{\overset{H}{|}}{C}}{}^* \dashrightarrow{\scriptstyle CH_2=CHY}
$$

$$
R\left[CH_2-\underset{\underset{Y}{|}}{\overset{\overset{H}{|}}{C}}\right]_m CH_2-\underset{\underset{Y}{|}}{\overset{\overset{H}{|}}{C}}{}^* \xrightarrow{\text{termination}} \left[CH_2-\underset{\underset{Y}{|}}{\overset{\overset{H}{|}}{C}}\right]_n \qquad (1\text{-}9)
$$

The growth of the polymer chain ceases when the reactive center is destroyed by one of a number of possible termination reactions.

One should not infer from the above discussion that chain polymerizations are faster than step polymerizations. The net rate at which monomer molecules disappear (i.e., the rate of polymerization) in step polymerization can be as great as or greater than that in chain polymerization. The difference between the two processes lies simply in the time required for the growth of each polymer molecule compared to the time required to achieve high conversions. Thus if we start out a chain polymerization and a step polymerization side by side we may observe a variety of situations with regard to their relative rates of polymerization. However, the molecular weights of the polymers produced at any time after the start of the reactions will always be very characteristically different for the two polymerizations. If the two polymerizations are stopped at 0.1% conversion, 1% conversion, 10% conversion, 40% conversion, 90% conversion, and so on, one will always observe the same behavior. The chain polymerization will show the presence of high-molecular-weight polymer molecules at all percents of conversion. There are no intermediate sized molecules in the reaction mixture—only monomer, high-polymer, and initiator species. The only change that occurs with conversion (i.e., reaction time) is the continuous increase in the number of polymer molecules (Fig. 1-1A). Polymer size is generally independent of percent conversion, although the amount of polymer certainly depends on it. On the other hand, high-molecular-weight polymer is obtained in step polymerizations only near the very end of the reaction (>98% conversion) (Fig. 1-1B). Thus both polymer size and the amount of polymer are dependent on conversion in step polymerization.

The classification of polymers according to polymerization mechanism, like that by structure and composition, is not without its ambiguities. Certain polymerizations

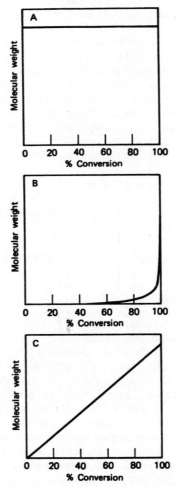

Fig. 1-1 Variation of molecular weight with conversion. A, chain polymerization; B, step polymerization; C, nonterminating ionic chain polymerization and protein synthesis.

show a linear increase of molecular weight with conversion (Fig. 1-1C) when the polymerization mechanism departs from the normal step or chain pathway. This is observed in certain ionic chain polymerizations, which involve a fast initiation process coupled with the absence of reactions that terminate the propagating reactive centers. Biological syntheses of proteins also show the behavior described by Fig. 1-1C because the various monomer molecules are directed to react in a very specific manner by an enzymatically controlled process.

The ring-opening polymerizations of cyclic monomers such as propylene oxide,

$$n\text{CH}_3\text{--CH}\overset{\displaystyle O}{\frown}\text{CH}_2 \rightarrow \left[\text{CH}_2\text{--}\underset{\displaystyle \text{CH}_3}{\text{CH}}\text{--O}\right]_n \qquad (1\text{-}10)$$

or ε-caprolactam,

$$n\begin{array}{c}\text{CH}_2\\ \diagup \quad \diagdown \\ \text{CH}_2 \qquad \text{CO}\\ \mid \qquad\quad \mid \\ \text{CH}_2 \qquad \text{NH}\\ \diagdown \quad \diagup \\ \text{CH}_2\text{--CH}_2\end{array} \rightarrow \text{--}(\text{NHCH}_2\text{CH}_2\text{CH}_2\text{CH}_2\text{CH}_2\text{CO}\text{--})_{\overline{n}} \qquad (1\text{-}11)$$

proceed either by step or chain mechanisms, depending on the particular monomer, reaction conditions, and initiator employed. The polymer obtained is the same regardless of the polymerization mechanism. Such polymerizations point out very clearly that one must distinguish between the classification of the polymerization mechanism and that of the polymer structure. The two classifications cannot always be used interchangeably. Polymers such as the polyethers and polyamides produced in Eqs. 1-10 and 1-11, as well as those from other cyclic monomers, must be separately classified as to polymerization mechanism and polymer structure. These polymers are structurally classified as condensation polymers, since they contain functional groups (e.g., ether, amide) in the polymer chain. They, like the polyurethanes, are not classified as addition polymers by the use of Carothers's original classification. The situation is even more complicated for a polymer such as that obtained from ε-caprolactam. The exact same polymer can be obtained by the step polymerization of the linear monomer ε-aminocaproic acid. It should suffice at this point to stress that the terms condensation and step polymer or polymerization are not synonymous nor are the terms addition and chain polymer or polymerization, even though these terms are often used interchangeably. The classification of polymers based only on polymer structure or only on polymerization mechanism is often an oversimplification that leads to ambiguity and error. Both structure and mechanism are usually needed in order to clearly classify a polymer.

1-2 NOMENCLATURE OF POLYMERS

Polymer nomenclature in general leaves much to be desired. A standard nomenclature system based on chemical structure as is used for small inorganic and organic compounds is most desirable. Unfortunately, the naming of polymers has not proceeded in a systematic manner until relatively recently. It is not at all unusual for a polymer to have several different names because of the use of different nomenclature systems. The nomenclature systems that have been used are based on either the structure of the polymer or the source of the polymer [i.e., the monomer(s) used in its synthesis] or trade names. Not only have there been several different nomenclature systems; their application has not always been rigorous. An important step toward standardization was made in the 1970s with the publication by the International Union of Pure and Applied Chemistry (IUPAC) of a detailed structure-based nomenclature system for single-strand organic polymers [IUPAC, 1973, 1976; Jenkins and Loening, 1989]. IUPAC nomenclature systems for inorganic and other types of polymers have since been developed.

1-2a Nomenclature Based on Source

The most simple and commonly used nomenclature system is probably that based on the source of the polymer. This system is applicable primarily to polymers synthesized from a single monomer as in addition and ring-opening polymerizations. Such polymers are named by adding the name of the monomer onto the prefix "poly" without a space or hyphen. Thus the polymers from ethylene and acetaldehyde are named polyethylene and polyacetaldehyde, respectively. When the monomer has a substituted parent name or a multiworded name or an abnormally long name, parentheses are placed around its name following the prefix "poly." The polymers from 3-methyl-1-pentene, vinyl chloride, propylene oxide, chlorotrifluoroethylene, and ε-caprolactam are named poly(3-methyl-1-pentene), poly(vinyl chloride), poly(propylene oxide), poly(chlorotrifluoroethylene), and poly(ε-caprolactam), respectively. Other examples are listed in Table 1-2. The parentheses are frequently omitted in common usage when naming polymers. Although this will often not present a problem, it is incorrect and in some cases the omission can lead to uncertainty as to the structure of the polymer named. Thus the use of polyethylene oxide instead of poly(ethylene oxide) can be ambiguous in denoting one of the following possible structures:

$$-\!\left(\!-CH_2CH_2-\!\right)_{\!n}\!O-\!\left(\!-CH_2CH_2-\!\right)_{\!n}\qquad\left(\!-CH_2CH_2-\!\right)_{\!n}\!O$$

$$\textbf{II}\qquad\qquad\qquad\qquad\textbf{III}$$

instead of the polymer, $-\!\left(CH_2CH_2-O-\right)_n$, from ethylene oxide.

Some polymers are named as being derived from hypothetical monomers. Thus poly(vinyl alcohol) is actually produced by the hydrolysis of poly(vinyl acetate)

$$\left[\begin{array}{c}CH_3COO\\|\\-CH_2-CH-\end{array}\right]_n + nH_2O \rightarrow \left[\begin{array}{c}HO\\|\\-CH_2-CH-\end{array}\right]_n + nCH_3COOCH \qquad (1\text{-}12)$$

It is, however, named as a product of the hypothetical monomer vinyl alcohol (which in reality exists exclusively as the tautomer–acetaldehyde).

Condensation polymers synthesized from single reactants are named in a similar manner. Examples are the polyamides and polyesters produced from amino acids and hydroxy acids, respectively. Thus, the polymer from 6-aminocaproic acid is named poly(6-aminocaproic acid)

$$n\text{H}_2\text{N}\!-\!\text{CH}_2\text{CH}_2\text{CH}_2\text{CH}_2\text{CH}_2\!-\!\text{COOH} \rightarrow$$

6-Aminocaproic acid

$$\text{+NH}\!-\!\text{CO}\!-\!\text{CH}_2\text{CH}_2\text{CH}_2\text{CH}_2\text{CH}_2\!\text{+}_n\!\text{-}$$ $$(1\text{-}13)$$

Poly(6-aminocaproic acid)

It should be noted that there is an ambiguity here in that poly(6-aminocaproic acid) and poly(ϵ-caprolactam) are one and the same polymer. The same polymer is produced from two different monomers—a not uncommonly encountered situation.

1-2b   Nomenclature Based on Structure (Non-IUPAC)

A number of the more common condensation polymers synthesized from two different monomers have been named by a semisystematic, structure-based nomenclature system other than the recent IUPAC system. The name of the polymer is obtained by following the prefix poly without a space or hyphen with parentheses enclosing the name of the structural grouping attached to the parent compound. The parent compound is the particular member of the class of the polymer—the particular ester, amide, urethane, and so on. Thus the polymer from hexamethylene diamine and sebacic acid is considered as the substituted amide derivative of the compound sebacic acid, $\text{HO}_2\text{C}(\text{CH}_2)_8\text{CO}_2\text{H}$, and is named poly(hexamethylene sebacamide). Poly(ethylene terephthalate) is the polymer from ethylene glycol and terephthalic acid, $p\text{-HO}_2\text{C}\!-\!\text{C}_6\text{H}_4\!-\!\text{CO}_2\text{H}$. The polymer from trimethylene glycol and ethylene diisocyanate is poly(trimethylene ethylene-urethane)

$$\text{+HN}\!-\!(\text{CH}_2)_6\!-\!\text{NHCO}\!-\!(\text{CH}_2)_8\!-\!\text{CO+}_n$$

Poly(hexamethylene sebacamide)

IV

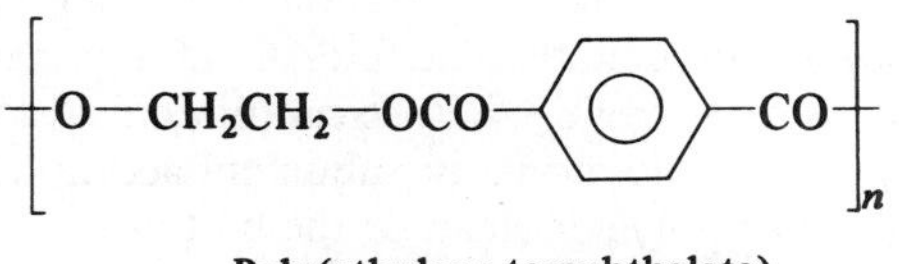

$$\left[\text{O}\!-\!\text{CH}_2\text{CH}_2\!-\!\text{OCO}\!-\!\!\langle\bigcirc\rangle\!\!-\!\text{CO}\right]_n$$

Poly(ethylene terephthalate)

V

$$\text{+O}\!-\!\text{CH}_2\text{CH}_2\text{CH}_2\!-\!\text{OCONH}\!-\!\text{CH}_2\text{CH}_2\!-\!\text{NHCO+}_n$$

Poly(trimethylene ethylene-urethane)

VI

A suggestion was made to name condensation polymers synthesized from two different monomers by following the prefix "poly" with parentheses enclosing the names of the two reactants, with the names of the reactants separated by the term

-*co*-. Thus, the polymer in Eq. 1-7 would be named poly(phenol-*co*-formaldehyde). This suggestion did not gain acceptance.

1-2c IUPAC Structure-Based Nomenclature System

The inadequacy of the preceding nomenclature systems was apparent as the polymer structures being synthesized became increasingly complex. The IUPAC rules allow one to name *single-strand* organic polymers in a systematic manner based on polymer structure. Single-strand organic polymers have any pair of adjacent repeat units interconnected through only one atom. All the polymers discussed up to this point and the overwhelming majority of polymers to be considered in this text are single-strand polymers. *Double-strand* polymers or *ladder* polymers have any pair of adjacent repeat units interconnected through more than one atom. An example of a double-strand polymer is

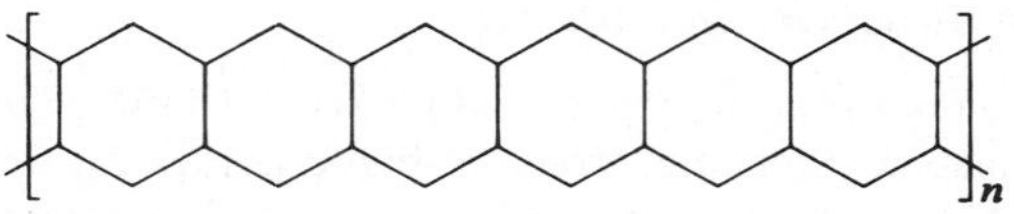

VII

Some aspects of double-strand polymers are considered in Sec. 2-14a, but there are no nomenclature rules for such polymers.

The basis of the IUPAC polymer nomenclature system is the selection of a preferred *constitutional repeating unit* (abbreviated as CRU). The CRU is also referred to as the *structural repeating unit*. The CRU is the smallest possible repeating unit of the polymer. It is a bivalent unit for a single-strand polymer. The name of the polymer is the name of the CRU in parentheses or brackets prefixed by "poly." The CRU is synonymous with the repeating unit defined in Sec. 1-1a except when the repeating unit consists of two symmetrical halves, as in the polymers $-(-CH_2CH_2-)_n-$ and $-(-CF_2CF_2-)_n-$. The CRU is CH_2 and CF_2, respectively, for polyethylene and polytetrafluoroethylene, while the repeating unit is CH_2CH_2 and CF_2CF_2, respectively.

The constitutional repeating unit is named as much as possible according to the IUPAC nomenclature rules for small organic compounds. The IUPAC rules for naming single-strand polymers dictate the choice of a single CRU so as to yield a unique name, by specifying both the *seniority* among the atoms or subunits making up the CRU and the direction to proceed along the polymer chain to the end of the CRU. A CRU is composed of two or more *subunits* when it cannot be named as a single unit. The following is a summary of the most important of the IUPAC rules for naming single-strand organic polymers:

1 The name of a polymer is the prefix "poly" followed in parentheses or brackets by the name of the CRU. The CRU is named by naming its subunits. Subunits are defined as the largest subunits that can be named by the IUPAC rules for small organic compounds.

2 The CRU is written from left to right beginning with the subunit of highest seniority and proceeding in the direction involving the shortest route to the subunit next in seniority.

3 The seniority of different types of subunits is heterocyclic rings > heteroatoms or acyclic subunits containing heteroatoms > carbocyclic rings > acyclic subunits containing only carbon. The presence of various types of atoms, groups of atoms, or rings that are not part of the main polymer chain but are substituents on the CRU do not affect this order of seniority.

4 For heterocyclic rings the seniority is a ring system with nitrogen in the ring > a ring system containing nitrogen and a heteroatom other than nitrogen as high as possible in the order of seniority defined by rule 5 below > a ring system having the largest number of rings > a ring system having the largest individual ring > a ring system having the greatest number of heteroatoms > a ring system having the greatest variety of heteroatoms > a ring system having the greatest number of heteroatoms highest in the order given in rule 3. When two heterocyclic subunits differ only in degree of unsaturation, seniority increases with the degree of unsaturation.

5 For heteroatom(s) or acyclic subunits containing heteroatom(s), the order of decreasing seniority is O, S, Se, Te, N, P, As, Sb, Bi, Si, Ge, Sn, Pb, B, Hg. (Any heteroatom has higher seniority than carbon—rule 3.) The seniority of other heteroatoms within this order is determined from their positions in the periodic table.

6 For carbocyclic rings, the seniority is the ring system containing the largest number of rings > the ring system having the largest individual ring at the first point of difference > the ring system having the largest number of atoms common to all its rings > the ring system having the lowest location number (referred to as *locant*), which designates the first point of difference for ring junctions > the ring system with the most unsaturation.

7 The above orders of seniority are superseded by the requirement of minimizing the number of free valences in the CRU, that is, the CRU should be a bivalent group wherever possible.

8 Where there is a choice subunits should be oriented so that the lowest locant results for substituents or groupings.

Let us illustrate some of these rules by naming a few polymers. For the polymer

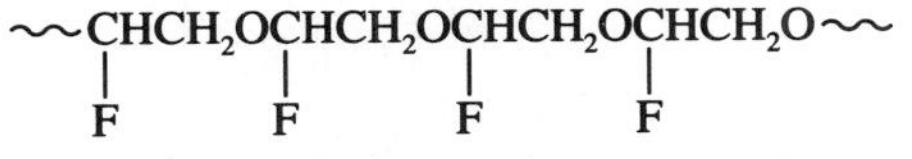

$$\sim\!\!\sim\! CHCH_2OCHCH_2OCHCH_2OCHCH_2O \sim\!\!\sim$$
$$\;\;\;|\qquad\;\;|\qquad\;\;|\qquad\;\;|$$
$$\;\;\;F\qquad F\qquad F\qquad F$$

VIII

the possible CRUs are

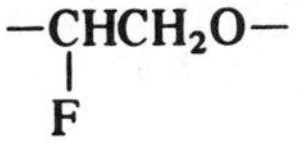

$$-CHCH_2O-\qquad -CH_2OCH-\qquad -OCHCH_2-$$
$$\;\;|\qquad\qquad\qquad\quad|\qquad\qquad\quad\;\;|$$
$$\;\;F\qquad\qquad\qquad\quad F\qquad\qquad\quad F$$

IX 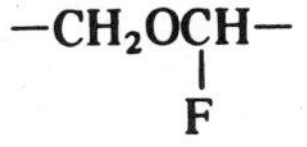   **X**    **XI**

$$-OCH_2CH-\qquad -CHOCH_2-\qquad -CH_2CHO-$$
$$\qquad\quad\;|\qquad\qquad\;|\qquad\qquad\qquad\quad|$$
$$\qquad\quad\;F\qquad\qquad\;F\qquad\qquad\qquad\;\;F$$

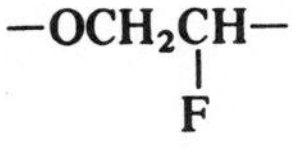

XII    **XIII** 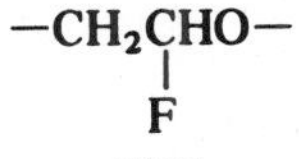   **XIV**

Note that CRUs **XII–XIV** are simply the reverse of CRUs **IX–XI**. Application of the nomenclature rules dictates the choice of only one of these as the CRU. That oxygen has higher seniority than carbon (rule 5) eliminates all except **XI** and **XII** as the CRU. Application of rule 8 results in **XI** as the CRU and the name poly[oxy(1-fluoroethylene)]. Choosing **XII** as the CRU would result in the name poly[oxy(2-fluoroethylene)] which gives the higher locant for the fluorine substituent. The name poly[oxy(fluoromethylene-methylene)] is incorrect because it does not define the largest possible subunit (which is CHFCH$_2$ vs CHF plus CH$_2$).

Rule 7 specifies —CH=CH— as the correct CRU in preference to =CH—CH=, since the former is bivalent, while the latter is tetravalent. The polymer —(—CH=CH—)$_n$ is poly(vinylene).

The higher seniority of heterocyclic rings over carbocyclic rings (rule 3) yields the CRU

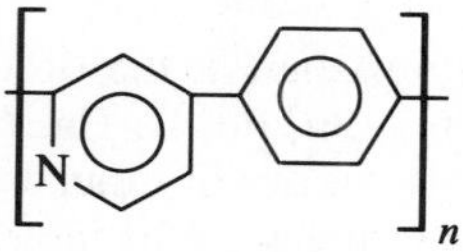

XV

with the name poly(2,4-pyridinediyl-1,4-phenylene).

The higher seniority with higher unsaturation for cyclic subunits (rules 4 and 6) and the seniority of carbocyclic subunits over acyclic carbon subunits (rule 3) yields the CRU

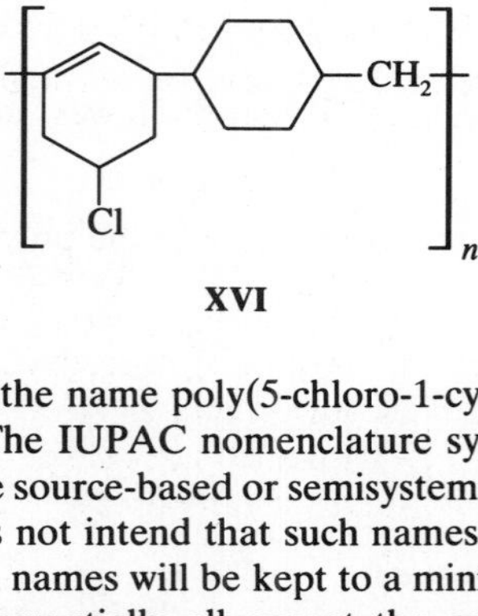

XVI

and the name poly(5-chloro-1-cyclohexen-1,3-ylene-1,4-cyclohexylene-methylene).

The IUPAC nomenclature system recognizes that most of the common polymers have source-based or semisystematic names that are well established by usage. IUPAC does not intend that such names be supplanted by the IUPAC names but hopes that such names will be kept to a minimum. It is clear that the IUPAC system will be used for essentially all except the common (commercial) polymers. For the latter it is doubtful that the IUPAC names will displace the presently established names. The IUPAC names for various of the common polymers are indicated below the more established source or semisystematic name in the following:

—(—CH$_2$CH$_2$—)$_n$

XVII

Polyethylene
Poly(methylene)

—(—CHCH$_2$—)$_n$
   |
  CH$_3$

XVIII

Polypropylene
Poly(propylene)

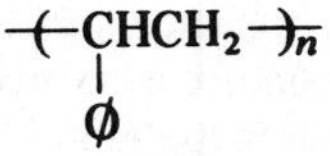

$$-\!\!\left(\!\!\begin{array}{c}CHCH_2\\|\\\phi\end{array}\!\!\right)_{\!\!\overline{n}}$$

XIX

Polystyrene
Poly(1-phenylethylene)

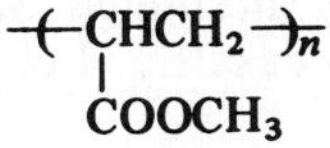

$$-\!\!\left(\!\!\begin{array}{c}CHCH_2\\|\\COOCH_3\end{array}\!\!\right)_{\!\!\overline{n}}$$

XX

Poly(methyl acrylate)
Poly[1-(methoxycarbonyl)ethylene]

$$-\!\!\left(OCH_2\right)_{\!\overline{n}}$$

XXI

Polyformaldehyde
Poly(oxymethylene)

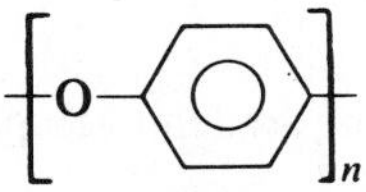

$$\left[O\!-\!\!\left\langle\bigcirc\right\rangle\right]_n$$

XXII

Poly(phenylene oxide)
Poly(oxy-1,4-phenylene)

$$-\!\!\left(NH(CH_2)_6NHCO(CH_2)_4CO\right)_{\!\overline{n}}$$

XIII

Poly(hexamethylene adipamide)
Poly(iminohexamethyleneiminoadipoyl)

$$-\!\!\left(NHCO(CH_2)_5\right)_{\!\overline{n}}$$

XXIV

Poly(ϵ-caprolactam) or poly(ϵ-
aminocaproic acid) (depending on
source of polymer)
Poly[imino(1-oxohexamethylene)]

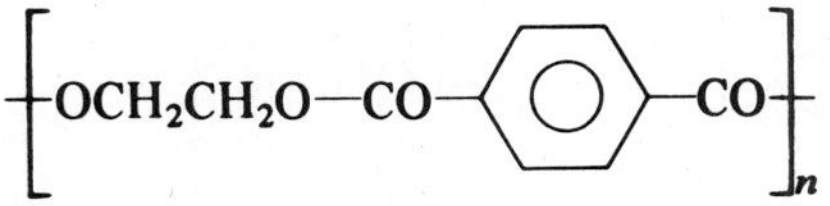

$$\left[OCH_2CH_2O\!-\!CO\!-\!\!\left\langle\bigcirc\right\rangle\!-\!CO\right]_n$$

XXV

Poly(ethylene terephthalate)
Poly(oxyethyleneoxyterephthaloyl)

The IUPAC nomenclature system will generally be used throughout this text with a few exceptions. One is the use of well-established names of the common polymers. Another exception will be in not always following rule 2 for writing the constitutional repeating unit (although the correct IUPAC name will usually be employed). Using the IUPAC choice of the CRU leads in some cases to structures that are longer and appear more complicated. Thus the IUPAC structure for the polymer in Eq. 1-3 is

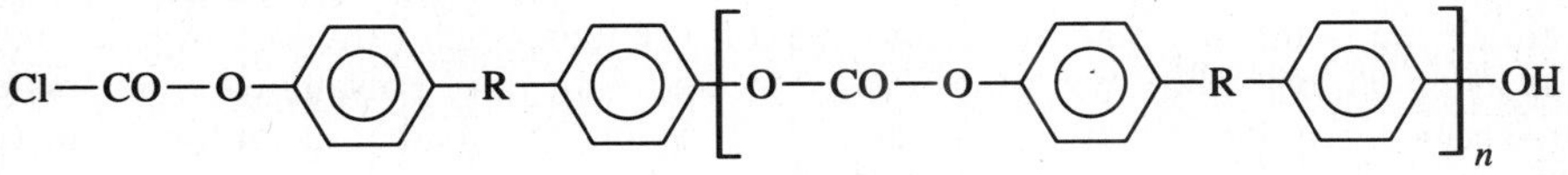

$$Cl\!-\!CO\!-\!O\!-\!\!\left\langle\bigcirc\right\rangle\!-\!R\!-\!\!\left\langle\bigcirc\right\rangle\!\!\left[O\!-\!CO\!-\!O\!-\!\!\left\langle\bigcirc\right\rangle\!-\!R\!-\!\!\left\langle\bigcirc\right\rangle\right]_n\!\!-\!OH$$

XXVI

which is clearly not as simple as

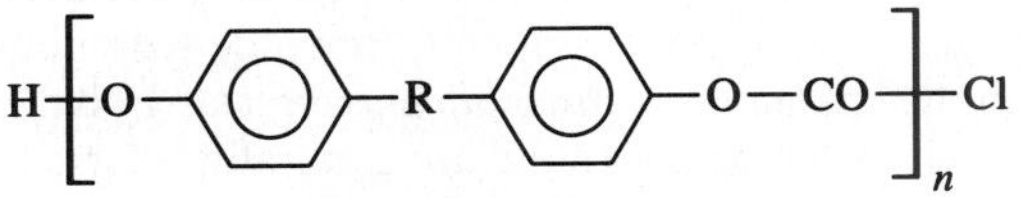

$$H\!\!\left[O\!-\!\!\left\langle\bigcirc\right\rangle\!-\!R\!-\!\!\left\langle\bigcirc\right\rangle\!-\!O\!-\!CO\right]_n\!\!-\!Cl$$

XXVII

although both **XXVI** and **XXVII** denote the exact same structure. This type of problem arises only with certain polymers and then only when the drawn structure is to include the ends of the polymer chain instead of simply the repeating unit or the CRU. The CRU will also generally not be used in equations such as Eq. 1-9. The polymerization mechanism in such reactions involves the propagating center on the substituted carbon atom of the monomer. Using the CRU would yield for the last part of Eq. 1-9,

$$
R \left[CH_2 - \underset{\underset{Y}{|}}{\overset{\overset{H}{|}}{C}} \right]_m CH_2 - \underset{\underset{Y}{|}}{\overset{\overset{H}{|}}{C^*}} \xrightarrow{\text{termination}} \left[\underset{\underset{Y}{|}}{\overset{\overset{H}{|}}{C}} - CH_2 \right]_n \tag{1-14}
$$

which appears unbalanced and confusing, at least to the beginning student, compared to

$$
R \left[CH_2 - \underset{\underset{Y}{|}}{\overset{\overset{H}{|}}{C}} \right]_m CH_2 - \underset{\underset{Y}{|}}{\overset{\overset{H}{|}}{C^*}} \xrightarrow{\text{termination}} \left[CH_2 - \underset{\underset{Y}{|}}{\overset{\overset{H}{|}}{C}} \right]_n \tag{1-9}
$$

Equation 1-9 has the repeating unit written the same way on both sides, while Eq. 1-14 has the repeating unit reversed on the right side relative to what it is on the left side.

1-2d Trade Names and Nonnames

Special terminology based on trade names has been employed for some polymers. Although trade names should be avoided, one must be familiar with those that are firmly established and commonly used. An outstanding example of trade-name nomenclature is the use of the name nylon for the polyamides from unsubstituted, nonbranched aliphatic monomers. Two numbers are added onto the word "nylon" with the first number indicating the number of methylene groups in the diamine portion of the polyamide and the second number the number of carbon atoms in the diacyl portion. Thus poly(hexamethylene adipamide) and poly(hexamethylene sebacamide) are nylon-6,6 and nylon-6,10, respectively. Unfortunately, variants of these names are all too frequently employed. The literature contains such variations of nylon-6,6 as nylon-66, 66-nylon, nylon 6/6, 6,6-nylon, and 6-6-nylon. Polyamides from single monomers are denoted by a single number to denote the number of carbon atoms in the repeating unit. Poly(ε-caprolactam) or poly(6-aminocaproic acid) is nylon-6.

In far too many instances trade-name polymer nomenclature conveys very little meaning regarding the structure of a polymer. Many condensation polymers, in fact, seem not to have names. Thus the polymer obtained by the step polymerization of formaldehyde and phenol is variously referred to as phenol-formaldehyde polymer, phenol-formaldehyde resin, phenolic, phenolic resin, and phenoplast. Polymers of formaldehyde or other aldehydes with urea or melamine are generally referred to as amino resins or aminoplasts without any more specific names. It is often extremely difficult to determine which aldehyde and which amino monomers have been used to synthesize a particular polymer being referred to as an amino resin. More specific nomenclature, if it can be called that, is afforded by indicating the two reactants as in names such as urea-formaldehyde resin or melamine-formaldehyde resin.

A similar situation exists with the naming of many other polymers. Thus the polymer

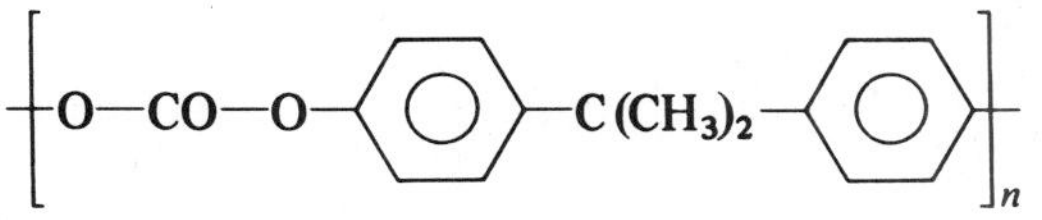

$$\left[\!\!-O\!-\!CO\!-\!O\!-\!\!\bigcirc\!\!-\!C(CH_3)_2\!-\!\!\bigcirc\!\!-\!\right]_n$$

XXVII

is usually referred to as "the polycarbonate from bisphenol A" or "polycarbonate." The IUPAC name for this polymer is poly(oxycarbonyloxy-1,4-phenylene-isopropylidene-1,4-phenylene).

1-3 LINEAR, BRANCHED, AND CROSSLINKED POLYMERS

Polymers can be classified as *linear*, *branched*, or *crosslinked* polymers depending on their structure. In the previous discussion on the different types of polymers and polymerizations we have considered only those polymers in which the monomer molecules have been linked together in one continuous length to form the polymer molecule. Such polymers are termed linear polymers. Under certain reaction conditions or with certain kinds of monomers the polymers can be quite different.

Branched polymers are often obtained in both step and chain polymerizations, although for reasons that are usually quite different in the two cases. Branched polymer molecules are those in which there are side branches of linked monomer molecules protruding from various central branch points along the main polymer chain. The difference between the shapes of linear and branched polymer molecules can be seen from the structural representations in Fig. 1-2. The branch points are indicated by heavy dots. The illustrations show that there are several different kinds of branched polymers. The branched polymer can be comblike in structure with either long (A) or short (B) branches. When there is extensive branching, the polymer can have a dendritic structure in which there are branches protruding from other branches, that is, branched branches (C). The presence of branching in a polymer usually has a large effect on many important polymer properties. The most significant property change brought about by branching is the decrease in crystallinity. Branched polymers do not pack as easily into a crystal lattice as do linear polymers.

It is important to point out that the term "branched polymer" does not refer to linear polymers containing side groups that are part of the monomer structure. Only those polymers that contain side branches composed of complete monomer units are termed "branched polymers." Thus polystyrene **XXVIII**

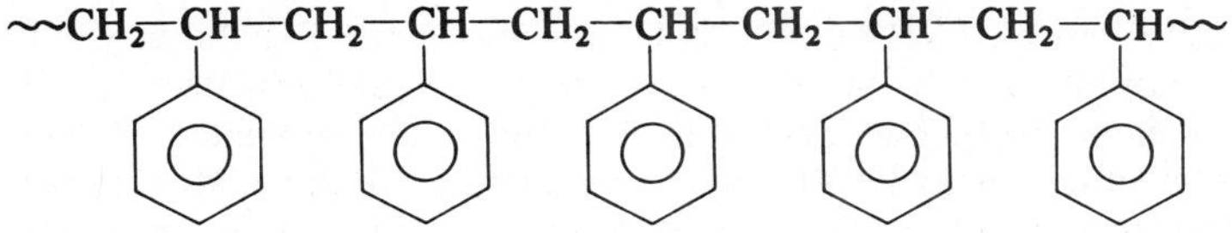

$$\sim\!CH_2\!-\!CH\!-\!CH_2\!-\!CH\!-\!CH_2\!-\!CH\!-\!CH_2\!-\!CH\!-\!CH_2\!-\!CH\!\sim$$

XXVIII

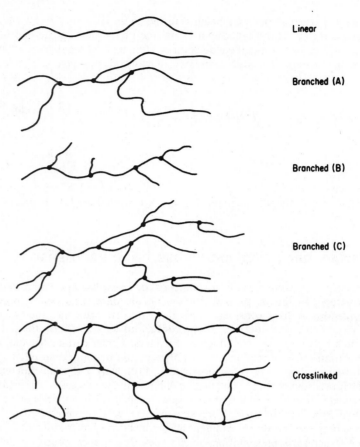

Linear

Branched (A)

Branched (B)

Branched (C)

Crosslinked

Fig. 1-2 Structure of linear, branched, and crosslinked polymers.

is classified as a linear polymer, and not as a branched polymer, for the reason that the phenyl groups are part of the monomer unit and are not considered as branches. Branched polystyrene would be the polymer **XXIX** in which one has one or more polystyrene branches protruding from the main linear polystyrene chain.

When polymers are produced in which the polymer molecules are linked to each other at points other than their ends, the polymers are said to be *crosslinked* (Fig. 1-2). Crosslinking can be made to occur during the polymerization process by the use of appropriate monomers. It can also be brought about after the polymerization by various chemical reactions. The crosslinks between polymer chains can be of different lengths depending on the crosslinking method and the specific conditions employed. One can also vary the number of crosslinks so as to obtain lightly or highly crosslinked polymers. When the number of crosslinks is sufficiently high, a *three-dimensional* or *space network* polymer is produced in which all the polymer chains in a sample have been linked together to form one giant molecule. Light crosslinking is used to impart good recovery (elastic) properties to polymers to be used as rubbers. High degrees of crosslinking are used to impart high rigidity and dimensional stability

$\sim$CH$_2$—CH—CH$_2$—CH—CH$_2$—C—CH$_2$—CH—CH$_2$—CH$\sim$

XXIX

(under conditions of heat and stress) to polymers such as the phenol-formaldehyde
and urea-formaldehyde polymers.

1-4 MOLECULAR WEIGHT

The molecular weight of a polymer is of prime importance in its synthesis and application. The interesting and useful mechanical properties that are uniquely associated
with polymeric materials are a consequence of their high molecular weight. Most
important mechanical properties depend on and vary considerably with molecular
weight as seen in Fig. 1-3. There is a minimum polymer molecular weight (A), usually
a thousand or so, to produce any significant mechanical strength at all. Above A,
strength increases rapidly with molecular weight until a critical point (B) is reached.
Mechanical strength increases more slowly above B and eventually reaches a limiting
value (C). The critical point B generally corresponds to the minimum molecular weight
for a polymer to begin to exhibit sufficient strength to be useful. Most practical applications of polymers require higher molecular weights to obtain higher strengths.

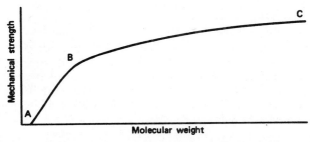

Fig. 1-3 Dependence of mechanical strength on polymer molecular weight.

The minimum useful molecular weight (B), usually in the range 5,000–10,000, differs for different polymers. The molecular weights at A and C also vary for different polymers. The plot in Fig. 1-3 generally shifts to the right as the magnitude of the intermolecular forces decreases. Polymer chains with stronger intermolecular forces, for example, polyamides and polyesters, develop sufficient strength to be useful at lower molecular weights than polymers having weaker intermolecular forces, for example, polyethylene.

Properties other than strength also show a significant dependence on molecular weight. However, most properties show different quantitative dependencies on molecular weight. Different polymer properties usually reach their optimum values at different molecular weights. Further, a few properties may increase with molecular weight to a maximum value and then decrease with further increase in molecular weight. An example is the ability to process polymers into useful articles and forms (e.g., film, sheet, pipe, fiber). Processability begins to decrease past some molecular weight as the viscosity becomes too high and melt flow too difficult. Thus the practical aspect of a polymerization requires one to carry out the process to obtain a compromise molecular weight—a molecular weight sufficiently high to obtain the required strength for a particular application without overly sacrificing other properties. Synthesizing the highest possible molecular weight is not necessarily the objective of a typical polymerization. Instead, one often aims to obtain a high but specified, compromise molecular weight. The utility of a polymerization is greatly reduced unless the process can be carried out to yield the specified molecular weight. The control of molecular weight is essential for the practical application of a polymerization process.

When one speaks of the molecular weight of a polymer, one means something quite different from that which applies to small-sized compounds. Polymers differ from the small-sized compounds in that they are polydisperse or heterogeneous in molecular weight. Even if a polymer is synthesized free from contaminants and impurities, it is still not a pure substance in the usually accepted sense. Polymers, in their purest form, are mixtures of molecules of different molecular weight. The reason for the polydispersity of polymers lies in the statistical variations present in the polymerization processes. When one discusses the molecular weight of a polymer, one is actually involved with its average molecular weight. Both the average molecular weight and the exact distribution of different molecular weights within a polymer are required in order to fully characterize it. The control of molecular weight and molecular weight distribution (MWD) is often used to obtain and improve certain desired physical properties in a polymer product.

Various methods based on solution properties are used to determine the average molecular weight of a polymer sample. These include methods based on colligative properties, light scattering, and viscosity [Billingham, 1977; Bohdanecky and Kovar, 1982; Collins et al., 1973; Morawetz, 1975; Slade, 1975]. The various methods do not yield the same average molecular weight. Different average molecular weights are obtained because the properties being measured are biased differently toward the different-sized polymer molecules in a polymer sample. Some methods are biased toward the larger-sized polymer molecules, while other methods are biased toward the smaller-sized molecules. The result is that the average molecular weights obtained are correspondingly biased toward the larger- or smaller-sized molecules. The following average molecular weights are determined:

1 The *number-average molecular weight* $\overline{M}_n$ is determined by experimental methods that count the number of polymer molecules in a sample of the polymer. The methods for measuring $\overline{M}_n$ are those that measure the colligative properties of solutions—vapor pressure lowering (vapor pressure osmometry), freezing point depression (cryoscopy), boiling point elevation (ebulliometry), and osmotic pressure (membrane osmometry). The colligative properties are the same for small and large molecules when comparing solutions at the same molal (or mole fraction) concentration. For example, a 1-molal solution of a polymer of molecular weight 10^5 has the same vapor pressure, freezing point, boiling point, and osmotic pressure as a 1-molal solution of a polymer of molecular weight 10^3 or a 1-molal solution of a small molecule such as hexane. $\overline{M}_n$ is defined as the total weight w of all the molecules in a polymer sample divided by the total number of moles present. Thus the number-average molecular is

$$\overline{M}_n = \frac{w}{\Sigma N_x} = \frac{\Sigma N_x M_x}{\Sigma N_x} \tag{1-15}$$

where the summations are over all the different sizes of polymer molecules from $x = 1$ to $x = \infty$ and N_x is the number of moles whose weight is M_x. Equation 1-15 can also be written as

$$\overline{M}_n = \Sigma \underline{N}_x M_x \tag{1-16}$$

where $\underline{N}_x$ is the mole-fraction (or the number-fraction) of molecules of size M_x. The most common methods for measuring $\overline{M}_n$ are membrane osmometry and vapor pressure osmometry since reasonably reliable commercial instruments are available for those methods. Vapor pressure osmometry, which measures vapor pressure indirectly by measuring the change in temperature of a polymer solution on dilution by solvent vapor, is generally useful for polymers with $\overline{M}_n$ below 10,000–15,000. Above that molecular-weight limit, the quantity being measured becomes too small to detect by the available instruments. Membrane osmometry is limited to polymers with $\overline{M}_n$ above about 20,000–30,000 and below 500,000. The lower limit is a consequence of the partial permeability of available membranes to smaller-sized polymer molecules. Above molecular weights of 500,000, the osmotic pressure of a polymer solution becomes too small to measure accurately. End-group analysis is also useful for measurements of $\overline{M}_n$ for certain polymers. For example, the carboxyl end groups of a polyester can be analyzed by titration with base and carbon–carbon double bond end groups can be analyzed by ^{1}H NMR. Accurate end-group analysis becomes difficult for polymers with $\overline{M}_n$ values above 20,000–30,000.

2   Light scattering by polymer solutions, unlike coligative properties, is greater for larger-sized molecules than for smaller-sized molecules. The average molecular weight obtained from light scattering measurements is the *weight-average molecular weight* $\overline{M}_w$ defined as

$$\overline{M}_w = \Sigma w_x M_x \tag{1-17}$$

where w_x is the weight-fraction of molecules whose weight is M_x. $\overline{M}_w$ can also be defined as

$$\overline{M}_w = \frac{\Sigma c_x M_x}{\Sigma c_x} = \frac{\Sigma c_x M_x}{c} = \frac{\Sigma N_x M_x^2}{\Sigma N_x M_x} \tag{1-18}$$

where c_x is the weight concentration of M_x molecules, c is the total weight concentration of all the polymer molecules, and the following relationships hold:

$$w_x = \frac{c_x}{c} \tag{1-19}$$

$$c_x = N_x M_x \tag{1-20}$$

$$c = \Sigma c_x = \Sigma N_x M_x \tag{1-21}$$

Since the amount of light scattered by a polymer solution increases with molecular weight, this method becomes more accurate for higher polymer molecular weights. There is no upper limit to the molecular weight that can be accurately measured except the limit imposed by insolubility of the polymer. The lower limit of $\overline{M}_w$ by the light scattering method is close to 5000–10,000. Below this molecular weight, the amount of scattered light is too small to measure accurately.

3 Solution viscosity is also useful for molecular-weight measurements. Viscosity, like light scattering, is greater for the larger-sized polymer molecules than for smaller ones. However, solution viscosity does not measure $\overline{M}_w$ since the exact dependence of solution viscosity on molecular weight is not exactly the same as light scattering. Solution viscosity measures the *viscosity-average molecular weight* $\overline{M}_v$ defined by

$$\overline{M}_v = [\Sigma w_x M_x^a]^{1/a} = \left[\frac{\Sigma N_x M_x^{a+1}}{\Sigma N_x M_x}\right]^{1/a} \tag{1-22}$$

where a is a constant. The viscosity- and weight-average molecular weights are equal when a is unity. $\overline{M}_v$ is less than $\overline{M}_w$ for most polymers, since a is usually in the range 0.5–0.9. However, $\overline{M}_v$ is much closer to $\overline{M}_w$ than to $\overline{M}_n$, usually within 20% of $\overline{M}_w$. The value of a is dependent on the hydrodynamic volume of the polymer and the effective volume of the solvated polymer molecule in solution, and varies with polymer, solvent, and temperature.

More than one average molecular weight is required to reasonably characterize a polymer sample. There is no such need for a monodisperse product (i.e., one composed of molecules whose molecular weights are all the same) for which all three average molecular weights are the same. The situation is quite different for a polydisperse polymer where all three molecular weights are different if the constant a in Eq. 1-22 is less than unity, as is the usual case. A careful consideration of Eqs. 1-15 through 1-22 shows that the number-, viscosity-, and weight-average molecular weights, in that order, are increasingly biased toward the higher-molecular-weight fractions in a polymer sample. For a polydisperse polymer

$$\overline{M}_w > \overline{M}_v > \overline{M}_n$$

with the differences between the various average molecular weights increasing as the molecular-weight distribution broadens. A typical polymer sample will have the molecular-weight distribution shown in Fig. 1-4. The approximate positions of the different average molecular weights are indicated on this distribution curve.

For most practical purposes, one usually characterizes the molecular weight of a polymer sample by measuring $\overline{M}_n$ and either $\overline{M}_w$ or $\overline{M}_v$. $\overline{M}_v$ is commonly used as a

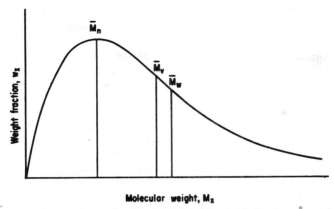

Fig. 1-4 Distribution of molecular weights in a typical polymer sample.

close approximation of $\overline{M}_w$, since the two are usually quite close (within 10–20%). Thus in most instances, one is concerned with the $\overline{M}_n$ and $\overline{M}_w$ of a polymer sample. The former is biased toward the lower-molecular-weight fractions, while the latter is biased toward the higher-molecular-weight fractions. The ratio of the two average molecular weights $\overline{M}_w/\overline{M}_n$ depends on the breadth of the distribution curve (Fig. 1-4) and is often useful as a measure of the polydispersity in a polymer. The value of $\overline{M}_w/\overline{M}_n$ would be unity for a perfectly monodisperse polymer. The ratio is greater than unity for all actual polymers and increases with increasing polydispersity.

The characterization of a polymer by $\overline{M}_n$ alone, without regard to the polydispersity, can be extremely misleading, since most polymer properties such as strength and melt viscosity are determined primarily by the size of the molecules that make up the bulk of the sample by weight. Polymer properties are much more dependent on the larger-sized molecules in a sample than on the smaller ones. Thus, for example, consider a hypothetical mixture containing 95% by weight of molecules of molecular weight 10,000, and 5% of molecules of molecular weight 100. (The low-molecular-weight fraction might be monomer, a low-molecular-weight polymer, or simply some impurity.) The $\overline{M}_n$ and $\overline{M}_w$ are calculated from Eqs. 1-15 and 1-17 as 1680 and 9505, respectively. The use of the $\overline{M}_n$ value of 1680 gives an inaccurate indication of the properties of this polymer. The properties of the polymer are determined primarily by the 10,000-molecular-weight molecules that comprise 95% of the weight of the mixture. The weight-average molecular weight is a much better indicator of the properties to be expected in a polymer. The utility of $\overline{M}_n$ resides primarily in its use to obtain an indication of polydispersity in a sample by measuring the ratio $\overline{M}_w/\overline{M}_n$.

In addition to the different average molecular weights of a polymer sample, it is frequently desirable and necessary to know the exact distribution of molecular weights. As indicated previously, there is usually a molecular-weight range for which any given polymer property will be optimum for a particular application. The polymer sample containing the greatest percentage of polymer molecules of that size is the one that will have the optimum value of the desired property. Since samples with the same average molecular weight may possess different molecular-weight distributions, information regarding the distribution allows the proper choice of a polymer for optimum performance. Various methods have been used in the past to determine the molecular-weight distribution of a polymer sample, including fractional extraction and fractional precipitation. These methods are laborious and determinations of molecular-weight distributions were not routinely performed. However, the development of *size exclusion chromatography* (SEC), also referred to as *gel permeation chromatography* (GPC) and the availability of automated commercial instruments have changed the situation. Molecular-weight distributions are now routinely performed in most laboratories using SEC.

Size exclusion chromatography involves the permeation of a polymer solution through a column packed with microporous beads of crosslinked polystyrene [Janca, 1984; Yau et al., 1979]. The packing contains beads of different-sized pore diameters. Molecules pass through the column by a combination of transport into and through the beads and through the interstitial volume (the volume between beads). Molecules that penetrate the beads are slowed down more in moving through the column than molecules that do not penetrate the beads; in other words, transport through the interstitial volume is faster than through the pores. The smaller-sized polymer molecules penetrate all the beads in the column since their molecular size (actually their hydrodynamic volume) is smaller than the pore size of the beads with the smallest-sized pores. A larger-sized polymer molecule does not penetrate all the beads since its molecular size is larger than the pore size of some of the beads. The larger the polymer molecular weight, the fewer beads that are penetrated and the greater is the extent of transport through the interstitial volume. The time for passage of polymer molecules through the column decreases with increasing molecular weight. The use of an appropriate detector (refractive index, viscosity, light scattering) measures the amount of polymer passing through the column as a function of time. This information and a calibration of the column with standard polymer samples of known molecular weight allow one to obtain the molecular weight distribution in the form of a plot such as that in Fig. 1-4. Not only does SEC yield the molecular weight distribution, but $\overline{M}_n$ and $\overline{M}_w$ (and also $\overline{M}_v$ if a is known) are also calculated automatically. SEC is now the method of choice for measurement of $\overline{M}_n$ and $\overline{M}_w$ since the SEC instrument is far easier to use compared to methods such as osmometry and light scattering.

1-5 PHYSICAL STATE

1-5a Crystalline and Amorphous Behavior

Solid polymers differ from ordinary, low-molecular-weight compounds in the nature of their physical state or *morphology*. Most polymers show simultaneously the characteristics of both crystalline solids and highly viscous liquids [Bassett, 1981; Keller, 1977; Mandelkern, 1976; Mark et al., 1984; Sharples, 1972; Sperling, 1986; Woodward, 1989; Wunderlich, 1973]. X-Ray and electron diffraction patterns often show the sharp features typical of three-dimensionally ordered, crystalline materials as well as the diffuse features characteristic of liquids. The terms *crystalline* and *amorphous* are used to indicate the ordered and unordered polymer regions, respectively. Different polymers show different degrees of crystalline behavior. The known polymers constitute a spectrum of materials from those that are completely amorphous to others that possess low to moderate to high crystallinity. The term *semicrystalline* is used to refer to polymers that are partially crystalline. Completely crystalline polymers are rarely encountered.

The exact nature of polymer crystallinity has been the subject of considerable controversy. The *fringed-micelle* theory, developed in the 1930s, considers polymers to consist of small-sized, ordered crystalline regions—termed *crystallites*—imbedded in an unordered, amorphous polymer matrix. Polymer molecules are considered to pass through several different crystalline regions with crystallites being formed when extended-chain segments from different polymer chains are precisely aligned together and undergo crystallization. Each polymer chain can contribute ordered segments to several crystallites. The segments of the chain in between the crystallites make up the unordered amorphous matrix. This concept of polymer crystallinity is shown in Fig. 1-5.

The *folded-chain lamella* theory arose in the last 1950s when polymer single crystals in the form of thin platelets termed *lamella*, measuring about 10,000 Å × 100 Å, were grown from polymer solutions. Contrary to previous expectations, X-ray diffraction patterns showed the polymer chain axes to be parallel to the smaller dimension of the platelet. Since polymer molecules are much longer than 100 Å, the polymer molecules are presumed to fold back and forth on themselves in an accordionlike manner in the process of crystallization. Chain folding was unexpected, since the most thermodynamically stable crystal is the one involving completely extended chains. The latter is kinetically difficult to achieve and chain folding is apparently the system's compromise for achieving a highly stable crystal structure under normal crystallization conditions. Two models of chain folding can be visualized. Chain folding is regular and sharp with a uniform fold period in the *adjacent-reentry* model (Fig. 1-6). In the *nonadjacent-reentry* or *switchboard* model (Fig. 1-7) molecules wander through the nonregular surface of a lamella before reentering the lamella or a neighboring lamella. In the chain-folded lamella picture of polymer crystallinity less than 100% crystallinity

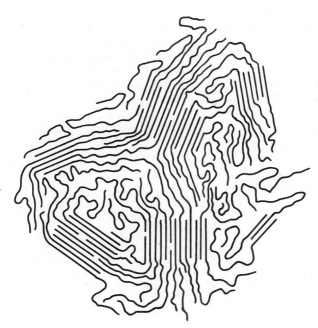

Fig. 1-5 Fringed-micelle picture of polymer crystallinity.

Fig. 1-6 Adjacent-reentry model of single crystal.

is attributed to defects in the chain-folding process. The defects may be imperfect folds, irregularities in packing, chain entanglements, loose chain ends, dislocations, occluded impurities, or numerous other imperfections. The adjacent reentry and switchboard models differ in the details of what constitutes the chain-folding defects. The switchboard model indicates that most defects are at the crystal surfaces, while the adjacent-reentry model indicates that defects are located as much within the crystal as at the crystal surfaces.

Folded-chain lamella represent the morphology not only for single crystals grown from solution but also polymers crystallized from the melt—which is how almost all commercial and other synthetic polymers are obtained. Melt-crystallized polymers have the most prominent structural feature of polymer crystals—the chains are oriented perpendicular to the lamella face so that chain folding must occur. Chain folding is maximum for polymers crystallized slowly near the crystalline melting temperature. Fast cooling (quenching) gives a more chaotic crystallization with less chain folding. Melt crystallization often develops as a spherical or spherulitic growth as seen under the microscope. Nucleation of crystal growth occurs at various nuclei and crystal growth proceeds in a radial fashion from each nucleus until the growth fronts from neighboring structures impinge on each other. These sperical structures, termed *spherulites*, completely fill the volume of a crystallized polymer sample. Spherulites have different sizes and degrees of perfection depending on the specific polymer and crystallization conditions.

A spherulite is a complex, polycrystalline structure (Fig. 1-8). The *nucleus* for spherulitic growth is the single crystal in which a multilayered stack is formed, and

Fig. 1-7 Switchboard model of single crystal.

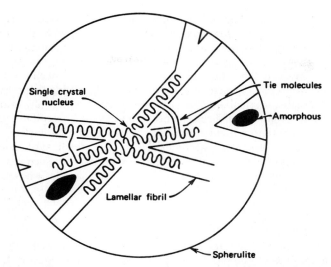

Fig. 1-8   Structural organization within a spherulite in melt-crystallized polymer.

each lamella extends to form a *lamellar fibril*. The flat ribbonlike lamellar fibrils diverge, twist, and branch as they grow outward from the nucleus. Growth occurs by chain folding with the polymer chain axes being perpendicular to the length of the lamellar fibril. The strength of polymers indicates that more than van der Waals forces hold lamellae together. There are *interlamellar* or *intercrystalline fibrils* (also termed *tie molecules*) between the lamellar fibrils within a spherulite and between fibrils of different spherulites. Some polymer molecules simultaneously participate in the growth of two or more adjacent lamellae and provide molecular links that reinforce the crystalline structure. The chain axes of tie molecules lie parallel to the long axes of the link—each link between lamellae is an extended-chain type of single crystal. The tie molecules are the main component of the modern picture of polymer crystallinity, which is a carryover from the fringe-micelle theory. The amorphous content of a semicrystalline, melt-crystallized polymer sample consists of the defects in the chain-folding structure, tie molecules, and the material that is either, because of entanglements, not included in the growing lamellar fibril or is rejected from it owing to its unacceptable nature; low molecular weight chains and nonregular polymer chain segments, for example, are excluded.

Some natural polymers such as cotton, silk, and cellulose have the extended-chain morphology, but their morphologies are determined by enzymatically controlled synthesis and crystallization processes. Extended-chain morphology is obtained in some synthetic polymers under certain circumstances. These include crystallization from the melt (or annealing for long time periods) under pressure or other applied stress and crystallization of polymers from the liquid crystalline state. The former has been observed with several polymers, including polyethylene and polytetrafluoroethylene. The latter is observed with polymers containing stiff or rigid-rod chains, such as poly(*p*-phenyleneterephthalamide) (Secs. 2–8f, 2–14g). Extended chain morphology is also obtained in certain polymerizations involving conversion of crystalline monomer to crystalline polymer, for example, polymerization of diacetylenes (Sec. 3–13c).

A variety of techniques have been used to determine the extent of crystallinity in a polymer, including X-ray diffraction, density, IR, NMR, and heat of fusion [Wunderlich, 1973]. X-Ray diffraction is the most direct method but requires the somewhat difficult separation of the crystalline and amorphous scattering envelops. The other methods are indirect methods but are easier to use since one need not be an expert in the field as with X-ray diffraction. Heat of fusion is probably the most often used method since reliable thermal analysis instruments are commercially available [Turi, 1981; Wendlandt, 1986]. The difficulty in using thermal analysis (differential scanning calorimetry and differential thermal analysis) or any of the indirect methods is the uncertainty in the values of the quantity measured (e.g., the heat of fusion per gram of sample or density) for 0 and 100% crystalline samples since such samples seldom exist. The best technique is to calibrate the method with samples whose crystallinites have been determined by X-ray diffraction.

1-5b   Determinants of Polymer Crystallinity

Regardless of the precise picture of order and disorder in polymers, the prime consideration that should be emphasized is that polymers have a tendency to crystallize. The extent of this crystallization tendency plays a most significant role in the practical ways in which polymers are used. This is a consequence of the large effect of crystallinity on the thermal, mechanical, and other important properties of polymers. Different polymers have different properties and are synthesized and used differently because of varying degrees of crystallinity. The extent of crystallinity developed in a polymer sample is a consequence of both thermodynamic and kinetic factors. In this discussion we will note the general tendency to crystallize under moderate crystallization conditions (that is, conditions that exclude extremes of time, temperature, and pressure). Thermodynamically crystallizable polymers generally must crystallize at reasonable rates if crystallinity is to be employed from a practical viewpoint. The extent to which a polymer crystallizes depends on whether its structure is conducive to packing into the crystalline state and on the magnitude of the secondary forces of the polymer chains. Packing is facilitated for polymer chains that have structural regularity, compactness, streamlining, and some degree of flexibility. The stronger the secondary forces, the greater will be the driving force for the ordering and crystallization of polymer chains.

Some polymers are highly crystalline primarily because their structure is conducive to packing, while others are crystalline primarily because of strong secondary forces. For still other polymers both factors may be favorable for crystallization. Polyethylene, for example, has essentially the best structure in terms of its ability to pack into the crystalline state. Its very simple and perfectly regular structure allows chains to pack tightly and without any restrictions as to which segment of one chain need line up next to which other segment of the same chain or of another chain. The flexibility of the polyethylene chains is also conducive to crystallization in that the comformations required for packing can be easily achieved. Even though its secondary forces are small, polyethylene crystallizes easily and to a high degree because of its simple and regular structure.

Polymers other than polyethylene have less simple and regular chains. Poly(ϵ-caprolactam) can be considered as a modified polyethylene chain containing the amide group in between every five methylenes. Poly(ϵ-caprolactam) and other polyamides are highly crystalline polymers. The amide group is a polar one and leads to much larger secondary forces in polyamides (due to hydrogen bonding) compared to polyethylene; this is most favorable for crystallization. However, the polyamide chains are not as simple as those of polyethylene and packing requires that chain segments be brought together so that the amide groups are aligned. This restriction leads to a somewhat lessened degree of crystallization in polyamides than expected, based only on a consideration of the high secondary forces. Crystallinity in a polymer such as a polyamide can be significantly increased by mechanically stretching it to facilitate the ordering and alignment of polymer chains.

Polymers such as polystyrene, poly(vinyl chloride), and poly(methyl methacrylate) show very poor crystallization tendencies. Loss of structural simplicity (compared to polyethylene) results in a marked decrease in the tendency toward crystallization. Fluorocarbon polymers such as poly(vinyl fluoride), poly(vinylidene fluoride), and polytetrafluoroethylene are exceptions. These polymers show considerable crystallinity since the small size of fluorine does not preclude packing into a crystal lattice. Crystallization is also aided by the high secondary forces. High secondary forces coupled with symmetry account for the presence of significant crystallinity in poly(vinylidene chloride). Symmetry alone without significant polarity, as in polyisobutylene, is insufficient for the development of crystallinity. (The effect of stereoregularity of polymer structure on crystallinity is postponed to Sec. 8-2a.)

Polymers with rigid, cyclic structures in the polymer chain, as in cellulose and poly(ethyleneterephthalate), are difficult to crystallize. Moderate crystallization does occur in these cases, as a result of the polar polymer chains. Additional crystallization can be induced by mechanical stretching. Cellulose is interesting in that native cellulose in the form of cotton is much more crystalline than cellulose that is obtained by precipitation of cellulose from solution (Sec. 9-3a). The biosynthesis of cotton proceeds with an enzymatic ordering of the polymer chains in spite of the rigid polymer chains. Excess chain rigidity in polymers due to extensive crosslinking, as in phenol-formaldehyde and urea-formaldehyde polymers, completely prevents crystallization.

Chain flexibility also effects the ability of a polymer to crystallize. Excessive flexibility in a polymer chain as in polysiloxanes and natural rubber leads to an inability of the chains to pack. The chain conformations required for packing cannot be maintained due to the high flexibility of the chains. The flexibility in the cases of the polysiloxanes and natural rubber is due to the bulky Si—O and *cis*-olefin groups, respectively. Such polymers remain as almost completely amorphous materials, which, however, show the important property of elastic behavior.

1-5c Thermal Transitions

Polymeric materials are characterized by two major types of transition temperatures—the *crystalline melting temperature* T_m and the *glass transition temperature* T_g. The crystalline melting temperature is the melting temperature of the crystalline domains of a polymer sample. The glass transition temperature is the temperature at which the amorphous domains of a polymer take on the characteristic properties of the glassy state—brittleness, stiffness, and rigidity. The difference between the two thermal transitions can be understood more clearly by considering the changes that occur in a liquid polymer as it is cooled. The translational, rotational, and vibrational energies of the polymer molecules decrease on cooling. When the total energies of the molecules have fallen to the point where the translational and rotational energies are essentially zero, crystallization is possible. If certain symmetry requirements are met, the molecules are able to pack into an ordered, lattice arrangement and crystallization occurs. The temperature at which this occurs is T_m. However, not all polymers meet the necessary symmetry requirements for crystallization. If the symmetry requirements are not met, crystallization does not take place, but the energies of the molecules continue to decrease as the temperature decreases. A temperature is finally reached—the T_g—at which long-range motions of the polymer chains stop. Long-range motion, also referred to as *segmental motion*, refers to the motion of a segment of a polymer chain by the concerted rotation of bonds at the ends of the segment. [Bond rotations about side chains, e.g., the C—CH$_3$ and C—COOCH$_3$ bonds in poly(methyl methacrylate), do not cease at T_g.]

Whether a polymer sample exhibits both thermal transitions or only one depends on its morphology. Completely amorphous polymers show only a T_g. A completely crystalline polymer shows only a T_m. Semicrystalline polymers exhibit both the crystalline melting and glass transition temperatures. Changes in properties such as specific volume and heat capacity occur as a polymer undergoes each of the thermal transitions. Figure 1–9 shows the changes in specific volume with temperature for completely amorphous and completely crystalline polymers (the solid lined plots). T_m is a first-order transition with a discontinuous change in the specific volume at the transition temperature. T_g is a second-order transition involving only a change in the temperature coefficient of the specific volume. (A plot of the temperature coefficient of the specific volume versus temperature shows a discontinuity.) The corresponding plot for a semicrystalline polymer consists of the plot for the crystalline polymer plus the dotted

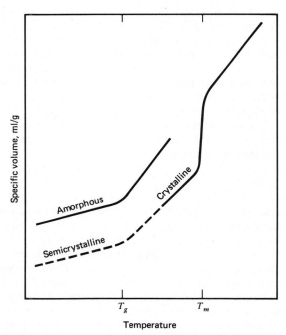

Fig. 1-9 Determination of glass transition and crystalline melting temperatures by changes in specific volume.

portion corresponding to the glass transition. A variety of methods have been used to determine T_g and T_m, including dilatometry (specific volume), thermal analysis, dynamic mechanical behavior, dielectric loss, and broad-line NMR. The most commonly used method is differential scanning calorimetry (DSC). DSC reflects the change in heat capacity of a sample as a function of temperature by measuring the heat flow required to maintain a zero temperature differential between an inert reference material and the polymer sample.

The melting of a polymer takes place over a wider temperature range than that observed for small organic molecules such as benzoic acid due to the presence of different-sized crystalline regions and the more complicated process for melting of large molecules. T_m, generally reported as the temperature for the onset of melting, is determined as the intersection from extrapolation of the two linear regions of Fig. 1-9 (before and after the onset). T_g also occurs over a wide temperature range and is determined by extrapolation of the two linear regions, before and after T_g. The glass transition is a less well understood process than melting. There are indications that it is at least partially a kinetic phenomenon. The experimentally determined value of T_g varies significantly with the time scale of the measurement. Faster cooling rates result in higher T_g values. Further, significant densification still takes place below T_g with the amount dependent on the cooling rate. Perhaps the best visualization of T_g involves the existence of a modest range of temperatures at which there is cessation of segmental motion for polymer chain segments of different lengths ($\sim$5–20 chain atoms).

Some polymers undergo other thermal transitions in addition to T_g and T_m. These include crystal–crystal transitions (i.e., transition from one crystalline form to another) and crystalline–liquid crystal transitions.

The values of T_g and T_m for a polymer affect its mechanical properties at any particular temperature and determine the temperature range in which that polymer can be employed. The T_g and T_m values for some of the common polymers are shown in Table 1-3 [Brandrup and Immergut, 1975, 1989]. (These are the values at 1 atm pressure.) Consider the manner in which T_g and T_m vary from one polymer to another. One can discuss the two transitions simultaneously since both are affected similarly by considerations of polymer structure. Polymers with low T_g values usually have low T_m values; high T_g and high T_m values are usually found together. Polymer chains that do not easily undergo bond rotation so as to pass through the glass transition would also be expected to melt with difficulty. This is reasonable, since similar considerations of polymer structure are operating in both instances. The two thermal transitions are generally affected in the same manner by the molecular symmetry, structural rigidity, and secondary forces of polymer chains [Billmeyer, 1984; Mark et al., 1984; Sperling, 1986; Williams, 1971]. High secondary forces (due to high polarity or hydrogen bonding) lead to strong crystalline forces requiring high temperatures for melting. High secondary forces also decrease the mobility of amorphous polymer chains, leading to high T_g. Decreased mobility of polymer chains, increased chain rigidity, and high T_g are found where the chains are substituted with several substituents as in poly(methyl methacrylate) and polytetrafluoroethylene or with bulky substituents as in polystyrene. The T_m values of crystalline polymers produced from such rigid chains would also be high. The effects of substituents are not always easy to understand. A comparison of polypropylene, poly(vinyl chloride), and poly(vinyl fluoride) with polyisobutylene, poly(vinylidene chloride), and poly(vinylidene fluoride), respectively, shows the polymers from 1,1-disubstituted ethylenes have lower T_g

TABLE 1-3 Thermal Transitions of Polymers[a]

Polymer	Repeating Unit	T_g (°C)	T_m (°C)
Polydimethylsiloxane	—OSi(CH$_3$)$_2$—	−127	−40
Polyethylene	—CH$_2$CH$_2$—	−125	137
Polyoxymethylene	—CH$_2$O—	−82	181
Natural rubber (polyisoprene)	—CH$_2$C(CH$_3$)=CHCH$_2$—	−73	28
Polyisobutylene	—CH$_2$C(CH$_3$)$_2$—	−73	44
Poly(ethylene oxide)	—CH$_2$CH$_2$O—	−41	66
Poly(vinylidene fluoride)	—CH$_2$CF$_2$—	−40	185
Polypropylene	—CH$_2$CH(CH$_3$)—	−8	176
Poly(vinyl fluoride)	—CH$_2$CHF—	41	200
Poly(vinylidene chloride)	—CH$_2$CCl$_2$—	−18	200
Poly(vinyl acetate)	—CH$_2$CH(OCOCH$_3$)—	32	
Poly(chlorotrifluoroethylene)	—CF$_2$CFCl—		220
Poly(ε-caprolactam)	—(CH$_2$)$_5$CONH—	52	223
Poly(hexamethylene adipamide)	—NH(CH$_2$)$_6$NHCO(CH$_2$)$_4$CO—	50	265
Poly(ethylene terephthalate)	—OCH$_2$CH$_2$OCO—⟨○⟩—CO—	61	270
Poly(vinyl chloride)	—CH$_2$CHCl—	81	273
Polystyrene	—CH$_2$CHφ—	100	250
Poly(methyl methacrylate)	—CH$_2$C(CH$_3$)(CO$_2$CH$_3$)—	105	200
Cellulose triacetate	(see structure)		306
Polytetrafluoroethylene	—CF$_2$CF$_2$—	117	327

[a]Data from Brandrup and Immergut [1975, 1989].

and T_m values than those from the monosubstituted ethylenes. One might have predicted the opposite result due to the greater polarity and molecular symmetry of the polymers from 1,1-disubstituted ethylenes. Apparently, the presence of two side groups instead of one separates polymer chains from each other and results in more flexible polymer chains. Thus, the effects of substituents on T_g and T_m depend on their number and identity.

The rigidity of polymer chains is especially high when there are cyclic structures in the main polymer chains. Polymers such as cellulose have high T_g and T_m values. On the other hand, the highly flexible polysiloxane chain (a consequence of the large size of Si) results in very low values of T_g and T_m.

Although T_g and T_m depend similarly on molecular structure, the variations in the two transition temperatures do not always quantitatively parallel each other. Table 1-3 shows the various polymers listed in order of increasing T_g values. The T_m values are seen to generally increase in the same order, but there are many polymers whose T_m values do not follow in the same exact order. Molecular symmetry, chain rigidity, and secondary forces do not affect T_g and T_m in the same quantitative manner. Thus polyethylene and polyoxymethylene have low T_g values because of their highly flexible chains; however, their simple and regular structures yield tightly packed crystal structures with high T_m values. An empirical consideration of the ratio T_g/T_m (Kelvin temperatures) for various polymers aids this discussion. The T_g/T_m ratio is approximately 1/2 for symmetrical polymers [e.g., poly(vinylidene chloride)], but the ratio is closer to 3/4 for unsymmetrical polymers (e.g., poly[vinyl chloride]). This result indicates that T_m is more dependent on molecular symmetry while T_g is more dependent on secondary forces and chain flexibility.

It should be evident that some of the factors that decrease the crystallization tendency of a polymer also lead to increased values of T_m (and also T_g). The reason for this is that the extent of crystallinity developed in a polymer is both kinetically and thermodynamically controlled, while the melting temperature is only thermodynamically controlled. Polymers with rigid chains are difficult or slow to crystallize, but the portion that does crystallize will have a high melting temperature. (The extent of crystallinity can be significantly increased in such polymers by mechanical stretching to align and crystallize the polymer chains.) Thus compare the differences between polyethylene and poly(hexamethylene adipamide). Polyethylene tends to crystallize easier and faster than the polyamide because of its simple and highly regular structure and is usually obtained with greater degrees of crystallinity. On the other hand, the T_m of the polyamide is much higher (by ~130°C) than that of polyethylene because of the much greater secondary forces.

1-6 APPLICATIONS OF POLYMERS

1-6a Mechanical Properties

Many polymer properties such as solvent, chemical, and electrical resistance and gas permeability are important in determining the use of a specific polymer in a specific application. However, the prime consideration in determining the general utility of a polymer is its mechanical behavior, that is, its deformation and flow characteristics under stress. The mechanical behavior of a polymer can be characterized by its *stress–strain* properties [Billmeyer, 1984; Nielsen, 1974]. This often involves observing the behavior of a polymer as one applies tension stress to it in order to elongate (strain) it to the point where it ruptures (pulls apart). The results are usually shown as a plot of the stress versus elongation (strain). The stress is usually expressed in newtons per square centimeter (N/cm^2) or megapascals (MPa) where 1 MPa = 100 N/cm^2. The strain is the fractional increase in the length of the polymer sample (i.e., $\Delta L/L$, where L is the original, unstretched sample length). The strain can also be expressed as the percent elongation, $\Delta L/L \times 100\%$. Although N/cm^2 is the SI unit for stress, psi (pounds per square inch) is found extensively in the literature. The conversion factor is 1 N/cm^2 = 1.450 psi. SI units will be used throughout this text with other commonly used units also indicated.

Several stress–strain plots are shown in Fig. 1-10. Four important quantities characterize the stress–strain behavior of a polymer:

1 *Modulus.* The resistance to deformation as measured by the initial stress divided by $\Delta L/L$.

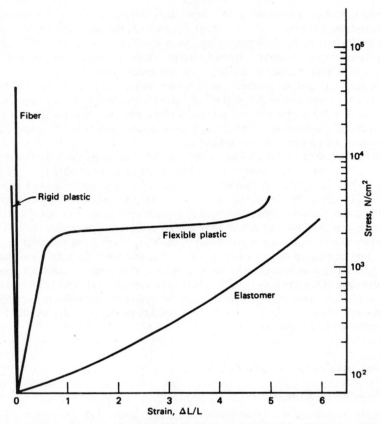

Fig. 1-10 Stress–strain plots for a typical elastomer, flexible plastic, rigid plastic, and fiber.

2 *Ultimate Strength or Tensile Strength*. The stress required to rupture the sample.

3 *Ultimate Elongation*. The extent of elongation at the point where the sample ruptures.

4 *Elastic Elongation*. The elasticity as measured by the extent of reversible elongation.

Polymers vary widely in their mechanical behavior depending on the degree of crystallinity, degree of crosslinking, and the values of T_g and T_m. High strength and low extensibility are obtained in polymers by having various combinations of high degrees of crystallinity or crosslinking or rigid chains (characterized by high T_g). High extensibility and low strength in polymers are synonymous with low degrees of crystallinity and crosslinking and low T_g values. The temperature limits of utility of a polymer are governed by its T_g and/or T_m. Strength is lost at or near T_g for an amorphous polymer and at or near T_m for a crystalline polymer.

An almost infinite variety of polymeric materials can be produced. The polymer scientist must have an awareness of the properties desired in the final polymer in order to make a decision about the polymer to be synthesized. Different polymers are synthesized to yield various mechanical behaviors by the appropriate combinations of crystallinity, crosslinking, T_g, and T_m. Depending on the particular combination, a specific polymer will be used as a *fiber*, *flexible plastic*, *rigid plastic*, or *elastomer* (rubber). Commonly encountered articles that typify these uses of polymers are clothing and rope (fiber), packaging films and seat covers (flexible plastic), eyeglass lenses and housings for appliances (rigid plastic), and rubber bands and tires (elastomer). Table 1-4 shows the uses of many of the common polymers. Some polymers are used in more than one category because certain mechanical properties can be manipulated by appropriate chemical or physical means, such as by altering the crystallinity or adding plasticizers (Sec. 3-13b-3-a) or copolymerization (Sec. 2-13, Chap. 6). Some polymers are used as both plastics and fibers, others as both elastomers and plastics.

1-6b   Elastomer, Fibers, Plastics

The differences between fibers, plastics, and elastomers can be seen in the stress–strain plots in Fig. 1-10. The modulus of a polymer is the initial slope of such a plot; the tensile strength and ultimate elongation are the highest stress and elongation values, respectively. Elastomers are the group of polymers that can easily undergo very large, reversible elongations (≤ 500–1000%) at relatively low stresses. This requires that the polymer be completely (or almost completely) amorphous with a low glass transition temperature and low secondary forces so as to obtain high polymer chain mobility. Some degree of crosslinking is needed so that the deformation is rapidly and completely reversible (elastic). The initial modulus of an elastomer should be very low (<100 N/cm^2), but this should increase fairly rapidly with increasing elongation; otherwise, it would have no overall strength and resistance to rupture at low strains. Most elastomers obtain the needed strength via crosslinking and the incorporation of reinforcing inorganic fillers (e.g., carbon black, silica). Some elastomers undergo a small amount of crystallization during elongation, especially at very high elongations, and this acts as an additional strengthening mechanism. The T_m of the crystalline regions must be below or not significantly above the use temperature of the elastomer

TABLE 1-4   Uses of Polymers

Elastomers	Plastics	Fibers
Polyisoprene	Polyethylene	
Polyisobutylene	Polytetrafluoroethylene	
	Poly(methyl methacrylate)	
	Phenol-formaldehyde	
	Urea-formaldehyde	
	Melamine-formaldehyde	
←——— Polystyrene ————→		
←——— Poly(vinyl chloride) ————→		
←——— Polyurethane ————→		
←——— Polysiloxane ————→		
	←——— Polyamide ————→	
	←——— Polyester ————→	
	←——— Cellulosics ————→	
	←——— Polypropylene ————→	
		Polyacrylonitrile

in order that the crystals melt and deformation be reversible when the stress is removed. Polyisoprene (natural rubber) is a typical elastomer—it is amorphous, is easily crosslinked, has a low T_g ($-73°C$), and has a low T_m ($28°C$). Crosslinked (moderately) polyisoprene has a modulus that is initially less than 70 N/cm^2; however, its strength increases to about 1500 N/cm^2 at 400% elongation and about 2000 N/cm^2 at 500% elongation. Its elongation is reversible over the whole elongation range, that is, up to just prior to the rupture point. The extent of crosslinking and the resulting strength and elongation characteristics of an elastomer cover a considerable range depending on the specific end use. The use of an elastomer to produce an automobile tire requires much more crosslinking and reinforcing fillers than does the elastomer used for producing rubber bands. The former application requires a stronger rubber with less tendency to elongate than the latter application. Extensive crosslinking of a rubber converts the polymer to a rigid plastic.

Fibers are polymers that have very high resistance to deformation—they undergo only low elongations (<10–50%) and have very high moduli ($>35,000$ N/cm^2) and tensile strengths ($>35,000$ N/cm^2). A polymer must be very highly crystalline and contain polar chains with strong secondary forces in order to be useful as a fiber. Mechanical stretching is used to impart very high crystallinity to a fiber. The crystalline melting temperature of a fiber must be above 200°C so that it will maintain its physical integrity during the use temperatures encountered in cleaning and ironing. However, T_m should not be excessively high—not higher than 300°C—otherwise, fabrication of the fiber by melt spinning may not be possible. The polymer should be soluble in solvents used for solution spinning of the fiber but not in dry-cleaning solvents. The glass transition temperature should have an intermediate value; too high a T_g interferes with the stretching operation as well as with ironing, while too low a T_g would not allow crease retention in fabrics. Poly(hexamethylene adipamide) is a typical fiber. It is stretched to high crystallinity and its amide groups yield very strong secondary forces due to hydrogen bonding; the result is very high tensile strength (70,000 N/cm^2), very high modulus (500,000 N/cm^2), and low elongation ($<20\%$). The T_m and T_g have optimal values of 265 and 50°C, respectively. [The use of polypropylene as a fiber is an exception to the generalization that polar polymers are required for fiber applications. The polypropylene used as a fiber has a highly stereoregular structure and can be mechanically stretched to yield a highly oriented polymer with the strength characteristics required of a fiber (see Sec. 8-4g-4).]

Plastics comprise a large group of polymers that have a wide range of mechanical behaviors in between those of the elastomers and fibers. There are two types of plastics—flexible plastics and rigid plastics. The flexible plastics possess moderate to high degrees of crystallinity and a wide range of T_m and T_g values. They have moderate to high moduli (15,000–350,000 N/cm^2), tensile strengths (1500–7000 N/cm^2), and ultimate elongations (20–800%). The more typical members of this subgroup have moduli and tensile strengths in the low ends of the indicated ranges with elongations in the high end. Thus polyethylene is a typical flexible plastic with a tensile strength of 2500 N/cm^2, a modulus of 20,000 N/cm^2, and an ultimate elongation of 500%. Other flexible plastics include polypropylene and poly(hexamethylene adipamide). Poly(hexamethylene adipamide) is used as both a fiber and a flexible plastic. It is a plastic when it has moderate crystallinity, while stretching converts it into a fiber. Many flexible plastics undergo large ultimate elongations—some as large as those of elastomers. However, they differ from elastomers in that only a small portion (approximately $<20\%$) of the ultimate elongation is reversible. The elongation of a plastic past the reversible region results in its permanent deformation, that is, the plastic will retain its elongated shape when the stress is removed.

The rigid plastics are quite different from the flexible plastics. The rigid plastics are characterized by high rigidity and high resistance to deformation. They have high moduli (70,000–350,000 N/cm^2) and moderate to high tensile strengths (3000–8500 N/cm^2), but more significantly, they undergo very small elongations (<0.5–3%) before rupturing. The polymers in this category are amorphous polymers with very rigid chains. The high chain rigidity is achieved in some cases by extensive crosslinking, for example, phenol-formaldehyde, urea-formaldehyde, and melamine-formaldehyde polymers. In other polymers the high rigidity is due to bulky side groups on the polymer chains resulting in high T_g values, for example, polystyrene ($T_g = 100°C$) and poly(methyl methacrylate) ($T_g = 105°C$).

REFERENCES

Bassett, D. C., "Principles of Polymer Morphology," Cambridge University Press, Cambridge, 1981.

Billingham, N. C., "Molar Mass Measurements in Polymer Science," Kogan Page, London, 1977.

Billmeyer, F. W., Jr., "Textbook of Polymer Science," Chaps. 11, 12, Wiley-Interscience, New York, 1984.

Bohdanecky, M. and J. Kovar, "Viscosity of Polymer Solutions," Elsevier, Amsterdam, 1982.

Brandrup, J. and E. H. Immergut, Eds., "Polymer Handbook," 2nd, 3rd eds., Wiley-Interscience, New York, 1975, 1989.

Carothers, W. H., *J. Am. Chem. Soc.*, **51**, 2548 (1929).

Collins, E. A., J. Bares, and F. W. Billmeyer, Jr., "Experiments in Polymer Science," Wiley, New York, 1973.

Flory, P. J., "Principles of Polymer Chemistry," Chap. 2, Cornell University Press, Ithaca, N.Y., 1953.

IUPAC, *Macromolecules*, **6**, 149 (1973); *Pure Appl. Chem.*, **48**, 375 (1976).

Janca, J., "Steric Exclusion Liquid Chromatography of Polymers," Marcel Dekker, New York, 1984.

Jenkins, A. D. and K. L. Loening, "Nomenclature," Chap. 2 in "Comprehensive Polymer Science," Vol. 1, Eds., C. Booth and C. Price, Pergamon Press, Oxford, 1989.

Keller, A., *J. Polym. Sci. Symp.*, **51**, 7 (1975), **59**, 1 (1977).

Mandelkern, L., *Acct. Chem. Res.*, **9**, 81 (1976).

Mark, J. E., A. Eisenberg, W. W. Graessley, L. Mandelkern, and J. L. Koenig, "Physical Properties of Polymers, American Chemical Society, Washington, D.C., 1984.

Morawetz, H., "Macromolecules in Solution," Chaps. IV–VI, Wiley-Interscience, New York, 1975.

Nielsen, L. E., "Mechanical Properties of Polymers and Composites," Vols. 1, 2, Marcel Dekker, New York, 1974.

Sharples, A., "Crystallinity," Chap. 4 in "Polymer Science," Vol. 1, A. D. Jenkins, Ed., North-Holland, Amsterdam, 1972.

Slade, P. E., Jr., Ed., "Polymer Molecular Weights," Parts I and II, Marcel Dekker, New York, 1975.

Sperling, L. H., "Introduction to Physical Polymer Science," Wiley, New York, 1986.

Turi, E. A., Ed., "Thermal Characterization of Polymeric Materials," Academic Press, New York, 1981.

Wendlandt, W. W., "Thermal Analysis," 3rd ed., Wiley, New York, 1986.

Williams, D. J., "Polymer Science and Engineering," Prentice-Hall, Englewood Cliffs, N.J., 1971.

Woodward, A. E., Atlas of Polymer Morphology," Hanser Verlag, Munich, 1989.

Wunderlich, B., "Macromolecular Physics," Vols. 1–3, Academic Press, New York, 1973.

Yau, W. W., J. J. Kirkland, and D. D. Bly, "Modern Size Exclusion Liquid Chromatography. Practice of Gel Permeation and Gel Filtration Chromatography," Wiley, New York, 1979.

PROBLEMS

1-1 Show by equations the overall chemical reactions involved in the synthesis of polymers from

  a. $CH_2{=}CH{-}CO_2H$

  b.

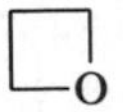

  c. $H_2N{-}(CH_2)_5{-}NH_2 + ClCO{-}(CH_2)_5{-}COCl$

  d. $HO{-}(CH_2)_5{-}CO_2H$

  e.

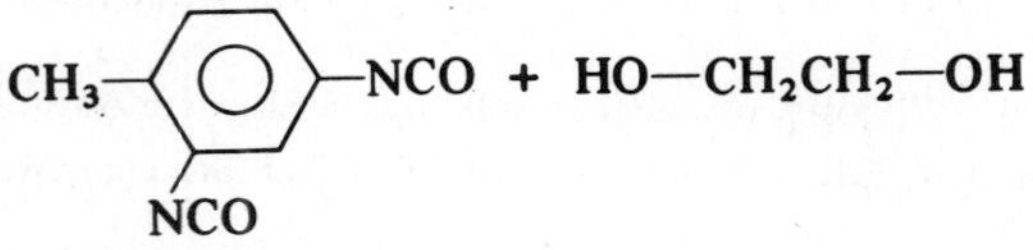

  f. $CH_2{=}CH{-}F$

1-2 What is the structure of the repeating unit in each of the polymers in Question 1? Can any other monomer(s) be used to obtain the same polymer structure for any of these polymers?

1-3 Classify the polymers as to whether they are condensation or addition polymers. Classify the polymerizations as to whether they are step, chain, or ring-opening polymerizations.

1-4 How would you experimentally determine whether the polymerization of an unknown monomer X was proceeding by a step or a chain mechanism?

1-5 Name each of the polymers in Question 1 by the IUPAC system. Indicate alternate names where applicable based on the polymer source, non-IUPAC structure system, or trade names.

1-6 Name each of the following polymers by the IUPAC system

  a.

  b.

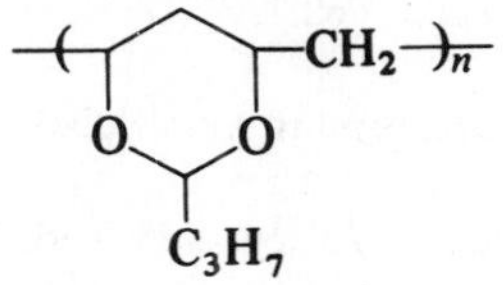

c.

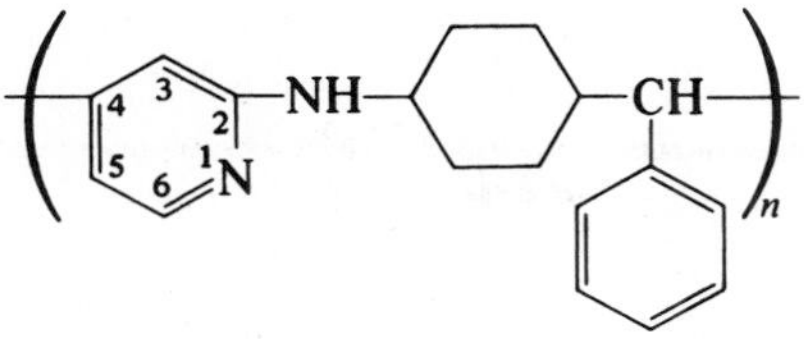

d.

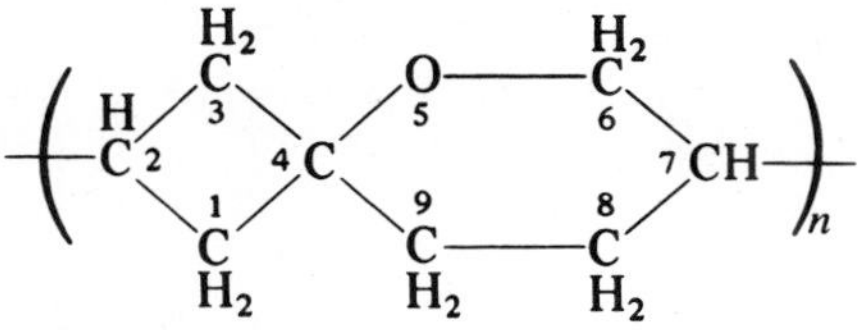

1-7 A sample of polystyrene is composed of a series of fractions of different-sized molecules:

Fraction	Weight Fraction	Molecular Weight
A	0.10	12,000
B	0.19	21,000
C	0.24	35,000
D	0.18	49,000
E	0.11	73,000
F	0.08	102,000
G	0.06	122,000
H	0.04	146,000

Calculate the number-average and weight-average molecular weights of this polymer sample. Draw a molecular weight distribution curve analogous to Fig. 1-4.

1-8 Indicate how the extent of polymer crystallinity is affected by chemical structure.

1-9 Define T_m and T_g and indicate how they are affected by chemical structure.

1-10 Describe the differences in the properties and uses of flexible plastics, rigid plastics, fibers, and elastomers. What types of chemical structures are typical of each?

CHAPTER 2

STEP POLYMERIZATION

Many of the common condensation polymers are listed in Table 1-1. In all instances the polymerization reactions shown are those proceeding by the step polymerization mechanism. This chapter will consider the characteristics of step polymerization in detail. The synthesis of condensation polymers by ring-opening polymerization will be subsequently treated in Chap. 7. A number of different chemical reactions may be used to synthesize polymeric materials by step polymerization. These include esterification, amidation, the formation of urethanes, aromatic substitution, and others. Polymerization usually proceeds by the reactions between two different functional groups, for example, hydroxyl and carboxyl groups, or isocyanate and hydroxyl groups.

All step polymerizations fall into two groups depending on the type of monomer(s) employed. The first involves two different bifunctional and/or polyfunctional monomers in which each monomer possesses only one type of functional group. (A *bifunctional monomer* is a monomer containing two functional groups per molecule. A *polyfunctional monomer* is one with three or more functional groups per molecule.) The second involves a single monomer containing both types of functional groups. The synthesis of polyamides illustrates both groups of polymerization reactions. Thus polyamides can be obtained from the reaction of diamines with diacids,

$$n\text{H}_2\text{N—R—NH}_2 + n\text{HO}_2\text{C—R}'\text{—CO}_2\text{H} \rightarrow$$

$$\text{H—(—NH—R—NHCO—R}'\text{—CO—)}_n\text{—OH} + (2n - 1)\text{H}_2\text{O} \qquad (2\text{-}1a)$$

or from the reaction of amino acids with themselves,

$$n\text{H}_2\text{N—R—CO}_2\text{H} \rightarrow \text{H—(—NH—R—CO—)}_n\text{—OH} + (n - 1)\text{H}_2\text{O} \qquad (2\text{-}1b)$$

The two groups of reactions can be represented in a general manner by the equations

$$n\text{A—A} + n\text{B—B} \rightarrow \text{—(—A—AB—B—)}_n \qquad (2\text{-}2)$$

$$nA—B \rightarrow —(\!-A—B\!-)_n— \tag{2-3}$$

where A and B are the two different types of functional groups. The characteristics of these two polymerization reactions are very similar. The successful synthesis of high polymers (i.e., polymer of sufficiently high molecular weight to be useful from the practical viewpoint) using any step polymerization reaction generally is more difficult than the corresponding small molecule reaction, since high polymer can only be achieved at very high conversions (>98–99%). A conversion of 90%, which would be considered excellent for the synthesis of ethyl acetate or methyl benzamide, is a disaster for the synthesis of the corresponding polyester or polyamide. The need for very high conversions to synthesize high polymer places several stringent requirements on any reaction to be used for polymerization—a favorable equilibrium and the absence of cyclization and other side reactions. These stringent requirements are met by a relatively small fraction of the reactions familiar to and used by chemists to synthesize small molecules.

2-1   REACTIVITY OF FUNCTIONAL GROUPS

2-1a   Basis for Analysis of Polymerization Kinetics

The kinetics of polymerization are of prime interest from two viewpoints. The practical synthesis of high polymers requires a knowledge of the kinetics of the polymerization reaction. From the theoretical viewpoint the significant differences between step and chain polymerizations reside in large part in their respective kinetic features.

Step polymerization proceeds by a relatively slow increase in molecular weight of the polymer. Consider the synthesis of a polyester from a diol and a diacid. The first step is the reaction of the diol and diacid monomers to form dimer,

$$HO—R—OH + HO_2C—R'—CO_2H \rightarrow$$
$$HO—R—OCO—R'—CO_2H + H_2O \tag{2-4}$$

The dimer then forms trimer by reaction with diol monomer,

$$HO—R—OCO—R'—CO_2H + HO—R—OH \rightarrow$$
$$HO—R—OCO—R'—COO—R—OH + H_2O \tag{2-5}$$

and also with diacid monomer,

$$HO—R—OCO—R'—CO_2H + HO_2C—R'—CO_2H \rightarrow$$
$$HO_2C—R'—COO—R—OCO—R'—CO_2H + H_2O \tag{2-6}$$

Dimer reacts with itself to form tetramer,

$$2HO—R—OCO—R'—CO_2H \rightarrow$$
$$HO—R—OCO—R'—COO—R—OCO—R'—CO_2H + H_2O \tag{2-7}$$

The tetramer and trimer proceed to react with themselves, with each other, and with monomer and dimer. The polymerization proceeds in this stepwise manner with the molecular weight of the polymer continuously increasing with time (conversion). Step polymerizations are characterized by the disappearance of monomer very early in the reaction far before the production of any polymer of sufficiently high molecular weight (approximately >5000–10,000) to be of practical utility. Thus for most step polymerizations there is less than 1% of the original monomer remaining at a point where the average polymer chain contains only ~10 monomer units. As will be seen in Chap. 3, the situation is quite different in the case of chain polymerization.

The rate of a step polymerization is, therefore, the sum of the rates of reaction between molecules of various sizes, that is, the sum of the rates for reactions such as

$$
\begin{aligned}
&\text{monomer} + \text{monomer} \rightarrow \text{dimer} \\
&\text{dimer} + \text{monomer} \rightarrow \text{trimer} \\
&\text{dimer} + \text{dimer} \rightarrow \text{tetramer} \\
&\text{trimer} + \text{monomer} \rightarrow \text{tetramer} \\
&\text{trimer} + \text{dimer} \rightarrow \text{pentamer} \\
&\text{trimer} + \text{trimer} \rightarrow \text{hexamer} \\
&\text{tetramer} + \text{monomer} \rightarrow \text{pentamer} \\
&\text{tetramer} + \text{dimer} \rightarrow \text{hexamer} \\
&\text{tetramer} + \text{trimer} \rightarrow \text{heptamer} \\
&\text{tetramer} + \text{tetramer} \rightarrow \text{octamer} \\
&\text{pentamer} + \text{trimer} \rightarrow \text{octamer} \\
&\text{pentamer} + \text{tetramer} \rightarrow \text{nonamer} \\
&\text{etc.} \\
&\text{etc.}
\end{aligned}
\tag{2-8a}
$$

which can be expressed as the general reaction

$$
n\text{-mer} + m\text{-mer} \rightarrow (n + m)\text{-mer} \tag{2-8b}
$$

The reaction mixture at any instance consists of various-sized diol, diacid, and hydroxy acid molecules. Any HO-containing molecule can react with any COOH-containing molecule. This is a general characteristic of step polymerization.

The kinetics of such a situation with innumerable separate reactions would normally be difficult to analyze. However, kinetic analysis is greatly simplified if one assumes that the reactivities of both functional groups of a bifunctional monomer (e.g., both hydroxyls of a diol) are the same, the reactivity of one functional group of a bifunctional reactant is the same irrespective of whether the other functional group has reacted, and the reactivity of a functional group is independent of the size of the molecule to which it is attached (i.e., independent of the values of n and m). These simplifying assumptions, often referred to as the concept of *equal reactivity of functional groups*, make the kinetics of step polymerization identical to those for the analogous small molecule reaction. As will be seen shortly, the kinetics of a polyesterification, for example, become essentially the same as that for the esterification of acetic acid with ethanol.

2-1b Experimental Evidence

These simplifying assumptions are justified on the basis that many step polymerizations have reaction rate constants that are independent of the reaction time or polymer molecular weight. It is, however, useful to examine in more detail the experimental and theoretical justifications for these assumptions. Studies of the reactions of certain nonpolymeric molecules are especially useful for understanding polymerization kinetics [Flory, 1953]. The independence of the reactivity of a functional group on molecular size can be observed from the reaction rates in a homologous series of compounds differing from each other only in molecular weight. Consider, for example, the rate constant data in the first column in Table 2-1 for the esterification of a series of homologous carboxylic acids [Bhide and Sudborough, 1925]:

$$H(CH_2)_xCO_2H + C_2H_5OH \xrightarrow{HCl} H(CH_2)_xCO_2C_2H_5 + H_2O \qquad (2\text{-}9)$$

It is evident that, although there is a decrease in reactivity with increased molecular size, the effect is only significant at a very small size. The reaction rate constant very quickly reaches a limiting value at $x = 3$, which remains constant and independent of molecular size. Analogous results are found for the polyesterification of sebacoyl chloride with α,ω-alkane diols [Ueberreiter and Engel, 1977]:

$$HO(CH_2)_xOH + Cl\text{—}OC(CH_2)_8CO\text{—}Cl \xrightarrow{-HCl}$$

$$\text{—}\!\!\left[\text{—}O(CH_2)_xOCO(CH_2)_8CO\text{—}\right]_n\!\!\text{—} \qquad (2\text{-}10)$$

The rate constant for esterification is independent of x for the compounds studied (Table 2-2). Examples of similar behavior can be found for other reactions [Rand et al., 1965]. The results for sebacoyl chloride offer direct evidence of the concept of functional group reactivity being independent of molecular size, since the rate constant is independent of n as well as x.

The independence of functional group reactivity of molecular size is contrary to

TABLE 2-1   Rate Constants for Esterification (25°C) in Homologous Compounds[a,b]

Molecular Size (x)	$k \times 10^4$ for $H(CH_2)_xCO_2H$	$k \times 10^4$ for $(CH_2)_x(CO_2H)_2$
1	22.1	
2	15.3	6.0
3	7.5	8.7
4	7.5	8.4
5	7.4	7.8
6		7.3
8	7.5	
9	7.4	
11	7.6	
13	7.5	
15	7.7	
17	7.7	

[a]Rate constants are in units of liters/mole-sec.
[b]Data from Bhide and Sudborough [1925].

TABLE 2-2   Rate Constants for
Polyesterification (26.9°C) of Sebacoyl Chloride
with α,ω-Alkane Diols in Dioxane[a,b]

Molecular Size (x)	$k \times 10^3$ for HO(CH$_2$)$_x$OH
5	0.60
6	0.63
7	0.65
8	0.62
9	0.65
10	0.62

[a]Rate constants are in units of liters/mole-sec.
[b]Data from Ueberreiter and Engel [1977].

the general impression that was widely held in the early years of polymer science. There was a general misconception of decreased reactivity with increased molecular size. This was due in large part to the fact that reactions with the larger-sized species were often attempted without adjusting concentrations so as to have equivalent concentrations of the reacting functional groups. In many instances the low or difficult solubility of the higher molecular weight reactants in a homologous series was responsible for the observed, apparent low reactivity. These effects are still pitfalls to be avoided.

2-1c Theoretical Considerations

There is theoretical justification for the independence of the reactivity of a functional group of molecular size [Rabinowitch, 1937]. The common misconception regarding the alleged low reactivity of groups attached to large molecules comes from the lower diffusion rates of the latter. However, the observed reactivity of a functional group is dependent on the collision frequency of the group, not on the diffusion rate of the whole molecule. The collision frequency is the number of collisions one functional group makes with other functional groups per unit of time. A terminal functional group attached to a growing polymer has a much greater mobility than would be expected from the mobility of the polymer molecule as a whole. The functional group has appreciable mobility due to the conformational rearrangements that occur in nearby segments of the polymer chain. The collision rate of such a functional group with neighboring groups will be about the same as for small molecules.

The lowered mobility of the polymer molecule as a whole alters the time distribution of collisions. A lower diffusion rate means that any two functional groups will undergo more total collisions before diffusing apart. Thus in any particular time interval (which is long compared to the interval required for diffusion of a group from one collision partner to the next) the number of partners with which a functional group undergoes collisions is less for a group attached to a polymer chain than for one attached to a small molecule. But, most significantly, their overall collision frequencies are the same in the two cases. If the diffusion of a functional group from one collision partner to the next were too slow to maintain the concentration of the reactive pairs of functional groups at equilibrium, one would observe decreased reactivity for the functional group. The opposite is the case, since reaction between two functional groups in a step polymerization occurs only about once in every 10^{13} collisions [Flory, 1953]. During the time interval required for this many collisions, sufficient diffusion of the molecule and of the functional group occurs to maintain the equilibrium concentration of collision pairs of functional groups. The net result of these considerations is that the reactivity of a functional group will be independent of the size of the molecule to which it is attached. Exceptions to this occur when the reactivities of the groups are very high and/or the molecular weights of the polymer are very high. The polymerization becomes diffusion-controlled in these cases because mobility is too low to allow maintenance of the equilibrium concentration of reactive pairs and of their collision frequencies.

2-1d   Equivalence of Groups in Bifunctional Reactants

The equivalence of the reactivities of the two functional groups in bifunctional reactants has also been demonstrated in many systems. The second column in Table 2-1 shows data for the esterification of a homologous series of dibasic acids:

$$HO_2C—(CH_2)_x—CO_2H + 2C_2H_5OH \rightarrow$$
$$C_2H_5O_2C—(CH_2)_x—CO_2C_2H_5 + 2H_2O \qquad (2\text{-}11)$$

The rate constants tabulated are the averages of those for the two carboxyl groups in each dibasic acid. The rate constants for the two carboxyls were the same within experimental error. Further, a very important observation is that the reactivity of one of the functional groups is not dependent on whether the other has reacted. As in the case of the monocarboxylic acids, the reactivity of the carboxyl group quickly reaches a limiting value. Similar results were observed for the polyesterification of sebacoyl chloride (Table 2-2).

It should not be concluded that the reactivity of a functional group is not altered by the presence of a second group. The data show that at small values of x up to 4 or 5, there is an effect of the second group. In most instances the effect disappears fairly rapidly with increasing molecular size as the two functional groups become isolated from each other. Thus the rate constant for the esterification of the dibasic acid with $x = 6$ is the same as that for the monobasic acids. It might appear at first glance that polymerizations involving bifunctional reagents with low x values would present a problem. However, the difference in reactivities of the functional groups of a bifunctional reagent compared to the functional group in a monofunctional reagent is not an important consideration in handling the kinetics of step polymerization. The most significant considerations are that the reactivity of a functional group in a bifunctional monomer is not altered by the reaction of the other group, and that the reactivities of the two functional groups are the same and do not change during the course of the polymerization.

2-2   KINETICS OF STEP POLYMERIZATION

Consider the polyesterification of a diacid and a diol to illustrate the general form of the kinetics of a typical step polymerization. Simple esterification is a well-known acid-catalyzed reaction and polyesterification, no doubt, follows the same course [Otton and Ratton, 1988; Vancso-Szmercsanyi and Makay-Bodi, 1969]. The reaction involves protonation of the carboxylic acid,

$$
\underset{\text{O}}{\overset{\text{O}}{\underset{\|}{\sim\sim\sim\text{C}}}}\text{—OH} + \text{HA} \underset{k_2}{\overset{k_1}{\rightleftharpoons}} \sim\sim\sim\underset{+}{\overset{\text{OH}}{\underset{|}{\text{C}}}}\text{—OH (A}^-) \qquad (2\text{-}12)
$$

I

followed by reaction of the protonated species **I** with the alcohol to yield the ester

$$
\sim\sim\sim\underset{+}{\overset{\text{OH}}{\underset{|}{\text{C}}}}\text{—OH} + \sim\sim\text{OH} \underset{k_4}{\overset{k_3}{\rightleftharpoons}} \sim\sim\sim\overset{\text{OH}}{\underset{\underset{+}{\overset{|}{\sim\sim\text{OH}}}}{\underset{|}{\text{C}}}}\text{—OH} \qquad (2\text{-}13)
$$

(A^-) (A^-)

II

$$
\sim\sim\sim\overset{\text{OH}}{\underset{\underset{+}{\overset{|}{\text{OH}}}}{\underset{|}{\text{C}}}}\text{—OH} \overset{k_5}{\rightleftharpoons} \sim\sim\sim\overset{\text{O}}{\overset{\|}{\text{C}}}\text{—O}\sim\sim + \text{H}_2\text{O} + \text{HA} \qquad (2\text{-}14)
$$

(A^-)

In the above equations $\sim\sim\sim$ and $\sim\sim$ are used to indicate all acid or alcohol species in the reaction mixture (i.e., monomer, dimer, trimer, . . . , n-mer). Species **I** and **II** are shown in the form of their associated ion pairs since polymerization often takes places in organic media of low polarity. (A^-) is the negative counterion derived from the acid HA. Polyesterifications, like many other step polymerizations, are equilibrium reactions. However, from the practical viewpoint of obtaining high yields of high-molecular-weight product such polymerizations are run in a manner so as to continuously shift the equilibrium in the direction of the polymer. In the case of a polyesterification this is easily accomplished by removal of the water that is a by-product of the reaction species **II** (Eq. 2-14). Under these conditions the kinetics of polymerization can be handled by considering the reactions in Eqs. 2-13 and 2-14 to be irreversible. (A consideration of the kinetics of a reversible step polymerization is given in Sec. 2-4c.)

   The rate of a step polymerization is conveniently expressed in terms of the concentrations of the reacting functional groups. Thus the polyesterification can be experimentally followed by titrating for the unreacted carboxyl groups with a base. The *rate of polymerization* R_p can then be expressed as the *rate of disappearance of carboxyl groups* $-d[\text{COOH}]/dt$. For the usual polyesterification, the polymerization rate is synonomous with the rate of formation of species **II**; that is, k_4 is vanishingly small (since the reaction is run under conditions that drive Eqs. 2-13 and 2-14 to the right), and k_1, k_2, and k_5 are large compared to k_3. An expression for the reaction rate can be obtained following the general procedures for handling a reaction scheme with the characteristics described [Moore and Pearson, 1981]. The rate of polymerization is given by

$$
R = \frac{-d[\text{COOH}]}{dt} = k_3[\overset{+}{\text{C}}(\text{OH})_2][\text{OH}] \qquad (2\text{-}15)
$$

where [COOH], [OH], and [$\overset{+}{C}(OH)_2$] represent the concentrations of carboxyl, hy-droxyl, and protonated carboxyl (**I**) groups, respectively. The concentration terms are in units of moles of the particular functional group per liter of solution.

Equation 2-15 is inconvenient in that the concentration of protonated carboxylic groups is not easily determined experimentally. One can obtain a more convenient expression for R_p by substituting for $\overset{+}{C}(OH)_2$ from the equilibrium expression

$$K = \frac{k_1}{k_2} = \frac{[\overset{+}{C}(OH)_2]}{[COOH][HA]} \tag{2-16}$$

for the protonation reaction (Eq. 2-12). Combination of Eqs. 2-15 and 2-16 yields

$$\frac{-d[COOH]}{dt} = k_3K[COOH][OH][HA] \tag{2-17}$$

Two quite different kinetic situations arise from Eq. 2-17 depending on the identity of HA, that is, on whether a strong acid such as sulfuric acid or p-toluenesulfonic acid is added as an exernal catalyst.

2-2a Self-Catalyzed Polymerization

In the absence of an externally added strong acid the diacid monomer acts as its own catalyst for the esterification reaction. For this case [HA] is replaced by [COOH] and Eq. 2-17 can be written in the usual form [Flory, 1953]

$$\frac{-d[COOH]}{dt} = k[COOH]^2[OH] \tag{2-18}$$

where K and k_3 have been combined into the experimentally determined rate constant k. Equation 2-18 shows the important characteristic of the self-catalyzed polymeri-zation—the reaction is third-order overall with a second-order dependence on the carboxyl concentration. The second-order dependence on the carboxyl concentration is comprised of two first-order dependencies—one for the carboxyl as the reactant and one as the catalyst.

For most polymerizations the concentrations of the two functional groups are very nearly stoichiometric, and Eq. 2-18 can be written as

$$\frac{-d[M]}{dt} = k[M]^3 \tag{2-19a}$$

or

$$\frac{-d[M]}{[M]^3} = k\,dt \tag{2-19b}$$

where [M] is the concentration of hydroxyl groups or carboxyl groups. Integration of Eq. 2-19b yields

$$2kt = \frac{1}{[M]^2} - \frac{1}{[M]_0^2} \tag{2-20}$$

where $[M]_0$ is the initial (at $t = 0$) concentration of hydroxyl or carboxyl groups. It is convenient at this point to write Eq. 2-20 in terms of the *extent* or *fraction of reaction* p defined as the fraction of the hydroxyl or carboxyl functional groups that has reacted at time t. p is also referred to as the *extent* or *fraction of conversion*. (The value of p is calculated from a determination of the amount of unreacted carboxyl groups.) The concentration $[M]$ at time t of either hydroxyl or carboxyl groups is then given by

$$[M] = [M]_0 - [M]_0 p = [M]_0(1 - p) \tag{2-21}$$

Combination of Eqs. 2-20 and 2-21 yields

$$\frac{1}{(1 - p)^2} = 2[M]_0^2 kt + 1 \tag{2-22}$$

2-2a-1   Experimental Observations

Equation 2-22 indicates that a plot of $1/(1 - p)^2$ versus t should be linear. This behavior has been generally observed in polyesterifications. Figure 2-1 shows the results for the polymerization of diethylene glycol, $(HOCH_2CH_2)_2O$, and adipic acid [Flory, 1939; Solomon, 1967, 1972]. The results are typical of the behavior observed for polyesterifications. Although various authors agree on the experimental data, there has been considerable disagreement on the interpretation of the results. At first glance the plot does not appear to exactly follow the relationship. The experimental points deviate from the third-order plot in the initial region below 80% conversion and in the later stages above 93% conversion. These deviations led various workers [Amass, 1979; Fradet and Marechal, 1982a, 1982b; Solomon, 1967, 1972] to suggest alternate kinetic expressions based on either 1- or $\frac{3}{2}$-order dependencies of the reaction rate on the carboxyl concentration, that is, 2- and $2\frac{1}{2}$-order dependencies according to

$$\frac{-d[COOH]}{dt} = k[COOH][OH] \tag{2-23a}$$

and

$$\frac{-d[COOH]}{dt} = k[COOH]^{3/2}[OH] \tag{2-23b}$$

However, a critical evaluation shows that both kinetic possibilities leave much to be desired. A plot (Fig. 2-2) of the experimental rate data according to Eq. 2-23a fits the experimental data well only in the region between 50 and 86% conversion with an excessively poor fit above 86% conversion. On the other hand, a plot (Fig. 2-3) according to Eq. 2-23b fits reasonably well up to about 80% conversion but deviates badly above that point. Neither of the two alternate kinetic plots comes close to being as useful as the third-order plot in Fig. 2-1. The third-order plot fits the experimental data much better than does either of the others at the higher conversions. The fit of the data to the third-order plot is reasonably good over a much greater range of the higher conversion region. The region of high conversion is of prime importance, since, as will be shown in Sec. 2-2a-3, high-molecular-weight polymer is only obtained at

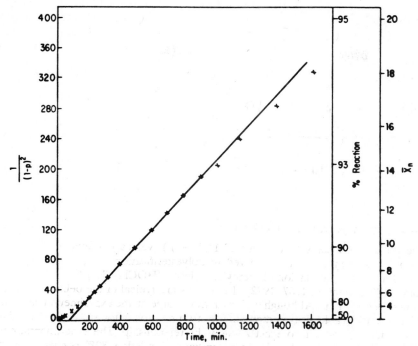

Fig. 2-1 Third-order plot of the self-catalyzed polyesterification of adipic acid with diethylene glycol at 166°C. After Solomon [1967] (by permission of Marcel Dekker, Inc., New York) from the data of Flory [1939] (by permission of American Chemical Society, Washington, D.C.).

high conversions. From the practical viewpoint, the low conversion region of the kinetic plot is of little significance.

2-2a-2   Reasons for Nonlinearity in Third-Order Plot

The failure to fit the data over the complete conversion range from 0 to 100% to a third-order plot has sometimes been ascribed to failure of the assumption of equal functional group reactivity, but this is an invalid conclusion. The nonlinearities are not inherent characteristics of the polymerization reaction. Similar nonlinearities have been observed for nonpolymerization esterification reactions such as esterifications of lauryl alcohol with lauric or adipic acid and diethylene glycol with caproic acid [Flory, 1939; Fradet and Marechal, 1982b].

2-2a-2-a   Low Conversion Region. Flory attributed the nonlinearity in the low conversion region to the large changes that take place in the reaction medium. The solvent for the reaction changes from an initial mixture of carboxylic acid and alcohol to an ester. There is a large decrease in the polarity of the reaction system as the polar alcohol and acid groups are replaced by the less polar ester groups with the simultaneous removal of water. This effect may exert itself in two ways—a change in the reaction rate constant or order of reaction. Either possibility is compatible with the experimental observations. The low conversion region of the third-order plot in Fig.

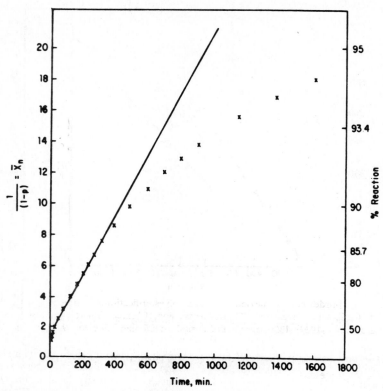

Fig. 2-2 Second-order plot of the self-catalyzed polyesterification of adipic acid with diethylene glycol at 166°C. After Solomon [1967] (by permission of Marcel Dekker, Inc., New York) from the data of Flory [1939] (by permission of American Chemical Society, Washington, D.C.).

2-1 is approximately linear but with a smaller slope (rate constant) than in the region of higher conversion. The direction of this change corresponds to that expected for a reaction involving a charged substrate (protonated carboxyl) and neutral nucleophile (alcohol). The reaction rate constant increases with decreased solvent polarity since the transition state is less charged than the reactants [Lowry and Richardson, 1987]. It has also been suggested that extensive association of both the diol and diacid reactants in the low conversion region lowers the reaction rate by effectively decreasing the concentrations of the reactive species, that is, free, unassociated OH and COOH groups [Ueberreiter and Hager, 1979].

The low conversion region fits an overall $2\frac{1}{2}$ reaction order (Fig. 2-3) better than it does 3-order. A change from $2\frac{1}{2}$- to 3-order as the reaction medium becomes less polar is compatible with a change from specific acid (i.e., H^+) catalysis to general-acid catalysis. The proton concentration is relatively high and $[H^+]$ is a more effective catalyst than the unionized carboxylic acid in the polar, low conversion region. HA in Eqs. 2-12, 216, and 2-17 is replaced by H^+. The reaction rate becomes first-order each in carboxyl and hydroxyl groups and H^+. The proton concentration is given by

$$[H^+] = (K_{HA}[HA])^{1/2} \tag{2-24}$$

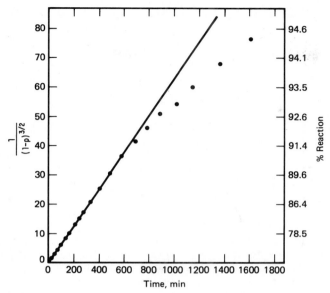

Fig. 2-3 A 2½-order plot of the self-catalyzed polyesterification of adipic acid with diethylene glycol at 166°C. After Solomon [1967] (by permission of Marcel Dekker, Inc., New York) from the data of Flory [1939] (by permission of American Chemical Society, Washington, D.C.).

where K_{HA} is the ionization constant for the carboxylic acid. Combination of Eqs. 2-17 and 2-24 yields

$$\frac{-d[\text{COOH}]}{dt} = k_3 K K_{HA}^{1/2}[\text{COOH}]^{3/2}[\text{OH}] \tag{2-25}$$

for the polymerization rate. The reaction medium becomes less polar after the low conversion (>80%) region and the major catalyst in the reaction system is the un-ionized carboxylic acid—the reaction is second-order in carboxylic acid and third-order overall (Eqs. 2-18 and 2-19).

Another problem is the very high concentrations of reactants present in the low conversion region. The correct derivation of any rate expression such as Eqs. 2-20 and 2-22 requires the use of activities instead of concentrations. The use of concentrations instead of activities assumes a direct proportionality between concentration and activity. This assumption is usually valid at the dilute and moderate concentrations where kinetic studies on small molecules are typically performed. However, the assumption often fails at high concentrations and those are the reaction conditions for the typical step polymerization that proceeds with neat reactants. A related problem is that neither concentration nor activity may be the appropriate measure of the ability of the reaction system to donate a proton to the carboxyl group. The acidity function h_0 is often the more appropriate measure of acidity for nonaqueous systems or systems containing high acid concentrations [Lowry and Richardson, 1987]. Unfortunately, the appropriate h_0 values are not available for polymerization systems.

Yet another possibility for the nonlinearity in the low conversion region is the decrease in the volume of the reaction mixture with conversion due to loss of one of the products of reaction (water in the case of esterification). This presents no problem if concentration is plotted against time as in Eq. 2-20. However, a plot of $1/(1 - p)^2$ against time (Eq. 2-22) has an inherent error since the formulation of Eq. 2-21 assumes a constant reaction volume (and mass) [Szabo-Rethy, 1971]. Elias [1985] derived the kinetics of step polymerization with correction for loss of water, but the results have not been tested. It is unclear whether this effect alone can account for the nonlinearity in the low conversion region of esterification and polyesterification.

2-2a-2-b High-Conversion Region. The nonlinearity observed in the third-order plot in the final stages of the polyesterification (Fig. 2-1) is probably not due to any of the above reasons, since the reaction system is fairly dilute and of relatively low polarity. Further, it would be difficult to conjecture that the factors responsible for nonlinearity at low conversions are present at high conversion but absent in between. It is more likely that other factors are responsible for the nonlinear region above 93% conversion.

Polyesterifications, like many step polymerizations, are carried out at moderate to high temperatures not only to achieve fast reaction rates but also to aid in removal of the small molecule by-product (often H_2O). The polymerization is an equilibrium reaction and the equilibrium must be displaced to the right (toward the polymer) to achieve high conversions (synonymous with high molecular weights). Partial vacuum (often coupled with purging of the reaction system by nitrogen gas) is usually also employed to drive the system toward high molecular weight. Under these conditions small amounts of one or the other or both reactants may be lost by degradation or volatilization. In the case of polyesterification, small degradative losses might arise from dehydration of the diol, decarboxylation of the diacid, or other side reactions (Sec. 2-8d). Although such losses may not be important initially, they can become very significant during the later stages of reaction. Thus a loss of only 0.3% of one reactant can lead to an error of almost 5% in the concentration of that reactant at 93% conversion. Kinetic studies on esterification and polyesterification have been performed under conditions which minimized the loss of reactants by volatilization or side reactions [Hamann et al., 1968]. An ester or polyester was synthesized in a first stage, purified, and then used as the solvent for a second-stage esterification or poly-esterification. The initial reactant concentrations of the second-stage reactions cor-responded to 80% conversion relative to the situation if the first-stage reaction had not been stopped but with an important difference—the reactant concentrations are accurately known. The second-stage reaction showed third-order behavior up to past 98–99% conversion (higher conversions were not studied). This compares with the loss of linearity in the third-order plot at about 93% conversion if the first-stage reactions were carried out without interruption.

Another possible reason for the observed nonlinearity is an increase in the rate of the reverse reaction. The polyesterification reaction (as well as many other step poly-merizations) is an equilibrium reaction. It often becomes progressively more difficult to displace the equilibrium to the right (toward the polymer) as the conversion in-creases. This is due in large part to the greatly increased viscosity of the reaction medium at high conversions. The viscosity in the adipic acid–diethylene glycol polym-erization increases from 0.015 to 0.30 poise during the course of the reaction [Flory, 1939]. This large viscosity increase decreases the efficiency of water removal and may lead to the observed decrease in the reaction rate with increasing conversion. High viscosity in the high conversion region may also lead to failure of the assumption of equal reactivity of functional groups—specifically to a decrease in functional group reactivity at very large molecular size if there is too large a decrease in molecular mobility.

2-2a-3 Molecular Weight of Polymer

The molecular weight of a polymer is of prime concern from the practical viewpoint, for unless a polymer is of sufficiently high molecular weight (approximately >5000–10,000) it will not have the desirable strength characteristics. It is therefore important to consider the change in polymer molecular weight with reaction time. For the case at hand of stoichiometric amounts of diol and diacid the number of unreacted carboxyl groups N is equal to the total number of molecules present in the system at some time t. This is so because each molecule larger than a monomer will on the average have a hydroxyl at one end and a carboxyl at the other end, while each diacid monomer molecule contains two carboxyls and each diol monomer contains no carboxyls.

The residue from each diol and each diacid (separately, not together) in the polymer chain is termed a *structural unit* (or a *monomer unit*). The *repeating unit* of the chain consists of two structural units, one each of the diol and diacid. The total number of structural units in any particular system equals the total number of bifunctional monomers initially present. The *number-average degree of polymerization* $\overline{X}_n$ is defined as the average number of structural units per polymer chain. (The symbols $\overline{P}$ and $\overline{DP}$ are also employed to signify the number-average degree of polymerization.) $\overline{X}_n$ is simply given as the total number of monomer molecules initially present divided by the total number of molecules present at time t,

$$\overline{X}_n = \frac{N_0}{N} = \frac{[M]_0}{[M]} \tag{2-26}$$

Combining Eqs. 2-21 and 2-26, one obtains

$$\overline{X}_n = \frac{1}{(1 - p)} \tag{2-27}$$

This equation relating the degree of polymerization to the extent of reaction was originally set forth by Carothers [1936] and is sometimes referred to as the *Carothers equation*.

The *number-average molecular weight* $\overline{M}_n$, defined as the total weight of a polymer sample divided by the total number of moles in it (Eqs. 1-15 and 1-16), is given by

$$\overline{M}_n = M_o \overline{X}_n + M_{eg} = \frac{M_o}{(1 - p)} + M_{eg} \tag{2-28}$$

where M_o is the mean of the molecular weights of the two structural units, and M_{eg} is the molecular weight of the end groups. For the polyesterification of adipic acid, $HO_2C(CH_2)_4CO_2H$, and ethylene glycol, $HOCH_2CH_2OH$, the repeating unit is

$$-OCH_2CH_2OCO(CH_2)_4CO-$$

III

and one half of its weight or 86 is the value of M_o. The end groups are H— and —OH and M_{eg} is 18. For even a modest molecular weight polymer the contribution of M_{eg} to $\overline{M}_n$ is negligibly small, and Eq. 2-28 becomes

$$\overline{M}_n = M_o\overline{X}_n = \frac{M_o}{(1 - p)} \tag{2-29}$$

Combination of Eqs. 2-22 and 2-27 yields

$$\overline{X}_n^2 = 1 + 2[M]_0^2 kt \tag{2-30}$$

Since the reaction time and degree of polymerization appear as the first and second powers, respectively, the polymer molecular weight will increase very slowly with reaction time except in the early stages of the reaction. This means that very long reaction times are needed to obtain a high-molecular-weight polymer product. The right-hand ordinate of Fig. 2-1 shows the variation of $\overline{X}_n$ with t. The slow increase of the molecular weight of the polymer with time is clearly apparent. The rate of increase of $\overline{X}_n$ with time decreases as the reaction proceeds. The production of high polymers requires reaction times that are too long from the practical viewpoint.

2-2b External Catalysis of Polymerization

The slow increase in molecular weight was mistakenly thought originally to be due to the low reactivity of functional groups attached to large molecules. It is, however, simply a consequence of the third-order kinetics of the direct polyesterification reaction. The realization of this kinetic situation led to the achievement of high-molecular-weight products in reasonable reaction times by employing small amounts of externally added strong acids (such as sulfuric acid or p-toluenesulfonic acid) as catalysts. Under these conditions, [HA] in Eq. 2-17 is the concentration of the catalyst. Since this remains constant throughout the course of the polymerization, Eq. 2-17 can be written as

$$\frac{-d[M]}{dt} = k'[M]^2 \tag{2-31}$$

where the various constant terms in Eq. 2-17 have been collected into the experimentally determinable rate constant k'. Equation 2-31 applies to reactions between stoichiometric concentrations of the diol and diacid.

Integration of Eq. 2-31 yields

$$k't = \frac{1}{[M]} - \frac{1}{[M]_0} \tag{2-32}$$

Combining Eqs. 2-32 and 2-21 yields the dependence of the degree of polymerization on reaction time as

$$[M]_0 k't = \frac{1}{(1 - p)} - 1 \tag{2-33a}$$

or

$$\overline{X}_n = 1 + [M]_0 k't \qquad (2\text{-}33b)$$

Data for the polymerization of diethylene glycol with adipic acid catalyzed by p-toluenesulfonic acid are shown in Fig. 2-4. The plot follows Eq. 2-33 with the degree of polymerization increasing linearly with reaction time. The much greater rate of increase of $\overline{X}_n$ with reaction time in the catalyzed polyesterification (Fig. 2-4) relative to the uncatalyzed reaction (Fig. 2-1) is a general and most significant phenomenon. The polyesterification becomes a much more economically feasible reaction when it is catalyzed by an external acid. The self-catalyzed polymerization is not a useful reaction from the practical viewpoint of producing high polymer in reasonable reaction times.

The nonlinearity in the initial region of Fig. 2-4 is, like that in Fig. 2.1, a characteristic of esterifications in general and not of the polymerization reaction. The

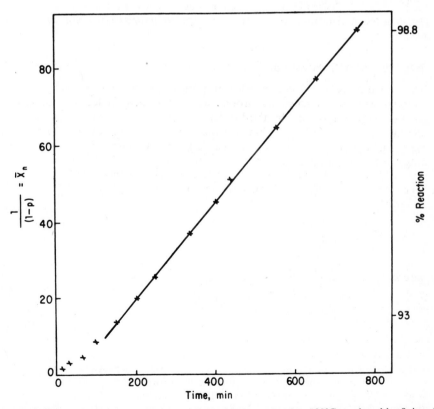

Fig. 2-4 Polyesterification of adipic acid with diethylene glycol at 109°C catalyzed by 0.4 mole % p-toluenesulfonic acid. After Solomon [1967] (by permission of Marcel Dekker, Inc., New York) from the data of Flory [1939] (by permission of American Chemical Society, Washington, D.C.).

general linearity of the plot in the higher conversion region is a strong confirmation of the concept of functional group reactivity independent of molecular size. Figure 2-4 shows that the polyesterification continues its second-order behavior at least up to a degree of polymerization of 90 corresponding to a molecular weight of ~10,000. There is no change in the reactivities of the hydroxyl and carboxyl groups in spite of the large increase in molecular size (and the accompanying large viscosity increase of the medium). Similar results have been observed in many other polymerizations. Data on the degradation of polymers also show the same effect. Thus in the acid hydrolysis of cellulose there is no effect of molecular size on hydrolytic reactivity up to a degree of polymerization of 1500 (molecular weight 250,000) [Flory, 1953]. The concept of functional group reactivity independent of molecular size has been highly successful in allowing the kinetic analysis of a wide range of polymerizations and reactions of polymers. Its validity, however, may not always be quite rigorous at very low or very high conversions.

2-2c   Step Polymerizations Other Than Polyesterification: Catalyzed versus Uncatalyzed

The kinetics of step polymerizations other than polyesterification follow easily from those considered for the latter. The number of different general kinetic schemes encountered in actual polymerization situations is rather small. Polymerizations by reactions between the A and B functional groups of appropriate monomers proceed by one of the following situations [Saunders and Dobinson, 1976]:

1   Some polymerizations, such as the formation of polyamides, proceed at reasonable rates as uncatalyzed reactions.

2   Other polymerizations such as those of urea, melamine, or phenol with formaldehyde (see Table 1-1) require an externally added acid or base catalyst to achieve the desired rates of reaction.

3   A few polymerizations can be reasonably employed either in a catalyzed or an uncatalyzed process. Polyurethane formation is an example of this type of behavior. The reaction between diols and diisocyanates is subject to base catalysis. However, the polymerization is often carried out as an uncatalyzed reaction to avoid various undesirable side reactions.

Regardless of the situation into which a particular polymerization falls, the observed overall kinetic features will be the same. The polymerization rates will be dependent on both the A and B groups. For the usual case where one has stoichiometric amounts of the two functional groups, the kinetics will be governed by Eq. 2-30 or 2-33. The observed kinetics will also be the same whether the polymerization is carried by the reaction of A—A and B—B monomers or by the self-reaction of an A—B monomer.

The derivation (Sec. 2-2b) of the kinetics of catalyzed polyesterification assumes that the catalyzed reaction is much faster than the uncatalyzed reaction, that is, $k' >> k$. This assumption is usually valid and therefore one can ignore the contribution by the uncatalyzed polyesterification to the total polymerization rate. For example, k' is close to two orders of magnitude larger than k for a typical polyesterification. For the atypical situation where k is not negligible relative to k', the kinetic expression for [M] or $\overline{X}_n$ as a function of reaction time must be derived [Hamann et al., 1968] starting with a statement of the polymerization rate as the sum of the rates of the catalyzed and uncatalyzed polymerizations,

$$\frac{-dM}{dt} = k[M]^3 + k'[M]^2 \tag{2-34}$$

Integration of Eq. 2-34 yields

$$k't = \frac{k}{k'} \ln \frac{[M](k[M]_0 + k')}{(k[M] + k')[M]_0} + \frac{1}{[M]} - \frac{1}{[M]_0} \tag{2-35}$$

The natural log term on the right side of Eq. 2-35 is the contribution of the uncatalyzed reaction. Its relative importance increases as k/k' increases. (When k/k' is very small, Eq. 2-35 converts to Eq. 2-32.)

2-2d   Nonequivalence of Functional Groups in Polyfunctional Reagents

2-2d-1   Examples of Nonequivalence

Before proceeding, it is useful to point out that there are instances where some or all parts of the concept of equal reactivity of functional groups are invalid [Kronstadt et al., 1978; Lovering and Laidler, 1962]. The assumption of equal reactivities of all functional groups in a polyfunctional monomer may often be incorrect. The same is true for the assumption that the reactivity of a functional group is independent of whether the other functional groups in the monomer have reacted. Neither of these assumptions is valid for 2,4-tolylene diisocyanate (**IV**)

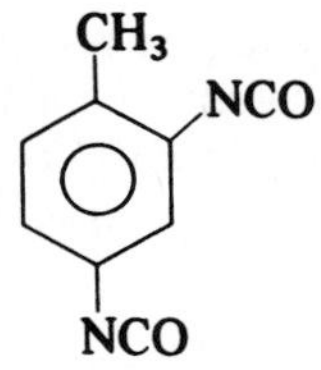

which is the diisocyanate monomer often employed in the synthesis of polyurethanes. Urethane formation can be depicted as occurring by the sequence

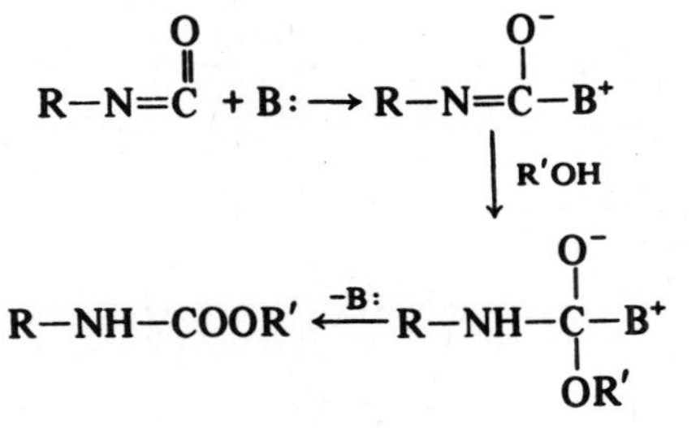

$$\tag{2-36}$$

where B: is a base catalyst (e.g., a tertiary amine). The reaction involves a rate-determining nucleophilic attack by the alcohol on the electrophilic carbon–nitrogen linkage of the isocyanate. This is substantiated by the observation that the reactivity of the isocyanate group increases with the electron-pulling ability of the substituent attached to it [Kaplan, 1961].

In 2,4-tolylene diisocyanate several factors cause the reactivities of the two functional groups to differ. These can be discussed by considering the data in Table 2-3 on the reactivities of various isocyanate groups compared to that in phenyl isocyanate toward reaction with n-butanol at 39.7°C in toluene solution with triethylamine as the catalyst [Brock, 1959, 1961]. It is clear that a methyl substituent deactivates the isocyanate group by decreasing its electronegativity; the effect is greater when the substituent is at the nearby ortho position. For the diisocyanates in Table 2-3 k_1 is the rate constant for the more reactive isocyanate group, while k_2 is the rate constant for the reaction of the less reactive isocyanate group after the more reactive one has been converted to a urethane group. One isocyanate group activates the other by electron withdrawal, as shown by the increased reactivity in the diisocyanates relative to the monoisocyanates. However, once the first isocyanate group has reacted the reactivity of the second group decreases, since the urethane group is a much weaker electron-pulling substituent than the isocyanate group. For 2,4-tolylene diisocyanate analysis of the experimental data at low conversion shows the para isocyanate group to be 2.7 times as reactive as the ortho groups. However, the effective difference in reactivity of the two functional groups is much greater ($k_1/k_2 = 11.9$), since the reactivity of the ortho group drops an additional factor of approximately 4 after reaction of the para isocyanate group.

The change in reactivity of one functional group upon reaction of the other has been noted in several systems as a difference in the reactivity of monomer compared to the other sized species, although all other species from dimer on up had the same reactivity [Hodkin, 1976; Yu et al., 1986]. This behavior usually occurs with monomers having two functional groups in close proximity and where polymerization involves a significant change in the electron-donating or electron-withdrawing ability of the functional group. Thus the reactivity of the hydroxyl group in ethylene glycol (**V**) toward esterification is considerably higher than the hydroxyl of a half-esterified glycol (**VI**) [Yamanis and Adelman, 1976].

$$HO—CH_2CH_2OH \qquad HO—CH_2CH_2—OCO—R—OCO\sim$$

$$\textbf{V} \qquad\qquad\qquad\qquad \textbf{VI}$$

TABLE 2-3   Reactivity of Isocyanate Group with n-C$_4$H$_9$OH^a

Isocyanate Reactant	Rate Constants^b		
	k_1	k_2	k_1/k_2
Monoisocyanate			
Phenyl isocyanate	0.406		
p-Tolyl isocyanate	0.210		
o-Tolyl isocyanate	0.0655		
Diisocyanates			
m-Phenylene diisocyanate	4.34	0.517	8.39
p-Phenylene diisocyanate	3.15	0.343	9.18
2,6-Tolylene diisocyanate	0.884	0.143	6.18
2,4-Tolylene diisocyanate	1.98	0.166	11.9

^{a}Data from Brock [1959].
^{b}k_1 and k_2 are units of liters/mole-min.

The nucleophilicity of a hydroxyl group is enhanced more by an adjacent hydroxyl relative to an adjacent ester group. Similarly, in the polymerization of sodium p-fluorothiophenoxide (an A—B reactant), nucleophilic substitution at the aromatic C—F bond occurs faster at fluorophenyl end groups of a growing polymer compared to monomer [Lenz et al., 1962]. The electron-donating effect of the para S$^-$ group in the monomer decreases the electron deficiency of the para carbon more than does a $\emptyset$S group in species larger than monomer. Polymerization of terephthalic acid with 4,6-diamino-1,3-benzenediol via oxazole formation (Eq. 2-224) proceeds with a sharp and continuous decrease in reaction rate with increasing polymer molecular weight [Cotts and Berry, 1981]. Reaction becomes progressively more diffusion-controlled with increasing molecular size due to the increasing rigid-rod structure of the growing polymer.

Trifunctional monomers constitute an important class of monomers. One often encounters such reactants in which the various functional groups have different reactivities. Thus, the polyesterification of glycerol (**VII**) with phthalic anhydride proceeds

$$
\begin{array}{l}
CH_2\!-\!OH \\
| \\
CH\!-\!OH \\
| \\
CH_2\!-\!OH
\end{array}
$$

VII

with incomplete utilization of the hydroxyl groups [Kienle et al., 1939]. This has been attributed to the lowered reactivity of the secondary hydroxyl group compared to the two primary hydroxyls. However, the explanation may be more complicated, since systems containing trifunctional reagents often reach a point in the reaction at which they cease being homogeneous solutions. Some systems have been found to contain dispersed polymer particles 0.05 to 1 μm in diameter [Solomon and Hopwood, 1966]. These particles have been termed *microgel particles* and are the result of crosslinking reactions (Sec. 2-10). Functional group reactivity in trifunctional monomers is considered further in Sec. 2-12.

2-2d-2 *Kinetics*

Kinetic analysis of a step polymerization becomes complicated when all functional groups in a reactant do not have the same reactivity. Consider the polymerization of A—A with B—B' where the reactivities of the two functional groups in the B—B' reactant are initially of different reactivities and, further, the reactivities of B and B' each change upon reaction of the other group. Even if the reactivities of the two functional groups in the A—A reactant are the same and independent of whether either group has reacted, the polymerization still involves four different rate constants. Any specific-sized polymer species larger than dimer is formed by two simultaneous routes. For example, the trimer A—AB—B'A—A is formed by

$$\text{B--B}' \xrightarrow[k_2]{\text{A--A}} \text{B--B}'\text{A--A} \tag{2-37a}$$

$$\text{A--A} \downarrow k_1 \qquad\qquad k_4 \downarrow \text{A--A}$$

$$\text{A--AB--B}' \xrightarrow[k_3]{\text{A--A}} \text{A--AB--B}'\text{A--A} \tag{2-37c}$$

$$\qquad (2\text{-}37b) \qquad\qquad\qquad (2\text{-}37d)$$

The two routes (one is Eqs. 2-37b and 2-37c; the other is Eqs. 2-37a and 2-37d) together comprise a complex reaction system that consists simultaneously of competitive, consecutive and competitive, parallel reactions.

Obtaining an expression for the concentration of A (or B or B') groups or the extent of conversion or $\overline{X}_n$ as a function of reaction time becomes much more difficult than for the case where the equal reactivity postulate holds, that is, where $k_1 = k_2 = k_3 = k_4$. As a general approach, one writes an expression for the total rate of disappearance of A groups in terms of the rates of the four reactions 2-37a, b, c, and d and then integrates that expression to find the time-dependent change in [A]. The difficulty arises because the differential equations that must be integrated are not linear equations and do not have exact solutions except in very particular cases. Numerical methods are then needed to obtain an approximate solution. Although this approach has been successfully used to describe some reactions of small molecules, there are few successful applications to polymerization systems [Moore and Pearson, 1981].

2-2d-2-a Polymerization of A—A with B—B'. The kinetics of the reaction system described in Eq. 2-37 is relatively difficult to treat because there are four different rate constants. Special (and simpler) cases involving only two rate constants can be more easily treated. One such case is where the two functional groups B and B' have different reactivities but their individual reactivities are decoupled in that neither the reactivity of B nor of B' changes upon reaction of the other group, that is,

$$k_1 \neq k_2 \tag{2-38a}$$

$$k_1 = k_4 \tag{2-38b}$$

$$k_2 = k_3 \tag{2-38c}$$

The reaction system converts from one (Eq. 2-37) that is, from the kinetic viewpoint, simultaneously competitive, consecutive (series) and competitive, simultaneous (parallel) to one (Eq. 2-38) that is only competitive, simultaneous. The polymerization consists of the B and B' functional groups reacting independently with A groups. The rates of disappearance of A, B, and B' functional groups are given by

$$\frac{-d[\text{B}]}{dt} = k_1[\text{A}][\text{B}] \tag{2-39}$$

$$\frac{-d[\text{B}']}{dt} = k_2[\text{A}][\text{B}'] \tag{2-40}$$

$$\frac{-d[\text{A}]}{dt} = k_1[\text{A}][\text{B}] + k_2[\text{A}][\text{B}'] \tag{2-41}$$

The polymerization rate is synonymous with the rate of disappearance of A groups (or the sum of the rates of disappearance of B and B' groups).

In the typical polymerization one has equimolar concentrations of the A—A and B—B' reactants at the start of polymerization. The initial concentrations of A, B, and B' groups are

$$[\text{A}]_0 = 2[\text{B}]_0 = 2[\text{B}']_0 \tag{2-42}$$

and the relationship between the concentrations of A, B, and B' at any time during polymerization is

$$[A] = [B] + [B']$$ (2-43)

Combination of Eqs. 2-41 and 2-43 yields the polymerization rate as

$$\frac{-d[A]}{dt} = (k_2 - k_1)[A][B] - k_2[A]^2$$ (2-44)

Introduction of the dimensionless variables α, β, γ, and τ and the parameter s defined by

$$\alpha = \frac{[A]}{[A]_0}$$ (2-45a)

$$\beta = \frac{[B]}{[B]_0}$$ (2-45b)

$$\gamma = \frac{[B']}{[B']_0}$$ (2-45c)

$$\tau = [B]_0 k_2 t$$ (2-45d)

$$s = \frac{k_2}{k_1}$$ (2-45e)

simplifies the mathematical solution of Eq. 2-44. α, β, and γ are the fractions of A, B, and B' groups, respectively, which remain unreacted at any time. τ is dimensionless time and s is the ratio of two rate constants. It is then possible to solve the coupled system of Eqs. 2-39 through 2-44 to give

$$\tau = \int_\beta^1 \frac{s\,d\beta}{\beta^2[1 + \beta^{(s-1)}]}$$ (2-46)

$$\gamma = \beta^s$$ (2-47)

where

$$\alpha = \frac{\beta + \gamma}{2} = \frac{\beta + \beta^s}{2}$$ (2-48)

The integral in Eq. 2-46 yields simple integrals for s values of 0, 1, and ∞. The dependence of α, β, and γ on τ in these instances are

1 For $s = 1$, that is, $k_2 = k_1$,

$$\beta = \frac{s}{s + 2\tau} = \gamma = \alpha$$ (2-49)

2 For $s = 0$, that is, $k_1 >> k_2$, where k_1 has some value and $k_2 \rightarrow 0$: B' groups do not react ($\gamma = 1$) and

$$\ln \left[\frac{(1 + \beta)}{2\beta} \right] = \tau \qquad (2\text{-}50\text{a})$$

$$\alpha = \frac{(1 + \beta)}{2} \qquad (2\text{-}50\text{b})$$

3 For $s = \infty$, that is, $k_2 >> k_1$, where k_2 has some value and $k_1 \rightarrow 0$: B groups do not react ($\beta = 1$) and

$$\ln \left[\frac{(1 + \gamma)}{2\gamma} \right] = \tau \qquad (2\text{-}51\text{a})$$

$$\alpha = \frac{(1 + \gamma)}{2} \qquad (2\text{-}51\text{b})$$

For almost all other values of s the integral in Eq. 2-46 must be numerically evaluated. The results of such calculations are presented in graphical form in Figs. 2-5 and 2-6 as plots of α, β, and γ vs log τ for s values of 1 and larger [Ozizmir and Odian, 1980a, 1980b]. There is no need for plots for $s < 1$, since the B and B' groups have interchangeable roles due to the symmetry of the reaction system. α at any time is exactly the same for $s = z$ or $s = 1/z$, where z is any number. The only difference for an s value of z or $1/z$ is whether it is the B or B' type of group, which reacts rapidly while the other reacts slowly. Figure 2-5 gives β and γ as a function of τ. When $s = 1$ both β and γ decay at the same rate according to Eq. 2-49. As s increases the β curves

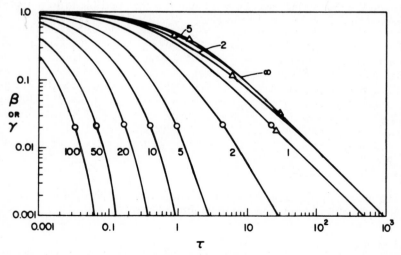

Fig. 2-5 Decay in β and γ with time in the polymerization of A—A with B—B'. β plots are indicated by $\triangle$; γ plots by $\bigcirc$. Values of s are noted for each plot. After Ozizmir and Odian [1980a, 1980b] (by permission of Wiley Interscience, New York).

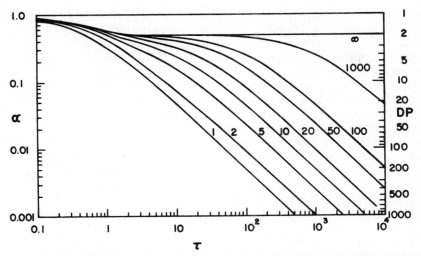

Fig. 2-6 Decay in α and increase in degree of polymerization with time in polymerization of A—A with B—B′. Values of s are noted for each plot. After Ozizmir and Odian [1980a, 1980b] (by permission of Wiley Interscience, New York).

move sharply above the curve for $s = 1$ since B groups react more slowly with A. Although k_2 is fixed, increasing values of s help move the γ curves gradually below the curve for $s = 1$, since B groups reacting more slowly with A leaves comparably more [A] for B′ groups to react with. As $s \to \infty$ the condition is reached in which [B] remains constant at $[B]_0$ during the complete decay of B′ according to Eq. 2-51b.

Figure 2-6 shows the decay in α and the corresponding increase in $\overline{X}_n$ with τ (since $\overline{X}_n = 1/\alpha$). It becomes progressively more difficult to achieve high degrees of polymerization in reasonable times for reaction systems of $s > 1$. At sufficiently large s (small k_1) the reaction time to obtain high polymer ($\overline{X}_n > 50$–100) becomes impractical. The extreme situation occurs at $s \to \infty$ ($k_1 \to 0$), where the maximum degree of polymerization is 2. Reaction systems with very small k_1 values must be avoided in order to achieve high degrees of polymerization. The desirable system from the viewpoint of polymerization rate and $\overline{X}_n$ is that with $s = 1$. (The exception to this generalization is for systems of $s > 1$ where both k_1 and k_2 are large but k_2 is larger.)

The A—A plus B—B′ system with $k_1 \neq k_4$ and $k_2 \neq k_3$ has been treated in a similar manner [Ozizmir and Odian, 1981]. The parameters α, β, γ, and s as defined by Eqs. 2-45a, 2-45b, 2-45c, and 2-45e are retained in the treatment. τ is redefined as

$$\tau = (k_1 + k_2)[B]_0 t \tag{2-52}$$

and the parameters u_1 and u_2

$$u_1 = \frac{k_4}{k_1} \tag{2-53}$$

$$u_2 = \frac{k_3}{k_2} \tag{2-54}$$

are introduced to describe the changes in the rate constants for the B and B' functional groups. The results are shown in Fig. 2-7 as a plot of α and the degree of polymerization versus τ for a range of values of s, u_1, and u_2. For the case where the two functional groups B and B' have the same initial reactivity ($s = 1$) (plots 1–3), the polymerization rate decreases if the reactivity of either B or B' decreases on reaction of the other group (i.e., if $u_1 < 1$ or $u_2 < 1$). The polymerization rate increases if the reactivity of either B or B' increases on reaction of the other group ($u_1 > 1$ or $u_2 > 1$). The effects reinforce each other when both u_1 and u_2 change in the same direction, but tend to cancel each other when u_1 and u_2 change in opposite directions. When the B and B' groups have different initial reactivities, for example, $s = 5$ (B' five times more reactive than B), the polymerization rate is decreased (compare plots 1 and 4) unless compensated for by an increase in the reactivity of B on reaction of B' ($u_1 > 1$). There is an increase in polymerization rate when the increased reactivity of B is large (compare plots 4 and 6). The polymerization rate becomes more depressed when B groups decrease in reactivity upon reaction of B' ($u_1 < 1$) (compare plots 4 and 5).

2-2d-2-b   Variation of Rate Constant with Size. The usual kinetic analysis of step polymerization, including that described for the A—A plus B—B' system, assumes the rate constants are independent of molecular size. The effect of a rate constant dependence on molecular size has been analyzed for A—B or A—A plus B—B polymerizations where a functional group on a monomer molecule has a different reactivity from the same group in species larger than monomer (i.e., for dimer and larger) [Goel et al., 1977; Gupta et al., 1979a, 1979b]. The following terms are defined:

$$u = \frac{k_m}{k_p} \tag{2-55}$$

$$\tau = [A]_0 k_p t \tag{2-56}$$

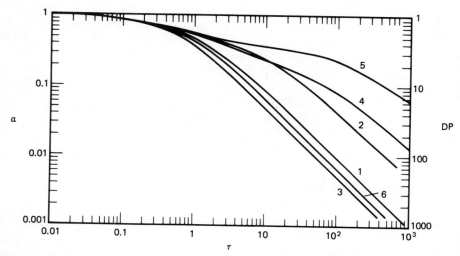

Fig. 2-7 Decay in α and increase in degree of polymerization of A—A with B—B' for $k_1 \neq k_4$ and $k_2 \neq k_3$. Values of s, μ_1, μ_2 are 1, 1, 1 (plot 1); 1, 0.2, 0.2 (plot 2); 1, 5, 5 (plot 3); 5, 1, 1 (plot 4); 5, 0.2, 1 (plot 5); 5, 5, 1 (plot 6). After Ozizmir and Odian [1981].

where k_m is the rate constant for A and B groups that are part of monomer molecules and k_p is the corresponding rate constant for A and B groups that are part of all species larger than monomer.

Figure 2-8 shows the results plotted as the polymer degree of polymerization versus τ for various values of u. The mathematical solution of this case shows that the initial limiting slope (at $\tau \to 0$) is $u/2$ while the final limiting slope (at large values of τ) is $\frac{1}{2}$ independent of the value of u. Most of these features are evident in Fig. 2-8. Dimer is produced more rapidly when monomer is more reactive than the larger sized species (that is, the larger the value of u). There is a subsequent slower rate of increase in $\overline{X}_n$ with the limiting slope of $\frac{1}{2}$ being reached more rapidly for larger u values. The initial rate of increase of $\overline{X}_n$ is slower for $u < 1$ (monomer less reactive than larger sized species), and it takes longer to reach the limiting slope of $\frac{1}{2}$. More significantly, it takes much longer reaction times to reach a particular degree of polymerization when $u < 1$. This is a kinetic characteristic of consecutive (series) reaction systems—

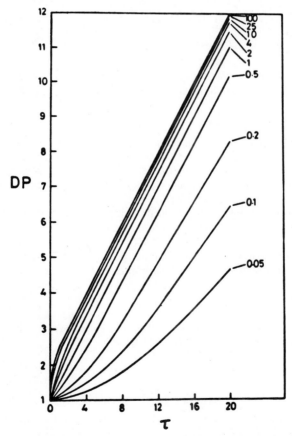

Fig. 2-8 Variation of degree of polymerization with time when monomer has a different reactivity from large-sized species. Values of u are noted for each plot; the plots for $u = 25$ and 100 are indistinguishable from each other. After Gupta et al. [1979a, 1979b] (by permission of Pergamon Press).

the overall rate of production of the final product (high polymer) is faster when the initial reaction (reaction of monomer) is faster and the later reaction (reaction of dimer and larger sized species) is slow than vice versa. The trends described in this section would also apply (qualitatively at least) for the case where k is a continuously varying function of molecular size. If k increases with molecular size, the situation is qualitatively analogous to the above system for $u < 1$. The situation is analogous to $u > 1$ when k decreases with molecular size.

2-3 ACCESSIBILITY OF FUNCTIONAL GROUPS

In order for a polymerization to yield high polymer, the polymer must not precipitate from the reaction mixture before the desired molecular weight is reached. Premature precipitation effectively removes the growing polymer molecules from the reaction; further growth is prevented because the polymer's functional end groups are no longer accessible to each other. The effect can be seen in Table 2-4 for the polymerization of bis(4-isocyanatophenyl)methane with ethylene glycol [Lyman, 1960].

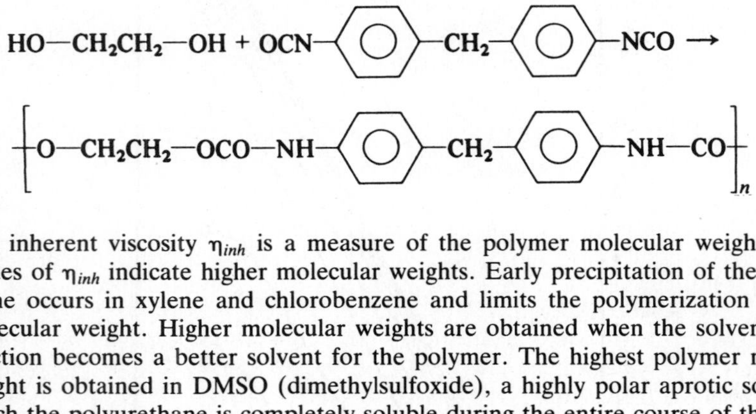

$$HO{-}CH_2CH_2{-}OH + OCN{-}\left\langle\bigcirc\right\rangle{-}CH_2{-}\left\langle\bigcirc\right\rangle{-}NCO \longrightarrow$$

$$\left[O{-}CH_2CH_2{-}OCO{-}NH{-}\left\langle\bigcirc\right\rangle{-}CH_2{-}\left\langle\bigcirc\right\rangle{-}NH{-}CO\right]_n \qquad (2\text{-}57)$$

The inherent viscosity η_{inh} is a measure of the polymer molecular weight. Larger values of η_{inh} indicate higher molecular weights. Early precipitation of the polyurethane occurs in xylene and chlorobenzene and limits the polymerization to a low molecular weight. Higher molecular weights are obtained when the solvent for the reaction becomes a better solvent for the polymer. The highest polymer molecular weight is obtained in DMSO (dimethylsulfoxide), a highly polar aprotic solvent, in which the polyurethane is completely soluble during the entire course of the polymerization.

Figure 2-9 shows similar behavior for the polymerization between terephthaloyl chloride and *trans*-2,5-dimethylpiperazine in mixtures of chloroform with carbon tet-

TABLE 2-4 Effect of Solvent on Molecular Weight in Polymerization of Bis(4-isocyanatophenyl)methane with Ethylene Glycol[a,b]

Solvent	η_{inh}[c]	Solubility of Polymer
Xylene	0.06	Precipitates at once
Chlorobenzene	0.17	Precipitates at once
Nitrobenzene	0.36	Precipitates after $\frac{1}{2}$ hour
Dimethyl sulfoxide	0.69	Polymer is soluble

[a]Data from Lyman [1960].
[b]Polymerization temperature: 115°C.
[c]Measured in dimethylformamide at room temperature.

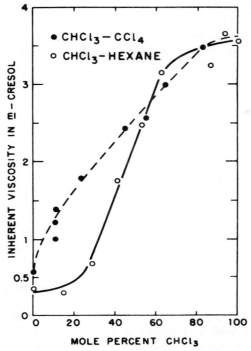

Fig. 2-9 Polymerization of terephthalic acid and *trans*-2,5-dimethylpiperazine in mixed solvents. After Morgan [1963, 1965] (by permission of Wiley-Interscience, New York).

rachloride or *n*-hexane [Morgan, 1963, 1965]. Chloroform is a good solvent for the polymer, while carbon tetrachloride and *n*-hexane are poor solvents. The inherent viscosity of the polymer (measured at 30°C in *m*-cresol) increases as the reaction mixture contains a larger proportion of chloroform. The better the reaction medium as a solvent for the polymer the longer the polymer stays in solution and the larger the polymer molecular weight. With a solvent medium that is a poor solvent for polymer, the molecular weight is limited by precipitation. Other examples of this behavior are described in Secs. 2-8c and 2-14.

    In addition to the effect of a solvent on the course of a polymerization due to the solvent being a poor or good solvent for the polymer, solvents affect polymerization rates and molecular weights due to preferential solvation or other specific interactions with either the reactants or transition state of the reaction or both. The direction of the solvation effect is generally the same in polymerization as in the corresponding small molecule reaction and will not be considered in detail. Thus polar solvents enhance the rate of a polymerization with a transition state more polar than the reactants. Polar solvents are not desirable for reactions involving transition states that are less polar than the reactants. The course of a polymerization can be dramatically affected by specific interactions of a solvent with the functional groups of a reactant. The reactivity of a functional group can be altered by specific interaction with solvent. Thus solvents markedly affect the polymer molecular weight in the polymerization of adipic acid and hexamethylene diamine with certain ketone solvents yielding the highest molecular weights [Ogata, 1973]. The molecular weight enhancement by ketones has been ascribed to an enhancement of the diamine nucleophilicity due possibly to a polar interaction between ketone and amine. Alternately, the intermediate formation of an imine may be responsible, since imines are formed from the diamine and ketone. The imine would be expected to be more reactive than the amine toward the carboxylic acid.

2-4 EQUILIBRIUM CONSIDERATIONS

2-4a Closed System

Many, if not most, step polymerizations involve equilibrium reactions, and it becomes important to analyze how the equilibrium affects the extent of conversion and, more importantly, the polymer molecular weight. A polymerization in which the monomer(s) and polymer are in equilibrium is referred to as an *equilibrium polymerization* or *reversible polymerization*. A first consideration is whether an equilibrium polymerization will yield high molecular weight polymer if carried out in a *closed system*. By a closed system is meant one where none of the products of the forward reaction are removed. Nothing is done to push or drive the equilibrium point for the reaction system toward the polymer side. Under these conditions the concentrations of products (polymer and usually a small molecule such as water) build up until the rate of the reverse reaction becomes equal to the polymerization rate. The reverse reaction is referred to generally as a *depolymerization* reaction; other terms such as *hydrolysis* or *glycolysis* may be used as applicable in specific systems. The polymer molecular weight is determined by the extent to which the forward reaction has proceeded when equilibrium is established.

Consider an external acid-catalyzed polyesterification

$$\sim\sim\sim COOH + \sim\sim\sim OH \overset{K}{\rightleftharpoons} \sim\sim\sim CO-O\sim\sim\sim + H_2O \tag{2-58}$$

in which the initial hydroxyl group and carboxyl group concentrations are both $[M]_0$. The concentration of ester groups $[COO]$ at equilibrium is $p[M]_0$ where p represents the extent of reaction at equilibrium. $p[M]_0$ also represents the water concentration at equilibrium. The concentrations of hydroxyl and carboxyl groups at equilibrium are each $([M]_0 - p[M]_0)$. The equilibrium constant for the polymerization is given by

$$K = \frac{[COO][H_2O]}{[COOH][OH]} = \frac{(p[M]_0)^2}{([M]_0 - p[M]_0)^2} \tag{2-59}$$

which simplifies to

$$K = \frac{p^2}{(1 - p)^2} \tag{2-60}$$

Solving for p yields

$$p = \frac{K^{1/2}}{1 + K^{1/2}} \tag{2-61}$$

Equation 2-61 yields the extent of conversion as a function of the equilibrium constant. To obtain an expression for the degree of polymerization as a function of K, Eq. 2-61 is combined with Eq. 2-27 to yield

$$\overline{X}_n = 1 + K^{1/2} \tag{2-62}$$

Table 2-5 shows p and $\overline{X}_n$ values calculated for various K values. These calculations clearly indicate the limitation imposed by equilibrium on the synthesis of even a modest molecular weight polymer. A degree of polymerization of 100 (corresponding to a molecular weight of approximately 10^4 in most systems) can be obtained in a closed system only if the equilibrium constant is almost 10^4. The higher molecular weights that are typically required for practical applications would require even larger equilibrium constants. A consideration of the equilibrium constants for various step polymerizations or the corresponding small molecule reactions quickly shows that polymerizations cannot be carried out as closed systems [Allen and Patrick, 1974; Saunders and Dobinson, 1976; Zimmerman, 1988]. For example, the equilibrium constant for a polyesterification is typically no larger than 1–10, K for a transesterification is in the range 0.1-1 and K for polyamidation is in the range 10^2–10^3. Although the equilibrium constant for a polyamidation is very high as K values go, it is still too low to allow the synthesis of high molecular weight polymer. Further, even for what appear to be essentially irreversible polymerizations, reversal of polymerization is a potential problem if the by-product small molecule is not removed (or, alternately, the polymer is not removed).

It is worth mentioning that K values reported in the literature for any specific step polymerization often differ considerably. Thus, K values for the polymerization of adipic acid and hexamethylene diamine range from a low of 250 to a high of 900. There are several reasons for these differences, not the least of which is the experimental difficulty in carrying out measurements on polymerizations involving highly concentrated systems (often containing only the monomers, without any solvent) at moderately high temperatures (200–300°C). Other reasons for the variation in K values are the effects of temperature and the concentration of the small molecule by-product on K. Most step polymerizations are exothermic and K decreases with increasing temperature (see Sec. 2-8a). The common practice of extrapolating values of K determined at certain temperatures to other temperatures can involve considerable error

TABLE 2-5 Effect of Equilibrium Constant on Extent of Reaction and Degree of Polymerization in Closed System

Equilibrium Constant (K)	p	$\overline{X}_n$
0.0001	0.0099	1.01
0.01	0.0909	1.10
1	0.500	2
16	0.800	5
81	0.900	10
361	0.950	20
2,401	0.980	50
9,801	0.990	100
39,601	0.995	200
249,001	0.998	500

if ΔH is not accurately known. The variation of K with the concentration of the small molecule by-product has been established in the polyamidation reaction but the quantitative effect has not been generally studied. The effect may be due to a change of K with polarity of the reaction medium.

2-4b Open, Driven System

The inescapable conclusion is that except in a minority of systems a step polymerization must be carried out as an *open, driven* system. That is, we must remove at least one of the products of the forward (polymerization) reaction so as to drive the equilibrium toward high molecular weights. It is usually more convenient to remove the small molecule by-product rather than the polymer. When water is the by-product it can be removed by a combination of temperature, reduced pressure, and purging with inert gas. Conveniently one often carries out step polymerizations at temperatures near or above the boiling point of water. This is usually done for purposes of obtaining desired reaction rates, but it has the added advantage of facilitating water removal. A small molecule by-product such as HCl can be removed in the same manner or by having a base present in the reaction system to neutralize the hydrogen chloride. Driving an equilibrium toward polymer requires considerable effort, since the water or hydrogen chloride or other small molecule must diffuse through and out of the reaction mixture. Diffusion is not so easy since the typical step polymerization system is fairly viscous at very high conversions. The polymerization can become diffusion-controlled under these conditions with the polymerization being controlled by the rate of diffusion of the small molecule by-product [Campbell et al., 1970].

The extent to which one must drive the system in the forward direction can be seen by calculating the lowering of the small molecule concentration, which is necessary to achieve a particular molecular weight. For the polyesterification (Eq. 2-58) one can rewrite Eq. 2-59 as

$$K = \frac{p[H_2O]}{[M]_0(1 - p)^2} \tag{2-63}$$

which is combined with Eq. 2-27 to yield

$$K = \frac{p[H_2O]\overline{X}_n^2}{[M]_0} \tag{2-64}$$

and this result combined with Eq. 2-27 to give

$$[H_2O] = \frac{K[M]_0}{\overline{X}_n(\overline{X}_n - 1)} \tag{2-65}$$

Equation 2-65, which applies equally to A—B polymerizations, indicates that $[H_2O]$ must be greatly reduced to obtain high $\overline{X}_n$ values. $[H_2O]$ is inversely dependent on essentially the square of $\overline{X}_n$ since $(\overline{X}_n - 1)$ is close to $\overline{X}_n$ for large values of $\overline{X}_n$. It is also seen that the level to which the water concentration must be lowered to achieve a particular degree of polymerization increases with increasing K and increasing initial concentration of the reactants. Table 2-6 shows the calculated $[H_2O]$ values for selected values of K and $\overline{X}_n$ at $[M]_0 = 5 M$. A concentration of 5 M is fairly typical of a step

TABLE 2-6 Effect of Water Concentration on Degree of Polymerization in Open, Driven System

K	$\overline{X}_n$	$[H_2O]^a$ (moles/liter)
0.1	1.32^b	1.18^b
	20	1.32×10^{-3}
	50	2.04×10^{-4}
	100	5.05×10^{-5}
	200	1.26×10^{-5}
	500	2.00×10^{-6}
1	2^b	2.50^b
	20	1.32×10^{-2}
	50	2.04×10^{-3}
	100	5.05×10^{-4}
	200	1.26×10^{-4}
	500	2.01×10^{-5}
16	5^b	4.00^b
	20	0.211
	50	3.27×10^{-2}
	100	8.10×10^{-3}
	200	2.01×10^{-3}
	500	3.21×10^{-4}
81	10^b	4.50^b
	20	1.07
	50	0.166
	100	4.09×10^{-2}
	200	1.02×10^{-2}
	500	1.63×10^{-3}
361	20^b	4.75^b
	50	0.735
	100	0.183
	200	4.54×10^{-2}
	500	7.25×10^{-3}

^{a}[H$_2$O] values are for [M]$_0$ = 5.
^{b}These values are for a closed reaction system at equilibrium.

polymerization which is often carried out with only the reactant(s) present (without solvent). The lowering of [H$_2$O] to achieve a particular $\overline{X}_n$ is less the more favorable the equilibirum (that is, the larger the K value). Thus the synthesis of polyamides (with typical K values $> 10^2$) is clearly easier from the equilibrium viewpoint than polyester synthesis ($K \sim 0.1$-1). Polyesterification requires a greater lowering of [H$_2$O] than does polyamidation. It should be understood that simply to lower [H$_2$O] as much as possible is not the desired approach. One needs to control the [H$_2$O] so as to obtain the desired degree of polymerization.

2-4c Kinetics of Reversible Polymerization

Although reversible or equilibrium polymerizations would almost always be carried out in an irreversible manner, it is interesting to consider the kinetics of polymerization for the case in which the reaction was allowed to proceed in a reversible manner.

(The kinetics of reversible ring-opening polymerizations are discussed in Sec. 7-2b-5.) The kinetics of the reversible ester interchange polymerization

$$n\text{HOCH}_2\text{CHO}-\text{CO}-\text{R}-\text{CO}-\text{OCH}_2\text{CH}_2\text{OH} \rightleftharpoons$$

VIII

$$\text{HOCH}_2\text{CH}_2\text{O}-[-\text{CO}-\text{R}-\text{CO-OCH}_2\text{CH}_2\text{O}-]_n\text{-H} + (n-1)\text{HOCH}_2\text{CH}_2\text{OH}$$

IX

$$(2\text{-}66)$$

have been studied in considerable detail [Challa, 1960; Chegolya et al., 1979]. The polymerization occurs by reaction between 2-hydroxyethyl ester end groups (more specifically, by reaction between hydroxyl and ester groups of 2-hydroxyethyl ester endgroups) and can be depicted as

$$2\sim\sim\text{RCOOCH}_2\text{CH}_2\text{OH} \underset{k_2}{\overset{k_1}{\rightleftharpoons}}$$

VIII

$$\sim\sim\text{RCOOCH}_2\text{CH}_2\text{OOCR}\sim\sim + \text{HOCH}_2\text{CH}_2\text{OH} \qquad (2\text{-}67)$$

IX

The rate of polymerization, denoted by the net rate of ethylene glycol formation $d[\text{G}]/dt$, is given by the difference in the rates of the forward and back reactions

$$R_p = \frac{d[\text{G}]}{dt} = k_1[\text{M}]^2 - 4k_2[\text{P}][\text{G}] \qquad (2\text{-}68)$$

where [M], [P], and [G] refer to the concentrations of the 2-hydroxyethyl ester end groups (**VIII**), the ethylene diester groups (**IX**), and ethylene glycol, respectively. The factor of 4 in Eq. 2-68 takes into account the statistical favoring of the reverse action. Reaction between any bifunctional ethylene diester group and any bifunctional glycol molecule can proceed in four equivalent ways; the forward reaction between any hydroxyl group and any ester linkage belonging to different 2-hydroxyethyl ester end groups can only go in one way.

The concentrations of the various species can be conveniently expressed in terms of the extent of reaction p at time t in the reaction by

$$[\text{M}] = 2(1 - p)[\text{R}]_0 \qquad (2\text{-}69)$$

$$[\text{P}] = p[\text{R}]_0 \qquad (2\text{-}70)$$

$$[\text{G}] = (q + p - p_0)[\text{R}]_0 \qquad (2\text{-}71)$$

where $[\text{R}]_0$ is the initial concentration of the repeating unit X

$$-\text{CH}_2\text{OOC}-\text{R}-\text{COOCH}_2-$$

X

and

$$q = \frac{[G]_0}{[R]_0} \tag{2-72}$$

$$p = \frac{2[R]_0 - [M]}{2[R]_0} \tag{2-73}$$

where $[G]_0$ is the initial ethylene glycol concentration. $[R]$ remains constant in the system at its initial value $[R]_0$. As p increases, the concentration of **IX** in the polymer increases by the amount of its decrease in **VIII**.

One should note that the various terms have been defined so that the kinetic treatment applies equally to polyesterification systems containing any initial combination of **VIII**, **IX**, and glycol. For the case in which only equivalent concentrations of polymer and glycol are initially present one has $p_0 = 1$, $q = 1$, $[M]_0 = 0$, $[P]_0 = [R]_0 = [G]_0$.

Combination of Eqs. 2-68 through 2-71 gives the rate expression

$$\frac{dp}{dt} = 4[R]_0\{k_1(1 - p)^2 - k_2 p(q + p - p_0)\} \tag{2-74}$$

which can be integrated between $t = 0$ ($p = 0$) and $t = t(p = p)$ to yield

$$k_1 t = \frac{K}{4B[R]_0} \ln \left\{ \frac{(p_0 - p_e)[(1 - K)(p - p_e) + B]}{(p - p_e)[(1 - K)(p_0 - p_e) + B]} \right\} \tag{2-75}$$

where p_e is the extent of reaction at equilibrium, $K = k_1/k_2$, and

$$B = [(2K + q - p_0)^2 - 4K(K - 1)]^{1/2} \tag{2-76}$$

Equation 2-75 is used to describe the polymerization rate in terms of the extent of reaction as a function of time once the equilibrium quantities K and p_e are known for a system.

2-5 CYCLIZATION VERSUS LINEAR POLYMERIZATION

2-5a Possible Cyclization Reactions

The production of linear polymers by the step polymerization of polyfunctional monomers is sometimes complicated by the competitive occurrence of *cyclization reactions*. Ring formation is a possibility in the polymerizations of both the A—B and A—A plus B—B types. Reactants of the A—B type such as amino or hydroxy acids may undergo intramolecular cyclization instead of linear polymerization

$$H_2N—R—COOH \rightarrow HN—R—CO \tag{2-77}$$

$$HO—R—COOH \rightarrow O—R—CO \tag{2-78}$$

Reactants of the A—A type are not likely to undergo direct cyclization instead of linear polymerization. The functional groups in a typical A—A reactant are incapable of reacting with each other under the conditions of step polymerization. Thus, there is usually no possibility of anhydride formation from reaction of the carboxyl groups of a diacid reactant under the reaction conditions of a polyesterification. Similarly, cyclization does not occur between hydroxyl groups of a diol, amine groups of a diamine, isocyanate groups of a diisocyanate, and so on.

Once linear polymerization has reached the dimer size, intramolecular cyclization is a possibility throughout any A—B

$$
\text{H} + \text{O} - \text{R} \rightarrow_n \text{COOH} \rightarrow + \text{O} - \text{R} \rightarrow_n \text{COO} \tag{2-79}
$$

or A—A + B—B polymerization

$$
\text{H} + \text{O} - \text{R} - \text{OCO} - \text{R}' - \text{CO} \rightarrow_n \text{OH} \rightarrow + \text{O} - \text{R} - \text{OCO} - \text{R}' - \text{CO} \rightarrow_n
$$

$$
\tag{2-80}
$$

2-5b Thermodynamic and Kinetic Considerations

Whether cyclization is competitive with linear polymerization for a particular reactant or pair of reactants depends on thermodynamic and kinetic considerations of the size of the ring structure which may be formed. An understanding of the relative ease of cyclization or linear polymerization comes from a variety of sources. These include direct studies with various bifunctional monomers in cyclization reactions (such as those in Eqs. 2-77 through 2-80) as well as ring-opening polymerization studies (Chap. 7) and data such as the heats of combustion of cyclic compounds [Carothers and Hill, 1933; Eliel, 1962; Sawada, 1976]. Consider first the thermodynamic stability of different sized ring structures. Some of the most useful data on the effect of ring size on thermodynamic stability is that on the heats of combustion of cycloalkanes (Table 2-7) (1kJ = 0.2388 kcal). A comparison of the heats of combustion per methylene group in these ring compounds with that in an open chain alkane yields a general measure of the thermodynamic stabilities of different sized rings. More precisely, thermodynamic stability decreases with increasing strain in the ring structure as measured by the differences in the heats of combustion per methylene group of the cycloalkane and the *n*-alkane. The strain in cyclic structures is very high for the 3- and 4-membered rings, decreases sharply for 5-, 6-, and 7-membered rings, increases for 8- to 11-membered rings, and then decreases again for larger rings.

The strain in ring structures is of two types—*angle strain* and *conformational strain*. Ring structures of less than five atoms are highly strained due to the high degree of angle strain, that is, the large distortion of their bond angles from the normal tetrahedral bond angle. Bond angle distortion is virtually absent in rings of five or more members. For rings larger than five atoms the strain due to bond angle distortion would be excessive for planar rings. For this reason rings larger than five atoms exist in more stable, nonplanar (puckered) forms. The differences in strain among rings of five members and larger are due to differences in conformational strain. The 5- and 7-membered rings are somewhat strained in comparison to the 6-membered ring due to the *torsional strain* arising from eclipsed conformations on adjacent atoms of the ring. Rings of 8 or more members have *transannular strain* arising from repulsive interactions between hydrogens or other groups which are forced to crowd positions

TABLE 2-7   Heats of Combustion and Strains of Cycloalkanes per Methylene Group[a]

$(CH_2)_n$ n	Heat of Combustion per Methylene Group (kJ/mole)	Strain per Methylene Group[b] (kJ/mole)
3	697.6	38.6
4	686.7	27.7
5	664.5	5.5
6	659.0	0.0
7	662.8	3.8
8	664.1	5.1
9	664.9	5.9
10	664.1	5.1
11	663.2	4.2
12	660.3	1.3
13	660.7	1.7
14	659.0	0.0
15	659.5	0.5
16	659.5	0.5
17	658.2	−0.8
n-Alkane	659.0	0.0

[a]Data from Eliel [1962].
[b]Calculated as the heat of combustion per methylene group minus the value (659.0) for the n-alkane methylene group.

in the interior of the ring structure. Transannular strain virtually disappears for rings larger than 11 members; the ring becomes sufficiently large to accommodate substituents without transannular repulsions.

The general order of thermodynamic stability of different sized ring structures is, thus, given by

3,4 << 5,7-11 < 12 and larger, 6

This same order of stability is generally observed for a variety of ring structures, including those containing atoms or groups other than methylene. Although data are not as extensive for ring structures such as ethers, lactones, or lactams, the general expectation is borne out. The substitution of an oxygen, carbonyl, nitrogen, or other group for methylene does not appreciably alter the bond angles in the ring structure. Replacement of a methylene group by a less bulky oxygen atom or carbonyl group may, however, slightly increase the stability of the ring structure by a decrease in the conformational strain. It has also been observed that substituents, other than hydrogen, on a ring structure generally increase its stability relative to the linear structure as repulsive interactions between substituents are less severe in the ring structure.

In addition to thermodynamic stability kinetic feasibility is important in determining the competitive position of cyclization relative to linear polymerization. Kinetic feasibility for the cyclization reaction depends on the probability of having the functional end groups of the reactant molecules approach each other. As the potential ring size increases, the monomers which would give rise to ring structures have many conformations, very few of which involve the two ends being adjacent. The probability of ring formation decreases as the probability of the two functional groups encountering each other decreases. The effect is reflected in an increasingly unfavorable entropy of activation.

The overall ease of cyclization is thus dependent on the interplay of two factors: (1) the continuous decrease in kinetic feasibility with ring size, and (2) the thermodynamic stability, which increases as the ring size increases to the 6-membered ring, then decreases as the ring size increases up to the 9- to 11-membered ring and then increases again for larger rings. The net result of the two factors is that one usually observes the ease of ring formation to be low for the 3-membered ring and very low for the 4-membered ring. The 3-membered ring is formed easier than the 4-membered ring due to the more favorable kinetic factor. There is a sharp increase for the 5-membered ring because of a highly favorable thermodynamic factor. The 6-membered ring is slightly less easily formed than the 5-membered because the slightly more favorable thermodynamic factor is negated by a decrease in the kinetic factor. There is a sharper decrease for the 7-membered ring and then a smaller decrease for the 8- to 10-membered rings as both the kinetic and thermodynamic factors become less favorable. The ease of ring formation increases for the 11- and 12-membered rings due to their slightly greater thermodynamic stabilities. The ease of cyclization decreases slowly with ring size for rings larger than the 12-membered ring. These larger-sized rings are comparable in thermodynamic stability but the kinetic factor is progressively less favorable. Specifically, the kinetic factor becomes less favorable for cyclization in proportion to the $\frac{3}{2}$ power of ring size [Jacobson and Stockmayer, 1950; Yuan et al., 1988].

These conclusions are qualitatively summarized in Fig. 2-10. From the practical viewpoint of obtaining linear polymerization, it turns out that ring formation in A—B polymerizations is only a problem when the monomer can form 5-, 6-, or 7-membered rings, that is, when R in Eqs. 2-77 and 2-78 contributes 3-5 atoms to the ring. Such A—B monomers are avoided. Also, one can use an A—A + B—B system instead of the A—B system. Cyclization (Eqs. 2-79 and 2-80) of the growing linear polymer in both A—B and A—A + B—B systems occurs to a very low extent for two reasons. First, one avoids the use of reactants with R and R' groups that would yield 5-, 6- or 7-membered rings at the dimer stage ($n = 2$ and 1, respectively, in Eqs. 2-79 and 2-80). Second, the cyclization of linear species for the typical monomers would yield large-sized rings and such rings are not easily formed. However, the extent of cycli-

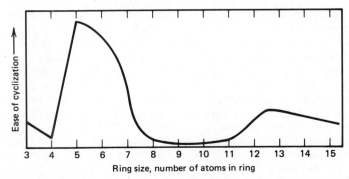

Fig. 2-10 Dependence of the ease of cyclization on the size of the ring.

zation of dimer, trimer, tetramer, and so on to large-sized cyclic oligomers is not zero. Although data are not available for a wide range of polymerization systems, the total cyclic content in some linear polymerizations (e.g., polyesterification and polyamidation) has been found to be 1–3% [Goodman, 1988; Semlyen, 1986; Zimmerman, 1988]. (The cyclic content in polysiloxanes is considerably higher; see Sec. 2-12f.) The presence of even these small amounts of cyclic oligomers materials can be detrimental to the commercial utilization of a polymer if the cyclics migrate out of the product during its utilization. Many commercial processes remove cyclics by extraction (e.g., with steam in polyamide production) or thermal devolatilization (polysiloxane production).

2-5c   Other Considerations

Step polymerizations are almost always carried out using high concentrations of the reactant(s). This is highly favorable for linear polymerization. Cyclization is a unimolecular (intramolecular) reaction, while linear polymerization is a bimolecular (intermolecular) reaction. The rate of the latter increases much faster than the rate of the former as the concentration of reactant(s) increase. Thus this factor of reactant concentration is superimposed on the previous thermodynamic and kinetic considerations. The concentration factor increases the overall competitive position of linear polymerization relative to cyclization. This is the reason that cyclization is even less of a problem than expected based on the previous discussions. Another factor that is often used to shift the competitive balance between linear polymerization and cyclization toward the former is the interconvertibility of the linear and cyclic structures under appropriate reaction conditions. In these cases it is possible to shift the equilibrium between the two structures in the direction of the linear polymer.

The previous discussions have concerned rings containing carbon, carbon and oxygen, or carbon and nitrogen atoms. The situation as regards the competition between cyclization and linear polymerization as a function of ring size may be altered when other atoms make up the ring structure. Thus in the case of polysiloxanes where the ring structure **XI** contains alternating oxygen and silicon atoms, rings of less than eight

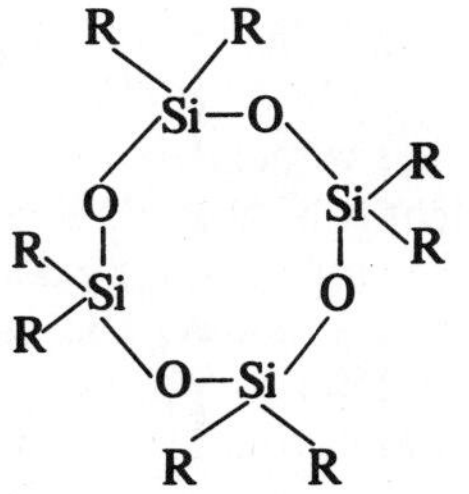

XI

atoms are quite strained because of the large size of the silicon atom, the greater length of the Si—O bond, and the large Si—O—Si bond angle. The optimum ring size is the 8-membered ring, although the preference is not overwhelming because of the kinetic factor. Larger-sized rings are not as favored for the reasons previously discussed.

2-6 MOLECULAR-WEIGHT CONTROL IN LINEAR POLYMERIZATION

2-6a Need for Stoichiometric Control

There are two important aspects with regard to the control of molecular weight in polymerizations. In the synthesis of polymers, one is usually interested in obtaining a product of very specific molecular weight, since the properties of the polymer will usually be highly dependent on molecular weight. Molecular weights higher or lower than the desired weight are equally undesirable. Since the degree of polymerization is a function of reaction time, the desired molecular weight can be obtained by quenching the reaction (e.g., by cooling) at the appropriate time. However, the polymer obtained in this manner is unstable in that subsequent heating leads to changes in molecular weight because the ends of the polymer molecules contain functional groups (referred to as *end groups*) that can react further with each other.

This situation is avoided by adjusting the concentrations of the two monomers (e.g., diol and diacid) so that they are slightly nonstoichiometric. One of the reactants is present in slight excess. The polymerization then proceeds to a point at which one reactant is completely used up and all the chain ends possess the same functional group—the group that is in excess. Further polymerization is not possible, and the polymer is stable to subsequent molecular-weight changes. Thus the use of excess diamine in the polymerization of a diamine with a diacid (Eq. 1-1) yields a polyamide (**XII**) with amine end groups

$$\text{excess } H_2N\text{—R—}NH_2 + HO_2C\text{—R}'\text{—}CO_2H \rightarrow$$

$$H\text{—}(\text{—NH—R—NHCO—R}'\text{—CO—})_n\text{—NH—R—}NH_2 \qquad (2\text{-}81)$$
$$\textbf{XII}$$

in the absence of any diacid for further polymerization. The use of excess diacid accomplishes the same result; the polyamide (**XIII**) in this case has carboxyl end groups

$$\text{excess } HO_2C\text{—R}'\text{—}CO_2H + H_2N\text{—R—}NH_2 \rightarrow$$

$$HO\text{—}(\text{—CO—R}'\text{—CONH—R—NH—})_n\text{—CO—R}'\text{—COOH} \qquad (2\text{-}82)$$
$$\textbf{XIII}$$

which are incapable of further reaction, since the diamine has completely reacted.

Another method of achieving the desired molecular weight is by the addition of a small amount of a *monofunctional monomer*. Acetic acid or lauric acid, for example, are often used to achieve molecular weight stabilization of polyamides. The monofunctional monomer controls and limits the polymerization of bifunctional monomers because its reaction with the growing polymer yields chain ends devoid of a functional group and therefore incapable of further reaction. Thus, the use of benzoic acid in Reaction 2-1a yields a polyamide (**XIV**) with phenyl end groups

$$H_2N\text{—R—}NH_2 + HO_2C\text{—R}'\text{—}CO_2H + \phi CO_2H \rightarrow$$

$$\phi\text{—CO—}(\text{—NH—R—NHCO—R}'\text{—CO—})_n\text{—NHRNHOCO}\phi \qquad (2\text{-}83)$$
$$\textbf{XIV}$$

which are unreactive toward polymerization.

2-6b Quantitative Aspects

In order to properly control the polymer molecular weight, one must precisely adjust the stoichiometric imbalance of the bifunctional monomers or of the monofunctional monomer. If the nonstoichiometry is too large, the polymer molecular weight will be too low. It is therefore important to understand the quantitative effect of the stoichiometric imbalance of reactants on the molecular weight. This is also necessary in order to know the quantitative effect of any reactive impurities that may be present in the reaction mixture either initially or that are formed by undesirable side reactions. Impurities with A or B functional groups may drastically lower the polymer molecular weight unless one can quantitatively take their presence into account. Consider now the various different reactant systems which are employed in step polymerizations:

TYPE 1. For the polymerization of the bifunctional monomers A—A and B—B (e.g., diol and diacid or diamine and diacid) where B—B is present in excess, the numbers of A and B functional groups are given by N_A and N_B, respectively. N_A and N_B are equal to twice the number of A—A and B—B molecules, respectively, that are present. The *stoichiometric imbalance r* of the two functional groups is given by $r = N_A/N_B$. (The ratio r is always defined so as to have a value equal to or less than unity but never greater than unity, i.e., the B groups are those in excess). The total number of monomer molecules is given by $(N_A + N_B)/2$ or $N_A(1 + 1/r)/2$.

The extent of reaction p is introduced here and defined as the fraction of A groups that have reacted at a particular time. The fraction of B groups that have reacted is given by rp. The fractions of unreacted A and B groups are $(1 - p)$ and $(1 - rp)$, respectively. The total numbers of unreacted A and B groups are $N_A(1 - p)$ and $N_B(1 - rp)$, respectively. The total number of polymer chain ends is given by the sum of the total number of unreacted A and B groups. Since each polymer chain has two chain ends, the total number of polymer molecules is one half the total number of chain ends or $[N_A(1 - p) + N_B(1 - rp)]/2$.

The number-average degree of polymerization $\overline{X}_n$ is the total number of A—A and B—B molecules initially present divided by the total number of polymer molecules

$$\overline{X}_n = \frac{N_A(1 + 1/r)/2}{[N_A(1 - p) + N_B(1 - rp)]/2} = \frac{1 + r}{1 + r - 2rp} \qquad (2\text{-}84)$$

Equation 2-84 shows the variation of $\overline{X}_n$ with the stoichiometric imbalance r and the extent of reaction p. There are two interesting limiting forms of this relationship. When the two bifunctional monomers are present in stoichiometric amounts (that is, $r = 1.000$), Eq. 2-84 reduces to the previously discussed Eq. 2-27

$$\overline{X}_n = \frac{1}{(1 - p)} \qquad (2\text{-}27)$$

On the other hand, when the polymerization is 100% complete (that is, $p = 1.000$), Eq. 2-84 becomes

$$\overline{X}_n = \frac{(1 + r)}{(1 - r)} \qquad (2\text{-}85)$$

In actual practice, p may approach but never becomes equal to unity.

Figure 2-11 shows plots of $\overline{X}_n$ vs the stoichiometric imbalance for several values of p in accordance with Eq. 2-84. The stoichiometric imbalance is expressed both as the ratio r and also as the mole percent excess of the B—B reactant over the A—A reactant. The various plots show how r and p must be controlled so as to obtain a particular degree of polymerization. However, one does not usually have complete freedom of choice of the r and p values in a polymerization. Complete control of the stoichiometric ratio is not always possible, since reasons of economy and difficulties in the purification of reactants may prevent one from obtaining r values very close to 1.000. Similarly, many polymerizations are carried out to less than 100% completion

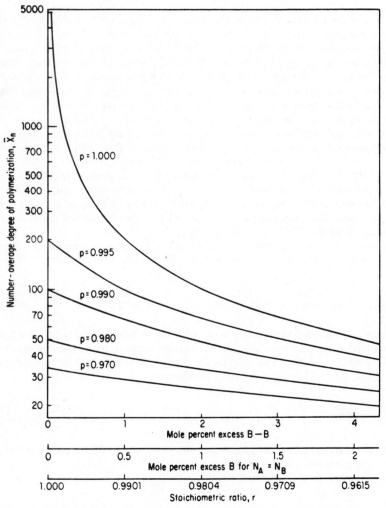

Fig. 2-11 Dependence of the number-average degree of polymerization $\overline{X}_n$ on the stoichiometric ratio r for different extents of reaction p in the polymerization of A—A with B—B.

(i.e., to $p < 1.000$) for reasons of time and economy. The time required to achieve each of the last few percent of reaction is close to that required for the first 97-98 or so percent of reaction. Thus a detailed consideration of Fig. 2-4 shows that the time required to go from $p = 0.97$ ($\overline{X}_n = 33.3$) to $p = 0.98$ ($\overline{X}_n = 50$) is approximately the same as that to reach $p = 0.97$ from the start of reaction.

Consider a few examples to illustrate the use of Eq. 2-84 and Fig. 2-11. For stoichiometric imbalances of 0.1 and 1 mole percent (r values of 1000/1001 and 100/101, respectively) at 100% reaction, the values of $\overline{X}_n$ are 2001 and 201, respectively. The degree of polymerization decreases to 96 and 66, respectively, at 99% reaction and to 49 and 40 at 98% reaction. It is clear that step polymerizations will almost always be carried out to at least 98% reaction, since a degree of polymerization of at least approximately 50–100 is usually required for a useful polymer. Higher conversions and the appropriate stoichiometric ratio are required to obtain higher degrees of polymerization. The exact combination of p and r values necessary to obtain any particular degree of polymerization is obtained from Fig. 2-11 and Eq. 2-84. One can also calculate the effect of losses of one reactant or both during the polymerization through volatilization, or side reactions. The precision required in the control of the stoichiometric ratio in a polymerization is easily found from Eq. 2-84. An error in the experimentally employed r value yields a corresponding error in $\overline{X}_n$. The shape of the plots in Fig. 2-11 shows that the effect of an error in r, however, is progressively greater at higher degrees of polymerization. Progressively greater control is required to synthesize the higher-molecular-weight polymer.

TYPE 2. The control of the degree of polymerization in the polymerization of an equimolar mixture A—A and B—B by the addition of small amounts of a monofunctional reactant, for example, B, has been described above. The same equations that apply to a Type 1 polymerization are also applicable here except that r must be redefined as

$$r = \frac{N_A}{N_B + 2N_{B'}} \tag{2-86}$$

where $N_{B'}$ is the number of B molecules present and $N_A = N_B$. The coefficient 2 in front of $N_{B'}$ is required since one B molecule has the same quantitative effect as one excess B—B molecule in limiting the growth of a polymer chain. Equations 2-84 to 2-86 do not apply to Type 2 systems unless equimolar amounts of A—A and B—B are present. Other situations are correctly analyzed only by Eqs. 2-142 and 2-144 in Sec. 2-10a.

TYPE 3. Polymerizations of A—B type monomers such as hydroxy and amino acids automatically take place with internally supplied stoichiometry. For such a polymerization Eqs. 2-84 and 2-85 apply with r equal to 1. This leads to a polymer product that is subsequently unstable toward molecular-weight changes because the end groups of the polymer molecules can react with each other. Molecular weight stabilization is usually accomplished by using a monofunctional B reactant. In this latter case the same equations apply with r redefined as

$$r = \frac{N_A}{N_B + 2N_{B'}} \tag{2-87}$$

where $2N_{B'}$ has the same meaning as in a Type 2 polymerization, and $N_A = N_B =$ the number of A—B molecules. (Bifunctional A—A or B—B monomers can also be employed to control molecular weight in this polymerization.)

The plots in Fig. 2-11 apply equally well to polymerizations of Types 1, 2, and 3, although the scale of the x-axis may be different. When the x-axis is expressed as the stoichiometric ratio r the scale is exactly the same for all three types of polymerization. Different scales will be applicable when the x-axis is expressed in terms of the mole percent excess of the molecular weight controlling reactant. Thus, the x-axis is shown as the mole percent excess of the B—B reactant for Type 1 polymerizations. For polymerizations of Type 2 when $N_A = N_B$ and those of Type 3, the x-axis is shown as the mole percent excess of B groups. The two x-axes differ by a factor of 2 because one B—B molecule is needed to give the same effect as one B molecule. The relationship between the degree of polymerization and the stoichiometric ratio (Eqs. 2-84 to 2-87 and Fig. 2-11) has been verified in a large number of step polymerizations. Its verification and use for molecular weight control has been extensively reviewed in several polymer systems, including polyamides, polyesters, and polybenzimidazoles [Korshak, 1966]. The effect of excess A—A or B—B reactants as well as A or B molecules follows the expected behavior.

The discussion in this section, including the derivations of Eqs. 2-84 through 2-86 assumes that the initial stoichiometric ratio of reactants is the effective stoichiometric ratio throughout the polymerization from start to finish. However, this is often not the case, as there may be losses of one or all reactants as polymerization proceeds. Losses are of two types. Monomer loss due to volatilization is not uncommon, since moderately high reaction temperatures are often used. For example, volatilization losses of the diamine reactant are a problem in polyamidation because of the much higher vapor pressure of diamine compared to the diacid. The extent of this loss depends on the particular diamine used, the specific reaction conditions (temperature, pressure) and whether the polymerization reactor has provision for preventing or minimizing the losses. Aside from volatilization losses, the other pathway by which reactant losses occur is by side reactions. Many polymerization systems involve reactants that can undergo some reaction(s) other than polymerization. Specific examples of such side reactions are described in Secs. 2-8, 2-12, and 2-14. Polymerization conditions usually involve a compromise between conditions that yield the highest reaction rates and those that minimize side reactions and volatilization losses. Because of the very large effect of r on polymer molecular weight this compromise may be much closer to reaction conditions that minimize any change in r. An alternate and/or simultaneous approach involves adjusting the stoichiometric imbalance by a continuous or batchwise replenishment of the "lost" reactant. The amount of added reactant must be precisely calculated based on a chemical analysis of the r value for the reaction system as a function of conversion. Also the additional amount of the "lost" reactant must be added at the appropriate time. Premature addition or the addition of an incorrect amount results in performing the polymerization at other than the required stoichiometric ratio.

2-6c Kinetics of Nonstoichiometric Polymerization

The kinetics of polymerizations involving nonstoichiometric amounts of A and B functional groups can be handled in a straightforward manner. Consider the external acid-catalyzed A—A plus B—B polymerization with $r < 1$. The polymerization rate, defined as the rate of disappearance of the functional groups present in deficient amount, is given by

$$\frac{-d[A]}{dt} = k[A][B] \qquad (2\text{-}88)$$

The following stoichiometry holds:

$$[A]_0 - [A] = [B]_0 - [B] \qquad (2\text{-}89)$$

where $[A]_0$ and $[B]_0$ are the initial concentrations of A and B groups. Combination of Eqs. 2-88 and 2-89 followed by integration [Moore and Pearson, 1981] yields

$$\frac{1}{[B]_0 - [A]_0} \ln \left[\frac{[A]_0[B]}{[B]_0[A]} \right] = kt \qquad (2\text{-}90)$$

which is combined with $r = [A]_0/[B]_0$ to give

$$\ln \frac{[B]}{[A]} = -\ln r + [B]_0(1 - r)kt \qquad (2\text{-}91)$$

A plot of ln ([B]/[A]) versus t is linear with a positive slope of $[B]_0(1 - r)k$ and an intercept of $-\ln r$.

When r is close to unity the polymerization rate is adequately described by the expressions in Secs. 2-2a and 2b for the case of $r = 1$. Only when r is considerably different from unity does it become necessary to employ Eq. 2-91 or its equivalent. Most step polymerizations are carried out with close to stoichiometric amounts of the two reacting functional groups. The main exception to this generalization are some of the reaction systems containing polyfunctional reactants (Secs. 2-10 and 2-12).

2-7 MOLECULAR-WEIGHT DISTRIBUTION IN LINEAR POLYMERIZATION

The product of a polymerization is a mixture of polymer molecules of different molecular weights. For theoretical and practical reasons it is of interest to discuss the distribution of molecular weights in a polymerization. The *molecular-weight distribution* (MWD) has been derived by Flory by a statistical approach based on the concept of equal reactivity of functional groups [Flory, 1953; Howard, 1961; Peebles, 1971]. The derivation which follows is essentially that of Flory and applies equally to A—B and stoichiometric A—A plus B—B types of step polymerizations.

2-7a Derivation of Size Distributions

Consider the probability of finding a polymer molecule containing x structural units. This is synonymous with the probability of finding a molecule with $(x - 1)$ A groups reacted and one A group unreacted. The probability that an A group has reacted at time t is defined as the extent of reaction p. The probability that $(x - 1)$ A groups have reacted is the product of $(x - 1)$ separate probabilities or p^{x-1}. Since the probability of an A group being unreacted is $(1 - p)$, the probability $\underline{N}_x$ of finding the molecule in question, with x structural units, is given by

$$\underline{N}_x = p^{x-1}(1 - p) \tag{2-92}$$

Since $\underline{N}_x$ is synonymous with the *mole-* or *number-fraction* of molecules in the polymer mixture that are x-mers (i.e., that contain x structural units), then

$$N_x = Np^{x-1}(1 - p) \tag{2-93}$$

where N is the total number of polymer molecules and N_x is the number that are x-mers. If the total number of structural units present initially is N_0, then $N = N_0(1 - p)$ and Eq. 2-93 becomes

$$N_x = N_0(1 - p)^2 p^{x-1} \tag{2-94}$$

Neglecting the weights of the end groups, the *weight-fraction* w_x of x-mers (i.e., the weight fraction of the molecules that contains x structural units) is given by $w_x = xN_x/N_0$ and Eq. 2-94 becomes

$$w_x = x(1 - p)^2 p^{x-1} \tag{2-95}$$

Equations 2-92 and 2-95 give the *number-* and *weight-distribution functions*, respectively, for linear step polymerizations at the extent of reaction p. These distributions are usually referred to as the *most probable* or *Flory* or *Flory–Schulz distributions*. Plots of the two distribution functions for several values of p are shown in Figs. 2-12 and 2-13. It is seen that on a number basis there are more monomer molecules than any polymer species regardless of the extent of reaction. Although the number of monomer molecules decreases as the extent of reaction increases, they are still the most plentiful species. The situation is quite different for the weight distribution of molecular weights. On a weight basis, the proportion of low-molecular-weight species is very small and decreases as the extent of reaction increases. The maxima

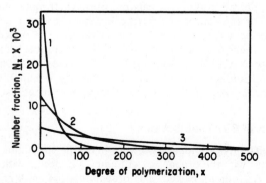

Fig. 2-12 Number-fraction distributions for linear polymerization. Plot 1—$p = 0.9600$; plot 2—$p = 0.9875$; plot 3—$p = 0.9950$. After Howard [1961] (by permission of Iliffe Books, Ltd., London).

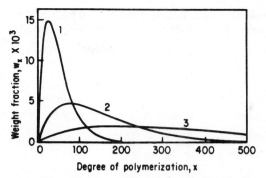

Fig. 2-13 Weight-fraction distributions for linear polmerization. Plot 1—$p = 0.9600$; plot 2— $p = 0.9875$; plot 3—$p = 0.9950$. After Howard [1961] (by permission of Iliffe Books, Ltd., London).

in Fig. 2-13 occur at $x = -(1/\ln p)$, which is very near the number-average degree of polymerization given by Eq. 2-27.

Experimental determinations of molecular-weight distribution are frequently obtained in an integral form in which the *combined* or *cumulative weight-fraction* I_x of all polymer molecules having degrees of polymerization up to and including x are plotted against x. For this reason it is useful to express the Flory distribution function in terms of I_x. This is done by integrating Eq. 2-95 to yield

$$I_x = (1/p) - [1 + (1 - p)x]p^{(x-1)} \tag{2-96}$$

The integral weight-distribution function is graphically illustrated in Fig. 2-14.

2-7b Breadth of Molecular-Weight Distribution

The number- and weight-average degree of polymerization $\overline{X}_n$ and $\overline{X}_w$ can be derived from the number- and weight-distribution functions, respectively. The number- and

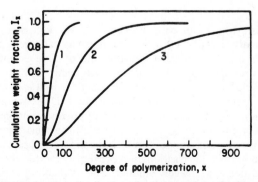

Fig. 2-14 Integral weight-fraction distributions for linear polymerization. Plot 1—$p = 0.9600$; plot 2—$p = 0.9875$; plot 3—$p = 0.9950$. After Howard [1961] (by permission of Iliffe Books, Ltd., London).

weight-average molecular weights have been defined by Eqs. 1-15 and 1-17. Dividing Eq. 1-15 by the weight M_o of a structural unit yields the number-average degree of polymerization as

$$\overline{X}_n = \frac{\Sigma x N_x}{\Sigma N_x} = \Sigma x \underline{N}_x \qquad (2\text{-}97)$$

where the summations are over all values of x. Combination of Eqs. 2-92 and 2-97 gives

$$\overline{X}_n = \Sigma x p^{x-1}(1 - p) \qquad (2\text{-}98)$$

Evaluation of this summation yields

$$\overline{X}_n = \frac{1}{(1 - p)}$$

which is the same result obtained earlier. Dividing Eq. 1-17 by M_o yields

$$\overline{X}_w = \Sigma x w_x \qquad (2\text{-}99)$$

Combination of Eqs. 2-95 and 2-99 gives

$$\overline{X}_w = \Sigma x^2 p^{x-1}(1 - p)^2 \qquad (2\text{-}100)$$

which is evaluated as

$$\overline{X}_w = \frac{(1 + p)}{(1 - p)} \qquad (2\text{-}101)$$

The *breadth of the molecular-weight distribution* is then given by

$$\frac{\overline{X}_w}{\overline{X}_n} = (1 + p) \qquad (2\text{-}102)$$

The ratio $\overline{X}_w/\overline{X}_n$ is synonymous with the ratio $\overline{M}_w/\overline{M}_n$ discussed in Sec. 1-4. It is a measure of the polydispersity of a polymer sample. The value of $\overline{X}_w/\overline{X}_n$ increases with the extent of reaction and approaches 2 in the limit of large extents of reaction. The ratio $\overline{X}_w/\overline{X}_n$ is also referred to as the *polydispersity index* (*PDI*).

The most probable distribution of Flory is generally well established, although its experimental verification has been somewhat limited. Direct evidence for the most probable distribution requires the fractionation of polymer samples followed by molecular-weight measurements on the fractions to allow the construction of experimental plots of $\underline{N}_x$, w_x, I_x versus x for comparison with the theoretical plots. The experimental difficulties involved in polymer fractionation previously limited the number of polymerizations that had been extensively studied. The availability of automated *size exclusion chromatography* (SEC), also referred to as *gel permeation chromatography* (GPC), has significantly increased the available data on molecular-weight distributions of polymers. (The availability of calibration standards for only certain polymers is a limiting factor for size exclusion chromatography. However, the combination of SEC with osmometry and light scattering measurements of molecular weight offers a powerful combination for examining polymer size distributions.) The Flory distribution has been experimentally verified for a number of step polymerizations, including polyamides and polyesters. Less direct verification of the most probable distribution has been made in many instances by determining the ratio $\overline{X}_w/\overline{X}_n$. For many different step polymerizations, this ratio has been found to be close to 2 as required by Eq. 2-102.

2-7c Interchange Reactions

Some polymers (polyesters, polyamides, polysulfides, and others) undergo *interchange reactions* under appropriate conditions. Interchange involves reaction between the terminal functional group of one polymer molecule with the interunit repeating linkage of another polymer molecule, for example, between the terminal —NH$_2$ and interunit —CONH— groups of polyamide molecules. Two polymer chains may react to yield one shorter and one longer chain,

$$\text{H}\!-\!(\!-\text{NH}\!-\!\text{R}\!-\!\text{CO}\!-\!)_n\text{OH} + \text{H}\!-\!(\!-\text{NH}\!-\!\text{R}\!-\!\text{CO}\!-\!)_m\text{OH} \rightarrow$$

$$\text{H}\!-\!(\!-\text{NH}\!-\!\text{R}\!-\!\text{CO}\!-\!)_w\text{OH} + \text{H}\!-\!(\!-\text{NH}\!-\!\text{R}\!-\!\text{CO}\!-\!)_z\text{OH} \qquad (2\text{-}103)$$

If free interchange occurs, the molecular weight distribution will be the Flory distribution described by Eqs. 2-92 and 2-95. Free interchange corresponds to all interunit linkages in all polymer molecules having equal probabilities of interchange. This is analogous to the concept of functional group reactivity independent of molecular size as applied to the interchange reaction. It is apparent that the presence of interchange during a polymerization will not affect the size distribution from that expected for the random polymerization. The Flory or most probable distribution is also that expected for the random scission of the interunit linkages of polymer chains, for example, in the hydrolysis of cellulose.

2-7d Alternate Approaches for Molecular-Weight Distribution

A number of treatments other than that by Flory have been given for the molecular weight distributions in linear step polymerizations [Burchard, 1979; Durand and Bruneau, 1979a, 1979b]. However, a knowledge of the average properties ($\overline{M}_n$, $\overline{M}_w$, and *PDI*) is often sufficient for many practical purposes. Macosko and Miller [1976] developed a useful statistical approach for obtaining the average properties without the need to calculate the molecular-weight distributions. This approach will be described for a polymerization system composed of A—A, B—B, and B'—B' where A groups can react with B groups and with B' groups [Ozizmir and Odian, 1980a, 1980b]. To keep the system as simple as possible, the molecular weights of the three structural units are taken as equal (and denoted by M_o) and the initial system contains equimolar amounts of B—B and B'—B' monomers with the total moles of A—A being equal to the sum of B—B and B'—B'.

The expected masses of polymers which contain a randomly selected A, B, or B' group in the system are denoted by w_A, w_B, and $w_{B'}$. In order to obtain these quantities it is convenient to introduce the quantities w_A^i, w_B^i, $w_{B'}^i$, and w_A^o, w_B^o, $w_{B'}^o$. Here w_A^i represents the expected mass attached to a randomly selected A group in the system looking inward toward the other A group of the A—A structural unit of which it is a part and w_A^o represents the expected mass attached looking outward from the randomly selected A group as shown in **XV**. Similarly, w_B^i, w_B^o, $w_{B'}^i$, and $w_{B'}^o$ represent the expected inward and outward masses attached to B and B' groups.

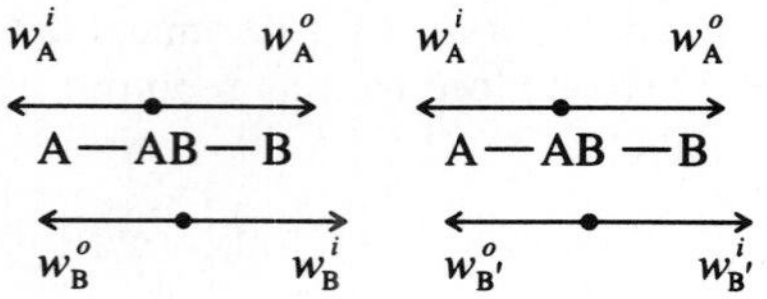

$$w_A^i \qquad w_A^o \qquad w_A^i \qquad w_A^o$$

$$\overleftrightarrow{\quad\bullet\quad} \qquad \overleftrightarrow{\quad\bullet\quad}$$

$$A—AB—B \qquad A—AB—B$$

$$\overleftrightarrow{\quad\bullet\quad} \qquad \overleftrightarrow{\quad\bullet\quad}$$

$$w_B^o \qquad w_B^i \qquad w_{B'}^o \qquad w_{B'}^i$$

XV

The following relationships hold for this system:

$$w_A = w_A^i + w_A^o \tag{2-104}$$

$$w_B = w_B^i + w_B^o \tag{2-105}$$

$$w_{B'} = w_{B'}^i + w_{B'}^o \tag{2-106}$$

$$w_A^i = M_o + w_A^o \tag{2-107}$$

$$w_A^o = p_A \left\{ \frac{p_B w_B^i}{(p_B + p_{B'})} + \frac{p_{B'} w_{B'}^i}{(p_B + p_{B'})} \right\} = \frac{(p_B w_B^i + p_{B'} w_{B'}^i)}{2} \tag{2-108}$$

$$w_B^i = M_o + w_B^o \tag{2-109}$$

$$w_B^o = p_B w_A^i \tag{2-110}$$

$$w_{B'}^i = M_o + w_{B'}^o \tag{2-111}$$

$$w_{B'}^o = p_{B'} w_A^i \tag{2-112}$$

$$p_A = \frac{(p_B + p_{B'})}{2} \tag{2-113}$$

Most of the above relationships are simple statements of material balance. Equations 2-104 through 2-106 state that the total mass of polymer attached to an A, B, or B' group is the sum of the inward and outward masses attached to that group. Equations 2-107, 2-109, and 2-111 state that the difference between the inward and outward masses attached to a group is M_o. Equation 2-112 indicates that the polymer mass attached to a B' group looking outward from the B' group equals the probability of that group having reacted with an A group multiplied by the mass of polymer attached to the A group looking inward from that A group. The corresponding descriptions for w_A^o and w_B^o are equations 2-108 and 2-110.

Equations 2-107 through 2-113 can be solved and their results combined with Eqs. 2-104 through 2-106 to yield

$$w_A = U - M_o \tag{2-114a}$$

$$w_B = M_o + p_B U \tag{2-114b}$$

$$w_{B'} = M_o + p_{B'} U \tag{2-114c}$$

where

$$U = 2w_A^i = \frac{2M_o + (p_B M_o + p_{B'} M_o)}{1 - (p_B^2 + p_{B'}^2)/2} \tag{2-115}$$

The weight-average molecular weight, obtained by a weight-averaging of w_A, w_B, and $w_{B'}$, is given by

$$\overline{M}_w = \frac{w_A}{2} + \frac{w_B}{4} + \frac{w_{B'}}{4} \tag{2-116}$$

which yields

$$\overline{M}_w = \frac{M_o}{4} \left\{ \frac{(2 + p_B + p_{B'})^2}{1 - (p_B^2 + p_{B'}^2)/2} \right\} \tag{2-117}$$

upon substitution for w_A, w_B, and $w_{B'}$. The number-average molecular weight, obtained by dividing the total number of moles of reactants initially present divided by the total present at any time, yields

$$\overline{M}_n = \frac{M_o}{(1 - p_A)} \tag{2-29}$$

It is useful to introduce the fractions of unreacted A, B, and B' functional groups

$$\alpha = 1 - p_A \tag{2-118a}$$

$$\beta = 1 - p_B \tag{2-118b}$$

$$\gamma = 1 - p_{B'} \tag{2-118c}$$

For the case where B and B' groups have the same reactivity

$$\gamma = \beta \tag{2-119}$$

Combination of Eqs. 2-113, 2-117 through 2-119, and 2-29 yields the polydispersity index as

$$PDI = \frac{(2 - \alpha)^2}{2 - (\beta^2 + \gamma^2)/2\alpha} \tag{2-120}$$

Equation 2-120 yields *PDI* as a function of conversion in a straightforward manner without having to solve differential equations to obtain the number- and weight-average molecular-weight distributions. One need only take a set of β values and then calculate the corresponding γ, α, and *PDI* values from Eqs. 2-119 and 2-120. The limit of *PDI* at complete conversion ($p_A = 1$, $\alpha = 0$) is 2 as for the Flory distribution. The Macosko–Miller method has also been applied to other polymerizations, including the A—B plus A'—B' system. In addition to being a simpler method for obtaining the average properties compared to the Flory and similar methods, it more readily allows an evaluation of the effect of various reaction parameters such as unequal group reactivity or nonstoichiometric amounts of reactants on $\overline{M}_n$, $\overline{M}_w$, and *PDI* (see Sec. 2-7e).

2-7e   Effect of Reaction Variables on MWD

2-7e-1   Unequal Reactivity of Functional Groups

The molecular-weight distribution and/or *PDI* has been described for several cases where the assumption of equal reactivity of functional groups is not valid. Unequal reactivity is easily handled by the Macosko–Miller method. For the A—A + B—B + B'—B' system described in the previous section, we simply redefine the relationship between β and γ by

$$\gamma = \beta^s \tag{2-47}$$

(which was derived previously in Sec. 2-2d-2) where s is the ratio of rate constants for reaction of B' and B groups [Ozizmir and Odian, 1980a, 1980b].

The polydispersity index increases (for conversions <100%) when the reactivities of B and B' groups are different compared to the case where the reactivities are the same. The opposite trend is found for the A—A + B—B' system [Gandhi and Babu, 1979, 1980]. There is a basic difference between the two- and the three-species reaction systems. In the three-species system, when B' groups have higher reactivity than B groups, the B'—B' monomer continues polymerization with A—A monomer, independently of the B—B monomer, to form large-sized polymer chains. In the two-species system, long chains cannot be formed without participation of B groups in the polymerization process. The extents of the broadening and narrowing effects in the two systems increase as the difference in reactivities of B and B' become larger. The effects are significant at low conversions; for instance, *PDI* is 2.10 at 50% conversion for the three-species system when B' is 20-fold more reactive than B compared to a *PDI* of 1.50 for the equal reactivity case. However, the PDI at complete conversion equals 2 in both the two- and three-species systems, which is the same limit as for the equal reactivity case.

2-7e-2   Change in Reactivity on Reaction

An increase in the reactivity of one functional group in a bifunctional reactant upon reaction of the other functional group results in an increase in *PDI* [Gandhi and Babu, 1979, 1980]. The *PDI* at complete conversion exceeds 2 when there is a greater than twofold increase in reactivity. *PDI* decreases when there is a decrease in reactivity of one functional group on reaction of the other group but the *PDI* at complete conversion is 2. The same trends are found but are more exaggerated when functional group reactivity varies continuously with molecular size [Gupta et al., 1979a, 1979b; Nanda and Jain, 1968].

2-7e-3   Nonstoichiometry of Functional Groups

Using the Macosko–Miller approach for the A—A + B—B + B'—B' system, we introduce r_1 and r_2

$$r_1 = \frac{[\text{B}]_0}{[\text{A}]_0} \tag{2-121a}$$

$$r_2 = \frac{[\text{B}']_0}{[\text{A}]_0} \tag{2-121b}$$

$$p_\text{A} = r_1 p_\text{B} + r_2 p_{\text{B}'} \tag{2-121c}$$

to describe the relative amounts and extents of reaction of the three reactants and reformulate Eqs. 2-108 and 2-115 as

$$w^o_A = (p_B r_1 w^i_B + p_{B'} r_2 w^i_{B'})$$ (2-122)

$$U = 2w^i_A = \frac{2M_o + (p_B r_1 M_o + p_{B'} r_2 M_o)}{1 - (p^2_B r_1 + p^2_{B'} r_2)/2}$$ (2-123)

The expressions for $\overline{M}_w$, $\overline{M}_n$, and *PDI* are complicated but can be solved numerically [Ozizmir and Odian, 1990]. When B and B' groups together are in excess over A groups ($r_1 + r_2 > 1$) and the reactivities of B and B' are the same ($s = 1$), *PDI* is very slightly decreased at all conversions in comparison to the stoichiometric case ($r_1 + r_2 = 1$). The final *PDI* is 1.98. For $s > 1$ (B' more reactive than B), *PDI* is initially decreased but increases with conversion and the final *PDI* is above 2. The trends are more exaggerated when B' is in excess over B and less exaggerated when B is in excess over B'. For example, for the case $r_1 + r_2 = 1.25$, $s = 20$, the final *PDI* is 2.63, 6.50, and 2.16, respectively, when r_1/r_2 is 1, 3, and 1/3. When A groups are in excess over B and B', *PDI* is very slightly decreased at all conversions in comparison to the stoichiometric case. The final *PDI* is 1.98 independent of s, r_1/r_2, and ($r_1 + r_2$).

Nonstoichiometric amounts of reactants decrease *PDI* at all conversions for A—A + B—B' polymerizations [Gandi and Babu, 1979, 1980]. The decrease in *PDI* is greater the larger the difference in the reactivities of B and B' groups when B—B' is in excess over A—A. The decrease in *PDI* is independent of the difference in reactivities of B and B' when A—A is in excess. The trend is the same with A—A in excess when B and B' have the same initial reactivity but the reactivity of each group changes on reaction of the other. However, a different trend is seen when B—B' is in excess. *PDI* increases with a limiting value at complete conversion of 2 or greater than 2 depending on whether the difference in reactivity of B and B' is less than or greater than 2.

2-8 PROCESS CONDITIONS

2-8a Physical Nature of Polymerization Systems

Several considerations are common to all processes for step polymerizations in order to achieve high molecular weights. One needs to employ a reaction with an absence or at least a minimum of side reactions, which would limit high conversions. Polymerizations are carried out at high concentrations to minimize cyclization and maximize the reaction rate. High purity reactants in stoichiometric or near-stoichiometric amounts are required. The molecular weight is controlled by the presence of controlled amounts of monofunctional reagents or an excess of one of the bifunctional reagents. Equilibrium considerations are also of prime importance. Since many step polymerizations are equilibrium reactions, appropriate means must be employed to displace the equilibrium in the direction of the polymer product. Distillation of water or other small molecule products from the reaction mixture by suitable reaction temperatures and reduced pressures are often employed for this purpose.

Table 2-8 shows values of some kinetic and thermodynamic characteristics of typical step polymerizations [Bekhli et al., 1967; Chelnokova et al., 1949; Fukumoto, 1956; Hamann et al., 1968; Malhotra and Avinash, 1975, 1976; Ravens and Ward, 1961;

TABLE 2-8 Values of Reaction Parameters in Typical Polymerizations

Reactants[a]	T (°C)	$k \times 10^3$ (liters/mole-sec)	E_a (kJ/mole)[b]	ΔH (kJ/mole)[b]
Polyester				
HO(CH₂)₁₀OH + HOOC(CH₂)₄COOH	161	7.5×10^{-2}	59.4	
HO(CH₂)₁₀OH + HOOC(CH₂)₄COOH[c]	161	1.6		
HOCH₂CH₂OH + p-HOOC—φ—COOH	150			−10.9
HO(CH₂)₆OH + ClOC(CH₂)₈COCl	58.8	2.9	41	
p-HOCH₂CH₂OOC—φ—COOCH₂CH₂OH	275	0.5	188	
p-HOCH₂CH₂OOC—φ—COOCH₂CH₂OH[d]	275	10	58.6	
Polyamide				
H₂N(CH₂)₆NH₂ + HOOC(CH₂)₈COOH	185	1.0	100.4	
Piperazine + p-Cl—CO—φ—CO—Cl		$10^7 - 10^8$		
H₂N(CH₂)₅COOH	235			≈24
Polyurethane				
m-OCN—φ—NCO + HOCH₂CH₂OCO(CH₂)₄COOCH₂CH₂OH	60	0.40[e]	31.4	
m-OCN—φ—NCO + HOCH₂CH₂OCO(CH₂)₄COOCH₂CH₂OH	60	0.23[f]	35.0	
Phenol-Formaldehyde Polymer				
φ—OH + H₂CO[c]	75	1.1[g]	77.4	
φ—OH + H₂CO[h]	75	0.048[g]	76.6	

[a]Uncatalyzed unless otherwise noted.
[b]1 cal = 4.184 J.
[c]Acid-catalyzed.
[d]Catalyzed by Sb₂O₃.
[e]k_1 value.
[f]k_2 value.
[g]Average k for all functional groups.
[h]Base-catalyzed.

Saunders and Dobinson, 1976; Stevenson, 1969; Ueberreiter and Engel, 1977]. These data have implications on the temperature at which polymerization is carried out. Most step polymerizations proceed at relatively slow rates at ordinary temperatures. High temperatures in the range of 150–200°C and higher are frequently used to obtain reasonable polymerization rates. Table 2-8 shows that the rate constants are not large even at these temperatures. Typical rate constants are of the order of 10^{-3} liter/mole-sec. There are a few exceptions of step polymerizations with significantly larger k values, for example, the polymerization reaction between acid halides and alcohols. The need to use higher temperatures can present several problems, including loss of one or the other reactant by degradation or volatilization. Oxidative degradation of polymer is also a potential problem in some cases. The use of an inert atmosphere (N_2, CO_2) can minimize oxidative degradation.

   Bulk or *mass polymerization* is the simplest process for step polymerizations, since it involves only the reactants and whatever catalyst, if any, which is required [Menikheim, 1988]. There is a minimum of potentialities for contamination, and product separation is simple. Bulk polymerization is particularly well suited for step polymerization because high-molecular-weight polymer is not produced until the very last stages of reaction. This means that the viscosity is relatively low throughout most of the course of the polymerization and mixing of the reaction mixture is not overly difficult. Thermal control is also relatively easy, since the typical reaction has both a modest activation energy E_a and enthalpy of polymerization ΔH. Although some step polymerizations have moderately high activation energies, for example, 100.4 kJ/mole for the polymerization of sebacic acid and hexamethylene diamine (Table 2-8), the ΔH is still only modestly exothermic. The exact opposite is the case for chain polymerizations, which are generally highly exothermic with high activation energies and where the viscosity increases much more rapidly. Thermal control and mixing present much greater problems in chain polymerizations.

   Bulk polymerization is widely used for step polymerizations. Many polymerizations, however, are carried out in *solution* with a solvent present to solubilize the reactants, or to allow higher reaction temperatures to be employed, or as a convenience in moderating the reaction and acting as a carrier.

2-8b   Different Reactant Systems

For many step polymerizations there are different combinations of reactants that can be employed to produce the same type of polymer (Table 1-1). Thus the polymerization of a hydroxy acid yields a polymer very similar to (but not the same as) that obtained by reacting a diol and diacid.

$$n\text{HO—R—CO}_2\text{H} \rightarrow \text{H} \text{—} (\text{O—R—CO})_{\overline{n}} \text{OH} + (n - 1)\text{H}_2\text{O} \qquad (2\text{-}124)$$

$$n\text{HO—R—OH} + n\text{HO}_2\text{C—R}'\text{—CO}_2\text{H} \rightarrow$$

$$\text{HO} \text{—} (\text{R—OCO—R}'\text{—COO})_{\overline{n}} \text{H} + (2n - 1)\text{H}_2\text{O} \qquad (2\text{-}125)$$

On the other hand, it is apparent that there are different reactant systems that can yield the exact same polymer. Thus the use of the diacid chloride or anhydride instead of the diacid in Eq. 2-125 would give exactly the same polymer product. The organic chemical aspects of the synthesis of various different polymers by different step polymerization processes have been discussed [Elias, 1984; Lenz, 1967; Morgan, 1965].

Whether one particular reaction or another is employed to produce a specific polymer depends on several factors. These include the availability, ease of purification, and properties (both chemical and physical) of the different reactants and whether one or another reaction is more devoid of destructive side reactions.

The ability to obtain high-molecular-weight polymer from a reaction depends on whether the equilibrium is favorable. If the equilibrium is unfavorable as it is in many instances, success depends on the ease with which the polymerization can be driven close to completion. The need for and the ease of obtaining and maintaining stoichiometry in a polymerization is an important consideration. The various requirements for producing a high polymer may be resolved in quite different ways for different polymers. One must completely understand each type of polymerization reaction so as to appropriately meet the stringent requirements for the synthesis of high-molecular-weight polymer. Various step polymerizations are described below and serve to illustrate how the characteristics of a polymerization reaction are controlled so as to obtain high polymer.

2-8c   Interfacial Polymerization

Many of the polymers that are produced by the usual high-temperature reactions could be produced at lower temperatures by using the faster Schotten–Baumann reactions of acid chlorides. Thus polyesters and polyamides could be produced by replacing the diacid or diester reactant by the corresponding diacyl chloride

$$n\text{ClCO—R—COCl} + n\text{HO—R'—OH} \rightarrow$$

$$-(\text{CO—R—COO—R'—O})_{n} + 2n\text{HCl} \tag{2-126}$$

$$n\text{ClCO—R—COCl} + n\text{H}_2\text{N—R'—NH}_2 \rightarrow$$

$$-(\text{CO—R—CONH—R'—NH})_{n} + 2n\text{HCl} \tag{2-127}$$

2-8c-1   Description of Process

The rate constants for these reactions are orders of magnitude greater than those for the corresponding reactions of the diacid or diester reactants (Table 2-8). The use of such reactants in a novel low-temperature polymerization technique called *interfacial polymerization* has been extensively studied [Morgan and Kwolek, 1959a, 1959b; Nikonov and Savinov, 1977]. Temperatures in the range 0–50°C are usually employed. Polymerization of two reactants is carried out at the interface between two liquid phases, each containing one of the reactants (Fig. 2-15). Polyamidation is performed at room temperature by placing an aqueous solution of the diamine on top of an organic phase containing the acid chloride. The reactants diffuse to and undergo polymerization at the interface. The polymer product precipitates and is continuously withdrawn in the form of a continuous film or filament if it has sufficient mechanical strength. Mechanically weak polymers that cannot be removed impede the transport of reactants to the reaction site and the polymerization rate decreases with time. The polymerization rate is usually diffusion-controlled, since the rates of diffusion of reactants to the interface are slower than the rate of reaction of the two functional groups. (This may not be the situation when reactions with small rate constants are employed.)

Interfacial polymerization is mechanistically different from the usual step polymerization in that the monomers diffusing to the interface will react only with polymer chain ends. The reaction rates are so high that diacid chloride and diamine monomer

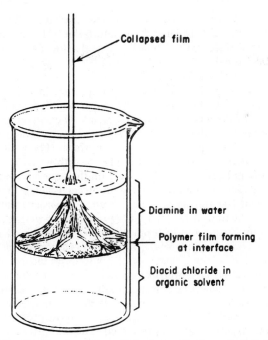

Collapsed film

Diamine in water

Polymer film forming at interface

Diacid chloride in organic solvent

Fig. 2-15 Interfacial polymerization; removal of polymer film from the interface. From Morgan and Kwolek [1959a, 1959b] (by permission of Division of Chemical Education, American Chemical Society, Washington, D.C. and Wiley-Interscience, New York); an original photograph, from which this figure was drawn, was kindly supplied by Dr. P. W. Morgan.

molecules will react with the growing polymer chain ends before they can penetrate through the polymer film to start the growth of new chains. There is thus a strong tendency to produce higher-molecular-weight polymer in the interfacial process compared to the usual processes. Also, interfacial polymerization does not require overall bulk stoichiometry of the reactants in the two phases. Stoichiometry automatically exists at the interface where polymerization proceeds. There is always a supply of both reactants at the interface due to diffusion from the organic and aqueous phases. Furthermore, high-molecular-weight polymer is formed at the interface regardless of the overall percent conversion based on the bulk amounts of the two reactants. The overall percent conversion can be increased by employing a stirred system as a means of increasing the total area of reacting interface.

Several reaction parameters must be controlled in order for interfacial polymerization to proceed successfully. An inorganic base must be present in the aqueous phase to neutralize the by-product hydrogen chloride. If it were not neutralized the hydrogen chloride would tie up the diamine as its unreactive amine hydrochloride salt leading to greatly lowered reaction rates. The acid chloride may undergo hydrolysis to the unreactive acid at high concentrations of the inorganic base or at low polymerization rates. Hydrolysis decreases not only the polymerization rate but greatly limits the polymer molecular weight, since it converts the diacid chloride into the diacid, which is unreactive at the temperatures employed in interfacial polymerization. The slower the polymerization rate, the greater the problem of hydrolysis as the acid chloride will have more time to diffuse through the interface and into the water layer. Thus acid hydrolysis prevents the use of the interfacial technique for the synthesis of polyesters from diols, since the reaction is relatively slow ($k \sim 10^{-3}$ liter/mole-sec). The reaction of diacid chlorides and diamines is so fast ($k \sim 10^4 - 10^5$ liters/mole-sec) that hydrolysis is usually completely absent.

The choice of the organic solvent is very important in controlling the polymer molecular weight, since it appears that the polymerization actually occurs on the organic solvent side of the interface in most systems. The reason for this is the greater tendency of the diamine to diffuse into the organic solvent compared to the diffusion of diacid chloride into the aqueous side of the interface. (For some systems, e.g., the reaction of the disodium salt of a dihydric phenol with a diacid chloride, the exact opposite is the case and polymerization occurs on the aqueous side of the interface.) An organic solvent that precipitates the high-molecular-weight polymer but not the low-molecular-weight fractions is desirable. Premature precipitation of the polymer will prevent the production of the desired high-molecular-weight product. Thus, for example, xylene and carbon tetrachloride are precipitants for all molecular weights of poly(hexamethylene sebacate), while chloroform is a precipitant only for the high-molecular-weight polymer. Interfacial polymerization with the former organic solvents would yield only low-molecular-weight polymer. The molecular-weight distributions observed in interfacial polymerizations are usually quite different from the most probable distribution [Arai et al., 1985; Korshak, 1966; Morgan, 1965]. Most interfacial polymerizations yield distributions broader than the most probable distribution, but narrower distributions have also been observed. The differences are probably due to fractionation when the polymer undergoes precipitation. The effect is dependent on the organic solvent used and the solubility characteristics of the polymer.

The organic solvent can also effect the polymerization by affecting the diffusion characteristics of the reaction system. A solvent that swells the precipitated polymer is desirable to maximize the diffusion of reactants through it to the reaction site. However, the swelling should not decrease the mechanical strength of the polymer below the level that allows it to be continuously removed from the interface. It has been found that the optimum molar ratio of the two reactants in terms of producing the highest yield and/or highest molecular weight is not always 1:1 and often varies with the organic solvent. The lower the tendency of the water-soluble reactant to diffuse into the organic phase, the greater must be its concentration relative to the other reactant's concentration. The optimum ratio of concentrations of the two reactants is that which results in approximately equalizing the rates of diffusion of the two reactants to the interface.

2-8c-2 *Utility*

The interfacial technique has several advantages. Bulk stoichiometry is not needed to produce high-molecular-weight polymers and fast reactions are used. The low temperatures allow the synthesis of polymers that may be unstable at the high temperatures required in the typical step polymerization. The interfacial technique has been extended to many different polymerizations, including the formation of polyamides, polyesters, polyurethanes, polysulfonamides, polycarbonates, and polyureas. However, there are disadvantages to the process, which have limited its commercial utility. These include the high cost of acid chloride reactants and the large amounts of solvents that must be used and recovered. Commercial utilization has been limited to the synthesis of polycarbonates, aliphatic polysulfides, and aromatic polyamides.

2-8d Polyesters

Various combinations of reactant(s) and process conditions are potentially available to synthesize polyesters [Goodman, 1988]. Polyesters can be produced by direct esterification of a diacid with a diol (Eq. 2-125) or self-condensation of a hydroxy carboxylic acid (Eq. 2-124). Since polyesterification, like many step polymerizations, is an equilibrium reaction, water must be continuously removed to achieve high conversions and high molecular weights. Control of the reaction temperature is important to minimize side reactions such as dehydration of the diol to form diethylene glycol

$$HOCH_2CH_2OH \rightarrow HOCH_2CH_2OCH_2CH_2OH + H_2O \qquad (2\text{-}128)$$

and β-scission of the polyester to form acid and alkene end groups that subsequently react to form an anhydride plus acetaldehyde

$$\sim\sim R - COOCH_2CH_2OCO - R \sim\sim \rightarrow$$

$$\sim\sim R - COOH + CH_2{=}CH - OCO - R \sim\sim \rightarrow$$

$$CH_3CHO + \sim\sim R - COOCO - R \sim\sim \qquad (2\text{-}129)$$

Other reported side reactions include dehydration between alcohol end groups, decarboxylation of diacid monomer, dehydration between carboxyl end groups, and scission and polymerization of the alkene end groups formed in Eq. 2-129. Side reactions directly interfere with the polymerization by altering the stoichiometric ratio of the reacting functional groups, and this affects the polymer molecular weight. Additionally, side reactions can have deleterious effects on polymer properties. For example, diethylene glycol formed by dehydration of ethylene glycol (Eq. 2-128) takes part in the polymerization. For poly(ethylene terephthalate), the T_m is decreased by the introduction of diethylene glycol units in place of ethylene glycol units in the polymer chain. The presence of acetaldehyde as an impurity causes problems when poly(ethylene terephthalate) is used to produce food and beverage containers. Acetaldehyde also results in discoloration in the final polymer product due to the formation of aldol-type products.

In industrial practice relatively few linear polyesters are synthesized by direct reactions of diacids and diols because of the high temperatures required to completely eliminate water. The reactions of acid anhydrides and diols suffer from the same problem. However, these reactions are used to synthesize low-molecular-weight and crosslinked polyesters. Commercial products based on phthalic and maleic anhydrides are produced (Sec. 2-12a). Polyesters have been synthesized by the reaction of diacylchlorides with diols. The reaction rates are generally too slow with aliphatic diols, but dihydric phenols are sufficiently reactive to yield high polymer (Sec. 2-8e). Ester interchange is often the most practical reaction for polyester synthesis. The ester interchange reaction, typically using a dimethylester, is usually a faster reaction than direct esterification of the diacid. Also, the diester often is more easily purified and has better solubility characteristics. Various weak bases such as manganese acetate, antimony(III) oxide, and titanium alkoxides are used to catalyze the polymerization [Shah et al., 1984].

The most important commercial polyester is poly(ethylene terephthalate), often referred to as PET. The IUPAC name is poly(oxyethyleneoxyterephthaloyl). Poly(ethylene terephthalate) is synthesized from dimethyl terephthalate and ethylene glycol by a two-stage ester interchange process. The first stage involves an ester interchange reaction to produce bis(2-hydroxyethyl)terephthalate along with small amounts of

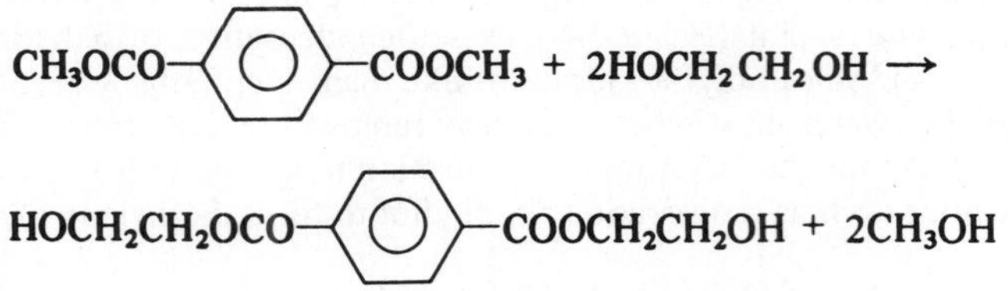

$$CH_3OCO-\left\langle\bigcirc\right\rangle-COOCH_3 \ + \ 2HOCH_2CH_2OH \ \longrightarrow$$

$$HOCH_2CH_2OCO-\left\langle\bigcirc\right\rangle-COOCH_2CH_2OH \ + \ 2CH_3OH \qquad (2\text{-}130)$$

larger-sized oligomers. The reactants are heated at temperatures increasing from 150 to 210°C and the methanol is continuously distilled off.

In the second-stage the temperature is raised to 270–280°C and polymerization proceeds with the removal of ethylene glycol being facilitated by using a partial vacuum

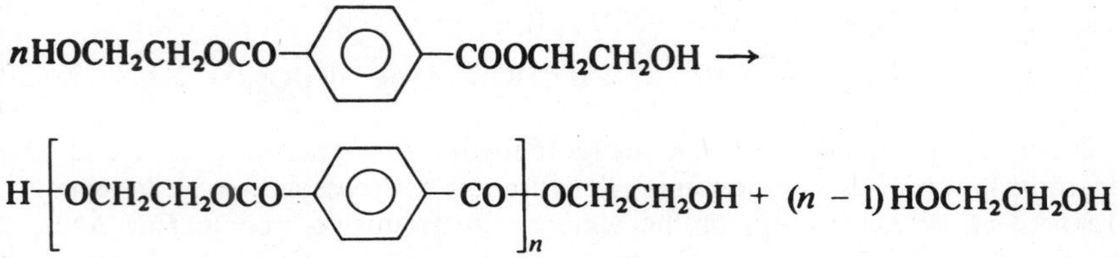

$$nHOCH_2CH_2OCO-\left\langle\bigcirc\right\rangle-COOCH_2CH_2OH \ \longrightarrow$$

$$H\left[OCH_2CH_2OCO-\left\langle\bigcirc\right\rangle-CO\right]_n OCH_2CH_2OH \ + \ (n-1)HOCH_2CH_2OH$$

$$(2\text{-}131)$$

of 0.5–1 torr (66–133 Pa). The first stage of the polymerization is a solution polymerization. The second stage is a melt polymerization since the reaction temperature is above the crystalline melting temperature of the polymer. Figure 2-16 illustrates a commercial process for this polymerization [Ellwood, 1967].

The properties and usefulness of the final polymer depends on controlling its structure by appropriate control of process variables during polymerization and subsequent processing into product. Temperature control and the choice of catalysts are critical in minimizing deleterious side reactions. A dual catalyst system is used in PET synthesis. The first-stage catalyst is an acetate of manganese, zinc, calcium, cobalt, or magnesium. Antimony(III) oxide is usually added as the second-stage catalyst; it is ineffective alone for the first-stage reaction. The first-stage catalyst is often inactivated by the addition of an alkyl or aryl phosphite or phosphate.

The production of high-molecular-weight polymer requires the complete removal of ethylene glycol because of the unfavorable equilibrium that would otherwise exist. If ethylene glycol were not removed equilibrium would be established at too low an extent of reaction (approximately $p < 0.7$) and the product would be of very low molecular weight (Fig. 2-11). A unique feature of the ester interchange process is the absence of the need for stoichiometric balance of the two functional groups at the start of the polymerization. Stoichiometric balance is inherently achieved in the last stages of the second step of the process. In fact, an excess of ethylene glycol is used in the initial stage to increase the rate of formation of bis(2-hydroxyethyl)terephthalate.

The availability of terephthalic acid in higher purity than previously available has led to its use in a modification of the two-stage process. Terephthalic acid and an

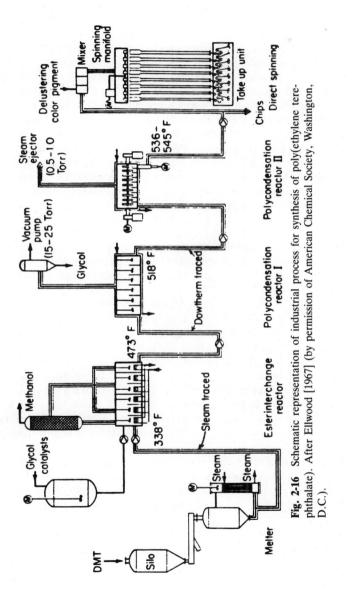

Fig. 2-16 Schematic representation of industrial process for synthesis of poly(ethylene tere-phthalate). After Ellwood [1967] (by permission of American Chemical Society, Washington, D.C.).

excess of ethylene glycol (in the form of a paste) are used to produce the bis(2-hydroxyethyl)terephthalate, which is then polymerized as described above.

Poly(ethylene terephthalate), known by the trade names *Mylar, Dacron*, and *Terylene*, is a very high volume polymer—over 4.5 billion pounds per year of fiber and plastic are produced in the United States. The world-wide production is about twice that of the United States. These figures are especially impressive when we note that PET was not introduced as a commercial product until 1953. Because of its high crystalline melting temperature (270°C) and stiff polymer chains, PET has good mechanical strength, toughness, and fatigue resistance up to 150–175°C as well as good chemical, hydrolytic, and solvent resistance. Fiber applications account for about 80% of the total PET production. Poly(ethylene terephthalate) fiber has outstanding crease resistance, has good abrasion resistance, can be treated with crosslinking resins to impart permanent-press (wash and wear) properties, and can be blended with cotton and other cellulosic fibers to give better feel and moisture permeation. Fiber applications include wearing apparel, curtain, upholstery, thread, tire cord, and fabrics for industrial filtration. Large amounts of PET find applications as plastics. Film applications include photographic, magnetic, and X-ray films or tapes, boil-in-bag food pouches, metallized films, and electrical insulation. The outstanding barrier properties of PET has led to a rapid growth in the use of blow molded bottles for soft drinks, beers, spirits, and other food products. Poly(ethylene terephthalate) also finds use as an *engineering plastic* where it replaces steel, aluminum, and other metals in the manufacture of precision moldings for electrical and electronic devices, domestic and office appliances, and automobile parts. In these engineering applications PET is often reinforced with glass fiber or compounded with silicones, graphite, or Teflon to improve strength and rigidity. The glass reinforced grades of PET are rated for continuous use at temperatures up to 140–155°C.

The use of PET as an engineering plastic has been somewhat limited by its relatively low rate of crystallization. This results in increased processing costs due to long mold recycle times. (The addition of nucleating agents such as talc, MgO, calcium silicate, zinc stearate, or plasticizers such as long-chain fatty esters helps overcome this problem.) Poly(butylene terephthalate) (PBT), produced by substituting 1,4-butanediol for ethylene glycol, crystallizes much faster than PET. Its maximum use temperature is 120–140°C, slightly lower than that of PET. The volume of PBT used for engineering applications is increasing faster than that of PET. PBT is not, however, a competitor in other applications of poly(ethylene terephthalate). More than 200 million pounds of PET and PBT are used annually in the United States in engineering plastics applications.

Other polyesters of commercial importance are polycarbonates, liquid crystal polyesters, and unsaturated polyesters (Secs. 2-8e, 2-14g, 2-12a).

Completely aliphatic polyesters, made from aliphatic diacid and aliphatic diol components), are not of major industrial importance because of their low melting temperatures and poor hydrolytic stability. (Low-molecular-weight aliphatic polyesters are used as plasticizers and prepolymer reactants in the synthesis of polyurethanes; see Secs. 2-12e, 2-13c-2).

2-8e Polycarbonates

Polycarbonates are polyesters of carbonic acid. The most important commercial polycarbonate is that based on 2,2′-bis(4-hydroxyphenyl)propane (*bisphenol A*) [Freitag et al., 1988]. It has been synthesized by the reaction of the dihydric phenol with phosgene or by ester interchange with diphenyl carbonate:

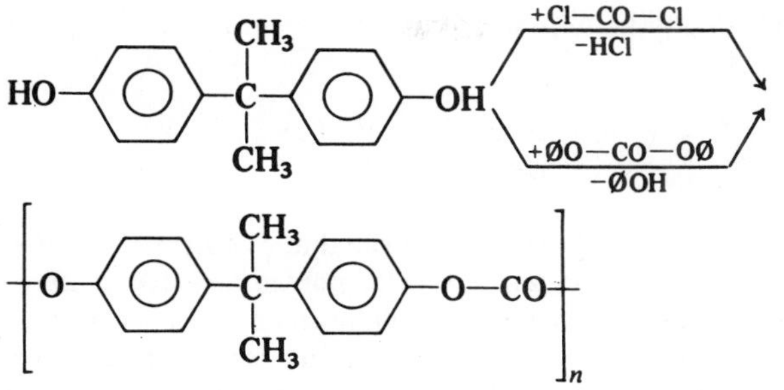

$$\hspace{12cm} (2\text{-}132)$$

Polymerization by the ester interchange route is carried out as a two-stage melt polymerization very similar to that described for poly(ethylene terephthalate). However, most industrial processes involve the phosgene reaction in a stirred interfacial polymerization. Overall economics and easier control of polymer molecular weight favor the phosgene process over ester interchange. Organic solvents such as chlorobenzene, 1, 2-dichloroethane, tetrahydrofuran (THF), anisole, and dioxane are used. The bisphenol A is usually dissolved in aqueous alkali to form the phenolate salt and then the organic solvent added followed by phosgene. The organic solvent prevents the loss of phosgene by hydrolysis and precipitation of the polymer before it has reached the desired molecular weight. Reaction temperatures in the range 0–50°C are used. Phase-transfer catalysts, such as quaternary ammonium and sulfonium salts and crown ethers, may be added to enhance the transfer of the phenolate salt across the interfacial boundary into the organic phase. The polymerization is usually a two-stage reaction. Oligomers are formed in the first stage. Tertiary amines are added in the second stage to catalyze the further reaction to high-molecular-weight polymer. (An alternate route to polycarbonates, the ring-opening polymerization of cyclic polycarbonate oligomers, has some potential advantages over the step polymerization route; see Sec. 7-5c.)

The polycarbonate based on bisphenol A has the IUPAC name poly(carbonyldioxy-1,4-phenyleneisopropylidene-1,4-phenylene) (trade names: *Lexan, Merlon, Calibre*); it is usually referred to as *polycarbonate* or *PC*. Although it can be crystallized ($T_m = 270°C$), most polycarbonates are amorphous ($T_g = 150°C$). Chain stiffening due to a combination of the benzene rings and bulky tetrasubstituted carbons in the polymer chain and the high T_g result in a good combination of mechanical properties over a considerable temperature range (15–130°C). PC has excellent resistance to acids and oxidants, better than PET, but is somewhat less resistant to bases compared to PET. Polycarbonate is comparable to PET in resistance to organic solvents at ambient temperature. At higher temperatures, PC is more resistant to aliphatic and aromatic solvents but less resistant to polar organic solvents.

Although its upper temperature limit (120°C for reinforced grades) and resistance to some solvents are low compared to other engineering plastics, polycarbonate finds many uses because of its exceptional transparency and impact resistance (toughness) as well as good dimensional resistance and creep resistance. Applications include food contact (baby bottles, microwave ovenware, food storage containers), medical (packaging for radiation sterilization, components for blood filters, sterilizers, dialyzers), glazing and optical (windows, traffic light lenses, motorcycle and aircraft windshields, safety helmets and glasses, eyeglasses), and other household, consumer, and automotive uses (ski poles, compact disks, battery cases, kitchen sinks, appliance and power tool housings). More than 400 million pounds of polycarbonate are produced annually in the United States.

2-8f Polyamides

The synthesis of polyamides follows a different route from that of polyesters. Although several different polymerization reactions are possible, polyamides are usually produced either by direct amidation of a diacid with a diamine or the self-amidation of an amino acid. The polymerization of amino acids is not as useful due to a greater tendency toward cyclization (Sec. 2-5b). Ring-opening polymerization of lactams is also employed to synthesize polyamides (Chap. 7). Poly(hexamethylene adipamide) or poly(iminohexamethyleneiminoadipoyl) (also referred to as *nylon 6/6*) is synthesized from hexamethylene diamine and adipic acid [Zimmerman, 1988]. A stoichiometric balance of amine and carboxyl groups is readily obtained by the preliminary formation of a 1:1 ammonium salt (**XVI**) in aqueous solution at a concentration of 50%. The salt is often referred to as a *nylon salt*. Stoichiometric balance can be controlled by adjusting the pH of the solution by appropriate addition of either diamine or diacid. The aqueous salt solution is concentrated to a slurry of approximately 60% or higher salt content by heating above 100°C. Polymerization is carried out by raising the temperature to about 210°C. Reaction proceeds under a steam pressure of about 250 psi (1.7 MPa), which effectively excludes oxygen. The pressure also prevents salt precipation and subsequent polymerization on heat-transfer surfaces.

$$n\text{H}_2\text{N}(\text{CH}_2)_6\text{NH}_2 + n\text{HO}_2\text{C}(\text{CH}_2)_4\text{CO}_2\text{H} \longrightarrow n\begin{bmatrix} {}^-\text{O}_2\text{C}(\text{CH}_2)_4\text{CO}_2^- \\ {}^+\text{H}_3\text{N}(\text{CH}_2)_6\text{NH}_3^+ \end{bmatrix} \longrightarrow$$

$$\textbf{XVI}$$

$$\text{H}\!\!-\!\!\!\left[\text{NH}-(\text{CH}_2)_6-\text{NHCO}-(\text{CH}_2)_4-\text{CO}\right]_{\overline{n}}\!\text{OH} + (2n-1)\text{H}_2\text{O} \qquad (2\text{-}133)$$

Unlike polyester synthesis, polyamidation is carried out without an external strong acid, since the reaction rate is sufficiently high without it. Further, amidation may not be an acid-catalyzed reaction. The equilibrium for polyamidation is much more favorable than that for the synthesis of polyesters. For this reason the amidation is carried out without concern for shifting the equilibrium until the last stages of reaction. Steam is released to maintain the pressure at about 1.7 MPa, while the temperature is continuously increased to 275°C. When 275°C is reached the pressure is slowly reduced to atmospheric pressure and heating continued to drive the equilibrium to the right. The later stage of the reaction is a melt polymerization since the reaction temperature is above the T_m. Figure 2-17 represents a commercial process for nylon 6/6 synthesis [Taylor, 1944].

Molecular-weight control and stabilization are accomplished by addition of a calculated amount of a monofunctional acid such as acetic acid. Diamine loss during polymerization is unavoidable because of its volatility. The loss must be quantitatively taken into account by careful control of process conditions and initial charges of reactants.

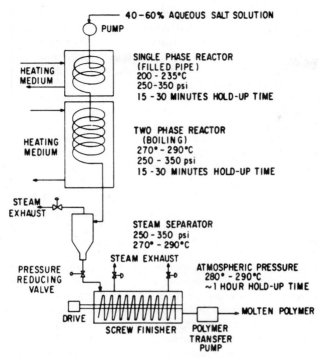

Fig. 2-17 Schematic representation of industrial process for synthesis of poly(hexamethylene adipamide). After Taylor [1944] (by permission of Wiley-Interscience, New York).

More than 3.25 billion pounds of polyamides are produced annually in the United States. Most of that is nylon 6/6; the second most important polyamide is nylon 6, which is produced by ring-opening polymerization of ε-caprolactam. Poly(hexamethylene adipamide) is an excellent fiber and plastic with a high crystalline melting point (265°C). Nylon 6/6 is moderately crystalline (~50%) as normally produced, but this is further increased for fiber applications by orientation via mechanical stretching. It has a very good combination of high strength, flexibility, toughness, abrasion resistance, dyeability, low coefficient of friction (self-lubricating), low creep, and resistance to solvents, oils, bases, fungi, and body fluids. The main limitation is its moisture pickup with resulting changes in dimensional and mechanical properties. Polyamides are more resistant to alkaline hydrolysis than polyesters but not as resistant to acid hydrolysis. Polyamides have better resistance to a range of organic solvents compared to PET and PC.

Nylon fiber uses include apparel, carpets, upholstery, tire reinforcements, ropes, seat belts, parachutes, fishing nets, and substrates for industrial coated fabrics. Nylon fiber has been losing market share to PET fiber for apparel applications since the latter is better for producing wrinkle-resistant apparel. However, nylon is still the fiber of choice for hosiery, stretch fabrics, and women's undergarments. Nylons are the largest volume engineering plastic—more than 500 million pounds per year in the United States. Fiber and mineral reinforcements are widely used for engineering plastic uses. The upper temperature for continuous use is 65–75°C for unreinforced grades and 100–115°C for glass and mineral reinforced grades. Applications include almost every industry and market—transportation (auto fender extensions, engine fans, brake and power steering reservoirs, valve covers, lamp housings, roof-rack components, light-duty gears for windshield wipers and speedometers), electrical/electronics (toggle switches, plugs, sockets, antenna-mounting devices, terminal blocks), industrial (self-lubricated gears and bearings, antifriction and snap-fit parts, parts for food and textile-processing equipment, valves, vending machines, pumps), film (meat and cheese packaging, cook-in-pouches, multilayer nylon-polyolefin protective barrier materials for oil and moisture resistance), and consumer (ski boots, racquet frames, kitchen utensils, toys, power tool housings).

Nylons 6/6 and 6 comprise more than 90% of the polyamide market. The two have similar properties but nylon 6 has a lower T_m (223°C). Small amounts of nylons 6/9, 6/10, 6/12, 11, 12, 12/12, and 4/6 are produced as specialty materials. Those with more methylene groups than nylons 6/6 and 6 have better moisture resistance, dimensional stability, and electrical properties but the degree of crystallinity, T_m, and mechanical properties are lower. Specialty nylons made from dimerized fatty acids find applications as hot-melt adhesives, crosslinking agents for epoxy resins, and thermographic inks.

The synthesis of aromatic polyamides (referred to as *aramid* polymers) is difficult to carry out using diacid and diamine reactants due to the lower reactivity of aromatic amines compared to aliphatic amines [Lenk, 1978]. The aromatic ring decreases the electron density of nitrogen through resonance interaction. The elevated temperatures required to achieve polymerization are generally too high, resulting in extensive side reactions that limit polymer molecular weight.

Aromatic polyamides are produced by using the faster reaction of a diamine with a diacid chloride, for example, for poly(*m*-phenylene isophthalamide) or poly(imino-1,3-phenyleneiminoisophthaloyl) (trade name: *Nomex*). The polymerization is carried out in solution at temperatures under 100°C with a tertiary base present to react with

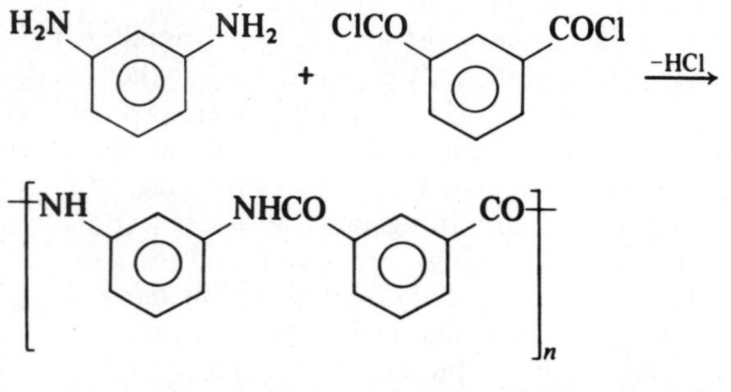

$$\text{(2-134)}$$

the liberated hydrogen chloride. Highly polar aprotic solvents such as dimethylacetamide (DMA), *N*-methylpyrrolidinone (NMP), and hexamethylphosphoramide (HMP) have been used to prevent premature precipitation of the growing polyamide chains. This is a significant problem in the synthesis of aromatic polyamides, since the reaction temperature is much lower than for aliphatic amines. The presence of lithium or calcium chloride promotes solubilization of the polymer, probably as a result of coordination of the metal ion with the carbonyl groups of the polyamide and/or solvent [Kwolek and Morgan, 1977]. The resulting solutions of these polymerizations are often used directly in spinning operations to produce fiber.

There is an effort to find reaction conditions suitable for carrying out the polyamidation using the diacid, instead of the diacyl chloride, at moderate temperatures. Various phosphorus compounds, such as triphenyl phosphite (ϕ_3PO) and phosphorus oxychloride ($POCl_3$), often in the presence of a base such as pyridine or imidazole and LiCl or $CaCl_2$, show promise for enhancing the polymerization reaction between diacid and diamine. Enhanced reaction rates, polymer yields, and polymer molecular weights have been reported in a number of systems [Aharoni et al., 1984; Arai et al., 1985; Higashi and Kobayashi, 1989; Higashi et al., 1988; Krigbaum et al., 1985]. The mechanism responsible for enhancing the reactivity of an acid toward amidation by an aromatic amine is not clear. One suggestion is that phosphorylation of the acid results in a phosphate ester that undergoes faster amidation compared to the acid since phosphate is a good leaving group. (There are similar efforts to achieve polyesterifications at moderate temperatures [Higashi et al., 1985; Moore and Stupp, 1990.].)

Poly(imino-1,4-phenyleneiminoterephthaloyl) (**XVII**) (trade names: *Kevlar, Twaron*) is synthesized from the corresponding *para*-substituted diamine and diacid chloride. Poly(iminocarbonyl-1,4-phenylene) (**XVIII**) (also known as *Kevlar*) is based on *p*-aminobenzoic acid.

idation using the diacid, instead of the diacyl chloride, at moderate temperatures.

The totally aromatic structures of the aramid polymers give them exceptional heat and flame resistance, very high melting points (generally above their decomposition temperature, which are $\geq 500°C$), ultrahigh strength, and better resistance to solvents, chemicals, and oxidizing agents compared to aliphatic polyamides. *Kevlar* has higher strength and modulus properties compared to *Nomex*. The para-structure of the former gives a rodlike extended-chain structure that forms liquid crystal solutions. (*Liquid crystal solutions* are solutions in which there is a high degree of ordering of solute molecules; see Sec. 2-14g.) Polymer crystallization from liquid crystal solutions results in a highly oriented, extended-chain morphology in the bulk polymer, resulting in high strength and high modulus.

Aramid polymers are much more expensive than the aliphatic polyamides. The use of aramid polymers is limited to those applications that justify the high cost. The present U.S. market is about 20 million pounds per year. The applications are those where one needs very high flame resistance (clothing for fire fighters and welders, welder's protective shield, upholstery and drapes), heat resistance (ironing board covers, insulation film for electrical motors and transformers, aerospace and military), dimensional stability (fire hose, V- and conveyor belts), or strength and modulus (circuit boards, bullet-proof vests, fiber optic and power lines, ship mooring ropes, auto tire cord).

2-8g Historical Aspects

It is worth exploring some of the historical aspects of polyamide and polyester chemistry at this point [Hounshell and Smith, 1988]. Carothers and co-workers at DuPont synthesized a large number of aliphatic polyesters in the 1930s. These had melting points below 100°C, which made them unsuitable for fiber use. Carothers then turned successfully to polyamides, based on the theoretical consideration that amides melt higher than esters. Polyamides were the first synthetic fibers to be produced commercially. The polyester and polyamide research at DuPont had a major impact on all of polymer science. Carothers laid the foundation for much of our understanding of how to synthesize polymeric materials. Out of that work came other discoveries in the late 1930s, including neoprene, an elastomer produced from chloroprene, and teflon, produced from tetrafluoroethylene. The initial commercial application for nylon 6/6 was women's hosiery, but this was short-lived with the intrusion of World War II. The entire nylon 6/6 production was allocated to the war effort in applications for parachutes, tire cord, sewing thread, and rope. The civilian applications for nylon products burst forth and expanded rapidly after the war.

Not only did Carothers's work secure the future of the DuPont chemical empire but he launched the synthetic fiber industry and changed the agricultural patterns of the Southern cotton states. Subsequent to the success of nylon, workers in the United Kingdom in the early 1950s achieved success in producing a polyester fiber by using terephthalate as the acid component. Cotton, no longer King of Fibers, accounts for less than 25% of the U.S. fiber market. Nylon, PET, and rayon (regenerated cellulose) account for the remainder.

2-9 MULTICHAIN POLYMERIZATION

2-9a Branching

The discussions until this point have been concerned with the polymerization of bifunctional monomers to form *linear polymers*. When one or more monomers with more than two functional groups per molecule are present the resulting polymer will be *branched* instead of linear. With certain monomers *crosslinking* will also take place with the formation of *network structures* in which a branch or branches from one polymer molecule become attached to other molecules. The structures of linear, branched, and crosslinked polymers are compared in Fig. 1-2.

Consider the polymerization of an A—B reactant in the presence of a small amount of a monomer A_f containing f functional groups per molecule. The value of f is termed the *functionality of the monomer*. The product of this polymerization will be a branched polymer in which f chains are attached to a central branch point (i.e., an A_f species).

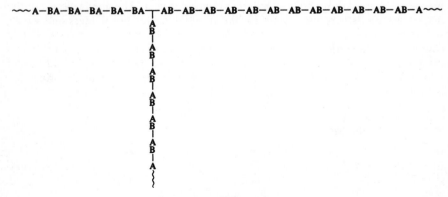

For the specific case of $f = 3$, polymerization of A—B in the presence of A$\underset{\displaystyle A}{\overset{\displaystyle |}{}}$A

leads to the structure **XIX**. A careful consideration of this structure shows that there can be only one A_f reactant molecule incorporated into each polymer molecule. Further, crosslinked species will not be formed. Reactions between two polymer molecules of the above type cannot occur since all growing branches possess A functional groups at their ends. Branch chains from one molecule cannot react with those from another. (This would not be true if A groups were capable of reacting with each other. However, that is not the usual situation.)

2-9b   Molecular-Weight Distribution

The molecular-weight distribution in this type of nonlinear polymerization will be much narrower than for a linear polymerization. Molecules of sizes very much different from the average are less likely than in linear polymerization, since this would require having the statistically determined f branches making up a molecule all very long or all very short. The distribution functions for this polymerization have been derived statistically [Peebles, 1971; Schaefgen and Flory, 1948], and the results are given as

$$\overline{X}_n = \frac{(frp + 1 - rp)}{(1 - rp)} \tag{2-135}$$

$$\overline{X}_w = \frac{(f - 1)^2(rp)^2 + (3f - 2)rp + 1}{(frp + 1 - rp)(1 - rp)} \tag{2-136}$$

The breadth of the distribution is characterized by

$$\frac{\overline{X}_w}{\overline{X}_n} = 1 + \frac{frp}{(frp + 1 - rp)^2} \tag{2-137}$$

which becomes

$$\frac{\overline{X}_w}{\overline{X}_n} = 1 + \frac{1}{f} \tag{2-138}$$

in the limit of $p = r = 1$.

The weight distribution (Eq. 2-136) is shown in Fig. 2-18 for several values of f. The extents of reaction have been adjusted to maintain a constant number-average degree of polymerization of 80 in all four cases. The size distribution becomes progressively narrower with increasing functionality of A_f. This is also evident from Eq. 2-138 where $\overline{X}_w/\overline{X}_n$ decreases from 2 for $f = 1$ to 1.25 for $f = 4$ (at $p = 1$). Linear polymers are formed for f values of 1 or 2, while branched polymers are formed when f is greater than 2. The case of $f = 1$ corresponds to the Type 3 polymerization discussed in Sec. 2-6. The distribution is the exact same as the most probable distribution for linear polymerization (Sec. 2-7). Linear polymerization with $f = 2$ is of interest in that the distribution is narrower than that for the usual linear polymerization. The linking of two statistically independent A—B type polymer chains into one polymer molecule via an A—A molecule leads to a significantly narrower distribution that in the usual polymerizations of the A—B or A—A plus B—B types.

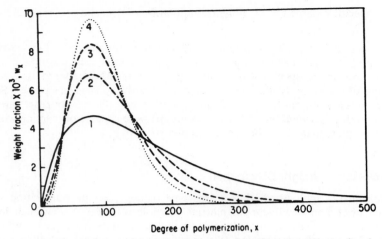

Fig. 2-18 Weight-fraction distributions for multichain polymerization. Plot 1—$f = 1$; plot 2—$f = 2$; plot 3—$f = 3$; plot 4—$f = 4$. After Howard [1961] (by permission of Iliffe Brooks, Ltd., London).

2-10 CROSSLINKING

Polymerization of the A—B plus A_f system (with $f > 2$) in the presence of B—B will lead not only to branching but also to a *crosslinked* polymer structure. Branches from one polymer molecule will be capable of reacting with those of another polymer molecule because of the presence of the B—B reactant. Crosslinking can be pictured as leading to structure **XX**, in which two polymer chains have been joined together (crosslinked) by a branch. The branch joining the two chains is referred to as a *crosslink*. A crosslink can be formed whenever there are two branches (e.g., those indicated in **XX** by the arrows) that have different functional groups at their ends, that is, one has an A group and the other a B group. Crosslinking will also occur in other polymerizations involving reactants with functionalities greater than two. These include the polymerizations

$$A—A + B_f \rightarrow$$
$$A—A + B—B + B_f \rightarrow$$
$$A_f + B_f \rightarrow$$

Crosslinking is distinguished by the occurrence of *gelation* at some point in the polymerization. At this point, termed the *gel point*, one first observes the visible formation of a *gel* or *insoluble polymer fraction*. (The gel point is alternately taken as the point at which the system loses fluidity as measured by the failure of an air bubble to rise in it.) The gel is insoluble in all solvents at elevated temperatures under conditions where polymer degradation does not occur. The gel corresponds to the formation of an infinite network in which polymer molecules have been crosslinked to each other to form a macroscopic molecule. The gel is, in fact, considered as one molecule. The nongel portion of the polymer remains soluble in solvents and is referred

```
~~~A—BA—BA┬AB—AB—AB—AB—AB—AB—BA—BA┬A~~~
          A                        A
          B                        B
          |                        |
          A                        B
          B                        A
          |                        |
          A                        B
          B                        A
          |                        |
     ⎧    A                        B ←
Crosslink⎨ B
     ⎩    |
          A                  ⟋AB—A ←
          B                  B
          |                  A
          B                  A
          A                  B
B—AB—BA—┴—AB—AB—AB—BA—BA—BA—┴A
A                           B
B                           A
A                           B
B                           A
A                           |
├—AB—A~~~
A
B
|
A
B
|
B
⟨
```

XX

to as *sol*. As the polymerization and gelation proceed beyond the gel point, the amount of gel increases at the expense of the sol as more and more polymer chains in the sol are crosslinked to the gel. There is a dramatic physical change that occurs during the process of gelation. The reaction mixture is transformed into a polymer of infinite viscosity.

The crosslinking reaction is an extremely important one from the commercial standpoint. Crosslinked plastics are increasingly used as engineering materials because of their excellent stability toward elevated temperatures and physical stress. They are dimensionally stable under a wide variety of conditions due to their rigid network structure. Such polymers will not flow when heated and are termed *thermosetting* polymers or simply *thermosets*. (Plastics that do soften and flow when heated, that is, uncrosslinked plastics, are called *thermoplastics*. Almost all of the polymers produced by chain polymerization are thermoplastics. There are only a relatively few such polymers which are crosslinked.) The commercial importance of the thermosets is seen from the fact that almost 20% of the over 60 billion pounds of plastics produced annually in the United States were crosslinked plastics.

In order to control the crosslinking reaction so that it can be used properly it is important to understand the relationship between gelation and the extent of reaction. Two general approaches have been used to relate the extent of reaction at the gel point to the composition of the polymerization system—based on calculating when $\overline{X}_n$ and $\overline{X}_w$, respectively, reach the limit of infinite size.

2-10a   Carothers Equation: $\overline{X}_n \to \infty$

2-10a-1   *Stoichiometric Amounts of Reactants*

Carothers derived a relationship between the extent of reaction at the gel point and the *average functionality* f_{avg} of the polymerization system for the case where the two functional groups A and B are present in equivalent amounts [Carothers, 1936]. The derivation follows in a manner very close to that for Eq. 2-84. The average functionality of a mixture of monomers is the average number of functional groups per monomer molecule for all types of monomer moelcules. It is defined by

$$f_{avg} = \frac{\Sigma N_i f_i}{\Sigma N_i} \tag{2-139}$$

where N_i is the number of molecules of monomer i with functionality f_i, and the summations are over all the monomers present in the system. Thus for a system consisting of 2 moles of glycerol (a triol) and 3 moles of phthalic acid (a diacid) there is a total of 12 functional groups per five monomer molecules, and f_{avg} is 12/5 or 2.4. For a system consisting of a equimolar amounts of glycerol, phthalic anhydride, and a monobasic acid, there is a total of six functional groups per three molecules and f_{avg} is 6/3 or 2.

In a system containing equivalent numbers of A and B groups, the number of monomer molecules present initially is N_0 and the corresponding total number of functional groups is $N_0 f_{avg}$. If N is the number of molecules after reaction has occurred, then $2(N_0 - N)$ is the number of functional groups that have reacted. The extent of reaction p is the fraction of functional groups lost,

$$p = \frac{2(N_0 - N)}{N_0 f_{avg}} \tag{2-140}$$

while the degree of polymerization is

$$\overline{X}_n = \frac{N_0}{N} \tag{2-141}$$

Combination of Eqs. 2-140 and 2-141 yields

$$\overline{X}_n = \frac{2}{2 - pf_{avg}} \tag{2-142}$$

which can be rearranged to

$$p = \frac{2}{f_{avg}} - \frac{2}{\overline{X}_n f_{avg}} \tag{2-143}$$

Equation 2-143, often referred to as the *Carothers equation*, relates the extent of reaction and degree of polymerization to the average functionality of the system.

An important consequence of Eq. 2-143 is its limiting form at the gel point where the number-average degree of polymerization becomes infinite. The *critical extent of reaction* p_c at the gel point is given by

$$p_c = \frac{2}{f_{avg}} \tag{2-144}$$

since the second term of the right side of Eq. 2-143 vanishes. Equation 2-144 can be used to calculate the extent of reaction required to reach the onset of gelation in a mixture of reacting monomers from its average functionality. Thus the glycerol-phthalic acid (2:3 molar ratio) system mentioned above has a calculated critical extent of reaction of 0.833.

2-10a-2 Extension of Nonstoichiometric Reactant Mixtures

Equations 2-143 and 2-144 apply only to systems containing stoichiometric numbers of the two different functional groups. For nonequivalent numbers of the functional groups, the average functionality calculated from Eq. 2-139 is too high. Thus consider the extreme example of a mixture of 1 mole of glycerol and 5 moles of phthalic acid. Using Eq. 2-139, one calculates a value of 13/6 or 2.17 for f_{avg}. This indicates that high polymer will be obtained. Further, one would predict from Eq. 2-144 that cross-linking will occur at $p_c = 0.922$. Both of these conclusions are grossly in error. It is apparent from previous discussions (Sec. 2-6) that the gross stoichiometric imbalance between the A and B functional groups in this system ($r = 0.3$) precludes the formation of any but extremely low molecular weight species. The polymerization will stop with the large portion of the acid functional groups being unreacted.

The average functionality of nonstoichiometric mixtures has been deduced [Pinner, 1956] as being equal to twice the total number of functional groups that are not in excess divided by the total number of all molecules present. This simply takes into account the fact that the extent of polymerization (and crosslinking, if it can occur) depends on the deficient reactant. The excess of the other reactant is not useful; in fact, it results in a lowering of the functionality of the system. For the above nonstoichiometric mixture of 1 mole of glycerol and 5 moles of phthalic acid, the f_{avg} value is correctly calculated as 6/6 or 1.00. This low value of f_{avg} is indicative of the low degree of polymerization that will occur in the system.

In a similar manner the average functionality of nonstoichiometric mixtures containing more than two monomers has been obtained. Consider a ternary mixture of N_A moles of A_{f_A}, N_C moles of A_{f_C}, and N_B moles of Bf_B with functionalities f_A, f_C, and f_B, respectively. In this system A_{f_A} and A_{f_C} contain the same functional groups (i.e., A groups) and the total number of A functional groups is less than the number of B groups (i.e., B groups are in excess). The average functionality in such a system is given by

$$f_{avg} = \frac{2(N_A f_A + N_C f_C)}{N_A + N_C + N_B} \tag{2-145}$$

or

$$f_{avg} = \frac{2r f_A f_B f_C}{f_A f_C + r\rho f_A f_B + r(1 - \rho) f_B f_C} \tag{2-146}$$

where

$$r = \frac{N_A f_A + N_C f_C}{N_B f_B} \tag{2-147}$$

$$\rho = \frac{N_C f_C}{N_A f_A + N_C f_C} \tag{2-148}$$

The term r is the ratio of A groups to B groups and has a value equal to or less than unity. r is comparable to the previously discussed stoichiometric imbalance. The term ρ is the fraction of all A functional groups that belong to the reactant with $f > 2$.

One can easily show by substitution of different values of r, ρ, f_A, f_B, and f_C into Eqs. 2-144 through 2-148 that crosslinking is accelerated (i.e., p_c decreases) for systems that contain closer to stoichiometric amounts of A and B functional groups (r closer to 1), systems with larger amounts of polyfunctional reactants (ρ closer to 1) and systems containing reactants of higher functionality (higher values of f_A, f_B, and f_C).

The Carothers equation and Eq. 2-84 overlap but are not equivalent. Recall the brief note in Sec. 2-6b that Eq. 2-84 does not apply to systems containing a monofunctional reactant unless it is present in addition to stoichiometric amounts of bifunctional reactants (either A—A + B—B or A—B). That is, for example, Eq. 2-84 applies to a reaction system of 10 moles each of A—A and B—B plus 1 mole B but does not apply to a system of 10 moles A—A, 9 moles B—B, and 1 mole B. In reality, the two reaction systems yield very nearly the same degree of polymerization at any particular extent of reaction. The Carothers equation yields $\overline{X}_n$ values of 1.909 and 1.905, respectively, for the two systems at complete conversion. The use of Eq. 2-84 for the second reaction system yields a gross distortion. The value of r obtained from Eq. 2-86 would be unity (the same as for a perfectly stoichiometric system of 10 moles each of A—A and B—B!), and this leads to a $\overline{X}_n$ value of infinity at complete conversion. This is correct for the stoichiometric system but clearly is incorrect for the reaction system containing the large amount of monofunctional B reactant. The use of Eq. 2-84 must be limited to situations in which there are stoichiometric amounts of bifunctional reactants (or A—B). Also, Eq. 2-84 is inapplicable to any systems containing a reactant with functionality greater than 2. The Carothers equation, without such limitations, has two general uses. First, it can be used in the form of Eq. 2-142 to calculate the degree of polymerization of a reaction system using the appropriate value of f_{avg} (whose calculation depends on whether stoichiometric or nonstoichiometric amounts are present). If one obtains a negative value of $\overline{X}_n$ in such a calculation, it means that the system is past the gel point at that conversion. Second, the use of Eq. 2-144 allows one to calculate the critical extent of reaction at the gel point. It should be kept clearly in mind that the extent of reaction calculated from Eq. 2-142 or 2-144 refers to the extent of reaction of the A functional groups (defined as those groups not in excess). The extent of reaction of the B functional groups is rp or rp_c.

2-10b   Statistical Approach to Gelation: $\overline{X}_w \rightarrow \infty$

Flory [1941, 1953] and Stockmayer [1943, 1952, 1953] used a statistical approach to derive an expression for predicting the extent of reaction at the gel point by calculating when $\overline{X}_w$ approaches infinite size. The statistical approach in its simplest form assumes that the reactivity of all functional groups of the same type is the same and independent of molecular size. It is further assumed that there are no intramolecular reactions between functional groups on the same molecule. (These assumptions are also inherent in the use of Eqs. 2-142 through 2-148. For this derivation the *branching coefficient* α is defined as the probability that a given functional group of a branch unit at the end of a polymer chain segment leads to another branch unit. For the polymerization of A—A with B—B and A_f this corresponds to obtaining a chain segment of the type

$$A—A + B—B + A_f \rightarrow A_{(f-1)}—A(B—BA—A)_nB—BA—A_{(f-1)} \tag{2-149}$$

where n may have any value from zero to infinity. The multifunctional monomer A_f is considered a *branch unit*, while the segments between branch units are defined as *chain segments*. The branch units occur on a polymer chain at the branch points (Fig. 1-2). Infinite networks are formed when n number of chains or chain segments give rise to more than n chains through branching of some of them. The criterion for gelation in a system containing a reactant of functionality f is that at least one of the $(f - 1)$ chain segments radiating from a branch unit will in turn be connected to another branch unit. The probability for this occurring is simply $1/(f - 1)$ and the *critical branching coefficient* α_c for gel formation is

$$\alpha_c = \frac{1}{(f - 1)} \tag{2-150}$$

The f in Eq. 2-150 is the functionality of the branch units, that is, of the monomer with functionality greater than 2. It is not the average functionality f_{avg} from the Carothers equation. If more than one type of multifunctional branch unit is present an average f value of all the monomer molecules with functionality greater than 2 is used in Eq. 2-150.

When $\alpha(f - 1)$ equals 1 a chain segment will, on the average, be succeeded by $\alpha(f - 1)$ chains. Of these $\alpha(f - 1)$ chains a portion α will each end in a branch point so that $\alpha^2(f - 1)^2$ more chains are created. The branching process continues with the number of succeeding chains becoming progressively greater through each succeeding branching reaction. The growth of the polymer is limited only by the boundaries of the reaction vessel. If, on the other hand, $\alpha(f - 1)$ is less than 1, chain segments will not be likely to end in branch units. For a trifunctional reactant ($f = 3$) the critical value of α is $\frac{1}{2}$.

The probability α is now related to the extent of reaction by determining the probability of obtaining a chain segment of the type shown in Eq. 2-149. The extents of reaction of A and B functional groups are p_A and p_B, respectively. The ratio of all A groups, both reacted and unreacted, that are part of branch units, to the total number of all A groups in the mixture is defined by ρ. (This corresponds to the same definition of ρ as in Eq. 2-148. The probability that a B group has reacted with a branch unit is $p_B\rho$. The probability that a B group has reacted with a nonbranch A—A unit is $p_B(1 - \rho)$. Therefore the probability of obtaining a segment of the type in Eq. 2-149 is given by $p_A[p_B(1 - \rho)p_A]^n p_B\rho$. Summation of this over all values of n and then evaluation of the summation yields

$$\alpha = \frac{p_A p_B \rho}{1 - p_A p_B(1 - \rho)} \tag{2-151}$$

Either p_A or p_B can be eliminated by using the ratio r of all A groups to all B groups to substitute $p_B = rp_A$ into Eq. 2-151 to yield

$$\alpha = \frac{rp_A^2\rho}{1 - rp_A^2(1 - \rho)} = \frac{p_B^2\rho}{r - p_B^2(1 - \rho)} \tag{2-152}$$

Equation 2-152 can be combined with Eq. 2-150 to yield a useful expression for the extent of reaction (of the A functional groups) at the gel point

$$p_c = \frac{1}{\{r[1 + \rho(f - 2)]\}^{1/2}} \tag{2-153}$$

Several special cases of Eqs. 2-153 and 2-152 are of interest. When the two functional groups are present in equivalent numbers, $r = 1$ and $p_A = p_B = p$, Eqs. 2-152 and 2-153 become

$$\alpha = \frac{p^2\rho}{1 - p^2(1 - \rho)} \tag{2-154}$$

and

$$p_c = \frac{1}{[1 + \rho(f - 2)]^{1/2}} \tag{2-155}$$

When the polymerization is carried out without any A—A molecules ($\rho = 1$) with $r < 1$ the equations reduce to

$$\alpha = rp_A^2 = \frac{p_B^2}{r} \tag{2-156}$$

and

$$p_c = \frac{1}{\{r[1 + (f - 2)]\}^{1/2}} \tag{2/157}$$

If both of the above conditions are present, $r = \rho = 1$, Eqs. 2-152 and 2-153 become

$$\alpha = p^2 \tag{2-158}$$

and

$$p_c = \frac{1}{[1 + (f - 2)]^{1/2}} \tag{2-159}$$

These equations do not apply for reaction systems containing monofunctional reactants and/or both A and B type of branch units. Consider the more general case of the system

$$A + A_2 + A_3 \ldots + A_i + B + B_2 + B_3 \ldots + B_j \rightarrow$$
$$\text{Crosslinked polymer} \tag{2-160}$$

where we have reactants ranging from monofunctional to ith functional for A functional groups and monofunctional to jth functional for B functional groups [Durand and Bruneau, 1982a, 1982b; Miller et al., 1979; Stafford, 1981; Stockmayer, 1952, 1953]. The extent of reaction at the gel point is given by

$$p_c = \frac{1}{\{r(f_{w,A} - 1)(f_{w,B} - 1)\}^{1/2}} \tag{2-161}$$

where $f_{w,A}$ and $f_{w,B}$ are weight-average functionalities of A and B functional groups, respectively, and r is the stoichiometric imbalance (Sec. 2-6b). The functionalities $f_{w,A}$ and $f_{w,B}$ are defined by

$$f_{w,A} = \frac{\Sigma f_{A_i}^2 N_{A_i}}{\Sigma f_{A_i} N_{A_i}} \tag{2-162}$$

$$f_{w,B} = \frac{\Sigma f_{B_j}^2 N_{B_j}}{\Sigma f_{B_j} N_{B_j}} \tag{2-163}$$

The summations in Eq. 2-162 are for all molecules containing A functional groups with N_{A_i} representing the number of moles of reactant A_i containing f_{A_i} number of A functional groups per molecule. The summations in Eq. 2-163 are for all molecules containing B functional groups with N_{B_j} representing the number of moles of reactant B_j containing f_{B_j} number of B functional groups per molecule.

The use of Eq. 2-161 can be illustrated for the following system:

4 moles A	2 moles B
51 moles A_2	50 moles B_2
2 moles A_3	3 moles B_3
3 moles A_4	3 moles B_5

r, $f_{w,A}$, $f_{w,B}$, and p_c are calculated as

$$r = \frac{1(4) + 2(51) + 3(2) + 4(3)}{1(2) + 2(50) + 3(3) + 5(3)} = 0.9841 \tag{2-164}$$

$$f_{w,A} = \frac{1^2(4) + 2^2(51) + 3^2(2) + 4^2(3)}{1(4) + 2(51) + 3(2) + 4(3)} = 2.2097 \tag{2-165}$$

$$f_{w,B} = \frac{1^2(2) + 2^2(50) + 3^2(3) + 5^2(3)}{1(2) + 2(50) + 3(3) + 5(3)} = 2.4127 \tag{2-166}$$

$$p_c = \frac{1}{\{0.9841(1.2097)(1.4127)\}^{1/2}} = 0.7711 \tag{2-167}$$

2-10c   Experimental Gel Points

The two approaches to the problem of predicting the extent of reaction at the onset of gelation differ appreciably in their predictions of p_c for the same system of reactants. The Carothers equation predicts the extent of reaction at which the number-average degree of polymerization becomes infinite. This must obviously yield a value of p_c

that is too large because polymer molecules larger than $\overline{X}_n$ are present and will reach the gel point earlier than those of size $\overline{X}_n$. The statistical treatment theoretically overcomes this error, since it predicts the extent of reaction at which the polymer size distribution curve first extends into the region of infinite size.

The gel point is usually determined experimentally as that point in the reaction at which the reacting mixture loses fluidity as indicated by the failure of bubbles to rise in it. Experimental observations of the gel point in a number of systems have confirmed the general utility of the Carothers and statistical approaches. Thus in the reactions of glycerol (a triol) with equivalent amounts of several diacids, the gel point was observed at an extent of reaction of 0.765 [Kienle and Petke, 1940, 1941]. The predicted values of p_c are 0.709 and 0.833 from Eqs. 2-153 (statistical) and 2-144 (Carothers), respectively. Flory [1941] studied several systems composed of diethylene glycol ($f = 2$), 1,2,3-propanetricarboxylic acid ($f = 3$), and either succinic or adipic acid ($f = 2$) with both stoichiometric and nonstoichiometric amounts of hydroxyl and carboxyl groups. Some of the experimentally observed p_c values are shown in Table 2-9 along with the corresponding theoretical values calculated by both the Carothers and statistical equations.

The observed p_c values in Table 2-9 as in many other similar systems fall approximately midway between the two calculated values. The Carothers equation gives a high value of p_c for reasons mentioned above. The experimental p_c values are close to but always higher than those calculated from Eq. 2-153. There are two reasons for this difference: the occurrence of intramolecular cyclization and unequal functional group reactivity. Both factors were ignored in the theoretical derivations of p_c. Intramolecular cyclization reactions between functional groups are wasteful of the reactants and require the polymerization to be carried out to a greater extent of reaction to reach the gel point. Stockmayer [1945] showed this to be the case for the reaction of pentaerythritol, $C(CH_2OH)_4$, ($f = 4$) with adipic acid. Gelation was studied as a function of concentration and the results extrapolated to infinite concentration where intramolecular reactions would be expected to be nil. The experimental p_c value of 0.578 ± 0.005 compared exceptionally well with the theoretical value of 0.577. The value of p_c calculated by the Carothers equation is 0.75 for this system.

In many reaction systems the difference between the observed p_c values and those calculated from Eq. 2-153 are at least partially ascribed to the failure of the assumption of equal reactivity of all functional groups of the same type. An example is the glycerol–phthalic acid system previously mentioned. The difference between the calculated and observed values of p_c (0.709 vs 0.765) would be decreased, but not eliminated, if the calculation were corrected for the known lower reactivity of the secondary hydroxyl group of glycerol.

TABLE 2-9 Gel Point Determinations for Mixtures of 1,2,3-Propanetricarboxylic Acid, Diethylene Glycol, and Either Adipic or Succinic Acid[a]

| $r = \dfrac{[CO_2H]}{[OH]}$ | ρ | Extent of Reaction at Gel Point (p_c) | | |
		Calculated from Eq. 2-144	Calculated from Eq. 2-153	Observed[a]
1.000	0.293	0.951	0.879	0.911
1.000	0.194	0.968	0.916	0.939
1.002	0.404	0.933	0.843	0.894
0.800	0.375	1.063	0.955	0.991

[a]Data from Flory [1941].

It is difficult to find crosslinking systems that are ideal in that all functional groups are of equal reactivity and intramolecular cyclization is negligible. The crosslinking of vinyl terminated poly(dimethylsiloxane) polymers with tri- and tetrafunctional silanes appears to be an exception. Thus the calculated and experimental p_c values were 0.578 and 0.583, respectively, for the tetrafunctional silane and 0.708 and 0.703, respectively, for the trifunctional silane (with $r = 0.999$) [Valles and Macosko, 1979].

Although both the Carothers and statistical approaches are used for the practical predication of gel points, the statistical approach is the more frequently employed. The statistical method is preferred, since it theoretically gives the gel point for the largest-sized molecules in a size distribution. From the practical viewpoint the use of the Carothers approach alone can be disastrous, since the Carothers approach always predicts a higher p_c than observed. Crosslinked polymer becomes intractable above p_c; the reaction must be stopped at an extent of reaction below p_c before starting the technological step of fabricating the polymer into the shape of final desired product (see Sec. 2-12). Further, the statistical approach offers a greater possibility of adaption to various systems of reactants as well as the inclusion of corrections for the non-applicability of the equal reactivity assumption and the occurrence of intramolecular reactions. Equation 2-153 and its modifications have been used extensively in the technology of crosslinked polymers.

2-10d Extensions of Statistical Approach

The statistical approach has been applied to systems containing reactants with functional groups of unequal reactivity [Case, 1957; Macosko and Miller, 1976; Miller and Macosko, 1978; Miller et al., 1979]. In this section we will consider some of the results for such systems. Figure 2-19 shows a plot of $\overline{M}_w$ vs extent of reaction for the various

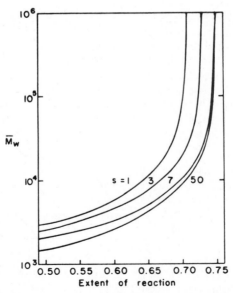

Fig. 2-19 $\overline{M}_w$ versus extent of reaction for polymerization of A_3 with B—B' for different s values. After Miller and Macosko [1978] (by permission of American Chemical Society, Washington, D.C.).

values of s at $r = 1$ for the system

$$A\underset{\underset{A}{|}}{}A + B\text{—}B'$$

where the two functional groups in the B—B' reactant have a difference in reactivity by a factor of s. s is the ratio of the rate constants for B and B' groups reacting with A groups. p_c is the value of p at which the curve becomes essentially vertical (i.e., $\overline{M}_w \to \infty$). The curves shift to the right with increasing s, although the shift becomes progressively smaller with increasing s. The extent of reaction at the gel point increases as the B and B' groups differ in reactivity. The more reactive B groups react to a greater extent earlier in the overall process, but gelation does not occur without reaction of the less reactive B' groups. Overall there is a wastage of the more reactive B groups and p_c is increased. There is an upper limit to this wastage with increasing s and p_c levels off with increasing s.

Figure 2-20 shows a p_c contour map for the system

$$A\underset{\underset{A''}{|}}{}A' + B\text{—}B$$

at $r = 1$ for various values of s_1 and s_2 with

$$s_1 = \frac{k}{k'} \tag{2-168a}$$

$$s_2 = \frac{k}{k''} \tag{2-168b}$$

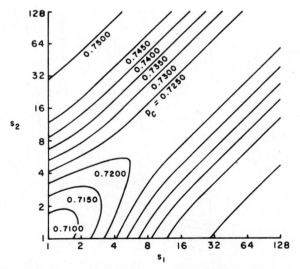

Fig. 2-20   Effect of s_1 and s_2 on p_c for polymerization of B—B with A—A'. p_c values are shown on the plot. After Miller and Macosko [1978] (by permission of American Chemical Society, Washington, D.C.).

where k, k', and k'' are the rate constants for the reactions of A, A', and A" groups, respectively, with B groups. p_c increases as s_1 or s_2 becomes larger than 1. If both s_1 and s_2 are larger than 1, the increase in p_c is greater the larger is the difference between s_1 and s_2. Plots such as Figs. 2-19 and 2-20 can be used to calculate the expected p_c for a system where the reactivities of the functional groups are known or the relative reactivities (s in the first system, s_1 and s_2 in the second system) in a particular reaction system can be determined from the measurement of p_c.

Some theoretical evaluations of the effect of intramolecular cyclization on gelation have been carried out [Gordon and Ross-Murphy, 1975; Kilb, 1958; Kumar et al., 1986; Stafford, 1981]. The main conclusion is that, although high reactant concentrations decrease the tendency toward cyclization, there is at least some cyclization occurring even in bulk polymerizations. Thus, even after correcting for unequal reactivity of functional groups, one can expect the actual p_c in a crosslinking system to be higher than the calculated value.

Various relationships have been derived to describe the crosslinking density of a system past p_c [Argyropoulos et al., 1987; Durand and Bruneau, 1982a, 1982b; Dusek, 1979a, 1979b; Dusek et al., 1987; Flory, 1953; Miller and Macosko, 1976] (see also Sec. 2-11). The density of crosslinks in a reaction system increases with conversion and with increasing functionality of reactants. For a system such as

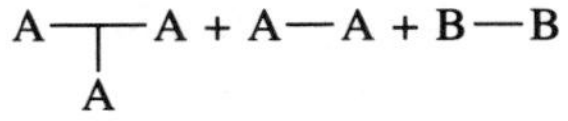

$$A{-}{\overset{\displaystyle |}{\underset{\displaystyle A}{}}}{-}A + A{-}A + B{-}B$$

the crosslink density increases as ρ increases. Varying the crosslink density by changing ρ is of practical interest in that it allows one to control the flexibility or rigidity of the final crosslinked polymer.

2-11 MOLECULAR-WEIGHT DISTRIBUTIONS IN NONLINEAR POLYMERIZATIONS

The molecular-size-distribution functions for three-dimensional polymers are derived in a manner analogous to those for linear polymers, but with more difficulty. The derivations have been discussed elsewhere [Flory, 1946, 1953; Stockmayer, 1943, 1952, 1953], and only their results will be considered here. The number N_x, number- or mole-fraction $\underline{N}_x$, and weight-fraction w_x of x-mer molecules in a system containing monomer(s) with $f > 2$ are given, respectively, by

$$N_x = N_0 \left[\frac{(fx - x)! f}{x!(fx - 2x + 2)!} \right] \alpha^{x-1}(1 - \alpha)^{fx-2x+2} \tag{2-169}$$

$$\underline{N}_x = \left[\frac{(fx - x)! f}{x!(fx - 2x + 2)!(1 - \alpha f/2)} \right] \alpha^{x-1}(1 - \alpha)^{fx-2x+2} \tag{2-170}$$

$$w_x = \left[\frac{(fx - x)! f}{(x - 1)!(fx - 2x + 2)!} \right] \alpha^{x-1}(1 - \alpha)^{fx-2x+2} \tag{2-171}$$

The number- and weight-average degrees of polymerization are given by

$$\overline{X}_n = \frac{1}{1 - (\alpha f/2)} \tag{2-172}$$

$$\overline{X}_w = \frac{(1 + \alpha)}{1 - (f - 1)\alpha} \tag{2-173}$$

$$\frac{\overline{X}_w}{\overline{X}_n} = \frac{(1 + \alpha)(1 - \alpha f/2)}{1 - (f - 1)\alpha} \tag{2-174}$$

These equations are general and apply equally for multifunctional reactions such as that of A_f with B_f or that of A_f with A—A and B—B. Depending on which of these reactant combinations is involved, the value of α will be appropriately determined by the parameters r, f, p, and ρ. For convenience the size distributions in the reaction of equivalent amounts of trifunctional reactants alone, that is, where $\alpha = p$, will be considered. A comparison of Eqs. 2-95 and 2-171 shows that the weight distribution of branched polymers is broader than that of linear polymers at equivalent extents of reaction. Furthermore, the distribution for the branched polymers becomes increasingly broader as the functionality of the multifunctional reactant increases. The distributions also broaden with increasing values of α. This is seen in Fig. 2-21, which shows the weight-fraction of x-mers as a function of α for the polymerization involving only trifunctional reactants.

Figure 2-22 shows a plot of the weight-fraction of different x-mers vs α or p for the trifunctional polymerization. The weight-fraction w_{gel} of the gel or ∞-mer is also plotted. A comparison of Figs. 2-21 and 2-22 with Fig. 2-12 for linear polymerization shows that the weight-fraction of monomer is always greater than that of any one of the other species (up to the point where the curves for w_1 and w_{gel} intersect). As reaction proceeds larger species are built up at the expense of the smaller ones with a maximum being reached at a value of p less than $\frac{1}{2}$. The maximum shifts to higher values of p for the larger species. The distribution broadens and reaches maximum

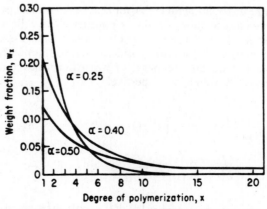

Fig. 2-21 Molecular-weight distribution as a function of the extent of reaction for the polymerization of trifunctional reactants where $\alpha = p$. After Flory [1946] (by permission of American Chemical Society, Washington, D.C.).

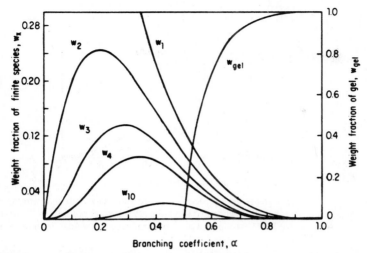

Fig. 2-22 Weight-fractions of various finite species and of gel in a trifunctional polymerization where $\alpha = p$. After Flory [1946] (by permission of American Chemical Society, Washington, D.C.).

heterogeneity at the gel point ($\alpha = \frac{1}{2}$), where the fraction of species that are highly branched is still small. The infinite network polymer is first formed at the gel point, and its weight-fraction rapidly increases. The species in the sol (consisting of all species other than gel) decrease in average size because the larger, branched species are preferentially tied into the gel. Past the point of intersection for the w_1 and w_{gel} curves gel is the most abundant species on a weight basis.

The broadening of the distribution with increasing α can also be noted by the $\overline{X}_w/\overline{X}_n$ value. Equations 2-172 to 2-174 show that the difference between the number- and weight-average degrees of polymerization increases very rapidly with increasing extent of reaction. At the gel point the breadth of the distribution $\overline{X}_w/\overline{X}_n$ is enormous, since $\overline{X}_w$ is infinite, while $\overline{X}_n$ has a finite value of 4 (Fig. 2-23). Past the gel point the value of $\overline{X}_w/\overline{X}_n$ for the sol fraction decreases. Finally at $\alpha = p = 1$ the whole system has been converted to gel (i.e., one giant molecule) and $\overline{X}_w/\overline{X}_n$ equals 1.

As previously mentioned, the behavior of systems containing bifunctional as well as trifunctional reactants is also governed by the equations developed above. The variation of w_x for the polymerization of bifunctional monomers, where the branching coefficient α is varied by using appropriate amounts of a trifunctional monomer, is similar to that observed for the polymerization of trifunctional reactants alone. The distribution broadens with increasing extent of reaction. The effect of unequal reactivity of functional groups and intramolecular is to broaden the molecular-weight distribution [Case, 1957; Macosko and Miller, 1976; Miller and Macosko, 1980; Muller and Burchard, 1978].

2-12 CROSSLINKING TECHNOLOGY

The theory of the crosslinking process and the parameters that control it has been discussed. Control of the process is extremely important in the commercial processing

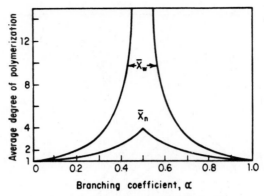

Fig. 2-23 Number- and weight-average degrees of polymerization as a function of α for a trifunctional polymerization. The portions of the plots after the gel point ($\alpha = \frac{1}{2}$) are for the sol fraction only. After Flory [1946] (by permission of American Chemical Society, Washington, D.C.).

of thermosetting plastics. This applies both to the reaction prior to the gel point and that subsequent to it. The period after the gel point is usually referred to as the curing period. Too slow or too rapid crosslinking can be detrimental to the properties of a desired product. Thus in the production of a thermoset foamed product the foam structure may collapse if gelation occurs too slowly. On the other hand, for reinforced and laminated products the bond strength of the components may be low if crosslinking occurs too quickly.

The fabrication techniques for producing objects from thermosetting and thermoplastic polymers are vastly different. In the case of the thermoplastics the polymerization reaction is completed by the plastics manufacturer. The fabricator takes the polymer and applies heat and pressure to produce a flowable material that can be shaped into the desired finished object. The situation is quite different for thermosetting plastics. A completely polymerized thermoset is no longer capable of flow and cannot be processed into an object. Instead, the fabricator receives an incompletely polymerized polymer—termed a *prepolymer*—from the plastics manufacturer and polymerization (and crosslinking) is completed during the fabrication step. For example, a prepolymer can be poured into an appropriate mold and then solidified to form the desired finished object by completing the polymerization. The prepolymers are usually in the molecular weight range 500–5000 and may be either liquid or solid.

Thermoset polymers are classified as A-, B-, and C-stage polymers according to their extent of reaction p compared to the extent of reaction at gelation p_c. The polymer is an A-stage polymer if $p < p_c$, B-stage if the system is close to the gel point (p_c), and C-stage when it is well past p_c. The A-stage polymer is soluble and fusible. The B-stage polymer is still fusible but is barely soluble. The C-stage polymer is highly crosslinked and both infusible and insoluble. The prepolymer employed by a fabricator is usually a B-stage polymer, although it may be an A-stage one. The subsequent polymerization of the prepolymer takes it through to the C-stage.

Thermoset plastics (often referred to as *resins*) can generally be grouped into two types depending on whether the same chemical reaction is used to both synthesize and crosslink the prepolymer. Generally, the older thermosetting materials involve the random crosslinking of bifunctional monomers with monomers of functionality greater than 2. The prepolymer is obtained in a first step by stopping the reaction, usually by cooling, at the desired extent of conversion (either A- or B-stage). Polymerization with crosslinking is completed during a second-step fabrication process, usually by heating. The newer thermosets involve prepolymers of a more specially designed structure. These prepolymers undergo crosslinking in the second step by a different chemical reaction than that for the first step. Because there is different chemistry in the two steps, one is not involved in any p_c considerations in the first step. The reactants are bifunctional with respect to the first-step chemistry. However, one or more of the reactants has some functional group capable of reaction under other conditions. Such prepolymer systems are advantageous because they generally offer greater control of the polymerization and crosslinking reactions, and very importantly, of the structure of the final product. The uses and markets for these newer thermosets are growing more rapidly than those of the older systems.

2-12a Polyesters, Unsaturated Polyesters, and Alkyds

Polyesters formed from phthalic anhydride and glycerol were among the first com-

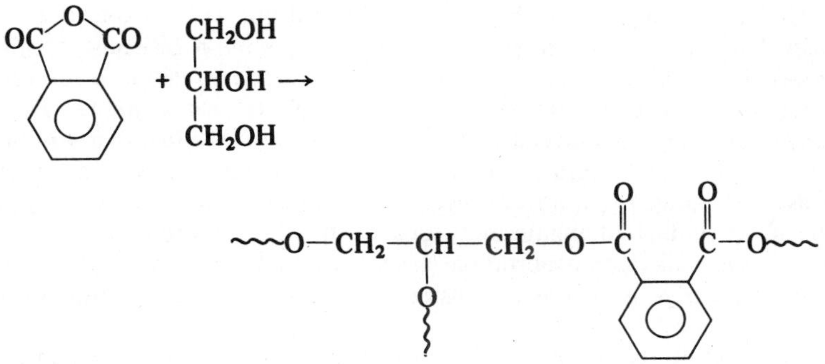

(2-175)

mercial crosslinked polyesters. Linear polyesters seldom are synthesized by the direct reactions of acids or acid anhydrides with alcohols because the higher temperatures required for high conversions lead to side reactions, which interfere with obtaining high molecular weights. This consideration is not overwhelmingly important for crosslinking systems, since crosslinking is achieved at far lower extents of reaction than are needed to obtain high polymer in a linear polymerization. Crosslinked polyesters are typically synthesized by direct esterification of acid or acid anhydride with alcohol.

Simple polyesters of the type described by Eq. 2-175 are too limited to be of commercial interest. Almost all crosslinked polyesters are either unsaturated polyesters or alkyd polyesters. These offer a greater ability to vary the final product properties to suit a targeted market. Also, they offer greater process control since different chemical reactions are involved in the polymerization and crosslinking reactions. A typical unsaturated polyester is that obtained by the polymerization of maleic anhydride and ethylene glycol. Maleic anhydride is only a bifunctional reactant

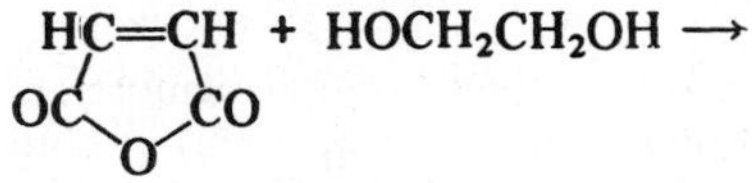

(2-176)

in the polyesterification reaction, but it has the potential for a higher functionality under the appropriate reaction conditions. The alkene double bond is reactive under radical chain polymerization conditions. Crosslinking is accomplished in a separate step by radical copolymerization with alkene monomers such as styrene, vinyl toluene, methyl methacrylate, triallyl cyanurate and diallyl phthalate (Sec. 6-6a).

Fumaric and itaconic acids are also used as the diacid component. Most reaction formulations involve a mixture of a saturated diacid (iso- and terephthalic, adipic) with the unsaturated diacid or anhydride in appropriate proportions to control the density of crosslinking (which depends on the carbon–carbon double bond content of the prepolymer) for specific applications [Parker and Peffer, 1977; Selley, 1988]. Propylene glycol, 1,4-butanediol, neopentyl glycol, diethylene glycol, and bisphenol A are also used in place of ethylene glycol as the diol component. Aromatic reactants are used in the formulation to improve the hardness, rigidity, and heat resistance of the crosslinked product. Halogenated reactants are used to impart flame resistance.

More than 1.3 billion pounds per year of unsaturated polyesters are used annually in the United States. Almost all unsaturated polyesters are used with fibrous reinforcements or filler. More than 80% of the market consists of structural applications that require the strengthening imparted by fibrous (usually fiber glass) reinforcement. The remainder is used without fibrous reinforcement but with inexpensive fillers to lower costs. Unsaturated polyesters have a good combination of resistance to softening and deformation at high temperature, electrical properties, resistance to corrosion, weak alkalies, and strong acids and possess very good weatherability. The liquid polyester prepolymers are especially easy to fabricate into infusible thermoset objects by casting in open molds, spray techniques as well as compression, hand layup, and resin-transfer molding. Unsaturated polyesters are used extensively in the construction (tub and shower units, building facades, specialty flooring, cultured onyx and marble, chemical storage tanks), transportation (truck cabs, auto body repair), and marine (boat hulls) industries as well as for business machine and electric handtool-molded parts.

Alkyds are unsaturated polyesters in which the unsaturation is located at chain ends instead of within the polymer chain [Lanson, 1986]. This is accomplished by using an unsaturated monocarboxylic acid in the polymerization. Various unsaturated

$$\text{phthalic anhydride} + \underset{\substack{|\\ \text{CH}_2\text{OH}}}{\overset{\substack{\text{CH}_2\text{OH}\\|}}{\text{CHOH}}} + \text{RCH}{=}\text{CH}{-}\text{R}'{-}\text{CO}_2\text{H} \longrightarrow$$

$$\text{RCH}{=}\text{CH}{-}\text{R}'{-}\text{CO}{-}\text{O}{-}\text{CH}_2{-}\underset{\substack{|\\\text{O}\\\}}{\text{CH}}{-}\text{CH}_2{-}\text{O}{-}\overset{\text{O}}{\overset{\|}{\text{C}}}\text{—(benzene)—}\overset{\text{O}}{\overset{\|}{\text{C}}}{-}\text{O}\sim\sim \qquad (2\text{-}177)$$

fatty acids are used. Some contain a single double bond (oleic, ricinoleic); others contain two or more double bonds, either isolated (linoleic, linolenic) or conjugated (eleostearic). The unsaturated fatty acid is typically used together with a saturated monoacid (typically a fatty acid such as lauric, stearic, palmitic but also benzoic and *p-t*-butylbenzoic). (The unsaturated and saturated fatty acids are often referred to as *drying* and *nondrying oils*, respectively.) The fatty acid content is in the range 30–60% or higher and the alkyd is referred to as a *short*, *medium*, or *long oil alkyd* depending on the fatty acid content. Although glycerol is the major polyhydric alcohol in alkyd formulations, a variety of other alcohols with functionality from 2 to 8 (diethylene glycol, sorbitol, pentaerythritol, tripentaerythritol) are used either alone or in combination. Alkyd prepolymers are crosslinked via oxidation with atmospheric oxygen (Sec. 9-2a). The crosslinking process is referred to as *drying* and is directly dependent on the content of unsaturated fatty acid. The crosslinking reaction involves chemical reactions different from those involved in prepolymer synthesis.

More than 200 million pounds of alkyds are produced annually in the United States. Almost all of this is used in coating applications. Alkyds undergo rapid drying and possess good adhesion, flexibility, strength, and durability. The upper temperature limit for continuous use for both alkyds and unsaturated polyesters is about 130°C. Considerable variations are possible in properties by manipulation of the alkyd formulation. For example, the use of higher functionality alcohols allows an increase in the fatty acid content, which imparts faster drying as well as increased hardness, better gloss retention, and improved moisture resistance. Alkyds are used as architectural enamels, exterior paints, top-side marine paints, and various metal primers and paints. Alkyds with carboxyl end groups are modified by reaction with hydroxy groups of nitrocellulose to produce lacquers. Modifications with amino resins (Sec. 2-12b-2) yields resins suitable as baked-on enamels for metal cabinets, appliances, venetian blinds, and toys. Blends of chlorinated rubber with alkyds are used as paints for highway marking, concrete, and swimming pools.

Water-based alkyds now represent a significant part of the total market. Alkyd prepolymers are rendered water-soluble or soluble in mixtures of water and alcohol (2-butoxyethanol, 2-butanol) by various changes in prepolymer synthesis. One approach is to introduce unesterified carboxyl groups into the polymer chain, for example, by using trimellitic anhydride (**XXI**) or dimethylolpropionic acid (**XXII**). The carboxyl groups are neutralized with base to form carboxylate groups along the poly-

| **XXI** | **XXII** | **XXIII** |

mer chain to promote water solubility. Another approach is to use poly(ethylene oxide) (**XXIII**) with hydroxyl end groups as an alcohol in the prepolymer synthesis.

2-12b Phenolic Polymers

2-12b-1 *Resole Phenolics*

Phenolic polymers are obtained by the polymerization of phenol ($f = 3$) with formaldehyde ($f = 2$) [Brydson, 1982; Kopf, 1988, Lenz, 1967]. The polymerization rate is pH-dependent, with the highest reaction rates occurring at both high and low pH. The strong base-catalyzed polymerization yields mixtures, referred to as *resoles*, *resole prepolymers*, or *resole phenolics*, of mononuclear methylolphenols (**XXIV**) and various dinuclear and polynuclear compounds such as **XXV** and **XXVI**. The specific com-

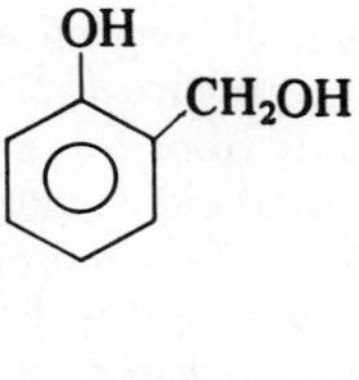

XXIVa 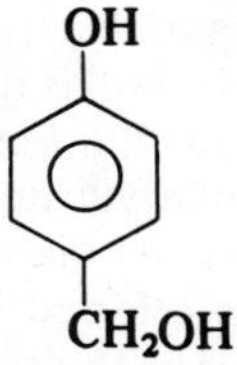   XXIVb

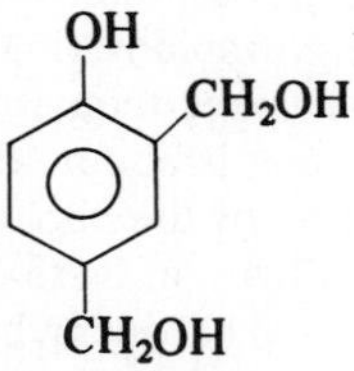

XXIVc 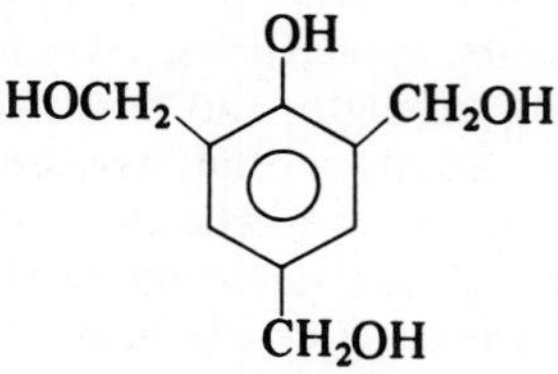   XXIVd

pounds shown as **XXV** and **XXVI** are only some of the possible products. The other

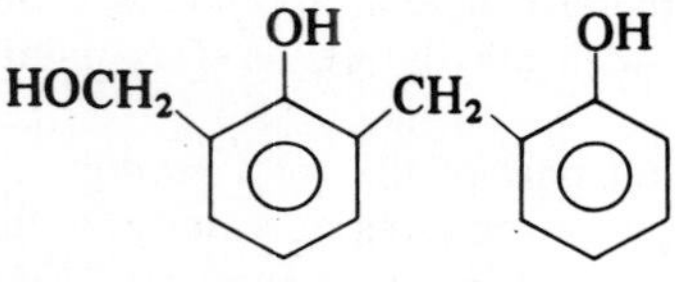

XXVa

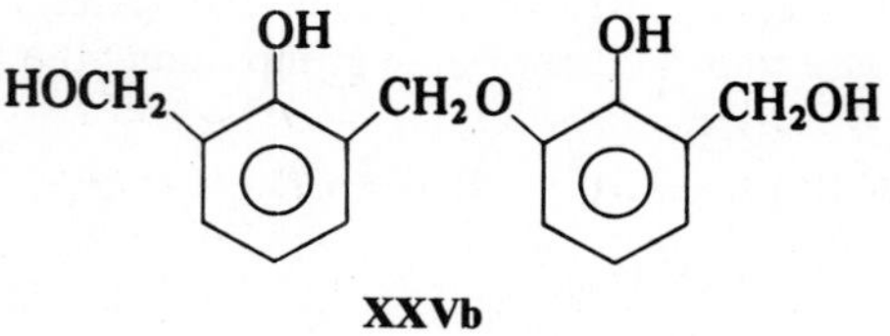

XXVb

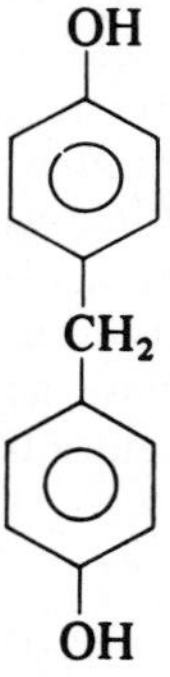

XXVc

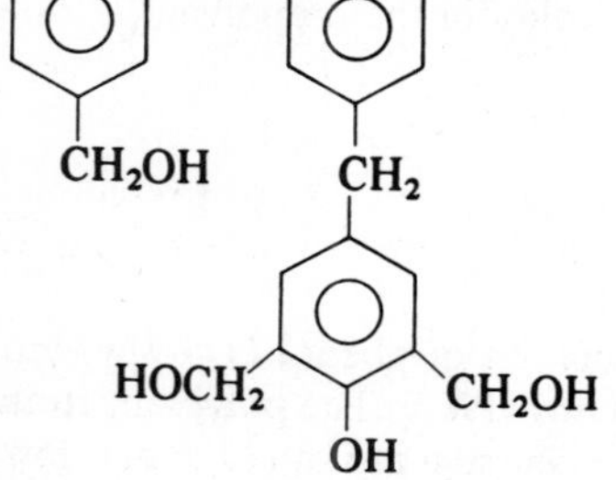

XXVI

products are those differing in the position of ring substitution (i.e., ortho versus para) and the type of bridge between rings (methylene versus ether). The polymerization is carried out using a molar ratio of formaldehyde to phenol of 1.2–3.0:1. A 1.2:1 ratio is the typical formulation. The formaldehyde is supplied as a 36–50% aqueous solution (referred to as *formalin*). The catalyst is 1–5% sodium, calcium, or barium hydroxide. Heating at 80–95°C for 1–3 hr is sufficient to form the prepolymer. Vacuum dehydration is carried out quickly to prevent overreaction and gelation.

Although phenol itself is used in the largest volume, various substituted phenols such as cresol (*o*-, *m*-, *p*-), *p*-butylphenol, resorcinol, and bisphenol A are used for specialty applications. Some use is also made of aldehydes other than formaldehyde— acetaldehyde, glyoxal, 2-furaldehyde.

The exact composition and molecular weight of the resole depend on the formaldehyde to phenol ratio, pH, temperature and other reaction conditions [Maciel et al., 1984; Werstler, 1986]. For example, higher formaldehyde compositions will yield resoles containing more of the mononuclear compounds. The resole prepolymer may be either a solid or liquid polymer and soluble or insoluble in water depending on composition and molecular weight. Molecular weights are in the range 500–5000, with most being below 2000. The resoles are usually neutralized or made slightly acidic. The pH of the prepolymer together with its composition determine whether it will be slow curing or highly reactive. Crosslinking (curing) is carried out by heating at temperatures up to 180°C. The crosslinking process involves the same chemical reactions as for synthesizing the prepolymer—the formation of methylene and ether bridges between benzene rings to yield a network structure of type **XXVII**. The relative

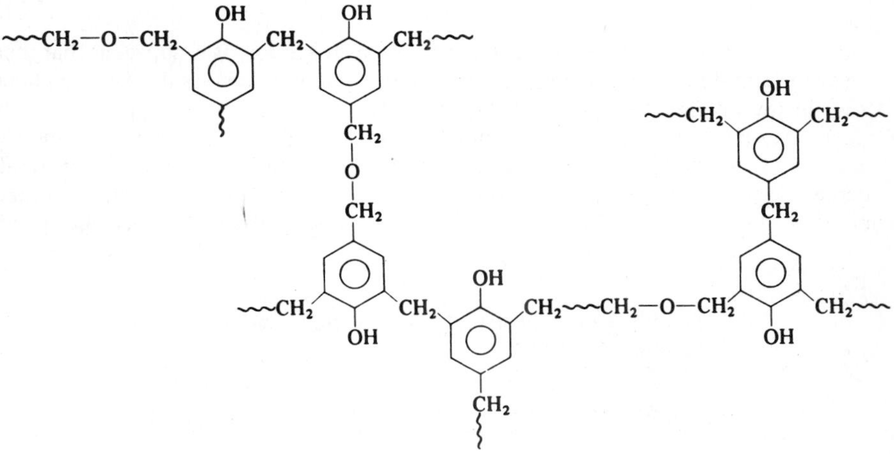

XXVII

importance of methylene and ether bridges is dependent on temperature [Lenz, 1967]. Higher curing temperatures favor the formation of methylene bridges but not to the exclusion of ether bridges.

The polymerization and crosslinking of phenol-formaldehyde is a highly useful industrial process. However, the reactions that take place are quite difficult to handle in a quantitative manner for a number of reasons. The assumption of equal reactivity of all functional groups in a monomer, independent of the other functional groups in the molecule and of whether the others are reacted, is dubious in this polymerization. Consider, for example, the routes by which trimethylolphenol (**XXIVd**) can be produced in this system:

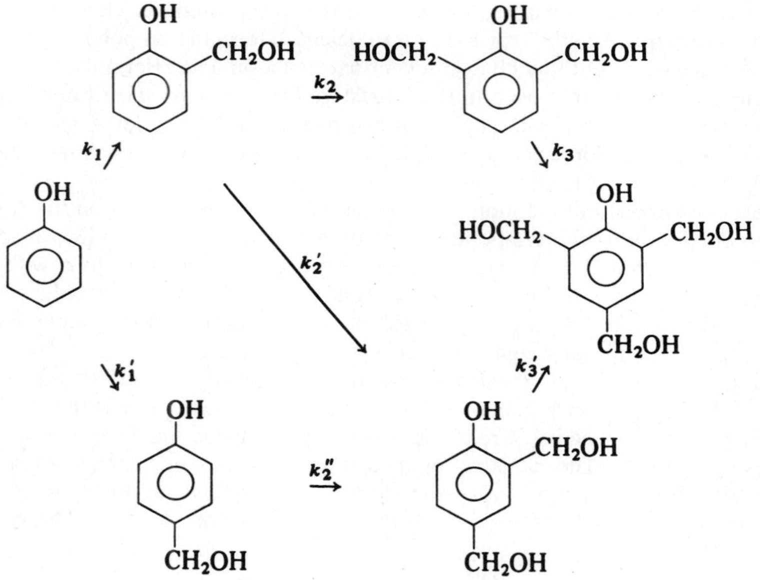

$$(2\text{-}178)$$

Table 2-10 shows the rate constants for the various reactions. It is apparent that there are significant differences in the reactivities of the different functional groups in phenol (i.e., in the different positions on the ring). The reaction between phenol and formaldehyde involves a nucleophilic attack by the phenolate anion on formaldehyde (Eq. 2-179) (or its hydrated form). The electron-pushing methylol groups generally increase the reactivity of phenol toward further substitution although steric hindrance modifies this effect. It is difficult to quantitatively discuss the reasons for the specific

TABLE 2-10 Reaction of Phenol with Formaldehyde Catalyzed by Base[a,b]

Reaction	Rate Constant $\times$ 10^4 (liters/mole-sec)
k_1	14.63
k_1'	7.81
k_2	13.50
k_2'	10.21
k_2''	13.45
k_3	21.34
k_3'	8.43

[a]Data from Zavitsas [1968]; Zavitsas et al. [1968].
[b]Reaction conditions: pH = 8.3, T = 57°C.

differences in reactivity, since there is no general agreement on even the relative values of the seven different rate constants. The evaluation of the different rate constants is difficult from both the viewpoints of chemical analysis of the mono-, di-, and trimethylolphenols and the mathematical analysis [Katz and Zwei, 1978] of the kinetic data.

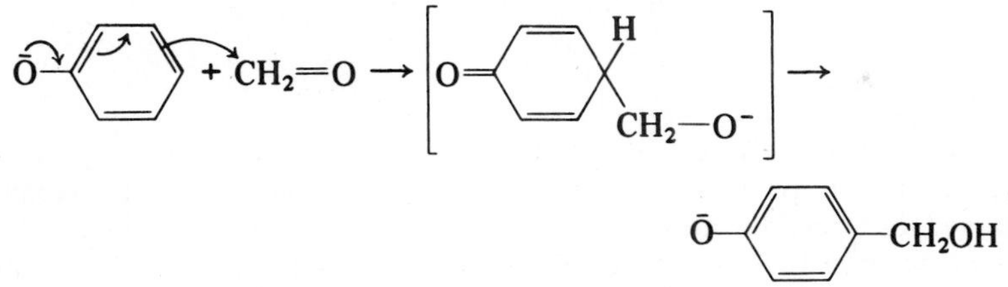

$$\bar{O} \!-\!\!\left\langle \right\rangle \!+\! CH_2 \!=\! O \longrightarrow \left[O \!=\!\!\left\langle \right\rangle_{CH_2 - O^-}^{H} \right] \longrightarrow$$

$$\bar{O} \!-\!\!\left\langle \right\rangle \!-\! CH_2OH \quad (2\text{-}179)$$

Not only are the different ring positions on phenol of different reactivity, but one expects that the two functional groups of formaldehyde would also differ. The second functional group of formaldehyde actually corresponds to the methylol group, since the reaction of a methylolphenol with a phenol molecule (or with a second methylolphenol) probably proceeds by a sequence such as

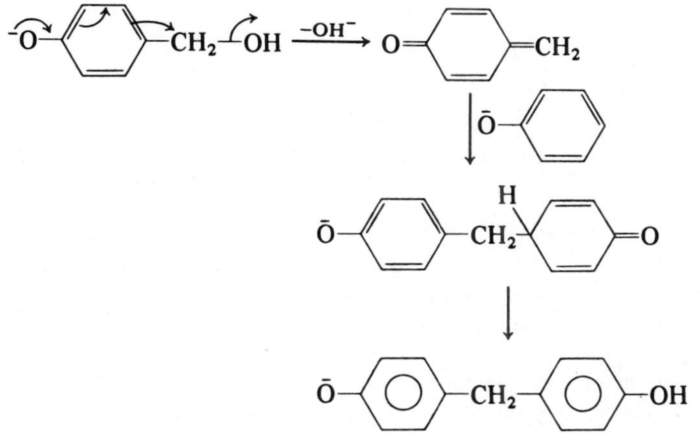

$$\bar{O} \!-\!\!\left\langle \right\rangle \!-\! CH_2 \!-\! OH \xrightarrow{-OH^-} O \!=\!\!\left\langle \right\rangle \!=\! CH_2$$

$$\Big\downarrow \bar{O} \!-\!\!\left\langle \right\rangle$$

$$\bar{O} \!-\!\!\left\langle \right\rangle \!-\! CH_2 \!-\!\!\left\langle \right\rangle_{H} \!=\! O$$

$$\Big\downarrow$$

$$\bar{O} \!-\!\!\left\langle \bigcirc \right\rangle \!-\! CH_2 \!-\!\!\left\langle \bigcirc \right\rangle \!-\! OH \quad (2\text{-}180)$$

Direct kinetic measurements are not available to show the reactivity of the methylol group in this reaction compared to the initial reaction of formaldehyde. However, the general observation that the amounts of di- and polynuclear compounds present in resole prepolymers differ widely depending on the reaction condition (temperature, pH, specific catalyst used, concentrations of reactants) indicates that the two functional groups of formaldehyde differ in reactivity.

A further complication in the phenol-formaldehyde polymerization is that it may involve a decrease in functional group reactivity with molecular size. This can easily happen in systems that undergo extensive crosslinking. Such systems may cease to be homogeneous solutions before the experimentally determined gel point. The gel point may be preceded by the formation of microgel particles (Sec. 2-2d) which are too small to be visible to the naked eye [Katz and Zwei, 1978]. Functional groups in the microgel particles would be relatively unreactive due to their physical unavailability. Similar considerations would apply for the reaction period subsequent to the gel point. A decrease in reactivity with size may not be due to molecular size but a consequence of steric shielding of ortho or para positions of benzene rings within a chain compared to those positions on rings at the chain ends [Kumar et al., 1980].

2-12b-2 Novolac Phenolics

Phenol-formaldehyde prepolymers, referred to as *novolacs*, are obtained by using a ratio of formaldehyde to phenol of 0.75–0.85:1, sometimes lower. Since the reaction system is starved for formaldehyde, only low molecular weight polymers can be formed and there is a much narrower range of products compared to the resoles. The reaction is accomplished by heating for 2–4 hr at or near reflux temperature in the presence of an acid catalyst. Oxalic and sulfuric acids are used in amounts of 1-2 and <1 part, respectively, per 100 parts phenol. The polymerization involves electrophilic aromatic substitution, first by hydroxymethyl carbocation and subsequently by benzyl carbo-

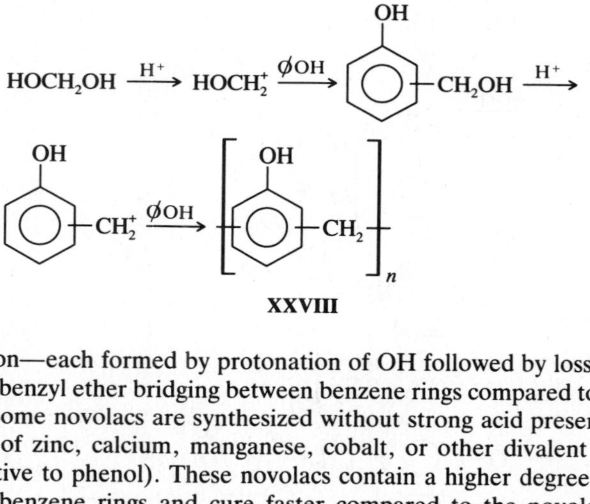

$$(2\text{-}181)$$

XXVIII

cation—each formed by protonation of OH followed by loss of water. There is much less benzyl ether bridging between benzene rings compared to the resole prepolymers.

Some novolacs are synthesized without strong acid present by using a carboxylate salt of zinc, calcium, manganese, cobalt, or other divalent metal ion (2% or more relative to phenol). These novolacs contain a higher degree of ortho substitution on the benzene rings and cure faster compared to the novolacs synthesized with the stronger acids. The mechanism for this ortho effect probably involves simultaneous complexation of the divalent metal ion with the phenol and carbocation hydroxyl groups. Similar effects of divalent ions on the relative amounts of ortho and para substitution are observed in the formation of resole prepolymers on comparison of catalysis by divalent metal hydroxides versus monovalent metal hydroxides.

The reaction mixture is dehydrated at temperatures as high as 160°C (higher temperatures can be tolerated than with resoles). The prepolymer is cooled, crushed, blended with 5–15% hexamethylenetetramine, $(CH_2)_6N_4$, and sold to the fabricator. Hexamethylenetetramine, referred to as *hexa*, is the product of the reaction of 6 moles of formaldehyde and 4 moles of ammonia. Curing occurs rapidly on heating with the formation of both methylene and benzylamine crosslinking bridges between benzene rings. The crosslinked network is pictured as **XXIX**.

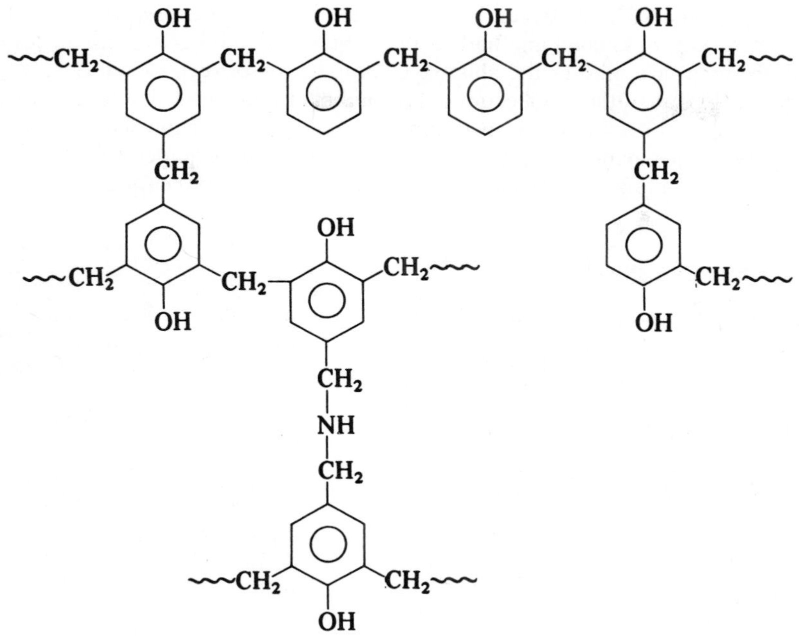

XXIX

2-12b-3 Applications

Phenolic polymers, the first commercial synthetic plastics, were introduced by Baekeland in 1909 through his Bakelite Company. Bakelite dominated this product until 1926 when its patent expired. Phenolic polymers are the largest-volume thermosetting plastics. More than 3 billion pounds are produced annually in the United States. Phenolics have high strength and dimensional stability combined with good resistance to impact, creep, solvents, and moisture. Most phenolics contain fillers or reinforcements. General-grade phenolics are filled with mica, clay, wood or mineral flour, cellulose, and chopped fabric. Engineering-grade phenolics, reinforced with glass or organic fibers, elastomers, graphite, and polytetrafluoroethylene, are rated for continuous use at 150–170°C.

The largest-volume application of phenolic polymers is the adhesive bonding material for the manufacture of plywood, particle board, wafer board, and fiberboard. The phenolic constitutes up to one-third of the weight of such products. Other adhesive applications include the binder for aluminum oxide or silicon carbide in abrasion wheels and disks and contact adhesives, usually blends of rubber and phenolic, used extensively in the furniture, automotive, construction, and footwear industries. Phenolics are used in a variety of coatings applications—baked-on coatings, coatings on cans, drums, metal pipe, electrical insulation, varnishes, and metal primers. The phenolic resins are generally too hard and brittle to be used alone for the coatings applications but blends with alkyds, polyethylene, and epoxy resins perform well. Laminates of cellulose paper, cotton fabric, glass and asbestos cloths, and wood veneer with phenolics are made in various forms (sheet, rod, tubes) that can be machined. These find wide uses as printed-circuit boards, gears, cams and protective and decorative surfaces for tables, furniture, kitchen and bathroom countertops, and walls. Composites containing phenolics are used as the ablative coating for space reentry vehicles, aircraft interiors and brakes. Other applications of phenolics include molded objects (switches, pump housings, cutlery handles, oven and toaster parts), friction materials (brake lining, transmission bands), foam insulation, carbonless copy paper, and the binder for sand in the manufacture of foundry molds for molden-metal casing.

2-12c Amino Plastics

The amino resins or plastics, closely related to the phenolics both in synthesis and applications, are obtained by the polymerization of formaldehyde with urea (**XXX**) ($f = 4$) or melamine (**XXXI**) ($f = 6$). Synthesis of the amino plastics can be carried

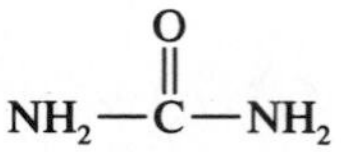

XXX	**XXXI**

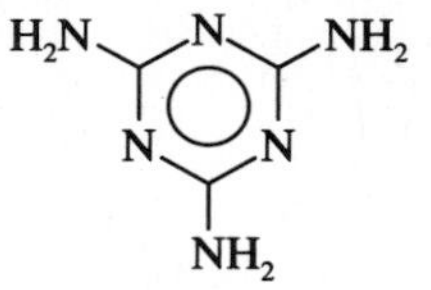

out in either alkaline or acidic conditions [Drumm and LeBlanc, 1972; Nair and Francis, 1983; Updegraff, 1985]. Control of the extent of reaction is achieved by pH and temperature control. The prepolymer can be made at various pH levels depending on the reaction temperature chosen. Polymerization is stopped by cooling and bringing the pH close to neutral. Curing of the prepolymer involves heating, usually in the presence of an added acid catalyst.

Polymerization of urea and formaldehyde yields various methylolureas (**XXXII**) as well as a mixture of higher oligomers as was the case for the phenolic prepolymers.

$$HOCH_2-NH-CO-NH_2 \qquad HOCH_2-NH-CO-NH-CH_2OH$$

XXXIIa   **XXXIIb**

$$(HOCH_2)_2N-CO-NH_2 \qquad (HOCH_2)_2N-CO-NH-CH_2OH$$

XXXIIc   **XXXIId**

$$(HOCH_2)_2N-CO-N(CH_2OH)_2$$

XXXIIe

The relative amounts of the various components is dependent on reaction conditions. The crosslinking reaction involves methylene (**XXXIII**), methylene ether (**XXXIV**),

$$\sim\sim CH_2-NH-CO-N-CH_2-NH-CO-NH\sim\sim$$
$$\qquad\qquad\qquad\qquad | $$
$$\qquad\qquad\qquad\quad CH_2$$
$$\qquad\qquad\qquad\qquad |$$
$$\sim\sim NH-CO-NH-CH_2-N-CO-NH-CH_2\sim\sim$$

XXXIII

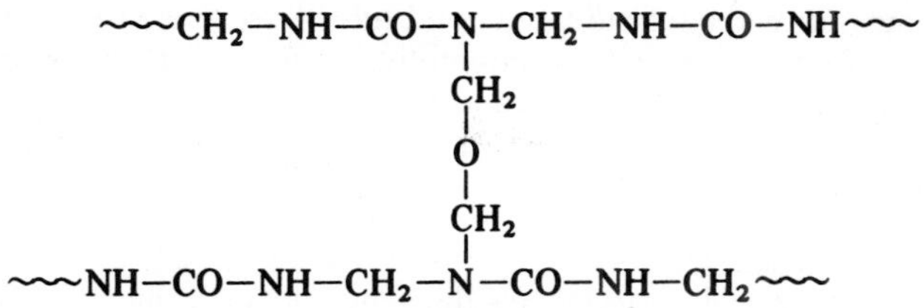

$$\sim\sim CH_2-NH-CO-N-CH_2-NH-CO-NH\sim\sim$$
$$\underset{\displaystyle CH_2}{|}$$
$$\underset{\displaystyle O}{|}$$
$$\underset{\displaystyle CH_2}{|}$$
$$\sim\sim NH-CO-NH-CH_2-N-CO-NH-CH_2\sim\sim$$
$$\textbf{XXXIV}$$

and cyclic (**XXXV**) bridges between urea units [Ebdon and Heaton, 1977; Tomita and Ono, 1979]. As in the phenol-formaldehyde system, there is considerable evidence that the assumption of equal reactivity of functional groups is invalid for both urea and formaldehyde [Tomita and Hirose, 1976]. The formation and curing of prepolymers from formaldehyde and melamine (2,4,6-triamino-*s*-triazine) follows in a similar manner.

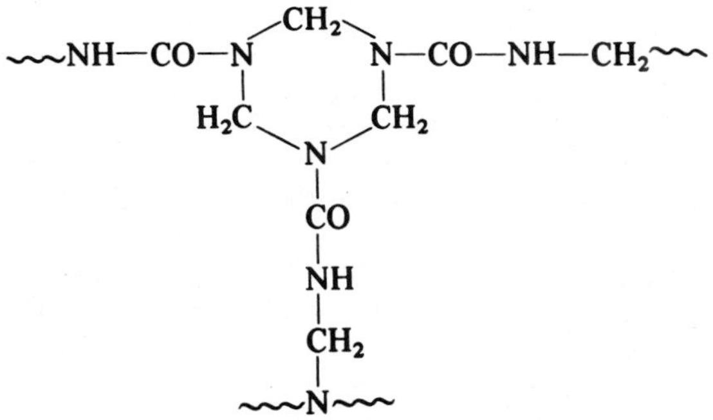

$$\textbf{XXXV}$$

More than 1.6 billion pounds of amino plastics are produced annually in the United States. The urea-formaldehyde polymers account for slightly more than 85% of the total. The amino plastics are similar in properties to the phenolics but are clearer and colorless. They are also harder but have somewhat lower impact strength and resistance to heat and moisture. The melamine resins are better than the ureas in hardness and resistance to heat and moisture. The melamine and urea resins are rated for continuous use at temperatures of 130–150°C and 100°C, respectively. The general applications of the amino and phenolic plastics are the same but there are uses where the amino plastics are superior. The melamine resins find an important niche due to their combination of clarity and lack of color compared to the phenolics and their superior hardness and heat and moisture resistance compared to urea resins. Molding and laminating techniques are used to produce colored dinnerware and dinnerware with decorative patterns. A combination of phenolic and melamine plastics are used to produce decorative laminated plastic sheets for tabletop and counters. The combination gives a superior product—a phenolic backing imparts superior mechanical properties while a top melamine layer imparts outstanding clarity and hardness. Amino resins are used to impart crease resistance and wash-and-wear properties to cellulosic fabrics. A garment with its appropriate creases and containing the low-molecular-weight amino resin is heated to achieve curing. Curing involves reaction of the amino resin with hydroxyl groups of the cellulosic fabric, and this "sets" the garment into its shape. Amino resins also find extensive as automotive paints; the cheaper ureas are used as primers while the melamines are used as topcoats.

2-12d Epoxy Resins

Epoxy resins or plastics are typically formed by the reaction of bisphenol A and epichlorhydrin [McAdams and Gannon, 1986]. The reaction actually involves the

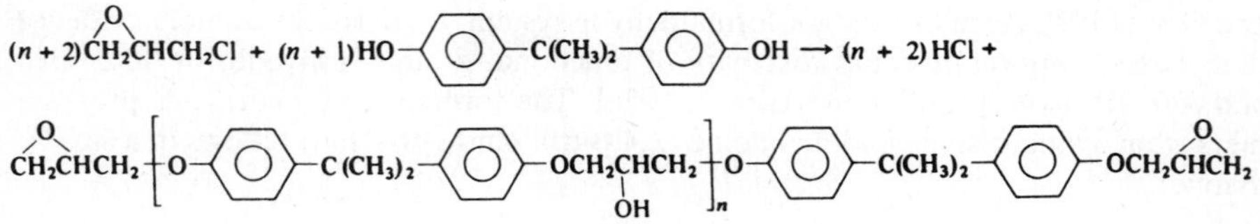

$$(n + 2)\text{CH}_2\text{CHCH}_2\text{Cl} + (n + 1)\text{HO}-\!\!\langle\bigcirc\rangle\!\!-\text{C(CH}_3)_2-\!\!\langle\bigcirc\rangle\!\!-\text{OH} \longrightarrow (n + 2)\,\text{HCl} +$$

$$\text{CH}_2\text{CHCH}_2\!\!\left[\text{O}-\!\!\langle\bigcirc\rangle\!\!-\text{C(CH}_3)_2-\!\!\langle\bigcirc\rangle\!\!-\text{OCH}_2\underset{\overset{|}{\text{OH}}}{\text{CHCH}_2}\right]_n\!\!\text{O}-\!\!\langle\bigcirc\rangle\!\!-\text{C(CH}_3)_2-\!\!\langle\bigcirc\rangle\!\!-\text{OCH}_2\text{CHCH}_2$$

$$(2\text{-}182)$$

sodium salt of bisphenol A since polymerization is carried out in the presence of an equivalent of sodium hydroxide. Reaction temperatures are in the range 50–95°C. Side reactions (hydrolysis of epichlorohydrin, reaction of epichlorohydrin with hydroxyl groups of polymer or impurities) as well as the stoichiometric ratio need to be controlled to produce a prepolymer with two epoxide end groups. Either liquid or solid prepolymers are produced by control of molecular weight; typical values of n are less than 1 for liquid prepolymers and in the range 2–30 for solid prepolymers.

Epichlorohydrin is reacted with a variety of hydroxy, carboxy, and amino compounds to form monomers with two or more epoxide groups, and these monomers are then used in the reaction with bisphenol A [Lohse, 1987]. Examples are the diglycidyl derivative of cyclohexane-1,2-dicarboxylic acid, the triglycidyl derivatives of p-aminophenol and cyanuric acid, and the polyglycidyl derivative of phenolic prepolymers. Epoxidized diolefins are also employed.

A variety of coreactants are used to cure epoxy resins either through the epoxide or hydroxyl groups. Polyamines are the most common curing agent with reaction involving ring-opening addition of amine.

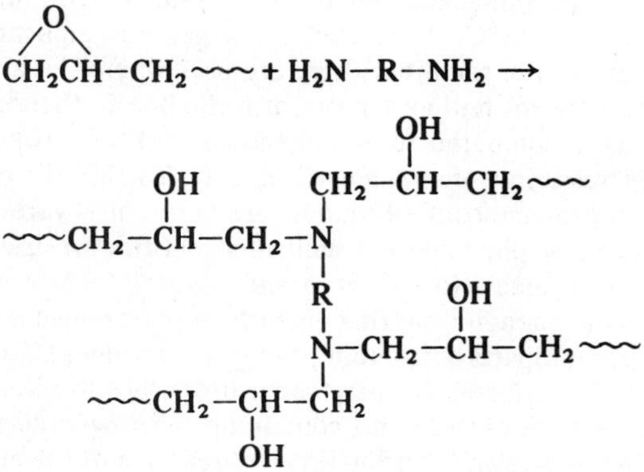

$$\text{CH}_2\text{CH}-\text{CH}_2\sim\sim + \text{H}_2\text{N}-\text{R}-\text{NH}_2 \longrightarrow$$

$$\sim\sim\text{CH}_2-\underset{\overset{|}{\text{OH}}}{\text{CH}}-\text{CH}_2-\text{N}\!\!\begin{array}{l}\text{CH}_2-\underset{\overset{|}{\text{OH}}}{\text{CH}}-\text{CH}_2\sim\sim\\[4pt]\text{R}\\[2pt]\text{N}-\text{CH}_2-\underset{\overset{|}{\text{OH}}}{\text{CH}}-\text{CH}_2\sim\sim\\[2pt]\sim\sim\text{CH}_2-\underset{\overset{|}{\text{OH}}}{\text{CH}}-\text{CH}_2\end{array}$$

$$(2\text{-}183)$$

Both primary and secondary amines are used with primary amines being more reactive than secondary amines [Glover et al., 1988]. Since each nitrogen–hydrogen bond is reactive in the curing reaction, primary and secondary amine groups are bi- and monofunctional, respectively. A variety of amines are used as crosslinking agents, including diethylene triamine ($f = 5$), triethylene tetramine ($f = 6$), 4,4′-diamino-diphenylmethane ($f = 4$), and polyaminoamides (e.g., the diamide formed from diethylene triamine and a dimerized or trimerized fatty acid). Other types of compounds have also been used to cure epoxy resins via the epoxide groups, including polythiols, dicyandiamide (cyanoguanidine), diisocyanates, and phenolic prepolymers. Some of these agents require weak bases such as tertiary amines and imidazole derivatives to accelerate the curing process.

Crosslinking of epoxy plastics through the hydroxyl groups of the repeat unit is used for prepolymers with low epoxide group contents. The most common curing agent is phthalic anhydride although other acid anhydrides such as tetrahydrophthalic, nadic methyl, and chloroendic anhydrides are used in specialty applications.

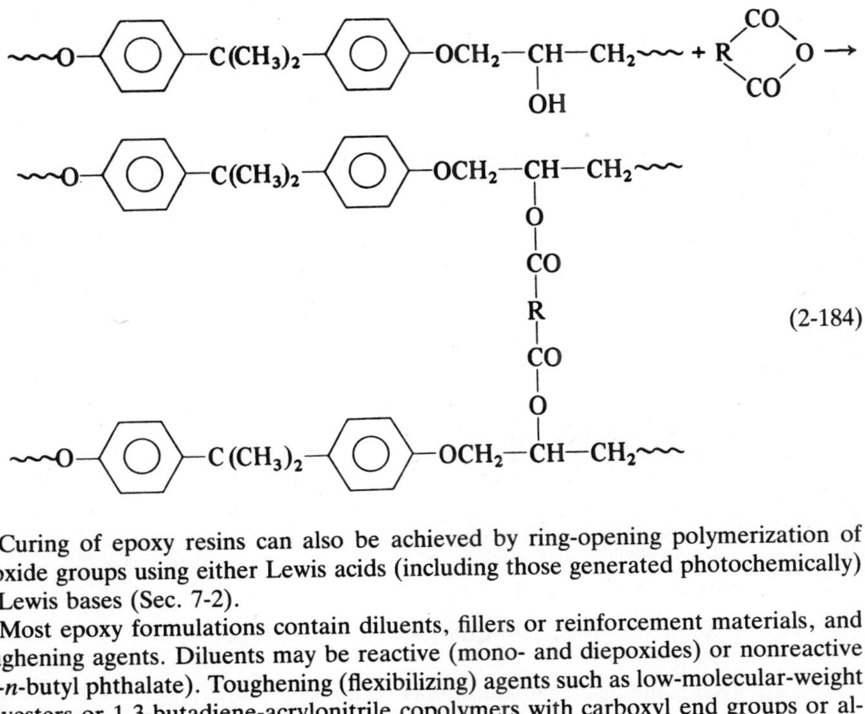

(2-184)

Curing of epoxy resins can also be achieved by ring-opening polymerization of epoxide groups using either Lewis acids (including those generated photochemically) or Lewis bases (Sec. 7-2).

Most epoxy formulations contain diluents, fillers or reinforcement materials, and toughening agents. Diluents may be reactive (mono- and diepoxides) or nonreactive (di-n-butyl phthalate). Toughening (flexibilizing) agents such as low-molecular-weight polyesters or 1,3-butadiene-acrylonitrile copolymers with carboxyl end groups or aliphatic diepoxides participate in the crosslinking process.

Epoxy resins possess high chemical and corrosion resistance, toughness and flexibility, good mechanical and electrical behavior, and outstanding adhesion to many different substrates. A wide range of products varying in properties and curing temperature are obtained for specific applications by proper selection of the monomers, additives, and curing agents. Use temperatures are in the range 90–130°C. More than 500 million pounds of epoxy plastics are produced annually in the United States. The applications for epoxies fall into two categories—coatings and structural. Coatings application include marine, maintenance, drum, and can coatings. Automotive primer coatings involve epoxy resins with ionic charges (either carboxylate or quaternary ammonium ion) that allow for electrodeposition of the resin. Waterborne epoxy coatings have been developed for beer and beverage containers. Structural composites are used in the military (missile casings, tanks), aircraft (rudders, wing skins, flaps), automobiles (leaf springs, drive shafts), and pipe in the oil, gas, chemical, and mining industries. Laminates are used in the electrical and electronics industries (circuit-board substrate, encapsulation of devices such as transistors, semiconductors, coils). Other applications include industrial and terrazzo flooring, repair of highway concrete cracks, coatings for roads and bridges, and a wide range of adhesive applications.

2-12e Polyurethanes

The synthesis of polyurethanes is usually presented as proceeding via the formation of carbamate (urethane) linkages by the reaction of isocyanates and alcohols.

$$n\text{HO}-\text{R}-\text{OH} + n\text{OCN}-\text{R}'-\text{NCO} \rightarrow$$

$$\text{HO}-(\text{R}-\text{OCONH}-\text{R}'-\text{NHCO}-\text{O})_{(n-1)}\ \text{R}-\text{OCONH}-\text{R}'-\text{NCO}$$

$$(2\text{-}185)$$

However, this is an oversimplification since other reactions are also usually involved [Backus, 1977; Backus et al., 1988]. Foamed products such as seat cushions and bedding are the largest applications of polyurethanes. Water is often deliberately added in the production of flexible polyurethane foams. Isocyanate groups react with water to form urea linkages in the polymer chain with the evolution of carbon dioxide. The

$$2\text{\small\char"007E\char"007E\char"007E}\text{NCO} + \text{H}_2\text{O} \rightarrow \text{\small\char"007E\char"007E\char"007E}\text{NH}-\text{CO}-\text{NH}\text{\small\char"007E\char"007E\char"007E} + \text{CO}_2 \qquad (2\text{-}186)$$

carbon dioxide acts as the *blowing agent* to form the foamed structure of the final product. (Low-boiling solvents such as fluorotrichloromethane are usually added to act as blowing agents in the synthesis of rigid polyurethane foamed products.) Many polyurethanes are synthesized using mixtures of diols and diamines. The diamines react with isocyanate groups to introduce additional urea linkages into the polymer. Thus, the typical polyurethane actually contains both urethane and urea repeat units.

$$2\text{\small\char"007E\char"007E\char"007E}\text{NCO} + \text{H}_2\text{N}-\text{R}''-\text{NH}_2 \rightarrow$$

$$\text{\small\char"007E\char"007E\char"007E}\text{NH}-\text{CO}-\text{NH}-\text{R}''-\text{NH}-\text{CO}-\text{NH}\text{\small\char"007E\char"007E\char"007E} \qquad (2\text{-}187)$$

(There are even some instances where only diamines are used in the synthesis. The polymer is then a polyurea although the manufacturer typically refers to it as a polyurethane.)

The situation is even more complex since the N—H bonds of both urethane and urea linkages add to isocyanate groups to form allophanate and biuret linkages, respectively. These reactions result in branching and crosslinking of the polymer. Equation 2-188 shows how a diisocyanate crosslinks the urethane and urea groups from two different polymer chains. The relative amounts of allophanate and biuret crosslinks in the polymer depend on the relative amounts of urea and urethane groups (which, in turn, depends on the relative amounts of diamine and diol in the reaction system) and reaction conditions. There is a greater tendency toward biuret linkages since the urea N—H is more reactive than the urethane N—H.

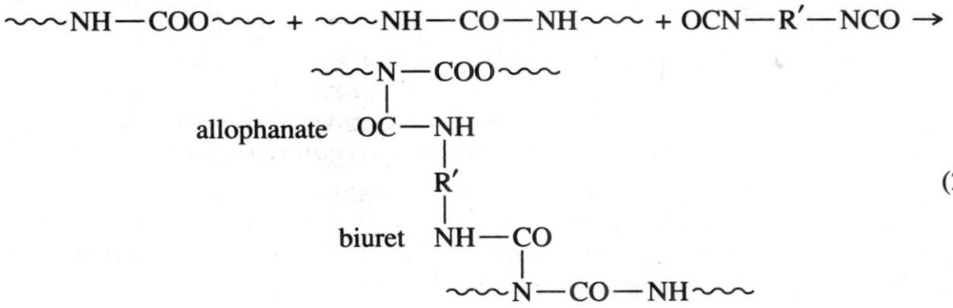

$$\sim\!\!\sim\!\!NH\!-\!COO\!\sim\!\!\sim + \sim\!\!\sim\!\!NH\!-\!CO\!-\!NH\!\sim\!\!\sim + OCN\!-\!R'\!-\!NCO \rightarrow$$

$\sim\!\!\sim\!\!N\!-\!COO\!\sim\!\!\sim$

allophanate $\quad \overset{|}{OC}\!-\!NH$

$\overset{|}{R'}$ $\qquad\qquad\qquad$ (2-188)

biuret $\quad NH\!-\!CO$

$\sim\!\!\sim\!\!N\!-\!CO\!-\!NH\!\sim\!\!\sim$

$$3\sim\!\!\sim\!\!R'\!-\!NCO \rightarrow \qquad\qquad\qquad\qquad (2\text{-}189)$$

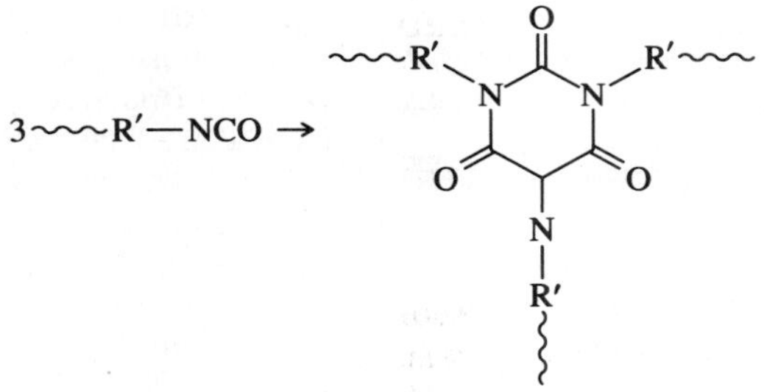

Trimerization of isocyanate groups to form isocyanurates also occurs and serves as an additional source of branching and crosslinking (Eq. 2-189).

These reactions present a complexity in carrying out polymerization. Simultaneously, we have the ability to vary polymer properties over a wide range by control of the relative amounts of reactants and the polymerization conditions. Further control of the final product is achieved by choice of monomers. Most polyurethanes involve a macroglycol (a low-molecular-weight polymer with hydroxyl end groups), diol or diamine, and diisocyanate. The macroglycol (also referred to as a *polyol*) is a polyether or polyester synthesized under conditions that result in two or more hydroxyl end groups. (Details of polyurethane synthesis involving macroglycols are described in Sec. 2-13c.) The diol monomers include ethylene glycol, 1,4-butanediol, 1,6-hexanediol, and *p*-di(2-hydroxyethoxy)benzene. The diamine monomers include diethyltoluene-diamine, methylenebis(*p*-aminobenzene), and 3,3'-dichloro-4,4'-diaminodiphenyl-methane. The diisocyanate monomers include hexamethylene diisocyanate, toluene 2,4- and 2,6-diisocyanates, and naphthalene 1,5-diisocyanate. Alcohol and isocyanate reactants with functionality greater than 2 are also employed.

The extent of crosslinking in polyurethanes depends on a combination of the amount of polyfunctional monomers present and the extent of biuret, allophanate, and trimerization reactions [Dusek, 1987]. The latter reactions are controlled by the overall stoichiometry and the specific catalyst present. Stannous and other metal carboxylates as well as tertiary amines are catalysts for the various reactions. Proper choice of the specific catalyst results in differences in the relative amounts of each reaction. Temperature also affects the extents of the different reactions. Polymerization temperatures are moderate, often near ambient and usually no higher than 100–120°C. Significantly higher temperatures are avoided because polyurethanes undergo several different types of degradation reactions, such as

$$\sim\sim NH-COO-CH_2CH_2\sim\sim \rightarrow \sim\sim NH_2 + CO_2 + CH_2=CH\sim\sim \qquad (2\text{-}190)$$

$$\sim\sim NH-COO-CH_2CH_2\sim\sim \rightarrow CO_2 + \sim\sim NH-CH_2CH_2\sim\sim \qquad (2\text{-}191)$$

as well as decomposition back to the alcohol and isocyanate monomers. Overall control in the synthesis of polyurethane foamed products also requires a balance between the polymerization–crosslinking and blowing processes. An imbalance between the chemical and physical processes can result in a collapse of the foamed structures (before solidification by crosslinking and/or cooling) or imperfections in the foam structures, which yields poor mechanical strength and performance.

The wide variations possible in synthesis give rise to a wide range of polyurethane products including flexible and rigid foams and solid elastomers, extrusions, coatings, and adhesives. Polyurethanes possess good abrasion, tear, and impact resistance coupled with oil and grease resistance. More than 2 billion pounds of polyurethane products are produced annually in the United States. Flexible foamed products include upholstered furniture and auto parts (cushions, backs, and arms), mattresses, and carpet underlay. Rigid foamed products with a closed-cell morphology possess excellent insulating properties and find extensive use in commercial roofing, residential sheathing, and insulation for water heaters, tanks, pipes, refrigerators, and freezers. Solid elastomeric products include forklift tires, skateboard wheels, automobile parts (bumpers, fascia, fenders, door panels, gaskets for trunk, windows, windshield, steering wheel, instrument panel), and sporting goods (golf balls, ski boots, football cleats). Many of these foam and solid products are made by reaction injection molding (RIM), a process in which a mixture of the monomers is injected into a mold cavity where polymerization and crosslinking take place to form the product. Reaction injection molding of polyurethanes, involving low-viscosity reaction mixtures and moderate reaction temperatures, is well suited for the economical molding of large objects such as automobile fenders. (Many of the elastomeric polyurethane products are *thermoplastic elastomers*; see Sec. 2-13c.)

2-12f Polysiloxanes

Polysiloxanes, also referred to as *silicones*, possess an unusual combination of properties that are retained over a wide temperature range (-100 to $250°C$). They have very good low temperature flexibility because of the low T_g value ($-127°C$). Silicones are very stable to high temperature, oxidation, chemical and biological environments, and weathering and possess good dielectric strength, and water repellency. Almost 0.5 billion pounds of polysiloxanes are produced annually in the United States in the form of fluids, resins, and elastomers. Fluid applications include fluids for hydraulics, antifoaming, water-repellent finishes for textiles, surfactants, greases, lubricants, and heating baths. Resins are used as varnishes, paints, molding compounds, electrical insulation, adhesives, laminates, and release coatings. Elastomer applications include sealants, caulks, adhesives, gaskets, tubing, hoses, belts, electrical insulation such as automobile ignition cable, encapsulating and molding applications, fabric coatings, encapsulants, and a variety of medical applications (antiflatulents, heart valves, encasing of pacemakers, prosthetic parts, contact lenses, coating of plasma bottles to avoid blood coagulation). Silicone elastomers differ markedly from other organic elastomers in the much greater effect of reinforcing fillers in increasing strength properties.

Polysiloxane fluids and resins are obtained by hydrolysis of chlorosilanes such as dichlorodimethyl-, dichloromethylphenyl-, and dichlorodiphenylsilanes [Brydson, 1982; Hardman and Torkelson, 1989]. The chlorosilane is hydrolyzed with water to a mixture of chlorohydroxy and dihydroxysilanes (referred to as *silanols*), which react with each other by dehydration and dehydrochlorination. The product is an equili-

$$
\underset{\underset{CH_3}{|}}{\overset{\overset{CH_3}{|}}{Cl-Si-Cl}} \xrightarrow[-HCl]{H_2O} \underset{\underset{CH_3}{|}}{\overset{\overset{CH_3}{|}}{Cl-Si-OH}} + \underset{\underset{CH_3}{|}}{\overset{\overset{CH_3}{|}}{HO-Si-OH}} \xrightarrow[-HCl]{-H_2O}
$$

$$
\left[-O - \underset{\underset{CH_3}{|}}{\overset{\overset{CH_3}{|}}{Si}} - \right]_n \tag{2-192}
$$

brated mixture of approximately equal amounts of cyclic oligomers and linear polysiloxanes. The amount of cyclics can vary from 20 to 80% depending on reaction conditions. The major cyclic oligomer is the tetramer with progressively decreasing amounts of higher cyclics. After the initial equilibration, a disiloxane terminating agent such as $[(CH_3)_3Si]_2O$ is added to stabilize the reaction mixture by termination of the linear species. The process may be carried out under either acidic or basic conditions depending on the desired product molecular weight. Basic conditions favor the production of higher molecular weight. Mixtures of cyclic oligomers and linear polymer may be employed directly as silicone fluids, or the cyclic content may be decreased prior to use by devolitilization (heating under vacuum). The synthesis of silicone resins proceeds in a similar manner except that the reaction mixture includes trichlorosilanes to bring about more extensive polymerization with crosslinking. Typically, the polymer product will be separated from an aqueous layer after the hydrolytic step, heated in the presence of a basic catalyst such as zinc octanoate to increase the polymer molecular weight and decrease the cyclic content, cooled, and stored. The final end-use application of this product involves further heating with a basic catalyst to bring about more extensive crosslinking.

Silicone elastomers are either *room-temperature vulcanization* (RTV) or *heat-cured silicone rubbers* depending on whether crosslinking is accomplished at ambient or elevated temperature. (The term *vulcanization* is a synonym for crosslinking. *Curing* is typically also used as a synonym for crosslinking but often refers to a combination of additional polymerization plus crosslinking.) RTV and heat-cured silicone rubbers typically involve polysiloxanes with degrees of polymerizations of 200–1500 and 2500–11,000, respectively. The higher-molecular-weight polysiloxanes cannot be synthesized by the hydrolytic step polymerization process. This is accomplished by ring-opening polymerization using ionic initiators (Sec. 7-11a). "One-component" RTV rubbers consist of an airtight package containing silanol-terminated polysiloxane, crosslinking agent (methyltriacetoxysilane, and catalyst (e.g., dibutyltin laurate). Moisture from the atmosphere converts the crosslinking agent to the corresponding silanol (acetic acid is a by-product), $CH_3Si(OH)_3$, which brings about further polymerization combined with crosslinking of the polysiloxane.

$$3 \sim\!\sim\!\sim SiR_2\!-\!OH + CH_3Si(OH)_3 \xrightarrow{-H_2O} \sim\!\sim\!\sim SiR_2\!-\!O\!-\!\underset{\underset{SiR_2}{\overset{|}{O}}}{\overset{\overset{CH_3}{|}}{Si}}\!-\!O\!-\!SiR_2\!\sim\!\sim\!\sim$$

$$(2\text{-}193)$$

Two-component RTV formulations involve separate packages for the polysiloxane and crosslinking agent. Hydrosilation curing involves the addition reaction between a polysiloxane containing vinyl groups (obtained by including methylvinyldichlorosilane in the original reaction mixture for synthesis of the polysiloxane) and a siloxane crosslinking agent that contains Si—H functional groups, such as 1,1,3,3,5,5,7,7-octamethyltetrasiloxane. The reaction is catalyzed by chloroplatinic acid or other soluble

$$\sim\!\sim\!\sim \underset{\underset{CH=CH_2}{|}}{SiR}\!-\!O\!\sim\!\sim\!\sim + Si[OSi(CH_3)_2H]_4 \rightarrow$$

$$Si\!\left[\!OSi(CH_3)_2\!-\!CH_2CH_2SiR\!-\!O\!\sim\!\sim\!\sim\!\right]_4 \quad (2\text{-}194)$$

Pt compound. Hydride-functional siloxanes can also crosslink silanol-terminated polysiloxanes. The reaction is catalyzed by tin salts and involves H_2 loss between Si—H and Si—O—H groups. Heat-curing of silicone rubbers usually involves free-radical initiators such as benzoyl peroxide (Sec. 9-2c). Hydrosilation at 50–100°C is also practiced.

2-12g Polysulfides

Polysulfide elastomers are produced by the reaction of an aliphatic dihalide, usually bis(2-chloroethyl)formal, with sodium polysulfide under alkaline conditions [Brydson, 1982; Ellerstein, 1988].

$$ClCH_2CH_2OCH_2OCH_2CH_2Cl + Na_2S_x \rightarrow$$

$$\underset{}{\left[\!-CH_2CH_2OCH_2OCH_2CH_2S_x\!-\!\right]_n} + NaCl \quad (2\text{-}195a)$$

The reaction is carried out with the dihalide suspended in an aqueous magnesium hydroxide phase. The value of x is slightly above 2. The typical polymerization system includes up to 2% 1,2,3-trichloropropane. The polymerization readily yields a polymer with a very high molecular weight, but this is not desirable until its end-use application. The molecular weight is lowered and the *polysulfide rank* (value of x) is simultaneously brought close to 2 by reductive treatment with NaSH and Na_2SO_3 followed by acidification. The result is a liquid, branched polysulfide with terminal thiol end groups and a molecular weight in the range 1000–8000. Curing to a solid elastomer is accomplished by oxidation of thiol to disulfide links by oxidants such as lead dioxide and *p*-quinone dioxime.

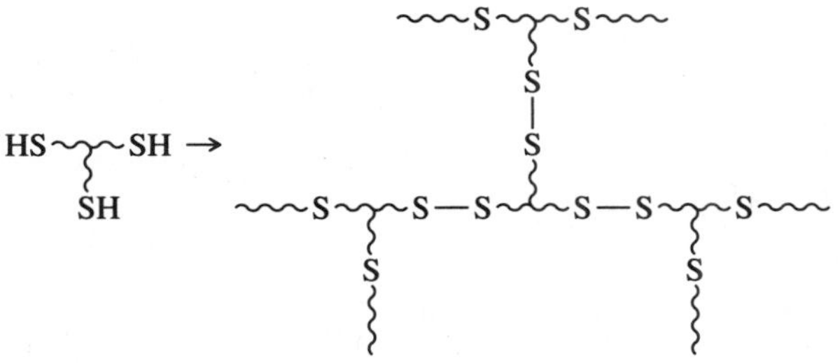

$$\text{HS} \sim \curlywedge \sim \text{SH} \rightarrow \qquad (2\text{-}195\text{b})$$

Polysulfides, often referred to as *thiokols*, are not produced at high volumes. These are specialty materials geared toward a narrow market. The annual production of polysulfides in the United States exceeds 40 million pounds. The advantages and disadvantages of polysulfides both reside in the disulfide linkage. They possess low temperature flexibility and very good resistance to ozone, weathering, oil, solvent (hydrocarbons as well as polar solvents such as alcohols, ketones, esters), and moisture. However, polysulfides have poor thermal stability and creep resistance, have low resilience, and are malodorous. (Aromatic polysulfides do not have the same deficiencies; see Sec. 2-14d.) The major applications of polysulfides are as sealants, caulks, gaskets, O-rings, and cements for insulating glass and fuel tanks, in marine applications, and in gasoline and fuel hose.

2-13 STEP COPOLYMERIZATION

It should be apparent that step polymerization is a versatile means of synthesizing a host of different polymers. The chemical structure of a polymer can be varied over a wide range in order to obtain a product with a particular combination of desirable properties. One can vary the functional group to produce a polyester, polyamide, or some other class of polymer as desired. Further, for any specific class of polymer, there is a considerable choice in the range of structures which can be produced. Consider, for example, the structure of polyamides. Polyamides with either of the general structures

—CO—R—CONH—R'—NH—   —NH—R—CO—

        XXXVI   **XXXVII**

can be synthesized depending on whether one uses the reaction of a diamine with a diacid or the self-reaction of an amino acid. A range of different polyamides can be obtained by varying the choice of the R and R' groups in structure **XXXVI** and the R group in structure **XXXVII**. Thus, for example, one can produce nylon 6/6 and nylon 6/10 by the reactions of hexamethylene diamine with adipic and sebacic acids, respectively. Nylon 6/10 is more flexible and moisture-resistant than nylon 6/6 as a result of the longer hydrocarbon segment derived from sebacic acid.

2-13a Types of Copolymers

Further variation is possible in the polymer structure and properties by using mixtures of the appropriate reactants such that the polymer chain can have different R and R' groups. Thus polyamide structures of types **XXXVIII** and **XXXIX** are possible variations on structures

—CO—R—CONH—R'—NHCO—R″—CONH—R‴—NH—

XXXVIII

—NH—R—CO—NH—R″—CO—

XXXIX

XXXVI and **XXXVII**, respectively. A polymer such as **XXXVIII** or **XXXIX** has two different repeat units and is referred to as a *copolymer*; the process by which it is synthesized is referred to as *copolymerization*. Polymers with structures **XXXVI** and **XXXVII**, each containing a single repeat unit, may be referred to as *homopolymers* to distinguish them from copolymers.

Different types of copolymers are possible with regard to sequencing of the two repeating units relative to each other. Thus a copolymer with an overall composition indicated by **XXXVIII** could have the *alternating copolymer* structure shown in **XXXVIII** in which the R, R', R″, and R‴ groups alternate in that order over and over again along the polymer chain, or the *block copolymer* structure **XL** in which

$$\text{---} (\text{---CO—R—CONH—R'—NH---})_m (\text{---CO—R″—CONH—R‴—NH---})_p$$

XL

blocks of one type of homopolymer structure are attached to blocks of another type of homopolymer structure, or the *statistical copolymer* structure in which there is an irregular (statistical) distribution of R and R″ groups as well as R' and R‴ groups along the copolymer chain. Similarly, one can have alternating, block, and statistical copolymers for the overall composition **XXXIX**.

For the statistical copolymer the distribution may follow different statistical laws, such as Bernoullian (zero-order Markov), first- or second-order Markov, depending on the specific reactants and the method of synthesis. This is discussed further in Secs. 6-2 and 6-5. Many statistical copolymers are produced via Bernoullian processes wherein the various groups are randomly distributed along the copolymer chain; such copolymers are *random copolymers*. The terminology used in this book is that recommended by IUPAC [Ring et al., 1985]. However, most literature references use the term *random copolymer* independent of the type of statistical distribution (which seldom is known).

The alternating and statistical copolymer structures can be symbolized as

〜〜ABABABABABABABABABABABAB〜〜

Alternating copolymer

〜〜AABABBBABAABAAABBABBBAAB〜〜

Statistical copolymer

where A and B represent two different repeating units. Different block copolymers are distinguished by the number of blocks per molecule, for example,

A_mB_p	$A_mB_pA_m$	$A_mB_pA_mB_p$	$(A_mB_p)_n$
Diblock	Triblock	Tetrablock	Multiblock

which are referred to as AB diblock, ABA triblock, ABAB tetrablock, and AB multiblock copolymers, respectively. For the various block copolymers the values of *m* and *p* as well as *n* are average values; thus, there is a distribution of block lengths and number of blocks along the copolymer chain. There is considerable structural versatility possible for statistical and block copolymers in terms of the relative amounts of A and B in a copolymer. For block copolymers there is the additional variation possible in the number of blocks of A and B and their block lengths (values of *m* and *p*).

Alternating, statistical, and random copolymers are named by following the prefix "poly" with the names of the two repeating units. The specific type of copolymer is noted by inserting *-alt-*, *-stat-*, or *-ran-* in between the names of the two repeating units; *-co-* is used when the type of copolymer is not specified, i.e., poly(A-*co*-B), poly(A-*alt*-B), poly(A-*stat*-B), poly(A-*ran*-B). Block copolymers are named by inserting *-block-* in between the names of the homopolymers corresponding to each of the blocks. The di-, tri-, tetra-, and multiblock copolymers are named as polyA-*block*-polyB, polyA-*block*-polyB-*block*-polyA, polyA-*block*-polyB-*block*-polyA-*block*-polyB, and poly(polyA-*block*-polyB), respectively. Adoption in the literature of some of these IUPAC recommendations for naming copolymers has been slow.

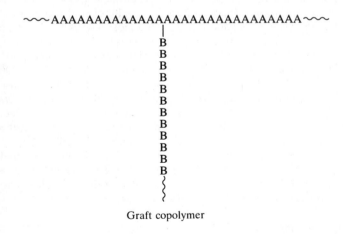

Graft copolymer

A fourth type of copolymer is the *graft copolymer* in which one or more blocks of homopolymer B are grafted as branches onto a main chain of homopolymer A. Graft copolymers are named by inserting *-graft-* in between the names of the corresponding homopolymers with the main chain being named first (e.g., polyA-*graft*-polyB). Graft copolymers are relatively unimportant for step polymerizations because of difficulties in synthesis. Graft copolymers are considered further in Sec. 9-8.

The discussion to this point has involved copolymers in which both repeating units have the same functional group. A second category of copolymer involves different functional groups in the two repeat units, for example, an amide–ester copolymer such as

$$-NH-R-CO-O-R''-CO-$$

XLI

instead of the amide–amide copolymer **XXXIX**. Both categories of copolymers are important, although not in the same manner.

2-13b Methods of Synthesizing Copolymers

2-13b-1 Statistical Copolymers

The copolymerization of a mixture of monomers offers a route to statistical copolymers; for instance, a copolymer of overall composition **XXXVIII** is synthesized by copolymerizing a mixture of the four monomers

$$\left.\begin{array}{l} HO_2C-R-CO_2H \\ HO_2C-R''-CO_2H \\ H_2N-R'-NH_2 \\ H_2N-R'''-NH_2 \end{array}\right\} \rightarrow \text{Copolymer of XXXVIII} \qquad (2\text{-}196)$$

It is highly unlikely that the reactivities of the various monomers would be such as to yield either block or alternating copolymers. The quantitative dependence of copolymer composition on monomer reactivities has been described [Korshak et al., 1976; Mackey et al., 1978; Russell et al., 1981]. The treatment is the same as that described in Chap. 6 for chain copolymerization (Secs. 6-2 and 6-5). The overall composition of the copolymer obtained in a step polymerization will almost always be the same as the composition of the monomer mixture since these reactions are carried out to essentially 100% conversion (a necessity for obtaining high-molecular-weight polymer). Further, for step copolymerizations of monomer mixtures such as in Eq. 2-196 one often observes the formation of random copolymers. This occurs either because there are no differences in the reactivities of the various monomers or the polymerization proceeds under reaction conditions where there is extensive interchange (Sec. 2-7c). The use of only one diacid or one diamine would produce a variation on the copolymer structure with either $R = R''$ or $R' = R'''$ [Jackson and Morris, 1988].

Statistical copolymers containing repeating units each with a different functional group can be obtained using appropriate mixtures of monomers. For example, a polyesteramide can be synthesized from a ternary mixture of a diol, diamine, and diacid or a binary mixture of a diacid and amine-alcohol [East et al., 1989].

2-13b-2 Alternating Copolymers

The alternating copolymer of composition **XXXVIII** cannot be synthesized. However, it is possible to synthesize an alternating copolymer in which $R'' = R$ by using a two-stage process. In the first stage a diamine is reacted with an excess of diacid to form a trimer

$$2nHO_2C-R-CO_2H + nH_2N-R'-NH_2 \rightarrow$$

$$nHO_2C-R-CONH-R'-NHCO-R-CO_2H \qquad (2\text{-}197)$$

The trimer is then reacted with an equimolar amount of a second diamine in the second stage

$$n\text{HO}_2\text{C}-\text{R}-\text{CONH}-\text{R}'-\text{NHCO}-\text{R}-\text{CO}_2\text{H} + n\text{H}_2\text{N}-\text{R}'''-\text{NH}_2 \rightarrow$$

$$\text{HO}-(-\text{CO}-\text{R}-\text{CONH}-\text{R}'-\text{NHCO}-\text{R}-\text{CONH}-\text{R}'''-\text{NH}-)_{\overline{n}}\text{H} +$$

$$(2n - 1)\text{H}_2\text{O} \tag{2-198}$$

Alternating copolymers with two different functional groups are similarly synthesized by using preformed reactants [Adduci and Amone, 1989; Gopal and Srinivasan, 1986; Mormann et al., 1989]. For example, polyamideurethanes analogous to **XXXIX** and **XXXVIII** are obtained by the polymerizations

$$\text{OCN}-\text{R}-\text{CONH}-\text{R}'-\text{OSi}(\text{CH}_3)_3 \xrightarrow[-(\text{CH}_3)_3\text{SiF}]{\text{HF}}$$

$$-(-\text{CO}-\text{NH}-\text{R}-\text{CO}-\text{NH}-\text{R}'-\text{O}-)_{\overline{n}} \tag{2-199}$$

$$\text{OCN}-\text{R}-\text{CONH}-\text{R}'-\text{NHCO}-\text{R}-\text{NCO} + \text{HO}-\text{R}''-\text{OH} \xrightarrow{\text{HF}}$$

$$-(-\text{CONH}-\text{R}-\text{CONH}-\text{R}'-\text{NHCO}-\text{R}-\text{NHCOO}-\text{R}''-\text{O}-)_{\overline{n}} \tag{2-200}$$

The silyl ether derivative of the alcohol is used in Eq. 2-199 since the corresponding alcohol $\text{OCN}-\text{R}-\text{CONH}-\text{R}'-\text{OH}$ cannot be isolated because of the high degree of reactivity of isocyanate and alcohol groups toward each other.

2-13b-3 Block Copolymers

There are two general methods for synthesizing block copolymers [Gaymans et al., 1989; Hedrick et al., 1989; Klein et al., 1988; Leung and Koberstein, 1986; Reiss et al., 1985; Speckhard et al., 1986]. The two methods, referred to here as the *one-prepolymer* and *two-prepolymer* methods, are described below for block copolymers containing different functional groups in the two repeat units. They are equally applicable to block copolymers containing the same functional group in the two repeating units. The two-prepolymer method involves the separate synthesis of two different prepolymers, each containing appropriate end groups, followed by polymerization of the two prepolymers via reaction of their end groups. Consider the synthesis of a polyester-*block*-polyurethane. A hydroxy-terminated polyester prepolymer is synthesized from $\text{HO}-\text{R}-\text{OH}$ and $\text{HOOC}-\text{R}'-\text{COOH}$ using an excess of the diol reactant. An isocyanate-terminated polyurethane prepolymer is synthesized from $\text{OCN}-\text{R}''-\text{NCO}$ and $\text{HO}-\text{R}'''-\text{OH}$ using an excess of the diisocyanate reactant. The α,ω-dihydroxypolyester and α,ω-diisocyanatopolyurethane prepolymers, referred to as *macrodiol and macrodiisocyanate*, respectively, are subsequently polymerized with each other to form the block copolymer.

$$p\text{H}-(-\text{O}-\text{R}-\text{OOC}-\text{R}'-\text{CO})_n\text{OR}-\text{OH} +$$

$$p\text{OCN}-(-\text{R}''-\text{NHCOO}-\text{R}'''-\text{OOCNH})_m\text{R}''-\text{NCO} \rightarrow$$

$$-[(-\text{O}-\text{R}-\text{OOC}-\text{R}'-\text{CO})_n\text{OR}-\text{OOCNH}-(-\text{R}''$$

$$-\text{NHCOO}-\text{R}'''-\text{OOCNH})_m\text{R}''-\text{NHCO}-]_{\overline{p}} \tag{2-201}$$

The block lengths n and m can be varied separately by adjusting the stoichiometric ratio r of reactants and conversion in each prepolymer synthesis. In typical systems the prepolymers have molecular weights in the range 500–6000. The molecular weight of the block copolymer is varied by adjusting r and conversion in the final polymerization. A variation of the two-prepolymer method involves the use of a coupling agent to join the prepolymers. For example, a diacyl chloride could be used to join together two different macrodiols or two different macrodiamines or a macrodiol with a macrodiamine.

The one-prepolymer method involves one of the above prepolymers with two "small" reactants. The macrodiol is reacted with a diol and diisocyanate

$$p\text{H} \overline{}(\text{O—R—OOC—R}'\text{—CO})_n\text{OR—OH} + mp\text{HO—R}'''\text{—OH} +$$

$$p(m + 1)\text{OCN—R}''\text{—NCO} \rightarrow$$

$$\overline{}[(\text{O—R—OOC—R}'\text{—CO})_n\text{OR—OOCNH} \overline{}(\text{R}''$$

$$\text{—NHCOO—R}'''\text{—OOCNH})_m\text{R}''\text{—NHCO} \overline{}_p \qquad (2\text{-}202)$$

The block lengths and the final polymer molecular weight are again determined by the details of the prepolymer synthesis and its subsequent polymerization. An often-used variation of the one-prepolymer method is to react the macrodiol with excess diisocyanate to form an isocyanate-terminated prepolymer. The latter is then *chain-extended* (i.e., increased in molecular weight) by reaction with a diol. The one- and two-prepolymer methods can in principle yield exactly the same final block copolymer. However, the dispersity of the polyurethane block length (m is an average value as are n and p) is usually narrower when the two-prepolymer method is used.

The prepolymers described above are one type of *telechelic polymer*. A telechelic polymer is one containing one or more functional end groups that have the capacity for selective reaction to form bonds with another molecule. The functionality of a telechelic polymer or prepolymer is equal to the number of such end groups. The macrodiol and macrodiisocyanate telechelic prepolymers have functionalities of two. Many other telechelic prepolymers were discussed in Sec. 2-12. (The term *functional polymer* has also been used to describe a polymer with one or more functional end groups.)

2-13c Utility of Copolymerization

There has been enormous commercial success in the synthesis of various copolymers. Intense research activity continues in this area of polymer science with enormous numbers of different combinations of repeating units being studied. Copolymer synthesis offers the ability to alter the properties of a homopolymer in a desired direction by the introduction of an appropriately chosen second repeating unit. One has the potential to combine the desirable properties of two different homopolymers into a single copolymer. Copolymerization is used to alter such polymer properties as crystallinity, flexibility, T_m, and T_g. The magnitudes, and sometimes even the directions, of the property alterations differ depending on whether statistical, alternating, or block copolymers are involved. The crystallinity of statistical copolymers is lower than that of either of the respective homopolymers (i.e., the homopolymers corresponding to the two different repeating units) because of the decrease in structural regularity. The melting temperature of any crystalline material formed is usually lower than that of either homopolymer. The T_g value will be in between those for the two homopolymers. Alternating copolymers have a regular structure, and their crystallinity may not be significantly affected unless one of the repeating units contains rigid, bulky, or excessively flexible chain segments. The T_m and T_g values of an alternating copolymer are in between the corresponding values for the homopolymers. Block copolymers often show significantly different behavior compared to alternating and statistical copolymers. Each type of block in a block copolymer shows the behavior (crystallinity, T_m, T_g) present in the corresponding homopolymer as long as the block lengths are not too short. This behavior is typical since A blocks from different polymer molecules aggregate with each other and, separately, B blocks from different polymer molecules aggregate with each other. This offers the ability to combine the properties of two very different polymers into one block copolymer. The exception to this behavior occurs infrequently—when the tendency for cross-aggregation between A and B blocks is the same as for self-aggregation of A blocks with A blocks and B blocks with B blocks.

Most commercial utilizations of copolymerization fall into one of two groups. One group consists of various statistical (usually random) copolymers in which the two repeating units possess the same functional group. The other group of commercial copolymers consists of block copolymers in which the two repeating units have different functional groups. There are very few commercial statistical copolymers in which the two repeating units have different functional groups. The reasons for this situation probably reside in the difficulty of finding one set of reaction conditions for simultaneously performing two different reactions, such as amidation simultaneously with ether formation. Also, the property enhancements available through this copolymerization route are apparently not significant enough in most systems to motivate one to overcome the synthetic problem by the use of specially designed monomers. The same synthetic and property barriers probably account for the lack of commercial alternating copolymers in which the two repeating units have different functional groups. Similar reasons account for the general lack of commercial block copolymers in which the two repeating units have the same functional group. The synthetic problem here is quite different. Joining together blocks of the same type of repeat unit is not a problem, but preventing interchange (Sec. 2-7c) between the different blocks is often not possible. Scrambling of repeating units via interchange is also a limitation on producing alternating copolymers in which both repeating units have the same functional group. With all the present research activity in copolymer synthesis, one might expect the situation to be significantly different in the future.

2-13c-1 *Statistical Copolymers*

Statistical copolymerization is practiced in many of the previously discussed polymerization systems in Secs. 2-8 and 2-12. Mixtures of phenol with an alkylated phenol such as cresol or bisphenol A are used for producing specialty phenolic resins. Trifunctional reactants are diluted with bifunctional reactants to decrease the crosslink density in various thermosetting systems. Methylphenyl-, diphenyl-, and methylvinyl-chlorosilanes are used in copolymerization with dimethyldichlorosilane to modify the properties of the polysiloxane. Phenyl groups impart better thermal and oxidative stability and improved compatibility with organic solvents. Vinyl groups introduce sites for crosslinking via peroxides (Sec. 9-2c). Polysiloxanes with good fuel and solvent resistance are obtained by including methyltrifluoropropylchlorosilane. Flame retardancy is imparted to the polycarbonate from bisphenol A (Eq. 2-132) by statistical or block copolymerization with a tetrabromo derivative of bisphenol A.

Statistical copolymerization of ethylene glycol and 1,4-butanediol with dimethyl terephthalate results in products with improved crystallization and processing rates compared to poly(ethylene terephthalate). *Polyarylates* (trade names: *Ardel*, *Arylon*, *Durel*), copolymers of bisphenol A with iso- and terephthalate units, combine the toughness, clarity, and processibility of polycarbonate with the chemical and heat resistance of poly(ethylene terephthalate). The homopolymer containing only terephthalate units is crystalline, insoluble, sometimes infusible, and difficult to process. The more useful copolymers, containing both tere- and isophthalate units, are amorphous, clear, and easy to process. Polyarylates are used in automotive and appliance hardware and printed-circuit boards. Similar considerations in the copolymerization of iso- and terephthalates with 1,4-cyclohexanedimethanol or hexamethylene diamine yield clear, amorphous, easy-to-process copolyesters or copolyamides, respectively, which are used as packaging film for meats, poultry, and other foods. The use of a mixture of unsaturated and saturated diacids or dianhydrides allows one to vary the crosslink density (which controls hardness and impact strength) in unsaturated polyesters (Sec. 2-12a). The inclusion of phthalic anhydride improves hydrolytic stability for marine applications.

An interesting copolyester synthesis is that carried out biologically by various bacteria such as *Alcaligenes eutrophus* and *Pseudomonas oleovorans* (Ballistreri et al., 1990; Doi et al., 1988; Gross et al., 1989]. The identity and relative amounts of the repeating units in the copolyester vary to an extent depending on the microorganism and its nutrient supply. *Biopol*, a commercial product in the United Kingdom, is reported to be a copolyester with 3-hydroxybutyrate and 3-hydroxyvalerate repeating units produced by *Alcaligenes eutrophus*. Such polymers have attracted some industrial interest because they are environmentally degradable. Degradable polymers offer an alternative to recycling of the worldwide mountains of nondegradable polymer wastes.

2-13c-2 Block Copolymers

Polyurethane multiblock copolymers of the type described by Eqs. 2-201 and 2-202 constitute an important segment of the commercial polyurethane market. The annual production is about 100 million pounds in the United States. These polyurethanes are referred to as *thermoplastic polyurethanes* (TPU) (trade names: *Estane*, *Texin*). They are among a broader group of elastomeric block copolymers referred to as *thermoplastic elastomers* (TPE). Crosslinking is a requirement to obtain the resilience associated with a rubber. The presence of a crosslinked network prevents polymer chains from irreversibly slipping past one another on deformation and allows for rapid and complete recovery from deformation.

Crosslinking in conventional elastomers is chemical; it is achieved by the formation of chemical bonds between copolymer chains (Sec. 9-2). Crosslinking of thermoplastic elastomers occurs by a physical process as a result of their microheterogeneous, two-phase morphology. This is a consequence of morphological differences between the two different blocks in the multiblock copolymer [Manson and Sperling, 1976]. One of the blocks, the polyester or polyether, is flexible (*soft*) and long while the other block, the polyurethane, is rigid (*hard*) and short. The rigidity of the polyurethane blocks is due to crystallinity (and hydrogen-bonding). (For other thermoplastic elastomers such as those based on styrene and 1,3-dienes, the rigidity of the hard polystyrene blocks is due to their glassy nature; see Sec. 5-4a.) The hard blocks from different copolymer molecules aggregate together to form rigid domains at ambient temperature. These rigid domains comprise a minor, discontinuous phase dispersed in the major, continuous phase composed of the rubbery blocks from different copolymer chains. The rigid domains act as *physical crosslinks* to hold together the soft, rubbery domains in a network structure. Physical crosslinking unlike chemical crosslinking is thermally reversible since heating over the T_m of the hard blocks softens the rigid domains and the copolymer flows. Cooling reestablishes the rigid domains, and the copolymer again behaves as a crosslinked elastomer.

Thermoplastic elastomers have the important practical advantage over conventional elastomers that there is no need for the additional chemical crosslinking reaction, which means that fabrication of objects is achieved on conventional equipment for thermoplastics. The transition from a molten polymer to a solid, rubbery product is much faster and takes place simply on cooling. There are significant advantages in recycle times, the ability to recycle scrap, and overall production costs. However, thermoplastic elastomers generally are not as effective as chemically crosslinked elastomers in solvent resistance and resistance to deformation at high temperature. TPE is not used in applications such as automobile tires, where these properties are important. Thermoplastic polyurethanes are used in applications such as wire insulation, automobile fascia, footwear (lifts, ski boots, football cleats), wheels (industrial, skateboard), and adhesives, where such properties are not important. Two types of macrodiols are used in the synthesis of TPU—polyesters, as described in Eqs. 2-201 and 2-202, and α,ω-dihydroxypolyethers obtained by ring-opening polymerizations (Chap. 7) of ethylene and propylene oxides and tetrahydrofuran (THF). The polyester-based TPU is tougher and has better oil resistance while the polyether-based TPU shows better low temperature flexibility and better hydrolytic stability. *Spandex* (trade name: *Lycra*) fibers are multiblock polyurethanes similar to TPU. Spandex typically differs from TPU in containing urea linkages instead of urethane in the hard blocks. These are obtained by using a diamine (instead of diol) to chain-extend isocyanate-terminated polyurethane prepolymers (synthesized from macroglycol and excess diisocyanate). Spandex fibers can undergo large, reversible deformations and are used in the manufacture of undergarments. The various multiblock polyurethanes are often referred to as *segmented polyurethanes*.

Other important thermoplastic elastomers are the multiblock polyetheresters (trade names: *Hytrel, Lomad, Gaflex*) and polyetheramides (trade names: *Pebax, Estamid, Grilamid*).

2-13c-3 *Polymer Blends and Interpenetrating Polymer Networks*

Polymer blends and *interpenetrating polymer networks* (IPN) are different from copolymers but like copolymers are used to bring together the properties of different polymers [Paul et al., 1988]. The total of all polymer blends (produced by both step and chain reactions) is estimated at about 3% of the total polymer production—about 2 billion pounds per year in the United States. There is considerable activity in this area since "new" products can be obtained and markets expanded by the physical mixing together of existing products. No new polymer need be synthesized.

Polycarbonate is blended with a number of polymers including PET, PBT, acrylonitrile–butadiene–styrene terpolymer (ABS) rubber, and styrene–maleic anhydride (SMA) copolymer. The blends have lower costs compared to polycarbonate and, in addition, show some property improvement. PET and PBT impart better chemical resistance and processability, ABS imparts improved processability, and SMA imparts better retention of properties on aging at high temperature. Polyphenylene oxide blended with high-impact polystyrene (HIPS) (polybutadiene-*graft*-polystyrene) has improved toughness and processability. The impact strength of polyamides is improved by blending with an ethylene copolymer or ABS rubber.

The interpenetrating polymer network is a blend of two different polymer networks without covalent bonds between the two networks. An IPN is obtained by the simultaneous or sequential crosslinking of two different polymer systems [Kim et al., 1986; Klempner and Berkowski, 1987; Sperling, 1981, 1986]. IPN synthesis is the only way of achieving the equivalent of a physical blend for systems containing crosslinked polymers. A simultaneous IPN, referred to as *SIN*, is produced by reacting a mixture of the monomers, crosslinking reagents, and catalysts for the two crosslinking systems. A sequential IPN (SIPN) is produced by reacting a mixture of one crosslinked polymer and the ingredients for the other crosslinking system. *Semi*-IPN and *pseudo*-IPN refer to sequential and simultaneous synthesis, respectively, of interpenetrating polymer networks in which one of the polymers is not crosslinked. Interpenetrating polymer networks are possible for a pair of step polymerization systems, a pair of chain polymerization systems, or the combination of step and chain polymerization systems.

There are a number of commercial interpenetrating polymer networks, although they seldom are identified as IPNs. The inclusion of a thermoplastic with an unsaturated polyester decreases the amount of shrinkage of the latter on crosslinking. The inclusion of a polyurethane makes the unsaturated polyester tougher and more resilient. Other examples of useful IPNs are epoxy resin–polysulfide, epoxy resin–polyester, epoxy resin–polyurethane, polyurethane–poly(methyl methacrylate), polysiloxane–polyamide, and epoxy resin–poly(diallyl phthalate). Many of these are not "pure" IPNs as defined above because of the presence of grafting and crosslinking between the two components. This is usually an advantage in producing an IPN with minimal phase separation.

2-14 NEWER POLYMERS AND POLYMERIZATIONS

The driving force in polymer synthesis is the search for polymers to replace other materials of construction especially metals. Polymers are lightweight and can be processed more easily and economically into a wide range of shapes and forms. The major synthetic efforts at present are in the areas of *high-temperature*, *liquid crystal*, and *conducting polymers*. There is an interrelationship between these three efforts, as will become apparent later.

2-14a Requirements for High-Temperature Polymers

There has been a continuing and strong effort since the late 1950s to synthesize *high-temperature polymers*. The terms *heat-resistant* and *thermally stable polymer*, used synonymously with *high-temperature polymer*, refer to a *high-performance polymer* that can be utilized at higher use temperatures; that is, its mechanical strength and modulus, stability to various environments (chemical, solvent, UV, oxygen), and dimensional stability at higher temperatures match those of other polymers at lower temperatures. The impetus for heat-resistant polymers comes from the needs in such technological areas as advanced air- and spacecraft, electronics, and defense as well as consumer applications. The advantages of heat-resistant polymers are the weight savings in replacing metal items and the ease of processing polymeric materials into various configurations. Lightweight polymers possessing high strength, solvent and chemical resistance, and serviceability at temperatures in excess of 250°C would find a variety of potential uses, such as automotive and aircraft components (including electrical and engine parts), nonstick and decorative coatings on cookware, structural components for aircraft, space vehicles, and missiles (including adhesives, gaskets, composite and molded parts, ablative shields), electronic and microelectronic components (including coatings, circuit boards, insulation), and components such as pipes, exhaust filter stacks, and other structural parts for the chemical and energy generating (nuclear, geothermal) plants. The synthetic routes studied have involved inorganic and semiinorganic as well as organic systems. The efforts to date have been much more fruitful in the organic systems, which will be discussed in this section. Inorganic and semiinorganic systems will be considered separately in Sec. 2-15.

Both chemical and physical factors determine the heat resistance of polymers [Cassidy, 1980; Critchley et al., 1983; Hergenrother, 1987; Marvel, 1975]. The strengths of the primary bonds in a polymer are the single most important determinant of the heat resistance of a polymer structure. This is especially critical with respect to the bonds in the polymer chain. Breakage of those bonds results in a deterioration of mechanical strength due to the drop in molecular weight. Bond breakages in pendant (side) groups on the polymer chain may not be as disastrous (unless it subsequently results in main-chain breakage). Aromatic ring systems (carbocyclic and heterocyclic) possess the highest bond strengths due to resonance stabilization and form the basis of almost all heat-resistant polymers. The inclusion of other functional groups in the polymer chain requires careful choice to avoid introducing weak links into an otherwise strong chain. Certain functional groups (ether, sulfone, imide, amide, CF_2) are much more heat resistant than others (alkylene, alicyclic, unsaturated, NH, OH).

A number of other factors weaken or strengthen the inherent heat resistance of a polymer chain. Polymer chains based on aromatic rings are desirable not only because of the high primary bond strengths but also because their rigid (stiff) polymer chains offer increased resistance to deformation and thermal softening. *Ladder* or *semiladder* polymer structures are possible for chains constructed of ring structures. A ladder or *two-strand* polymer chain has an uninterrupted sequence of rings joined one to another at two connecting atoms (see structure **VII** in Sec. 1-2c). A semiladder structure has single bonds interconnecting some of the rings. The ladder polymer is more desirable from the viewpoint of obtaining rigid polymer chains. Also, the ladder polymer may be more heat-resistant since two bond cleavages (compared to only one bond cleavage for the semiladder structure) in the same vicinity are required before there is a large drop in chain length and mechanical strength. Ladder polymers have been synthesized but have no practical utility because of a complete lack of processability. High molecular weight and crosslinking are desirable for the same reason. Strong secondary attractive forces (including dipole–dipole and hydrogen-bond interactions) improve heat resistance. Crystallinity increases heat resistance by serving as physical crosslinks that increase polymer chain rigidity and the effective secondary attractions. Branching lowers heat resistance by preventing close packing of polymer chains.

The factors that lead to increased heat resistance also present problems with respect to the synthesis of polymers and their utilization. Rigid polymer chains lead to decreased polymer solubility, and this may present a problem in obtaining polymer molecular weights sufficiently high to possess the desired mechanical strength. Low-molecular-weight polymers may precipitate from the reaction mixture and prevent further polymerization. Polymers with highly rigid chains may also be infusible and intractable, which makes it difficult to process them by the usual techniques into various shapes, forms, and objects. The synthesis of heat-resistant polymers may then require a compromise away from polymer chains with maximum rigidity in order to achieve better solubility and processing properties. There are two general approaches to this compromise. One approach involves the introduction of some flexibilizing linkages, such as isopopylidene, C=O, and SO$_2$, into the rigid polymer chain by using an appropriate monomer or comonomer. Such linkages decrease polymer chain rigidity while increasing solubility and processability. The other approach involves the synthesis of reactive telechelic oligomers containing functional end groups capable of reacting with each other. The oligomer, possessing a molecular weight of 500–4000 and two functional end groups, is formed into the desired end-use object by the usual polymer processing techniques. Subsequent heating of the oligomer results in reaction of the functional end groups with each other. The oligomer undergoes polymerizaton to higher molecular weight (referred to as *chain extension*). Crosslinking may also occur depending on the functionality of the A groups.

A number of the polymers considered previously—polycarbonate, aramid, and polyarylate—were among the first commercial successes in the efforts to synthesize polymers with increasingly high use temperatures. In the following sections we will discuss some of the other commercially available heat-resistant polymers followed by a consideration of research efforts to move further up in the temperature scale.

2-14b   Aromatic Polyethers by Oxidative Coupling

The *oxidative coupling* polymerization of many 2,6-disubstituted phenols to form aromatic polyethers is accomplished by bubbling oxygen through a solution of the

$$n\text{HO}-\!\!\left\langle\bigcirc\right\rangle\!\!\begin{smallmatrix}R\\ \\R'\end{smallmatrix} + \frac{n}{2}\,\text{O}_2 \xrightarrow[\text{amine}]{\text{Cu}^+} \left[\text{O}-\!\!\left\langle\bigcirc\right\rangle\!\!\begin{smallmatrix}R\\ \\R'\end{smallmatrix}\right]_n + n\text{H}_2\text{O} \qquad (2\text{-}203)$$

phenol in an organic solvent (toluene) containing a catalytic complex of a cuprous salt and amine [Aycock et al., 1988; Bartmann and Kowalczik, 1988; Finkbeiner et al., 1977; Hay and Dana, 1989; Mobley, 1984].

Amines such as diethylamine, morpholine, pyridine, and N,N,N',N'-tetramethyl-ethylenediamine are used to solubilize the metal salt and increase the pH of the reaction system so as to lower the oxidation potential of the phenol reactant. The polymerization does not proceed if one uses an amine that forms an insoluble metal complex. Some copper–amine catalysts are inactivated by hydrolysis via the water formed as a by-product of polymerization. The presence of a desiccant such as anhydrous magnesium sulfate or 4-Å molecular sieve in the reaction mixture prevents this inactivation. Polymerization is terminated by sweeping the reaction system with nitrogen and the catalyst is inactivated and removed by using an aqueous chelating agent.

Polymerization proceeds rapidly under mild conditions (25–50°C) for phenols containing small substituents. Phenols with one or more bulky o-substituents, such as i-propyl or t-butyl, undergo dimerization instead of polymerization. Dimerization is also the major reaction for phenols with *o*-methoxy substituents. The ratio of amine-to-cuprous ion and the reaction temperature determine the extent of carbon–oxygen coupling (polymerization) relative to carbon–carbon coupling (dimerization), probably

$$2HO-\underset{R}{\overset{R}{\bigcirc}} \rightarrow O=\underset{R}{\overset{R}{\bigcirc}}=\underset{R}{\overset{R}{\bigcirc}}=O \qquad (2\text{-}204)$$

by affecting the nature of the complex formed among cuprous ion, amine, and reaction intermediate. Higher amine:cuprous ratios and lower reaction temperatures favor polymerization while dimerization is favored by higher temperatures and lower amine:cuprous ratios. Phenols with only one ortho substituent undergo branching during polymerization.

The mechanism of oxidative coupling polymerization of phenols involves the reaction sequence in Eqs. 2-205 through 2-209. The copper-amine complex functions as a catalyst for the oxidation by oxidation by oxygen of the phenol to the monomeric phenoxy radical **XLII**. The phenoxy radical then undergoes carbon–oxygen coupling to yield the dimer (Eq. 2-205). Polymerization proceeds by oxidation of the dimer to the corresponding phenoxy radical (Eq. 2-206) followed by carbon–oxygen coupling reactions with monomeric phenoxy radical (Eq. 2-207) and with itself (Eq. 2-208). The trimer continues its growth by repeating the oxidation–coupling sequence; the tetramer continues its growth by dissociation (Eq. 2-209) into two phenoxy radicals (a monomer and a trimer) that can undergo coupling reactions. Growth to high-molecular-weight polymer proceeds by successive oxidation–coupling sequences.

The polymer from 2,6-dimethylphenol is the commercial product referred to as *poly(p-phenylene oxide)* or PPO. The IUPAC name is poly(oxy-2,6-dimethyl-1,4-phenylene). PPO possesses very little measurable crystallinity ($T_m = 262$–$267°C$, $T_g = 205$–$210°C$). There is very little neat PPO used commercially because its high melt viscosity makes processing too difficult. The commercially available products are blends of PPO with high-impact polystyrene (HIPS) (polybutadiene-*graft*-polystyrene). (Some blends with other rubbery polymers are also available.) Blending overcomes the processing problem by lowering the melt viscosity and also increases the impact strength. However, the continuous use temperature for the blend, referred to as *modified* PPO (trade name: *Noryl*), is decreased to the 80–110°C range with the specific use temperature depending on the relative amount of polystyrene. Modified PPO has close to the lowest water absorption of all engineering plastics, and this imparts excellent electrical properties over a considerable humidity range. It has excellent resistance to aqueous environments, including acids, bases, and oxidants, but this is not matched by its resistance to organic solvents. The resistance of PPO to organic solvents is not as good as that of PET and is much poorer than the resistance of polyamides such as nylon 6/6. PPO has better resistance to aromatic, halogenated, and polar organic solvents compared to polycarbonate.

About 250 million pounds of modified PPO are produced annually in the United States. As with all engineering plastics, the applications for modified PPO include both filled (including reinforced) and unfilled grades. The continuous use temperatures of 80–110°C indicated above are the maximum use temperatures and these require

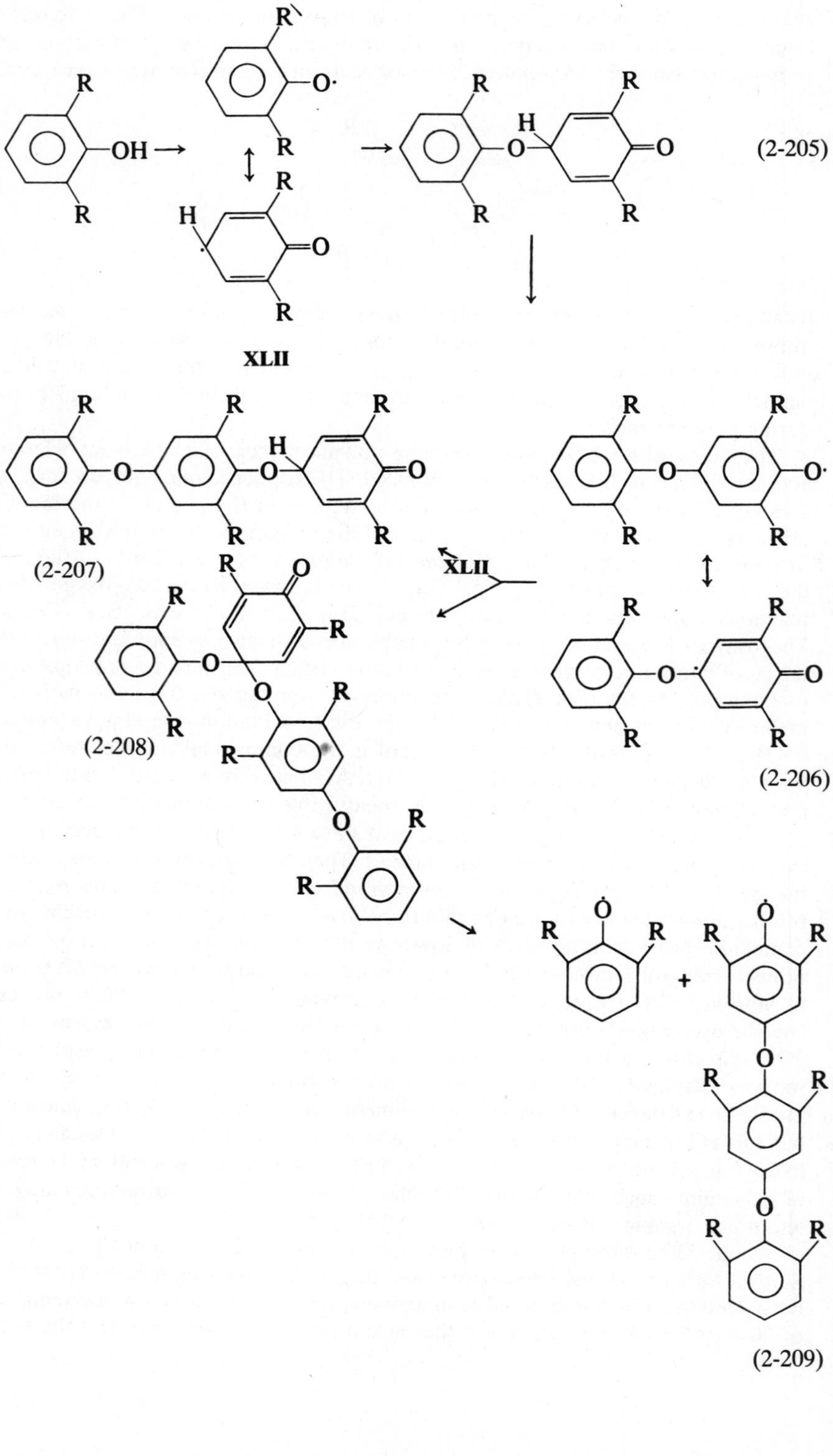

(2-205)

XLII

(2-207)

(2-208)

XLII

(2-206)

(2-209)

reinforced grades (e.g., reinforced with fiber glass or carbon-black-filled). A wide range of applications are found for modified PPO: automotive (dashboard, wheel covers, metalized grilles, and trim), electrical (radomes, fuse boxes, wiring splice devices), consumer and business (refrigerator door liners, computer housing, keyboard frame, steam iron), and liquid handling (pipe, valves, pump housing, impellors).

2-14c Aromatic Polyethers by Nucleophilic Substitution

Polyetherketones and *polyethersulfones*, also referred to as *polyketones* and *polysulfones*, are synthesized by nucleophilic aromatic substitution between aromatic dihalides and bisphenolate salts [Cassidy, 1980; Clagett, 1986; Critchley et al., 1983; Durvasula et al., 1989; Harris and Johnson, 1989; May, 1988]. This is shown in Eq. 2-210, where X is halogen and Y is C=O or SO_2.

$$X\!\!-\!\!\langle\bigcirc\rangle\!\!-\!\!Y\!\!-\!\!\langle\bigcirc\rangle\!\!-\!\!X + NaO\!-\!Ar\!-\!ONa \xrightarrow{-NaCl}$$

$$\left[\!\langle\bigcirc\rangle\!\!-\!\!Y\!\!-\!\!\langle\bigcirc\rangle\!\!-\!\!O\!-\!Ar\!-\!O\right]_n \quad (2\text{-}210)$$

Although aromatic halides are typically not reactive toward nucleophilic substitution, polymerization is rather facile in these systems because of the electron-withdrawing sulfone and carbonyl groups. The bisphenolate salt is formed *in situ* by the addition of the bisphenol and sodium or other alkali metal carbonate or hydroxide. Polysulfones are typically synthesized from aromatic dichlorides in a polar aprotic solvent such as 1-methyl-2-pyrrolidinone (NMP) or dimethyl sulfoxide. The polar aprotic solvent increases the nucleophilicity of the phenoxide anion by preferential solvation of cations but not anions. Reaction temperatures of 130–160°C are used mainly because of the poor solubility of the diphenolate salt. An excess of monohydric phenol or monochloroalkane is used to control the polymer molecular weight. A dry reaction system is required for polymerization to yield high-molecular-weight polymer. The presence of water hydrolyzes the phenolate salt, generating sodium hydroxide, which reacts with the aromatic dichloride and alters the stoichiometric ratio of reacting functional groups. Polymerization is carried out in the absence of oxygen to prevent oxidation of the bisphenolate. Polyketones are typically synthesized from aromatic difluorides in diphenyl sulfone at temperatures of 200–350°C. The higher reaction temperatures compared to polysulfone synthesis are needed to keep the polymer from premature precipitation, which would limit the polymer molecular weight. There are efforts to use variations on the synthetic scheme to allow lower polymerization temperatures. For example, use of the ketal derivative instead of the ketone allows lower temperatures since the ketal polymer has better solubility characteristics [Kelsey et al., 1987]. The ketone polymer would be generated from the ketal by hydrolysis after achieving high molecular weights at lower polymerization temperatures.

Electrophilic aromatic substitution of acyl and sulfonyl halides on aromatic reactants has been studied as an alternate method of synthesizing polysulfones and polyketones. This approach is not competitive with nucleophilic substitution probably because it yields mixtures of *o*- and *p*-catenation.

Polymers **XLIII–XLVII** are commercially available. **XLIII** and **XLIV** are referred to as *polyetheretherketone* (PEEK) (trade name: *Victrex*) and *polyetherketone* (PEK) (trade name: *Victrex, Hostatec, Ultrapek*), respectively. **XLV**, **XLVI**, and **XLVII** are

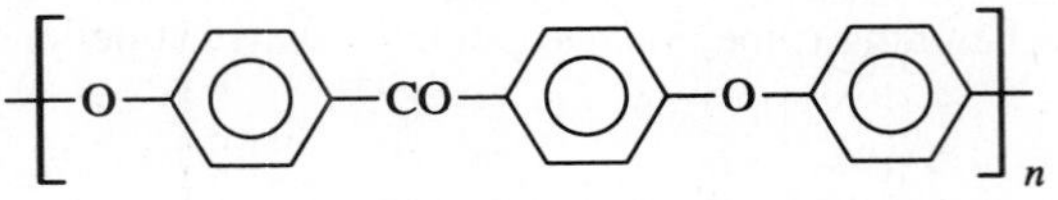

XLIII

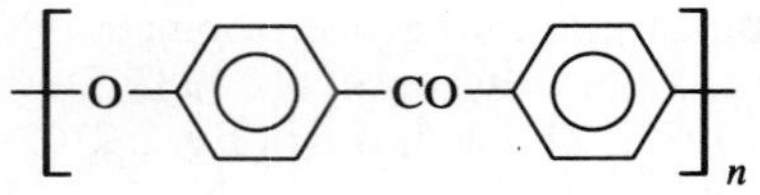

XLIV

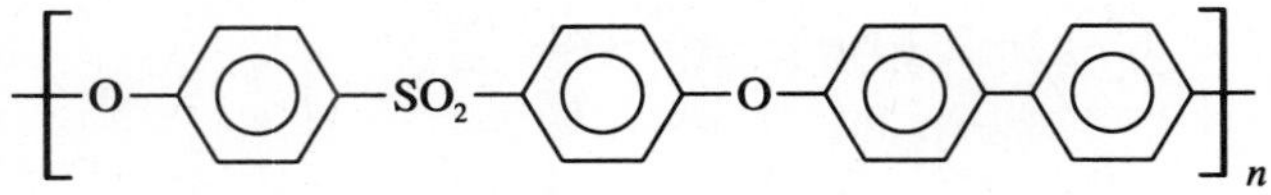

XLV

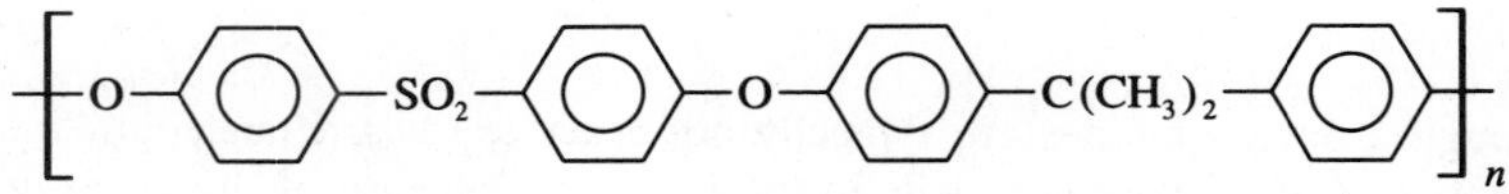

XLVI

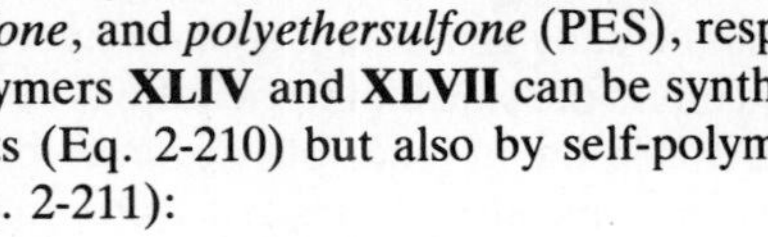

XLVII

known by the relatively nondescriptive names *polyphenylsulfone, bisphenol A polysulfone*, and *polyethersulfone* (PES), respectively (trade names: *Radel, Udel, Victrex*). Polymers **XLIV** and **XLVII** can be synthesized not only using A—A and B—B reactants (Eq. 2-210) but also by self-polymerization of the appropriate A—B reactant (Eq. 2-211):

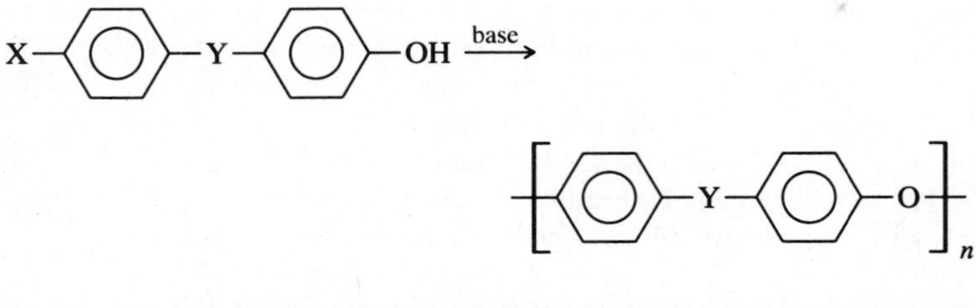

$$(2\text{-}211)$$

The polysulfones are transparent, amorphous polymers but possess good mechanical properties due to the rigid polymer chains. The glass transition temperatures are in the 180–230°C range, and polysulfones are rated for continuous use in the 150–200°C range. The resistance of polysulfones to aqueous environments, including acids, bases, and oxidants, is excellent. Their resistance to aromatic and halogenated organic solvents is comparable to PPO; PPO is better toward polar solvents but poorer toward aliphatics. Bisphenol A polysulfone differs from the other members of the polysulfone family because of the isopropylidene group. It has the lower glass transition and use temperatures and is less resistant to organic solvents. More than 10 million pounds of polysulfones are produced annually in the United States. The very good thermal and hydrolytic stability of polysulfones makes them useful for microwave cookware and in medical, biological, and food processing and service applications that require repeated cleaning with hot water or steam. Other applications include circuit breakers, electrical connectors, switch and relay bases, radomes, battery cases, camera bodies, aircraft interior parts, and membrane supports for reverse osmosis. Outdoor use is somewhat limited because of relatively low UV stability.

The polyketones are partially crystalline (approximately 35%). PEEK and PEK, respectively, have glass transition temperatures of 143 and 165°C and melting temperatures of 334 and 365°C. Both polyketones have excellent resistance to a wide range of aqueous and organic environments. Their resistance to organics matches that of the polyamides, while their resistance to aqueous environments matches that of the polysulfones. PEEK and PEK are rated for continuous service at temperatures of 240–280°C. These relatively new engineering plastics have an annual production in the United States of about one million pounds. The polyketones are finding applications in parts subjected to high temperature and aggressive environments: automotive (bearings, piston parts), aerospace (structural components), oil and chemical (pumps, compressor valve plates), and electrical–electronic (cable, insulation) applications.

2-14d   Aromatic Polysulfides

Poly(p-phenylene sulfide) or poly(thio-1,4-phenylene) (trade name: *Ryton*) (PPS) is synthesized by the reaction of sodium sulfide with p-dichlorobenzene in a polar solvent

$$Cl\text{--}\underset{}{\bigcirc}\text{--}Cl + Na_2S \xrightarrow{-NaCl} \left[\underset{}{\bigcirc}\text{--}S\right]_n \tag{2-212}$$

such as 1-methyl-2-pyrrolidinone (NMP) at about 250°C and 1.1 MPa (160 psi) [Cassidy, 1980; Hill and Brady, 1988; Lopez and Wilkes, 1989; Rajan et al., 1988]. The reaction mechanism may be more complicated than a simple nucleophilic aromatic substitution [Koch and Heitz, 1983]. Copolymers of *m*- and *p*-dichlorobenzenes have been synthesized. PPS undergoes a slow curing process when heated at 315–425°C. The curing reaction involves chain extension, oxidation, and crosslinking but is poorly understood. There is spectroscopic evidence for crosslinking via sulfur, oxygen, and aromatic bridges between polymer chains [Hawkins, 1976]. The presence of 1,2,4-trichlorobenzene during polymerization allows the production of a high-molecular-weight PPS without the thermal curing step.

PPS is a highly (60–65%) crystalline polymer with $T_m = 285°C$ and $T_g = 85°C$. It is rated for continuous service at 200–240°C, placing PPS between the polysulfones and the polyketones. It has inherent flame resistance, and its stability toward both organic and aqueous environments is excellent. PPS is comparable to polysulfones and polyketones in resistance to acids and bases but is somewhat less resistant to oxidants. The resistance of PPS to organic solvents is comparable to that of polyamides and polyketones. More than 15 million pounds of PPS are produced annually in the United States. Applications of PPS include automotive (components requiring heat and fluid resistance, sockets and reflectors for lights), consumer (microwave oven components, hair-dryer grille), industrial (oilfield downhole components, motor insulation, pumps and valves), blends with fluorocarbon polymers (release coating for cookware, appliances, molds), and protective coatings (valves, pipe, heat exchangers, electromotive cells). PPS is an alternative to traditional thermosetting plastics in some of these applications.

2-14e Aromatic Polyimides

Aromatic polyimides are synthesized by the reactions of dianhydrides with diamines, for example, the polymerization of pyromellitic anhydride with p-phenylenediamine

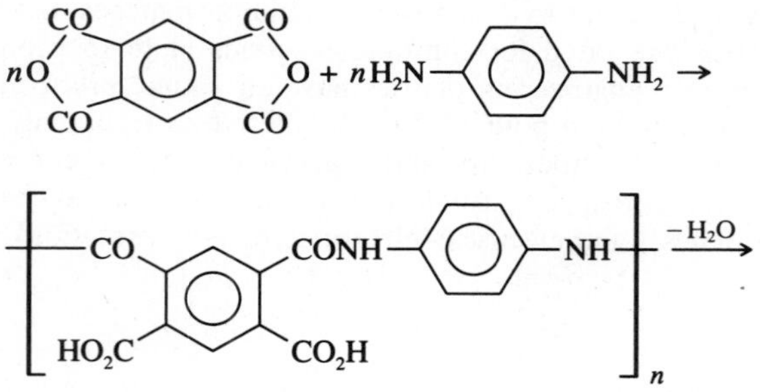

XLVIII

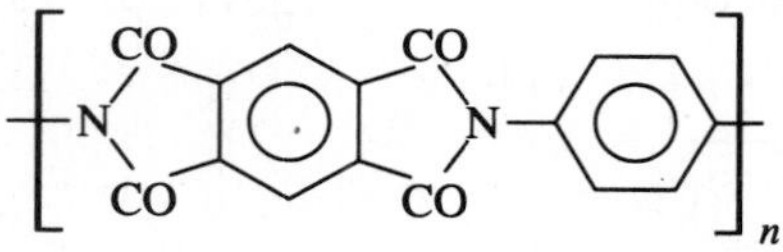

XLIX

$$(2\text{-}213a)$$

to form poly(pyromellitimido-1,4-phenylene) (Eq. 2-213a) [de Abajo, 1988; Cassidy, 1980; Hergenrother, 1987; Johnston et al., 1987; Mittal, 1984; Pyun et al., 1989; Scola and Vontell, 1989]. Solubility considerations sometimes result in using the half acid–half ester of the dianhydride instead of the dianhydride. Polymerization can also be accomplished using diisocyanates in place of diamines.

The direct production of high-molecular-weight aromatic polyimides in a one-stage polymerization cannot be accomplished because the polyimides are insoluble and infusible. The polymer chains precipitate from the reaction media before high molecular weights are obtained. Polymerization is accomplished by a two-stage process. The first stage involves an amidation reaction carried out in a polar aprotic solvent such as NMP or N,N-dimethylacetamide (DMAC) to produce a high-molecular-weight poly(amic acid) (**XLVIII**). Processing of the polymer can be accomplished only prior to the second stage, at which point it is still soluble and fusible. It is insoluble and infusible after the second stage of the process. The poly(amic acid) is formed into the desired physical form of the final polymer product (e.g., film, fiber, laminate, coating), and then the second stage of the process is performed. The poly(amic acid) is cyclo-dehydrated to the polyimide (**XLIX**) by heating at temperatures above 150°C.

The poly(amic acid) is kept in solution during the amidation by using mild temperatures, no higher than 70°C and usually lower than 30°C, to keep the extent of cyclization to a minimum. Less polar, more volatile solvents than NMP and DMAC are sometimes used to facilitate solvent evaporation in the second-stage reaction. This requires the use of an acidic or basic catalyst since the reaction rate is decreased in a less polar solvent. The poly(amic acid) is prone to hydrolytic cleavage and has relatively poor storage stability unless kept cold and dry during storage. Temperatures as high as 300°C are used for the second-stage solid-state cyclization reaction although the use of vacuum or dehydrating agents such as acetic anhydride–pyridine allow lower cyclization temperatures.

Polyimides (PI) are amorphous materials with a range of glass transition temperatures depending on the specific structure. The resistance of PI to organic solvents is excellent. Polyimides show good oxidation resistance and hydrolytic stability toward acidic environments, comparable to PET and better than nylon 6/6. However, PI undergoes hydrolytic degradation in strongly alkaline environments, comparable to polycarbonate and poorer than PET and nylon 6/6. The high temperature resistance of polyimides is excellent, with continuous use temperatures of 300–350°C possible, especially for structures containing only aromatic rings. Wholly aromatic polyimides are generally too stiff for most applications. A variety of different dianhydrides and diamines have been used to introduce less rigid units, such as biphenyl, alicyclic, methylene, carbonyl, sulfone, isopropylidene, perfluoroisopropylidene, ether, and thioether, into the polyimide chain.

One of the first commercial polyimides was based on the polymerization of 4,4'-diaminodiphenyl ether with pyromellitic dianhydride (trade names: *Kapton*, *Vespel*, *Pyralin*). This and other similar polyimides are available in the form of poly(amic acid) solutions that are used as high-temperature wire enamels and insulating varnishes and for coating fiber glass and other fabrics. Polyimide films find applications as insulation for electric motors and missile and aircraft wire and cable. There is also some processing of PI by powder-technology (low-temperature forming followed by sintering) and compression molding (requiring higher pressures than normal) to form such automotive and aircraft engine parts as bushings, seals, piston rings, and bearings.

The lack of easy processability of polyimides has impeded the practical utilization of this high-performance material until relatively recently. Several different modifications of the polyimide system have successfully produced materials that are readily processable. *Polyetherimides* (PEI) are polyimides containing sufficient ether as well as other flexibilizing structural units to impart melt processability by conventional techniques, such as injection molding and extrusion. The commercially available PEI (trade name: *Ultem*) is the polymer synthesized by nucleophilic aromatic substitution between the disodium salt of bisphenol A and 1,3-bis(4-nitrophthalimido)benzene (Eq. 2-213b) [Clagett, 1986]. This is the same reaction as that used to synthesize polyethersulfones and polyetherketones (Eq. 2-210) except that nitrate ion is displaced instead of halide. Polymerization is carried out at 80–130°C in a polar solvent (NMP, DMAC). It is also possible to synthesize the same polymer by using the diamine–

dianhydride reaction. Everything being equal (cost and availability of pure reactants), the nucleophilic substitution reaction is the preferred route because of the more moderate reaction conditions.

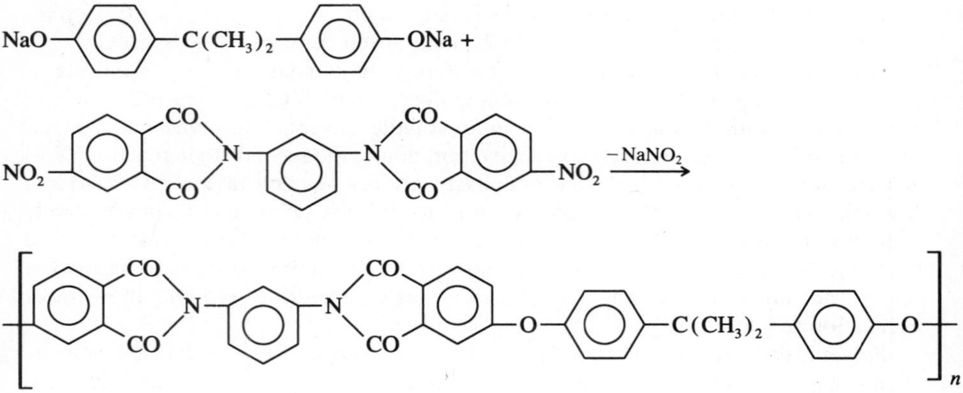

$$\text{(2-213b)}$$

PEI is amorphous with a glass transition temperature of 215°C and continuous use temperature of 170–180°C. The solvent and chemical resistance of PEI is comparable to that of polyimides as described above, except that it is soluble in partially halogenated organic solvents. Processability by conventional techniques coupled with a wide range of desirable properties has resulted in rapid growth for this new polymer. PEI also has the advantage of not having any volatiles (water) produced during processing as occurs during the second stage cyclization of poly(amic acids). Voids and strains can be generated as volatiles are lost during processing. Between 5 and 10 million pounds of PEI are produced annually in the United States. PEI presently accounts for more than two-thirds of all polyimide polymers. PEI resins are used in many industries: electrical–electronic (circuit boards, radomes, high-temperature switches), transportation (fuel system components, transmission and jet engine components, lamp sockets), medical (surgical instrument handles, trays, and other items requiring sterilization by steam, ethylene oxide, chemical means, or γ-radiation), and consumer and business (microwave cookware, curling irons, food packaging, memory disks, fans for computer and other equipment).

Polyamideimides (PAI), (trade name: *Torlon*) containing both amide and imide functional groups in the polymer chain, are produced by the reaction of trimellitic anhydride (or a derivative) with various diamines (Eq. 2-214). PAI resins are amorphous polymers with glass transition temperatures of 270–285°C and continuous use temperatures of 220–230°C. Polyamideimides have excellent resistance to organic solvents. PAI has good resistance to acids, bases, and oxidants but is attacked by bases at high temperature. Polyamideimides are used in various aircraft (jet engine, generator, and compressor components), automotive (universal joints, transmission seal rings and washers), and industrial (machine gears and other mechanical components, nonstick and low-friction coatings) applications.

Bismaleimide (BMI) polymers are produced from the combination of a diamine and bismaleimide (Eq. 2-215) [de Abajo, 1988; Cassidy, 1980]. Polymerization is carried out with the bismaleimide in excess to produce maleimide end-capped telechelic

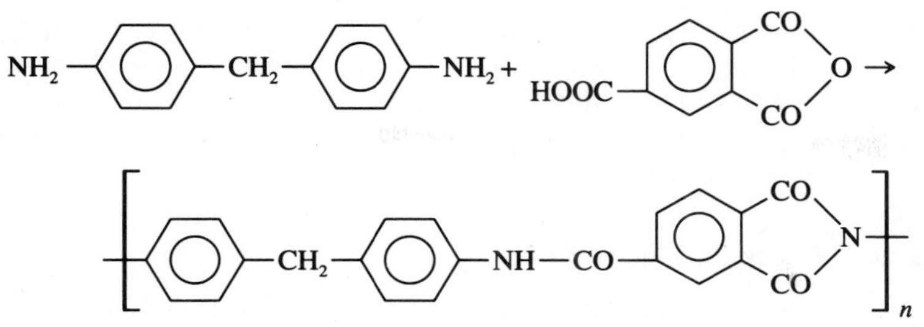

$$\text{(2-214)}$$

oligomers (L). Heating at temperatures of 180°C and higher results in crosslinking via radical chain polymerization (Chap. 3) of the maleimide carbon–carbon double bonds (Eq. 2-216). The properties of the thermoset product can be varied by changing the Ar and Ar′ groups and the crosslink density. The latter depends on the bismaleim-

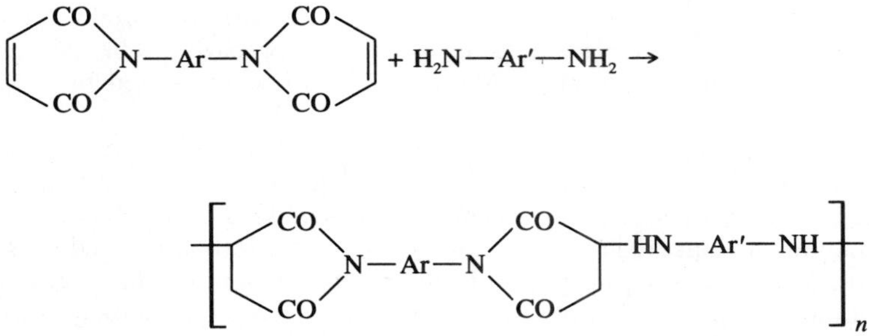

$$\text{(2-215)}$$

ide:diamine ratio. [Further variations are possible by radical copolymerization (Chap. 6) of L with alkene monomers.] Some newer BMI polymer formulations use o,o'-diallylbisphenols as coreactants in place of diamine [Zahir et al., 1989]. Chain extension and crosslinking involves a repetitive sequence of two reactions. The ene reaction forms a styrene-type adduct by addition of the allyl group to a maleimide double bond. The adduct subsequently undergoes a Diels–Alder reaction in which the adduct and a second maleimide double bond act as diene and dienophile, respectively.

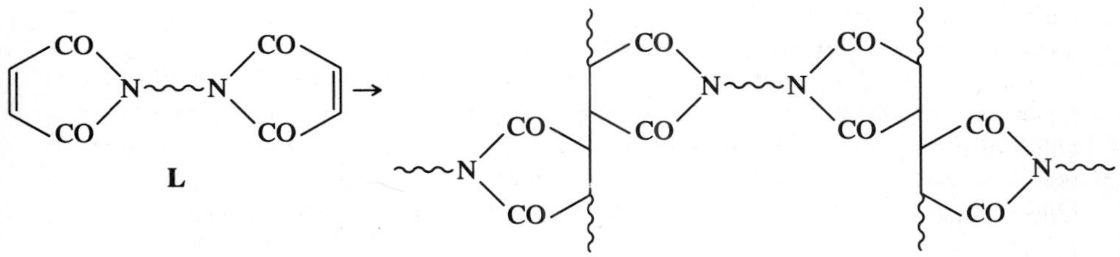

$$\text{(2-216)}$$

BMI polymers have glass transition temperatures in excess of 260°C and continuous use temperatures of 200–230°C. BMI polymers lend themselves to processing by the same techniques used for epoxy polymers. They are finding applications in high-performance structural composites and adhesives (e.g., for aircraft, aerospace, and defense applications) used at temperatures beyond the 150–180°C range for the epoxies.

Polyimidosulfides have been synthesized by the corresponding addition reaction between bismaleimides and dithiols, although commercial products are not available [White and Scaia, 1984].

2-14f Reactive Telechelic Oligomer Approach

We have noted several successful alternatives—PEI, PAI, and BMI—used to overcome the inherent lack of processability of polyimides. The BMI approach can be considered a precursor of the *reactive telechelic oligomer approach* for overcoming processability problems in many of the more advanced heat-resistant polymer systems. The polymer chains in those materials, consisting of heterocyclic and carbocyclic rings, often show poor processability. Even for systems where processability is not a problem—including previously discussed polymers such as polyketone and polysulfone—the reactive telechelic oligomer approach introduces the ability to form thermoset products in addition to thermoplastics. The reactive telechelic oligomer approach will be specifically described in this section for the polyimides, but it has been practiced for a large number of different polymers.

The reactive telechelic oligomer approach is relatively simple. Polymerization is carried out in the usual manner except that one includes a monofunctional reactant to stop reaction at the oligomer stage, generally in the 500–3000 molecular weight range. The monofunctional reactant not only limits polymerization but end-caps the oligomer with functional groups capable of subsequent reaction to achieve curing of the oligomer (i.e., chain extension as well as crosslinking in most cases). Many different functional groups have been used for this purpose—alkyne, 5-norbornene, maleimide, biphenylene, nitrile, [2.2]paracyclophane, and cyanate [Harrys and Spinelli, 1985]. Maleimide, 5-norbornene (bicyclo[2.2.1]hept-4-ene), and alkyne end-capped polyimide oligomers are commercially available [Mittal, 1984]. Maleimide end-capped oligomers (trade name: *Kerimid*) are cured by heating as described above for the BMI polymers. Maleimide end groups can be introduced into polyimide oligomers simply by carrying out polymerization with a mixture of diamine, dianhydride, and maleic anhydride. An alternate approach is to polymerize the dianhydride in the presence of excess diamine to form an amine end-capped polyimide that is then reacted with maleic anhydride [Lyle et al., 1989].

5-Norbornene end-capped polyimide oligomers (trade name: *PMR*, *LARC*) are obtained by including 5-norbornene-2,3-dicarboxylic anhydride (*nadic anhydride*) (**LI**) in the polymerization reaction between a dianhydride and diamine [de Abajo, 1988; Hergenrother, 1987]. Heating the oligomer at 270–320°C results in a thermoset product. The main reaction responsible for crosslinking probably involves a reverse (*retro*) Diels–Alder reaction of the 5-norbornene system to yield cyclopentadiene and maleimide end-capped oligomer (Eq. 2-217). This is followed by radical copolymerization of cyclopentadiene with the carbon–carbon double bond of maleimide (Eq. 2-218).

LI

Crosslinking appears to be more complex with additional reactions occurring simultaneously and to different extents depending on the cure temperature—copolymerization of cyclopentadiene with **LII** as well as **LIII**, Diels–Alder reaction of cyclopentadiene and **LII** to form an adduct that copolymerizes with **LII** and/or **LIII** [Hay et al., 1989].

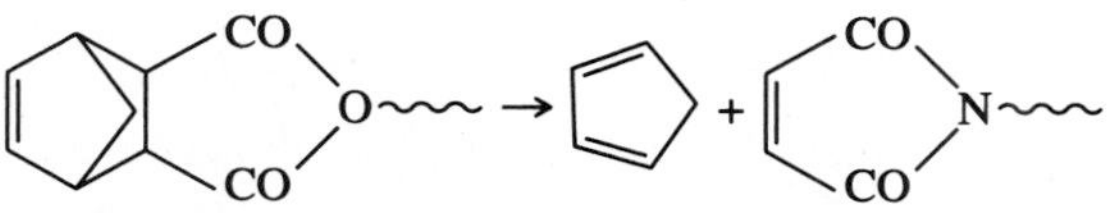

(2-217)

LII   **LIII**

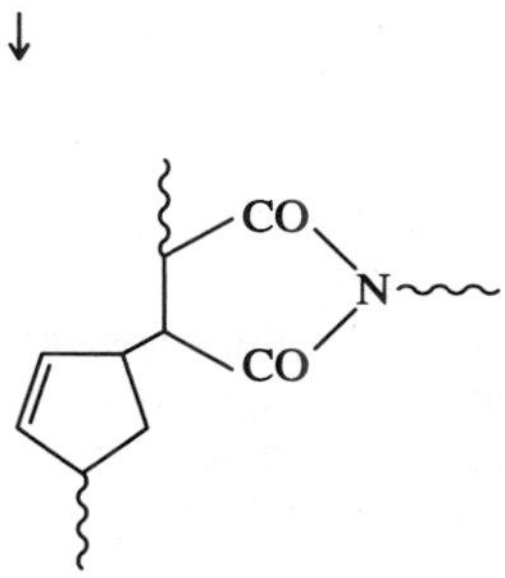

(2-218)

Alkyne end-capped polyimide oligomers (trade name: *Thermid*, *ATPI*) are obtained by including *p*-aminophenylacetylene in the polymerization mixture of a diamine and a dianhydride. Curing to a thermoset product takes place on heating at 250–400°C. The lower cure temperatures require the presence of a catalyst such as $Ni(P\emptyset_3)_2(CO)_2$. Alkyne end-capped oligomers were developed with the idea that cyclotrimerization would yield aromatic structures (Eq. 2-219) as the crosslinking unit and the resulting thermoset would have better heat resistance than the thermosets derived from the maleimide and 5-norbornene end-capped oligomers. The latter ther-

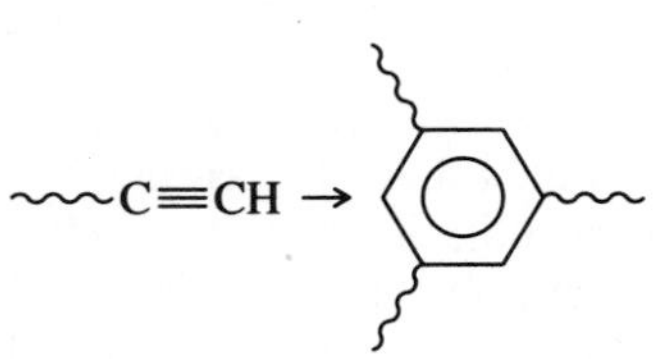

$$\sim\sim\sim C\equiv CH \rightarrow$$

(2-219)

mosets did not realize the full potential of the polyimides system because of the presence of aliphatic groupings introduced through the end groups and their curing reactions. The alkyne end-capped oligomers have the same problem as cyclotrimerization accounts for only 30% of the total reaction [Hergenrother, 1985; Takeichi and Stille, 1986]. The formation of carbon–carbon double bonds and other functionalities act as weak links in the overall polymer structure. Nitrile (—C≡N) and cyanate (—O—C≡N) groups also undergo cyclotrimerization, but the thermoset structures are not as stable as those formed using alkyne functional groups [Harrys and Spinelli, 1985; Temin, 1982–1983].

Many other functional groups are under consideration. Biphenylene end groups react with each other to form tetrabenzocyclooctatetraene linkages, but the process involves only chain extension without crosslinking [Recca and Stille, 1978]. Cross-linking is achieved when biphenylene groups are incorporated as internal units within the polymer main chain [Droske and Stille, 1984]. Incorporation of functional groups

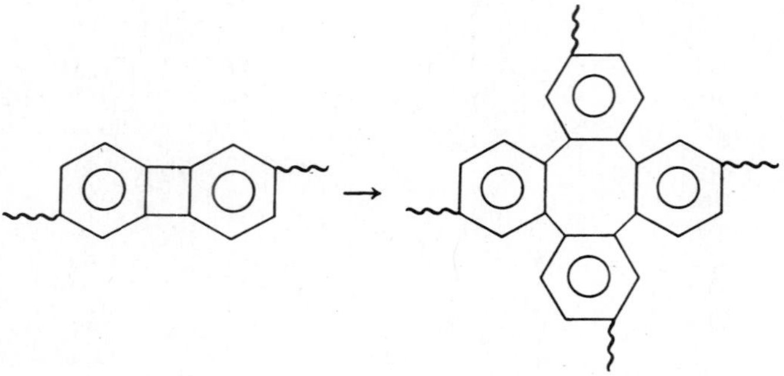

$$(2\text{-}220)$$

as internal units instead of end groups has been studied for other functional groups. For example, biphenylene end-capped oligomers have been crosslinked by reaction with oligomers containing alkyne functions in the main chain [Cannizzo et al., 1989; Takeichi and Stille, 1986]. Phenanthrene units are formed by reaction of alkyne and biphenylene functions.

Other groups being studied include [2.2]paracyclophane (**LIV**) and allyl 5-norbor-nene (**LV**) [Renner and Kramer, 1989; Tan and Arnold, 1988; Upshaw and Stille,

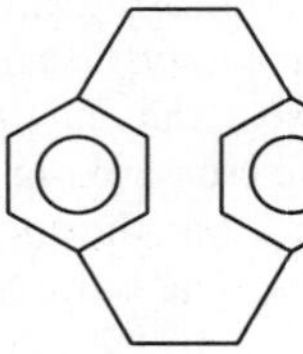

LIV 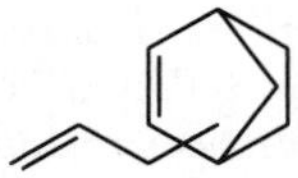   LV

1988]. The reactions occurring during curing are not established. The curing process due to [2.2] paracyclophane units probably arise from an initial homolytic breakage at the bond connecting the two methylene groups followed by various radical–radical coupling reactions. Curing due to allyl 5-norbornene units involves both Diels–Alder and ene reactions.

2-14g Liquid Crystal Polymers

Liquid crystal polymers (LCP) are polymers that exhibit liquid crystal characteristics either in solution (*lyotropic* liquid crystal) or in the melt (*thermotropic* liquid crystal) [Ballauf, 1989; Brown and Crooker, 1983; Finkelmann, 1987; Morgan, 1987]. We need to define the *liquid crystal* state before proceeding. In general, solids (crystals) have three-dimensional, long-range ordering of molecules. The molecules are said to be ordered or oriented with respect to their centers of mass and their molecular axes. The physical properties (e.g., refractive index, electrical conductivity, coefficient of thermal expansion) of a wide variety of crystalline substances vary in different directions. Such substances are referred to as *anisotropic* substances. Substances that have the same properties in all directions are referred to as *isotropic* substances. For example, liquids that possess no long-range molecular order in any dimension are described as isotropic.

Not all crystalline substances lose all long-range molecular order on melting. Some crystalline subtances retain one- or two-dimensional long-range molecular order above the crystalline melting point. Such substances are then described as being in the liquid crystal state. They show the flow behavior of isotropic liquids but appear to possess some kind of long-range molecular order greater than the isotropic liquid state but less than the true crystal state. Heating a thermotropic liquid crystal eventually results in its transformation to the liquid state at some temperature, $T_{lc,i}$, referred to as the *liquid crystal–isotropic phase transformation temperature for thermotropic liquid crystals*.

Our understanding of lyotropic liquid crystals follows in a similar manner. The action of solvent on a crystalline substance disrupts the lattice structure and most compounds pass into solution. However, some compounds yield liquid crystal solutions that possess long-range ordering intermediate between solutions and crystals. The lyotropic liquid crystal can pass into the solution state by the addition of more solvent and/or heating to a higher temperature. Thermotropic and lyotropic liquid crystals, both turbid in appearance, become clear when they pass into the liquid and solution states, respectively.

Liquid crystal behavior is due chiefly to molecular rigidity that excludes more than one molecule occupying a particular volume, rather than intermolecular attractive forces. The requirement for a substance to exhibit liquid crystal behavior is molecular shape anisotropy, which is not affected by conformational changes. Some biological molecules (e.g., tobacco mosaic virus, polypeptides, and cellulose) are liquid crystals because of their rigid helical rod conformations. This is not the case for synthetic polymers. The typical polymer liquid crystal is a rigid-rod molecule. (Theoretical studies indicate the molecular length must exceed the diameter by a factor greater than 6 [Flory and Ronca, 1979].) Other anisotropic molecular shapes (e.g., disks) also show liquid crystal behavior but these are much more rare for polymers. Polymer liquid crystals typically show either *smectic* or *nematic* liquid crystal behavior. These are shown in Fig. 2-24, where each ▬ represents a polymer molecule. There is a parallel arrangement of the long molecular axes in both nematic and smectic liquid crystals. Smectic liquid crystals are more ordered than nematic liquid crystals as a result of differences in the orientation of their molecular chain ends. Molecular chain ends are coterminal (i.e., lined up next to each other) in smectics but are out of register

Liquid

Nematic LC

Smectic LC

Fig. 2-24 Molecular ordering in nematic and smectic liquid crystals compared to liquid. After Brown and Crooker [1983] (by permission of American Chemical Society).

(i.e., have no particular orientation) in nematics. Another difference between smectic and nematic liquid crystals is that smectics are layered while nematics are usually not layered. (A third type of liquid crystal, *cholesteric*, involves nematic packing in planes with the molecular axes in each plane turned at an angle relative to those in the next plane, but is rarely observed in polymers.) Smectic liquid crystals, being more ordered, are more viscous than nematics (everything else being equal). Microscopic observations are of prime importance in distinguishing nematic and smectic structures [Demus and Richter, 1978].

Rigid-rod liquid crystal polymer molecules result from having rigid groups either within the polymer chain or as side groups on the polymer chain. These two types of LC polymers, referred to as *main-chain* and *side-chain* LC polymers, respectively, are depicted in Fig. 2-25. The rigid groups are referred to as *mesogens* or *mesogenic* groups. The mesogens are shown as rigid-rod groups, but other shapes (e.g., disks) are also possible. The collection of mesogenic unit results in a molecule which is a rigid rod with molecular shape anisotropy and liquid crystal behavior.

Liquid crystal behavior in polymers has practical consequences. Liquid crystal melts or solutions have lower viscosities than melts or solutions of random-coil polymers and are easier to process. More importantly, the extension and orientation of polymer chains during processing yields highly crystalline products with high modulus and high strength. The moduli and strengths of chain-extended LC polymers is considerably higher than those of the chain-folded crystalline polymers formed from non-LC polymers. LC polymers also have significantly increased crystalline melting temperatures as a result of the extended-chain morphology. Thus, the synthetic guidelines are the same for synthesis of heat resistant and LC polymers—build rigid polymer chains.

Poly(1,4-oxybenzoyl) (**LVI**), obtained by self-reaction of *p*-hydroxybenzoic acid, and the various aramids (Sec. 2-8f) were among the first LC polymers studied. The experience in commercializing poly(1,4-oxybenzoyl) and the aramids illustrates that one can have "too much of a good thing." None of these polymers can be melt

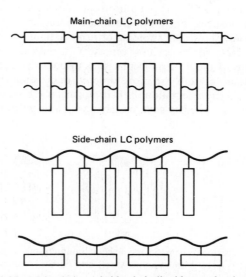

Fig. 2-25   Main-chain and side-chain liquid crystal polymers.

processed since they undergo decomposition before melting ($T_m > 500°$). The aramids are spin-processed into fibers using lyotropic solutions in sulfuric acid. It would be preferable to melt-process the polymers or, at least, to use a less agressive and/or dangerous solvent, but these options do not exist. Poly(1,4-oxybenzoyl) (trade name: *Ekonol*) has found limited utility because of a lack of easy processability. It does not

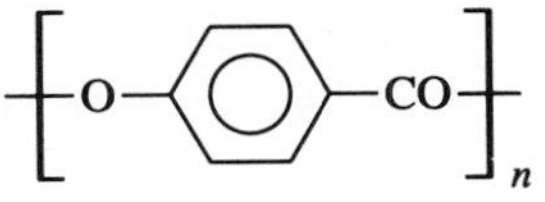

LVI

have sufficient solubility even in an aggressive solvent to allow solution processing. It can be processed by plasma spraying or powder sintering.

There is a need to decrease the crystalline melting temperature in such very-high melting LC polymers to achieve a degree of processability. Copolymerization is used to alter the polymer chains in the direction of decreased rigidity, but one needs to do it carefully to retain as high modulus, strength, and T_m as possible. Chain rigidity can be lowered by reducing molecular linearity by incorporating a less symmetrical comonomer unit (*o* and *m* instead of *p* or naphthalene ring, other than 1,4-, instead of benzene ring), a comonomer containing a flexible spacer (single bond, oxygen, or methylene between mesogenic monomer units) or a comonomer with flexible side groups. This approach has been successfully applied to poly(1,4-oxybenzoyl). Two copolyesters are commercial products. *Vecta* is a copolymer of *p*-hydroxybenzoic acid with 6-hydroxy-2-naphthoic acid; *Xydar* is a copolymer of *p*-hydroxybenzoic acid with *p*,*p*′-biphenol and terephthalic acid. These polymers have continuous service temperatures of up to 240°C and find applications for microwave ovenware, automotive components, components for chemical pumps and distillation towers, and electronic devices.

Many of the polymers discussed in the following sections exhibit liquid crystal behavior.

2-14h 5-Membered Ring Heterocyclic Polymers

Polyimides are obtained from amine and carboxyl reactants when the ratio of amine to acid functional groups is 1:2. If reactants with the reverse ratio of amine to acid functional groups are employed, polybenzimidazoles (PBI) are producecd; for instance, polymerization of 3,3′diaminobenzidine and diphenyl isophthalate yields

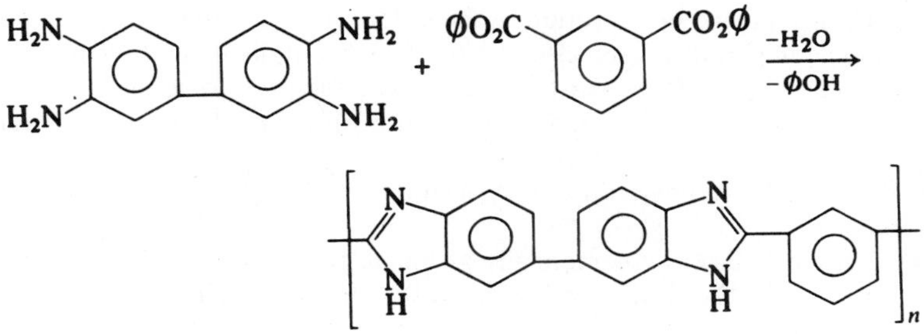

(2-221)

LVIIa

poly(5,5'-bibenzimidazole-2,2-'-diyl-1,3-phenylene) [Buckley et al., 1988; Hergen-rother, 1987; Marvel, 1975; Ueda et al., 1985]. The reaction probably proceeds by a sequence of two nucleophilic reaction—a nucleophilic substitution to form an amine–amide (Eq. 2-222) followed by cyclization via nucleophilic addition (Eq. 2-223).

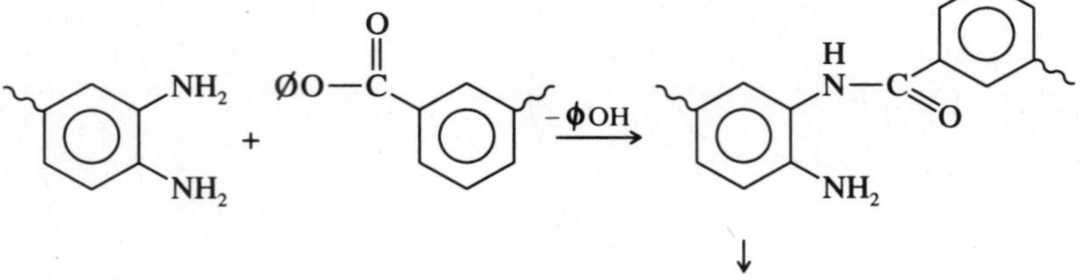

(2-222)

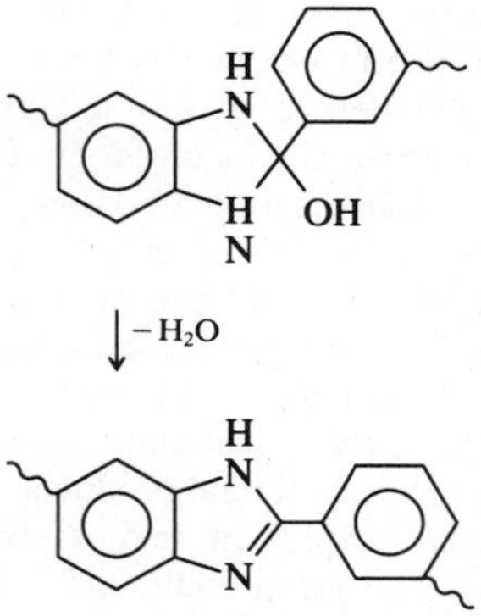

(2-223)

Polymerization is often carried out as a two-stage melt polymerization. Oxygen is removed from the reaction system by vacuum and the system purged with nitrogen to avoid loss of stoichiometry by oxidation of the tetramine reactant. The reaction mixture is heated to about 290°C in the first stage with the reaction starting soon after the reactants form a melt (150°C). The high-volume foam produced in this stage is removed, cooled, and crushed to a fine powder. The second-stage reaction involves heating of the powder at 370–390°C under nitrogen.

Considerable efforts have centered on carrying out the synthesis of polybenzimi-dazoles at more moderate temperatures. Polymerization of the isophthalic acid or its diphenyl ester have been successfully carried out in polyphosphoric acid or methane-sulfonic acid–phosphorus pentoxide at 140–180°C, but the reaction is limited by the very low solubilities (<5%) of the reactants in that solvent. The lower reaction temperature is a consequence of activation of the carboxyl reactant via phosphorylation. Lower reaction temperatures are also achieved in hot molten nonsolvents such as sulfolane and diphenyl sulfone, but the need to remove such solvents by a filtration or solvent extraction is a disadvantage. Alternate routes to the polybenzimidazole system have also been explored, including the reaction of tetramines with dialdehydes (imine formation followed by reductive cyclization) [Korshak et al., 1984; Neuse and Loonat, 1983].

The PBI in Eq. 2-221 is commercially available in the form of fiber, composite resin, and formed objects (trade name: *Celazole*). It has good stability and mechanical behavior up to 300°C and higher (about 25°C higher than the most stable polyimides), does not burn and is self-extinguishing. PBI has better hydrolytic stability than aromatic polyamides and polyimides. It has been used to manufacture flight suits for Apollo and Skylab astronauts and escape suits for Space Shuttle astronauts. Fiber is spun from PBI using lyotropic solutions of the polymer in solvents such as N,N-dimethylacetamide, in which it is reasonably soluble at the processing temperature of 250°C. Sulfonated PBI has potential for protective clothing for fire fighters. PBI itself is not appropriate for such use since it undergoes extensive shrinkage when exposed to flames. Sulfonation of PBI (on the benzene rings) greatly reduces this shrinkage problem. Other potential applications being considered for PBI include hot-melt adhesives and hollow fibers and flat membranes for reverse osmosis.

It should be noted that structure **LVIIa** (Eq. 2-221) is an oversimplification for the polymer formed from 3,3-diaminobenzidene and diphenyl isophthalate. Dehydration after ring closure can occur toward either of the two nitrogens, and one would expect more or less random placements of the carbon–nitrogen double bonds. Thus, the PBI structure is a random copolymer of repeating units **LVIIa**, **LVIIb**, and **LVIIc**.

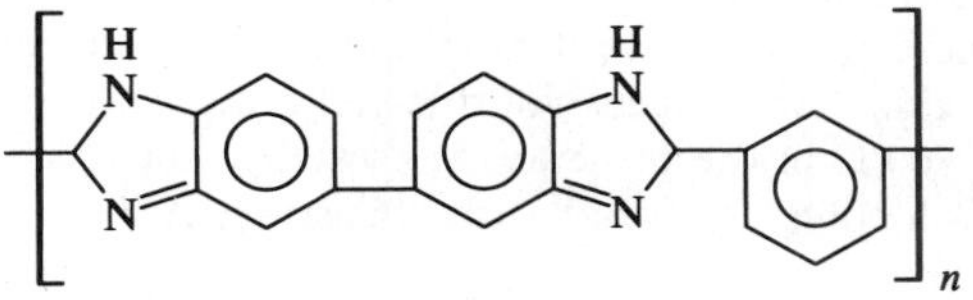

LVIIb

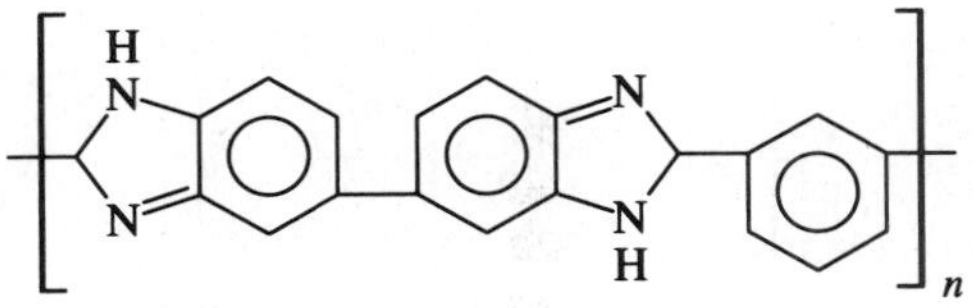

LVIIc

Polybenzoxazoles (PBO) and polybenzthiazoles (PBT) are related to and synthesized in a manner similar to polybenzimidazoles [Chow et al., 1989; Evers et al., 1981; Hergenrother, 1987; Krause et al., 1988; Maruyama et al., 1988; Wolfe, 1988; Wolfe et al., 1981]. Synthesis of PBO and PBT is accomplished by the reactions of a dicarboxyl

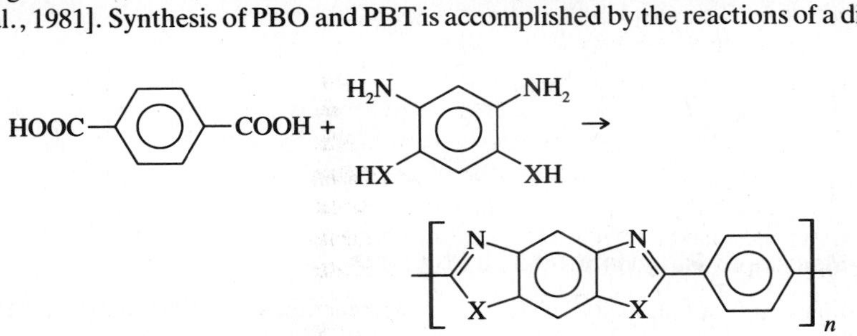

(2-224)

LVIII

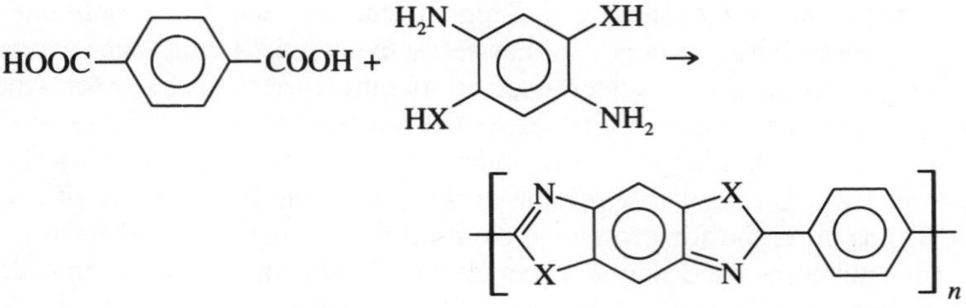

$$\text{HOOC}-\underset{}{\underset{}{\bigcirc}}-\text{COOH} + \underset{HX}{\overset{H_2N}{\underset{}{\bigcirc}}}\overset{XH}{\underset{NH_2}{}} \rightarrow$$

$$\left[\underset{X}{\overset{N}{\underset{}{\bigcirc}}}\overset{}{\underset{N}{\bigcirc}}-\underset{}{\bigcirc}\right]_n \tag{2-225}$$

LIX

reactant with a bis(*o*-aminophenol) (X = O) and a bis(*o*-aminothiophenol) (X = S), respectively. Polymers **LVIII** and **LIX** for X = O are named poly(benzo[1,2-*d*:5,4-*d'*]bisoxazole-2,6-diyl-1,4-phenylene] and poly(benzo[1,2-*d*:4,5-*d'*]bisoxazole-2,6-diyl-1,4-phenylene], respectively. The names of the corresponding polymers for X = S are obtained by replacing bisoxazole by bisthiazole. Structures **LVIII** and **LIX** have been referred to as cis-PBO or -PBT and trans-PBO or -PBT, respectively. The properties of PBO and PBT polymers are similar to those of polybenzimidazoles.

The polymerization of a dihydrazide with a diacyl chloride yields a polyhydrazide. Heating the polyhydrazide at 100–200°C in the presence of a catalyst such as a carbodiimide or thionyl chloride yields a poly(1,3,4-oxadiazole) [Cassidy, 1980; Gebben

$$\text{H}_2\text{NNH}-\text{CO}-\text{Ar}-\text{CO}-\text{NHNH}_2 + \text{Cl}-\text{CO}-\text{Ar}'-\text{CO}-\text{Cl} \xrightarrow{-\text{HCl}}$$

$$-[\text{Ar}-\text{CO}-\text{NHNH}-\text{CO}-\text{Ar}'-\text{CO}-\text{NHNH}-\text{CO}]_n \xrightarrow{-\text{H}_2\text{O}}$$

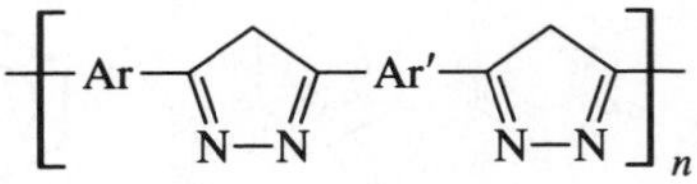

$$\left[\text{Ar}-\underset{N-N}{\overset{}{\bigcirc}}-\text{Ar}'-\underset{N-N}{\overset{}{\bigcirc}}\right]_n \tag{2-226}$$

et al., 1988]. The poly(1,3,4-oxadiazole) structure can also be obtained by reaction of hydrazine with a diacyl chloride or diacid [Ueda and Sugita, 1988].

A wide range of polymers based on other heterocyclic 5-membered rings have been investigated, including pyrazole, triazole, tetrazole, 1,3,4-thiazidazole, and thiophene [Cassidy, 1980; Critchley et al., 1983].

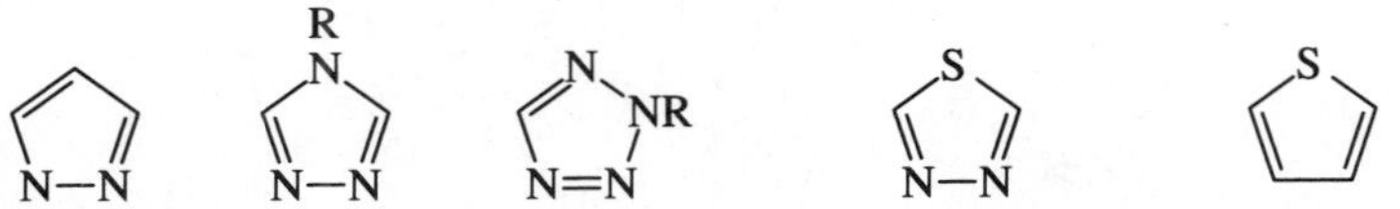

Pyrazole Triazole Tetrazole 1,3,4-Thiazidazole Thiophene

2-14i 6-Membered Ring Heterocyclic Polymers

Polyquinolines are obtained by the Friedlander reaction of a bis(*o*-aminoaromatic aldehyde) (or ketone) with an aromatic bis(ketomethylene) reactant [Stille, 1981]. The quinoline ring is formed by a combination of an aldol condensation and imine formation. Polymerization is carried out at 135°C in *m*-cresol with poly(phosphoric acid) as the catalyst. The reaction also proceeds under base catalysis, but there are more side reactions. There has been a considerable effort to use the reactive telechelic oligomer approach to producing crosslinked polyquinolines [Sutherlin and Stille, 1986].

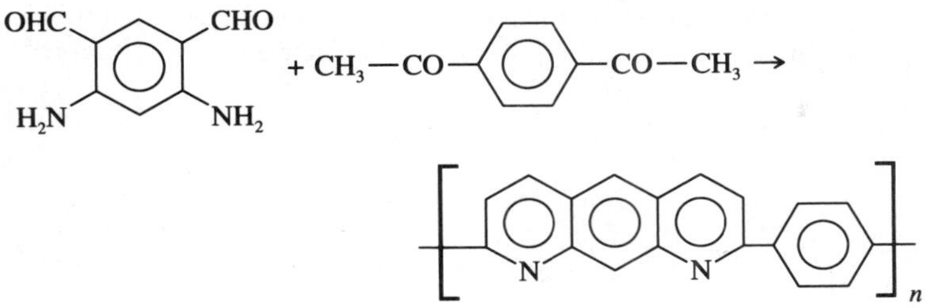

$$\tag{2-227}$$

Polymers with quinoline rings in the repeating unit have been obtained by reactions other than the Friedlander reaction [Moore and Robello, 1986; Ruan and Litt, 1988].

Polyquinoxalines are obtained by the polymerization of aromatic bis(o-diamine) and bis(α-ketoaldehyde) reactants in m-cresol at about 80°C [Hergenrother, 1988].

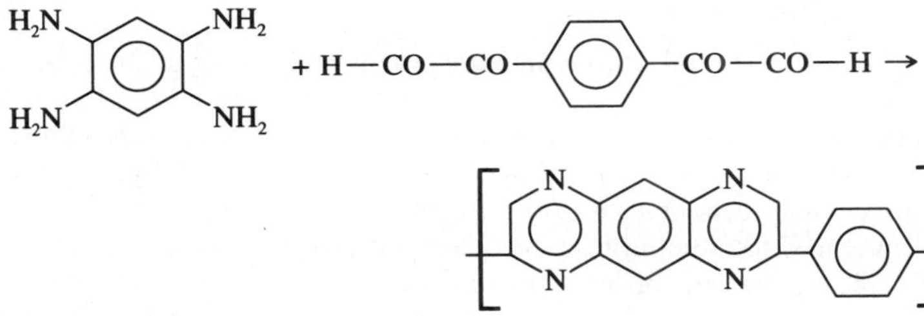

$$\tag{2-228}$$

Reaction temperatures above 100°C are avoided since the result is branching. The corresponding polyphenylquinoxalines (**LX**) are synthesized by using a bis(phenyl-α-diketone) instead of the bis(α-ketoaldehyde) reactant.

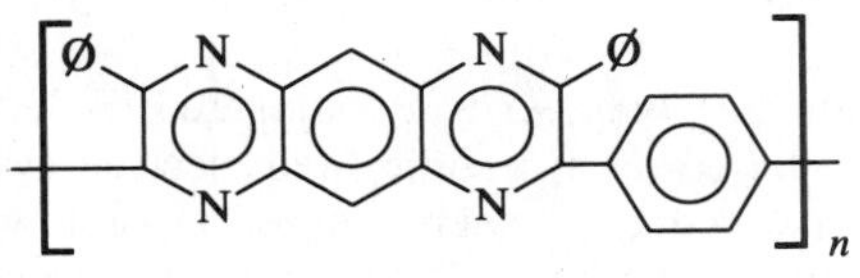

LX

Polyquinoxalines and polyphenylquinoxalines are often referred to as PQ and PPQ, respectively. The large majority of these polymers are amorphous, although a few show some crystalline behavior. PQ and PPQ are promising materials as they possess a very good combination of properties—excellent thermal and oxidative stability, resistance to acids and bases, high glass transition temperature, and high strength and modulus. These properties are coupled with a significant ability to be processed. This is especially the case for PPQ polymers. There is a considerable range between the glass transition temperature and the decomposition temperature, which allows for thermoplastic processing. Also, PPQ is soluble in phenolic and chlorinated solvents although PQ is soluble only in phenolic solvents. The favorable combination of properties has resulted in considerable efforts to explore the synthetic and practical limits of the PQ and PPQ systems. A number of ladder PQ and PPQ polymers have been studied, but the results have been disappointing, as is the case with other ladder polymers. Nonladder PQ and PPQ polymers in the form of films, composites, and adhesives show good high-temperature properties, but their utilization is impeded by high costs.

Many other 6-membered ring systems have been explored as the basis for high-

| Anthrazoline | Quinazoline | Benzoxazinone | s-Triazine |

temperature polymers, including anthrazoline, quinazolone, benzooxazinone, and s-triazine [Cassidy, 1980; Critchley et al., 1983].

2-14j   Conducting Polymers

There has been a considerable effort to synthesize organic polymers that are electrically conducting [Aldissi, 1989; Fromer and Chance, 1986; Potember et al., 1987]. Conducting polymers offer two potential advantages over the traditional inorganic materials used as conductors. First, processing of conducting polymers by molding and other plastics processing techniques into various electrical and electronic devices and forms is easy compared to metallurgical processes used for inorganic conducting materials. Second, the lightweight property of polymeric materials would make certain types of applications more practical and economical. The biggest and most immediate potential application for conducting polymers is lightweight rechargeable batteries for portable devices (e.g., tools) and vehicles. Conducting polymers would serve both current carrying and ion-conduction functions by replacing traditional electrode and electrolyte substances. The other application area for conducting polymers is their use to build circuitry elements, both passive (conducting circuits) and active (p–n and Schottky junctions).

The structural requirement for a conducting polymer is a conjugated π-electron system. The typical polymers studied are based on a π-conjugated electron system constituting the polymer backbone, but conducting polymers are also possible when the π-conjugated electron system is present as side chains. Polymers with conjugated π-electron systems display unusual electronic properties including high electron affinities and low ionization potentials. Such polymers are easily oxidized or reduced by charge-transfer agents (*dopants*) that act as electron acceptors or electron donors, respectively. This results in poly(radical–cation)- or poly(radical–anion)-doped conducting materials (p-type and n-type, respectively) with near-metallic conductivity in many cases. Electrons are removed from or added to the conjugated π-electron system to form extra holes or electrons, respectively, which carry current by wandering through the polymer chain. The dopant plays an additional role other than bringing about oxidation or reduction. It acts as the bridge or connection for carrying current between different polymer chains. Oxidative dopants include AsF_5, I_2, $AlCl_3$, and $MoCl_5$. Sodium, potassium, and lithium naphthalides are reductive dopants. The dopant, necessary to impart electrical conduction, often decreases but does not eliminate the processability of a polymer.

A number of other characteristics are required for a viable polymeric conductor. Chain orientation is needed to enhance the conducting properties of a polymeric material, especially the intermolecular conduction (i.e., conduction of current from one polymer molecule to another). This is a problem with many of the polymers that are amorphous and show poor orientation. For moderately crystalline or oriented polymers, there is the possibility of achieving the required orientation by mechanical stretching. Liquid crystal polymers would be especially advantageous for electrical conduction because of the high degree of chain orientation that can be achieved. A problem encountered with some doped polymers is a lack of stability. These materials are either oxidants or reductants relative to other compounds, especially water and oxygen.

The first polymer shown to have conducting behavior was polyacetylene (Sec. 8-4d-2). Some of the polymers investigated as heat-resistant polymers have conjugated π-electron systems and offer potential as conducting polymers. Among the polymers we have covered up to this point, poly(p-phenylene sulfide), polyquinoline, and polyquinoxaline have shown promise as conducting polymers. We now consider a number of other polymers that are at least as promising as conducting polymers.

2-14j-1 Electrochemical Polymerization

Passage of current through a solution results in the loss of electrons, and compounds are oxidized at the anode. Electrons are gained and compounds reduced at the cathode. This process is referred to as *electrochemical polymerization* when polymer is formed. Polypyrrole is obtained by the electrochemical polymerization of pyrrole in a solvent such as acetonitrile, tetrahydrofuran, or propylene carbonate [Genies et al., 1983; Lowen and Van Dyke, 1990; Naarmann, 1987; Samuelson and Druy, 1986]. An electrolyte such as tetrabutylammonium tosylate or tetrafluoroborate or lithium perchlorate is present, and polymerization is carried out either at constant voltage or constant current.

The mechanism for polymerization involves oxidation of pyrrole at the α-position to form a radical–cation (**LXI**), which undergoes radical coupling to yield the dimer–dication (**LXII**). The latter loses two protons to yield the dimer (**LXIII**). The dimer repeats the same reaction sequence—loss of an electron to form a dimer radical–cation, coupling with itself and **LXI** to form the tetramer–dication and trimer–dication, respectively, followed by a two proton loss to yield tetramer and trimer. Propagation to polymer proceeds via repetition of the same sequence—one-electron loss, coupling of different-sized radical–cations, deprotonation. This polymerization mechanism bears considerable resemblance to that for the oxidative polymerization of 2,6-disubstituted phenols (Sec. 2-14b).

Electrochemical polymerization, as usually carried out, does not yield the neutral, nonconducting polypyrrole shown in Eq. 2-231 but the oxidized (doped), conducting form. (One can cycle back and forth between the conducting and nonconducting forms, colored black and light yellow, respectively, by reversing polarity.) The doped polymer would have a structure such as **LXIV**, where A^- is the anion of the electrolyte. The doped polymer precipitates out and coats the surface of the anode during polymerization. The polymerization reaction and polymer properties (conductivity and mechanical strength) are dependent on such parameters as identity and concentration of electrolyte, reaction temperature, and current density.

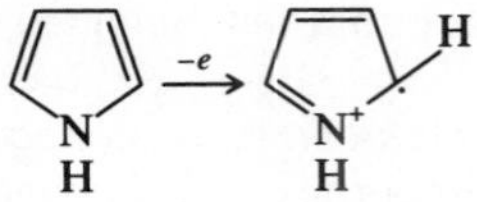

$$(2\text{-}229)$$

LXI

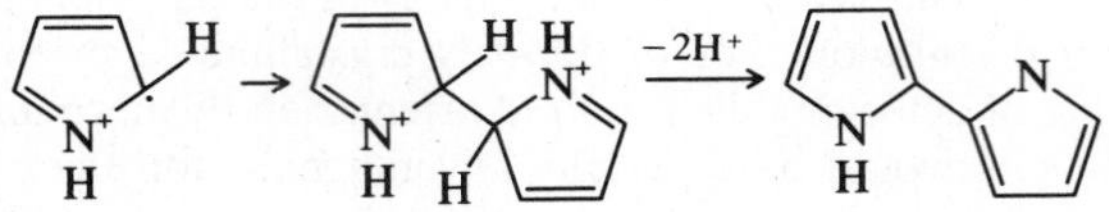

$$(2\text{-}230)$$

LXII   **LXIII**

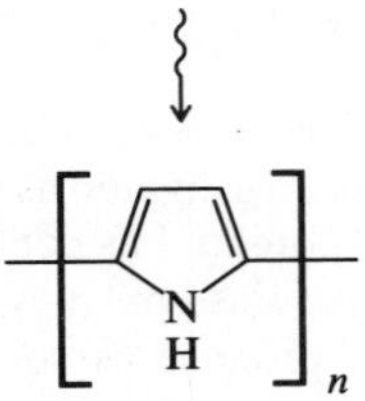

$$(2\text{-}231)$$

Electrochemical polymerization also takes place with many other aromatic and heterocyclic compounds, including thiophene, aniline, furan, carbazole, and azulene.

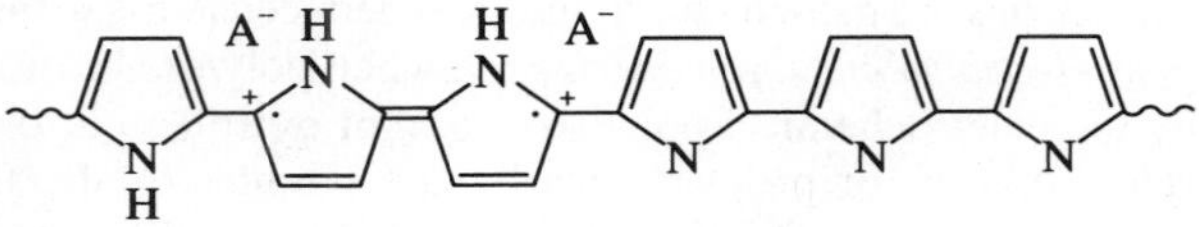

LXIV

The polymers from aniline and thiophene, together with polypyrrole, are the most promising as conducting polymers. Polymerization of thiophene proceeds in a manner similar to pyrrole except that there is a combination of propagation at both the α- and β-positions, which results in the formation of a branched product [Roncali et al., 1989]. Polymerization of aniline proceeds with coupling of nitrogen to carbon to yield polyaniline (**LXV**), which undergoes oxidation to the doped radical–cation (**LXVI**) [Johannsen et al., 1989; Leclerc et al., 1989; Macdiarmid et al., 1987; Wei et al., 1989,

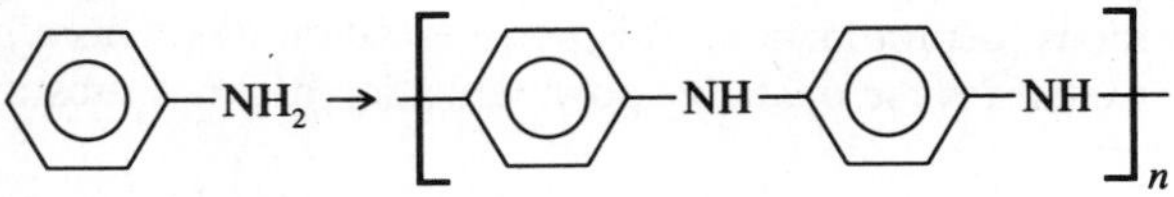

LXV

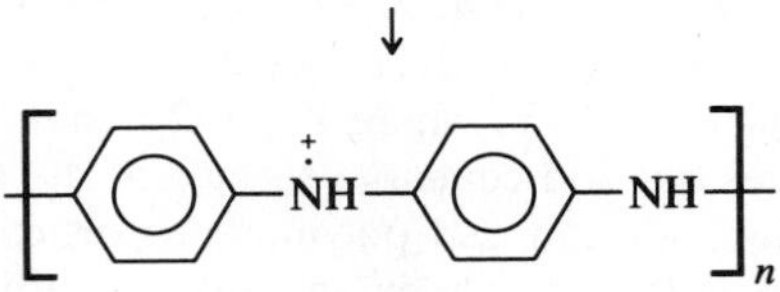

$$(2\text{-}232)$$

LXVI

1990]. Aniline as well as pyrrole, thiophene, and other aromatics are also polymerized by chemical oxidants, e.g., pyrrole is polymerized by ferrous or silver ions and aniline by persulfate [Chao and March, 1988].

2-14j-2 Poly(p-phenylene) via Oxidative Polymerization

One of the earliest attempts to synthesize a heat-resistant polymer was the oxidative-coupling polymerization of benzene to poly(p-phenylene) [Jones and Kovacic, 1987; Kovacic and Jones, 1987; Milosevich et al., 1983]. The reaction requires the presence

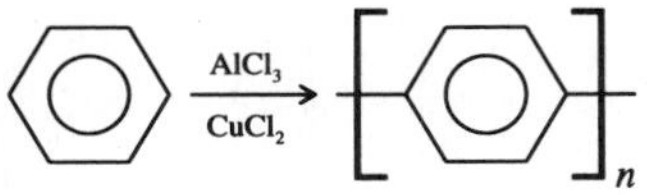

$$(2\text{-}233)$$

of both a Lewis acid catalyst and an oxidant. The most widely used combination is aluminum chloride–cupric chloride. The detailed mechanism of the polymerization is not well established, but it bears resemblance to the mechanism described above for the electrochemical polymerization of pyrrole.

Poly(p-phenylene) is of interest not only for heat resistance but also as a conducting polymer. Unfortunately, the oxidative-coupling reaction leads to an insoluble and intractable product of low molecular weight ($\overline{X}_n \leq 10$–15) and irregular structure (due to a mixture of para and ortho substitution). Similar results have been found for polymerization of biphenyl, p-terphenyl, naphthalene, and other aromatics. Alternate approaches to the poly(p-phenylene) system have been studied [McKean and Stille, 1988]. Poly(p-2,5-di-n-hexylphenylene), degree of polymerization of approximately 30, was obtained by the palladium catalyzed self-coupling of 4-bromo-2,5-di-n-hexylbenzeneboronic acid [Rehahn et al., 1990].

Oxidative coupling polymerizations of dialkynes and dithiols to polyalkynes and polydisulfides have also been studied [Jones and Kovacic, 1987].

2-14j-3 Poly(p-phenylene Vinylene)

Poly(p-phenylene vinylene) (IUPAC name: poly[1,4-phenylene-1,2-ethenediyl]) is obtained in a two-step sequence (Eq. 2-234) involving successive base- and thermally

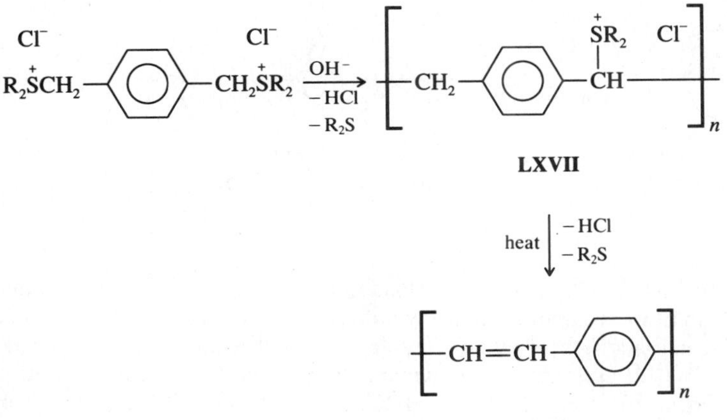

LXVII

$$(2\text{-}234)$$

LXVIII

induced eliminations of dialkyl sulfide and hydrogen halide [Lenz et al., 1988; Lenz et al., 1989; Memeger, 1989]. Attempts to synthesize **LXVIII** via the Wittig reaction of a dialdehyde and bistriphenylphosphonium salt were unsuccessful in that only very low-molecular-weight, insoluble, and intractable product was obtained [Horhold and Helbig, 1987]. The synthesis from the bissulfonium salt achieves a high-molecular-weight product since the precursor polymer **LXVII** is soluble.

The corresponding azomethine polymers such as **LXIX** have been synthesized by

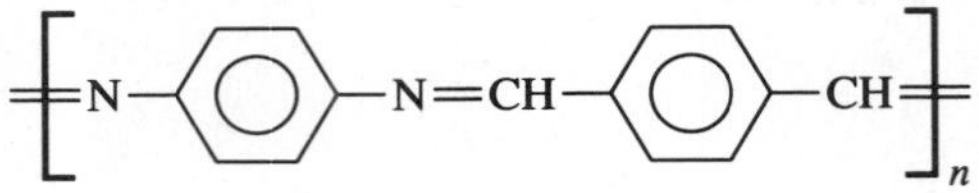

LXIX

the reaction of the appropriate diamine and dialdehyde [Cheng et al., 1989; Morgan et al., 1987] but are not sufficiently stable toward hydrolysis for conducting polymer applications. Both **LXVIII** and **LXIX** show liquid crystal behavior.

2-14k Miscellaneous Polymerizations

2-14k-1 Cycloaddition or Four-Center Polymerization

The Diels–Alder reaction, involving the [4 + 2] cycloaddition of an unsaturated group (dienophile) to a 1,3-diene, has been studied for the synthesis of ladder polymers, such as the reaction of 2-vinyl-1,3-butadiene with benzoquinone [Bailey, 1972; Bailey et al., 1962]. Related polymerizations are those utilizing the [2 + 2] cycloaddition

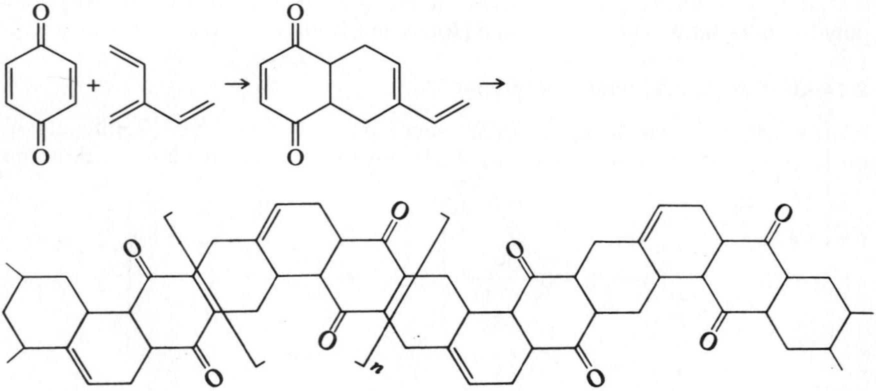

$$(2\text{-}235)$$

reaction [Dilling, 1983]. While [4 + 2] cycloaddition reactions are thermally induced, [2 + 2] cycloadditions are photochemically induced (as is the case for small molecule reactions). The self-addition of alkene double bonds is the most extensively studied example of a [2 + 2] cycloaddition polymerizations, such as the polymerization of 2,5-distyrylpyrazine [Braun and Wegner, 1983; Hasegawa et al., 1988]. This polymerization is a solid-state reaction involving irradiation of crystalline monomer with ultraviolet or ionizing radiation. The reaction is an example of a *topochemical* or *lattice-controlled* polymerization in which reaction proceeds either inside the monomer crystal where the product structure and symmetry are controlled by the packing of monomer in the lattice or at defect sites where product structure and symmetry are determined by the packing of monomer at these sites.

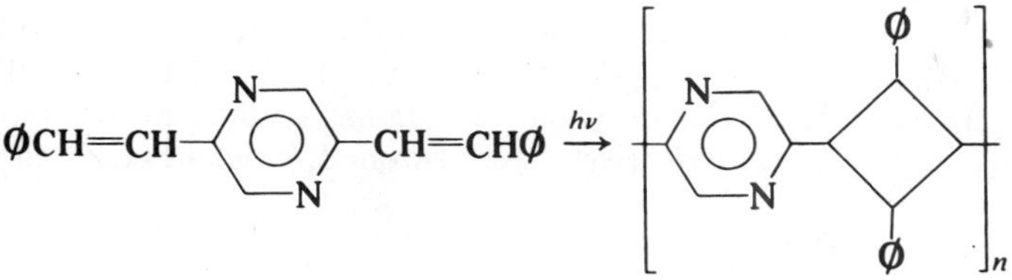

$$\phi CH{=}CH{-}\left(\begin{array}{c} N{-} \\ \bigcirc \\ {-}N \end{array}\right){-}CH{=}CH\phi \xrightarrow{h\nu} \left[\begin{array}{c} \phi \\ \left(\begin{array}{c} N{-} \\ \bigcirc \\ {-}N \end{array}\right){-}\diamond{-} \\ \phi \end{array}\right]_n \tag{2-236}$$

A number of other cycloaddition polymerizations have been reported, including the [3 + 2] cycloaddition reaction between hexafluoroacetone azine and electron-poor alkenes [Nuyken et al., 1988; 1990] and the [4 + 4] cycloaddition between anthracene derivatives [Dilling, 1983].

2-14k-2   *Starburst Dendrimer Polymers*

Starburst dendrimer polymers possess a radially symmetrical star-shaped architecture in which there are successive cascades of branched polymer structures. The sequence of reactions used in the synthesis of one such polymer involves a repetitive sequence of Michael additions of amine to α,β-unsaturated ester followed by nucleophilic substitution of ester by amine [Tomalia et al., 1985, 1986]. The first step involves Michael addition of ammonia with excess methyl acrylate followed by reaction with excess ethylenediamine to yield **LXX**—a star molecule with three arms. Structure **LXX** is reacted with excess methyl acrylate followed by excess ethylenediamine to yield the star molecule **LXXI** containing six arms. The two-step sequence is repeated to yield star polymers, referred to as *starburst polymers*, with successive doubling of the number

$$NH_3 \xrightarrow[\text{2. } H_2NCH_2CH_2NH_2]{\text{1. } CH_2{=}CH{-}COOCH_3} H_2NCH_2CH_2NHCOCH_2CH_2{-}N{\overset{\displaystyle CH_2CH_2CONHCH_2CH_2NH_2}{\underset{\displaystyle CH_2CH_2CONHCH_2CH_2NH_2}{\Big\langle}}} \tag{2-237}$$

LXX

$\downarrow$ repeat

$$\begin{array}{c} H_2NCH_2CH_2NHCOCH_2CH_2 \\ \searrow \\ NCH_2CH_2NHCOCH_2CH_2{-}N \\ \nearrow \\ H_2NCH_2CH_2NHCOCH_2CH_2 \end{array}\quad \begin{array}{c} CH_2CH_2CONHCH_2CH_2N{\overset{\displaystyle CH_2CH_2CONHCH_2CH_2NH_2}{\underset{\displaystyle CH_2CH_2CONHCH_2CH_2NH_2}{\Big\langle}}} \\ \\ CH_2CH_2CONHCH_2CH_2N{\overset{\displaystyle CH_2CH_2CONHCH_2CH_2NH_2}{\underset{\displaystyle CH_2CH_2CONHCH_2CH_2NH_2}{\Big\langle}}} \end{array}\cdot$$

LXXI   (2-238)

of arms per molecule. The terms *arboral* or *cauliflower polymer* have also been used to describe these types of polymers, and a number of different reactions have been used for their synthesis [Hawker and Frechet, 1990; Newkome et al., 1986; Wilson and Tomalia, 1989]. Work is in progress to ascertain whether there are unique properties associated with the starburst architecture.

2-14k-3   Spiro Structures

Polymer chains based on *spiro structures* have been studied as another route to heat-resistant polymers [Bailey and Volpe, 1976; Kurita et al., 1979]. These are polymer chains built of spiro rings; that is, there is only one atom common to two adjacent rings. An example is the polyspiroketal synthesized from 1,4-cyclohexanedione and

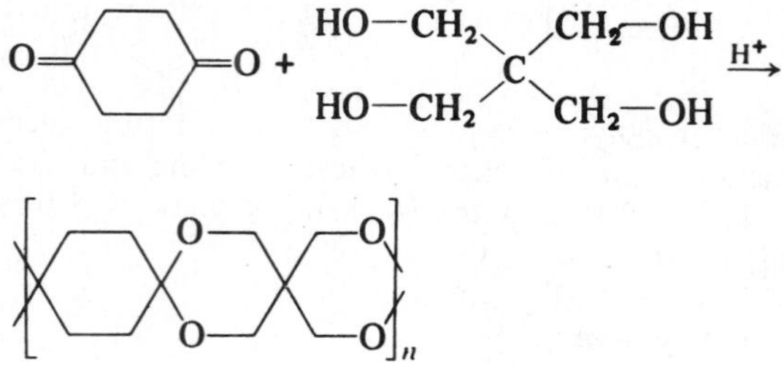

$$
\text{(chemical reaction scheme)} \tag{2-239}
$$

pentaerythritol. Unfortunately, the low solubility and intractability of spiro polymers makes it difficult to synthesis or utilize high-molecular-weight polymer.

2-15   INORGANIC AND ORGANOMETALLIC POLYMERS

The drive for heat-resistant polymers has led to an exploration of polymers based on inorganic elements since bond energy considerations indicate that such polymers should be superior to organics in thermal stability [Allcock and Lampe, 1981; Ray, 1978; Stone and Graham, 1962]. Both *inorganic* and *organometallic* polymers have been studied. Inorganic polymers are polymers containing no organic groups, while organometallic polymers contain a combination of inorganic and organic groups. Inorganic and organometallic polymers have potential for a variety of uses—as partial or complete substitutes for organic fibers, elastomers, and plastics where flame and heat resistance is important; as marine antifoulants, bactericides, medicinals, fungicides, adhesives, photoresists, photosensitizers, and photostabilizers; and as conducting polymers. In general, much of this potential is unrealized, although significant exceptions exist—the polysiloxanes (Secs. 2-12f, 7-11a), poly(*p*-phenylene sulfide) (Sec. 2-14d), polyphosphazenes (Sec. 7-11b), and polysilanes (Sec. 2-15b-3) are commercial polymers.

2-15a   Inorganic Polymers

With essentially no exceptions attempts to synthesize inorganic polymers that can directly substitute for organic polymers have been unsuccessful. Although the intrinsic thermal stability is often good, a variety of difficulties must be overcome before a usable polymer is obtained. Inorganic polymers typically suffer from various combinations of poor hydrolytic stability, low polymer molecular weights, and low chain flexibility. For example, inorganic polymers with high molecular weights and good hydrolytic stability are often highly inflexible. A large deficiency for many polymers is their intractability to current techniques of fabricating polymers into products such as film, fiber, tubing, and other objects. This does not mean that inorganic polymers are useless. Inorganic polymers comprise many of the materials that are employed at home, industry, and elsewhere. Let us briefly consider some of those materials. [Several wholly inorganic polymers, polysulfur, polyselenium, polythiazyl, and poly(dichlorophosphazene), are discussed in Chap. 7.]

2-15a-1 Minerals

A variety of mineral-type materials are inorganic polymers [Gimblett, 1963; Ray, 1978]. Silica [$(SiO_2)_n$] is found in nature in various crystalline forms, including sand, quartz, and agate. The various crystalline forms of silica consist of three-dimensional, highly crosslinked polymer chains composed of SiO_4 tetrahedra where each oxygen atom is bonded to two silicon atoms and each silicon atom is bonded to four oxygen atoms. *Silicates*, found in most clays, rocks, and soils, are also based on SiO_4 tetrahedra. However, they differ from silica in having Si:O ratios under 1:2 (compared to 1:2 for silica) and contain Si—O$^-$ groups with associated metal cations. Single-strand, double-strand (ladder), sheet (composed of multiple-double-strand sheets analogous to graphite), and three-dimensional polymer chains occur depending on the Si:O ratio, cation:Si ratio, and charge on the cation. Talc contains magnesium and possesses a sheet silicate structure. Tremolite, an asbestos mineral, has a double-strand structure and contains calcium and magnesium. The silicates, similar to silica itself, are highly rigid materials because of their ladder, sheet, and three-dimensional structure. Even the single-strand chains are rigid since the cations act to hold together adjacent polymer chains.

2-15a-2 Glasses

Silicate *glasses* are produced by melting and rapidly cooling silica or a mixture of silica with other materials [Thornton and Colangelo, 1985]. The product is an amorphous glass since molten silica has a strong tendency to supercool. (Special procedures with regard to heating and cooling rates are required to achieve crystallization.) Silica glass differs from crystalline silica in that there is less than full coordination of all silicon and oxygen atoms; that is, not every silicon is coordinated to four oxygens and not every oxygen is coordinated to two silicons. *Fused silica glass*, consisting of virtually pure silica, is the most chemically resistant of glasses and exhibits the maximum continuous service temperature (900°C). Its relatively high cost limits its use to special applications such as the fiber optics used in information and image transmission and medical fiberscopes for internal examination of humans and animals. The most commonly encountered glass, referred to as *soda-lime glass*, is made by incorporating various amounts of calcium, sodium, and potassium into the silicate by adding the appropriate compounds, such as sodium and calcium carbonates or oxides, to molten silica. The properties of the glass (e.g., hardness, softening temperature) are varied by varying the relative amounts of the different cations. This is the glass used for windows, lightbulbs, bottles, and jars. Optical glass is similar to soda-lime glass but is much harder because it contains less sodium and more potassium. Colored glasses are obtained by the addition of appropriate compounds; for example, chromium(III) and cobalt(II) oxides yield green and blue glasses, respectively. Photochromic glasses, which darken reversibly on exposure to light, are obtained by including silver halide in the glass formulation. Vitreous enamels on metal objects and glazes on pottery are glass coatings obtained by covering the item with a paste of the appropriate oxides and heating to a high temperature.

Borosilicate and *aluminosilicate* glasses are produced by adding B_2O_3 (borax) and Al_2O_3 (alumina), respectively, to molten silica. This produces a structure where boron and aluminum atoms, respectively, replace some silicon atoms in the silicate polymer chain. Laboratory glassware is manufactured from borosilicate glass, which additionally contains sodium and calcium (trade name: *Pyrex*). The very low coefficient of thermal expansion of borosilicate glasses makes them far less prone to breakage on heating and cooling and especially useful for volumetric glassware. This glass also has very good chemical resistance. Greater chemical resistance, when required, is obtained by using borosilicate glasses with a very high (99.6%) silica content. Pyrex-type borosilicate glasses contain 70–80% silica. (Quartz glassware is used for very special applications.) Aluminosilicate glasses containing calcium and magnesium are used for cookware. There are many naturally occurring aluminosilicate minerals, such as feldspars and zeolites, which contain various combinations of sodium, potassium, and calcium.

A variety of other glassy inorganic polymers are known [Ray, 1978]. *Polymetaphosphates* are linear polymers produced by heating an alkali dihydrogen phosphate (Eq. 2-240). *Polyultraphosphates*, the corresponding branched and crosslinked ana-

$$\text{NaH}_2\text{PO}_4 \xrightarrow{-\text{H}_2\text{O}} \left[\begin{array}{c} \text{O} \\ \| \\ \text{--P--O--} \\ | \\ \text{ONa} \end{array}\right]_n \tag{2-240}$$

logs, are obtained by having phosphoric acid present in the reaction mixture. *Chalcogenide glasses* are crosslinked polymers formed by the fusion of a chalcogen (S, Se, Te) with one or more of various polyvalent elements (e.g., As, Cd, Ge, P, Si, Sn). Some of these polymers are useful as infrared transparent windows in applications such as infrared detectors and for encapsulating photosensitive transistors.

2-15a-3 Ceramics

Ceramic materials have higher heat resistance than do their corresponding glass compositions. The *traditional ceramics*, such as brick, pottery, and porcelain, are aluminosilicates derived from clay as the main raw material [Jastrzebski, 1976; Thornton and Colangelo, 1985]. These are heterogeneous materials in which microcrystals are embedded in a glass matrix. Clays are produced by the weathering of the mineral feldspar, an aluminosilicate containing sodium and potassium ions. Weathering of feldspar produces kaolinite, empirical formula $Al_2Si_2O_5(OH)_4$, in the form of small thin platelets. Clays are mixtures of kaolinite with small amounts of feldspar, sand, and other minerals. Most clays are reddish because of the presence of iron(III) oxide as an impurity. Ceramic objects are made by working a mixture of clay with water. Other silicate materials are often added depending on the desired properties in the final product. The presence of water renders the mixture pliable as kaolinite platelets can slide over one another. *Firing* (heating) of the shaped object results in a complex set of reactions. Water is lost during the process and the kaolinite platelets can no longer slip past one another. In addition, an aluminosilicate glassy network with its associated cations is formed. The resulting ceramic consists of kaolinite platelets distributed in a glassy aluminosilicate matrix. The partially crystalline nature of these ceramics results in denser materials with higher strength, heat conduction, and heat resistance compared to glasses. Traditional ceramics find a range of applications: pottery, construction bricks, whitewares (porcelain coatings on various household items and spark plugs, ovenware), and refractory (furnace bricks, mortar, liners, coatings). Most of these applications require the heat-resistant properties inherent in the ceramic materials. However, the strength requirements are not very high in such applications. A second group of ceramic materials, often referred to as *new ceramics*, have been developed in more recent years. These include various pure and mixed oxides, carbides, nitrides, borides, and silicides (e.g., BeO, ZrO, α-Al$_2$O$_3$, α-SiC, TiC, BN, AlN, Si$_3$N$_4$, ZrB$_2$, TaB$_2$; empirical formulas). The new ceramics are used in applications requiring high strength at high temperature, such as parts for gas and jet engines, nuclear plants, high-temperature chemical plants and abrasives, cutting tools, and dies for high-speed machining of metals. Silicon carbide is used for the heat-resistant tiles that protect the space shuttle during reentry into the earth's atmosphere. Many of the new ceramics are close to being completely crystalline materials. Others consist of crystalline material embedded in a glassy matrix, but the amount of crystalline phase is typically greater than that for the traditional ceramics. Synthesis of the various materials used for the newer ceramics typically requires high-temperature processes; for instance, SiC is produced by the reaction of silica and anthracite coal or coke in an electric resistance furnace. The production of objects from the newer ceramics follows in the same manner as the traditional ceramics but with higher firing temperatures (800–1800°C). The firing process at these higher temperatures is referred to as *sintering*.

The formation of objects from glass and ceramic (especially the new ceramic) materials is achieved only at very high temperatures where there are considerable limitations on the manipulative techniques available to form desired shapes and forms. Two new methods hold considerable promise for producing glass and ceramic objects more easily and at much more moderate temperatures. The key to both methods is the use of very different chemical reactions, based on organic derivatives of the inorganic element, to form the glass or ceramic compound(s). The *preceramic polymer* method is described in Sec. 2-15b-3. The *sol–gel* method involves the *in situ* generation of silica by the base- or acid-catalyzed hydrolysis of tetramethoxy- or tetraethoxysilane in an alcohol–water mixture (Eq. 2-241) [Brinker et al., 1984; Hench and West, 1990; Klein, 1985; Nogami and Moriya, 1980].

$$Si(OR)_4 \xrightarrow[-ROH]{H_2O} \left(SiO_2 \right)_{\overline{n}} \tag{2-241}$$

The polymerization conditions (temperature, reactant, and catalyst concentrations) may be varied to yield reaction times from minutes to hours or days for completion. Polymerization generally involves initial reaction at ambient temperature followed by an aging process at 50–80°C to develop a controlled network structure. The acid-catalyzed process is slower and easier to control than the base-catalyzed process. Tetraacetoxysilanes have also been used as the silica precursor.

The reaction mixture consists of a silica gel swollen with solvent (alcohol and water) after aging. Densification occurs by a drying step in which solvent is evaporated from the reaction mixture. The product after densification is a microporous silica referred to as a *xerogel* or *aerogel* depending on the conditions of solvent removal. Normal solvent evaporation of the open (to the atmosphere) reaction mixture results in a high-density xerogel. Bleeding solvents slowly from a closed reaction mixture kept under conditions where the solvents are above their critical conditions results in an aerogel. The aerogel is generally of low density and useful as a high-temperature thermal insulator. The last phase of densification may involve temperatures considerably in excess of 100°C to complete the polymerization reaction and drive off the last amounts of solvent. Fabrication of the silica gel (e.g., by coating onto some object or device or pouring into a mold) can be carried out at various times during the overall process depending on the specific application. A variety of mixed silicate glasses can be obtained by including other compounds in the reaction mixture. Borosilicate, aluminosilicate, and mixed boroaluminosilicate glasses containing various additional cations (e.g., sodium, barium, potassium) are produced by polymerizing a mixture of tetraalkoxysilane with the appropriate alkoxides, acetates, or nitrates of the other metals.

A polyacyloxyaluminoxane coordinated with a carboxylic acid (**LXXII**) is a pre-

$$\left[\begin{array}{c} \text{HOCOR} \\ | \\ -\text{Al}-\text{O}- \\ | \\ \text{OCOR} \end{array}\right]_n$$

LXXII

cursor for alumina fibers [Kimura et al., 1989]. The polymer, synthesized by reaction of trialkylaluminum with a carboxylic acid and water, can be melt spun into fibers at 200°C. Pyrolysis in stages at temperatures of up to 1400°C produces alumina fibers.

2-15b Organometallic Polymers

Although inorganic polymers find important applications as glasses and ceramics, their high rigidity make them unsuitable for replacement of organic polymers. Organometallic polymers, a compromise between inorganic and organic polymers, have been studied in the effort to achieve the goal of heat-resistant polymers processable by the techniques used for organic polymers. The general approach involves decreasing the extent of ladder, sheet, and three-dimensional structures as well as bridging between different chains through cation–oxygen anion interactions by replacing some of the oxygen linkages to metal atoms by organic linkages. The model for this approach is the polysiloxane system, where the presence of two organic groups attached to each silicon atom greatly changes chain rigidity compared to silica. Although many systems have and are being investigated, commercial success in this field is limited at present.

2-15b-1 *Polymerization via Reaction at Metal Bond*

Many polymerizations used to synthesize organometallic polymers involve cleavage of a bond to the metal atom of one reactant followed by attachment at that metal of a fragment derived from a second reactant [Carraher et al., 1978; Pittman et al., 1987; Sheets et al., 1985]. Polyphosphonates and polyphosphoramides are obtained by reacting an aryl (or alkyl) phosphonic dichloride with a diol and diamine, respectively

[Kricheldorf et al., 1988; Percec et al., 1979; Pretula and Penczek, 1990]. These and other phosphorus-containing polymers have received attention because of their flame-retardant properties, either as a flame-retardant polymer directly or as an additive to some other polymer.

$$
\text{Cl}-\underset{\underset{\text{O}}{\|}}{\overset{\overset{\text{Ar}}{|}}{\text{P}}}-\text{Cl} + \text{HO}-\text{Ar}'-\text{OH} \xrightarrow{-\text{HCl}} \left[\underset{\underset{\text{O}}{\|}}{\overset{\overset{\text{Ar}}{|}}{\text{P}}}-\text{O}-\text{Ar}'-\text{O}\right]_n \qquad (2\text{-}242)
$$

$$
\text{Cl}-\underset{\underset{\text{O}}{\|}}{\overset{\overset{\text{Ar}}{|}}{\text{P}}}-\text{Cl} + \text{H}_2\text{N}-\text{Ar}'-\text{NH}_2 \xrightarrow{-\text{HCl}} \left[\underset{\underset{\text{O}}{\|}}{\overset{\overset{\text{Ar}}{|}}{\text{P}}}-\text{NH}-\text{Ar}'-\text{NH}\right]_n \qquad (2\text{-}243)
$$

Carboxylate and sulfide groups have been similarly used to obtain organometallic polymers (Eqs. 2-244 and 2-245). Tin-containing polymers have been of interest for

$$
\text{R}_2\text{SnCl}_2 + \text{NaOOC}-\text{R}'-\text{COONa} \xrightarrow{-\text{NaCl}} \left[\underset{\underset{\text{R}}{|}}{\overset{\overset{\text{R}}{|}}{\text{Sn}}}-\text{OCO}-\text{R}'-\text{COO}\right]_n \qquad (2\text{-}244)
$$

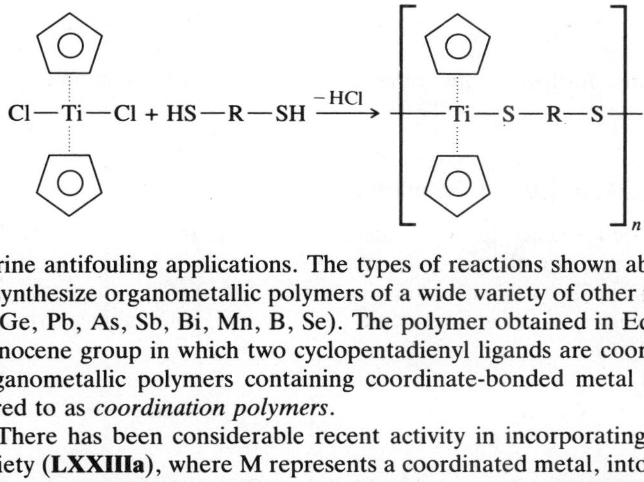

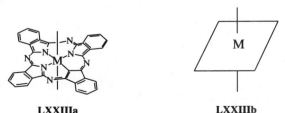

$$
\qquad (2\text{-}245)
$$

marine antifouling applications. The types of reactions shown above have been used to synthesize organometallic polymers of a wide variety of other metals (e.g., Zr, Hf, Si, Ge, Pb, As, Sb, Bi, Mn, B, Se). The polymer obtained in Eq. 2-245 contains the titanocene group in which two cyclopentadienyl ligands are coordinated to titanium. Organometallic polymers containing coordinate-bonded metal atoms are often referred to as *coordination polymers*.

There has been considerable recent activity in incorporating the phthalocyanine moiety (**LXXIIIa**), where M represents a coordinated metal, into polymer structures. Structure **LXXIIIa** is symbolized in brief by structure **LXXIIIb**. High-temperature

LXXIIIa	**LXXIIIb**

dehydration of the phthalocyanine diol or nucleophilic displacements of the dichloride have been successful [Marks, 1985; Snow and Griffith, 1988]. The phthalocyanine

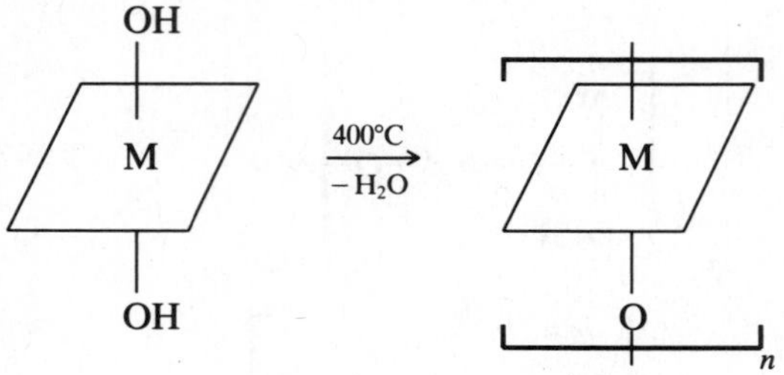

$$
\begin{array}{c}
\text{OH} \\
| \\
\boxed{M} \\
| \\
\text{OH}
\end{array}
\quad \xrightarrow[-\,H_2O]{400°C} \quad
\begin{array}{c}
| \\
\boxed{M} \\
| \\
O \\
\end{array}_n
\qquad (2\text{-}246)
$$

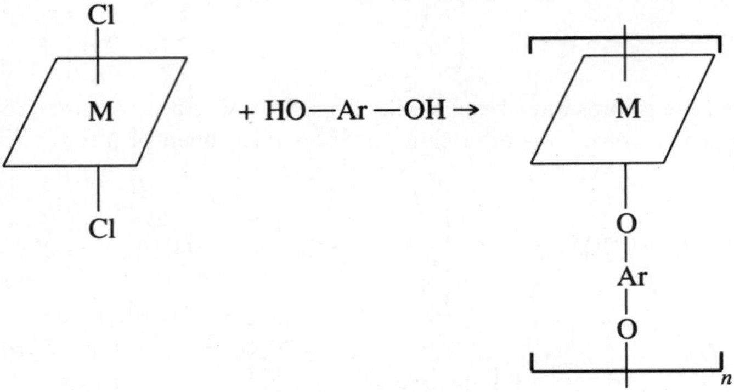

$$
\begin{array}{c}
\text{Cl} \\
| \\
\boxed{M} \\
| \\
\text{Cl}
\end{array}
+ \; HO{-}Ar{-}OH \rightarrow
\begin{array}{c}
| \\
\boxed{M} \\
| \\
O \\
| \\
Ar \\
| \\
O \\
\end{array}_n
\qquad (2\text{-}247)
$$

polymers show promise for high-temperature, catalytic, and electrical conduction applications.

2-15b-2 *Polymerization without Reaction at Metal Atom*

An alternate synthetic approach to organometallic polymers is to perform reactions where the metal atom is not the reaction site. An example is a polyesterification between an organometallic monomer containing two carboxyl groups and a diol. Reaction of the diacid with a diamine would yield an organometallic polymer via

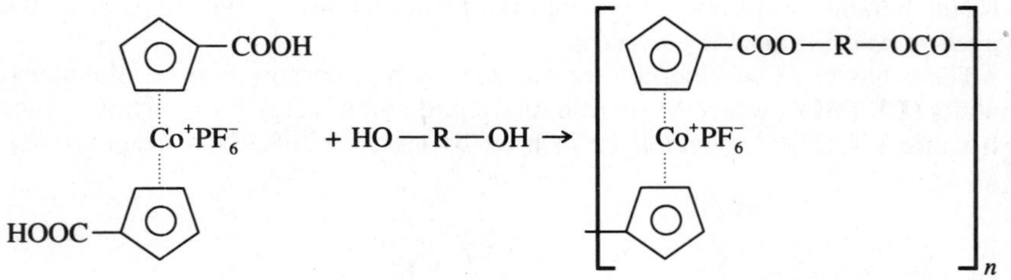

$$
\begin{array}{c}
\text{COOH} \\
Co^+PF_6^- \\
HOOC
\end{array}
+ \; HO{-}R{-}OH \rightarrow
\left[
\begin{array}{c}
\text{COO}{-}R{-}OCO \\
Co^+PF_6^-
\end{array}
\right]_n
$$

$$(2\text{-}248)$$

polyamidation. A wide range of other reactions covered in this chapter can be used to synthesize organometallic polymers.

2-15b-3 Polysilanes

Polysilanes are polymers in which there is catenation of silicon, that is, where silicon atoms are bonded to each other in a continuous manner. Synthesis of polysilanes involves the *Wurtz coupling* of diorganodichlorosilanes with sodium metal [Miller and Michl, 1989; West, 1986; West and Maxka, 1988; Yajima et al., 1978]. The reaction

$$RR'SiCl_2 \xrightarrow[-NaCl]{Na} \left[\begin{array}{c} R \\ | \\ Si \\ | \\ R' \end{array} \right]_n \qquad (2\text{-}249)$$

is typically carried out in a hydrocarbon solvent such as toluene, xylene, or octane at temperatures above 100°C. Polymerization can be achieved at ambient temperature in the presence of ultrasound, which produces high temperature and pressures locally for short bursts [Kim and Matyjaszewski, 1988]. The mechanism for this reaction is not established, but most evidence indicates polymerization is a complex process involving some combination of radical, anionic, and silylene (the Si analog of carbene) intermediates [Gauthier and Worsfold, 1989; Matyjaszewski et al., 1988]. Furthermore, although included in this chapter, the polymerization is probably a chain reaction. It is included here because of its technological importance in complementing the sol–gel process for producing ceramics.

Polysilanes have been synthesized with various combinations of alkyl and aryl substituents. Polysilanes, such as polydimethylsilane or polydiphenylsilane, with symmetrical substitution are highly crystalline and show little or no solubility in a range of organic solvents. (The two polymers are also known as *polydimethylsilylene* and *polydiphenylsilylene*, respectively.) Crystallinity is decreased and solubility increased when R and R' are different or for copolymers derived from two different symmetrically substituted dichlorosilanes. There is considerable interest in polysilanes from several viewpoints. Many of the interesting properties of polysilanes result from the relative ease of delocalization of the electrons in the catenated Si—Si σ-bonds as evidenced by the strong ultraviolet absorption at 290–365 nm. Polysilanes undergo photolytic radical cleavage with a high quantum yield and offer potential as radical initiators and positive photoresists. (In a positive photoresist application, the portions of a polymer not protected by a mask are degraded by irradiation and then dissolved by solvent or photovolatilized.) Polysilanes also offer potential as semiconductor, photoconductor, and nonlinear optical materials.

The largest interest in polysilanes arises from their use as *preceramic polymers*. The normal powder metallurgy techniques for processing ceramic materials limits the complexity of the objects that can be produced. Polysilane chemistry offers an alternate with good potential for producing a variety of objects, including fiber. Thermolysis of a polysilane in an inert atmosphere at 450°C yields a polycarbosilane through a complex rearrangement process. For example, polydimethylsilylene yields polymethylsilamethylene (**LXXIV**). A soluble portion of the polycarbosilane is isolated by fractional precipitation from *n*-hexane and used as a ceramic precursor. The soluble polycarbosilane can be formed into objects (including fibers) and then pyrolyzed at 1300°C to yield the corresponding crystalline β-silicon carbide ceramic objects. Other organometallic polymers are being studied for use as precursor polymers for other ceramic systems, such as polysilazanes for silicon nitride [Baney and Chandra, 1988].

$$\left[\begin{array}{c} CH_3 \\ | \\ -Si- \\ | \\ CH_3 \end{array}\right]_n \xrightarrow{450°C} \left[\begin{array}{c} H \\ | \\ -Si-CH_2- \\ | \\ CH_3 \end{array}\right]_n \xrightarrow{1300°C} \beta\text{-SiC} \qquad (2\text{-}250)$$

LXXIV

The use of a monoalkyltrichlorosilane in the Wurtz-type polymerization is reported to yield $(RSi)_n$ referred to as *polyalkylsilyne* [Bianconi et al., 1989]. The elemental composition of the polymer would suggest either a three-dimensional crosslinked, linear conjugated, or aromatic structure, but none of these possibilities have been verified experimentally.

REFERENCES

de Abajo, J., *Makromol. Chem. Macromol. Symp.*, **22** 141 (1988).

Adduci, J. M. and M. J. Amone, *J. Polym. Sci. Polym. Chem. Ed.*, **27**, 1115 (1989).

Aharoni, S. M., W. B. Hammond, J. S. Szobota, and D. Masilamani, *J. Polym. Sci. Polym. Chem. Ed.*, **22**, 2579 (1984).

Aldissi, M., *Makromol. Chem. Macromol. Symp.*, **24**, 1 (1989).

Allcock, H. and F. W. Lampe, "Contemporary Polymer Chemistry," Chap. 7, Prentice-Hall, Englewood Cliffs, N.J., 1981.

Allen, P. E. M. and C. R. Patrick, "Kinetics and Mechanisms of Polymerization Reactions," Chap. 5, Halsted Press (Wiley), New York, 1974.

Amass, A. J., *Polymer*, **20**, 515 (1979).

Arai, Y., M. Watanabe, K. Sanui, and N. Ogata, *J. Polym. Sci. Polym. Chem. Ed.*, **23**, 3081 (1985).

Argyropoulos, D. S., R. M. Berry, and H. I. Bolker, *Makromol. Chem.*, **188**, 1985 (1987).

Aycock, D., V. Abolins, and D. M. White, "Poly(phenylene ether)," pp. 1–30 in "Encyclopedia of Polymer Science and Engineering," Vol. 13, H. F. Mark, N. M. Bikales, C. G. Overberger, and G. Menges, Eds., Wiley-Interscience, New York, 1988.

Backus, J. K., "Polyurethanes," Chap. 17 in "Polymerization Processes," C. E. Schildknecht, Ed. (with I. Skeist), Wiley-Interscience, New York, 1977.

Backus, J. K., C. D. Blue, P. M. Boyd, F. J. Cama, J. H. Chapman, J. L. Eakin, S. J. Harasin, E. R. McAfee, D. G. McCarty, J. N. Rieck, H. G. Schmelzer, and E. P. Squiller, "Poly-urethanes," pp. 243–303 in "Encyclopedia of Polymer Science and Engineering," Vol. 13, H. F. Mark, N. M. Bikales, C. G. Overberger, and G. Menges, Eds., Wiley-Interscience, New York, 1988.

Bailey, W. J., "Diels–Alder Polymerization," Chap. 6 in "Step-Growth Polymerizations," D. H. Solomon, Ed., Marcel Dekker, New York, 1972.

Bailey, W. J., J. Economy, and M. E. Hermes, *J. Org. Chem.*, **27**, 3295 (1962).

Bailey, W. J. and A. A. Volpe, *Polym. Prepr.*, **8**(1), 292 (1976).

Ballauff, M., *Angew. Chem. Int. Ed. Engl.*, **28**, 253 (1989).

Ballistreri, A., G. Montaudo, G. Impallomeni, R. W. Lenz, Y. B. Kim, and R. C. Fuller, *Macromolecules*, **23**, 5059 (1990).

Baney, R. and G. Chandra, "Preceramic Polymers," pp. 312–344 in "Encyclopedia of Polymer Science and Engineering," Vol. 13, H. F. Mark, N. M. Bikales, C. G. Overberger, and G. Menges, Eds., Wiley-Interscience, New York, 1988.

Bartmann, M. and U. Kowalczik, *Makromol. Chem.*, **189**, 2285 (1988).

Bekhli, E. Yu., O. V. Nesterov, and S. G. Entelis, *J. Polym. Sci.*, **C16**, 209 (1967).

Bhide, B. V. and J. J. Sudborough, *J. Indian Inst. Sci.*, **8A**, 89 (1925).

Bianconi, P. A., F. C. Schilling, and T. W. Weidman, *Macromolecules*, **22**, 1697 (1989).

Braun, H.-G. and G. Wegner, *Makromol. Chem.*, **184**, 1103 (1983).

Brinker, C. J., K. D. Keefer, D. W. Schaefer, R. A. Assink, B. D. Kay, and C. S. Ashley, *J. Non-Cryst. Solids*, **63**, 45 (1984).

Brock, F. H., *J. Org. Chem.*, **24**, 1802 (1959); *J. Phys. Chem.*, **65**, 1638 (1961).

Brown, G. H. and P. P. Crooker, *Chem. Eng. News*, **61**, 24 (Jan. 31, 1983).

Brydson, J. A., "Plastics Materials," 4th ed., Chaps. 19 and 23, Butterworth Scientific, London, 1982.

Buckley, A., D. E. Stuetz, and G. A. Serad, "Polybenzimidazoles," pp. 572–601 in "Encyclopedia of Polymer Science and Engineering," Vol. 11, H. F. Mark, N. M. Bikales, C. G. Overberger, and G. Menges, Eds., Wiley-Interscience, New York, 1988.

Burchard, W., *Polymer*, **20**, 589 (1979).

Campbell, G. A., E. F. Elton, and E. G. Bobalek, *J. Polym. Sci.*, **14**, 1025 (1970).

Cannizzo, L. F., T. Hagiwara, and J. K. Stille, *Makromol. Chem. Suppl.*, **15**, 85 (1989).

Carothers, W. H., *Trans. Faraday Soc.*, **32**, 39 (1936).

Carothers, W. H. and J. W. Hill, *J. Am. Chem. Soc.*, **55**, 5043 (1933).

Carraher, C. E., Jr., J. E. Sheats, and C. U. Pittman, Jr., Eds., "Organometallic Polymers," Academic Press, New York, 1978.

Case, L. C., *J. Polym. Sci.*, **26**, 333 (1957); **29**, 455 (1958); **48**, 27 (1960).

Cassidy, P. E., "Thermally Stable Polymers," Marcel Dekker, New York, 1980.

Challa, G., *Makromol. Chem.*, **38**, 105 (1960).

Chao, T. H. and J. March, *J. Polym. Sci. Polym. Chem. Ed.*, **26**, 743 (1988).

Chegolya, A. S., V. V. Shevchenko, and G. D. Miklailov, *J. Polym. Sci. Polym. Chem. Ed.*, **17**, 889 (1979).

Chelnokova, G. I., S. R. Rafikov, and V. V. Korshak, *Dokl. Akad. Nauk SSSR*, **64**, 353 (1949).

Cheng, S. Z. D., J. J. Janimak, K. Sridhar, and F. W. Harris, *Polymer*, **30**, 494 (1989).

Chow, A. W., S. P. Bitler, P. E. Penwell, D. J. Osborne, and J. F. Wolfe, *Macromolecules*, **22**, 3514 (1989).

Clagett, D. C., "Engineering Plastics," pp. 94–131 in "Encyclopedia of Polymer Science and Engineering," Vol. 6, H. F. Mark, N. M. Bikales, C. G. Overberger, and G. Menges, Eds., Wiley-Interscience, New York, 1986.

Cotts, D. B. and G. C. Berry, *Macromolecules*, **14**, 930 (1981).

Critchley, J. P., G. J. Knight, and W. W. Wright, "Heat-Resistant Polymers," Plenum Press, New York, 1983.

Demus, D. and L. Richter, "Textures of Liquid Crystal Polymers," Verlag Chemie, Weinheim, 1978.

Dilling, W. L., *Chem. Rev.*, **83**, 1 (1983).

Doi, Y., M. Kunioka, Y. Nakamura, and K. Soga, *Macromolecules*, **21**, 2722 (1988).

Droske, J. P. and J. K. Stille, *Macromolecules*, **17**, 1 (1984).

Drumm, M. F. and J. R. LeBlanc, "The Reactions of Formaldehyde with Phenols, Melamine, Aniline and Urea," Chap. 5 in "Step-Growth Polymerizations," D. H. Solomon, Ed., Marcel Dekker, New York, 1972.

Durand, D. and C-M. Bruneau, *Eur. Polym. J.*, **13**, 463 (1977a); *Makromol. Chem.*, **178**, 3237 (1977b) and **179**, 147 (1978); *J. Polym. Sci. Polym. Phys. Ed.*, **17**, 273, 295 (1979a); *Makromol. Chem.*, **180**, 2947 (1979b); *Polymer*, **23**, 69 (1982a); *Makromol. Chem.*, **183**, 1021 (1982b).

Durvasula, V. R., F. A. Stuber, and D. Bhattacharjee, *J. Polym. Sci. Polym. Chem. Ed.*, **27**, 661 (1989).

Dusek, K., *Makromol. Chem. Suppl.*, **2**, 35 (1979a); *Polym. Bull.*, **1**, 523 (1979b); **17**, 481 (1987).

Dusek, K., M. Ilansky, and J. Somkovarsky, *Polym. Bull.*, **18**, 209 (1987).

East, G. C., V. Kalyvas, J. E. McIntyre, and A. H. Milburn, *Polymer*, **30**, 558 (1989).

Ebdon, J. R. and P. E. Heaton, *Polymer*, **18**, 971 (1977).

Elias, H.-G., "Macromolecules," Vol. 2, 2nd ed., Plenum Press, New York, 1984.

Elias, H.-G., *Makromol. Chem.*, **186**, 847 (1985).

Eliel, E. L., "Stereochemistry of Carbon Compounds," Chap. 7, McGraw-Hill, New York, 1962.

Ellerstein, S., "Polysulfides," pp. 186–196 in "Encyclopedia of Polymer Science and Engineering," Vol. 13, H. F. Mark, N. M. Bikales, C. G. Overberger, and G. Menges, Eds., Wiley-Interscience, New York, 1988.

Ellwood, P., *Chem. Eng.*, **74**, 98 (Nov. 20, 1967).

Evers, R. C., F. E. Arnold, and T. E. Helminiak, *Macromolecules*, **14**, 925 (1981).

Finkbeiner, H. L., A. S. Hay, and D. M. White, "Polymerizations by Oxidative Coupling," Chap. 15 in "Polymerization Processes," C. E. Schildknecht, Ed. (with I. Skeist), Wiley-Interscience, New York, 1977.

Finkelmann, H., *Angew. Chem. Int. Ed. Engl.*, **26**, 816 (1987).

Flory, P. J., *J. Am. Chem. Soc.*, **61**, 3334 (1939); **63**, 3083, 3091, 3096 (1941); *Chem. Rev.*, **39**, 137 (1946).

Flory, P. J., "Principles of Polymer Chemistry," Chaps. 3, 8, 9, Cornell University Press, Ithaca, N.Y., 1953.

Flory, P. J. and G. Ronca, *Mol. Cryst, Liq. Cryst.*, **54**, 289 (1979).

Fradet, A. and E. Marechal, *Adv. Polym. Sci.*, **43**, 53 (1982a); *J. Macromol. Sci. Chem.*, **A17**, 859 (1982b).

Freitag, D., U. Grigo, P. R. Muller, and W. Nouvertne, "Polycarbonates," pp. 648–718 in "Encyclopedia of Polymer Science and Engineering," Vol. 11, H. F. Mark, N. M. Bikales, C. G. Overberger, and G. Menges, Eds., Wiley-Interscience, New York, 1988.

Frommer, J. E. and R. R. Chance, "Electrically Conducting Polymers," pp. 462–507 in "Encyclopedia of Polymer Science and Engineering," Vol. 5, H. F. Mark, N. M. Bikales, C. G. Overberger, and G. Menges, Eds., Wiley-Interscience, New York, 1986.

Fukumoto, O., *J. Polym. Sci.*, **22**, 263 (1956).

Gandhi, K. S. and S. V. Babu, *AIChE J.*, **25**, 266 (1979); *Macromolecules*, **13**, 791 (1980).

Gauthier, S. and D. J. Worsfold, *Macromolecules*, **22**, 2213 (1989).

Gaymans, R. J., P. Schwering, and J. L. deHaan, *Polymer*, **30**, 974 (1989).

Gebben, B., M. H. V. Mulder, and C. A. Smolders, *Makromol. Chem. Macromol. Symp.*, **20/21**, 37 (1988).

Genies, E. M., G. Bidan, and A. F. Diaz, *J. Electroanal. Chem.*, **149**, 101 (1983).

Gimblett, F. G. R., "Inorganic Polymer Chemistry," Butterworths, London, 1963.

Glover, D. J., J. V. Duffy, and B. Hartmann, *J. Polym. Sci. Poly. Chem. Ed.*, **26**, 79 (1988).

Goel, R., S. K. Gupta, and A. Kumar, *Polymer*, **18**, 851 (1977).

Goodman, I, "Polyesters," pp. 1–75 in "Encyclopedia of Polymer Science and Engineering," Vol. 12, H. F. Mark, N. M. Bikales, C. G. Overberger, and G. Menges, Eds., Wiley-Interscience, New York, 1988.

Gopal, J. and M. Srinivasan, *Makromol. Chem.*, **187**, 1 (1986).

Gordon, M. and S. B. Ross-Murphy, *Pure Appl. Chem.*, **43**, 1 (1975).

Gross, R. A., C. DeMello, and R. W. Lenz, *Macromolecules*, **22**, 1106 (1989).

Gupta, S. K., A. Kumar, and A. Bhargova, *Polymer*, **20**, 305 (1979a); *Eur. Polym. J.*, **15**, 557 (1979b).

Hamann, S. D., D. H. Solomon, and J. D. Swift, *J. Macromol. Sci. Chem.*, **A2**, 153 (1968).

Hardman, B. and A. Torkelson, "Silicones," pp. 204–308 in "Encyclopedia of Polymer Science and Engineering," Vol. 12, H. F. Mark, N. M. Bikales, C. G. Overberger, and G. Menges, Eds., Wiley-Interscience, New York, 1989.

Harris, J. E. and R. N. Johnson, "Polysulfones," pp. 196–211 in "Encyclopedia of Polymer Science and Engineering," Vol. 13, H. F. Mark, N. M. Bikales, C. G. Overberger, and G. Menges, Eds., Wiley-Interscience, New York, 1989.

Harrys, F. W. and H. J. Spinelli, Eds., "Reactive Oligomers," ACS Symp. Ser., **282**, Washington, D.C., 1985.

Hasegawa, M., T. Katsumata, Y. Ito, K. Saigo, and Y. Iitaka, *Macromolecules*, **21**, 3134 (1988).

Hawker, C. J. and J. M. J. Frechet, *Macromolecules*, **23**, 4726 (1990).

Hawkins, R. T., *Macromolecules*, **9**, 189 (1976).

Hay, A. S. and D. D. Dana, *J. Polym. Sci. Polym. Chem. Ed.*, **27**, 873 (1989).

Hay, J. N., J. D. Boyle, S. F. Parker, and D. Wilson, *Polymer*, **30**, 1032 (1989).

Hedrick, J. L., H. R. Brown, D. C. Hofer, and R. D. Johnson, *Macromolecules*, **22**, 2048 (1989).

Hench, L. L. and J. K. West, *Chem. Rev.*, **90**, 33 (1990).

Hergenrother, P. M., "Acetylene-Terminated Prepolymers," pp. 61–86 in "Encyclopedia of Polymer Science and Engineering," Vol. 1, H. F. Mark, N. M. Bikales, C. G. Overberger, and G. Menges, Eds., Wiley-Interscience, New York, 1985.

Hergenrother, P. M., "Heat-Resistant Polymers," pp. 639–665 in "Encyclopedia of Polymer Science and Engineering," Vol. 7, H. F. Mark, N. M. Bikales, C. G. Overberger, and G. Menges, Eds., Wiley-Interscience, New York, 1987.

Hergenrother, P. M., "Polyquinoxalines," pp. 55–87 in "Encyclopedia of Polymer Science and Engineering," Vol. 13, H. F. Mark, N. M. Bikales, C. G. Overberger, and G. Menges, Eds., Wiley-Interscience, New York, 1988.

Hergenrother, P. M., B. J. Jensen, and S. J. Havens, *Polymer*, **29**, 358 (1988).

Higashi, F., A. Hoshio, Y. Yamada, and M. Ozawa, *J. Polym. Sci. Polym. Chem. Ed.*, **23**, 69 (1985).

Higashi, F. and A. Kobayashi, *J. Polym. Sci. Polym. Chem. Ed.*, **27**, 507 (1989).

Higashi, F., Y.-N. Lee, and A. Kobayashi, *J. Polym. Sci. Polym. Chem. Ed.*, **26**, 2077 (1988).

Hill, Jr., H. W. and D. G. Brady, "Poly(arylene sulfide)s," pp. 531–557 in Encyclopedia of Polymer Science and Engineering," Vol. 11, H. F. Mark, N. M. Bikales, C. G. Overberger, and G. Menges, Eds., Wiley-Interscience, New York, 1988.

Hodkin, J. H., *J. Polym. Sci. Polym. Chem. Ed.*, **14**, 409 (1976).

Horhold, H.-H. and M. Helbig, *Makromol. Chem. Macromol. Symp.*, **12**, 229 (1987).

Hounshell, D. A. and J. K. Smith, Jr., "Science and Corporate Strategy: DuPont R & D, 1902–1980," Cambridge, New York, 1988.

Howard, G. J., "The Molecular Weight Distribution of Condensation Polymers," Vol. I, pp. 185–231 in "Progress in High Polymers," J. C. Robb and F. W. Peaker, Eds., Iliffe Books, London, 1961.

Jackson, W. J., Jr., and J. C. Morris, *J. Polym. Sci. Polym. Chem. Ed.*, **26**, 835 (1988).

Jacobson, H. and W. H. Stockmayer, *J. Chem. Phys.*, **18**, 1600 (1950).

Jastrzebski, Z. D., "Nature and Properties of Engineering Materials," 2nd ed., Chap. 9, Wiley, New York, 1976.

Johannsen, I., J. B. Torrance, and A. Nazzal, *Macromolecules*, **22**, 566 (1989).

Johnston, J. C., M. A. B. Meador, and W. B. Alston, *J. Polym. Sci. Polym. Chem. Ed.*, **25**, 2175 (1987).

Jones, M. B. and P. Kovacic, "Oxidative Polymerization," pp. 670–683 in "Encyclopedia of Polymer Science and Engineering," Vol. 10, H. F. Mark, N. M. Bikales, C. G. Overberger, and G. Menges, Eds., Wiley-Interscience, New York, 1987.

Kaplan, M., *J. Chem. Eng. Data*, **6**, 272 (1961).

Katz, D. and I. G. Zwei, *J. Polym. Sci. Polym. Chem. Ed.*, **16**, 597 (1978).

Kelsey, D. R., L. M. Robeson, R. A. Clendinnig, and C. S. Blackwell, *Macromolecules*, **20**, 1205 (1987).

Kienle, R. H., P. A. Van der Meulen, and F. E. Petke, *J. Am. Chem. Soc.*, **61**, 2258 (1939).

Kienle, R. H. and F. E. Petke, *J. Am. Chem. Soc.*, **62**, 1053 (1940); **63**, 481 (1941).

Kilb, R. W., *J. Phys. Chem.*, **62**, 969 (1958).

Kim, B. S., D. S. Lee, and S. C. Kim, *Macromolecules*, **19**, 2589 (1986).

Kim, H. K., and K. Matyjaszewski, *J. Am. Chem. Soc.*, **110**, 3321 (1988).

Kimura, Y., M. Furukawa, H. Yamane, and T. Kitao, *Macromolecules*, **22**, 79 (1989).

Klein, L. C., *Ann. Rev. Mater. Sci.*, **15**, 227 (1985).

Klein, P. G., J. R. Ebdon, and D. J. Hourston, *Polymer*, **29**, 1079 (1988).

Klempner, D. and L. Berkowski, "Interpenetrating Polymer Networks," pp. 279–341 in "Encyclopedia of Polymer Science and Engineering," Vol. 8, H. F. Mark, N. M. Bikales, C. G. Overberger, and G. Menges, Eds., Wiley-Interscience, New York, 1987.

Koch, W. and W. Heitz, *Makromol. Chem.*, **184**, 779 (1983).

Kopf, P. W., "Phenolic Resins," pp. 45–95 in "Encyclopedia of Polymer Science and Engineering," Vol. 12, H. F. Mark, N. M. Bikales, C. G. Overberger, and G. Menges, Eds., Wiley-Interscience, New York, 1988.

Korshak, V. V., *Pure Appl. Chem.*, **12**, 101 (1966).

Korshak, V. V., A. L. Rusanov, and D. S. Tugushi, *Polymer*, **25**, 1539 (1984).

Korshak, V. V., S. V. Vinogradova, S. I. Kuchanov, and V. A. Vasnev, *J. Macromol. Sci. Rev. Macromol. Chem.*, **C14**, 27 (1976).

Kovacic, P. and M. B. Jones, *Chem. Rev.*, **87**, 357 (1987).

Krause, S. J., T. B. Haddock, D. L. Vezie, P. G. Lenhert, W.-F. Hwang, G. E. Price, T. E. Helminak, J. F. O'Brien, and W. W. Adams, *Polymer*, **29**, 1354 (1988).

Kricheldorf, H. R., H. Koziel, and E. Witek, *Makromol. Chem. Rapid. Commun.*, **9**, 217 (1988).

Krigbaum, W. R., R. Kotek, Y. Mihara, and J. Preston, *J. Poly. Sci. Polym. Chem. Ed.*, **23**, 1907 (1985).

Kronstadt, M., P. L. Dubin, and J. A. Tyburczy, *Macromolecules*, **11**, 37 (1978).

Kumar, A., A. K. Kulshreshtha, and S. K. Gupta, *Polymer*, **21**, 317 (1980).

Kumar, A., S. Wahal, S. Sastri, and S. K. Gupta, *Polymer*, **27**, 583 (1986).

Kurita, K., N. Hirakawa, T. Dobashi, and Y. Iwakura, *J. Polym. Sci. Polym. Chem. Ed.*, **17**, 2567 (1979).

Kwolek, S. L. and P. W. Morgan, *Macromolecules*, **10**, 1390 (1977).

Kwolek, S. L., P. W. Morgan, and J. R. Schaefgen, "Liquid Crystalline Polymers," pp. 1–61 in "Encyclopedia of Polymer Science and Engineering," Vol. 9, H. F. Mark, N. M. Bikales, C. G. Overberger, and G. Menges, Eds., Wiley-Interscience, New York, 1987.

Lanson, H. J., "Alkyd Resins," pp. 644–679 in "Encyclopedia of Polymer Science and Engineering," Vol. 1, H. F. Mark, N. M. Bikales, C. G. Overberger, and G. Menges, Eds., Wiley-Interscience, New York, 1986.

Leclerc, M., J. Guay, and L. H. Dao, *Macromolecules*, **22**, 649 (1989).

Lenk, R. S., *J. Polym. Sci. Macromol. Rev.*, **13**, 355 (1978).

Lenz, R. W., "Organic Chemistry of Synthetic High Polymers," Chaps. 4–8, Wiley-Interscience, New York, 1967.

Lenz, R. W., C.-C. Han, J. Stenger-Smith, and F. E. Karasz, *J. Polym. Sci. Polym. Chem. Ed.*, **26**, 3241 (1988).

Lenz, R. W., C.-C. Han, and M. Lux, *Polymer*, **30**, 1041 (1989).

Lenz, R. W., C. E. Handlovits, and H. A. Smith, *J. Polym. Sci.*, **58**, 351 (1962).

Leung, L. M. and J. T. Koberstein, *Macromolecules*, **19**, 707 (1986).

Lopez, L. C. and G. L. Wilkes, *J. Macromol. Sci. Rev. Macromol. Chem. Phys.*, **C29**, 83 (1989).

Lohse, F., *Makromol. Chem. Macromol. Symp.*, **7**, 1 (1987).

Lovering, E. G. and K. J. Laidler, *Can. J. Chem.*, **40**, 31 (1962).

Lowen, S. V. and J. D. Van Dyke, *J. Polym. Sci. Polym. Chem. Ed.*, **28**, 451 (1990).

Lowry, T. H. and K. S. Richardson, "Mechanism and Theory in Organic Chemistry," 3rd ed., Chaps. 3, 4, Harper and Row, New York, 1987.

Lyle, G. D., J. S. Senger, D. H. Chen, S. Kilis, S. D. Wu, D. K. Mohanty, and J. E. McGrath, *Polymer*, **30**, 978 (1989).

Lyman, D. J., *J. Polym. Sci.*, **45**, 49 (1960).

Macdiarmid, A. G., J. C. Chiang, A. F. Richter, and A. J. Epstein, *Synth, Met.*, **18**, 285 (1987).

Maciel, G. E., I-S. Chuang, and L. Gollob, *Macromolecules*, **17**, 1081 (1984).

Mackey, J. H., V. A. Pattison, and J. A. Pawlak, *J. Polym. Sci. Polym. Chem. Ed.*, **16**, 2849 (1978).

Macosko, C. W. and D. R. Miller, *Macromolecules*, **9**, 199 (1976).

Malhotra, H. C. and (Mrs.) Avinash, *Indian J. Chem.*, **13**, 1159 (1975); *J. Appl. Polym. Sci.*, **20**, 2461 (1976).

Manson, J. A. and L. H. Sperling, "Polymer Blends and Composites," Plenum Press, New York, 1976.

Marks, T. J., *Science*, **227**, 881 (1985).

Maruyama, Y., Y. Oishi, M. Kakimoto, and Y. Imai, *Macromolecules*, **21**, 2305 (1988).

Marvel, C. S., *J. Macromol. Sci. Rev. Macromol. Chem.*, **C13**, 219 (1975).

Matyjaszewski, K., Y. L. Chen, and H. K. Kim, "New Synthetic Routes to Polysilanes," Chap. 6 in "Inorganic and Organometallic Polymers," M. Zeldin, K. J. Wynne, and H. R. Allcock, Eds., ACS Symp. Ser., **360**, Washington, 1988.

May, R., "Polyetheretherketones," pp. 313–320 in "Encyclopedia of Polymer Science and Engineering," Vol. 12, H. F. Mark, N. M. Bikales, C. G. Overberger, and G. Menges, Eds., Wiley-Interscience, New York, 1988.

McAdams, L. V. and J. A. Gannon, "Epoxy Resins," pp. 322–382 in "Encyclopedia of Polymer Science and Engineering," Vol. 6, H. F. Mark, N. M. Bikales, C. G. Overberger, and G. Menges, Eds., Wiley-Interscience, New York, 1986.

McKean, D. R. and J. K. Stille, *Macromolecules*, **21**, 294 (1988).

Memeger, Jr., W., *Macromolecules*, **22**, 1577 (1989).

Menikheim, "Polymerization Procedures, Industrial," pp. 504–541 in "Encyclopedia of Polymer Science and Engineering," Vol. 12, H. F. Mark, N. M. Bikales, C. G. Overberger, and G. Menges, Eds., Wiley-Interscience, New York, 1988.

Miller, D. R. and C. W. Macosko, *Macromolecules*, **9**, 206 (1976); **11**, 656 (1978); **13**, 1063 (1980).

Miller, D. R., E. M. Valles, and C. W. Macosko, *Polym. Eng.Sci.*, **19**, 272 (1979).

Miller, R. D. and J. Michl, *Chem. Rev.*, **89**, 1359 (1989).

Milosevich, S. A., K. Saichek, L. Hinchey, W. B. England, and P. Kovacic, *J. Am. Chem. Soc.*, **105**, 1088 (1983).

Mittal, K. L., "Polyimides: Synthesis, Characterization, and Applications," Vols. 1, 2, Plenum, New York, 1984.

Mobley, D. P., *J. Polym. Sci. Poly. Chem. Ed.*, **22**, 3203 (1984).

Moore, J. A. and D. R. Robello, *Macromolecules*, **19**, 2669 (1986).

Moore, J. S. and S. I. Stupp, *Macromolecules*, **23**, 65 (1990).

Moore, J. W. and R. G. Pearson, "Kinetics and Mechanism," 3rd ed., Chaps. 2, 3, 8, 9, Wiley, New York, 1981.

Morgan, P. W., *J. Polym. Sci.*, **C4**, 1075 (1963).

Morgan, P. W., "Condensation Polymers: By Interfacial and Solutions Methods," Chap. IV, Wiley-Interscience, New York, 1965.

Morgan, P. W. and S. L. Kwolek, *J. Chem. Ed.*, **36**, Cover and p. 182 (1959a); *J. Polym. Sci.*, **40**, 299 (1959b).

Morgan, P. W., S. L. Kwolek, and T. C. Pletcher, *Macromolecules*, **20**, 729 (1987).

Mormann, W., N. Tiemann, and E. Turuskan, *Polymer*, **30**, 1127 (1989).

Muller, M. and W. Burchard, *Makromol. Chem.*, **179**, 1821 (1978).

Naarmann, H., *Makromol. Chem. Macromol. Symp.*, **8**, 1 (1987).

Nair, R. and D. J. Francis, *Polymer*, **24**, 626 (1983).

Nanda, V. S. and S. C. Jain, *J. Chem. Phys.*, **49**, 1318 (1968).

Neuse, E. W. and M. S. Loonat, *Macromolecules*, **16**, 128 (1983).

Newkome, G. R., G. R. Baker, M. J. Saunders, P. S. Russo, V. G. Gupta, Z. Yao, J. E. Miller, and K. Bouillion, *J. Chem. Soc. Chem. Commun.*, 752 (1986).

Nikonov, V. Z. and V. M. Savinov, "Polyamides," Chap. 15 in "Interfacial Synthesis," Vol. II, F. Millich and C. E. Carraher, Jr., Eds., Marcel Dekker, New York, 1977.

Nogami, M. and Y. Moriya, *J. Non-Cryst. Solids*, **37**, 191 (1980).

Nuyken, O., G. Maier, and K. Burger, *Makromol. Chem.*, **189**, 2245 (1988); **191**, 2455 (1990).

Ogata, N., "Synthesis of Condensation Polymers," pp. 1–50 in "Progress in Polymer Science Japan," Vol. 6, S. Onogi and K. Uno, Eds., Halsted Press (Wiley), New York, 1973.

Otton, J. and S. Ratton, *J. Polym. Sci. Polym. Chem. Ed.*, **26**, 2183 (1988).

Ozizmir, E. and G. Odian, *J. Polym. Sci. Polym. Chem. Ed.*, **18**, 1089 (1980a); **18**, 2281 (1980b); unpublished results, 1981, 1990.

Parker, E. E. and J. R. Peffer, "Unsaturated Polyester Resins," Chap. 3 in "Polymerization Processes," C. E. Schildknecht, Ed. (with I. Skeist), Wiley-Interscience, New York, 1977.

Paul, D. R., J. W. Barlow, and H. Keskkula, "Polymer Blends," pp. 399–461 in "Encyclopedia of Polymer Science and Engineering," Vol. 12, H. F. Mark, N. M. Bikales, C. G. Overberger, and G. Menges, Eds., Wiley-Interscience, New York, 1988.

Peebles, L. H., Jr., "Molecular Weight Distribution in Polymers," Chaps. 1, 4, 5, Wiley-Interscience, New York, 1971.

Percec, S. A., Natansohn, and M. Dima, *Angew. Makromol. Chem.*, **80**, 143 (1979).

Pinner, S. H., *J. Polym. Sci.*, **21**, 153 (1956).

Pittman, C. U., Jr., C. E. Carraher, Jr., and J. R. Reynolds, "Organometallic Polymers," pp. 541–594 in "Encyclopedia of Polymer Science and Engineering," Vol. 10, H. F. Mark, N. M. Bikales, C. G. Overberger, and G. Menges, Eds., Wiley-Interscience, New York, 1987.

Potember, R. S., R. C. Hoffman, H. S. Hu, J. E. Cocchiaro, C. A. Viands, R. A. Murphy, and T. O. Poehler, *Polymer*, **28**, 574 (1987).

Pyun, E., R. J. Mathisen, and C. S. P. Sung, *Macromolecules*, **22**, 1174 (1989).

Pretula, J. and S. Penczek, *Makromol. Chem.*, **191**, 671 (1990).

Rabinowitch, E., *Trans. Faraday Soc.*, **33**, 1225 (1937).

Rajan, C. R., V. M. Nadkarni, and S. Ponrathnam, *J Polym. Sci. Polym. Chem. Ed.*, **26**, 2581 (1988).

Rand, L., B. Thir, S. L. Reegen, and K. C. Frisch, *J. Appl. Polym. Sci.*, **9**, 1787 (1965).

Ravens, D. A. S. and J. M. Ward, *Trans. Faraday Soc.*, **57**, 150 (1961).

Ray, N. H., "Inorganic Polymers," Academic Press, New York, 1978.

Recca, A., and Stille, J. K., *Macromolecules*, **11**, 479 (1978).

Rehahn, M., A.-D. Schluter, and G. Wegner, *Makromol. Chem.*, **191**, 1991 (1990).

Reiss, G., G. Hurtrez, and P. Bahadur, "Block Copolymers," pp. 324–434 in "Encyclopedia of Polymer Science and Engineering," Vol. 2, H. F. Mark, N. M. Bikales, C. G. Overberger, and G. Menges, Eds., Wiley-Interscience, New York, 1985.

Renner, A. and A. Kramer, *J. Poly. Sci. Polym. Chem. Ed.*, **27**, 1301 (1989).

Ring, W., I. Mita, A. D. Jenkins, and N. M. Bikales, *Pure Appl. Chem.*, **57**, 1427 (1985).

Roncali, J., R. Garreau, D. Delkabouglise, F. Garnier, and M. Lemaire, *Makromol. Chem. Macromol. Symp.*, **24**, 77 (1989).

Ruan, J. Z. and M. H. Litt, *Macromolecules*, **21**, 876 (1988).

Russell, G. A., P. M. Henrichs, J. M. Hewitt, H. R. Grashof, and M. A. Sandhu, *Macromolecules*, **14**, 1764 (1981).

Samuelson, L. A. and M. A. Druy, *Macromolecules*, **19**, 824 (1986).

Saunders, J. H. and F. Dobinson, "The Kinetics of Polycondensation Reactions," Chap. 7 in "Comprehensive Chemical Kinetics," Vol. 15, C. H. Bamford and C. F. H. Tipper, Eds., American Elsevier, New York, 1976.

Sawada, H., "Thermodynamics of Polymerization," Chap. 6, Marcel Dekker, New York, 1976.

Schaefgen, J. R. and P. J. Flory, *J. Am. Chem. Soc.*, **70**, 2709 (1948).

Scola, D.A. and J. H. Vontell, *Chemtech*, 112 (1989).

Selley, J., "Polyesters, Unsaturated," pp. 256–290 in "Encyclopedia of Polymer Science and Engineering," Vol. 12, H. F. Mark, N. M. Bikales, C. G. Overberger, and G. Menges, Eds., Wiley-Interscience, New York, 1988.

Semlyen, J. A., "Cyclic Polymers," Chaps. 1, 6, Elsevier, London, 1986.

Shah, T. H., J. I. Bhatty, G. A. Gamien, and D. Dollimore, *Polymer*, **25**, 1333 (1984).

Sheets, J. E., C. E. Carraher, Jr., and C. U. Pittman, Jr., "Metal-containing Polymer Systems," Plenum Press, New York, 1985.

Snow, A. W. and J. R. Griffith, "Phthalocyanine Polymers," pp. 212–225 in "Encyclopedia of Polymer Science and Engineering," Vol. 11, H. F. Mark, N. M. Bikales, C. G. Overberger, and G. Menges, Eds., Wiley-Interscience, New York, 1988.

Solomon, D. H., *J. Macromol. Sci. Rev. Macromol. Chem.*, **Cl**(1), 179 (1967).

Solomon, D. H., "Polyesterification," Chap. 1 in "Step-Growth Polymerizations," D. H. Solomon, Ed., Marcel Dekker, New York, 1972.

Solomon, D. H. and J. J. Hopwood, *J. Appl. Polym. Sci.*, **10**, 1431 (1966).

Speckhard, T. A., J. A. Miller, and S. L. Cooper, *Macromolecules*, **19**, 1558 (1986).

Sperling, L. H., "Interpenetrating Polymer Networks and Related Materials," Plenum Press, New York, 1981.

Sperling, L. H., "Recent Developments in Interpenetrating Polymer Networks and Related Materials," Chap. 2 in "Multicomponent Polymer Materials," D. R. Paul and L. H. Sperling, Eds., Am. Chem. Soc. Adv. Ser., **211**, Washington, D. C., 1986.

Stafford, J. W., *J. Polym. Sci. Polym. Chem. Ed.*, **17**, 3375 (1979); **19**, 3219 (1981).

Stevenson, R. W., *J. Polym. Sci.*, **A-1**(7), 375 (1969).

Stille, J. K., *Macromolecules*, **14**, 870 (1981).

Stockmayer, W. H., *J. Chem. Phys.*, **11**, 45 (1943); *J. Polym. Soc.*, **9**, 69 (1952); **11**, 424 (1953).

Stockmayer, W. H., in "Advancing Fronts in Chemistry," Chap. 6, S. B. Twiss, Ed., Van Nostrand Reinhold, New York, 1945.

Stone, F. G. A. and W. A. G. Graham, Eds., "Inorganic Polymers," Academic Press, New York, 1962.

Sutherlin, D. M. and J. K. Stille, *Macromolecules*, **19**, 251 (1986)

Szabo-Rethy, E., *Eur. Polym. J.*, **7**, 1485 (1971).

Takeichi, T., and J. K. Stille, *Macromolecules*, **19**, 2093, 2103, 2108 (1986).

Tan, L.-S., and F. E. Arnold, *J. Polym. Chem. Polym. Chem. Ed.*, **26**, 1819 (1988).

Taylor, G. B., U. S. Patent 2,361,717 (DuPont), October 31, 1944.

Temin, S. C., *J. Macromol. Sci. Rev. Macromol. Chem. Phys.*, **C22**, 131 (1982–1983).

Thornton, P. A. and V. J. Colangelo, "Fundamentals of Engineering Materials," Chap. 15, Prentice-Hall, Englewood-Cliffs, N.J., 1985.

Tomalia, D. A., H. Baker, J. Dewald, M. Hall, G. Kallos, S. Martin, J. Roeck, J. Ryder, and P. Smith, *Polym. J.*, **17**, 117 (1985); *Macromolecules*, **19**, 2466 (1986).

Tomita, B. and Y. Hirose, *J. Polym. Sci. Polym. Chem. Ed.*, **14**, 387 (1976).

Tomita, B. and H. Ono, *J. Polym. Sci. Polym. Chem. Ed.*, **17**, 3205 (1979).

Ueberreiter, K. and M. Engel, *Makromol. Chem.*, **178**, 2257 (1977).

Ueberreiter, K. and W. Hager, *Makromol. Chem.*, **180**, 1697 (1979).

Ueda, M., M. Sato, and A. Mochizuki, *Macromolecules*, **18**, 2723 (1985).

Ueda, M. and H. Sugita, *J. Polym. Sci. Polym. Chem. Ed.*, **26**, 159 (1988).

Updegraff, I. H., "Amino Resins," pp. 752–789 in "Encyclopedia of Polymer Science and Technology," Vol. 1, H. F. Mark, N. M. Bikales, C. G. Overberger, and G. Menges, Eds., Wiley-Interscience, New York, 1985.

Upshaw, T. A., and J. K. Stille, *Macromolecules*, **21**, 2010 (1988).

Valles, E. M. and C. W. Macosko, *Macromolecules*, **12**, 521, 673 (1979).

Vansco-Szmercsanyi, I. and E. Makay-Bodi, *Eur. Polym. J.*, **5**, 145, 155 (1969).

Wei, Y., R. Hariharan, and S. A. Patel, *Macromolecules*, **23**, 758 (1990).

Wei, Y., X. Tang, Y. Sun, and W. Focke, *J. Polym. Sci. Polym. Chem. Ed.*, **27**, 2385 (1989).

Werstler, D. D., *Polymer*, **27**, 750 (1986).

West, R., *J. Organomet. Chem.*, **300**, 327 (1986).

West, R., and J. Maxka, "Polysilane High Polymers: An Overview," Chap. 2 in "Inorganic and Organometallic Polymers," M. Zeldin, K. J. Wynne, and H. R. Allcock, Eds., ACS Symp. Ser., **360**, Washington, 1988.

White, J. E. and M. D. Scaia, *Polymer*, **25**, 850 (1984).

Wilson, L. R. and D. A. Tomalia, *Polym. Prep.*, **30**(1), 115 (1989).

Wolfe, J. F., "Polybenzothiazoles and Polybenzoxazoles," pp. 601–635 in "Encyclopedia of Polymer Science and Engineering," Vol. 11, H. F. Mark, N. M. Bikales, C. G. Overberger, and G. Menges, Eds., Wiley-Interscience, New York, 1988.

Wolfe, J. F., B. H. Loo, and F. E. Arnold, *Macromolecules*, **14**, 915 (1981).

Yajima, S., Y. Hasegawa, J. Hayashi, and M. Iimura, *J. Mater. Sci.*, **13**, 2569 (1978).

Yamanis, J. and M. Adelman, *J. Polym. Sci. Polym. Chem. Ed.*, **14**, 1945, 1961 (1976).

Yu, T.-y., S.-k. Fu, C.-y., Jiang, W.-z. Cheng, and R.-y. Xu, *Polymer*, **27**, 1111 (1986).

Yuan, S.-F., A. J. Masters, C. V. Nicholas, and C. Booth, *Makromol. Chem.*, **189**, 823 (1988).

Zahir, S., M. A. Chaudari, and J. King, *Macromol. Chem. Macromol. Symp.*, **25**, 141 (1989).

Zavitsas, A. A., *J. Polym. Sci.*, **A-1**(6), 2533 (1968).

Zavitsas, A. A., R. D. Beaulieu, and J. R. LeBlanc, *J. Polym. Sci.*, **A-1**(6), 2541 (1968).

Zimmerman, J., "Polyamides," pp. 315–381 in "Encyclopedia of Polymer Science and Engineering," Vol. 11, H. F. Mark, N. M. Bikales, C. G. Overberger, and G. Menges, Eds., Wiley-Interscience, New York, 1988.

PROBLEMS

2-1 Derive an expression for the rate of polymerization of stoichiometric amounts of adipic acid and hexamethylene diamine. Indicate the assumptions inherent in the derivation. Derive an expression for the rate of polymerization of non-stoichiometric amounts of the two reactants.

2-2 A 21.3-g sample of poly(hexamethylene adipamide) is found to contain 2.50×10^{-3} moles of carboxyl groups by both titration with base and infrared spectroscopy. From these data the polymer is calculated to have a number-average molecular weight of 8520. What assumption is made in the calculation? How can one experimentally obtain the correct value of $\overline{M}_n$?

2-3 Describe and draw the structure of the polyester obtained in each of the following polymerizations:

a. $HO_2C-R-CO_2H + HO-R'-OH$
b. $HO_2C-R-CO_2H + HO-R''-OH$
$$\qquad\qquad\qquad\qquad\qquad\quad |$$
$$\qquad\qquad\qquad\qquad\qquad OH$$
c. $HO_2C-R-CO_2H + HO-R''-OH + HO-R'-OH$
$$\qquad\qquad\qquad\qquad\qquad\quad |$$
$$\qquad\qquad\qquad\qquad\qquad OH$$

Will the structure of the polymer produced in each case depend on the relative amounts of the reactants? If so, describe the differences.

2-4 Describe and draw the structure of the polyester obtained in each of the following polymerizations:

a. $HO-R-CO_2H$
b. $HO-R-CO_2H + HO-R'-OH$
c. $HO-R-CO_2H + HO-R''-OH$
$$\qquad\qquad\qquad\qquad\qquad\quad |$$
$$\qquad\qquad\qquad\qquad\qquad OH$$
d. $HO-R-CO_2H + HO-R''-OH + HO-R''-OH$
$$\qquad\qquad\qquad\qquad\qquad\qquad\qquad\qquad\quad |$$
$$\qquad\qquad\qquad\qquad\qquad\qquad\qquad\qquad OH$$

Will the structure of the polymer produced in each of cases b, c, and d depend on the relative amounts of the reactants? If so, describe the differences.

2-5 Compare the molecular weight distributions that are expected for the polymerizations in Questions 2-3a, 2-4a, 2-4b, and 2-4c.

2-6 Discuss the possibility of cyclization in the polymerization of
a. $H_2N-(CH_2)_m-CO_2H$
b. $HO-(CH_2)_2-OH + HO_2C-(CH_2)_m-CO_2H$
for the cases where m has values from 2 to 10? At what stage(s) in the reaction is cyclization possible? What factors determine whether cyclization or linear polymerization is the predominant reaction?

2-7 Show that the time required to go from $p = 0.98$ to $p = 0.99$ is very close to the time to reach $p = 0.98$ from the start of polymerization for the external acid-catalyzed polymerization of an equimolar mixture of a diol and diacid.

2-8   The polymerization between equimolar amounts of a diol and diacid proceeds with an equilibrium constant of 200. What will be the expected degree of polymerization and extent of reaction if the reaction is carried out in a closed system without removal of the by-product water? To what level must $[H_2O]$ be lowered in order to obtain a degree of polymerization of 200 if the initial concentration of carboxyl groups is $2\ M$?

2-9   A heat-resistant polymer *Nomex* has a number-average molecular weight of 24,116. Hydrolysis of the polymer yields 39.31% by weight *p*-aminoaniline, 59.81% terephthalic acid, and 0.88% benzoic acid. Write the formula for this polymer. Calculate the degree of polymerization and the extent of reaction. Calculate the effect on the degree of polymerization if the polymerization had been carried out with twice the amount of benzoic acid.

2-10   Calculate the number-average degree of polymerization of an equimolar mixture of adipic acid and hexamethylene diamine for extents of reaction 0.500, 0.800, 0.900, 0.950, 0.970, 0.990, 0.995.

2-11   Calculate the feed ratio of adipic acid and hexamethylene diamine that should be employed to obtain a polyamide of approximately 15,000 molecular weight at 99.5% conversion. What is the identity of the end groups of this product? Do the same calculation for a 19,000-molecular-weight polymer.

2-12   What is the proportion of benzoic acid that should be used with an equimolar mixture of adipic acid and hexamethylene diamine to produce a polymer of 10,000 molecular weight at 99.5% conversion? Do the same calculation for 19,000 and 28,000 molecular weight products.

2-13   Calculate the extent of reaction at which gelation occurs for the following mixtures:
   a. Phthalic anhydride and glycerol in stoichiometric amounts
   b. Phthalic anhydride and glycerol in the molar ratio 1.500:0.980
   c. Phthalic anhydride, glycerol, and ethylene glycol in the molar ratio 1.500:0.990:0.002
   d. Phthalic anhydride, glycerol, and ethylene glycol in the molar ratio 1.500:0.500:0.700

Compare the gel points calculated from the Carothers equation (and its modifications) with those using the statistical approach. Describe the effect of unequal functional groups reactivity (e.g., for the hydroxyl groups in glycerol) on the extent of reaction at the gel point.

2-14   Show by equations the polymerization of melamine and formaldehyde to form a crosslinked structure.

2-15   Describe by means of equations how random and block copolymers having the following compositions could be synthesized:

   a.

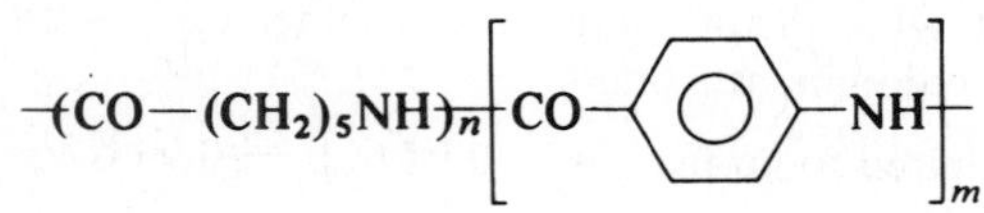

b. 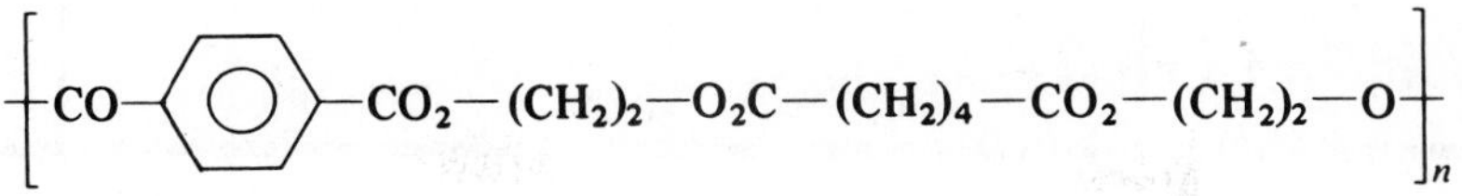 $\left[CO\text{-}\left\langle\bigcirc\right\rangle\text{-}CO_2\text{-}(CH_2)_2\text{-}O_2C\text{-}(CH_2)_4\text{-}CO_2\text{-}(CH_2)_2\text{-}O \right]_n$

2-16 How would you synthesize a block copolymer having segments of the following structures?

$+CH_2CH_2CH_2\text{-}O+_m$ and $\left[O\text{-}CH_2CH_2\text{-}OCONH\text{-}\underset{\overset{\displaystyle CH_3}{}}{\left\langle\bigcirc\right\rangle}\text{-}NHCO \right]_p$

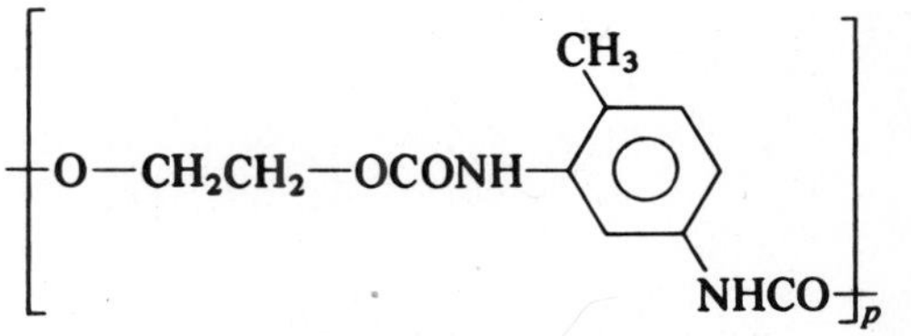

2-17 Distinguish between spiro, ladder, and semiladder polymers. Give examples of each.

CHAPTER 3

RADICAL CHAIN POLYMERIZATION

In the previous chapter, the synthesis of polymers by step polymerization was considered. Polymerization of unsaturated monomers by chain polymerization will be discussed in this and several of the subsequent chapters. Chain polymerization is initiated by a reactive species R^* produced from some compound I termed an *initiator*.

$$I \rightarrow R^* \tag{3-1}$$

The reactive species, which may be a free radical, cation, or anion, adds to a monomer molecule by opening the π-bond to form a new radical, cation, or anion center, as the case may be. The process is repeated as many more monomer molecules are successively added to continuously propagate the reactive center

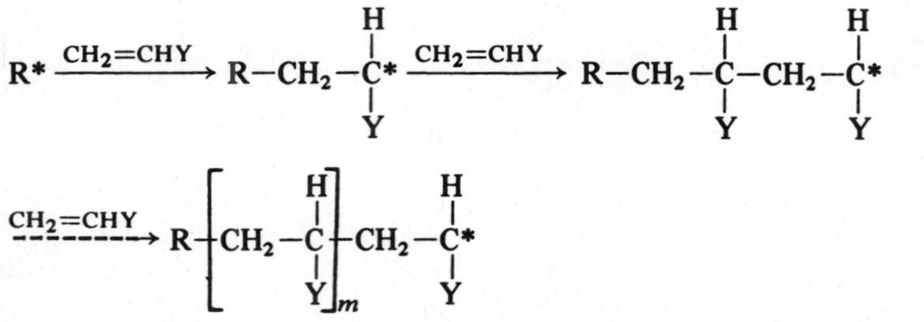

$$R^* \xrightarrow{CH_2=CHY} R-CH_2-\underset{\underset{Y}{|}}{\overset{\overset{H}{|}}{C}}* \xrightarrow{CH_2=CHY} R-CH_2-\underset{\underset{Y}{|}}{\overset{\overset{H}{|}}{C}}-CH_2-\underset{\underset{Y}{|}}{\overset{\overset{H}{|}}{C}}*$$

$$\xrightarrow{CH_2=CHY} R \left[CH_2-\underset{\underset{Y}{|}}{\overset{\overset{H}{|}}{C}} \right]_m CH_2-\underset{\underset{Y}{|}}{\overset{\overset{H}{|}}{C}}* \tag{3-2}$$

Polymer growth is terminated at some point by destruction of the reactive center by an appropriate reaction depending on the type of reactive center and the particular reaction conditions.

3-1 NATURE OF RADICAL CHAIN POLYMERIZATION

3-1a Comparison of Chain and Step Polymerization

Chain polymerization proceeds by a distinctly different mechanism from step polymerization. The most significant difference is that high-molecular-weight polymer is formed immediately in a chain polymerization. A radical, anionic, or cationic reactive center, once produced, adds many monomer units in a chain reaction and grows rapidly to a large size. The monomer concentration decreases throughout the course of the reaction as the number of high-polymer molecules increases. At any instant the reaction mixture contains only monomer, high polymer, and the growing chains. The molecular weight of the polymer is relatively unchanged during the polymerization, although the overall percent conversion of monomer to polymer increases with reaction time.

The situation is quite different for a step polymerization. Whereas only monomer and the propagating species can react with each other in chain polymerization, any two molecular species present can react in step polymerization. Monomer disappears much faster in step polymerization as one proceeds slowly to produce dimer, trimer, tetramer, and so on. The molecular weight increases throughout the course of the reaction and high-molecular-weight polymer is not obtained until the end of the polymerization. Long reaction times are necessary for both high percent conversion and high molecular weights.

3-1b Radical versus Ionic Chain Polymerizations

3-1b-1 General Considerations of Polymerizability

Whether a particular monomer can be converted to polymer depends on both thermodynamic and kinetic considerations. The polymerization will be impossible under any and all reaction conditions if it does not pass the test of thermodynamic feasibility. Polymerization is possible only if the free energy difference ΔG between monomer and polymer is negative (Sec. 3-9b). A negative ΔG does not, however, mean that polymerization will be observed under a particular set of reaction conditions (type of initiation, temperature, etc.). The ability to carry out a thermodynamically feasible polymerization depends on its kinetic feasibility—on whether the process proceeds at a reasonable rate under a proposed set of reaction conditions. Thus, whereas the polymerization of a wide variety of unsaturated monomers is thermodynamically feasible, very specific reaction conditions are often required to achieve kinetic feasibility in order to accomplish a particular polymerization.

Although radical, cationic, and anionic initiators are used in chain polymerizations, they cannot be used indiscriminately, since all three types of initiation do not work for all monomers. Monomers show varying degrees of selectivity with regard to the type of reactive center that will cause their polymerization. Most monomers will undergo polymerization with a radical initiator, although at varying rates. However, monomers show high selectivity toward ionic initiators. Some monomers may not polymerize with cationic initiators, while others may not polymerize with anionic initiators. The variety of behaviors can be seen in Table 3-1. The types of initiation which bring about the polymerization of various monomers to high-molecular-weight polymer are indicated. Thus, although the polymerization of all the monomers in Table 3-1 is thermodynamically feasible, kinetic feasibility is achieved in many cases only with a specific type of initiation.

The carbon–carbon double bond in vinyl monomers and the carbon–oxygen double bond in aldehydes and ketones are the two main types of linkages that undergo chain polymerization. The polymerization of the carbon–carbon double bond is by far the most important of the two types of monomers. The carbonyl group is not prone to polymerization by radical initiators because of its polarized nature.

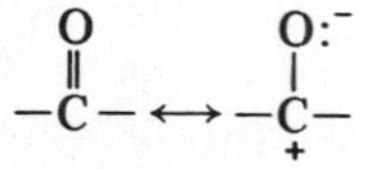

$$
\begin{matrix} O \\ \parallel \\ -C- \end{matrix} \longleftrightarrow \begin{matrix} O{:}^{-} \\ \mid \\ -\overset{+}{C}- \end{matrix}
\tag{3-3}
$$

Aldehydes and ketones are polymerized by both anionic and cationic initiators (Chap. 5).

3-1b-2 Effects of Substituents

Unlike the carbonyl linkage, the carbon–carbon double bond undergoes polymerization by both radical and ionic initiators. The difference arises because the π-bond of a vinyl monomer can respond appropriately to the initiator species by either homolytic or heterolytic bond breakage

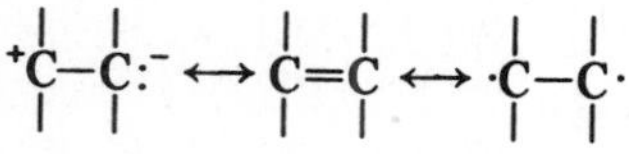

$$
\overset{+}{C}-\overset{\mid}{\underset{\mid}{C}}{:}^{-} \longleftrightarrow \overset{\mid}{\underset{\mid}{C}}{=}\overset{\mid}{\underset{\mid}{C}} \longleftrightarrow \cdot\overset{\mid}{\underset{\mid}{C}}-\overset{\mid}{\underset{\mid}{C}}\cdot
\tag{3-4}
$$

A wide range of carbon–carbon double bonds undergo chain polymerization. Table 3-1 shows monomers with alkyl, alkenyl, aryl, halogen, alkoxy, ester, amide, nitrile, and heterocyclic substituents on the alkene double bond.

TABLE 3-1 Types of Chain Polymerization Undergone by Various Unsaturated Monomers

Monomers	Type of Initiation		
	Radical	Cationic	Anionic
Ethylene	+	−	+
1-Alkyl olefins (α-olefins)	−	+	−
1,1-Dialkyl olefins	−	+	−
1,3-Dienes	+	+	+
Styrene, α-methyl styrene	+	+	+
Halogenated olefins	+	−	−
Vinyl esters ($CH_2{=}CHOCOR$)	+	−	−
Acrylates, methacrylates	+	−	+
Acrylonitrile, methacrylonitrile	+	−	+
Acrylamide, methacrylamide	+	−	+
Vinyl ethers	−	+	−
N-Vinyl carbazole	+	+	−
N-Vinyl pyrrolidone	+	+	−
Aldehydes, ketones	−	+	+

Whether a vinyl monomer polymerizes by radical, anionic, or cationic initiators depends on the inductive and resonance characteristics of the substituent(s) present. The effect of the substituent manifests itself by its alteration of the electron-cloud density on the double bond and its ability to stabilize the possible radical, anion, or cation formed. Electron-donating substituents such as alkoxy, alkyl, alkenyl, and phenyl increase the electron density on the carbon–carbon double bond

$$CH_2^{\delta-}{=}CH{\leftarrow}\overset{\delta+}{Y}$$

and facilitate its bonding to a cationic species. Further, these substituents stabilize the cationic propagating species by resonance, for example, in the polymerization of vinyl ethers,

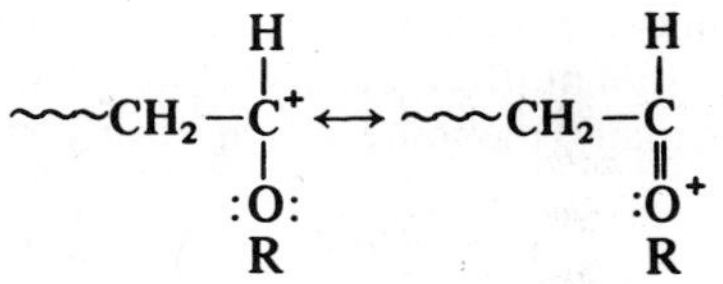

(3-5a)

The alkoxy substituent allows a delocalization of the positive charge. If the substituent were not present (e.g., in ethylene), the positive charge would be localized on the single α-carbon atom. The presence of the alkoxy group leads to stabilization of the carbocation by delocalization of the positive charge over two atoms—the carbon and the oxygen. Similar delocalization effects occur with phenyl, vinyl, and alkyl substituents, for example, for styrene polymerization,

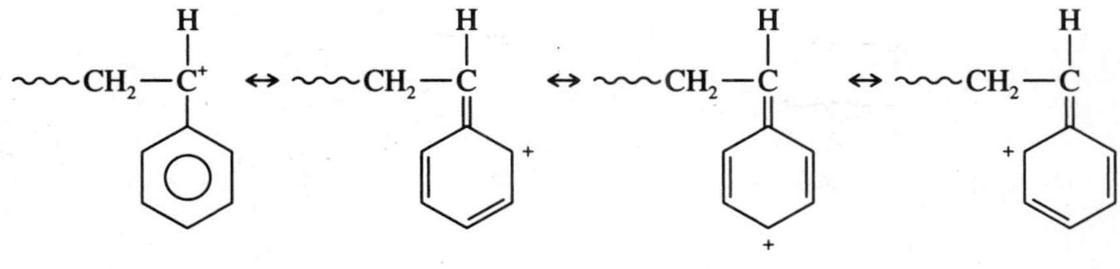

(3-5b)

Thus monomers such as isobutylene, styrene, methyl vinyl ether, and isoprene undergo polymerization by cationic initiators. The effect of alkyl groups in facilitating cationic polymerization is weak, and it is only the 1,1-dialkyl alkenes that undergo cationic polymerization.

Electron-withdrawing substituents such as cyano and carbonyl (aldehyde, ketone, acid, or ester) facilitate the attack of an anionic species by decreasing the electron density on the double bond

$$CH_2^{\delta+}{=}CH{\rightarrow}\overset{\delta-}{Y}$$

They stabilize the propagating anionic species by resonance, for example, for acrylonitrile polymerization

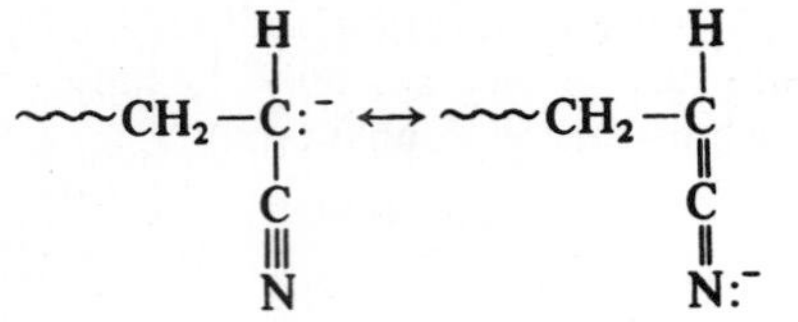

$$\text{~~CH}_2-\overset{\overset{\displaystyle H}{|}}{\underset{\underset{\displaystyle N}{\overset{\displaystyle |||}{\underset{}{C}}}}{C}}\text{:}^- \longleftrightarrow \text{~~CH}_2-\overset{\overset{\displaystyle H}{|}}{\underset{\underset{\displaystyle N\text{:}^-}{\overset{\displaystyle ||}{\underset{}{C}}}}{C}} \tag{3-6}$$

Stabilization of the propagating carbanion occurs by delocalization of the negative charge over the α-carbon and the nitrogen of the nitrile group. Alkenyl and phenyl substituents, although electron-withdrawing inductively, can resonance stabilize the anionic propagating species in the same manner as a cyano group. Monomers such as styrene and 1,3-butadiene can, therefore, undergo anionic as well as cationic polymerization. Halogens withdraw electrons inductively and push electrons by resonance, but both effects are relatively weak and neither anionic nor cationic polymerization is appreciably facilitated for halogenated monomers such as vinyl chloride.

Contrary to the high selectivity shown in cationic and anionic polymerization, radical initiators bring about the polymerization of almost any carbon–carbon double bond. Radical species are neutral and do not have stringent requirements for attacking the π-bond or for the stabilization of the propagating radical species. Resonance stabilization of the propagating radical occurs with almost all substituents, for example,

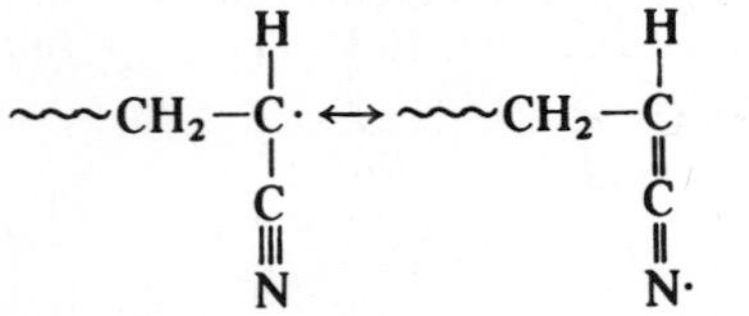

$$\text{~~CH}_2-\overset{\overset{\displaystyle H}{|}}{\underset{\underset{\displaystyle N}{\overset{\displaystyle |||}{\underset{}{C}}}}{C}}\cdot \longleftrightarrow \text{~~CH}_2-\overset{\overset{\displaystyle H}{|}}{\underset{\underset{\displaystyle N\cdot}{\overset{\displaystyle ||}{\underset{}{C}}}}{C}} \tag{3-7a}$$

$$\text{~~CH}_2-\overset{\displaystyle H}{\underset{\displaystyle}{C}}\cdot \longleftrightarrow \text{~~CH}_2-\overset{\displaystyle H}{\underset{\displaystyle}{C}} \longleftrightarrow \text{~~CH}_2-\overset{\displaystyle H}{\underset{\displaystyle}{C}} \longleftrightarrow \text{~~CH}_2-\overset{\displaystyle H}{\underset{\displaystyle}{C}} \tag{3-7b}$$

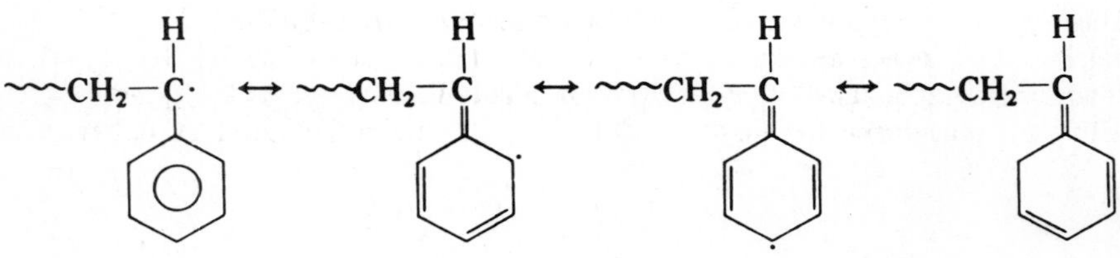

$$\text{~~CH}_2-\overset{\overset{\displaystyle H}{|}}{\underset{\underset{\displaystyle :\ddot{C}l:}{}}{C}}\cdot \longleftrightarrow \text{~~CH}_2-\overset{\overset{\displaystyle H}{|}}{\underset{\underset{\displaystyle :\ddot{C}l\cdot}{\overset{\displaystyle ||}{}}}{C}} \tag{3-7c}$$

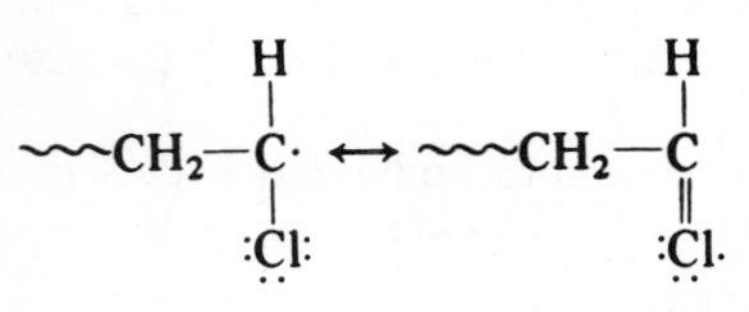

Thus almost all substituents are able to stabilize the propagating radical by delocalization of the radical over two or more atoms. The remainder of this chapter will be concerned with the detailed characteristics of radical chain polymerization. Ionic chain polymerizations will be considered in Chap. 5.

3-2 STRUCTURAL ARRANGEMENT OF MONOMER UNITS

3-2a Possible Modes of Propagation

There are two possible points of attachment on monosubstituted ($X = H$) or 1,1-disubstituted monomers for a propagating radical—either on carbon 1,

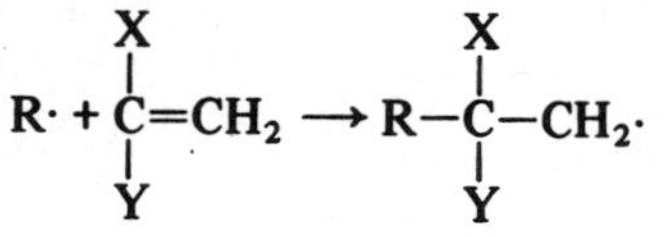

$$R\cdot + \underset{\underset{Y}{|}}{\overset{\overset{X}{|}}{C}}{=}CH_2 \rightarrow R{-}\underset{\underset{Y}{|}}{\overset{\overset{X}{|}}{C}}{-}CH_2\cdot \tag{3-8}$$

$$\mathbf{I}$$

or carbon 2,

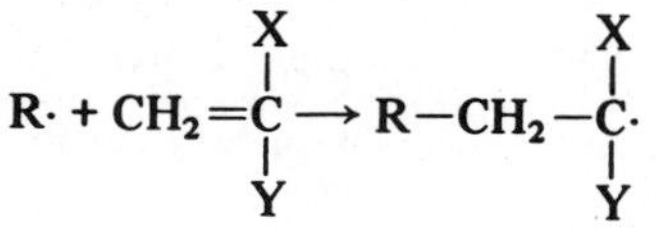

$$R\cdot + CH_2{=}\underset{\underset{Y}{|}}{\overset{\overset{X}{|}}{C}} \rightarrow R{-}CH_2{-}\underset{\underset{Y}{|}}{\overset{\overset{X}{|}}{C}}\cdot \tag{3-9}$$

$$\mathbf{II}$$

If each successive addition of monomer molecules to the propagating radical occurs in the same manner as Eq. 3-9 or Eq. 3-8, the final polymer product will have an arrangement of monomer units in which the substituents are on alternate carbon atoms:

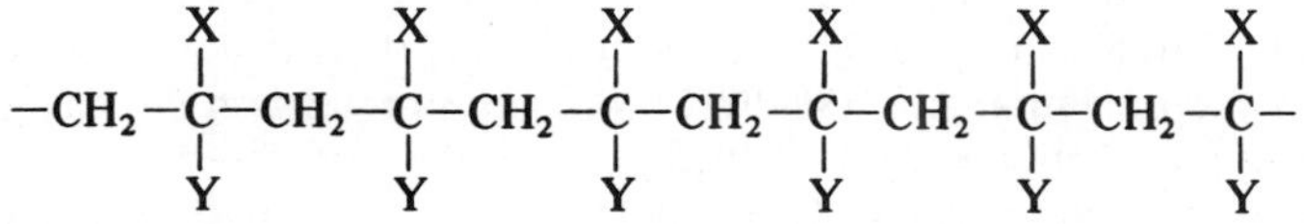

$$-CH_2{-}\underset{\underset{Y}{|}}{\overset{\overset{X}{|}}{C}}{-}CH_2{-}\underset{\underset{Y}{|}}{\overset{\overset{X}{|}}{C}}{-}CH_2{-}\underset{\underset{Y}{|}}{\overset{\overset{X}{|}}{C}}{-}CH_2{-}\underset{\underset{Y}{|}}{\overset{\overset{X}{|}}{C}}{-}CH_2{-}\underset{\underset{Y}{|}}{\overset{\overset{X}{|}}{C}}{-}CH_2{-}\underset{\underset{Y}{|}}{\overset{\overset{X}{|}}{C}}{-}$$

$$\mathbf{III}$$

This type of arrangement (III) is usually referred to as a *head-to-tail* or *H-T* or *1,3-placement* of monomer units. An inversion of this mode of addition by the polymer chain propagating alternately via Eqs. 3-9 and 3-8 would lead to a polymer structure with a *1,2*-placement of substituents at one or more places in the final polymer chain

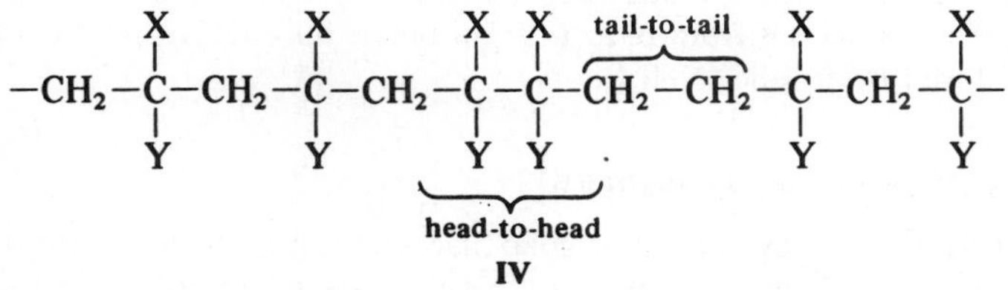

$$-CH_2{-}\underset{\underset{Y}{|}}{\overset{\overset{X}{|}}{C}}{-}CH_2{-}\underset{\underset{Y}{|}}{\overset{\overset{X}{|}}{C}}{-}CH_2{-}\underset{\underset{Y}{|}}{\overset{\overset{X}{|}}{C}}{-}\underset{\underset{Y}{|}}{\overset{\overset{X}{|}}{C}}{-}CH_2{-}CH_2{-}\underset{\underset{Y}{|}}{\overset{\overset{X}{|}}{C}}{-}CH_2{-}\underset{\underset{Y}{|}}{\overset{\overset{X}{|}}{C}}{-}$$

tail-to-tail / head-to-head

$$\mathbf{IV}$$

1,2-Placement is usually referred to as *head-to-head* or *H–H* placement.

The head-to-tail placement would be expected to be overwhelmingly predominant, since successive propagations by Eq. 3-9 are favored on both steric and resonance grounds. The propagating radical (radical II) formed by attachment of a radical at carbon 2 is the more stable one. Radical II can be stabilized by the resonance effects of the X and Y substituents. The substituents cannot stabilize radical I, since they are not attached to the carbon bearing the unpaired electron. Further, the approach (and subsequent attachment) of a propagating radical at the unsubstituted carbon (carbon 2) of a monomer molecule is much less sterically hindered compared to the approach at the substituted carbon (carbon 1). A propagation proceeding with predominantly H-T placement is a *regioselective* process, that is, one orientation (H–T) is favored over another (H–H). The term *isoregic* has been used to indicate a polymer structure with exclusive head-to-tail placements. The terms *syndioregic* and *aregic* would be used for polymer structures with alternating and random arrangements, respectively, of H–T and H–H placements.

3-2b Experimental Evidence

These theoretical predictions have been experimentally verified for a number of polymers. The presence of no more than 1–2% head-to-head placement in various polymers of vinyl esters such as poly(vinyl acetate) has been determined by hydrolysis of the polymer to poly(vinyl alcohol) and the periodate oxidation of the 1,2-glycol units [Hayashi and Ostu, 1969].

$$\sim\sim\sim CH_2-\underset{\underset{OH}{|}}{CH}-\underset{\underset{OH}{|}}{CH}-CH_2\sim\sim\sim \xrightarrow{IO_4^-} \sim\sim\sim CH_2-\underset{\overset{\|}{O}}{CH} + \underset{\overset{\|}{O}}{CH}-CH_2\sim\sim\sim \quad (3\text{-}10)$$

Although appropriate chemical reactions do not exist to detect the occurrence of head-to-head propagation in all but a few polymers, the use of high-resolution nuclear magnetic resonance spectroscopy (including ^1H, ^{13}C, and F^{19}) has greatly increased our knowledge in this area [Bovey, 1972; Bovey et al., 1977]. It is now clear that H–T propagation is the predominant (>98–99%) mode of propagation in chain polymerization. The only exceptions occur when the substituents on the double bond are small (and do not offer appreciable steric hindrance to the approaching radical) and do not have a large resonance-stabilizing effect, specifically when fluorine is the substituent. Thus the extents of head-to-head placements in poly(vinyl fluoride), poly(vinylidene fluoride), polytrifluoroethylene, and polychlorotrifluoroethylene are about 10, 5, 12, and 2%, respectively [Cais and Kometani, 1984, 1988].

The effect of increasing the polymerization temperature is an increase in the extent of H–H placement, but the effect is small. Thus the H–H content in poly(vinyl acetate) increases from 1 to 2% when the temperature is increased from 40 to 100°C [Moritani and Iwasaki, 1978]; the increase is from 10 to 14% for polytrifluoroethylene when the temperature is increased from −80 to 80°C.

3-2c Synthesis of Head-to-Head Polymers

Some polymers consisting entirely of head-to-head placements have been deliberately synthesized to determine if significant property differences exist compared to the head-to-tail polymers. The synthetic approach involves an appropriate choice of monomer for the particular H–H polymer. For example, H–H poly(vinyl chloride) is obtained by chlorination of 1,4-poly-1,3-butadiene,

$$\underset{}{+CH_2-CH{=}CH-CH_2+_n} \xrightarrow{Cl_2} +CH_2-\underset{\underset{Cl}{|}}{CH}-\underset{\underset{Cl}{|}}{CH}-CH_2+_n \quad (3\text{-}11)$$

and H–H polystyrene by hydrogenation of 1,4-poly-2,3-diphenylbutadiene,

$$\text{---}CH_2-\underset{\phi}{\underset{|}{C}}=\underset{\phi}{\underset{|}{C}}-CH_2\text{---}_n \xrightarrow{H_2} \text{---}CH_2-\underset{\phi}{\underset{|}{CH}}-\underset{\phi}{\underset{|}{CH}}-CH_2\text{---}_n \tag{3-12}$$

[Inoue et al., 1977; Kawaguchi et al., 1985].

3-3 RATE OF RADICAL CHAIN POLYMERIZATION

3-3a Sequence of Events

Radical chain polymerization is a chain reaction consisting of a sequence of three steps—*initiation*, *propagation*, and *termination*. The initiation step is considered to involve two reactions. The first is the production of free radicals by any one of a number of reactions. The usual case is the homolytic dissociation of an initiator or catalyst species I to yield a pair of radicals $R\cdot$

$$I \xrightarrow{k_d} 2R\cdot \tag{3-13}$$

where k_d is the rate constant for the catalyst dissociation. The second part of the initiation involves the addition of this radical to the first monomer molecule to produce the chain initiating species $M_1\cdot$

$$R\cdot + M \xrightarrow{k_i} M_1\cdot \tag{3-14a}$$

where M represents a monomer molecule and k_i is the rate constant for the initiation step (Eq. 3-14a). For the polymerization of $CH_2{=}CHY$, Eq. 3-14a takes the form

$$R\cdot + CH_2{=}CHY \longrightarrow R-CH_2-\underset{Y}{\overset{H}{\underset{|}{\overset{|}{C}}}}\cdot \tag{3-14b}$$

(The radical $R\cdot$ is often referred to as an *initiator radical* or a *primary radical*.)

Propagation consists of the growth of $M_1\cdot$ by the successive additions of large numbers (hundreds and perhaps thousands) of monomer molecules according to Eq. 3-2. Each addition creates a new radical which has the same identity as the one previously, except that it is larger by one monomer unit. The successive additions may be presented by

$$M_1\cdot + M \xrightarrow{k_p} M_2\cdot \tag{3-15a}$$

$$M_2\cdot + M \xrightarrow{k_p} M_3\cdot \tag{3-15b}$$

$$M_3\cdot + M \xrightarrow{k_p} M_4\cdot \tag{3-15c}$$

etc., etc.

or in general terms

$$M_n\cdot + M \xrightarrow{k_p} M_{n+1}\cdot \tag{3-15d}$$

where k_p is the rate constant for propagation. Propagation with growth of the chain to high-polymer proportions takes place very rapidly. The value of k_p for most monomers is in the range 10^2–10^4 liters/mole-sec. This is a large rate constant—much larger than those usually encountered in step polymerizations (see Table 2-8).

At some point, the propagating polymer chain stops growing and terminates. Termination with the annihilation of the radical centers occurs by bimolecular reaction between radicals. Two radicals react with each other by *combination* (*coupling*),

$$\sim\sim\text{CH}_2-\underset{\underset{\text{Y}}{|}}{\overset{\overset{\text{H}}{|}}{\text{C}}}\cdot + \cdot\underset{\underset{\text{Y}}{|}}{\overset{\overset{\text{H}}{|}}{\text{C}}}-\text{CH}_2\sim\sim \xrightarrow{k_{tc}} \sim\sim\text{CH}_2-\underset{\underset{\text{Y}}{|}}{\overset{\overset{\text{H}}{|}}{\text{C}}}-\underset{\underset{\text{Y}}{|}}{\overset{\overset{\text{H}}{|}}{\text{C}}}-\text{CH}_2\sim\sim \tag{3-16a}$$

or, more rarely, by *disproportionation* in which a hydrogen radical that is *beta* to one radical center is transferred to another radical center. This results in the formation of two polymer molecules—one saturated and one unsaturated.

$$\sim\sim\text{CH}_2-\underset{\underset{\text{Y}}{|}}{\overset{\overset{\text{H}}{|}}{\text{C}}}\cdot + \cdot\underset{\underset{\text{H}}{|}}{\overset{\overset{\text{H}}{|}}{\text{C}}}-\underset{\underset{\text{H}}{|}}{\overset{\overset{\text{H}}{|}}{\text{C}}}\sim\sim \xrightarrow{k_{td}} \sim\sim\text{CH}_2-\underset{\underset{\text{Y}}{|}}{\overset{\overset{\text{H}}{|}}{\text{C}}}\text{H} + \underset{\underset{\text{Y}}{|}}{\overset{\overset{\text{H}}{|}}{\text{C}}}=\overset{\overset{\text{H}}{|}}{\text{C}}\sim\sim \tag{3-16b}$$

Termination can also occur by a combination of coupling and disproportionation. The two different modes of termination can be represented in general terms by

$$\text{M}_n\cdot + \text{M}_m\cdot \xrightarrow{k_{tc}} \text{M}_{n+m} \tag{3-17a}$$

$$\text{M}_n\cdot + \text{M}_m\cdot \xrightarrow{k_{td}} \text{M}_n + \text{M}_m \tag{3-17b}$$

where k_{tc} and k_{td} are the rate constants for termination by coupling and disproportionation, respectively. One can also express the termination step by

$$\text{M}_n\cdot + \text{M}_m\cdot \xrightarrow{k_t} \text{dead polymer} \tag{3-18}$$

where the particular mode of termination is not specified and

$$k_t = k_{tc} + k_{td} \tag{3-19}$$

The term *dead polymer* signifies the cessation of growth for the propagating radical. The propagation reaction would proceed indefinitely until all the monomer in a reaction system were exhausted if it were not for the strong tendency toward termination. Typical termination rate constants are in the range of 10^6–10^8 liters/mole-sec or orders of magnitude greater than the propagation rate constants. The much greater value of k_t (whether k_{tc} or k_{td}) compared to k_p does not prevent propagation because the radical species are present in very low concentrations and because the polymerization rate is dependent on only the one-half power of k_t (as will be discussed in Sec. 3-4a-2).

3-3b Rate Expression

Equations 3-13 through 3-19 constitute the detailed mechanism of a free-radical-initiated chain polymerization. The chain nature of the process resides in the propagation step (Eq. 3-15) in which large numbers of monomer molecules are converted to polymer for each initial radical species produced in the first step (Eq. 3-13). In order to obtain a kinetic expression for the overall rate of polymerization, it is necessary to assume that k_p and k_t are independent of the size of the radical. (This assumption is inherent in the notation used in Eqs. 3-15 through 3-18.) This is exactly the same type of assumption that was employed in deriving the kinetics of step polymerization. There is experimental evidence that indicates that, although radical reactivity depends on molecular size, the effect of size vanishes after the dimer or trimer [Kerr, 1973].

Monomer disappears by the initiation reaction (Eq. 3-14 as well as by the propagation reactions (Eq. 3-15). The *rate of monomer disappearance*, which is synonymous with the *rate of polymerization*, is given by

$$\frac{-d[M]}{dt} = R_i + R_p \tag{3-20}$$

where R_i and R_p are the rates of initiation and propagation, respectively. However, the number of monomer molecules reacting in the initiation step is far less than the number in the propagation step for a process producing high polymer. To a very close approximation the former can be neglected and the polymerization rate is given simply by the rate of propagation

$$\frac{-d[M]}{dt} = R_p \tag{3-21}$$

The rate of propagation, and therefore the rate of polymerization, is the sum of many individual propagation steps. Since the rate constants for all the propagation steps are the same, one can express the polymerization rate by

$$R_p = k_p[M\cdot][M] \tag{3-22}$$

where [M] is the monomer concentration and $[M\cdot]$ is the total concentration of all chain radicals, that is, all radicals of size $M_1\cdot$ and larger.

Equation 3-22 for the polymerization rate is not directly usable because it contains a term for the concentration of radicals. Radical concentrations are difficult to measure quantitatively, since they are very low ($\sim 10^{-8}$ M), and it is therefore desirable to eliminate $[M\cdot]$ from Eq. 3-22. In order to do this, the *steady-state* assumption is made that the concentration of radicals increases initially, but almost instantaneously reaches a constant, steady-state value. The rate of change of the concentration of radicals quickly becomes and remains zero during the course of the polymerization. This is equivalent to stating that the rates of initiation R_i and termination R_t of radicals are equal or

$$R_i = 2k_t[M\cdot]^2 \tag{3-23}$$

The steady-state assumption is not unique to polymerization kinetics. It is often used in developing the kinetics of many small molecule reactions that involve highly reactive intermediates present at very low concentrations—conditions that are met in radical chain polymerizations. The theoretical validity of the steady-state assumption has been discussed [Kondratiev, 1969] and its experimental validity shown in many polymerizations. As is developed in Sec. 3-8b, typical polymerizations achieve a steady-state after a period, which may be at most a minute.

The right side of Eq. 3-23 represents the rate of termination. There is no specification as to whether termination is by coupling or disproportionation since both follow the same kinetic expression. This use of the factor of 2 in the termination rate equation follows the generally accepted convention for reactions destroying radicals in pairs. It is also generally employed for reactions creating radicals in pairs as in Eq. 3-13. (In using the polymer literature one should be aware that the factor of 2 is not universally employed.) Rearrangement of Eq. 3-23 to

$$[\text{M·}] = \left(\frac{R_i}{2k_t}\right)^{1/2} \tag{3-24}$$

and substitution into Eq. 3-22 yields

$$R_p = k_p[\text{M}] \left(\frac{R_i}{2k_t}\right)^{1/2} \tag{3-25}$$

for the rate of polymerization. It is seen that Eq. 3-25 has the significant conclusion of the dependence of the polymerization rate on the square root of the initiation rate. Doubling the rate of initiation does not double the polymerization rate; the polymerization rate is increased only by the factor $\sqrt{2}$. This behavior is a consequence of the bimolecular termination reaction between radicals.

3-3c Experimental Determination of R_p

It is appropriate at this point to briefly discuss the experimental procedures used to determine polymerization rates for both step and radical chain polymerizations. R_p can be experimentally followed by measuring the change in any property that differs for the monomer(s) and polymer, for example, solubility, density, refractive index, and spectral absorption [Collins et al., 1973; McCaffery, 1970; Stickler, 1987]. Some techniques are equally useful for step and chain polymerizations, while others are more appropriate for only one or the other. Techniques useful for radical chain polymerizations are generally applicable to ionic chain polymerizations. The utility of any particular technique also depends on its precisions and accuracy at low, medium, and high percents of conversion. Some of the techniques have the inherent advantage of not needing to stop the polymerization to determine the percent conversion, that is, conversion can be followed versus time on the same reaction sample.

3-3c-1 *Physical Separation and Isolation of Reaction Product*

This technique involves isolating (followed by drying and weighing) the polymer from the reaction mixture in either of two ways: precipitation of the polymer by addition of a nonsolvent or distillation of monomer. The technique is not generally useful for step polymerization systems, since they consist of a range of different (small)-sized species that are difficult to separate from each other. Isolation of the polymer is most applicable to chain polymerizations, which consist of monomer and high polymer, differing greatly in their physical properties. The technique is, however, time-consuming and requires termination of the polymerization and great care to obtain accurate results.

A modification of this approach, which is applicable to many step polymerizations, involves measuring the rate of formation of the small molecule by-product. For example, the apparatus in Fig. 3-1 can be used to follow the rate of a polyesterification. Water distills out of the reaction vessel, is condensed and then collected in a calibrated trap (k). An advantage of this approach is that one can follow the entire course of a polymerization from 0 to 100% reaction on one reaction sample.

3-3c-2 Chemical and Spectroscopic Analysis

Chemical analysis of the unreacted monomer functional groups as a function of time is useful for step polymerizations. For example, polyesterification can be followed accurately by titration of the carboxyl group concentration with standard base or analysis of hydroxyl groups by reaction with acetic anhydride. The rate of chain polymerization of vinyl monomers can be followed by titration of the unreacted double bonds with bromine.

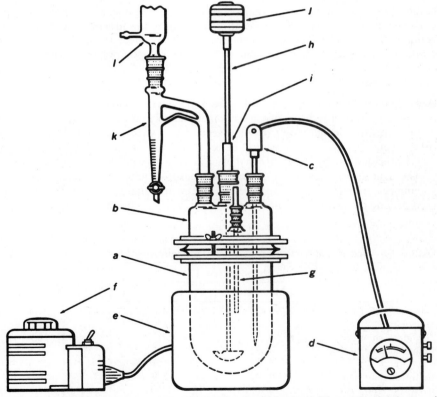

Fig. 3-1 Polyesterification apparatus: (a) reaction kettle bottom, (b) reaction kettle top, (c) thermocouple, (d) pyrometer, (e) heating mantle, (f) variable transformer, (g) nitrogen-inlet tube, (h) stainless steel stirrer, (i) stirrer bushing, (j) air motor, (k) graduated distillation trap, (i) condenser. After McCaffery [1970] (by permission of McGraw-Hill, New York).

The disappearance of monomer or appearance of polymer can also be followed by infrared (IR), ultraviolet (UV), and nuclear magnetic resonance (NMR) spectroscopy. One can follow the decrease in absorption signal(s) due to monomer and/or increase in absorption signal(s) due to polymer. For example, for styrene polymerization, NMR signals characteristic of the monomer, which disappear upon reaction, include signals for $CH_2=C$ (5.23 and 5.73 ppm) and $CH-C$ (6.71 ppm) protons; simultaneously, signals characteristic of the CH_2 and CH protons of polymer appear at 1.44 and 1.84 ppm, respectively. The accuracy of the spectroscopic technique can be high when the monomer and polymer signals do not overlap each other.

3-3c-3 Dilatometry

Dilatometry utilizes the volume change that occurs upon polymerization to follow conversion time. It is often an accurate method for chain polymerizations because of the large difference in density between monomer and polymer (Table 3-2). The conversion is conveniently followed in a *dilatometer* (Fig. 3-2) whose volume has been calibrated. Purified monomer is added to the dilatometer so that the volume comes into the B region. After purging the system of oxygen by freezing the monomer and evacuation, the apparatus is sealed off above portion B. The dilatometer is placed in a constant temperature bath and the volume change of the polymerizing system, which is quantitatively related to the percent conversion, followed with time. Dilatometry is not useful for the usual step polymerization where there is a small molecule by-product that results in no appreciable volume change upon polymerization.

3-3c-4 Other Methods

Most chain polymerizations (but not step polymerizations) involve considerable changes in refractive index between monomer and polymer. For example, the changes in the refractive indices (n_D^{20}) of vinyl acetate and styrene are 0.0711 and 0.0477, respectively. Temperature control to better than 0.02°C is needed to obtain the necessary precision (in the fourth decimal place of refractive index) to use this technique.

The heat of polymerization ΔH can be measured quite accurately by differential scanning calorimetry and is a measure of conversion. Other techniques that have been used include dielectric constant measurements, and gas chromatography.

TABLE 3-2 Densities of Some Monomers and Polymers[a]

| Monomer | Density (g/cm^3) at 25°C | | Volume Change (%) |
	Monomer	Polymer	
Vinyl chloride	0.919	1.406	34.4
Acrylonitrile	0.800	1.17	31.0
Vinylidene chloride	1.213	1.71	28.7
Methyl acrylate	0.952	1.223	22.1
Vinyl acetate[b]	0.934	1.191	21.6
Methyl methacrylate	0.940	1.179	20.6
Styrene	0.905	1.062	14.5

[a]Data from Collins et al. [1973].
[b]Density values are for 20°C.

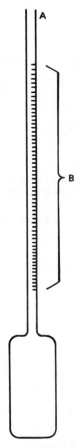

Fig. 3-2 Dilatometer.

3-4 INITIATION

The derivation of Eq. 3-25 is general in that the reaction for the production of radicals (Eq. 3-13) is not specified. The initiation rate is simply shown as R_i. A variety of initiator systems can be used to bring about the polymerization. (The term *catalyst* is often used synonomously with initiator, but it is incorrect in the classical sense, since the initiator is consumed. The use of the term catalyst may be condoned since very large numbers of monomer molecules are converted to polymer for each initiator molecule which is consumed.) Radicals can be produced by a variety of thermal, photochemical, and redox methods [Bamford, 1988; Eastmond, 1976a, 1976b, 1976c]. In order to function as a useful source of radicals an initiator system should be readily available, stable under ambient or refrigerated conditions, and possess a practical rate of radical generation at temperatures which are not excessive (approximately <150°C).

3-4a Thermal Decomposition of Initiators

3-4a-1 Types of Initiators

The *thermal, homolytic dissociation* of initiators is the most widely used mode of generating radicals to initiate polymerization—for both commercial polymerizations and theoretical studies. Polymerizations initiated in this manner are often referred to as *thermal initiated* or *thermal catalyzed* polymerizations. The number of different types of compounds that can be used as thermal initiators is rather limited. One is usually limited to compounds with bond dissociation energies in the range 100–170 kJ/mole. Compounds with higher or lower dissociation energies will dissociate too slowly or too rapidly. Only a few classes of compounds—including those with O—O, S—S, N—O bonds—possess the desired range of dissociation energies. However, it is only the peroxides which find extensive use as radical sources. The other classes of compounds are usually either not readily available or not stable enough. Several different types of peroxy compounds are widely used [Sheppard, 1988]. These are acyl peroxides such as acetyl and benzoyl peroxides,

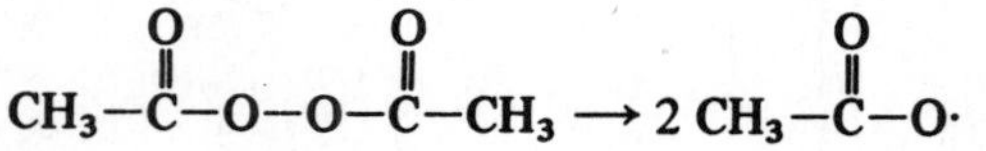

$$CH_3-\overset{\overset{\displaystyle O}{\|}}{C}-O-O-\overset{\overset{\displaystyle O}{\|}}{C}-CH_3 \longrightarrow 2\ CH_3-\overset{\overset{\displaystyle O}{\|}}{C}-O\cdot \tag{3-26a}$$

$$\phi-\overset{\overset{\displaystyle O}{\|}}{C}-O-O-\overset{\overset{\displaystyle O}{\|}}{C}-\phi \longrightarrow 2\ \phi-\overset{\overset{\displaystyle O}{\|}}{C}-O\cdot \tag{3-26b}$$

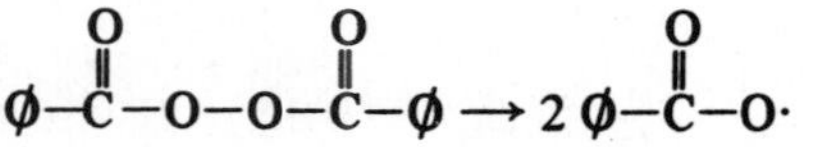

alkyl peroxides such as cumyl and *t*-butyl peroxides,

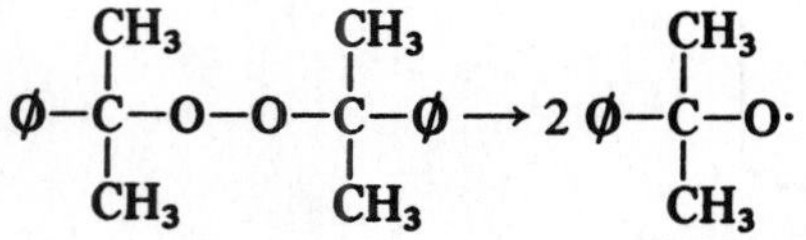

$$\phi-\overset{\overset{\displaystyle CH_3}{|}}{\underset{\underset{\displaystyle CH_3}{|}}{C}}-O-O-\overset{\overset{\displaystyle CH_3}{|}}{\underset{\underset{\displaystyle CH_3}{|}}{C}}-\phi \longrightarrow 2\ \phi-\overset{\overset{\displaystyle CH_3}{|}}{\underset{\underset{\displaystyle CH_3}{|}}{C}}-O\cdot \tag{3-26c}$$

$$CH_3-\overset{\overset{\displaystyle CH_3}{|}}{\underset{\underset{\displaystyle CH_3}{|}}{C}}-O-O-\overset{\overset{\displaystyle CH_3}{|}}{\underset{\underset{\displaystyle CH_3}{|}}{C}}-CH_3 \longrightarrow 2\ CH_3-\overset{\overset{\displaystyle CH_3}{|}}{\underset{\underset{\displaystyle CH_3}{|}}{C}}-O\cdot \tag{3-26d}$$

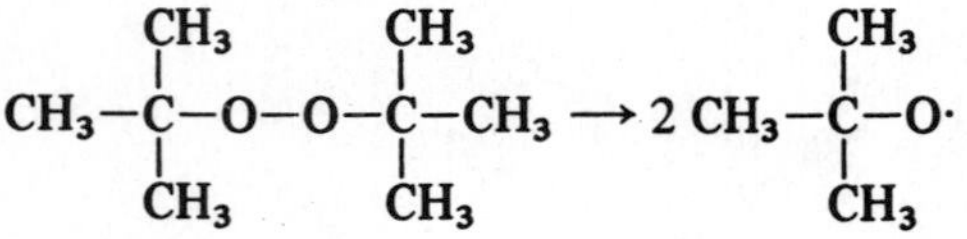

hydroperoxides such as *t*-butyl and cumyl hydroperoxides,

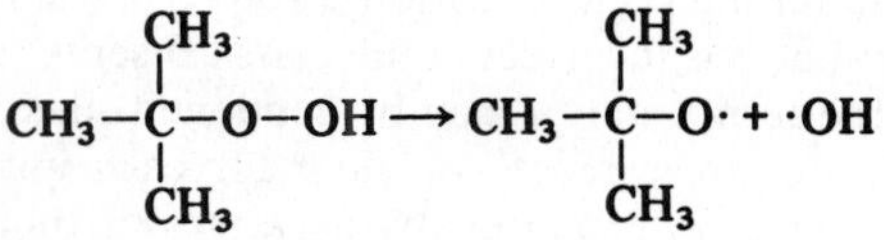

$$CH_3-\overset{\overset{\displaystyle CH_3}{|}}{\underset{\underset{\displaystyle CH_3}{|}}{C}}-O-OH \longrightarrow CH_3-\overset{\overset{\displaystyle CH_3}{|}}{\underset{\underset{\displaystyle CH_3}{|}}{C}}-O\cdot + \cdot OH \tag{3-26e}$$

$$\phi-\overset{\overset{\displaystyle CH_3}{|}}{\underset{\underset{\displaystyle CH_3}{|}}{C}}-O-OH \longrightarrow \phi-\overset{\overset{\displaystyle CH_3}{|}}{\underset{\underset{\displaystyle CH_3}{|}}{C}}-O\cdot + \cdot OH \tag{3-26f}$$

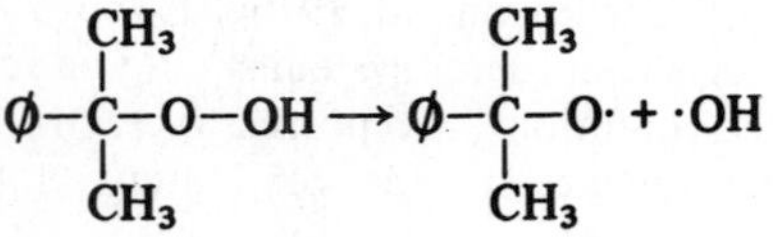

and peresters such as *t*-butyl perbenzoate,

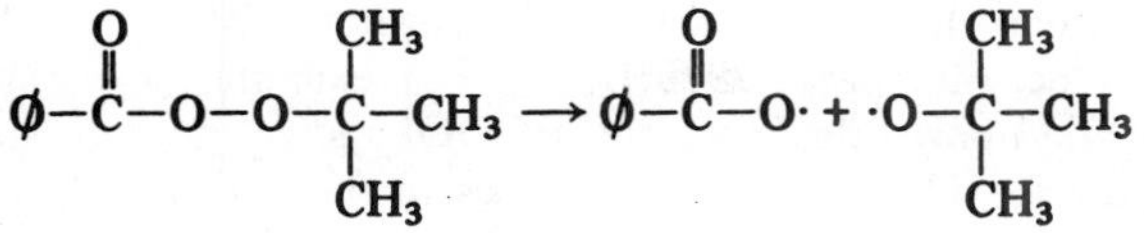

$$\phi-\overset{\overset{\text{O}}{\|}}{\text{C}}-\text{O}-\text{O}-\overset{\overset{\text{CH}_3}{|}}{\underset{\underset{\text{CH}_3}{|}}{\text{C}}}-\text{CH}_3 \rightarrow \phi-\overset{\overset{\text{O}}{\|}}{\text{C}}-\text{O}\cdot + \cdot\text{O}-\overset{\overset{\text{CH}_3}{|}}{\underset{\underset{\text{CH}_3}{|}}{\text{C}}}-\text{CH}_3 \qquad (3\text{-}26g)$$

Other peroxides used to initiate polymerization are acyl alkylsulfonyl peroxides (**Va**), dialkyl peroxydicarbonates (**Vb**), diperoxyketals (**Vc**), and ketone peroxides (**Vd**).

$$\text{R}-\text{SO}_2-\text{OO}-\text{CO}-\text{R}' \qquad \text{RO}-\text{CO}-\text{OO}-\text{CO}-\text{OR}$$

Va   **Vb**

$$(\text{ROO})_2\text{C}(\text{R}')_2 \qquad \text{RR}'\text{C}(\text{OOH})_2$$

Vc   **Vd**

Aside from the various peroxy compounds, the main other class of compound used extensively as initiators are the azo compounds. 2,2'-Azobisisobutyronitrile (AIBN) is the most important member of this class of initiators, although other azo com-

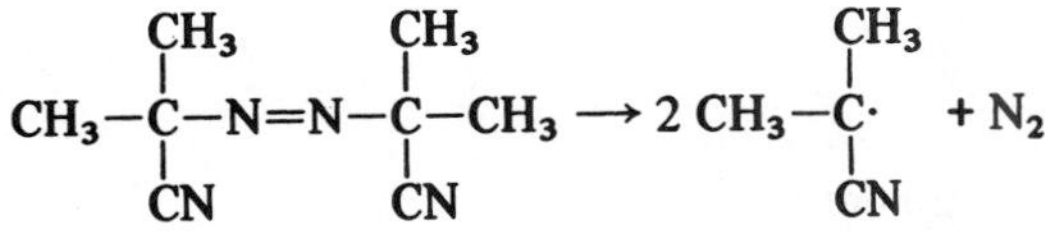

$$\text{CH}_3-\overset{\overset{\text{CH}_3}{|}}{\underset{\underset{\text{CN}}{|}}{\text{C}}}-\text{N}{=}\text{N}-\overset{\overset{\text{CH}_3}{|}}{\underset{\underset{\text{CN}}{|}}{\text{C}}}-\text{CH}_3 \rightarrow 2\,\text{CH}_3-\overset{\overset{\text{CH}_3}{|}}{\underset{\underset{\text{CN}}{|}}{\text{C}}}\cdot \;\; + \text{N}_2 \qquad (3\text{-}26h)$$

pounds such as 2,2'-azobis(2,4-dimethylpentanenitrile) and 1,1'-azobis(cyclohexane-carbonitrile) are also used [Sheppard, 1985]. The facile dissociation of azo compounds is not due to the presence of a weak bond as is the case with the peroxy compounds. The C—N bond dissociation energy is high (ca. 290 kJ/mole), but the driving force for homolysis is the formation of the highly stable nitrogen molecule.

Among other initiators that have been studied are disulfides

$$\text{RS}-\text{SR} \rightarrow 2\text{RS}\cdot \qquad (3\text{-}26i)$$

and tetrazenes

$$\text{R}_2\text{N}-\text{N}{=}\text{N}-\text{NR}_2 \rightarrow 2\text{R}_2\text{N}\cdot + \text{N}_2 \qquad (3\text{-}26j)$$

[Oda et al., 1978; Sato et al., 1979a, 1979b; Sugiyama, 1982].

The various initiators are used at different temperatures depending on their rates of decomposition. Thus azobisisobutyronitrile (AIBN) is commonly used at 50–70°C, acetyl peroxide at 70–90°C, benzoyl peroxide at 80–95°C, and dicumyl or di-*t*-butyl peroxide at 120–140°C. The value of the decomposition rate constant k_d varies in the range 10^{-4}–10^{-9} sec^{-1} depending on the initiator and temperature [Eastmond, 1976a, 1976b, 1976c]. Most initiators are used at temperatures where k_d is usually 10^{-4}–10^{-6} sec^{-1}. The great utility of the peroxides as a class of initiators arises from the availability in stable form of many different compounds with a wide variety of use temperatures.

The differences in the rates of decomposition of the various initiators are related to differences in the structures of the initiators and of the radicals produced. The effects of structure on initiator reactivity have been discussed elsewhere [Bamford, 1988; Eastmond, 1976a, 1976b, 1976c; Sheppard, 1985, 1988]. For example, k_d is larger for acyl peroxides than for alkyl peroxides since the RCOO· radical is more stable than the RO· radical and for R—N=N—R, k_d increases in the order R = allyl, benzyl > tertiary > secondary > primary [Koenig, 1973].

The differences in the decomposition rates of various initiators can be conveniently expressed in terms of the *initiator half-life* $t_{1/2}$ defined as the time for the concentration of I to decrease to one half its original value. The rate of initiator disappearance by Eq. 3-13 is

$$\frac{-d[I]}{dt} = k_d[I] \tag{3-27}$$

which on integration yields

$$[I] = [I]_0 e^{-k_d t} \tag{3-28a}$$

or

$$\log \frac{[I]_0}{[I]} = k_d t \tag{3-28b}$$

where $[I]_0$ is the initiator concentration at the start of polymerization. $t_{1/2}$ is obtained as

$$t_{1/2} = \frac{0.693}{k_d} \tag{3-29}$$

by setting $[I] = [I]_0/2$. Table 3-3 lists the initiator half-lives for several common initiators at various temperatures.

3-4a-2 Kinetics of Initiation and Polymerization

The rate of producing primary radicals by thermal homolysis of an initiator R_d (Eqs. 3-13 and 3-26) is given by

$$R_d = 2fk_d[I] \tag{3-30}$$

where $[I]$ is the concentration of the initiator and f is the *initiator efficiency*. The initiator efficiency is defined as the fraction of the radicals produced in the homolysis reaction that initiate polymer chains. The value of f is usually less than unity due to wastage reactions. The factor of 2 in Eq. 3-30 follows the convention previously discussed for Eq. 3-23.

The initiation reaction in polymerization is composed of two steps (Eqs. 3-13 and 3-14) as discussed previously. In most polymerizations, the second step (the addition of the primary radical to monomer) is much faster than the first step. The homolysis

TABLE 3-3 Half-Lives of Initiators[a,b]

Initiator	50°C	60°C	70°C	85°C	100°C	115°C	130°C	145°C	155°C	175°C
					Half-Life at					
Azobisisobutyronitrile	74 hr		4.8 hr		7.2 min					
Benzoyl peroxide			7.3 hr	1.4 hr	19.8 min					
Acetyl peroxide	158 hr		8.1 hr	1.1 hr						
Lauryl peroxide	47.7 hr	12.8 hr	3.5 hr	31 min						
t-Butyl peracetate				88 hr	12.5 hr	1.9 hr	18 min			
Cumyl peroxide						13 hr	1.7 hr	16.8 min		
t-Butyl peroxide					218 hr	34 hr	6.4 hr	1.38 hr		
t-Butyl hydroperoxide					338 hr				44.9 hr	4.81 hr

[a]Data from Brandrup and Immergut [1989] and Huyser [1970].
[b]$t_{1/2}$ values are for benzene or toluene solutions of the initiators.

of the initiator is the rate determining step in the initiation sequence and the rate of initiation is then given by

$$R_i = 2fk_d[I] \tag{3-31}$$

Substitution of Eq. 3-31 into Eq. 3-25 yields

$$R_p = k_p[M] \left(\frac{fk_d[I]}{k_t}\right)^{1/2} \tag{3-32}$$

3-4a-3 Dependence of Polymerization Rate on Initiator

Equation 3-32 describes the most common case of radical chain polymerization. It shows the polymerization rate to be dependent on the square root of the initiator concentration. This dependence has been abundantly confirmed for many different monomer-initiator combinations over wide ranges of monomer and initiator concentrations [Eastmond, 1976a, 1976b, 1976c; Kamachi et al., 1978; Santee et al., 1964; Schulz and Blaschke, 1942; Vrancken and Smets, 1959]. Figure 3-3 shows typical data illustrating the square-root dependence on [I]. Deviations from this behavior are found under certain conditions. The order of dependence of R_p on [I] may be observed to be less than one-half at very high initiator concentrations. However, such an effect is not truly a deviation from Eq. 3-32. It may be due to a decrease in f with increasing initiator concentration (Sec. 3-4g-2).

Alternately, the termination mode may change from the normal bimolecular termination between propagating radicals to *primary termination*, which involves propagating radicals reacting with primary radicals [Berger et al., 1977; Deb and Kapoor, 1979; Ito, 1980; Mahabadi and Meyerhoff, 1979]:

$$M_n\cdot + R\cdot \xrightarrow{k_{tp}} M_n\text{—}R \tag{3-33a}$$

This occurs if primary radicals are produced at too high a concentration and/or in the presence of two low a monomer concentration to be completely and rapidly scavenged by monomer (by Eq. 3-14a). If termination occurs exclusively by primary termination, the polymerization rate is given by

$$R_p = \frac{k_p k_i[M]^2}{k_{tp}} \tag{3-33b}$$

Eq. 3-33b, derived by combining the rate expressions for Eqs. 3-14a, 3-15d, and 3-33a, shows that the polymerization rate becomes independent of the initiator concentration (but not k_i) and second-order in monomer concentration.

Primary termination and the accompanying change in the order of dependence of R_p on [I] may also be found in the Trommsdorff polymerization region (Sec. 3-10). Situations also arise where the order of dependence of R_p on [I] will be greater than one-half. This behavior may be observed in the Trommsdorff region if the polymer radicals do not undergo termination or under certain conditions of chain transfer or inhibition (Sec. 3-7).

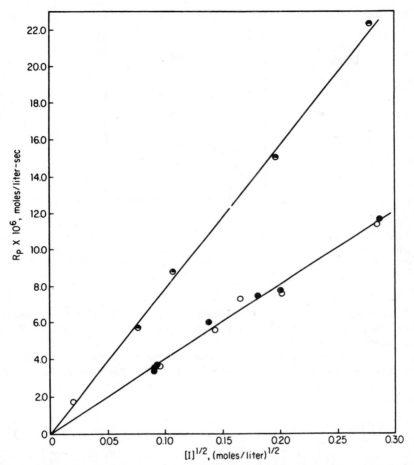

Fig. 3-3 Square-root dependence of the polymerization rate R_p on the initiator concentration [I]. ⊖ = Methyl methacrylate, benzoyl peroxide, 50°C. After Schulz and Blaschke [1942] (by permission of Akademische Verlagsgesellschaft, Geest and Portig K.-G., Leipzig). ○, ● = Vinyl benzoate, azobisisobutyronitrile, 60°C. After Santee et al. [1964] and Vrancken and Smets [1959] (by permission of Huthig and Wepf Verlag, Basel).

3-4a-4 Dependence of Polymerization Rate on Monomer

The rate expression Eq. 3-32 requires a first-order dependence of the polymerization rate on the monomer concentration. This is indeed found to be the general behavior for many polymerizations [Kamachi et al., 1978]. Figure 3-4 shows the first-order relationship for the polymerization of methyl methacrylate [Sugimura and Minoura, 1966]. However, there are many polymerizations where R_p shows a higher than first-order dependence on [M]. Thus the rate of polymerization depends on the $\frac{3}{2}$-power of the monomer concentration in the polymerization of styrene in chlorobenzene solution at 120°C initiated by t-butyl peresters [Misra and Mathiu, 1967]. The benzoyl

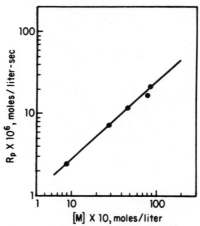

Fig. 3-4 First-order dependence of the polymerization rate R_p of methyl methacrylate on the monomer concentration [M]. The initiator is the *t*-butyl perbenzoate–diphenylthiourea redox system. After Sugimura and Minoura [1966] (by permission of Wiley-Interscience, New York).

peroxide initiated polymerization of styrene in toluene at 80°C shows an increasing order of dependence of R_p on [M] as [M] decreases [Horikx and Hermans, 1953]. The dependence is 1.18-order at [M] = 1.8 and increases to 1.36-order at [M] = 0.4. These effects may be caused by a dependence of the initiation rate on the monomer concentration. Equation 3-28 was derived on the assumption that R_i is independent of [M]. The initiation rate can be monomer-dependent in several ways. The initiator efficiency f may vary directly with the monomer concentration

$$f = f'[\text{M}] \tag{3-34}$$

which would lead (by substitution of Eq. 3-34 into Eqs. 3-31 and 3-32) to first-order dependence of R_i on [M] and $\frac{3}{2}$-order dependence of R_p on [M]. (This effect has been observed and is discussed in Sec. 3-4f.) The equivalent result arises if the second step of the initiation reaction (Eq. 3-14 were to become the rate-determining step instead of the first step (Eq. 3-13). This occurs when $k_d > k_i$ or when [M] is low. It is also frequently encountered in polymerizations initiated photolytically or by ionizing radiation (Secs. 3-4c and 3-4d) and in some redox-initiated polymerizations (Sec. 3-4b).

In some systems it appears that the initiation step differs from the usual two-step sequence of Eqs. 3-13 and 3-14. Thus in the *t*-butyl hydroperoxide-styrene system only a minor part of the initiation occurs by the first-order homolysis reaction (Eq. 3-26f), which accounts for the complete decomposition of *t*-butyl hydroperoxide in the absence of styrene. Homolysis of the hydroperoxide occurs at a much faster rate in the presence of styrene than in its absence. The increased decomposition rate in the *t*-butyl hydroperoxide–styrene system occurs by a *molecule-induced homolysis* reaction which is first-order in both styrene and hydroperoxide [Walling and Heaton, 1965]. The initiation reaction may be written as

$$\text{M} + \text{I} \rightarrow \text{M} \cdot + \text{R} \cdot \tag{3-35}$$

and will result in a $\frac{3}{2}$-order dependence of R_p on [M]. This initiation is probably best considered as an example of redox initiation (Sec. 3-4b).

Other exceptions to the first-order dependence of the polymerization rate on the monomer concentration occur when termination is not by bimolecular reaction of propagating radicals. Second-order dependence of R_p on [M] occurs for primary termination (Eq. 3-33a) and certain redox-initiated polymerizations (Sec. 3-4b-2). Less than first-order dependence of R_p on [M] has been observed for polymerizations (Sec. 9-8a-2) taking place inside a solid under conditions where monomer diffusion into the solid is slower than the normal propagation rate [Odian et al., 1980] and also in some redox polymerizations (Sec. 3-4b-2) [Mapunda-Vlckova and Barton, 1978].

3-4b Redox Initiation

Many oxidation–reduction reactions produce radicals that can be used to initiate polymerization. This type of initiation is referred to as *redox initiation*, *redox catalysis*, or *redox activation*. A prime advantage of redox initiation is that radical production occurs at reasonable rates over a very wide range of temperatures, depending on the particular redox system, including initiation at moderate temperatures of 0–50°C and even lower. This allows a greater freedom of choice of the polymerization temperature than is possible with the thermal homolysis of initiators. Some redox polymerizations can be initiated photolytically as well as thermally. A wide range of redox reactions, including both inorganic and organic components either wholly or in part, may be employed for this purpose. Some redox systems involve direct electron transfer between reductant and oxidant, while others involve the intermediate formation of reductant-oxidant complexes; the latter are charge-transfer complexes in some cases.

3-4b-1 Types of Redox Initiators

1. Peroxides in combination with a reducing agent are a common source of radicals, for example, the reaction of hydrogen peroxide with ferrous ion

$$H_2O_2 + Fe^{2+} \rightarrow HO^- + HO\cdot + Fe^{3+} \tag{3-36a}$$

Ferrous ion also promotes the decomposition of a variety of other compounds including various types of organic peroxides [Bamford, 1988].

$$ROOR \xrightarrow{Fe^{2+}} RO^- + RO\cdot \tag{3-36b}$$

$$ROOH \xrightarrow{Fe^{2+}} HO^- + RO\cdot \tag{3-36c}$$

$$\underset{ROOCR'}{\overset{O}{\|}} \xrightarrow{Fe^{2+}} R'\underset{}{\overset{O}{\|}}CO^- + RO\cdot \tag{3-36d}$$

Other reductants such as Cr^{2+}, V^{2+}, Ti^{3+}, Co^{2+}, and Cu^+ can be employed in place of ferrous ion in many instances. Most of these redox systems are aqueous or emulsion systems. Redox initiation with acyl peroxides can be carried out in organic media by using amines as the reductant [Morsi et al., 1977; O'Driscoll et al., 1965]. An interesting system is the combination of benzoyl peroxide and an *N,N*-dialkylaniline. The difference in the rates of decomposition between such a redox system and the simple thermal homolysis of the peroxide alone is very striking. The decomposition rate constant k_d for pure benzoyl peroxide in styrene polymerizations is 1.33×10^{-4} sec^{-1} at 90°C, while that for the benzoyl peroixde-N,N-diethylaniline redox system is 1.25×10^{-2} liters/mole-sec at 60°C and 2.29×10^{-3} liters/mole-sec at 30°C. The redox system has a much larger decomposition rate. Radical production in this redox system appears to proceed via an initial ionic displacement by the nitrogen of the aniline on the peroxide linkage

$$\phi-\overset{R}{\underset{R}{N}}-R + \phi-\overset{O}{\overset{\|}{C}}-O-O-\overset{O}{\overset{\|}{C}}-\phi \longrightarrow \left[\phi-\overset{R}{\underset{R}{N}}-O-\overset{O}{\overset{\|}{C}}-\phi\right]^{+} \phi-\overset{O}{\overset{\|}{C}}-O^{-} \longrightarrow$$

$$\underset{\text{V}}{}$$

$$\phi-\overset{+}{\underset{R}{N}}-R + \phi-\overset{O}{\overset{\|}{C}}-O\cdot + \phi-\overset{O}{\overset{\|}{C}}-O^{-} \qquad (3\text{-}37)$$

In support of this mechanism, the rate of initiation increases with increasing nucleophilicity of the amine. Initiation is by the ϕCOO· radical. The amino cation–radical is generally not an effective initiator as shown by the absence of nitrogen in the polymer; it apparently disappears by some unknown side reactions such as dimerization and/or deprotonation. (However, there is some evidence that the cation-radical produced by the interaction of benzoyl peroxide and N-vinylcarbazole (NVC) initiates cationic polymerization of NVC simultaneous with the radical polymerization [Bevington et al., 1978].) Other types of peroxides also appear susceptible to this type of acceleration. Peroxide decomposition is also accelerated by transition metal ion complexes such as copper (II) acetylacetonate [Ghosh and Maity, 1978; Shahani and Indictor, 1978]. Zinc chloride accelerates the rate of decomposition of AIBN in methyl methacrylate by a factor of 8 at 50°C, with a corresponding increase in polymerization rate [Lachinov et al., 1977]. The mechanism may be analogous to the amine–peroxide system. The effect of ZnCl$_2$ and other metal ions on accelerating R_p is well established, although the mechanism in most cases appears to involve changes in k_p and/or due to complexation of the metal ions with the monomer and/or propagating radicals.

   2. The combination of a variety of inorganic reductants and inorganic oxidants initiates radical polymerization, for example,

$$^{-}O_3S-O-O-SO_3^{-} + Fe^{2+} \rightarrow Fe^{3+} + SO_4^{2-} + SO_4^{-}\cdot \qquad (3\text{-}38a)$$

$$^{-}O_3S-O-O-SO_3^{-} + S_2O_3^{2-} \rightarrow SO_4^{2-} + SO_4^{-}\cdot + \cdot S_2O_3^{-} \qquad (3\text{-}38b)$$

Other redox systems include reductants such as HSO_3^{-}, SO_3^{2-}, $S_2O_3^{2-}$, and $S_2O_5^{2-}$ in combination with oxidants such as Ag^{+}, Cu^{2+}, Fe^{3+}, ClO_3^{-}, and H_2O_2.

   3. Organic-inorganic redox pairs initiate polymerization, usually but not always by oxidation of the organic component, for example, the oxidation of an alcohol by Ce^{4+},

$$R-CH_2-OH + Ce^{4+} \xrightarrow{k_d} Ce^{3+} + H^{+} + R-\dot{C}H-OH \qquad (3\text{-}39)$$

or by V^{5+}, Cr^{6+}, Mn^{3+} [Fernandez and Guzman, 1989; Misra and Bajpai, 1982; Nayak and Lenka, 1980]. Other redox pairs include

a. Oxidation of thiol compounds such as thiourea and thioglycollic acid by CrO_4^-, $S_2O_8^-$, Mn^{3+}, and Fe^{3+} [Balakrishnan and Subbu, 1988; Lenka and Nayak, 1987; Pramanick and Chatterjee, 1980].

b. Oxidation of oxalic and citric acids by permanganate, Mn^{3+} and Ce^{4+} [Kaliyamurthy et al., 1979; Misra et al., 1984].

c. Organometallic derivatives of transition metals in low (usually 0) oxidation state such as $Mo(CO)_6$, $Mn_2(CO)_{10}$, ferrocene, and cobaltocene in combination with an organic halide (usually CCl_4) initiate polymerization by electron transfer from the metal to the halide, for example,

$$Cp_2Fe + CCl_4 \rightleftharpoons Cp_2Fe \overset{\delta+}{\cdots} \overset{\delta-}{CCl_4} \rightarrow Cp_2Fe^+Cl^- + CCl_3 \cdot \qquad (3\text{-}40)$$
$$\textbf{VII}$$

where Cp represents cyclopentadienyl anion [Ouchi et al., 1978].

4. There are some initiator systems in which the monomer itself acts as one component of the redox pair. Examples are thiosulfate plus acrylamide or methacrylic acid and N,N-dimethylaniline plus methyl methacrylate [Manickam et al., 1978; Tsuda et al., 1984].

3-4b-2   Rate of Redox Polymerization

The kinetics of redox initiated polymerizations generally fall into two categories depending on the termination mode. Many of these polymerizations proceed in the same manner as other polymerizations in terms of the propagation and termination steps, the only difference being the source of radicals for the initiation step. For these polymerizations where termination is by bimolecular reaction of propagating radicals, the initiation and polymerization rates will be given by appropriate expressions that are very similar to those developed previously:

$$R_i = k_d[\text{reductant}][\text{oxidant}] \qquad (3\text{-}41)$$

$$R_p = k_p[\text{M}] \left(\frac{k_d[\text{reductant}][\text{oxidant}]}{2k_t} \right)^{1/2} \qquad (3\text{-}42)$$

Equations 3-41 and 3-42 differ from those (Eqs. 3-31 and 3-32) discussed previously in that the factor of 2 is absent from the expression for R_i, since typically only one radical is produced per oxidant-reductant pair.

Some redox polymerizations involve a change in the termination step from the usual bimolecular reaction to monomolecular termination involving the reaction between the propagating radicals and a component of the redox system. This leads to kinetics which are appreciably different than those previously encountered. Thus, in the alcohol-Ce^{4+} system (Eq. 3-39), termination occurs according to

$$M_n \cdot + Ce^{4+} \rightarrow Ce^{3+} + H^+ + \text{dead polymer} \qquad (3\text{-}43)$$

at high cerric ion concentrations. The propagating radical loses a hydrogen to form a dead polymer molecule with an olefinic endgroup. The rates of initiation and termination are given by

$$R_i = k_d[Ce^{4+}][alcohol] \tag{3-44}$$

$$R_t = k_t[Ce^{4+}][M\cdot] \tag{3-45}$$

By making the usual steady-state assumption (i.e., $R_i = R_t$), one obtains the polymerization rate as

$$R_p = \frac{k_d k_p[M][alcohol]}{k_t} \tag{3-46}$$

In many redox polymerizations, monomer may actually be involved in the initiation process. Although not indicated above, this is the case for initiations described under item 3b; and, of course, for item 4. R_p will show a higher dependence on [M] in these cases than indicated by Eqs. 3-42 and 3-46. First-order dependence of R_i on [M] results in $\frac{3}{2}$- and 2-order dependencies of R_p on [M] for bimolecular and monomolecular terminations, respectively. For those redox initiations involving the equilibrium formation of intermediate complexes that lead to radical formation (for example, Eq. 3-40), derivation of the kinetics follows in a straight-forward manner. Thus R_i for initiation by Eq. 3-40 is

$$R_i = K k_d[Cp_2Fe][CCl_4] \tag{3-47}$$

where K is the equilibrium constant for formation of the complex between ferrocene and CCl_4.

3-4c Photochemical Initiation

Photochemical or *photoinitiated* polymerizations occur when radicals are produced by ultraviolet and visible light irradiation of a reaction system [Oster and Yang, 1968; Pappas, 1988]. In general, light absorption results in radical production by either of two pathways:

1. Some compound in the system undergoes excitation by energy absorption and subsequent decomposition into radicals.
2. Some compound undergoes excitation and the excited species interacts with a second compound (by either energy transfer or redox reaction) to form radicals derived from the latter and/or former compound(s).

The term *photosensitizer* was originally used to refer to the second pathway, especially when it involved energy transfer, but that distinction has become blurred. The mechanism for photoinitiation in a reaction system is not always clear-cut and may involve both pathways. *Photosensitizer* is now used to refer to any substance that either increases the rate of photoinitiated polymerization or shifts the wavelength at which polymerization occurs.

Photoinitiation of polymerization offers significant practical advantages. Radical production and polymerization can be spatially directed (i.e., confined to particular regions) and turned on or off simply by turning the light source on or off. Additionally, the initiation rates can be controlled by a combination of the source of radicals, light intensity, and temperature. Extensive use is made of these advantages in the printing and coatings industries [Pappas, 1988; Pelgrims, 1978]. Photochemical polymerization has found applications in decorative and protective coatings and inks for metal, paper, wood, and plastics; in photolithography for producing integrated and printed circuits; and in curing dental materials. Many of these applications involve a combination of polymerization and crosslinking. Crosslinking is often achieved by the use of monomers with two or more double bonds per molecule (Sec. 6-6a). Acrylate systems are the most common, although unsaturated polyester and styrene systems are also employed. (Curing of epoxy resins via ionic photoinitiators is also commercially important; see Sec. 7-2b-7). Another approach for producing photocrosslinked coatings is the crosslinking of preformed polymers. This requires the use of specially designed polymers with reactive functional groups (e.g., double bonds) and/or appropriate crosslinking agents. Photopolymerizations are of special interest in applications where economic and/or environmental considerations require the use of solvent-free systems. The significant drawback to photopolymerization is that the penetration through a thickness of material is low, limiting use to the surface-type application.

Photopolymerization is used in the photoimaging industry. The reaction system for photoimaging applications is referred to as a *photoresist*. The development of a printed circuit involves coating a copper–laminate substrate with the photoresist followed by irradiation through a mask with transparent areas corresponding to the desired copper circuitry. The unexposed areas are uncured and easily dissolved by solvent. Copper is then selectively etched away from below the unexposed areas. The copper below the exposed areas is protected by a polymer coating during the etching process. Finally, that polymer coating is removed to yield the desired copper printed circuit.

3-4c-1 Bulk Monomer

The irradiation of some monomers results in the formation of an excited state M^* by the absorption of light photons (quanta)

$$M + h\nu \rightarrow M^* \tag{3-48}$$

The excited species undergoes homolysis to produce radicals

$$M^* \rightarrow R\cdot + R'\cdot \tag{3-49}$$

capable of initiating the polymerization of the monomer. The identities of the radicals $R\cdot$ and $R'\cdot$ are not usually well established. Nor is it clear that photolysis of a bulk monomer always results in the simple homolysis to yield two radicals per one monomer molecule as described by Eq. 3-49. Initiation by photolysis of a monomer is limited to those monomers where the double bond is conjugated with other groups (e.g., styrene, methyl methacrylate) such that absorption will occur above the vacuum UV region (200 nm) where light sources are readily available. However, unless absorption occurs above 300–325 nm, there is the practical limitation of the need for quartz reaction vessels, since glass does not transmit appreciable amounts of energy at wavelengths below 300–325 nm. A further practical limitation for most monomers is that the efficiency of initiation (*quantum yield*) by photolysis of bulk monomer is almost always considerably less than the initiator systems described below.

3-4c-2 *Irradiation of Thermal and Redox Initiators*

The various thermal and redox initiators described in Secs. 3-4a and 3-4b can also be used in photoinitiation. In the usual case irradiation yields the same radicals as described previously for thermal or redox initiation. However, not all thermal and redox initiators are equally useful as photoinitiators, since many do not absorb light in the practical wavelength region. On the other hand, the photochemical method allows the use of a wider range of compounds as initiators due to the higher selectivity of photolytic homolysis. For example, for compounds other than the thermal and redox initiators discussed previously, homolysis occurs at too high a temperature and usually results in the production of a wide spectrum of different radicals (and ions) as various bonds randomly break. Thus photochemical initiation can be used with carbonyl compounds such as ketones. Although both aliphatic and aromatic ketones have been studied [Ledwith, 1977; Pappas, 1988], the aromatic ketones are more useful in commercial practice, since their absorptions occur at longer wavelength and their quantum yields are higher. Benzophenone and acetophenone and their derivatives are the most commonly encountered simple ketones. Ketones undergo homolysis by one or both (often simultaneously) of two processes—fragmentation and hydrogen abstraction. Fragmentation involves

$$\phi-\overset{\overset{\displaystyle O}{\|}}{C}-\phi' \xrightarrow{h\nu} \left(\phi-\overset{\overset{\displaystyle O}{\|}}{C}-\phi'\right)^{*} \longrightarrow \phi-\overset{\overset{\displaystyle O}{\|}}{C}\cdot + \cdot\phi' \tag{3-50}$$

while hydrogen abstraction occurs only in the presence of a hydrogen donor (RH).

$$\phi-\overset{\overset{\displaystyle O}{\|}}{C}-\phi' \xrightarrow{h\nu} \left(\phi-\overset{\overset{\displaystyle O}{\|}}{C}-\phi'\right)^{*} \xrightarrow{RH} \phi-\overset{\overset{\displaystyle OH}{|}}{\underset{\displaystyle \cdot}{C}}-\phi' + R\cdot \tag{3-51}$$

Amines with abstractable α-hydrogens are the most efficient and extensively used hydrogen donors; less efficient donors include alcohols and ethers. The hydrogen abstraction reaction is generally more efficient than fragmentation and occurs at higher wavelengths (lower-energy photons).

Similar photolytic reactions occur with benzoin (**VIII**), benzoin ethers (**IX**), benzil (**X**), and benzil ketals (**XI**). These and related compounds are extensively used as

$$\phi-\overset{\overset{\displaystyle O}{\|}}{C}-\overset{\overset{\displaystyle OH}{|}}{C}H-\phi \qquad \phi-\overset{\overset{\displaystyle O}{\|}}{C}-\overset{\overset{\displaystyle OR}{|}}{C}H-\phi$$
$$\textbf{VIII} \qquad\qquad\qquad \textbf{IX}$$

$$\phi-\overset{\overset{\displaystyle O}{\|}}{C}-\overset{\overset{\displaystyle O}{\|}}{C}-\phi \qquad \phi-\overset{\overset{\displaystyle O}{\|}}{C}-\overset{\overset{\displaystyle OR}{|}}{\underset{\underset{\displaystyle OR}{|}}{C}}-\phi$$
$$\textbf{X} \qquad\qquad\qquad \textbf{XI}$$

commercial photoinitiators, usually in conjunction with hydrogen donors [Encinas et al., 1989a, 1989b; Groenenboom et al., 1982; Ledwith, 1976; Lipscomb and Tarshiani, 1988; Pappas, 1988]. The relative amounts of the fragmentation and hydrogen abstraction reactions vary depending on the type of initiator (which determines the relative stabilities of the radicals formed from the two reactions) and also on the hydrogen donor used.

A variety of other photoinitiators have been studied, including chelate and other organometallic derivatives of transition metals, pyrene–triethylamine, acylphosphine oxides, thioxanthones, and fluorene [Baxter and Davidson, 1988; Lougnot et al., 1989].

Dye-sensitized photopolymerizations are of interest in that the spectral range of photoinitiation can be extended into the visible region. A variety of dyes such as methylene blue, thionine, fluorescein, and eosin undergo excitation and can interact with an appropriate substance to yield radicals capable of initiating polymerization. Dye-sensitized polymerizations often comprise redox systems involving electron or hydrogen transfer between the excited dye and some other substance. An example is the methylene blue-p-toluenesulfinate ion system. The p-toluenesulfinate ion is photooxidized to the corresponding radical, which initiates polymerization.

$$CH_3-\langle\bigcirc\rangle-SO_2^- \longrightarrow CH_3-\langle\bigcirc\rangle-SO_2 \cdot \tag{3-52}$$

Some dye-sensitized polymerizations appear to proceed by energy transfer from excited dye Z^* to another compound C,

$$Z^* + C \rightarrow Z + C^* \tag{3-53}$$

Radicals form by decomposition of the excited state of C,

$$C^* \rightarrow \text{radicals} \tag{3-54}$$

The overall result is that, whereas C^* cannot be produced by direct irradiation of C with frequency ν, the excitation of C is accomplished because Z^* is able to transfer the energy to C at an appropriate frequency ν', which can be absorbed by C.

The detailed mechanism of photoexcitation is discussed elsewhere [Holden, 1988]. Although beyond the scope of the present text, a few considerations are presented. Most excitations of ketones involve $n \rightarrow \pi^*$ excited states, although $\pi \rightarrow \pi^*$ may also be important depending on the particular photoinitiator and solvent. ($n \rightarrow \pi^*$ transitions are usually lower in energy than $\pi \rightarrow \pi^*$.) Photoinitiation in a number of redox systems including some dye-sensitized, organometallic, and ketone–amine systems has been postulated as occurring through the formation of exciplex charge transfer complexes. An exciplex (**XIII** or **XV**), which is a charge transfer complex, is formed by complexation (either **XII** or **XIV**) of a donor (D) and acceptor (A) after photoexcitation of only one or the other component (not both) in systems which give no evidence of ground-state complexation:

$$D \xrightarrow{h\nu} D^* \xrightarrow{A} (D^*,A) \longrightarrow (\dot{D}^+,\dot{A}^-)^* \longrightarrow \text{radicals} \tag{3-55}$$
$$\qquad\qquad\quad \textbf{XII} \qquad\quad \textbf{XIII}$$

$$A \xrightarrow{h\nu} A^* \xrightarrow{D} (D,A^*) \longrightarrow (\dot{D}^+,\dot{A}^-)^* \longrightarrow \text{radicals} \tag{3-56}$$
$$\qquad\qquad\quad \textbf{XIV} \qquad\quad \textbf{XV}$$

Formation of the exciplex is a different process than occurs in systems in which the donor and acceptor form a ground-state complex (**XVI**), which undergoes excitation (**XVII**),

$$A + D \longrightarrow (A,D) \overset{h\nu}{\longrightarrow} (\dot{A}^+,\dot{D}^-)^* \longrightarrow \text{radicals} \qquad (3\text{-}57)$$
$$\textbf{XVI} \qquad \textbf{XVII}$$

The ground-state complex **XVI** may or may not be a charge-transfer complex. Exciplex interaction occurs readily because D* is a better donor than D and A* is a better acceptor than A. Thus exciplex formation expands the range of acceptor and donor species that undergo complexation.

3-4c-3   Rate of Photopolymerization

The rate of photochemical initiation is given by

$$R_i = 2\phi I_a \qquad (3\text{-}58)$$

where I_a is the intensity of absorbed light in moles (called *Einsteins* in photochemistry) of light quanta per liter-second and ϕ is the number of propagating chains initiated per light photon absorbed. ϕ is referred to as the *quantum yield for initiation*. The factor of 2 in Eq. 3-58 is used to indicate that two radicals are produced per molecule undergoing photolysis. The factor of 2 is not used for those initiating systems that yield only one radical instead of two. Thus the maximum value of ϕ is 1 for all photoinitiating systems. ϕ is synonymous with f in that both describe the efficiency of radicals in initiating polymerization. (The reader is cautioned that the above definition of ϕ is not always used. An alternate approach is to replace ϕ by $f\phi'$, where ϕ' is the number of radicals produced per light photon absorbed and f is the previously defined initiator efficiency. Also, quantum yields can be defined in terms of the quantum yield for reaction of the initiator species or quantum yield for monomer polymerized. These quantum yields are not synonymous with the quantum yield for initiation.)

An expression for the polymerization rate is obtained by combining Eqs. 3-58 and 3-25 to yield

$$R_p = k_p[\text{M}] \left(\frac{\phi I_a}{k_t} \right)^{1/2} \qquad (3\text{-}59)$$

It is often convenient to express the absorbed light intensity by

$$I_a = \epsilon I_o[\text{A}]b \qquad (3\text{-}60)$$

where I_o is the incident light intensity, A is the species that undergoes photo-excitation, ϵ is the molar absorptivity (extinction coefficient) of A at the particular frequency of radiation absorbed, and b is the thickness of reaction system being irradiated. (ϵ usually has units of liter/mole-cm.) Combination of Eq. 3-60 with Eqs. 3-58 and 3-25 allows R_i and R_p to be expressed as

$$R_i = 2\phi\epsilon I_o[\text{A}]b \qquad (3\text{-}61)$$

$$R_p = k_p[M] \left(\frac{\phi \epsilon I_o[A]b}{k_t} \right)^{1/2} \tag{3-62}$$

The light intensities delivered by various light sources are usually known in units such as kcal/sec, kJ/sec, or erg/sec, and it is necessary to convert them into the appropriate units of Einsteins (moles) of light quanta per liter-sec before use in the above equations. This is accomplished by a knowledge of the wavelength λ or frequency v of the light energy employed. The energy in an Einstein of light is Nhv or Nhc/λ where h is Planck's constant, c is the speed of light, and N is Avogadro's number.

The use of Eqs. 3-61 and 3-62 assumes that the incident light intensity does not vary appreciably throughout the thickness of the reaction vessel. This will be true only when the absorption of light is quite low or very thin reaction vessels are employed. For most polymerizations, the light absorption will not be negligible and I_o and I_a will vary with thickness. Under these conditions an expression for I_a can be obtained from Lambert–Beer's law

$$I = I_o(10^{-\epsilon[A]b'}) \tag{3-63}$$

where I is the light intensity at a distance b' into the reaction vessel. The light intensity absorbed by the reaction system is then given by

$$I_a = I_o(1 - 10^{-\epsilon[A]b}) \tag{3-64}$$

where b is the total thickness of the reaction vessel. The polymerization rate can then be analyzed by combining Eqs. 3-59 and 3-64 to yield

$$R_p = k_p[M] \left(\frac{\phi I_o[1 - 10^{-\epsilon[A]b}]}{k_t} \right)^{1/2} \tag{3-65}$$

3-4c-3-a Measurement of Absorbed Light. The use of Eq. 3-65 may be avoided and Eq. 3-59 (with I_a defined by Eq. 3-64) used instead by directly measuring I_a in the particular polymerization system. Measurement of light intensity is generally referred to as *actinometry*. Chemical, thermal, and electrical actinometers are used [Ranby and Rabek, 1975]. Thermal and electrical types include photomultipliers, semiconductor photodetectors, and thermocouples which operate on the principle of converting photon energy to either electrical or thermal energy. The thermal and electrical actinometers are generally more convenient to use than chemical actinometers. I_a is measured by placing the thermal or electrical actinometer directly behind the reaction vessel and measuring the differences in I_o when the vessel is empty compared to when it holds the reaction system. A chemical actinometer involves using a chemical reaction whose quantum yield is known. The most-used chemical actinometer is probably potassium ferrioxalate. Irradiation of $K_3Fe(C_2O_4)_3$ as an aqueous acidified solution results in reduction of ferric ions to ferrous. Ferrous ions can be followed spectrophotometrically as the red-colored complex with 1,10-phenanthroline. The quantum yield for ferric to ferrous conversion varies with wavelength and temperature. Other chemical actinometers include uranyl oxalate (involving oxidation of oxalate ion) and benzophenone (reduction to benzhydrol). Chemical actinometers are placed directly into the reaction vessel to be used for polymerization studies.

3-4c-3-b   General Observations. Inspection of the rate equations developed above for photochemical polymerizations indicates R_p is first-order in [M], $\frac{1}{2}$-order in light intensity and $\frac{1}{2}$-order in [A] for the case where there is negligible attenuation of the light intensity in traversing the reaction vessel, that is, Eqs. 3-59 and 3-62 apply. When the photoexcitation involves monomer (A = M), the dependence of R_p on [M] becomes $\frac{3}{2}$-order. Abnormal orders in [M] have been found under certain circumstances. An extreme example is the iodine bromide photopolymerization of methyl methacrylate where R_p was found to be 3-order in monomer and 0-order in both iodine bromide and I_o at high intensities [Ghosh and Banerjee, 1978]. This was explained by initiation involving the formation of primary radicals from a monomer-iodine bromide complex together with a slow reaction of primary radicals with monomer and coupled with primary termination. At lower light intensities bimolecular termination occurs and R_p is 2-order in [M] and $\frac{1}{2}$-order in both iodine bromide and I_o. On the other extreme, the dependence of R_p on [M] drops from first-order to $\frac{1}{2}$-order as [M] increases in the photopolymerization of methyl methacrylate with ketone photoinitiators [Lissi et al., 1979]. With increasing monomer concentration, monomer quenches the excited state for photoinitiation. Quenching is a generally observed phenomenon in photopolymerizations—leading to a lower than expected dependence of R_p on some component in the system that acts as the *quencher*. In simple terms a *quencher* (which can be A or M or solvent) undergoes energy transfer with the photoexcited species to dissipate the excitation energy. Quenching can be observed if the concentration of A or M is too high.

For photopolymerizations where appreciable attenuation of light intensity occurs, the dependence of R_p on [M], [A], and I_o will be more complex—being exactly defined by Eq. 3-65. Equation 3-65 is the useful equation for describing the practical aspects of photopolymerization. It allows one to calculate the maximum thickness of a reaction system that can be photopolymerized. That maximum thickness is the value of b for which

$$10^{-\epsilon[A]b} \rightarrow 0 \tag{3-66}$$

The maximum thickness decreases with increasing ϵ and [A]. But from the practical viewpoint, one almost never works with reaction systems of the maximum thickness. As a light beam penetrates through a reaction system, I_o and I_a decrease continuously as described by Eqs. 3-63 and 3-64. Thus, although the total R_p and R_i increase with thickness, there is a gradient in R_i and R_p with thickness. If one attempts to polymerize any particular thickness of reaction system it is more important to calculate the R_p at various points in that thickness (e.g., the front surface, middle, and rear surface) than simply to calculate the total R_p for that complete sample thickness. If the reaction systems's thickness is too large, the polymerization rate in the rear (deeper) thicknesses will be too low and those portions will be only partially polymerized—the result would be highly negative in terms of uniformity of product properties. A consideration of Eqs. 3-64 and 3-65 indicates that this problem is more severe for reactions carried out at high values of [A] and ϵ because the variation of light intensity and R_i with thickness is greater for larger values of [A] and ϵ.

Many commercial photopolymerizations are carried out in air. This results in a major complication due to inhibition of radicals in the surface layers of a reaction system by reaction with oxygen (Sec. 3-7b). The problem of nonuniformity of initiation rates can be minimized (but not completely overcome) by using a photoinitiator with absorption bands at two different wave lengths, one of which absorbs strongly (large ϵ) and the other weakly (small ϵ). The high ϵ absorption band gives a high intensity at the surface of the reaction system to overcome the loss of radicals by reaction with oxygen, but the intensity rapidly decreases as a function of penetration. The intensity of the low ϵ absorption band is less effected by penetration and provides a more uniform intensity as a function of thickness.

3-4d   Initiation by Ionizing Radiations

Radioactive sources and particle accelerators are used to initiate chain polymerizations. Electrons (β-rays), neutrons, and α-particles (He^{2+}) are particulate radiations while γ- and x-rays are electromagnetic radiations. The interactions of these radiations with matter are more complex than those of light [Chapiro, 1962; Wilson, 1974]. The chemical effects of the different types of radiation are qualitatively the same, although there are quantitative differences. Molecular excitation may occur with the subsequent formation of radicals in the same manner as in photolysis, but ionization of a compound C by ejection of an electron

$$C + radiation \rightarrow C^{\ddagger} + e^{-} \tag{3-67}$$

is more probable because of the higher energies of these radiations compared to visible or ultraviolet light energy. (Ionizing radiations have particle or photon energies in the range 10 keV–100 meV (1 fJ–16,000 fJ) compared to 1–6 eV for visible-ultraviolet photons.) For this reason such radiations are termed *ionizing radiations*. The cation formed in Eq. 3-67 is a radical–cation $C^{\ddagger}$ formed by loss of a π-electron and has both radical and positive centers (Sec. 5-2a-3-d). The radical–cation can propagate at the radical and/or cationic centers depending on reaction conditions. The radical–cation can also dissociate to form separate radical and cationic species

$$C^{\ddagger} \rightarrow A\cdot + B^{+} \tag{3-68}$$

The initially ejected electron may be attracted to the cation B^{+} with the formation of another radical,

$$B^{+} + e^{-} \rightarrow B\cdot \tag{3-69}$$

Radicals may also be produced by a sequence of reactions initiated by the capture of an ejected electron by C

$$C + e^{-} \rightarrow C^{-} \tag{3-70}$$

$$C^{-} \rightarrow B\cdot + A^{-} \tag{3-71}$$

$$A^{-} \rightarrow A\cdot + e^{-} \tag{3-72}$$

where C^{-} may or may not be an excited species depending on the energy of the electron.

The radiolysis of olefinic monomers results in the formation of cations, anions, and free radicals as described above. It is then possible for these species to initiate chain polymerizations. Whether a polymerization is initiated by the radicals, cations, or anions depends on the monomer and reaction conditions. Most radiation-initiated polymerizations are radical polymerizations. It is usually only at low temperatures that the ionic species are stable enough to initiate polymerization. At ambient temperatures or higher, the ionic species usually are not stable and dissociate to yield radicals. Radiolytic initiation can also be carried out using initiators or other compounds which are prone to undergo decomposition on irradiation.

Radiation-initiated ionic polymerizations, however, have been established in a number of systems. This occurs under conditions where ionic species have reasonable stabilities and the monomer is susceptable to ionic propagation. Cationic polymerizations of isobutylene, vinyl ethers, and styrene and anionic polymerization of acrylonitrile have been observed [Chapiro, 1962; Kubota et al., 1978; Wilson, 1974]. Ionic polymerizations have been observed at temperatures as high as 30–50°C, but lower temperatures are generally required, usually not because the ionic rates increase with decreasing temperature but because the radical contribution decreases. The role of reaction conditions has been well-established for styrene polymerization [Machi et al., 1972; Squire et al., 1972; Takezaki et al., 1978]. The dryness of the reaction system and the absence of any other species which can terminate ions (either those formed in the radiolytic reactions (Eqs. 3-67 through 3-72) or the ionic propagating species) is critical. Water and other similar compounds terminate ions by transferring a proton or negative fragment (see Chap. 5). Styrene monomer as normally purified and "dried" often has sufficient water present (ca. 10^{-2}–10^{-4} M) to prevent ionic polymerization— only radical polymerization occurs. With *superdried* monomer ($[H_2O] < 10^{-5}$–10^{-7} M), one observes a bimodal-molecular-weight distribution. This indicates the simultaneous occurrence of cationic and radical polymerizations since the two are known to yield different-molecular-weight polymers. With decreasing water concentration ($\sim < 10^{-7}$–10^{-10} M) the polymerization proceeds entirely by cationic propagation as evidenced by a unimodal MWD, which is characteristic of the cationic MWD. The extent of the ionic contribution relative to radical polymerization also increases with increasing radiation intensity, perhaps due to the rapid depletion of water at the higher intensities. Other evidence for the ionic contributions include kinetic and activation energy data.

The kinetics of radiation-initiated polymerizations follow in a relatively straightforward manner those of photolytic polymerization. The initiation rate is determined by the radiation intensity and the concentration and susceptibility of the compound that radiolyzes to yield the initiating species (ionic and/or radical). The final expression for R_p is determined by the exact details of the initiation, propagation and termination steps.

Radiation-initiated polymerization has achieved some commercial success in the coatings area but far less than photochemical polymerization in spite of the greater penetrating power of ionizing radiation (except for α-particles). The reason is the higher costs and safety problems of ionizing radiation sources compared to photochemical sources.

3-4e Pure Thermal Initiation

Many monomers appear to undergo a spontaneous polymerization when heated in the apparent absence of catalysts. On careful investigation few of these involve the thermal production of radicals from the monomer. In most cases the observed polymerizations are initiated by the thermal homolysis of impurities (including peroxides or hydroperoxides formed due to O_2) present in the monomer. Most monomers when exhaustively purified (and in exhaustively purified reaction vessels) do not undergo a *purely thermal, self-initiated polymerization* in the dark. Only styrene has been unequivocally shown to undergo self-initiated polymerization. Methyl methacrylate was also considered to undergo self-initiated polymerization but recent work indicates that most, but not all, of the previously reported self-initiated polymerization was caused by adventitious peroxides that were difficult to exclude by the usual purification techniques [Lehrle and Shortland, 1988]. Other monomers that may be susceptible to self-initiated polymerization are substituted styrenes, acenaphthylene, 2-vinylthiophene, and 2-vinylfuran as well as certain pairs of monomers [Brand et al., 1980; Hall, 1983; Lingnau et al., 1980; Sato et al., 1977]. The rate of self-initiated polymerization is much slower than the corresponding polymerization initiated by thermal homolysis of an initiator such as AIBN, but far from negligible; for example, the self-initiated polymerization rate for neat styrene at 60°C is 1.98×10^{-6} mole/liter-sec [Graham et al., 1979].

The initiation mechanism for styrene polymerization has been established [Barr et al., 1978; Graham et al., 1979; Husain and Hamielec, 1978; Kaufman, 1979; Olaj et al., 1976, 1977a, 1977b]. It involves the formation of a Diels–Alder dimer (**XVIII**) of styrene followed by transfer of a hydrogen atom from the dimer to a styrene molecule

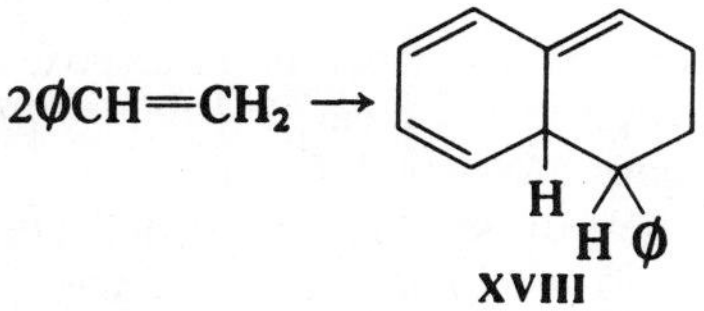

$$(3\text{-}73a)$$

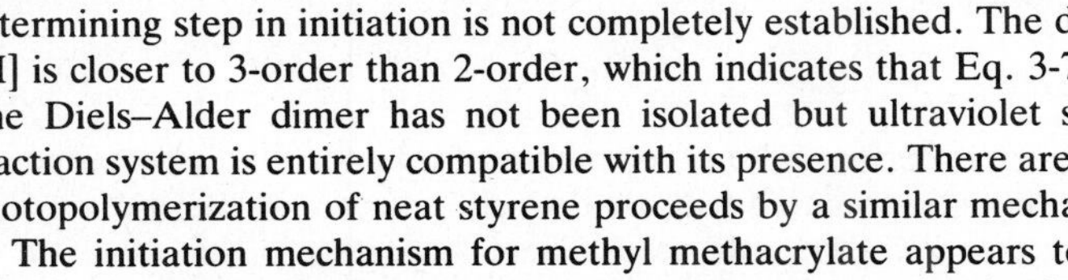

$$(3\text{-}73b)$$

Whether formation of the Diels–Alder dimer or its reaction with styrene is the rate-determining step in initiation is not completely established. The dependence of R_i on [M] is closer to 3-order than 2-order, which indicates that Eq. 3-73b is the slow step. The Diels–Alder dimer has not been isolated but ultraviolet spectroscopy of the reaction system is entirely compatible with its presence. There are indications that the photopolymerization of neat styrene proceeds by a similar mechanism.

The initiation mechanism for methyl methacrylate appears to involve the initial formation of a biradical by reaction of two monomer molecules followed by hydrogen

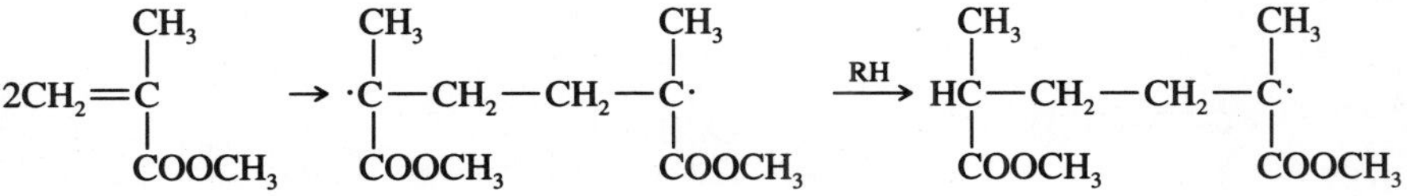

$$(3\text{-}74)$$

transfer from some species in the reaction system to convert the biradical to a mono-radical [Lingnau and Meyerhoff, 1984a, 1984b].

3-4f Other Methods of Initiation

3-4f-1 Electroinitiation

Electrolysis of monomer solutions has been studied as a means of initiating polymerization [Olaj, 1987; Otero and Mugarz, 1987]. Aqueous and various organic solvents (e.g., DMF, methylene chloride, methanol) have been employed in the presence of various inorganic compounds. The latter are used either to conduct current or participate in the initiation process. Depending on the components present, radicals and/or ions are formed when current is passed through the reaction system, and these initiate polymerization. Such polymerizations are referred to as *electroinitiated* or *electrolytic polymerizations*. An interesting example is the electrolysis of an acetonitrile solution of acrylamide containing tetrabutylammonium perchlorate [Samal and Nayak, 1988]. Radical polymerization occurs at the anode initiated by $ClO_4\cdot$ formed by oxidation of ClO_4^-. Anionic polymerization occurs at the cathode initiated by direct transfer of electrons to the monomer (Sec. 5-3a-2).

3-4f-2 Plasma

Plasma polymerization occurs when a gaseous monomer is placed in an electric discharge at low pressures under conditions where a plasma is created [Boenig, 1988; Yasuda, 1986]. A plasma consists of ionized gaseous molecules. In some cases, the system is heated and/or placed in a radiofrequency field to assist in creating the plasma. A variety of organic molecules including alkenes, alkynes, and alkanes undergo polymerization to high-molecular-weight products under these conditions. The propagation mechanisms appear to involve both ionic and radical species. Plasma polymerization offers a potentially unique method of forming thin polymer films for uses such as thin-film capacitors, antireflection coatings, and various types of thin membranes.

3-4g Initiator Efficiency

3-4g-1 Definition of ƒ

When a material balance is performed on the amount of initiator that is decomposed during a polymerization and compared with that which initiates polymerization, it is apparent that the initiator is inefficiently used. There is wastage of initiator due to two reactions. One is the *induced decomposition of initiator* by the attack of propagating radicals on the initiator, for example,

$$M_n\cdot + \phi\overset{O}{\overset{\|}{C}}O-O\overset{O}{\overset{\|}{C}}\phi \longrightarrow M_n-O\overset{O}{\overset{\|}{C}}\phi + \phi\overset{O}{\overset{\|}{C}}O\cdot \tag{3-75}$$

This reaction is termed *chain transfer to initiator* and is considered further in Sec. 3-6b. The induced decomposition of initiator does not change the radical concentration during the polymerization, since the newly formed radical ($\phi COO\cdot$) will initiate a new polymer chain. However, the reaction does result in a wastage of initiator. A molecule of initiator is decomposed without an increase in the number of propagating radicals or the amount of monomer being converted to polymer.

The second wastage reaction is that involving the side reaction(s) of the radicals formed in the primary step of initiator decomposition. Some of the radicals formed in the primary decomposition step, for example, in the reaction

$$\phi \overset{O}{\underset{\|}{C}} O - O \overset{O}{\underset{\|}{C}} \phi \longrightarrow 2 \ \phi COO \cdot \tag{3-76}$$

undergo reactions to form neutral molecules instead of initiating polymerization. It is this wastage reaction that is referred to when discussing the initiator efficiency f. The initiator efficiency f is defined as the fraction of radicals formed in the primary step of initiator decomposition, which are successful in initiating polymerization. The initiator efficiency is considered exclusive of any initiator wastage by induced decomposition. The reader is cautioned, however, that reported literature values of f do not always make this distinction. Very frequently, the calculations of f neglect and do not correct for the occurrence of induced decomposition. Such f values may be considered as the "effective" or "practical" initiator efficiency in that they give the net or overall initiation efficiency of the initial catalyst concentration. The preferred method of quantitatively handling induced decomposition of initiator, however, is that which is discussed in Sec. 3-6b-3.

3-4g-2 Mechanism of $f < 1$: Cage Effect

The values of f for most initiators lie in the range 0.3–0.8. To understand why the initiator efficiency will be less than unity, consider the following reactions that occur in polymerizations initiated by benzoyl peroxide:

$$\phi COO - OOC\phi \rightleftharpoons [2\phi COO \cdot] \tag{3-77}$$

$$[2\phi COO \cdot] \longrightarrow [\phi COO\phi + CO_2] \tag{3-78}$$

$$[2\phi COO \cdot] + M \longrightarrow \phi CO_2 \cdot + \phi COOM \cdot \tag{3-79}$$

$$[2\phi COO \cdot] \longrightarrow 2\phi COO \cdot \tag{3-80}$$

$$\phi COO \cdot + M \longrightarrow \phi COOM \cdot \tag{3-81}$$

$$\phi COO \cdot \longrightarrow \phi \cdot + CO_2 \tag{3-82}$$

$$\phi \cdot + M \longrightarrow \phi M \cdot \tag{3-83}$$

$$\phi \cdot + \phi COO \cdot \longrightarrow \phi COO\phi \tag{3-84}$$

$$2\phi \cdot \longrightarrow \phi - \phi \tag{3-85}$$

The brackets around any species indicates the presence of a solvent cage, which traps the radicals for some period before they diffuse apart. Equation 3-77 represents the primary step of initiator decomposition into two radicals, which are held within the solvent cage. The radicals in the solvent cage may undergo recombination (the reverse of Eq. 3-77), reaction with each other (Eq. 3-78), reaction with monomer (Eq. 3-79), or diffusion out of the solvent cage (Eq. 3-80). Once outside the solvent cage the radicals may react with monomer (Eq. 3-81) or decompose according to Eq. 3-82 to yield a radical which may undergo various reactions (Eqs. 3-83 to 3-85). Reaction 3-82 probably also occurs within the solvent cage followed by diffusion of the two species.

Recombination of the primary radicals (Eq. 3-77) has no effect on the initiator efficiency. Initiation of the polymerization occurs by Eqs. 3-79, 3-81, and 3-83. The initiator efficiency is decreased by the reactions indicated by Eqs. 3-78, 3-84, and 3-85, since their products are stable and cannot give rise to radicals. Of these reactions the decomposition within the solvent cage (Eq. 3-78) is usually much more significant than the others in decreasing the value of f. That this reaction is competitive with those leading to initiation of polymerization can be seen by considering the time scale [Koenig and Fischer, 1973; Noyes, 1961] of the various events. The average life of neighboring radicals is perhaps 10^{-10}–10^{-9} sec. Since rate constants for radical–radical reactions are in the range 10^7 liters/mole-sec and higher, and the concentration of radicals in the solvent cage is ~ 10 M, there is a reasonable probability that Reaction 3-78 will occur. Reaction 3-79 cannot compete effectively with Reaction 3-78, since radical addition reactions have much lower rate constants (10–10^5 liters/mole-sec).

Secondary-type decay of radicals within the solvent cage can also occur

$$[2\phi COO\cdot] \longrightarrow [\phi COO\cdot + CO_2 + \phi\cdot] \tag{3-86}$$

$$[\phi COO\cdot + CO_2 + \phi\cdot] \longrightarrow [CO_2 + \phi COO\phi] \tag{3-87}$$

$$[\phi COO\cdot + CO_2 + \phi\cdot] \longrightarrow \phi COO\cdot + CO_2 + \phi\cdot \tag{3-88}$$

$$[2\phi COO\cdot] \longrightarrow [2CO_2 + 2\phi\cdot] \tag{3-89}$$

$$[2CO_2 + 2\phi\cdot] \longrightarrow [2CO_2 + \phi{-}\phi] \tag{3-90}$$

$$[2CO_2 + 2\phi\cdot] \longrightarrow 2CO_2 + 2\phi\cdot \tag{3-91}$$

Reactions 3-86 and 3-89 coupled, respectively, with Eqs. 3-88 and 3-91 do not lower f but do lower f when coupled, respectively, with Eqs. 3-87 and 3-90. The radicals that escape the solvent cage (Eqs. 3-88 and 3-91) then react as per Eqs. 3-81 through 3-85.

Lowering of the initiator efficiency by reactions analogous to Eq. 3-78 is referred to as the *cage effect* [Bamford, 1988; Koenig and Fischer, 1973; Martin, 1973]. It is a general phenomenon observed in almost all initiation systems. About the only exceptions are the initiation systems such as

$$Fe^{2+} + H_2O_2 \rightarrow Fe^{3+} + HO^- + HO\cdot \tag{3-92}$$

which do not produce radicals in pairs. However, even in some such instances the radical may be capable of reacting with a nonradical species in the solvent cage.

Acetoxy radicals from acetyl peroxide undergo partial decarboxylation and radical combination within the solvent cage leads to stable products incapable of producing radicals

$$CH_3COO{-}OOCCH_3 \longrightarrow [2\ CH_3COO\cdot] \nearrow^{[CH_3COOCH_3 + CO_2]} \searrow_{[CH_3CH_3 + 2\ CO_2]}$$

$$[CH_3COOCH_3 + CO_2] \tag{3-93a}$$

$$[CH_3CH_3 + 2\ CO_2] \tag{3-93b}$$

The homolysis of AIBN occurs in a concerted process with simultaneous breakage of the two C—N bonds to yield nitrogen and 2-cyano-2-propyl radicals. Reaction of the radicals with each other can occur in two ways to yield tetramethylsuccinodinitrile and dimethyl-*N*-(2-cyano-2-isopropyl)ketenimine.

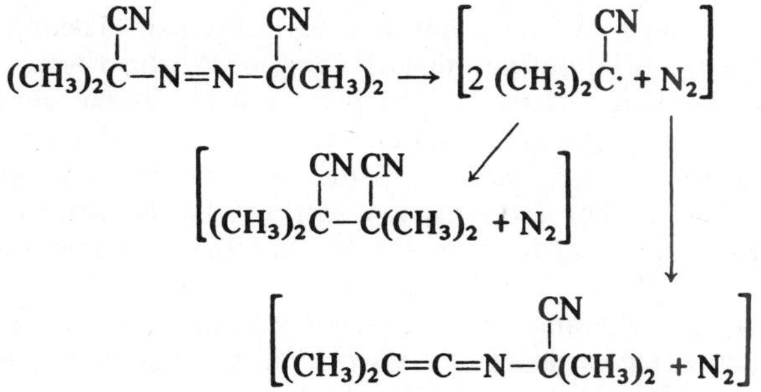

$$
\underset{\substack{|\\ \text{CN}}}{(CH_3)_2C}-N{=}N-\underset{\substack{|\\ \text{CN}}}{C(CH_3)_2} \rightarrow \left[2\ \underset{\substack{|\\ \text{CN}}}{(CH_3)_2C\cdot} + N_2\right]
$$

$$
\left[\underset{\substack{|\ \ |\\ \text{CN CN}}}{(CH_3)_2C{-}C(CH_3)_2} + N_2\right] \tag{3-94a}
$$

$$
\left[(CH_3)_2C{=}C{=}N-\underset{\substack{|\\ \text{CN}}}{C(CH_3)_2} + N_2\right] \tag{3-94b}
$$

A minor reaction of 2-cyano-2-propyl radicals is disproportionation to methacrylonitrile and isobutyronitrile [Moad et al., 1984; Starnes et al., 1984]. This presents a complication for polymerizations carried out to high conversions where the methacrylonitrile concentration is significant since methacrylonitrile undergoes copolymerization with many monomers.

The cage effect has also been well-documented as the cause of lowered ϕ values in photoinitiation [Berner et al., 1978; Braun and Studenroth, 1979]. Benzoin photolysis yields benzaldehyde,

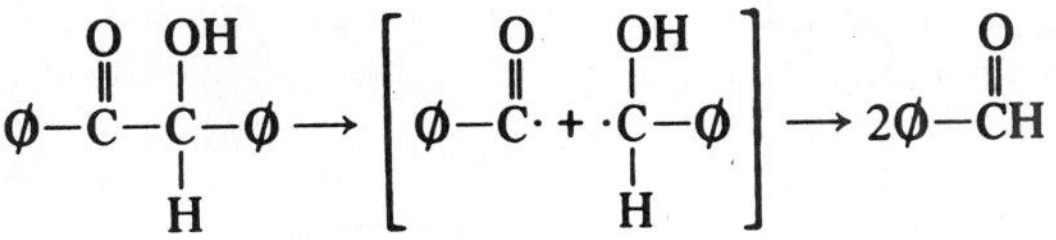

$$
\phi{-}\underset{\substack{||\\ O}}{C}{-}\underset{\substack{|\\ H}}{\overset{OH}{C}}{-}\phi \rightarrow \left[\phi{-}\underset{\substack{||\\ O}}{C}\cdot + \cdot\underset{\substack{|\\ H}}{\overset{OH}{C}}{-}\phi\right] \rightarrow 2\phi{-}\underset{\substack{||\\ O}}{C}H \tag{3-95}
$$

The photolysis of benzildimethylketal yields benzoyl and benzylketal radicals in the primary cleavage. The latter radical undergoes decay within the cage to yield methyl benzoate and methyl radical. Coupling of the methyl and benzoyl radicals decrease f.

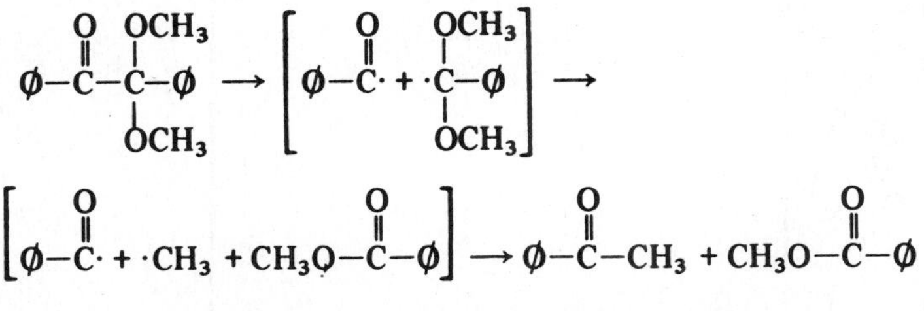

$$
\phi{-}\underset{\substack{||\\ O}}{C}{-}\underset{\substack{|\\ OCH_3}}{\overset{OCH_3}{C}}{-}\phi \rightarrow \left[\phi{-}\underset{\substack{||\\ O}}{C}\cdot + \cdot\underset{\substack{|\\ OCH_3}}{\overset{OCH_3}{C}}{-}\phi\right] \rightarrow
$$

$$
\left[\phi{-}\underset{\substack{||\\ O}}{C}\cdot + \cdot CH_3 + CH_3O{-}\underset{\substack{||\\ O}}{C}{-}\phi\right] \rightarrow \phi{-}\underset{\substack{||\\ O}}{C}{-}CH_3 + CH_3O{-}\underset{\substack{||\\ O}}{C}{-}\phi \tag{3-96}
$$

The initiator efficiency and quantum yield for initiation are generally independent of monomer concentration and of the radical coupling reactions occurring outside the solvent cage (Eqs. 3-84, 3-85). The prime reason for $f < 1$ is the reactions occurring within the solvent cage. Once a radical has diffused out of the solvent cage, the reaction with monomer (Eq. 3-81) occurs predominantly in preference to other reactions [Fink, 1983]. Even if Reaction 3-82 occurs this will be followed by Reaction 3-83 in preference to the two radical combination reactions (Eqs. 3-84 and 3-85). The preference for initiation of polymerization arises from the much greater monomer concentrations $(10^{-1}-10\ M)$ compared to the radical concentrations $(10^{-7}-10^{-9}\ M)$ that are normally present in polymerization systems. However, one can observe a variation of f with the monomer concentration at low monomer concentrations. Figure 3-5 shows this effect in the AIBN initiation of styrene polymerization. The initiator efficiency increases very rapidly with [M] and a limiting value is quickly reached. Theoretical predictions [Koenig and Fischer, 1973] indicate that an effect of [M] on f would be expected at monomer concentrations below $10^{-1}-10^{-2}\ M$, and this is the general observation. f is affected at lower concentrations as one increases the initiation rate, since the radical–radical reactions become more probable. An example is the photopolymerization of methyl methacrylate using benzoin methyl ether [Carlblom and Pappas, 1977], where ϕ decreases from 0.65 to 0.30 as the concentration of the photosensitizer increases from $10^{-3}\ M$ to $4 \times 10^{-2}\ M$.

The initiator efficiency varies to differing extents depending on the identities of the monomer, solvent, and initiator. f decreases as the viscosity of the reaction medium increases. With increasing viscosity, the lifetimes of radicals in the solvent cage are increased—leading to greater extents of radical–radical reactions within the solvent cage. The effect of viscosity has been observed when viscosity was altered by using different solvents. The effect can be quite large, for example, f for bis-azo-1,1'-cyclohexane nitrile at 0°C decreases from 0.22 in benzene to 0.086 in diethyl phthalate [Fischer et al., 1969]. f has also been observed to decrease during the course of a polymerization, since both [M] and viscosity increase with conversion [Russell et al., 1988]. For example, f for AIBN in the polymerization of styrene decreases from 0.75 at conversions up to 30% to 0.20 at 90–95% conversion [Solomon and Moad, 1987]. In some cases f varies with solvent as a result of the solvent reacting with radicals before the radicals can initiate polymerization. The initiator efficiencies of AIBN, benzoyl peroxide, and t-butyl hydroperoxide are lower in carbon tetrachloride or 1-

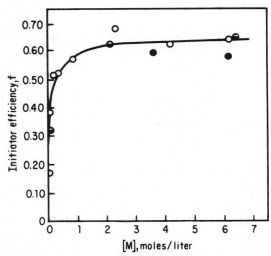

Fig. 3-5   Effect of styrene concentration on the initiator efficiency of azobisisobutyronitrile. Symbols ●, ○, and ◓ refer to experiments with initiator concentrations of 0.20, 0.50 and 1.00 g/liter, respectively. After Bevington [1955] (by permission of The Faraday Society, London).

chlorooctane than in aromatic solvents [Nozaki and Bartlett, 1946; Shahani and Indictor, 1978; Walling, 1954]. Reactions of radicals with solvents are discussed further in Sec. 3-6. The change in initiator efficiency with solvent may in a few instances be due to a solvation effect, for example, the specific solvation of electrophilic radicals by electron-donor solvents [Martin, 1973]. The initiator efficiency for any particular initiator may also vary for different monomers. Thus the value of f for AIBN in the polymerizations of methyl methacrylate, vinyl acetate, styrene, vinyl chloride, and acrylonitrile ranged from 0.6 to 1.0, increasing in that order [Arnett and Peterson, 1952]. This order is a consequence of the relative rates with which radicals add to the different monomers.

3-4g-3 *Experimental Determination of f*

Various methods are employed for the experimental evaluation of the initiator efficiency. (These methods are generally also applicable to determining ϕ for photopolymerization). One method involves the determination and comparison of both the initiator decomposition and the production of polymer molecules. The initiator decomposition is best determined during an actual polymerization. Independent measurement of initiator decomposition in the absence of monomer can give erroneous results. The rates of decomposition of initiator can be quite different for pure initiator as compared to initiator in the presence of monomer, since f varies with monomer concentration. Also in some cases monomer is involved in the initiation step [e.g., molecule-induced homolysis (Eq. 3-35), certain redox initiations described in Sec. 3-4b-1]. AIBN decomposition can be followed relatively easily by following the evolution of nitrogen. Measurement of the polymer number-average molecular weight allows a determination of f by comparison of the number of radicals produced with the number of polymer molecules obtained. This method requires a knowledge of whether termination occurs by coupling or by disproportionation, since the former results in two initiator fragments per polymer molecule and the latter in only one. The occurrence of induced decomposition must be taken into account in calculating the number of radicals produced. Induced decomposition is negligible for azonitriles but occurs to an appreciable extent with many peroxides.

The second method is a variation of the first in which the number of initiator fragments in the polymer is determined by direct analysis of the polymer end groups. The analysis is relatively difficult to perform accurately because of the low concentration of the end groups. For a 50,000-molecular-weight polymer, the end groups comprise only about 0.1% of the total weight of the sample. The use of isotopically labeled initiators such as ^{14}C-labeled AIBN and benzoyl peroxide and ^{35}S-labeled potassium persulfate is a sensitive method for determining the number of initiator fragments [Bonta et al., 1976]. Isotopically labeled initiators also allow a quantitative analysis of the initiator radicals in systems where their identities are not clearcut. For example, the use of benzoyl peroxide with ^{14}C in the carbonyl carbon shows that ϕCOO· is responsible for 98% of initiation of styrene polymerization at 80°C, but this drops to about 60% (with ϕ· initiating 40% of the propagating chains) at 30°C [Berger et al., 1977].

A third method involves the use of radical scavengers, which count the number of radicals in a system by rapidly stopping their growth. Three groups of radical scavengers are used [Bamford, 1988; Koenig and Fischer, 1973]. The first group consists of stable radicals such as 2,2-diphenylpicrylhydrazyl (DPPH) obtained by oxidation of diphenylpicrylhydrazine:

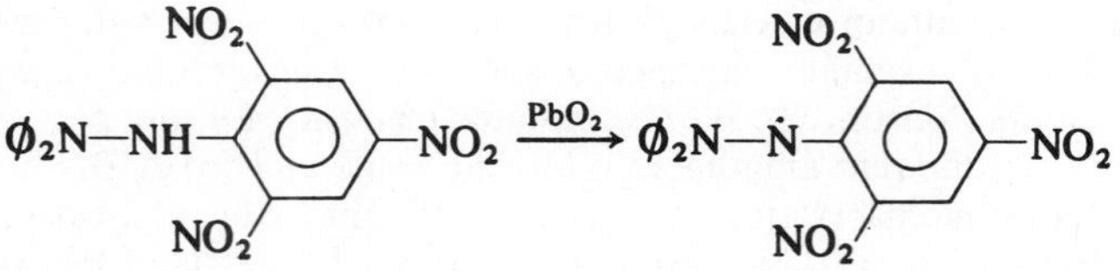

$$(3\text{-}97)$$

The DPPH radical terminates other radicals, probably by the reaction

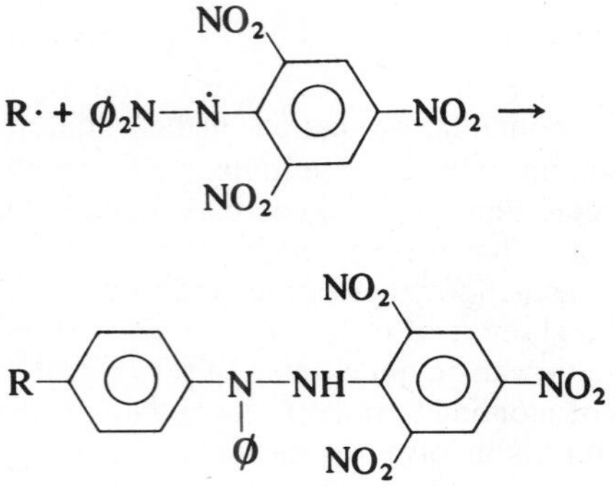

$$(3\text{-}98)$$

The reaction can be easily followed spectrophotometrically, since the DPPH radical is deep violet and the product is usually light yellow or colorless. Other stable radicals include nitroxides, α, γ-bis(diphenylene)-β-phenylallyl, oxygen, triphenylverdazyl, and galvinoxyl [Areizaga and Guzman, 1988; Braun and Czerwinski, 1987; Stickler, 1987]. A second group reacts by fast hydrogen or halogen transfer and consists of thiols, iodine, bromine, and dihydroanthracene. The third group, consisting of molecules that react with radicals, include p-benzoquinone and duroquinone (2,3,5,6-tetramethylbenzoquinone) [Janzen, 1971]. Nitroso and nitrone compounds have been found to be highly useful [Evans, 1979; Sato et al., 1977]. These compounds, referred to as *spin traps*, react with radicals to form stable radicals. 2-Methyl-2-nitrosopropane (**XIX**) and phenyl-N-t-butylnitrone (**XX**) have been used extensively:

$$(CH_3)_3C-N=O + R\cdot \longrightarrow (CH_3)_3C-\overset{\displaystyle |}{\underset{\displaystyle R}{N}}-\dot{O} \qquad (3\text{-}99)$$

$$\textbf{XIX} \qquad\qquad\qquad\qquad \textbf{XXI}$$

$$\phi-CH=\overset{\displaystyle O^-}{\underset{\displaystyle +}{N}}-C(CH_3)_3 + R\cdot \longrightarrow \phi-\overset{\displaystyle |}{\underset{\displaystyle R}{CH}}-\overset{\displaystyle |}{\underset{\displaystyle \dot O}{N}}-C(CH_3)_3 \qquad (3\text{-}100)$$

$$\textbf{XX} \qquad\qquad\qquad\qquad\qquad \textbf{XXII}$$

Electron spin resonance (ESR) spectroscopy can be advantageously used to measure the radical concentrations of the nitroxide radicals (**XXI** and **XXII**) produced, since these are much more stable than the R$\cdot$ radicals. Of greater importance, ESR can be used to determine the structure of R$\cdot$, since the ESR of the nitroxide radical is quite sensitive to the structure of R. (For this purpose, nitroso spin traps are more useful, since the R group in the nitroxide radical is nearer to the lone electron.) This can allow a determination of the structures of radicals first-formed in initiator decomposition, the radicals that actually initiate polymerization (if they are not identical with the former) as well as the propagating radicals [Rizzardo and Solomon, 1979; Sato et al., 1975].

The radical scavenging methods require the general caution that the stoichiometry of the reaction between scavenger and radical must be established. A problem with some scavengers is that their reaction with radicals may not be quantitative. The DPPH radical is an extremely efficient scavenger in many systems. It completely stops vinyl acetate and styrene polymerizations even at concentrations below $10^{-4}\ M$ [Bartlett and Kwart, 1950]. However, the scavenging effect of DPPH is not universally quantitative for all monomers.

A fourth method is the *dead-end* polymerization technique, which allows the simultaneous determination of k_d [Catalgil and Baysal, 1987; Reimschuessel and Creasy, 1978; Stickler, 1979]. Dead-end polymerization refers to a polymerization in which the initiator concentration decreases to such a low value that the half-life of the propagating polymer chains approximates that of the initiator. Under such circumstances the polymerization stops short of completion, and one observes a limiting conversion of monomer to polymer at infinite reaction time. Consider a dead-end polymerization initiated by the thermal homolysis of an initiator. Equations 3-28a and 3-32 can be combined and rearranged to yield

$$\frac{-d[M]}{[M]} = k_p \left(\frac{fk_d[I]_0}{k_t}\right)^{1/2} e^{-k_d t/2}\ dt \tag{3-101}$$

which on integration leads to

$$-\ln\frac{[M]}{[M]_0} = -\ln(1 - p) = 2k_p \left(\frac{f[I]_0}{k_t k_d}\right)^{1/2} (1 - e^{-k_d t/2}) \tag{3-102}$$

where p is the extent of conversion of monomer to polymer and is defined by $([M]_0 - [M])/[M]_0$. At long reaction times ($t \to \infty$), [M] and p reach the limiting values of $[M]_\infty$ and p_∞, respectively, and Eq. 3-102 becomes

$$-\ln\frac{[M]_\infty}{[M]_0} = -\ln(1 - p_\infty) = 2k_p \left(\frac{f[I]_0}{k_t k_d}\right)^{1/2} \tag{3-103}$$

Dividing Eq. 3-102 by Eq. 3-103, rearranging, and then taking logarithms of both sides leads to the useful expression

$$-\ln\left[1 - \frac{\ln(1 - p)}{\ln(1 - p_\infty)}\right] = \frac{k_d t}{2} \tag{3-104}$$

The value of k_d is then easily found from the slope of a plot of the left side of Eq. 3-104 vs time. Figure 3-6 shows the plot for the AIBN initiated polymerization of isoprene at three different temperatures [Gobran et al., 1960]. Since k_d is determined, f can be obtained from either Eq. 3-32 or 3-103 if the ratio $k_p/k_t^{1/2}$ is known from other studies.

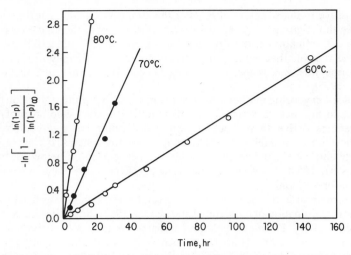

Fig. 3-6 Dead-end polymerization of isoprene initiated by azobisisobutyronitrile. After Gobran et al. [1960] (by permission of Wiley-Interscience, New York).

Equation 3-102 is also useful for determining the time needed to reach different extents of conversion for actual polymerization systems where both [M] and [I] decrease with time.

3-4h Other Aspects of Initiation

There are two possible modes of addition of a primary radical from initiator to the double bond of a monomer: *head* and *tail additions*, corresponding to Eqs. 3.8 and 3-9, respectively. Tail addition is favored over head addition for the same reasons head-to-tail propagation is favored over head-to-head propagation (Sec. 3-2a). This expectation is generally verified in the few systems which have been experimentally examined [Bevington, 1988; Moad et al., 1982, 1984; Solomon and Moad, 1987]. Tail addition occurs exclusively in the AIBN and *t*-butyl peroxide initiated polymerizations of styrene and the AIBN initiated polymerization of vinyl chloride; there is no detectable head addition. For example, the ratio of tail:head additions varies from 20:1 to 13:1, depending on the reaction conditions, for the benzoyl-peroxide-initiated polymerization of styrene. The extent of head addition increases for monomers where there is a smaller difference in stability between the two possible radicals. The tail:head ratio is about 5–6 for the *t*-butyl peroxide initiated polymerization of vinyl acetate. The various results, especially those with vinyl acetate, show that the selectivity between tail and head additions in the initiation step is not as great as the selectivity between head-to-tail and head-to-head additions in propagation, presumably because the typical primary radical is more reactive than the typical propagating radical.

The initiation process appears more complicated than described above, although data are not available in more than a few systems. The benzoyl peroxide initiated polymerization of styrene involves considerable substitution of initiator radicals on the benzene ring for polymerizations carried out at high conversions and high initiator concentrations. About one-third of the initiator radicals from *t*-butyl peroxide abstract hydrogen atoms from the α-methyl groups of methyl methacrylate, while there is no such abstraction for initiator radicals from benzoyl peroxide or AIBN.

3-5 MOLECULAR WEIGHT

3-5a Kinetic Chain Length

The *kinetic chain length* v of a radical chain polymerization is defined as the average number of monomer molecules consumed (polymerized) per each radical, which initiates a polymer chain. This quantity will obviously be given by the ratio of the polymerization rate to the initiation rate or to the termination rate, since the latter two rates are equal.

$$v = \frac{R_p}{R_i} = \frac{R_p}{R_t} \tag{3-105}$$

Combination of Eqs. 3-22, 3-23, and 3-105 yields

$$v = \frac{k_p[M]}{2k_t[M\cdot]} \tag{3-106}$$

or

$$v = \frac{k_p^2[M]^2}{2k_t R_p} \tag{3-107}$$

For polymerization initiated by the thermal homolysis of an initiator, Eq. 3-32 can be substituted into Eq. 3-107 to yield

$$v = \frac{k_p[M]}{2(f k_d k_t[I])^{1/2}} \tag{3-108}$$

Equations 3-106 through 3-108 show a very significant characteristic of radical chain polymerizations. The kinetic chain length is inversely dependent on the radical concentration or the polymerization rate. This is of great practical significance—any attempt to increase the polymerization rate by increasing the radical concentration comes at the expense of producing small-sized polymer molecules. Increasing the polymerization rates and radical concentrations in radical polymerizations leads to small-sized polymer molecules. The kinetic chain length at constant polymerization rate is a characteristic of the particular monomer and independent of the method of initiation. Thus for any monomer the kinetic chain length will be independent of whether the polymerization is initiated by thermal, redox, or photochemical means, whether initiators are used, or of the particular initiator used, if the [M·] or R_p is the same.

3-5b Mode of Termination

The *number-average degree of polymerization* $\overline{X}_n$, defined as the average number of monomer molecules contained in a polymer molecule, is related to the kinetic chain length. If the propagating radicals terminate by coupling (Eq. 3-16a), a dead polymer molecule is composed of two kinetic chain lengths and

$$\overline{X}_n = 2\nu \tag{3-109a}$$

For termination by disproportionation (Eq. 3-16b) the kinetic chain length is synonymous with the number-average degree of polymerization

$$\overline{X}_n = \nu \tag{3-109b}$$

The number-average molecular weight of a polymer is given by

$$\overline{M}_n = M_o\overline{X}_n \tag{3-110}$$

where M_o is the molecular weight of the monomer.

The mode of termination is experimentally determined from the observation of the number of initiator fragments per polymer molecule. This requires the analysis of the molecular weight of a polymer sample as well as the total number of initiator fragments contained in that sample. Termination by coupling results in two initiator fragments per polymer molecule, while disproportionation results in one initiator fragment per polymer molecule. The fractions of propagating chains, a and $(1 - a)$, respectively, which undergo termination by coupling and disproportionation can be related to b, the average number of initiator fragments per polymer molecule. For a reaction system composed of n propagating chains coupling yields an initiator fragments and $an/2$ polymer molecules, while disproportionation yields $(1 - a)n$ initiator fragments and $(1 - a)n$ polymer molecules. The average number of initiator fragments per polymer molecule, defined as the total initiator fragments divided by the total number of polymer molecules, is given as

$$b = \frac{an + (1 - a)n}{an/2 + (1 - a)n} = \frac{2}{2 - a} \tag{3-111}$$

from which the fractions of coupling and disproportionation are obtained as

$$a = \frac{2b - 2}{b} \tag{3-112a}$$

$$1 - a = \frac{2 - b}{b} \tag{3-112b}$$

Although experimental data are not available for all monomers, most polymer radicals appear to terminate predominately or entirely by coupling (except where chain transfer predominates). Studies with small, aliphatic radicals clearly predict this tendency [Tedder, 1974]. However, varying extents of disproportionation are observed depending on the reaction system. Disproportionation increases when the propagating radical is sterically hindered or has many β-hydrogens available for transfer. Thus, whereas styrene, methyl acrylate, and acrylonitrile undergo termination almost exclusively by coupling, methyl methacrylate undergoes termination by a combination of coupling and disproportionation [Ayrey et al., 1977; Bamford, 1988; Bonta et al., 1976]. Increased temperature increases the extent of disproportionation, with the most significant effect for sterically hindered radicals. The extent of disproportionation in methyl methacrylate increases from 67% at 25°C to 80% at 80°C [Bamford et al., 1969b]. There appears to be a tendency toward disproportionation for highly reactive radicals such as those in vinyl acetate and ethylene polymerizations. However, the effect is relatively small as the extent of disproportionation in these two polymerizations is no greater than 10% at most [Bamford et al., 1969a, 1969b].

3-6 CHAIN TRANSFER

3-6a Effect of Chain Transfer

In many polymerization systems the polymer molecular weight is observed to be lower than predicted on the basis of the experimentally observed extents of termination by coupling and disproportionation. This effect is due to the premature termination of a growing polymer by the transfer of a hydrogen or other atom or species to it from some compound present in the system—the monomer, initiator, or solvent, as the case may be. These radical displacement reactions are termed *chain-transfer* reactions and may be depicted as

$$M_n \cdot + XA \xrightarrow{k_{tr}} M_n\text{---}X + A \cdot \tag{3-113}$$

where XA may be monomer, initiator, solvent, or other substance and X is the atom or species transferred. Chain transfer to initiator was referred to earlier as induced initiator decomposition (Sec. 3-4g-1).

The rate of a chain-transfer reaction is given by

$$R_{tr} = k_{tr}[M\cdot][XA] \tag{3-114}$$

where k_{tr} is the chain-transfer rate constant. Chain transfer results in the production of a new radical $A\cdot$ which then reinitiates polymerization

$$A\cdot + M \xrightarrow{k_a} M\cdot \tag{3-115}$$

Chain transfer is a *chain-breaking* reaction; it results in a decrease in the size of the propagating polymer chain. The effect of chain transfer on the polymerization rate is dependent on whether the rate of reinitiation is comparable to that of the original propagating radical. Table 3-4 shows the four main possible situations that may be encountered. Reinitiation is rapid in Cases 1 and 2 and one observes no change in the polymerization rate. The same number of monomer molecules are consumed per unit time with the formation of larger numbers of smaller-sized polymer molecules. The relative decrease in $\overline{X}_n$ depends on the magnitude of the transfer constant. When the transfer rate constant k_{tr} is much larger than that for propagation (Case 2) the result is a very small-sized polymer ($\overline{X}_n \simeq 1\text{--}5$)—referred to as a *telomer*. When reinitiation is slow compared to propagation (Cases 3 and 4) one observes a decrease in R_p as well as in $\overline{X}_n$. The magnitude of the decrease in R_p is determined by the relative values of k_p and k_{tr}. The remainder of this section will be concerned with Case 1. Cases 3 and 4 will be considered further in Sec. 3-7. Case 2 (telomerization) is not within the scope of this text.

TABLE 3-4   Effect of Chain Transfer on R_p and $\overline{X}_n$

Case	Relative Rate Constants for Transfer, Propagation, and Reinitiation		Type of Effect	Effect on R_p	Effect on $\overline{X}_n$
1	$k_p \gg k_{tr}$	$k_a \simeq k_p$	Normal chain transfer	None	Decrease
2	$k_p \ll k_{tr}$	$k_a \simeq k_p$	Telomerization	None	Large decrease
3	$k_p \gg k_{tr}$	$k_a < k_p$	Retardation	Decrease	Decrease
4	$k_p \ll k_{tr}$	$k_a < k_p$	Degradative chain transfer	Large decrease	Large decrease

For chain transfer by Case 1 the kinetic chain length remains unchanged, but the number of polymer molecules produced per kinetic chain length is altered. The number-average degree of polymerization is no longer given by v or $2v$ for disproportionation and coupling, respectively. Chain transfer is important in that it may alter the molecular weight of the polymer product in an undesirable manner. On the other hand, controlled chain transfer may be employed to advantage in the control of molecular weight at a specified level.

The degree of polymerization must now be redefined as the polymerization rate divided by the sum of the rates of all chain-breaking reactions (i.e., the normal termination mode plus all chain-transfer reactions). For the general case of a polymerization initiated by the thermal homolysis of a catalyst and involving termination by coupling and chain transfer to monomer, initiator, and the compound S (referred to as a *chain-transfer agent*), the number-average degree of polymerization follows from Eq. 3-114 as

$$\overline{X}_n = \frac{R_p}{(R_t/2) + k_{tr,M}[M\cdot][M] + k_{tr,S}[M\cdot][S] + k_{tr,I}[M\cdot][I]} \tag{3-116}$$

The first term in the denominator denotes coupling and the other three denote chain transfer by monomer, chain-transfer agent, and initiator, respectively. A *chain-transfer constant C* for a substance is defined as the ratio of the rate constant k_{tr} for the chain transfer of a propagating radical with that substance to the rate constant k_p for propagation of the radical. The chain-transfer constants for monomer, chain transfer agent, and initiator are then given by

$$C_M = \frac{k_{tr,M}}{k_p} \qquad C_S = \frac{k_{tr,S}}{k_p} \qquad C_I = \frac{k_{tr,I}}{k_p} \tag{3-117}$$

Combining Eqs. 3-22, 3-31, 3-32, 3-116, and 3-117 with $R_i = R_t$ yields

$$\frac{1}{\overline{X}_n} = \frac{2R_p}{R_i} + C_M + C_S \frac{[S]}{[M]} + C_I \frac{[I]}{[M]} \tag{3-118a}$$

or

$$\frac{1}{\overline{X}_n} = \frac{k_t R_p}{k_p^2 [M]^2} + C_M + C_S \frac{[S]}{[M]} + C_I \frac{k_t R_p^2}{k_p^2, f k_d [M]^3} \tag{3-118b}$$

which show the quantitative effect of the various transfer reactions on the number-average degree of polymerization. Various methods can be employed to determine the values of the chain transfer constants. Equation 3-118 is often referred to as the *Mayo equation*.

3-6b Transfer to Monomer and Initiator

3-6b-1 Determination of C_M and C_I

Two special cases of Eq. 3-118 are of interest. When a chain-transfer agent is absent, the term in [S] disappears and

$$\frac{1}{\overline{X}_n} = \frac{k_t R_p}{k_p^2 [M]^2} + C_M + C_I \frac{k_t R_p^2}{k_p^2 f k_d [M]^3} \tag{3-119a}$$

or

$$\frac{1}{\overline{X}_n} = \frac{k_t R_p}{k_p^2 [M]^2} + C_M + C_I \frac{[I]}{[M]} \tag{3-119b}$$

Equation 3-119a is quadratic in R_p and the appropriate plot of $1/\overline{X}_n$ vs R_p in Fig. 3-7 for styrene polymerization shows the effect to varying degrees depending on the initiator [Baysal and Tobolsky, 1952]. The initial portion of the plot is linear, but at higher concentrations of initiator and therefore high values of R_p the plot deviates from linearity as the contribution of transfer to initiator increases. The intercept of the linear portion yields the value of C_M. The slope of the linear portion is given by $k_t/k_p^2 [M]^2$ from which the important quantity k_p^2/k_t can be determined since the monomer concentration is known. For systems in which chain transfer to initiator is negligible a plot of $1/\overline{X}_n$ versus R_p will be linear over the whole range.

Several methods are available for the determination of C_I. Equation 3-119a can be rearranged and divided through by R_p to yield

$$\left[\frac{1}{\overline{X}_n} - C_M \right] \frac{1}{R_p} = \frac{k_t}{k_p^2 [M]^2} + \frac{C_I k_t R_p}{k_p^2 f k_d [M]^3} \tag{3-120}$$

A plot of experimental data as the left side of Eq. 3-120 versus R_p yields a straight line whose slope is $(C_I k_t/k_p^2 f k_d [M]^3)$. The initiator transfer constant can be determined from the slope because the various other quantities are known or can be related to known quantities through Eq. 3-32. When chain transfer to monomer is negligible, one can rearrange Eq. 3-119b to yield

$$\left[\frac{1}{\overline{X}_n} - \frac{k_t R_p}{k_p^2 [M]^2} \right] = C_I \frac{[I]}{[M]} \tag{3-121}$$

A plot of the left side of Eq. 3-121 versus [I]/[M] yields a straight line whose slope is C_I. Figure 3-8 shows the appropriate plot for the *t*-butyl hydroperoxide polymerization of styrene [Walling and Heaton, 1965].

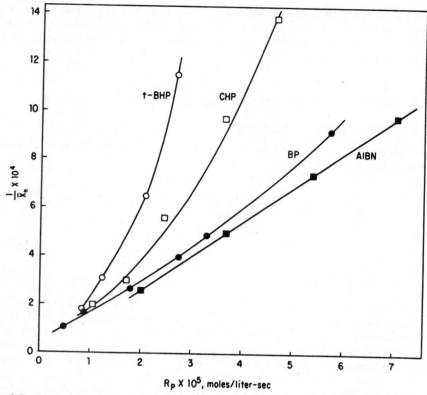

Fig. 3-7 Dependence of the degree of polymerization of styrene on the polymerization rate. The effect of chain transfer to initiator is shown for *t*-butyl hydroperoxide (○), cumyl hydroperoxide (□), benzoyl peroxide (●), and azobisisobutyronitrile (■) at 60°C. After Baysal and Tobolsky [1952] (by permission of Wiley-Interscience, New York).

3-6b-2 Monomer Transfer Constants

Using the methods described the values of C_M and C_I in the benzoyl peroxide polymerization of styrene have been found to be 0.00006 and 0.055, respectively [Mayo et al., 1951]. The amount of chain transfer to monomer which occurs is negligible in this polymerization. The chain-transfer constant for benzoyl peroxide is appreciable, and chain transfer with initiator becomes increasingly important as the initiator concentration increases. These effects are shown in Fig. 3-9, where the contributions of the various sources of chain ends are indicated. The topmost curve shows the total number of polymer molecules per 10^5 syrene monomer units. The difference between successive curves gives the number of polymer molecules terminated by normal coupling termination, transfer to benzoyl peroxide, and transfer to styrene.

The monomer chain-transfer constants are generally small for most monomers—being in the range 10^{-5} to 10^{-4} (Table 3-5). Chain transfer to monomer places the upper limit to the polymer molecular weight than can be obtained, assuming the absence of all other transfer reactions. Transfer to monomer does not, however,

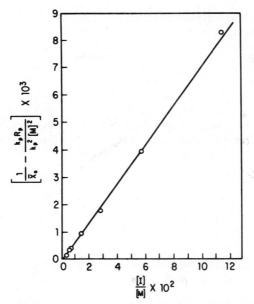

Fig. 3-8 Determination of initiator chain-transfer constants in the *t*-butyl hydroperoxide initiated polymerization of styrene in benzene solution at 70°C. After Walling and Heaton [1965] (by permission of American Chemical Society, Washington, D.C.).

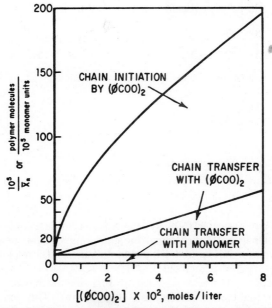

Fig. 3-9 Contribution of various sources of chain termination in the benzoyl peroxide initiated polymerization of styrene at 60°C. After Mayo et al. [1951] (by permission of American Chemical Society, Washington, D.C.).

prevent the synthesis of polymers of sufficiently high molecular weight to be of practical importance. C_M is generally low because the reaction

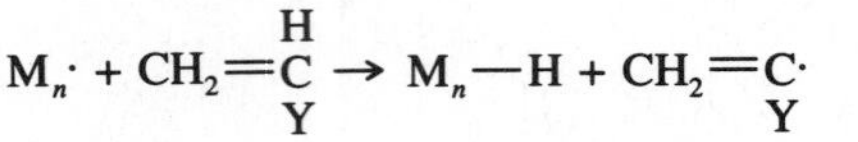

$$M_n\cdot + CH_2{=}\underset{Y}{\overset{H}{C}} \rightarrow M_n{-}H + CH_2{=}\underset{Y}{C\cdot} \qquad (3\text{-}122)$$

involves breaking the strong vinyl C—H bond. The largest monomer transfer constants are generally observed when the propagating radicals have very high reactivities, for example, ethylene, vinyl acetate, and vinyl chloride. Chain transfer to monomer for vinyl acetate had been attributed to transfer from the acetoxy methyl group [Nozakura et al., 1972]

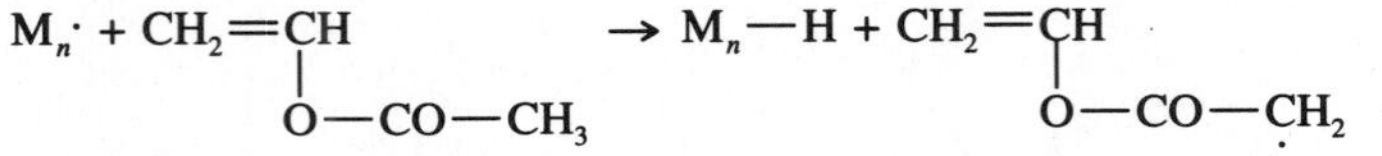

$$M_n\cdot + \underset{O-CO-CH_3}{CH_2{=}CH} \qquad \rightarrow M_n{-}H + \underset{O-CO-CH_2}{\underset{\cdot}{CH_2{=}CH}} \qquad (3\text{-}123)$$

$$\textbf{XXIII}$$

although experiments with vinyl trideuteroacetate and trideuterovinyl acetate indicate that more than 90% of the transfer occurs at the vinyl hydrogens [Litt and Chang, 1981].

The very high value of C_M for vinyl chloride is attributed to a reaction sequence involving the propagating center **XXIVa** formed by head-to-head addition [Hjertberg and Sorvik, 1983; Llauro-Darricades et al., 1989; Starnes, 1985; Starnes et al., 1983; Tornell, 1988]. Intramolecular migration of a chlorine atom (Eq. 3-124) yields the secondary radical **XXIVb**, which subsequently transfers the chlorine atom to monomer (Eq. 3-125) to yield poly(vinyl chloride) (referred to as PVC) with double bond end

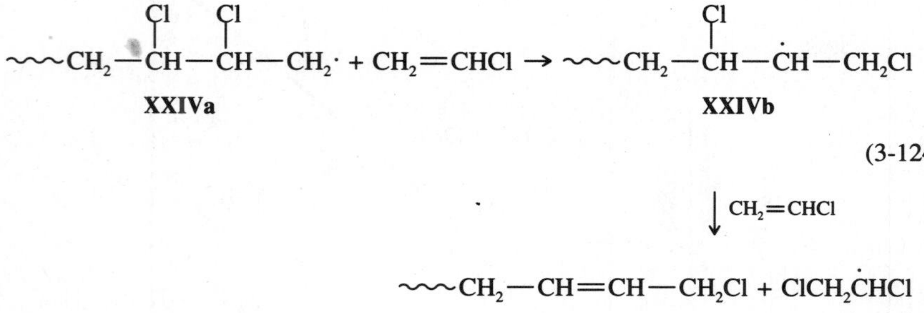

$$\overset{Cl}{\underset{}{}}\overset{Cl}{\underset{}{}}$$
$$\sim\sim CH_2{-}CH{-}CH{-}CH_2\cdot + CH_2{=}CHCl \rightarrow \sim\sim CH_2{-}\overset{Cl}{CH}{-}\overset{\cdot}{CH}{-}CH_2Cl$$

$$\textbf{XXIVa} \qquad\qquad\qquad\qquad\qquad\qquad\qquad \textbf{XXIVb}$$

$$(3\text{-}124)$$

$$\Big\downarrow CH_2{=}CHCl$$

$$\sim\sim CH_2{-}CH{=}CH{-}CH_2Cl + ClCH_2\overset{\cdot}{CH}Cl$$

$$\textbf{XXVa} \qquad\qquad\qquad \textbf{XXVb}$$

$$(3\text{-}125)$$

groups (**XXVa**) and radical **XXVb**. Radical **XXVb** reinitiates polymerization to yield polymer molecules with 1,2-dichloroalkane end groups. Consistent with this mechanism, the amounts of **XXVa** and 1,2-dichloroalkane end groups are approximately equal (0.7–0.8 per polymer molecule).

The C_M value of vinyl chloride is sufficiently high that the maximum number-average molecular weight that can be achieved is 50,000–100,000. This limit is still reasonable from the practical viewpoint–PVC of 50,000 molecular weight is a very useful product.

Not only the case of vinyl chloride but also styrene shows that the observed chain transfer to monomer is not the simple reaction described by Eq. 3-122. Considerable

TABLE 3-5   Monomer Chain-Transfer Constants[a]

Monomer	$C_M \times 10^4$
Acrylamide	0.6, 0.12[b]
Acrylonitrile	0.26–0.3
Ethylene	0.4–4.2
Methyl acrylate	0.036–0.325
Methyl methacrylate	0.07–0.25
Styrene	0.30–0.60
Vinyl acetate	1.75–2.8
Vinyl chloride	10.8–16

[a]All C_M values are for 60°C except where otherwise noted.
[b]Value at 40°C.
[c]Braks and Huang [1978]; all other data from Brandrup and Immergut [1989].

evidence [Olaj et al., 1977a, 1977b] indicates that the experimentally observed C_M for styrene may be due in large part to the Diels–Alder dimer **XVIII** transferring a hydrogen (probably the same hydrogen transferred in the thermal initiation process) to monomer.

3-6b-3   Initiator Transfer Constants

Different initiators have varying transfer constants (Table 3-6). Further, the value of C_I for a particular initiator also varies with the reactivity of the propagating radical. Thus there is a fivefold difference in C_I for cumyl hydroperoxide toward poly(methyl methacrylate) radical compared to polystyryl radical. The latter is the less reactive radical; see Sec. 6-3b.

Azonitriles have generally been considered to be "cleaner" initiators in the sense of being devoid of transfer, but recent work [Braks and Huang, 1978] clearly indicates this is not true. The transfer with azonitriles probably occurs by the displacement

TABLE 3-6   Initiator Chain-Transfer Constants[a,b]

Initiator	C_I for Polymerization of Styrene	Methyl Methacrylate	Acrylamide
2,2'-Azobisisobutyronitrile[c]	0.091–0.14	0.02	—
t-Butyl peroxide	0.00076–0.00092	—	—
Cumyl peroxide (50°C)	0.01	—	—
Lauroyl peroxide (70°C)	0.024	—	—
Benzoyl peroxide	0.048–0.10	0.02	—
t-Butyl hydroperoxide	0.035	—	—
Cumyl hydroperoxide	0.063	0.33	—
Persulfate (40°C)[d]	—	—	0.0026

[a]Data from Brandrup and Immergut [1989], unless otherwise noted.
[b]All C_I values are for 60°C except where otherwise noted.
[c]Ayrey and Haynes [1974]; Braks and Huang [1978].
[d]Shawki and Hamielec, 1979.

reaction

$$M_n\cdot + RN{=}NR \rightarrow M_n{-}R + N_2 + R\cdot \tag{3-126}$$

Many peroxides have significant chain-transfer constants. Dialkyl and diacyl peroxides undergo transfer by

$$M_n\cdot + RO{-}OR \rightarrow M_n{-}OR + RO\cdot \tag{3-127}$$

where R = alkyl or acyl. The acyl peroxides have higher transfer constants than the aklyl peroxides due to the weaker $O{-}O$ bond of the former. The hydroperoxides are usually the strongest transfer agents among the initiators. Transfer probably involves hydrogen atom abstraction

$$M_n\cdot + ROO{-}H \rightarrow M_n{-}H + ROO\cdot \tag{3-128}$$

The typical effect of initiator chain transfer [Baysal and Tobolsky, 1952] can be seen graphically in Fig. 3-8. The decrease of polymer size due to chain transfer to initiator is much less than indicated from the C_I values because it is the quantity $C_I[I]/[M]$, which effects $\overline{X}_n$ (Eq. 3-119b). The initiator concentrations are quite low (10^{-4}–10^{-2} M) in polymerization and the ratio $[I]/[M]$ is typically in the range 10^{-3}–10^{-5}.

3-6c   Transfer to Chain-Transfer Agent

3-6c-1   Determination of C_S

The second special case of Eq. 3-118 consists of the situation where transfer with the chain-transfer agent is most important. In some instances the chain-transfer agent is the solvent, while in others it is an added compound. In such a case the third term on the right side of Eq. 3-118 makes the biggest contribution to the determination of the degree of polymerization. By the appropriate choice of polymerization conditions, one can determine the value of C_S for various chain-transfer agents [Gregg and Mayo, 1948]. By using low concentrations of initiators or initiators with negligible C_I values (e.g., AIBN), the last term in Eq. 3-118 becomes negligible. The first term on the right side of the equation may be kept constant by keeping $R_p/[M]^2$ constant by appropriately adjusting the initiator concentration throughout a series of separate polymerizations. Under these conditions, Eq. 3-118 takes the form

$$\frac{1}{\overline{X}_n} = \left(\frac{1}{\overline{X}_n}\right)_0 + C_S\frac{[S]}{[M]} \tag{3-129}$$

where $(1/\overline{X}_n)_0$ is the value of $1/\overline{X}_n$ in the absence of the chain-transfer agent. $(1/\overline{X}_n)_0$ is the sum of the first, second, and fourth terms on the right side of Eq. 3-118. C_S is then determined as the slope of the linear plot of $1/\overline{X}_n$ vs $[S]/[M]$. Such plots are shown in Fig. 3-10 for several chain-transfer agents in styrene polymerization. This method has been found to be of general utility in determining values of C_S.

   An alternate method includes a modification of the first method by plotting $[1/\overline{X}_n - (1/\overline{X}_n)_0]$ instead of $1/\overline{X}_n$ versus $[S]/[M]$. Under conditions where $R_p/[M]^2$ is

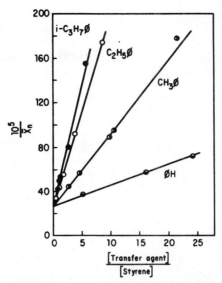

Fig. 3-10   The effect of various chain-transfer agents on the degree of polymerization of styrene at 100°C. After Gregg and Mayo [1948] (by permission of American Chemical Society, Washington, D.C.).

not constant one may plot the quantity $(1/\overline{X}_n - k_t R_p/k_p^2[M]^2)$ against $[S]/[M]$. This yields a straight line with slope C_S. Another method involves dividing the rate expression for transfer (Eq. 3-114) by that for propagation (Eq. 3-122) to yield

$$\frac{d[S]/dt}{d[M]/dt} = \frac{k_{tr,S}[S]}{k_p[M]} = C_S \frac{[S]}{[M]} \tag{3-130}$$

The value of C_S is obtained as the slope of the line obtained by plotting the ratio of the rates of disappearance of transfer agent and monomer $d[S]/d[M]$ versus $[S]/[M]$.

3-6c-2   Structure and Reactivity

The transfer constants for various compounds are shown in Table 3-7. These data are useful for the information they yield regarding the relationship between structure and reactivity in radical displacement reactions. Aliphatic hydrocarbons such as cyclohexane with strong C—H bonds show low transfer constants. Benzene has an even lower C_S value because of the stronger C—H bonds. Transfer to benzene appears not to involve hydrogen abstraction but addition of the propagating radical [Deb and Ray, 1978] to the benzene ring:

$$\mathbf{M}_n \cdot + \langle\bigcirc\rangle \longrightarrow \mathbf{M}_n\text{---}\langle\rangle \tag{3-131}$$

TABLE 3-7 **Transfer Constants for Chain-Transfer Agents**[a,b]

	$C_S \times 10^4$ for Polymerization of	
Transfer Agent	Styrene	Vinyl Acetate
Benzene	0.023	1.2
Cyclohexane	0.031	7.0
Heptane	0.42	17.0 (50°C)
Toluene	0.125	21.6
Ethylbenzene	0.67	55.2
Isopropylbenzene	0.82	89.9
t-Butylbenzene	0.06	3.6
n-Butyl chloride	0.04	10
n-Butyl bromide	0.06	50
2-Chlorobutane	1.2	—
Acetone	4.1	11.7
Acetic acid	2.0	1.1
n-Butyl alcohol	1.6	20
Ethyl ether	5.6	45.3
Chloroform	3.4	150
n-Butyl iodide	1.85	800
Butylamine	7.0	—
Triethylamine	7.1	370
Di-n-butyl sulfide	22	260
Di-n-butyl disulfide	24	10,000
Carbon tetrachloride	110	10,700
Carbon tetrabromide	22,000	390,000
n-Butyl mercaptan	210,000	480,000

[a]Data from Brandrup and Immergut [1989] and Eastmond [1976a, 1976b, 1976c].
[b]All values are for 60°C unless otherwise noted.

The presence of the weaker benzylic hydrogens in toluene, ethylbenzene, and iso-propylbenzene leads to higher C_S values relative to benzene. The benzylic C—H is abstracted easily because the resultant radical is resonance stabilized.

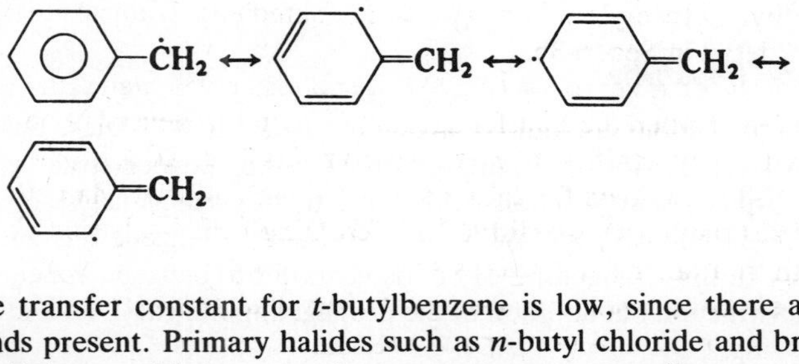

$$(3\text{-}132)$$

The transfer constant for t-butylbenzene is low, since there are no benzylic C—H bonds present. Primary halides such as n-butyl chloride and bromide behave similar to aliphatics with low transfer constants, corresponding to a combination of either aliphatic C—H bond breakage or the low stability of a primary alkyl radical on abstraction of Cl or Br. The iodide, on the other hand, transfers an iodide atom and shows a much higher C_S value due to the weakness of the C—I bond. Secondary halides have somewhat higher C_S values due to the greater stability of the secondary radical formed.

Acids, carbonyl compounds, ethers, amines, and alcohols have transfer constants higher than those of aliphatic hydrocarbons, corresponding to C—H breakage and stabilization of the radical by an adjacent O, N, or carbonyl group.

The weak S—S bond leads to high transfer constants for disulfides,

$$M_n \cdot + RS{-}SR \longrightarrow M_n{-}SR + RS \cdot \tag{3-133}$$

The high C_S values for carbon tetrachloride and carbon tetrabromide are due to the weak carbon–halogen bonds. These bonds are especially weak because of the excellent stabilization of the trihalocarbon radicals formed by resonance involving the halogen free pairs of electrons:

$$|\overline{\underline{Cl}}{-}\dot{C}{-}\overline{\underline{Cl}}| \leftrightarrow |\dot{\underline{Cl}}{=}C{-}\overline{\underline{Cl}}| \leftrightarrow |\overline{\underline{Cl}}{-}C{=}\dot{\underline{Cl}}| \leftrightarrow |\overline{\underline{Cl}}{-}C{-}\overline{\underline{Cl}}| \tag{3-134}$$

The greater transfer constant for carbon tetrabromide compared to the tetrachloride is due to the weaker C—Br bond. The low C_S value for chloroform compared to carbon tetrachloride is explained by C—H bond breakage in the former. The thiols have the largest transfer constants of any known compounds due to the weak S—H bond.

Two interesting observations are made when the C_S values for various compounds are compared in the polymerization of various different monomers. The absolute value of the transfer constant for any one compound may change very significantly depending on the monomer being polymerized. This is clearly seen in Table 3-7 where many of the transfer agents are 1–2 orders of magnitude more active in vinyl acetate polymerization compared to styrene polymerization. This effect is a consequence of the greater reactivity of the vinyl acetate propagating radical. The chain-transfer constant for any one compound generally increases in the order of increasing radical reactivity. The order of radical reactivity is vinyl chloride > vinyl acetate > acrylonitrile > methyl acrylate > methyl methacrylate > styrene > 1,3-butadiene. Radical reactivity is discussed in greater detail in Sec. 6-3b.

The order of reactivity of a series of transfer agents usually remains the same irrespective of the monomer when the transfer agents are relatively neutral in polarity. However, there are many very significant deviations from this generalization for polar transfer agents. Table 3-8 shows data for carbon tetrachloride and triethylamine with several monomers. The monomers are listed in decreasing order of reactivity for transfer reactions with neutral transfer agents such as hydrocarbons. It is apparent that the electron-rich (electron-donor) transfer agent triethylamine has enhanced reactivity with the electron-poor (electron-acceptor) monomers, especially acrylonitrile. The electron-poor (electron-acceptor) transfer agent carbon tetrachloride has enhanced reactivity with the electron-rich (electron-donor) monomers vinyl acetate and styrene and markedly lowered reactivity with the electron-poor monomers.

The enhancement of chain-transfer reactivity has been postulated [Patnaik et al.,

TABLE 3-8 Polar Effects in Chain Transfer[a,b,c]

| | Chain-Transfer Agent | | | |
| | CCl$_4$ | | (C$_2$H$_5$)$_3$N | |
Monomer	$C_S \times 10^4$	k_{tr}	$C_S \times 10^4$	k_{tr}
Vinyl acetate	10,700	2400	370	85
Acrylonitrile	0.85	0.17	3800	760
Methyl acrylate	1.25[c]	0.26[c]	400	84
Methyl methacrylate	2.4	0.12	1900	98
Styrene	110	1.8	7.1	0.12

[a]C_S values are taken from Brandrup and Immergut [1989]; Eastmond [1976a, 1976b, 1976c] and are for 60°C unless otherwise noted.
[b]k_{tr} values were calculated from Eq. 3-117 using the k_p values from Table 3-10.
[c]C_S value is at 80°C. The k_{tr} was calculated using the k_p value for 60°C.

1979; Walling, 1957] as occurring by stabilization of the respective transition states for the transfer reactions by contributions from polar structures such as

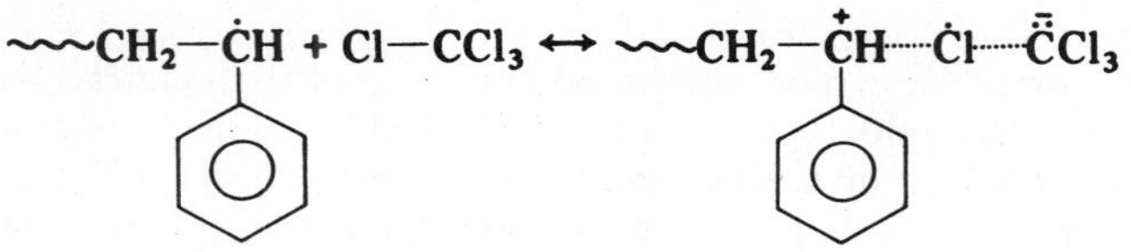

$$\sim\sim CH_2-\overset{\cdot}{C}H + Cl-CCl_3 \longleftrightarrow \sim\sim CH_2-\overset{+}{C}H\cdots\overset{-}{C}l\cdots\overset{..}{C}Cl_3 \tag{3-135a}$$

$$\sim\sim CH_2-\underset{CN}{\overset{\cdot}{C}H} + H-\underset{CH_3}{CH}-N(CH_3)_2 \longleftrightarrow$$

$$\sim\sim CH_2-\underset{CN}{\overset{\overset{..}{-}}{C}H}\cdots\overset{..}{H}\cdots\underset{CH_3}{\overset{+}{C}H}-N(CH_3)_2 \tag{3-135b}$$

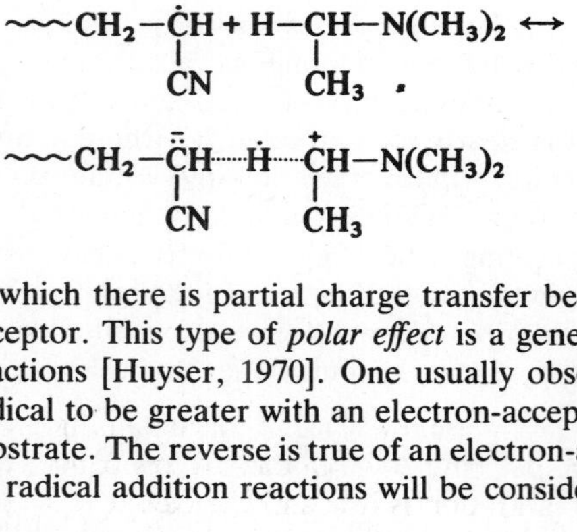

in which there is partial charge transfer between an electron donor and an electron acceptor. This type of *polar effect* is a general one often encountered in free radical reactions [Huyser, 1970]. One usually observes the reactivity of an electron-donor radical to be greater with an electron-acceptor substrate than with an electron-donor substrate. The reverse is true of an electron-acceptor radical. The effect of polar effects on radical addition reactions will be considered in Sec. 6-3b-3.

3-6c-3 Applications of Chain-Transfer Agents

The use of Eqs. 3-118, 3-119, and 3-129 allows one to quantitatively determine the effects of transfer to monomer, initiator, and solvent on the molecular weight of the product that can be obtained from a reaction system. Of prime importance is the control of the concentrations of monomer, initiator, and solvent and/or the use of deliberately added chain-transfer agents to control the molecular weight of a polymerization. Equation 3-129 can be used to determine the concentration of transfer agent needed to obtain a specifically desired molecular weight. When used in this manner, transfer agents are called *regulators* or *modifiers*. Transfer agents with transfer constants of 1 or greater are especially useful, since they can be used in small concentrations. Thus mercaptans such as n-dodecyl mercaptan are used in the industrial emulsion copolymerization of styrene and butadiene for SBR rubbers. The production of very low-molecular-weight polymers by chain transfer (telomerization) is an industrially useful reaction. Ethylene polymerized in chloroform yields, after fluorination of its end groups, a starting material for fluorinated lubricants. Low-molecular-weight acrylic ester polymers have been used as plasticizers. On the other hand, benzene is used as the solvent for the production of high-molecular-weight polyethylene because of its low transfer constant.

It is also useful to point out that chain-transfer studies yield further corroboration of the concept of functional group reactivity independent of molecular size. Thus one can vary the degree of polymerization for a monomer by using different chain-transfer agents or different concentrations of the same transfer agent. Under these conditions the propagation rate constant k_p is found to be independent of $\overline{X}_n$. Further, the transfer constant k_{tr} for a particular transfer agent is also independent of the size of the propagating radical.

3-6d Chain Transfer to Polymer

The previous discussion has ignored the possibility of chain transfer to polymer molecules. Transfer to polymer results in the formation of a radical site on a polymer chain. The polymerization of monomer at this site leads to the production of a branched polymer, for example,

$$\text{M}_n\text{·} + \text{\char`\~\char`\~CH}_2-\underset{\underset{\text{H}}{|}}{\overset{\overset{\text{Y}}{|}}{\text{C}}}\text{\char`\~\char`\~} \rightarrow \text{M}_n-\text{H} + \text{\char`\~\char`\~CH}_2-\underset{\text{·}}{\overset{\overset{\text{Y}}{|}}{\text{C}}}\text{\char`\~\char`\~}$$

$$\text{M}\downarrow$$

$$\text{\char`\~\char`\~CH}_2-\underset{\underset{\text{M}_m}{|}}{\overset{\overset{\text{Y}}{|}}{\text{C}}}\text{\char`\~\char`\~} \qquad (3\text{-}136)$$

Ignoring chain transfer to polymer does not present a difficulty in obtaining precise values of C_I, C_M, and C_S, since these are determined from data at low conversions. Under these conditions the polymer concentration is low and the extent of transfer to polymer is negligible.

Transfer to polymer cannot, however, be neglected for the practical situation where polymerization is carried to complete or high conversion. The effect of chain transfer to polymer plays a very significant role in determining the physical properties and the ultimate applications of a polymer [Small, 1975]. As indicated in Chap. 1, branching drastically decreases the crystallinity of a polymer.

The transfer constant C_P for chain transfer to polymer is not easily obtained [Yamamoto and Sugimoto, 1979]. C_P cannot be simply determined by introducing the term $C_P[P]/[M]$ into Eq. 3-118 as is indicated in many introductory polymer texts. Transfer to polymer does not necessarily lead to a decrease in the overall degree of polymerization. Each act of transfer produces a branched polymer molecule of larger size than initially present in addition to prematurely terminating a propagating polymer chain.

The evaluation of C_P involves the difficult determination of the number of branches produced in a polymerization relative to the number of monomer molecules polymerized. This can be done by polymerizing a monomer in the presence of a known concentration of polymer of known molecular weight. The product of such an experiment consists of three different types of molecules:

Type 1. Unbranched molecules of the initial polymer.
Type 2. Unbranched molecules produced by polymerization of the monomer.
Type 3. Branched molecules arising from transfer of Type 2 radicals to Type 1 molecules.

The number of new polymer molecules produced in the system yields the number of Type 2 molecules. The total number of branches is obtained by performing a mass balance on the system and assuming the size of a branch is the same as a Type 2 molecule. This experimental analysis is inherently difficult and is additionally complicated if chain transfer to initiator, monomer, or some other species is occurring simultaneously. Other methods of determining the polymer transfer constant have also been employed but are usually not without ambiguity. Thus, for example, C_P for poly(vinyl acetate) has been determined by degradative hydrolysis [Lindeman, 1967]. This method assumes that transfer occurs at the acetoxy methyl group leading to polymer branches,

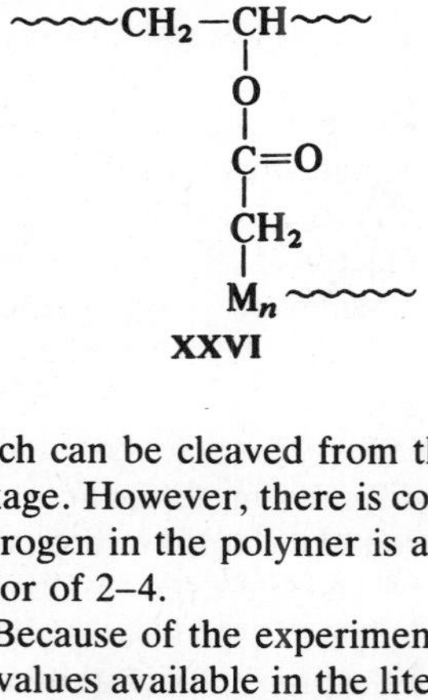

$$
\begin{array}{c}
\sim\!\!\sim\!\!\sim\!\text{CH}_2-\underset{\displaystyle |}{\text{CH}}\!\sim\!\!\sim \\
\text{O} \\
| \\
\text{C}=\text{O} \\
| \\
\text{CH}_2 \\
| \\
\text{M}_n \sim\!\!\sim\!\!\sim
\end{array}
$$

XXVI

which can be cleaved from the original polymer molecule by ester hydrolysis of the linkage. However, there is considerable evidence [Clarke et al., 1961] that the tertiary hydrogen in the polymer is abstracted easier than the acetoxy methyl hydrogen by a factor of 2–4.

Because of the experimental difficulties involved, there are relatively few reliable C_P values available in the literature. The values that are available [Eastmond, 1976a, 1976b, 1976c; Ham, 1967] for any one polymer often vary considerably from each other. It is often most useful to consider the small model compound analog of a polymer (e.g., ethylbenzene or isopropylbenzene for polystyrene) to gain a correct perspective of the importance of polymer chain transfer. A consideration of the best available C_P values and those of the appropriate small model compounds indicates that the amount of transfer to polymer will not be high in most cases even at high conversion. C_P values are about 10^{-4} or slightly higher for many polymers such as polystyrene and poly(methyl methacrylate).

Flory [1947] derived the equation

$$\rho = -C_P \left[1 + \left(\frac{1}{p} \right) \ln (1 - p) \right] \tag{3-137}$$

to express the *branching density* ρ as a function of the polymer transfer constant C_P and the extent of reaction p. The branching density ρ is the number of branches per monomer molecule polymerized [Fanood and George, 1988]. Equation 3-137 indicates that branching increases rapidly during the later stages of a polymerization. Using a C_P value of 1×10^{-4}, one calculates from Eq. 3-137 that there will be 1.0, 1.6, and 2.2 branches for every 10^4 monomer units polymerized at 80, 90, and 95% conversion, respectively. Experimental data verify this result quite well. For styrene polymerization at 80% conversion, there is about one branch for every $4-10 \times 10^3$ monomer units for polymer molecular weights of 10^5-10^6. This corresponds to about 1 polymer chain in 10 containing a branch [Bevington et al., 1947].

The extent of branching is greater in polymers, such as poly(vinyl acetate), poly(vinyl chloride), and polyethylene, which have very reactive propagating radicals. Poly(vinyl acetate) has a C_P value that is probably in the range $2-5 \times 10^{-4}$. Further, vinyl acetate monomer was noted earlier (Table 3-5) as having a large C_M value. Transfer to monomer yields a species that initiates polymerization at its radical center and, at its double bond, enters the propagation reaction. The result is extensive branching in poly(vinyl acetate) since a branch is formed for each act of transfer to either monomer or polymer [Stein, 1964; Wolf and Burchard, 1976].

The extent of branching in polyethylene varies considerably depending on the polymerization temperature and other reaction conditions (Sec. 3-13b-1), but may reach as high as 15–30 branches per 500 monomer units. The branches in polyethylene are of two types: short branches (less than 7 carbons) and long branches. The long branches, formed by the "normal" chain transfer to polymer reaction (Eq. 3-137], affect the melt flow (viscosity) properties of the polymer and thus greatly influence its processing characteristics [Jacovic et al., 1979; Starck and Lindberg, 1979]. The short branches, which outnumber the long branches by a factor of 20–50, have a very significant effect on the polymer crystallinity. Radical-polymerized polyethylene has a maximum crystallinity of 60–70% due to the presence of the short branches. The identity of the short branches has been established by high-resolution ^{13}C NMR spectroscopy as well as infrared spectroscopy, and radiolytic and pyrolytic fragmentation studies [Baker et al., 1979; Bovey et al., 1976; Bowmer and O'Donnell, 1977; Hay et al., 1986; Randall, 1978; Sugimura et al., 1981; Usami and Takayama, 1984]. The presence of ethyl, n-butyl, n-amyl, and n-hexyl branches has been clearly established. Although the relative amounts of the different branches varies considerably depending on the polymerization conditions, n-butyl branches are the most abundant for most polyethylenes [Axelson et al., 1979]. Most polyethylenes contain 5–15 n-butyl branches and 1–2 each of ethyl, n-amyl, and n-hexyl branches per 1000 carbon atoms. Some polyethylenes possess more ethyl than n-butyl branches. Small amounts of methyl, n-propyl, and other branches have been detected in some studies.

The generally accepted mechanism for the formation of short branching in polyethylene involves a *backbiting* intramolecular transfer reaction in which the propagating radical **XXVII** abstracts hydrogens from the fifth, sixth, and seventh methylene groups from the radical end (Eqs. 3-138a,b,c). The resultant radicals **XXVIIIa,b,c** propagate with monomer with the resultant formation of *n*-hexyl, *n*-amyl, and *n*-butyl branches, respectively. The general predominance of *n*-butyl branches is ascribed to its being formed through a six-membered transition state (consisting of the five carbons and the hydrogen being abstracted). Ethyl branches arise from radical **XXVIIIc** undergoing a second intramolecular transfer reaction after the addition of one ethylene

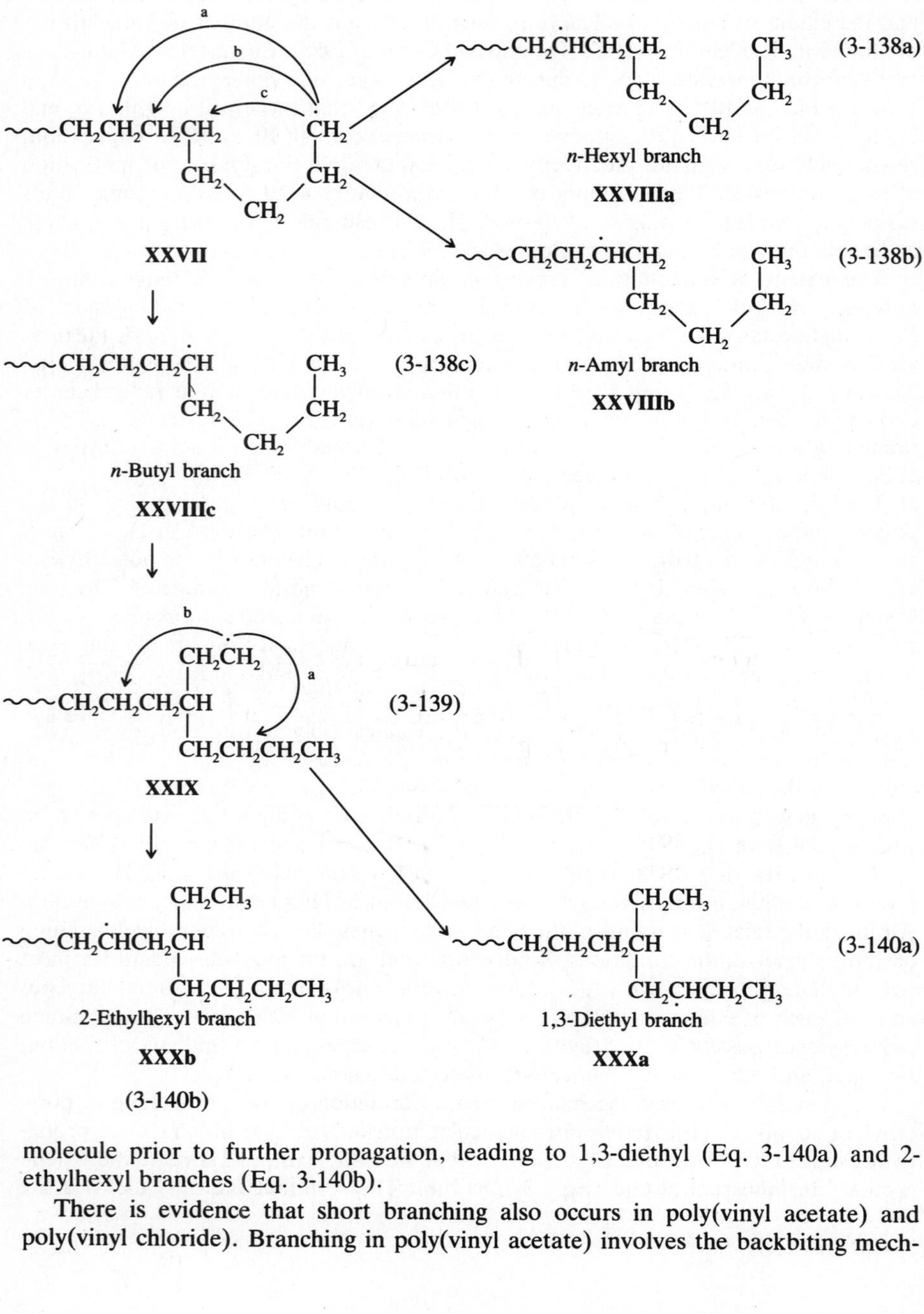

molecule prior to further propagation, leading to 1,3-diethyl (Eq. 3-140a) and 2-ethylhexyl branches (Eq. 3-140b).

There is evidence that short branching also occurs in poly(vinyl acetate) and poly(vinyl chloride). Branching in poly(vinyl acetate) involves the backbiting mechanism as in polyethylene, although the exact identity and distribution of different-sized branches is not established [Adelman and Ferguson, 1975; Morishima and Nozakura, 1976; Nozakura et al., 1976]. Short branching in poly(vinyl chloride) is a more complex process [Hjertberg and Sorvik, 1983; Starnes, 1985; Starnes et al., 1983; Tornell, 1988]. 2-Chloroethyl and 2,4-dichloro-*n*-butyl branches are formed via the backbiting route with the later more abundant (ca. 1 per 1000 monomer units) than the former. However, the most abundant branches are chloromethyl groups—up to 4–6 per 1000 monomer units. Chloromethyl groups occur when radical **XXIVa** (Eq. 3-124) undergoes propagation:

$$\sim\sim\text{CH}_2\text{CHCl}-\overset{|}{\underset{\underset{\text{XXIVa}}{\text{CH}_2\text{Cl}}}{\text{CH}_2}}\cdot + \text{CH}_2=\text{CHCl} \longrightarrow \sim\sim\text{CH}_2\text{CHClCHCH}_2\text{CHCl} \qquad (3\text{-}141)$$
$$\underset{\text{CH}_2\text{Cl}}{|}$$

Long branches in PVC, up to about one branch per 2000 monomer units, arise from hydrogen abstraction at the CHCl group in the polymer chain.

3-7 INHIBITION AND RETARDATION

The addition of certain substances suppresses the polymerization of monomers. These substances act by reacting with the initiating and propagating radicals and converting them either to nonradical species or radicals of reactivity too low to undergo propagation. Such polymerization suppressors are classified according to their effectiveness. *Inhibitors* stop every radical and polymerization is completely halted until they are consumed. *Retarders* are less efficient and stop only a portion of the radicals. In this case, polymerization occurs, but at a slower rate. The difference between inhibitors and retarders is simply one of degree and not kind. Figure 3-11 shows these effects in the thermal polymerization of styrene [Schulz, 1947]. Polymerization is completely stopped by benzoquinone, a typical inhibitor, during an *induction* or *inhibition period* (curve 2). At the end of this period, when the benzoquinone has been consumed, polymerization proceeds at the same rate as in the absence of inhibitor (curve 1). Nitrobenzene, a retarder, lowers the polymerization rate without an inhibition period

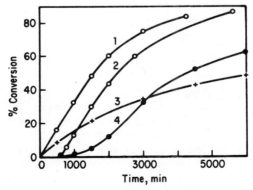

Fig. 3-11 Inhibition and retardation in the thermal, self-initiated polymerization of styrene at 100°C. Plot 1—no inhibitor; plot 2—0.1% benzoquinone; plot 3—0.5% nitrobenzene; plot 4—0.2% nitrosobenzene. After Schulz [1947] (by permission of Verlag Chemie GmbH, Weinheim).

(curve 3). The behavior of nitrosobenzene, $\emptyset$NO, is more complex (curve 4). It is initially an inhibitor but is apparently converted to a product which acts as a retarder after the inhibition period. This latter behavior is not at all uncommon. Inhibition and retardation are usually the cause of the irreproducible polymerization rates observed with insufficiently purified monomers. Impurities present in the monomer may act as inhibitors or retarders. On the other hand, inhibitors are invariably added to commercial monomers to prevent premature thermal polymerization during storage and shipment. These inhibitors are usually removed prior to polymerization or, alternatively, an appropriate excess of initiator may be used to compensate for their presence.

3-7a   Kinetics of Inhibition or Retardation

The kinetics of retarded or inhibited polymerization can be analyzed using a scheme consisting of the usual initiation (Eq. 3-14), propagation (Eq. 3-15), and termination (Eq. 3-16) reactions in addition to the inhibition reaction

$$M_n\cdot \; + \; Z \xrightarrow{\;k_z\;} M_n \; + \; Z\cdot \quad \text{and/or} \quad M_nZ\cdot \tag{3-142}$$

where Z is the inhibitor or retarder. Z acts either by adding to the propagating radical to form $M_nZ\cdot$ or by chain transfer of hydrogen or other radical to yield $Z\cdot$ and polymer. The kinetics are simplified if one assumes that $Z\cdot$ and $M_nZ\cdot$ do not reinitiate polymerization and also they terminate without regeneration of Z.

The steady-state assumption for the radical concentration leads to

$$\frac{d[M\cdot]}{dt} = R_i - 2k_t[M\cdot]^2 - k_z[Z][M\cdot] = 0 \tag{3-143}$$

which can be combined with Eq. 3-22 to yield

$$\frac{2R_p^2 k_t}{k_p^2[M]^2} + \frac{R_p[Z]k_z}{k_p[M]} - R_i = 0 \tag{3-144}$$

Equation 3-144 has been used to correlate rate data in inhibited polymerizations. A consideration of Eq. 3-144 shows that R_p is inversely proportional to the ratio k_z/k_p of the rate constants for inhibition and propagation. This ratio is often referred to as the *inhibition constant z*, that is,

$$z = \frac{k_z}{k_p} \tag{3-145}$$

It is further seen that R_p depends on R_i to a power between one-half and unity depending on the relative magnitudes of the first two terms in Eq. 3-144. Two limiting cases of Eq. 3-144 exist. When the second term is negligible compared to the first the polymerization is not retarded, and Eq. 3-144 simplifies to Eq. 3-25.

For the case where the retardation is strong ($k_z/k_p \gg 1$) normal bimolecular termination will be negligible. Under these conditions the first term in Eq. 3-144 is negligible and one has

$$\frac{R_p[Z]k_z}{k_p[M]} - R_i = 0 \tag{3-146a}$$

or

$$R_p = \frac{k_p[M]R_i}{k_z[Z]} = \frac{-d[M]}{dt} \tag{3-146b}$$

Equation 3-146 shows the rate of retarded polymerization to be dependent on the first power of the initiation rate. Further, R_p is inversely dependent on the inhibitor concentration. The induction period observed for inhibited polymerization is directly proportional to the inhibitor concentration.

The inhibitor concentration will decrease with time and $[Z]$ at any time is given by

$$[Z] = [Z]_0 - \frac{R_i t}{y} \tag{3-147}$$

where $[Z]_0$ is the initial concentration of Z, t is time, and y is the number of radicals terminated per inhibitor molecule. Combination of Eqs. 3-145, 3-146b, and 3-147 yields

$$\frac{-d[M]}{dt} = \frac{R_i[M]}{z([Z]_0 - R_i t/y)} \tag{3-148a}$$

or, by rearrangement,

$$\frac{-1}{d \ln [M]/dt} = \frac{z[Z]_0}{R_i} - \frac{zt}{y} \tag{3-148b}$$

A plot of the left side of Eq. 3-148b versus time is linear and the values of z and y can be obtained from the intercept and slope, respectively, if $[Z]_0$ and R_i are known. The method involves difficult experimentation, since the polymerization rates being measured are quite small, especially if z is large. Figure 3-12 shows Eq. 3-148b plotted for vinyl acetate polymerization inhibited by 2,3,5,6-tetramethylbenzoquinone(duroquinone) [Bartlett and Kwart, 1950]. Table 3-9 shows selected z values for various systems.

A careful consideration of Eq. 3-148b shows that for a strong retarder ($z \gg 1$), the polymerization rate will be negligible until the inhibitor concentration is markedly reduced. When the inhibitor concentration becomes sufficiently low, propagation can become competitive with the inhibition reaction. This is more readily seen by considering the equation

$$\frac{d[Z]}{d[M]} = \frac{z[Z]}{[M]} \tag{3-149}$$

obtained by dividing the rate expression for the disappearance of inhibitor (Eq. 3-142) by that for monomer disappearance (Eq. 3-16). Integration of Eq. 3-149 yields

$$\log \left(\frac{[Z]}{[Z]_0} \right) = z \log \left(\frac{[M]}{[M]_0} \right) \tag{3-150}$$

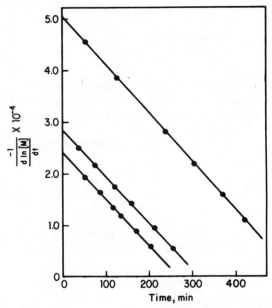

Fig. 3-12 The inhibition of the benzoyl peroxide initiated polymerization of vinyl acetate by duroquinone at 45°C. The three lines are for different concentrations of duroquinone. After Bartlett and Kwart [1950] (by permission of American Chemical Society, Washington, D.C.).

where $[Z]_0$ and $[M]_0$ are initial concentrations. It is apparent from this expression that the inhibitor must be almost completely consumed before the monomer can be polymerized. Equation 3-150 can also be used to determine the inhibition constant from the slope of a plot of log $[Z]$ versus log $[M]$.

3-7b Types of Inhibitors and Retarders

As with chain-transfer agents, one observes that the inhibition constant for any particular compound varies considerably depending on the reactivity and polarity of the propagating radical (Table 3-9). Various types of compounds act as inhibitors and retarders. Stable free radicals that are too stable to initiate polymerization but that can still react with radicals are one type of radical terminator. Diphenylpicrylhydrazyl (DPPH) is such a radical and its use as a radical scavenger has been discussed (Sec. 3-4g-3). The stoichiometry between the number of kinetic chains terminated and the number of DPPH radicals consumed is 1:1.

The most useful class of inhibitors are molecules that react with chain radicals to yield radicals of low reactivity. Quinones such as benzoquinone and chloranil (2,3,5,6-tetrachlorobenzoquinone) are an important class of inhibitor. The behavior of quinones is quite complex [Eastmond, 1976a, 1976b, 1976c; George, 1967; Small, 1975; Yamamoto and Sugimoto, 1979; Yassin and Risk, 1978a, 1978b]. Two major types of products are obtained—quinone and ether—formed by reaction at the C and O atoms of a quinone, respectively. Attack of a propagating radical at oxygen

$$M_n^{\cdot} + O = \!\!\!\left\langle \!\!\!\bigcirc\!\!\! \right\rangle \!\!\!= O \longrightarrow M_n - O - \!\!\!\left\langle \!\!\!\bigcirc\!\!\! \right\rangle \!\!\!- O\cdot \tag{3-151}$$

XXXI

TABLE 3-9   Inhibitor Constants[a]

Inhibitor	Monomer[b]	$z = k_z/k_p$
Nitrobenzene	Methyl acrylate	0.00464
	Styrene	0.326
	Vinyl acetate	11.2
1,3,5-Trinitrobenzene	Methyl acrylate	0.204
	Styrene	64.2
	Vinyl acetate	404
p-Benzoquinone	Acrylonitrile	0.91
	Methyl methacrylate	5.7
	Styrene	518
Chloranil	Methyl methacrylate (44°C)	0.26
	Styrene	2,040
DPPH	Methyl methacrylate (44°C)	2,000
FeCl$_3$	Acrylonitrile (60°C)	3.3
	Styrene (60°C)	536
CuCl$_2$	Acrylonitrile (60°C)	100
	Methyl methacrylate (60°C)	1,027
	Styrene	~11,000
Oxygen	Methyl methacrylate	33,000
	Styrene	14,600
Sulfur	Methyl methacrylate (44°C)	0.075
	Vinyl acetate (44°C)	470
Aniline	Methyl acrylate	0.0001
	Vinyl acetate	0.015
Phenol	Methyl acrylate	0.0002
	Vinyl acetate	0.012
p-Dihydroxybenzene	Vinyl acetate	0.7
1,2,3-Trihydroxybenzene	Vinyl acetate	5.0
2,4,6-Trimethylphenol	Vinyl acetate	0.5

[a]Data from Brandrup and Immergut [1989].
[b]All data are for 50°C unless otherwise noted.

yields the aryloxy radical **XXXI**, which can terminate by coupling and/or disproportionation with another propagating radical (or itself) or add monomer. Attack on the
ring carbon yields radical **XXXII**, which can react with another propagating radical
(Eq. 3-152) to form the quinone **XXXIII**. The latter may itself be an inhibitor. An

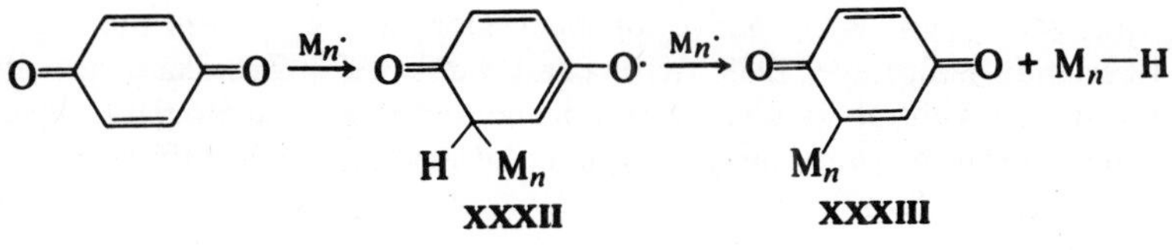

$$(3\text{-}152)$$

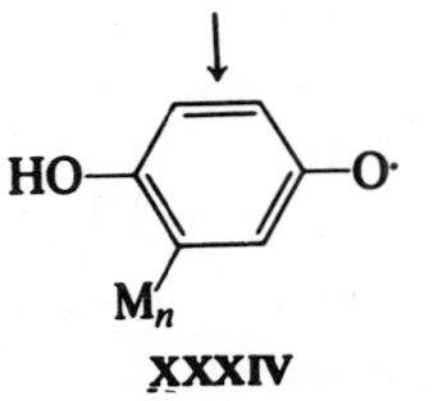

$$(3\text{-}153)$$

alternate route for radical **XXXII** is rearrangement to **XXXIV** followed by coupling or termination with other radicals. The overall stoichiometry (i.e., the value of y in Eqs. 3-147 and 3-148) between the number of kinetic chains terminated and the number of the original quinone molecules consumed varies considerably depending on the particular quinone and radical. Although y often is 1, both pathways (Eqs. 3-151 and 3-152/3-153) can yield y values larger than 1 (up to 2) depending on the reactivities of **XXXI** and **XXXIII**.

Polar effects appear to be of prime importance in determining the effect of quinones. p-Benzoquinone and chloranil (which are electron-poor) act as inhibitors toward electron-rich propagating radicals (vinyl acetate and styrene) but only as retarders toward the electron-poor acrylonitrile and methyl methacrylate propagating radicals. A further observation is that the inhibiting ability of a quinone toward electron-poor monomers can be increased by the addition of an electron-rich third component such as an amine. Thus the presence of triethylamine converts chloranil from a very weak retarder to an inhibitor toward methyl methacrylate.

Oxygen is a powerful inhibitor, as seen from the very large z values. Oxygen reacts with radicals to form the relatively unreactive peroxy radical

$$M_n\cdot + O_2 \rightarrow M_n\text{—}OO\cdot \tag{3-154}$$

that reacts with itself or another propagating radical by coupling and disproportionation reactions to form inactive products (probably peroxides and hydroperoxides) [George and Ghosh, 1978; Koenig and Fischer, 1973; Kishore and Bhanu, 1988; Maybod and George, 1977]. Peroxy radicals can also slowly add monomer to yield an alternating copolymer [Kishore et al., 1981]. The action of oxygen is anomolous in that it is known to initiate some polymerizations. Some commercial processes for ethylene polymerization involve initiation by oxygen [Tatsukami et al., 1980]. Initiation probably occurs by thermal decomposition of the peroxides and hydroperoxides formed from the monomer (or from impurities present in the system). Whether oxygen is an inhibitor or, more rarely, an initiator will be highly temperature-dependent. Initiation will occur at higher temperatures where the peroxides and hydroperoxides are unstable.

Contrary to the general impression, phenol and aniline are poor retarders even toward highly reactive radicals such as the poly(vinyl acetate) propagating radical (Table 3-9). Phenols with electron-donating groups (and, to some extent, similarly substituted anilines) act as more powerful retarders; for example, z is 5.0 for 2,4,6-trimethylphenol compared to 0.012 for phenol toward vinyl acetate polymerization. The presence of an electron-withdrawing group on the phenol ring decreases its activity as an inhibitor. Most phenols are active, or much more active, only in the presence of oxygen [Kurland, 1980; Levy, 1985; Prabha and Nandi, 1977; Zahalka et al., 1988]. The mechanism for inhibition by phenols has been attributed to hydrogen transfer to the propagating radical, but this does not explain the synergistic effect of oxygen. The

$$M_n\cdot + ArOH \rightarrow M_n\text{—}H + ArO\cdot \tag{3-155}$$

inhibiting effect of phenols in the presence of oxygen is probably a consequence of the transfer reaction

$$M_n\text{—}OO\cdot + ArOH \rightarrow M_n\text{—}OOH + ArO\cdot \tag{3-156}$$

being faster than the reaction in Eq. 3-154; that is, phenols scavenge peroxy radicals more effectively than do alkyl (carbon) radicals. Most phenols have y values close to 2, indicating that the phenoxy radical formed in Eq. 3-156 reacts with propagating radicals (carbon and/or peroxy). This probably involves the propagating radical adding to the ring or abstracting hydrogen from a methyl group on the ring for methyl-substituted phenols. The inhibiting action of di- and trihydroxybenzenes such as hydroquinone, t-butylcatcehol (1,2-dihydroxy-4-t-butylbenzene), and pyrogallol (1,2,3-trihydroxybenzene) in the presence of oxygen may be due to their oxidation to quinones instead of or in addition to the reactions described by Eqs. 3-155 and 3-156 [Georgieff, 1965].

Substituted anilines behave similarly to the phenols, although relatively little data are available. N-phenyl-N'-isopropyl-p-phenylenediamine is an efficient inhibitor in the polymerization of styrene only in the presence of oxygen [Winkler and Nauman, 1988]. However, the effectiveness of phenothiazine as an inhibitor in the polymerization of acrylic acid is independent of oxygen [Levy, 1985].

Aromatic nitro compounds terminate propagating chains, with a greater effect for more reactive and electron-rich radicals. Nitro compounds inhibit vinyl acetate, retard styrene but have little effect on methyl acrylate and methyl methacrylate [Eastmond, 1976a, 1976b, 1976c; Schulz, 1947; Tabata et al., 1964]. The effectiveness increases with the number of nitro groups on the ring. 1,3,5-Trinitrobenzene has z values that are 1–2 orders of magnitude greater than nitrobenzene (Table 3-9). The mechanism of radical termination involves attack on both the aromatic ring and the nitro group. Attack on the aromatic ring yields radical **XXXV**, which reacts with another propagating radical (Eq. 3-157) or with monomer (Eq. 3-158).

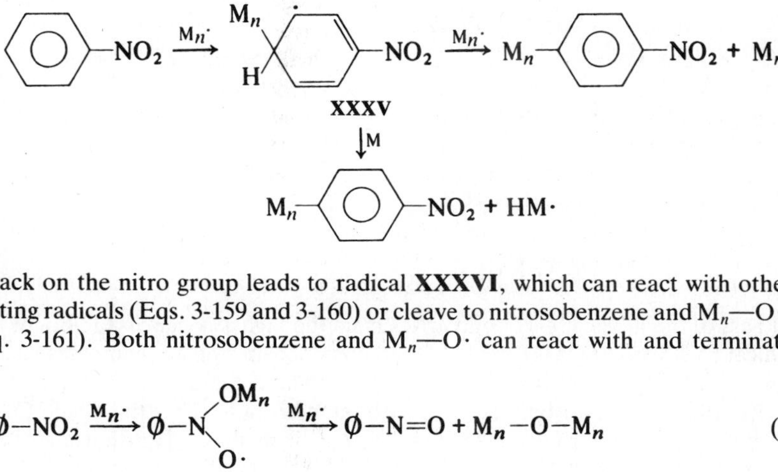

$$(3\text{-}157)$$

$$(3\text{-}158)$$

Attack on the nitro group leads to radical **XXXVI**, which can react with other propagating radicals (Eqs. 3-159 and 3-160) or cleave to nitrosobenzene and M_n—O· radical (Eq. 3-161). Both nitrosobenzene and M_n—O· can react with and terminate other

$$(3\text{-}159)$$

$$(3\text{-}160)$$

$$M_n\text{—O·} + \emptyset\text{—NO} \qquad\qquad (3\text{-}161)$$

propagating radicals. The reaction in Eq. 3-160 probably also involves disproportionation—hydrogen transfer from the propagating radical to **XXXV**. Overall, one observes high y values—up to 5 or 6 for 1,3,5-trinitrobenzene.

Oxidants such as $FeCl_3$ and $CuCl_2$ are strong inhibitors, terminating radicals by the reactions [Billingham et al., 1980; Chetia and Dass, 1976; Matsuo et al., 1977]:

$$\sim\sim CH_2\dot{C}H\phi + FeCl_3 \nearrow \sim\sim CH_2CHCl\phi + FeCl_2 \quad \text{(3-162a)}$$

$$\searrow \sim\sim CH=CH\phi + HCl + FeCl_2 \quad \text{(3-162b)}$$

Reductants also terminate propagating radicals, but much less effectively.

Other inhibitors include sulfur, carbon, aromatic azo compounds, and chlorophosphines [Nigenda et al., 1977; Uemura et al., 1977].

3-7c Autoinhibition of Allylic Monomers

An especially interesting case of inhibition is the internal or *autoinhibition* of allylic monomers (CH_2=CH—CH_2Y). Allylic monomers such as allyl acetate polymerize at abnormally low rates with the unexpected dependence of the rate on the first power of the initiator concentration. Further, the degree of polymerization, which is independent of the polymerization rate, is very low—being only 14 for allyl acetate. These effects are the consequence of *degradative chain transfer* (Case 4 in Table 3-4). The propagating radical in such a polymerization is very reactive, while the allylic C—H (the C—H bond alpha to the double bond) in the monomer is quite weak—resulting in facile chain transfer to monomer

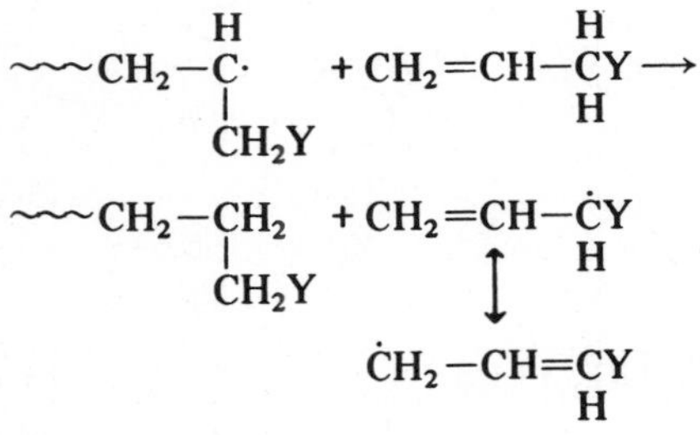

$$\sim\sim CH_2-\underset{\underset{CH_2Y}{|}}{\overset{\overset{H}{|}}{C}}\cdot + CH_2=CH-\underset{\underset{H}{|}}{\overset{\overset{H}{|}}{C}}Y \longrightarrow$$

$$\sim\sim CH_2-\underset{\underset{CH_2Y}{|}}{CH_2} + CH_2=CH-\underset{\underset{H}{|}}{\dot{C}}Y \quad \text{(3-163)}$$

$$\downarrow$$

$$\dot{C}H_2-CH=\underset{\underset{H}{|}}{C}Y$$

XXXVII

The weakness of the allylic C—H bond arises from the high resonance stability of the allylic radical (**XXXVII**) that is formed. Degradative chain transfer competes exceptionally well with normal propagation and the polymer chains are terminated by transfer after the addition of only a very few monomer units. That the allylic C—H bond is the one broken in the transfer reaction has been shown [Bartlett and Tate, 1953] by experiments with CH_2=CH—CD_2OCOCH_3. The deuterated allyl acetate polymerizes 1.9–2.9 times as fast as normal allyl acetate and has a degree of polymerization 2.4 times as large under the same conditions. This is what would be expected since the C—D bond is stronger than the C—H bond due to its lower zero point energy and degradative chain transfer would therefore be decreased in the deuterated monomer.

The allylic radicals which are formed are too stable to initiate polymerization and the kinetic chain also terminates when the transfer occurs. The allylic radicals undergo termination by reaction with each other or, more likely, with propagating radicals

[Litt and Eirich, 1960]. Reaction 3-163 is equivalent to termination by an inhibitor, which is the monomer itself in this case. In this polymerization the propagation and termination reactions will have the same general kinetic expression with first-order dependencies on initiator and monomer concentrations, since the same reactants and stoichiometry are involved. The degree of polymerization is simply the ratio of the rate constants for propagation and termination and is independent of the initiator concentration.

The low reactivity of α-olefins such as propylene or of 1,1-dialkyl olefins such as isobutylene toward radical polymerization is probably a consequence of degradative chain transfer with the allylic hydrogens. It should be pointed out, however, that other monomers such as methyl methacrylate and methacrylonitrile, which also contain allylic C—H bonds, do not undergo extensive degradative chain transfer. This is due to the lowered reactivity of the propagating radicals in these monomers. The ester and nitrile substituents stabilize the radicals and decrease their reactivity toward transfer. Simultaneously the reactivity of the monomer toward propagation is enhanced. These monomers, unlike the α-olefins and 1,1-dialkyl olefins, yield high polymers in radical polymerizations.

3-8 DETERMINATION OF ABSOLUTE RATE CONSTANTS

3-8a Non-Steady-State Kinetics

There are five different types of rate constants that are of concern in radical chain polymerization—those for initiation, propagation, termination, chain transfer, and inhibition. The use of polymerization data under steady-state conditions allows the evaluation of only the initiation rate constant k_d (or k_i for thermal initiation). The ratio $k_p/k_t^{1/2}$ or k_p^2/k_t can be obtained from Eq. 3-25, since R_p, R_i, and [M] are measurable. Similarly, the chain transfer constant k_{tr}/k_p and the inhibition constant k_z/k_p can be obtained by any one of several methods discussed. However, the evaluation of the individual k_p, k_t, k_{tr}, and k_z values under steady-state conditions requires the accurate determination of the propagating radical concentration. This would allow the determination of k_p from Eq. 3-22 followed by the calculation of k_t, k_{tr}, and k_z from the ratios $k_p/k_t^{1/2}$, k_{tr}/k_p, and k_z/k_p. Electron spin resonance (ESR) spectroscopy is clearly the technique to use for measurement of the propagating radical concentration. However, the propagating radical concentrations are too low in typical steady-state polymerizations to determine accurately using commercial ESR instruments. Specially modified ESR spectrometers have been used in a few instances, but this approach awaits major instrument advances before it can be used routinely [Bresler et al., 1974; Kamachi et al., 1985; Shen et al., 1987]. Until then, it is necessary to employ non-steady-state experiments in combination with steady-state experiments to determine the individual reaction rate constants. The treatment described below for a photochemical polymerization under non-steady-state conditions is essentially that of Flory [1953] and Walling [1957].

Radicals are generated abruptly by "putting on" the light source. The rates of initiation and termination are given by Eqs. 3-58 and 3-23, respectively. The rate of change of the radical concentration is given by the difference of their rates of production and termination

$$\frac{d[\text{M}\cdot]}{dt} = 2\phi I_a - 2k_t[\text{M}\cdot]^2 \tag{3-164}$$

For steady-state conditions $d[\text{M}\cdot]/dt$ is zero and

$$2\phi I_a = 2k_t[\text{M}\cdot]_s \tag{3-165}$$

The subscript s in this and other equations denotes steady-state quantities. Combining Eqs. 3-164 and 3-165 yields

$$\frac{d[\text{M}\cdot]}{dt} = 2k_t([\text{M}\cdot]_s^2 - [\text{M}\cdot]^2) \tag{3-166}$$

Equation 3-166 gives the rate of change of the radical concentration for any particular condition as the difference in the rates of termination under steady-state conditions and that particular condition.

It is convenient and indeed necessary at this point to define the parameter τ_s as the *average lifetime of a growing radical* under steady-state conditions. The radical lifetime is given by the steady-state radical concentration divided by its steady-state rate of disappearance

$$\tau_s = \frac{[\text{M}\cdot]_s}{2k_t[\text{M}\cdot]_s^2} = \frac{1}{2k_t[\text{M}\cdot]_s} \tag{3-167}$$

Combination of Eq. 3-167 with Eq. 3-22 at steady-state yields

$$\tau_s = \frac{k_p[\text{M}]}{2k_t(R_p)_s} \tag{3-168}$$

The individual constants k_p and k_t could be determined from Eqs. 3-168 and 3-25 if τ_s were known. It is the objective of non-steady-state experiments to determine τ_s for just this purpose.

Integration of Eq. 3-166 yields

$$\ln\left[\frac{(1 + [\text{M}\cdot]/[\text{M}\cdot]_s)}{(1 - [\text{M}\cdot]/[\text{M}\cdot]_s)}\right] = 4k_t[\text{M}\cdot]_s(t - t_0) \tag{3-169}$$

where t_0 is the integration constant such that $[\text{M}\cdot] = 0$ at time t_0. Combining Eqs. 3-167 and 3-169 gives

$$\tanh^{-1}\left(\frac{[\text{M}\cdot]}{[\text{M}\cdot]_s}\right) = \frac{(t - t_0)}{\tau_s} \tag{3-170}$$

or

$$\frac{[\text{M}\cdot]}{[\text{M}\cdot]_s} = \frac{R_p}{(R_p)_s} = \tanh\left(\frac{(t - t_0)}{\tau_s}\right) \tag{3-171}$$

(The reader should be familiar with the hyperbolic tangent and inverse hyperbolic tangent so as to derive Eqs. 3-170 and 3-171.) The t in these equations is the time of illumination. Now consider this polymerization as proceeding with intermittent illumination, that is, with alternate light and dark periods. At the very beginning the radical concentration is zero, but at any time following the radical concentration never falls to zero again. It has instead some value $[M\cdot]_2 > 0$ at the start of a period of illumination or light period (which corresponds to the end of a dark period). Therefore, Eq. 3-170 becomes

$$\frac{t_0}{\tau_s} = -\tanh^{-1}\left(\frac{[M\cdot]_2}{[M\cdot]_s}\right) \tag{3-172}$$

and

$$\tanh^{-1}\left(\frac{[M\cdot]_1}{[M\cdot]_s}\right) - \tanh^{-1}\left(\frac{[M\cdot]_2}{[M\cdot]_s}\right) = \frac{t}{\tau_s} \tag{3-173}$$

where $[M\cdot]_1$ is the radical concentration after a light period of time t. $[M\cdot]_1$ also corresponds to the radical concentration at the start of a dark period. The curve OAE in Fig. 3-13 shows the buildup in radical concentration during a light period according to Eq. 3-173 when the initial radical concentration is zero at $t_0 = 0$.

The light is now turned off at time t and radical decay occurs during the dark period according to

$$\frac{d[M\cdot]}{dt} = -2k_t[M\cdot]^2 \tag{3-174}$$

which on integration yields

$$\frac{1}{[M\cdot]} - \frac{1}{[M\cdot]_1} = 2k_t t' \tag{3-175}$$

where $[M\cdot]$ is the radical concentration after a further time interval t' corresponding to the time of darkness. Multiplying Eq. 3-175 through by $[M\cdot]_s$ and combining with

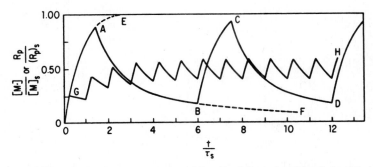

Fig. 3-13 Plot of $[M]/[M]_s$ versus time during alternating light and dark periods. Adapted from P. J. Flory, "Principles of Polymer Chemistry." Copyright 1953 by Cornell University. Used by permission of Cornell University Press, Ithaca, N.Y.

Eq. 3-167 yields

$$\frac{[M\cdot]_s}{[M\cdot]} - \frac{[M\cdot]_s}{[M\cdot]_1} = \frac{t'}{\tau_s} \tag{3-176}$$

or

$$\frac{(R_p)_s}{R_p} - \frac{(R_p)_s}{(R_p)_1} = \frac{t'}{\tau_s} \tag{3-177}$$

The radical decay according to Eqs. 3-176 and 3-177 is shown as curve ABF in Fig. 3-14.

One method of determining τ_s is the *after-effect technique*, which employs Eq. 3-177 or its equivalent. The polymerization rate $(R_p)_s$ under steady-state conditions (also referred to as *stationary conditions*) is observed using constant illumination. The illumination is stopped and the subsequent rate R_p is determined as a function of time t' after the start of the dark period. The ratio $(R_p)_s/R_p$ is plotted against t' according to Eq. 3-177 and the slope is $1/\tau_s$. The experimental procedures for obtaining τ_s by this technique, although quite difficult because very slow reactions of very short duration must be accurately measured, have been successfully used [Bamford and Hirooka, 1984; Stickler, 1987]. A complementary technique, the *preeffect technique*, follows the polymerization period during illumination before steady-state conditions have been reached. The accuracy of both techniques have been improved by the use of laser sources that offer higher light intensities and better control of the pulse rate and light intensity compared to conventional light sources [Norrish and Thrush, 1956].

3-8b Rotating Sector Method

The more common technique employed to determine τ_s is the *rotating sector method* involving the alternating of light and dark periods. This method requires a study of the polymerization rate as a function of cycle time (the time for one light and one

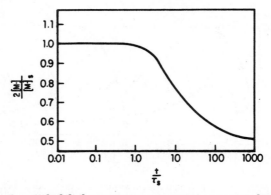

Fig. 3-14 Semilog plot of $2[M]/[M]_s$ versus t/τ_s. After Matheson et al. [1949] (by permission of American Chemical Society, Washington, D.C.).

dark period). Consider the specific case where the ratio r of the length t' of the dark period to the length t of a light period is 3. If the cycle time is very long compared to τ_s (slow flickering), the R_p will be equal to $(R_p)_s$ during the entire light period and zero during the entire dark period. The reason for this is that the times for the radical concentration to reach steady-state or to decay to zero are small compared to times t and t', respectively. The average polymerization rate $\overline{R}_p$ over one complete cycle will be one fourth that of the steady-state value, since the system is illuminated only one fourth of the time, that is

$$\overline{R}_p = \frac{(R_p)_s}{4} \tag{3-178}$$

If the cycle time is reduced (fast flickering), radical decay during the dark period is incomplete and the steady-state radical concentration is not reached during the light period (Fig. 3-14). The radical concentration averaged over a cycle will be greater than $(R_p)_s/4$ because the decay during a dark period is less than the buildup during a light period. A very fast cycle time maintains the radical concentration at approximately a constant level (curve OGH in Fig. 3-14) and is equivalent to polymerization under constant illumination at an intensity $1/(1 + r)$ times that actually used. The ratio of the average polymerization rate $(\overline{R}_p)_\infty$ at very high or infinite speed of sector rotation to the steady-state rate is then given by

$$\frac{(\overline{R}_p)_\infty}{(R_p)_s} = \frac{1}{(1 + r)^{1/2}} \tag{3-179}$$

which equals $\frac{1}{2}$ for the case of $r = 3$. Thus the average rate increases from one fourth to one half of $(R_p)_s$ as the cycle or flickering frequency $1/(t + rt)$ increases from a very low to a very high value compared with $1/\tau_s$.

The mathematical treatment of flickering illumination has been described [Briers et al., 1926]. After a number of cycles, the radical concentration oscillates uniformly with a constant radical concentration $[M\cdot]_1$ at the end of each light period of duration t and a constant radical concentration $[M\cdot]_2$ at the end of each dark period of duration $t' = rt$. Applying Eqs. 3-173 and 3-177a, one obtains

$$\tanh^{-1}\left(\frac{[M\cdot]_1}{[M\cdot]_s}\right) - \tanh^{-1}\left(\frac{[M\cdot]_2}{[M\cdot]_s}\right) = \frac{t}{\tau_s} \tag{3-180}$$

and

$$\frac{[M\cdot]_s}{[M\cdot]_2} - \frac{[M\cdot]_s}{[M\cdot]_1} = \frac{rt}{\tau_s} \tag{3-181}$$

The maximum and minimum values of the ratios $[M\cdot]_1/[M\cdot]_s$ and $[M\cdot]_2/[M\cdot]_s$ can be calculated from Eqs. 3-180 and 3-181 for given values of r and t/τ_s. The average radical concentration $\overline{[M\cdot]}$ over a cycle or several cycles is given by

$$\overline{[M\cdot]}(t + rt) = \int_0^t [M\cdot]\, dt + \int_0^{t'} [M\cdot]\, dt' \tag{3-182}$$

where the first integral (for the light period) is given by Eq. 3-180 and the second integral (for the dark period) is given by Eq. 3-181. Evaluation of the integrals in Eq. 3-182 yields

$$\frac{\overline{[M\cdot]}}{[M\cdot]_s} = (r + 1)^{-1} \left[1 + \frac{\tau_s}{t} \ln \left(\frac{[M\cdot]_1/[M\cdot]_2 + [M\cdot]_1/[M\cdot]_s}{1 + [M\cdot]_1/[M\cdot]_s} \right) \right] \tag{3-183}$$

Using Eq. 3-183 with Eqs. 3-180 and 3-181, one can calculate the ratio $\overline{[M\cdot]}/[M\cdot]_s$ as a function of t/τ_s for a fixed value of r. A semilog plot of such data for $r = 3$ is shown in Fig. 3-14. In accordance with an earlier conclusion, the radical concentration falls from one-half of the steady-state value for fast flickering to one-fourth for slow flickering.

In order to experimentally determine the τ_s value for a particular system, one interposes a rotating sector or disk in between the system and the light source [Nagy et al., 1983]. The sector has a portion cut out, which determines the value of r. The steady-state polymerization rate is first measured without the sector present. Then the average rate $\overline{R}_p$ is measured with the sector present at increasing rates of sector rotation. The cycle time as well as t and t' are determined by the rate of sector rotation. The data are plotted at the rate ratio $\overline{R}_p/(R_p)_s$ versus log t. Alternately, one can plot the data as the rate ratio $\overline{R}_p/(\overline{R}_p)_\infty$ since this ratio is related to $\overline{R}_p/(R_p)_s$ through Eq. 3-179. The theoretical curve (e.g., Fig. 3-14) for the same r value is placed on top of the experimental curve and shifted on the abscissa until a best fit is obtained. The displacement on the abscissa of one curve from the other yields log τ_s since the abscissa for the theoretical curve is log t − log τ_s. Figure 3-15 shows such a determination for the polymerization of methacrylamide in water solution initiated by the photochemical decomposition of hydrogen peroxide [Dainton and Sisley, 1963].

The experimental determination of τ_s then allows the calculation of k_p, k_t, k_{tr}, and k_z. The ratios $k_p/k_t^{1/2}$ and k_p/k_t are obtained from Eqs. 3-25 and 3-168, respectively. Combination of these ratios yields the individual values of k_p and k_t. Quantities such

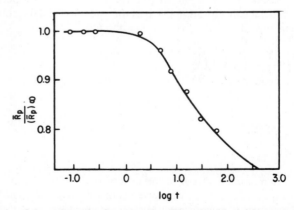

Fig. 3-15 The ratio of the polymerization rate of methacrylamide for intermittent illumination with light periods of t sec to the value at very high (infinite) sector speed plotted against t. The circles are the experimental points; the line is the theoretical plot. After O'Driscoll and Mahabadi [1976] (by permission of The Faraday Society, London).

as $[M \cdot]_s$ and R_t can be calculated from Eqs. 3-22 and 3-23, respectively. k_{tr} and k_z can be obtained from the values of the chain transfer and inhibition constants. Determination of τ_s also allows one to evaluate the validity of the usual steady-state assumption in polymerization using Eq. 3-171. The time required for $[M \cdot]$ and R_p to reach their steady-state values is calculated as 65, 6.5, and 0.65 sec, respectively, for τ_s values of 10, 1, and 0.1 sec. Thus in the typical polymerization study the steady-state assumption is valid after a couple of minutes at most.

An experimental variation on the rotating sector is *spatially intermittent polymerization*, which uses a glass tubular reactor surrounded by a metal cylinder having narrow, regularly spaced slots through which light shines. Monomer is pumped through the reactor and passes alternatively through light and dark regions. There is a time-varying rate of radical production (and polymerization rate) that depends on the slot width, spacing between slots, and flow rate. The mathematical treatment is very similar to that described for the rotating sector method [O'Driscoll and Mahabadi, 1976].

Most nonsteady-state experiments are presently performed using a *pulsed laser* for the light source. The use of a laser source is advantageous for the reasons mentioned in the discussion on the after- (post-) and preeffect techniques. Also, there is no need for a rotating sector. The reaction system is subjected to pulsed irradiation by a laser source and the ratio r is defined accurately by the pulse width (time) and pulse frequency of the laser.

A recent advance in the pulsed laser technique involves the determination of k_p from measurements of the number-average degree of polymerization under non-steady-state conditions [Buback et al., 1987; Davis et al., 1989; Olaj and Bitai, 1987; Olaj et al., 1987; Schnoll-Bitai and Olaj, 1990]. Under pulsed laser irradiation, primary radicals are formed in very short times compared both to the average lifetimes of propagating radicals and also the times for conversion of primary radicals to propagating radicals. Consider the propagation that proceeds during a dark period (laser off). The subsequent flash of light (laser on) produces a high concentration of primary radicals in the presence of the propagating radicals. A large fraction of the primary radicals terminate the propagating radicals. The remaining fraction of primary radicals initiate new propagating radicals that propagate during the remaining portion of the very brief light period and then during the subsequent dark period. When the next light period occurs, the cycle of events is repeated. Bimolecular termination is present throughout the light and dark periods. However, primary termination is the major termination process because of the high radical concentrations formed during each brief "laser on" light period. Determination of the number-average degree of polymerization of the major component $(\overline{X}_n)_0$ of the formed polymer allows a determination of k_p from

$$(\overline{X}_n)_0 = t_d k_p [M] \tag{3-184}$$

where t_d is the duration of the dark period.

This approach is promising because it allows one to obtain the propagation rate constant from a second measurement, that is, from $(\overline{X}_n)_0$ as well as from R_p. The general applicability of the measurements based on the degree of polymerization have yet to be firmly established. The major problem is identification of $(\overline{X}_n)_0$ since there is (as expected) a distribution of different-sized propagating radicals terminated by primary termination, on top of which there is a distribution of propagating radicals terminated by bimolecular coupling and disproportionation.

3-8c   Typical Values of Reaction Parameters

Table 3-10 shows the values of the various concentrations, rates, and rate constants involved in the photopolymerization of methacrylamide as well as the range of values that are generally encountered in radical chain polymerizations. For the methacrylamide case, the experimentally determined quantities were R_i, $(R_p)_s$, [M], [I], $k_p/k_t^{1/2}$, τ_s, and k_p/k_t. All of the other parameters were then calculated in the appropriate manner. These values are typical of radical polymerizations. The k_p value ($\sim 10^3$) is larger by many orders of magnitude than the usual reaction rate constant (for example, Table 2-1, 2-2, and 2-10 show rate constants of approximately 10^{-3}, 10^{-2}, and 10^{-6} for esterification, urethane formation, and phenol-formaldehyde polymerization, respectively). Propagation is therefore rapid and high polymer is formed essentially instantaneously. However, the even larger k_t value ($\sim 10^7$) leads to quick termination, low radical concentrations ($\sim 10^{-8}$ M), and short radical lifetimes. The radical lifetime for methacrylamide in these experiments was 2.62 sec, but it can be shorter or longer under other conditions or for other monomers. It is interesting to compare the experimental value (8.75×10^{-9}) of R_i with the calculated value (8.73×10^{-9}) of R_t. The excellent agreement of the two indicates the validity of the steady-state assumption.

The rate constants for propagation and termination have been determined for many monomers. The values for some of the common monomers are shown in Table 3-11. The monomers have been listed in order of decreasing k_p values (which does not necessarily correspond to the exact order of decreasing k_t values). The order of k_p values will be discussed in Sec. 6-3b, since k_p is a function of monomer reactivity and radical reactivity.

The reader is cautioned that there is often a considerable divergence in the literature for values of rate constants [Buback et al., 1988]. One needs to examine the experimental details of literature reports to choose appropriately the values to be used for any needed calculations. Apparently different values of a rate constant may be a consequence of experimental error, experimental conditions (e.g., differences in conversion, solvent viscosity), or method of calculation (e.g., different workers using different literature values of k_d for calculating R_i, which is subsequently used to calculate $k_p/k_t^{1/2}$ from an experimental determination of R_p).

TABLE 3-10   Reaction Parameters in Radical Chain Polymerization

Quantity	Units	General Range of Values	Methacrylamide Photopolymerization[a]
R_i	moles/liter-sec	10^{-8}–10^{-10}	8.75×10^{-9}
k_d	sec^{-1}	10^{-4}–10^{-6}	—
[I]	moles/liter	10^{-2}–10^{-4}	3.97×10^{-2}
$[M\cdot]_s$	moles/liter	10^{-7}–10^{-9}	2.30×10^{-8}
$(R_p)_s$	moles/liter-sec	10^{-4}–10^{-6}	3.65×10^{-6}
[M]	moles/liter	10–10^{-1}	0.20
k_p	liters/mole-sec	10^2–10^4	7.96×10^2
R_t	moles/liter-sec	10^{-8}–10^{-10}	8.73×10^{-9}
k_t	liters/mole-sec	10^6–10^8	8.25×10^6
τ_s	sec	10^{-1}–10	2.62
k_p/k_t	none	10^{-4}–10^{-6}	9.64×10^{-5}
$k_p/k_t^{1/2}$	$(liters/mole-sec)^{1/2}$	1–10^{-2}	2.77×10^{-1}

[a]Values are taken directly or recalculated from Dainton and Sisley [1963].

TABLE 3-11   Kinetic Parameters in Radical Chain Polymerization[a,b,c,d]

Monomer	$k_p \times 10^{-3}$	E_p	$A_p \times 10^{-7}$	$k_t \times 10^{-7}$	E_t	$A_t \times 10^{-9}$
Vinyl chloride (50°C)	11.0	16	0.33	210	17.6	600
Tetrafluoroethylene (83°C)	9.10	17.4	—	—	—	—
Vinyl acetate	2.30	18	3.2	2.9	21.9	3.7
Acrylonitrile	1.96	16.2	—	7.8	15.5	—
Methyl acrylate	2.09	29.7	10	0.95	22.2	15
Methyl methacrylate	0.515	26.4	0.087	2.55	11.9	0.11
2-Vinylpyridine	0.186	33	—	3.3	21	—
Styrene	0.165	26	0.45	6.0	8.0	0.058
Ethylene	0.242	18.4	—	54.0	1.3	—
1,3-Butadiene	0.100	24.3	12	—	—	—

[a] k_p and k_t values are for 60°C unless otherwise noted. The units of k_p and k_t are liters/mole-sec.
[b] E_p values are in kJ/mole of polymerizing monomer; E_t values are in kJ/mole of propagating radicals.
[c] A_p and A_t values are in liters/mole-sec.
[d] Data are from Brandrup and Immergut [1989], Eastmond [1976a, 1976b, 1976c], and Walling [1957].

3-9   ENERGETIC CHARACTERISTICS

3-9a   Activation Energy and Frequency Factor

The effect of temperature on the rate and degree of polymerization is of prime importance in determining the manner of performing a polymerization. Increasing the reaction temperature usually increases the polymerization rate and decreases the polymer molecular weight. Figure 3-16 shows this effect for the thermal, self-initiated polymerization of styrene. However, the quantitative effect of temperature is complex since R_p and $\overline{X}_n$ depend on a combination of three rate constants—k_d, k_p, and k_t. Each of the rate constants for initiation, propagation, and termination can be expressed by an Arrhenius-type relationship

$$k = Ae^{-E/RT} \tag{3-185a}$$

or

$$\ln k = \ln A - \frac{E}{RT} \tag{3-185b}$$

where A is the *collision frequency factor*, E the *Arrhenius activation energy*, and T the Kelvin temperature. A plot of $\ln k$ versus $1/T$ allows the determination of both E and A from the slope and intercept, respectively. Values of E_p, the activation energy for propagation, and E_t, the activation energy for termination, for several monomers are shown in Table 3-10. It is interesting to note that the variations in the values of A_p, the frequency factor for propagation, are much greater than those in E_p—indicating that steric effects are probably the more important factor in determining the absolute value of k_p. Thus the more hindered monomers (e.g., methyl methacrylate) have lower k_p and A_p values than the less hindered ones. The A_p values in general are much lower than the usual value (10^{11}–10^{13}) of the frequency factor for a bimolecular reaction—probably due to a large decrease in entropy on polymerization. The variations in the values of A_t, the frequency factor for termination, generally follow

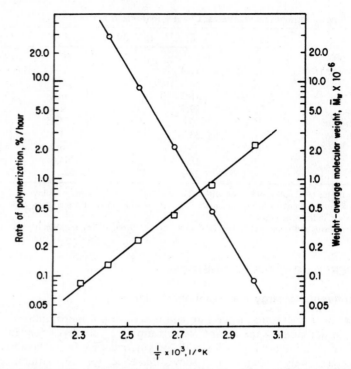

Fig. 3-16   Dependence of the polymerization rate (○) and polymer molecular weight (□) on the temperature for the thermal self-initiated polymerization of styrene. After Roche and Price [1952] (by permission of Dow Chemical Co., Midland, Mich.).

along the same line as the A_p values. The A_t values in general are larger than the A_p values.

3-9a-1   Rate of Polymerization

Consider the activation energy for various radical chain polymerizations. For a polymerization initiated by the thermal decomposition of an initiator, the polymerization rate depends on the ratio of three rate constants $k_p (k_d/k_t)^{1/2}$ in accordance with Eq. 3-32. The temperature dependence of this ratio, obtained by combining three separate Arrhenius-type equations, is given by

$$\ln \left[k_p \left(\frac{k_d}{k_t} \right)^{1/2} \right] = \ln \left[A_p \left(\frac{A_d}{A_t} \right)^{1/2} \right] - \frac{[E_p + (E_d/2) - (E_t/2)]}{RT} \tag{3-186}$$

The *composite* or *overall activation energy for the rate of polymerization* E_R is $[E_p + (E_d/2) - (E_t/2)]$. Since R_p is given by Eq. 3-32, one can write Eq. 3-186 as

$$\ln R_p = \ln \left[A_p \left(\frac{A_d}{A_t} \right)^{1/2} \right] + \ln \left[(f[\mathrm{I}])^{1/2}[\mathrm{M}] \right] - \frac{E_R}{RT} \tag{3-187}$$

E_R and $A_p \, (A_d/A_t)^{1/2}$ can then be obtained from the slope and intercept, respectively, of a plot of $\ln R_p$ versus $1/T$ (similar to Fig. 3-16).

E_d, the activation energy for initiator decomposition, is in the range 120–150 kJ/mole for most of the commonly used initiators (Table 3-12). The E_p and E_t values for most monomers are in the ranges 20–40 kJ/mole and 8–20 kJ/mole, respectively (Tables 3-11 and Brandrup and Immergut, 1989). The overall activation energy E_R for most polymerizations initiated by thermal initiator decomposition is about 80–90 kJ/mole. This corresponds to a two- or threefold rate increase for a 10°C temperature increase. The situation is different for other modes of initiation. Thus redox initiation (for example, Fe^{2+} with thiosulfate or cumene hydroperoxide) has been discussed as taking place at lower temperatures compared to the thermal polymerizations. One indication of the difference between the two different initiation modes is the difference in activation energies. Redox initiation will have an E_d value of only about 40–60 kJ/mole—or about 80 kJ/mole less than that for thermal initiation [Barb et al., 1951]. This leads to an E_R for redox polymerization of about 40 kJ/mole—or about one half the value for nonredox initiators.

For a purely photochemical polymerization, the initiation step is temperature-independent ($E_d = 0$), since the energy for initiator decomposition is supplied by light quanta. The overall activation energy for photochemical polymerization is then only about 20 kJ/mole. This low value of E_R indicates that R_p for photochemical polymerizations will be relatively insensitive to temperature compared to other polymerizations. The effect of temperature on photochemical polymerizations is complicated, however, since most photochemical initiators can also decompose thermally. At higher temperatures the initiators may undergo appreciable thermal decomposition in addition to the photochemical decomposition. In such cases, one must take into account both the thermal and photochemical initiations. The initiation and overall activation energies for a purely thermal self-initiated polymerization are approximately the same as for initiation by the thermal decomposition of an initiator. For the thermal, self-initiated polymerization of styrene the activation energy for initiation is 121 kJ/mole and E_R is 86 kJ/mole [Barr et al., 1978; Hui and Hamielec, 1972]. However, purely thermal polymerizations proceed at very slow rates because of the low probability of the initiation process due to the very low values (10^4–10^6) of the frequency factor.

3-9a-2 Degree of Polymerization

To determine the effect of temperature on the molecular weight of the polymer produced in a thermally catalyzed polymerization where transfer reactions are negligible,

TABLE 3-12 Thermal Decomposition of Initiators[a,b,c]

Initiator	$k_d \times 10^5$	T (°C)	E_d
2,2′-Azobisisobutyronitrile	0.845	60	123.4
Acetyl peroxide	2.39	70	136.0
Benzoyl peroxide	5.50	85	124.3
Cumyl peroxide	1.56	115	170.3
t-Butyl peroxide	3.00	130	146.9
t-Butyl hydroperoxide	0.429	155	170.7

[a] All data are for decompositions in benzene solution.
[b] Data from Brandrup and Immergut [1989].
[c] The units of k_d are sec^{-1}; the units of E_d are kJ/mole.

one must consider the ratio $k_p/(k_d k_t)^{1/2}$, since it determines the degree of polymerization (Eq. 3-108). The variation of this ratio with temperature is given by

$$\ln\left[\frac{k_p}{(k_d k_t)^{1/2}}\right] = \ln\left[\frac{A_p}{(A_d A_t)^{1/2}}\right] - \frac{[E_p - (E_d/2) - (E_t/2)]}{RT} \tag{3-188}$$

where the energy term $[E_p - (E_d/2) - (E_t/2)]$ is the *composite* or *overall activation energy for the degree of polymerization* $E_{\overline{X}_n}$. For bimolecular termination, $\overline{X}_n$ is governed by Eqs. 3-108 and 3-109a and one can write

$$\ln \overline{X}_n = \ln\left[\frac{A_p}{(A_d A_t)^{1/2}}\right] + \ln\left[\frac{[M]}{(f[I])^{1/2}}\right] - \frac{E_{\overline{X}_n}}{RT} \tag{3-189}$$

$E_{\overline{X}_n}$ has a value of about -60 kJ/mole in typical cases and $\overline{X}_n$ decreases rapidly with increasing temperature. $E_{\overline{X}_n}$ is about the same for a purely thermal, self-initiated polymerization (Fig. 3-16). For a pure photochemical polymerization $E_{\overline{X}_n}$ is positive by approximately 20 kJ/mole, since E_d is zero, and $\overline{X}_n$ increases moderately with temperature. For all other cases, $\overline{X}_n$ decreases with temperature.

When chain transfer occurs in the polymerization, $\overline{X}_n$ is given by an appropriate form of Eq. 3-118. The temperature dependence of $\overline{X}_n$ can be quite complex depending on the relative importance of the various terms in Eq. 3-118. For the case where chain transfer to compound S is controlling (Eq. 3-129), one obtains

$$-\ln\left[\frac{[M]}{[S]}\left(\frac{1}{\overline{X}_n} - \frac{1}{(\overline{X}_n)_0}\right)\right] = \ln\frac{k_p}{k_{tr,S}} = \ln\frac{A_p}{A_{tr,S}} - \frac{(E_p - E_{tr,S})}{RT} \tag{3-190}$$

The quantity $(E_p - E_{tr,S})$ is now $E_{\overline{X}_n}$ and can be obtained from a plot of either of the two forms of the left side of Eq. 3-190 versus $1/T$. $E_{tr,S}$ usually exceeds E_p by 20–65 kJ/mole with the more active transfer agents having lower values. The term $(E_p - E_{tr,S})$ is usually -20 to -65 kJ/mole (Table 3-13) and the molecular weight therefore decreases with increasing temperature. The frequency factors for transfer reactions are usually greater than those for propagations, and the low transfer constant of a particular transfer agent is a consequence of the high activation energy only.

TABLE 3-13   Activation Parameters for Chain Transfer in Styrene Polymerization (60°C)[a]

Transfer Agent	$-(E_p - E_{tr,S})$	$\log(A_{tr,S}/A_p)$
Cyclohexane	56.1	3.1
Benzene	62.0	3.9
Toluene	42.3	1.7
Ethylbenzene	23.0	-0.55
Isopropylbenzene	23.0	-0.47
t-Butylbenzene	57.4	3.8
n-Butyl chloride	58.6	4
n-Butyl bromide	46.1	2
n-Butyl iodide	29.3	1
Carbon tetrachloride	20.9	1

[a]Data from Gregg and Mayo [1948, 1953a, 1953b].

3-9b Thermodynamics of Polymerization

3-9b-1 *Significance of* ΔG, ΔH, *and* ΔS

The thermodynamic characteristics $(\Delta G, \Delta H, \Delta S)$ of polymerization are important to an understanding of the effect of monomer structure on polymerization. Further, knowledge of ΔH allows one to maintain the desired R_p and $\overline{X}_n$ by appropriate thermal control of the process. The ΔG, ΔH, and ΔS for a polymerization are the differences in *free energy*, *enthalpy*, and *entropy*, respectively, between 1 mole of monomer and 1 mole of repeating units in the polymer product. The thermodynamic properties of a polymerization relate only to the propagation step, since polymerization consists of single acts of initiation and termination and a large number of propagation steps.

Chain polymerizations of alkenes are exothermic (negative ΔH) and exoentropic (negative ΔS). The exothermic nature of polymerization arises because the process involves the exothermic conversion of π-bonds in monomer molecules into σ-bonds in the polymer. The negative ΔS for polymerization arises from the decreased degrees of freedom (randomness) for the polymer relative to the monomer. Thus, polymerization is favorable from the enthalpy viewpoint but unfavorable from the entropy viewpoint. Table 3-14 shows the wide range of ΔH values for various monomers. The ΔS values fall in a narrower range of values. The methods of evaluating ΔH and ΔS have been reviewed [Dainton and Ivin, 1950, 1958]. These include direct calorimetric measurements of ΔH for the polymerization, determination by the difference between

TABLE 3-14   Enthalpy and Entropy of Polymerization at 25°C[a,b]

Monomer	$-\Delta H$	$-\Delta S$
Ethylene[c]	93	155
Propene	84	116
1-Butene	83.5	113
Isobutylene	48	121
1,3-Butadiene	73	89
Isoprene	75	101
Styrene	73	104
α-Methylstyrene	35	110
Vinyl chloride	72	—
Vinylidene chloride	73	89
Tetrafluoroethylene	163	112
Acrylic acid	67	—
Acrylonitrile	76.5	109
Maleic anhydride	59	—
Vinyl acetate	88	110
Methyl acrylate	78	—
Methyl methacrylate	56	117

[a]Data from Brandrup and Immergut [1989]; Sawada [1976].
[b]ΔH refers to the conversion of liquid monomer to amorphous or (slightly) crystalline polymer. ΔS refers to the conversion of monomer (at a concentration of 1 M) to amorphous or slightly crystalline polymer. The subscripts *lc* are often used with ΔH and ΔS to show the initial and final states (that is, ΔH_{lc} and ΔS_{lc}). The units of ΔH are kJ/mole of polymerized monomer; the units of ΔS are J/K-mole.
[c]Data are for conversion of gaseous monomer to crystalline polymer.

the heats of combustion of monomer and polymer, and measurements of the equilibrium constant for the polymerization. The overall thermodynamics of the polymerization of alkenes is quite favorable. The value of ΔG given by

$$\Delta G = \Delta H - T \Delta S \qquad (3\text{-}191)$$

is negative because the negative $T \Delta S$ term is outweighed by the negative ΔH term.

One should recall the earlier discussion (Sec. 3-1b) on the thermodynamic and kinetic feasibilities of polymerization. The data in Table 3-14 clearly show the general thermodynamic feasibility for any carbon–carbon double bond. Although the relative thermodynamic feasibility of any one monomer varies depending on the substituents present in the monomer, ΔG is negative in all cases and polymerization is favored. However, thermodynamic feasibility does not indicate the experimental conditions that may be required to bring about the polymerization. Thus Table 3-1 showed that the kinetic feasibility of polymerization varies considerably from one monomer to another in terms of whether radical, cationic, or anionic initiation can be used for the reaction. In some instances thermodynamically feasible polymerizations may require very specific catalyst systems. This is the case with the α-olefins, which cannot be polymerized to high molecular polymers by any of the conventional radical or ionic initiators. The polymerization of these monomers was not achieved until the discovery of the Ziegler–Natta or coordination type initiators (Chap. 8).

3-9b-2 Effect of Monomer Structure

Consider the effect of monomer structure on the enthalpy of polymerization. The ΔH values for ethylene, propene, and 1-butene are very close to the difference (82–90 kJ/mole) between the bond energies of the π-bond in an alkene and the σ-bond in an alkane. The ΔH values for the other monomers vary considerably. The variations in ΔH for differently substituted ethylenes arise from any of the following effects:

1 Differences in the resonance stabilization of monomer and polymer due to differences in conjugation or hyperconjugation.
2 Steric strain differences in the monomer and polymer arising from bond angle deformation, bond stretching, or interactions between nonbonded atoms.
3 Differences in hydrogen bonding or dipole interactions in the monomer and polymer.

Many substituents stabilize the monomer but have no appreciable effect on polymer stability, since resonance is only possible with the former. The net effect is to decrease the exothermicity of the polymerization. Thus hyperconjugation of alkyl groups with the C=C lowers ΔH for propylene and 1-butene polymerizations. Conjugation of the C=C with substituents such as the benzene ring (styrene and α-methylstyrene), and alkene double bond (butadiene and isoprene), the carbonyl linkage (acrylic acid, methyl acrylate, methyl methacrylate), and the nitrile group (acrylonitrile) similarly leads to stabilization of the monomer and decreases enthalpies of polymerization. When the substituent is poorly conjugating as in vinyl acetate, the ΔH is close to the value for ethylene.

The effect of 1,1-disubstitution manifests itself by decreased ΔH values. This is a consequence of steric strain in the polymer due to interactions between substituents on alternating carbon atoms of the polymer chain

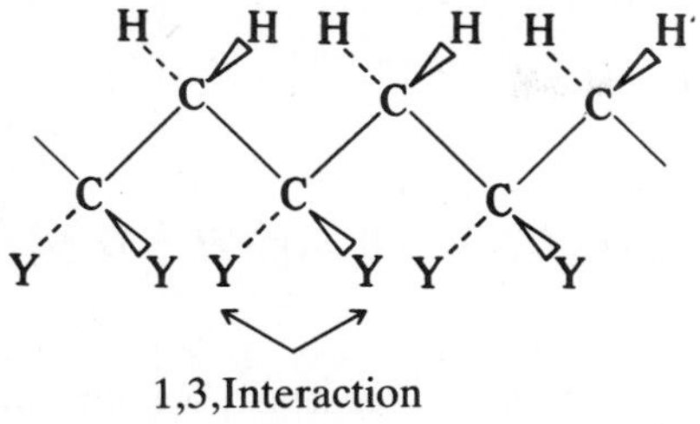

1,3,Interaction

XXXVIII

In the picture above the main polymer chain is drawn in the plane of this text with the H and Y substituents placed above and below the plane of the text. The dotted and triangular lines indicate substituents below and above this plane, respectively. Such interactions are referred to as *1,3-interactions* and are responsible for the decreased ΔH values in monomers such as isobutylene, α-methylstyrene, methyl methacrylate, and vinylidene chloride. The effect in α-methylstyrene is especially significant. The ΔH value of -35 kJ/mole is essentially the smallest heat of polymerization of any monomer.

A contributing factor to the lowering of ΔH in some cases is a decrease in hydrogen bonding or dipole interactions on polymerization. Monomers such as acrylic acid and acylamide are significantly stabilized by strong intermolecular associations. The intermolecular associations are not as important in the polymer because its steric constraints prevent the required lining up of substituents.

The ΔH value for vinyl chloride is lowered relative to that for ethylene due to increased steric strain in the polymer and increased resonance stabilization of the monomer. However, it is not clear why ΔH for vinylidene chloride is not lower than that for vinyl chloride. The abnormally high ΔH for tetrafluoroethylene is difficult to understand. A possible explanation may involve increased stabilization of the polymer due to the presence of intermolecular association (dipole interaction).

While the ΔH values vary over a wide range for different monomers, the ΔS values are less sensitive to monomer structure, being relatively constant within the range of 100–120 J/K-mole. The $T \Delta S$ contribution to the ΔG for polymerization will be small as indicated earlier and will vary only within a narrow range. Thus the variation in the $T \Delta S$ term at 50°C for all monomers is in the range 30–40 kJ/mole. The entropy changes which occur upon polymerization have been analyzed for several monomers [Dainton and Ivin, 1950, 1958]. The ΔS of polymerization arises primarily from the loss of the translational entropy of the monomer. Losses in the rotational and vibrational entropies of the monomer are essentially balanced by gains in the rotational and vibrational entropies of the polymer. Thus ΔS for polymerization is essentially the translational entropy of the monomer, which is relatively insensitive to the structure of the monomer.

3-9b-3 *Polymerization of 1,2-Disubstituted Ethylenes*

With few exceptions, 1,2-disubstituted ethylenes containing substituents larger than fluorine such as maleic anhydride, stilbene, 1,2-dichlroethylene, and benzalacetophenone exhibit very little or no tendency to undergo polymerization [Bacskai, 1976; Kellou and Jenner, 1976; Seno et al., 1976]. Steric inhibition is the cause of this behavior, but the effect is different than that responsible for the difficulty with which 1,1-disubstituted ethylenes polymerize.

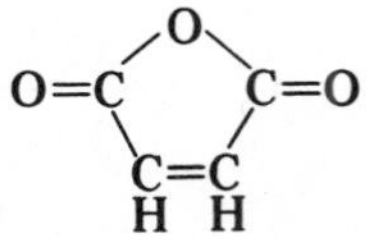

$\phi CH=CH\phi$ $ClCH=CHCl$ $\phi CH=CH-CO-\phi$

Polymers from 1,2-disubstituted ethylenes (**XXXIX**) possess 1,3-interactions, but

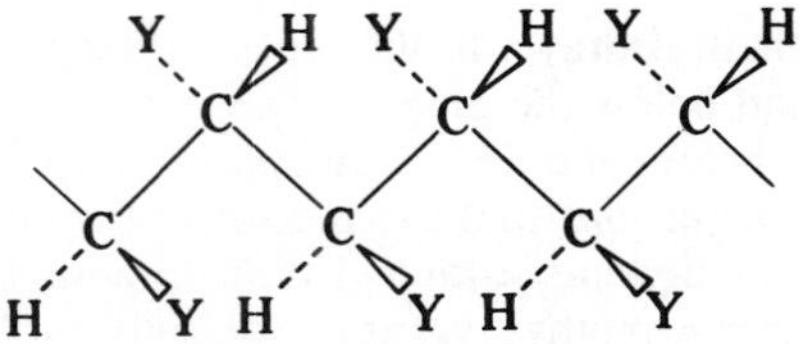

XXXIX

the steric strain is not as severe as in **XXXVIII**. Both **XXXVIII** and **XXXIX** possess the same number of 1,3-interactions but the distribution of the interactions is different. For **XXXVIII**, pairs of 1,3-carbons each have a pair of 1,3-interactions. No pair of 1,3-carbons in **XXXIX** has more than a single 1,3-interaction. Thus the ΔH value for maleic anhydride is -59 kJ/mole, which is favorable for polymerization compared to the value for some 1,1-disubstituted ethylenes.

The very low tendency of 1,2-disubstituted ethylenes to polymerize is due to kinetic considerations superimposed on the thermodynamic factor. The approach of the propagating radical to a monomer molecule is sterically hindered. The propagation step is extremely slow due to steric interactions between the β-substituent of the propagating species and the two substituents of the incoming monomer molecule:

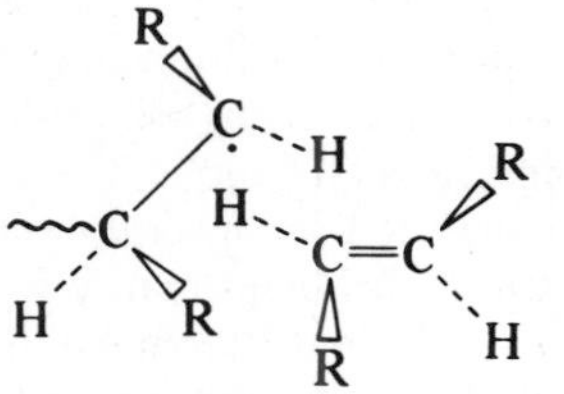

XL

Some success has been achieved in polymerizing maleic anhydride and maleimides, although the molecular weights are generally low and high initiator concentrations are required [Cubertson, 1987; Gaylord, 1975; Gaylord and Mehta, 1988; Zott and Heusinger, 1978]. The polymerization of maleic anhydride is complex—reaction proceeds with some loss of carbon dioxide and ionic or radical–ionic intermediates may be involved. Some dialkyl fumarates are polymerized by radical initiators while there are no reports of radical polymerization of dialkyl maleates (unless a base is present to isomerize the maleate to fumarate) [Otsu et al., 1988]. Vinylene carbonate (**XLI**) has been polymerized to high molecular weights [Field and Schaefgen, 1962; Reim-schuessel and Creasy, 1978].

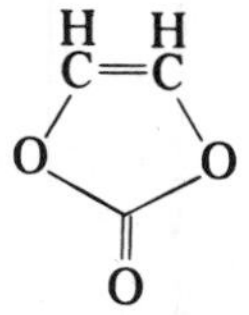

$$
\begin{array}{c}
\text{H} \quad \text{H} \\
\text{C} = \text{C} \\
\text{O} \qquad \text{O} \\
\text{O}
\end{array}
$$

XLI

3-9c   Polymerization–Depolymerization Equilibria

3-9c-1   Ceiling Temperature

For most chain polymerizations there is some temperature at which the reaction becomes a reversible one, that is, the propagation step (Eq. 3-15) should be written as an equilibrium reaction [Joshi and Zwolinski, 1967; Sawada, 1976],

$$\text{M}_n\cdot + \text{M} \underset{k_{dp}}{\overset{k_p}{\rightleftharpoons}} \text{M}_{n+1}\cdot \tag{3-192}$$

where k_{dp} is the rate constant of the reverse reaction—termed *depolymerization* or *depropagation*. The overall effect of temperature on polymerization is complex due to the presence of this propagation–depropagation equilibrium. When the temperature is initially increased for the polymerization of a monomer the polymerization rate increases as k_p increases (Sec. 3-9a-1). However, at higher temperatures the depropagation rate constant k_{dp}, which was initially zero, increases and becomes significant with increasing temperature. Finally, a temperature–the *ceiling temperature* T_c—is reached at which the propagation and depropagation rates are equal. These effects are shown in Fig. 3-17 for styrene. At the ceiling temperature the net rate of polymer production is zero.

The equilibrium position for the monomer–polymer equilibrium in Eq. 3-192 will be dependent on the temperature with increased temperature resulting in a shift to the left, since the forward reaction is exothermic. The reaction isotherm

$$\Delta G = \Delta G^0 + RT \ln K \tag{3-193}$$

is applicable to an analysis of polymerization–depolymerization equilibria. ΔG^0 is the ΔG of polymerization for the monomer and polymer in appropriate standard states [Dainton and Ivin, 1950, 1958]. The standard state for monomer is often taken as the pure liquid or a 1-M solution. The standard state for polymer is usually the solid amorphous or slightly crystalline polymer or a solution that is 1 M in the repeating unit of the polymer. For an equilibrium situation $\Delta G = 0$ by definition and Eq. 3-193 may be combined with Eq. 3-191 to yield

$$\Delta G^0 = \Delta H^0 - T\Delta S^0 = -RT \ln K \tag{3-194}$$

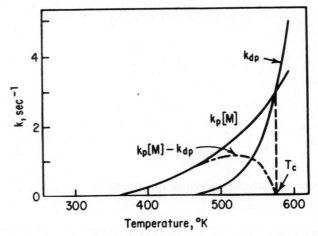

Fig. 3-17   The variation of $k_p[M]$ and k_{dp} with temperature for styrene. After Dainton and Ivin [1958] (by permission of The Chemical Society, Burlington House, London).

The equilibrium constant is defined by k_p/k_{dp} or more conveniently by

$$K = \frac{[M_{n+1}\cdot]}{[M_n\cdot][M]} = \frac{1}{[M]} \tag{3-195}$$

Combination of Eqs. 3-194 and 3-195 yields

$$T_c = \frac{\Delta H^0}{\Delta S^0 + R \ln [M]_c} \tag{3-196a}$$

or

$$\ln [M]_c = \frac{\Delta H^0}{RT_c} - \frac{\Delta S^0}{R} \tag{3-196b}$$

   Equation 3-196b shows the *equilibrium monomer concentration* $[M]_c$ as a function of the reaction or ceiling temperature T_c. Since ΔH^0 is a negative quantity, the monomer concentration in equilibrium with polymer increases with increasing temperature, that is, a plot of $\ln [M]_c$ versus $1/T$ is linear with a negative slope of $\Delta H^0/R$ and an intercept of $-\Delta S^0/R$. This means that there is a series of ceiling temperatures corresponding to different equilibrium monomer concentrations. For any monomer solution of concentration $[M]_c$ there is a temperature T_c at which polymerization does not occur. (For each $[M]_c$ there is a corresponding plot analogous to Fig. 3-17, in which $k_{dp} = k_p[M]$ at its T_c.) Stated another way, the polymerization of a particular monomer solution at a particular temperature proceeds until equilibrium is established, that is, until the monomer concentration decreases to the $[M]_c$ value corresponding to that T_c temperature. Thus higher initial monomer concentrations are required with increasing temperature in order to observe a net production of polymer before equilibrium is established. There is an upper temperature limit above which polymer cannot be obtained even from pure monomer. The reader is cautioned to note that the literature often appears to refer only to a singular T_c value—"the ceiling temperature." It is clear from the discussion above that each monomer concentration has its own T_c value. The apparent designation of a singular T_c value usually refers to the T_c for the pure monomer or in some cases to that for the monomer at unit molarity.

For many of the alkene monomers, the equilibrium position for the propagation–depropagation equilibrium is far to the right under the usual reaction temperatures employed, that is, there is essentially complete conversion of monomer to polymer for all practical purposes. Table 3-15 shows the monomer concentrations at 25°C for a few monomers [Cook et al., 1958; McCormick, 1957; Wall, 1960; Worsfold and Bywater, 1957]. Data are also shown for the ceiling temperatures of the pure monomers. The data do indicate that the polymer obtained in any polymerization will contain some concentration of residual monomer as determined by Eq. 3-196. Further, there are some monomers for which the equilibrium is not particularly favorable for polymerization, for example, α-methylstyrene. Thus at 25°C a 2.2-M solution of α-methylstyrene will not undergo polymerization. Pure α-methylstyrene will not polymerize at 61°C. Methyl methacrylate is a borderline case in that the pure monomer can be polymerized below 220°C, but the conversion will be appreciably less than complete. Thus, for example, the value of $[M]_c$ at 110°C is 0.139 M [Brandrup and Immergut, 1989]. Equations 3-196a and 3-196b apply equally well to ionic chain and ring-opening polymerizations as will be seen in subsequent chapters. The lower temperatures of ionic polymerizations offer a useful route to the polymerization of many monomers that cannot be polymerized by radical initiation because of their low ceiling temperatures. The successful polymerization of a previously unpolymerizable monomer is often simply a matter of carrying out the reaction at a temperature below its ceiling temperature.

Interestingly, it should not be assumed that a polymer will be useless above its ceiling temperature. A dead polymer that has been removed from the reaction media will be stable and will not depolymerize unless an active end is produced by bond cleavage of an end group or at some point along the polymer chain. When such an active site is produced by thermal, chemical, photolytic, or other means, depolymerization will follow until the monomer concentration becomes equal to $[M]_c$ for the particular temperature. The thermal behavior of many polymers, however, is much more complex. Degradative reactions other than depolymerization will often occur at temperatures below the ceiling temperature.

TABLE 3-15 Polymerization–Depolymerization Equilibria[a]

Monomer	$[M]_c$ at 25°C	T_c for Pure Monomer (°C)
Vinyl acetate	1×10^{-9}	—
Methyl acrylate	1×10^{-9}	—
Ethylene	—	400
Styrene	1×10^{-6}	310
Methyl methacrylate	1×10^{-3}	220
α-Methylstyrene	2.2	61
Isobutylene	—	50

[a]Data from Cook et al. [1958]; McCormick [1957]; Wall [1960]; Worsfold and Bywater [1957].

3-9c-2 *Floor Temperature*

The ceiling temperature phenomenon is observed because ΔH is highly exothermic, while ΔS is mildly exoentropic. The opposite type of phenomenon occurs in rare instances where ΔS is endoentropic ($\Delta S = +$) and ΔH is very small (either $+$ or $-$) or zero. Under these conditions, there will be a *floor temperature* T_f below which polymerization is not possible. This behavior has been observed in only three cases— the polymerizations of cyclic sulfur and selenium octamers and octamethylcyclotetrasiloxane to the corresponding linear polymers (Secs. 7-6a,b). ΔH is 9.5, 13.5, and nearly 0 kJ/mole, respectively, and ΔS is 27, 31, and 6.7 J/K-mole, respectively, for the three systems [Brandrup and Immergut, 1989; Lee and Johannson, 1966, 1976].

3-10 AUTOACCELERATION

3-10a Course of Polymerization

Radical chain polymerizations are characterized by the presence of an *autoacceleration* in the polymerization rate as the reaction proceeds [North, 1974]. One would normally expect a reaction rate to fall with time (i.e., the extent of conversion), since the monomer and initiator concentrations decrease with time. However, the exact opposite behavior is observed in many polymerizations—the reaction rate increases with conversion. A typical example is shown in Fig. 3-18 for the polymerization of methyl methacrylate in benzene solution [Schulz and Haborth, 1948]. The curve for the 10% methyl methacrylate solution shows the behavior that would generally be expected. The curve for neat (pure) monomer shows a dramatic autoacceleration in the polymerization rate. Such behavior is referred to as the *gel effect*. (The term *gel* as used here is different from its usage in Sec. 2-10; it does not refer to the formation of a crosslinked polymer.) The terms *Trommsdorf effect* and *Norrish–Smith effect* are also used in recognition of the early workers in the field. Similar behavior has been observed for a variety of monomers, including styrene, vinyl acetate, and methyl methacrylate [Abuin and Lissi, 1977; Balke and Hamielec, 1973; Cardenas and O'Driscoll, 1976, 1977; Small, 1975; Turner, 1977; Yamamoto and Sugimoto, 1979]. It turns out that

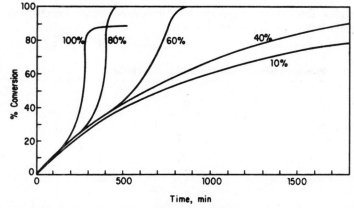

Fig. 3-18 Autoacceleration in the benzoyl peroxide initiated polymerization of methyl methacrylate in benzene at 50°C. The different plots represent various concentrations of monomer in solvent. After Schulz and Haborth [1948] (by permission of Huthig and Wepf Verlag, Basel).

the gel effect is the "normal" behavior for most polymerizations. The gel effect should not be confused with the autoacceleration that would be observed if a polymerization were carried out under nonisothermal conditions such that the reaction temperature increased with conversion (since ΔH is negative). The gel effect is observed under isothermal reaction conditions.

A more critical analysis of accurate polymerization data indicates that the situation is complicated; three stages can be distinguished in some polymerizations when $R_p/[M][I]^{1/2}$ is plotted against time (or conversion) (Fig. 3-19) [Dionisio et al., 1979; Dionisio and O'Driscoll, 1980; Sack et al., 1988]. Plotting $R_p/[M][I]^{1/2}$ instead of percent conversion takes into account the concentration changes in monomer and initiator with time. *Stage I* involves either a constant rate (IA) or declining rate (IB) with time. *Stage II* constitutes the autoaccelerative gel effect region. *Stage III* involves either a constant (IIIA) or delining (IIIB) rate.

3-10b Diffusion-Controlled Termination

An understanding of this behavior requires that we appreciate termination is a diffusion-controlled reaction best described as proceeding by the three-step process [Mahabadi and O'Driscoll, 1977a, 1977b; North, 1974]:

1 *Translational diffusion* of two propagating radicals (i.e., movement of the whole radicals) until they are in close proximity to each other:

$$M_n{}^{\cdot} + M_m{}^{\cdot} \underset{k_2}{\overset{k_1}{\rightleftharpoons}} [M_n{}^{\cdot} \text{---} M_m{}^{\cdot}] \tag{3-197}$$
$$\mathbf{XLIIa}$$

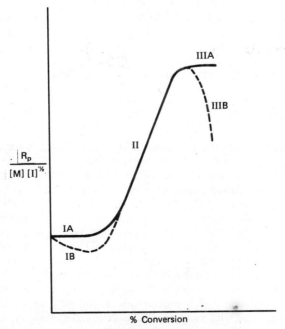

Fig. 3-19 Effect of conversion on polymerization rate.

2   Rearrangement of the two chains so that the two radical ends are sufficiently close for chemical reaction, which occurs by *segmental diffusion* of the chains, that is, by the movement of segments of a polymer chain relative to other segments

$$[M_n{}^\cdot \text{ --- } M_m{}^\cdot] \underset{k_4}{\overset{k_3}{\rightleftharpoons}} [M_n{}^\cdot / M_m{}^\cdot]$$

   XLIIa   **XLIIb**

$$(3\text{-}198)$$

3   Chemical reaction of two radical ends

$$[M_n{}^\cdot / M_m{}^\cdot] \xrightarrow{k_c} \text{dead polymer}$$

   XLIIb

$$(3\text{-}199)$$

Theoretical considerations indicate that k_c would be very large, about 8×10^9 liters/mole-sec, in low viscosity media (such as bulk monomer) for the reaction between two radicals. The rate constants for reactions of small radicals (e.g., methyl, ethyl, propyl) are close to this value (being about 2×10^9 liters/mole-sec) [Ingold, 1973]. Experimentally determined k_t values for radical polymerizations, however, are considerably lower, usually by two orders of magnitude or more (see Table 3-11). Thus diffusion is the rate-determining process for termination, $k_c \gg k_4$ and one obtains

$$R_t = \frac{k_1 k_3 [M\cdot]^2}{k_2 + k_3} \tag{3-200}$$

by assuming steady-state concentrations of both **XLIIa** and **XLIIb**. Two limiting cases of termination arise. For the case of slow translational diffusion, $k_3 \gg k_2$, and

$$R_t = k_1 [M\cdot]^2 \tag{3-201a}$$

For the case of slow segmental diffusion, $k_2 \gg k_3$, and

$$R_t = \frac{k_1 k_3 [M\cdot]^2}{k_2} \tag{3-201b}$$

Thus the experimentally observed termination rate constant k_t corresponds to k_1 and $k_1 k_3 / k_2$, respectively, for the two limiting situations.

Recent work has shown that segmental diffusion and translational diffusion are expected to be affected differently with conversion [Dionisio and O'Driscoll, 1980; Mahabadi and O'Driscoll, 1977a, 1977b; Mahabadi and Rudin, 1979]. With increasing conversion the polymerization medium becomes a poorer solvent due to the increased polymer concentration. The size of the randomly coiled up propagating radical in solution (referred to as *coil*) becomes smaller and there is an effective higher concentration gradient across the coil. Segmental diffusion of the radical end out of the coil to encounter another radical is increased. Simultaneously, the increasing polymer concentration decreases translational diffusion as the reaction medium becomes more viscous and, at sufficiently high concentrations, the polymer radicals become more crowded and entangled with each other. Chain entanglement leads to a faster decrease in translational diffusion relative to the decrease with increasing viscosity. Stage IA behavior corresponds to the situation observed for many monomers where the increase in segmental diffusion is apparently exactly counterbalanced by the decrease in translational diffusion (i.e., k_t remains constant). When the initial increase in segmental diffusion is greater than the decrease in translational diffusion, k_t increases and the polymerization rate decreases (Stage IB). Moderate Stage IB behavior has been observed in several polymerizations (styrene, methyl methacrylate [Abuin et al., 1978; High et al., 1979]). At this time, no system has been observed with more than a moderate decrease in $R_p/[M][I]^{1/2}$ with conversion.

At some point, translational diffusion decreases faster than the increase in segmental diffusion and rapid autoacceleration occurs—Stage II—the gel effect. As the polymerization proceeds the viscosity of the system increases with subsequent chain entanglement [Lachinov et al., 1979] and termination becomes increasingly slower. Although propagation is also hindered, the effect is much smaller, since k_p values are smaller than k_t values by a factor of 10^4–10^5. Termination involves the reaction of two large polymer radicals, while propagation involves the reaction of small monomer molecules and only one large radical. High viscosity affects the former much more than the latter. Therefore, the quantity $k_p/k_t^{1/2}$ increases and the result in accordance with Eq. 3-25 is an increase in R_p with conversion. A second consequence of this effect is an increase in molecular weight with conversion as required by Eq. 3-105. These conclusions have been verified by the quantitative evaluation of the k_p and k_t values as a function of the percent conversion. Thus, Table 3-16 shows data on the polymerization of methyl methacrylate [Hayden and Melville, 1960]. It is seen that k_p is relatively unaffected until 50% conversion (of monomer to polymer) has been reached, whereas k_t has decreased by almost two orders of magnitude in the same span. The $k_p/k_t^{1/2}$ ratio and the polymerization rate simultaneously increase rapidly at first and then taper off as k_p is also affected in the later stages of reaction. The attention of the reader is also called to the data showing the increase in the radical lifetime with increasing conversion.

At very high (e.g., above 50% conversion for the system in Table 3-16), k_p becomes sufficiently affected that the $R_p/[M][I]^{1/2}$ begins to level off (Stage IIIA behavior) or decrease (Stage IIIB behavior). Stage IIIB behavior is much more common than Stage IIIA. The decrease in rate during Stage IIIB, sometimes referred to as the *glass or vitrification effect*, can be extremely pronounced depending on the reaction temperature. The glass transition temperature of a polymerization reaction mixture increases

TABLE 3-16   Effect of Conversion on the Polymerization of Methyl Methacrylate (22.5°C)[a]

% Conversion	Rate (%/hr)	τ (sec)	k_p	$k_t \times 10^{-5}$	$(k_p/k_t^{1/2}) \times 10^2$
0	3.5	0.89	384	442	5.78
10	2.7	1.14	234	273	4.58
20	6.0	2.21	267	72.6	8.81
30	15.4	5.0	303	14.2	25.5
40	23.4	6.3	368	8.93	38.9
50	24.5	9.4	258	4.03	40.6
60	20.0	26.7	74	0.498	33.2
70	13.1	79.3	16	0.0564	21.3
80	2.8	216	1	0.0076	3.59

[a]Data from Hayden and Melville [1960].

with conversion of monomer to polymer. Polymerization can stop appreciably short of full conversion if the reaction system has a percent conversion whose glass transition temperature exceeds the reaction temperature [Friis and Hamielec, 1976; Mita and Horie, 1987; Sundberg and James, 1978]. For example, this occurs in the polymerization of pure methyl methacrylate in Fig. 3-18 since the polymerization temperature (50°C) is considerably below the glass transition temperature of poly(methyl methacrylate (105°C).

Recent work indicates that variations in initiator efficiency f are also important in understanding the effect of conversion on rate and degree of polymerization. The calculations of k_p and k_t in Table 3-16 were carried out with the assumption of a constant f independent of conversion, but it is known that f varies with conversion. The initiator efficiency decreases with conversion in an approximately linear manner until the high conversions of the Stage IIIB region are reached; thereafter, there is a much steeper drop in f with conversion [Russell et al., 1988; Sack et al., 1988]. Thus, the behavior in Stage IIIB is a consequence of decreases in both k_p and f, not just a decrease in k_p. If one corrects the calculations of rate constants in previous work (e.g., Table 3-16) for the variation in f with conversion, the trends for k_t and k_p are essentially unchanged.

3-10c   Effect of Reaction Conditions

The rate of segmental diffusion is predicted to increase (with a corresponding increase in k_t and decreases in rate and degree of polymerization) by any factor that decreases the coil size of the propagating radical. Thus segmental diffusion will increase with decreasing polymer molecular weight and goodness of reaction solvent and with increasing conversion [Mahabadi, 1987; Olaj and Zifferer, 1987]. The various factors affecting segmental diffusion are interrelated. At zero or very low conversion, k_t is larger in a poor solvent compared to k_t in a good solvent. It is possible to observe a "crossover" at somewhat higher conversion whereby the k_t in the good solvent is larger than k_t in the poor solvent. This occurs because coil size decreases with increasing polymer concentration, but the decrease is steeper in better solvents [Mahabadi and Rudin, 1979]. Many of these predictions for the course of polymerization in Stage I have been verified [Abuin et al., 1978; Dionisio and O'Driscoll, 1980; Ludwico and Rosen, 1975, 1976].

The behavior in Stage II is dominated by the decreased rate of translational diffusion caused by the increasing viscosity of the reaction system. Viscosity increases with increasing polymer molecular weight and solvent goodness. Increased polymer molecular weight also leads to earlier chain entanglements. These expectations have been verified in a number of studies. Lower polymer molecular weights moderate the gel effect (Stage II behavior) by shifting the conversion at which the gel effect begins and the steepness of the subsequent increase in rate [Abuin and Lissi, 1977; Cardenas and O'Driscoll, 1976, 1977]. Vinyl acetate yields a lower-molecular-weight polymer due to chain transfer to monomer and does not show as dramatic a gel effect as styrene or methyl methacrylate. Temperature also plays a large role—higher temperatures decrease the viscosity of the reaction medium. This delays the gel effect and the autoacceleration may not be as pronounced. Similar effects are observed in the presence of solvents and chain transfer agents. Solvents lower the viscosity of the medium directly, while chain transfer agents do so by lowering the polymer molecular weight. The effect of solvent in methyl methacrylate polymerization is seen in Fig. 3-19. The effect of viscosity on the gel effect has also been examined by adding polymer to a monomer prior to initiating polymerization; R_p increases with viscosity (which increases with polymer concentration) [Kulkarni et al., 1979].

The percent conversion for the onset of the gel effect varies considerably depending on the monomer and the reaction conditions (which determine coil size, viscosity, entanglements). In some systems the gel effect has been reported as occurring at only a few percent conversion, in others, not until 60–70% conversion. Further, there is considerable difficulty in obtaining sufficiently accurate data to allow a precise evaluation of the onset of the gel effect, especially if the onset occurs gradually.

3-10d   Related Phenomena

3-10d-1   Occlusion (Heterogeneous) Polymerization

Autoacceleration of the gel effect type is observed for the polymerization of monomers whose polymers are insoluble or weakly soluble in their own monomers. Examples of this type of behavior are acrylonitrile, vinyl chloride, trifluorochloroethylene, and vinylidene chloride [Billingham and Jenkins, 1976; Gromov et al., 1980; Guyot, 1987; Jenkins, 1967; Olaj et al., 1977a, 1977b; Talamini and Peggion, 1967]. Similar effects are observed in other polymerizations when one uses solvents which are nonsolvents for the polymer [e.g., methanol for polystyrene, hexane for poly(methyl methacrylate)]. The accelerative effects observed in these instances of heterogeneous polymerization are similar to the gel effect and are caused by a decrease in k_t relative to k_p. The growing polymeric radicals become coiled up, since they are essentially insoluble or on the verge of insolubility in the solvent or in their own monomer. Termination between radicals again becomes progressively more difficult, while propagation may still proceed reasonably well. The reason for the decrease in *occlusion* or *heterogeneous* polymerization is probably due more to a decrease in segmental diffusion than to decreased translational diffusion. Although the propagating radical coil size is decreased in the poor reaction medium, unlike homogeneous systems, segmental diffusion is not enhanced, since diffusion must take place in a poor medium.

3-10d-2   Template Polymerization

Template (matrix) polymerization involves the polymerization of a monomer in the presence of a polymer, often a polymer derived from a different monomer [Gons et al., 1978; Matuszewska-Czerwik and Polowinski, 1990; Srivstava et al., 1987; Tan and Challa, 1987, 1988; Tewari and Srivastava, 1989; van de Grampel et al., 1988]. The original polymer and formed polymer are often referred to as *parent* and *daughter polymer*, respectively. Examples include acrylic acid with polyethyleneimine or poly(*N*-vinylpyrrolidinone), *N*-vinylimidazole with poly(methacrylic acid), acrylonitrile with poly(vinyl acetate), and methyl methacrylate with poly(methyl methacrylate). Many of these polymerizations proceed with rate enhancements relative to the corresponding polymerizations in the absence of polymer. The rate enhancements are typically by a factor no larger than 2–5, although a few larger enhancements have been observed. Two types of mechanisms appear to be operative depending on how strongly monomer is absorbed to polymer. A *zip* mechanism occurs when monomer is strongly complexed with the parent polymer. The complexed monomer molecules are lined up next to each other, which facilitates rapid propagation when an initiating radical from the surrounding solution adds to one of the absorbed monomer molecules.

For monomers that only weakly complex with the parent polymer, polymerization occurs by a *pickup* mechanism. Initiation occurs in the surrounding solution to form a propagating oligomer that then complexes with the parent polymer. Subsequent propagation proceeds as monomer is picked up at the propagating center. Rate enhancement for zip propagation probably results from an increase in k_p due to alignment of monomer molecules on the parent polymer coupled with a decrease in k_t due to slower translational and segmental diffusion of propagating radicals. Only the latter effect is operative for template polymerizations involving the pickup mechanism. Template polymerizations have been of interest from the viewpoint of synthesizing stereoregular polymers (Chap. 8) by using a stereoregular parent polymer but this approach has been successful in very few instances.

3-10e Dependence of Polymerization Rate on Initiator and Monomer

The occurrence of the three stages of homogeneous polymerization and the occlusion and template polymerizations give rise to observed deviations of the usual kinetic expressions (e.g., Eqs. 3-32 and 3-108). It is not unusual to observe R_p dependence on [M] of greater than first-order and R_p dependence on [I] of either less than or greater than $\frac{1}{2}$-order. (A difficulty in evaluating literature data is that it is not always clear what reaction conditions (Stage I, II, or III, occlusion, template?) exist in a particular study.) Such results are almost always artifacts. For both homogeneous and heterogeneous polymerizations, one may observe a higher than first-order dependence on [M] due to the $k_p/(2k_t)^{1/2}$ ratio varying with [M] to some power, which coupled with the normal first-order monomer dependence of propagation gives a higher than first-order dependence of R_p on [M]. The $k_p/(2k_t)^{1/2}$ may vary with [M] because changing [M] may change the viscosity and/or goodness of the solvent (which affect the rates of segmental and/or translational diffusion). An increased dependence of R_p on [M] may also result from the second step (Eq. 3-14a) of the initiation sequence becoming the rate-determining step due to decreased mobility of the primary radicals.

A lower than $\frac{1}{2}$-order dependence of R_p on [I] may be the indirect result of the corresponding change in polymer molecular weight. The termination rate constant would increase with polymer molecular weight, which in turn decreases with initiator concentration, leading to a dependence of R_p on [I] less than $\frac{1}{2}$-order, even 0-order in some instances. On the other hand, the dependence of R_p on [I] becomes close to first-order in some heterogeneous polymerizations because the propagating radicals become so coiled up and inaccessible as to be incapable of undergoing termination under the usual reaction conditions [Ueda et al., 1984]. Termination becomes first-order in radical concentration. The presence of buried or trapped nonterminated radicals has been shown by ESR and the ability of photochemically produced reaction mixtures to continue polymerizing for days after the discontinuation of illumination [Sato et al., 1984]. Such long-lived radicals have been studied as a means of producing block copolymers [Sato et al., 1983; Seymour and Stahl, 1977]. Thus, vinyl acetate has been polymerized in a poor and/or viscous solvent to produce occluded macroradicals, which then initiate the polymerization of methyl methacrylate, acrylic acid, acrylonitrile, styrene, and N-vinylpyrrolidinone to produce various block copolymers.

Under conditions where bimolecular termination between propagating radicals becomes difficult because of the increased viscosity or heterogeneity, primary termination may become important or even the only mode of termination. The latter leads to R_p being second order in [M] and zero order in [I].

3-10f Other Accelerative Phenomena

Various other rate and molecular-weight accelerations have been observed. The polymerizations of acrylonitrile, N-vinylpyrrolidinone, and acrylic acid are faster in the presence of water or hydrogen-bonding solvents even though the reaction medium becomes homogeneous instead of heterogeneous as in nonpolar solvents [Burillo et al., 1980; Chapiro and Dulieu, 1977; Laborie, 1977; Olive and Olive, 1978; Senogles and Thomas, 1978]. The rate acceleration is due to an increase in the propagation step. The effect is very large in some cases; for example, k_p for acrylonitrile at 25°C is 2.8×10^4 liters/mole-sec in water compared to 400 liters/mole-sec in dimethylformamide. Water simultaneously hydrogen-bonds with both monomer and the propagating radical chain end (through the CN group) to increase the effective local monomer concentration at the radical site and also to increase the reactivity of the monomer and/or radical by a polar effect. This effect of water has been referred to as a *template* effect. (This template effect is different from that discussed in Sec. 3-10d-2.) An alternate mechanism proposes that the solvent binds monomer molecules into linear aggregates, which undergo faster propagation due to a favorable orientation (referred to as *zip propagation*). This autoacceleration sometimes occurs only after some polymer has formed, the formed polymer assisting in stabilizing the linear monomer aggregates (similar to template polymerization).

Carboxylic acid monomers such as acrylic and methacrylic acids and *trans*-butadiene-1-carboxylic acid have lower polymerization rates in good solvents and/or at higher pH where monomer exists in ionized form. Repulsions between the carboxylate anion groups of the propagating chain end and monomer result in lowered reactivity in the propagation step [Bando and Minoura, 1976; Ponratnam and Kapur, 1977]. Monomers containing β-diketone and β-ketoester groups (e.g., ethyl 4-methyl-3-oxo-4-pentenoate) exhibit higher reaction rates in benzene compared to acetonitrile. The enol tautomer, more plentiful in the less polar solvent, is more reactive than the keto tautomer [Masuda et al., 1987, 1989].

The addition of Lewis acids such as $ZnCl_2$ and $AlCl_3$ often increases the polymerization rate of monomers with electron-withdrawing substituents, such as acrylamide, methyl methacrylate, and acrylonitrile [Gromov et al., 1980a, 1980b; Haeringer and Reiss, 1978; Kabanov, 1987; Liaw and Chung, 1983; Madruga and Sanroman, 1981; Maekawa et al., 1978; Sugiyama and Lee, 1977]. The mechanism of acceleration is not well established. In some instances the metal salt increases the initiation rate by complexing with the initiator. Increased reactivity toward propagation by complexation of the propagating radical and/or monomer with the Lewis acid appears to be the predominant effect in other systems. Simultaneous effects on both initiation and propagation (and possibly also termination) are indicated in many systems, since the polymer molecular weight often shows a maximum and a minimum with increasing concentration of Lewis acid. Some reports of complexation between a monomer or propagating radical with solvent—leading to changes in reactivity—have also been reported [Ghosh and Mukhopadhyay, 1980; Kamachi et al., 1979].

3-11 MOLECULAR-WEIGHT DISTRIBUTION

3-11a Low-Conversion Polymerization

The molecular-weight distribution in radical chain polymerizations is more complex than those in step polymerization. Radical chain polymerization involves several possible modes by which propagation is terminated—disproportionation, coupling, and various transfer reactions. The situation is further complicated, since the molecular weight of the polymer produced at any instant varies with the overall percent conversion due to changes in the monomer and catalyst concentrations and the propagation and termination rate constants. Molecular-weight distributions can be easily calculated for polymerizations restricted to low conversions where all of the kinetic parameters ($[M]$, $[I]$, k_d, k_p, k_p) are approximately constant [Flory, 1953; Macosko and Miller; 1976; Peebles, 1971; Tompa, 1976; Vollmert, 1973]. Under these conditions, the polymer molecular weight does not change with conversion.

Consider the situation where one polymer molecule is produced from each kinetic chain. This is the case for termination by disproportionation or chain transfer or a combination of the two, but without combination. The molecular-weight distributions are derived in this case in exactly the same manner as for linear step polymerization (Sec. 2-7). Equations 2-92, 2-94, 2-95, 2-27, 2-101, and 2-102 describe the number-

$$\underline{N}_x = (1 - p)p^{x-1} \tag{2-92}$$

$$N_x = N_0(1 - p)^2 p^{x-1} \tag{2-94}$$

$$w_x = x(1 - p)^2 p^{x-1} \tag{2-95}$$

$$\overline{X}_n = \frac{1}{(1 - p)} \tag{2-27}$$

$$\overline{X}_w = \frac{(1 + p)}{(1 - p)} \tag{2-101}$$

$$\frac{\overline{X}_w}{\overline{X}_n} = (1 + p) \tag{2-102}$$

fraction, number, and weight-fraction distributions, the number- and weight-average degrees of polymerization, and the breadth of the distribution, respectively.

One difference in the use of these equations for radical chain polymerizations compared to step polymerizations is the redefinition of p as the probability that a propagating radical will continue to propagate instead of terminating. The value of p is given as the rate of propagation divided by the sum of the rates of all reactions that a propagating radical may undergo

$$p = \frac{R_p}{R_p + R_t + R_{tr}} \tag{3-202}$$

where R_p, R_t, and R_{tr} are the rates of propagation, termination by disproportionation, and chain transfer, respectively. There is a very important second difference in the use of the above equations for radical chain polymerization compared to step polymerizations. The equations apply to the whole reaction mixture for a step polymerization but only to the polymer fraction of the reaction mixture for a radical chain polymerization.

A consideration of the above equations indicates that high polymer (i.e., large values of $\overline{X}_n$ and $\overline{X}_w$) will only be produced if p is close to unity. This is certainly what one expects based on the previous discussions in Sec. 3-5. The distributions described by Eqs. 2-92, 2-94, and 2-95 have been shown in Figs. 2-12, 2-13, and 2-14. The breadth of the size distribution $\overline{X}_w/\overline{X}_n$ (also referred to as *PDI*, the polydispersity index) has a limiting value of two as p approaches unity.

For termination by coupling ($R_{tr} = 0$ and R_t is the rate of coupling) where a polymer arises from the combination of two kinetic chains, the size distribution is narrower. The situation is analogous to that for step polymerizations with branching (Sec. 2-9). Polymer molecules of sizes much different from the average are less likely, since the probability for coupling between same-sized propagating radicals is the same as that between different-sized propagating radicals. The size distributions can be derived in a manner analogous to that used in Sec. 2-7a. Consider the probability $\underline{N}_{y+z}$ of formation of an x-sized polymer molecule by the coupling of y- and z-sized propagating radicals. $\underline{N}_{y+z}$ is the product of the probabilities, $\underline{N}_y$ and $\underline{N}_z$, of forming the y- and z-sized propagating radicals. $\underline{N}_y$ and $\underline{N}_z$ are given by

$$\underline{N}_y = (1 - p)p^{y-1} \tag{3-203}$$

$$\underline{N}_z = (1 - p)p^{z-1} \tag{3-204}$$

which are obtained in the same manner as Eq. 2-92. Equation 3-204 can be rewritten as

$$\underline{N}_z = (1 - p)p^{x-y-1} \tag{3-205}$$

since $y + z = x$ and $\underline{N}_{y+z}$ is then given by

$$\underline{N}_{y+z} = (1 - p)^2 p^{x-2} \tag{3-206}$$

Equation 3-206 gives the probability of forming an x-sized polymer molecule by only one of many equally probable pathways. An x-sized polymer molecule can be obtained by couplings of y- and z-sized radicals, z- and y-sized radicals, $(y + 1)$- and $(z - 1)$-sized radicals, $(z - 1)$- and $(y + 1)$-sized radicals, $(y + 2)$- and $(z - 2)$-sized radicals, $(z - 2)$- and $(y + 2)$-sized radicals, $(y - 1)$- and $(z + 1)$-sized radicals, $(z + 1)$- and $(y - 1)$-sized radicals, and so on. There are $(x - 1)$ possible pathways of producing an x-sized polymer molecule when x is an even number. Each of the pathways has the same probability—that given by Eq. 3-206—and the total probability $\underline{N}_x$ of forming an x-sized polymer molecule is given by

$$\underline{N}_x = (x - 1)(1 - p)^2 p^{x-2} \tag{3-207}$$

$\underline{N}_x$ is synonymous with the mole- or number-fraction distribution. When x is an odd number, there are x pathways and the $(x - 1)$ term in Eq. 3-207 should be replaced by x. For polymerizations yielding high polymer, the difference between x and $(x + 1)$ is negligible and can be ignored. [Some derivations of size distributions show exponents of y and z, respectively, in Eqs. 3-203 and 3-204 instead of $(y - 1)$ and $(z - 1)$, which results in an exponent of x instead of $(x - 2)$ in Eqs. 3-206 and 3-207. The differences are, again, unimportant for systems that yield high polymer.]

Using the same approach as in Secs. 2-7a and 2-7b, the following equations are derived for the weight-fraction distribution, number- and weight-average degrees of polymerization, and $\overline{X}_x/\overline{X}_n$

$$w_x = \tfrac{1}{2}x(1 - p)^3(x - 1)p^{x-2} \tag{3-208}$$

$$\overline{X}_n = \frac{2}{1 - p} \tag{3-209}$$

$$\overline{X}_w = \frac{2 + p}{1 - p} \tag{3-210}$$

$$\frac{\overline{X}_w}{\overline{X}_n} = \frac{2 + p}{2} \tag{3-211}$$

For polymerizations where termination occurs by a combination of coupling, disproportionation, and chain transfer, one can obtain the size distribution by a weighted combination of the above two sets of distribution functions. Thus the weight distribution can be obtained as

$$w_x = Ax(1 - p)^2 p^{x-1} + \tfrac{1}{2}(1 - A)x(1 - p)^3(x - 1)p^{x-2} \tag{3-212}$$

where A is the fraction of polymer molecules formed by disproportionation and chain-transfer reactions [Smith et al., 1966].

3-11b High-Conversion Polymerization

For many practical radical chain polymerizations, the reaction is carried out to high or complete conversion. The size distributions for high-conversion polymerizations become much broader than those described above or those in step polymerizations. Both the monomer and initiator concentrations decrease as a polymerization proceeds. The polymer molecular weight depends on the ratio $[M]/[I]^{1/2}$ according to Eq. 3-108. In the usual situation $[I]$ decreases faster than $[M]$ and the molecular weight of the polymer produced at any instant increases with the conversion. The overall molecular-weight distributions for high or complete conversion polymerizations are quite broad, with the $\overline{X}_w/\overline{X}_n$ ratio being in the range 2–5. Broad molecular-weight distributions are usually not desirable from the practical viewpoint, since most polymer properties show optimum values at specific molecular weights. Commercial polymerization processes often involve the addition of multiple charges of initiator and/or monomer during the course of the reaction to minimize the molecular weight broadening.

When autoacceleration occurs in polymerization there is even larger broadening of the size distribution. The large increases in the $k_p/k_t^{1/2}$ ratio (Table 3-16) that accompany the gel effect lead to large increases in the sizes of the polymers being produced as the polymerization proceeds. $\overline{X}_w/\overline{X}_n$ values as high as 5–10 may be observed when the gel effect is present. Excessive molecular-weight broadening occurs when branched polymers are produced by chain transfer to polymer. Chain transfer to polymer can lead to $\overline{X}_w/\overline{X}_n$ ratios as high as 20–50. This very extensive size broadening occurs because chain transfer to polymer increases as the polymer size increases. Thus branching leads to even more branching as a polymerization proceeds. One usually attempts to minimize the molecular-weight broadening due to the get effect and chain transfer to polymer, but it is quite difficult to do so successfully. Thus, for example, low temperatures minimize chain transfer to polymer but maximize the gel effect (Sec. 3-10a).

3-12 EFFECT OF PRESSURE

The effect of pressure on polymerization, although not extensively studied, is important from the practical viewpoint since several monomers are polymerized at pressures above atmospheric. Pressure affects polymerization through changes in concentrations, rate constants, and equilibrium constants [Ogo, 1984; Weale, 1974; Zutty and Burkhart, 1962]. The commercial polymerizations of most gaseous monomers (e.g., vinyl chloride, vinylidene chloride, tetrafluoroethylene, vinyl fluoride) are carried out at very moderate pressures of about 5–10 MPa (1 MPa = 145 psi), where the primary effect is one of increased concentration (often in a solvent) leading to increased rates. Significant changes in rate and equilibrium constants occur only at higher pressures such as the 100–300-MPa pressures used to produce the large-volume polymer—low-density polyethylene (Sec. 3-13b-1).

3-12a Effect on Rate Constants

High pressure can have appreciable effects on polymerization rates and polymer molecular weights. Increased pressure usually results in increased polymerization rates and molecular weights. Figure 3-20 shows these effects for the radiation-initiated polymerization of styrene [Moore et al., 1977].

3-12a-1 *Volume of Activation*

The effect of pressure on R_p and $\overline{X}_n$, like that of temperature, manifests itself by changes in the three rate constants—initiation, propagation, and termination. The

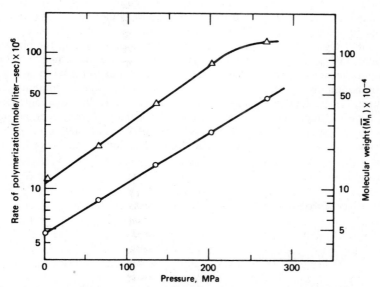

Fig. 3-20 The effect of pressure on the polymerization rate (○) and polymer molecule weight (△) for the radiation-initiated polymerization of styrene at 25°C. After Moore et al. [1977] (by permission of Wiley-Interscience, New York.)

quantitative effect of P (at constant temperature) on a rate constant is given by

$$\frac{d \ln k}{dP} = \frac{-\Delta V^{\ddagger}}{RT} \tag{3-213}$$

where $\Delta V^{\ddagger}$ is the *volume of activation*, that is, the change in volume $(cm^3/mole)$ in going from the reactant(s) to the transition state [Asano and LeNoble, 1978]. A negative value of $\Delta V^{\ddagger}$ corresponds to the volume of the transition state being smaller than that of the reactants; the result is that increased pressure leads to an increase in the reaction rate constant. A $\Delta V^{\ddagger}$ of -10 $cm^3/mole$ corresponds to 1.5- and 60-fold increases in rate constant at 100 and 1000 MPa, respectively, at 25°C. The increases are 2.2- and 3500-fold if $\Delta V^{\ddagger}$ is -20 $cm^3/mole$. Positive values of $\Delta V^{\ddagger}$ correspond to a decrease in the reaction rate constant with increasing pressure as the volume of the transition state is larger than that of the reactants.

The $\Delta V^{\ddagger}$ term for initiation by the thermal decomposition of an initiator $\Delta V_d^{\ddagger}$ is positive, since the reaction is a unimolecular decomposition involving a volume expansion for the initiator going to the transition state. The initiation rate decreases with increasing pressure. $\Delta V_d^{\ddagger}$ is in the range of $+4$ to $+13$ $cm^3/mole$, with most initiators being in the narrower range of $+5$ to $+10$ $cm^3/mole$ [Luft et al., 1978; Neumann, 1972; Walling and Metzger, 1959]. The measured $\Delta V_d^{\ddagger}$ for a particular initiator often varies with solvent viscosity indicating that a large portion of $\Delta V_d^{\ddagger}$ is due to increased pressure decreasing the ability of radicals to diffuse from the solvent cage. This is further verified by the observation that $\Delta V_d^{\ddagger}$ determined by the disappearance of initiator is lower than that measured using radical scavengers, which only detect radicals outside the solvent cage. It is estimated that $\Delta V_d^{\ddagger}$ would be $+4$ to $+5$ $cm^3/mole$ for most initiators if there were no cage effect [Martin, 1973]. The volume of activation for propagation $\Delta V_p^{\ddagger}$ is negative, since the reaction involves two species coming together (i.e., a volume decrease occurs) in going to the transition state. $\Delta V_p^{\ddagger}$ is in the range -15 to -24 $cm^3/mole$ for most monomers [Brandrup and Immergut, 1989; Ogo, 1984; Weale, 1974]. The volume of activation for bimolecular termination $\Delta V_t^{\ddagger}$ is positive with values that are generally in the range $+13$ to $+25$ $cm^3/mole$ for most monomers. At first glance one would expect $\Delta V_t^{\ddagger}$ to be negative, since termination involves a decrease in volume when two radicals come together in the transition state. However, since the termination step is diffusion-controlled, increased pressure decreases k_t by increasing the viscosity of the reaction medium [Nicholson and Norrish, 1956; O'Driscoll, 1977, 1979]. (Another result of the increased viscosity is that the gel effect is observed at lower conversions at higher pressures.)

3-12a-2 *Rate of Polymerization*

The variation of the polymerization rate with pressure will depend on the variation of the ratio $k_p(k_d/k_t)^{1/2}$ with pressure in accordance with Eqs. 3-32 and 3-213. This latter variation will be given by

$$\frac{d \ln [k_p(k_d/k_t)^{1/2}]}{dP} = \frac{-\Delta V_R^{\ddagger}}{RT} \tag{3-214}$$

where $\Delta V_R^{\ddagger}$ is the *overall volume of activation for the rate of polymerization* and is given by

$$\Delta V_R^{\ddagger} = \frac{\Delta V_d^{\ddagger}}{2} + \Delta V_p^{\ddagger} - \frac{\Delta V_t^{\ddagger}}{2} \tag{3-215}$$

The various terms are such that the value of $\Delta V_R^{\ddagger}$ is negative by 15 to 20 cm^3/mole for most monomers (as high as 25 cm^3/mole in some cases) and the rate of polymerization is increased by high pressures (see Fig. 3-20). Thus, $\Delta V_R^{\ddagger}$ is -17.1, -17.2, -15.6, -26.3, and -17.5 cm^3/mole, respectively, for styrene, vinyl acetate, methyl methacrylate, butyl methacrylate, and ethylene [Buback and Lendle, 1983; Ogo, 1984]. It is worth noting that the relative effect of pressure on R_p is less than that of temperature from the practical viewpoint. For a polymerization with $\Delta V_R^{\ddagger} = -25$ cm^3/mole and $E_R = 80$ kJ/mole an increase in pressure from 0.1 to 400 MPa at 50°C yields about the same increase in R_p as does an increase in temperature from 50 to 105°C. The increase in temperature is considerably easier to achieve than the pressure increase. Commercial high-pressure polymerization is usually carried out only when the polymerization temperature is relatively high to begin with (see the discussion on ethylene polymerization in Sec. 3-13b-1).

3-12a-3 Degree of Polymerization

The variation of the degree of polymerization with pressure depends on the variation of the ratio $k_p/(k_d k_t)^{1/2}$ with pressure as given by

$$\frac{d \ln [k_p/(k_d k_t)^{1/2}]}{dP} = \frac{-\Delta V_{\overline{X}_n}^{\ddagger}}{RT} \tag{3-216}$$

where $\Delta V_{\overline{X}_n}^{\ddagger}$ is the *overall volume of activation for the degree of polymerization* and is given by

$$\Delta V_{\overline{X}_n}^{\ddagger} = \Delta V_p^{\ddagger} - \frac{\Delta V_d^{\ddagger}}{2} - \frac{\Delta V_t^{\ddagger}}{2} \tag{3-217}$$

$\Delta V_{\overline{X}_n}^{\ddagger}$ is negative by about 20–25 cm^3/mole in most cases and the polymer molecular weight increases with increasing pressure (Fig. 3-21). $\Delta V_{\overline{X}_n}^{\ddagger}$ is more negative than $\Delta V_R^{\ddagger}$, since $\Delta V_d^{\ddagger}$ appears as $-\Delta V_d^{\ddagger}/2$ in the former. Thus $\overline{X}_n$ changes more steeply with pressure than does R_p. However, one often observes as in Fig. 3-21 that the increase in molecular weight with pressure is not continuous but reaches a limiting value at some pressure. The explanation for this effect is that chain transfer to monomer becomes progressively more important compared to bimolecular termination with increasing pressure. At the highest pressures bimolecular termination becomes insignificant and chain transfer to monomer is the major determinant of molecular weight.

Chain-transfer reactions would be expected to increase in rate with increasing pressure since transfer is a bimolecular reaction with a negative volume of activation. The variation of chain-transfer constants with pressure, however, differ depending on the relative effects of pressure on the propagation and transfer rate constants. For the case where only transfer to chain-transfer agent S is important, C_S varies with pressure according to

$$\frac{d \ln C_S}{dP} = \frac{d \ln (k_{tr,S}/k_p)}{dP} = \frac{-(\Delta V_{tr,S}^\ddagger - \Delta V_p^\ddagger)}{RT} \tag{3-218}$$

where $\Delta V_{tr,S}^\ddagger$ is the volume of activation for chain transfer. For some transfer agents, $\Delta V_{tr,S}^\ddagger$ is smaller than $\Delta V_p^\ddagger$ and C_S increases with pressure; the opposite occurs with other transfer agents. However, the difference in activation volumes for propagation and transfer is generally small (1–7 cm^3/mole) and the change in C_S with pressure is generally small. Thus, C_S for carbon tetrachloride in styrene polymerization is decreased by only 15% in going from 0.1 to 440 MPa [Toohey and Weale, 1962]. An interesting observation is that the extent of the back-biting self-transfer reaction in polyethylene decreases with increasing pressure [Woodbrey and Ehrlich, 1963]. This means that the amount of branching in polyethylene decreases as the polymerization pressure increases.

3-12b Thermodynamics of Polymerization

The variation of the equilibrium constant for a reaction with pressure (at constant temperature) takes the same form as for the variation of a rate constant

$$\frac{d \ln K}{dP} = \frac{-\Delta V}{RT} \tag{3-219}$$

where ΔV is the reaction volume (the difference in volume between reactants and products). K increases with increasing pressure since ΔV is negative. ΔV values are generally in the range -15 to -25 cm^3/mole. Polymerization is thermodynamically more favored at high pressure. Analysis shows that the effect of high pressure is exerted primarily through the change in ΔS. Increased pressure makes both ΔH and ΔS less negative, but the quantitative effect is much greater on ΔS, resulting in a more favorable (more negative) ΔG with increasing pressure. (The changes in ΔH and $T\Delta S$ are ~5 and 25 kJ/mole, respectively, at 1000 MPa relative to 0.1 MPa at 25°C [Allen and Patrick, 1974].)

Increased pressure increases the ceiling temperature and decreases the equilibrium monomer concentration according to

$$\frac{dT_c}{dP} = \frac{T_c \Delta V}{\Delta H} \tag{3-220}$$

$$\frac{d \ln [M]_c}{dP} = \frac{\Delta V}{RT} \tag{3-221}$$

The strong dependence of T_c and $[M]_c$ on pressure can lead to the observation of a *threshold pressure*—a pressure below which polymerization does not occur for a particular monomer concentration at a particular temperature.

3-12c Other Effects of Pressure

High pressures also affect the course of polymerization through changes in the physical properties of a reaction system. (The effect of increased viscosity has previously been mentioned.) The freezing point of a liquid usually rises with increasing pressure, about 15 to 25°C per 100 MPa. Thus increasing pressure may result in changing the state of a reaction system from liquid to solid or the exact nature of the liquid state with corresponding complications in the rate and/or thermodynamics of polymerization [Sasuga and Takehisa, 1978; Weale, 1974]. Dielectric constant also increases with increasing pressure [Ogo et al., 1978]. These changes often lead to discontinuities or changes in the quantitative effect of pressure on rate and degree of polymerization and/or equilibrium constant.

3-13 PROCESS CONDITIONS

The conditions under which radical polymerizations are carried out are both of the homogeneous and heterogeneous types. This classification is usually based on whether the initial reaction mixture is homogeneous or heterogeneous. Some homogeneous systems may become heterogeneous as polymerization proceeds due to insolubility of the polymer in the reaction media. Mass and solution polymerizations are homogeneous processes; suspension and emulsion polymerizations are heterogeneous processes. Emulsion polymerization will be discussed separately in Chap. 4. The other processes will be considered here. All monomers can be polymerized by any of the various processes. However, it is usually found that the commercial polymerization of any one monomer is best carried out by one or two of the processes. The laboratory techniques for carrying out polymerizations have been discussed [Moore, 1977–1985; Sandler and Karo, 1974, 1980; Takemoto and Mujata, 1980]. Up-to-data technological details of commercial polymerization processes are difficult to ascertain, although general descriptions are available [Barrett, 1975; Brydson, 1982; Grulke, 1989; Matsumoto et al., 1977; Menikheim, 1989; Munzer and Trommsdorff, 1977; Rodriguez, 1982; Saunders, 1973; Schildknecht, 1977].

3-13a General Considerations

3-13a-1 Bulk (Mass) Polymerization

Bulk or *mass polymerization* of a pure monomer offers the simplest process with a minimum of contamination of the product. However, bulk polymerization is difficult to control due to the characteristics of radical chain polymerization. Their highly exothermic nature, the high activation energies involved, and the tendency toward the gel effect combine to make heat dissipation difficult. Bulk polymerization requires careful temperature control. Further, there is also the need for strong and elaborate stirring equipment since the viscosity of the reaction system increases rapidly at relatively low conversion. The viscosity and exotherm effects make temperature control difficult. Local hot spots may occur–resulting in degradation and discoloration of the polymer product and a broadened molecular-weight distribution due to chain transfer to polymer. In the extreme case, uncontrolled acceleration of the polymerization rate can lead to disastrous "runaway" reactions [Sebastian and Biesenberger, 1979]. Bulk polymerization is not used commercially for chain polymerizations nearly as much as for step polymerizations because of the difficulties indicated. It is, however, used in the polymerizations of ethylene, styrene, and methyl methacrylate. The heat dissipation and viscosity problems are circumvented by carrying out the polymerizations to low conversions with separation and recycling of unreacted monomer. An alternative is to carry out polymerization in stages—to low conversion in a large reactor and to final conversion in thin layers (either on supports or free-falling streams).

3-13a-2 Solution Polymerization

Polymerization of a monomer in a solvent overcomes many of the disadvantages of the bulk process. The solvent acts as diluent and aids in the transfer of the heat of polymerization. The solvent also allows easier stirring, since the viscosity of the reaction mixture is decreased. Thermal control is much easier in *solution polymerization* compared to bulk polymerization. On the other hand, the presence of solvent may present new difficulties. Unless the solvent is chosen with appropriate consideration, chain transfer to solvent can become a problem. Further, the purity of the polymer may be affected if there are difficulties in removal of the solvent. Vinyl acetate, acrylonitrile, and esters of acrylic acid are polymerized in solution.

3-13a-3 Heterogeneous Polymerization

Heterogeneous polymerization is used extensively to control the thermal and viscosity problems. Two types of heterogeneous polymerization—*precipitation* and *suspension*—can be distinguished. Precipitation polymerizations begin as homogeneous polymerizations but are quickly converted to heterogeneous polymerizations. This occurs in polymerization of a monomer either in bulk or in solution (usually aqueous but sometimes organic) where the polymer formed is insoluble in the reaction medium. Bulk polymerization of vinyl chloride and solution polymerization of acrylonitrile in water are examples of precipitation polymerization. Precipitation polymerizations are often referred to as *powder* or *granular* polymerizations because of the forms in which the final polymer products are obtained. The initiators used in precipitation polymerization are soluble in the initial reaction medium. Polymerization proceeds after precipitation by absorption of monomer and initiator (and/or initiating radicals) into the polymer particles.

Suspension polymerization (also referred to as *bead* or *pearl* polymerization) is carried out by suspending the monomer (discontinuous phase) as droplets (50–500 μm in diameter) in water (continuous phase). Styrene, acrylic and methacrylic esters, vinyl chloride, vinyl acetate, and tetrafluoroethylene are polymerized by the suspension method. The water:monomer weight ratio varies from 1:1 to 4:1 in most polymerizations. The monomer droplets (subsequently converted to polymer particles) are prevented from coalescing by agitation and the presence of *suspension stabilizers* (also referred to as *dispersants* or *surfactants*). Two types of stabilizers are used—water-soluble polymers (often in the presence of electrolyte or buffer) and water-insoluble inorganic powders. The former type includes poly(vinyl alcohol), hydroxypropyl cellulose, sodium poly(styrene sulfonate), and sodium salt of acrylic acid–acrylate ester copolymer; the latter type includes talc, hydroxyapatite, barium sulfate, kaolin, magnesium carbonate and hydroxide, calcium phosphate, and aluminum hydroxide. The levels of suspension stabilizers are typically less than 0.1 weight % (wt %) of the aqueous phase (although the water-soluble polymers are sometimes used at higher concentrations). This is much lower than the surfactant concentrations used in emulsion polymerizations (typically as high as 1–5%) (Chap. 4) and accounts for the higher monomer droplet sizes in suspension polymerization. Also, unlike emulsion polymerization, the two-phase system cannot be maintained in suspension polymerization without agitation. Another difference is that the dispersants used in suspension polymerization seldom form colloidal micelles as in emulsion polymerization.

The initiators used in suspension polymerization are soluble in the monomer droplets. Such initiators are often referred to as *oil-soluble initiators*. Each monomer droplet in a suspension polymerization is considered to be a mini bulk polymerization system. The kinetics of polymerization within each droplet are the same as those for the corresponding bulk polymerization.

Suspension polymerizations in the presence of high concentrations ($>1\%$) of the water-soluble stabilizers (and usually initiated by water-soluble initiators) are used to produce latexlike dispersions of particles having small particle size in the range 0.5–10 μm. These suspension polymerizations are sometimes (but not always) referred to as *dispersion polymerizations*. The term *microsuspension polymerization* has also been used, especially when the monomer droplet size is 1 μm or smaller [Hunkeler et al., 1989]. *Inverse microsuspension* systems involve an organic solvent as the continuous phase with droplets of a water-soluble monomer (e.g., acrylamide), either neat or dissolved in water. The term *inverse dispersion* has been used to describe systems involving the organic solvent as continuous phase with dissolved monomer and initiator that yield insoluble polymer.

Dispersion and emulsion polymerizations are used to produce finely divided, stable latices and dispersions that can be used directly as coatings, paints (the commonly encountered "water-based paints"), adhesives, and other products but are not as useful for producing high-purity products free of the dispersants and other additives. The other heterogeneous polymerizations are more useful for this purpose since larger particles are produced and these can be much more easily isolated and purified by methods such as filtration and centrifugation.

3-13a-4 Other Processes

Several other techniques for carrying out radical chain polymerizations have been studied. Polymerization of *clathrate* (or *inclusion* or *canal*) complexes of monomer (guest molecule) with host molecules such as urea, thiourea, perhydrotriphenylene, apocholic and deoxycholic acids, or β-cyclodextrin offers the possibility of obtaining stereoregular polymers (Chap. 8) or polymers with extended-chain morphology by the proper choice of the host–guest pair and reaction conditions [Farina and Di Silvestro, 1988; Miyata et al., 1988]. (Clathrates differ from other adducts such as hydrates and coordination complexes in that the latter involve significant intermolecular interactions, e.g., hydrogen-bonding, ion–dipole, while the former involve much weaker forces that are related to the steric fit between host and guest moelcules.) Polymerizations of monomers in the form of vesicles, monolayers, and multilayers have been studied with a view toward obtaining polymers with functions that mimic organized biological systems such as lipid bilayers [Day and Ringsdorf, 1979; Ohno et al., 1989; Okahata et al., 1987; Uchida et al., 1989; Yuan et al., 1989]. Radical polymerizations have also been carried out with monomers in the solid state [Chapiro, 1962; Chatani, 1974] and liquid crystal state [Hsu and Blumstein, 1977].

3-13b Specific Commercial Polymers

3-13b-1 Polyethylene

Radical chain polymerization of ethylene to polyethylene

$$CH_2{=}CH_2 \rightarrow {-\!\left(\!-CH_2CH_2\!-\right)_{\!n}\!-} \tag{3-222}$$

is carried out at high pressures of 120–300 MPa and at temperatures above the T_m of polyethylene (Fig. 3-21) [Doak, 1986]. Batch processes are not useful since the long residence time gives relatively poor control of product properties. Long-chain branching due to intermolecular chain transfer becomes excessive with deleterious effects on the physical properties. Continuous processes allow better control of the polymerization.

Tubular reactors are most often used, although autoclave reactors are also employed. Tubular reactors consist of a number of sections, each with an inner diameter of 2–6 cm and length of 0.5–1.5 km, arranged in the shape of an elongated coil. The polymerization mixture has very high linear velocities (>10 m/sec) with short reaction times (0.25–2 min). Trace amounts of oxygen ($\leq$ 300 ppm) are typically used as the initiator, often in combination with an alkyl or acyl peroxide or hydroperoxide. Ethylene is compressed in stages with cooling between stages and introduced into the reactor. Initiator and chain-transfer agent are added during a late stage of compression or simultaneously with introduction of monomer into the reactor. Propane, n-butane, cyclohexane, propene, 1-butene, i-butene, acetone, 2-propanol, and propionaldehyde have been used as chain-transfer agents. The initial reaction temperature is typically 140–180°C, but this increases along the length of the reactor to peak temperatures as high as 300–325°C before decreasing to about 250–275°C due to the presence of cooling jackets. Polymerization occurs in the highly compressed gaseous state where ethylene behaves much as a liquid (even though ethylene is above its critical temperature). The reaction system after the start of polymerization is homogeneous (polymer swollen

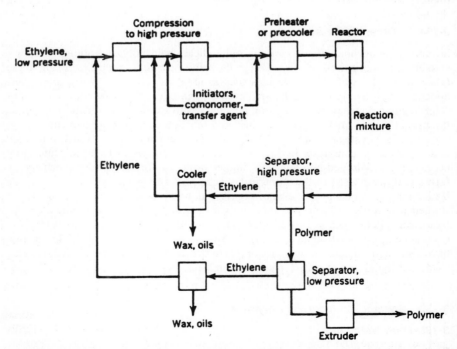

Fig. 3-21 Flow diagram of high-pressure polyethylene process. After Doak [1986] (by permission of Wiley-Interscience, New York).

by monomer) if the pressure is above 200 MPa. Some processes involve multizone reactors where there are multiple injections of initiator (and, often, monomer and chain-transfer agent) along the length of the tubular reactor.

The conversion per pass is usually 15–20% and 20–30% per pass for single-zone (no additional injections of reactants) and multizone reactors, respectively. After leaving the reactor, polyethylene is removed from the reaction mixture by reducing the pressure, usually in two stages in series at 20–30 MPa and 0.5 MPa or less. Molten polyethylene is extruded, pelletized, and cooled. The recovered ethylene is cooled to allow waxes and oils (i.e., very-low-molecular-weight polymer) to be separated prior to recycling of ethylene to the reactor. Overall, ethylene polymerization is carried out as a bulk polymerization (i.e., no added solvent or diluent) in spite of the inherent thermal and viscosity problems. Control is achieved by limiting conversion to no more than 30% per pass. The polymerization is effectively carried out as a solution or suspension process (with unreacted monomer acting as solvent or diluent) depending on whether the pressure is above or below 200 MPa.

Autoclave reactors differ from tubular reactors in two respects. Autoclave reactors have much smaller length-to-diameter ratios (2–4 for single-zone and up to 15–18 for multizone reactors) and operate at a much narrower reaction temperature range.

The polyethylene produced by radical polymerization is referred to as *low-density polyethylene* (LDPE) or *high-pressure polyethylene* to distinguish it from the polyethylene synthesized using coordination catalysts (Sec. 8-4g-2). The latter polyethylene is referred to as *high-density polyethylene* (HDPE) or *low-pressure polyethylene*. Low-density polyethylene is more highly branched (both short and long branches) than high-density polyethylene and is therefore lower in crystallinity (40–60% vs. 70–90%) and density (0.91–0.93 g/cm^3 vs. 0.94–0.96 g/cm^3).

Low-density polyethylene has a wide range and combination of desirable properties. Its very low T_g of about $-120°C$ and moderately high crystallinity and T_m of 105–115°C give it flexibility and utility over a wide temperature range. It has a good combination of strength, flexibility, impact resistance, and melt flow behavior. The alkane nature of polyethylene imparts a very good combination of properties. Polyethylene is highly resistant to water and many aqueous solutions even at high temperatures. It has good solvent, chemical, and oxidation resistance, although it is slowly attacked by oxidizing agents and swollen by hydrocarbon and chlorinated solvents at room temperature. Polyethylene is a very good electrical insulator. Commercial low-density polyethylenes have number-average molecular weights in the range 20,000–100,000, with $\overline{X}_w/\overline{X}_n$ in the range 3–20. A wide range of LDPE products are produced with different molecular weights, molecular-weight distributions, and extents of branching (which affects crystallinity, density, and strength properties), depending on the polymerization temperature, pressure (i.e., ethylene concentration), and reactor type. Long-chain branching increases with increasing temperature and conversion but decreases with increasing pressure. Short-chain branching decreases with increasing temperature and conversion but increases with pressure. Autoclave processes generally yield polyethylenes with narrower moelcular-weight distributions but increased long-chain branching compared to tubular processes.

The wide range and combination of properties, including the ease of fabrication by a variety of techniques, makes LDPE the largest-volume polymer. Over 9 billion pounds are produced annually in the United States; worldwide production is probably more than three times that amount. These amounts are especially significant when one realizes that commercial quantities of polyethylene did not become available until after World War II in 1945. Polyethylene was discovered in the early 1930s at the ICI laboratories in England and played an important role in the outcome of World War II. The combination of good electrical properties, mechanical strength, and processability made possible the development of radar, which was critical to turning back Germany's submarine threat and air attacks over England.

Film applications account for over 60% of polyethylene consumption. Most of this is extrusion-blown film for packaging and household uses (bags, pouches and wrap for food, clothes, trash, and dry cleaning) and agricultural and construction applications (greenhouses, industrial tank and other liners, moisture and protective barriers). Injection molding of toys, housewares, paint-can lids, and containers accounts for another 10–15%. About 5% of the LDPE produced is used as electrical wire and cable insulation (extrusion coating and jacketing) for power transmission and communication. Extrusion coating of paper to produce milk, juice, and other food cartons and multiwall bags accounts for another 10%. Other applications include blow-molded squeeze bottles for glue and personal-care products.

Trade names for polyethylene include *Alathon*, *Alkathene*, *Fertene*, *Grex*, *Hostalen*, *Marlex*, *Nipolon*, and *Petrothene*.

Low- and high-density polyethylene, polypropylene, and polymers of other alkene (olefin) monomers constitute the *polyolefin* family of polymers. All except LDPE are produced by coordination catalysts. Coordination catalysts are also used to produce linear low-density polyethylene (LLDPE), which is essentially equivalent to LDPE in structure, properties, and applications (Sec. 8-4g-3). The production figures given above for LDPE includes LLDPE. The latter makes up about 30% of the total. (Copolymers constitute 20–25% of all low density polyethylenes; see Sec. 6-8b.)

3-13b-2   Polystyrene

Continuous solution polymerization is the most important method for the commercial production of polystyrene although suspension polymerization is also used [Moore,

$$CH_2{=}CH \rightarrow {-}(CH_2CH{)}_n \qquad\qquad (3\text{-}223)$$
$$\quad\; \overset{|}{\varnothing} \qquad\qquad\quad \overset{|}{\varnothing}$$

1989]. Emulsion polymerization is important for ABS production (Chap. 4) but not for polystyrene itself.

Figure 3-22 shows a generalized solution process for polymerization of styrene. Actual processes may have up to five reactors in series; processes with only one reactor are sometimes used. Styrene, solvent (usually ethylbenzene in amounts of 2–30%), and occasionally initiator are fed to the first reactor. Solvent is used primarily for viscosity control with the amount determined by the exact configuration of the reactor and the polymer molecular weight desired. A secondary function of solvent is control of molecular weight by chain transfer, although more effective chain transfer agents are also used. The reactors are run at successively increasing temperatures with 180°C in the last reactor. For thermal, self-initiated polymerization, the first reactor is run at 120°C. The temperature in the first reactor is 90°C when initiators are used. Both single- and two-initiator systems are used. Final conversions of 60–90% are achieved in the last reactor. The reaction mixture is passed through a vacuum devolatilizer to remove solvent and unreacted monomer that are condensed and recycled to the first reactor. The devolatilized polystyrene (at 220–260°C) is fed to an extruder and pel-

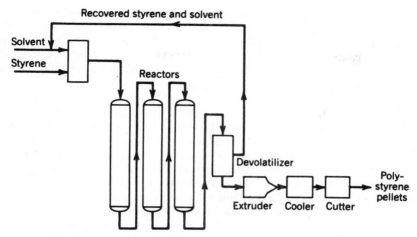

Fig. 3-22   Continuous solution polymerization of styrene. After Moore [1989] (by permission of Wiley-Interscience, New York).

letized. Control of the heat dissipation and viscosity problems is achieved by the use of a solvent, limiting conversion to less than 100%, and the sequential increasing of reaction temperature.

Commercial polystyrenes (PS) have number-average molecular weights in the range 50,000–150,000 with $\overline{X}_w/\overline{X}_n$ values of 2–4. Although completely amorphous (T_g = 85°C), its bulky rigid chains (due to phenyl–phenyl interactions) impart good strength with high-dimensional stability (only 1–3% elongation); polystyrene is a typical rigid plastic. PS is a very good electrical insulator, has excellent optical clarity due to the lack of crystallinity, possesses good resistance to aqueous acids and bases, and is easy to fabricate into products since only T_g must be exceeded for the polymer to flow. However, polystyrene has some limitations. It is attacked by hydrocarbon solvents, has poor weatherability (UV, oxygen, and ozone attack) due to the labile benzylic hydrogens, is somewhat brittle, and has poor impact strength due to the stiff polymer chains. The upper temperature limit for using polystyrene is low because of the lack of crystallinity and low T_g. In spite of these problems, styrene polymers are used extensively, over 7 billion pounds of plastics and over 2 billion pounds of elastomers per year in the United States. Weathering problems of styrene products are significantly decreased by compounding with appropriate stabilizers (UV absorbers and/or antioxidants). Solvent resistance can be improved somewhat by compounding with glass fiber and other reinforcing agents. Copolymerization and polymer blends (Sec. 2-13a-3b) are used extensively to increase the utility of styrene products. *Copolymerization* involves polymerizing a mixture of two monomers. The product, referred to as a *copolymer*, contains both monomers in the polymer chain—in the alternating, statistical, block, or graft arrangement depending on the detailed chemistry of the specific monomers and reactions conditions (Sec. 2-13a, Chap. 6). Polymer blends are physical mixtures of two different materials (either homopolymers or copolymers). All the elastomeric styrene products are copolymers or blends; none are homopolystyrene. No more than about one-third of the styrene plastic products are homopolystyrene. The structural details of the copolymers and blends will be considered in Sec. 6-8a.

About two billion pounds of styrene homopolymer are produced annually in the United States in the form of a product referred to as *crystal polystryene*. Although the term *crystal* is used to describe the great optical clarity of the product, crystal polystyrene is not crystalline; it is completely amorphous. A variety of products are formed by injection molding, including tumblers, dining utensils, hairbrush handles, housewares, toys, cosmetic containers, camera parts, audiotape cassettes, computer disk reels, stereo dust covers, and office fixtures. Medical applications include various items sterilized by ionizing radiation (pipettes, Petri dishes, medicine containers). Extruded sheet is used for lighting and decoration applications. Biaxially oriented sheet is thermoformed into various shapes, such as blister packaging and food packaging trays. *Expandable polystyrene*, either crystal polystyrene or styrene copolymers impregnated with a blowing agent (usually pentane), is used to produce various foamed product. Among the products are disposable drinking cups (especially coffee cups), cushioned packaging, and thermal insulation used in the construction industry. Extruded foam sheets are converted by thermoforming into egg cartons, meat and poultry trays, and packaging for fast-food takeouts. The production of expandable polystyrene and styrene copolymers probably exceeds one billion pounds annually in the United States.

Trade names for PS include *Carinex*, *Cellofoam*, *Dylene*, *Fostarene*, *Hostyren*, *Lustrex*, *Styron*, and *Styrofoam*.

3-13b-3 Vinyl Family

The *vinyl family* of polymers consists of poly(vinyl chloride), poly(vinylidene chloride), poly(vinyl acetate), and their copolymers and derived polymers.

3-13b-3-a Poly(vinyl chloride)   Most poly(vinyl chloride) (PVC) is commercially produced by suspension polymerization [Brydson, 1982; Tornell, 1988]. Bulk and

$$CH_2{=}CH \rightarrow -(CH_2CH)_n \qquad (3\text{-}224)$$
$$\quad | \qquad\qquad | $$
$$\quad Cl \qquad\qquad Cl$$

emulsion polymerizations are used to a much lesser extent, and solution polymerization is seldom used. Suspension polymerization of vinyl chloride is generally carried out in a batch reactor such as that shown in Fig. 3-23. A typical recipe includes 180 parts water and 100 parts vinyl chloride plus small amounts of dispersants (<1 part), monomer-soluble initiator, and chain-transfer agent (trichloroethylene). All components except monomer are charged into the reactor, which is then partially evacuated. Vinyl chloride is drawn in, sometimes by using pressurized oxygen-free nitrogen to force monomer into the reactor. The reactants are then heated in the closed system to about 50°C and the pressure rises to about 0.5 MPa. The temperature is maintained at about 50°C as polymerization proceeds. When the pressure is about 0.05 MPa, corresponding to about 90% conversion, excess monomer is vented off to be recycled. Removal of residual monomer typically involves passing the reaction mixture through a countercurrent of steam. The reaction mixture is then cooled, and the polymer separated, dried in hot air at about 100°C, sieved to remove any oversized particles, and stored. Typical number-average molecular weights for commercial PVC are in the range 30,000–80,000.

Poly(vinyl chloride) has very low crystallinity but achieves strength because of the

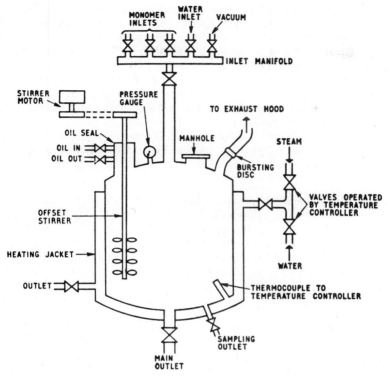

Fig. 3-23 Typical polymerization vessel for suspension polymerization of vinyl chloride. After Brydson [1982] (by permission of Butterworth Scientific, London).

bulky polymer chains (a consequence of the large C1 groups on every other carbon). This is apparent in the high T_g of 81°C, although T_g is not so high that processing by a variety of techniques (injection molding, extrusion, blow molding, calendering) is impaired. Poly(vinyl chloride) is relatively unstable to light and heat with the evolution of hydrogen chloride, which can have deleterious effects on the properties of nearby objects (e.g., electrical components) as well as physiological effects on humans. The commercial importance of PVC would be greatly reduced were it not for the fact that this instability can be controlled by blending with appropriate additives such as metal oxides and carbonates as well as fatty acid salts. These additives stabilize PVC by slowing down the dehydrochlorination reaction and absorption of the evolved hydrogen chloride.

Poly(vinyl chloride) is a very tough and rigid material with extensive applications. Its range of utilization is significantly expanded by *plasticization*, which converts *rigid PVC* to *flexible PVC*. Plasticization involves blending PVC with plasticizers (e.g., di-*i*-octyl phthalate, tritolyl phosphate, epoxidized oils). The plasticizer imparts flexibility by acting effectively as an "internal lubricant" between PVC chains. (Plasticization accomplishes for PVC what copolymerization and polymer blending accomplished for polystyrene—changing a rigid material into a flexible one. The plasticization route is not useful for polystyrene as PS is not so compatible with plasticizers.) Blending of

PVC with rubbery polymers to improve impact strength is also practiced. About 8.5 billion pounds of PVC, about evenly divided between rigid and flexible grades, are produced annually in the United States. Rigid pipe for home and other construction accounts for more than 40% of the total PVC market. *Vinyl* siding, window frames, rain gutters, and downspouts account for another 10–15%. Packaging applications (bottles, box lids, blister packaging) account for about 8%. More than 10% of PVC is used for flooring (large sheet and tile). Other applications include wire and cable insulation for buildings, telecommunications and appliance wiring, garden hose, surgical and other protective gloves, and coated products.

Trade names for PVC include *Carina, Corvic, Darvic, Elvic, Exon, Geon, Koroseal, Marvinol, Nipeon, Opalon, Vybak, Vygen,* and *Vyram.* The generic term *vinyl* is often used for PVC products.

3-13b-3-b   Other Members of Vinyl Family. More than 0.5 billion pounds of poly(vinyl acetate) (PVAc), poly(vinylidene chloride), and their copolymers and polymers derived from them are produced annually in the United States. PVAc does not have sufficient strength for producing the types of products obtained from polyethylene, polystyrene, and poly(vinyl chloride) since it is noncrystalline and has a T_g of only 28°C. However, poly(vinyl acetate) (**XLIII**) and its copolymers find uses as water-

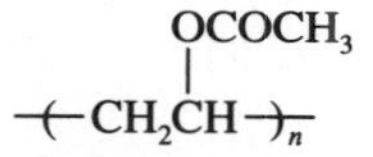

$$\begin{array}{c} \text{OCOCH}_3 \\ | \\ -\!\!\left(\!\text{CH}_2\text{CH}\!\right)\!\overline{}_n \end{array}$$

XLIII

based paints; as adhesives for paper, textiles, and wood (labeling, spackling compound, white glue); and as a sizing or coating compound for paper and textiles. Emulsion polymerization is the predominant commercial process with the resulting latices used directly for most applications. PVAc is also used to produce poly(vinyl alcohol) and acetals of the latter (Sec. 9-4). Poly(vinyl alcohol) (PVA) is used as thickening agents, coatings, adhesives, and for water-soluble packages for bath salts, bleaches, detergents, and other materials. The formal and butyral derivatives of PVA are used, respectively, as heat-resistant wire enamel insulation and the interlayer in safety glass laminates.

Poly(vinylidene chloride) (**XLIV**) and its copolymers with vinyl chloride, acrylo-

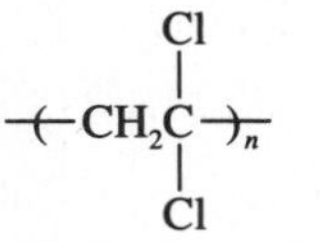

$$\begin{array}{c} \text{Cl} \\ | \\ -\!\!\left(\!\text{CH}_2\text{C}\!\right)\!\overline{}_n \\ | \\ \text{Cl} \end{array}$$

XLIV

nitrile, and acrylates, usually produced by the suspension or emulsion process, are useful as oil, fat, oxygen, and moisture-resistant packaging films (*Saran wrap*), containers, coatings, tank liners, and monofilaments in drapery fabrics and industrial filter cloths.

3-13b-4 Acrylic Family

The *acrylic* family of polymers includes polymers and copolymers of acrylic and meth-acrylic acids and esters, acrylonitrile, and acrylamide [Kine and Novak, 1985; Nemec and Bauer, 1985; Peng, 1985; Thomas and Wang, 1985].

3-13b-4-a Acrylate and Methacrylate Products. Close to 2 billion pounds of poly-meric products are produced annually based on acrylic and methacrylic esters, about evenly divided between acrylates and methacrylates. A substantial fraction of the methacrylate products are copolymers. Most of the acrylate products are copolymers. The copolymers contain various combinations of acrylate and/or methacrylate mon-omers, including combinations of ester and acid monomers. Methyl methacrylate (MMA) is by far the most important methacrylate ester monomer, accounting for 90% of the volume of methacrylic ester monomers. Ethyl and *n*-butyl acrylates account for about 80% of the total volume of acrylate ester monomers.

Poly(methyl methacrylate) (PMMA) is polymerized by solution, suspension, and

$$
\begin{array}{ccc}
\text{COOCH}_3 & & \text{COOCH}_3 \\
| & & | \\
\text{CH}_2\!=\!\text{C} & \rightarrow & \text{+}\!\text{CH}_2\text{C}\text{+}_n \\
| & & | \\
\text{CH}_3 & & \text{CH}_3
\end{array}
\qquad (3\text{-}225)
$$

emulsion processes. PMMA is completely amorphous but has high strength and ex-cellent dimensional stability due to the rigid polymer chains ($T_g = 105°C$). It has exceptional optical clarity, very good weatherability, and impact resistance (can be machined) and is resistant to many chemicals, although attacked by organic solvents. Polyacrylates differ considerably from the polymethacrylates. They are much softer since the polymer chains are not nearly as rigid [e.g., poly(ethyl acrylate), $T_g = -24°C$] due to the absence of the methyl groups on alternating carbons of the polymer chains. Polymethacrylates tend to be used as shaped objects, while polyacrylates find applications requiring flexibility and extensibility.

Various rigid PMMA products, such as sheet, rod, and tube, are produced by bulk polymerization in a *casting* process. Polymerization is carried out in stages to allow easier dissipation of heat and control of product dimensions since there is a very large (21%) volume contraction on polymerization. Partially polymerized monomer (about 20% conversion) is produced by heating for about 10 min at 90°C with a peroxide. [Uncatalyzed thermal polymerization for much longer periods (about 2 hr) is also used to achieve greater control.] The syrup is cooled to ambient temperature and poured into a mold, and then the assembly is heated in a water or air bath to progressively higher temperatures. The maximum temperature used is 90°C, since higher temper-atures can lead to bubble formation in the product as the boiling point of methyl methacrylate is 100.5°C. The mold often involves spacer units, which, together with clamps, exert pressure on the polymerizing system to accommodate shrinkage during polymerization. Continuous cast polymerization processes are also used extensively. Bulk polymerization and suspension polymerization are used to produce poly-methacrylate molding powders.

Applications for rigid methacrylate polymer products include signs and glazing substitute for safety glass and to prevent vandalism (school, home, and factory win-dows, patio window doors, aircraft windows and canopies, panels around hockey rinks, bank teller windows, bath and shower enclosures, skylights), indoor and outdoor lighting, lenses, diffusers, and louvers, architectural structures (domes over pools and stadia, archways between buildings), bathtubs and sanitary fixtures, optical fibers for light transmission, plastic eyeglass lenses, dentures and dental filling materials, and contact lenses (hard and soft).

Solution and emulsion polymerizations are used to produce acrylate and methacrylate polymer products for nonrigid applications. These applications include coatings (interior and exterior paints, including automotive paints), textiles (finishes for improving abrasion resistance, binding pigments, reducing shrinkage of wool, drapery and carpet backing), additives for engine oils and fluids, and caulks and sealants.

Trade names for acrylate and methacrylate polymer products include *Acrylite*, *Dicalite*, *Lucite*, *Plexiglas*, and *Rhoplex*.

3-13b-4-b Polyacrylonitrile. The most important commercial processes for polyacrylonitrile (**XLV**) are solution and suspension polymerizations. Almost all of the

$$\begin{array}{c} CN \\ | \\ -\!\!\left(CH_2CH\right)_{\!n} \\ \textbf{XLV} \end{array}$$

products containing acrylonitrile are copolymers. Styrene-acrylonitrile copolymers (SAN) are useful as plastics (Sec. 6-8a).

About 600 million pounds of *modacrylic* and *acrylic fibers* are produced annually in the United States [trade names: *Acrilon* (now *Monsanto fiber*), *Dynel*, *Orlon*, *Verel*]. Acrylic fibers contain at least 85% by weight of acrylonitrile, while modacrylic fibers contain less than 85% acrylonitrile but no less than 35%. Halogen-containing monomers, such as vinyl chloride, vinylidene chloride, and vinyl bromide, are copolymerized with acrylonitrile to impart flame retardancy. Methyl acrylate and vinyl acetate are used to increase the solubility of the polymer in solvents used for spinning of fibers. Monomers with acidic or basic groups are used to improve dyeability with basic and acid dyes, respectively. Acrylic and modacrylic fibers are the only fibers made from a monomer with a carbon–carbon double bond. Acrylonitrile is the only such monomer with a sufficiently polar group (nitrile) to yield a polymer with the high secondary forces required in a fiber. All other fibers are made from condensation polymers such as polyesters and polyamides. (The only other exception to this generalization is polypropylene; see Sec. 8-4g-4.)

Acrylic and modacrylic fibers have a wool-like appearance and feel, and excellent resistance to heat, ultraviolet radiation, and chemicals [Bajaj and Kumari, 1987]. These fibers have replaced wool in many applications, such as socks, pullovers, sweaters, and craft yarns. Other applications include tenting, awning fabric, and sandbags for rivershore stabilization. The use of acrylic and modacrylic fibers in carpets is low since these materials do not hold up well to recycling through hot–humid conditions. This also prevents its use in the easy-care garment market.

3-13b-4-c Other Members of Acrylic Family. From the viewpoint of sales volume, all other members of the acrylic family constitute a small fraction of the total. However, many of them are useful specialty polymers. Polyacrylamide (**XLVI**), poly(acrylic acid) (**XLVII**), and poly(methacrylic acid) (**XLVIII**) and some of their copolymers

$$\underset{\textbf{XLVI}}{\text{--}(\text{CH}_2\text{CH})_{\overline{n}}} \quad \quad \underset{\textbf{XLVII}}{\text{--}(\text{CH}_2\text{CH})_{\overline{n}}} \quad \quad \underset{\textbf{XLVIII}}{\text{--}(\text{CH}_2\text{C})_{\overline{n}}}$$

with CONH₂, COOH on XLVI/XLVII and COOH, CH₃ on XLVIII.

are used in various applications that take advantage of their solubility in water. Poly(acrylic acid) and poly(methacrylic acid) are used as thickening agents (water used in secondary recovery of oil, rubber emulsions applied to fabrics to produce nonslip backing for floor covering), adhesives, dispersants for inorganic pigments in paints, flocculants (to aggregate suspended particles in metal recovery, clarification of waste and potable waters), and crosslinked ion-exchange resins. Polyacrylamides, often containing ionic comonomers, are used as flocculants in industrial waste and municipal drinking water treatments, papermaking, and mining; crosslinked polyacrylamide is used in gel electrophoresis.

3-13b-5 *Fluoropolymers*

The *fluoropolymer* family consists of polymers produced from alkenes in which one or more hydrogens have been replaced by fluorine. The most important members of this family are polytetrafluoroethylene (PTFE) (**XLIX**), polychlorotrifluoroethylene (PCTFE) (**L**), poly(vinyl fluoride) (PVF) (**LI**), poly(vinylidene fluoride) (PVDF) (**LII**); copolymers of tetrafluoroethylene with ethylene, perfluoropropyl vinyl ether,

$$\underset{\textbf{XLIX}}{\text{--}(\text{CF}_2\text{CF}_2)_{\overline{n}}} \quad \underset{\textbf{L}}{\text{--}(\text{CF}_2\text{CFCl})_{\overline{n}}} \quad \underset{\textbf{LI}}{\text{--}(\text{CH}_2\text{CHF})_{\overline{n}}} \quad \underset{\textbf{LII}}{\text{--}(\text{CH}_2\text{CF}_2)_{\overline{n}}}$$

and perfluoropropylene; and the copolymer of ethylene and chlorotrofluoroethylene. The fluoropolymers are obtained mainly by suspension polymerization; emulsion polymerization is also practiced. The molecular weights of the polymers are high, ranging up to 10^5–10^6 for PTFE, apparently due to the lack of appreciable chain-transfer reactions and the precipitation of growing radicals (leading to greatly decreased termination).

The commercial production of fluoropolymers is very small compared to commodity polymers such as polyethylene or PVC. But the production is still significant on an absolute basis, about 20 million pounds annually in the United States, and it finds a wide range of important specialized applications [Gangal, 1989]. PTFE is the largest-volume fluoropolymer. This family of polymers offer unique performance. Within the fluoropolymers are polymers that withstand a wide variety of chemical environments and are useful at temperatures as low as $-200°C$ and as high as $260°C$. Most fluoropolymers are totally insoluble in common organic solvents and unaffected by hot concentrated acids and bases. They have outstanding electrical resistance, low coefficient of friction (self-lubricating and nonstick parts), and will not support combustion. The properties of polytetrafluoroethylene, high crystallinity, T_m of 327°C, and very high melt viscosity, preclude its processing by the usual techniques for plastics. Fabrication requires the powder and cold extrusion techniques of metal forming. The various copolymers of tetrafluoroethylene and the other fluoropolymers, possessing lower T_m and crystallinity, were developed to overcome the lack of melt processability of PTFE.

The applications for fluoropolymers are specialty applications, limited by their high costs (about \$12–\$15 per pound compared to \$0.40 for LDPE, \$0.60 for PS, \$0.75–\$1.00 for phenolic, urea–formaldehyde, alkyds resins, \$1.10 for PMMA, \$2.70 for nylon 6/6 and polycarbonate, \$20–\$40 for PEEK).

Fluoropolymer applications include electrical (coaxial cable, tape, spaghetti tubing), mechanical (seals, piston rings, bearings, antistick coatings for cookware, self-lubricating parts), chemical (pipe liners, overbraided hose liners, gaskets, thread sealant tapes), and micropowders used in plastics, inks, lubricants, lacquer finishes. A unique application of fluoropolymers is protection of the Statue of Liberty against corrosion.

Trade names for fluoropolymers include *Aflon, Fluon, Foraflon, Halar, Halon, Hostaflon, Kel-F, Kynar, Polyflon, Tedlar, Teflon,* and *Tefzel.*

3-13b-6 *Polymerization of Dienes*

The polymerization of *dienes* (i.e., monomers with two double bonds per molecule) can proceed in a variety of ways depending on the chemical structure of the diene. Dienes are classified as *conjugated* or *nonconjugated dienes* depending on whether the two double bonds are conjugated with each other. Conjugated dienes are also referred to as *1,3-dienes*. The polymerization of nonconjugated dienes often yields branched and crosslinked products. Nonconjugated dienes are often used to bring about crosslinking in other alkene monomers by the technique of copolymerization (Sec. 6-6a). Certain nonconjugated dienes undergo a special type of cyclization polymerization (cyclopolymerization) by an alternating intramolecular reaction mechanism (Sec. 6-6b).

The polymerization of conjugated dienes is of special interest. Two different types of polymerization reactions occur with 1,3-dienes such as 1,3-butadiene, isoprene(2-methyl-1,3-butadiene), and chloroprene (2-chloro-1,3-butadiene)

$$CH_2{=}CH{-}CH{=}CH_2 \qquad CH_2{=}\underset{\underset{CH_3}{|}}{C}{-}CH{=}CH_2 \qquad CH_2{=}\underset{\underset{Cl}{|}}{C}{-}CH{=}CH_2$$

1,3-Butadiene   Isoprene   Chloroprene

One of the polymerization routes involves polymerization of one or the other of the double bonds in the usual manner. The other route involves the two double bonds acting in a unique and concerted manner. Thus addition of an initiating radical to a 1,3-diene such as 1,3-butadiene yields an allylic radical with the two equivalent resonance forms **LIII** and **LIV**

$$R\cdot + CH_2{=}CH{-}CH{=}CH_2 \rightarrow \quad \begin{matrix} R{-}CH_2{-}\overset{\cdot}{C}H{-}CH{=}CH_2 & \textbf{LIII} \\ \updownarrow & \\ R{-}CH_2{-}CH{=}CH{-}\overset{\cdot}{C}H_2 & \textbf{LIV} \end{matrix} \qquad (3\text{-}226)$$

Propagation may then occur by attachment of the successive monomer units either at carbon 2 (propagation via **LIII**) or at carbon 4 (propagation via **LIV**). The two modes of polymerization are referred to as *1,2*-polymerization and *1,4-polymerization*, respectively. The polymers obtained have the repeating structures

$$\left[\begin{matrix} CH_2{-}CH \\ \quad\quad | \\ \quad CH{=}CH_2 \end{matrix}\right]_n \qquad\qquad {\left(CH_2{-}CH{=}CH{-}CH_2\right)}_n$$

1,2-Polymerization   1,4-Polymerization

and are named 1,2-poly-1,3-butadiene and 1,4-poly-1,3-butadiene, respectively. Both polymers are unsaturated—the *1,2-polymer* has pendant unsaturation while there is unsaturation along the polymen chain in the *1,4-polymer*. The polymerization of 1,3-dienes is considered in greater detail in Sec. 8-6.

The polymerization and copolymerization of 1,3-dienes is of commercial importance in the annual production of over 4 billion pounds of elastomers and about 2 billion pounds of plastics in the United States (Secs. 6-8a, 8-6e).

3-13b-7   *Miscellaneous Commercial Polymers*

3-13b-7-a   Poly(p-xylylene). Pyrolysis of *p*-xylene yields poly(*p*-xylylene) [Errede and Szwarc, 1958; Humphrey, 1984; Lee, 1977–1978; Lee and Wunderlich, 1975;

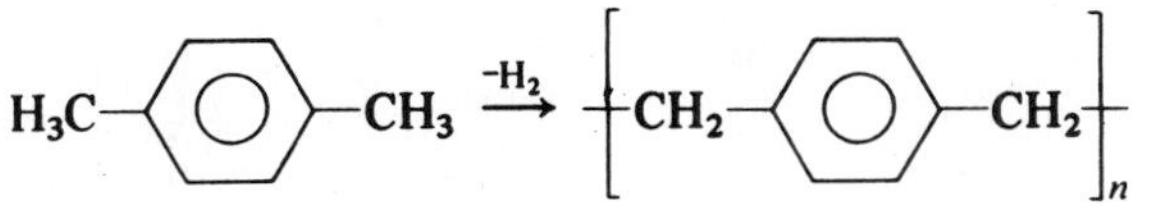

$$H_3C-\underset{}{\bigcirc}-CH_3 \xrightarrow{-H_2} \left[CH_2-\underset{}{\bigcirc}-CH_2\right]_n \qquad (3\text{-}227)$$

Surendran et al., 1987]. The reaction is carried out by pyrolyzing *p*-xylene at temperatures of up to 950°C (in the absence of oxygen) to yield di-*p*-xylylene (**LV**) (also named [2,2]-*paracyclophane*), which is isolated as a stable solid. Subsequent polymerization of di-*p*-xylylene to poly(*p*-xylylene) involves vaporization of di-*p*-xylylene by heating to about 160°C at 1 torr (1.33×10^2 Pa) followed by cleavage to *p*-xylylene diradical (**LVI**) by heating at 680°C at 0.5 torr. The diradical nature of *p*-xylylene was

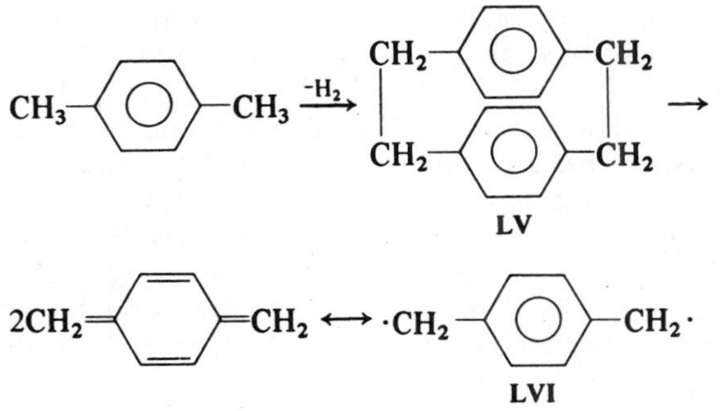

$$CH_3-\underset{}{\bigcirc}-CH_3 \xrightarrow{-H_2} \begin{array}{c} CH_2-\underset{}{\bigcirc}-CH_2 \\ | \qquad\qquad | \\ CH_2-\underset{}{\bigcirc}-CH_2 \end{array} \longrightarrow$$

LV

$$2CH_2=\underset{}{\bigcirc}=CH_2 \longleftrightarrow \cdot CH_2-\underset{}{\bigcirc}-CH_2\cdot \qquad (3\text{-}228)$$

LVI

shown by ESR and trapping of α,α'-diiodo-*p*-xylene with iodine. *p*-Xylylene is very reactive and can be trapped only in dilute solution at low temperature; the half-life at $-78°C$ and 0.12 M is about 21 hr. *p*-Xylylene spontaneously polymerized with crystallization when condensed on surfaces kept at temperatures below about 30°C at 0.1 torr. The reaction is extremely fast; quantitative conversion of di-*p*-xylylene to poly(*p*-xylylene) takes place in a fraction of a second. The overall polymerization process, often referred to as *transport polymerization*, involves transport of the monomer (*p*-xylylene) in the gas phase under vacuum from its place of synthesis to the solid surface on which it polymerizes. This physical process makes poly(*p*-xylylene) useful commercially for forming thin coatings on objects with true conformality of the coating to all surfaces including deep penetration into small spaces. Poly(*p*-xylylene) (trade name: *Parylene N*; IUPAC name: poly[1,4-phenyleneethylene]) and the corresponding polymers derived from 2-chloro-*p*-xylene and 2,5-dichloro-*p*-xylene (*Par-*

ylenes C and D) have a modest but growing importance in the electronics industry for coating and encapsulating delicate microcircuitry for protection from hostile environments. The polymers are crystalline (T_m near or above 400°C) with excellent mechanical and electrical insulating characteristics at temperatures as low as -200°C and as high as 200°C. Other potential uses include coating of orthopedic parts (bone pins and joint prosthesis) to make them biocompatible and conservation of archival and artifact objects.

Polymerization is initiated by coupling of two *p*-xylylene radicals to form a diradical (**LVII**), which then grows by the addition of *p*-xylylene at both of its radical centers.

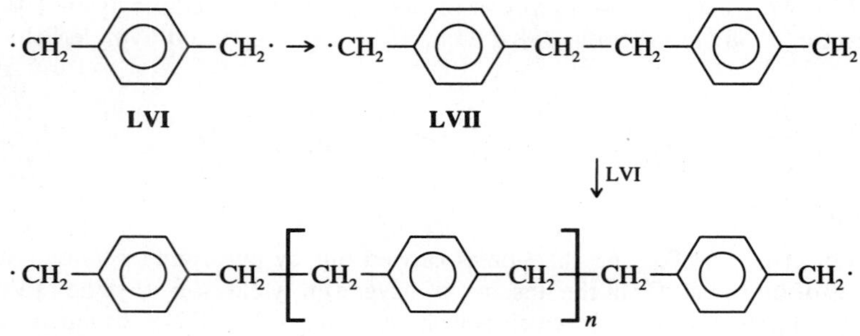

Whether the formation of poly(*p*-xylylene) should be included in this chapter is not clear. Decisive data are not available to indicate the classification of this polymerization as a step or chain reaction. The formation of high polymer occurs instantaneously when *p*-xylylene contacts the cool surface, precluding the evaluation of polymer molecular weight versus conversion. Also, the mode of termination for this reaction is unknown.

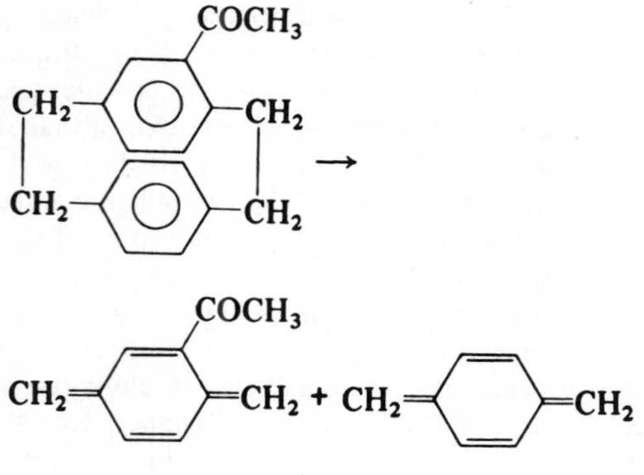

One aspect of the polymerization that is well established is the initiation step when di-*p*-xylylene is pyrolyzed. An alternate initiation mode involving the direct formation of the diradical **LVII** from **LV** by cleavage of only one of the two CH_2—CH_2 bonds is ruled out from experiments with monosubstituted di-*p*-xylylenes. When acetyl-di-*p*-xylylene is pyrolyzed and the pyrolysis vapor led through successive condensation surfaces at temperatures of 90 and 25°C, respectively, the result is the formation of two different polymers neither of which is poly(acetyl-di-*p*-xylylene). Pyrolysis yields acetyl-*p*-xylylene and *p*-xylylene which undergo separate polymerizations at different condensation temperatures. Poly(acetyl-p-xyxylene) is produced at the 90°C surface and poly(p-xylylene) at the 25°C surface.

Transport polymerization has also been studied with other monomers, including methylene

$$CH_2{=}CO \text{ or } CH_2{=}NN \xrightarrow{-CO \text{ or } N_2} :CH_2 \rightarrow -\!\!\left(\!CH_2\!\right)_{\!n}\!- \qquad (3\text{-}231)$$

and other carbenes (phenylcarbene, 1,4-phenylenecarbene), silylenes such as :Siϕ_2 and germylenes such as :GeCl$_2$ and :Geϕ_2 [Lee, 1977–1978; Lee and Wunderlich, 1978].

3-13b-7-b   Poly(N-vinylcarbazole). *Poly(N-vinylcarbazole)* (PVCB) is produced by the polymerization of N-vinylcarbazole (**LVIII**). PVCB was originally used as a ca-

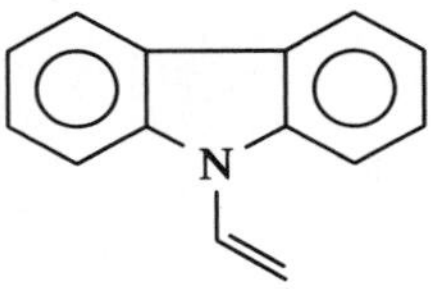

LVIII

pacitor dielectric, since it has very good electrical resistance over a range of temperatures and frequencies. Other electrical applications included switch parts and coaxial cable spacers. These applications are now very limited. The main application of PVCB today is in electrostatic dry copying (*xerography*) machines, a consequence of its photoconductivity.

3-13b-7-c   Poly(N-vinylpyrrolidinone). *Poly(N-vinylpyrrolidinone)* (PVP) is obtained by polymerization of N-vinylpyrrolidinone (**LIX**). The largest use is in hair

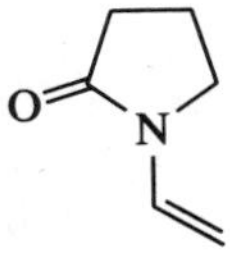

LIX

sprays and wave sets because of its property as a film-former, with good adhesion to hair, film luster, and ease of removal by washing with water. PVP is also used in hand creams and liquid makeups. It finds applications in the textile industry because of its affinity for many dyestuffs; it is used for dye stripping, in the formulation of sizes and finishes, and to assist in dye-leveling processes. PVP was used extensively as a blood plasma substitute in Germany during World War II. It is still used as a blood plasma substitute or stockpiled for emergency use in some countries.

3-13c   Other Polymerizations

The radical chain polymerizations of a variety of other monomers have been studied for purposes of obtaining specific properties in the polymer products (e.g., heat resistance, electrical conductivity). Some of these polymerizations are described in this section. With very few exceptions, none of these polymers are of commercial importance at this time.

3-13c-1 Organometallic Polymers

Various organometallic compounds containing polymerizable carbon–carbon double bonds have been synthesized and their polymerization studied [Carraher, 1981; Carraher and Sheats, 1978; Pittman et al., 1987]. Much of the interest in these polymerizations has been to obtain polymers with improved thermal stability or electrical (semi)conductivity, but such objectives have not been realized. Among the organometallic monomers studied are vinylferrocene (**LX**), tricarbonylchromiumstyrene (**LXI**), and trialkyltin methacrylate (**LXII**). Polymerization of most organometallic

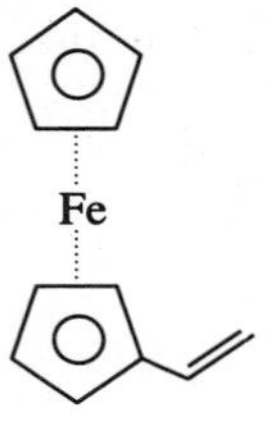

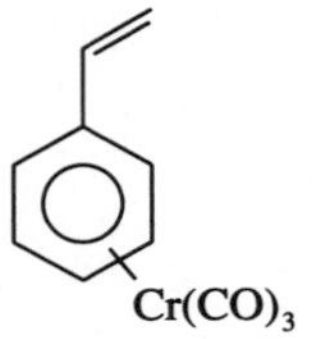

monomers proceeds in a straightforward manner, but there are complications in some cases. Polymerization of vinylferrocene proceeds with termination by *intramolecular electron transfer* (Eq. 3-232 where F_c represents the ferrocene moiety) superimposed

$$\sim\!\!\sim\!\!\text{CH}_2\!-\!\overset{\displaystyle |}{\underset{\displaystyle \text{Fc}}{\text{CH}}}\!\cdot \;\rightarrow\; \sim\!\!\sim\!\!\text{CH}_2\!-\!\overset{\displaystyle |}{\underset{\displaystyle \text{Fc}^+}{\text{CH}^-}} \;\xrightarrow{\text{H}^+}\; \sim\!\!\sim\!\!\text{CH}_2\!-\!\overset{\displaystyle |}{\underset{\displaystyle \text{Fc}^+}{\text{CH}_2}} \qquad (3\text{-}232)$$

$$\textbf{LXIII} \qquad\qquad\qquad \textbf{LXIV}$$

on termination by coupling and disproportionation. This involves electron transfer from iron to the radical center to form carbanion **LXIII**. Carbanion **LXIII** abstracts a proton from solvent to form **LXIV**, which subsequently yields a paramagnetic high-spin iron complex of unknown structure [Willis and Sheats, 1984].

Copolymers containing trialkyltin methacrylate or acrylate as one of the comonomers are used to formulate antifouling marine paints to prevent the growth of barnacles and fungi on shore installations and ship bottoms [Yaeger and Castelli, 1978]. The "killing" action of these paints depends on hydrolysis of the trialkyltin moiety, which is toxic to marine organisms, from the polymer.

3-13c-2 Functional Polymers

The antifouling paints described above are an example of a *functional polymer*, a polymer that contains some specific moiety (trialkyltin in the particular example) with a specific property. There is considerable activity in the synthesis of a variety of functional polymers. For example, chloroamphenicol, a broad-spectrum antibiotic, has been esterified with acrylic acid to yield **LXV**. Polymerization or copolymerization of **LXV** yields a *polymeric drug*, which has the advantage of an *in vivo* slow release

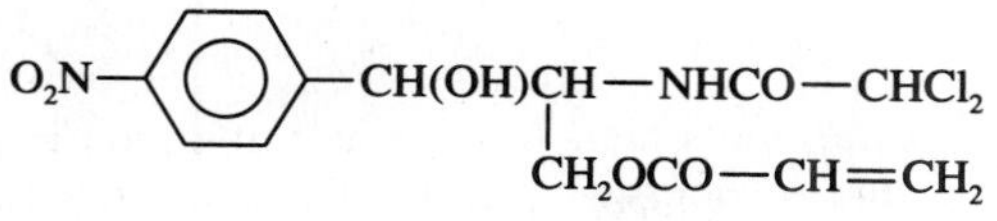

$$O_2N-\underset{}{\bigcirc}-CH(OH)\underset{\underset{CH_2OCO-CH=CH_2}{|}}{CH}-NHCO-CHCl_2$$

LXV

$$(n\text{-}C_{18}H_{37})_2\underset{\underset{CH_3}{|}}{\overset{Br^-}{N^+}}-CH_2CH_2OCO-\underset{}{\bigcirc}-CH=CH_2$$

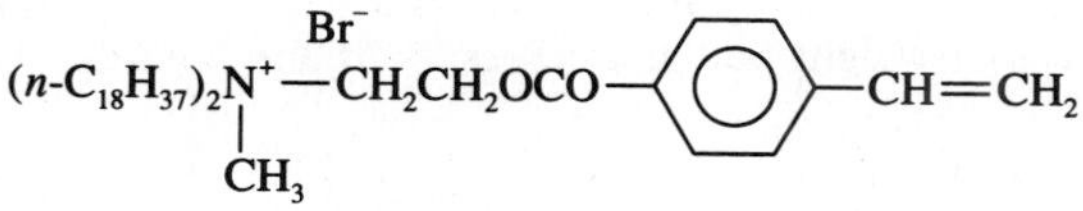

LXVI

of chloroamphenicol via ester hydrolysis [Meslard et al., 1986]. The result would be a more constant *in vivo* concentration of the drug.

The polymerization of **LXVI** has been studied for synthesizing *liposomes* (lipid vesicles) [Yuan et al., 1989]. The polymer from 1-(*p*-vinylphenyl)-2-tetrazoline-5-thione (**LXVII**) is of interest in terms of the known ability of its heterocyclic moiety

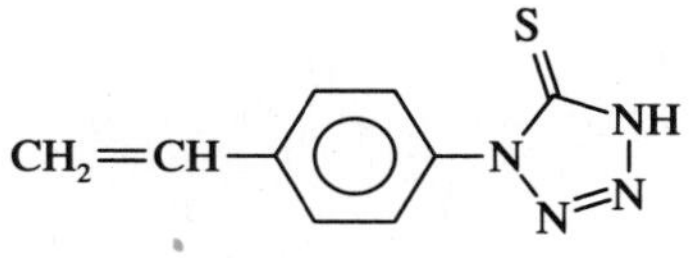

$$CH_2=CH-\underset{}{\bigcirc}-\underset{\underset{N\underset{\displaystyle N}{\diagdown}}{\overset{S}{\underset{||}{\underset{}{N}}}}-\overset{NH}{\underset{N}{\diagup}}$$

LXVII

to form insoluble compounds with heavy-metal ions such as silver and lead [Grasshoff et al., 1978]. Monomers such as **LXVIII** have been studied for synthesizing liquid

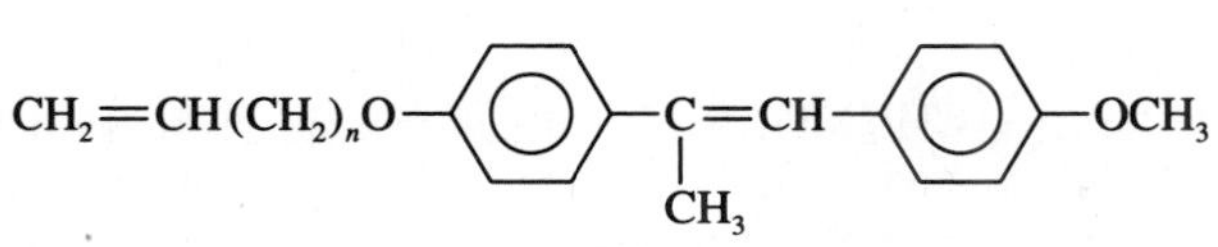

$$CH_2=CH(CH_2)_nO-\underset{}{\bigcirc}-\underset{\underset{CH_3}{|}}{C}=CH-\underset{}{\bigcirc}-OCH_3$$

LXVIII

crystal polymers (Sec. 2-14g) [Finkelmann, 1987; Percec and Tomazos, 1989; Zhou et al., 1987]. (The location of the mesogenic 4′-methoxy-α-methylstilbene moiety relative to the polymer chain can be varied by varying *n*.) Other systems studied include monomers with saccharide (sugar), crown ether, and nucleic acid moieties [Akashi et al., 1979; Klein and Herzog, 1987; Nakai et al., 1977; Peramunage et al., 1989].

A different type of functional polymer is the *telechelic polymer* (Sec. 2-13b). Telechelic polymers containing a functional group such as OH or COOH at each end are useful for synthesizing block copolymers by step polymerization. Telechelic polymers containing particular functional end groups are obtained by using initiators that carry those functional groups [Heitz, 1987]. Telechelics with hydroxyl end groups are obtained in a radical polymerization by using H_2O_2 or 4,4′-azobis(4-cyano-*n*-pentanol) as the initiator. The use of 4,4′-azobis(4-cyanovaleric acid) as initiator results in telechelics with carboxyl end groups.

3-13c-3 *Acetylenic Monomers*

The polymerization of acetylene (alkyne) monomers has received attention in terms of the potential for producing conjugated polymers with electrical conductivity. Simple alkynes such as phenylacetylene do undergo radical polymerization but the molecular weights are low ($\overline{X}_n < 25$) [Amdur et al., 1978]. Ionic and coordination polymerizations of alkynes result in high-molecular-weight polymers (Secs. 5-7d and 8-4d-2).

$$\phi C\equiv CH \rightarrow -(-\phi C{=}CH-)_{\overline{n}} \tag{3-233}$$

The radical 1,4-polymerization of certain diacetylenes (conjugated dialkynes) proceeds to yield high-molecular-weight polymer (10^5) when carried out in the solid state. Diacetylene itself has not been successfully polymerized; it is too reactive to be polymerized in a controlled manner. 1,4-Disubstituted diacetylenes are polymerized by ionizing radiation, UV, or heat [Baughman and Yee, 1978; Chance, 1986; Eckhardt et al., 1983; Patel et al., 1978; Wegner, 1980]. The reaction proceeds via a dicarbene propagating species (**LXIX**) with the final polymer product possessing an alternating ene–yne-conjugated structure (**LXX**). The most studied diacetylene is

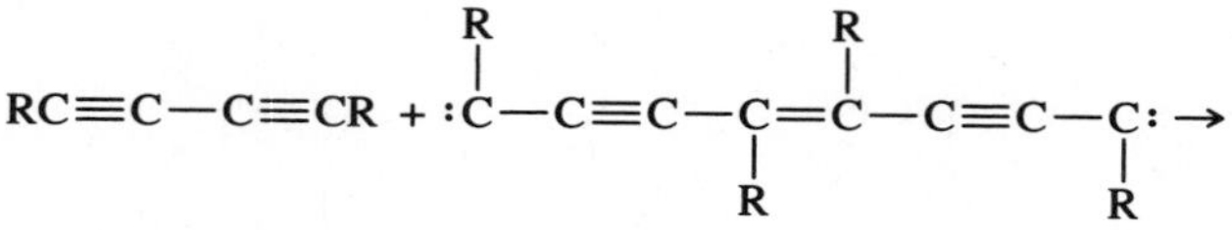

$$RC\equiv C-C\equiv CR + \ :\!\overset{\textstyle R}{\underset{\textstyle R}{C}}-C\equiv C-\overset{\textstyle R}{\underset{\textstyle R}{C}}{=}\overset{}{C}-C\equiv C-\overset{}{C}\!: \ \rightarrow$$

LXIX

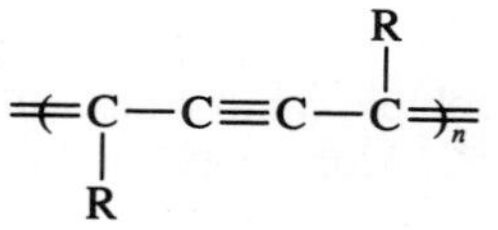

$$=\!\!\!\overset{}{C}-C\equiv C-\overset{\textstyle R}{\underset{\textstyle R}{C}}\!\!=\!\!\!\!\!\!\frac{}{}_{n} \tag{3-234}$$

LXX

that with the R groups being —CH$_2$OSO$_2\phi$CH$_3$; others include those with R = —(CH$_2$)$_4$OCONHCH$_2$COOC$_4$H$_9$ and —(CH$_2$)$_4$OCONHϕ. High-molecular-weight polymer is obtained only when the reaction proceeds as a lattice-controlled (topochemical) solid-state polymerization. This occurs only when the R groups have the appropriate bulkiness such that the diacetylene monomer and corresponding polymer have essentially the same crystal structure. The topochemical reaction transforms a monomer crystal to a polymer crystal with similar dimensions. Polymer single crystals of macroscopic dimensions, several centimeters in some cases, have been synthesized. The conjugated π-electron system of the polydiacetylenes imparts intense light absorption in the visible region as well as interesting (photo)conductivity and nonlinear optical properties.

REFERENCES

Abuin, E. and E. A. Lissi, *J. Macromol. Sci. Chem.*, **A11**, 287 (1977).

Abuin, E. B., E. A. Lissi, and A. Marquez, *J. Polym. Sci. Polym. Chem. Ed.*, **16**, 3003 (1978).

Adelman, R. L. and R. C. Ferguson, *J. Polym. Sci. Polym. Chem. Ed.*, **13**, 891 (1975).

Akashi, M., H. Takada, Y. Inaki, and K. Takemoto, *J. Polym. Sci. Polym. Chem. Ed.*, **17**, 747 (1979).

Allen, P. E. M. and C. R. Patrick, "Kinetics and Mechanisms of Polymerization Reactions," Chaps. 2, 3, 4, 7, Wiley, New York, 1974.

Amdur, S., A. T. Y. Cheng, C. J. Wong, P. Ehrlich, and R. D. Allendoefer, *J. Polym. Sci. Polym. Chem. Ed.*, **16**, 407 (1978).

Areizaga, J. F. and G. M. Guzman, *Makromol. Chem. Macromol. Symp.* **20/21**, 77 (1988).

Arnett, L. M. and J. H. Peterson, *J. Am. Chem. Soc.*, **74**, 2031 (1952).

Asano, T. and W. J. Le Noble, *Chem. Rev.*, **78**, 407 (1978).

Axelson, D. E., G. C. Levy, and L. Mandelkern, *Macromolecules*, **12**, 41 (1979).

Ayrey, G. and A. C. Haynes, *Makromol. Chem.*, **175**, 1463 (1974).

Ayrey, G., M. J. Humphrey, and R. C. Poller, *Polymer*, **18**, 840 (1977).

Backsai, R., *J. Polym. Sci. Polym. Chem. Ed.*, **14**, 1797 (1976).

Bajaj, P. and S. Kumari, *J. Macromol. Sci.-Rev. Macromol. Chem. Phys.*, **C27**, 181 (1987).

Baker, C., P. David, and W. F. Maddams, *Makromol. Chem.*, **180**, 975 (1979).

Balakrishnan, T. and S. Subbu, *J. Polym. Sci. Polym. Chem. Ed.*, **26**, 355 (1988).

Balke, S. T. and A. E. Hamielec, *J. Appl. Polym. Sci.*, **17**, 905 (1973).

Bamford, C. H., "Radical Polymerization," pp. 708–867 in "Encyclopedia of Polymer Science and Engineering," Vol. 13, H. F. Mark, N. M. Bikales, C. G. Overberger, and G. Menges, Eds., Wiley-Interscience, New York, 1988.

Bamford, C. H., R. W. Dyson, and G. C. Eastward, *Polymer*, **10**, 885 (1969a).

Bamford, C. H., G. C. Eastmond, and D. Whittle, *Polymer*, **10**, 771 (1969b).

Bamford, C. H. and M. Hirooka, *Polymer*, **25**, 1791 (1984).

Bando, Y. and Y. Minoura, *J. Polym. Sci. Polym. Chem. Ed.*, **14**, 1183, 1195 (1976).

Barb, W. G., J. H. Baxendale, P. George, and K. R. Hargrave, *Trans. Faraday Soc.*, **47**, 462, 591 (1951).

Barr, N. J., W. I. Bengough, G. Beveridge, and G. B. Park, *Eur. Polym. J.*, **14**, 245 (1978).

Barrett, K. E. J., Ed., "Dispersion Polymerization in Organic Media," Wiley, New York, 1975.

Bartlett, P. D. and H. Kwart, *J. Am. Chem. Soc.*, **72**, 1051 (1950).

Bartlett, P. D. and F. A. Tate, *J. Am. Chem. Soc.*, **75**, 91 (1953).

Baughman, R. H. and K. C. Yee, *J. Polym. Sci. Macromol. Rev.*, **13**, 219 (1978).

Baxter, J. E., R. S. Davidson, H. J. Hageman, and T. Overeem, *Makromol. Chem.*, **189**, 2769 (1988).

Baysal, B. and A. V. Tobolsky, *J. Polym. Sci.*, **8**, 529 (1952).

Berger, K. C., P. C. Deb, and G. Meyerhoff, *Macromolecules*, **10**, 1075 (1977).

Berner, G., R. Kirchmayer, and G. Rist, *J. Oil Col. Chem. Assoc.*, **61**, 105 (1978).

Bevington, J. C., *Trans. Faraday Soc.*, **51**, 1392 (1955); *Makromol. Chem. Macromol. Symp.*, **20/21**, 59 (1988).

Bevington, J. C., C. J. Dyball, B. J. Hunt, and J. Leech, *Polymer*, **19**, 991 (1978).

Bevington, J. C., G. M. Guzman, and H. W. Melville, *Proc. Roy. Soc.* (*London*), **A221**, 453, 547 (1947).

Billingham, N. C. and A. D. Jenkins, "Free Radical Polymerization in Heterogeneous Systems," Chap. 6 in "Comprehensive Chemical Kinetics," Vol. 14A, C. H. Bamford and C. F. H. Tipper, Eds., American Elsevier, New York, 1976.

Billingham, N. C., A. J. Chapman, and A. D. Jenkins, *J. Polym. Sci. Polym. Chem. Ed.*, **18**, 827 (1980).

Boenig, H. V., "Plasma Polymerization," pp. 248–261 in "Encyclopedia of Polymer Science and Engineering," Vol. 11, H. F. Mark, N. M. Bikales, C. G. Overberger, and G. Menges, Eds., Wiley-Interscience, New York, 1988.

Bonta, G., B. M. Gallo, S. Russo, and C. Uliana, *Polymer*, **17**, 217 (1976).

Bovey, F. A., "High Resolution NMR of Macromolecules," Chap. 8, Academic Press, New York, 1972.

Bovey, F. A., F. C. Schilling, T. K. Kwei, and H. L. Frisch, *Macromolecules*, **10**, 559 (1977).

Bovey, F. A., F. C. Schilling, F. L. McCracken, and H. L. Wagner, *Macromolecules*, **9**, 76 (1976).

Bowmer, T. N. and J. H. O'Donnell, *Polymer*, **18**, 1032 (1977).

Braks, J. G. and R. Y. M. Huang, *J. Appl. Polym. Sci.*, **22**, 3111 (1978).

Brand, E., M. Stickler, and G. Meyerhoff, *Makromol. Chem.*, **181**, 913 (1980).

Brandrup, J. and E. H. Immergut, Eds. (with W. McDowell), "Polymer Handbook," Wiley-Interscience, New York, 1989.

Braun, D. and W. K. Czerwinski, *Makromol. Chem.*, **188**, 2371 (1987).

Braun, D. and R. Studenroth, *Angew. Makromol. Chem.*, **79**, 79 (1979).

Bresler, S. E., E. N. Kosbekov, V. N. Fornichev, and V. N. Shadrin, *Makromol. Chem.*, **175**, 2875 (1974).

Briers, F., D. L. Chapman, and E. Walters, *J. Chem. Soc.*, 562 (1926).

Brydson, J. A., "Plastics Materials," 4th ed., Chaps. 10, 12–17, 21, Butterworths, London, 1982.

Buback, M., L. H. Garcia-Rubio, R. G. Gilbert, D. H. Napper, J. Guillot, A. E. Hamielec, D. Hill, K. F. O'Driscoll, O. F. Olaj, J. Shen, D. Solomon, G. Moad, M. Stickler, M. Tirrell, and M. A. Winnik, *J. Polym. Sci. Polym. Chem. Ed.*, **26**, 293 (1988).

Buback, M., B. Huckestein, and U. Leinhos, *Makromol. Chem. Rapid. Commun.*, **8**, 473 (1987).

Buback, M. and H. Lendle, *Makromol. Chem.*, **184**, 193 (1983).

Burillo, G., A. Chapiro, and Z. Makowski, *J. Polym. Sci. Polym. Chem. Ed.*, **18**, 327 (1980).

Cais, R. E. and J. M. Kometani, *Macromolecules*, **17**, 1932 (1984); *Polymer*, **29**, 168 (1988).

Cardenas, J. N. and K. F. O'Driscoll, *J. Polym. Sci. Polym. Chem. Ed.*, **14**, 883 (1976); **15**, 1883, 2097 (1977).

Carlblom, L. H. and S. P. Pappas, *J. Polym. Sci. Polym. Chem. Ed.*, **15**, 1381 (1977).

Carraher, C. E., Jr., *J. Chem. Ed.*, **58**, 921 (1981).

Carraher, C. E., Jr. and J. E. Sheats, Eds., "Organometallic Polymers," Academic Press, New York, 1978.

Catalgil, H. H. and B. M. Baysal, *Makromol. Chem.*, **188**, 495 (1987).

Chance, R. R., "Diacetylene Polymers," pp. 767–779 in "Encyclopedia of Polymer Science and Engineering," Vol. 4, H. F. Mark, N. M. Bikales, C. G. Overberger, and G. Menges, Eds., Wiley-Interscience, New York, 1986.

Chapiro, A., "Radiation Chemistry of Polymer Systems," Chap. IV, Wiley-Interscience, New York, 1962.

Chapiro, A. and J. Dulieu, *Eur. Polym. J.*, **13**, 563 (1977).

Chatani, Y., "Structural Features of Solid-State Polymerization," in "Progress in Polymer Science Japan," Vol. 7, K. Imahori and T. Higashimura, Eds., Halsted Press (Wiley), New York, 1974, p. 149.

Chetia, P. D. and N. N. Dass, *Eur. Polym. J.*, **12**, 165 (1976).

Clarke, J. T., R. O. Howard, and W. H. Stockmayer, *Makromol. Chem.* **44–46**, 427 (1961).

Collins, E. A., J. Bares, and F. W. Billlmeyer, Jr., "Experiments in Polymer Science," Chap. 5, Wiley-Interscience, New York, 1973.

Cook, R. E., F. S. Dainton, and K. J. Ivin, *J. Polym. Sci.*, **29**, 549 (1958).

Culbertson, B. M., "Maleic and Fumaric Polymers," pp. 225–294 in "Encyclopedia of Polymer Science and Engineering," Vol. 9, H. F. Mark, N. M. Bikales, C. G. Overberger, and G. Menges, Eds., Wiley-Interscience, New York, 1987.

Dainton, F. S. and K. J. Ivin, *Trans. Faraday Soc.*, **46**, 331 (1950); *Quart. Rev. (London).*, **12**, 61 (1958).

Dainton, F. S. and W. D. Sisley, *Trans. Faraday Soc.*, **59**, 1369 (1963).

Davis, T. P., D. F. O'Driscoll, M. C. Piton, and M. A. Winnik, *Macromolecules*, **22**, 2785 (1989).

Day, D. R. and H. Ringsdorf, *Makromol. Chem.*, **180**, 1059 (1979).

Deb, P. C. and S. K. Kapoor, *Eur. Polym. J.*, **15**, 477, 961 (1979).

Deb, P. C. and S. Ray, *Eur. Polym. J.*, **14**, 607 (1978).

Dionisio, J. M., H. K. Mahabadi, J. F. O'Driscoll, E. Abuin, and E. A. Lissi, *J. Polym. Sci. Polym. Chem. Ed.*, **17**, 1891 (1979).

Dionisio, J. M. and K. F. O'Driscoll, *J. Polym. Sci. Polym. Chem. Ed.*, **18**, 241 (1980).

Doak, K. W., "Low Density Polyethylene (High Pressure)," pp. 386–429 in "Encyclopedia of Polymer Science and Engineering," Vol. 6, 2nd ed., H. F. Mark, N. M. Bikales, C. G. Overberger, and G. Menges, Eds., Wiley-Interscience, New York, 1986.

Eastmond, G. C., "The Kinetics of Free Radical Polymerization of Vinyl Monomers in Homogeneous Solutions," Chap. 1 in "Comprehensive Chemical Kinetics," Vol. 14A, C. H. Bamford and C. F. H. Tipper, Eds., American Elsevier, New York, 1976a.

Eastmond, G. C., "Kinetic Data for Homogeneous Free Radical Polymerizations of Various Monomers," Chap. 3 in "Comprehensive Chemical Kinetics," Vol. 14A, C. H. Bamford and C. F. H. Tipper, Eds., American Elsevier, New York, 1976b.

Eastmond, G. C., "Chain Transfer, Inhibition and Retardation," Chap. 2 in "Comprehensive Chemical Kinetics," Vol. 14A, C. H. Bamford and C. F. H. Tipper, Eds., American Elsevier, New York, 1976c.

Eckhardt, H., T. Prusik, and R. R. Chance, *Macromolecules*, **16**, 732 (1983).

Encinas, M. V., J. Garrido, and E. A. Lissi, *J. Polym. Sci. Polym. Chem. Ed.*, **27**, 139 (1989a).

Encinas, M. V., C. Majmud, J. Garrido, and E. A. Lissi, *Macromolecules*, **22**, 563 (1989b).

Errede, L. A. and M. Szwarc, *Quart. Rev.*, **12**, 301 (1958).

Evans, C. A., *Aldrichim. Acta*, **12(2)**, 23 (1979).

Fanood, M. H. R., and M. H. George, *Polymer*, **29**, 134 (1988).

Farina, M. and G. Di Silvestro, "Polymerization in Clathrates," pp. 486–504 in "Encyclopedia of Polymer Science and Engineering," Vol. 12, H. F. Mark, N. M. Bikales, C. G. Overberger, and G. Menges, Eds., Wiley-Interscience, New York, 1987.

Fernandez, M. D. and G. M. Guzman, *J. Polym. Sci. Polym. Chem. Ed.*, **27**, 2427 (1989).

Field, N. D. and J. R. Schaefgen, *J. Polym. Sci.*, **58**, 533 (1962).

Fink, J. K., *J. Polym. Sci. Polym. Chem. Ed.*, **21**, 1445 (1983).

Finkelmann, H., *Angew. Chem. Int. Ed. Engl.*, **26**, 816 (1987).

Fischer, J. P., G. Mucke, and G. V. Schulz, *Ber. Bunsenges. Phys. Chem.*, **73**, 154 (1969).

Flory, P. J., *J. Am. Chem. Soc.*, **69**, 2893 (1947).

Flory, P. J., "Principles of Polymer Chemistry," Chap. IV, Cornell University Press, Ithaca, N.Y., 1953.

Friis, N. and A. E. Hamielec, *ACS Symp. Ser.*, **24**, 82 (1976).

Gangal, S. V., "Tetrafluoroethylene Polymers," pp. 577–649 in "Encyclopedia of Polymer Science and Engineering," Vol. 16, H. F. Mark, N. M. Bikales, C. G. Overberger, and G. Menges, Eds., Wiley-Interscience, New York, 1989.

Gaylord, N. G., *J. Macromol. Sci. Rev. Macromol. Chem.*, **C13**, 235 (1975).

Gaylord, N. G. and R. Mehta, *J. Polym. Sci. Polym. Chem. Ed.*, **26**, 1903 (1988).

George, M. H., "Styrene," Chap. 3 in "Vinyl Polymerization," Vol. 1, Part I, G. E. Ham, Ed., Marcel Dekker, New York, 1967.

George, M. H. and A. Ghosh, *J. Polym. Sci. Polym. Chem. Ed.*, **16**, 981 (1978).

Georgieff, K. K., *J. Appl. Polym. Sci.*, **9**, 2009 (1965).

Ghosh, P. and H. Banerjee, *J. Polym. Sci. Polym. Chem. Ed.*, **16**, 633 (1978).

Ghosh, P. and S. N. Maity, *Eur. Polym. J.*, **14**, 855 (1978).

Ghosh, P. and G. Mukhopadhyay, *Makromol. Chem.*, **180**, 2253 (1979); *J. Polym. Sci. Polym. Chem. Ed.*, **18**, 283 (1980).

Gobran, R. H., M. B. Berenbaum, and A. V. Tobolsky, *J. Polym. Sci.*, **46**, 431 (1960).

Gons, J., L. J. P. Straatnam, and G. Challa, *J. Polym. Sci. Polym. Chem. Ed.*, **16**, 427 (1978).

Graham, W. D., J. G. Green, and W. A. Pryor, *J. Org. Chem.*, **44**, 907 (1979).

Grasshoff, J. M., J. L. Reid, and L. D. Taylor, *J. Polym. Sci. Polym. Chem. Ed.*, **16**, 2401 (1978).

Gregg, R. A. and F. R. Mayo, *J. Am. Chem. Soc.*, **70**, 2373 (1948); *Disc. Faraday Soc.*, **2**, 6133 (1953a); *J. Am. Chem. Soc.*, **75**, 3530 (1953b).

Groenenboom, C. J., H. J. Hageman, T. Overeem, and A. J. M. Weber, *Makromol. Chem.*, **183**, 281 (1982).

Gromov, V. F., N. I. Galperina, T. O. Osmanov, P. M. Khomikovskii, and A. D. Abkin, *Eur. Polym. J.*, **16**, 529 (1980a).

Gromov, V. F., T. Osmanov, P. M. Khomikovskii, and A. D. Abkin, *Eur. Polym. J.*, **16**, 803 (1980b).

Grulke, E. A., "Suspension Polymerization," pp. 443–473 in "Encyclopedia of Polymer Science and Engineering," Vol. 16, H. F. Mark, N. M. Bikales, C. G. Overberger, and G. Menges, Eds., Wiley-Interscience, New York, 1989.

Guyot, A., *Makromol. Chem. Macromol. Symp.*, **10/11**, 461 (1987).

Haeringer, A. and G. Reiss, *Eur. Polym. J.*, **14**, 117 (1978).

Hall, H. K., Jr., *Angew. Chem. Int. Ed. Engl.*, **22**, 440 (1983).

Ham, G. E., "General Aspects of Free Radical Polymerization," Chap. 1 in "Vinyl Polymerization," Vol. 1, Part I, G. E. Ham, Ed., Marcel Dekker, New York, 1967.

Hay, J. N., P. J. Mills, and R. Ognjanovic, *Polymer*, **27**, 677 (1986).

Hayashi, K. and T. Ostu, *Makromol. Chem.*, **127**, 54 (1969).

Hayden, P. and H. W. Melville, *J. Polym. Sci.*, **43**, 201 (1960).

Heitz, W., *Makromol. Chem. Macromol. Symp.*, **10/11**, 297 (1987).

High, K. A., H. B. Lee, and D. T. Turner, *Macromolecules*, **12**, 332 (1979).

Hill, A. and K. W. Doak "Mechanism of Free Radical Polymerization of Ethylene," Chap. 7 in "Crystalline Olefin Polymers," Part I, R. A. V. Raff and K. W. Doak, Eds., Wiley-Interscience, New York, 1965.

Hjertberg, T. and E. M. Sorvik, *Polymer*, **24**, 673, 685 (1983).

Holden, D. A., "Photochemistry," pp. 126–154 in "Encyclopedia of Polymer Science and Engineering," Vol. 11, H. F. Mark, N. M. Bikales, C. G. Overberger, and G. Menges, Eds., Wiley-Interscience, New York, 1988.

Horikx, M. W. and J. J. Hermans, *J. Polym. Sci.*, **11**, 325 (1953).

Hsu, E. C. and A. Blumstein, *J. Polym. Sci. Polym. Chem. Ed.*, **15**, 129 (1977).

Hui, A. W. and A. E. Hamielec, *J. Appl. Polym. Sci.*, **16**, 749 (1972).

Humphrey, B. J., *Studies Conservation*, **29**, 117 (1984).

Hunkeler, D., A. E. Hamielec, and W. Baade, *Polymer*, **30**, 127 (1989).

Husain, A. and A. E. Hamielec, *J. Appl. Polym. Sci.*, **22**, 1207 (1978).

Huyser, E. S., "Free Radical Chain Reactions," Chap. 10 and pp. 314–330, Wiley, New York, 1970.

Ingold, K. U., "Rate Constants for Free Radical Reactions in Solution," Chap. 2 in "Free Radicals," Vol. I, J. K. Kochi, Ed., Wiley, New York, 1973.

Inoue, H., M. Helbid, and O. Vogl, *Macromolecules*, 10, 1331 (1977).

Ito, K., *J. Polym. Sci. Polym. Chem. Ed.*, **18**, 701 (1980); *Macromolecules*, **13**, 193 (1980).

Jacovic, M. S., D. Pollock, and R. S. Porter, *J. Appl. Polym. Sci.*, **23**, 517 (1979).

Janzen, E. G., *Acct. Chem. Res.*, **4**, 31 (1971).

Jenkins, A. D., "Occlusion Phenomenon in the Polymerization of Acrylonitrile and Other Monomers," Chap. 6 in "Vinyl Polymerization," Vol. 1, Part I, G. E. Ham, Ed., Marcel Dekker, New York, 1967.

Joshi, R. M. and B. J. Zwolinski, "Heats of Polymerization and Their Structural and Mechanistic Implications," in "Vinyl Polymerization," Vol. 1, Part I, Chap. 8, G. E. Ham, Ed., Marcel Dekker, New York, 1967.

Kabanov, V. A., *Makromol. Chem. Macromol. Symp.*, **10/11**, 193 (1987).

Kaliyamurthy, K., P. Elayaperumal, T. Balakrishnan, and M. Santappa, *Makromol. Chem.*, **180**, 1575 (1979).

Kamachi, M., M. Kohno, Y. Kuwae, and S. Nozakura, *Polymer J.*, 17, 541 (1985).

Kamachi, M., J. Satoh, and D. J. Liaw, *Polym. Bull.*, **1**, 581 (1979).

Kamachi, M., J. Satoh, and S.-I. Nozakura, *J. Polym. Sci. Polym. Chem. Ed.*, **16**, 1789 (1978).

Kaufmann, H. F., *Makromol. Chem.*, **180**, 2649, 2665, 2681 (1979).

Kawaguchi, H., P. Loeffler, and O. Vogl, *Polymer*, **26**, 1257 (1985).

Kellou, M. S. and G. Jenner, *Eur. Polym. J.*, **12**, 883 (1976).

Kerr, J. A., "Rate Processes in the Gas Phase," Chap. 1 in "Free Radicals," Vol. I, J. K. Kochi, Ed., Wiley, New York, 1973.

Kine, B. B. and R. W. Novak, "Acrylic and Methacrylic Ester Polymers," pp. 234–299 in "Encyclopedia of Polymer Science and Engineering," Vol. 1, H. F. Mark, N. M. Bikales, C. G. Overberger, and G. Menges, Eds., Wiley-Interscience, New York, 1985.

Kishore, K. and V. A. Bhanu, *J. Polym. Sci. Polym. Chem. Ed.*, **26**, 2832 (1988).

Kishore, K., V. Gayathri, and K. Ravindran, *J. Macromol. Sci. Chem.*, **16**, 1359 (1981).

Klein, J. and D. Herzog, *Makromol. Chem.*, **188**, 1217 (1987).

Koenig, T., "The Decomposition of Peroxides and Azoalkanes," Chap. 3 in "Free Radicals," Vol. I, J. K. Kochi, Ed., Wiley, New York, 1973.

Koenig, T. and H. Fischer, "Cage Effects," Chap. 4 in "Free Radicals." Vol. I, J. K. Kochi, Ed., Wiley, New York, 1973.

Kondratiev, V. N., "Chain Reactions," Chap. 2 in "Comprehensive Chemical Kinetics," Vol. 2, C. H. Bamford and C. F. H. Tipper, Eds., American Elsevier, New York, 1969.

Kubota, H., V. Ya. Kabanov, D. R. Squire, and V. Stannett, *J. Macromol. Sci. Chem.*, **A12**, 1299 (1978).

Kulkarni, M. G., R. A. Mashelkar, and L. S. Doraiswamy, *J. Polym. Sci. Polym. Chem. Ed.*, **17**, 713 (1979).

Kurland, J. J., *J. Polym. Sci. Polym. Chem. Ed.*, **18**, 1139 (1980).

Laborie, F., *J. Polym. Sci. Polym. Chem. Ed.*, **15**, 1255, 1275 (1977).

Lachinov, M. P., R. A. Simonian, T. G. Georgieva, V. P. Zubov, and V. A. Kabonov, *J. Polym. Sci. Polym. Chem. Ed.*, **17**, 613 (1979).

Lachinov, M. B., V. P. Zubov, and V. A. Kabanov, *J. Polym. Sci. Polym. Chem. Ed.*, **15**, 1777 (1977).

Ledwith, A., *J. Oil Col. Chem. Assoc.*, **59**, 157 (1976); *Pure Appl. Chem.*, **49**, 431 (1977).

Lee, C. J., *J. Macromol. Sci. Rev. Macromol. Chem.*, **C16**, 79 (1977–1978).

Lee, C. J. and B. Wunderlich, *J. Appl. Polym. Sci. Appl. Polym. Symp.*, **26**, 291 (1975); *Makromol. Chem.*, **179**, 561 (1978).

Lee, C. L. and O. K. Johannson, *J. Polym. Sci.*, **A-1(4)**, 3013 (1966); *J. Polym. Sci. Polym. Chem. Ed.*, **14**, 729 (1976).

Lehrle, R. S. and A. Shortland, *Eur. Polym. J.*, **24**, 425 (1988).

Lenka, S. and P. L. Nayak, *J. Polym. Sci. Polym. Chem. Ed.*, **25**, 1563 (1987).

Levy, L. B., *J. Polym. Sci. Polym. Chem. Ed.*, **23**, 1505 (1985).

Liaw, D.-J. and K-C. Chung, *Makromol. Chem.*, **184**, 29 (1983).

Lindeman, M. K., "The Mechanism of Vinyl Acetate Polymerization," Chap. 4 in "Vinyl Polymerization," Vol. 1, Part I, G. E. Ham, Ed., Marcel Dekker, New York, 1967.

Lingnau, J. and G. Meyerhoff, *Macromolecules*, **17**, 941 (1984a); *Makromol. Chem.*, **185**, 587 (1984b).

Lingnau, J., M. Stickler, and G. Meyerhoff, *Eur. Polym. J.*, **16**, 785 (1980).

Lipscomb, N. T. and Y. Tarshiani, *J. Polym. Sci. Polym. Chem. Ed.*, **26**, 529 (1988).

Lissi, E. A., M. V. Encina, and M. T. Abarca, *J. Polym. Sci. Polym. Chem. Ed.*, **17**, 19 (1979).

Litt, M. H. and K. H. S. Chang, *ACS Symp. Ser.*, **165**, 455 (1981).

Litt, M. and F. R. Eirich, *J. Polym. Sci.*, **45**, 379 (1960).

Llauro-Darricades, M-F., N. Bensemra, A. Guyot, and R. Petiaud, *Makromol. Chem. Macromol. Symp.*, **29**, 171 (1989).

Lougnot, D. J., C. Turck, and J. P. Fouassier, *Macromolecules*, **22**, 108 (1989).

Ludwico, W. A. and S. L. Rosen, *J. Appl. Polym. Sci.*, **19**, 757 (1975); *J. Polym. Sci. Polym. Chem. Ed.*, **14**, 2121 (1976).

Luft, G., P. Mehrling, and H. Seidl, *Angew. Makromol. Chem.*, **73**, 95 (1978).

Machi, S., J. Silverman, and D. J. Metz, *J. Phys. Chem.*, **76**, 930 (1972).

Macosko, C. W. and D. R. Miller, *Macromolecules*, **9**, 199 (1976).

Madruga, E. L. and J. Sanroman, *J. Polym. Sci. Polym. Chem. Ed.*, **19**, 1101 (1981).

Maekawa, T., S. Mah, and S. Okamura, *J. Macromol. Sci. Chem.*, **A12**, 1 (1978).

Mahabadi, H. K., *Makromol. Chem. Macromol. Symp.*, **10/11**, 127 (1987).

Mahabadi, H. K. and G. Meyerhoff, *Eur. Polym. J.*, **15**, 607 (1979).

Mahabadi, H. K. and K. F. O'Driscoll, *J. Polym. Sci. Polym. Chem. Ed.*, **15**, 283 (1977a); *Macromolecules*, **10**, 55 (1977b).

Mahabadi, H. K. and A. Rudin, *J. Polym. Sci.*, **17**, 1801 (1979).

Manickam, S. P., K. Venkatarao, U. C. Singh, and N. R. Subbaratnam, *J. Polym. Sci. Polym. Chem. Ed.*, **16**, 2701 (1978).

Mapunda-Vlckova, J. and J. Barton, *Makromol. Chem.*, **179**, 113 (1978).

Martin, J. C., "Solvation and Association," Chap. 20 in "Free Radicals," Vol. II, J. K. Kochi, Ed., Wiley, New York, 1973.

Masuda, S., M. Tanaka, and T. Ota, *Polymer*, **28**, 1945 (1987); *J. Polym. Sci. Polym. Chem. Ed.*, **27**, 855 (1989).

Matheson, M. S., E. E. Auer, E. B. Bevilacqua, and E. J. Hart, *J. Am. Chem. Soc.*, **71**, 497 (1949).

Matsumoto, M., K. Takakura, and T. Okaya, "Radical Polymerizations in Solution," Chap. 7 in "Polymerization Processes," C. E. Schildknecht, Ed. (with I. Skeist), Wiley-Interscience, New York, 1977.

Matsuo, K., G. W. Nelb, R. G. Nelb, and W. H. Stockmayer, *Macromolecules*, **10**, 654 (1977).

Matuszewski-Czerwik, J. and S. Polowinski, *Eur. Polym. J.*, **26**, 549 (1990).

Maybod, H. and M. H. George, *J. Polym. Sci. Polym. Lett. Ed.*, **15**, 693 (1977).

Mayo, F. R., R. A. Gregg, and M. S. Matheson, *J. Am. Chem. Soc.*, **73**, 1691 (1951).

McCaffery, E. M., "Laboratory Preparations for Macromolecular Chemistry," Experiments 4, 16, 17, and 27, McGraw-Hill, New York, 1970.

McCormick, H., *J. Polym. Sci.*, **25**, 488 (1957).

Menikheim, V., "Polymerization Procedures, Industrial," pp. 504–541 in "Encyclopedia of Polymer Science and Engineering," Vol. 12, H. F. Mark, N. M. Bikales, C. G. Overberger, and G. Menges, Eds., Wiley-Interscience, New York, 1988.

Meslard, J.-C., L. Yean, F. Subira, and J.-P. Vairon, *Makromol. Chem.*, **187**, 787 (1986).

Misra, G. S. and U. D. N. Bajpai, *Prog. Polym. Sci.*, **8**, 61 (1982).

Misra, G. S. and V. R. B. Mathiu, *Makromol. Chem.*, **100**, 5 (1967).

Misra, G. S., P. S. Bassi, and S. L. Abrol, *J. Polym. Sci. Polym. Chem. Ed.*, **22**, 1883 (1984).

Mita, I. and K. Horie, *J. Macromol. Sci.-Rev. Macromol. Chem. Phys.*, **C27**, 91 (1987).

Miyata, M., T. Tsuzuki, F. Noma, K. Takemot, and M. Kamachi, *Makromol. Chem. Rapid Commun.*, **9**, 45 (1988).

Moad, G., D. H. Solomon, S. R. Johns, and R. I. Willing, *Macromolecules*, **15**, 1188 (1982); **17**, 1094 (1984).

Moore, E. R., Ed., "Styrene Polymers," pp. 1–246 in "Encyclopedia of Polymer Science and Engineering," Vol. 16, 2nd ed., H. F. Mark, N. M. Bikales, C. G. Overberger, and G. Menges, Eds., Wiley-Interscience, New York, 1989.

Moore, J. A., Ed., "Macromolecular Syntheses," Vols. 1–9, Wiley-Interscience, New York, 1977–1985.

Moore, P. W., F. W. Ayscough, and J. G. Clouston, *J. Polym. Sci. Polym. Chem. Ed.*, **15**, 1291 (1977).

Morishima, Y. and S.-I. Nozakura, *J. Polym. Sci. Polym. Chem. Ed.*, **14**, 1277 (1976).

Moritani, T. and H. Iwasaki, *Macromolecules*, **11**, 1251 (1978).

Morsi, S. E., A. B. Zaki, and M. A. El-Khyami, *Eur. Polym. J.*, **13**, 851 (1977).

Munzer, M. and E. Trommsdorff, "Polymerizations in Suspension," Chap. 5 in "Polymerization Processes," C. E. Schildknect, Ed. (with I. Skeist), Wiley-Interscience, New York, 1977.

Nagy, A., D. Szalay, T. Foldes-Berezsnich and F. Tudos, *Eur. Polym. J.*, **19**, 1047 (1983).

Nakai, S., T. Nakaya, and M. Imoto, *Makromol. Chem.*, **178**, 2963 (1977).

Nayak, P. L. and S. Lenka, *J. Macromol. Sci. Rev. Macromol. Chem.*, **C19**, 83 (1980).

Nemec, J. W. and W. Bauer, Jr., "Acrylic and Methacrylic Polymers," pp. 211–234 in "Encyclopedia of Polymer Science and Engineering," Vol. 1, H. F. Mark, N. M. Bikales, C. G. Overberger, and G. Menges, Eds., Wiley-Interscience, New York, 1985.

Neumann, R. C., *Acct. Chem. Res.*, **5**, 381 (1972).

Nicolson, A. E. and R. G. W. Norrish, *Disc. Faraday Soc.*, **22**, 104 (1956).

Nigenda, S. E., D. Caballero, and T. Ogawa, *Makromol. Chem.*, **178**, 2989 (1977).

Norrish, R. G. W. and B. A. Thrush, *Quart. Rev. (London)*, **10**, 149 (1956).

North, A. M., "The Influence of Chain Structure on the Free Radical Termination Reaction," Chap. 5 in "Reactivity, Mechanism and Structure in Polymer Chemistry," A. D. Jenkins and A. Ledwith, Eds., Wiley-Interscience, New York, 1974.

Noyes, R. M., "Effects of Diffusion Rates on Chemical Kinetics," Chap. 5 in "Progress in Reaction Kinetics," Vol. 1, G. Porter, Ed., Pergamon, New York, 1961.

Nozaki, K. and P. D. Bartlett, *J. Am. Chem. Soc.*, **68**, 1686 (1946).

Nozakura, S.-K., Y. Morishima, H. Iimura, and Y. Irie, *J. Polym. Sci. Polym. Chem. Ed.*, **14**, 759 (1976).

Nozakura, S.-K., Y. Morishima, and S. Murahashi, *J. Polym. Sci. Polym. Chem. Ed.*, **10**, 2781, 2853 (1972).

Oda, T., T. Maeshima, and K. Sugiyama, *Makromol. Chem.*, **179**, 2331 (1978).

Odian, G., A. Derman, and K. Imre, *J. Polym. Sci. Polym. Chem. Ed.*, **18**, 737 (1980).

O'Driscoll, K. F., *Makromol. Chem.*, **178**, 899 (1977); **180**, 2053 (1979).

O'Driscoll, K. F., P. F. Lyons, and R. Patsiga, *J. Polym. Sci.*, **A3**, 1567 (1965).

O'Driscoll, K. F. and H. K. Mahabadi, *J. Polym. Sci. Polym. Chem. Ed.*, **14**, 869 (1976).

Ogo, Y., *J. Macromol. Sci. Rev. Macromol. Chem. Phys.*, **C24**, 1 (1984).

Ogo, Y., M. Yoshikawa, S. Ohtani, and T. Imoto, *J. Polym. Sci. Polym. Chem. Ed.*, **16**, 1413 (1978).

Ohno, H., S. Takeoka, H. Iwai, and E. Tsuchida, *Macromolecules*, **22**, 61 (1989).

Okahata, Y., H. Noguchi, and T. Seki, *Macromolecules*, **20**, 15 (1987).

Olaj, O. F., *Makromol. Chem. Macromol. Symp.*, **8**, 235 (1987); **10/11**, 165 (1988).

Olaj, O. F. and I. Bitai, *Angew. Makromol. Chem.*, **155**, 177 (1987).

Olaj, O. F., I. Bitai, and F. Hinkelmann, *Makromol. Chem.*, **188**, 1689 (1987).

Olaj, O. F., J. W. Breitenbach, K. J. Parth, and N. Philippovich, *J. Macromol. Sci. Chem.*, **A11**, 1319 (1977a).

Olaj, O. F., H. F. Kaufmann, and J. W. Breitenbach, *Makromol. Chem.*, **177**, 3065 (1976), **178**, 2707 (1977b).

Olaj, O. F. and G. Ziffer, *Makromol. Chem. Macromol. Symp.*, **10/11**, 165 (1987).

Olive, G. H. and S. H. Olive, *Polym. Bull.*, **1**, 47 (1978).

Oster, G. and N.-L. Yang, *Chem. Rev.*, **68**, 125 (1968).

Otero, T. F. and M. A. Mugarz, *Makromol. Chem.*, **188**, 2885 (1987).

Otsu, T., T. Yasuhara, and A. Matsumoto, *J. Macromol. Sci. Chem.*, **A25**, 537 (1988).

Ouchi, T., H. Taguchi, and M. Imoto, *J. Macromol. Sci. Chem.*, **A12**, 719 (1978).

Pappas, S. P., "Photopolymerization," pp. 186–212 in "Encyclopedia of Polymer Science and Engineering," Vol. 11, H. F. Mark, N. M. Bikales, C. G. Overberger, and G. Menges, Eds., Wiley-Interscience, New York, 1988.

Patel, G. N., R. R. Chance, E. A. Turi, and Y. P. Khanna, *J. Am. Chem. Soc.*, **100**, 6644 (1978).

Patnaik, L. N., M. K. Rout, S. P. Rout, and A. Rout, *Eur. Polym. J.*, **15**, 509 (1979).

Peebles, L. H., Jr., "Molecular Weight Distributions in Polymers," Chap. 2, Wiley-Interscience, New York, 1971.

Pelgrims, J., *J. Oil Col. Chem. Assoc.*, **61**, 114 (1978).

Peng, F. M., "Acrylonitrile Polymers," pp. 426–470 in "Encyclopedia of Polymer Science and Engineering," Vol. 1, H. F. Mark, N. M. Bikales, C. G. Overberger, and G. Menges, Eds., Wiley-Interscience, New York, 1985.

Peramunage, D., J. E. Fernandez, and L. H. Garcia-Rubio, *Macromolecules*, **22**, 2845 (1989).

Percec, V. and D. Tomazos, *Macromolecules*, **22**, 2062 (1989).

Pittman, C. U., Jr., C. E. Carraher, Jr., and J. R. Reynolds, "Organometallic Polymers," pp. 541–594 in "Encyclopedia of Polymer Science and Engineering," Vol. 10, H. F. Mark, N. M. Bikales, C. G. Overberger, and G. Menges, Eds., Wiley-Interscience, New York, 1987.

Ponratnam, S. and S. L. Kapur, *Makromol. Chem.*, **178**, 1029 (1977).

Prabha, R. and U. S. Nandi, *J. Polym. Sci. Polym. Chem. Ed.*, **15**, 1973 (1977).

Pramanick, D. and A. K. Chatterjee, *J. Polym. Sci. Polym. Chem. Ed.*, **18**, 311 (1980).

Ranby, B. and J. F. Rabek, "Photodegradation, Photo-oxidation and Photostabilization of Polymers," Chaps. 1 and 12, Wiley-Interscience, New York, 1975.

Randall, J. C., *J. Appl. Polym. Sci.*, **22**, 585 (1978).

Reimschuessel, H. K. and W. S. Creasy, *J. Polym. Sci. Polym. Chem. Ed.*, **16**, 845 (1978).

Rizzardo, E. and D. H. Solomon, *Polym. Bull.*, **1**, 529 (1979).

Roche, A. and C. C. Price in "Styrene: Its Polymers, Copolymers, and Derivatives," R. H. Boundy, R. F. Boyer, and S. M. Stoesser, Eds., unpublished data in Table 7-1, p. 216, Van Nostrand Reinhold, New York, 1952.

Rodriguez, F., "Principles of Polymer Systems," Chap. 13, McGraw-Hill, New York, 1982.

Russell, G. T., D. H. Napper, and R. G. Gilbert, *Macromolecules*, **21**, 2133, 2141 (1988).

Sack, R., G. V. Schulz, and G. Meyerhoff, *Macromolecules*, **21**, 3345 (1988).

Samal, S. K. and B. Nayak, *J. Polym. Sci. Polym. Chem. Ed.*, **26**, 1035 (1988).

Sandler, S. R. and W. Karo, "Polymer Syntheses," Academic Press, New York, Vols. 1 and 2 (1974), Vol. 3 (1980).

Santee, G. F., R. H. Marchessault, H. G. Clark, J. J. Kearny, and V. Stannett, *Makromol. Chem.*, **73**, 177 (1964).

Sasuga, T. and M. Takehisa, *J. Macromol. Sci. Chem.*, **A12**, 1307, 1343 (1978).

Sato, T., M. Abe, and T. Otsu, *Makromol. Chem.*, **178**, 1267, 1951 (1977); **180**, 1165 (1979a).

Sato, T., S. Kita and T. Otsu, *Makromol. Chem.*, **176**, 561 (1975); **180**, 1911 (1979b).

Sato, T., T. Iwaki, S. Mori, T. Otsu, *J. Poly. Sci. Polym. Chem. Ed.*, **21**, 819 (1983).

Sato, T., J. Miyamoto, and T. Otsu, *J. Polym. Sci. Polym. Chem. Ed.*, **22**, 3921 (1984).

Saunders, K. J., "Organic Polymer Chemistry," Chaps. 2–7, 18, 19, Chapman and Hall, London, 1973.

Sawada, H., "Thermodynamics of Polymerization," Chaps. 1, 2, 5, Marcel Dekker, New York, 1976.

Schildknecht, C. E. "Cast Polymerizations," Chap. 2 and "Other Bulk Polymerizations," Chap. 4 in "Polymerization Processes," C. E. Schildknecht, Ed. (with I. Skeist), Wiley-Interscience, New York, 1977.

Schnoll-Bitai, I. and O. F. Olaj, *Makromol. Chem.*, **191**, 2491 (1990).

Schulz, G. V., *Chem. Ber.*, **80**, 232 (1947).

Schulz, G. V. and F. Blaschke, *Z. Physik Chem.* (*Leipzig*), **B51**, 75 (1942).

Schulz, G. V. and G. Haborth, *Makromol. Chem.*, **1**, 106 (1948).

Sebastian, D. H. and J. A. Biesenberger, *Polym. Eng. Sci.*, **19**, 190 (1979).

Seno, M., M. Ishii, M. Ibonai, T. Kuramochi, and M. Miyashita, *J. Polym. Sci. Polym. Chem. Ed.*, **14**, 1292 (1976).

Senogles, E. and R. A. Thomas, *J. Polym. Sci. Polym. Lett. Ed.*, **16**, 555 (1978).

Seymour, R. B. and G. A. Stahl, *J. Macromol. Sci. Chem.*, **A11**, 53 (1977).

Shahani, C. J. and N. Indictor, *J. Polym. Sci. Polym. Chem. Ed.*, **16**, 2683, 2997 (1978).

Shawki, S. M. and A. E. Hamielec, *J. Appl. Polym. Sci.*, **23**, 334 (1979).

Shen, J., Y. Tian, Y. Zeng, and Z. Qiu, *Makromol. Chem. Rapid. Commun.*, **8**, 615 (1987).

Sheppard, C. S., "Azo Compounds," pp. 143–157 in "Encyclopedia of Polymer Science and Engineering," Vol. 2, H. F. Mark, N. M. Bikales, C. G. Overberger, and G. Menges, Eds.., Wiley-Interscience, New York, 1985.

Sheppard, C. S., "Peroxy Compounds," pp. 1–21 in "Encyclopedia of Polymer Science and Engineering," Vol. 11, H. F. Mark, N. M. Bikales, C. G. Overberger, and G. Menges, Eds.., Wiley-Interscience, New York, 1988.

Small, P. A., *Adv. Polym. Sci.*, **18**, 1 (1975).

Smith, W. B., J. A. May, and C. W. Kim, *J. Polym. Sci.*, **A2**, 365 (1966).

Solomon, D. H. and G. Moad, *Makromol. Chem. Macromol. Symp.*, **10/11**, 109 (1987).

Squire, D. R., J. A. Cleaveland, T. M. A. Hossain, W. Oraby, E. P. Stahel, and V. T. Stannett, *J. Appl. Polym. Sci.*, **16**, 645 (1972).

Srivastava, A. K., S. K. Nigam, A. K. Shukla, S. Saini, P. Kumar, and N. Tewari, *J. Macromol. Sci. Rev. Macromol. Chem. Phys.*, **C27**, 171 (1987).

Starck, P. and J. J. Lindberg, *Angew. Makromol. Chem.*, **75**, 1 (1979).

Starnes, W. H., Jr., *Pure Appl. Chem.*, **57**, 1001 (1985).

Starnes, W. H., Jr., I. M. Plitz, F. C. Schilling, G. M. Villacorta, G. S. Park, and A. H. Saremi, *Macromolecules*, **17**, 2507 (1984).

Starnes, W. H., Jr., F. C. Schilling, I. M. Plitz, R. E. Cais, D. J. Freed, R. L. Hartless, and F. A. Bovey, *Macromolecules*, **16**, 790 (1983).

Stein, D. J., *Makromol. Chem.*, **76**, 157, 169 (1964).

Stickler, M., *Makromol. Chem.*, **180**, 2615 (1979).

Stickler, M., *Makromol. Chem. Macromol. Symp.*, **10/11**, 17 (1987).

Sugimura, T. and Y. Minoura, *J. Polym. Sci.*, **A-1**(4), 2735 (1966).

Sugimura, Y., T. Usami, T. Nagaya, and S. Tsuge, *Macromolecules*, **14**, 1787 (1981).

Sugiyama, K., T. Oda, and T. Maeshima, *Makromol. Chem.*, **183**, 1 (1982).

Sugiyama, K. and S.-W. Lee, *Makromol. Chem.*, **178**, 421 (1977).

Sundberg, D. C. and D. R. James, *J. Polym. Sci. Polym. Chem. Ed.*, **16**, 523 (1978).

Surendran, G., M. Gazicki, W. J. James, and H. Yasuda, *J. Polym. Sci. Polym. Chem. Ed.*, **25**, 1481 (1987).

Tabata, Y., K. Ishigure, K. Oshima, and H. Sobue, *J. Polym. Sci.*, **A2**, 2445 (1964).

Takemoto, K. and M. Mujata, *J. Macromol. Sci. Rev. Macromol. Chem.*, **C18**, 83 (1980).

Takezaki, J., T. Okada, and I. Sakurada, *J. Appl. Polym. Sci.*, **22**, 2683, 3311 (1978).

Talamini, G. and E. Peggion, "Polymerization of Vinyl Chloride and Vinylidene Chloride," Chap. 5 in "Vinyl Polymerization," Vol. 1, Part I, G. E. Ham, Ed., Marcel Dekker, New York, 1967.

Tan, Y. Y. and G. Challa, *Makromol. Chem. Macromol. Symp.*, **10/11**, 214 (1987); "Template Polymerization," pp. 544–569 in "Encyclopedia of Polymer Science and Engineering," Vol. 16, H. F. Mark, N. M. Bikales, C. G. Overberger, and G. Menges, Eds., Wiley-Interscience, New York, 1988.

Tatsukami, Y., T. Takahashi, and H. Yoshioka, *Makromol. Chem.*, **181**, 1107 (1980).

Tedder, J. M., "The Reactivity of Free Radicals," Chap. 2 in "Reactivity, Mechanism and Structure in Polymer Chemistry," A. D. Jenkins and A. Ledwith, Eds., Wiley-Interscience, New York, 1974.

Tewari, N. and A. K. Srivastava, *J. Polym. Sci. Polym. Chem. Ed.*, **27**, 1065 (1989).

Thomas, W. M. and D. W. Wang, "Acrylamide Polymers," pp. 169–211 in "Encyclopedia of Polymer Science and Engineering," Vol. 1, H. F. Mark, N. M. Bikales, C. G. Overberger, and G. Menges, Eds., Wiley-Interscience, New York, 1985.

Tompa, H., "The Calculation of Mole-Weight Distributions from Kinetic Schemes," Chap. 7 in "Comprehensive Chemical Kinetics," Vol. 14A, C. H. Bamford and C. F. H. Tipper, Eds., American Elsevier, New York, 1976.

Toohey, A. C. and K. E. Weale, *Trans. Faraday Soc.*, **58**, 2446 (1962).

Tornell, B., *Polym.-Plast. Technol. Eng.*, **27**, 1 (1988).

Tsuda, K., S. Kondo, K. Yamashita, and K. Ito, *Makromol. Chem.*, **185**, 81 (1984).

Turner, D. T., *Macromolecules*, **10**, 221 (1977).

Uchida, M., T. Tanizaki, T. Kunatike, and T. Kajiyama, *Macromolecules*, **22**, 2381 (1989).

Ueda, M., S. Shouji, T. Ogata, M. Kamachi, and C. U. Pittman, Jr., *Macromolecules*, **17**, 2800 (1984).

Uemura, H., T. Taninaka, and Y. Minoura, *J. Polym. Sci. Polym. Lett. Ed.*, **15**, 493 (1977).

Usami, T. and S. Takayama, *Macromolecules*, **17**, 1756 (1984).

van de Grampel, H. T., Y. Y. Ton, and G. Challa, *Makromol. Chem. Macromol. Symp.*, **20/21**, 83 (1988).

Vollmert, B. (translated by E. H. Immergut), "Polymer Chemistry," Springer-Verlag, New York, 1973.

Vrancken, A. and G. Smets, *Makromol. Chem.*, **30**, 197 (1959).

Wall, L. A., *Soc. Plastic Eng. J.*, **16**, 1 (1960).

Walling, C., *J. Polym. Sci.*, **14**, 214 (1954).

Walling, C., "Free Radicals in Solution," Chaps. 3–5, Wiley-Interscience, New York, 1957.

Walling, C. and L. Heaton, *J. Am. Chem. Soc.*, **87**, 38 (1965).

Walling, C. and G. Metzger, *J. Am. Chem. Soc.*, **81**, 5365 (1959).

Weale, K. E., "The Influence of Pressure on Polymerization Reactions," Chap. 6 in "Reactivity, Mechanism and Structure in Polymer Chemistry," A. D. Jenkins and A. Ledwith, Eds., Wiley-Interscience, New York, 1974.

Wegner, G., *Disc. Faraday Soc.*, **68**, 494 (1980).

Willis, T. C. and J. E. Sheats, *J. Polym. Sci. Polym. Chem. Ed.*, **22**, 1077 (1984).

Wilson, J. E., "Radiation Chemistry of Monomers, Polymers and Plastics," Chaps. 1–5, Marcel Dekker, New York, 1974.

Winkler, R. E., and E. B. Nauman, *J. Polym. Sci. Polym. Chem. Ed.*, **26**, 2857 (1988).

Wolf, C. and W. Burchard, *Makromol. Chem.*, **177**, 2519 (1976).

Woodbrey, J. C. and P. Ehrlich, *J. Am. Chem. Soc.*, **85**, 1580 (1963).

Worsfold, D. J. and S. Bywater, *J. Polym. Sci.*, **26**, 299 (1957).

Yaeger, W. L. and V. J. Castelli, "Antifouling Applications of Various Tin-Containing Organometallic Polymers," pp. 175–180 in "Organometallic Polymers," C. E. Carraher, Jr. and J. E. Sheats, Eds., Academic Press, New York, 1978.

Yamamoto, K. and M. Sugimoto, *J. Macromol. Sci. Chem.*, **A13**, 1067 (1979).

Yassin, A. A. and N. A. Risk, *Polymer*, **19**, 57 (1978a); *J. Polym. Sci. Polym. Chem. Ed.*, **16**, 1475 (1978b).

Yasuda, H., "Plasma Polymerization," Academic Press, New York, 1986.

Yuan, Y., P. Tundo, and J. H. Fendler, *Macromolecules*, **22**, 29 (1989).

Zahalka, H. A., B. Robillard, L. Hughes, J. Lusztyk, G. W. Burton, E. G. Janzen, Y. Kotake, and K. U. Ingold, *J. Org. Chem.*, **53**, 3789 (1988).

Zhou, Q.-F., H.-M. Li, and X.-D. Feng, *Macromolecules*, **20**, 233 (1987).

Zott, H. and H. Heusinger, *Eur. Polym. J.*, **14**, 89 (1978).

Zutty, N. L. and R. D. Burkhart, "Polymer Synthesis at High Pressures," Chap. 3 in "Polymerization and Polycondensation Processes," N. A. J. Platzker, Ed., American Chemical Society, Van Nostrand Reinhold, New York, 1962.

PROBLEMS

3-1   When one considers the various polymers produced from carbon–carbon double bond monomers, the following generalizations are apparent:

  a. The polymers are produced almost exclusively from ethylene, monomers that have one substituent on the double bond, or monomers that have two substituents on the same carbon atom of the double bond. Monomers containing one substituent on each carbon of the double bond generally do not polymerize.

  b. Most of the chain polymerizations are carried out by radical initiation; relatively few are produced by ionic initiation.

  Why? Are there good reasons for these generalizations or are they simply a matter of chance? Discuss.

3-2   Show by chemical equations the polymerization of acrylonitrile initiated by the thermal decomposition of cumyl hydroperoxide.

3-3 Using carbon-14 labeled AIBN as an initiator, a sample of styrene is polymerized to an average degree of polymerization of 1.52×10^4. The AIBN has an activity of 9.81×10^7 counts/min-mole in a scintillation counter. If 3.22 g of the polystyrene has an activity of 203 counts/min, determine the mode of termination by appropriate calculations.

3-4 Poly(vinyl acetate) of number-average molecular weight 100,000 is hydrolyzed to poly(vinyl alcohol). Oxidation of the latter with periodic acid to cleave 1,2-diol linkages yields a poly(vinyl alcohol) with $\overline{X}_n = 200$. Calculate the percentages of head-to-tail and heat-to-head linkages in the poly(vinyl acetate).

3-5 The benzoyl peroxide initiated polymerization of a monomer follows the simplest kinetic scheme, that is, $R_p = k_p[M](fk_d[I]/k_t)^{1/2}$ with all rate constants and f being independent of conversion. For a polymerization system with $[M]_0 = 2\ M$ and $[I]_0 = 10^{-2}$ molar, the limiting conversion p_∞ is 10%. To increase p_∞ to 20%:

a. Would you increase or decrease $[M]_0$ and by what factor?

b. Would you increase or decrease $[I]_0$ and by what factor? How would the rate and degree of polymerization be affected by the proposed changes in $[I]_0$?

c. Would you increase or decrease the reaction temperature for the case of thermal initiated polymerization? For the case of photopolymerization (assuming that an increase in temperature does not cause thermal decomposition of initiator)?

E_d, E_p, and E_t are 64, 32, and 8 kJ/mole, respectively.

3-6 For a radical polymerization with bimolecular termination, the polymer produced contains 1.30 initiator fragments per polymer molecule. Calculate the relative extents of termination by disproportionation and coupling, assuming that no chain transfer reactions occur.

3-7 A solution 0.20 molar in monomer and $4.0 \times 10^{-3}\ M$ in a peroxide initiator is heated at 60°C. How long will it take to achieve conversion? $k_p = 145$ liters/mole-sec, $k_t = 7.0 \times 10^7$ liters/mole-sec, $f = 1$, and the initiator half-life is 44 hr.

3-8 Show that a plot of the fraction of monomer polymerized versus time t yields a straight line for $t \gg \tau_s$ and that the extension of this line cuts the time axis at $t = \tau_s \ln 2$. (Hint: Use $[M\cdot] = [M\cdot]_s \tanh(t/\tau_s)$ as a starting point and assume low conversion.)

3-9 The following data were obtained in the thermal initiated bulk polymerization of monomer Z ($[M] = 8.3\ M$) using radical initiator W at 60°C:

$R_p \times 10^3$ (mole/liter-sec)	$\overline{X}_n$
0.0050	8350
0.010	5550
0.020	3330
0.050	1317
0.10	592
0.15	358

Calculate C_M, $k_p/k_t^{1/2}$, and fk_d in this polymerization if it is experimentally observed that $R_p = 4.0 \times 10^{-4}[I]^{1/2}$. Is chain transfer to initiator important? If it is, describe how to calculate C_I.

3-10 Consider the polymerization of styrene initiated by di-*t*-butyl peroxide at 60°C. For a solution of 0.01 *M* peroxide and 1.0 *M* styrene in benzene, the initial rates of initiation and polymerization are 4.0×10^{-11} mole/liter-sec and 1.5×10^{-7} mole/liter-sec, respectively. Calculate the values of (fk_d), the initial kinetic chain length, and the initial degree of polymerization. Indicate how often on the average chain transfer occurs per each initiating radical from the peroxide. What is the breadth of the molecular weight distribution that is expected, that is, what is the value of $\overline{X}_w/\overline{X}_n$? Use the following chain transfer constants:

$$C_M = 8.0 \times 10^{-5}$$

$$C_I = 3.2 \times 10^{-4}$$

$$C_P = 1.9 \times 10^{-4}$$

$$C_S = 2.3 \times 10^{-6}$$

3-11 For a particular application the molecular weight of the polystyrene obtained in Question 3-10 is too high. What concentration of *n*-butyl mercaptan should be used to lower the molecular weight to 85,000? What will be the polymerization rate for the reaction in the presence of the mercaptan?

3-12 Consider the polymerization reaction in Question 3-10. Aside from increasing the monomer concentration, what means are available for increasing the polymerization rate? Compare the alternate possibilities with respect to any changes that are expected in the molecular weight of the product.

3-13 Show by chemical equations the reactions involved in chain transfer by hexane, benzene, isopropylbenzene, propanol, butyl iodide, carbon tetrabromide, *n*-butyl mercaptan, and di-*n*-butyl sulfide. Compare and discuss the differences in the transfer constants of these agents for vinyl acetate polymerization (Table 3-7).

3-14 The polymerization of methyl acrylate (one molar in benzene) is carried out using a photosensitizer and 3130 Å light from a filtered mercury arc lamp. Light is absorbed by the system at the rate of 1.0×10^5 ergs/liter-sec. If the quantum yield for radical production in this system is 0.50, calculate the rates of initiation and polymerization.

3-15 Consider the polymerization of bulk styrene by ultraviolet irradiation. The initial polymerization rate and degree of polymerization are 1.0×10^{-3} moles/liter-sec and 200, respectively, at 27°C. What will be the corresponding values for polymerization at 77°C.?

3-16 The same initial polymerization rate and degree of polymerization as in Question 3-15 are obtained at 27°C for a particular AIBN thermal-initiated polymerization of styrene. Calculate the R_p and $\overline{X}_n$ values at 77°C.

3-17 A radical chain polymerization following $R_p = [M]k_p(fk_d[I]/k_t)^{1/2}$ shows the indicated conversions for specified initial monomers and initiator concentrations and reaction times:

Experiment	Temperature (°C)	[M] (moles/liter)	[I] × 10³ (moles/liter)	Reaction Time (min)	Conversion (%)
1	60	1.00	2.5	500	50
2	80	0.50	1.0	700	75
3	60	0.80	1.0	600	40
4	60	0.25	10.0	?	50

Calculate the reaction time for 50% conversion in Experiment 4. Calculate the overall activation energy for the rate of polymerization.

3-18 Calculate the equilibrium monomer concentration $[M]_c$ for radical polymerization of 1,3-butadiene at 27°C. assuming that ΔH^0 and ΔS^0 are given by the values in Table 3-14. Repeat the calculations for 77°C and 127°C.

3-19 Most radical chain polymerizations show a one-half-order dependence of the polymerization rate on the initiation rate R_i (or the initiator concentration [I]). Describe and explain under what reaction conditions (i.e., what type(s) of initiation and/or termination) radical chain polymerizations will show the following dependencies:

a. First-order.
b. Zero-order.

Explain clearly the polymerization mechanisms which give rise to these different kinetic orders. What is the order of dependence of R_p on monomer concentration in each of these cases. Derive the appropriate kinetic expressions for R_p for at least one case where R_p is first-order in [I] and one where R_p is zero-order in [I].

3-20 What is the breadth of the size distribution to be expected for a low conversion polymerization where termination is entirely by coupling. Discuss the manner in which each of the following situations alters the size distribution:

a. Chain transfer to n-butyl mercaptan.
b. High conversion.
c. Chain transfer to polymer.
d. Autoacceleration.

For those situations where there is a tendency toward a broadening of the size distribution, discuss the possible process conditions which may be used to decrease this tendency.

3-21 Calculate the rate and degree of polymerization of methyl methacrylate initiated by AIBN at 50°C at 2500 atm relative to the corresponding quantities at 1 atm if $\Delta V_R^{\ddagger} = -19.0 \text{ cm}^3/\text{mole}$ and $\Delta V_d^{\ddagger} = 3.8 \text{ cm}^3/\text{mole}$.

CHAPTER 4

EMULSION POLYMERIZATION

Emulsion polymerization refers to a unique process employed for some radical chain polymerizations. It involves the polymerization of monomers in the form of emulsions (i.e., colloidal dispersions). The process bears a superficial resemblance to suspension polymerization (Sec. 3-13a-3) but is quite different in mechanism and reaction characteristics. Emulsion polymerization differs from suspension polymerization in the type and smaller size of the particles in which polymerization occurs, in the kind of initiator employed, and in the dependence of polymer molecular weight on reaction parameters.

4-1 DESCRIPTION OF PROCESS

4-1a Utility

Emulsion polymerization was first employed during World War II for producing synthetic rubbers from 1,3-butadiene and styrene. This was the start of the synthetic rubber industry in the United States. It was a dramatic development because the Japanese naval forces threatened access to the natural-rubber (NR) sources, which were necessary for the war effort. Synthetic rubber has advanced significantly from the first days of "balloon" tires, which had a useful life of 5000 mi to present-day tires, which are good for 50,000 mi. Emulsion polymerization is presently the predominant process for the commercial polymerizations of vinyl acetate, chloroprene, various acrylate copolymerizations, and copolymerizations of butadiene with styrene and acrylonitrile. It is also used for methacrylates, vinyl chloride, acrylamide, and some fluorinated ethylenes.

The emulsion polymerization process has several distinct advantages. The physical state of the emulsion (colloidal) system makes it easy to control the process. Thermal and viscosity problems are much less significant than in bulk polymerization. The product of an emulsion polymerization, referred to as a *latex*, can in many instances be used directly without further separations. (However, there may be the need for appropriate blending operations, e.g., for the addition of pigments.) Such applications include paints, coatings, finishes, and floor polishes. Aside from the physical difference between the emulsion and other polymerization processes, there is one very significant kinetic difference. For the other processes there is an inverse relationship (Eq. 3-107) between the polymerization rate and the polymer molecular weight. This drastically limits one's ability to make large changes in the molecular weight of a polymer, from 25,000 to 100,000 or from 100,000 to 25,000. Large decreases in the molecular weight of a polymer can be made without altering the polymerization rate by using chain-transfer agents. However, large increases in molecular weight can be made only by decreasing the polymerization rate by lowering the initiator concentration or lowering the reaction temperature. Emulsion polymerization is a unique process in that it affords the means of increasing the polymer molecular weight without decreasing the polymerization rate. Because of a different reaction mechanism, emulsion polymerization has the advantage of being able to simultaneously attain both high molecular weights and high reaction rates.

4-1b   Qualitative Picture

4-1b-1   *Components and Their Locations*

The physical picture of emulsion polymerization is based on the original qualitative picture of Harkins [1947] and the quantitative treatment of Smith and Ewart [1948] with subsequent contributions by other workers [Blackley, 1975; Casey et al., 1990; Fitch, 1980; Gardon, 1977; Hawkett et al., 1977, 1980; Piirma, 1982; Poehlein, 1986; Ugelstad and Hansen, 1976]. Table 4-1 shows a typical recipe for an emulsion polymerization [Vandenberg and Hulse, 1948]. This formulation, one of the early ones employed for the production of styrene-1,3-butadiene rubber (trade name: *GR-S*), is typical of all emulsion polymerization systems. The main components are the monomer(s), *dispersing medium*, *emulsifier*, and water-soluble initiator. The dispersing medium is the liquid, usually water, in which the various components are dispersed by means of the emulsifier. The ratio of water to monomer(s) is generally in the range 70/30 to 40/60 (by weight). The action of the emulsifier (also referred to as *surfactant*

TABLE 4-1   Composition of a GR-S Recipe for Emulsion Polymerization of Styrene-Butadiene[a]

Component	Parts by Weight
Styrene	25
Butadiene	75
Water	180
Emulsifier (Dresinate 731)	5
n-Dodecyl mercaptan	0.5
NaOH	0.061
Cumene hydroperoxide	0.17
FeSO$_4$	0.017
Na$_4$P$_2$O$_7$·10 H$_2$O	1.5
Fructose	0.5

[a]Data from Vandenberg and Hulse [1948].

or *soap*) is due to its molecules having both hydrophilic and hydrophobic segments. Various other components may also be present in the emulsion system. Thus, a mercaptan is used in the above formulation as a chain transfer agent to control the polymer molecular weight. The initiator is the hydroperoxide-ferrous ion redox system and the function of fructose is probably to regenerate ferrous ion by reducing the ferric ion produced in the initiation reaction (Eq. 3-36c). The sodium pyrophosphate acts to solubilize the iron salts in the strongly alkaline reaction medium. The emulsion system is usually kept in a well-agitated state during reaction.

The locations of the various components in an emulsion system will now be considered. When the concentration of a surfactant exceeds its *critical micelle concentration* (CMC), the excess surfactant molecules aggregate together to form small colloidal clusters referred to as *micelles*. The transformation of a solution to the colloidal state as the surfactant concentration exceeds the CMC occurs to minimize the free energy of solution (heat is liberated) and is accompanied by a sharp drop in the surface tension of the solution. Electrical conductivity, ion activities, viscosity, and other solution properties also show marked changes at CMC. CMC values are in the range 0.001–0.1 mole/liter, with most surfactants having values in the lower end of the range. Since surfactant concentrations in most emulsion polymerizations (0.1–3 wt % based on the aqueous phase) exceed CMC by one or more orders of magnitude, the bulk of the surfactant is in the micelles. Typical micelles have dimensions of 2–10 nm (1 nm = 10 Å = 10^{-3} μm) with each micelle containing 50–150 surfactant molecules. Most authors show the shape of micelles as being spherical, but this is not always the case. Both spherical and rodlike micelles are observed depending on the surfactant and its concentration. The surfactant molecules are arranged in a micelle with their hydrocarbon portions pointed toward the interior of the micelle and their ionic ends outward toward the aqueous phase. The number of micelles and their size depends on the amount of emulsifier. Large amounts of emulsifier yield larger numbers of smaller-sized particles.

When a water-insoluble or slightly water-soluble monomer is added, a very small fraction dissolves in the continuous aqueous phase. The water solubilities of most monomers are quite low, although the spread is large; for example, styrene, butadiene, vinyl chloride, methyl methacrylate, and vinyl acetate are soluble to the extent of 0.07, 0.8, 7, 16, 25 g/liter, respectively, at 25°C [Gardon, 1977]. An additional but still small portion of the monomer enters the interior hydrocarbon portions of the micelles. This is evidenced by X-ray and light-scattering measurements showing that the micelles increase in size as monomer is added. The amount of monomer in micelles compared to that in solution is much greater for the water-insoluble, nonpolar monomers. For example, the amount of micellar monomer is 2-, 5-, and 40-fold larger for methyl methacrylate, butadiene, and styrene, respectively, than the amount in solution [Bovey et al., 1955]. For vinyl acetate, the amount of micellar monomer is only a few percent of that in solution [Dunn, 1985].

The largest portion of the monomer (>95%) is dispersed as *monomer droplets* whose size depends on the stirring rate. The monomer droplets are stabilized by surfactant molecules absorbed on their surfaces. Monomer droplets have diameters in the range 1–10 μm (10^3–10^4 nm) or larger. Thus, in a typical emulsion polymerization system, the monomer droplets are much larger than the monomer-containing micelles. Consequently, while the concentration of micelles is 10^{17}–10^{18} per milliliter, there are at most 10^{10}–10^{11} monomer droplets per milliliter. A further difference between micelles and monomer droplets is that the total surface area of the micelles is larger than that of the droplets by more than two orders of magnitude. The size, shape, and concentration of each of the various types of particles in the emulsion system are obtained from electron microscopy, light scattering, ultracentrifugation, photon correlation spectroscopy, and other techniques [Debye and Anacker, 1951; Kratohvil, 1964; Munro et al., 1979].

4-1b-2 Site of Polymerization

The initiator is present in the water phase and this is where the initiating radicals are produced. The rate of radical production R_i is typically of the order of 10^{13} radicals per milliliter per second. (The symbol ρ is often used instead of R_i in emulsion polymerization terminology.) The locus of polymerization is now of prime concern. The site of polymerization is not the monomer droplets since the initiators employed are insoluble in the organic monomer. Such initiators are referred to as *oil-insoluble initiators*. This situation distinguishes emulsion polymerization from suspension polymerization. Oil-soluble initiators are used in suspension polymerization and reaction occurs in the monomer droplets. The absence of polymerization in the monomer droplets in emulsion polymerization has been experimentally verified. If one halts an emulsion polymerization at an appropriate point before complete conversion is achieved, the monomer droplets can be separated and analyzed. An insignificant amount (approximately <0.1%) of polymer is found in the monomer droplets in such experiments. Polymerization takes place almost exclusively in the micelles. Monomer droplets do not compete effectively with micelles in capturing radicals produced in solution because of the much smaller total surface area of the droplets.

Polymerization of the monomer in solution undoubtedly takes place but does not contribute significantly, since the monomer concentration is low and propagating radicals would precipitate out of aqueous solution at very small (oligomeric) size. The micelles act as a meeting place for the organic (oil-soluble) monomer and the water-soluble initiator. The micelles are favored as the reaction site because of their high monomer concentration (similar to bulk monomer concentration) compared to the monomer in solution. As polymerization proceeds, the micelles grow by the addition of monomer from the aqueous solution whose concentration is replenished by dissolution of monomer from the monomer droplets. A simplified schematic representation of an emulsion polymerization system is shown in Fig. 4-1. The system consists of three types of particles: monomer droplets, inactive micelles in which polymerization is not occurring, and active micelles in which polymerization is occurring. The latter are no longer considered as micelles but are referred to as *polymer particles*. An emulsifier molecule is shown as ○− to indicate one end (○) is polar or ionic and the other end (−) nonpolar.

The mechanism for *particle nucleation* (i.e, formation of polymer particles) proceeds by two simultaneous processes. One is the entry of radicals (either primary radicals or, more likely, oligomeric radicals formed by solution polymerization) from the aqueous phase into the micelles (*micellar nucleation*). The other, *homogeneous nucleation*, involves solution-polymerized oligomeric radicals becoming insoluble and precipitating onto themselves (or onto dead oligomer) [Fitch et al., 1969; Hansen and Ugelstad, 1978]. The precipitated species become stabilized by absorbing surfactant (from solution, monomer droplets, and micelles) and on subsequent absorption of monomer are the equivalent of polymer particles formed by micellar nucleation. The relative extents of micellar and homogeneous nucleation are expected to vary with the water solubility of the monomer and the surfactant concentration. Higher water solubility and low surfactant concentration favor homogeneous nucleation; micellar

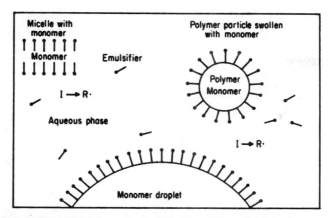

Fig. 4-1 Simplified representation of an emulsion polymerization system.

nucleation is favored by low water solubility and high surfactant concentration. (That homogeneous nucleation occurs is evidenced by the occurrence of emulsion polymerization of systems where the surfactant concentration is below CMC [Roe, 1968].) Micellar nucleation is probably the predominant mechanism for a highly water-insoluble monomer such as styrene [Hansen and Ugelstad, 1979a, 1979b] with homogeneous nucleation the predominant mechanism for a water-soluble monomer such as vinyl acetate [Zollars, 1979].

It has been suggested that the major growth process for the first-formed polymer particles, sometimes referred to as *precursor particles*, is coagulation with other particles and not polymerization of monomer. This coagulation, referred to as *coagulative nucleation*, is then considered as part of the overall nucleation sequence for the formation of *mature* polymer particles whose subsequent growth occurs entirely by polymerization. The experimental evidence for the coagulative nucleation process is the positive skewness of the polymer particle size distribution determined at short reaction times. This indicates that the rate of formation of polymer particles is an increasing function of time that is incompatible with a one-step mechanism but compatible with the two-step process of micellar and/or homogeneous nucleation followed by coagulative nucleation [Feeney et al., 1984]. The driving force for coagulation of precursor particles is their relative instability compared to larger-sized particles. The small size of a precursor particle (several nanometers) with its high curvature of the electrical double layer precludes the high surface charge density required for high colloidal stability. Once the particles reach a larger size with high colloidal stability, there is no longer a driving force for coagulation and further growth occurs only by polymerization.

4-1b-3 *Progress of Polymerization*

A variety of behaviors are observed for the polymerization rate versus conversion depending on the relative rates of initiation, propagation, and termination, which are in turn dependent on the monomer and reaction conditions (Fig. 4-2). Irrespective of the particular behavior observed, three *intervals* (I, II, III) can be discerned in all emulsion polymerizations based on the *particle number N* (the concentration of poly-

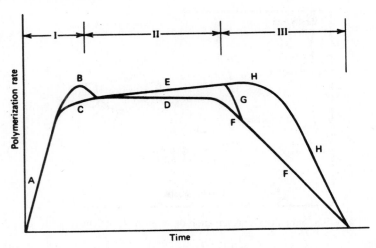

Fig. 4-2 Different rate behaviors observed in emulsion polymerization. After Gardon [1977] (by permission of Wiley-Interscience, New York).

mer particles in units of number of particles per milliliter) and the existence of a separate monomer phase (i.e., monomer droplets). There is a separate monomer phase in Intervals I and II but not in III. The particle number increases with time in Interval I and then remains constant during Intervals II and III. Particle nucleation occurs in Interval I with the polymerization rate increasing with time as the particle number builds up. Monomer diffuses into the polymer particles to replace that which has reacted. The reaction system undergoes a very significant change during Interval I. The particle number stabilizes at some value which is only a small fraction, typically about 0.1%, of the concentration of micelles initially present. (N is in range 10^{13}–10^{15} particles per milliliter.) As the polymer particles grow in size and contain polymer as well as monomer, they absorb more and more surfactant (in order to maintain stability) from that which is in solution. The point is quickly reached at which the surfactant concentration in solution falls below its CMC, the inactive micelles become unstable and disappear with dissolution of micellar surfactant. By the end of Interval I or very early in Interval II all or almost all of the surfactant in the system has been absorbed by the polymer particles. As a consequence the monomer droplets are relatively unstable and will coalesce if agitation is stopped. Interval I is generally the shortest of the three intervals, its duration varying in the range 2–15% conversion. Interval I is longer for low initiation rates as more time is needed to attain the steady-state particle number. The more water-soluble monomers such as vinyl acetate tend to complete Interval I faster than the less water-soluble monomers. This is probably a consequence of the significant extent of homogeneous nucleation occurring simultaneously with micellar nucleation, resulting in achieving the steady-state particle number sooner. The predicted maximum in Fig. 4-2 (curve AB), arising from a transient high particle number and/or high proportion of particles containing propagating radicals, is often not distinguishable experimentally, since it is not a high maximum. The maximum is observed for many monomers when the initiation rates are sufficiently high.

Polymerization proceeds in the polymer particles as the monomer concentration in the particles is maintained at the equilibrium (saturation) level by diffusion of monomer from solution, which in turn is maintained at the saturation level by dissolution of monomer from the monomer droplets. The monomer concentration in the polymer particles is high; the volume fraction of monomer ϕ_m is 0.2, 0.3, 0.5, 0.6, 0.71, and 0.85 for ethylene, vinyl chloride, butadiene, styrene, methyl methacrylate, and vinyl acetate, respectively [Gardon, 1977]. The polymerization rate either is constant (behavior D) or increases slightly with time (E) during Interval II. The latter behavior, which may begin immediately as shown in Fig. 4-2 or after a constant rate period, is a consequence of the gel or Trommsdorff effect (Sec. 3-10a). The polymer particles increase in size as the monomer droplets decrease. Interval II ends when the monomer droplets disappear. The transition from Interval II to III occurs at lower conversions as the water solubility of the monomer increases and the extent of swelling of the polymer particles by monomer increases. For monomers (e.g., vinyl chloride) with low water solubility and low ϕ_m, the transition occurs at about 70–80% conversion. The transition occurs at progressively lower conversion as the proportion of the total monomer in the system that is contained in the droplets decreases: styrene and butadiene at 40–50% conversion, methyl methacrylate at 25%, and vinyl acetate at 15% [Zollars, 1979].

The particle number remains the same in Interval III as in Interval II but the monomer concentration decreases with time, since monomer droplets are no longer present. The decrease in ϕ_m is slower with the more water-soluble monomers as the monomer in solution acts as a reservoir. The presence of a gel effect continues in Interval III. The quantitative interplay of a decreasing monomer concentration with the gel effect determines the exact behavior observed in this interval (GF or H). Polymerization continues at a steadily decreasing rate as the monomer concentration in the polymer particles decreases. Final conversions of essentially 100% are usually achieved. The final polymer particles, spherical in shape, usually have diameters of 50–200 nm, which places them intermediate in size between the initial micelles and monomer droplets. (Polymer particles as small as 10 nm and as high as several μm have been produced in emulsion polymerization.)

4-2 QUANTITATIVE ASPECTS

4-2a Rate of Polymerization

An expression for the rate of polymerization can be obtained by considering first the rate in a single polymer particle in which propagation is occurring (i.e., a particle containing a radical) and then the number of such particles. At the start of polymerization in a typical system where the concentration of micelles is 10^{18} per milliliter and the initiation rate is 10^{13} radicals per milliliter-second, a radical diffuses into a micelle every 10^5 sec at the start of Interval I. As the system progresses through Interval I, this time period decreases sharply, since the concentration of micelles is decreasing. A radical enters each particle on an average of every 10 sec during Intervals II and III, where N is typically 10^{14} particles per milliliter. Once inside the micelle or polymer particle, a radical propagates in the usual manner at a rate r_p dependent on the propagation rate constant k_p and the monomer concentration [M] in the particle.

$$r_p = k_p[\text{M}] \tag{4-1}$$

The monomer concentration is usually quite high since in many cases the equilibrium swelling of the particle by monomer is of the order 50–85% by volume. Values of [M] as high as 5 M are common. [M] varies only weakly with the size of the polymer particles.

Consider now what occurs on the entry of a radical into a particle that already has a radical. For most reaction systems, the radical concentration in a polymer particle is 10^{-6} M or higher. This is a higher radical concentration than in the homogeneous polymerization systems and the radical lifetime here is only a few thousandths of a second. The entry of a second radical into the polymer particle results in immediate bimolecular termination. Thus the polymer particle will have either one or zero radicals. The presence of two radicals in one particle is synonymous with zero radicals, since termination occurs so quickly. The particle is then dormant until another (the third) radical arrives. The particle is again activated and propagation proceeds until the next radical. The cycle of alternate growth and inactivity of the polymer particle continues until the monomer conversion is essentially complete.

The rate of polymerization R_p at any instant is given by the product of the concentration of active particles [P·] and the rate of propagation in a particle.

$$R_p = k_p[M][P·] \tag{4-2}$$

[P·] is conveniently expressed by

$$[P·] = \frac{10^3 N' \bar{n}}{N_A} \tag{4-3}$$

where N' is the concentration of micelles plus particles, $\bar{n}$ is the average number of radicals per micelle plus particle, and N_A is the Avogadro number. The use of $10^3/N_a$ in Eq. 4-3 and in the subsequent equations expresses [P·] in moles/liter and R_p in moles/liter-sec. Combination of Eqs. 4-2 and 4-3 yields the polymerization rate as

$$R_p = \frac{10^3 N' \bar{n} k_p[M]}{N_A} \tag{4-4}$$

$N' \bar{n}$ is zero at the start of Interval I, since $\bar{n} = 0$. N' decreases, $\bar{n}$ increases and the product $N' \bar{n}$ increases with time during Interval I. At the start of Interval II, N' has reached its steady-state value N. $\bar{n}$ may or may not reach an absolutely constant value. Behavior D in Interval II usually involves a steady-state $\bar{n}$ value, while behavior E usually involves a slow increase in $\bar{n}$ with conversion. $\bar{n}$ will remain approximately constant or increase in Interval III although a decrease will occur if the initiation rate decreases sharply due to exhaustion of the initiator concentration. Most texts show

$$R_p = \frac{10^3 N \bar{n} k_p[M]}{N_A} \tag{4-5}$$

for the polymerization rate instead of the more general Eq. 4-4. Equation 4-5 applies to Intervals II and III where only polymer particles exist (no micelles). It is during Intervals II and III that the overwhelming percent of monomer conversion to polymer takes place. In the remainder of Section 4-2, the discussions will be concerned only with these Intervals.

The value of $\bar{n}$ during Intervals II and III is of critical importance in determining R_p and has been the subject of much theoretical and experimental work. Three *cases* can be distinguished—*Cases* 1, 2, and 3. The major differences between the three cases are the occurence of radical diffusion out of the polymer particles (*desorption*), the particle size, modes of termination, and the rates of initiation and termination relative to each other and to the other reaction parameters. The quantitative interplay of these factors leading to Case 1, 2, or 3 behavior has been discussed [Nomura, 1982; Ugelstad and Hansen, 1976]. Our discussion will be in qualitative terms.

CASE 2: $\bar{n} = 0.5$. This is the case usually described in texts as applicable to most emulsion polymerizations. It occurs when desorption of radicals does not occur or is negligible compared to the rate of radicals entering particles (*absorption*) and the particle size is too small, relative to the bimolecular termination rate constant, to accommodate more than one radical. Under these conditions, a radical entering a polymer particle is trapped within that particle and undergoes propagation until another radical enters, at which point there is essentially instantaneous termination. Any polymer particle will be active half of the time and dormant the other half of the time. In other words, at any given moment half of the polymer particles contain one radical and are growing while the other half are dormant. The number of radicals per particle $\bar{n}$ averaged over all the particles is 0.5. Case 2 behavior also requires the initiation rate not be excessively low and negligible termination of radicals in the aqueous phase.

CASE 1: $\bar{n} < 0.5$. The average number of radicals per particle can drop below 0.5 if radical desorption from particles and termination in the aqueous phase are not negligible. The decrease in $\bar{n}$ is larger for small particle sizes and low initiation rates.

CASE 3: $\bar{n} > 0.5$. Some fraction of the polymer particles must contain two or more radicals per particle in order for $\bar{n}$ to be larger than 0.5, since there will always be a fraction (a very significant fraction) that has zero radical per particle. This occurs if the particle size is large or the termination rate constant is low while termination in the aqueous phase and desorption are not important and the initiation rate is not too low.

Although most texts indicate that Case 2 is the predominant behavior for all monomers, this is not true. Certain monomers, especially vinyl acetate and vinyl chloride, follow Case 1 behavior under a variety of reaction conditions [Blackley, 1975; Nomura and Harada, 1981]. For example, $\bar{n}$ is approximately 0.1 or lower for vinyl acetate and vinyl chloride [Gilbert and Napper, 1974; Litt et al., 1970]. Values of $\bar{n}$ are calculated from Eq. 4-5 using the k_p value from bulk polymerization at the appropriate percent conversion, that is, at the conversion corresponding to the volume fraction of monomer in the polymer particles. The monomers which show strong Case 1 behavior are those with high monomer chain-transfer constants. Chain transfer to monomer results in a small-sized monomer radical that can desorb from the polymer particle much more readily than the large-sized propagating radical. This was verified by carrying out emulsion polymerizations with intermittent ionizing radiation. The polymerization rate decays to zero after irradiation ceases but before all of the monomer has polymerized [Lansdowne et al., 1980; Ley et al., 1969; Sundari, 1979]. If desorption of monomer radicals did not occur, polymerization should continue until monomer would be exhausted. The polymerization rate decays for all monomers but at very different rates. The decay rate, which follows the desorption rate, increased as the monomer chain transfer constant increased.

The effect of reaction conditions on $\bar{n}$ (and R_p, of course) can be observed even with styrene, which shows a very strong tendency toward Case 2 behavior under a wide range of reaction conditions [Brooks and Qureshi, 1976; Hawkett et al., 1980]. *Seed* polymerization [Hayashi et al., 1989], involving the addition of monomer and initiator to a previously prepared emulsion of polymer particles, is especially useful for this purpose since it allows the variation of certain reaction parameters while holding N constant. Thus, $\bar{n}$ in seeded styrene polymerization drops from 0.5 to 0.2 when the initiator concentration decreases from 10^{-2} to 10^{-5} M. At sufficiently low R_i, the rate of radical absorption is not sufficiently high to counterbalanace the rate of desorption. One also observes that above a particular initiation rate ([I] $= 10^{-2}$ M in this case), the system maintains Case 2 behavior with $\bar{n}$ constant at 0.5 and R_p independent of R_i. A change in R_i simply results in an increased rate of alternation of activity and inactivity in each polymer particle. Similar experiments show that $\bar{n}$ drops below 0.5 for styrene when the particle size becomes sufficiently small. The extent of radical desorption increases with decreasing particle size since the travel distance for radical diffusion from a particle decreases.

Case 3 behavior occurs when the particle size is sufficiently large (about 0.1–1 μm) relative to k_t such that two or more radicals can coexist in a polymer particle without instantaneous termination. This effect is more pronounced as the particle size and percent conversion increase. At high conversion the particle size increases and k_t decreases, leading to an increase in $\bar{n}$. The increase in $\bar{n}$ occurs at lower conversions for the larger-sized particles. Thus for styrene polymerization $\bar{n}$ increases from 0.5 to only 0.6 at 90% conversion for 0.7-μm particles. On the other hand, for 1.4-μm particles, $\bar{n}$ increases to about 1 at 80% conversion and more than 2 at 90% conversion [Chatterjee et al., 1979; Gerrens, 1959]. Much higher values of $\bar{n}$ have been reported in other emulsion polymerizations [Ballard et al., 1986; Mallya and Plamthottam, 1989].

Consider now the implications of Eq. 4-5. The values of k_p, [M] and, to a large extent, $\bar{n}$ are specified for any particular monomer. The polymerization rate is then determined by the value of N. Increasing the surfactant concentration and increasing R_i increases N (Sec. 4-2c) and, therefore, R_p. These trends are shown in Figs. 4-3 and 4-4 [Hansen and Ugelstad, 1979a, 1979b; Vidotto et al., 1970]. It should be noted that the polymerization rate is unaffected by changes in R_i once particle nucleation

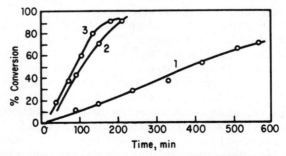

Fig. 4-3 Plot of percent conversion vs time for emulsion polymerizations of styrene with different concentrations of potassium laurate at 60°C. The moles of emulsifier per polymerization charge (containing 180 g H_2O, 100 g styrene, 0.5 g $K_2S_2O_8$) are 0.0035 (plot 1), 0.007 (plot 2), and 0.014 (plot 3). After Williams and Bobalek [1966] (by permission of Wiley-Interscience, New York).

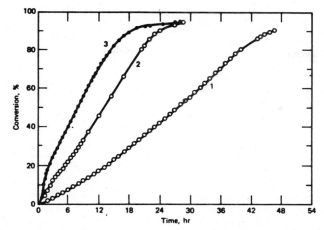

Fig. 4-4 Plot of percent conversion versus time for emulsion polymerization of vinyl chloride at 50°C for monomer/water ratio of 26/74 and 0.883% surfactant. The initiator concentrations are 0.0012% (plot 1), 0.0057% (plot 2), and 0.023% (plot 3). After Vidotto et al. [1970] (by permission of Huthig and Wepf Verlag. Basel).

has ceased at the end of Interval I. Such changes would only result in changing the rate of alternation of activity and inactivity in each polymer particle.

4-2b   Degree of Polymerization

The number-average degree of polymerization in an emulsion polymerization can be obtained by considering what occurs in a single polymer particle. The rate r_i at which primary radicals enter a polymer particle is given by

$$r_i = \frac{R_i}{N} \tag{4-6}$$

This is the same as the rate of termination r_t of a polymer chain for Case 2 behavior, since termination occurs immediately on the entry of a radical into a polymer particle in which a polymer chain is propagating. The degree of polymerization is then the rate of growth of a polymer chain divided by the rate at which primary radicals enter the polymer particle, that is, Eq. 4-1 divided by Eq. 4-5.

$$\overline{X}_n = \frac{r_p}{r_i} = \frac{Nk_p[\mathrm{M}]}{R_i} \tag{4-7}$$

Figure 4-5 shows the viscosity-average molecular weights in the emulsion polymerizations of styrene of Fig. 4-3. The results are in line with Eq. 4-7 in that the polymer size increases with the emulsifier concentration.

It should be noted that the degree of polymerization in an emulsion polymerization is synonymous with the kinetic chain length. Although termination is by bimolecular coupling, one of the radicals is a primary (or oligomeric) radical, which does not significantly contribute to the size of a dead polymer molecule. The derivation of Eq.

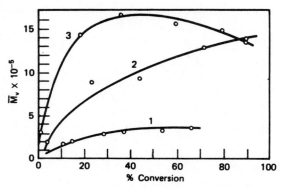

Fig. 4-5 Plot of viscosity-average molecular weight versus percent conversion for emulsion polymerizations of styrene with different concentrations of potassium laurate at 60°C. The moles of emulsifier per polymerization charge (containing 180 g H_2O, 100 g styrene, 0.5 g $K_2S_2O_8$) are 0.0035 (plot 1), 0.007 (plot 2), and 0.0014 (plot 3). After Williams and Bobalek [1966] (by permission of Wiley-Interscience, New York).

4-7 assumes the absence of any termination by chain transfer. If chain transfer occurs the degree of polymerization will be given by

$$\overline{X}_n = \frac{r_p}{r_i + \Sigma r_{tr}} \tag{4-8}$$

where Σr_{tr} is the sum of the rates of all transfer reactions. The rate of a chain transfer reaction in a polymer particle would be given by an equation of the type

$$r_{tr} = k_{tr}[XA] \tag{4-9}$$

analogous to the case of transfer in homogeneous polymerization (Eq. 3-113).

The degree of polymerization, like the polymerization rate, varies directly with N, but the degree of polymerization also varies indirectly with R_i. A consideration of Eqs. 4-5 and 4-7 with their analogues for homogeneous, radical chain polymerization (Eqs. 3-25 and 3-109) shows the significant characteristic of the emulsion process. In homogeneous polymerization, one can increase the polymerization rate by increasing the rate of initiation, but the result is a simultaneous lowering of the polymer molecular weight. No experimental variable is available to increase R_p without decreasing $\overline{X}_n$. The situation is quite different in emulsion polymerization. The rate and degree of polymerization can be simultaneously increased by increasing the number of polymer particles at a constant initiation rate.

Equations 4-7 and 4-8 require modification to be applicable to Case 3 behavior where a significant fraction of the polymer particles have 2 or more radicals per particle. For such particles, one still has $r_i = r_t$ (assuming a steady-state $\overline{n}$) but the degree of polymerization will be twice that for Case 2, since termination is by coupling between propagating radicals instead of propagating and primary (or oligomeric) radicals. Thus, the overall degree of polymerization for Case 3 behavior will be between $\overline{X}_n$ as calculated from Eq. 4-7 and twice that value, the exact value being the average between the two weighted in proportion to the fraction of particles which contain more than one propagating radical.

4-2c Number of Polymer Particles

The number of polymer particles is the prime determinant of the rate and degree of polymerization since it appears as the first power in boths Eqs. 4-5 and 4-7. The formation (and stabilization) of polymer particles by both micellar nucleation and homogeneous nucleation involves the adsorption of surfactant from the micelles, solution, and monomer droplets. The number of polymer particles that can be stabilized is dependent on the total surface area of surfactant present in the system $a_s S$, where a_s is the interfacial surface area occupied by a surfactant molecule and S is the total concentration of surfactant in the system (micelles, solution, monomer droplets). However, N is also directly dependent on the rate of radical generation. The quantitative dependence of N on $a_s S$ and R_i has been derived as

$$N = k \left(\frac{R_i}{\mu}\right)^{2/5} (a_s S)^{3/5} \tag{4-10}$$

where μ is the rate of volume increase of a polymer particle (which can be determined from r_p and geometric considerations). The same result has been derived for both micellar and homogeneous nucleations—each in the absence of coagulative nucleation [Roe, 1968; Smith and Ewart, 1948]. The value of k is between 0.37 and 0.53 depending on the assumptions made regarding the relative efficiencies of radical capture by micelles versus polymer particles and which geometric parameter of the particle (radius, surface area or volume) determines the rate at which polymer particles capture radicals. We should note that high particle numbers are associated with small particle size and low particle numbers with large particle size. Equation 4-10 leads to the prediction that the particle radius will be inversely dependent on the 0.20- and 0.13- order of S and R_i, respectively.

A consideration of Eq. 4-10 together with Eqs. 4-5 and 4-7 shows that both R_p and $\overline{X}_n$ depend on the $\frac{3}{5}$-power of the total surfactant concentrations. The polymerization rate varies with the $\frac{2}{5}$-power of R_i while the degree of polymerization varies inversely with the $\frac{2}{5}$-power of R_i. The dependence of R_p on R_i does not contradict the earlier conclusion regarding the independence of the polymerization rate on the rate of radical production. The rate of radical generation affects the number of polymer particles formed, which in turn determines the polymerization rate. However, once an emulsion polymerization system has reached a steady state with regard to N, the rate of radical generation no longer has any effect on the polymerization rate as long as initiation is taking place. Further and very significantly, it should be noted that the number of polymer particles can be increased by increasing the emulsifier concentration while maintaining a constant rate of radical generation. Thus from the practical viewpoint one can simultaneously increase R_p and $\overline{X}_n$ by increasing N. Increasing N by increasing R_i increases R_p but at the expense of decreasing $\overline{X}_n$.

The predicted dependence of N on S and R_i for the formation of polymer particles by micellar and homogeneous nucleation followed by coagulative nucleation is given by Eq. 4-11 [Feeney et al., 1984].

$$N \propto R_i^{2/5} S^{0.4\text{-}1.2} \tag{4-11}$$

The occurrence of coagulative nucleation does not alter the $\frac{2}{5}$-power dependence of N on R_i. However, the coagulative nucleation mechanism indicates a more complex dependence of N on S. The exponent of S decreases monotonically from 1.2 to 0.4 with increasing S. The concentration of polymer particles is higher and the nucleation time is longer for systems with high surfactant concentrations. Polymer particle formation becomes less efficient at longer times as there is a greater tendency for capture of precursor particles by polymer particles when the latter concentrations are high. Within the overall behavior predicted by Eq. 4-11, there is compatibility with the $\frac{2}{5}$-power dependence of N on R_i predicted by the Eq. 4-10.

Nonpolar monomers such as styrene, with little tendency toward radical desorption, generally show $\frac{3}{5}$- and $\frac{2}{5}$-power dependencies of N on S and R_i, respectively. This result, however, cannot be taken to exclude coagulative nucleation since one cannot preclude the exponent of the dependence of N on S being larger and smaller, respectively, than $\frac{3}{5}$ at lower and higher concentrations of surfactant than those studied. Monomers such as vinyl acetate and vinyl chloride, which show Case 1 behavior, tend to show a dependence of N on S in line with that predicted by Eq. 4-11, indicating the presence of coagulative nucleation. Simultaneously, the dependence of N on R_i deviates markedly from Eq. 4-11. When extensive radical desorption occurs, the large fraction of nucleation is initiated by desorbed radicals with the result that N is little affected by R_i. Thus, the order of dependence of N on S is 0.64 for styrene, 0.86 for methyl methacrylate, 1.0 for vinyl chloride, and 1.0 for vinyl acetate, while the orders of dependence on R_i are 0.36, 0.20, 0, and 0, respectively [Hansen and Ugelstad, 1979a, 1979b]. The emulsion copolymerization of acrylonitrile and butyl acrylate shows a decrease in the exponent of the dependence of N and S from 0.67 to 0.40 with increasing surfactant concentration when an anionic surfactant was used [Capek et al., 1988]. The exponent was close to one for polymerization in the presence of a cationic surfactant.

Anomolous results have been observed in some emulsion polymerizations—inverse dependencies of N, R_p, and $\overline{X}_n$ on surfactant concentration. Some surfactants act as inhibitors or retarders of polymerization, especially of the more highly reactive radicals from vinyl acetate and vinyl chloride [Okamura and Motoyama, 1962; Stryker et al., 1967]. This is most apparent with surfactants possessing unsaturation (e.g., certain fatty acid soaps). Degradative chain transfer through allyl hydrogens is probably quite extensive.

The polymer particles decrease in stability during Intervals II and III since the total polymer particle surface area increases and the coverage of the surface with surfactant decreases. The relative decrease in particle stability appears to be insufficient to cause coalescence as long as stirring is maintained since N is generally observed to be constant. In some systems, however, the stability decreases sufficiently to cause the particles to coalesce and N decreases with conversion [Blackley, 1975].

4-3 OTHER CHARACTERISTICS OF EMULSION POLYMERIZATION

4-3a Initiators

The initiators used in emulsion polymerization are water-soluble initiators such as potassium or ammonium persulfate, hydrogen peroxide, and 2,2′-azobis(2-amidino-propane) dihydrochloride. Partially water-soluble peroxides such as succinic acid peroxide and t-butyl hydroperoxide and azo compounds such as 4,4′-azobis(4-cyanopentanoic acid) have also been used. Redox systems such as persulfate with ferrous ion

OTHER CHARACTERISTICS OF EMULSION POLYMERIZATION

(Eq. 3-38a) are commonly used. Redox systems are advantageous in yielding desirable initiation rates at temperatures below 50°C. Other useful redox systems include cumyl hydroperoxide or hydrogen peroxide with ferrous, sulfite, or bisulfite ion.

4-3b Surfactants

Anionic surfactants are the most commonly used surfactants in emulsion polymerization [Blackley, 1975; Gardon, 1977]. These include fatty acid soaps (sodium or potassium stearate, laurate, palmitate), sulfates, and sulfonates (sodium lauryl sulfate and sodium dodecylbenzene sulfonate). The sulfates and sulfonates are useful for polymerization in acidic medium where fatty acid soaps are unstable or where the final product must be stable toward either acid or heavy-metal ions. Nonionic surfactants such as poly(ethylene oxide), poly(vinyl alcohol) and hydroxyethyl cellulose are sometimes used in conjunction with anionic surfactants for improving the freeze–thaw and shear stability of the polymer or to aid in controlling particle size and size distribution. The presence of the nonionic surfactant imparts a second mode of colloidal stabilization, in addition to electrostatic stabilization by the anionic surfactant, via steric interference with the van der Waals attraction between polymer particles. Nonionic surfactants are also of use where the final polymer latex should be insensitive to changes in pH over a wide range. Nonionic surfactants are only infrequently used alone, since their efficiency in producing stable emulsions is less than that of the anionic surfactants. Anionic surfactants are generally used at a level of 0.2–3 wt % based on the amount of water; nonionic surfactants are used at the 2–10% level. Cationic surfactants such as dodecylammonium chloride and cetyltrimethylammonium bromide are much less frequently used than anionic surfactants because of their inefficient emulsifying action or adverse effects on initiator decomposition. Also, cationic surfactants are more expensive than anionic surfactants.

Surfactants increase particle number and decrease particle size as their concentration in the initial reaction charge is increased. However, one can use delayed addition of surfactant after nucleation is complete to improve particle stability, without affecting the particle number, size, and size distribution.

4-3c Other Components

The quality of the water used in emulsion polymerization is important. Deionized water may be used since the presence of foreign ions or ions in uncontrolled concentrations can interfere with both the initiation process and the action of the emulsifier. Antifreeze additives are used to allow polymerization at temperatures below 0°C. These include inorganic electrolytes as well as organics such as ethylene glycol, glycerol, methanol, and monoalkyl ethers of ethylene glycol. The addition of inorganic electrolytes often affects the polymerization rate and stability of the emulsion. Sequestering agents such as ethylenediamine tetraacetic acid or its alkali metal salts may be added to help solubilize a component of the initiator system or to deactivate traces of calcium and magnesium ions present in the water. Buffers such as phosphate or citrate salts may be used to stabilize the latex toward pH changes.

4-3d Propagation and Termination Rate Constants

Emulsion polymerization proceeds in a polymer particle where the concentration of polymer is quite high throughout the reaction. This type of system is then similar to a bulk polymerization in the later stages of reaction and one would anticipate the occurrence of the Trommsdorff effect. The propagation rate constant for an emulsion polymerization can be obtained for Case 2 systems from the polymerization rate using Eq. 4-5, where $\bar{n} = 0.5$. One can ascertain that Case 2 behavior is present by observing whether the polymerization rate in Interval II is insensitive to changes in the initiation rate. The value of k_p can also be obtained from the degree of polymerization using Eq. 4-7. This is often a more reliable measure of k_p since there is no need to make any assumption on the value of $\bar{n}$. The propagation rate constant is generally found to have the same value in emulsion polymerization as in the corresponding bulk polymerization at high conversion—more specifically, at a conversion corresponding to the volume fraction of polymer in monomer that exists in the emulsion system.

4-3e Energetics

The heat of an emulsion polymerization is the same as that for the corresponding bulk or solution polymerization, since ΔH is essentially the enthalpy change of the propagation step. Thus, the heats of emulsion polymerization for acrylic acid, methyl acrylate, and methyl methacrylate are -67, -77, and -58 kJ/mole, respectively [McCurdy and Laidler, 1964], in excellent agreement with the ΔH values for the corresponding homogeneous polymerizations (Table 3-14).

The effect of temperature on the rate of emulsion polymerization, although not extensively studied, is generally similar to that on homogeneous polymerization with a few modifications. The overall rate of polymerization increases with an increase in temperature. Temperature increases the rate by increasing both k_p and N. The increase in N is due to the increased rate of radical generation at higher temperatures. Opposing this trend to a slight extent is the small decrease in the concentration of monomer in the particles at higher temperatures. Thus, the value of [M] for styrene decreases ~15% in going from 30 to 90°C [Smith and Ewart, 1948]. The overall activation energy for emulsion polymerization is, thus, a combination of the activation energies for propagation, radical production, and [M]. For the few systems studied, the overall activation energies for emulsion polymerization are approximately the same or less than those for the corresponding homogeneous polymerization [Stavrova et al., 1965].

Carrying out an emulsion polymerization requires an awareness of the *Krafft point* of an ionic surfactant and the *cloud point* of a nonionic surfactant. Micelles are formed only at temperatures above the Krafft point of an ionic surfactant. For a nonionic surfactant, micelles are formed only at temperatures below the cloud point. Emulsion polymerization is carried out below the cloud temperature of a nonionic surfactant and above the Krafft temperature of an ionic surfactant.

4-3f Molecular Weight and Particle Size Distribution

Theoretical considerations indicate that compartmentalization of radicals in polymer particles does not change the polydispersity index *PDI* ($= \overline{X}_w/\overline{X}_n$) in emulsion polymerization from its value of 2 in homogeneous polymerization when termination takes place by transfer to monomer, chain-transfer agent, or other substance [Giannetti et al., 1988; Katz et al., 1969; Lichti et al., 1980, 1982]. However, emulsion polymerization results in molecular-weight broadening when termination involves the bimolecular reaction between radicals. While short propagating chains are likely to couple or disproportionate with longer chains in homogeneous polymerization (*PDI* = 1.5 and 2 for coupling and disproportionation, respectively), any two chains that undergo bimolecular termination in emulsion polymerization are not random. The broadening of *PDI* in emulsion polymerization is greater for disproportionation than for coupling. For Case 2 behavior, coupling of the propagating chain in a polymer particle with the low-molecular-weight entering radical does not greatly affect *PDI*. Such coupling is equivalent to termination by chain transfer and *PDI* has a value of 2 compared to 1.5 for homogeneous polymerization. When termination is by disproportionation, *PDI* has a value of 4 at $\bar{n} = 0.5$ compared to 2 for homogeneous polymerization. Low-molecular-weight radicals entering the polymer particles disproportionate with propagating radicals and increase the number of low-molecular-weight molecules; $\overline{X}_n$ is decreased while $\overline{X}_w$ is essentially unchanged and $\overline{X}_w/\overline{X}_n$ increases. When $\bar{n} > 0.5$ (Case 3), the tendency toward molecular-weight broadening decreases as the sizes of the radicals undergoing coupling or disproportionation become more nearly the same size. *PDI* tends toward the values in homogeneous polymerization (1.5 and 2 for coupling and disproportionation, respectively) as $\bar{n}$ increases from 0.5 to 2.

Although the preceding discussion indicates that *PDI* is generally larger in emulsion polymerization compared to homogeneous polymerization, the opposite is usually observed because of different trends in *PDI* as a function of conversion [Cooper, 1974; Lin and Chiu, 1979]. The molecular weight of the polymer produced during a batch emulsion polymerization remains reasonably constant during a large part (i.e., Interval II) of the overall reaction, since N, [M], k_p, and R_i are relatively constant. This is different from homogeneous batch polymerization where substantial changes occur in the molecular weight and *PDI* with conversion throughout the whole conversion range.

In addition to the molecular-weight distribution, there is a particle size distribution in emulsion polymerization [Chen and Wu, 1988; Gardon, 1977; Lichti et al., 1982]. The particle size distribution *PSD* is expressed, analogously to the molecular-weight distribution, as the ratio of the weight-average particle size to number-average particle size. (Different particle sizes are calculated depending on whether one uses the particle radius, diameter, or volume as the measure of particle size.) The particle-size distribution is a consequence of the distribution of times at which different polymer particles are nucleated. The polydispersity is maximum during Interval I and narrows considerably during the subsequent period. There has been an effort to produce narrow-particle-size distributions (*PSD*) by controlling the nucleation process, choice and amount of surfactant, and the use of seed emulsion polymerization, temperature, and other reaction variables. Narrow-particle-size distributions are useful in applications such as calibration of electron microscope, ultracentrifuge, aerosol counting, and light-scattering instruments and the measurement of pore sizes of filters and membranes. Narrow particle distributions, with *PSD* values of 1.1 and lower, have been obtained by choosing reaction conditions with short nucleation times (short Interval I relative to Intervals II and III), increased latex stability (to prevent coagulation), and decreased Interval III times.

4-3g Surfactant-Free Emulsion Polymerization

The presence of surfactant is a disadvantage for certain applications of emulsion polymers such as those involving instrument calibration and pore size determination. The presence of adsorbed surfactant gives rise to somewhat variable properties since the amount of adsorbed surfactant can vary with the polymerization and application conditions. Removal of the surfactant, on the other hand, can lead to coagulation or flocculation of the destabilized latex. *Surfactant-free emulsion polymerization*, involving no added surfactant, is a useful approach to solving this problem [Chainey et al., 1987; Hearn et al., 1985]. The technique requires the use of an initiator yielding initiating species that impart surface-active properties to the polymer particles. Persulfate is a useful initiator for this purpose. Latices prepared by the surfactant-free technique are stabilized by chemically bound sulfate groups of the $SO_4^-\cdot$ initiating species derived from persulfate ion. Since the surface-active groups are chemically bound, the latices can be purified (freed of unreacted monomer, initiator, etc.) without loss of stability, and their stability is retained over a wider range of use conditions than the corresponding latices produced using surfactants. A characteristic of surfactant-free emulsion polymerization is that the particle number is generally lower by up to about two orders of magnitude compared to the typical emulsion polymerization, typically 10^{12} versus 10^{14} particles per milliliter. This is simply a consequence of the lower total particle surface area that can be stabilized by the sulfate groups alone relative to that when added surfactant is present.

4-3h Core–Shell Model

Several groups have suggested that emulsion polymerization does not occur homogeneously throughout a polymer particle but follows a *core–shell model* in which polymerization takes place within the outer periphery (shell) of a particle. The strongest evidence for this model comes from experiments in which radioactive styrene was added at various times after the start of a styrene emulsion polymerization. Analysis of the resulting latex particles showed that radioactive (new) polymer was formed in the outer shell of the polymer particles [Grancio and Williams, 1970]. Two different mechanisms have been proposed to explain preferential polymerization in the shell layer. One mechanism invokes a nonuniform distribution of both monomer and polymer in the polymer particles—an outer monomer-rich shell surrounding an inner polymer-rich core [Williams, 1971]. The other mechanism suggests that radicals derived from water-soluble initiators, such as $SO_4^-\cdot$, are not able to penetrate into the central core of polymer particles because of their hydrophilic character [Chern and Poehlein, 1987].

4-3i Inverse Emulsion Polymerization

In the conventional emulsion polymerization, a hydrophobic monomer is emulsified in water and polymerization initiated with a water-soluble initiator. Emulson polymerization can also be carried out as an *inverse emulsion polymerization* [Poehlein, 1986]. Here, an aqueous solution of a hydrophilic monomer is emulsified in a nonpolar organic solvent such as xylene or paraffin and polymerization initiated with an oil-soluble initiator. The two types of emulsion polymerizations are referred to as *oil-in-water* (*o/w*) and *water-in-oil* (*w/o*) emulsions, respectively. Inverse emulsion polymerization is used in various commercial polymerizations and copolymerizations of acrylamide as well as other water-soluble monomers. The end use of the reverse latices often involves their addition to water at the point of application. The polymer dissolves readily in water, and the aqueous solution is used in applications such as secondary oil recovery and flocculation (clarification of waste water, metal recovery).

Nonionic surfactants such as sorbitan monooleate yield more stable emulsions than do ionic surfactants. However, the latices from inverse emulsion polymerizations are generally less stable than those from conventional emulsion polymerizations and flocculation is a problem. Stable latices can be obtained by *inverse microemulsion polymerization* [Candau, 1987]. Inverse microemulsion polymerization involves systems devoid of droplets, in which only micelles are present. Microemulsions are obtained by using larger amounts of surfactant and continuous phase relative to monomer compared to the amounts used for conventional emulsions. The term *macroemulsion* is sometimes used to refer to a "conventional-sized" emulsion system. (The micelle size is the same in macroemulsions and microemulsions). Both water-soluble and oil-soluble initiators have been used in inverse microemulsion polymerizations. The polymer particles produced from inverse microemulsion polymerization have much smaller sizes than do those from the corresponding macroemulsion polymerization and, in the presence of a considerable amount of surfactant, are well stabilized against flocculation. Oil-in-water microemulsion polymerizations have also been studied [Kuo et al., 1987]. *Miniemulsion polymerization* systems are somewhere in between macro and micro systems [Choi et al., 1985; Delgado et al., 1989]. They contain both micelles and monomer droplets, but the monomer droplets are smaller than in macro systems. Such systems, although labeled "emulsion," may involve a combination of emulsion and suspension polymerizations. Also, the reader is cautioned that some systems referred to in the literature as microemulsion polymerizations may actually be micro-suspension polymerizations. This confusion results when the reaction system is inadequately characterized in terms of the presence or absence of micelles and monomer droplets and their relative numbers and the size of monomer droplets. For both micro and mini reaction systems in which the initiator is soluble in the continuous phase, the mechanism for polymerization is determined by the relative surface areas of micelles versus monomer droplets (if both types of particles are present). Emulsion polymerization occurs if only micelles are present; suspension polymerization occurs if only monomer droplets are present.

REFERENCES

Ballard, M. J., R. G. Gilbert, D. H. Napper, P. J. Pomery, P. W. O'Sullivan, and J. H. O'Donnell, *Macromolecules*, **19**, 1303 (1986).

Blackley, D. C., "Emulsion Polymerization," Applied Science Publishers, London, 1975.

Bovey, F. A., I. M. Kolthoff, A. I. Medalia, and E. J. Meehan, "Emulsion Polymerization," Interscience, New York, 1955.

Brooks, B. W. and M. K. Qureshi, *Polymer*, **17**, 740 (1976).

Candau, F. "Microemulsion Polymerizations," pp. 718–724 in "Encyclopedia of Polymer Science and Engineering," Vol. 9, H. F. Mark, N. M. Bikales, C. G. Overberger, and G. Menges, Eds., Wiley-Interscience, New York, 1987.

Capek, I., M. Mlynarova, and J. Barton, *Makromol. Chem.*, **189**, 341 (1988).

Casey, B. S., I. A. Maxwell, B. R. Morrison, and R. G. Gilbert, *Makromol. Chem. Macromol. Symp.*, **31**, 1 (1990).

Chainey, M., J. Hearn, and M. C. Wilkinson, *J. Polym. Sci. Polym. Chem. Ed.*, **25**, 505 (1987).

Chatterjee, S. P., M. Banerjee, and R. S. Konar, *J. Polym. Sci. Polym. Chem. Ed.*, **17**, 2193 (1979).

Chen, S.-A. and K.-W. Wu, *J. Polym. Sci. Polym. Chem. Ed.*, **26**, 1143 (1988).

Chern, C.-S. and G. W. Poehlein, *J. Polym. Sci. Polym. Chem. Ed.*, **25**, 617 (1987).

Choi, Y. T., M. S. El-Aasser, E. D. Sudol, and J. W. Vanderhoff, *J. Polym. Sci. Polym. Chem. Ed.*, **23**, 2973 (1985).

Cooper, W., "Emulsion Polymerization," Chap. 7 in "Reactivity Mechanism and Structure in Polymer Chemistry," A. D. Jenkins and A. Ledwith, Eds., Wiley-Interscience, New York, 1974.

Debye, P. and E. W. Anacker, *J. Phys. Coll. Chem.*, **55**, 644 (1951).

Delgado, J., M. S. El-Aaser, C. A. Silebi, and J. W. Vanderhoff, *J. Polym. Sci. Polym. Chem. Ed.*, **27**, 193 (1989).

Dunn, A. S., *Makromol. Chem. Suppl.*, **10/11**, 1 (1985).

Feeney, P. J., D. H. Napper, and R. G. Gilbert, *Macromolecules*, **17**, 2520 (1984).

Fitch, R. M. Ed., "Polymer Colloids II," Plenum Press, New York, 1980.

Fitch, R. M., M. B. Prenosil, and K. J. Sprick, *J. Polym. Sci.*, **C27**, 95 (1969).

Gardon, J. L., "Emulsion Polymerization," Chap. 6 in "Polymerization Processes," C. E. Schild-knecht, Ed. (with I. Skeist), Wiley-Interscience, New York, 1977.

Gerrens, H., *Adv. Polym. Sci.*, **1**, 234 (1959).

Giannetti, E., G. Storti, and M. Morbidelli, *J. Polym. Sci. Polym. Chem. Ed.*, **26**, 1865 (1988).

Gilbert, R. G. and D. H. Napper, *J. Chem. Soc. Faraday I*, **71**, 391 (1974).

Grancio, M. R. and D. J. Williams, *J. Polym. Sci.*, **A–1**(8), 2617 (1970).

Hansen, F. K. and J. Ugelstad, *J. Polym. Sci. Polym. Chem. Ed.*, **16**, 1953 (1978); **17**, 3033, 3047, 3069 (1979a); *Makromol. Chem.*, **180**, 2423 (1979b).

Harkins, W. D., *J. Am. Chem. Soc.*, **69**, 1428 (1947).

Hawkett, B. S., D. H. Napper, and R. G. Gilbert, *J. Chem. Soc.*, *Faraday I*, **73**, 690 (1977); *J. Chem. Soc.*, *Faraday Trans. I*, **76**, 1323 (1980).

Hayashi, S., A. Komatsu, and T. Hirai, *J. Polym. Sci. Polym. Chem. Ed.*, **27**, 157 (1989).

Hearn, J., M. C. Wilkinson, A. R. Goodall, and M. Chainey, *J. Polym. Sci. Polym. Chem. Ed.*, **23**, 1869 (1985).

Katz, S., R. Shinnar, and G. M. Saidel, *Adv. Chem. Ser.*, **91**, 145 (1969).

Kratohvil, J. P., *Anal. Chem.*, **36**, 485R (1964).

Kuo, P.-L., N. J. Turro, C.-M. Tseng, M. S. El-Aasser, and J. W. Vanderhoff, *Macromolecules*, **20**, 1216 (1987).

Lansdowne, S. W., R. G. Gilbert, D. H. Napper, and D. F. Sangster, *J. Chem. Soc.*, *Faraday Trans. I*, **76**, 1344 (1980).

Ley, G. J. M., C. Schneider, and D. O. Hummel, *J. Polym. Sci.*, **C27**, 119 (1969).

Lichti, G., R. G. Gilbert, and D. H. Napper, *J. Polym. Sci. Polym. Chem. Ed.*, **18**, 1297 (1980); "Theoretical Predictions of the Particle Size and Molecular Weight Distributions in Emulsion Polymerization," Chap. 3 in "Emulsion Polymerization," I. Piirma, Ed., Academic Press, New York, 1982.

Lin, C.-C. and W.-Y. Chiu, *J. Appl. Polym. Sci.*, **23**, 2049 (1979).

Lin, C.-C., W.-Y. Chiu, and L.-C. Huang, *J. Appl. Polym. Sci.*, **25**, 565 (1980).

Litt, M., R. Patsiga, and V. Stannett, *J. Polym. Sci.*, **A–1**(8), 3607 (1970).

Mallya, P. and S. S. Plamthottam, *Polym. Bull.*, **21**, 497 (1989).

McCurdy, K. G. and K. J. Laidler, *Can. J. Chem.*, **42**, 818 (1964).

Munro, D., A. R. Goodall, M. C. Wilkinson, K. Randle, and J. Hearn, *J. Colloid Interface Sci.*, **68**, 1 (1979).

Nomura, M., "Desorption and Reabsorption of Free Radicals in Emulsion Polymerization," Chap. 5 in "Emulsion Polymerization," I. Piirma, Ed., Academic Press, New York, 1982.

Nomura, M. and M. Harada, *J. Appl. Polym. Sci.*, **26**, 17 (1981).

Okamura, S. and T. Motoyama, *J. Polym. Sci.*, **58**, 221 (1962).

Piirma, I., Ed., "Emulsion Polymerization," Academic Press, New York, 1982.

Poehlein, G. W., "Emulsion Polymerization," pp. 1–51 in "Encyclopedia of Polymer Science and Engineering," Vol. 6, H. F. Mark, N. M. Bikales, C. G. Overberger, and G. Menges, Eds., Wiley-Interscience, New York, 1986.

Roe, C. P., *Ind. Eng. Chem.*, **60**, 20 (1968).

Smith, W. V. and R. W. Ewart, *J. Chem. Phys.*, **16**, 592 (1948).

Stavrova, S. D., M. F. Margaritova, and S. S. Medvedev, *Vysokomol. Soyedin.*, **7**, 792 (1965).

Stryker, H. K., G. J. Mantell, and A. F. Helin, *J. Polym. Sci.*, **11**, 1 (1967).

Sundari, F., *J. Appl. Polym. Sci.*, **24**, 1031 (1979).

Ugelstad, J. and F. K. Hansen, *Rubber Chem. Technol.*, **49**, 536 (1976).

Vandenberg, E. J. and G. E. Hulse, *Ind. Eng. Chem.*, **40**, 932 (1948).

Vidotto, G., A. C. Arrialdi, and G. Talamini, *Makromol. Chem.*, **134**, 41 (1970).

Williams, D. J., "Polymer Science and Engineering," Chap. 4, Prentice-Hall, Englewood Cliffs, N.J., 1971.

Williams, D. J. and E. G. Bobalek, *J. Polym. Sci.*, A–1(4), 3065 (1966).

Zollars, R. L., *J. Appl. Polym. Sci.*, **24**, 1353 (1979).

PROBLEMS

4-1 Describe the components of an emulsion polymerization system on a macroscopic level. Compare the pros and cons of emulsion polymerization as a process condition in comparison to bulk and solution polymerization.

4-2 Describe the microscopic picture of emulsion polymerization according to Harkins, Smith, and Ewart. Where are the monomer, initiator, and emulsifier located? Describe the changes which take place as the reaction proceeds to 100% conversion.

4-3 What are the characteristic overall features that distinguish emulsion polymerization from homogeneous polymerization? Compare the two with regard to the heat of polymerization and the effect of temperature on the polymerization rate.

4-4 Quantitatively compare the rate and degree of polymerization of styrene polymerized in bulk at 60°C with an emulsion polymerization (Case 2 behavior: $\bar{n} = 0.5$) containing 1.0×10^{15} polymer particles per milliliter. Assume that $[M] = 5.0$ molar, $R_i = 5.0 \times 10^{12}$ radicals per milliliter per second and all rate constants are the same for both systems. For each polymerization system, indicate the various ways (if any) by which the polymerization rate can be affected without affecting the degree of polymerization.

4-5 Describe the reaction conditions under which deviations from Case 2 behavior are observed; that is, when does $\bar{n} > 0.5$ (Case 3) and $\bar{n} < 0.5$ (Case 1)?

CHAPTER 5

IONIC CHAIN POLYMERIZATION

The ionic chain polymerization of the carbon-carbon and carbon-oxygen double bonds was briefly described in Chap. 3. Ionic chain polymerizations will now be considered in detail. The bulk of the discussion will involve the polymerization of the carbon-carbon double bond by cationic and anionic initiators (Secs. 5-2 and 5-3). The last part of the chapter will consider the polymerization of other unsaturated linkages. A few of the polymerizations in this chapter, although not classified as ionic chain polymerizations, are included for the sake of convenience. Polymerizations initiated by coordination compounds and metal oxides are usually also ionic in nature. These are termed coordination polymerizations and will be considered separately in Chap. 8. The polymerizations of dienes will also not be considered in this chapter. Cyclopolymerization of nonconjugated dienes and the 1,2- and 1,4-polymerizations of conjugated dienes will be discussed in Chaps. 6 and 8.

5-1 COMPARISON OF RADICAL AND IONIC POLYMERIZATIONS

Almost all monomers containing the carbon-carbon double bond undergo radical polymerization, while ionic polymerizations are highly selective (Table 3-1). Cationic polymerization is essentially limited to those monomers with electron-releasing substituents such as alkoxy, phenyl, vinyl, and 1,1-dialkyl. Anionic polymerization takes place with monomers possessing electron-withdrawing groups such as nitrile, carboxyl, phenyl, and vinyl. The selectivity of ionic polymerization is due to the very strict requirements for stabilization of anionic and cationic propagating species (Sec. 3-1b-2). The commercial utilization of cationic and anionic polymerizations is rather limited because of this high selectivity of ionic polymerizations compared to radical polymerization (and the greater importance of coordination polymerization compared to ionic polymerization).

Ionic polymerizations are not as well understood as radical polymerizations because of experimental difficulties involved in their study. The nature of the reaction media in ionic polymerizations is often not clear since heterogeneous inorganic initiators are often involved. Further, it is extremely difficult in most instances to obtain reproducible kinetic data because ionic polymerizations proceed at very rapid rates and are extremely sensitive to the presence of small concentrations of impurities and other adventitious materials. The rates of ionic polymerizations are usually greater than those of radical polymerizations.

Cationic and anionic polymerizations have many similar characteristics. Both depend on the formation and propagation of ionic species, a positive one in one case and a negative one in the other. The formation of ions with sufficiently long lifetimes for propagation to yield high-molecular-weight products generally requires stabilization of the propagating centers by solvation. Relatively low or moderate temperatures are also needed to suppress termination, transfer, and other chain-breaking reactions which destroy propagating centers.

Although solvents of high polarity are desirable to solvate the ions, they cannot be employed for several reasons. The highly polar hydroxylic solvents (water, alcohols) react with and destroy most ionic initiators. Other polar solvents such as ketones prevent initiation of polymerization by forming highly stable complexes with the initiators. Ionic polymerizations are, therefore, usually carried out in solvents of low or moderate polarity such as methyl chloride, ethylene dichloride, and pentane, although moderately high polarity solvents such as nitrobenzene are also used. In such solvents one usually does not have only a single type of propagating species. For any propagating species such as $\sim\!\sim\!\sim$BA in cationic polymerization, one can visualize the range of behaviors from one extreme of a completely *covalent* species (**I**) to the other of a completely *free* (and highly solvated) *ion* (**IV**)

$$\sim\!\sim\!\sim\text{BA} \qquad \sim\!\sim\!\sim\text{B}^+\text{A}^- \qquad \sim\!\sim\!\sim\text{B}^+\|\text{A}^- \qquad \sim\!\sim\!\sim\text{B}^+ + \text{A}^-$$

$$\textbf{I} \qquad\qquad \textbf{II} \qquad\qquad \textbf{III} \qquad\qquad \textbf{IV}$$

The intermediate species include the *tight* or *contact ion pair* (**II**) (also referred to as the *intimate ion pair*) and the *solvent-separated* or *loose ion pair* (**III**). The intimate ion pair has a *counter-* or *gegenion* of opposite charge close to the propagating center (unseparated by solvent). The solvent-separated ion pair involves ions that are partially separated by solvent molecules. The propagating cationic chain end has a negative counterion. For an anionic polymerization the charges in species **II–IV** are reversed; that is, B carries the negative charge and A the positive charge. There is a propagating anionic chain end with a positive counterion.

Most ionic polymerizations involve two types of propagating species, an ion pair and a free ion **IV**, coexisting in equilibrium with each other. The identity of the ion pair (i.e., whether the ion pair is best described as species **II** or **III**) depends on the particular reaction conditions, especially the solvent employed. Increased solvent polarity favors the loose ion pair while the tight ion pair predominates in solvents of low polarity. The ion pairs in cationic polymerization tend to be loose ion pairs even in solvents of low or moderate polarity since the counterions (e.g., bisulfate, $SbCl_6^-$, perchlorate) are typically large ions. The lower charge density of a large counterion results in smaller electrostatic attractive forces between the propagating center and counterion. The nature of the ion pairs is much more solvent-dependent in anionic polymerizations where the typical counterion (e.g., Li^+, Na^+) is small. The covalent species **I** is generally ignored since it is usually unreactive or much lower in reactivity compared to the other species. (However, see Sec. 5-2f-1 for systems in which the covalent species is important.) Free-ion concentrations are generally much smaller than ion-pair concentrations but the relative concentrations are greatly affected by the reaction conditions. Increased solvent polarity results in a shift from ion pairs to free ions. The nature of the solvent has a large effect in ionic polymerization since the different types of propagating species have different reactivities. Loose ion pairs are more reactive than tight ion pairs. Free ions are orders of magnitude higher in reactivity than ion pairs in anionic polymerization. Ion pairs are generally no more than an order of magnitude lower in reactivity compared to free ions in cationic polymerization.

Ionic polymerizations are characterized by a wide variety of modes of initiation and termination. Unlike radical polymerization, termination in ionic polymerization never involves the bimolecular reaction between two propagating polymer chains of like charge. Termination of a propagating chain occurs by its reaction with the counterion, solvent, or other species present in the reaction system.

5-2 CATIONIC POLYMERIZATION OF THE CARBON–CARBON DOUBLE BOND

5-2a Initiation

Various initiators can be used to bring about the polymerization of monomers with electron-releasing substituents [Gandini and Cheradame, 1980, 1985; Kennedy and Marechal, 1982; Sauvet and Sigwalt, 1989] as described below. Additional information on cationic initiators can be found in Chap. 7.

5-2a-1 *Protonic Acids*

Protonic (Brønsted) acids initiate cationic polymerization by protonation of the olefin. The method depends on the use of an acid that is strong enough to produce a reasonable concentration of the protonated species

$$HA + RR'C{=}CH_2 \rightarrow RR'\overset{+}{C}(A)^-_{\underset{|}{CH_3}} \tag{5-1}$$

but the anion of the acid should not be highly nucleophilic, otherwise it will terminate the protonated olefin by combination (i.e., by covalent bond formation)

$$RR'\overset{+}{\underset{\underset{CH_3}{|}}{C}}(A)^- \rightarrow RR'\overset{\overset{A}{|}}{C}{-}CH_3 \tag{5-2}$$

The method used in drawing the ionic species in the above two equations and throughout this chapter is meant to show that the ionic species usually do not exist as free ions but as ion pairs. The parentheses around the anionic fragment is used to indicate that the negative counterion remains close to the positive fragment. (Even when the counterion is not shown, one should understand it is present.) However, keep in mind that free ions coexist with the ion pairs, and it is the relative concentrations which determine the overall polymerization rate.

The nomenclature for positively charged organic ions has undergone some change. The older term, no longer used, for the trivalent, trigonal sp^2-hybridized species such as those in Eqs. 5-1 and 5-2 is *carbonium ion*. Olah [1972, 1988] proposed that *carbenium ion* be used instead with the term *carbonium ion* being reserved for pentavalent charged carbon ions (e.g., nonclassical ions) and the term *carbocation* encompassing both carbenium and carbonium ions. The term carbenium ion for the trivalent carbon ion has not taken firm hold. Most text and journal references use the term *carbocation*, and so will this text. The term *carbocation polymerization* is used synonymously with *cationic polymerization* in the literature.

The requirement for the anion not to be excessively nucleophilic generally limits the utility of most strong acids as cationic initiators. Hydrogen halides are ineffective as initiators of cationic polymerization because of the highly nucleophilic character of halide ions. One forms only the 1:1 addition product of alkene and hydrogen halide under most conditions. Hydrogen iodide shows some tendency to polymerize vinyl ethers, but the polymer yields and molecular weights are low (but see Sec. 5-2a-3-a). Other strong acids with less nucleophilic anions, such as perchloric, sulfuric, phosphoric, fluoro- and chlorosulfonic, methanesulfonic, and trifluoromethanesulfonic (triflic) acids, initiate cationic polymerization. However, the polymer molecular weights rarely exceed a few thousand.

5-2a-2 *Lewis Acids*

Various Lewis acids are used to initiate cationic polymerization, generally at low temperatures, with the formation of high-molecular-weight polymers in high yield. These include metal halides (e.g., $AlCl_3$, BF_3, $SnCl_4$, $SbCl_5$, $ZnCl_2$, $TiCl_4$, PCl_5), organometallic derivatives (e.g., $RAlCl_2$, RAl_2Cl, R_3Al), and oxyhalides (e.g., $POCl_3$, CrO_2Cl). (Many of the metal halides are familiar to chemists as Friedel–Crafts catalysts.) Lewis acids are by far the most important means of initiating cationic polymerizations. $AlCl_3$ is the most important Lewis acid for industrial cationic polymerizations.

Initiation by Lewis acids requires and/or proceeds faster in the presence of either a proton donor (*protogen*) such as water, hydrogen halide, alcohol, and carboxylic acid or a carbocation donor (*cationogen*) such as *t*-butyl chloride or triphenylmethyl chloride. Thus dry isobutylene is unaffected by dry boron trifluoride but polymerization occurs immediately when trace amounts of water are added [Evans and Meadows, 1950]. The terminology of Kennedy and Marechal [1982] is used in this book; the protogen or cationogen is referred to as the *initiator* while the Lewis acid is the *coinitiator*. The reader is cautioned that much of the published literature until recently used the reverse terminology. The protogen or cationogen is referred to as the initiator since it supplies the proton or cation which ultimately adds to monomer to initiate polymerization. The initiator and coinitiator, comprising an *initiating system*, react to form an *initiator–coinitiator complex* (or *syncatalyst system*) which then proceeds to donate a proton or carbocation to monomer and, thus, to initiate propagation [Dunn, 1979; Kennedy, 1976]. The initiation process for boron trifluoride and water is

$$BF_3 + H_2O \rightleftharpoons BF_3 \cdot OH_2 \tag{5-3a}$$

$$BF_3 \cdot OH_2 + (CH_3)_2C{=}CH_2 \rightarrow (CH_3)_3C^+(BF_3OH)^- \tag{5-4a}$$

(The initiator–coinitiator complex $BF_3 \cdot OH_2$ is often shown as $H^+(BF_3OH)^-$ but there is no evidence to support the latter type of species for any Lewis acids.) Initiation by aluminum chloride and t-butyl chloride is described by

$$AlCl_3 + (CH_3)_3CCl \rightleftharpoons (CH_3)_3C^+(AlCl_4)^- \qquad (5\text{-}3b)$$

$$(CH_3)_3C^+(AlCl_4)^- + \emptyset CH{=}CH_2 \rightarrow (CH_3)_3CCH_2\overset{+}{C}H\emptyset(AlCl_4)^- \qquad (5\text{-}4b)$$

The initiation process can be generalized as

$$I + ZY \overset{K}{\rightleftharpoons} Y^+(IZ)^- \qquad (5\text{-}3c)$$

$$Y^+(IZ)^- + M \overset{k_i}{\longrightarrow} YM^+(IZ)^- \qquad (5\text{-}4c)$$

where I, ZY, and M represent the coinitiator, initiator, and monomer, respectively. Initiation by the combination of a Lewis acid and protogen or cationogen has the advantage over initiation by a Brønsted acid that the anion IZ^- is far less nucleophilic than A^-. This prolongs the life of the propagating carbocation and allows propagation to proceed to higher molecular weight.

A considerable body of evidence indicates that a number of Lewis acids, especially those with higher acid strength such as $AlCl_3$ and $TiCl_4$, initiate polymerization by a *self-ionization* process in addition to the coinitiation process [Gandini and Cheradame, 1980, 1985; Grattan and Plesch, 1977, 1980; Masure et al., 1978; 1980]. The generally accepted mechanism for initiation by self-ionization involves bimolecular ionization

$$2AlBr_3 \rightleftharpoons AlBr_2^+(AlBr_4)^- \qquad (5\text{-}3d)$$

followed by reaction with monomer,

$$AlBr_2^+(AlBr_4)^- + M \rightarrow AlBr_2M^+(AlBr_4)^- \qquad (5\text{-}4d)$$

The Lewis acid acts as both the initiator and coinitiator in this process. An alternate self-ionization mechanism is the direct addition of initiator to monomer,

$$TiCl_4 + M \rightarrow TiCl_3M^+Cl^- \qquad (5\text{-}5)$$

Closely related to the self-ionization process is the increased initiation which occurs with certain mixtures of Lewis acids, such as $TiCl_4$ and $AlCl_3$ or $FeCl_3$ and BCl_3 [Marek et al., 1988]. Initiation involves a process analogous to Eq. 5-3d

$$FeCl_3 + BCl_3 \rightleftharpoons FeCl_2^+(BCl_4^-) \qquad (5\text{-}3e)$$

Most of the experimental evidence to support the self-ionization process is indirect, consisting of kinetic, conductance, and spectrophotometric data for reaction systems at different levels of dryness and purity. In the typical experiment, one concludes that self-ionization occurs if polymerization is achieved in reaction systems subjected to the most stringent purification and drying procedures. The major problem in ascertaining whether self-initiation occurs in a particular reaction system, and, if it does, its extent relative to the initiation process involving a coinitiator is the large effect exerted by small amounts of proton or carbocation donors. Thus water concentrations of 10^{-3} M are sufficient to increase the initiation rate by a factor of 10^3 for $TiCl_4$ and $AlCl_3$ in CH_2Cl_2 [Masure et al., 1978, 1980; Sauvet et al., 1978]. Polymerizations in the presence of sterically hindered pyridines such as 2,6-di-t-butylpyridine and 2,4,6-tri-t-butylpyridine offer further evidence for the presence of a self-ionization initiation process [Gandini and Martinez, 1988]. Sterically hindered pyridines (SHP) are generally active proton scavengers but do not react with Lewis acids or carbocations; in other words, steric hindrance cuts down on the reactivity of the nitrogen toward electrophilic species larger than protons. The presence of SHP results in complete inhibition of polymerization in some systems but only lowered reaction rates in other systems. For the latter, one observes a continuous decrease in polymerization rate with increasing concentration of SHP up to some critical SHP concentration. Thereafter, there is a residual (self-initiation) polymerization rate that is unaffected by increasing SHP concentration. The SHP method is not completely without ambiguity since SHP may not scavenge protons efficiently in a heterogeneous reaction system or in systems where protons never appear as such but are transferred in a concerted process. These and other data [Grattan and Plesch, 1977, 1980; Kennedy, 1976] do, however, show that when self-ionization occurs, its contribution to the overall initiation process is small in the presence of a proton or carbocation donor. For polymerizations carried out under most reaction conditions, the moisture content (and/or level of other carbocation or proton donors) is often sufficient so that self-ionization constitutes only a minor proportion of the total initiation process. (Conventional "dry-box" conditions usually do not involve moisture levels lower than 10^{-3} M [Kennedy, 1976].)

The activity of an initiator-coinitiator complex is dependent on its ability to donate a proton or carbocation to the monomer which, in turn, depends on the initiator, coinitiator, solvent, and monomer. The extent of formation of the initiator-coinitiator complex (i.e., the value of K in Eq. 5-3c) and its rate of addition to monomer (i.e., the value of k_i in Eq. 5-4c) generally increase with increasing acidity of the Lewis acid coinitiator. Thus the general order of activity of aluminum coinitiators corresponds to their order of acidity: $AlCl_3 > AlRCl_2 > AlR_2Cl > AlR_3$ and $AlR_2I > AlR_2Br > AlR_2Cl$ [DiMaina et al., 1977; Kennedy and Marechal, 1982; Magagnini et al., 1977]. The activity of the initiator-coinitiator complex also increases with increasing acidity of the initiator, for example, hydrogen chloride $>$ acetic acid $>$ nitroethane $>$ phenol $>$ water $>$ methanol $>$ acetone in the polymerization of isobutylene with tin (IV) chloride [Kennedy, 1968; Plesch, 1963]. A word of caution regarding these generalizations: the order of activity of a series of initiators or coinitiators may differ depending on the identity of the other component, monomer, solvent, or the presence of competing reactions. For example, the activity of boron halides in isobutylene polymerization, $BF_3 > BCl_3 > BBr_3$, with water as the initiator is the opposite of their acidities. Hydrolysis of the boron halides to inactive products, increasing in the order $BBr_3 > BCl_3 > BF_3$, is responsible for the observed polymerization results [Kennedy et al., 1977, Kennedy and Feinberg, 1978].

The reactivity of organic halide cationogens in initiation depends on carbocation stability in a complex manner. Increased carbocation stability results in the formation of higher concentrations of carbocations from the cationogen but the carbocations have lower reactivity. Differences in the stability of the carbocation formed from the cationogen compared to the propagating carbocation are also important in determining the effectiveness of a cationogen. Primary and secondary alkyl halides are generally ineffective as initiators of cationic polymerization. Primary and secondary carbocations are formed too slowly and/or in extremely low concentrations. (There are a few reports of initiation by primary or secondary halides [Colclough and Dainton, 1958; Toman et al., 1989a, 1989b], but initiation more likely involves self-ionization of the Lewis acid or the presence of adventitious water.) Tertiary carbocations such as *t*-butyl and 2-phenyl-*i*-propyl (cumyl) are sufficiently stable to form but are not more stable than the carbocations derived from their additions to monomers such as isobutylene, styrene, or *N*-vinylcarbazole, so that polymerizations of those monomers occur. Cumyl

N-Vinylcarbazole

and *t*-butyl carbocations have been generated from cumyl and *t*-butyl esters and ethers as well as from the halides [Faust and Kennedy, 1987; Mishra and Kennedy, 1987]. Highly stable carbocations such as trityl, $\phi_3 C^+$ and cycloheptatrienyl (tropylium), $C_7 H_7^+$ are generally too stable to be very efficient in polymerizing the less reactive monomers such as isobutylene and styrene but polymerization of the more reactive monomers (*p*-methoxystyrene, vinyl ethers, indene, *N*-vinylcarbazole) proceeds rapidly [Ledwith, 1979a, 1979b; Rooney, 1978]. (Trityl and tropylium carbocations are sufficiently stable that their salts with stable anions such as hexafluoroantimonate, SbF_6^-, can be purchased in pure crystalline form from chemical vendors.) Acylium ions (oxocarbocations) have also been used to initiate cationic polymerization [Sawamoto et al., 1977, 1978]. These can be formed *in situ* or separately prepared as the solid salt, which is subsequently added to the monomer solution. Acetyl perchlorate,

$$ R-\overset{\overset{\displaystyle O}{\|}}{C}-F + SbF_5 \rightarrow R-\overset{\overset{\displaystyle O}{\|}}{C^+}(SbF_6^-) \tag{5-6} $$

$CH_3CO-OClO_3$, is covalent but is assumed to generate acetyl ions via a dissociation equilibrium.

Many polymerizations exhibit a maximum polymerization rate at some ratio of initiator to coinitiator [Biswas and Kabir, 1978, 1979; Colclough and Dainton, 1958; Taninaka and Minoura, 1976]. The polymerization rate increases with increasing [initiator]/[coinitiator], reaches a maximum, and then either decreases or levels off. Figure 5-1 shows this behavior for the polymerization of styrene initiated by tin(IV) chloride-water in carbon tetrachloride. The decrease in rate at higher initiator concentration is usually ascribed to inactivation of the coinitiator by initiator. The inactivation process in a system such as $SnCl_4$-H_2O may involve hydrolysis of Sn—Cl bonds to Sn—OH. There is experimental evidence for such reactions when comparable concentrations of coinitiator and initiator are present. However, the rate maxima as in Fig. 5-1 are observed at quite low [initiator]/[coinitiator] ratios where corresponding experimental evidence is lacking. An alternate mechanism for the behavior in Fig. 5-1 is that initiator, above a particular concentration, competes successfully with monomer for the initiator-coinitiator complex (**V**),

$$ \underset{\textbf{V}}{SnCl_4 + H_2O \rightleftharpoons SnCl_4 \cdot OH_2} \overset{H_2O}{\underset{}{\rightleftharpoons}} \underset{\textbf{VI}}{(H_3O^+)(SnCl_4OH^-)} \tag{5-7} $$

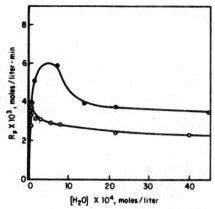

Fig. 5-1 Effect of water concentration on the $SnCl_4$ initiated polymerization rate of styrene in carbon tetrachloride at 25°C. Symbols $\circ$ and $\bullet$ refer to initiator concentrations of 0.08 and 0.12 M, respectively. After Colclough and Dainton [1958] (by permission of the Faraday Society, London).

to yield the oxonium salt (**VI**) which is too unreactive in protonating olefins because the basicity of the carbon-carbon double bond is far less than that of water [Ledwith and Sherrington, 1974] (see also Sec. 5-2c-5). The optimum initiator to coinitiator ratio varies considerably depending on the initiator, coinitiator, monomer, solvent, and temperature, since these factors affect the balance between the competing processes of initiation and inactivation.

5-2a-3 Other Initiators

5-2a-3-a Iodine. Chlorine, bromine, and iodine act as cationogens in the presence of the more active Lewis acids such as trialkyl aluminum or dialkylaluminum halide [DiMaina et al., 1977; Magagnini et al., 1977]. The initiating species is the halonium ion X^+ present in low concentration via the equilibrium reaction between Lewis acid and halogen.

Iodine is unique among the halogens in that it initiates polymerization of the more reactive monomers (styrene, vinyl ether, acenaphthylene. N-vinylcarbazole) even in

Acenaphthylene

the absence of a Lewis acid [Johnson and Young, 1976; Sauvet and Sigwalt, 1989]. Iodine is not the actual initiator in this system. Iodine adds to the double bond to form a diiodide that eliminates hydrogen iodide. The hydrogen iodide generated by this process acts as the cationogen with iodine acting as a Lewis acid to form the initiating system. It was noted earlier (Sec. 5-2a-1) that hydrogen iodide is not an

$$I_2 + CH_2{=}CH \longrightarrow ICH_2{-}CHI \xrightarrow{-HI} ICH{=}CH \qquad (5\text{-}8)$$
$$\qquad\;\; | \qquad\qquad\qquad | \qquad\qquad\qquad |$$
$$\qquad\;\; OR \qquad\qquad\quad OR \qquad\qquad\quad OR$$

efficient initiator because iodide ion is too nucleophilic. The presence of iodine "activates" (polarizes or dissociates) the C—I bond sufficiently to allow propagation to proceed. (See Sec. 5-2g-1 for details of the propagation mechanism.) This initiation route is more efficiently utilized by directly adding a mixture of hydrogen iodide and either iodine or a metal halide such as ZnX_2 or SnX_2 to the reaction system [Highashimura et al., 1988].

5-2a-3-b *Photoinitiation.* Aryldiazonium ($ArN_2^+Z^-$), diaryliodonium ($Ar_2I^+Z^-$), and triarylsulfonium ($Ar_3S^+Z^-$) salts, where Z^- is a nonnucleophilic and photostable anion such as tetrafluoroborate (BF_4^-), hexafluoroantimonate (SbF_6^-), and hexafluorophosphate (PF_6^-), are effective photoinitiators of cationic polymerization [Crivello, 1984; Crivello et al., 1988; Ledwith, 1977, 1979; Yagci and Schnabel, 1988]. Aryldiazonium salts have limited practical utility because of their inherent thermal stability. Diaryliodonium and triarylsulfonium salts are very stable—so stable that their mixtures with highly reactive monomers do not undergo polymerization on long-term storage. Some of these initiators have found commercial application in the photocrosslinking of epoxy resins through cationic polymerization.

Diaryliodonium and triarylsulfonium salts act as photoinitiators of cationic polymerization. Photolytic cleavage of an Ar—I or Ar—S bond yields a radical–cation (Eq. 5-9) that reacts with HY to yield an initiator–coinitiator complex that acts as a proton

$$Ar_2I^+(PF_6^-) \xrightarrow{h\nu} ArI^{+\cdot}(PF_6^-) + Ar\cdot \qquad (5\text{-}9a)$$

$$\downarrow HY$$

$$ArI + Y\cdot + H^+ (PF_6^-) \qquad (5\text{-}10a)$$

$$Ar_3S^+(SbF_6^-) \xrightarrow{h\nu} Ar_2S^{+\cdot}(SbF_6^-) + Ar\cdot \qquad (5\text{-}9b)$$

$$\downarrow HY$$

$$Ar_2S + Y\cdot + H^+ (SbF_6^-) \qquad (5\text{-}10b)$$

donor to initiate cationic polymerization. HY may be solvent or some other deliberately added substance such as an alcohol (or adventitious impurity, including water) with a labile hydrogen. Overall, the process is a photolytically induced redox reaction between the cation–radical and HY. Interestingly, the same result has been achieved thermally without photolysis by coupling the appropriate reducing agent (e.g., ascorbic acid or copper(I) benzoate) with a diaryliodonium salt [Crivello et al., 1983]. The spectral response for photoinitiation by the diaryliodonium and triarylsulfonium salts can be extended to longer wavelengths by either appropriate alterations in the aromatic groups and/or the use of photosensitizers.

5-2a-3-c *Electroinitiation.* *Electrolytic* or *electroinitiated* polymerization involves initiation by cations formed via electrolysis of some component of the reaction system (monomer, solvent, electrolyte, or other deliberately added substance) [Cerrai et al., 1976, 1979; Funt et al., 1976; Oberrauch et al., 1978; Olaj, 1987]. Thus initiation in the presence of perchlorate ion proceeds by oxidation of perchlorate followed by hydrogen abstraction

$$ClO_4^- \xrightarrow{-e} ClO_4 \cdot \xrightarrow{HY} HClO_4 \tag{5-11}$$

where HY is some hydrogen donor in the system. Perchloric acid is the actual initiating species.

Electrolytic polymerization in the presence of polynuclear aromatic compounds such as perylene proceeds via oxidation to the stable radical–cation **VII** which initiates

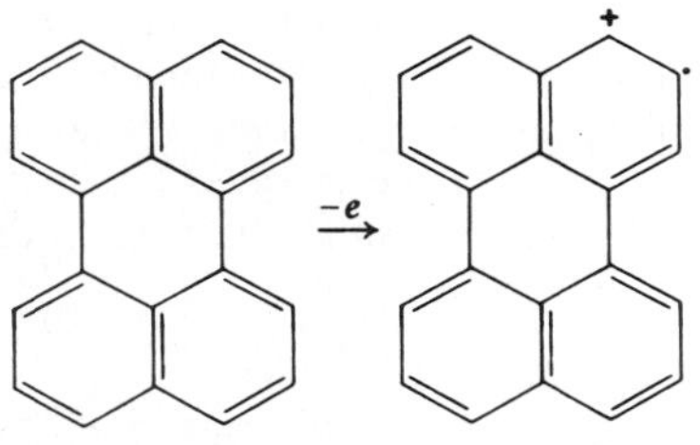

$$\tag{5-12}$$

Perylene   **VII**

polymerization by *electron transfer* from monomer to **VII** to form the monomer radical–cation **VIII**.

$$VII + CH_2{=}CH \rightarrow Perylene + \overset{\cdot}{C}H_2{=}\overset{+}{C}H \rightarrow \overset{+}{C}H{-}CH_2CH_2{-}\overset{+}{C}H \tag{5-13}$$

$$\hspace{2.5cm} | \hspace{4.2cm} | \hspace{1.0cm} | \hspace{2.5cm} |$$
$$\hspace{2.5cm} OR \hspace{4.0cm} OR \hspace{0.7cm} OR \hspace{2.2cm} OR$$

            VIII   **IX**

Propagation is considered to proceed through the dicarbocation **IX** formed by dimerization of **VIII**, although direct evidence for the dicarbocation is lacking.

Some electroinitiated polymerizations appear to proceed via monomer radical–cations (**VIII**) formed directly by electron transfer from monomer to the anode. Monomer radical–cations have also been postulated as the initiating species in some polymerizations when monomers are heated or photolyzed in the presence of electron acceptors such as metal halides, 1,2,4,5-tetracyanobenzene, and pyromellitic anhydride [Halaska et al., 1986; Irie and Hayashi, 1975; Marek and Toman, 1980]. The monomer–electron acceptor complex (which may be a charge-transfer complex) undergoes electron transfer to produce the monomer radical–cation, which is then the initiator, for example,

$$VCl_4 + M \rightarrow complex \xrightarrow[\text{or } h\nu]{\text{heat}} VCl_4^- + M^{\ddagger} \tag{5-14}$$

5-2a-3-d Ionizing Radiation. Ionizing radiations (electrons, neutrons, γ- and β-rays, etc.) initiate cationic polymerization [Deffieux et al., 1983; Stannett et al., 1989]. The mechanism for initiation is not well established, although it is generally accepted that the first event involves the formation of a radical–cation such as **VIII** by the ejection of a π-electron. The radical–cation can undergo subsequent reactions to form other radical, anionic, and cationic species (Sec. 3-4d). Whether one observes radical, cationic, or anionic polymerization depends on the monomer and reaction conditions. Styrene can undergo polymerization by all three mechanisms. For superdry reaction systems at 25°C, the overwhelming mechanism for polymerization is cationic with about 2.5% anionic and negligible radical reaction. Radical polymerization becomes the dominant process with increasing reaction temperature. The presence of water or other protogens markedly decreases the extent of ionic polymerization relative to radical polymerization. Isobutylene shows negligible tendency to undergo any polymerization except cationic, and that occurs only at lower temperatures (usually considerably below 0°C). At higher temperatures, no polymerization occurs since the cationic reaction is not favored.

The actual species responsible for cationic polymerizations initiated by ionizing radiation is not established. The most often described mechanism postulates reaction between radical–cation and monomer to form separate cationic and radical species; subsequently, the cationic species propagates rapidly while the radical species propagates very slowly. The proposed mechanism for isobutylene involves transfer of a hydrogen radical from monomer to the radical–cation to form the t-butyl carbocation and an unreactive allyl-type radical:

$$(CH_3)_2C{=}CH_2 \xrightarrow{\text{radiation}} (CH_3)_2\overset{+}{C}{-}\overset{\cdot}{C}H_2$$

$$\downarrow {\scriptstyle (CH_3)_2C{=}CH_2}$$

$$(CH_3)_3C^+ + CH_3{-}\underset{\underset{CH_2\cdot}{|}}{C}{=}CH_2 \qquad (5\text{-}15)$$

The evidence for this mechanism is based on mass spectroscopy of the gas phase radiolysis of isobutylene, which may not be applicable to the typical liquid phase polymerization system. Initiation in condensed systems probably follows the same course as electroinitiation and UV or thermal initiations of monomer–electron acceptor systems (Sec. 5-2a-3-d), specifically, coupling of radical–cations to form dicarbocations (**IX**).

Cationic polymerization initiated by ionizing radiation is markedly different from other cationic polymerizations in that the propagating species is a free ion remote from a counterion. Overall electrical neutrality is maintained by electrons trapped by the monomer.

5-2b Propagation

The initiator ion pair (consisting of the carbocation and its negative counterion) produced in the initiation step (Eq. 5-4) proceeds to propagate by successive additions of monomer molecules

$$H{-}[CH_2C(CH_3)_2{-}]_n^+ (BF_3OH)^- + (CH_3)_2C{=}CH_2 \rightarrow$$

$$H{-}[CH_2C(CH_3)_2{-}]_n{-}CH_2\overset{+}{C}(CH_3)_2(BF_3OH)^- \qquad (5\text{-}16a)$$

or

$$HM_n^+(IZ)^- + M \xrightarrow{k_p} HM_nM^+(IZ)^- \tag{5-16b}$$

This addition proceeds by insertion of monomer between the carbocation and its negative counterion.

The propagation reaction can be complicated in some cases by the occurrence of intramolecular rearrangements due to 1,2-hydride ion ($H:^-$) or 1,2-methide ($CH_3:^-$) shifts. Polymerizations proceeding with rearrangement are referred to as *isomerization polymerizations*. The extent of rearrangement during cationic propagation will depend on the relative stabilities of the propagating and rearranged carbocations and the relative rates of propagation and rearrangement. Both factors favor propagation without rearrangement for monomers such as styrene, indene, acenaphthylene, coumarone, vinyl ethers, and isobutylene. Not only do these monomers propagate via

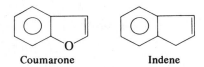

Coumarone   Indene

reasonably stable carbocations such as tertiary, benzyl, and oxycarbocations; the carbocations have no routes available for rearrangement to more stable carbocations. Extensive rearrangement during propagation occurs for a variety of 1-alkenes (α-olefins) [Cesca, 1985]. Propylene, 1-butene, and higher 1-alkenes yield oligomers (DP no higher than 10–20) with highly irregular structures due to various combinations of 1,2-hydride and 1,2-methide shifts. For example, propylene polymerization proceeds to give an extremely complicated oligomer structure with methyl, ethyl, *n*- and *i*-propyl, and other groups. (Hydride transfer to polymer is also involved; see Sec. 5-2c-4.) The propagating secondary carbocations are insufficiently stable to propagate without extensive rearrangement. Simultaneously, only oligomers are formed since none of the rearrangement pathways are favorable for rapid propagation (relative to a variety of chain-transfer and termination reactions).

Isomerization polymerizations yield high-molecular-weight products when reaction proceeds through relatively simple rearrangement routes involving stable carbocations. Thus polymerization of 3-methyl-1-butene yields high polymer containing both the first-formed (**Xa**) and rearranged (**Xb**) repeating units in varying amounts depending

$$-CH_2-\underset{\underset{\displaystyle CH(CH_3)_2}{|}}{CH}- \qquad -CH_2-CH_2-C(CH_3)_2-$$

Xa   **Xb**

on the reaction temperature [Kennedy et al., 1964]. Isomerization occurs by a 1,2-hydride ion shift in the first-formed propagating carbocation (**XIa**) prior to the addition of the next monomer unit. The rearranged ion (**XIb**) is a tertiary carbocation and is more stable than the first-formed carbocation, which is a secondary carbocation. The product contains mostly the rearranged repeating unit but some normal propagation occurs at higher temperatures due to kinetic reasons. The product contains about 70

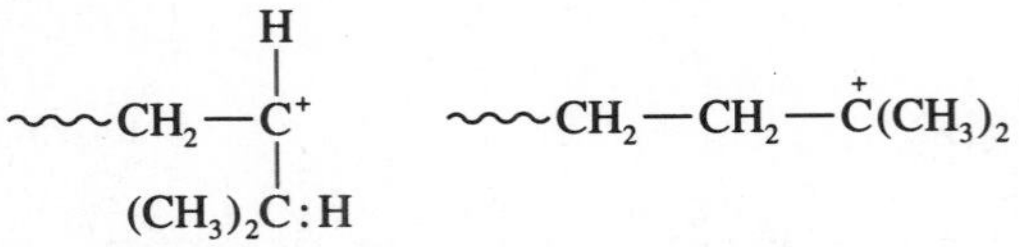

$$\begin{array}{c} \text{H} \\ | \\ \sim\!\!\sim\!\!\text{CH}_2\!-\!\overset{+}{\text{C}} \\ | \\ (\text{CH}_3)_2\text{C}\!:\!\text{H} \end{array} \qquad\qquad \sim\!\!\sim\!\!\text{CH}_2\!-\!\text{CH}_2\!-\!\overset{+}{\text{C}}(\text{CH}_3)_2$$

$$\qquad\qquad \textbf{XIa} \qquad\qquad\qquad\qquad\qquad \textbf{XIb}$$

and 100% of the rearranged repeating unit at polymerization temperatures of -130 and $-100°C$, respectively.

The use of high resolution ^{1}H and C^{13} NMR has made clear that there is great complexity to the rearrangements occurring for certain monomers. Five different repeating units, derived from carbocations **XII–XVI**, are found in the polymer from 4-methyl-1-butene [Ferraris et al., 1977; Kennedy and Johnston, 1975]. The first-formed carbocation **XII** undergoes hydride shifts to form carbocations **XIII**, **XIV**, and **XV**;

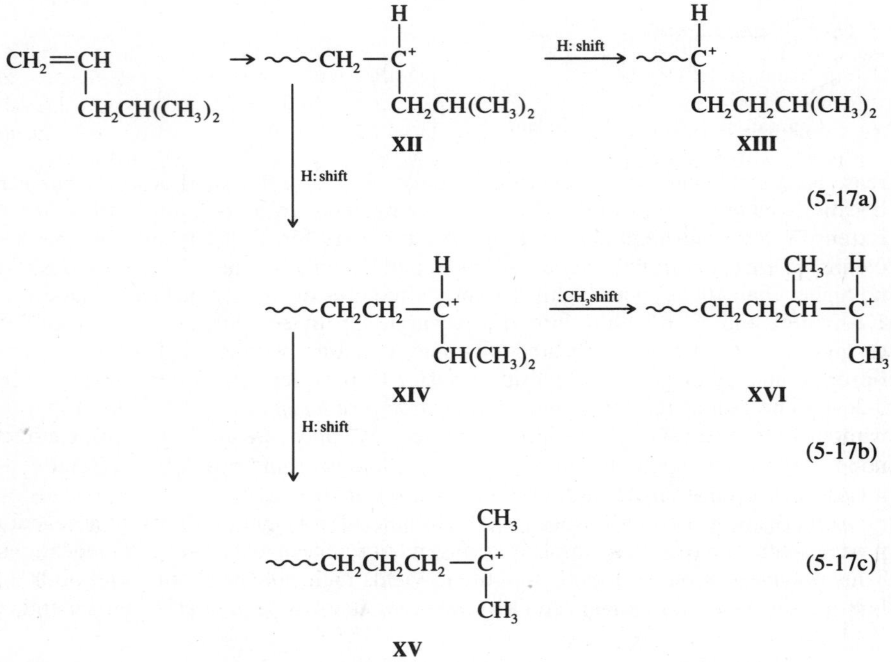

$$\begin{array}{ccc} \text{CH}_2\!=\!\text{CH} & & \\ | & \rightarrow & \sim\!\!\sim\!\!\text{CH}_2\!-\!\overset{+}{\text{C}} \\ \text{CH}_2\text{CH}(\text{CH}_3)_2 & & | \\ & & \text{CH}_2\text{CH}(\text{CH}_3)_2 \\ & & \textbf{XII} \end{array} \xrightarrow{\text{H: shift}} \begin{array}{c} \text{H} \\ | \\ \sim\!\!\sim\!\!\overset{+}{\text{C}} \\ | \\ \text{CH}_2\text{CH}_2\text{CH}(\text{CH}_3)_2 \\ \textbf{XIII} \end{array}$$

$$\text{(5-17a)}$$

(H: shift, downward arrow)

$$\begin{array}{c} \text{H} \\ | \\ \sim\!\!\sim\!\!\text{CH}_2\text{CH}_2\!-\!\overset{+}{\text{C}} \\ | \\ \text{CH}(\text{CH}_3)_2 \\ \textbf{XIV} \end{array} \xrightarrow{:\text{CH}_3\text{shift}} \begin{array}{c} \text{CH}_3 \quad \text{H} \\ | \qquad | \\ \sim\!\!\sim\!\!\text{CH}_2\text{CH}_2\text{CH}\!-\!\overset{+}{\text{C}} \\ | \\ \text{CH}_3 \\ \textbf{XVI} \end{array}$$

$$\text{(5-17b)}$$

(H: shift, downward arrow)

$$\begin{array}{c} \text{CH}_3 \\ | \\ \sim\!\!\sim\!\!\text{CH}_2\text{CH}_2\text{CH}_2\!-\!\overset{+}{\text{C}} \\ | \\ \text{CH}_3 \\ \textbf{XV} \end{array} \qquad\qquad \text{(5-17c)}$$

XIV rearranges to **XVI** by a methide shift. The repeating unit derived from **XV**, the most stable carbocation, is present in the greatest abundance (42–51%). The other carbocations are of comparable stability and the repeating units derived from them are found in comparable amounts.

The driving force in some isomerization polymerizations is relief of steric strain. Polymerization of β-pinene proceeds by the first-formed carbocation **XVII** rearranging to **XVIII** via cleavage of the strained four-membered ring and migration of the resulting *gem*-dimethyl carbanion center [Kennedy and Marechal, 1982].

Other monomers that undergo isomerization polymerization include 5-methyl-1-hexene, 4,4-dimethyl-1-pentene, 6-methyl-1-heptene, α-pinene, and vinylcyclopropane [Cesca, 1985; Corno et al., 1979].

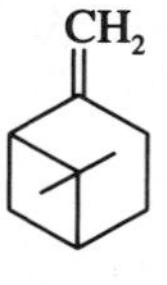

β-Pinene 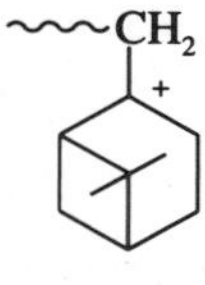   XVII 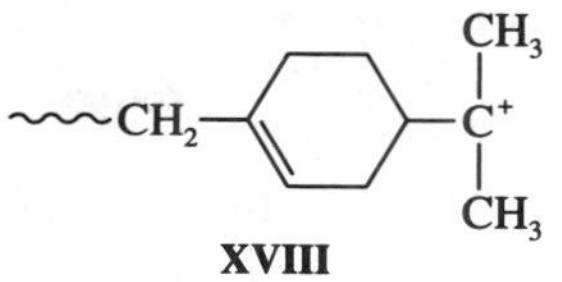   XVIII

5-2c Chain Transfer and Termination

Various reactions lead to termination of chain growth in cationic polymerization [Dunn, 1979; Gandini and Cheradame, 1985; Kennedy, 1975; Kennedy and Marechal, 1982; Ledwith and Sherrington, 1974; Lenz, 1967]. Many of the reactions that terminate the growth of a propagating chain do not, however, terminate the kinetic chain because a new propagating species is generated in the process.

5-2c-1 Chain Transfer to Monomer

Chain transfer to monomer is probably the most important chain-breaking reaction for most monomers. There are few polymerizations where it does not occur. Transfer to monomer involves transfer of a β-proton from the carbocation to a monomer molecule with the formation of terminal unsaturation in the polymer molecule

$$H \text{---} [\text{---} CH_2C(CH_3)_2 \text{---}]_n \text{---} CH_2\overset{+}{C}(CH_3)_2(BF_3OH)^- + CH_2 \text{=} C(CH_3)_2 \rightarrow$$

$$(CH_3)_3C^+(BF_3OH)^- + H \text{---} [\text{---} CH_2C(CH_3)_2 \text{---}]_n \text{---} CH_2C(CH_3) \text{=} CH_2$$

$$+ H \text{---} [\text{---} CH_2C(CH_3)_2 \text{---}]_n \text{---} CH \text{=} C(CH_3)_2 \tag{5-18a}$$

or

$$HM_nM^+(IZ)^- + M \xrightarrow{k_{tr,M}} M_{n+1} + HM^+(IZ)^- \tag{5-18b}$$

There are two different types of β-protons, and, therefore, two different unsaturated end groups are possible for isobutylene as well as some other monomers such as indene and α-methylstyrene. The relative amounts of the two end groups depend on the counterion, identity of the propagating center, and other reaction conditions. Only one type of unsaturated end group (internal) is possible for other monomers such as styrene, ethyl vinyl ether, and coumarone.

It should be noted that the kinetic chain is not terminated by this reaction since a new propagating species is regenerated. Many polymer molecules are usually produced for each initiator-coinitiator species present. Chain transfer to monomer is on much more favorable terms with propagation in many cationic polymerizations compared to radical polymerization. Since it is kinetically indistinguishable from propagation, the relative rates of transfer and propagation are given by the ratio $k_{tr,M}/k_p$, which is the chain-transfer constant for monomer C_M. The value of C_M determines the molecular weight of the polymer if other chain-breaking processes are not significant. The larger the value of C_M the lower will be the molecular weight.

Chain transfer to monomer is the principal reaction that limits polymer molecular weight for most monomers, especially at reaction temperatures higher than about 20°C. Since chain transfer to monomer generally has a higher activation energy than propagation, it is usually suppressed by working at lower reaction temperatures.

Another type of chain transfer to monomer reaction is that involving hydride ion transfer from monomer to the propagating center [Kennedy and Squires, 1967].

$$H\text{---}[\text{---}CH_2C(CH_3)_2\text{---}]_n\text{---}CH_2\overset{+}{C}(CH_3)_2(BF_3OH)^- + CH_2\text{=}C(CH_3)_2 \rightarrow$$
$$CH_2\text{=}C(CH_3)\text{---}\overset{+}{C}H_2(BF_3OH)^- + H\text{---}[\text{---}CH_2C(CH_3)_2\text{---}]_n\text{---}CH_2CH(CH_3)_2$$

$$(5\text{-}19)$$

This reaction may account in part for the oligomers obtained in the polymerization of propylene, 1-butene, and other 1-alkenes where the propagation reaction is not highly favorable (due to the low stability of the propagating carbocation). Unreactive 1-alkenes and 2-alkenes have been used to control polymer molecular weight in cationic polymerization of reactive monomers, presumably by hydride transfer to the unreactive monomer. The importance of hydride ion transfer from monomer is not established for the more reactive monomers. For example, hydride transfer by monomer is less likely a mode of chain termination compared to proton transfer to monomer for isobutylene polymerization since the tertiary carbocation formed by proton transfer is more stable than the allyl carbocation formed by hydride transfer. Similar considerations apply to the polymerizations of other reactive monomers. Hydride transfer is not a possibility for those monomers without easily transferable hydrogens, such as N-vinylcarbazole, styrene, vinyl ethers, and coumarone.

The two types of chain transfer to monomer are kinetically indistinguishable, but one (Eq. 5-18) results in unsaturated end groups, while the other (Eq. 5-19) results in saturated end groups.

5-2c-2 *Spontaneous Termination*

Termination can also take place by rearrangement of the propagating ion pair. *Spontaneous termination* involves regeneration of the initiator-coinitiator complex by expulsion from the propagating ion pair with the polymer molecule left with terminal unsaturation.

$$H\text{---}[\text{---}CH_2C(CH_3)_2\text{---}]_n\text{---}CH_2\overset{+}{C}(CH_3)_2(BF_3OH)^- \rightarrow$$
$$BF_3\cdot OH_2 + H\text{---}[\text{---}CH_2C(CH_3)_2\text{---}]_n\text{---}CH_2C(CH_3)\text{=}CH_2 \qquad (5\text{-}20a)$$

or, in more general terms,

$$HM_nM^+(IZ)^- \xrightarrow{k_{ts}} M_{n+1} + H^+(IZ)^- \qquad (5\text{-}20b)$$

This type of termination, also referred to as *chain transfer to counterion*, differs kinetically from chain transfer to monomer in that the rate of chain transfer to monomer has a first-order dependence on monomer while chain transfer to counterion does not depend on monomer. Chain transfer to counterion is almost never a dominant termination reaction compared to chain transfer to monomer. Chain transfer to counterion is similar to chain transfer to monomer in not terminating the kinetic chain.

5-2c-3 *Combination with Counterion*

Termination by combination of the propagating center with the counterion

$$HM_nM^+(IZ)^- \xrightarrow{k_t} HM_nMIZ \qquad (5\text{-}21a)$$

occurs, for example, in the trifluoroacetic acid initiated polymerization of styrene [Throssell et al., 1956]

$$H \underbrace{\left(-CH_2CH\phi \right)_n} CH_2\overset{+}{C}H\phi(OCOCF_3)^- \rightarrow$$

$$H \underbrace{\left(-CH_2CH\phi \right)_n} CH_2CH\phi - OCOCF_3 \qquad (5\text{-}21b)$$

Alternately, the propagating ion may combine with an anionic fragment from the counterion, for example

$$H \underbrace{\left[-CH_2C(CH_3)_2 \right]_n} CH_2\overset{+}{C}(CH_3)_2(BF_3OH)^- \rightarrow$$

$$H \underbrace{\left[-CH_2C(CH_3)_2 \right]_n} CH_2C(CH_3)_2OH + BF_3 \qquad (5\text{-}22)$$

or

$$H \underbrace{\left[-CH_2C(CH_3)_2 \right]_n} CH_2\overset{+}{C}(CH_3)_2(BCl_3OH)^- \rightarrow$$

$$H \underbrace{\left[-CH_2C(CH_3)_2 \right]_n} CH_2C(CH_3)_2Cl + BCl_2OH \qquad (5\text{-}23)$$

Termination by combination differs from the other modes of termination in that the kinetic chain is usually terminated, since the concentration of the initiator-coinitiator complex decreases.

Equations 5-22 and 5-23 indicate the complexity of cationic polymerization even when seemingly similar initiators such as BCl_3 and BF_3 are used. Termination in the BCl_3-initiated polymerizations of isobutylene and styrene occurs almost exclusively by combination with chloride [Kennedy and Feinberg, 1978; Kennedy et al., 1977]. For BF_3 initiated polymerization, chain transfer to monomer is the major mode of chain breaking with a minor contribution by combination with OH. The differences are explained by the order of bond strengths: B—F > B—O > B—Cl [Jolly, 1984]. A similar situation is found in the polymerization of styrene by trityl salts. Polymerization occurs readily when the counterion is SbF_6^- but poorly with the corresponding hexachloroantimonate counterion [Johnson and Pearce, 1976]. Chloride ion easily transfers from the counterion to terminate the propagating center, while fluoride is inactive toward transfer.

Combination with counterion is also important when aluminum alkyl–alkyl halide initiating systems are used [DiMaina et al., 1977; Kennedy, 1976; Reibel et al., 1979]. Termination occurs either by *alkylation*

$$\sim\sim\sim CH_2 - \overset{+}{C}(CH_3)_2(R_3AlCl)^- \rightarrow \sim\sim\sim CH_2 - C(CH_3)_2R + R_2AlCl \qquad (5\text{-}24)$$

or *hydridation*

$$\sim\sim\sim CH_2 - \overset{+}{C}(CH_3)_2([CH_3CH_2]_3AlCl)^- \rightarrow$$

$$\sim\sim\sim CH_2CH(CH_3)_2 + CH_2{=}CH_2 + (CH_3CH_2)_2AlCl \qquad (5\text{-}25)$$

Alkylation involves transfer of an alkyl anion to the propagating center. Hydridation involves transfer of a hydride ion from the alkyl anion to the propagating center. Hydridation occurs in preference to alkylation when the alkyl aluminum contains β-hydrogens.

5-2c-4 Chain Transfer to Polymer

Several *chain transfer to polymer* reactions are possible in cationic polymerization. Transfer of the cationic propagating center can occur either by electrophilic aromatic substitution or hydride transfer. *Intramolecular electrophilic aromatic substitution* (or *backbiting*) occurs in the polymerization of styrene as well as other aromatic monomers

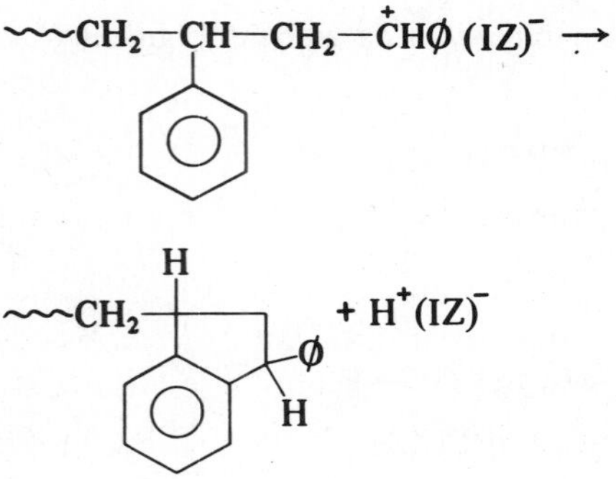

$$\text{~~~CH}_2\text{—CH—CH}_2\text{—}\overset{+}{\text{C}}\text{H}\phi\ (IZ)^- \rightarrow$$

$$\text{~~~CH}_2 \overset{H}{\underset{\phi}{\diagdown}} + H^+(IZ)^- \qquad\qquad (5\text{-}26)$$

with the formation of terminal indanyl structures and regeneration of the initiator-coinitiator complex [Hatada et al., 1980; Rooney, 1976, 1978].

Some branching has been detected in the polymerizations of styrene and anethole (β-methyl-*p*-methoxystyrene), indicating *intermolecular aromatic substitution* by a propagating carbocation on the aromatic ring of another polymer chain [Hatada et al., 1980; Kennedy and Marechal, 1982].

Intermolecular hydride transfer to polymer probably accounts for the short-chain branching found in the polymerizations of 1-alkenes such as propylene. The propagating carbocations are reactive secondary carbocations that can abstract tertiary hy-

$$\text{~~~CH}_2\text{—}\underset{H}{\overset{+}{\text{C}}}\text{R} + \text{~~~CH}_2\text{—}\underset{R}{\overset{H}{\text{C}}}\text{~~~} \longrightarrow$$

$$\text{~~~CH}_2\text{—}\underset{H}{\overset{H}{\text{C}}}\text{R} + \text{~~~CH}_2\text{—}\underset{R}{\overset{+}{\text{C}}}\text{~~~} \qquad\qquad (5\text{-}27)$$

drogens from the polymer [Plesch, 1953]. This reaction and the corresponding intramolecular transfers (Sec. 5-2b) are responsible for the production of only low molecular products from 1-alkenes.

5-2c-5 Other Transfer and Termination Reactions

Various transfer agents (denoted by S or XA as in Chap. 3), present as solvent, impurity, or deliberately added to the reaction system, can terminate the growing polymer chain by transfer of a negative fragment A^-

$$HM_nM^+(IZ)^- + XA \xrightarrow{k_{tr.S}} HM_nMA + X^+(IZ)^- \qquad\qquad (5\text{-}28)$$

Water, alcohols, acids, anhydrides, and esters have varying chain-transfer properties [Mathieson, 1963]. The presence of any of these transfer agents in sufficient concentrations results in Reaction 5-28 becoming the dominant mode of termination. Termination by these compounds involves transfer of HO, RO, or RCOO anion to the propagating carbocation. Aromatics, ethers, and alkyl halides are relatively weak chain-transfer agents. Transfer to aromatics occurs by alkylation of the aromatic ring.

Although a chain-transfer agent decreases the degree of polymerization in proportion to its concentration (Sec. 5-2d), it is not expected to affect the polymerization rate since the initiator–coinitiator complex should be regenerated on transfer. However, one finds that the more active transfer agents such as water, alcohols, and acids do decrease the polymerization rate; that is, they function as inhibitors or retarders. The decrease in polymerization rate is caused by inactivation of the coinitiator by reaction with the chain-transfer agent, such as hydrolysis of $SnCl_4$. An alternate mechanism is inactivation of the proton or X^+ species (generated via Reaction 5-28) by solvation. For example, in the presence of considerable amounts of water, the preferred reaction for protons is probably not addition to alkene but reaction with water to form hydronium ion; thus, water is a stronger base than an alkene. As indicated in Sec. 5-2a-2, the polymerization rate is usually observed to increase with increasing initiator concentration (at constant coinitiator concentration), reach a maximum, and then decrease or level off.

Compounds such as amines, triaryl or trialkylphosphines, and thiophene act as inhibitors or retarders by converting propagating chains to stable cations that are unreactive to propagation [Biswas and Kamannarayana, 1976], for example,

$$Hm_nM^+(IZ)^- + :NR_3 \rightarrow HM_nM\overset{+}{N}R_3(IZ)^- \tag{5-29}$$

Phosphines have been advantageously used to convert propagating carbocations to highly stable phosphonium ions that can be studied with ^{31}P NMR [Brzezinska et al., 1977].

p-Benzoquinone acts as an inhibitor in cationic polymerization by proton transfer from the propagating carbocation and/or initiator–coinitiator complex to form p-hydroquinone. For styrene polymerization, copolymerization between p-benzoquinone and styrene is also important in the inhibiting action of p-benzoquinone [Ragimov et al., 1980].

The most nucleophilic of the reagents discussed, such as water, alcohol (often with KOH), ammonia, and amines, are often used in excess to *quench* a cationic polymerization. This is typically carried out after complete (or at least maximum) conversion has been reached in order to inactivate the coinitiator by the processes described above.

5-2d Kinetics

5-2d-1 Different Kinetic Situations

The overall kinetics vary considerably depending largely on the mode of termination in a particular system. Consider the case of termination by combination of the propagating center with the counterion (Eq. 5-21). The kinetic scheme of initiation, propagation, and termination consists of Eqs. 5-3, 5-4, 5-16, and 5-21, respectively. The derivation of the rate expression for this polymerization under steady-state conditions ($R_i = R_t$) follows in a manner analogous to that used in radical polymerization (Sec.

3-3b). The rates of initiation, propagation, and termination are given by

$$R_i = Kk_i[I][ZY][M] \tag{5-30}$$

$$R_p = k_p[YM^+(IZ)^-][M] \tag{5-31}$$

$$R_t = k_t[YM^+(IZ^-)] \tag{5-32}$$

where $[YM^+(IZ)^-]$ is the total concentration of all-sized propagating centers,

$$[YM^+(IZ)^-] = \frac{Kk_i[I][ZY][M]}{k_t} \tag{5-33}$$

Combining Eqs. 5-31 and 5-33 yields the rate of polymerization as

$$R_p = \frac{R_i k_p[M]}{k_t} = \frac{Kk_i k_p[I][ZY][M]^2}{k_t} \tag{5-34}$$

The number-average degree of polymerization is obtained as the propagation rate divided by the termination rate

$$\overline{X}_n = \frac{R_p}{R_t} = \frac{k_p[M]}{k_t} \tag{5-35}$$

When chain breaking involves chain transfer to monomer (Eqs. 5-18 and 5-19), spontaneous termination (Eq. 5-20), and chain transfer to chain transfer agent S in addition to combination with the counterion, the concentration of the propagating species remains unchanged (assuming relatively small amounts of S such that the coinitiator is not inactivated), and the polymerization rate is again given by Eq. 5-34. However, the degree of polymerization is decreased by these other chain-breaking reactions and is given by the polymerization rate divided by the sum of all chain-breaking reactions

$$\overline{X}_n = \frac{R_p}{R_t + R_{ts} + R_{tr,M} + R_{tr,S}} \tag{5-36}$$

The rates of spontaneous termination and the two transfer reactions are given by

$$R_{ts} = k_{ts}[YM^+(IZ)^-] \tag{5-37a}$$

$$R_{tr,M} = k_{tr,M}[YM^+(IZ)^-][M] \tag{5-37b}$$

$$R_{tr,S} = k_{tr,S}[YM^+(IZ)^-][S] \tag{5-37c}$$

Combination of Eq. 5-36 with Eqs. 5-31, 5-32, and 5-37 yields

$$\overline{X}_n = \frac{k_p[M]}{k_t + k_{ts} + k_{tr,M} + k_{tr,S}[S]} \tag{5-38}$$

or

$$\frac{1}{\overline{X}_n} = \frac{k_t}{k_p[M]} + \frac{k_{ts}}{k_p[M]} + C_M + C_S \frac{[S]}{[M]} \tag{5-39}$$

where C_M and C_S are the chain-transfer constants for monomer and chain-transfer agent S defined by $k_{tr,M}/k_p$ and $k_{tr,S}/k_p$, respectively. Equation 5-39 is the cationic polymerization equivalent of the previously described Mayo equation (Eq. 3-118) for radical polymerization.

For the case where chain transfer to S (Eq. 5-29) terminates the kinetic chain, the polymerization rate is decreased and is given by

$$R_p = \frac{Kk_ik_p[I][ZY][M]^2}{k_t + k_{tr,S}[S]} \tag{5-40}$$

The various rate expressions were derived on the assumption that the rate-determining step in the initiation process is Reaction 5-4. If this is not the situation, the forward reaction in Eq. 5-3 is rate-determining. The initiation rate becomes independent of monomer concentration and is expressed by

$$R_i = k_1[I][ZY] \tag{5-41}$$

The polymerization rate expressions (Eqs. 5-34 and 5-40) will then be modified by replacing Kk_i by k_1, and there will be a one order lower dependence of R_p on [M]. The degree of polymerization is unchanged and still described by Eq. 5-39.

The expressions (Eqs. 5-34 and 5-40) for R_p in cationic polymerization point out one very significant difference between cationic and radical polymerizations. Radical polymerizations show a one-half-order dependence of R_p on R_i, while cationic polymerizations show a first-order dependence of R_p on R_i. The difference is a consequence of their basically different modes of termination. Termination is second-order in the propagating species in radical polymerization but only first-order in cationic polymerization. The one exception to this generalization is certain cationic polymerizations initiated by ionizing radiation (Secs. 5-2a-3-d and 3-4d). Initiation consists of the formation of radical–cations from monomer on radiolysis followed by dimerization to form a dicarbocation similar to **IX** in Eq. 5-13. An alternate proposal is reaction of

$$M \xrightarrow{\text{radiation}} \cdot M^+ + e^- \tag{5-42}$$

$$\cdot M^+ \rightarrow {}^+M\!\!-\!\!M^+ \tag{5-43}$$

the radical–cation with monomer to form a monocarbocation species (Eq. 5-15). In either case, the carbocation centers propagate by successive additions of monomer with radical propagation not favored at low temperatures in superpure and dry systems.

In the usual situation the radiolytic reaction (Eq. 5-42) determines the rate of initiation and R_i is given by

$$R_i = IG[M] \tag{5-44}$$

where I is the radiation intensity and G is the number of radical–cations formed per 100 eV of energy absorbed. In sufficiently pure systems (concentrations of water and other terminating agents $<10^{-7} - 10^{-10}$ M), termination of the propagating carbocation centers occurs by combination with a negative fragment Y^-

$$\sim\sim M^+ + Y^- \xrightarrow{k_t} \sim\sim M-Y \qquad (5\text{-}45)$$

Y^- is either a solvated electron (displaced electron formed during the radiolytic reaction) or the product of the electron having reacted with some compound in the reaction system [Allen et al., 1974; Hayashi et al., 1967; Kubota et al., 1978; Williams et al., 1967]. If Y^- is an electron, the propagating carbocation centers are converted to radical centers that subsequently undergo reaction with some species in the reaction system to form molecular species. The termination rate is given by

$$R_{t'} = k_{t'}[M^+][Y^-] = k_{t'}[M^+]^2 \qquad (5\text{-}46)$$

where $[M^+] = [Y^-]$ due to the stoichiometry in the radiolytic reaction. Setting $R_i = R_{t'}$ and combining the result with Eq. 5-31 yields the polymerization rate as

$$R_p = k_p[M] \left(\frac{R_i}{k_{t'}}\right)^{1/2} = k_p[M]^{3/2} \left(\frac{GI}{k_{t'}}\right)^{1/2} \qquad (5\text{-}47)$$

which is quite different from other cationic polymerizations in that the polymerization rate is $\frac{1}{2}$-order in R_i. The $\frac{1}{2}$-order dependence is then the same as observed in radical polymerization. For cationic polymerization initiated by radiation in reaction systems where the major termination occurs by reaction with some species other than Y^-, the termination process is first-order in propagating centers (not second-order as in Eq. 5-47) and the polymerization rate will show first-order dependence on R_i.

5-2d-2   *Validity of Steady-State Assumption*

The assumption of a steady-state for $[YM^+(IZ)^-]$ is not valid in many, if not most, cationic polymerizations, which proceed so rapidly that steady state is not achieved [Kennedy and Feinberg, 1978; Kennedy and Marechal, 1982]. Some of these reactions (e.g., isobutylene polymerization by $AlCl_3$ at $-100°C$) are essentially complete in a matter of seconds or minutes. Even in slower polymerizations, the steady state may not be achieved if $R_i > R_t$ [Villesange et al., 1977]. The concentration of propagating centers slowly increases throughout the polymerization, reaching a maximum late in the reaction, and then decreases. When initiation is fast, steady state may still not be achieved if the termination rate is lower or higher. The expressions in Sec. 5-2d-1 for R_p can only be employed if there is assurance that steady-state conditions exist, at least during some portion of the overall reaction. The existence of a steady state can be ascertained by measuring $[YM^+(IZ)^-]$ as a function of time. Since this is relatively difficult in most systems, it is more convenient to observe the polymerization rate as a function of time. Steady state is implied if R_p is constant with conversion except for changes due to decreased monomer and initiator concentrations. A more rapid decline in R_p with time than indicated by the decreases in $[M]$ and $[ZY]$ signifies a non-steady state. The absence of a steady state would also be indicated by an increase in R_p with time. The reader is cautioned that many of the experimental expressions reported in the literature to describe the kinetics of specific cationic polymerizations are invalid, since they are based on data where steady-state conditions do not apply [e.g., Biswas and Kabir, 1978, 1979]. Contrary to these considerations for R_p, the derivations of the expressions for $\overline{X}_n$ do not assume steady-state conditions.

Another consideration in the application of the various kinetic expressions is the uncertainty in some reaction systems as to whether the initiator-coinitiator complex is soluble. Failure of the usual kinetic expressions to describe a cationic polymerization may indicate that the reaction system is actually heterogeneous. The method of handling the kinetics of heterogeneous polymerizations is described in Sec. 8-4c.

5-2d-3   Molecular-Weight Distribution

The theoretical molecular-weight distributions for cationic chain polymerizations are the same as those described in Sec. 3-11 for radical chain polymerizations terminating by reactions in which each propagating chain is converted to one dead polymer molecule, that is, not including the formation of a dead polymer molecule by bimolecular coupling of two propagating chains. Equations 2-92 through 2-95, 2-27, 2-101, and 2-102 with p defined by Eq. 3-202 are applicable to cationic chain polymerizations carried out to low conversions. The polydispersity index ($PDI = \overline{X}_w/\overline{X}_n$) has a limit of 2. Many cationic polymerizations proceed with rapid initiation (in some cases, essentially instantaneous initiation), which narrows the molecular-weight distribution. In the extreme case where termination and transfer reactions are very slow or nonexistent, this would yield a very narrow molecular-weight distribution with PDI close to one (Secs. 5-2g-2, 5-3b-1). The polydispersity index is greater than one when chain breaking reactions are operative with values generally between 1 and 2 but also greater than 2 depending on the chain-breaking reactions and their rates relative to propagation [Cai et al., 1988; Yan and Yuan, 1986, 1987; Yuan and Yan, 1988].

For polymerizations carried out to high conversions where the concentrations of propagating centers, monomer, and transfer agent as well as rate constants change, the polydispersity index increases considerably. Relatively broad molecular-weight distributions are generally encountered in cationic polymerizations.

5-2e   Absolute Rate Constants

5-2e-1   Experimental Methods

The determination of the various rate constants (k_i, k_p, k_t, k_{ts}, k_{tr}) for cationic chain polymerization is much more difficult than in radical polymerization (or in anionic polymerization). R_p data from experiments under steady-state conditions is convenient for calculating rate constants, since the concentration of propagating species is not required. R_p data from non-steady-state experiments can be used, but only when the concentration of the propagating species is known. Unfortunately, most cationic polymerizations proceed neither under steady-state conditions nor with a knowledge of the concentration of the propagating species. Because the expressions for $\overline{X}_n$ do not depend on either steady-state reaction conditions or a knowledge of the concentration of propagating species, it is more convenient to calculate the ratios of various rate constants from $\overline{X}_n$ data than from R_p data. The use of $\overline{X}_n$ data, like the use of R_p data, does require that one employ data obtained at low conversions where rate constants and reactant concentrations have not changed appreciably. Further, the techniques used for measurement of $\overline{X}_n$, size exclusion chromatography (SEC), membrane and vapor pressure osmometry, require careful utilization to avoid their inherent limitations.

The degree of polymerization under various reaction conditions is used to obtain the k_t/k_p, k_{ts}/k_p, $k_{tr,M}/k_p$ ($= C_M$), and $k_{tr,S}/k_p$ ($= C_S$) ratios from Eq. 5-39. Experiments with varying [M] in the absence of chain-transfer agents yield a linear plot of $1/\overline{X}_n$ versus $1/[M]$ with intercept equal to C_M. The slope of the plot is given by ($k_t/k_p + k_{ts}/k_p$). The two ratios can be separated from each other by chemical analysis of the polymer end groups. Spontaneous termination and chain transfer to monomer both yield polymers with unsaturated end groups, while combination with counterion yields polymer end groups derived from the counterion. The end-group analysis combined with the calculated values of C_M and ($k_t/k_p + k_{ts}/k_p$) allow the separation of the latter two ratios. The value of C_S is obtained by carrying out experiments with varying amounts of chain-transfer agent. A plot of the data according to Eq. 3-129 as $1/\overline{X}_n$ versus $[S]/[M]$ is linear with a slope of C_S. $(1/\overline{X}_n)_0$ represents the value of

$$\frac{1}{\overline{X}_n} = \left(\frac{1}{\overline{X}_n}\right)_0 + C_S \frac{[S]}{[M]} \tag{3-129}$$

$1/\overline{X}_n$ in the absence of chain-transfer agent and is given by the sum of the first three terms on the right side of Eq. 5-39.

The determination of the individual rate constants requires the determination of k_p, a difficult task and one that has not often been performed well [Dunn, 1979; Kennedy and Marechal, 1982; Plesch, 1971, 1984, 1988]. The value of k_p is obtained directly from Eq. 5-31 from a determination of the polymerization rate. However, this requires the critical evaluation of the concentration of propagating species. The literature contains too many instances where the propagating species concentration is taken as equal to the concentration of initiator without experimental verification. Such an assumption holds only if $R_p < R_i$ and all the initiator is active, that is, the initiator is not associated or consumed by side reactions.

There are two general methods for the experimental evaluation of the propagating species concentration, neither method being experimentally simple nor unambiguous. One involves *short-stopping* a polymerizing system by adding a highly efficient terminating agent. All propagating centers are quickly terminated with incorporation into the polymer of an end group derived from the terminating agent. The end groups in the polymer are analyzed after separation of the polymer from the other components of the reaction system. This method is limited by the general difficulty of end-group analysis since the concentration of end groups can be quite low and by the need to assume that the terminating agent terminates all propagating centers. 2-Bromothiophene has been used to short-stop styrene polymerization with the bromine content of the polymer being analyzed by neutron activation [Higashimura et al., 1971]. Termination of isobutylene polymerization by 2,6-di-t-butylphenol involves aromatic alkylation in the para position of the phenol by the propagating carbocations. Analysis of the 2,6-di-t-butylphenol end groups is accomplished by UV spectroscopy [Russell and Vail, 1976].

The second method for determining the propagating species concentration involves direct UV-visible spectroscopic analysis of the propagating species during polymerization. The high extinction coefficients of some aromatic carbocation propagating species coupled with the availability of highly accurate spectrophotometers has resulted in extensive use of this method. Measurement of the UV absorbance of the polymerizing system as a function of time allows one to determine k_i, R_i, the concentration of propagating species, and k_p. The UV method has also been used to study very fast polymerizations. In *stopped-flow, rapid-scan spectroscopy*, separate monomer and initiator solutions are rapidly forced through a mixing chamber (where instantaneous mixing occurs) and then into a capillary tube located in a spectrophotometer, flow is stopped and the progress of reaction followed by measuring the change in absorbance with time (Chance, 1974; Sawamoto and Higashimura, 1979; Szwarc, 1968].

5-2e-2   *Difficulty in Interpreting Rate Constants*

Most reported values of k_p and other rate constants and kinetic parameters are questionable for several reasons. First, as discussed above, there is ambiguity about the concentration of propagating species. Second, the calculations of various rate constants and kinetic parameters are often based on inadequately substantiated reaction kinetics and mechanisms [Kennedy and Marechal, 1982; Plesch, 1971, 1990]. Even if the propagating species concentration is accurately known, the use of Eq. 5-31 without verification can lead to incorrect results. This occurs when the kinetics of cationic chain polymerization deviate from Eq. 5-31 and the scheme described in Sec. 5-2d-1 under certain reaction conditions (Sec. 5-2f). Third, there is the large question of how to interpret any obtained rate constants in view of the known multiplicity of propagating carbocation propagating species. The kinetic expressions in Sec. 5-2d are written in terms of only one type of propagating species—usually shown as the ion pair. This is incorrect, since both ion pairs and free ions are simultaneously present in most polymerization systems, usually in equilibrium with each other (Sec. 5-1). Thus the correct expression for the rate of any step (initiation, propagation, termination, transfer) in the polymerization should include separate terms for the respective contributions of the two types of propagating species. As an example, the propagation rate should be written as

$$R_p = k_p^+[YM^+][M] + k_p^{\pm}[YM^+(IZ)^-][M] \tag{5-48}$$

where $[YM^+]$ and $[YM^+(IZ)^-]$ are the concentrations of free ions and ion pairs, respectively, and k_p^+ and $k_p^{\pm}$ are the corresponding propagation rate constants. Most reported k_p values are only *apparent* or *pseudo* or *global* rate constants, k_p^{app}, obtained from the polymerization rate using the expression

$$R_p = k_p^{app}[M^*][M] \tag{5-49}$$

where $[M^*]$ is the total concentration of both types of propagating species. The apparent rate constant is thus not really a rate constant but a combination of rate constants and concentrations:

$$k_p^{app} = \frac{k_p^+[YM^+] + k_p^{\pm}[YM^+(IZ)^-]}{[YM^+] + [YM^+(IZ^-)]} \tag{5-50}$$

There are two approaches to the separation of k_p^{app} into the individual k_p^+ and $k_p^{\pm}$ values. One approach involves the experimental determination of the individual concentrations of free ions and ion pairs by a combination of conductivity with short-stop experiments or UV–visible spectroscopy. Conductivity directly yields the concentration of free ions; that is, only free ions conduct. Short-stop experiments yield the total of the ion-pair and free-ion concentrations. UV–visible spectroscopy for those monomers (mostly aromatic) where it is applicable is also used to obtain the total of the free-ion and ion-pair concentrations. It is usually assumed that ion pairs show the same UV–visible absorption as free ions since the ion pairs in cationic systems are loose ion pairs (due to the large size of the negative counterions; see Sec. 5-1). This approach is limited by the assumptions and/or experimental difficulties inherent in the various measurements. Conductivity measurements on systems containing low concentrations of ions are difficult to perform, and impurities can easily lead to erroneous results. The short-stop experiments do not distinguish between ion pairs and free ions, and the assumption of the equivalence of free ions and ion pairs in the spectroscopic method is not firmly established. The second approach involves determination of the polymerization rate at various concentrations of coinitiator and initiator and in both presence and absence of an added common ion salt. The latter, containing the counterion of the propagating carbocation, accentuates propagation by ion pairs by depressing their ionization to free ions. This approach, used extensively in anionic polymerization, is detailed in Sec. 5-3. It is equally applicable to cationic polymerization but has not been used extensively. In fact, neither approach has been used extensively in cationic polymerization.

One of the most often encountered errors in reported k_p^{app} values is their assignment as $k_p^{\pm}$. This can be erroneous since relatively small concentrations of free ions can have a significant effect on k_p^{app}. For the equilibrium between ion pairs and free ions

$$YM^+(IZ)^- \overset{K}{\rightleftharpoons} YM^+ + IZ^- \tag{5-51}$$

it can be shown [Plesch, 1973, 1977, 1984] that the ratio of concentrations of free ions and ion pairs is

$$\frac{[YM^+]}{[YM^+(IZ)^-]} = \frac{(1 + 4C/K)^{1/2} - 1}{2C/K + 1 - (1 + 4C/K)^{1/2}} \tag{5-52}$$

where $C = [YM^+] + [YM^+(IZ)^-]$. The relative concentrations of free ions and ion pairs depend on the ratio C/K according to Eq. 5-52. Free ions constitute approximately 99, 90, 62, 27, 9, 2, 1, and 0.3% of the propagating species at C/K values of 0.01, 0.1, 1, 10, 10^2, 10^3, 10^4, and 10^5, respectively. Cationic polymerizations are carried out over a wide range of C/K values, typically anywhere from 10^3 to 10^{-2}, depending on the specifics of the reaction system. Consider styrene polymerization by triflic (trifluoromethanesulfonic) acid in 1,2-dichloroethane at 20°C where K is 2.8×10^{-7} mole/liter [Kunitake and Takarabe, 1979]. Experiments performed at acid concentrations of 2.8×10^{-5} M ($C/K = 10^2$) would involve 91% propagation by ion pairs and 9% propagation by free ions. The relative contribution of free ions to the overall propagation increases for lower acid concentrations. There is considerable error if the k_p^{app} value is simply taken to be $k_p^{\pm}$ (as is often done since K is typically unknown and one cannot make the appropriate corrections). The ratio $k_p^+/k_p^{\pm}$ is probably in the range 5–50 for most systems. The calculated $k_p^{\pm}$ values are high by factors of 1.5 and 6 for $k_p/k_p^{\pm}$ values of 5 and 50, respectively. One can safely equate k_p^{app} with $k_p^{\pm}$ only for systems where $C/K = 10^3 - 10^4$ or larger. This can be done by using higher concentrations of triflic acid as long as there are no solubility problems or the reaction rates are not excessively high.

For most of the systems reported in the literature, C/K is not known—very often, neither K nor C is known. For two-component initiator–coinitiator systems, C is usually taken to be the initiator concentration [YZ] when the coinitiator is in excess or the coinitiator concentration [I] when the initiator is in excess. C may be lower than [YZ] or [I] due to association; that is, only a fraction of [YZ] or [I] may be active in polymerization. This may also be the case for one-component initiators such as triflic acid. It would be prudent to determine the actual value of C in any polymerization system—usually a difficult task and seldom achieved. Experimental difficulties have also limited our knowledge of K values, which are obtained most directly from conductivity measurements or, indirectly, from kinetic data. A comparison of polymerizations in the absence and presence of a common ion salt (e.g., tetra-n-butylammonium triflate for the triflic acid initiated polymerization) are useful for ascertaining whether significant amounts of free ions are present in a reaction system.

Polymerizations initiated by ionizing radiation or stable carbocation salts such as trityl or tropylium hexachloroantimonate are useful for evaluating the free-ion propagation rate constant. Ionizing radiation yields free ions (in the absence of ion pairs) whose concentrations can be obtained by conductivity measurements [Deffieux et al., 1980; Hayashi et al., 1971; Hsieh et al., 1980; Stannett et al., 1976]. Many carbocation salts are sufficiently stable to be isolated, purified, and characterized as crystalline products. Thus their concentration in a polymerization system is known, unlike the situation with other initiation systems such as boron trifluoride–water. It is often assumed that polymerizations initiated by these salts proceed with propagation carried entirely by free ions and the k_p^{app} is taken to be k_p^+. This is a valid assumption only under certain conditions. K is probably $10^{-4} - 10^{-5}$ for such systems [Gandini and Cheradame, 1985; Subira et al., 1988]. If C/K is no larger than 0.1 (e.g., C no higher than 10^{-5} for $K = 10^{-4}$), the system consists of 90% free ions and 10% ion pairs. The presence of 10% of the less reactive (ion pair) propagating species results in no more than a 10% error in the value of k_p^+ obtained by equating k_p^{app} with k_p^+.

Throughout the remainder of the chapter it should be understood that any rate constants that are presented are apparent rate constants unless otherwise indicated to be those for the free ion or ion pair. In general, the available data are used to point out certain trends (e.g., the effect of solvent on reactivity) without necessarily accepting the exact value of any reported rate constant as that for the ion pair or free ion. Comparison of data from different investigators should be done with caution.

5-2e-3 Comparison of Rate Constants

Table 5-1 shows the various kinetic parameters, including k_p^+ and $k_p^\pm$, in the polymerization of styrene initiated by triflic acid in methylene chloride at 20°C. Data for the polymerization of i-butyl vinyl ether initiated by trityl hexachloroantimonate in methylene chloride at 0°C are shown in Table 5-2. Table 5-3 shows k_p^+ values for several polymerizations initiated by ionizing radiation or trityl salts. Values of $k_p^\pm$ for two of the reaction systems are also included. A comparison of the k_p^+ and $k_p^\pm$ values for the styrene, p-methoxystyrene, and N-vinylcarbazole polymerizations shows the free-ion propagation rate to be an order of magnitude higher than the ion-pair propagation rate constant. Although there are relatively few other systems in which reasonably accurate measures of both k_p^+ and $k_p^\pm$ are available, it is generally considered that no more than the one order of magnitude difference in reactivity exists for free ions and ion pairs in cationic polymerization [Gandini and Cheradame, 1985; Matyjaszewski, 1989; Mayr et al., 1988; Sauvet and Sigwalt, 1989]. The counterion is

TABLE 5-1 Kinetic Parameters in CF_3SO_3H Polymerization of Styrene at 20°C in $ClCH_2CH_2Cl^a$

Parameter	Value
[Styrene]	0.27–0.40 M
[CF_3SO_3H]	3.8–7.1 × 10^{-3} M
k_i	10–23 liters/mole-sec
K_d	4.2 × 10^{-7} mole/liter
k_p^+	1.2 × 10^6 liters/mole-sec
$k_p^{\pm}$	1.0 × 10^5 liters/mole-sec
k_{ts}	170–280 sec^{-1}
k_t	<0.01k_{ts}
$k_{tr,M}$	1–4 × 10^3 liters/mole-sec

^{a}Data from Kunitake and Takarabe [1979].

typically quite large for cationic polymerization (e.g., $SbCl_6^-$, $CF_3SO_3^-$); the ion pair is a very loose ion pair with little difference in availability of the positive charge center for reaction compared to the free ion.

A comparison of Tables 5-1 to 5-3 with corresponding data for radical chain polymerization (see Tables 3-10 and 3-11) allows us to understand why cationic polymerizations are generally faster than radical polymerizations. The propagation rate constants in cationic polymerization are similar to or greater than those for radical polymerization. However, the termination rate constants are considerably lower in cationic polymerization. The polymerization rate is determined by the ratio of rate constants—k_p/k_t in cationic polymerization and $k_p/k_t^{1/2}$ in radical polymerization. The former ratio is larger than the latter by up to four orders of magnitude depending on the monomers being compared. Cationic polymerization is further favored, since the concentration of propagating species is usually much higher than in a radical polymerization. Cationic polymerizations often proceed with propagating species concentrations as high as 10^{-5} M. The concentration of propagating radicals is much lower, typically 10^{-7}–10^{-9} M.

5-2e-4 C_M and C_S Values

Monomer chain-transfer constants are shown in Tables 5-4 and 5-5 for various polymerizations of styrene and isobutylene. It is apparent that the C_M values for styrene are, in general, larger than those for isobutylene by one to two orders of magnitude. This corresponds to the greater reactivity of isobutylene toward propagation. Although

TABLE 5-2 Kinetic Parameters in $\phi_3C^+SbCl_6^-$ Polymerization of i-Butyl Vinyl Ether in CH_2Cl_2 at 0°C^a

Parameter	Value
[$\phi_3C^+SbCl_6^-$]	6.0 × 10^{-5} M
k_i	5.4 liters/mole-sec
k_p^+	7.0 × 10^3 liters/mole-sec
$k_{tr,M}$	1.9 × 10^2 liters/mole-sec
$k_{ts} + k_t$	0.2 sec^{-1}

^{a}Data from Subira et al. [1976].

TABLE 5-3 Propagation Rate Constants

Monomer	Initiator	Solvent	Temperature (°C)	$k_p^+ \times 10^{-4}$ (liters/mole-sec)
Isobutylene[a]	Radiation	Bulk	0	15000
Styrene[b]	Radiation	Bulk	15	350
p-Methoxystyrene[c]	Radiation	Bulk	0	300
	$\phi_3C^+SbCl_6^-$	CH_2Cl_2	10	36[g]
N-Vinylcarbazole[d]	$\phi_3C^+SbF_6^-$	CH_2Cl_2	20	60[h]
i-Propyl vinyl ether[e]	$\phi_3C^+SbCl_6^-$	CH_2Cl_2	0	1.1
	Radiation	CH_2Cl_2	0	8.6
Isoprene[f]	Radiation	Bulk	0	0.2

[a]Data from Williams and Taylor [1969].
[b]Data from Williams et al. [1967].
[c]Data from Sauvet et al. [1986].
[d]Data from Rooney [1976, 1978].
[e]Data from Subira et al. [1988], Deffieux et al. [1983].
[f]Data from Williams et al. [1976].
[g]$k_p^{\pm} = 4.1 \times 10^4$ liters/mole-sec.
[h]$k_p^{\pm} = 5.0 \times 10^4$ liters/mole-sec.

extensive data are not available, one generally observes that the monomers with the larger k_p values usually possess lower C_M values. The subject of the relative reactivities of different monomers in cationic polymerization will be considered in Chap. 6. In general, one observes the order of reactivity: vinyl ether > isobutylene > styrene, isoprene.

The values of C_S for some compounds in the polymerization of styrene are shown in Table 5-6. The larger transfer constants are associated with transfer agents that

TABLE 5-4 Monomer Transfer Constant for Styrene

Initiator	Solvent	Temperature (°C)	$C_M \times 10^2$
$SnCl_4$	ϕH	30	1.9[a]
$SnCl_4$	CCl_4—ϕNO_2 (3:7)	0	0.51[b]
$SnCl_4$	C_2H_5Br	−63	0.02[c]
$TiCl_4$	ϕH	30	2.0[d]
$TiCl_4$	$(CH_2Cl)_2$—ϕH (3:7)	30	1.5[d]
$TiCl_4$	CH_2Cl_2	−60	0.04[e]
$TiCl_4$	CH_2Cl_2	−90	<0.005[e]
$FeCl_3$	ϕH	30	1.2[c]
BF_3	ϕH	30	0.82[c]
BF_3	CH_2Cl_2	−50	0.057[f]
CF_3SO_3H	CH_2Cl_2	20	1.5[g]

[a]Data from Okamura and Higashimura [1956a, 1956b, 1960].
[b]Data from Endres and Overberger [1955].
[c]Data from Biddulph and Plesch [1960].
[d]Data from Sakurda et al. [1958].
[e]Data from Mathieson [1963].
[f]Data from Kennedy and Feinberg [1978], Kennedy et al. [1977].
[g]Data from Kunitake and Takarabe [1979].

TABLE 5-5 Monomer Transfer Constants in Cationic Polymerization of Isobutylene in CH_2Cl_2

Coinitiator-Initiator	Temperature (°C)	$C_M \times 10^4$
$TiCl_4$-H_2O[a]	-20	21.2
	-50	6.60
	-78	1.52
$TiCl_4$-Cl_3CCO_2H[a]	-20	26.9[b]
	-50	5.68
	-78	2.44
$SnCl_4$-Cl_3CCO_2H[a]	-20	60.0
	-50	36.0
	-78	5.7
BF_3-H_2O[b]	-25	15
	-50	3.9

[a]Data from Imanishi et al. [1961].
[b]Data from Kennedy and Feinberg [1978], Kennedy et al. [1977].

possess a weakly bonded negative fragment (e.g., CH_3O^- in CH_3OH) or are readily alkylated ($CH_3O\varnothing$). However, the value of C_S for a transfer agent is seen to vary considerably when the initiator, coinitiator, or solvent is changed.

5-2f Effect of Reaction Medium

The nature of the reaction medium can play a significant role in cationic polymerization. Changes in rate and/or degree of polymerization often occur when the solvent

TABLE 5-6 Chain-Transfer Constants in Styrene Polymerization

Initiator or Coinitiator	Solvent	Temperature (°C)	Transfer Agent	$C_M \times 10^2$
$SnCl_4$	$\varnothing H$	30	$\varnothing H$	0.22[a]
H_2SO_4	$(CH_2Cl)_2$	25	i-$C_3H_7\varnothing$	4.5[b]
$SnCl_4$	C_6H_{12}	0	i-$C_3H_7\varnothing$	0.60[c]
$SnCl_4$	CCl_4—$\varnothing NO_2(3:7)$	0	$CH_3O\varnothing$	162[d]
$SnCl_4$	CCl_4	30	CCl_4	0.52[e]
$SnCl_4$	$\varnothing H$	30	CH_3OH	312[e]
$SnCl_4$	$(CH_2Cl)_2$	30	CH_3OH	90[e]
$SnCl_4$	$\varnothing H$	30	CH_3COOH	350[e]
$SnCl_4$	$(CH_2Cl)_2$	30	CH_3COOH	40[e]
BF_3	$\varnothing H$	30	CH_3COOH	144[e]
$SnCl_4$	$\varnothing H$	30	$(CH_3CO)_2O$	960[e]
$SnCl_4$	$\varnothing H$	30	$CH_3COOC_2H_5$	12[e]
H_2SO_4	$(CH_2Cl)_2$	25	$(C_2H_5)_2O$	19[b]
CCl_3COOH	None	25	Benzoquinone	24[f]

[a]Data from Biddulph and Plesch [1960].
[b]Data from Jenkinson and Pepper [1961].
[c]Data from Overberger et al. [1956].
[d]Data from Endres and Overberger [1955].
[e]Data from Okamura and Higashimura [1956a, 1956b].
[f]Data from Brown and Mathieson [1957, 1958].

or counterion are changed. The effect of reaction medium can affect the reaction in several ways. The equilibrium between free ions and ion pairs will be shifted toward free ions by high-polarity solvents. This ususally results in an increased polymerization rate since free ions propagate faster than ion pairs, but the effect is not huge since k_p^+ typically does not exceed $k_p^{\pm}$ by more than an order of magnitude. However, large solvent effects can be observed for systems in which the counterion has a significant tendency to form a covalent bond to the carbocation. Low-polarity solvents drive the reaction system toward the covalent species **I**, (see Sec. 5-1), which propagates more slowly than the free ion or ion pair. In addition, solvent can affect the rate and degree of polymerization by changing propagation rate constants for the various propagating species.

5-2f-1   *Propagation by Covalent Species*

There is considerable evidence that some cationic polymerizations proceed not only through ionic intermediates (free ions and ion pairs) but also through covalent intermediates. Polymerizations proceeding through covalent propagating species are referred to as *pseudocationic polymerizations*.

Pseudocationic polymerization has been extensively studied for the perchloric acid polymerization of styrene [Dunn et al., 1976; Gandini and Cheradame, 1980; Matyjaszewski, 1988; Plesch, 1971, 1988]. This polymerization proceeds with propagation by a combination of ionic (free ions and ion pairs) and covalent species, the relative amounts of which depend on the specific reaction conditions. Three successive stages are seen in polymerizations carried out at $-20°C$ in methylene chloride. Stage I involves a rapid, short-lived reaction involving ions followed by a much slower but also much longer Stage II reaction in which ions cannot be detected either by conductivity or spectroscopy. The·conclusion of Stage II leads to a rapid and simultaneous increase in conductivity and polymerization rate as well as spectroscopic evidence of ionic propagating species (Stage III). At higher temperatures (-20 to $30°C$), there is no Stage I and Stage III is shorter. At temperatures below $-80°C$, one observes only the rapid Stage I polymerization. Stages I and III are attributed to propagations by a combination of free ions and ion pairs. Free ions and ion pairs formed in Stage I are insufficiently stable at the higher reaction temperature and/or are destroyed by residual amounts of water and impurities. Ion combination yields the covalent perchlorate ester, which is stabilized via solvation by monomer. The Stage II reaction is attributed to pseudocationic (covalent) propagation involving monomer insertion into the C—O bond of a perchlorate ester:

$$\sim\sim CH_2\text{--}CH \quad ClO_2 \rightarrow \sim\sim CH_2\text{--}CH\text{--}CH_2\text{--}CHOClO_3 \qquad (5\text{-}53)$$

At high conversion, there is insufficient monomer to stablize the covalent ester and ionization occurs, leading to the rapid Stage III polymerization carried by ions. Stage III reaction proceeds with a composite propagation rate constant for free ions and ion pairs of 10^4–10^5 liters/mole-sec at -60 to $-80°C$. The propagation rate constant for the Stage II covalent propagation is much lower—0.1–20 liters/mole-sec depending on solvent and temperature.

Aside from the lack of conductivity and spectroscopic absorption, several other observations support the covalent propagating species. There is no effect of added water on the rate of the Stage II reaction, although Stages I and III are suppressed. Size exclusion chromatography has shown a bimodal molecular-weight distribution for polymerizations carried out under conditions where there are significant extents of the two types of propagations [Pepper, 1974]. A bimodal MWD consists of a mixture of two molecular-weight distributions that are sufficiently different to be distinguished by SEC. The presence of a bimodal molecular-weight distribution indicates that there are two different propagating species that are not in fast equilibrium with each other (fast relative to their propagation rates). There would be a unimodal MWD (with some broadening) if two different propagating species were interconverting at a rate greater than either of their propagation rates. This is what is observed in various polymerizations where only free ions and ion pairs are present; in other words, free ions and ion pairs interconvert faster than they propagate. The bimodal MWD observed in the perchloric acid polymerization of styrene is attributed to separate propagations by covalent ester and free ion/ion pair. Propagation by free ions–ion pairs yields polymers (molecular weight ca. 10^4), while covalent propagation yields oligomers (molecular weight 600–700). (The low molecular weight of the latter indicates that extensive transfer to monomer occurs in competition with the slow propagation of the ester.) The amount of the low-molecular-weight fraction relative to the high-molecular-weight fraction as observed in the SEC correlates with the fraction of the total reaction carried by the Stage II reaction as observed by rate measurements. The high-molecular-weight fraction can be suppressed by the addition of a common ion salt such as tetra-n-butylammonium perchlorate, which pushes the equilibrium between ionic and covalent propagating species toward the ester. The extent of the Stage II propagation by ester relative to the ionic propagations in Stages I and III depends not only on temperature and water level but also solvent polarity. Only Stage II behavior is observed in solvents of low polarity (ca. $\epsilon < 6$), while both covalent and ionic propagations occur in more polar solvents. Propagation by covalent ester is not observed in solvents more polar than the more polar chlorinated hydrocarbons ($\epsilon > 9$–10).

The pseudocationic mechanism of covalent propagation is not as dramatic a departure from the cationic mechanism as appears. It is the extreme of cationic polymerization under conditions (reaction medium, temperature, counterion) where free ions and ion pairs are insufficiently stabilized. [An alternative possibility to covalent propagation as an explanation for pseudocationic polymerization is propagation based on very tight ion pairs—so tight that they do not respond spectroscopically or chemically (e.g., reaction with water) as do loose ion pairs).]

Similar results have been observed for a number of other polymerizations—styrene (CH_3COClO_4, FSO_3H, $ClSO_3H$, CH_3SO_3H, CF_3SO_3H, CF_3COOH), p-methylstyrene (CF_3SO_3H, I_2, CH_3COClO_4), p-methoxystyrene (CF_3SO_3H, I_2), and p-chlorostyrene (CH_3COClO_4, CF_3SO_3H) [Higashimura et al., 1976; Sawamoto et al., 1977, 1978; Sawamoto and Higashimura, 1978, 1979; Tanizaki et al., 1986]. Polymerizations in low-polarity solvents yield bimodal molecular-weight distributions that respond to changes in reaction conditions in the same manner as described for the styrene-perchloric acid system. The high-molecular-weight fraction is formed faster than the low-molecular-weight fraction but is suppressed in the presence of a common ion salt or in solvents of lower polarity. Pseudocationic polymerizations are not observed for typical Lewis acid initiating systems (e.g., $BF_3 + H_2O$), since the counterions cannot form covalent bonds to carbon. Certain polymerizations of isobutylene, vinyl ethers, and N-vinylcarbazole, referred to as *living polymerizations*, have similarities to pseudocationic polymerization (Sec. 5-2g).

5-2f-2 *Solvent Effects*

The shift from covalent propagation to ionic propagation with increasing solvent polarity expresses itself in significant changes in the overall rate and degree of polymerization. For the perchloric acid polymerization of styrene, there is an increase in overall reaction rate by about three orders of magnitude when polymerization is carried out in 1,2-dichloroethane ($\epsilon = 9.72$) compared to carbon tetrachloride ($\epsilon = 2.24$) [Pepper and Reilly, 1962]. Table 5-7 shows data for the polymerization of p-methoxystyrene by iodine [Kanoh et al., 1965]. The apparent propagation rate constant increases by more than two orders of magnitude in going from carbon tetrachloride ($\epsilon = 2.24$) to methylene chloride ($\epsilon = 9.08$). (Although the dielectric constant ϵ is generally used as an indication of the solvating power of a solvent, it is not necessarily a quantitative measure. Specific solvation and polarization effects, which are not adequately described by ϵ, may be important in a particular system.)

Effects similar to those described above are also observed even for reaction systems where covalent propagation does not occur. Thus, $BCl_3 + H_2O$ do not polymerize neat isobutylene but polymerization occurs for solutions of isobutylene in solvents (e.g., CH_2Cl_2, CH_3Cl) more polar than isobutylene [Kennedy et al., 1977]. Ionic propagating species (either free ions or ion pairs) are not sufficiently stable and do not form unless the reaction medium possesses some minimum solvating ability.

The preceding discussion assumes that the effect of solvent polarity on reaction rate (and polymer molecular weight) manifests itself only by altering the concentration of propagating centers. However, solvent polarity is also expected to alter propagation rate constants, k_p^+, $k_p^\pm$, and k_p^c, the rate constant for propagation by covalent species. One expects k_p^+ to decrease with increasing solvent polarity [Ledwith and Sherrington, 1974]. The transition state for propagation by a free ion involves charge dispersal, so that solvents of higher polarity stabilize the reactants more than the transition state [Reichardt, 1988]. The opposite effect is expected for propagation by a covalent species since the transition state involves the development of charged centers from neutral reactants. The effect of solvent polarity on $k_p^\pm$ is more difficult to predict. Whether increased solvent polarity increases or decreases $k_p^\pm$ depends on whether the transition state has a higher or lower dipole moment than the ion pair. There are very few data

TABLE 5-7 Effect of Solvent on Cationic Polymerization of p-Methoxystyrene by Iodine at 30°C[a]

Solvent	k_p^{app} (liters/mole-sec)
CH_2Cl_2	17
CH_2Cl_2/CCl_4, 3/1	1.8
CH_2Cl_2/CCl_4, 1/1	0.31
CCl_4	0.12

[a]Data from Kanoh et al. [1965].

available on the effect of solvents on the different propagation rate constants. A comparison of reported k_p^+ values for styrene, ethyl vinyl ether, and i-propyl vinyl ether in different solvents shows the expected decrease in propagation rate constant with increasing solvent polarity [Deffieux et al., 1983; Sawamoto and Higashimura, 1978, 1979]. Table 5-8 shows data for i-propyl vinyl ether in radiation polymerization at 30°C. A plot of ln k_p^+ versus $1/\epsilon$ is reasonably linear with a positive slope as expected for the reaction of an ion and neutral molecule [Reichardt, 1988]. The available data for k_p^c are consistent with charge development in the transition state. k_p^c for styrene polymerization by CH_3SO_3H, CF_3SO_3H, and CH_3COOH either increases slightly or is approximately constant with increasing solvent polarity [Sawamoto et al., 1978]. Data for $k_p^{\pm}$ as a function of solvent polarity are not available, although Plesch [1984] suggests that $k_p^{\pm}$ increases with solvent polarity.

The need for solvation of ionic propagating species, and perhaps also covalent species, in cationic polymerization is clearly evident when reaction is carried out in media of low dielectric constant. In addition to lowered polymerization rates in poorly solvating media, one frequently encounters increased kinetic orders in one of the reactants. The polymerization rate may show an increased order of dependence on the monomer, initiator, or coinitiator. Thus, the rate for the tin(IV) chloride polymerization of styrene depends on $[M]^2$ in benzene solution and $[M]^3$ in carbon tetrachloride solution [Okamura and Higashimura, 1956a, 1956b, 1960]. Carbon tetrachloride is a poor solvating agent compared to benzene, and the higher order in styrene concentration is due to styrene taking part in solvation of propagating species. At high concentrations of styrene (and also in neat styrene) the order in styrene decreases to 2 as the reaction medium becomes equivalent to the benzene system. (That benzene is a better solvating medium than carbon tetrachloride is not entirely clear from a consideration of dielectric constants since the values are quite close—2.28 and 2.24 for benzene and carbon tetrachloride, respectively. However, other measures of solvating power such as the E_T^N scale show the difference between benzene and CCl_4. The E_T^N value for a solvent is derived from the energy of UV excitation for a pyridinium-N-phenoxide betaine dye in the solvent. Benzene and CCl_4 have E_T^N values of 0.111 and 0.052, respectively, with larger values being indicative of higher solvating power [Reichardt, 1988].)

Plesch [1989, 1990] has pointed out that complexation of propagating species (whether free ion or ion pair) with monomer introduces further ambiguities into the interpretation of rate constant data. For example, a comparison of free-ion propagation rate constants from two different reaction systems is less clear-cut when one system involves free ions and the other involves monomer-complexed free ions than when both systems propagate through the same type of species.

TABLE 5-8   Effect of Solvent on k_p^c in Radiation Polymerization of i-Propyl Vinyl Ether at 30°C^a

Solvent	ϵ^b	k_p^+ (liters/mole-sec)
ØH	2.7	57
Bulk	3.0	130
$(C_2H_5)_2O$	3.7	34
CH_2Cl_2	6.0	1.5
CH_3NO_2	19.5	0.02

^{a}Data from Deffieux et al. [1983].
^{b}Calculated from ϵ values of solvent and monomer assuming additivity of volumes and ϵ of mixture to vary with volume fraction of solvent.

The polymerization of styrene by trichloroacetic acid without solvent and in 1,2-dichloroethane and nitroethane solutions illustrates the situation where the initiator solvates ionic propagating species [Brown and Mathieson, 1958]. The kinetic order in the concentration of trichloroacetic acid increases from one in the highly polar nitroethane to two in the less polar 1,2-dichloroethane to three in neat styrene.

Another effect of reaction medium on polymerization is the observation that the polymerization rate in some systems (e.g., styrene–triflic acid and N-vinylcarbazole–n-butylmagnesium bromide) becomes independent of or inversely proportional to the monomer concentration at higher monomer concentrations [Biswas and John, 1978; Chmelir and Schulz, 1979; Gandini and Cheradame, 1980; Hatada et al., 1980; Sawamoto et al., 1977, 1978]. This phenomenon is probably due to the formation of a nonreactive complex between monomer and initiator–coinitiator. Although complex formation is relatively unimportant at lower concentrations, a significant portion of the initiator–coinitiator is tied up in the complex at high monomer concentrations and unavailable to initiate polymerization.

5-2f-3   Counterion Effects

Data for various reaction systems can be compared to observe an effect of counterion on cationic polymerization; for instance, the pseudo–propagation constant for styrene polymerization at 25°C in 1,2-dichloroethane increases from 0.003 for I_2 as the initiator to 0.42 and 17.0 for $SnCl_4$–H_2O and $HClO_4$, respectively [Kanoh et al., 1962, 1963; Pepper and Reilly, 1962]. Interpretation of these results in terms of an effect of counterion on propagation is of dubious value since the rate constants being compared are only pseudo–propagation rate constants, which are composites of concentration and rate constant terms for various propagating species—free ion, ion pair, covalent ester. All available information indicates that there is little effect of counterion on reactivity of ion pairs since the ion pairs in cationic polymerization are loose ion pairs. Clearly, there is no effect of counterion expected for free-ion reactivity. The main effect of counterion is in reaction systems where covalent species are easily formed. There is a drop in polymerization rate irrespective of whether the covalent species undergoes propagation. (Contrary to the situation in cationic polymerization, there is a significant effect of counterion on the ion pair propagation rate constant in anionic polymerization; see Sec. 5-3e-2-b.)

5-2g   Living Polymerization

Living polymerizations are polymerizations in which propagating centers do not undergo either termination or transfer. The living nature of such systems is apparent in several ways. For a highly purified reaction system devoid of impurities, the spectroscopic absorption for the propagating species persists throughout the polymerization and does not disappear or decrease in intensity at 100% conversion. A plot of $\overline{M}_n$ versus conversion is linear. Further, after 100% conversion is reached, additional polymerization takes place by adding more monomer to the reaction system. The added monomer is also polymerized quantitatively. The molecular weight of the living polymer increases as the number of propagating species does not change. Such polymerizations offer the potential for producing structures with defined end groups (by deliberate termination of a living system with appropriate reagents) and block copolymers (by sequential addition of two or more different monomers) (Sec. 5-4). The first living polymerizations were achieved in anionic polymerizations of styrene and 1,3-dienes by Szwarc in the mid 1950s (Sec. 5-3b-1). Some of the potential of living anionic polymerization has been realized with commercial processes for styrene–butadiene block copolymers and *cis*-1,4-polybutadiene.

It has been much more difficult to achieve living cationic polymerizations. Propagating carbocationic centers are very reactive but not necessarily more reactive than their anionic counterparts. The difference has been the difficulty in finding reaction conditions that stabilize (decrease the reactivity of) the carbocations sufficiently to prevent transfer and termination but that still allow propagation (as well as initiation) to occur. Both transfer and termination are facile reactions in cationic systems. An approximation to living polymerization, referred to as *quasiliving carbocationic polymerization*, was achieved for styrene, substituted sytrenes, isobutylene, and vinyl ethers under certain restrictive reaction conditions [Faust et al., 1982–1983; Puskas et al., 1982–1983; Sawamoto and Kennedy, 1982–1983; Toman et al., 1989a, 1989b]. For polymerizations taking place under conditions where transfer to monomer, spontaneous termination, and combination with counterion are reversible or suppressed and no other chain-breaking processes occur, the lifetimes of propagating species are extended by carrying out polymerization with a very slow but continuous addition of monomer. The incoming monomer is consumed mostly by propagation, and the polymerization has the characteristics of a living polymerization. However, these are quasiliving, not truly living, polymerizations since the transfer and termination reactions are far from completely absent. The quasiliving approach has been used to synthesize various telechelic polymers and block copolymers (Sec. 5-4).

5-2g-1 Reaction Characteristics

Major advances were made in the 1980s, when a number of living cationic polymerizations were achieved. These include various polymerizations of *p*-methylstyrene, *p*-methoxystyrene, *N*-vinylcarbazole, isobutylene, and styrene [Faust and Kennedy, 1987; Faust et al., 1986; Higashimura et al., 1988; Higashimura and Sawamoto, 1985; Kaszas et al., 1990; Kennedy, 1990; Kojima et al., 1989, 1990; Nuyken and Kroner, 1990; Sawamoto and Higashimura, 1986, 1990; Sawamoto et al., 1987; Tanizaki et al., 1986; Zsuga et al., 1990]. The key to living cationic polymerization is the use of a reaction system in which the propagating centers are of sufficiently low reactivity that transfer and termination reactions are suppressed but not so unreactive that propagation is suppressed. This requires an appropriate matching of the stability of the carbocationic center with counterion, solvent polarity, and polymerization temperature. Among the specific living systems are

1. *N*-vinylcarbazole: HI in toluene at −40°C.
2. Vinyl ethers: HI + ZnI_2 in toluene at −40°C to 25°C.
3. Vinyl ethers: $(C_2H_5)_2AlCl_2$ + 1-(*i*-butoxy)ethyl acetate in *n*-hexane with ether or ester at 0–70°C.
4. *p*-Methylstyrene: CH_3COClO_4 in 1:4 CH_2Cl_2:ϕCH_3 at −78°C.
5. Isobutylene: Cumyl acetate + BCl_3 in CH_2Cl_2 at −30°C.

Figure 5-2 shows results for the polymerization of *i*-butyl vinyl ether by HI/I_2 in methylene chloride at −15°C [Sawamoto and Higashimura, 1986]. These data are typical of cationic living polymerizations. The plot of $\overline{M}_n$ versus conversion is linear over the complete conversion range for successive additions of two batches of monomer. The fact that the $\overline{M}_n$ values for both batches of monomer fall on the same line shows that the system undergoes reaction with a constant concentration of propagating species; that is, termination and chain transfer are absent. The presence of termination

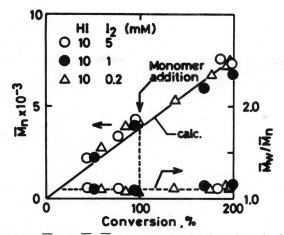

Fig. 5-2 Dependence of $\overline{M}_n$ and $\overline{M}_w/\overline{M}_n$ on conversion for the polymerization of *i*-butyl vinyl ether by HI/I$_2$ in CH$_2$Cl$_2$ at $-15°$C. [M] $= 0.38$ M at beginning of each batch; [HI] $= 0.01$ M; [I$_2$] $= 0.02$ M ($\triangle$), 0.001 M ($\bullet$), 0.005 M ($\circ$). After Sawamoto and Higashimura [1986] (by permission of Huthig and Wepf Verlag, Basel).

or transfer would result in the plot showing a declining slope (i.e., downward curvature) with increasing conversion.

The mechanistic descriptions of living cationic polymerization by various authors appear very similar to that for the pseudocationic polymerization of styrene by perchloric acid. It is generally proposed that a covalent (or near covalent, i.e., very tight ion pair) species is formed by addition of initiator to monomer. Subsequent propagation involves a more or less concerted process in which the bond between monomer and (covalently bound) counterion is stretched to generate the partially charged positive center and counterion simultaneously with the addition of monomer at the developing positive center. Whether the propagating center is best described as a covalent species or a very tight ion pair may be somewhat of a semantic problem. Whichever is the case, it is clear that the propagating centers in living systems are very different from those in nonliving (and nonpseudo) cationic systems. The latter involve either free ions or loose ion pairs. Living cationic and pseudocationic systems involve propagating centers with much lower charge densities.

The propagating mechanism responsible for living polymerizations initiated by HI can be pictured as

$$\text{(5-54)}$$

$$\text{(5-55)}$$

with propagation via stretching of the C—I bond with (Eq. 5-55) or without (Eq. 5-54) the presence of a Lewis acid such as zinc iodine to assist in stretching the C—I bond. Whether the Lewis acid is needed to achieve living polymerization depends on the strength of the C—I bond, which is determined primarily by the ability of R to stabilize the δ+ center but also by solvent, temperature, and Lewis acid. The balance of factors for N-vinylcarbazole is such that living polymerization occurs without a Lewis acid in toluene at −40°C. The C—I bond is stretched sufficiently for propagation to proceed but not stretched to an extent that gives rise to transfer and termination. The presence of a Lewis acid such as zinc iodide or iodine results in a nonliving (but more rapid) polymerization. The Lewis acid converts the counterion I^- to a more stable, less nucleophilic species (ZnI_3^- or I_3^-), and this results in greater reactivity for the δ+ propagating center. The polymerization in the absence of a Lewis acid also becomes nonliving when carried out in the more polar solvent methylene chloride at −78°C where ion pairs–free ions are more easily formed. Suppression of the ionization to ion pairs–free ions by the addition of a common ion salt, tetra-n-butylammonium iodide, reconverts the system to a living polymerization. A variety of other initiating systems achieve living polymerization of N-vinylcarbazole when the counterion is insufficiently nucleophilic to reduce the reactivity of the propagating centers (e.g., perchlorate ion). At the other extreme, there are counterions (e.g., Cl^-) too nucleophilic to allow either living or nonliving polymerization.

For vinyl ethers, there is less stabilization of the δ+ center by an ether oxygen compared to the carbazole nitrogen. I^- is too nucleophilic a counterion, and neither living nor nonliving polymerization occurs. The presence of I_2 or ZnI_2 decreases the nucleophilicity of the counterion (i.e., weakens the C—I bond sufficiently for living polymerization. Conditions for living polymerization are found in both toluene and methylene chloride. The temperature range over which living polymerization occurs is largest (−40 to 25°C) for ZnI_2 in toluene. The upper temperature limit for living polymerization is about 0°C when I_2 is used in either solvent or when ZnI_2 is used in methylene chloride. Polymerizations initiated by stronger Lewis acids such as BF_3 and $SnCl_4$ are not living.

The overall balance of reaction conditions needed to obtain the appropriate nucleophilicity of the counterion for the required weakening of the C—I bond is obtained not only by variations in counterion, solvent, and temperature but also by including an additional component. For example, i-butyl vinyl ether is polymerized at 0°C in n-hexane by $(C_2H_5)_2AlCl_2$ + 1-(i-butoxy)ethyl acetate but the polymerization is not living. The counterion in this system is acetate anion complexed with $(C_2H_5)_2AlCl_2$. (This system is similar to that used for the living polymerization of isobutylene where the initiating system is cumyl acetate and $TiCl_4$ or BCl_3 and the counterion is also acetate complexed with a Lewis acid [Faust and Kennedy, 1987; Kaszas et al., 1990]). The system is converted to a living polymerization over the temperature range 0 to +70°C by the addition of dioxane or ethyl acetate [Aoshima and Higashimura, 1989; Kishimoto et al., 1989]. The mechanism for the action of dioxane and ethyl acetate probably involves direct interaction with the δ+ propagating center (instead of altering the nucleophilicity of the counterion) since a comparison of the effects of various ethers and esters indicates the tendency toward living polymerization increases with the Lewis basicity (not acidity) of the additive. More recently, dimethyl sulfide has been used to stabilize δ+ propagating centers in polymerizations of i-butyl vinyl ether initiated by $(C_2H_5)_2AlCl_2$ or BF_3 with water [Cho et al., 1990].

5-2g-2   Rate and Degree of Polymerization

The rate of polymerization in living systems is expressed simply as the rate of propagation

$$R_p = k_p[M^*][M] \tag{5-56}$$

where $[M^*]$ is the total concentration of living propagating centers. The number-average degree of polymerization is given by the ratio of the concentrations of monomer and living ends

$$\overline{X}_n = \frac{[M]}{[M^*]} \tag{5-57}$$

The concentration of living centers $[M^*]$ is the concentration of initiator for an initiating system such as $HI + ZnI_2$ or $HI + I_2$. This has been verified by direct determination of $[M^*]$ by adding ethyl sodiomalonate to a polymerization mixture to cap the propagating centers followed by ^1H NMR analysis for the ethyl groups (Choi et al., 1990). The degree of polymerization is independent of the concentration of coinitiator (ZnI_2 or I_2), although the polymerization rate increases with increasing coinitiator concentration. This indicates that only a fraction of living propagating centers are actively propagating at any moment. There are active and inactive propagating centers in rapid equilibrium with each other. The living polymerizations are deliberately terminated when desired by the addition of methanol or other source of nucleophilic species.

A consequence of the absence of termination in a polymerization is that the polymer produced should be essentially monodisperse (i.e, $\overline{M}_w = \overline{M}_n$) under certain conditions. Initiation must be fast relative to propagation so that all of the propagating centers begin to propagate almost simultaneously; that is, all polymer molecules grow for the same length of time. Efficient mixing throughout the polymerization is required and depropagation must be slow relative to propagation. The size distribution will be given by the Poisson distribution [Flory, 1940; Peebles, 1971; Szwarc, 1968]

$$\frac{\overline{X}_w}{\overline{X}_n} = 1 + \frac{\overline{X}_n}{(\overline{X}_n + 1)^2} \tag{5-58}$$

which can be approximated by

$$\frac{\overline{X}_w}{\overline{X}_n} \simeq 1 + \frac{1}{\overline{X}_n} \tag{5-59}$$

Equation 5-59 shows that for any but a very low-molecular-weight polymer, the size distribution will be vary narrow with $\overline{X}_w/\overline{X}_n$ ($= PDI$) being close to unity. For a polymer of number-average degree of polymerization of 100 having a Poisson distribution, the PDI would be 1.01. Most of the living systems described above yield polymers having close to the Poisson distribution with *PDI* values of less than 1.1–1.2. Figure 5-2 shows data for the polymerization of *i*-butyl vinyl ether by the hydrogen iodide–iodine initiating system.

The occurrence of a narrow size distribution is often used as a diagnostic indication of living polymerization. Thus, in the reaction system described in Sec. 5-2g-1 where one passes from nonliving to living polymerizations by changes in coinitiator, temperature, or solvent, one observes a shift from a unimodal molecular-weight distribution (PDI considerably above 1; theoretical $PDI = 2$ as per Sec. 5-2d-3) for nonliving polymerization to a bimodal distribution for a mixture of living and nonliving polymerizations to a unimodal distribution (PDI close to 1) for living polymerization. The rates and molecular weights for the living polymerizations are lower than those for the nonliving polymerizations. The latter proceed by free-ion and ion-pair centers that are much more reactive than the propagating species in living systems. A range of molecular weights has been obtained in the various living systems. Some studies have concentrated on the synthesis of low-molecular-weight telechelic polymers with molecular weights of only a few thousand, while polymers of number-average molecular weights up to 50,000–100,000 have been synthesized in other studies.

The formation of Poisson distributions is corroboration of the fast equilibrium assumed between active and inactive propagating centers. For any propagating center, the fractions of its total lifetime spent as active and inactive centers are very close to those for other propagating centers.

None of the cationic living polymerization systems are as long-lived as the anionic systems to be described in Sec. 5-3b-1. The various cationic living systems show the characteristics of living systems as long as monomer is present, but decay of propagating species begins in the absence of monomer. The half-lives for propagating species in the absence of monomer vary from less than an hour to many hours at the higher temperatures but were as high as a day or more for lower temperatures depending on the particular system [Choi et al., 1990; Higashimura and Sawamoto, 1989]. The lifetimes for the isobutylene systems appear to be lower than those for other monomers, probably only minutes [Kennedy, 1989].

5-2h Energetics

Cationic polymerizations, like their radical counterparts, are quite exothermic, since the reaction involves the conversion of π-bonds to σ-bonds. The heat of polymerization for any particular monomer is essentially the same irrespective of the mode of initiation (if the monomer can be polymerized by both radical and cationic initiators).

From a consideration of Eqs. 5-34 and 5-35, the composite activation energies E_R and $E_{\overline{X}_n}$ for the rate and degree of polymerization, respectively, are obtained as

$$E_R = E_i + E_p - E_t \tag{5-60}$$

$$E_{\overline{X}_n} = E_p - E_t \tag{5-61}$$

where E_i, E_p, and E_t are the activation energies for the initiation, propagation, and chain termination steps, respectively. For the activation energy for $\overline{X}_n$, E_t will be replaced by an E_{tr} term or terms when transfer occurs. Propagation involves the addition of an ion or ion pair to a monomer in a medium of low polarity and does not require a large activation energy (ca. <20–25 kJ/mole). The values of E_i and E_t are greater than E_p in most cases. The net interaction of the various activation energies leads to values of E_R in the range of -20 to $+40$ kJ/mole. For many polymerization systems E_R is negative and one observes the rather unusual phenomenon of increasing polymerization rates when decreasing temperatures. The negative activation energy for the polymerization rate is not, however, a complete generalization. The sign and value of E_R vary from one monomer to another. Even for the same monomer, the value of E_R may vary considerably depending on the initiator, coinitiator, and solvent [Biswas and Kabir, 1978, 1979; Biswas and Kamannarayana, 1976; Brown and

Mathieson, 1957, 1958; Plesch, 1953]. Table 5-9 shows that E_R may vary from -35.5 to $+58.6$ kJ/mole for styrene polymerization. The variations in E_R are a consequence of the differences in E_i, E_p, and E_t caused by the differences in the initiator, coinitiator, and the solvating power of the reaction medium. It should be noted that, irrespective of sign, the values of E_R are generally smaller than in radical polymerizations. The rates of most cationic polymerizations do not change with temperature as much as those of radical polymerizations.

The reader should understand clearly that a negative value of E_R does not mean a negative activation energy for some reactant(s) going to a transition state. No simple, single step in any reaction pathway will have a negative activation energy. A negative E_R is simply a consequence of E_t being larger than $E_i + E_p$. Increased temperature decreases R_p because termination becomes more favored relative to initiation and propagation. What appears at first glance to be even more anomolous are negative values of E_p observed in a few polymerizations [Cotrel et al., 1976; Villesange et al., 1977]. The negative activation energy indicates that the measured E_p includes not only the activation energy for propagation but also a term for some other process. Since increased temperature decreases the dielectric constant of a liquid, the negative E_p may be indicative of a change in the relative concentrations of different propagating species (free ion, ion pair, covalent) or the effect of solvent polarity of k_p. The phenomenon may also be responsible for the changes in E_R with temperature that are sometimes observed [Biswas and Kabir, 1978, 1979; Plesch, 1971].

The composite activation energy for the degree of polymerization $E_{\overline{X}_n}$ is almost always negative because E_t or E_{tr} is generally greater than E_p irrespective of the mode of termination or transfer [Biswas and Kabir, 1978, 1979; DiMaina et al., 1977; Kennedy and Feinberg, 1978; Kennedy et al., 1977]. This means that the degree of polymerization decreases as the temperature increases. $E_{\overline{X}_n}$ is more negative when transfer reactions predominate over termination by combination or spontaneous termination since the transfer reactions have larger activation energies. As the polymerization temperature is increased, the principal mode of chain breaking shifts from termination to transfer [Biswas and Kamannarayana, 1976; Kennedy and Squires, 1965]. Further, there may be a shift from one transfer mode to another as the temperature is changed. Figure 5-3 shows a plot of log $E_{\overline{X}_n}$ versus $1/T$ for the polymerization of isobutylene by $AlCl_3$ in methylene chloride solution. There is a change in the slope of the plot at approximately $-100°C$ from an $E_{\overline{X}_n}$ value of -23.4 to -3.1 kJ/mole. This has been attributed to a change in the chain-breaking step from chain transfer to monomer below $-100°C$ to chain transfer to solvent above $-100°C$. The change in $E_{\overline{X}_n}$ may also be (at least partially) due to the effect of temperature on the relative amounts of the different types of propagating species.

TABLE 5-9 Activation Energy for Rate of Cationic Polymerization of Styrene

Initiating System	Solvent	E_R (kJ/mole)
$TiCl_4$-H_2O[a]	$(CH_2Cl)_2$	-35.5
$TiCl_4$-CCl_3CO_2H[a]	ϕCH_3	-6.3
CCl_3CO_2H	C_2H_5Br	12.6
$SnCl_4$-H_2O	ϕH	23.0
CCl_3CO_2H	$(CH_2Cl)_2$	33.5
CCl_3CO_2H	CH_3NO_2	58.6

[a]Data from Plesch [1953]; all other data from Brown and Mathieson [1957, 1958].

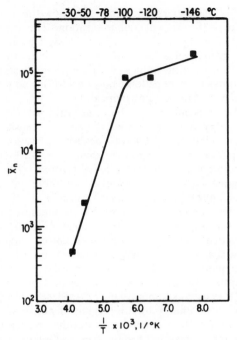

Fig. 5-3 Temperature dependence of $\overline{X}_n$ for the aluminum chloride polymerization of isobutylene. After Kennedy and Squires [1965] (by permission of Butterworth and Co. Ltd., London).

5-2i Commercial Applications of Cationic Polymerization

5-2i-1 *Polyisobutylene Products*

Various polymerizations and copolymerizations of isobutylene to produce polyisobutylene, *polybutene*, and *butyl rubber* constitute the most important commercial applications of cationic polymerization [Kennedy and Marechal, 1982; Kresge et al., 1987; Russell and Wilson, 1977]. The polybutenes are very low-molecular-weight copolymers ($\overline{M}_n < 3000$), containing about 80% isobutylene and 20% other butenes (mostly 1-butene), used in adhesives, caulking, sealants, lubricants, plasticizers, and additives for motor oils and transmission fluids (for viscosity improvements). Low-molecular-weight polyisobutylenes (up to $\overline{M}_v$ 5–10 $\times$ 10^4) range from viscous liquids to tacky semisolids and are used for sealant and caulking applications. High-molecular-weight polyisobutylenes ($\overline{M}_v > 10^5$) are rubbery solids used to make uncured (uncrosslinked) rubber products and as impact modifiers of thermoplastics. The molecular weights are controlled primarily by choice of polymerization temperature. Polymerization at -40 to 10°C with AlCl$_3$ (BF$_3$ or TiCl$_4$ is used in some processes) produces the lower-molecular-weight products. The high-molecular-weight polyisobutylenes are obtained at considerably lower reaction temperatures (-100 to -90°C) by using a process similar to that for butyl rubber.

Butyl rubber (BR) is a copolymer of isobutylene with small amounts of isoprene produced by aluminum chloride initiated polymerization. About 500 million pounds

$$CH_2\!=\!\underset{\underset{\displaystyle \text{Isoprene}}{|}}{\overset{\overset{\displaystyle CH_3}{|}}{C}}\!-\!CH\!=\!CH_2$$

of butyl rubber are produced annually in the United States. Polymerization is carried out in a continuous process using methyl chloride as diluent as shown in Fig. 5-4. Isobutylene and isoprene are purified and then mixed together in methyl chloride solution. The initiation system is produced by passing methyl chloride through beds of aluminum chloride at 30–45°C followed by dilution with methyl chloride and the addition of the initiator (protogen or cationogen). The initiation and monomer solutions are chilled separately to -100 to $-90°C$ by using boiling ethylene and propylene or propane as heat exchangers (referred to as C_2 and C_3 chillers in Fig. 5-4). The cold monomer and initiation solutions are injected rapidly and continuously into the reactor. Polymerization occurs almost instantaneously with the polymer precipitating as a fine slurry in methyl chloride. The slurry overflows into a flash tank containing steam and hot water. Solvent and unreacted monomers are vaporized and pass into a recovery system. Small amounts of zinc or aluminum stearate (to control the particle size of the butyl rubber latex) and antioxidant are added to the water in the flash tank. The aqueous slurry of butyl rubber passes into a stripping tank operating under vacuum to pull off residual amounts of methyl chloride and unreacted monomer. The stripped slurry is then pumped to a series of drying extruders to dewater the product and compressed into bales ready for shipment.

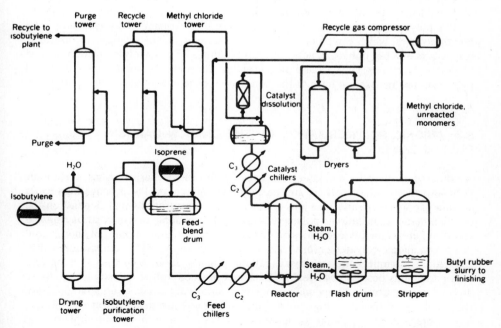

Fig. 5-4 Slurry process for production of butyl rubber. After Kresge et al. [1987] (by permission of Wiley-Interscience, New York).

The isoprene incorporates double bonds into the polymer chains, which makes subsequent vulcanization (crosslinking) possible (Sec. 9-2b). The amount of isoprene is in the range of 0.5–2.5% of the amount of isobutylene, the exact amount depending on the extent of crosslinking that is desired. Molecular weights of at least 200,000 are needed to obtain products that are not tacky. Control of molecular weight is achieved by working at the lower reaction temperatures and regulating the amounts of transfer and terminating agents present. Butyl rubber, unlike natural rubber, does not crystallize on cooling and hence remains flexible down to $-50°C$. The low degree of unsaturation imparts good resistance to aging, moisture, chemicals, and ozone. The low gas permeability of butyl rubber makes it well suited for innertubes for tires (which are still used in most nonpassenger automobile tires). The high-damping behavior of butyl rubber makes it well suited for engine and auto-body mounts. Other uses include electrical cable insulation, protective gloves, chemical tank liners, and pharmaceutical stoppers.

5-2i-2 Other Products

Hydrocarbon resins comprise a range of low-molecular-weight products ($\overline{M}_n < 3000$) used as adhesives, hot-melt coatings, tackifying agents, inks, and additives in rubber [Kennedy and Marechal, 1982]. These include products based on monomers derived from petroleum as well as plant sources. The petroleum-derived products include polymers produced from various alkenes, isoprene, piperylene, styrene, α-methylstyrene, vinyltoluene, and dicyclopentadiene. The plant-derived products include *polyterpenes* obtained by the polymerization of dipentene, α-pinene, and β-pinene. Low-molecular-weight coumarone–indene polymers, referred to as *coal-tar resins*, are used in surface coatings and floor tiles.

The polymerization of alkyl vinyl ethers is of some commercial importance. The homopolymers, which can be obtained only by cationic polymerization, are useful as plasticizers of other polymers, adhesives, and coatings. (The copolymerization of vinyl ethers with acrylates, vinyl acetate, maleic anhydride, and other monomers is achieved by radical polymerization but not the homopolymerizations of alkyl vinyl ethers.)

5-3 ANIONIC POLYMERIZATION OF THE CARBON–CARBON DOUBLE BOND

Anionic chain polymerizations show many of the same characteristics as cationic polymerizations, although there are some distinct differences. The propagating species are anionic ion pairs and free ions with relative concentrations that depend on the reaction media as in cationic polymerization. Unlike cationic polymerization, there is a large difference in the reactivities of ion pairs and free ions in anionic polymerization. Although anionic polymerizations generally proceed rapidly at low temperatures, they are seldom as temperature-sensitive as cationic polymerizations. Further, most anionic polymerizations possess positive E_R values and proceed well at and somewhat above ambient temperatures. Many anionic polymerizations are easier to understand, since the identities of the initiating species and counterions are much better established. The range of solvents useful for anionic polymerization is limited to aliphatic and aromatic hydrocarbons and ethers. Halogenated solvents, suitable for cationic polymerization, cannot be used for anionic polymerization because of their facile nucleophilic substitution reactions with carbanions. Other polar solvents such as esters and ketones are also excluded as a result of reaction with carbanions.

Termination occurs by transfer of a positive fragment, usually a proton, from the solvent or some transfer agent (often deliberately added), although other modes of termination are also known. Many anionic polymerizations are living polymerizations when the reaction components are appropriately chosen.

5-3a Initiation

5-3a-1 Nucleophilic Initiators

A variety of basic (nucleophilic) initiators have been used to initiate anionic polymerization [Bywater, 1975, 1976, 1985; Fontanille, 1989; Morton, 1983; Morton and Fetters, 1977; Richards, 1979; Szwarc, 1983; Young et al., 1984]. These include covalent or ionic metal amides such as $NaNH_2$ and $LiN(C_2H_5)_2$, alkoxides, hydroxides, cyanides, phosphines, amines, and organometallic compounds such as n-C_4H_9Li and $\emptyset MgBr$. Initiation involves the addition to monomer of a nucleophile (base), either a neutral (B:) or negative (B:$^-$) species.

Alkyllithium compounds are probably the most useful of these initiators, employed commercially in the polymerizations of 1,3-butadiene and isoprene. Initiation proceeds by addition of the metal alkyl to monomer

$$C_4H_9Li + CH_2{=}CHY \rightarrow C_4H_9-CH_2-\overset{\displaystyle Y}{\underset{\displaystyle H}{\overset{|}{\underset{|}{C}}}}{:}^-(Li^+) \tag{5-62}$$

followed by propagation

$$C_4H_9-CH_2-\overset{\displaystyle Y}{\underset{\displaystyle H}{\overset{|}{\underset{|}{C}}}}{:}^-(Li^+) + n\,CH_2{=}CHY \longrightarrow$$

$$C_4H_9{-}(\!-CH_2CHY{-}\!)_{\overline{n}}CH_2-\overset{\displaystyle Y}{\underset{\displaystyle H}{\overset{|}{\underset{|}{C}}}}{:}^-(Li^+) \tag{5-63}$$

The extensive use of alkyllithium initiators is due to their solubility in hydrocarbon solvents. Alkyls or aryls of the heavier alkali metals are poorly soluble in hydrocarbons, a consequence of their more ionic nature. The heavier alkali metal compounds, as well as alkyllithiums, are soluble in more polar solvents such as ethers. The use of most of the alkali metal compounds in ether solvents, especially the more ionic ones, is somewhat limited by their reactivity toward ethers. The problem is overcome by working below ambient temperatures and/or using less reactive (i.e., resonance-stabilized) anions as in benzylpotassium, cumylcesium, and diphenylmethyllithium.

Alkyl derivatives of the alkaline-earth metals have also been used to initiate anionic polymerization. Organomagnesium compounds are considerably less active than organolithiums, as a result of the much less polarized metal–carbon bond. They can only initiate polymerization of monomers more reactive than styrene and 1,3-dienes, such as 2- and 4-vinylpyridines, and acrylic and methacrylic esters. Organostrontium and organobarium compounds, possessing more polar metal–carbon bonds, are able to polymerize styrene and 1,3-dienes as well as the more reactive monomers.

In the relatively few anionic polymerizations initiated by neutral nucleophiles such as an amine or phosphine

$$R_3N: + CH_2{=}\underset{\underset{H}{|}}{\overset{\overset{Y}{|}}{C}} \longrightarrow R_3\overset{+}{N}{-}CH_2{-}\underset{\underset{H}{|}}{\overset{\overset{Y}{|}}{C}}{:}^{-} \qquad (5\text{-}64a)$$

$$\Big\downarrow CH_2{=}CHY$$

$$R_3\overset{+}{N}{-}\!\!\left(\!CH_2CHY\!\right)_{\!n}\!CH_2{-}\underset{\underset{H}{|}}{\overset{\overset{Y}{|}}{C}}{:}^{-} \qquad (5\text{-}64b)$$

the proposed propagating species is a *zwitterion* [Cronin and Pepper, 1988; Eromosele et al., 1989; Pepper and Ryan, 1983]. The zwitterion propagating species has the glaring deficiency of requiring increasing charge separation as propagation proceeds. Stabilization of zwitterion species may involve the positive end of one zwitterion propagating chain acting as the counterion of the carbanion end of another zwitterion propagating chain. The need for a zwitterion propagating species is avoided if initiation is proposed as occurring via hydroxide ion formed by the reaction of amine with adventious water [Donnelly et al., 1977; Ogawa and Romera, 1977].

The initiator required to polymerize a monomer depends on the reactivity of the monomer toward nucleophilic attack. Monomer reactivity increases with increasing ability to stabilize the carbanion charge. Very strong nucleophiles such as amide ion or alkyl carbanion are needed to polymerize monomers, such as styrene and 1,3-butadiene, with relatively weak electron-withdrawing substituents. Weaker nucleophiles, such as alkoxide and hydroxide ions, can polymerize monomers with strongly electron-withdrawing substituents, such as acrylonitrile, methyl methacrylate, and methyl vinyl ketone, although the efficiency is lower than that of the stronger nucleophiles. A monomer, such as methyl-α-cyanoacrylate, with two electron-withdrawing substituents can be polymerized with very mild nucleophiles such as Br^-, CN^-, amines, and phosphines. This monomer, used in many so-called superglues, polymerizes on contact with many surfaces. Polymerization probably involves initiation by water (or OH^- from water) [Donnelly et al., 1977].

5-3a-2 *Electron Transfer*

Szwarc and co-workers have studied the interesting and useful polymerizations initiated by aromatic radical-anions such as sodium naphthalene [Szwarc, 1968, 1974, 1983]. Initiation proceeds by the prior formation of the active initiator, the naphthalene radical–anion (**XIX**)

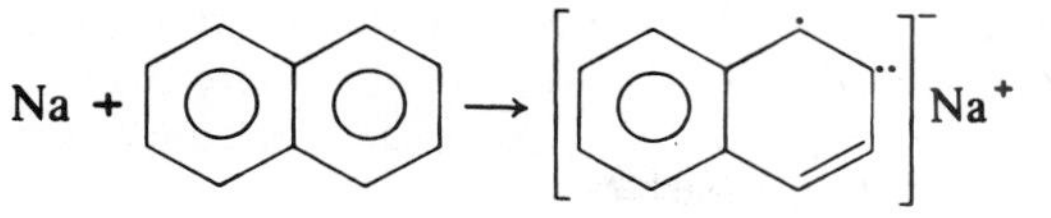

$$\text{Na} + \left[\bigcirc\bigcirc\right] \longrightarrow \left[\bigcirc\bigcirc\right]^{\overset{..}{\cdot}} \; \text{Na}^+ \tag{5-65a}$$

XIX

The reaction involves the transfer of an electron from the alkali metal to naphthalene. The radical nature of the anion–radical has been established from electron spin resonance spectroscopy and the carbanion nature by their reaction with carbon dioxide to form the carboxylic acid derivative. The equilibrium in Eq. 5-65a depends on the electron affinity of the hydrocarbon and the donor properties of the solvent. Biphenyl is less useful than naphthalene since its equilibrium is far less toward the anion–radical than for naphthalene. Anthracene is also less useful even though it easily forms the anion–radical. The anthracene anion–radical is too stable to initiate polymerization. Polar solvents are needed to stabilize the anion–radical, primarily via solvation of the cation. Sodium naphthalene is formed quantitatively in tetrahydrofuran (THF), but dilution with hydrocarbons results in precipitation of sodium and regeneration of naphthalene. For the less electropositive alkaline-earth metals, an even more polar solvent than THF [e.g., hexamethylphosphoramide (HMPA)] is needed.

The naphthalene anion–radical (which is colored greenish-blue) transfers an electron to a monomer such as styrene to form the styryl radical–anion (**XX**),

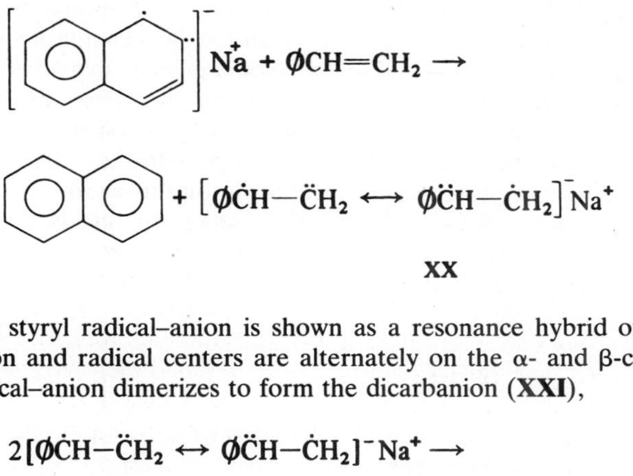

$$\left[\bigcirc\bigcirc\right]^{\overset{..}{\cdot}} \; \overset{+}{\text{Na}} + \emptyset\text{CH}{=}\text{CH}_2 \longrightarrow$$

$$\bigcirc\bigcirc + \left[\emptyset\overset{\cdot}{\text{C}}\text{H}{-}\overset{..}{\text{C}}\text{H}_2 \longleftrightarrow \emptyset\overset{..}{\text{C}}\text{H}{-}\overset{\cdot}{\text{C}}\text{H}_2\right]^- \text{Na}^+ \tag{5-65b}$$

XX

The styryl radical–anion is shown as a resonance hybrid of the forms wherein the anion and radical centers are alternately on the α- and β-carbon atoms. The styryl radical–anion dimerizes to form the dicarbanion (**XXI**),

$$2\left[\emptyset\overset{\cdot}{\text{C}}\text{H}{-}\overset{..}{\text{C}}\text{H}_2 \longleftrightarrow \emptyset\overset{..}{\text{C}}\text{H}{-}\overset{\cdot}{\text{C}}\text{H}_2\right]^- \text{Na}^+ \longrightarrow$$

$$\text{Na}^+ \left[^-{:}\underset{\underset{\text{H}}{|}}{\overset{\overset{\emptyset}{|}}{\text{C}}}{-}\text{CH}_2{-}\text{CH}_2{-}\underset{\underset{\text{H}}{|}}{\overset{\overset{\emptyset}{|}}{\text{C}}}{:}^-\right] \overset{..}{\text{Na}}^+ \tag{5-66}$$

XXI

That this reaction occurs is shown by electron spin resonance measurements, which indicate the complete disappearance of radicals in the system immediately after the addition of monomer. The dimerization occurs to form the styryl dicarbanion instead of $^-{:}\text{CH}_2\text{CH}\emptyset\text{CH}\emptyset\text{CH}_2{:}^-$, since the former is much more stable. The styryl dianions so-formed are colored red (the same as styryl monocarbanions formed via initiators such as *n*-butyllithium). Anionic propagation occurs at both carbanion ends of the styryl dianion

$$Na^+ \begin{bmatrix} \phi \\ ^-:C-CH_2-CH_2-C:^- \\ H \quad\quad H \end{bmatrix} Na^+ + (n + m)\ \phi CH{=}CH_2\ \longrightarrow$$

$$Na^+ \begin{bmatrix} \phi \\ ^-:C-CH_2\text{-}(CH\phi-CH_2\text{-})_n(CH_2-CH\phi\text{-})_m CH_2-C:^- \\ H \quad\quad\quad\quad\quad\quad\quad\quad\quad\quad\quad\quad\quad\quad H \end{bmatrix} Na^+ \quad (5\text{-}67)$$

Although the suggestion that the styryl anion-radical adds a few monomer molecules prior to dimerization has not been discounted, the reaction kinetics (Sec. 5-3d) clearly show that better than 99% of the propagation occurs through the dianion. Dimerization of radical centers is highly favored by their high concentrations, typically 10^{-3}–10^{-2} M (much higher than in a radical polymerization) and the large rate constants (10^6–10^8 liters/mole-sec) for radical coupling [Wang et al., 1978, 1979].

Electron-transfer initiation from other radical-anions, such as those formed by reaction of sodium with nonenolizable ketones, azomethines, nitriles, azo and azoxy compounds, has also been studied. In addition to radical–anions, initiation by electron transfer has been observed when one uses certain alkali metals in liquid ammonia. Polymerizations initiated by alkali metals in liquid ammonia proceed by two different mechanisms. In some systems, such as the polymerizations of styrene and meth-acrylonitrile by potassium, the initiation is due to amide ion formed in the system [Overberger et al., 1960]. Such polymerizations are analogous to those initiated by alkali amides. Polymerization in other systems cannot be due to amide ion. Thus polymerization of methacrylonitrile by lithium in liquid ammonia proceeds at a much faster rate than that initiated by lithium amide in liquid ammonia [Overberger et al., 1959]. The mechanism of polymerization is considered to involve the formation of a *solvated electron*,

$$Li + NH_3 \rightarrow Li^+(NH_3) + e^-(NH_3) \quad\quad\quad (5\text{-}68)$$

Such ammonia solutions are noted by their characteristic deep blue color. The solvated electron is then transferred to the monomer to form a radical–anion,

$$e^-(NH_3) + CH_2{=}CHY \rightarrow [\dot{C}H_2-\ddot{C}HY \leftrightarrow \ddot{C}H_2-\dot{C}HY]^-(NH_3) \quad (5\text{-}69)$$

The radical-anion proceeds to propagate in the same manner as discussed above for initiation by sodium naphthalene. (Polymerizations in liquid ammonia are very different from those in organic solvents in that free ions probably constitute the major portion of propagating species.)

Electron-transfer initiation also occurs in heterogeneous polymerizations involving dispersions of an alkali metal in monomer. Initiation involves electron transfer from the metal to monomer followed by dimerization of the monomer radical–anion to form the propagating dianion [Fontanille, 1989; Gaylord and Dixit, 1974; Morton and Fetters, 1977]. The rate of initiation is dependent on the surface area of the metal since the reaction is heterogeneous. Increased surface area of metal is achieved by condensing metal vapors directly into a reaction mixture or as a thin coating (mirror) of the metal on the inside walls of the reaction vessel or simply using small particle size of the metal.

Initiation by ionizing radiation occurs by electron transfer. Some component of the reaction system, either the solvent or monomer, undergoes radiolysis

$$S \rightarrow S^{+} + e_{solv} \tag{5-70}$$

to yield a cation–radical and solvated electron [Stannett et al., 1989]. If a monomer with an electron-withdrawing substituent is present, polymerization occurs by addition of the electron to monomer followed by dimerization to the dicarbanion and propagation (Sec. 5-2a-3-d).

Electroinitiated polymerization proceeds by direct electron addition to monomer to generate the monomer anion–radical, although initiation in some systems may involve the formation of an anionic species by electrolytic reaction of some component of the reaction system (often, the electrolyte) [Olaj, 1987].

5-3b   Termination

5-3b-1   Polymerizations without Termination

Termination of a propagating carbanion by combination with the counterion occurs in only a few instances, such as in electroinitiated polymerization when the contents of the anode and cathode chambers are mixed and in initiation by ionizing radiation. Termination by combination of the anion with a metal counterion does not take place. Many anionic polymerizations, especially of nonpolar monomers such as styrene and 1,3-butadiene, take place under conditions in which there are no effective termination reactions. Propagation occurs with complete consumption of monomer to form living polymers. The propagating anionic centers remain intact because transfer of proton or other positive species from the solvent does not occur. Living polymers are produced as long as one employs solvents, such as benzene, tetrahydrofuran, and 1,2-dimethoxyethane, which are relatively inactive in chain transfer with carbanions. The polymerization of styrene by amide ion in liquid ammonia, one of the first anionic systems to be studied in detail, is one of the few anionic polymerizations where chain transfer to solvent is extensive [Higginson and Wooding, 1952].

The nonterminating character of living anionic polymerizations is apparent in several different ways. Many of the propagating carbanions are colored. If a reaction system is highly purified so that impurities are absent, the color of the carbanions is observed to persist throughout the polymerization and does not disappear or change at 100% conversion. Further, after 100% conversion is reached, additional polymerization can be effected by adding more monomer, either the same monomer or a different monomer. The added monomer is also polymerized quantitatively and the molecular weight of the living polymer is increased.

5-3b-2   Termination by Impurities and Deliberately Added Transfer Agents

Most anionic (as well as cationic) polymerizations are carried out in an inert atmosphere with rigorously cleaned reagents and glassware since trace impurities lead to termination. Moisture adsorbed on the surface of glassware is usually removed by flaming under vacuum or washing with a living polymer solution. Oxygen and carbon dioxide from the atmosphere add to propagating carbanions to form peroxy and carboxyl anions. These are normally not reactive enough to continue propagation. (The peroxy and carboxyl anions usually are finally obtained as HO and HOOC groups when a proton donor is subsequently added to the polymerization system.) Any moisture present terminates propagating carbanions by proton transfer,

$$\sim\sim CH_2 - \underset{\underset{H}{|}}{\overset{\overset{\phi}{|}}{C}}{:}^- + H_2O \longrightarrow \sim\sim CH_2 - \underset{\underset{H}{|}}{\overset{\overset{\phi}{|}}{C}}H + HO^- \tag{5-71}$$

The hydroxide ion is usually not sufficiently nucleophilic to reinitiate polymerization and the kinetic chain is broken. Water has an especially negative effect on polymerization, since it is an active chain-transfer agent. For example, $C_{tr,S}$ is approximately 10 in the polymerization of styrene at 25°C with sodium naphthalene [Szwarc, 1960], and the presence of even small concentrations of water can greatly limit the polymer molecular weight and polymerization rate. The adventitious presence of other proton donors may not be as much of a problem. Ethanol has a transfer constant of about 10^{-3}. Its presence in small amounts would not prevent the formation of high polymer because transfer would be slow, although the polymer would not be living.

Living polymers are terminated by the deliberate addition of a chain-transfer agent such as water or alcohol to the reaction system after all of the monomer has reacted.

5-3b-3 *Spontaneous Termination*

Living polymers do not live forever. In the absence of terminating agents the concentration of carbanion centers decays with time [Fontanille, 1989; Glasse, 1983]. Polystyryl carbanions are the most stable of all living anionic systems, as they are stable for weeks in hydrocarbon solvents. (This is much more stable than any of the living cationic systems studied to date; see Sec. 5-2g). The mechanism for the decay of polystyryl carbanions on aging, referred to as *spontaneous termination*, is not completely established. The generally accepted mechanism, based on spectroscopy (IR, UV–visible) of the reaction system and final polymer after treatment with water, consists of *hydride elimination*

$$\sim\sim CH_2CH\phi CH_2 - \underset{\underset{H}{|}}{\overset{\overset{\phi}{|}}{C}}{:}^- Na^+ \longrightarrow \sim\sim CH_2CH\phi CH=CH\phi + H{:}^- Na^+ \tag{5-72}$$
$$\textbf{XXII}$$

followed by abstraction of an allylic hydrogen from **XXII** by a carbanion center to yield the unreactive 1,3-diphenylallyl anion (**XXIII**) [Spach et al., 1962]. The sodium

$$\sim\sim CH_2 - \underset{\underset{H}{|}}{\overset{\overset{\phi}{|}}{C}}{:}^- + \sim\sim CH_2CH\phi CH=CH\phi \longrightarrow$$

$$\sim\sim CH_2 - CH_2\phi + \sim\sim CH_2\overset{-}{C}\phi CH=CH\phi \tag{5-73}$$
$$\textbf{XXIII}$$

hydride eliminated in Reaction 5-72 may also participate in hydrogen abstraction from **XXII**.

The stability of polystyryl carbanions is considerably lower in polar solvents with decay occurring over the course of a few days at room temperature, but the stability is considerably better at lower temperatures. Termination in polar solvents (ethers such as THF and 1,2-dimethoxyethane) probably involves α-hydrogen abstraction and/ or nucleophilic attack on the C—O bond. The living polymers of 1,3-butadiene and isoprene decay much faster than polystyryl carbanions.

5-3b-4 Terminating Reactions of Polar Monomers

The anionic polymerizations of polar monomers, such as methyl methacrylate, methyl vinyl ketone, and acrylonitrile, are less understood than those for the nonpolar monomers. The polar monomers contain substituents that are reactive toward nucleophiles. This leads to termination and side reactions competitive with both initiation and propagation, resulting in complex polymer structures [Bywater, 1975, 1985; Hogen-Esch and Smid, 1987; Muller, 1989; Muller et al., 1986; Warzelhan et al., 1978]. Living polymerizations are much less frequently encountered.

Several different nucleophilic substitution reactions have been observed in the polymerization of methyl methacrylate. Attack of initiator on monomer

$$CH_2=\overset{\overset{\displaystyle CH_3}{|}}{C}-\overset{\overset{\displaystyle O}{\|}}{C}-OCH_3 + R^-Li^+ \longrightarrow CH_2=\overset{\overset{\displaystyle CH_3}{|}}{C}-\overset{\overset{\displaystyle O}{\|}}{C}-R + CH_3O^-Li^+ \qquad (5\text{-}74)$$

converts the active alkyllithium to the less active alkoxide initiator. Further, methyl methacrylate (MMA) is converted to *i*-propenyl alkyl ketone to the extent that this reaction occurs. The resulting polymerization is a copolymerization between the two monomers, not a homopolymerization of MMA. More importantly, this results in a slower reaction (and lower polymer molecular weight) since the carbanion derived from the ketone is not as reactive as the carbanion from MMA. Nucleophilic substitution by intramolecular backbiting attack of a propagating carbanion

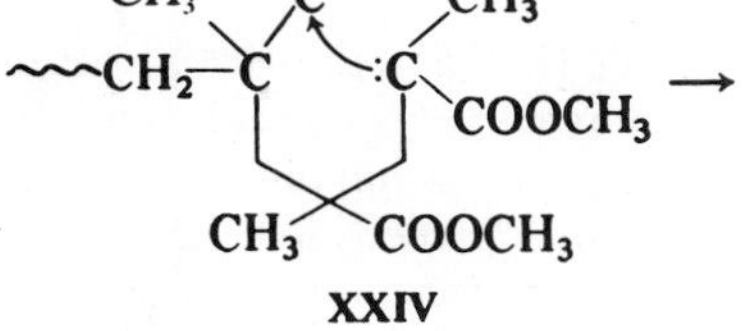

XXIV

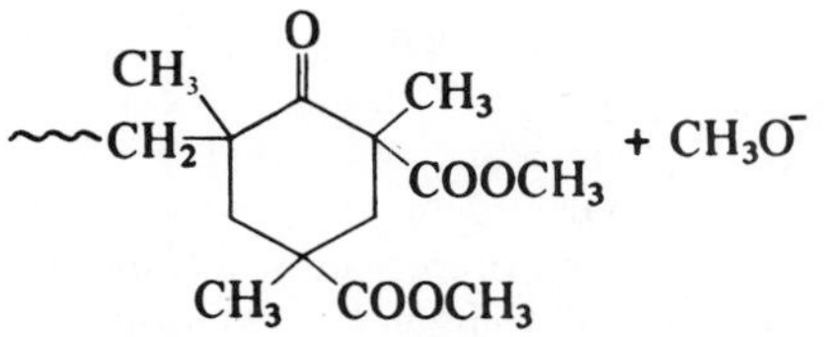

$$(5\text{-}75)$$

lowers the polymer molecular weight and decreases the polymerization rate since methoxide is a weak initiator. Other reactions have been proposed—nucleophilic attack by a propagating carbanion on monomer to displace methoxide and yield polymer with an i-propenyl keto end group and the intermolecular analogue of Reaction 5-75 to yield branched polymer—but there is little supportive data.

The side reactions, which not only lower the polymerization rate and polymer molecular weight but also broaden the molecular-weight distribution, predominate over "normal" polymerization in hydrocarbon solvents. Reaction 5-74 is the major reaction when butyllithium is used as the initiator with less than 1% of the initiator resulting in high-molecular-weight polymer. This reaction can be significantly minimized but not completely eliminated by using a less nucleophilic initiator, such as diphenylmethyllithium, cumylcesium, or polystyryl carbanions to which a few units of α-methylstyrene have been added [Wiles and Bywater, 1965]. However, Reaction 5-75 as well as the other side reactions can be sufficiently minimized by using the less nucleophilic initiator in a polar solvent such as THF at low temperatures (< -20 to $-70°C$) to obtain living polymerizations of acrylate and methacrylate monomers. The addition of a common ion (e.g., by using a mixture of alkyllithium and LiCl) to decrease the nucleophilicity of the propagating centers by forming tighter ion pairs also increases the tendency toward living polymerization [Fayt et al., 1987; Teyssie et al., 1990; Varshney et al., 1990].

Porphinatoaluminum compounds of structure **XXIV** ($Z = CH_3$, $S\emptyset$) are reported to initiate the living polymerizations of acrylate and methacrylate monomers at

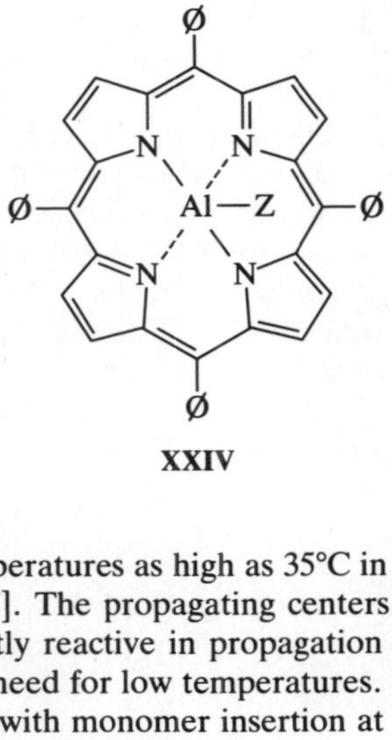

XXIV

temperatures as high as 35°C in methylene chloride [Inoue et al., 1990; Kuroki et al., 1987]. The propagating centers are sufficiently unreactive in side reactions but sufficiently reactive in propagation to allow polymerization in nonpolar solvent without the need for low temperatures. Propagation probably proceeds through covalent species with monomer insertion at Al—C bonds. Further discussion of these initiators is found in Sec. 7-2a-1.

Polymerizations of vinyl ketones such as methyl vinyl ketone are also complicated by nucleophilic attack of the initiator and propagating carbanion at the carbonyl group although few details have been established [Dotcheva and Tsvetanov, 1985; Hrdlovic et al., 1979; Nasrallah and Baylouzian, 1977]. Nucleophilic attack in these polymers results in addition, while that at the ester carbonyl of acrylates and methacrylates yields substitution. The major side reaction is an intramolecular Aldol-type condensation. Abstraction of an α-hydrogen from a methyl group of the polymer by either initiator or propagating carbanion yields an α-carbanion that attacks the carbonyl group of the adjacent repeat unit.

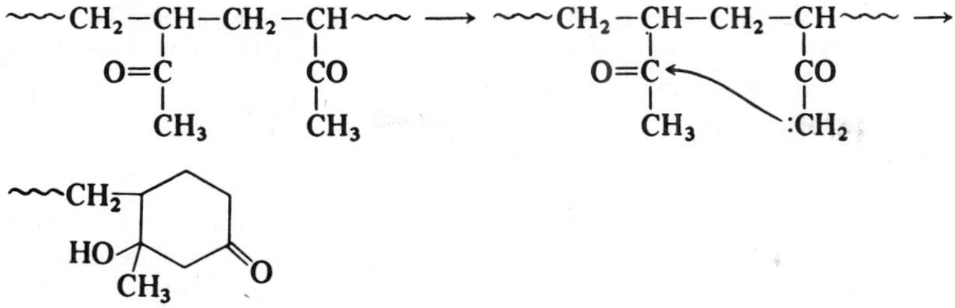

$$\sim\!\!\sim\!CH_2-CH-CH_2-CH\!\sim\!\!\sim \longrightarrow \sim\!\!\sim\!CH_2-CH-CH_2-CH\!\sim\!\!\sim \longrightarrow$$

$$\underset{\underset{CH_3}{|}}{O=C} \qquad \underset{\underset{CH_3}{|}}{CO} \qquad \underset{\underset{CH_3}{|}}{O=C} \qquad \underset{\underset{:CH_2}{|}}{CO}$$

$$\sim\!\!\sim\!CH_2$$

$$HO \qquad\qquad O$$

$$CH_3$$

$$\tag{5-76}$$

Acylonitrile polymerization is similarly complicated by addition reactions of the initiator and/or propagating nucleophiles with the nitrile group [Berger and Adler, 1986; Tsvetanov, 1979; Vankerckhoven and Van Beylen, 1978].

5-3c Group Transfer Polymerization

Although living anionic polymerizations of methacrylates and acrylates can be achieved at low temperatures, the low temperatures limit the practical utility of such processes. *Group transfer polymerization* yields living polymerizations without resorting to low temperatures [Muller, 1990; Reetz, 1988; Schubert and Bandermann, 1989; Schubert et al., 1989; Shen et al., 1989; Sogah et al., 1987, 1990; Webster, 1987]. The initiator is a silyl ketene acetal (**XXV**) that is typically synthesized from an ester enolate

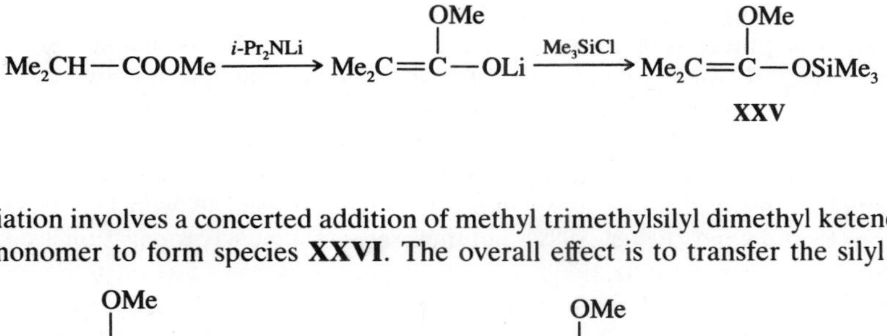

$$Me_2CH-COOMe \xrightarrow{\ i\text{-}Pr_2NLi\ } Me_2C\!\!=\!\!\overset{\overset{\displaystyle OMe}{|}}{C}-OLi \xrightarrow{\ Me_3SiCl\ } Me_2C\!\!=\!\!\overset{\overset{\displaystyle OMe}{|}}{C}-OSiMe_3$$

$$\textbf{XXV}$$

$$\tag{5-77}$$

Initiation involves a concerted addition of methyl trimethylsilyl dimethyl ketene acetal to monomer to form species **XXVI**. The overall effect is to transfer the silyl ketene

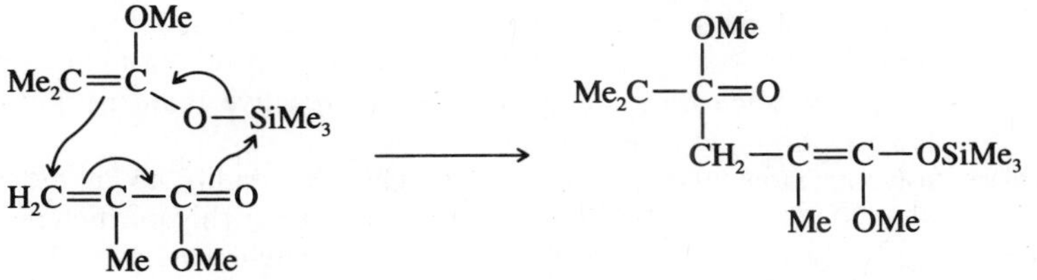

$$Me_2C\!\!=\!\!\overset{\overset{\displaystyle OMe}{|}}{C}\underset{O-SiMe_3}{\overset{}{\diagdown}}$$

$$H_2C\!\!=\!\!\overset{}{\underset{\underset{Me}{|}}{C}}-\overset{}{\underset{\underset{OMe}{|}}{C}}\!\!=\!\!O \qquad\longrightarrow\qquad Me_2C-\overset{\overset{\displaystyle OMe}{|}}{C}\!\!=\!\!O$$

$$\underset{\underset{Me\ \ OMe}{|\ \ \ |}}{CH_2-C\!\!=\!\!C-OSiMe_3} \qquad\tag{5-78}$$

$$\textbf{XXVI}$$

acetal center from initiator to monomer. Propagation proceeds in a similar manner

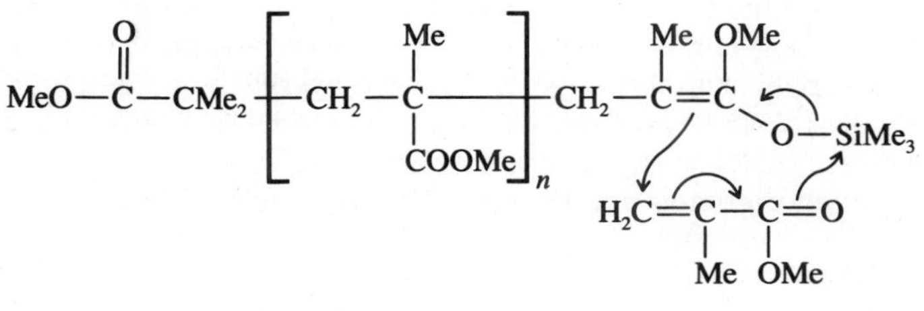

$$MeO-\overset{\overset{\displaystyle O}{\|}}{C}-CMe_2\!\!\left[\!CH_2-\overset{\overset{\displaystyle Me}{|}}{\underset{\underset{COOMe}{|}}{C}}\!\right]_{\!n}\!\!CH_2-\overset{\overset{\displaystyle Me\ \ OMe}{|\ \ \ \ |}}{C\!\!=\!\!C}\underset{O-SiMe_3}{\overset{}{\diagdown}}$$

$$H_2C\!\!=\!\!\overset{}{\underset{\underset{Me\ \ OMe}{|\ \ \ |}}{C}}-C\!\!=\!\!O \qquad\tag{5-79}$$

with the ketene acetal double bond acting as the propagating center. The proposed mechanism for both initiation and propagation involves nucleophilic attack of the π-electrons of the ketene acetal on monomer.

Group transfer polymerization (GTP) requires either a nucleophilic or Lewis acid catalyst. Bifluoride (HF_2^-) and fluoride ions, supplied by soluble reagents such as tris(dimethylamino)sulfonium bifluoride, $[(CH_3)_2N]_3SHF_2$, and $(n\text{-}C_4H_9)_4NF$, are the most effective nucleophilic catalysts, although other nucleophiles (CN^-, acetate, p-nitrophenolate) are also useful. Zinc chloride, bromide, and iodide, and dialkyl-aluminum chloride have been used as Lewis acid (electrophilic) catalysts. Nucleophilic catalysts function by assisting in the displacement of the trimethylsilyl group and are effective at low concentrations (<0.1 mole % relative to initiator). Lewis acid catalysts probably function by coordination with the carbonyl oxygen of monomer to increase the electrophilicity of the double bond, making the latter more susceptible to nucleophilic attack. Much higher concentrations of Lewis acid catalysts are needed, 10–20 mole % based on initiator for the aluminum catalysts but 10–20 mole % based on monomer for the zinc catalysts.

Group transfer polymerization requires the absence of materials such as water with active hydrogens, but oxygen does not interfere with the reaction. The range of solvents suitable for GTP is wider than for other anionic polymerizations. N,N-Dimethylformamide has been used for polymerizations with nucleophilic catalysts, and chlorinated hydrocarbons as well as acetonitrile for polymerizations with Lewis acid catalysts, although ethers such as tetrahydrofuran and aromatics such as toluene are probably the most commonly used solvents. The concerted nature of the reaction with the absence of full-fledged anionic propagating centers appears responsible for the ability to carry out polymerization in the more active solvents. Solvents such as chlorinated hydrocarbons and DMF cannot be used in anionic polymerizations initiated by sodium naphthalene, alkyllithium, and similar initiators because of nucleophilic reactions between solvent and initiator. The concerted reaction is also assumed to be responsible for the lowered sensitivity of GTP toward the various side reactions present in the usual anionic polymerization. The same side reactions occur as with anionic polymerization [Brittain and Dicker, 1989], but their rates relative to propagation are lower. This allows living polymerizations to be achieved at higher reaction temperatures. Living polymerizations of methacrylates have been successful at temperatures of 0–50°C. Lower reaction temperatures ($<0°C$) are required to obtain living systems with acrylate monomers since they are typically more reactive than methacrylates toward both polymerization and side reactions.

Control of polymer molecular weight can be achieved readily up to molecular weights of 10,000 to 20,000. Higher molecular weights require highly purified reagents as in other anionic polymerizations and the use of nucleophilic catalysts. The large amounts of Lewis acid catalysts required increases the extent of termination. Further, Lewis acid catalysts have a large disadvantage in terms of the purification required to obtain a product free of inorganic materials.

The concerted mechanism has been questioned for nucleophile-catalyzed group transfer polymerization [Quirk and Bidinger, 1989]. The alternate mechanism proposed for GTP involves a generation of the usual anionic propagating species in *low concentrations* by nucleophilic displacement of the trimethyl silyl group of initiator by the nucleophilic catalyst. The key to control of termination by side reactions in this mechanism involves a reversible complexation of the low concentration of propagating centers by the high concentration of silyl ketene acetal polymer chain ends. This decreases reactivity of propagating centers toward terminating side reactions relative to propagation.

A variation of GTP, referred to as *aldol GTP*, involves polymerization of a silyl vinyl ether initiated by an aldehyde [Sogah and Webster, 1986; Webster, 1987]. Both initiation and propagation involve nucleophilic addition of the vinyl ether to the aldehyde carbonyl group with transfer of the trialkyl silyl group from vinyl ether to the carbonyl oxygen (Eq. 5-80a). The reaction has similar characteristics as GTP. The

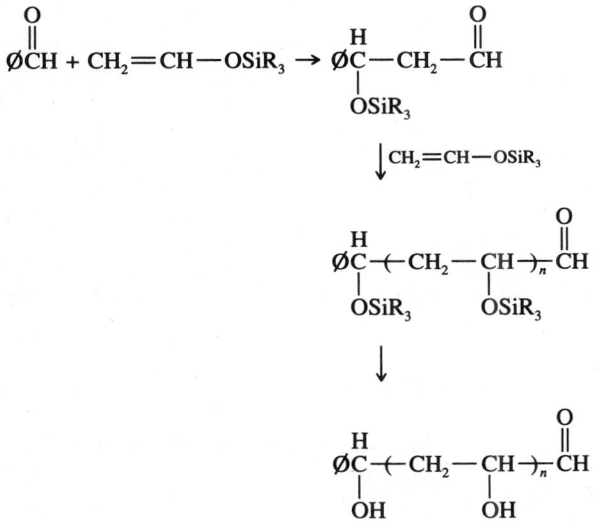

$$
\overset{O}{\overset{\|}{\varnothing CH}} + CH_2{=}CH{-}OSiR_3 \rightarrow \underset{OSiR_3}{\overset{H}{\underset{|}{\varnothing C}}}{-}CH_2{-}\overset{O}{\overset{\|}{CH}}
$$

$$\downarrow CH_2{=}CH{-}OSiR_3$$

$$
\underset{OSiR_3}{\overset{H}{\underset{|}{\varnothing C}}}{+}CH_2{-}\underset{OSiR_3}{\underset{|}{CH}}{\rightarrow}{}_n\overset{O}{\overset{\|}{CH}} \tag{5-80a}
$$

$$\downarrow$$

$$
\underset{OH}{\overset{H}{\underset{|}{\varnothing C}}}{+}CH_2{-}\underset{OH}{\underset{|}{CH}}{\rightarrow}{}_n\overset{O}{\overset{\|}{CH}} \tag{5-80b}
$$

product is a silated poly(vinyl alcohol) (PVA), which can be hydrolyzed by acid to PVA (Eq. 5-80b).

5-3d Kinetics of Polymerization with Termination

The polymerization rate for an anionic polymerization where termination occurs simultaneously with propagation follows in exactly the manner described for cationic polymerizations (Sec. 5-2d). For potassium amide initiated polymerization in liquid ammonia, initiation involves the dissociation of potassium amide followed by addition of amide ion to the first monomer unit.

$$
KNH_2 \overset{K}{\rightleftharpoons} K^+ + H_2N{:}^- \tag{5-81a}
$$

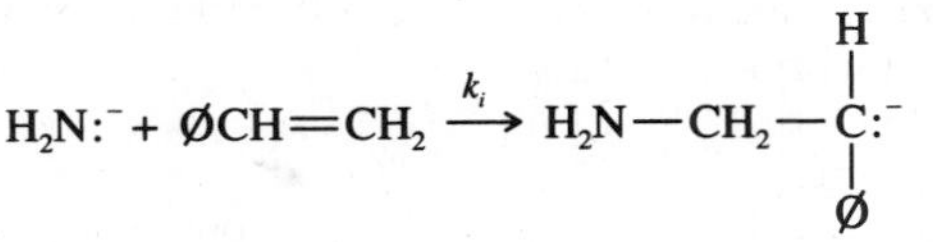

$$
H_2N{:}^- + \varnothing CH{=}CH_2 \overset{k_i}{\longrightarrow} H_2N{-}CH_2{-}\underset{\varnothing}{\overset{H}{\underset{|}{\overset{|}{C}}}}{:}^- \tag{5-81b}
$$

The rate of initiation is given by

$$
R_i = k_i[H_2N{:}^-][M] \tag{5-82}
$$

or

$$R_i = \frac{k_i K[\text{M}][\text{KNH}_2]}{[\text{K}^+]} \tag{5-83}$$

Propagation proceeds according to

$$\text{H}_2\text{N}-\text{M}_n^- + \text{M} \xrightarrow{k_p} \text{H}_2\text{N}-\text{M}_n\text{M}^- \tag{5-84}$$

with a rate given by

$$R_p = k_p[\text{M}^-][\text{M}] \tag{5-85}$$

where $[\text{M}^-]$ represents the total concentration of the propagating anionic centers.
    Chain transfer to solvent

$$\text{H}_2\text{N}-\text{M}_n^- + \text{NH}_3 \xrightarrow{k_{tr,\text{NH}_3}} \text{H}_2\text{N}-\text{M}_n-\text{H} + \text{NH}_2^- \tag{5-86}$$

is extensive but does not terminate the kinetic chain since amide ion is regenerated.
Termination occurs by transfer to adventitious water,

$$\text{H}_2\text{N}-\text{M}_n^- + \text{H}_2\text{O} \xrightarrow{k_{tr,\text{H}_2\text{O}}} \text{H}_2\text{N}-\text{M}_n-\text{H} + \text{HO}^- \tag{5-87}$$

or other impurity present. The rates of Reactions 5-86 and 5-87 are given by

$$R_{tr,\text{NH}_3} = k_{tr,\text{NH}_3}[\text{M}^-][\text{NH}_3] \tag{5-88}$$
$$R_{tr,\text{H}_2\text{O}} = k_{tr,\text{H}_2\text{O}}[\text{M}^-][\text{H}_2\text{O}] \tag{5-89}$$

The polymerization rate, derived in the usual manner by combining Eqs. 5-82, 5-83,
5-85, and 5-89 with the assumption of a steady state for $[\text{M}^-]$, is obtained as

$$R_p = \frac{Kk_i k_p[\text{M}]^2[\text{KNH}_2]}{k_{tr,\text{H}_2\text{O}}[\text{K}^+][\text{H}_2\text{O}]} \tag{5-90}$$

    The number-average degree of polymerization is given by

$$\frac{1}{\overline{X}_n} = \frac{C_{\text{NH}_3}[\text{NH}_3]}{[\text{M}]} + \frac{C_{\text{H}_2\text{O}}[\text{H}_2\text{O}]}{[\text{M}]} \tag{5-91}$$

Amide ion initiated polymerizations in liquid ammonia are about the only anionic
polymerizations studied that proceed with termination.

5-3e Kinetics of Living Polymerization

5-3e-1 Polymerization Rate

The rate of polymerization in nonterminating systems can be expressed as the rate of
propagation

$$R_p = k_p^{\text{app}}[\text{M}^-][\text{M}] \tag{5-92}$$

where [M$^-$] is the total concentration of all types of living anionic propagating centers (free ions and ion pairs). [M$^-$] can be determined by reacting the living polymer with a terminating agent such as methyl iodide, carbon dioxide, or other electrophilic substance followed by analysis of the amount of terminating agent incorporated into the polymer. The use of isotopically labeled terminating agents can increase the analytical sensitivity. Ultraviolet–visible spectroscopy is also useful for determining the total concentration of propagating species. Since free ions constitute a very small percentage of [M$^-$] for most systems, equating the concentration of ion pairs with [M$^-$] does not introduce a significant error. The living ends are monoanions in polymerizations initiated by butyllithium and similar initiators and dianions in polymerizations initiated by electron transfer. The two types of anions are referred to as *one-ended* (monofunctional) and *two-ended* (bifunctional) living anions.

Equation 5-92 applies for the case where initiation is rapid relative to propagation. This condition is met for polymerizations in polar solvents. However, polymerizations in nonpolar solvent frequently proceed with an initiation rate that is of the same order of magnitude or lower than propagation. More complex kinetic expressions analogous to those developed for radical and nonliving cationic polymerizations apply for such systems [Pepper, 1980; Szwarc et al., 1987].

As with certain cationic systems, many anionic living polymerizations proceed too rapidly to be followed by techniques such as dilatometry. The stopped-flow technique (Sec. 5-2e-1) is useful for studying these fast polymerizations. Ultraviolet–visible spectroscopy of the rapidly mixed reaction system contained in a capillary tube allows one to follow the initiation rate (observing the increase in optical density of propagating species) and/or the polymerization rate (by following loss of monomer). The polymerization rate can also be obtained by a modification of the apparatus in which polymerization is stopped by running the contents of the capillary tube into a solvent containing a terminating agent. The reaction time is given by the ratio of the capillary volume to the flow rate. Short reaction times from 0.005 to 2 sec can be accurately studied in this manner. The conversion and, hence, R_p and apparent propagation rate constant are obtained by analyzing the quenched reaction mixture for either polymer or unreacted monomer.

It is useful to understand the reasons for the faster reaction rates encountered in many anionic polymerizations compared to their radical counterparts. This can be done by comparing the kinetic parameters in appropriate rate equations: Eq. 3-22 for radical polymerization and Eq. 5-92 for anionic polymerization. The k_p values in radical polymerization are similar to the k_p^{app} values in anionic polymerization. Anionic k_p^{app} values may be 10- to 100-fold lower than in radical polymerization for polymerization in hydrocarbon solvents, while they may be 10- to 100-fold higher for polymerizations in ether solvents. The major difference in the rates of anionic and radical polymerizations resides in the lack of termination in anionic polymerization and the large difference in the concentrations of the propagating species. The concentration of propagating radicals is 10^{-9}–10^{-7} M, while that for propagating anions is often as high as 10^{-4}–10^{-2} M. Thus anionic polymerization rates are much higher than radical rates based only on the concentrations of propagating species.

5-3e-2 *Effects of Reaction Media*

The propagation rate constant and the polymerization rate for anionic polymerization are dramatically affected by the nature of both the solvent and the counterion. Thus the data in Table 5-10 show the pronounced effect of solvent in the polymerization of styrene by sodium naphthalene (3×10^3 M) at 25°C. The apparent propagation rate

TABLE 5-10   Effect of Solvent on Anionic Polymerization^a of Styrene

Solvent	Dielectric Constant (ϵ)	k_p^{app} liters/mole-sec
Benzene	2.2	2
Dioxane	2.2	5
Tetrahydrofuran	7.6	550
1,2-Dimethoxyethane	5.5	3,800

^{a}Data from Szwarc and Smid [1964].

constant is increased by two and three orders of magnitude in tetrahydrofuran and 1,2-dimethoxyethane, respectively, compared to the rate constants in benzene and dioxane. The polymerization is much faster in the more polar solvents. That the dielectric constant is not a quantitative measure of solvating power is shown by the higher rate in 1,2-dimethoxyethane (DME) compared to tetrahydrofuran (THF). The faster rate in DME may be due to a specific solvation effect arising from the presence of two ether functions in the same molecule.

The increase in k_p^{app} with increased solvating power of the reaction medium is due mainly to the increased fraction of free ions present relative to ion pairs. It would be more informative to obtain the individual propagation rate constants for the free ions and ion pairs as well as the relative amounts of the two types of propagating species.

5-3e-2-a Evaluation of Individual Propagation Rate Constants for Free Ions and Ion Pairs.
The rate of polymerization is appropriately expressed as the sum of the rates for the free propagating anion P^- and the ion pair $P^-(C^+)$:

$$R_p = k_p^-[P^-][M] + k_p^{\pm}[P^-(C^+)][M] \tag{5-93}$$

where k_p^- and $k_p^{\pm}$ are the propagation rate constants for the free ion and ion pair, respectively, $[P^-]$ and $[P^-(C^+)]$ are the concentrations of the free ion and ion pair, and $[M]$ is the monomer concentration. C^+ in the above notation is the positive counterion. Comparison of Eqs. 5-92 and 5-93 yields the apparent k_p as

$$k_p^{app} = \frac{k_p^-[P^-] + k_p^{\pm}[P^-(C^+)]}{[M^-]} \tag{5-94}$$

The two propagating species are in equilibrium according to

$$P^-(C^+) \overset{K}{\rightleftharpoons} P^- + C^+ \tag{5-95}$$

governed by the dissociation constant K given by

$$K = \frac{[P^-][C^+]}{[P^-(C^+)]} \tag{5-96}$$

For the case where $[P^-] = [C^+]$, that is, there is no source of either ion other than $P^-(C^+)$, the concentration of free ions is

$$[P^-] = (K[P^-(C^+)])^{1/2} \tag{5-97}$$

The extent of dissociation is small under most conditions, the concentration of ion pairs is close to the total concentration of living ends and Eq. 5-97 can be rewritten as

$$[P^-] = (K[M^-])^{1/2} \tag{5-98}$$

The concentration of ion pairs is given by

$$[P^-(C^+)] = [M^-] - (K[M^-])^{1/2} \tag{5-99}$$

Combination of Eqs. 5-94, 5-98, and 5-90 yields k_p^{app} as a function of $[M^-]$

$$k_p^{\text{app}} = k_p^{\mp} + \frac{(k_p^- - k_p^{\mp})K^{1/2}}{[M^-]^{1/2}} \tag{5-100}$$

Polymerizations can also be carried out in the presence of excess counterion by adding a strongly dissociating salt (e.g., NaBⵁ$_4$ to supply excess Na$^+$). The concentration of free ions, depressed by the common ion effect, is given by

$$[P^-] = \frac{K[M^-]}{[C^+]} \tag{5-101}$$

When the added salt is strongly dissociated and the ion pairs slightly dissociated, the counterion concentration is very close to that of the added salt [CZ]

$$[C^+] \simeq [CZ] \tag{5-102}$$

The concentrations of free anions and ion pairs are given by

$$[P^-] = \frac{K[M^-]}{[CZ]} \tag{5-103}$$

$$[P^-(C^+)] = [M^-] - \frac{K[M^-]}{[CZ]} \tag{5-104}$$

which are combined with Eq. 5-85 to yield

$$k_p^{\text{app}} = k_p^{\mp} + \frac{(k_p^- - k_p^{\mp})K}{[CZ]} \tag{5-105}$$

Equations 5-100 and 5-105 allow one to obtain k_p^-, $k_p^{\mp}$, and K from k_p^{app} values obtained in the absence and presence of added common ion. A plot of k_p^{app} obtained in the absence of added common ion versus $[M^-]^{1/2}$ yields a straight line whose slope and intercept are $(k_p^- - k_p^{\mp})K^{1/2}$ and $k_p^{\mp}$, respectively. A plot of k_p^{app} obtained in the presence of added common ion versus $[CZ]^{-1}$ yields a straight line whose slope and intercept are $(k_p^- - k_p^{\mp})K$ and $k_p^{\mp}$, respectively. Figures 5-5 and 5-6 show these plots for polystyryl sodium in 3-methyltetrahydrofuran at 20°C. The combination of the two slopes and two intercepts allows the individual calculation of k_p^-, $k_p^{\mp}$, and K. (*Note*: K as well as $[P^-]$ and $[P^-(C^+)]$ can also be independently determined from conductivity measurements.)

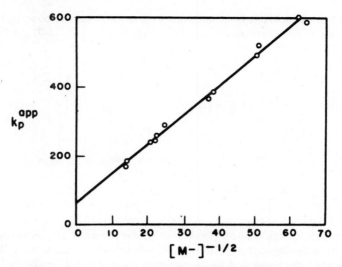

Fig. 5-5 Polymerization of styrene by sodium naphthalene in 3-methyltetrahydrofuran at 20°C. After Schmitt and Schulz [1975] (by permission of Pergamon Press Ltd., Oxford).

5-3e-2-b   Reactivity in Anionic Polymerization. Table 5-11 shows the values of K and the propagation rate constants for free ions and ion pairs in styrene polymerization in THF at 25°C with various alkali metal counterions [Bhattacharyya et al., 1965a,b]. The corresponding $k_p^{\pm}$ values in dioxane are also presented. The values of K and k_p^- in dioxane could not be obtained as conductivity measurements indicated no detectable dissociation of ion pairs to free ions in dioxane. A consideration of the data in Table

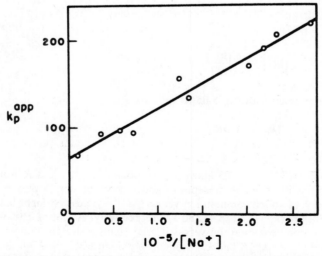

Fig. 5-6 Polymerization of styrene by sodium naphthalene in 3-methyltetrahydrofuran at 20°C in the presence of sodium tetraphenylborate. After Schmitt and Schulz [1975] (by permission of Pergamon Press Ltd., Oxford).

TABLE 5-11 Effect of Counterion on Anionic Polymerization of Styrene[a]

	Polymerization in Tetrahydrofuran			$k_p^{\mp}$
Counterion	$k_p^{\mp}$	$K \times 10^7$	k_p^-	for Dioxane
Li^+	160	2.2		0.94
Na^+	80	1.5		3.4
K^+	60–80	0.8	6.5×10^4	19.8
Rb^+	50–80	0.1		21.5
Cs^+	22	0.02		24.5

[a]Data from Bhattacharyya et al. [1965a, 1965b].

5-11 allows an understanding of reactivity in anionic polymerization. The polymerization data shows the much greater reactivity of the free ion compared to any of the ion pairs. The value of k_p^- is 6.5×10^4 liters/mole-sec which is larger by a factor of 10^2–10^3 than the $k_p^{\mp}$ values. Similar differences between k_p^- and $k_p^{\mp}$ have been observed in polymerizations of methyl methacrylate (MMA), 2-vinylpyridine (2VP), and isoprene (IP) [Bywater, 1985; Jeuk and Muller, 1982; Van Beylen et al., 1988]. For reference, k_p^- has values of 3.8×10^3, 1.0×10^5, and 2.0×10^3 liters/mole-sec for IP (25°C), 2VP (25°C), and MMA (-98°C), respectively, all measured in THF [Bywater, 1985].

The K values in Table 5-11 indicate that increased solvating power affects the reaction rate primarily through an increase in the concentration of free ions. When lithium is the counterion, one calculates from the equilibium constant that 1.48% of the propagating centers are free ions in THF (for a system where the total concentration of all propagating centers is 10^{-3} M) compared to zero in dioxane. Since free ions are so much more reactive than ion pairs, their small concentration has a very large effect on the observed polymerization rate. The majority of the propagation is carried by free ions; only about 10% of the observed reaction rate is due to ion pairs. It is worth mentioning that K values independently measured by conductivity are in excellent agreement with those obtained from the kinetic measurements. The values K from conductivity are 1.9, 1.5, 0.7, and 0.028×10^{-7}, respectively, for lithium, sodium, potassium, and cesium counterions [Geacintov et al., 1962; Shimomura et al., 1967a, 1967b; Szwarc, 1969]. Table 5-11 shows that the dissociation constant for the ion pair decreases in going from lithium to cesium as the counterion. The order of increasing K is the order of increasing solvation of the counterion. The smaller Li^+ is solvated to the greatest extent and the larger Cs^+ is the least solvated. The decrease in K has a very significant effect on the overall polymerization, since there is a very significant change in the concentration of the highly reactive free ions. Thus the free-ion concentration for polystyryl cesium ($K = 0.02 \times 10^{-7}$) is less than 10% that of polystyryl lithium ($K = 10^{-7}$).

The reactivities of the various ion pairs also increase in the same order as the K values: Li > Cs. The fraction of the ion pairs that are of the solvent-separated type increases with increasing solvation of the counterion. Solvent-separated ion pairs are much more reactive than contact ion pairs (Sec. 5-3e-4). The lower values of $k_p^{\mp}$ in dioxane relative to THF are also a consequence of the presence of a smaller fraction of the more reactive solvent-separated ion pairs. The order of reactivity for the different ion pairs in dioxane is the reverse of that in tetrahydrofuran. Solvation is not important in dioxane. The ion pair with the highest reactivity is that with the weakest bond between the carbanion center and counterion. The bond strength decreases and reactivity increases with increasing size of counterion. However, the effect of increasing counterion size levels off after K^+ as $k_p^{\mp}$ is approximately the same for potassium, rubidium, and cesium. The effect of counterion on ion-pair reactivity is different for methyl methacrylate (MMA) compared to styrene. The value of $k_p^{\mp}$ is 1 for lithium counterion and 30–33 for the larger alkali metal counterions for polymerization in tetrahydrofuran at $-98°C$ [Jeuk and Muller, 1982; Szwarc, 1983]. The difference between sodium counterion and the larger counterions for MMA in THF is similar to that observed for styrene in dioxane. These results have been interpreted as indicating the absence of solvation by THF for MMA polymerization due to the presence of intramolecular solvation. Intramolecular solvation involves electron donation from the carbonyl oxygen of the penultimate unit (i.e., the unit just before the end unit of the propagating chain) as shown in **XXVII** [Kraft et al., 1978, 1980]. This additional

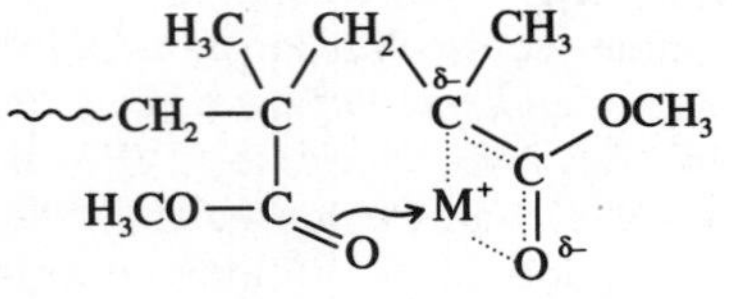

XXVII

binding of the counterion to the polymer accounts for the low dissociation constant $(K < 10^{-9})$ for polymethyl methacrylate and also poly(2-vinylpyridine) ion pairs [Tardi and Sigwalt, 1972; Van Beylen et al., 1988]. Intramolecular solvation in poly(2-vinylpyridine) involves electron donation from the nitrogen of a penultimate pyridine ring. This effect may also be responsible for the decrease in dissociation constant of the poly(2-vinylpyridine) ion pair with decreasing size of counterion in THF. Thus K is 1.1×10^{-9}, 2.5×10^{-9}, and 8.3×10^{-10} for Cs^+, K^+, and Na^+, respectively [Szwarc, 1983]. This is the reverse of the order observed for polystyryl ion pairs in THF. Apparently, smaller counterions "fit better" or "more tightly" into the intramolecular solvation sphere.

5-3e-3 Degree of Polymerization

The number-average degree of polymerization for a living polymerization is given by the ratio of the concentrations of monomer and living ends (Eq. 5-57). For the usual situation where all of the initiator I is converted into propagating living anionic ends, Eq. 5-57 becomes

$$\overline{X}_n = \frac{2[M]}{[I]} \tag{5-106}$$

or

$$\overline{X}_n = \frac{[M]}{[I]} \tag{5-107}$$

depending on the mode of initiation. Equation 5-106 applies to polymerizations initiated by electron transfer since each final polymer molecule originates from two initiator molecules (e.g., one dianionic propagating species is formed from two sodium naphthalenes). Initiation processes other than electron transfer (e.g., alkyllithium) involve one polymer molecule per initiator molecule and Eq. 5-107 is applicable.

Narrow molecular-weight distributions are obtained for systems with fast initiation and efficient mixing in the absence of depropagation, termination, and transfer reactions (Eq. 5-59). *PDI* values below 1.1–1.2 are found for many living polymerizations. The living polymer technique offers a unique method of synthesizing standard polymer samples of known and well-defined molecular weights. Commercially available molecular-weight standards are now available for a number of polymers—polystyrene, polyisoprene, poly(α-methylstyrene), poly(2-vinylpyridine), poly(methyl methacrylate), polyisobutylene, and poly(tetrahydrofuran). All except the last two polymers are synthesized by living anionic polymerization. The last two are obtained by living cationic polymerization with ring-opening polymerization (Chap. 7) used for poly(tetrahydrofuran). These polymer standards are useful as calibration standards in molecular weight measurements by size exclusion chromatography, membrane and vapor pressure osmometry, and viscometry.

The occurrence of any termination, transfer or side reactions result in broadening of the molecular-weight distribution. The termination reactions in methacrylate polymerizations (at other than low temperatures in polar solvents) and depropagation in α-methylstyrene polymerizations broaden *PDI* [Chaplin and Yaddehige, 1980; Malhotra et al., 1977; Malhotra, 1978]. Although the bulk of propagation is carried by a small fraction of the propagating species (i.e., the free ions), this does not significantly broaden the molecular-weight distribution since the free ions and ion pairs are in rapid equilibrium. Each polymer chain propagates as both free ion and ion pair over its lifetime and the average fractions of its lifetime spent as free ion and ion pair are not too different than for any other propagating chain.

5-3e-4 Energetics: Solvent-Separated and Contact Ion Pairs

The data available on the temperature dependence of the rates of living polymerization show the experimental activation energy E_R is generally relatively low and positive. One should note that E_R for living polymerization is the activation energy for propagation. The polymerization rates are relatively insensitive to temperature but increase with increasing temperature. Furthermore, the activation energy varies considerably depending on the solvent employed in the polymerization as was the case for cationic polymerization. Thus the activation energy for propagation in the system styrene–sodium naphthalene is 37.6 kJ/mole in dioxane and only 4.2 kJ/mole in tetrahydrofuran [Allen et al., 1960; Geacintov et al., 1962; Stretch and Allen, 1961]. The molecular weight of the polymer produced in a nonterminating polymerization is unaffected by temperature if transfer agents are absent. The situation can be different if transfer agents are initially present.

Most of the activation energy data reported in the literature are *apparent activation energies* corresponding to the values for the apparent propagation rate constant. The effect of temperature on propagation is complex—temperature simultaneously affects the relative concentrations of free ions and ion pairs and the individual rate constants for the free ions and ion pairs. Temperature affects the rate constants k_p^- and $k_p^\mp$ in the manner all rate constants are affected, increasing temperature increases the values of k_p^- and $k_p^\mp$. However, the effect of temperature on the concentration of free ions relative to ion pairs is in the opposite direction. The change with temperature of the equilibrium constant for dissociation of ion pairs into free ions is given by the relationship

$$\ln K = -\frac{\Delta H}{RT} + \frac{\Delta S}{R} \tag{5-108}$$

where ΔH is negative and K increases with decreasing temperature [Schmitt and Schultz, 1975; Shimomura et al., 1976b]. For example, ΔH is about -37 kJ/mole for polystyryl sodium in tetrahydrofuran, which corresponds to an increase in K by a factor of about 300 as the temperature changes from 25 to $-70°C$. The free-ion concentration is higher by a factor of about 20 at $-70°C$ compared to 25°C for a living end concentration of 10^{-3} M. The change in K with temperature is less for polystyryl cesium as ΔH is about -8 kJ/mole. The opposing effects of temperature on K and on the propagation rate constants results in apparent activation energies which are often low. Apparent activation energies for propagation in poorer solvating media will be higher than those in better solvating media. Little ionization to free ions takes place in the former and temperature has little effect on K. Significant ionization occurs in better solvents, K changes considerably with temperature, and the effect of T on K may come close to offsetting its effects on k_p^- and $k_p^\pm$. This clearly is the reason why the apparent activation energy for polystyryl sodium is 37.6 kJ/mole in dioxane but only 4.2 kJ/mole in tetrahydrofuran. Similarly, the effect of T on K in more polar solvents is greater for the smaller, better-solvated ions (Na^+) compared to the larger, more poorly solvated ions (Cs^+).

Evaluation of k_p^- and $k_p^\pm$ and the corresponding activation parameters has been carried out for polystyryl sodium and cesium in several different ether solvents (THF, tetrahydropyran, 1,2-dimethoxyethane) [Muller, 1989; Schmitt and Schulz, 1975; Szwarc, 1968, 1974, 1983]. k_p^- is independent of counterion, indicating that the observed value is for the free ion. Further, k_p^- is independent of the solvent, although one expects a decrease in the rate constant with increasing solvating power for reaction between an ionic species and a neutral molecule (Sec. 5-2f-2). Apparently, the range of solvents studied (various ethers) did not contain a large enough difference in solvating power to observe the expected effect even though there was a significant variation in the dielectric constants of the ethers. The various ethers are assumed to be similar in their specific solvation of the free anionic propagating centers. (One should keep in mind that the range of solvents appropriate to anionic polymerization is quite limited—mostly hydrocarbons and ethers.) The activation energy E_p^- and the frequency factor A_p^- for the free ion are 16.7 kJ/mole and 10^8 liters/mole-sec, respectively. It is useful to note that the propagation rate constant for the free polystyryl anion ($k_p^- \approx 10^5$) is larger than that for the free radical (165 liter/mole-sec) by three orders of magnitude. Propagation by the anion is favored by both a lower activation energy and a higher frequency factor ($E_p = 26$ and $A_p = 4.5 \times 10^6$ for radical propagation). The lower activation energy for anionic propagation is reasonable, since the interaction between anion and monomer should generate attractive forces (due to polarization) that reduce the potential energy barrier to addition. The more favorable frequency factor results from a decrease in order of the surrounding reaction medium when the negative charge of the propagating anion is dispersed in the transition state. From Table 5-3 we can note that the propagation rate constant for the free polystyryl carbocation is larger than for the carbanion by a factor of 10–100 as expected. Carbocation addition involves the use of vacant orbitals of the carbocation, while anionic propagation requires the use of antibonding orbitals, since all bonding orbitals on both monomer and carbanion are filled.

The calculation of $k_p^{\mp}$ and the corresponding activation parameters $E_p^{\mp}$ and $A_p^{\mp}$ proceeds in a reasonably straightforward manner for polymerizations in solvents of low polarity (no higher than dioxane). However, anomolous behavior is observed for polymerizations in solvents which are better solvating media than dioxane. $k_p^{\mp}$ is observed to increase with decreasing temperature in some systems, leading to negative activation energies. The activation energy for propagation by polystyryl sodium ion pairs in tetrahydrofuran is -6.2 kJ/mole over the temperature range -80 to $25°C$ [Shimomura et al., 1967a, 1967b]. $E_p^{\mp}$ for poly(α-methylstyryl) sodium in THF is -8.8 kJ/mole over the range -25 to $5°C$ [Hui and Ong, 1976]. More significantly, experiments carried out over a sufficiently wide temperature range showed that $E_p^{\mp}$ changed sign with temperature. Figure 5-7 shows the Arrhenius plot of $k_p^{\mp}$ versus $1/T$ for polystyryl sodium in tetrahydrofuran and 3-methyltetrahydrofuran [Schmitt and Schulz, 1975]. The plots are S-shaped with two inflection points. These anomolous results indicate that two different types of ion pairs are present and undergoing propagation, the contact ion pairs and solvent-separated ion pairs (corresponding to structures II and III in Sec. 5-1).

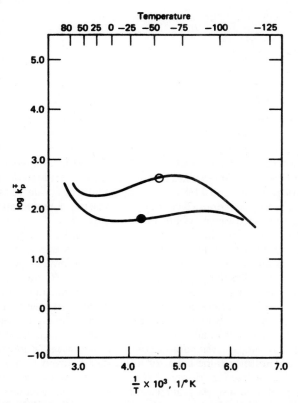

Fig. 5-7 Propagation rate constant for polystyryl sodium ion pairs in tetrahydrofuran (○) and 3-methyltetrahydrofuran (●). After Schmitt and Schulz [1975] (by permission of Pergamon Press Ltd., Oxford).

The observed ion pair propagation constant $k_p^{\mp}$ is an apparent rate constant. $k_p^{\mp}$ is a composite of the rate constants for the contact ion pair (k_c) and solvent-separated ion pair (k_s) according to

$$k_p^{\mp} = xk_s + (1 - x)k_c \tag{5-109}$$

or

$$k_p^{\mp} = \frac{(k_c + k_s K_{cs})}{(1 + K_{cs})} \tag{5-110}$$

where x and $(1 - x)$ are the fractions of solvent-separated and contact ion pairs, respectively, and K_{cs} is the equilibrium constant for interconversion between the two types of ion pairs

$$\sim\sim\sim P^-(C^+) \underset{}{\overset{K_{cs}}{\rightleftharpoons}} \sim\sim\sim P^-\| C^+ \tag{5-111}$$

The variation of $k_p^{\mp}$ with temperature depends on the interplay of the separate variations of k_c, k_s, and K_{cs} on temperature according to

$$\ln k_c = \ln A_c - \frac{E_c}{RT} \tag{5-112}$$

$$\ln k_s = \ln A_s - \frac{E_s}{RT} \tag{5-113}$$

$$\ln K_{cs} = -\frac{\Delta H_{cs}}{RT} + \frac{\Delta S_{cs}}{R} \tag{5-114}$$

A consideration of Eqs. 5-109, 5-110, 5-112, 5-113, and 5-114 indicates the reasons for the behavior observed in Fig. 5-7. There are no solvent-separated ion pairs at the highest temperature (ca. 70°C in THF). Decreasing temperature decreases $k_p^{\mp}$ because k_c is decreasing. As the temperature continues decreasing, a temperature is reached at which solvent-separated ion pairs are formed. Since $k_s > k_c$, $k_p^{\mp}$ goes through a minimum (at ca. 30°C), and then increases (provided $E_s < -\Delta H_{cs}$). However, a further decrease in temperature causes the rate to go through another inflection. There is so much conversion of contact ion pairs to solvent-separated ion pairs that any additional increase in the fraction of the latter is insufficient to counter the conventional effect of E_s. k_s decreases with decreasing T, $k_p^{\mp}$ reaches a maximum (at ca. −75°C) and then decreases. The overall effect of T on $k_p^{\mp}$ depends on the relative values of E_c, E_s, ΔH_{cs}, and ΔS_{cs}. In a poor solvent such as dioxane, there is a negligible fraction of solvent-separated ion pairs at all temperatures; k_c and $k_p^{\mp}$ decrease with decreasing temperature over the complete temperature range. In a moderately good solvent (THF and 3-methyltetrahydrofuran), the behavior in Fig. 5-7 is observed. For a sufficiently good solvent where the fraction of solvent-separated ion pairs is close to one, k_s and $k_p^{\mp}$ would decrease continuously with decreasing temperature. Data indicate that hexamethylphosphoramide is such a solvent for polystyryl sodium [Schmitt and Schulz, 1975].

Experimental data of $k_p^{\mp}$ versus $1/T$ can be fitted to the preceding equations to yield values of the various activation and thermodynamic parameters pertinent to ion pair propagation (Table 5-12). Corresponding values of the parameters for free-ion propagation are included in Table 5-12 for comparison. The solvent-separated ion pair is approximately half as reactive as the free ion, while the contact ion pair is more than three orders of magnitude less reactive. The fact that k_p^- is only slightly more than twice k_s clearly indicates that the solvent-separated ion pair is in an environment similar to that of the free ion. The frequency factors for the three types of propagating centers are very similar; reactivity differences are due to the differences in activation energies. The free ion and solvent-separated ion pair have comparable activation energies. The much higher value of E_c indicates the need to separate the anion and counterion in the transition state so that monomer can be inserted. The K_{cs} values show that a large fraction of the ion pairs in THF are solvent-separated ion pairs. Since $k_s \gg K_c$, any significant concentration of solvent-separated ion pairs will contribute heavily to the overall propagation rate. The wide variation in the relative amounts of solvent-separated and contact ion pairs with solvent is evident from the values of K_{cs} (0.13, 0.002, 0.0001, and $< 10^{-5}$ for 1,2-dimethoxyethane, tetrahydrofuran, tetrahydropyran, and dioxane, respectively, at 25°C) [Szwarc, 1983].

The reactivity of the solvent-separated ion pair is hardly affected by the counterion. This can be observed from the $k_p^\mp$ values in THF (Table 5-11). Most of the observed propagation in THF is due to solvent-separated ion pairs and $k_p^\mp$ is a good indication of k_s. The variation in $k_p^\mp$ from Li^+ to Cs^+ is relatively small and is probably due more to differences in the fractions of solvent-separated ion pairs than to differences in k_s. The reactivity of contact ion pairs is more sensitive to the counterion. The variation of $k_p^\mp$ in dioxane is by a factor of 25 between the different counterions. Since the fraction of solvent-separated ion pairs is extremely small in dioxane, $k_p^\mp$ is indicative of k_c. The larger, more loosely held cesium counterion results in a higher reactivity for the contact ion pair. The variation of k_c and k_s with solvating power of the reaction medium is not established. Some data indicate that k_s is insensitive to solvent while k_c increases with increasing solvating power, but these results are limited to ether solvents.

TABLE 5-12 Propagation of Polystyryl Sodium in Tetrahydrofuran[a]

k_p^- (20°C)	1.3×10^5 liters/mole-sec
K (20°C)	4.0×10^{-8} mole/liter (20°C)
	79.1×10^{-8} mole/liter ($-48°C$)
E_p^-	16.6 kJ/mole
A_p^-	1.0×10^8 liters/mole-sec
E_c	36.0 kJ/mole
A_c	6.3×10^7 liters/mole-sec
k_c (20°C)	24 liters/mole-sec
E_s	19.7 kJ/mole
A_s	2.08×10^8 liters/mole-sec
k_s (20°C)	5.5×10^4 liters/mole-sec
ΔH_{cs}	-19.7 kJ/mole
ΔS_{cs}	-142 J/K mole
K_{cs}	2.57×10^{-3} (20°C)
	2.57×10^{-2} ($-50°C$)

[a]Data for Schmitt and Schulz [1975].

5-3e-5   *Association Phenomena in Alkyllithium*

A complication for polymerizations initiated by organolithium compounds in nonpolar solvents, such as cyclohexane, *n*-hexane, and benzene, is *association* (*aggregation*) of the various organolithium species present in the reaction system. Association occurs with organolithium compounds because of the small size of lithium and its possessing more low energy orbitals than electrons. Each lithium associates and participates in bonding with two or more organic moieties. The phenomenon is important since the associated species are far less reactive than unassociated species in initiation and propagation. The effects of association have been extensively studied for styrene, isoprene, and 1,3-butadiene polymerizations initiated by *n*-, *s*-, and *t*-butyllithium, but there are considerable discrepancies between the results and mechanistic interpretations of different workers [Bywater, 1975, 1985; Morton, 1983; Van Beylen et al., 1988; Young et al., 1984].

The initiation and propagation reactions are typically found to show fractional orders of dependence of rate on alkyllithium. The situation is quite complex. The fractional orders for initiation and propagation are usually not the same and often vary depending on the monomer, solvent, and initiator and their absolute as well as relative concentrations. For styrene polymerization by *n*-butyllithium in aromatic solvents, the initiation and propagation rates are proportional to only the $\frac{1}{6}$- and $\frac{1}{2}$-powers of *n*-butyllithium concentration, respectively. These results have been interpreted in terms of the following association equilibria

$$(C_4H_9Li)_6 \overset{K_1}{\rightleftharpoons} 6C_4H_9Li \tag{5-115}$$

$$(C_4H_9\text{---}M_n^-Li^+)_2 \overset{K_2}{\rightleftharpoons} 2C_4H_9\text{---}M_n^-Li^+ \tag{5-116}$$

with initiator and propagating ion pairs in equilibria with the corresponding hexamer and dimer, respectively. (Equation 5-115 is an oversimplification. Dissociation of hexamer to monomeric *n*-butyllithium probably takes place in stages.) One derives the concentrations of unassociated initiator and propagating ion pairs as

$$[C_4H_9Li] = K_1^{1/6}[(C_4H_9Li)_6]^{1/6} \tag{5-117}$$

$$[C_4H_9\text{---}M_n^-Li^+] = K_2^{1/2}[(C_4H_9\text{---}M_n^-Li^+)_2]^{1/2} \tag{5-118}$$

which explain the dependence of R_i and R_p on the $\frac{1}{6}$- and $\frac{1}{2}$-power, respectively, of the *n*-butyllithium concentration. Molecular-weight measurements by viscometry and light scattering have been used to support the presence of the hexamer and dimer species, but the results from different workers are often contradictory [Van Beylen et al., 1988]. This interpretation of the fractional kinetic orders assumes that only monomeric initiator and propagating ion pairs are reactive. This assumption is not universally accepted for all reaction systems. It has been suggested that for some systems the fractional reaction orders are artifacts resulting from the complex manner in which concentration affects the reaction of monomer with associated initiator and propagating species [Young et al., 1984].

The results are slightly different with *s*- and *t*-butyllithium. R_p is still $\frac{1}{2}$-order in initiator but R_i is $\frac{1}{4}$- instead of $\frac{1}{6}$-order in initiator. *s*- and *t*-butyllithium exist as tetramers, not hexamers like *n*-butyllithium, because of the more sterically hindered alkyl groups. The propagating species still associates as dimers, since the initiating alkyl group becomes far removed from the propagating carbanion center. The rates of initiation by *s*- and *t*-butyllithium are greater than for *n*-butyllithium, a consequence of the greater nucleophilicity of tertiary and secondary carbanions over primary carbanions. The relative reactivities of *s*- and *t*-butyllithium are not always in the same order but depend on the monomer, indicating that steric effects are also important.

Polymerization in aliphatic hydrocarbons is considerably slower than in aromatic hydrocarbons because of decreased dissociation of initiator and propagating ion-pair aggregates. The course of reaction in aliphatic hydrocarbons is complex compared to that in aromatic solvents. Initiation is very slow at the start of reaction but proceeds with autoacceleration as *cross* or *mixed association* of initiator and propagating ion pairs replaces self-association of initiator. Cross association is weaker and results in an increased concentration of monomeric initiator. This effect may also explain the higher-order dependence of R_i on initiator (typically between $\frac{1}{2}$- and 1-order) in aliphatic solvents, especially for *s*- and *t*-butyllithium. R_p is still $\frac{1}{2}$-order in initiator independent of solvent.

The situation is similar qualitatively but differs quantitatively for isoprene and 1,3-butadiene. The dependence of R_p on initiator varies from $\frac{1}{6}$- to $\frac{1}{2}$-order depending on the specific reaction system. The reaction orders for all monomers are affected by the relative as well as absolute concentrations of initiator and monomer. Thus the dependence of R_p on initiator for the *n*-butyllithium polymerization of isoprene in benzene at 30°C is $\frac{1}{4}$-order at initiator concentrations above 10^{-4} M but $\frac{1}{2}$-order at initiator concentrations below 10^{-4} M [Van Beylen et al., 1988]. Higher initiator concentrations yield higher degrees of aggregation and lower kinetic orders. The excess of monomer over initiator is also important. Higher kinetic orders are often observed as the monomer:initiator ratio increases, apparently as a result of breakup of initiator and propagating ion-pair associations by monomer.

The association phenomena occurring with alkyllithium initiators in nonpolar solvents results in very low polymerization rates. A typical styrene or isoprene polymerization by butyllithium in benzene is orders of magnitude slower compared to the corresponding sodium naphthalene polymerization. When butyllithium polymerizations are carried out in polar solvents such as tetrahydrofuran, the association of initiator and propagating species vanishes completely and the polymerizations are much more rapid. The association phenomena can also be disrupted by adding a Lewis base, which can coordinate with the initiator (Alev et al., 1980; Cheminat et al., 1979; Dumas et al., 1978; Fontanille, 1989; Muller, 1989]. Polyamines such as N,N,N',N'-tetramethylethylene diamine (TMEDA) and polyethers such as tetraglyme, CH_3O-$(CH_2CH_2O)_4CH_3$, are more effective in breaking up the association than are simple ethers and amines such as diethyl ether or triethyl amine. The effectiveness of a Lewis base in promoting initiation and polymerization decreases if the base:initiator ratio is too high. One observes the polymerization rate to increase with increasing [base]/[RLi] and then level off at a maximum rate. With many Lewis bases, the polymerization rate decreases after reaching the maximum as [base]/[RLi] is increased further [Hay et al., 1976]. Complexation of base with initiator becomes so extensive as to decrease the reactivity of initiator.

The macrocyclic crown ethers such as 18-crown-6 (**XXVIII**) and cryptands such as 2.2.2-cryptand (**XXIX**) are extremely powerful for breaking up association of organolithium compounds. These macrocyclic ligands complex lithium and also other counterions so strongly as to greatly increase the concentration of free-ion propagating species, resulting in very large rate increases (Cheng and Halasa, 1976; Deffieux and Boileau, 1976].

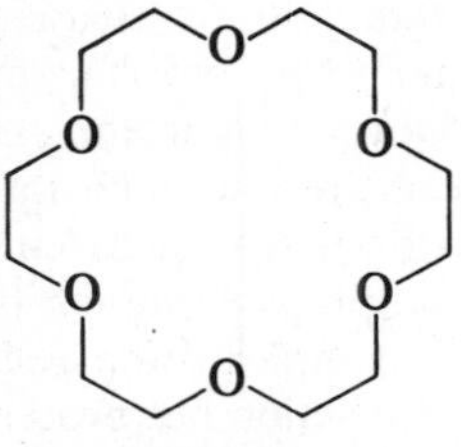

XXVIII 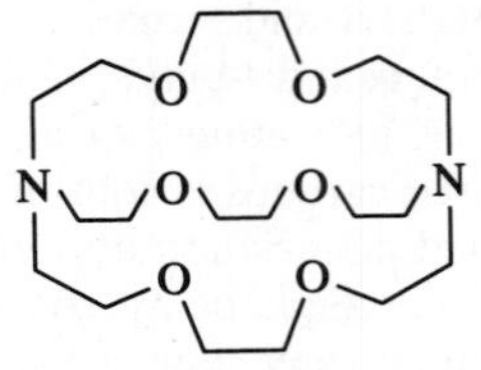   XXIX

5-3e-6 *Other Phenomena*

The need for solvation in anionic polymerization manifests itself in some instances by other deviations from the normal reaction rate expressions. Thus the butyllithium polymerization of methyl methacrylate in toluene at $-60°C$ shows a second-order dependence of R_p on monomer concentration [L'Abbe and Smets, 1967]. In the non-polar toluene, monomer is involved in solvating the propagating species [Busson and Van Beylen, 1978]. When polymerization is carried out in the mixed solvent dioxane–toluene (a more polar solvent than toluene), the normal first-order dependence of R_p on [M] is observed. The lithium diethylamide, $LiN(C_2H_5)_2$, polymerization of styrene at 25°C in THF–benzene similarly shows an increased order of dependence of R_p on [M] as the amount of tetrahydrofuran is decreased [Hurley and Tait, 1976].

Propagation of two-ended (bifunctional) propagating species often proceeds at a lower rate than the corresponding monoanion species as a result of *triple ion* formation [Bhattacharyya et al., 1964]. For example, ionic dissociation of the counterion from one end of the cesium salt of a two-ended propagating species **XXX** yields **XXXI** in which the newly dissociated anionic center remains near the ion-pair at the other end of **XXXI**. The result is **XXXII**, referred to as a triple ion.

$$ (5\text{-}119) $$

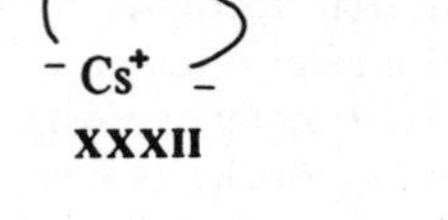

$$ (5\text{-}120) $$

XXX   XXXI

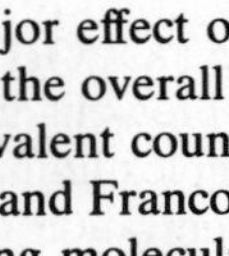

XXXII

The triple ion propagates faster than a simple ion pair but slower than a free ion [Muller, 1989]. The major effect of triple ion formation is a lowering of the concentration of free ions and the overall result is a decrease in polymerization rate. Similar results are found for divalent counterions such as barium and strontium [de Groof et al., 1975, 1977; Mathis and Francois, 1978; Mathis et al., 1978]. Triple ion formation decreases with increasing molecular weight of the two-ended dicarbanion; thus, R_p increases with conversion. As the length of the intramolecular chain between the two anion centers increases, the probability of cyclization to the triple ion decreases.

The higher reactivity of 2-vinylpyridine relative to styrene has been attributed to a combination of intramolecular solvation and triple ion formation [Sigwalt, 1975; Soum et al., 1977].

5-4 BLOCK COPOLYMERS

There has been increasing activity in synthesizing block copolymers (Sec. 2-13a) since these offer the potential for obtaining products that can incorporate the desirable properties of two or more homopolymers. There are several different approaches to synthesizing polymers with long blocks of two (or more) different monomers [Allport and James, 1973; Ceresa, 1973; Noshay and McGrath, 1977]. All the methods require considerable theoretical understanding and manipulative skills to obtain well-defined products.

5-4a Sequential Monomer Addition

Sequential addition of monomers to a living anionic polymerization system is at present the most useful method of synthesizing well-defined block copolymers [Morton, 1983; Morton and Fetters, 1977; Rempp et al., 1988]. An AB diblock copolymer is produced by polymerization of monomer A to completion using an initiator such as butyllithium. Monomer B is then added to the living polyA carbanions. When B has reacted completely a terminating agent such as water is added

$$A \xrightarrow{\text{RLi}} R \wedge\wedge\wedge AAA^- \xrightarrow{\text{B}} R \wedge\wedge\wedge AAA \wedge\wedge\wedge BBB^- \xrightarrow{\text{H}_2\text{O}}$$

$$R \wedge\wedge\wedge AAA \wedge\wedge\wedge BBBH \tag{5-121}$$

and then the AB block copolymer isolated [Fetters et al., 1978]. This method can be used to synthesize any of the various types of block copolymers (di, tri, tetra, penta, multiblocks) by employing the proper sequencing of additions of different monomers provided that each propagating carbanion can initiate the polymerization of the next monomer. For example, polystyryl carbanion will initiate polymerization of methyl methacrylate, but the reverse does not occur. Synthesis of a styrene–methyl methacrylate diblock copolymer requires that styrene be the first monomer. Further, it is useful to decrease the nucleophilicity of polystyryl carbanions by adding a small amount of 1,1-diphenylethylene to minimize attack at the ester function of methyl methacrylate (MMA). Sequencing is not a problem in synthesizing block copolymers of styrene with isoprene or 1,3-butadiene. The length of each segment in the block copolymer is determined by the amount of each monomer added relative to the amount of initiator. The overall properties of the product vary with the block lengths of the different monomers.

Difunctional initiators such as sodium naphthalene can be advantageously used to synthesize ABA, BABAB, CABAC, and other symmetrical block copolymers [Benson et al., 1985, Foss et al, 1977] more efficiently, that is, using fewer cycles of monomer additions. Difunctional initiators can also be prepared by reacting a diene (e.g., m-diisopropenylbenzene, 1,3-bis[1-phenylvinyl]benzene) with two equivalents of s- or t-butyllithium. Monomer B is initiated by a difunctional initiator followed by monomer A. A polymerizes at both ends of the B block to form an ABA triblock. BABAB or CABAC block copolymers are synthesized by the addition of monomer B or C to the ABA living polymer. The use of a difunctional initiator is the only way to synthesize a MMA–styrene–MMA triblock polymer since MMA carbanion does not initiate styrene polymerization.

Styrene-isoprene-styrene and styrene-1,3-butadiene-styrene ABA block copolymers, containing short styrene blocks on each side of a long diene block, are produced commercially. These are useful as thermoplastic elastomers (trade names: *Cariflex*, *Kraton*, *Soloprene*, *Stereon*), which behave as elastomers at ambient temperatures but are thermoplastic at elevated temperatures where they can be molded and remolded (Sec. 2-13c-2). The polystyrene blocks of the copolymer aggregate to form glassy (hard) domains at ambient temperatures. These effectively crosslink the rubbery (soft) polydiene segments and also act as reinforcing filler particles to increase strength. The physical crosslinking is reversible since heating above the glass transition temperature of polystyrene (90–100°C) softens the hard segments and the triblock copolymer flows under stress. Cooling reestablishes the hard polystyrene domains, and the material again acts as a crosslinked elastomer. Thermoplastic elastomers have the advantage that they can be processed as thermoplastics instead of thermosets. Diblock copolymers do not show good elastomer behavior. One needs the triblock structure with the hard (physical crosslinking) blocks at both ends of the polymer chain to obtain rubbery behavior. Higher-level block structures (tetra, penta, multiblocks) do not offer significant advantages over the triblock copolymer.

The range of monomers that can be incorporated into block copolymers by the living anionic route includes not only the carbon–carbon double bond monomers susceptible to anionic polymerization but also certain cyclic monomers, such as ethylene oxide, propylene sulfide, lactams, lactones, and cyclic siloxanes (Chap. 7). Thus one can synthesize block copolymers involving each of the two types of monomers. Some of these combinations require an appropriate adjustment of the propagating center prior to the addition of the cyclic monomer. For example, carbanions from monomers such as styrene or methyl methacrylate are not sufficiently nucleophilic to polymerize lactones. The block copolymer with a lactone can be synthesized if one adds a small amount of ethylene oxide to the living polystyryl system to convert propagating centers to alkoxide ions prior to adding the lactone monomer.

Group transfer polymerization allows the synthesis of block copolymers of different methacrylate or acrylate monomers, such as methyl methacrylate and allyl methacrylate [Webster and Sogah, 1989]. The synthesis of mixed methacrylate–acrylate block copolymers requires that the less reactive monomer (methacrylate) be polymerized first. The silyl dialkylketene acetal propagating center from methacrylate polymerization is more reactive for initiation of acrylate polymerization than the silyl monoalkylketene acetal propagating center from acrylate polymerization is for initiation of methacrylate polymerization. Bifunctional initiators such as 1,4-bis(methoxytrimethylsiloxymethylene)cyclohexane (**XXXIII**) are useful for synthes-

XXXIII

izing ABA block copolymers where the middle block is methacrylate [Steinbrecht and Bandermann, 1989; Yu et al., 1988].

AB and ABA block copolymers, where A is a vinyl ether and B is a different vinyl ether or *p*-methoxystyrene, have been synthesized by living cationic polymerization using the HI/I$_2$ initiating system [Higashimura and Sawamoto, 1985, 1989; Miyamoto et al., 1984, 1985].

5-4b Telechelic Polymers

Telechelic polymers, containing one or more end groups with the capacity to react with other molecules, are useful for synthesizing block and other types of copolymers [Nuyken and Pask, 1989]. Living anionic polymers can be terminated with a variety of electrophilic reagents to yield telechelic polymers [Fontanille, 1989; Quirk et al., 1989; Rempp et al., 1988]. For example, reaction with carbon dioxide, ethylene oxide, and allyl bromide yield polymers terminated with carboxyl, hydroxyl, and allyl groups, respectively. Functionalization with hydroxyl or carboxyl groups can also be achieved

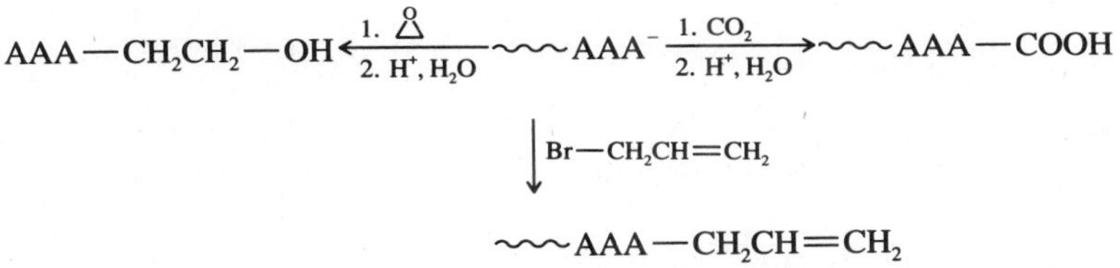

$$AAA-CH_2CH_2-OH \xleftarrow[\text{2. H}^+,\text{H}_2\text{O}]{1. \overset{O}{\triangle}} \rightsquigarrow AAA^- \xrightarrow[\text{2. H}^+,\text{H}_2\text{O}]{1. CO_2} \rightsquigarrow AAA-COOH$$

$$\downarrow Br-CH_2CH{=}CH_2 \qquad\qquad (5\text{-}122)$$

$$\rightsquigarrow AAA-CH_2CH{=}CH_2$$

by reaction with a lactone or anhydride, respectively. When the telechelic polymer contains a bifunctional group, such as the allyl-terminated telechelic above, it is referred to as a *macromonomer* or *macromer* since it is capable of participating in a polymerization reaction through that functional group. Polymers terminated with polymerizable carbon–carbon double bonds are also obtained by reacting the living polymer with *p*-vinylbenzyl chloride or a sequence of ethylene oxide followed by acryloyl chloride [Takaki et al., 1987].

Telechelic polymers are also obtained by using initiators containing the desired functional group. This often requires that the functional group be protected or blocked during initiation and polymerization since many functional groups act as facile terminating agents. Thus an amine-terminated telechelic is obtained by prior synthesis of an initiator containing an amine function protected by trimethylsilylation [Dickstein and Lillya, 1989]. The protected initiator **XXXIV** is used to initiate polymerization and subsequently the trimethylsilyl group is removed by acid hydrolysis.

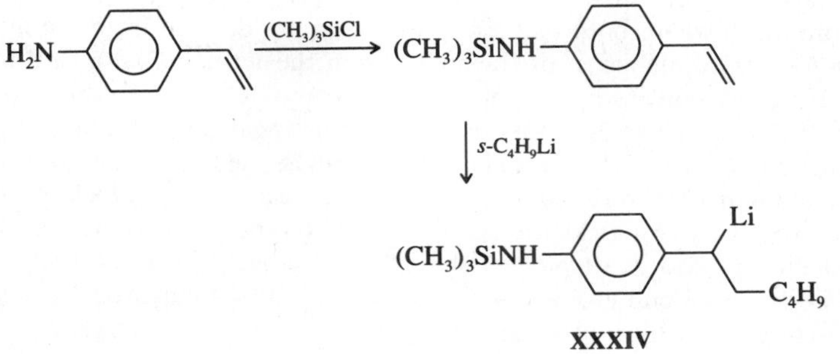

$$H_2N-\!\!\left\langle\bigcirc\right\rangle\!\!\diagdown \xrightarrow{(CH_3)_3SiCl} (CH_3)_3SiNH-\!\!\left\langle\bigcirc\right\rangle\!\!\diagdown$$

$$\downarrow s\text{-}C_4H_9Li$$

$$(CH_3)_3SiNH-\!\!\left\langle\bigcirc\right\rangle\!\!\begin{array}{c}\overset{\displaystyle Li}{|}\\ CH\\ |\\ CH_2\\ |\\ C_4H_9\end{array} \qquad (5\text{-}123)$$

$$\textbf{XXXIV}$$

Monofunctional telechelic polymers containing groups such as hydroxyl, amine, or carboxyl can be used to obtain diblock copolymers by reaction with other telechelic polymers. The latter includes telechelics obtained by the anionic route or telechelics obtained by other routes (e.g., step and ring-opening polymerizations). Telechelic polymers containing functional groups at each chain end are obtained by carrying out the above reactions with two-ended living anionic polymers. These bifunctional telechelics can be used to synthesize triblock and multiblock copolymers as well as telechelic prepolymers to be used in step polymerization (Sec. 2-13b) [Register et al., 1988].

Telechelic polymers are obtained by group transfer polymerization by using initiators (**XXXV**) containing protected functional groups [Webster and Sogah, 1989]. Hydroxyl- and carboxyl-terminated telechelic polymers are obtained by using initiator **XXXV** with R = —$CH_2CH_2OSi(CH_3)_3$ and —$Si(CH_3)_3$, respectively (R' = —CH_3). The HO and COOH groups are obtained by hydrolysis of the silyl ketene acetal end

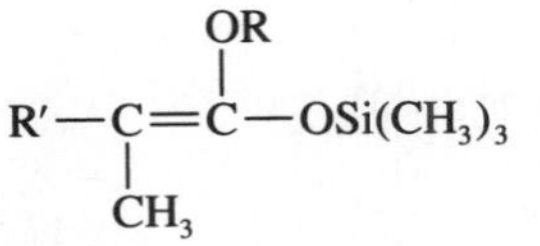

$$
\begin{array}{ccc}
\underset{\overset{|}{CH_3}}{\overset{\overset{OR}{|}}{R'-C}}=C-OSi(CH_3)_3 & (C_2H_5COO)_2\overset{..}{C}:^-Na^+ & CH_2=CH-COOCH_2CH_2\overset{\overset{CH_3}{|}}{C}H-I \\
\mathbf{XXXV} & \mathbf{XXXVI} & \mathbf{XXXVII}
\end{array}
$$

groups of the polymer. Another approach is reaction of the living silyl ketene acetal end group with an electrophilic reagent, such as benzaldehyde to yield a blocked hydroxyl function or p-bromomethylstyrene to yield a vinyl macromer.

Telechelic polymers are obtained from cationic living polymerizations initiated by HI/I_2 by using a terminating agent containing the desired functional group [Higashimura and Sawamoto, 1989]. For example, the use of sodiomalonic ester (**XXXVI**) yields a carboxyl-terminated telechelic after hydrolysis and decarboxylation for R = H and a vinyl-terminated telechelic is obtained for R = —$CH_2CH_2OCH=CH_2$. Functional groups can also be introduced by using appropriately substituted initiators; for instance, initiation by **XXXVII** plus I_2 or ZnI_2 yields a telechelic with a vinyl end group. Bifunctional telechelics with iodide end groups are obtained by employing a diiodide as initiator. This can be used to synthesize block copolymers by polymerization of a second monomer or other telechelics by use of appropriate terminating agents or directly by including the desired functional group as part of the diiodide initiator.

Similar approaches have been used for synthesizing polyisobutylene telechelics [Kennedy, 1985; Kennedy and Smith, 1980]. The *inifer* approach involves an initiator that is simultaneously an active transfer agent with the result that bifunctional telechelics are produced where both end group functions are derived from the initiator. The term *inifer* is a combination of fragments from the terms *ini*tiator and trans*fer* agent. For example, polymerization of isobutylene by BCl_3 and 1,4-bis(α,α-dimethylchloromethyl)benzene (**XXXVIII**) yields a bifunctional telechelic polymer with t-alkyl chloride end groups. This telechelic polymer can be used as the initiator together with diethylaluminum chloride as coinitiator to synthesize an ABA triblock copolymer, for example, where the A blocks are poly(α-methylstyrene) or polystyrene. The dichloro telechelic can also be employed to synthesize other telechelics, such as a telechelic with i-propenyl end groups via dehydrochlorination or hydroxyl end groups via hydroboration of the i-propenyl groups. Telechelic polymer **XXXIX** can also be

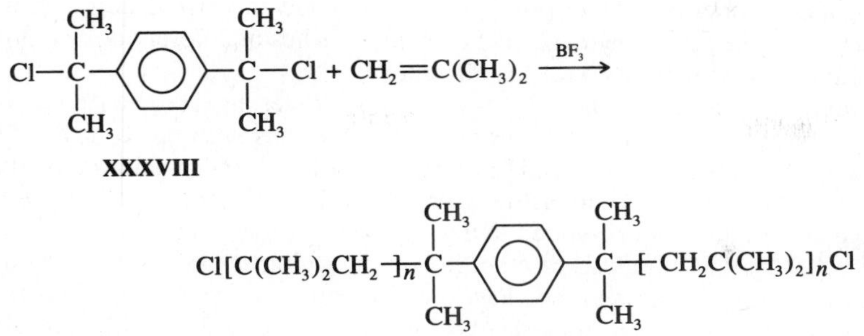

$$Cl[C(CH_3)_2CH_2\!+\!_n C\!-\!\langle\bigcirc\rangle\!-\!C\!+\!CH_2C(CH_3)_2]_nCl \quad (5\text{-}124)$$

XXXIX

synthesized by using a di-*t*-acetoxy initiator (**XL**) with BCl_3 [Faust et al., 1989; Zsuga et al., 1989].

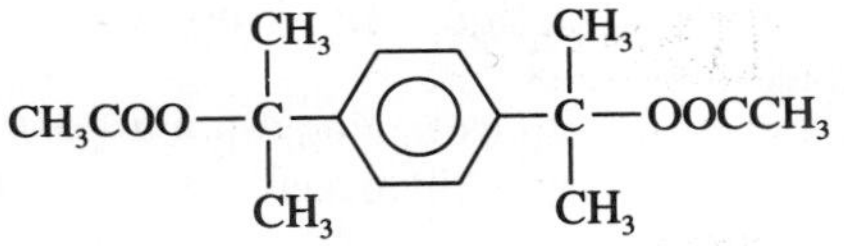

XL

5-4c Coupling Reactions

Living anionic block copolymers can be linked by coupling reactions; for example, a living AB diblock copolymer can be linked by 1,6-dibromohexane to yield an ABA triblock copolymer

$$\sim\!\!\sim\!A_nB_m^- \xrightarrow{Br(CH_2)_6Br} \sim\!\!\sim\!A_nB_m(CH_2)_6B_mA_n\!\sim\!\!\sim \quad (5\text{-}125)$$

Coupling reactions allow the synthesis of *star* polymers by using a multifunctional coupling agent such as $SiCl_4$, CH_3SiCl_3, $(ClCH_2)_4$, triallyloxy-*s*-triazine, and 1,2,4,5-tetrachloromethylbenzene [Alward et al., 1986; Lutz and Rempp, 1988; Rempp et al., 1988]; for example,

$$\sim\!\!\sim\!A_n^- + SiCl_4 \longrightarrow \sim\!\!\sim\!\!A_n\!-\!\underset{\underset{A_n}{|}}{\overset{\overset{A_n}{|}}{Si}}\!-\!A_n\!\sim\!\!\sim \quad (5\text{-}126)$$

An excess of living polymer is used to force the coupling reaction to completion. However, steric factors influence the extent and efficiency of coupling. Si—Cl bonds are more efficiently substituted compared to C—Cl bonds. Complete conversion is hindered when the coupling sites are closer together and for the more bulky living polymers. Polybutadienyl anions couple quantitatively with $SiCl_4$ to give the tetrasubstituted star polymer but polyisoprenyl anions yield only the trisubstituted product. Ths synthesis of mixed star polymers, that is, with branches of different polymers, has been attempted but structural control is difficult.

Much more extensively branched polymers, sometimes referred to as *comb* or *porcupine* polymers, are obtained by using coupling agents with much higher functionality [e.g., $Si(CH_2CH_2SiCl_3)_4$, $[CH_2Si(CH_2CH_2SiCl_3)_3]_2$, and *p*-chloromethylated polystyrene] or a diene (e.g., *p*-divinylbenzene), sometimes together with a two-ended living polymer instead of a one-ended living polymer. [Roovers et al., 1989; Taromi and Rempp, 1989]. The synthesis of other polymer topologies has also been studied. Cyclic and *eight-shaped* polystyrenes have been obtained by coupling a two-ended living anionic polymer with dimethyldichlorosilane and 1,2-bis(dichloromethyl-silyl)ethane, respectively, in dilute solution [Antonietti and Folsch, 1988; Antonietti et al., 1989; Roovers and Toporowski, 1983; Szwarc, 1989].

Coupling of GTP living polymers with halide terminating agents has also been achieved, and there are possibilities for various types of polymer structures in a manner analogous to that described above [Webster and Sogah, 1989]. In addition, star polymers have been synthesized by using polyfunctional initiators or by copolymerization of living GTP polymers with dimethacrylate monomers.

There is relatively little work reported on the coupling of living cationic polymers. However, 3- and 4-arm star polymers have been obtained by using tri and tetrafunctional initiators [Huang et al., 1988].

An interesting approach to block copolymers is the mutual termination of a living anionic polymer with a living cationic polymer. The only successful example of this approach is the coupling of living polystyrene with living polytetrahydrofuran [Richards et al., 1978]. This approach has the potential for synthesizing a wide variety of block structures, but the efficiency of the coupling reaction is generally low because of proton transfer from the carbocation to the carbanion, resulting in two homopolymers instead of the block copolymer.

5-4d Transformation Reactions

Another approach to block copolymers involves changing the type of propagating center part way through the synthesis via a *transformation reaction* [Burgess et al., 1977; Richards, 1980; Souel et al., 1977; Tung et al., 1985]. For example, after completion of the living anionic polymerization of monomer A, the carbanion centers are

$$\sim\sim\sim AAA^- Li^+ \xrightarrow{COCl_2} \sim\sim\sim AAA-CO-Cl \xrightarrow{AgSbF_6} \sim\sim\sim AAA^+ SbF_6^- \xrightarrow{B}$$

$$\sim\sim\sim AAA \sim\sim\sim BBB^+ SbF_6^-$$

$$(5\text{-}127)$$

transformed into carbocations by reaction with excess phosgene followed by silver hexafluoroantimonate. The carbocations are then used to polymerize monomer B.

Anionic propagating centers are transformed into radical centers in the presence of a second monomer that undergoes radical propagation, by reaction with trimethyl lead chloride followed by heating

$$\sim\sim\sim AAA^- Li^+ \xrightarrow[-LiCl]{ClPb(CH_3)_3} \sim\sim\sim AAA-Pb(CH_3)_3 \xrightarrow[-(CH_3)_3Pb\cdot]{heat}$$

$$\sim\sim\sim AAA\cdot \xrightarrow{B} \sim\sim\sim AAA\sim\sim\sim BBB\cdot$$

$$(5\text{-}128)$$

Anionic and cationic block copolymerizations are each limited at best to monomers that propagate through anionic and cationic centers, respectively. Transformation reactions offer the potential for synthesizing a wider range of block copolymers. This potential has not been realized since very few of the transformation reactions are quantitative or other side reactions occur. Thus coupling of two propagating carbanions by one phosgene molecule competes with the one-to-one transformation process in Eq. 5-127. The anionic-to-radical transformation in Eq. 5-128 involves the formation of trimethyllead radical, which initiates the homopolymerization of monomer B.

5-5 DISTINGUISHING BETWEEN RADICAL, CATIONIC, AND ANIONIC POLYMERIZATIONS

There is sometimes a question as to whether a particular initiator or initiator system initiates polymerization by radical, cationic, or anionic means. Such a question can easily arise, for example, in polymerizations initiated by ionizing radiation. The mode of initiation of a particular initiator can be distinguished by a consideration of its characteristics compared to those of known radical, cationic, and anionic initiators:

1 Ionic polymerizations usually proceed at lower temperatures than radical polymerizations. Although ionic reaction temperatures are usually below 0°C, there are numerous ionic polymerizations that proceed at temperatures somewhat above 0°C. Radical polymerizations, on the other hand, almost always proceed at temperatures appreciably above approximately 50°C. Furthermore, ionic polymerizations invariably have lower activation energies than their radical counterparts and, in many cases, they may actually possess negative activation energies.

2 Ionic polymerizations are distinguished by their marked sensitivity to changes in the polarity and solvating ability of the reaction media and counterion effects. Radical polymerizations do not show such effects.

3 The addition of known radical scavengers such as the DPPH radical to a polymerizing system will halt polymerization if it is a radical reaction. Ionic polymerizations will be unaffected by such additions. One must be careful, however, not to use a radical scavenger that also affects ionic polymerization. Thus, benzoquinone would be a poor choice as a radical scavenger, since it can also act as an inhibitor in ionic polymerization.

4 The chain transfer constant for an additive or solvent in the polymerization can be determined. This value can then be compared with the transfer constants for the same substance in the polymerization of the same monomer by known radical, cationic, and anionic initiators.

5 Copolymerization behavior can also be used to distinguish between radical and ionic polymerizations (see Chap. 6).

5-6 CARBONYL POLYMERIZATION

The polymerization of the carbonyl group in aldehydes yields polymers, referred to as *polyacetals*, since they contain the acetal repeating structure

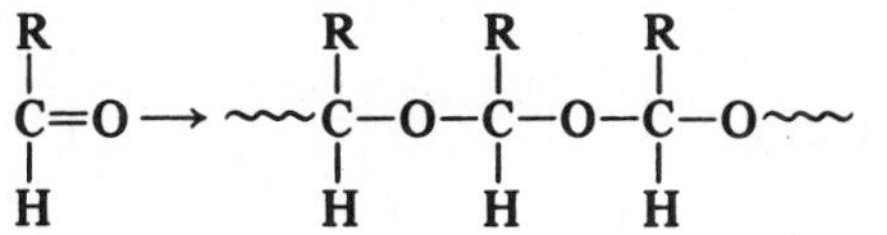

$$
\begin{array}{c}
\text{R} \\
| \\
\text{C}=\text{O} \\
| \\
\text{H}
\end{array}
\rightarrow
\begin{array}{c}
\text{R} \quad\quad \text{R} \quad\quad \text{R} \\
| \quad\quad\;\; | \quad\quad\;\; | \\
\sim\!\!\sim\!\!\sim\text{C}-\text{O}-\text{C}-\text{O}-\text{C}-\text{O}\sim\!\!\sim \\
| \quad\quad\;\; | \quad\quad\;\; | \\
\text{H} \quad\quad \text{H} \quad\quad \text{H}
\end{array}
\tag{5-129}
$$

Besides the carbon-carbon double bond, this is the only other unsaturated linkage whose polymerization has been successfully carried out to an appreciable extent [Furukawa and Saegusa, 1963; Kubisa et al., 1980; Vogl, 1967, 1976]. The polymerization of formaldehyde has been studied much more intensely than that of other aldehydes. Although formaldehyde was successfully polymerized over a hundred years ago, it was not until much later that high-molecular-weight polymers of other aldehydes were obtained.

Progress in the polymerization of the carbonyl linkage did not result until there was an understanding of the effect of ceiling temperature (T_c) on polymerization (Sec. 3-9c). With the major exception of formaldehyde and one or two other adehydes, carbonyl monomers have low ceiling temperatures (Table 5-13). Most carbonyl monomers have ceiling temperatures at or appreciably below room temperature. The low T_c values for carbonyl polymerizations are due primarily to the ΔH factor. The entropy of polymerization of the carbonyl double bond in aldehydes is approximately the same as that for the alkene double bond. The enthalpy of polymerization for the carbonyl double bond, however, is appreciably lower. Thus ΔH for acetaldehyde polymerization is only about 29 kJ/mole compared to the usual 80–90 kJ/mole for polymerization of the carbon–carbon double bond (Table 3-14) [Hashimoto et al., 1976, 1978].

Most of the early attempts to polymerize carbonyl compounds were carried out at temperatures that were, in retrospect, too high. The use of temperatures above T_c resulted in the absence of polymer formation due to the unfavorable equilibrium between monomer and polymer. Carbonyl monomers were successfully polymerized to high polymer when the polymerization reactions were carried out at temperatures below the ceiling temperatures of the monomers. Both anionic and cationic initiators, of the types used in alkene polymerizations, can be used to initiate the polymerization of the carbonyl double bond. These polymerizations are similar in their general characteristics to the ionic polymerizations of alkene monomers. The number of detailed mechanistic studies of carbonyl polymerization is, however, very small.

A characteristic of aldehyde polymerization is the precipitation, often with crystallization, of the polymer during polymerization. Depending on the solvent used, polymerization rate, state of agitation, and other reaction conditions, the polymerization can slow down or even stop due to occlusion of the propagating centers in the precipitated polymer. The physical state and surface area of the precipitated polymer influence polymerization by their effect on the availability of propagating centers and the diffusion of monomer to those centers.

TABLE 5-13   Ceiling Temperatures [a]

Monomer	T_c (°C)
Formaldehyde	119 [b]
Trifluoroacetaldehyde	81
Trichloroacetaldehyde	11
Propanal	−31
Acetaldehyde	−39
Pentanal	−42

[a]Data from Brandup and Immergut [1989], Kubisa et al. [1980], Vogl [1976].
[b]Monomer concentration = 1 atm (gaseous); all other data are for neat liquid monomer.

5-6a Anionic Polymerization

The carbonyl group is polymerized by a variety of anionic initiators. The strength of the base required to initiate polymerization depends on the substituent(s) attached to the carbonyl group.

5-6a-1 Formaldehyde

The carbonyl group of formaldehyde is highly susceptible to nucleophilic attack and this monomer can be polymerized with almost any base. Metal alkyls, alkoxides, phenolates, and carboxylates, hydrated alumina, amines, phosphines, and pyridine are effective in polymerizing formaldehyde.

The polymerization can be pictured as follows:

Initiation

$$A^-(G^+) + CH_2{=}O \rightarrow A{-}CH_2{-}O^-(G^+) \tag{5-130}$$

Propagation

$$HO{-}(\!\!{-}CH_2{-}O{-}\!)_{\overline{n}}\,CH_2{-}O^-(G^+) + CH_2{=}O \rightarrow$$
$$HO{-}(\!\!{-}CH_2{-}O{-}\!)_{\overline{(n+1)}}\,CH_2{-}O^-(G^+) \tag{5-131}$$

Termination by chain transfer

$$HO{-}(\!\!{-}CH_2{-}O{-}\!)_{\overline{n}}\,CH_2{-}O^-(G^+) + ZH \rightarrow$$
$$HO{-}(\!\!{-}CH_2{-}O{-}\!)_{\overline{n}}\,CH_2{-}OH + Z^-(G^+) \tag{5-132}$$

with initiation by anionic species A^- to form an alkoxide anion with nearby counterion G^+. Propagation proceeds in a like manner and termination occurs by transfer of a proton from ZH. The chain-transfer agent ZH may be any of a variety of compounds that can transfer a proton to the propagating alkoxide anion, such as water or an alcohol. The chain-transfer agent can have an effect on the polymerization rate if Z^- is not as effective as A^- in reinitiating polymerization.

5-6a-2 Other Carbonyl Monomers

The polymerization of carbonyl monomers other than formaldehyde follows in a similar manner, although the basicity of the initiator required may be quite different. Strong bases are required to initiate the polymerization of aliphatic aldehydes such as acetaldehyde and higher aldehydes [Starr and Vogl, 1978, 1979]. The inductive effect of an alkyl substituent destabilizes the propagating anion **XLI** by increasing the negative charge density on oxygen. The alkyl group also decreases reactivity for steric reasons.

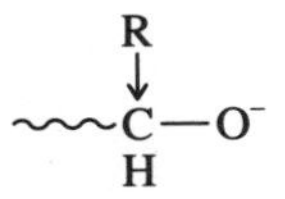

XLI 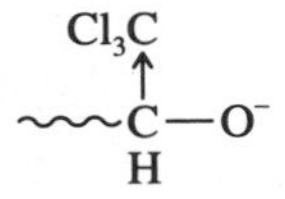   **XLII**

(Steric considerations are probably also responsible for the lower T_c values relative to formaldehyde.) Thus weak bases such as amines cannot be used to polymerize higher aldehydes than formaldehyde. Alkali metal alkyls and alkoxides are required. The presence of adventious water is detrimental, since the initiator reacts to form hydroxide ion, which is too weak to initiate polymerization. Ketones are unreactive toward polymerization because of the steric and inductive effects of two alkyl groups. Exceptions to this generalization are some copolymerizations of ketones with formaldehyde and the polymerization of thiocarbonyl monomers [Colomb et al., 1978]. Aromatic aldehydes are also unreactive.

A side reaction occurring with acetaldehyde and high aldehydes containing α-hydrogens is aldol condensation [Hashimoto et al., 1976, 1978; Yamamoto et al., 1978]. Aldol reaction can be extensive at ambient temperatures and higher but is avoided by polymerization at low temperature.

The substitution of halogens on the alkyl group of an aliphatic aldehyde greatly enhances its polymerizability (both from the kinetic and thermodynamic viewpoints [Kubisa and Vogl, 1980]. Trichloroacetaldehyde (chloral) is easily polymerized by such weak bases as pyridine, alkali thiocyanates, and even chloride ion. Further, the polymerization of chloral by butyllithium at $-78°C$ is complete in less than a second [Busfield and Whalley, 1965]. The electron-withdrawing inductive effect of the halogens acts to stabilize the propagating anion **XLII** by decreasing the charge density on the negative oxygen.

The effect of halogens on reactivity is also seen for fluorothiocarbonyl monomers. Thiocarbonyl fluoride is polymerized at $-78°C$ by a trace of a mild base such as

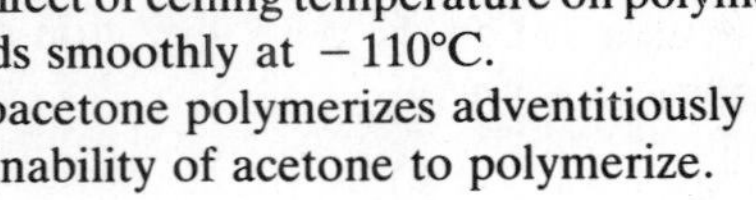

$$\overset{S}{\overset{\|}{F-C-F}} \qquad \overset{S}{\overset{\|}{CF_3-C-CF_3}}$$

Thiocarbonyl fluoride   Hexafluoroacetone

dimethylformamide. The polymerization of hexafluoroacetone is an extreme example of the effect of ceiling temperature on polymerization. T_c is very low but polymerization proceeds smoothly at $-110°C$.

Thioacetone polymerizes adventitiously even at very low temperature, in contrast to the inability of acetone to polymerize.

5-6b   Cationic Polymerization

Carbonyl monomers can be polymerized by acidic initiators, although their reactivity is lower than in anionic polymerization. Protonic acids such as hydrochloric and acetic acids and Lewis acids of the metal halide type are effective in initiating the cationic polymerization of carbonyl monomers. The initiation and propagation steps in polymerizations initiated with protonic acids can be pictured as

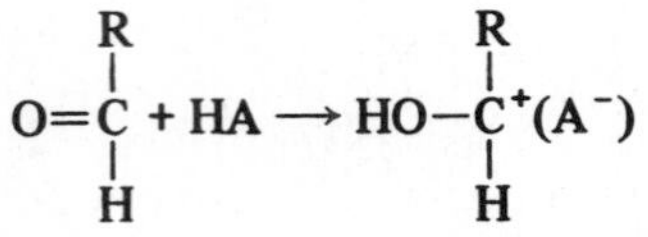

$$O=\overset{R}{\underset{H}{\overset{|}{C}}}+HA \longrightarrow HO-\overset{R}{\underset{H}{\overset{|}{C^+}}}(A^-) \qquad\qquad (5\text{-}133)$$

$$H\text{---}(O\text{---}CHR)_{\overline{n}}O\text{---}\underset{\underset{H}{|}}{\overset{\overset{R}{|}}{C}}{}^+(A^-) + O{=}\underset{\underset{H}{|}}{\overset{\overset{R}{|}}{C}} \longrightarrow H\text{---}(O\text{---}CHR)_{\overline{(n+1)}}O\text{---}\underset{\underset{H}{|}}{\overset{\overset{R}{|}}{C}}{}^+(A^-)$$

$$(5\text{-}134)$$

The termination reaction probably involves chain transfer with water or some other species present,

$$H\text{---}(O\text{---}CHR)_{\overline{(n+1)}}O\text{---}\underset{\underset{H}{|}}{\overset{\overset{R}{|}}{C}}{}^+(A^-) + H_2O \longrightarrow$$

$$H\text{---}(O\text{---}CHR)_{\overline{(n+1)}}O\text{---}\underset{\underset{H}{|}}{\overset{\overset{R}{|}}{C}}\text{---}OH + HA \qquad (5\text{-}132)$$

Competing side reactions in cationic polymerization of carbonyl monomers include cyclotrimerization and acetal interchange. Cyclotrimerization is minimized by low polarity solvents, low temperatures, and initiators of low acidity. Acetal interchange reactions among different polymer chains do not occur except at higher temperatures. Acetaldehyde and higher aldehydes are reasonably reactive in cationic polymerization compared to formaldehyde. Haloaldehydes are lower in reactivity compared to their nonhalogen counterparts.

5-6c Radical Polymerization

The carbonyl double bond has not generally been susceptible to polymerization by radical initiators. There are two reasons for this behavior. First, the carbonyl group is highly polarized and not prone to attack by a radical. Second, most radicals are produced at temperatures above the ceiling temperatures of carbonyl monomers. There are, however, a few recent and isolated cases of carbonyl polymerizations by radical initiators. Trifluoroacetaldehyde has been polymerized using benzoyl peroxide at 22°C. The polymerization is slow with 18 hr required to obtain 90% conversion [Busfield and Whalley, 1965]. However, fluorothiocarbonyl monomers such as thiocarbonyl fluoride have been polymerized at high rates by using a trialkylboron–oxygen redox system at −78°C [Sharkey, 1975]. Radicals are produced [Barney et al., 1966] by the sequence

$$R_3B + O_2 \rightarrow R_2BOOR \qquad (5\text{-}136)$$

$$R_2BOOR + 2\,R_3B \rightarrow R_2BOBR_2 + R_2BOR + 2\,R\cdot \qquad (5\text{-}137)$$

Thioacetone is also polymerized by this redox system.

Radical polymerizations are observed with these monomers because the electron-withdrawing substituents on the carbonyl and thiocarbonyl group decrease its polarity (and polarizability). The greater susceptibility of the thiocarbonyl double bond to radical polymerization is due to the lower electronegativity of sulfur compared to oxygen. The thiocarbonyl group is less polar than the carbonyl group and more prone to attack by a radical. The successful radical polymerization of thiocarbonyl monomers is also due in large part to the low temperature initiation process employed.

5-6d End-Capping

The polyacetals obtained by the chain polymerization of carbonyl monomers are generally unstable at ambient or moderate temperature due to the effect of ceiling temperature. Because of their low ceiling temperatures, they undergo facile depolymerization to monomer. Thus it would appear that this class of polymers would not be of any practical utility. However, the low ceiling temperatures of polyacetals have been circumvented by converting their reactive hydroxyl end groups into unreactive ester linkages by reaction with an anhydride.

$$\text{HO} \!-\!\!\left(\!-\text{CH}_2\text{O}\!-\!\right)_{\!n}\!\text{CH}_2\text{OH} \xrightarrow{\text{(RCO)}_2\text{O}} \text{RCOO} \!-\!\!\left(\!-\text{CH}_2\text{O}\!-\!\right)_{\!n}\!\text{CH}_2\text{OCOR}$$

$$(5\text{-}138)$$

This reaction is referred to as *end-capping* or *end-blocking*. The result is that a reactive site (i.e., an anion or cation) does not form at the chain end and depolymerization does not occur at the ceiling temperature of the polymer. In other words, the polymer chains are end-blocked from depolymerizing. Acetic anhydride is the most often used capping reagent. The decomposition temperatures of the polyacetals are increased by a couple of hundred degrees by the end-capping technique.

The polymerization of formaldehyde to polyoxymethylene (probably by anionic initiation), followed by end-capping with acetic anhydride is carried out on a commercial scale. Polyoxymethylene (referred to as POM; trade name: *Delrin*) is highly crystalline (60–77%) because of the ease of packing of the simple, polar polymer chains ($T_m = 175°C$). The commercial products have number-average molecular weights of 20,000–70,000 ($PDI \sim 2$). POM has a good combination of properties—high strength, toughness, resistance to creep, fatigue and abrasion, low coefficient of friction, low moisture absorption—and is used as an engineering plastic. It has excellent resistance to organic solvents, but the acetal structure is attacked by strong acids. The acetal structure is resistant to attack by base, but bases attack the ester end groups, and this results in depolymerization at higher temperatures. More than 150 million pounds of polyoxymethylene are produced annually in the United States. A large fraction of this is a copolymer (trade name: *Celcon*) produced by copolymerization of trioxane (the cyclic trimer of formaldehyde) with a small amount of a cyclic ether or acetal comonomer (Sec. 7-2b-7). Copolymerization accomplishes the same objective as end-capping. Depolymerization, when it occurs in a copolymer chain, proceeds only until the chain end is the other monomer unit. At that point further depolymerization does not take place. The melting temperature of the copolymer is about 10°C lower than the homopolymer but is rated for continuous use at higher temperatures (104°C versus 86°C) due to a decreased tendency toward depolymerization. The copolymer is also much more resistant to bases.

The applications of polyoxymethylene include plumbing and hardware (ballcock valves, shower heads, fittings and valves, pump and filter housings), machinery (gears, bearings, rollers, conveyor chains, airflow valves), transportion (fuel pump housing, cooling fan parts, fuel caps, window lift mechanisms and cranks, door handles, steering column–gear shift assemblies), appliances (can opener drivetrain, food mixer parts, pump and water sprinkler parts), and electronics (winders, reels, and guide rolls for audio and video cassettes, control disks, gears, transmission parts) [Dolce and Grates, 1985].

5-7 MISCELLANEOUS POLYMERIZATIONS

5-7a Monomers with Two Different Polymerizable Groups

Most monomers contain only one polymerizable group. There are some monomers with two polymerizable groups per molecule. Polymerization of such monomers can lead to more than one polymer structure. The polymerization of 1,3-dienes, of large industrial importance, is discussed in Sec. 8-6. 1-Substituted-1,2-dienes (*allenes*) undergo polymerization through both the substituted and unsubstituted double bond

$$RCH{=}C{=}CH_2 \longrightarrow \ {-}(CHR{-}\underset{\overset{\|}{CH_2}}{C}{-})_{\overline{n}} \ \text{and} \ {-}(\underset{\overset{\|}{CHR}}{C}{-}CH_2{-})_{\overline{n}} \tag{5-139}$$

$$\textbf{XLIII} \qquad\qquad \textbf{XLIV}$$

Various types of initiators have been studied, including ionic, radical, and coordination (Chap. 8) [Ghalamkar-Moazzam and Jacobs, 1978; Leland et al., 1977]. Polymerization through the unsubstituted double bond is the predominant reaction.

Some monomers such as acrolein and ketene contain two different types of polymerizable groups ($C{=}O$ and $C{=}C$). The polymerization of acrolein has been studied with radical, cationic, and anionic initiators [Calvaryrac et al., 1973; Gulino et al., 1981; Schulz, 1967; Yamashita et al., 1979]. Radical polymerization proceeds exclusively through the vinyl group

$$nCH_2{=}\underset{\overset{|}{CHO}}{CH} \longrightarrow \ {-}\left[CH_2{-}\underset{\overset{|}{CHO}}{CH}\right]_n \tag{5-140}$$

Anionic polymerization yields a polymer containing two types of repeating units cor-

$${-}CH_2{-}\underset{\overset{|}{CHO}}{CH}{-} \qquad {-}\underset{\overset{|}{CH{=}CH_2}}{CH}{-}O{-}$$

responding to reaction through both the alkene and carbonyl double bonds. However, polymerization through the alkene group is the dominant reaction. 1,4-Polymerization also occurs:

$$nCH_2{=}CH{-}CH{=}O \longrightarrow \ {-}(CH_2{-}CH{=}CH{-}O{-})_{\overline{n}} \tag{5-141}$$

Cationic polymerization of acrolein has been reported to yield polymers containing both types of repeating units, although one would not expect the alkene group to be reactive toward cationic initiation.

Dimethylketene has been polymerized by anionic initiators. Three different repeat units are found (**XLV, XLVI, XLVII**) corresponding to polymerization through the alkene double bond, carbonyl double bond, and the two double bonds in an alternating

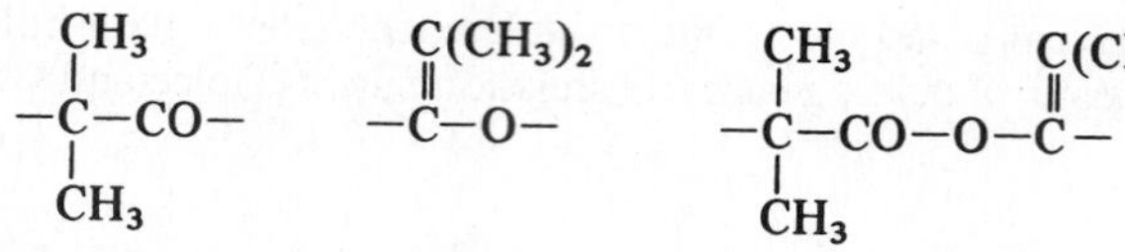

$$\underset{\textbf{XLV}}{\overset{\overset{\displaystyle CH_3}{|}}{\underset{\underset{\displaystyle CH_3}{|}}{-C-CO-}}} \qquad \underset{\textbf{XLVI}}{\overset{\overset{\displaystyle C(CH_3)_2}{\|}}{-C-O-}} \qquad \underset{\textbf{XLVII}}{\overset{\overset{\displaystyle CH_3}{|}}{\underset{\underset{\displaystyle CH_3}{|}}{-C-CO-O-}}\overset{\overset{\displaystyle C(CH_3)_2}{\|}}{C-}}$$

manner, respectively [Pregaglia et al., 1963; Yamashita and Nunamoto, 1962]. Polymerization through the more polar carbonyl group to form the polyacetal is favored by solvents of high polarity. Polymerization of the less polar alkene group, either completely to form the polyketone or partially to form the polyester, is favored by nonpolar solvents. Polyketone formation is favored by Li, Mg, and Al counterions, and polyester by Na and K.

5-7b   Hydrogen-Transfer Polymerization

The polymerization of acrylamide (or methacrylamide) by strong bases such as sodium, organolithium compounds, and alkoxides yields a polymer structure

$$n\text{CH}_2=\text{CH}-\text{CO}-\text{NH}_2 \rightarrow -(-\text{CH}_2-\text{CH}_2-\text{CO}-\text{NH}-)_n- \qquad (5\text{-}142)$$

which is quite unexpected [Bush and Breslow, 1968; Camino et al., 1970, 1977; Otsu et al., 1976]. The polyamide structure of the polymer has been confirmed by the hydrolysis of the polymer to 3-aminopropanoic acid.

The polymerization mechanism involves addition of the anionic initiator to the carbon–carbon double bond of the monomer

$$\text{B}^- + \text{CH}_2=\text{CHCONH}_2 \rightarrow \text{BCH}_2-\bar{\text{C}}\text{HCONH}_2 \qquad (5\text{-}143)$$

followed by proton abstraction from another monomer molecule to form the amide anion (**XLVIII**)

$$\text{BCH}_2-\bar{\text{C}}\text{HCONH}_2 + \text{CH}_2=\text{CHCONH}_2 \rightarrow$$

$$\underset{\textbf{XLVIII}}{\text{BCH}_2\text{CH}_2\text{CONH}_2 + \text{CH}_2=\text{CHCO}\bar{\text{N}}\text{H}} \qquad (5\text{-}144)$$

The amide anion initiates polymerization, via a similar reaction sequence, by addition to monomer

$$\text{CH}_2=\text{CHCO}\bar{\text{N}}\text{H} + \text{CH}_2=\text{CHCONH}_2 \rightarrow$$

$$\text{CH}_2=\text{CHCONHCH}_2\bar{\text{C}}\text{HCONH}_2 \qquad (5\text{-}145)$$

$$\text{CH}_2=\text{CHCONHCH}_2\bar{\text{C}}\text{HCONH}_2 + \text{CH}_2=\text{CHCONH}_2 \rightarrow$$

$$\text{CH}_2=\text{CHCONHCH}_2\text{CH}_2\text{CONH}_2 + \text{CH}_2=\text{CHCO}\bar{\text{N}}\text{H} \qquad (5\text{-}146)$$

Propagation follows in a like manner,

$$CH_2=CHCONH-(-CH_2CH_2CONH-)_{\overline{n}}H + CH_2=CHCON\bar{H} \longrightarrow$$
$$\textbf{XLIX}$$

$$CH_2=CHCONH-CH_2-\bar{C}HCONH-(-CH_2CH_2CONH-)_{\overline{n}}H \xrightarrow{monomer}$$

$$CH_2=CHCONH-(-CH_2CH_2CONH-)_{\overline{(n+1)}}H + CH_2=CHCONH \qquad (5\text{-}147)$$

This polymerization follows a path that is quite different than any of the previously discussed polymerizations. The propagating center is not an ion or a radical but is the carbon–carbon double bond at one end of the propagating species **XLIX**. In addition, it is not the monomer but the monomer anion **XLVIII** that adds to the propagating center. The evidence for this polymerization mechanism comes from analysis for the alkene end group and the observation that the reaction is a step polymerization as regard the buildup of polymer molecular weight. This mechanism is very similar to that for the anionic ring-opening polymerization of cyclic amides (Sec. 7-3a). By analogy to the latter, the monomer anion can be referred to as *activated monomer*. The overall polymerization reaction is often referred to as a *hydrogen-transfer polymerization*.

Polymerization of acrylamide does not proceed entirely through the hydrogen-transfer route under all reaction conditions. Polymers with mixed structures, resulting from concurrent hydrogen-transfer polymerization and normal polymerization through the carbon–carbon double bond, occur in some cases. The relative extents of the two modes of propagation are affected by the initiator, solvent, monomer concentration and temperature [Imanishi, 1989]. Polymerization of *p*-vinylbenzamide follows a similar course [Jung and Heitz, 1988]. Anionic polymerization of α- and β-aminopropionitrile

$$NH_2CH_2-CN \longrightarrow -(-NH-CH_2-\underset{\underset{H}{\overset{\|}{N}}}{C}-)_{\overline{n}} \qquad (5\text{-}148)$$

also proceeds through a hydrogen-transfer mechanism [Ree and Minoura, 1976, 1978].

5-7c Polymerization and Cyclotrimerization of Isocyanates

Isocyanates are polymerized through the carbon–nitrogen double bond to 1-nylons,

$$nR-N=C=O \longrightarrow \left[\underset{\overset{|}{R}}{N}-CO\right]_{\overline{n}} \qquad (5\text{-}149)$$

by anionic initiators such as metal alkyls, sodium naphthalene, and sodium cyanide [Bur and Fetters, 1976; Dabi and Zilka, 1980; Eromosele and Pepper, 1987; Kresta et al., 1979]. The polymerization has a low ceiling temperature and cyclotrimerization

to the isocyanurate

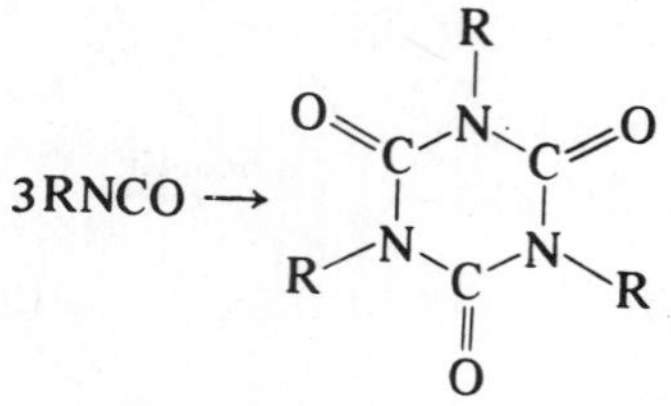

$$3\,RNCO \rightarrow$$

(5-150)

is the predominant reaction at ambient temperature and above.

5-7d Monomers with Triple Bonds

Alkynes have been polymerized using ionic and radical

$$n\,RC\equiv CH \longrightarrow \left[\begin{array}{c} R \\ | \\ -C=CH- \end{array} \right]_n$$

(5-151)

initiators, but the polymer molecular weights are low. High molecular weights are obtained by using Ziegler–Natta coordination catalysts (Sec. 8-4d-2) [Chien et al., 1980]. The polymers are of considerable interest in terms of their potential as (semi)conducting materials.

Nitriles have been polymerized by radical and ionic

$$n\,RC\equiv N \longrightarrow \left[\begin{array}{c} R \\ | \\ -C=N- \end{array} \right]_n$$

(5-152)

initiators. The polymer structures are often quite complex, although anionic polymerization yields the cleanest reaction [Wohrle, 1972, 1983; Wohrle and Knothe, 1988].

Polymerization of isocyanides to polyiminomethylenes is best achieved with cationic

$$R-NC \longrightarrow \left(\begin{array}{c} NR \\ \| \\ -C- \end{array} \right)_n$$

(5-153)

initiators although anionic and radical polymerizations also occur [Millich, 1988].

REFERENCES

Alev, S., A. Collet, M. Viguier, and F. Schue, *J. Polym. Sci. Polym. Chem. Ed.*, **18**, 1155 (1980).

Allen, C. C., W. Oraby, T. M. A. Hossain, E. P. Stahel, D. R. Squire, and V. T. Stannett, *J. Appl. Polym. Sci.*, **18**, 709 (1974).

Allen, G., G. Gee, and C. Stretch, *J. Polym. Sci.*, **48**, 189 (1960).

Allport, D. C. and W. H. James, Eds., "Block Copolymers," Halsted Press (Wiley), New York, 1973.

Alward, D. B., D. J. Kinning, E. L. Thomas and L. J. Fetters, *Macromolecules*, **19**, 215 (1986).

Antonietti, M., D. Ehlich, K. J. Folsch, H. Sillescu, M. Schmidt, and P. Lindner, *Macromolecules*, **22**, 2802 (1989).

Antonietti, M. and K. J. Folsch, *Makromol. Chem. Rapid. Commun.*, **9**, 423 (1988).

Aoshima, S. and T. Higashimura, *Macromolecules*, **22**, 1009 (1989).

Barney, A. L., J. M. Bruce, Jr., J. N. Coker, N. W. Jacobson, and W. H. Sharkey, *J. Polym. Sci.*, **A-1**(4), 2617 (1966).

Benson, R. S., Q. Wu, A. R. Ray, and D. J. Lyman, *J. Polym. Sci. Polym. Chem. Ed.*, **23**, 399 (1985).

Berger, W. and H.-J. Adler, *Makromol. Chem.*, **3**, 301 (1986).

Bhattacharyya, D. N., C. L. Lee, J. Smid, and M. Szwarc, *J. Phys. Chem.*, **69**, 612 (1965a).

Bhattacharyya, D. N., J. Smid, and M. Szwarc, *J. Phys. Chem.*, **69**, 624 (1965b).

Bhattacharyya, D. N., J. Smid, and M. Szwarc, *J. Am. Chem. Soc.*, **86**, 5024 (1964).

Biddulph, R. H. and P. H. Plesch, *J. Chem. Soc.*, 3913 (1960).

Biswas, M. and K. J. John, *J. Polym. Sci. Polym. Chem. Ed.*, **16**, 3025 (1978).

Biswas, M. and G. M. A. Kabir, *Eur Polym. J.*, **14**, 861 (1978); *J. Polym. Sci. Polym. Chem. Ed.*, **17**, 673 (1979).

Biswas, M. and P. Kamannarayana, *J. Polym. Sci. Polym. Chem. Ed.*, **14**, 2071 (1976).

Brandrup, J. and E. H. Immergut, Eds., "Polymer Handbook," 3rd ed., p. II-316, Wiley-Interscience, New York, 1989.

Brittain, W. J. and I. B. Dicker, *Macromolecules*, **22**, 1054 (1989).

Brown, C. P. and A. R. Mathieson, *J. Chem. Soc.*, 3608, 3612, 3620, 3631 (1957); 3445 (1958).

Brzezinska, K., W. Chivalkowska, P. Kubisa, K. Matyjaszewski, and S. Penczek, *Makromol. Chem.*, **178**, 2491 (1977).

Bur, A. J. and L. J. Fetters, *Chem. Rev.*, **76**, 727 (1976).

Burgess, F. J., A. V. Cunliffe, J. R. MacCallum, and D. H. Richards, *Polymer*, **18**, 719, 726, 733 (1977).

Busfield, W. K. and E. Whalley, *Can. J. Chem.*, **43**, 2289 (1965).

Bush, L. W. and D. S. Breslow, *Macromolecules*, **1**, 189 (1968).

Busson, R. and M. Van Beylen, *Macromolecules*, **10**, 1320 (1978).

Bywater, S., "Anionic Polymerization," Chap. 2 in "Progress in Polymer Science," Vol. 4, A. D. Jenkins, Ed., Pergamon Press, New York, 1975.

Bywater, S., "Anionic Polymerization of Olefins," Chap. 1 in "Comprehensive Chemical Kinetics," Vol. 15, "Non-Radical Polymerization," C. H. Bamford and C. F. H. Tipper, Eds., Elsevier, New York, 1976.

Bywater, S., "Anionic Polymerization," pp. 1–43 in "Encyclopedia of Polymer Science and Engineering," Vol. 2, H. F. Mark, N. M. Bikales, C. G. Overberger, and G. Menges, Eds., Wiley-Interscience, New York, 1985.

Cai, G.-F, D.-Y. Yan, and M. I itt, *Macromolecules*, **21**, 578 (1988).

Calvaryrac, H., P. Thivollet, and J. Gole, *J. Polym. Sci.*, **11**, 1631 (1973).

Camino, G., M. Guaita, and L. Trossarelli, *Makromol. Chem.*, **136**, 155 (1970); *J. Polym. Sci. Polym. Lett. Ed.*, **15**, 417 (1977).

Ceresa, R. J., Ed., "Block and Graft Copolymerization," Wiley-Interscience, New York, 1973.

Cerrai, P., G. Guerra, and M. Tricoli, *Eur. Polym. J.*, **12**, 247 (1976); *Eur. Polym. J.*, **15**, 153 (1979).

Cesca, S., "Isomerization Polymerization," pp. 463–487 in "Encyclopedia of Polymer Science and Engineering," Vol. 2, H. F. Mark, N. M. Bikales, C. G. Overberger, and G. Menges, Eds., Wiley-Interscience, New York, 1985.

Chance, B., "Rapid Flow Methods," Chap. II in "Techniques of Chemistry," Vol. VI, Part II, G. G. Hammes, Ed., Wiley, New York, 1974.

Chaplin, R. P. and S. Yaddehige, *J. Macromol. Sci. Chem.*, **A14**, 23 (1980).

Cheminat, A., G. Friedmann, and M. Brini, *J. Polym. Sci. Polym. Chem. Ed.*, **17**, 2865 (1979).

Cheng, T. C. and A. F. Halasa, *J. Polym. Sci. Polym. Chem. Ed.*, **14**, 583 (1976).

Chien, J. C. W., F. E. Karasz, G. E. Wnek, and A. G. MacDiarmid, *J. Polym. Sci. Polym. Chem. Ed.*, **18**, 45 (1980).

Chmelir, M. and G. V. Schulz, *Polym. Bull.*, **1**, 355 (1979).

Cho, C. G., B. A. Feit, and O. W. Webster, *Macromolecules*, **23**, 1918 (1990).

Choi, W. O., M. Sawamoto, and T. Higashimura, *Macromolecules*, **23**, 48 (1990).

Colclough, R. O. and F. S. Dainton, *Trans. Faraday Soc.*, **54**, 886, 894 (1958).

Colomb, H. O., F. E. Bailey, Jr., and R. D. Lundberg, *J. Polym. Sci. Polym. Lett. Ed.*, **16**, 507 (1978).

Corno, C., G. Ferraris, A. Priola, and S. Cesca, *Macromolecules*, **12**, 404 (1979).

Cotrel, R., G. Sauvet, J. P. Vairon, and P. Sigwalt, *Macromolecules*, **9**, 931 (1976).

Crivello, J. V., *Adv. Polym. Sci.*, **62**, 1 (1984).

Crivello, J. V., J. L. Lee, and D. A. Conlon, *Makromol. Chem. Macromol. Symp.*, **13/14**, 145 (1988).

Crivello, J. V., T. P. Lockhart, and J. L. Lee, *J. Polym. Sci. Polym. Chem. Ed.*, **21**, 97 (1983).

Cronin, J. P. and D. C. Pepper, *Makromol. Chem.*, **189**, 85 (1988).

Dabi, S. and A. Zilka, *Eur. Polym. J.*, **16**, 95, 471, 475 (1980).

Deffieux, A. and S. Boileau, *Macromolecules*, **9**, 369 (1976).

Deffieux, A., D. R. Squire, and V. Stannett, *Polym. Bull.*, **2**, 469 (1980).

Deffieux, A., J. A. Young, W. C. Hsieh, D. R. Squire, and V. Stannett, *Polymer*, **24**, 573 (1983).

de Groof, B., M. Van Beylen, and M. Szwarc, *Macromolecules*, **8**, 397 (1975).

de Groof, B., B. Mortier, M. Van Beylen, and M. Szwarc, *Macromolecules*, **10**, 598 (1977).

Dickstein, W. H. and C. P. Lillya, *Macromolecules*, **22**, 3882 (1989).

DiMaina, M., S. Cesca, P. Giusti, G. Ferrraris, and P. L. Magagnini, *Makromol. Chem.*, **178**, 2223 (1977).

Dolce, T. J. and J. A. Grates, "Acetal Resins," pp. 42–61 in "Encyclopedia of Polymer Science and Engineering," Vol. 1, H. F. Mark, N. M. Bikales, C. G. Overberger, and G. Menges, Eds., Wiley-Interscience, New York, 1985.

Donnelly, E. F., D. S. Johnston, D. C. Pepper, and D. J. Dunn, *J. Polym. Sci. Polym. Lett. Ed.*, **15**, 399 (1977).

Dotcheva, D. T. and C. B. Tsvetanov, *Makromol. Chem.*, **186**, 2103 (1985).

Dumas, S., V. Marti, J. Sledz, and F. Schue, *J. Polym. Sci. Polym. Lett. Ed.*, **16**, 81 (1978).

Dunn, D. J., "The Cationic Polymerization of Vinyl Monomers," Chap. 2 in "Developments in Polymerization-1," R. N. Haward, Ed., Applied Science Publishers, London, 1979.

Dunn, D. J., E. Mathias, and P. H. Plesch, *Eur. Polym. J.*, **12**, 1 (1976).

Endres, G. F. and C. G. Overberger, *J. Am. Chem. Soc.*, **77**, 2201 (1955).

Eromosele, I. C. and D. C. Pepper, *J. Polym. Sci. Poly. Chem. Ed.*, **25**, 3499 (1987).

Eromosele, I. C., D. C. Pepper, and B. Ryan, *Makromol. Chem.*, **190**, 1613 (1989).

Evans, A. G. and G. W. Meadows, *Trans. Faraday Soc.*, **46**, 327 (1950).

Faust, R. and J. P. Kennedy, *J. Polym. Sci. Polym. Chem. Ed.*, **25**, 1847 (1987).

Faust, R., A. Fehervari, and J. P. Kennedy, *J. Macromol. Sci. Chem.*, **A18**, 1209 (1982–1983); *Polym. Bull.*, **15**, 317 (1986).

Faust, R., M. Zsuga, and J. P. Kennedy, *Polym. Bull.*, **21**, 125 (1989).

Fayt, R., R. Forte, C. Jacobs, R. Jerome, T. Ouhadi, P. Teyssie, and S. K. Varshney, *Macromolecules*, **20**, 1444 (1987).

Ferraris, G., C. Corno, A. Priola, and S. Cesca, *Macromolecules*, **10**, 188 (1977).

Fetters, L. J., E. M. Firer, and M. Dafauti, *Macromolecules*, **10**, 1200 (1978).

Flory, P. J., *J. Am. Chem. Soc.*, **62**, 1561 (1940).

Fontanille, M., "Carbanionic Polymerization: General Aspects and Initiation," pp. 365–386 and "Carbanionic Polymerization: Termination and Functionalization," pp. 425–432 in "Comprehensive Polymer Science," Vol. 3, G. C. Eastmond, A. Ledwith, S. Russo, and P. Sigwalt, Eds., Pergamon Press, London, 1989.

Foss, R. P., H. A. Jacobson, and A. H. Sharkey, *Macromolecules*, **10**, 287 (1977).

Funt, B. L., N. Severs, and A. Glasel, *J. Polym. Sci. Polym. Chem. Ed.*, **14**, 2763 (1976).

Furukawa, J. and T. Saegusa, "Polymerization of Aldehydes and Oxides," Wiley-Interscience, New York, 1963.

Gandini, A. and H. Cheradame, *Adv. Polym. Sci.*, **34/35**, 1 (1980).

Gandini, A. and H. Cheradame, "Cationic Polymerization," pp. 729–814 in "Encyclopedia of Polymer Science and Engineering," Vol. 2, H. F. Mark, N. M. Bikales, C. G. Overberger, and G. Menges, Eds., Wiley-Interscience, New York, 1985.

Gandini, A. and A. Martinez, *Makromol. Chem. Macromol. Symp.*, **13/14**, 211 (1988).

Gaylord, N. G. and S. S. Dixit, *J. Polym. Sci. Macromol. Rev.*, **8**, 51 (1974).

Geacintov, C., J. Smid, and M. Szwarc, *J. Am. Chem. Soc.*, **84**, 2508 (1962).

Ghalamkar-Moazzam, M. and T. L. Jacobs, *J. Polym. Sci. Polym. Chem. Ed.*, **16**, 615, 701 (1978).

Glasse, M. D., "Spontaneous Termination in Living Polymers," pp. 133–196 in "Progress in Polymer Science," Vol. 9, A. D. Jenkins and V. T. Stannett, Eds., Pergamon Press, London, 1983.

Grattan, D. W. and P. H. Plesch, *J. Chem. Soc. Dalton Trans.*, 1734 (1977); *Makromol. Chem.*, **181**, 751 (1980).

Gulino, D., J. P. Pascault, and Q. T. Pham, *Makromol. Chem.*, **182**, 2321 (1981).

Halaska, V., J. Pecka, and M. Marek, *Makromol. Chem. Macromol. Symp.*, **3**, 3 (1986).

Hashimoto, K., H. Sumitomo, and S. Ohsawa, *J. Polym. Sci. Polym. Chem. Ed.*, **14**, 1221 (1976); **16**, 435 (1978).

Hatada, K., T. Kitayama, and H. Yuki, *Polym. Bull.*, **2**, 15 (1980).

Hay, J. N., D. S. Harris, and M. Wiley, *Polymer*, **17**, 613 (1976).

Hayashi, K., K. Hayashi, and S. Okamura, *J. Polym. Sci.*, **A-1**(9), 2305 (1971).

Hayashi, K. H. Yamazawa, T. Takagaki, F. Williams, K. Hayashi, and S. Okamura, *Trans. Faraday Soc.*, **63**, 1489 (1967).

Higashimura, T., S. Aoshima, and M. Sawamoto, *Makromol. Chem. Macromol. Symp.*, **13/14**, 457 (1988).

Higashimura, T., O. Kishiro, and T. Takeda, *J. Polym. Sci. Polym. Chem. Ed.*, **14**, 1089 (1976).

Higashimura, T., H. Kusano, T. Masuda, and S. Okamura, *J. Polym. Sci. Polym. Lett. Ed.*, **9**, 463 (1971).

Higashimura, T. and M. Sawamoto, *Makromol. Chem. Suppl.*, **12**, 153 (1985); private communication (1989).

Higginson, W. C. and N. S. Wooding, *J. Chem. Soc.*, 760, 1178 (1952).

Hogen-Esch, T. E. and J. Smid, Eds., "Recent Advances in Anionic Polymerization," Elsevier, New York, 1987.

Hrdlovic, P., J. Trekoval, and I. Lukac, *Eur. Polym. J.*, **15**, 229 (1979).

Hsieh, W. C., H. Kubota, D. R. Squire, and V. Stannett, *J. Polym. Sci. Polym. Chem. Ed.*, **18**, 2773 (1980).

Huang, K. J., M. Zsuga, and J. P. Kennedy, *Polym. Bull.*, **19**, 43 (1988).

Hui, K. M. and Y. K. Ong, *J. Polym. Sci. Polym. Chem. Ed.*, **14**, 1311 (1976).

Hurley, S. A. and P. J. T. Tait, *J. Polym. Sci. Polym. Chem. Ed.*, **14**, 1565 (1976).

Imanishi, Y., "Carbanionic Polymerization: Hydrogen Migration Polymerization," pp. 451–455 in "Comprehensive Polymer Science," Vol. 3, G. C. Eastmond, A. Ledwith, S. Russo, and P. Sigwalt, Eds., Pergamon Press, London, 1989.

Imanishi, Y., T. Higashimura, and S. Okamura, *Chem. High Polym.* (*Tokyo*), **18**, 333 (1961).

Inoue, S., T. Aida, M, Kuroki, and Y. Hosokawa, *Makromol. Chem. Macromol. Symp.*, **32**, 255 (1990).

Irie, M. and K. Hayashi, "Photoinduced Ionic Polymerizations," pp. 105–142 in "Progress in Polymer Science Japan," K. Imahori and T. Higashimura, Eds., Halsted Press (Wiley), New York, 1975.

Jenkinson, D. H. and D. C. Pepper, *Proc. Roy. Soc.* (London), **A263**, 82 (1961).

Jeuk, H. and A. H. E. Muller, *Makromol. Chem. Rapid. Commun.*, **3**, 121 (1982).

Johnson, A. F. and R. N. Young, *J. Polym. Sci. Symp.*, **56**, 211 (1976).

Johnson, A. F. and D. A. Pearce, *J. Polym. Sci. Polym. Symp.*, **56**, 57 (1976).

Jolly, W. L., "Modern Inorganic Chemistry," McGraw-Hill, New York, 1984.

Jung, H. and W. Heitz, *Makromol. Chem. Rapid Commun.*, **9**, 373 (1988).

Kanoh, N., A. Gotoh, T. Higashimura, and S. Okamura, *Makromol. Chem.*, **63**, 115 (1963).

Kanoh, N., T. Higashimura, and S. Okamura, *Makromol. Chem.*, **56**, 65 (1962).

Kanoh, N., K. Ikeda, A. Gotoh, T. Higashimura, and S. Okamura, *Makromol. Chem.*, **86**, 200 (1965).

Kaszas, G., J. E. Puskas, C. C. Chen, and J. P. Kennedy, *Macromolecules*, **23**, 3909 (1990).

Kennedy, J. P., *J. Polym. Sci.*, **A-1**(6), 3139 (1968); *J. Polym. Sci. Symp.*, **56**, 1 (1976); *Polym. J.*, **17**, 29 (1985); private communication (1989).

Kennedy, J. P., "Cationic Polymerization of Olefins: A Critical Inventory," Wiley-Interscience, New York, 1975.

Kennedy, J. P., *Makromol. Chem. Macromol. Symp.*, **32**, 119 (1990).

Kennedy, J. P. and S. C. Feinberg, *J. Polym. Sci. Polym. Chem. Ed.*, **16**, 2191 (1978).

Kennedy, J. P. and J. E. Johnston, *Adv. Polym. Sci.*, **19**, 57 (1975).

Kennedy, J. P. and E. Marechal, "Carbocationic Polymerization," Wiley-Interscience, New York, 1982.

Kennedy, J. P., S. Y. Huang, and S. C. Feinberg, *J. Polym. Sci. Polym. Chem. Ed.*, **15**, 2801, 2869 (1977).

Kennedy, J. P., L. S. Minckler, G. Wanless, and R. M. Thomas, *J. Polym. Sci.*, **A2**, 2093 (1964).

Kennedy, J. P. and R. A. Smith, *J. Polym. Sci. Polym. Chem. Ed.*, **18**, 1523, 1539 (1980).

Kennedy, J. P. and R. G. Squires, *Polymer*, **6**, 579 (1965); *J. Macromol. Sci. Chem.*, **A1**, 861 (1967).

Kishimoto, Y., S. Aoshima, and T. Higashimura, *Macromolecules*, **22**, 3877 (1989).

Kojima, K., M. Sawamoto, and T. Higashimura, *Macromolecules*, **22**, 1552 (1989); **23**, 948 (1990).

Kraft, R., A. H. E. Muller, H. Hocker, and G. V. Schulz, *Makromol. Chem. Rapid. Commun.*, **1**, 363 (1980).

Kraft, R., A. H. E. Muller, V. Warzelhan, H. Hocker, and G. V. Schulz, *Macromolecules*, **11**, 1093 (1978).

Kresge, E. N., R. H. Schatz, and H.-C. Wang, "Isobutylene Polymers," pp. 423–448 in "Encyclopedia of Polymer Science and Engineering," Vol. 8, H. F. Mark, N. M. Bikales, C. G. Overberger, and G. Menges, Eds., Wiley-Interscience, New York, 1987.

Kresta, J. E., R. J. Chang, S. Kathiriya, and K. C. Frisch, *Makromol. Chem.*, **180**, 1081 (1979).

Kubisa, P. and O. Vogl, *Polymer*, **21**, 525 (1980).

Kubisa, P., K. Neeld, J. Starr, and O. Vogl, *Polymer*, **21**, 1433 (1980).

Kubota, H., V. Ya. Kabanov, D. R. Squire, and V. T. Stannett, *J. Macromol. Sci. Chem.*, **A12**, 1299 (1978).

Kunitake, T. and K. Takarabe, *Macromolecules*, **12**, 1061, 1067 (1979).

Kuroki, M., T. Aida, and S. Inoue, *J. Am. Chem. Soc.*, **109**, 4737 (1990).

L'Abbe, G. and G. Smets, *J. Polym. Sci.*, **A-1**(5), 1359 (1967).

Ledwith, A., *Pure Appl. Chem.*, **49**, 431 (1977); **51**, 159 (1979a); *Makromol. Chem. Suppl.*, **3**, 348 (1979b).

Ledwith, A. and D. C. Sherrington, "Reactivity and Mechanism in Cationic Polymerization," Chap. 9 in "Reactivity, Mechanism and Structure in Polymer Chemistry," A. D. Jenkins and A. Ledwith, Eds., Wiley-Interscience, New York, 1974.

Leland, J., J. Boucher, and K. Anderson, *J. Polym. Sci. Polym. Chem. Ed.*, **15**, 2785 (1977).

Lenz, R. W., "Organic Chemistry of Synthetic High Polymers," Wiley-Interscience, New York, 1967, Chaps. 13, 14.

Lutz, P. and P. Rempp, *Makromol. Chem.*, **189**, 1051 (1988).

Magagnini, P. L., S. Cesca, P. Giusti, A. Priola, and M. DiMaina, *Makromol. Chem.*, **178**, 2235 (1977).

Malhotra, S. L., *J. Macromol. Sci. Chem.*, **A12**, 73 (1978).

Malhotra, S. L., J. Leonard, and M. Thomas, *J. Macromol. Sci. Chem.*, **A11**, 2213 (1977).

Marek, M. and L. Toman, *Makromol. Chem. Rapid Commun.*, **1**, 161 (1980).

Marek, M., J. Pecka, and V. Halaska, *Makromol. Chem. Macromol. Symp.*, **13/14**, 443 (1988).

Masure, M., G. Sauvet, and P. Sigwalt, *J. Polym. Sci. Polym. Chem., Ed.*, **16**, 3065 (1978); *Polym. Bull.*, **2**, 699 (1980).

Mathieson, A. R., "Styrene," Chap. 6 in "The Chemistry of Cationic Polymerization," P. H. Plesch, Ed., Macmillan, New York, 1963.

Mathis, C., L. Christmann, and B. Francois, *J. Polym. Sci. Polym. Chem. Ed.*, **16**, 1285 (1978).

Mathis, C. and B. Francois, *J. Polym. Sci. Polym. Chem. Ed.*, **16**, 1297 (1978).

Matyjaszewski, K., *Makromol. Chem. Macromol. Symp.*, **13/14**, 389 (1988).

Matyjaszewski, K., "Carbocationic Polymerization: Styrene and Substituted Styrenes," Chap. 41 in "Comprehensive Polymer Science," Vol. 3, G. C. Eastmond, A. Ledwith, S. Russo, and P. Sigwalt, Eds., Pergamon Press, Oxford, 1989.

Mayr, H., R. Schneider, and C. Schade, *Makromol. Chem. Macromol. Symp.*, **13/14**, 43 (1988).

Millich, F., "Polyisocyanides," pp. 383–398 in "Encyclopedia of Polymer Science and Engineering," Vol. 12, H. F. Mark, N. M. Bikales, C. G. Overberger, and G. Menges, Eds., Wiley-Interscience, New York, 1988.

Mishra, M. K. and J. P. Kennedy, *J. Macromol. Sci. Chem.*, **24**, 933 (1987).

Miyamoto, M., M. Sawamoto, and T. Higashimura, *Macromolecules*, **17**, 2228 (1984); **18**, 123 (1985).

Morton, M., "Anionic Polymerization: Principles and Practice," Academic Press, New York, 1983.

Morton, M. and L. J. Fetters, "Anionic Polymerizations and Block Copolymers," Chap. 9 in

"Polymerization Processes," C. E. Schildknecht, Ed. (with I. Skeist), Wiley-Interscience, New York, 1977.

Muller, A. H. E., "Carbanionic Polymerization: Kinetics and Thermodynamics," pp. 387–423 in "Comprehensive Polymer Science," Vol. 3, G. C. Eastmond, A. Ledwith, S. Russo, and P. Sigwalt, Eds., Pergamon Press, London, 1989.

Muller, A. H. E., *Makromol. Chem. Macromol. Symp.*, **32**, 87 (1990).

Muller, A. H. E., L. Lochman, and J. Trekoval, *Makromol. Chem.*, **187**, 1473 (1986).

Nasrallah, E. and S. Baylouzian, *Polymer*, **18**, 1173 (1977).

Noshay, A. and J. E. McGrath, "Block Copolymers," Academic Press, New York, 1977.

Nuyken, O. and H. Kroner, *Makromol. Chem.*, **191**, 1 (1990).

Nuyken, O. and S. Pask, "Telechelic Polymers," pp. 494–532 in "Encyclopedia of Polymer Science and Engineering," Vol. 16, H. F. Mark, N. M. Bikales, C. G. Overberger, and G. Menges, Eds., Wiley-Interscience, New York, 1989.

Oberrauch, E., T. Salvatori, and S. Cesca, *J. Polym. Sci. Polym. Lett. Ed.*, **16**, 345 (1978).

Ogawa, T. and J. Romero, *Eur. Polym. J.*, **13**, 419 (1977).

Okamura, S. and T. Higashimura, *J. Polym. Sci.*, **21**, 289 (1956a); *Chem. High. Polym. (Tokyo)*, **13**, 342, 397 (1956b) and **17**, 57 (1960).

Olah, G. A., *J. Am. Chem. Soc.*, **94**, 808 (1972); *Makromol. Chem. Macromol. Symp.*, **13/14**, 1 (1988).

Olaj, O. F., *Makromol. Chem. Macromol. Symp.*, **8**, 235 (1987).

Otsu, T., B. Yamada, M. Itahashi, and T. Mori, *J. Polym. Sci. Polym. Chem. Ed.*, **14**, 1347 (1976).

Overberger, C. G., G. F. Endres, and A. Monaci, *J. Am. Chem. Soc.*, **78**, 1969 (1956).

Overberger, C. G., E. M. Pearce, and N. Mayes, *J. Am. Chem. Soc.*, **34**, 109 (1959).

Overberger, C. G., H. Yuki, and N. Urakawa, *J. Polym. Sci.*, **45**, 127 (1960).

Peebles, L. H., Jr., "Molecular Weight Distributions in Polymers," Wiley-Interscience, New York, 1971, Chap. 3.

Pepper, D. C., *Makromol. Chem.*, **175**, 1077 (1974); *Eur. Polym. J.*, **16**, 407 (1980).

Pepper, D. C. and P. J. Reilly, *J. Polym. Sci.*, **58**, 639 (1962).

Pepper, D. C. and B. Ryan, *Makromol. Chem.*, **184**, 383, 395 (1983).

Plesch, P. H., *J. Chem. Soc.*, 1653 (1953); *Adv. Polym. Sci.*, **8**, 137 (1971); *Br. Polym. J.*, **5**, 1 (1973); *J. Polym. Sci. Symp.*, **56**, 373 (1977); *Makromol. Chem. Macromol. Symp.*, **13/14**, 375, 393 (1988); *Eur. Polym. J.*, **25**, 875 (1989); *Makromol. Chem. Macromol. Symp.*, **32**, 299 (1990).

Plesch, P. H., "Isobutene," Chap. 4 in "The Chemistry of Cationic Polymerizations," P. H. Plesch, Ed., Macmillan, New York, 1963.

Plesch, P. H., "Propagation Rate Constants in the Cationic Polymerization of Alkenes," pp. 1–16 in "Cationic Polymerization and Related Processes," E. J. Goethals, Ed., Academic Press, New York, 1984.

Pregaglia, G. F., M. Minaghi, and M. Cambini, *Makromol. Chem.*, **67**, 10 (1963).

Puskas, J., G. Kaszas, J. P. Kennedy, T. Kelen, and F. Tudos, *J. Macromol. Sci. Chem.*, **A18**, 1229, 1263, (1982–1983).

Quirk, R. P. and G. P. Bidinger, *Polym. Bull.*, **22**, 63 (1989).

Quirk, R. P., J. Yin, and L. J. Fetters, *Macromolecules*, **22**, 85 (1989).

Ragimov, A. V., A. Yu. Nagiev, B. I. Liogonky, and A. A. Berlin, *J. Polym. Sci. Polym. Chem. Ed.*, **18**, 713 (1980).

Ree, K. and Y. Minoura, *Makromol. Chem.*, **177**, 2897 (1976); **179**, 1145 (1978).

Reetz, M. T., *Angew. Chem. Int. Ed. Engl.*, **27**, 994 (1988).

Register, R. A., M. Foucart, R. Jerome, Y. S. Ding, and S. L. Cooper, *Macromolecules*, **21**, 1009 (1988).

Reibel, L., J. P. Kennedy, and D. Y. L. Chung, *J. Polym. Sci. Polym. Chem. Ed.*, **17**, 2757 (1979).

Reichardt, C., "Solvents and Solvent Effects in Organic Chemistry," 2nd ed., VCH, Weinheim, 1988, Chaps. 5, 7.

Rempp, R., E. Franta, and J.-E. Herz, *Adv. Polym. Sci.*, **86**, 145 (1988).

Richards, D. H., "Anionic Polymerization," Chap. 1 in "Developments in Polymerization-1," R. N. Haward, Ed., Applied Science Pubishers, Essex, U.K., 1979.

Richards, D. H., *Br. Polym. J.*, **12**, 537 (1980).

Richards, D. H., S. B. Kingston, and T. Souel, *Polymer*, **19**, 69, 806, 807 (1978).

Rooney, J. M., *J. Polym. Sci. Symp.*, **56**, 47 (1976); *Makromol. Chem.*, **179**, 165, 2419 (1978).

Roovers, J. P. and P. Toporowski, *Macromolecules*, **16**, 843 (1983).

Roovers, J., P. Toporowski, and J. Martin, *Macromolecules*, **22**, 1897 (1989).

Russell, K. E. and L. G. M. C. Vail, *J. Polym. Sci. Symp.*, **56**, 183 (1976).

Russell, K. E. and G. J. Wilson, "Cationic Polymerizations," Chap. 10 in "Polymerization Processes," C. E. Schildknecht, Ed. (with I. Skeist), Wiley-Interscience, New York, 1977.

Sakurda, Y., T. Higashimura, and S. Okamura, *J. Polym. Sci.*, **33**, 496 (1958).

Sauvet, G. and P. Sigwalt, "Carbocationic Polymerization: General Aspects and Initiation," Chap. 39 in "Comprehensive Polymer Science," Vol. 3, G. C. Eastmond, A. Ledwith, S. Russo, and P. Sigwalt, Eds., Pergamon Press, Oxford, 1989.

Sauvet, G., M. Moreau, and P. Sigwalt, *Makromol. Chem. Macromol. Symp.*, **3**, 33 (1986).

Sauvet, G., J. P. Vairon, and P. Sigwalt, *J. Polym. Sci. Polym. Chem. Ed.*, **16**, 3047 (1978).

Sawamoto, M., J. Fujimori, and T. Higashimura, *Macromolecules*, **20**, 916 (1987).

Sawamoto, M. and T. Higashimura, *Macromolecules*, **11**, 328, 501 (1978); **12**, 581 (1979); *Makromol. Chem. Macromol. Symp.*, **3**, 83 (1986); **32**, 131 (1990).

Sawamoto, M. and J. P. Kennedy, *J. Macromol. Sci. Chem.*, **A18**, 1275 (1982–1983).

Sawamoto, M., T. Masuda, and T. Higashimura, *Makromol. Chem.*, **178**, 1497 (1977); *J. Polym. Sci. Polym. Chem. Ed.*, **16**, 2675 (1978).

Schmitt, B. J. and G. V. Schulz, *Eur. Polym. J.*, **11**, 119 (1975).

Schubert, W. and F. Bandermann, *Makromol. Chem.*, **190**, 2721 (1989).

Schubert, W., H.-D. Sitz, and F. Bandermann, *Makromol. Chem.*, **190**, 2193 (1989).

Schulz, R. C., "Polymerization of Acrolein," Chap. 7 in "Vinyl Polymerization," Vol. 1, Part I, G. E. Ham, Ed., Marcel Dekker, New York, 1967.

Sharkey, W. H., *Adv. Polym. Sci.*, **17**, 73 (1975).

Shen, W.-P., W.-D. Zhu, M.-F. Yang, and L. Wang, *Makromol. Chem.*, **190**, 3061 (1989).

Shimomura, T., J. Smid, and M. Szwarc, *J. Am. Chem. Soc.*, **89**, 5743 (1967a).

Shimomura, T., K. J. Tolle, J. Smid, and M. Szwarc, *J. Am. Chem. Soc.*, **89**, 796 (1967b).

Sigwalt, P., *J. Polym. Sci. Symp.*, **50**, 95 (1975).

Sogah, D. Y. and O. W. Webster, *Macromolecules*, **19**, 1775 (1986).

Sogah, D. Y., W. R. Hertler, O. W. Webster, and G. M. Cohen, *Macromolecules*, **20**, 1473 (1987).

Sogah, D. Y., W. R. Hertler, I. B. Dicker, P. A. DePra, and J. R. Butera, *Makromol. Chem. Macromol. Symp.*, **32**, 75 (1990).

Souel, T., F. Schue, M. Abadie, and D. H. Richards, *Polymer*, **18**, 1293 (1977).

Soum, A., M. Fontanille, and P. Sigwalt, *J. Polym. Sci. Polym. Chem. Ed.*, **15**, 659 (1977).

Spach, G., M. Levy, and M. Szwarc, *J. Chem. Soc.*, 355 (1962).

Stannett, V. H. Garreau, C. C. Ma, J. M. Rooney, and D. R. Squire, *J. Polym. Sci. Symp.*, **56**, 233 (1976).

Stannett, V. T., J. Silverman, and J. L. Garnett, "Polymerization by High-Energy Radiation," pp. 317–336 in "Comprehensive Polymer Science," Vol. 4, G. C. Eastmond, A. Ledwith, S. Russo, and P. Sigwalt, Eds., Pergamon Press, London, 1989.

Starr, J. and O. Vogl, *Makromol. Chem.*, **179**, 2621 (1978); *J. Polym. Sci. Polym. Chem. Ed.*, **17**, 1923 (1979).

Steinbrecht, K. and F. Bandermann, *Makromol. Chem.*, **190**, 2183 (1989).

Stretch, C. and G. Allen, *Polymer*, **2**, 151 (1961).

Subira, F., G. Sauvet, J. P. Vairon, and P. Sigwalt, *J. Polym. Sci. Symp.*, **56**, 221 (1976).

Subira, F., J. P. Vairon, and P. Sigwalt, *Macromolecules*, **21**, 2339 (1988).

Szwarc, M., *Adv. Polym. Sci.*, **2**, 275 (1960); *Acct. Chem. Res.*, **2**, 87 (1969).

Szwarc, M., "Carbanions, Living Polymers, and Electron-Transfer Processes," Wiley-Interscience, New York, 1968.

Szwarc, M., "Ions and Ion Pairs in Ionic Polymerization," Chap. 4 in "Ions and Ion Pairs in Organic Reactions," Vol. 2, M. Szwarc, Ed., Wiley-Interscience, New York, 1974.

Szwarc, M., *Adv. Poly. Sci.*, **49**, 1 (1983); *Makromol. Chem.*, **190**, 567 (1989).

Szwarc, M. and J. Smid, "The Kinetics of Propagation of Anionic Polymerization and Copolymerization," Chap. 5 in "Progress in Reaction Kinetics," Vol. 2, G. Porter, Ed., Pergamon Press, Oxford, 1964.

Szwarc, M., M. Van Beylen, and D. Van Hoyweghen, *Macromolecules*, **20**, 445 (1987).

Takaki, M., R. Asami, K. Asano, and H. Hanahata, *Polym. Bull.*, **17**, 403 (1987).

Taninaka, T. and Y.Minoura, *J. Polym. Sci. Polym. Chem. Ed.*, **14**, 685 (1976).

Tanizaki, A., M. Sawamoto, and T. Higashimura, *J. Polym. Sci. Polym. Chem. Ed.*, **24**, 87 (1986).

Tardi, M. and P. Sigwalt, *Eur. Polym. J.*, **8**, 151 (1972).

Taromi, F. A. and P. Rempp, *Makromol. Chem.*, **190**, 1791 (1989).

Teyssie, Ph., R. Fayt, J. P. Hautekeer, C. Jacobs, R. Jerome, L. Leemans, and S. K.Varshney, *Makromol. Chem. Macromol. Symp.*, **32**, 61 (1990).

Throssell, J. J., S. P. Sood, M. Szwarc, and V. Stannett, *J. Am. Chem. Soc.*, **78**, 1122 (1956).

Toman, L., S. Pokorny, and J. Spevacek, *J. Polym. Sci. Polym. Chem. Ed.*, **27**, 2229 (1989a).

Toman, L., J. Spevacek, and S. Polorny, *J. Polym. Sci. Poly. Chem. Ed.*, **27**, 785 (1989b).

Tsvetanov, Ch. B., *Eur. Polym. J.*, **15**, 503 (1979).

Tung, L. H., G. Y.-S. Lo, and J. A. Griggs, *J. Polym. Sci. Polym. Chem. Ed.*, **23**, 1551 (1985).

Van Beylen, M., S. Bywater, G. Smets, M. Szwarc, and D. J. Worsfold, *Adv. Polym. Sci.*, **86**, 87 (1988).

Vankerckhoven, H. and M. Van Beylen, *Eur. Polym. J.*, **14**, 189, 273 (1978).

Varshney, S. K., J. P. Hautekeer, R. Fayt, R. Jerome, and Ph. Teyssie, *Macromolecules*, **23**, 2618 (1990).

Villesange, M., G. Sauvet, J. P. Vairon, and P. Sigwalt, *J. Macromol. Sci. Chem.*, **A11**, 391 (1977).

Vogl, O., "Polyaldehydes," Marcel Dekker, New York, 1967.

Vogl, O., "Kinetics of Aldehyde Polymerization," Chap. 5 in "Comprehensive Chemical Kinetics," Vol. 15, C.H. Bamford and C. F. H. Tipper, Eds., Elsevier, New York, 1976.

Wang, H. C., G. Levin, and M. Szwarc, *J. Am. Chem. Soc.*, **100**, 3969 (1978); *J. Phys. Chem.*, **83**, 785 (1979).

Warzelhan, V., G. Lohr, H. Hocker, and G. V. Schulz, *Makromol. Chem.*, **179**, 2211 (1978).

Webster, O. W., "Group-Transfer Polymerization," pp. 580–588 in "Encyclopedia of Polymer

Science and Engineering," Vol. 7, H. F. Mark, N. M. Bikales, C. G. Overberger, and G. Menges, Eds., Wiley-Interscience, New York, 1987.

Webster, O. W. and D. Y. Sogah, "Group Transfer and Aldol Group Transfer Polymerization," pp. 163–169 in "Comprehensive Polymer Science," Vol. 4, G. C. Eastmond, A. Ledwith, S. Russo, and P. Sigwalt, Eds., Pergamon Press, London, 1989.

Wiles, D. M. and S. Bywater, *Trans. Faraday Soc.*, **61**, 150 (1965).

Williams, F., K. Hayashi, K. Ueno, K. Hayashi, and S. Okamura, *Trans. Faraday Soc.*, **63**, 1501 (1967).

Williams, F., A. Shinkawa, and J. P. Kennedy, *J. Polym. Sci. Symp.*, **56**, 421 (1976).

Williams, F. and R. B. Taylor, *J. Am. Chem. Soc.*, **91**, 3728 (1969).

Wohrle, D., *Adv. Polym. Sci.*, **10**, 35 (1972); **50**, 45 (1983).

Wohrle, D. and G. Knothe, *J. Polym. Sci. Polym. Chem. Ed.*, **26**, 2435 (1988).

Yagci, Y. and W. Schnabel, *Makromol. Chem. Macromol. Chem.*, **13/14**, 167 (1988).

Yamamoto, T., S. Konagaya, and A. Yamamoto, *J. Polym. Sci. Polym. Lett. Ed.*, **16**, 7 (1978).

Yamashita, N., H. Inoue, and T. Maeshima, *J. Polym. Sci. Polym. Chem. Ed.*, **17**, 2739 (1979).

Yamashita, Y. and S. Nunamoto, *Makromol. Chem.*, **58**, 244 (1962).

Yan, D.-Y. and C.-M. Yuan, *Makromol. Chem.*, **187**, 2629, 2641 (1986); **188**, 333, 341 (1987).

Young, R. N., R. P. Quirk, and L. J. Fetters, *Adv. Polym. Sci.*, **56**, 1 (1984).

Yu, H.-S., W.-J. Choi, K.-T. Lim, and S.-K. Choi, *Macromolecules*, **21**, 2893 (1988).

Yuan, C. and D. Yan, *Polymer*, **29**, 24 (1988).

Zsuga, M., L. Balogh, T. Kelen, and J. Borbely, *Polym. Bull.*, **23**, 335 (1990).

Zsuga, M., R. Faust, and J. P. Kennedy, *Polym. Bull.*, **21**, 273 (1989).

PROBLEMS

5-1 Consider the following monomers and initiating systems:

Initiating Systems	Monomers
$(\phi CO_2)_2$	$\phi CH{=}CH_2$
$(CH_3)_3COOH + Fe^{2+}$	$CH_2{=}CHCN$
Na + naphthalene	$CH_2{=}C(CH_3)_2$
H_2SO_4	$CH_2{=}CH{-}O{-}n{-}C_4H_9$
$BF_3 + H_2O$	$CH_2{=}CH{-}Cl$
$n{-}C_4H_9Li$	$CH_2{=}C(CH_3){-}CO_2CH_3$
	$CH_2{=}O$
	$CF_2{=}S$

What is the actual initiating species that initiates polymerization for each of the initiating systems? Show equations. Which initiating system(s) can be used to polymerize each of the various monomers? Explain. What general reaction conditions (e.g., temperature, solvent) are required for each polymerization?

5-2 Isobutylene is polymerized under conditions where chain transfer to monomer is the predominant chain-breaking reaction. A 4.0-g sample of the polymer was found to decolorize 6.0 ml of an 0.01 *M* solution of bromine in carbon tetrachloride. Calculate the number-average molecular weight of the polyisobutylene.

5-3 The rates of most reactions increase with increasing temperature. For certain polymerizations, the rate decreases with temperature. Under what different conditions can this type of behavior occur? Explain. Consider step polymerizations as well as radical and ionic chain polymerizations in answering this question.

5-4 Consider the cationic polymerization of isobutylene using $SnCl_4$ as the coinitiator and water as the initiator. Under certain reaction conditions, the polymerization rate was found to be first-order in $SnCl_4$, first-order in water, and second-order in isobutylene. The number-average molecular weight of the initially formed polymer is 20,000. A 1.00-g sample of the polymer contains 3.0×10^{-5} moles of OH groups; it does not contain chlorine. Show the reaction sequence of initiation, propagation, and termination steps for this polymerization and derive the appropriate expressions for the rate and degree of polymerization. Indicate clearly any assumptions made in the derivations.

5-5 Under what reaction conditions might a cationic polymerization with $SnCl_4$ plus water show a dependence of the polymerization rate which is

a. Zero-order in water or $SnCl_4$.

b. Second-order in water or $SnCl_4$.

c. First-order in monomer.

5-6 A monomer Z is polymerized in the presence of an initiating system Y. The following experimental observations are made:

a. The degree of polymerization decreases as the reaction temperature increases.

b. The degree of polymerization is affected by the solvent used.

c. The degree of polymerization is first-order in monomer concentration.

d. The rate of polymerization increases as the reaction temperature increases.

Is this polymerization proceeding by a step, radical chain, cationic chain, or anionic chain mechanism? Discuss clearly how each experimental observation is consistent with your answer.

5-7 A 2.0 M solution of styrene in ethylene dichloride is polymerized at 25°C using 4.0×10^{-4} M sulfuric acid. Calculate the initial degree of polymerization. What would be the degree of polymerization if the monomer solution contains isopropylbenzene at a concentration of 8.0×10^{-5} M? Use the following data as needed: $k_p = 7.6$ liters/mole-sec, $k_{tr.M} = 1.2 \times 10^{-1}$ liters/mole-sec, $k_{ts} = 4.9 \times 10^{-2}$ sec^{-1}, k_t (combination) $= 6.7 \times 10^{-3}$ sec^{-1}, $C_s = 4.5 \times 10^{-2}$.

5-8 What experimental approaches are available for determining whether the polymerization of a particular monomer by ionizing radiation proceeds by a radical or ionic mechanism?

5-9 The sodium naphthalene polymerization of methyl methacrylate is carried out in benzene and tetrahydrofuran solutions. Which solution will yield the highest polymerization rate? Discuss the effect of solvent on the relative concentrations of the different types of propagating centers.

5-10 Assume that 1.0×10^{-3} mole of sodium naphthalene is dissolved in tetrahydrofuran and then 2.0 moles of styrene is introduced into the system by a rapid injection technique. The final total volume of the solution is 1 liter. Assume that the injection of styrene results in instantaneous homogeneous mixing. It is found that half of the monomer is polymerized in 2000 sec. Calculate the propagation rate constant. Calculate the degree of polymerization at 2000 sec and at 4000 sec of reaction time.

5-11 A 1.5 M solution of styrene in tetrahydrofuran is polymerized at 25°C by sodium naphthalene at a concentration of 3.2×10^{-5} M. Calculate the polymerization rate and degree of polymerization using appropriate data from Table 5-11. What fractions of the polymerization rate are due to free ions and ion pairs, respectively? Repeat the calculations for 3.2×10^{-2} M sodium naphthalene.

5-12 Consider propagation by polystyryl sodium ion pairs in a 1 M styrene solution in tetrahydrofuran. For an ion pair concentration of 2.0×10^{-3} M, calculate the relative contributions of contact and solvent-separated ion pairs to the propagation process. Use appropriate data from Table 5-12.

5-13 Show by equations the synthesis of the following types of block copolymers
   a. ABA
   b. CABAC
   where A, B, and C represent styrene, butadiene, and isoprene, respectively.

CHAPTER 6

CHAIN COPOLYMERIZATION

For most step polymerizations, for example, in the synthesis of poly(hexamethylene adipamide) or poly(ethylene terephthalate), two reactants or monomers are used in the process, and the polymer obtained contains two different kinds of structures in the chain. This is not the case for chain polymerizations where only one monomer need be used to produce a polymer. However, chain polymerizations can be carried out with mixtures of two monomers to form polymeric products with two different structures in the polymer chain. This type of chain polymerization process in which two monomers are simultaneously polymerized is termed a *copolymerization* and the product is a *copolymer*. It is important to stress that the copolymer is not an alloy of two homopolymers but contains units of both monomers incorporated into each co-polymer molecule. The process can be depicted as

$$M_1 + M_2 \rightarrow$$

$$\sim\sim M_1M_2M_2M_1M_2M_2M_2M_1M_1M_2M_2M_1M_1M_2M_1M_1M_1M_2M_2M_1M_1 \sim\sim \qquad (6\text{-}1)$$

The two monomers enter into the copolymer in overall amounts determined by their relative concentrations and reactivities. The simultaneous chain polymerization of different monomers can also be carried out with mixtures of three or more monomers. Such polymerizations are generally referred to as *multicomponent copolymerizations*; the term *terpolymerization* is specifically used for systems of three monomers.

Copolymerization is also important in step polymerization. Relatively few studies on step copolymerization have been carried out, although there are considerable commercial applications. Unlike the situation in chain copolymerization, the overall composition of the copolymer obtained in a step copolymerization is usually the same as the feed composition since step reactions must be carried out to close to 100% conversion for the synthesis of high-molecular-weight polymers. Further, most step polymerizations are equilibrium reactions and the initially formed copolymer com-

452

position is rapidly changed by equilibration. This chapter is concerned entirely with chain copolymerization. The main features of step copolymerization have been covered in Sec. 2-13.

6-1 GENERAL CONSIDERATIONS

6-1a Importance of Chain Copolymerization

Chain copolymerization is important from several considerations. Much of our knowledge of the reactivities of monomers, free radicals, carbocations, and carbanions in chain polymerization comes from copolymerization studies. The behavior of monomers in copolymerization reactions is especially useful for studying the effect of chemical structure on reactivity. Copolymerization is also very important from the technological viewpoint. It greatly increases the ability of the polymer scientist to tailor-make a polymer product with specifically desired properties. Polymerization of a single monomer is relatively limited as to the number of different products that are possible. The term *homopolymerization* is often used to distinguish the polymerization of a single monomer from the copolymerization process.

Copolymerization allows the synthesis of an almost unlimited number of different products by variations in the nature and relative amounts of the two monomer units in the copolymer product. A prime example of the versatility of the copolymerization process is the case of polystyrene. More than 8 billion pounds per year of polystyrene products are produced annually in the United States. Only about one-third of the total is styrene homopolymer. Polystyrene is a brittle plastic with low impact strength and low solvent resistance (Sec. 3-13b-2). Copolymerization as well as blending greatly increase the usefulness of polystyrene. Styrene copolymers and blends of copolymers are useful not only as plastics but also as elastomers. Thus copolymerization of styrene with acrylonitrile leads to increased impact and solvent resistance while copolymerization with 1,3-butadiene leads to elastomeric properties. Combinations of styrene, acrylonitrile, and 1,3-butadiene improve all three properties simultaneously. This and other technological applications of copolymerization are discussed further in Secs. 5-2h and 6-8.

6-1b Types of Copolymers

The copolymer described by Eq. 6-1, referred to as a *statistical copolymer*, has a distribution of the two monomer units along the copolymer chain that follows some statistical law, for example, Bernoullian (zero-order Markov) or first- or second-order Markov. Copolymers formed via Bernoullian processes have the two monomer units distributed randomly and are referred to as *random copolymers*. The reader is cautioned that the distinction between the terms statistical and random, recommended by IUPAC [Ring et al., 1985], has generally not been followed in the literature. Most references use the term *random copolymer* independent of the type of statistical process involved in synthesizing the copolymer. There are three other types of copolymer structures—*alternating*, *block*, and *graft*. The alternating copolymer contains the two monomer units in equimolar amounts in a regular alternating distribution,

$$\sim\sim\sim M_1M_2M_1M_2M_1M_2M_1M_2M_1M_2M_1M_2M_1M_2M_1M_2M_1M_2M_1M_2M_1M_2M_1M_2M_1\sim\sim\sim$$

I

Alternating, statistical, and random copolymers are named by following the prefix "poly" with the names of the two repeating units. The specific type of copolymer is noted by inserting *-alt-*, *-stat-*, or *-ran-* in between the names of the two repeating units with *-co-* used when the type of copolymer is not specified, for example, poly(styrene-*alt*-acrylonitrile), poly(styrene-*stat*-acrylonitrile), poly(styrene-*ran*-acrylonitrile), and poly(styrene-*co*-acrylonitrile).

Block and graft copolymers differ from the other copolymers in that there are long sequences of each monomer in the copolymer chain. A block copolymer is a linear copolymer with one or more long uninterrupted sequences of each polymeric species,

$$\sim\!\!\sim\!\!\sim M_1M_1M_1M_1M_1M_1M_1M_1M_1M_1M_1M_1M_1M_2M_2M_2M_2M_2M_2M_2M_2M_2M_2M_2M_2\sim\!\!\sim\!\!\sim$$

II

while a graft copolymer is a branched copolymer with a backbone of one monomer to which are attached one or more side chains of another monomer,

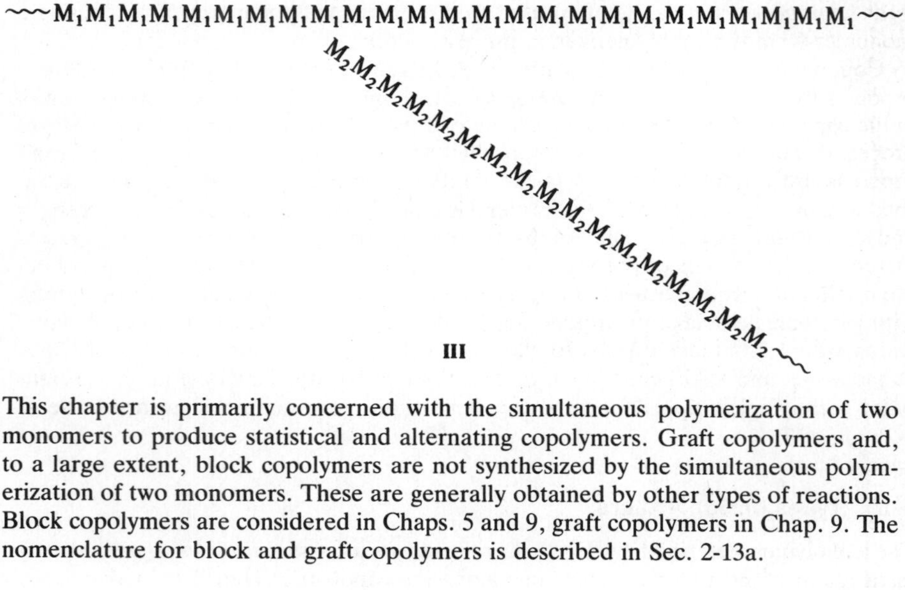

$$\sim\!\!\sim\!\!\sim M_1M_1M_1M_1M_1M_1M_1M_1M_1M_1M_1M_1M_1M_1M_1M_1M_1M_1M_1M_1M_1M_1M_1M_1M_1M_1\sim\!\!\sim$$

$$M_2M_2M_2M_2M_2M_2M_2M_2M_2M_2M_2M_2M_2M_2M_2M_2M_2M_2$$

III

This chapter is primarily concerned with the simultaneous polymerization of two monomers to produce statistical and alternating copolymers. Graft copolymers and, to a large extent, block copolymers are not synthesized by the simultaneous polymerization of two monomers. These are generally obtained by other types of reactions. Block copolymers are considered in Chaps. 5 and 9, graft copolymers in Chap. 9. The nomenclature for block and graft copolymers is described in Sec. 2-13a.

6-2 COPOLYMER COMPOSITION

6-2a Copolymerization Equation; Monomer Reactivity Ratios

The composition of a copolymer is usually different from the composition of the comonomer feed from which it is produced. In other words, different monomers have differing tendencies to undergo copolymerization. It was observed early that the relative copolymerization tendencies of monomers usually bore little resemblance to their relative rates of homopolymerization [Staudinger and Schneiders, 1939]. Some monomers are more reactive in copolymerization than indicated by their rates of homopolymerization; other monomers are less reactive. Further, and most dramatically, a few monomers, such as maleic anhydride, stilbene, and fumaric esters, undergo facile copolymerization with radical initiation, although they have very little or no tendency to undergo homopolymerization.

The composition of a copolymer thus cannot be determined simply from a knowledge of the homopolymerization rates of the two monomers. The determinants of copolymerization composition have been elucidated by several workers by assuming the chemical reactivity of the propagating chain (which may be a free radical, carbocation, or carbanion) in a copolymerization is dependent only on the identity of the monomer unit at the growing end and independent of the chain composition preceding the last monomer unit [Alfrey and Goldfinger, 1944; Mayo and Lewis, 1944; Wall, 1944; Walling, 1957]. This is referred to as the *first-order Markov* or *terminal model* of copolymerization. Consider the case for the copolymerization of the two monomers M_1 and M_2. Although radical copolymerization has been more extensively studied and is more important than ionic copolymerization, we will consider the general case without specification as to whether the mode of initiation is by a radical, anionic, or cationic species. Copolymerization of the two monomers leads to two types of propagating species—one with M_1 at the propagating end and the other with M_2. These can be represented by M_1^* and M_2^* where the asterisk represents either a radical, a carbocation ion, or a carbanion as the propagating species depending on the particular case. If it is assumed that the reactivity of the propagating species is dependent only on the monomer unit at the end of the chain (referred to as the *end* or *ultimate unit*), four propagation reactions are then possible. Monomers M_1 and M_2 can each add either to a propagating chain ending in M_1 or to one ending in M_2, that is,

$$M_1^* + M_1 \xrightarrow{k_{11}} M_1^* \qquad (6\text{-}2)$$

$$M_1^* + M_2 \xrightarrow{k_{12}} M_2^* \qquad (6\text{-}3)$$

$$M_2^* + M_1 \xrightarrow{k_{21}} M_1^* \qquad (6\text{-}4)$$

$$M_2^* + M_2 \xrightarrow{k_{22}} M_2^* \qquad (6\text{-}5)$$

where k_{11} is the rate constant for a propagating chain ending in M_1 adding to monomer M_1, k_{12} that for a propagating chain ending in M_1 adding to monomer M_2, and so on. The propagation of a reactive center by addition of the same monomer (i.e., Reactions 6-2 and 6-5) is often referred to as *homopropagation* or *self-propagation*; propagation of a reactive center by addition of the other monomer (Reactions 6-3 and 6-4) is referred to as *cross-propagation* or a *crossover* reaction. All propagation reactions are assumed to be irreversible.

Monomer M_1 disappears by Reactions 6-2 and 6-4, while monomer M_2 disappears by Reactions 6-3 and 6-5. The rates of disappearance of the two monomers, which are synonymous with their rates of entry into the copolymer, are given by

$$-\frac{d[M_1]}{dt} = k_{11}[M_1^*][M_1] + k_{21}[M_2^*][M_1] \qquad (6\text{-}6)$$

$$-\frac{d[M_2]}{dt} = k_{12}[M_1^*][M_2] + k_{22}[M_2^*][M_2] \qquad (6\text{-}7)$$

Dividing Eq. 6-6 by Eq. 6-7 yields the ratio of the rates at which the two monomers enter the copolymer, that is, the copolymer composition, as

$$\frac{d[M_1]}{d[M_2]} = \frac{k_{11}[M_1^*][M_1] + k_{21}[M_2^*][M_1]}{k_{12}[M_1^*][M_2] + k_{22}[M_2^*][M_2]} \tag{6-8}$$

In order to remove the concentration terms in M_1^* and M_2^* from Eq. 6-8, a steady-state concentration is assumed for each of the reactive species M_1^* and M_2^* separately. For the concentrations of M_1^* and M_2^* to remain constant their rates of interconversion must be equal. In other words, the rates of Reactions 6-3 and 6-4 are equal,

$$k_{21}[M_2^*][M_1] = k_{12}[M_1^*][M_2] \tag{6-9}$$

Equation 6-9 can be rearranged and combined with Eq. 6-8 to yield

$$\frac{d[M_1]}{d[M_2]} = \frac{\dfrac{k_{11}k_{21}[M_2^*][M_1]^2}{k_{12}[M_2]} + k_{21}[M_2^*][M_1]}{k_{22}[M_2^*][M_2] + k_{21}[M_2^*][M_1]} \tag{6-10}$$

Dividing the top and bottom of the right side of Eq. 6-10 by $k_{21}[M_2^*][M_1]$ and combining the result with the parameters r_1 and r_2, which are defined by

$$r_1 = \frac{k_{11}}{k_{12}} \quad \text{and} \quad r_2 = \frac{k_{22}}{k_{21}} \tag{6-11}$$

one finally obtains

$$\frac{d[M_1]}{d[M_2]} = \frac{[M_1](r_1[M_1] + [M_2])}{[M_2]([M_1] + r_2[M_2])} \tag{6-12}$$

Equation 6-12 is known as the *copolymerization equation* or the *copolymer composition equation*. The copolymer composition, $d[M_1]/d[M_2]$, is the molar ratio of the two monomer units in the copolymer. $d[M_1]/d[M_2]$ is expressed by Eq. 6-12 as being related to the concentrations of the two monomers in the feed, $[M_1]$ and $[M_2]$, and the parameters r_1 and r_2. The parameters r_1 and r_2 are termed the *monomer reactivity ratios*. Each r as defined above in Eq. 6-11 is the ratio of the rate constant for a reactive propagating species adding its own type of monomer to the rate constant for its addition of the other monomer. The tendency of two monomers to copolymerize is noted by r values between zero and unity. An r_1 value greater than unity means that M_1^* preferentially adds M_1 instead of M_2, while an r_1 value less than unity means that M_1^* preferentially adds M_2. An r_1 value of zero would mean that M_1 is incapable of undergoing homopolymerization.

The copolymerization equation can also be expressed in terms of mole fractions instead of concentrations. If f_1 and f_2 are the mole fractions of monomers M_1 and M_2 in the feed, and F_1 and F_2 are the mole fractions of M_1 and M_2 in the copolymer, then

$$f_1 = 1 - f_2 = \frac{[M_1]}{[M_1] + [M_2]} \tag{6-13}$$

and

$$F_1 = 1 - F_2 = \frac{d[M_1]}{d[M_1] + d[M_2]} \qquad (6\text{-}14)$$

Combining Eqs. 6-13 and 6-14 with Eq. 6-12 yields

$$F_1 = \frac{r_1 f_1^2 + f_1 f_2}{r_1 f_1^2 + 2f_1 f_2 + r_2 f_2^2} \qquad (6\text{-}15)$$

Equation 6-15 gives the copolymer composition as the mole fraction of monomer M_1 in the copolymer and is often more convenient to use than the previous form (Eq. 6-12) of the copolymerization equation.

6-2b   Statistical Derivation of Copolymerization Equation

Although the above derivation involves the steady-state assumption, the copolymerization equation can also be obtained by a statistical approach without invoking steady-state conditions [Farina, 1990; Goldfinger and Kane, 1948; Melville et al., 1947; Tirrell, 1986; Vollmert, 1973]. We proceed to determine the number-average sequence lengths, $\bar{n}_1$ and $\bar{n}_2$, of monomers 1 and 2, respectively. $\bar{n}_1$ is the average number of M_1 monomer units that follow each other consecutively in a sequence uninterrupted by M_2 units but bounded on each end of the sequence by M_2 units. $\bar{n}_2$ is the average number of M_2 monomer units in a sequence uninterrupted by M_1 units but bounded on each end by M_1 units.

The transition or conditional probability p_{11} of forming a $M_1 M_1$ dyad in the copolymer chain is given by the ratio of the rate for M_1^* adding M_1 to the sum of the rates for M_1^* adding M_1 and M_2, that is,

$$p_{11} = \frac{R_{11}}{R_{11} + R_{12}} \qquad (6\text{-}16)$$

where R_{11} and R_{12} are the rates of Reactions 6-2 and 6-3, respectively. Substitution of the expressions for R_{11} and R_{12} into Eq. 6-16 yields

$$p_{11} = \frac{r_1}{r_1 + ([M_2]/[M_1])} \qquad (6\text{-}17)$$

Similarly, the transition probabilities p_{12}, p_{21}, and p_{22} for forming the dyads $M_1 M_2$, $M_2 M_1$, and $M_2 M_2$, respectively, are given by

$$p_{12} = \frac{R_{12}}{R_{11} + R_{12}} = \frac{[M_2]}{r_1[M_1] + [M_2]} \qquad (6\text{-}18)$$

$$p_{21} = \frac{R_{21}}{R_{21} + R_{22}} = \frac{[M_1]}{r_2[M_2] + M_1} \qquad (6\text{-}19)$$

$$p_{22} = \frac{R_{22}}{R_{21} + R_{22}} = \frac{r_2[M_2]}{r_2[M_2] + [M_1]} \qquad (6\text{-}20)$$

where R_{21} and R_{22} are the rates of Reactions 6-4 and 6-5, respectively.

The sum of the transition probabilities of addition to M_1^* and M_2^* are each, separately, equal to 1:

$$p_{11} + p_{12} = 1 \tag{6-21a}$$

$$p_{21} + p_{22} = 1 \tag{6-21b}$$

The number-average sequence length $\bar{n}_1$ of monomer M_1 is

$$\bar{n}_1 = \sum_{x=1}^{x=\infty} x(\underline{N}_1)_x = (\underline{N}_1)_1 + 2(\underline{N}_1)_2 + 3(\underline{N}_1)_3 + 4(\underline{N}_1)_4 + \cdots \tag{6-22}$$

where $(\underline{N}_1)_x$ is the mole fraction of a sequence of M_1 units of length x and the summation is over all sized sequence lengths from 1 to infinity. $(\underline{N}_1)_x$ is the probability of forming such a sequence and is given by

$$(\underline{N}_1)_x = (p_{11})^{(x-1)}p_{12} \tag{6-23}$$

The meaning of Eq. 6-23 can be seen by considering, for example, the probability of forming a sequence $M_1M_1M_1$. The probability of forming such a sequence is the probability p_{11} of M_1^* adding M_1 multiplied by the probability p_{11} of a second addition of M_1 multiplied by the probability p_{11} of a third addition of M_1 multiplied by the probability p_{12} of addition of M_2 or $p_{11}^3 p_{12}$.

$$\bar{n}_1 = p_{12}(1 + 2p_{11} + 3p_{11}^2 + 4p_{11}^3 + \cdots) \tag{6-24}$$

For $p_{11} < 1$, which holds in a copolymerization, the expansion series in Eq. 6-24 is $1/(1 - p_{11})^2$ and Eq. 6-24 becomes

$$\bar{n}_1 = \frac{p_{12}}{(1 - p_{11})^2} = \frac{1}{p_{12}} = \frac{r_1[M_1] + [M_2]}{[M_2]} \tag{6-25}$$

In a similar manner one obtains

$$\bar{n}_2 = \frac{p_{21}}{(1 - p_{22})^2} = \frac{1}{p_{21}} = \frac{r_2[M_2] + [M_1]}{[M_1]} \tag{6-26}$$

The mole ratio of monomers M_1 and M_2 contained in the copolymer is given by the ratio of the two number-average sequence lengths,

$$\frac{\bar{n}_1}{\bar{n}_2} = \frac{d[M_1]}{d[M_2]} = \frac{[M_1](r_1[M_1] + [M_2])}{[M_2]([M_1] + r_2[M_2])} \tag{6-27}$$

which is exactly the same result as Eq. 6-12 (and Eq. 6-15). Thus the copolymerization equation holds for copolymerizations carried out under both steady-state and non-steady-state conditions provided that the reactivity of a propagating species is dependent only on the ultimate unit, depropagation does not occur, and high polymer is formed. Equation 6-27 (also 6-12 and 6-15) describes the *first-order Markov* or *terminal model* of copolymerization.

6-2c Range of Applicability of Copolymerization Equation

The copolymerization equation has been experimentally verified in innumerable comonomer systems. The copolymerization equation is equally applicable to radical, cationic, and anionic chain copolymerizations, although the r_1 and r_2 values for any particular comonomer pair can be drastically different depending on the mode of initiation. Thus the r_1 and r_2 values for the comonomer pair of styrene (M_1) and methyl methacrylate (M_2) are 0.52 and 0.46 in radical copolymerization, 10 and 0.1 in cationic polymerization, and 0.1 and 6 in anionic copolymerization [Landler, 1950; Pepper, 1954]. Figure 6-1 shows that these different r_1 and r_2 values give rise to large differences in the copolymer composition depending on the mode of initiation. The ionic copolymerizations are predictably much more selective than radical copolymerization. Methyl methacrylate as expected shows increased reactivity in anionic copolymerization and decreased reactivity in cationic copolymerization, while the opposite is observed for styrene. Thus for an equimolar styrene–methyl methacrylate feed the copolymer is approximately a 1:1 copolymer in the radical case but is mostly styrene in the cationic copolymerization and mostly methyl methacrylate in the anionic copolymerization. The high selectivity of ionic copolymerization limits its practical use. Since only a small number of monomers undergo ionic copolymerization, the range of copolymer products that can be obtained is limited. On the other hand, almost all monomers undergo radical copolymerization and a wide range of products can be synthesized.

For any specific type of initiation (i.e., radical, cationic, or anionic) the monomer reactivity ratios and therefore the copolymer composition equation are independent of many reaction parameters. Since termination and initiation rate constants are not

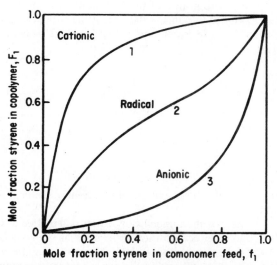

Fig. 6-1 Dependence of the instantaneous copolymer composition F_1 on the initial comonomer feed composition f_1 for styrene–methyl methacrylate in cationic (plot 1), radical (plot 2), and anionic (plot 3) copolymerizations initiated by SnCl$_4$, benzoyl peroxide, and Na/liquid NH$_3$, respectively. After Pepper [1954] (by permission of The Chemical Society, Burlington House, London) from data in Landler [1950] (by permission of Gauthier-Villars, Paris).

involved, the copolymer composition is independent of differences in the rates of initiation and termination or of the absence or presence of inhibitors or chain transfer agents. Under a wide range of conditions the copolymer composition is independent of the degree of polymerization. The only limitation on this generalization is that the copolymer be a high polymer. Further, the particular initiation system used in a radical copolymerization has no effect on copolymer composition. The same copolymer composition is obtained irrespective of whether initiation occurs by the thermal homolysis of initiators such as AIBN or peroxides, photolysis, radiolysis, or redox systems. However, the copolymer composition can be affected by the identity of the counterion in ionic copolymerizations (Sec. 6-4).

6-2d   Types of Copolymerization Behavior

Different types of copolymerization behavior are observed depending on the values of the monomer reactivity ratios. Copolymerizations can be classified into three types based on whether the product of the two monomer reactivity ratios $r_1 r_2$ is unity, less than unity, or greater than unity.

6-2d-1   Ideal Copolymerization: $r_1 r_2 = 1$

A copolymerization is termed *ideal* when the $r_1 r_2$ product is unity. Ideal copolymerization occurs when the two types of propagating species M_1^* and M_2^* show the same preference for adding one or the other of the two monomers. Under these conditions

$$\frac{k_{22}}{k_{21}} = \frac{k_{12}}{k_{11}} \quad \text{or} \quad r_2 = \frac{1}{r_1} \tag{6-28}$$

and the relative rates of incorporation of the two monomers into the copolymer are independent of the identity of the unit at the end of the propagating species. For an ideal copolymerization Eq. 6-28 is combined with Eq. 6-12 or 6-15 to yield the copolymerization equation as

$$\frac{d[M_1]}{d[M_2]} = \frac{r_1[M_1]}{[M_2]} \tag{6-29a}$$

or

$$F_1 = \frac{r_1 f_1}{r_1 f_1 + f_2} \tag{6-29b}$$

Most ionic copolymerizations (both anionic and cationic) are characterized by the ideal type of behavior.

When $r_1 = r_2 = 1$ the two monomers show equal reactivities toward both propagating species. The copolymer composition is the same as the comonomer feed with a random placement of the two monomers along the copolymer chain. Such behavior is referred to as *random* or *Bernoullian*. For the case where the two monomer reactivity ratios are different, that is, $r_1 > 1$ and $r_2 < 1$ or $r_1 < 1$ and $r_2 > 1$, one of the monomers is more reactive than the other toward both propagating species. The copolymer will contain a larger proportion of the more reactive monomer in random placement.

Figure 6-2 shows the variation in the copolymer composition as a function of the

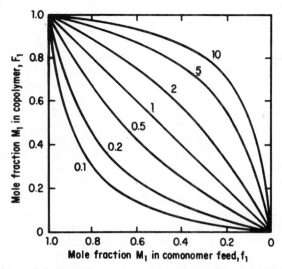

Fig. 6-2 Dependence of the instantaneous copolymer composition F_1 on the initial comonomer feed composition f_1 for the indicated values of r_1 where $r_1 r_2 = 1$. After Walling [1957] (by permission of Wiley, New York) from plot in Mayo and Walling [1950] (by permission of American Chemical Society, Washington, D.C.).

comonomer feed composition for different values of r_1 [Mayo and Walling, 1950]. The term *ideal copolymerization* is used to show the analogy between the curves in Fig. 6-2 and those for vapor-liquid equilibria in ideal liquid mixtures. The copolymer is richer in M_1 when $r_1 > 1$ and is poorer in M_1 when $r_1 < 1$. The term *ideal copolymerization* does not in any sense connote a desirable type of copolymerization. A very important practical consequence of ideal copolymerizations is that it becomes progressively more difficult to produce copolymers containing appreciable amounts of both monomers as the difference in reactivities of the two monomers increases. This is one of the reasons for the fact that ionic copolymerization is not of large practical significance. When, for example, $r_1 = 10$ and $r_2 = 0.1$, copolymers containing appreciable amounts of M_2 cannot be obtained. Thus a comonomer feed composition of 80 mole percent M_2 ($f_2 = 0.8$) would yield a copolymer containing only 18.5 mole percent M_2 ($F_2 = 0.185$). It is only when r_1 and r_2 do not differ markedly (for example, $r_1 = 0.5$-2) that there will exist a large range of comonomer feed compositions, which yield copolymers containing appreciable amounts of both monomers.

6-2d-2 Alternating Copolymerization: $r_1 = r_2 = 0$

When $r_1 = r_2 = 0$ (and $r_1 r_2 = 0$), the two monomers enter into the copolymer in equimolar amounts in a nonrandom, alternating arrangement along the copolymer chain. This type of copolymerization is referred to as *alternating copolymerization*. Each of the two types of propagating species preferentially adds the other monomer, that is, M_1^* adds only M_2 and M_2^* adds only M_1. The copolymerization equation reduces to

$$\frac{d[M_1]}{d[M_2]} = 1 \tag{6-30a}$$

or

$$F_1 = 0.5 \tag{6-30b}$$

The copolymer has the alternating structure **I** irrespective of the comonomer feed composition. Many radical copolymerizations show a tendency toward the alternation behavior.

The behavior of most comonomer systems lies between the two extremes of ideal and alternating copolymerization. As the $r_1 r_2$ product decreases from unity toward zero, there is an increasing *tendency toward alternation*. Perfect alternation occurs when r_1 and r_2 are both zero. The tendency toward alternation and the tendency away from ideal behavior increases as r_1 and r_2 become progressively less than unity. The range of behaviors can be seen by considering the situation where r_2 remains constant at 0.5 and r_1 varies between 2 and 0. Figure 6-3 shows the copolymer composition as a function of the feed composition in these cases. The curve for $r_1 = 2$ shows the ideal type of behavior described previously. As r_1 decreases below 2, there is an increasing tendency toward the alternating behavior with each type of propagating species preferring to add the other monomer. The increasing alternation tendency is measured by the tendency of the product $r_1 r_2$ to approach zero. Of great practical significance is the fact that a larger range of feed compositions will yield copolymers containing sizable amounts of both monomers. However, when $r_1 r_2$ is very small or zero the alternation tendency is too great and the range of copolymer compositions that can be obtained is again limited. In the extreme case where both r_1 and r_2 are zero only the 1:1 alternating copolymer can be produced. This would show in Fig. 6-3 as a horizontal line at $F_1 = 0.5$ (but the line would not touch either the left or right ordinates).

The curves in Fig. 6-3 illustrate an interesting characteristic of copolymerizations

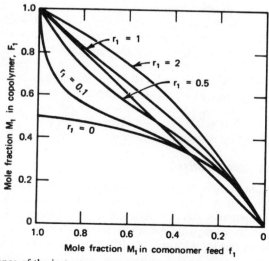

Fig. 6-3 Dependence of the instantaneous copolymer composition F_1 on the initial comonomer feed composition f_1 for the indicated values of r_1 with r_2 being constant at 0.5. After Walling [1957] (by permission of Wiley, New York) from plot in Mayo and Walling [1950] (by permission of American Chemical Society, Washington, D.C.).

with a tendency toward alternation. For values of r_1 and r_2 both less than unity the F_1 vs f_1 curves cross the line representing $F_1 = f_1$. At these intersections or *crossover points* the copolymer and feed compositions are the same and copolymerization occurs without a change in the feed composition. Such copolymerizations are termed *azeotropic copolymerizations*. The condition under which azeotropic copolymerization occurs, obtained by combination of Eq. 6-12 with $d[M_1]/d[M_2] = [M_1]/[M_2]$, is

$$\frac{[M_1]}{[M_2]} = \frac{(r_2 - 1)}{(r_1 - 1)} \tag{6-31a}$$

or

$$f_1 = \frac{(1 - r_2)}{(2 - r_1 - r_2)} \tag{6-31b}$$

A special situation arises when one of the monomer reactivity ratios is much larger than the other. For the case of $r_1 \gg r_2$ (that is, $r_1 \gg 1$ and $r_2 \ll 1$), both types of propagating species preferentially add monomer M_1. There is a tendency toward *consecutive homopolymerization* of the two monomers. Monomer M_1 tends to homopolymerize until it is consumed; monomer M_2 will subsequently homopolymerize. An extreme example of this type of behavior is shown by the radical polymerization of styrene–vinyl acetate with monomer reactivity ratios of 55 and 0.01. (See Secs. 6-3b-1 and 6-3c-1 for a further discussion of this commonomer system.)

6-2d-3 Block Copolymerization: $r_1 > 1$, $r_2 > 1$

If both r_1 and r_2 are greater than unity (and therefore, also $r_1 r_2 > 1$) there is a tendency to form a block copolymer (structure **II**) in which there are blocks of both monomers in the chain. This type of behavior has been encountered only in a few copolymerizations initiated by coordination catalysts (Sec. 8-4d-2). The extreme case of both r_1 and r_2 being much larger than unity—corresponding to the simultaneous and independent homopolymerizations of the two monomers—has not been observed except in one or two systems [Gumbs et al., 1969].

6-2e Variation of Copolymer Composition with Conversion

The various forms of the copolymerization equation (Eqs. 6-12 and 6-15) give the *instantaneous copolymer composition*—the composition of the copolymer formed from a particular feed composition at very low degrees of conversion (approximately < 5%) such that the composition of the comonomer feed is relatively unchanged from its initial value. For all copolymerizations except azeotropic copolymerizations the comonomer feed and copolymer compositions are different. The comonomer feed changes in composition as one of the monomers preferentially enters the copolymer. Thus there is a drift in the comonomer composition toward the less reactive monomer as the degree of conversion increases. This results in a similar variation of copolymer composition with conversion. In order to determine the instantaneous copolymer composition as a function of conversion for any given comonomer feed one must resort to an integrated form of the copolymerization equation. Direct integration of Eq. 6-12 for this purpose has resulted in solutions of varying utility [Kruse, 1967; Mayo and Lewis, 1944; Van der Meer et al., 1978; Wall, 1944].

The most generally useful method for analyzing copolymer composition as a function of conversion is that developed by Skeist [1946]. Consider a system initially containing a total of M moles of the two monomers and in which the copolymer formed is richer in monomer M_1 than is the feed (that is, $F_1 > f_1$). When dM moles of monomers have been copolymerized, the polymer will contain $F_1 d$M moles of monomer 1 and the feed will contain $(M - d\text{M})(f_1 - df_1)$ moles of monomer 1. A material balance for monomer 1 requires that the moles of M_1 copolymerized equal the difference in the moles of M_1 in the feed before and after reaction, or

$$M f_1 - (M - d\text{M})(f_1 - df_1) = F_1 d\text{M} \tag{6-32}$$

Equation 6-32 can be rearranged (neglecting the $df_1 d$M term, which is small) and converted to the integral form

$$\int_{M_0}^{M} \frac{d\text{M}}{M} = \ln \frac{M}{M_0} = \int_{(f_1)_0}^{f_1} \frac{df_1}{(F_1 - f_1)} \tag{6-33}$$

where M_0 and $(f_1)_0$ are the initial values of M and f_1.

Equation 6-15 allows the calculation of F_1 as a function of f_1 for a given set of r_1 and r_2 values. These can then be employed as $(F_1 - f_1)$ to allow the graphical or numerical integration of Eq. 6-33 between the limits of $(f_1)_0$ and f_1. In this manner one can obtain the variations in the feed and copolymer compositions with the degree of conversion (defined as $1 - M/M_0$).

Equation 6-33 has been integrated to the useful closed form

$$1 - \frac{M}{M_0} = 1 - \left[\frac{f_1}{(f_1)_0}\right]^\alpha \left[\frac{f_2}{(f_2)_0}\right]^\beta \left[\frac{(f_1)_0 - \delta}{f_1 - \delta}\right]^\gamma \tag{6-34}$$

which relates the degree of conversion to changes in the comonomer feed composition [Dionisio and O'Driscoll, 1979; Meyer and Chan, 1968; Meyer and Lowry, 1965]. The zero subscripts indicate initial quantities and the other symbols are given by

$$\alpha = \frac{r_2}{(1 - r_2)} \qquad \beta = \frac{r_1}{(1 - r_1)}$$

$$\gamma = \frac{(1 - r_1 r_2)}{(1 - r_1)(1 - r_2)} \qquad \delta = \frac{(1 - r_2)}{(2 - r_1 - r_2)} \tag{6-35}$$

Equation 6-34 or its equivalent has been used to correlate the drift in the feed and copolymer compositions with conversion for a number of different copolymerization systems [Capek et al., 1983; O'Driscoll et al., 1984; Stejskal et al., 1986; Teramachi et al., 1985].

A few examples will illustrate the utility of Eqs. 6-33 and 6-34. Figure 6-4 shows the behavior observed in the radical copolymerization of styrene and methyl methacrylate. F_1 and F_2 are the instantaneous copolymer compositions for the instantaneous feed compositions f_1 and f_2, respectively. The *average* or *cumulative composition* of the copolymer as a function of conversion is also shown. The copolymer produced is slightly richer in methyl methacrylate than the feed because methyl methacrylate has a slightly larger monomer reactivity ratio than styrene. The feed becomes richer in

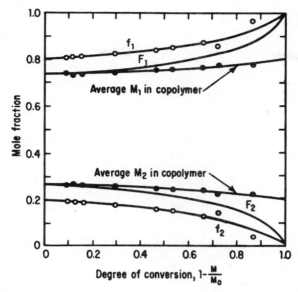

Fig. 6-4 Variations in feed and copolymer compositions with conversion for styrene (M_1)-methyl methacrylate (M_2) with $(f_1)_0 = 0.80$, $(f_2)_0 = 0.20$ and $r_1 = 0.53$, $r_2 = 0.56$. After Dionisio and O'Driscoll [1979] (by permission of Wiley, New York).

styrene with conversion—leading to an increase in the styrene content of the copolymer with conversion. Figure 6-4 also shows the average composition of all the copolymer formed up to some degree of conversion as a function of conversion. The average copolymer composition becomes richer in styrene than methyl methacrylate but less so than the instantaneous copolymer composition.

One can show the drift of copolymer composition with conversion for various comonomer feed compositions by a three-dimensional plot such as that in Fig. 6-5 for the radical copolymerization of styrene (M_1)-2-vinylthiophene (M_2). This is an ideal copolymerization with $r_1 = 0.35$ and $r_2 = 3.10$. The greater reactivity of the 2-vinylthiophene results in its being incorporated preferentially into the first-formed copolymer. As the reaction proceeds the feed and therefore the copolymer become progressively enriched in styrene. This is shown by Fig. 6-6, which describes the distribution of copolymer compositions at 100% conversion for several different initial feeds.

Corresponding data for the alternating radical copolymerization of styrene (M_1)-diethyl fumarate (M_2)($r_1 = 0.30$ and $r_2 = 0.07$) are shown in Figs. 6-7 and 6-8. This system undergoes azeotropic copolymerization at 57 mole percent styrene. Feed compositions near the azeotrope yield narrow distributions of copolymer composition except at high conversion where there is a drift to pure styrene or pure fumarate depending on whether the initial feed contains more or less than 57 mole percent styrene. The distribution of copolymer compositions becomes progressively wider as the initial feed composition differs more from the azeotropic composition.

In the commercial use of copolymerization it is desirable to obtain a copolymer with as narrow a distribution of compositions as possible, since polymer properties

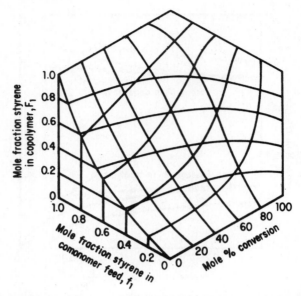

Fig. 6-5 Dependence of the instantaneous copolymer composition F_1 on the initial comonomer feed composition f_1 and the percent conversion of styrene (M_1)-2-vinylthiophene (M_2) with $r_1 = 0.35$ and $r_2 = 3.10$. After Walling [1957] (by permission of Wiley, New York) from plot in Mayo and Walling [1950] (by permission of American Chemical Society, Washington, D.C.).

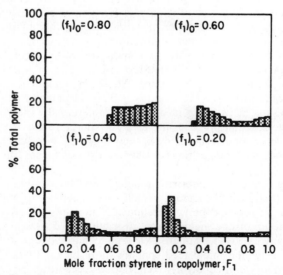

Fig. 6-6 Distribution of copolymer composition at 100% conversion for styrene–2-vinylthiophene at the indicated values of mole fraction styrene in the initial comonomer feed. After Billmeyer [1984] (by permission of Wiley-Interscience, New York) from plot in Mayo and Walling [1950] (by permission of American Chemical Society, Washington, D.C.).

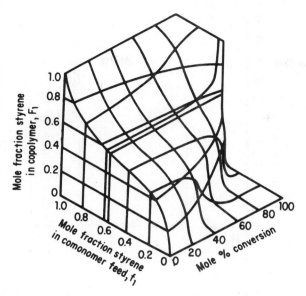

Fig. 6-7 Dependence of the instantaneous copolymer composition F_1 on the initial comonomer feed composition f_1 and the percent conversion of styrene (M_1)-diethyl fumarate (M_2) with r_1 = 0.30 and r_2 = 0.07. After Walling [1957] (by permission of Wiley, New York) from plot in Mayo and Walling [1950] (by permission of American Chemical Society, Washington, D.C.).

(and therefore utilization) are often highly dependent on copolymer composition [Athey, 1978]. Two approaches are simultaneously used to minimize heterogeneity in the copolymer composition. One is the choice of comonomers. Choosing a pair of monomers whose copolymerization behavior is such that F_1 is not too different from f_1 is highly desirable as long as that copolymer has the desired properties. The other approach is to maintain the feed concentration approximately constant by the batchwise or continuous addition of the more reactive monomer. The control necessary in maintaining f_1 constant depends on the extent to which the copolymer composition differs from the feed.

6-2f   Experimental Evaluation of Monomer Reactivity Ratios

Most procedures for evaluating r_1 and r_2 involve the experimental determination of the copolymer composition for several different comonomer feed compositions in conjunction with a differential form of the copolymerization equation (Eq. 6-12 or 6-15). Copolymerizations are carried out to as low degrees of conversion as possible (ca. < 5%) to minimize errors in the use of the differential equation. The copolymer composition is determined either directly by analysis of the copolymer or indirectly by analysis of comonomer feed. The techniques used for copolymer analysis include radioisotopic labeling and spectroscopy (IR, UV, NMR). Comonomer feed compositions are typically analyzed by high-pressure liquid chromatography (HPLC) or gas chromatography (GC). The techniques for analyzing comonomer compositions are inherently more sensitive than those for copolymer composition. However, this is

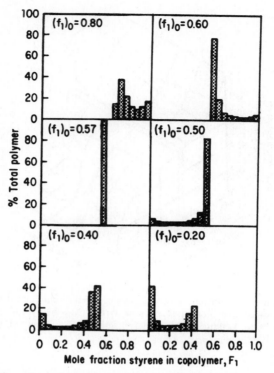

Fig. 6-8 Distribution of copolymer composition at 100% conversion for styrene–diethyl fumarate at the indicated values of mole fraction styrene in the initial comonomer feed. After Billmeyer [1984] (by permission of Wiley-Interscience, New York) from plot in Mayo and Walling [1950] (by permission of American Chemical Society, Washington, D.C.).

offset by the fact that the determination of copolymer composition by comonomer feed analysis requires the measurement of small differences between large numbers.

Various methods have been used to obtain monomer reactivity ratios from the copolymer composition data. The most often used method involves a rearrangement of the copolymer composition equation into a form linear in the monomer reactivity ratios. Mayo and Lewis [1944] rearranged Eq. 6-12 to

$$r_2 = \frac{[M_1]}{[M_2]}\left[\frac{d[M_2]}{d[M_1]}\left\{1 + \frac{r_1[M_1]}{[M_2]}\right\} - 1\right] \tag{6-36}$$

Data for the feed and copolymer compositions for each experiment with a given feed are substituted into Eq. 6-36 and r_2 is plotted as a function of various assumed values of r_1. Each experiment yields a straight line and the intersection of the lines for different feeds gives the best values of r_1 and r_2. Any variations observed in the points of intersection of various lines are a measure of the experimental errors in the composition data and the limitations of the mathematical treatment (see below). The composition data can also be treated by linear least-squares regression analysis instead of the graphical analysis.

Fineman and Ross [1950] rearranged Eq. 6-15 to

$$G = r_1 F - r_2 \tag{6-37}$$

where $G = X(Y - 1)/Y$, $F = X^2/Y$, $X = [M_1]/[M_2]$, and $Y = d[M_1]/d[M_2]$. G is plotted against F to yield a straight line with slope r_1 and intercept r_2.

The experimental composition data are unequally weighted by the Mayo–Lewis and Fineman–Ross plots with the data for the high or low compositions (depending on the equation used) having the greatest effect on the calculated values of r_1 and r_2 [Tidwell and Mortimer, 1965, 1970]. This often manifests itself by different values of r_1 and r_2 depending on which monomer is indexed as M_1.

Kelen and Tudos [1975] refined the linearization method by introducing an arbitrary positive constant α into Eq. 6-37 to spread the data more evenly so as to give equal weighing to all data points. Their results are expressed in the form

$$\eta = \left[r_1 + \frac{r_2}{\alpha} \right] \xi - \frac{r_2}{\alpha} \tag{6-38}$$

where

$$\eta = \frac{G}{\alpha + F} \tag{6-39a}$$

$$\xi = \frac{F}{\alpha + F} \tag{6-39b}$$

Plotting η against ξ gives a straight line that yields $-r_2/\alpha$ and r_1 as intercepts on extrapolation to $\xi = 0$ and $\xi = 1$, respectively. The value of α, chosen as $\alpha = (F_m F_M)^{1/2}$ where F_m and F_M are the lowest and highest F values, respectively, distributes the experimental data symmetrically on the plot.

Even with the Kelen–Tudos refinement there are statistical limitations inherent in the linearization method. The independent variable in any form of the linear equation is not really independent, while the dependent variable does not have a constant variance [O'Driscoll and Reilly, 1987]. The most statistically sound method of analyzing composition data is the nonlinear method, which involves plotting the instantaneous copolymer composition versus comonomer feed composition for various feeds and then determining which theoretical curve best fits the data by trial-and-error selection of r_1 and r_2 values. The pros and cons of the two methods have been discussed in detail, along with approaches for the best choice of feed compositions to maximize the accuracy of the r_1 and r_2 values [Bataille and Bourassa, 1989; Hautus et al., 1984; Kelen and Tudos, 1990; Leicht and Fuhrmann, 1983; Tudos and Kelen, 1981].

The use of a differential form of the copolymerization equation assumes that the feed composition does not change during the experiment, but this is not true. One holds the conversion as low as possible, but there are limitations since one must be able to isolate a sample of the copolymer for direct analysis (or, if copolymer analysis is via the change in feed composition, there must be a significantly measurable change in the feed composition). The use of an integrated form of the copolymerization equation does not have this limitation [Hamielec et al., 1989; O'Driscoll and Reilly, 1987; Plaumann and Branston, 1989]. The change in copolymer composition or feed composition with conversion is measured and the results curve fitted to an integrated form of the copolymerization equation. For example, one method uses the previously described Eq. 6-34 [Meyer, 1966]. Experimental data on the variation of feed composition with conversion are plotted as f_1 or f_2 versus $1\text{-}M/M_0$ to yield curves like those in Fig. 6-4. Using computational procedures, one then determines the best values of r_1 and r_2 that must be used with Eq. 6-34 to yield the same curve.

6-2g Microstructure of Copolymers

6-2g-1 Sequence-Length Distribution

The copolymerization equation describes the copolymer composition on a macroscopic scale, that is, the overall composition of a copolymer sample produced from a comonomer feed. This leaves unanswered two details concerning molecular-level composition of the copolymer. The first concerns the exact arrangement of the two monomers along the polymer chain. Although the copolymers produced by copolymerization are often referred to as *random copolymers*, this term is inappropriate, since it indicates that the two monomer units are randomly placed along the copolymer chain. This only occurs for the case $r_1 = r_2 = 1$. There is a definite trend toward a regular microstructure for all other cases. For example, if $r_1 > 1$, once a propagating species of type M_1^* is formed it will tend to a sequence of M_1 units. M_1^* will tend to add the other monomer unit if $r_1 < 1$. However, there is a random aspect to copolymerization due to the probabilistic nature of chemical reactions. Thus, an r_1 value of 2 does not imply that 100% of all M_1 units will be found as part of $M_1M_1M_1M_2$ sequences. A very large fraction, but not all, M_1 units will be found in such a sequence; a small fraction of M_1 units will be randomly distributed.

The microstructure of a copolymer is defined by the distributions of the various lengths of the M_1 and M_2 sequences, that is, the *sequence-length distributions*. The probabilities or mole fractions $(\underline{N}_1)_x$ and $(\underline{N}_2)_x$ of forming M_1 and M_2 sequences of length x are given (Sec. 6-2b) by

$$(\underline{N}_1)_x = (p_{11})^{(x-1)}p_{12} \tag{6-23}$$

$$(\underline{N}_2)_x = (p_{22})^{(x-1)}p_{21} \tag{6-40}$$

where the p values are defined by Eqs. 6-17 through 6-20. Equations 6-23 and 6-40 allow one to calculate the mole fractions of different lengths of M_1 and M_2 sequences. Distributions for a number of copolymerization systems have been described [Tirrell, 1986; Tosi, 1967–1968; Vollmert, 1973]. Figure 6-9 shows the sequence-length distribution for an ideal copolymerization with $r_1 = r_2 = 1$ for an equimolar feed composition ($f_1 = 0.5$). For this system $p_{11} = p_{12} = p_{22} = p_{21} = 0.50$. Although the most plentiful sequence is M_1 at 50%, there are considerable amounts of other sequences: 25%, 12.5%, 6.25%, 3.13%, 1.56%, and 0.78%, respectively, of dyad, triad, tetrad, pentad, hexad, and heptad sequences. The distribution of M_2 sequences is exactly the same as for M_1 sequences. For a feed composition other than equimolar, the distribution becomes narrower for the monomer present in lower amount and broader for the monomer in larger amount—a general phenomenon observed for all sequence-length distributions. It is clear that the copolymer, with an overall composition of $F_1 = F_2 = 0.50$, has a microstructure that is very different from that of a perfectly alternating copolymer.

The sequence-length distribution for an ideal copolymerization with $r_1 = 5, r_2 =$

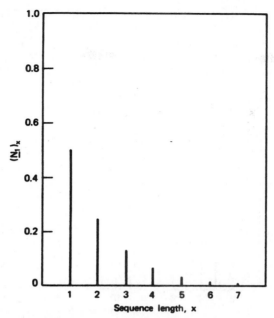

Fig. 6-9 Sequence-length distribution for an ideal copolymerization with $r_1 = r_2 = 1$ and $f_1 = f_2$. After Vollmert [1973] (by permission of Springer-Verlag New York, Inc., New York).

0.2 for an equimolar feed composition is shown in Fig. 6-10. This copolymerization has $p_{11} = p_{21} = 0.8333$ and $p_{12} = p_{22} = 0.1667$. Both M_1^* and M_2^* propagating centers have a 5:1 tendency to add M_1 over M_2, but M_1 pentad sequences are not the most plentiful, although they are among the most plentiful. The most plentiful sequence is M_1 at 16.7% with 14.0%, 11.6%, 9.7%, 8.1%, 6.7%, 5.6%, and 4.7%, respectively, of dyad, triad, tetrad, pentad, hexad, heptad, and octad M_1 sequences. There are smaller amounts of longer sequences: 3.2% of 10-unit, 1.3% of 15-unit, and 0.4% of 20-unit M_1 sequences. The sequence-length distribution is much narrower for the less reactive M_2 monomer. Single M_1 units are by far the most plentiful (83.3%) with 13.9% dyads, 2.3% triads, and 0.39% tetrads.

An alternating copolymerization with $r_1 = r_2 = 0.1$ and $f_1 = f_2$ has the sequence-length distribution shown in Fig. 6-11 ($p_{11} = p_{22} = 0.0910$; $p_{12} = p_{21} = 0.9090$). The sequence-length distributions for both monomer units are identical. The single M_1 and single M_2 sequences are overwhelmingly the most plentiful—the M_1M_2 comprises 90.9% of the copolymer structure. There are 8.3% and 0.75%, respectively, of dyad and triad sequences of both M_1 and M_2. Compare this distribution with that in Fig. 6-9 for the ideal copolymer having the identical overall composition. The large difference in the distributions for the two copolymers clearly indicates the difference between alternating and ideal behavior.

High-resolution nuclear magnetic resonance spectroscopy, especially ^{13}C NMR, is a powerful tool for analysis of copolymer microstructure [Bailey and Henrichs, 1978; Bovey, 1972; Randall, 1977, 1989; Randall and Ruff, 1988]. The predicted sequence-length distributions have been verified in a number of comonomer systems. Copolymer microstructure also gives an alternate method for evaluation of monomer reactivity

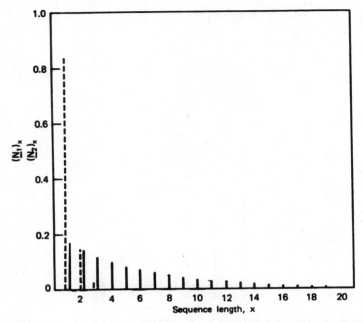

Fig. 6-10 Sequence-length distribution for an ideal copolymerization with $r_1 = 5$, $r_2 = 0.2$ and $f_1 = f_2$. $(\underline{N}_1)_x$ is represented by —; $(\underline{N}_2)_x$ by ---. The plots for $(\underline{N}_2)_x$ are shown slightly to the left of the actual sequence-length. After Vollmert [1973] (by permission of Springer-Verlag New York, Inc., New York).

ratios [Randall, 1977]. The method follows that described in Sec. 8-10 for stereochemical microstructure. For example, for the terminal model, the mathematical equations from Sec. 8-10-2 apply except that P_{mm}, P_{mr}, P_{rm}, and P_{rr} are replaced by p_{11}, p_{12}, p_{21}, and p_{22}.

6-2g-2 Copolymer Compositions of Different Molecules

A second uncertainty concerning copolymer composition is the distribution of composition from one copolymer molecule to another in a sample produced at any given degree of conversion [Galvan and Tirrell, 1986; Tacx et al., 1988]. Stockmayer [1945] indicated that the distribution of copolymer composition due to statistical fluctuations generally follows a very sharp Gaussian curve. Although the distribution is wider for low-molecular-weight copolymers and for ideal copolymerizations compared to alternating copolymerizations, it is relatively narrow in all practical cases. Thus it is calculated that for an ideal copolymer containing an average of 50 mole percent of each component, only 12% of the copolymer molecules contain less than 43% of either monomer for $\overline{X}_n = 100$, while only 12% contain less than 49% of either monomer at $\overline{X}_n = 10,000$. These theoretical conclusions of Stockmayer have been experimentally verified in a limited manner [Phillips and Carrick, 1962a, 1962b].

6-2h Multicomponent Copolymerization

Terpolymerization, the simultaneous polymerization of three monomers, has become increasingly important from the commercial viewpoint. The improvements that are

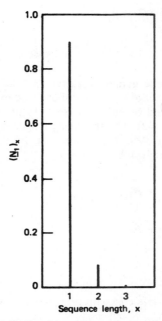

Fig. 6-11 Sequence-length distribution for an alternating copolymerization with $r_1 = r_2 = 0.1$ and $f_1 = f_2$. After Vollmert [1973] (by permission of Springer-Verlag New York, Inc., New York).

obtained by copolymerizing styrene with acrylonitrile or butadiene have been mentioned previously. The radical terpolymerization of styrene with acrylonitrile and butadiene increases even further the degree of variation in properties that can be built into the final product. Many other commercial uses of terpolymerization exist. In most of these the terpolymer has two of the monomers present in major amounts to obtain the gross properties desired, with the third monomer in a minor amount for modification of a special property. Thus the ethylene-propylene elastomers are terpolymerized with minor amounts of a diene in order to allow the product to be subsequently crosslinked.

The quantitative treatment of terpolymerization is quite complex, since nine propagation reactions,

Reaction	Rate
$M_1 \cdot + M_1 \rightarrow M_1 \cdot$	$R_{11} = k_{11}[M_1 \cdot][M_1]$
$M_1 \cdot + M_2 \rightarrow M_2 \cdot$	$R_{12} = k_{12}[M_1 \cdot][M_2]$
$M_1 \cdot + M_3 \rightarrow M_3 \cdot$	$R_{13} = k_{13}[M_1 \cdot][M_3]$
$M_2 \cdot + M_1 \rightarrow M_1 \cdot$	$R_{21} = k_{21}[M_2 \cdot][M_1]$
$M_2 \cdot + M_2 \rightarrow M_2 \cdot$	$R_{22} = k_{22}[M_2 \cdot][M_2]$
$M_2 \cdot + M_3 \rightarrow M_3 \cdot$	$R_{23} = k_{23}[M_2 \cdot][M_3]$
$M_3 \cdot + M_1 \rightarrow M_1 \cdot$	$R_{31} = k_{31}[M_3 \cdot][M_1]$
$M_3 \cdot + M_2 \rightarrow M_2 \cdot$	$R_{32} = k_{32}[M_3 \cdot][M_2]$
$M_3 \cdot + M_3 \rightarrow M_3 \cdot$	$R_{33} = k_{33}[M_3 \cdot][M_3]$

(6-41)

and six monomer reactivity ratios,

$$r_{12} = \frac{k_{11}}{k_{12}}, \; r_{13} = \frac{k_{11}}{k_{13}}, \; r_{21} = \frac{k_{22}}{k_{21}}, \; r_{23} = \frac{k_{22}}{k_{23}}, \; r_{31} = \frac{k_{33}}{k_{31}}, \; r_{32} = \frac{k_{33}}{k_{32}} \qquad (6\text{-}42)$$

are involved (as well as six termination reactions). The expression for the rate R of each of the propagation reactions is shown above. An expression for the terpolymer composition can be obtained by either the steady-state or statistical approach used for copolymerization (Sec. 6-2). The steady-state approach is described here [Alfrey and Goldfinger, 1944; Ham, 1964, 1966; Mirabella, 1977]. The rates of disappearance of the three monomers are given by

$$-\frac{d[M_1]}{dt} = R_{11} + R_{21} + R_{31} \qquad (6\text{-}43a)$$

$$-\frac{d[M_2]}{dt} = R_{12} + R_{22} + R_{32} \qquad (6\text{-}43b)$$

$$-\frac{d[M_3]}{dt} = R_{13} + R_{23} + R_{33} \qquad (6\text{-}43c)$$

The assumption of steady-state concentrations for $M_1\cdot$, $M_2\cdot$, and $M_3\cdot$ radicals can be expressed as

$$R_{12} + R_{13} = R_{21} + R_{31} \qquad (6\text{-}44a)$$

$$R_{21} + R_{23} = R_{12} + R_{32} \qquad (6\text{-}44b)$$

$$R_{31} + R_{32} = R_{13} + R_{23} \qquad (6\text{-}44c)$$

Combination of Eqs. 6-43 with 6-44, and the use of the appropriate rate expressions from Eq. 6-41 for each R term, yields the terpolymer composition as

$$d[M_1]:d[M_2]:d[M_3] =$$

$$[M_1] \left\{ \frac{[M_1]}{r_{31}r_{21}} + \frac{[M_2]}{r_{21}r_{32}} + \frac{[M_3]}{r_{31}r_{23}} \right\} \left\{ [M_1] + \frac{[M_2]}{r_{12}} + \frac{[M_3]}{r_{13}} \right\}$$

$$:[M_2] \left\{ \frac{[M_1]}{r_{12}r_{31}} + \frac{[M_2]}{r_{12}r_{32}} + \frac{[M_3]}{r_{32}r_{13}} \right\} \left\{ [M_2] + \frac{[M_1]}{r_{21}} + \frac{[M_3]}{r_{23}} \right\}$$

$$:[M_3] \left\{ \frac{[M_1]}{r_{13}r_{21}} + \frac{[M_2]}{r_{23}r_{12}} + \frac{[M_3]}{r_{13}r_{23}} \right\} \left\{ [M_3] + \frac{[M_1]}{r_{31}} + \frac{[M_2]}{r_{32}} \right\} \qquad (6\text{-}45)$$

A simpler expression for the terpolymer composition has been obtained by expressing the steady state with the relationships

$$R_{12} = R_{21} \qquad (6\text{-}46a)$$

$$R_{23} = R_{32} \qquad (6\text{-}46b)$$

$$R_{31} = R_{13} \qquad (6\text{-}46c)$$

instead of those in Eq. 6-44 [Valvassori and Sartori, 1967]. The combination of Eqs. 6-46 and 6-43 yields the terpolymer composition as

$$d[M_1]:d[M_2]:d[M_3] = [M_1]\left\{[M_1] + \frac{[M_2]}{r_{12}} + \frac{[M_3]}{r_{13}}\right\}$$

$$:[M_2]\frac{r_{21}}{r_{12}}\left\{\frac{[M_1]}{r_{21}} + [M_2] + \frac{[M_3]}{r_{23}}\right\}$$

$$:[M_3]\frac{r_{31}}{r_{13}}\left\{\frac{[M_1]}{r_{31}} + \frac{[M_2]}{r_{32}} + [M_3]\right\} \tag{6-47}$$

The conventional (Eq. 6-45) and simplified (Eq. 6-47) terpolymerization equations can be used to predict the composition of a terpolymer from the reactivity ratios in the two component systems M_1/M_2, M_1/M_3, and M_2/M_3 [Braun and Cei, 1987; Disselhoff, 1978; Janovic et al., 1983; Walling and Briggs, 1945]. Table 6-1 shows the predicted and experimental terpolymer compositions in several systems of three monomers. The compositions calculated by either of the terpolymerization equations show good agreement with the experimentally observed compositions. Neither equation is superior to the other in predicting terpolymer compositions. Both equations have been successfully extended to multicomponent copolymerizations of four or more monomers [Alfrey and Goldfinger, 1946; Ham, 1964; Tomescu and Simionescu, 1976; Valvassori and Sartori, 1967; Walling and Briggs, 1945]. The general agreement of the calculated and experimental compositions for a system of four monomers is also shown in Table 6-1.

TABLE 6-1   Predicted and Experimental Compositions in Radical Terpolymerization[a]

| | Feed Composition | | Terpolymer (mole percent) | | |
| | | | | Calculated from | |
System	Monomer	Mole percent	Found	Eq. 6-45	Eq. 6-47
1	Styrene	31.24	43.4	44.3	44.3
	Methyl methacrylate	31.12	39.4	41.2	42.7
	Vinylidene chloride	37.64	17.2	14.5	13.0
2	Methyl methacrylate	35.10	50.8	54.3	56.6
	Acrylonitrile	28.24	28.3	29.7	23.5
	Vinylidene chloride	36.66	20.9	16.0	19.9
3	Styrene	34.03	52.8	52.4	53.8
	Acrylonitrile	34.49	36.7	40.5	36.6
	Vinylidene chloride	31.48	10.5	7.1	9.6
4	Styrene	35.92	44.7	43.6	45.2
	Methyl methacrylate	36.03	26.1	29.2	33.8
	Acrylonitrile	28.05	29.2	26.2	21.0
5	Styrene	20.00	55.2	55.8	55.8
	Acrylonitrile	20.00	40.3	41.3	41.4
	Vinyl chloride	60.00	4.5	2.9	2.8
6	Styrene	25.21	40.7	41.0	41.0
	Methyl methacrylate	25.48	25.5	27.3	29.3
	Acrylonitrile	25.40	25.8	24.8	22.8
	Vinylidene chloride	23.91	8.0	6.9	6.9

[a]Data and calculations from Valvassori and Sartori [1967] and Walling and Briggs [1945].

The terpolymerization and multicomponent composition equations are generally valid only when all of the monomer reactivity ratios have finite values. When one or more of the monomers is incapable of homopolymerization the equations generally become indeterminate. Various modified expressions based on both the conventional and simplified equations have been derived for these and other special cases [Chien and Finkenaur, 1985; Quella, 1989].

6-3 RADICAL COPOLYMERIZATION

The discussions thus far have been quite general without any specification as to whether copolymerization occurs by radical or ionic propagation. Consider now some of the specific characteristics of radical copolymerization.

6-3a Effect of Reaction Conditions

6-3a-1 Reaction Medium

Monomer reactivity ratios are generally but not always independent of the reaction medium in radical copolymerization. There is a real problem here in that the accuracy of r values is often insufficient to allow one to reasonably conclude whether r_1 or r_2 vary with changes in reaction media. The more recent determinations of r values by high-resolution NMR are much more reliable than previous data for this purpose. It has been observed that the experimentally determined monomer reactivity ratios are affected by the reaction medium in certain systems. This occurs in some but not all polymerizations carried out under nonhomogeneous conditions. Emulsion and suspension polymerizations occasionally show copolymer compositions different from those in bulk or solution polymerization when the comonomer feed composition at the reaction site (monomer droplet, micelle) is different from that in the bulk of the reaction system [Doiuchi and Minoura, 1977; Plochocka, 1981]. This can occur in emulsion polymerization if the relative solubilities of the two monomers in the micelles are different from the feed composition in the dispersing medium or diffusion of one of the monomers into the micelles is too slow. Such behavior has been observed in several systems, such as styrene copolymerization with acrylonitrile or itaconic acid [Fordyce and Chapin, 1947; Fordyce and Ham, 1948]. The phenomenon also occurs in suspension polymerization if one of the monomers has appreciable solubility in the dispersing medium. The monomer reactivity ratios are unchanged in these systems. The apparent discrepancies are simply due to altered f_1 and f_2 values at the reaction sites. If the correct f_1 and f_2 values are determined and used in calculations of expected copolymer compositions, the discrepancies disappear [Smith, 1948].

Deviations are also observed in some copolymerizations where the copolymer formed is poorly soluble in the reaction medium [Pichot and Pham, 1979; Pichot et al., 1979; Suggate, 1978, 1979]. Under these conditions, altered copolymer compositions are observed if one of the monomers is preferentially adsorbed by the copolymer. Thus for methyl methacrylate (M_1)-N-vinylcarbazole (M_2) copolymerization, $r_1 = 1.80$, $r_2 = 0.06$ in benzene but $r_1 = 0.57$, $r_2 = 0.75$ in methanol [Ledwith et al., 1979]. The propagating copolymer chains are completely soluble in benzene but are microheterogeneous in methanol. N-vinylcarbazole (NVC) is preferentially adsorbed by the copolymer compared to methyl methacrylate. The comonomer composition in the domain of the propagating radical sites (trapped in the precipitating

copolymer) is richer in NVC than the comonomer feed composition in the bulk solution. NVC enters the copolymer to a greater extent than expected based on the feed
composition.

Some effect of viscosity on r has been observed [Kelen and Tudos, 1974; Rao et
al., 1976]. Copolymerization of styrene (M_1)-methyl methacrylate (M_2) in bulk leads
to a copolymer containing less styrene than when reaction is carried out in benzene
solution [Johnson et al., 1978]. The gel effect in bulk polymerization decreases the
mobility of styrene resulting in a decrease in r_1 and an increase in r_2.

The monomer reactivity ratio for an acidic or basic monomer shows a dependence
on pH since the identity of the monomer changes with pH. For example, acrylic acid
(M_1)-acrylamide (M_2) copolymerization shows $r_1 = 0.90$, $r_2 = 0.25$ at pH $= 2$ but
$r_1 = 0.30$, $r_2 = 0.95$ at pH $= 9$ [Ponratnam and Kapur, 1977; Truong et al., 1986].
Acrylic acid exists as the acrylate anion at high pH. Acrylate anion shows a decreased
tendency to homopropagate as well as add to propagating centers with electron-rich
substituents such as the amide group. This results in a decrease in r_1 and an increase
in r_2. A related phenomenon is the increase in the monomer reactivity ratio for ethyl
3-oxo-4-pentenoate when it copolymerizes with styrene in a nonpolar solvent compared
to a polar solvent [Masuda et al., 1987a, 1987b]. Ethyl 3-oxo-4-pentenoate exists in
a keto–enol equilibrium with the concentration of enol increasing with solvent polarity.
The enol has a higher reactivity compared to the keto form, and this results in a
copolymer richer in ethyl 3-oxo-4-pentenoate for copolymerization in nonpolar solvents.

Copolymerizations involving the combination of polar (M_1) and nonpolar (M_2)
monomers often show different behavior depending on the polarity of the reaction
medium. The copolymer composition is richer in the less polar monomer for reaction
in a polar (either aprotic or protic) solvent compared to nonpolar solvent. Calculations
of monomer reactivity ratios show a decrease in r_1 usually coupled with an increase
in r_2 for copolymerization in the polar solvent relative to values in the nonpolar solvent.
This behavior has been observed in systems such as styrene with acrylamide, acrylonitrile, 2-hydroxyethyl methacrylate, acrylic acid, or methacrylic acid, and methacrylic
acid with methyl methacrylate [Boudevska and Todorova, 1985; Harwood, 1987; Lebduska et al., 1986; Plochocka, 1981]. A variety of explanations have been put forth
to describe the results. These include complexation of polar solvent with polar monomer to decrease the latter's reactivity and ionization of acidic monomers to yield the
alkanoate monomers, which possess lower reactivity. Partitioning of the two monomers
between the bulk solution and the domain of a growing polymer radical is indicated
by the observation that the copolymers produced in different solvents have the same
microstructure when copolymers of the same composition are compared. Copolymers
are richer in the less polar monomer for reaction in polar solvents because of the feed
composition in the domain of the growing polymer radical being richer in the less
polar monomer.

6-3a-2 Temperature

An examination of various compilations of monomer reactivity ratios [Greenley, 1989a;
Young, 1975] shows that r_1 and r_2 are relatively insensitive to temperature provided
that propagation is irreversible. Thus the r_1 and r_2 values for styrene-1,3-butadiene
are 0.64 and 1.4 at 5°C, and 0.60 and 1.8 at 45°C. The r_1 and r_2 values for styrene–
methyl methacrylate are 0.52 and 0.46 at 60°C, and 0.59 and 0.54 at 131°C. The
monomer reactivity ratio is the ratio of two propagation rate constants, and its variation

with temperature will depend on the difference in propagation activation energies according to

$$r_1 = \frac{k_{11}}{k_{12}} = \frac{A_{11}}{A_{12}} \exp \left[\frac{(E_{12} - E_{11})}{RT} \right] \tag{6-48}$$

where E_{11} and A_{11} are the propagation activation energy and frequency factor for M_1 radical adding M_1 monomer, respectively, and E_{12} and A_{12} are the corresponding values for M_1 radical adding M_2 monomer.

The effect of temperature on r is not large, since activation energies for radical propagation are relatively small and, more significantly, fall in a narrow range such that $E_{12} - E_{11}$ is less than 10 kJ/mole for most pairs of monomers. However, temperature does have an effect, since $E_{12} - E_{11}$ is not zero. An increase in temperature results in a less selective copolymerization as the two monomer reactivity ratios of a comonomer pair each tend toward unity with decreasing preference of either radical for either monomer. Temperature has the greatest effect on those systems for which the r values deviate markedly from unity, behavior which is much more typical of ionic copolymerization than radical copolymerization.

6-3a-3 Pressure

The monomer reactivity ratio varies with pressure according to

$$\frac{d \ln r_1}{dP} = \frac{-(\Delta V^{\ddagger}_{11} - \Delta V^{\ddagger}_{12})}{RT} \tag{6-49}$$

where $\Delta V^{\ddagger}_{11}$ and $\Delta V^{\ddagger}_{12}$ are the propagation activation volumes for radical M_1 adding monomers M_1 and M_2, respectively [Burkhart and Zutty, 1962; Jenner, 1979; Jenner and Aieche, 1978; Van Der Meer et al., 1977a, 1977b]. Although propagation rates increase considerably with pressure, r is less sensitive to pressure, since $(\Delta V^{\ddagger}_{11} - \Delta V^{\ddagger}_{12})$ is smaller than either $\Delta V^{\ddagger}_{11}$ or $\Delta V^{\ddagger}_{12}$. The effect of pressure is in the same direction as that of temperature. Increasing pressure tends to decrease the selectivity of the copolymerization as the r values change in the direction of ideal copolymerization behavior. Thus the $r_1 r_2$ product for styrene-acrylonitrile changes from 0.026 at 1 atm to 0.077 at 1000 atm, while $r_1 r_2$ for methyl methacrylate-acrylonitrile changes from 0.16 to 0.91. Copolymerizations which are ideal at lower pressure remain so at higher pressure.

6-3b Reactivity

The monomer reactivity ratios for many of the most common monomers in radical copolymerization are shown in Table 6-2. These data are useful for a study of the relation between structure and reactivity in radical addition reactions. The reactivity of a monomer toward a radical depends on the reactivities of both the monomer and the radical. The relative reactivities of monomers and their corresponding radicals can be obtained from an analysis of the monomer reactivity ratios [Walling, 1957]. The reactivity of a monomer can be seen by considering the inverse of the monomer reactivity ratio $(1/r)$. The inverse of the monomer reactivity ratio gives the ratio of the rate of reaction of a radical with another monomer to its rate of reaction with its

TABLE 6-2 Monomer Reactivity Ratios in Radical Copolymerization

M_1	r_1	M_2	r_2	T (°C)
Acrylic acid	0.24	n-Butyl methacrylate	3.5	50
	0.25	Styrene	0.15	60
	8.7	Vinyl acetate	0.21	70
Acrylonitrile	0.86	Acrylamide	0.81	40
	0.046	1,3-Butadiene	0.36	40
	0.69	Ethyl vinyl ether	0.060	80
	0.98	Isobutylene	0.020	50
	1.5	Methyl acrylate	0.84	50
	0.14	Methyl methacrylate	1.3	70
	0.020	Styrene	0.29	60
	5.5	Vinyl acetate	0.060	70
	3.6	Vinyl chloride	0.044	50
	0.92	Vinylidene chloride	0.32	60
	0.020	2-Vinylpyridine	0.43	60
	0.11	4-Vinylpyridine	0.41	60
Allyl acetate	0	Methyl methacrylate	23	60
	0	Styrene	90	60
	0.70	Vinyl acetate	1.0	60
1,3-Butadiene	0.75	Methyl methacrylate	0.25	90
	1.4	Styrene	0.58	50
	8.8	Vinyl chloride	0.040	50
Diethyl fumarate	0	Acrylonitrile	8.0	60
	0.070	Styrene	0.30	60
	0.44	Vinyl acetate	0.011	60
	0.48	Vinyl chloride	0.13	60
Diethyl maleate	0	Acrylonitrile	12	60
	0	Methyl methacrylate	20	60
	0.010	Styrene	6.1	70
	0.040	Vinyl acetate	0.17	60
	0.046	Vinyl chloride	0.90	70
Ethylene	0	Acrylonitrile	7.0	20
	0.010	n-Butyl acrylate	14	150
	0.38	Tetrafluoroethylene	0.10	25
	0.79	Vinyl acetate	1.4	130
Fumaronitrile	0.019	n-Dodecyl vinyl ether[a]	0.004	60
	0	Methyl methacrylate	6.7	79
	0.006	Styrene	0.29	65
Maleic anhydride	0	Acrylonitrile	6.0	60
	0.045	n-Butyl vinyl ether[a]	0	50
	0.012	Methyl acrylate	2.8	75
	0.010	Methyl methacrylate	3.4	75
	0.08	cis-Stilbene[a]	0.07	60
	0.03	trans-Stilbene[a]	0.03	60
	0.005	Styrene	0.050	50
	0	Vinyl acetate	0.019	75
	0	Vinyl chloride	0.098	75
Methacrylic acid	2.4	Acrylonitrile	0.092	70
	0.60	Styrene	0.12	60
	24	Vinyl chloride	0.064	50
	0.58	2-Vinylpyridine	1.7	70

TABLE 6-2 (*Continued*)

M_1	r_1	M_2	r_2	T (°C)
Methacrylonitrile	0.46	Ethyl methacrylate	0.83	80
	0.25	Styrene	0.25	80
	12	Vinyl acetate	0.01	70
Methyl acrylate	0.84	Acrylonitrile	1.5	50
	0.070	1,3-Butadiene	1.1	5
	3.3	Ethyl vinyl ether	0	60
	2.8	Maleic anhydride	0.012	75
	0.40	Methyl methacrylate	2.2	50
	0.80	Styrene	0.19	60
	6.4	Vinyl acetate	0.030	60
	4.4	Vinyl chloride	0.093	50
	0.17	2-Vinylpyridine	1.7	60
	0.20	4-Vinylpyridine	1.7	60
Methyl methacrylate	0.36	Acenaphthylene	1.1	60
	0.46	Styrene	0.52	60
	9.0	Vinyl chloride	0.070	68
	2.4	Vinylidene chloride	0.36	60
α-Methylstyrene	0.14	Acrylonitrile	0.030	75
	0.040	Maleic anhydride	0.080	60
	0.27	Methyl methacrylate	0.48	60
	0.14	Styrene	1.2	60
Methyl vinyl ketone	0.35	Styrene	0.29	60
	8.3	Vinyl chloride	0.10	70
	1.8	Vinylidene chloride	0.55	70
Styrene	90	Ethyl vinyl ether	0	80
	42	Vinyl acetate	0	60
	15	Vinyl chloride	0.010	60
	1.8	Vinylidene chloride	0.087	60
Vinyl acetate	3.4	Ethyl vinyl ether	0.26	60
	0.24	Vinyl chloride	1.8	60
	0.030	Vinylidene chloride	4.7	68

[a]Data from Young [1975]; all other data from Greenley [1989a].

own monomer

$$\frac{1}{r_1} = \frac{k_{12}}{k_{11}} \tag{6-50}$$

Table 6-3 shows $1/r$ values calculated from the data in Table 6-2. The data in each vertical column show the *monomer reactivities* of a series of different monomers toward the same reference polymer radical. Thus the first column shows the reactivities of the monomers toward the butadiene radical, the second column shows the monomer reactivities toward the styrene radical, and so on. It is important to note that the data in each horizontal row in Table 6-3 cannot be compared; the data can only be compared in each vertical column.

TABLE 6-3   Relative Reactivities ($1/r$) of Monomers with Various Polymer Radicals[a]

Monomer	Butadiene	Styrene	Vinyl Acetate	Vinyl Chloride	Methyl Methacrylate	Methyl Acrylate	Acrylonitrile
				Polymer Radical			
Butadiene		1.7		29	4	20	50
Styrene	0.7		100	50	2.2	5.0	25
Methyl methacrylate	1.3	1.9	67	10		2	6.7
Methyl vinyl ketone		3.4	20	10			1.7
Acrylonitrile	3.3	2.5	20	25	0.82	1.2	
Methyl acrylate	1.3	1.3	10	17	0.52		0.67
Vinylidene chloride		0.54	10		0.39		1.1
Vinyl chloride	0.11	0.059	4.4		0.10	0.25	0.37
Vinyl acetate		0.019		0.59	0.050	0.11	0.24

[a]$1/r$ values calculated from data of Table 6-2.

6-3b-1   Resonance Effects

The monomers have been arranged in Table 6-3 in their general order of reactivity. The order of monomer reactivities is approximately the same in each vertical column irrespective of the reference radical. The exceptions that occur are due to the strong alternating tendency of certain comonomer pairs. Table 6-3 and other similar data show that substituents increase the reactivity of a monomer toward radical attack in the general order

$$-\phi, -CH{=}CH_2 > -CN, -COR > -COOH, -COOR >$$
$$-Cl > -OCOR, -R > -OR, -H$$

The order of monomer reactivities corresponds to the order of increased resonance stabilization by the particular substituent of the radical formed from the monomer. Substituents composed of unsaturated linkages are most effective in stabilizing the radicals because of the loosely held π-electrons, which are available for resonance stabilization. Substituents such as halogen, acetoxy, and ether are increasingly ineffective in stabilizing the radicals since only the nonbonding electrons on halogen or oxygen are available for interaction with a radical. The spread in the effectiveness of the various substituents in enhancing monomer reactivity is about 50–200-fold depending on the reactivity of the radical. The less reactive the attacking radical, the greater is the spread in reactivities of the different monomers. The effect of a second substituent in the 1-position as in vinylidene chloride is approximately additive.

The order of *radical reactivities* can be obtained by multiplying the $1/r$ values by the appropriate propagation rate constants for homopolymerization (k_{11}). This yields the values of k_{12} for the reactions of various radical–monomer combinations (Table 6-4). The k_{12} values in any vertical column in Table 6-4 give the order of monomer reactivities—as was the case for the data in Table 6-3. The data in any horizontal row give the order of radical reactivities toward a reference monomer. (The Q_1 and e_1 values in the last two vertical columns should be ignored at this point; they will be considered in Sec. 6-3b-4.)

TABLE 6-4 Rate Constants (k_{12}) for Radical–Monomer Reactions[a]

| | Polymer Radical | | | | | | | | |
Monomer (M_2)	Butadiene	Styrene	Methyl Methacrylate	Acrylonitrile	Methyl Acrylate	Vinyl Acetate	Vinyl Chloride	Q	e
Butadiene	100	280	2,060	98,000	41,800		319,000	1.70	−0.50
Styrene	70	165	1,130	49,000	10,045	230,000	550,000	1.00	−0.80
Methyl methacrylate	130	314	515	13,100	4,180	154,000	110,000	0.78	0.40
Acrylonitrile	330	413	422	1,960	2,510	46,000	225,000	0.48	1.23
Methyl acrylate	130	215	268	1,310	2,090	23,000	187,000	0.45	0.64
Vinyl chloride	11	9.7	52	720	520	10,100	11,000	0.056	0.16
Vinyl acetate		3.4	26	230	230	2,300	6,490	0.026	−0.88

[a] k_{12} values calculated from data in Tables 3-11 and 6-3.

As with monomer reactivities it is seen that the order of radical reactivities is essentially the same irrespective of the monomer used as reference. The order of substituents in enhancing radical reactivity is the opposite of their order in enhancing monomer reactivity. A substituent that increases monomer reactivity does so because it stabilizes and decreases the reactivity of the corresponding radical. A consideration of Table 6-4 shows that the effect of a substituent on radical reactivity is considerably larger than its effect on monomer reactivity. Thus vinyl acetate radical is about 100–1000 times more reactive than styrene radical toward a given monomer, while styrene monomer is only 50–100 times more reactive than vinyl acetate monomer toward a given radical. A comparison of the self-propagation rate constants (k_p) for vinyl acetate and styrene shows that these two effects very nearly compensate each other. The k_p for vinyl acetate is only 16 times that of styrene (Table 3-11).

The interaction of radical reactivity and monomer reactivity in determining the rate of a radical-monomer reaction can be more clearly seen by the use of the reaction coordinate diagram in Fig. 6-12. Figure 6-12 shows the potential energy changes accompanying the radical-monomer reaction as a function of the separation between the atoms forming the new bond. These energy changes are shown for the four possible reactions between resonance-stabilized and nonstabilized monomers and radicals,

$$R \cdot + M \rightarrow R \cdot \tag{6-51a}$$

$$R \cdot + M_s \rightarrow R_s \cdot \tag{6-51b}$$

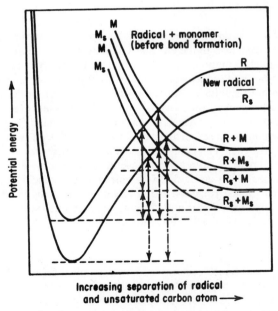

Fig. 6-12 Reaction coordination diagram for the reaction of a polymer radical with a monomer. The dependence of the potential energy of the system (radical + monomer) on the separation between the radical and the unsaturated carbon atom of the monomer is shown. The subscript s indicates the presence of a substituent that is capable of resonance stabilization. Activation energies are represented by the solid-line arrows; heats of reaction by the broken-line arrows. After Walling [1957] (by permission of Wiley, New York).

$$R_s \cdot + M_s \rightarrow R_s \cdot \qquad\qquad\qquad (6\text{-}51c)$$

$$R_s \cdot + M \rightarrow R \cdot \qquad\qquad\qquad (6\text{-}51d)$$

where the presence or absence of the subscript s indicates the presence or absence, respectively, of a substituent that is capable of resonance stabilization. Vinyl acetate and styrene monomers are examples of M and M_s, respectively; vinyl acetate and styrene radicals are examples of $R \cdot$ and $R_s \cdot$, respectively.

There are two sets of potential energy curves in Fig. 6-12. One set of four repulsion curves represents the energetics of the approach of a radical to a monomer; the other set of two Morse curves represents the stability of the bond (or of the polymer radical) finally formed. The intersections of the curves represent the transition states for the monomer-radical reactions (Eqs. 6-51a to 6-51d) where the unbonded and bonded states have the same energies. The various activation energies and heats of reaction are represented by the sold-line and broken-line arrows, respectively. The separation between the two Morse curves is significantly larger than that between either the top or bottom two repulsion curves, since substituents are much more effective in decreasing radical reactivity than in increasing monomer reactivity.

Figure 6-12 shows that the order of reaction rate constants for the various monomer-radical reactions is

$$R_s \cdot + M < R_s \cdot + M_s < R \cdot + M < R \cdot + M_s$$
$$(6\text{-}51d)(6\text{-}51c)(6\text{-}51a)(6\text{-}51b)$$

since the order of activation energies is the exact opposite. (This assumes that there are no appreciable differences in the entropies of activation—a reasonable assumption for sterically unhindered monomers.) This order of reactivity concisely summarizes the data in Tables 6-3 and 6-4 as well as many homopolymerization data. It is clear that monomers without stabilizing substituents (e.g., vinyl chloride or vinyl acetate) will self-propagate faster than those with stabilizing substituents (e.g., styrene) (Reaction 6-51a versus 6-51c). Copolymerization, on the other hand, will occur primarily between two monomers with stabilizing substituents or between two monomers without stabilizing substituents. The combination of a monomer with a stabilizing substituent and one without (e.g., styrene–vinyl acetate) yields a system in which a combination of Reactions 6-51b and 6-51d is required to have facile copolymerization. This does not occur, since Reaction 6-51d is very slow. Thus in the styrene–vinyl acetate system copolymerization is not efficient, since styrene radical is too unreactive to add to the unreactive vinyl acetate monomer.

6-3b-2 Steric Effects

The rates of radical–monomer reactions are also dependent on considerations of steric hindrance. This is easily observed by considering the reactivities of di-, tri-, and tetrasubstituted ethylenes in copolymerization. Table 6-5 shows the k_{12} values for the reactions of various chloroethylenes with vinyl acetate, styrene, and acrylonitrile radicals. The effect of a second substituent on monomer reactivity is approximately additive when both substituents are in the 1- or α-position. However, a second substituent when in the 2- or β-position of the monomer results in a decrease in reactivity due to steric hindrance between it and the radical to which it is adding. Thus 2–10-fold increases and 2–20-fold decreases in the reactivities of vinylidene chloride and 1,2-dichloroethylene, respectively, are observed compared to vinyl chloride.

TABLE 6-5   Rate Constants (k_{12}) for Radical–Monomer Reactions[a]

Monomer	Polymer Radical		
	Vinyl Acetate	Styrene	Acrylonitrile
Vinyl chloride	10,000	9.7	725
Vinylidene chloride	23,000	89	2,150
cis-1,2-Dichloroethylene	365	0.79	
trans-1,2-Dichloroethylene	2,320	4.5	
Trichloroethylene	3,480	10.3	29
Tetrachloroethylene	338	0.83	4.2

[a] k_{12} values calculated from data in Tables 3-11 and 6-2 and Eastman and Smith [1976].

Although the reactivity of 1,2-disubstituted ethylenes in copolymerization is low, it is still much greater than their reactivity in homopolymerization. It was observed in Sec. 3-9b-3 that the steric hinderance between a β-substituent on the attacking radical and a substituent on the monomer is responsible for the inability of 1,2-disubstituted ethylenes to homopolymerize. The reactivity of 1,2-disubstituted ethylenes toward copolymerization is due to the lack of β-substituents on the attacking radicals (e.g., the styrene, acrylonitrile, and vinyl acetate radicals).

A comparison of the cis- and trans-1,2-dichloroethylenes shows the trans isomer to be the more reactive by a factor of 6 [Dawson et al., 1969]. This is a general phenomenon observed in comparing the reactivities of cis- and trans-1,2-disubstituted ethylenes. The cis isomer, which is usually also the less stable isomer, is the less reactive one toward reaction with a radical. The difference in reactivity has been attributed to the inability of the cis isomer to achieve a completely coplanar conformation in the transition state—a requirement for resonance stabilization of the newly formed radical by the substituent.

The data on the reactivities of trichloroethylene and tetrachloroethylene further illustrate the competitive effects of substitutions on the 1- and 2-positions of ethylene. Trichloroethylene is more reactive than either of the 1,2-dichloroethylenes but less reactive than vinylidene chloride. Tetrachloroethylene is less reactive than trichloroethylene—analogous to the difference in reactivities between vinyl chloride and 1,2-dichloroethylene. The case of polyfluoroethylenes is an exception to the generally observed large decrease in reactivity with polysubstitution. Tetrafluoroethylene and chlorotrifluoroethylene show enhanced reactivity due apparently to the small size of the fluorine atoms.

6-3b-3   Alternation; Polar Effects and Complex Participation

It was noted earlier that the exact quantitative order of monomer reactivities is not the same when different reference radicals are considered (Tables 6-3 and 6-4). Analogously, the exact order of radical reactivities varies depending on the reference monomer. Monomer reactivity cannot be considered independent of radical reactivity and vice versa. One observes enhanced reactivities in certain pairs of monomers due apparently to subtle radical–monomer interactions. This effect is a very general one in radical copolymerization and corresponds to the alternating tendency of the co-monomer pairs. The deviation of the $r_1 r_2$ product from unity and its approach to zero is a measure of the alternating tendency. One can list monomers in order of their $r_1 r_2$ values with other monomers in such a manner that the further apart two monomers are, the greater is their tendency toward alternation (Table 6-6). (Ignore the e values

TABLE 6-6 Values of r_1r_2 in Radical Copolymerization[a,b]

	n-Butyl vinyl ether (−1.50)	Butadiene (−0.50)	Styrene (−0.80)	Vinyl acetate (−0.88)	Vinyl chloride (0.16)	Methyl methacrylate (0.40)	Vinylidene chloride (0.34)	Methyl vinyl ketone (1.06)	Acrylonitrile (1.23)	Diethyl fumarate (2.26)
Butadiene (−0.50)	0.78									
Styrene (−0.80)	0.55	0.31								
Vinyl acetate (−0.88)	0.39	0.19	0.34							
Vinyl chloride (0.16)	1.0	<0.1	0.24	0.30						
Methyl methacrylate (0.40)	0.61	0.006	0.16	0.6	0.96					
Vinylidene chloride (0.34)	0.99		0.10	0.35	0.83	0.18				
Methyl vinyl ketone (1.06)	1.1	0.0004	0.016	0.21	0.11		0.34			
Acrylonitrile (1.23)		~0	0.021	0.0049	0.056	0.13	0.56			
Diethyl fumarate (2.26)		~−0.002	0.006	0.00017	0.0024					
Maleic anhydride (3.69)										

[a] r_1r_2 values are calculated from data in Table 6-2 Greeley [1989a].
[b] values are shown in parentheses after each monomer.

until Sec. 6-3b-4). Thus acrylonitrile undergoes ideal copolymerization with methyl vinyl ketone ($r_1r_2 = 1.1$) and alternating copolymerization with butadiene ($r_1r_2 = 0.006$).

The order of monomers in Table 6-6 is one based on the polarity of the double bond. Monomers with electron-pushing substituents are located at the top (left) of the table and those with electron-pulling substituents at the bottom (right). The r_1r_2 value decreases progressively as one considers two monomers further apart in the table. The significant conclusion is that the tendency toward alternation increases as the difference in polarity between the two monomers increases. A most dramatic and useful aspect of alternating copolymerization is the copolymerization of monomers that show little or no tendency to homopolymerize. Maleic anhydride, diethyl fumarate, and fumaronitrile do not homopolymerize but will readily form (very strongly) alternating copolymers with electron-donor monomers such as styrene, vinyl ethers, and N-vinylcarbazole [Baldwin, 1965; Chiang et al., 1977; Yoshimura et al., 1978]. The copolymerization of

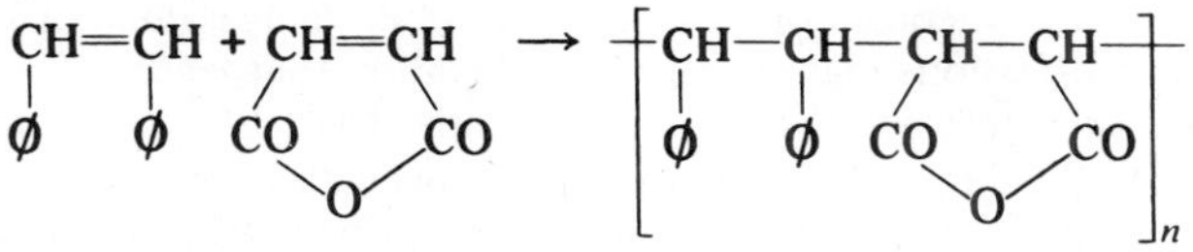

$$(6\text{-}52)$$

takes place even though neither monomer undergoes appreciable homopolymerization.

Two mechanisms have been proposed to explain the strong alternation tendency between electron-donor and electron-acceptor monomers. One mechanism, analogous to the polar effect in chain transfer (Sec. 3-6c-2), considers that interaction between an electron-acceptor radical and an electron-donor monomer or an electron-donor radical and an electron-acceptor monomer leads to a decrease in the activation energy for cross-propagation [Price, 1948; Walling et al., 1948]. The transition state is stabilized by partial electron transfer between the electron-donor and electron-acceptor species, for example, between styrene monomer and maleic anhydride radical

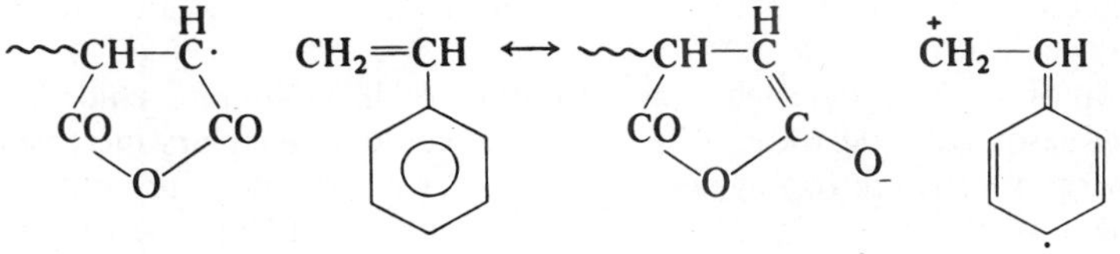

$$(6\text{-}53)$$

and similarly between styrene radical and maleic anhydride monomer,

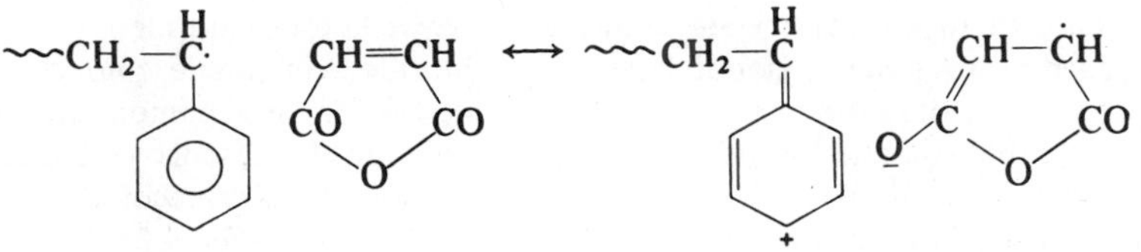

$$(6\text{-}54)$$

The second mechanism suggests [Cowie, 1989; Furukawa, 1986; Gaylord, 1972; Georgiev and Zubov, 1978] that alternating copolymerization results from homopolymerization of a 1:1 complex formed between donor and acceptor monomers

$$M_1 + M_2 \overset{K}{\rightleftharpoons} \underset{\text{complex}}{M_1M_2} \tag{6-55}$$

$$\sim\!\sim\!\sim M_1M_2\cdot + M_1M_2 \overset{k}{\longrightarrow} \sim\!\sim\!\sim M_1M_2M_1M_2\cdot \tag{6-56}$$

Spectroscopic (UV and NMR) evidence for the formation of charge transfer complexes between electron-donor and electron-acceptor monomers supports this mechanism in a number of systems [Hirai, 1976; Kuntz et al., 1978]. Additional evidence includes the marked tendency to alternation over a wide range of feed compositions, high reaction rate at or near the equimolar feed composition (corresponding to the maximum concentration of complex), the absence of a significant effect on molecular weight of added chain-transfer agents [Dodgson and Ebdon, 1977]. Some alternating copolymerizations proceed spontaneously without any added free-radical initiator [Hall et al., 1987]. Such copolymerizations may involve complexation that leads to radical formation by electron transfer between the monomers [Sato et al., 1971].

The stereochemistry of alkyl vinyl ether copolymerizations with fumaronitrile and maleonitrile supports the complex mechanism [Butler and Do, 1989; Olson and Butler, 1984]. The two copolymerizations yield different stereochemical results, indicating that the cis- and trans-arrangements of substituents in maleonitrile and fumaronitrile, respectively, are preserved in the copolymer. If maleonitrile and fumaronitrile entered the copolymer as individual monomer molecules, instead of as part of a complex with the alkyl vinyl ether, the two copolymers would possess the same stereochemistry. Contradictory evidence for the complex mechanism was found during radical trapping experiments in the copolymerization of N-phenylmaleimide (NPM) and 2-chloroethyl vinyl ether (CEVE) system [Jones and Tirrell, 1986, 1987; Prementine et al., 1989]. Only N-phenylmaleimide added to the radical trap, whereas both monomers should have added if reaction involved the 1:1 complex of NPM and CEVE. The validity of this experiment is unclear since alkyl mercury salts, used to generate radicals, would be expected to complex with CEVE and prevent its complexation with NPM [Butler and Do, 1989].

The addition of a Lewis acid such as zinc chloride, dialkylaluminum chloride, and alkylaluminum sesquichloride ($AlR_{1.5}Cl_{1.5}$) to some comonomer pairs increases the tendency to form alternating copolymers. Comonomer pairs which are mildly alternating become strongly alternating [Cowie, 1989; Furukawa, 1986]. Even more surprising, strongly alternating behavior occurs between monomers that normally do not copolymerize or whose alternation tendency is very low. Thus alternating copolymerization is achieved between electron-acceptor monomers such as acrylonitrile, methyl acrylate, methyl methacrylate, and methyl vinyl ketone and electron-donor monomers such as propylene, isobutylene, vinyl chloride, vinylidene chloride, and vinyl acetate. A consideration of the r values in Table 6-2 for these comonomer pairs clearly indicates that the alternating tendency is not high in the absence of the Lewis acid. In the presence of the Lewis acid these comonomer pairs become strongly alternating. Two mechanisms have been proposed to explain the effect of Lewis acids, both involving initial complexation of the Lewis acid with the electron-acceptor monomer to produce a binary complex with enhanced electron-acceptor properties. The binary complex participates in copolymerization either through a cross-propagation mechanism or via interaction with the electron-donor monomer to form a ternary complex that subsequently undergoes homopropagation. The latter mechanism is clearly indicated in those reaction systems where there is a quantitative relationship between the tendency toward alternation and the concentration of the ternary complex. Both mechanisms explain why copolymerization is possible with monomers such as propylene or isobutylene, which do not homopolymerize as a result of degradative chain transfer. Degradative chain transfer is less competitive with polymerization through the binary or ternary complex.

Both the cross-propagation and complex mechanisms are probably operative in alternating copolymerizations with the relative importance of each depending on the particular reaction system. The tendency toward alternation, with or without added Lewis acid, is temperature- and concentration-dependent. Alternation decreases with increasing temperature and decreasing total monomer concentration since the extent of complex formation decreases. When the alternation tendency is less than absolute because of high reaction temperature, low monomer concentration, absence of a Lewis acid, or an imbalance in the coordinating abilities of the two monomers, copolymerization proceeds simultaneously by the two mechanisms. The quantitative aspects of this situation are considered in Sec. 6–5.

6-3b-4   *Q-e Scheme*

Various attempts have been made to place the radical–monomer reaction on a quantitative basis in terms of correlating structure with reactivity. Success in this area would give a better understanding of copolymerization behavior and allow the prediction of the monomer reactivity ratios for comonomer pairs that have not yet been copolymerized. A useful correlation is the $Q-e$ scheme of Alfrey and Price [1947], who proposed that the rate constant for a radical–monomer reaction, for example, for the reaction of $M_1 \cdot$ radical with M_2 monomer, be written as

$$k_{12} = P_1 Q_2 \exp(-e_1 e_2) \tag{6-57}$$

where P_1 and Q_2 are measures of the resonance stabilization of $M_1 \cdot$ radical and M_2 monomer, respectively, and e_1 and e_2 are measures of their polar properties. By assuming that the same e value applies to both a monomer and its corresponding radical (that is, e_1 defines the polarities of M_1 and $M_1 \cdot$, while e_2 defines the polarities of M_2 and $M_2 \cdot$), one can write expressions for k_{11}, k_{22}, k_{21} analogous to Eq. 6-57. These can be appropriately combined to yield the monomer reactivity ratios in the forms

$$r_1 = \frac{Q_1}{Q_2} \exp[-e_1(e_1 - e_2)] \tag{6-58}$$

$$r_2 = \frac{Q_2}{Q_1} \exp[-e_2(e_2 - e_1)] \tag{6-59}$$

which correlate monomer–radical reactivity with the parameters Q_1, Q_2, e_1, and e_2. The basis of the $Q-e$ scheme (Eqs. 6-57 to 6-59) is the theoretically unsatisfactory suggestion that the alternating tendency is due to ground-state electrostatic interactions between permanent charges in the monomer and radical [Price, 1948]. Although there have been attempts to place it on a solid theoretical basis [Colthup, 1982; O'Driscoll and Yonezawa, 1966], the $Q-e$ scheme is best considered as an empirical approach to placing monomer reactivity on a quantitative basis. Monomer reactivity is separated into the parameter Q, which describes the resonance factor (and to a slight extent the steric factor) present in the monomer, and the parameter e, which describes the polar factor.

Consider now the use of the $Q-e$ scheme to predict monomer reactivity ratios. Values of Q and e have been assigned to monomers based on their r values and the arbitrarily chosen reference values of $Q = 1$ and $e = -0.80$ for styrene [Greenley, 1989b]. Table 6-7 shows the average Q and e values for some common monomers. The practical success of the $Q-e$ scheme in predicting the r_1 and r_2 values for co-monomer pairs not previously copolymerized has been limited in its quantitative aspects. The reason for this is that the Q and e values for a monomer are not unique values for both experimental and theoretical reasons. The precision of the calculated Q and e values is often poor as a result of inaccuracies in the experimentally determined r values. Further, the Q and e values for a monomer vary considerably depending on the monomer with which it is copolymerized as a result of inherent deficiencies of the $Q-e$ scheme. It does not explicitly take into account steric factors which may affect monomer reactivity ratios for certain radical–monomer combinations. The assumption of the same e value for a monomer and its corresponding radical is also inadequate. Attempts to refine the $Q-e$ scheme by using separate e values for monomer and radical have not been successful.

In spite of these deficiencies the $Q-e$ scheme is a reasonable qualitative and even semiquantitative approach to the effect of structure on monomer reactivity. It can be used to give a general idea of the behavior to be expected from a comonomer pair that has not been studied. The $Q-e$ values can be used to more quantitatively discuss reactivity data such as those in Table 6-4 and 6-6. It is clear that the monomers (with the exception of 1,3-butadiene) are lined up in Table 6-6 in order of their e values. This order defines the polarities of the various monomers. The relative importance of resonance and polar factors in determining monomer reactivity can be discussed by considering the data in Table 6-4 in terms of the Q and e values of the monomers. The various reference radicals can be divided into two groups: one composed of the relatively unreactive radicals (styrene and 1,3-butadiene) and the other composed of reactive radicals (all except styrene and 1,3-butadiene). For the reactive radicals, such as vinyl chloride and vinyl acetate, monomer reactivity depends on Q with k_{12} increasing with increasing Q and polar effects are too subtle to discern. The unreactive radicals from styrene and 1,3-butadiene show the same trends except that one can discern the polar effects. There is enhanced ractivity of these radicals (possessing negative e values) toward a monomer such as acrylonitrile with relatively high positive e values. The resonance factor is, however, more important than the polar factor; the former determines the magnitude of monomer reactivity. Thus monomer reactivities toward the butadiene and styrene radicals fall into two groups—one group of monomers with high Q values and high reactivities and another group with low Q values and low reactivities.

A number of useful generalizations are possible regarding which pairs of monomers will copolymerize and the behavior to be expected. Copolymerization proceeds poorly with monomers whose Q values are very different, since copolymer formation would require the energetically unfavorable conversion of a resonance-stabilized radical to a less stabilized radical and vice versa (Fig. 6-12). Thus vinyl chloride and vinyl acetate do not copolymerize well with styrene or butadiene. Copolymerization is more suitable between monomers of similar Q values, preferably high Q values, for example, styrene

TABLE 6-7 *Q* and *e* Values for Monomers*[a]**

Monomer	Q	e
Acenaphthalene	0.72	−1.88
Ethyl vinyl ether	0.018	−1.80
Propene	0.009	−1.69
N-Vinylpyrrolidone	0.088	−1.62
n-Butyl vinyl ether	0.038	−1.50
i-Butyl vinyl ether	0.030	−1.27
p-Methoxystyrene	1.53	−1.40
Isobutylene	0.023	−1.20
Allyl acetate	0.24	−1.07
Vinyl acetate	0.026	−0.88
α-Methylstyrene	0.97	−0.81
Styrene	1.00	−0.80
Indene	0.13	−0.71
p-Bromostyrene	1.30	−0.68
Allyl chloride	0.026	−0.60
Isoprene	1.00	−0.55
1,3-Butadiene	1.70	−0.50
2-Vinylpyridine	1.41	−0.42
Ethylene	0.016	0.05
Vinyl chloride	0.056	0.16
m-Nitrostyrene	2.19	0.20
Vinylidene chloride	0.31	0.34
Methyl methacrylate	0.78	0.40
Acrylamide	0.23	0.54
Methacrylic acid	0.98	0.62
Methyl acrylate	0.45	0.64
Methacrylonitrile	0.86	0.68
Vinyl fluoride	0.008	0.72
4-Vinylpyridine	2.47	0.84
n-Butyl acrylate	0.38	0.85
Acrylic acid	0.83	0.88
1-Hexene	0.035	0.92
Methyl vinyl ketone	0.66	1.06
Diethyl maleate	0.053	1.08
Acrylonitrile	0.48	1.23
Tetrafluoroethylene	0.032	1.63
o-Chlorostyrene	2.66	1.57
Diethyl fumarate	0.25	2.26
Fumaronitrile	0.29	2.73
Maleic anhydride	0.86	3.69

[a]Data from Greenley [1989b].

and butadiene. Ideal copolymerization occurs between two monomers having similar *Q* and *e* values, for example, styrene-butadiene, vinyl chloride–vinyl acetate, and acrylonitrile-methyl acrylate. The tendency toward alternation is greatest for monomers having the same *Q* values with high *e* values of opposite sign.

6-3b-5 Other Quantitative Approaches to Reactivity

Linear-free-energy relationships such as the Hammett and Taft equations [Lowry and Richardson, 1987] have been used to correlate copolymerization behavior with structure, but the approach is limited to considering a series of monomers which are similar in structure. Walling [1957] applied the Hammett equation to copolymerization among various meta- and para-substituted styrenes. The Taft equation in the form

$$\log\left(\frac{1}{r_1}\right) = \sigma^*\rho^* + \delta E_S \tag{6-60}$$

has been used to correlate the reactivity of various vinyl esters (M_2) toward the radical from ethylene (M_1) [Van der Meer et al., 1977b]. The variables ρ^* and δ are proportionality constants indicative of the susceptability of the radical–monomer reaction to the effect of polar and steric factors, respectively, present in the monomer. Variables σ^* and E_S are constants which indicate the polar and steric factors, respectively, due to the substituents on the double bond of the monomer. For the series of vinyl esters, $CH_2=CHOCOR$, σ^* and E_S are the substituent constants for the various R groups. Similar correlations have been successful for alkyl acrylates and methacrylates [Otsu, 1987].

Other attempts to correlate reactivity in copolymerization include the *product probability* and *patterns of reactivity* approaches [Bamford and Jenkins, 1965; Ham, 1967].

6-3c   Rate of Copolymerization

The rate of copolymerization, unlike the copolymer composition, depends on the initiation and termination steps as well as on the propagation steps. In the usual case both monomers combine efficiently with the initiator radicals and the initiation rate is independent of the feed composition. Two different approaches have been used to derive expressions for the rate of copolymerization.

6-3c-1   *Chemical-Controlled Termination*

One approach assumes the termination reaction to proceed by chemical control [Walling, 1949]. Copolymerization consists of four propagation reactions (Eqs. 6-2 through 6-5) and the three termination reactions

$$M_1\cdot \ +\ M_1\cdot \ \xrightarrow{\ k_{i11}\ } \tag{6-61}$$

$$M_2\cdot \ +\ M_2\cdot \ \xrightarrow{\ k_{i22}\ } \tag{6-62}$$

$$M_1\cdot \ +\ M_2\cdot \ \xrightarrow{\ k_{i12}\ } \tag{6-63}$$

corresponding to termination between like radicals (Eqs. 6-61 and 6-62) and cross-termination between unlike radicals (Eq. 6-63).

The overall rate of copolymerization is given by the sum of the four propagation rates,

$$R_p = -\frac{d[M_1] + d[M_2]}{dt} = k_{11}[M_1\cdot][M_1] + k_{12}[M_1\cdot][M_2]$$

$$+ k_{22}[M_2\cdot][M_2] + k_{21}[M_2\cdot][M_1] \tag{6-64}$$

In order to eliminate radical concentrations from Eq. 6-64, two steady-state assumptions are made. A steady-state concentration is assumed for each type of radical,

$$k_{21}[M_2\cdot][M_1] = k_{12}[M_1\cdot][M_2] \tag{6-65}$$

as was done via Eq. 6-9 in deriving the copolymer composition equation. Steady-state is also assumed for the total concentration of radicals,

$$R_i = 2k_{t11}[M_1\cdot]^2 + 2k_{t12}[M_1\cdot][M_2\cdot] + 2k_{t22}[M_2\cdot]^2 \tag{6-66}$$

Eliminating radical concentrations from Eq. 6-64 by combining it with Eqs. 6-65 and 6-66 and then using the definitions of r_1 and r_2, one obtains for the copolymerization rate,

$$R_p = \frac{(r_1[M_1]^2 + 2[M_1][M_2] + r_2[M_2]^2)R_i^{1/2}}{\{r_1^2\delta_1^2[M_1]^2 + 2\phi r_1 r_2\delta_2\delta_1[M_1][M_2] + r_2^2\delta_2^2[M_2]^2\}^{1/2}} \tag{6-67}$$

where

$$\delta_1 = \left(\frac{2k_{t11}}{k_{11}^2}\right)^{1/2} \tag{6-68a}$$

$$\delta_2 = \left(\frac{2k_{t22}}{k_{22}^2}\right)^{1/2} \tag{6-68b}$$

$$\phi = \frac{k_{t12}}{2(k_{t11}k_{t22})^{1/2}} \tag{6-68c}$$

The δ terms are simply the reciprocals of the familiar $k_p/(2k_t)^{1/2}$ ratios for the homopolymerizations of the individual monomers. The ϕ term represents the ratio of half the cross-termination rate constant to the geometric mean of the rate constants for self-termination of like radicals. The factor of 2 is present in the denominator of Eq. 6-68c since cross-termination is statistically favored over termination by like radicals by a factor of 2. A $\phi < 1$ means that cross-termination is not favored, while $\phi > 1$ means that cross-termination is favored.

The δ_1 and δ_2 values are obtained from homopolymerization and the r_1 and r_2 from copolymerization data. Experimental determination of the rate of copolymerization then allows the calculation of ϕ from Eq. 6-67. Values of ϕ for several monomer pairs are shown in Table 6-8 along with their $r_1 r_2$ values. There is a general trend of ϕ values greater than unity indicating that cross-termination is favored. The tendency toward cross-termination parallels the tendency toward cross-propagation (i.e., toward alternation) in that the ϕ increases as $r_1 r_2$ approaches zero. This has led to the conclusion that polar effects are responsible for the tendency toward cross-termination. The reaction between radicals of dissimilar polarity is enhanced because of stabilization of the transition state for termination by electron-transfer effects analogous to those used to describe enhanced chain transfer (Eq. 3-135) and alternation (Eqs. 6-53 and 6-54).

Alternating copolymerizations proceed with enhanced propagation (r_1 and r_2 both <1) and enhanced termination ($\phi > 1$). The interplay of these two opposing effects makes it difficult to predict the dependence of rate on feed composition. The overall

TABLE 6-8   Values of φ and $r_1 r_2$ in Radical Copolymerization[a]

Comonomer	φ	$r_1 r_2$
Styrene–butyl acrylate	150	0.07
Styrene–methyl acrylate	50	0.14
Methyl methacrylate-*p*-methoxystyrene	24	0.09
Styrene–methyl methacrylate	13	0.24
Styrene–methacrylonitrile	6[b]	0.16[c]
Styrene–methoxystyrene	1	0.95

[a]Data from Walling [1957] unless otherwise noted.
[b]Data from Ito [1978].
[c]Data from Greenley [1989a].

shape of a rate-composition plot will depend on the values of φ, r_1, and r_2. For an alternating copolymerization there will usually be at least some range of feed compositions over which the copolymerization rate will be lower than the homopolymerization rate of the monomer present in greater amount because of the enhanced termination offsetting enhanced propagation. The rate for other feed compositions may be similar to or higher than for homopolymerization of the other monomer. Figure 6-13 shows this complex behavior for styrene–methacrylonitrile. The two plots are the calculated rates for φ values of 1 and 6. The experimental data follows the plot for φ = 6 quite well.

Styrene-vinyl acetate presents an interesting case of an ideal copolymerization. The value of r_2 is very small and Eq. 6-67 is approximated by

$$R_p = \left([M_1] + \frac{2[M_2]}{r_1} \right) \frac{R_i^{1/2}}{\delta_1} \tag{6-69}$$

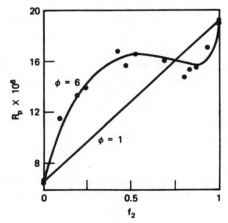

Fig. 6-13 Dependence of the rate of radical copolymerization of styrene–methacrylonitrile at 60°C on the comonomer feed composition. The two plots are the theoretical plots calculated for φ values of 1 and 6; the circles represent the experimental data. After Ito [1978] (by permission of Wiley-Interscience, New York).

from which it is observed that the addition of very small amounts of styrene to vinyl acetate results in very large decreases in the polymerization rate. Styrene essentially inhibits the polymerization of vinyl acetate. Vinyl acetate radicals are rapidly converted to styrene even at low styrene concentrations because of the high reactivity of both vinyl acetate radical and styrene monomer. The styrene radicals react very slowly with vinyl acetate monomer and are unable to react with styrene monomer because the latter is present in small amounts. The net effect is an almost complete cessation of polymerization.

6-3c-2 *Diffusion-Controlled Termination*

The preceding treatment of copolymerization with the ϕ factor can be ambiguous, since it is well established that termination in radical polymerization is generally diffusion-controlled [Atherton and North, 1962; North, 1963]. The value ϕ cannot be interpreted primarily in terms of the chemical effects of the radical chain ends. The dependence of the rate of termination on the translational and segmental diffusion of the polymer chains was discussed in Sec. 3-10b. A ϕ value different from unity should then be interpreted in terms of the changes that occur in the translational and segmental diffusion of the chains according to their composition. The observation that ϕ in several systems varies with copolymer composition corroborates this interpretation and makes the use of Eq. 6-67 with a single ϕ value of dubious utility [Barb, 1953; Braun and Czerwinski, 1987].

A kinetic expression for the rate of diffusion-controlled copolymerization is obtained by considering the termination reaction as the reaction

$$\left.\begin{array}{c} M_1\cdot \; + \; M_1\cdot \\ M_1\cdot \; + \; M_2\cdot \\ M_2\cdot \; + \; M_1\cdot \end{array}\right\} \xrightarrow{k_{t(12)}} \text{dead copolymer} \tag{6-70}$$

where the termination rate constant $k_{t(12)}$ is a function of the copolymer composition. The condition for the steady state for the total concentration of radicals then takes the form

$$R_i = 2k_{t(12)}([M_1\cdot] + [M_2\cdot])^2 \tag{6-71}$$

instead of Eq. 6-66. Combination of Eqs. 6-64, 6-65, and 6-71 with the definitions of r_1 and r_2 yields the rate of copolymerization as

$$R_p = \frac{(r_1[M_1]^2 + 2[M_1][M_2] + r_2[M]^2)R_i^{1/2}}{k_{t(12)}^{1/2}\left\{\left(\dfrac{r_1[M_1]}{k_{11}}\right) + \left(\dfrac{r_2[M_2]}{k_{22}}\right)\right\}} \tag{6-72}$$

One would expect the termination rate constant $k_{t(12)}$ to be a function of the termination rate constants for the corresponding two homopolymerizations. In an ideal situation this dependence might take the form

$$k_{t(12)} = F_1 k_{t11} + F_2 k_{t22} \tag{6-73}$$

where $k_{t(12)}$ is the average of k_{t11} and k_{t22} each weighted on the basis of the copolymer composition in mole fractions (F_1 and F_2).

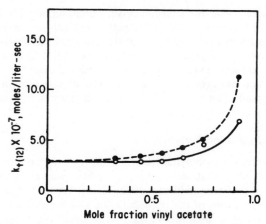

Fig. 6-14 Dependence of the termination rate constant $k_{t(12)}$ on the mole fraction of vinyl acetate in the radical copolymerization of vinyl acetate–methyl methacrylate. The solid-line plot and open circles represent calculations by Eq. 6-73; the broken-line plot and solid circles represent calculations by Eq. 6-72. After Atherton and North [1962] (by permission of the Faraday Society, London).

The utility of Eq. 6-72 in correlating copolymerization data has been established in several studies [O'Driscoll et al., 1967; Prochazka and Kratochvil, 1983]. Thus, Fig. 6-14 shows the experimentally determined $k_{t(12)}$ values (i.e., the values calculated from Eq. 6-72) for vinyl acetate–methyl methacrylate copolymerization (dotted curve). The expected variation of $k_{t(12)}$ with the comonomer (and copolymer) composition can be handled by Eq. 6-72, whereas Eq. 6-67 with a single ϕ value is not applicable. Figure 6-14 also shows the $k_{t(12)}$ values calculated via Eq. 6-73 from the copolymer composition and the termination rate constants for the two homopolymerizations (solid curve). Equation 6-73 appears to be qualitatively but not quantitatively valid. Evidently, the method of weighting the k_{t11} and k_{t22} values directly with the copolymer composition is not correct. Refinements of this treatment have included expressing $k_{t(12)}$ as a function of the viscosity of the reaction medium and introducing penultimate considerations into the termination process [Braun and Czerwinski, 1989].

6-4 IONIC COPOLYMERIZATION

Ionic copolymerizations are different from radical copolymerizations in several respects. Ionic copolymerizations are much more selective. The number of comonomer pairs that undergo either cationic or anionic copolymerization is relatively limited because of the wider range of monomer reactivities in ionic copolymerization [Bywater, 1976; Kennedy and Marechal, 1983; Morton, 1983]. Cationic copolymerization is limited to monomers with electron-donating substituents and anionic copolymerization to monomers with electron-withdrawing substituents. For comonomer pairs that undergo ionic copolymerization, the general tendency is toward the ideal type of behavior (Sec. 6-2c-1), with the $r_1 r_2$ product approaching unity, where the relative reactivities of the two monomers toward the two different ionic propagating centers are approximately the same. There is a general lack of any tendency toward alternation. Furthermore, quite a few copolymerizations proceed with $r_1 r_2$ values greater than unity. Thus there are relatively few monomer pairs that yield copolymers containing large proportions of both monomers.

Another characteristic feature of ionic copolymerizations is the sensitivity of the monomer reactivity ratios to changes in the initiator, reaction medium, or temperature. This is quite different from the behavior observed in radical copolymerization. Monomer reactivity ratios in radical copolymerization are far less dependent on reaction conditions.

6-4a Cationic Copolymerization

6-4a-1 Reactivity

The effect of a substituent on the reactivity of a monomer in cationic copolymerization depends on the extent to which it increases the electron density on the double bond and on its ability to resonance stabilize the carbocation that is formed. However, the order of monomer reactivities in cationic copolymerization (as in anionic copolymerization) is not nearly as well defined as in radical copolymerization. Reactivity is often influenced to a larger degree by the reaction conditions (solvent, counterion, temperature) than by the structure of the monomer. There are relatively few reports in the literature in which monomer reactivity has been studied for a wide range of different monomers under conditions of the same solvent, counterion, and reaction temperature.

Among the most extensive studies of monomer reactivity have been those involving the copolymerization of various meta- and para-substituted styrenes with other styrene monomers (styrene, α-methylstyrene, and p-chlorostyrene) as the reference monomer [Kennedy and Marechal, 1983]. The relative reactivities of the various substituted styrenes have been correlated by the Hammett sigma–rho relationship

$$\log \left(\frac{1}{r_1} \right) = \rho\sigma \tag{6-74}$$

For example, $\log (1/r_1)$ values for a series of meta- and para-substituted styrenes copolymerized with styrene were plotted against the sigma substituent constants to yield a straight line with slope ρ of negative sign. The sigma value of a substituent is a quantitative measure of that substituent's total electron-donating or electron-withdrawing effect by both resonance and induction. Electron-withdrawing and electron-donating substituents have positive and negative sigma constants, respectively. A negative value of ρ means $1/r_1$ is increased by electron-donating substituents as expected for cationic polymerization. (A positive value of ρ would mean $1/r_1$ is increased by electron-withdrawing substituents.) Substituents increase the reactivity of styrene in the approximate order

$$p\text{-OCH}_3 > p\text{-CH}_3 > p\text{-H} > p\text{-Cl} > m\text{-Cl} > m\text{-NO}_2$$
$$\ \ (-0.27) \qquad (-0.17) \qquad (0) \qquad (+0.23) \qquad (+0.37) \qquad (+0.71)$$

which follows the order of their electron-donating effect as indicated by the sigma values shown in parentheses. More recently $\log (1/r_1)$ for meta- and para-substituted styrenes has been correlated with the ^{13}C NMR chemical shifts of the β-carbon of the substituted styrenes [Hatada et al., 1977; Wood et al., 1989]. Similar correlations have been observed for the cationic copolymerizations of para-substituted benzyl vinyl ethers with benzyl vinyl ether. The correlation of log $(1/r_1)$ with chemical shift (δ) is analogous to the correlation with σ, since both δ and σ measure the electron-donating ability of the substituent.

Although the Hammett-type approach is most useful for the quantitative correlation of monomer reactivity with structure, it is applicable only to substituted styrenes. One is, however, usually more interested in the relative reactivities of the commonly encountered monomers such as isoprene, acrylonitrile, and isobutene. The appropriate quantitative data are relatively sparse for these monomers. The generally observed order of monomer reactivity is

vinyl ethers > isobutylene > styrene, isoprene

which is the order expected on the basis of the electron-pushing ability of the various substituents. Monomers with electron-pulling substituents such as acrylonitrile, methyl methacrylate, and vinyl chloride show negligible reactivity in cationic copolymerization. There has been some success in correlating log $(1/r_1)$ with the e values from the $Q-e$ scheme [Ham, 1977].

Steric effects similar to those in radical copolymerization are also operative in cationic copolymerizations. Table 6-9 shows the effect of methyl substituents in the α- and β-positions of styrene. Reactivity is increased by the α-methyl substituent due to its electron-pushing power. The decreased reactivity of β-methylstyrene relative to styrene indicates that the steric effect of the β-substituent outweighs its polar effect of increasing the electron density on the double bond. Furthermore, the trans-β-methylstyrene appears to be more reactive than the cis isomer, although the difference is much less than in radical copolymerization (Sec. 6-3b-2). It is worth noting that 1,2-disubstituted alkenes have finite r values in cationic copolymerization compared to the values of zero in radical copolymerization (Table 6-2). There is a tendency for 1,2-disubstituted alkenes to self-propagate in cationic copolymerization, although this tendency is nil in the radical reaction.

6-4a-2 Effect of Solvent and Counterion

It has previously been shown that large changes can occur in the rate of a cationic polymerization by using a different solvent and/or different counterion (Sec. 5-2f). The monomer reactivity ratios are also affected by changes in the solvent or counterion. The effects are often complex and difficult to predict since changes in solvent or counterion often result in alterations in the relative amounts of the different types of

TABLE 6-9 Steric Effects in Copolymerization of α- and β-Methylstyrenes (M₁) with p-Chlorostyrene (M₂)[a,b]

M₁	r_1	r_2
Styrene	2.31	0.21
α-Methylstyrene	9.44	0.11
trans-β-Methylstyrene	0.32	0.74
cis-β-Methylstyrene	0.32	1.0

[a]Data from Overberger et al. [1951, 1954, 1958].
[b]SnCl₄ in CCl₄ at 0°C.

propagating centers (free ion, ion pair, covalent), each of which may be differently affected by solvent. As many systems do not show an effect as do show an effect of solvent or counterion on r values [Kennedy and Marechal, 1983]. The dramatic effect that solvents can have on monomer reactivity ratios is illustrated by the data in Table 6-10 for isobutylene-p-chlorostyrene. The aluminum bromide-initiated copolymerization shows $r_1 = 1.01$, $r_2 = 1.02$ in n-hexane but $r_1 = 14.7$, $r_2 = 0.15$ in nitrobenzene. The variation in r values has been attributed to the preferential solvation of propagating centers in the nonpolar medium (n-hexane) by the more polar monomer (p-chlorostyrene). The increased concentration of p-chlorostyrene at the reaction site results in its greater incorporation into the copolymer than expected based on the composition of the comonomer feed in the bulk solution. Calculation of r values using the bulk comonomer feed composition results in a lower value of r_1 coupled with a higher value of r_2. In the polar nitrobenzene the propagating centers are completely solvated by the solvent without participation by p-chlorostyrene, and the more reactive isobutylene exhibits its greater reactivity.

The effect of solvent on monomer reactivity ratios cannot be considered independent of the counterion employed. Again, the situation is difficult to predict with some comonomer systems showing altered r values for different initiators and others showing no effects. Thus the isobutylene–p-chlorostyrene system (Table 6-10) shows different r_1 and r_2 for AlBr$_3$ and SnCl$_4$. The interdependence of the effects of solvent and counterion are shown in Table 6-11 for the copolymerization of styrene and p-methylstyrene. The initiators are listed in order of their strength as measured by their effectiveness in homopolymerization studies. Antimony pentachloride is the strongest initiator and iodine the weakest. The order is that based on the relative concentrations of different types of propagating centers. Polymerizations by iodine and trichloroacetic proceed predominantly through covalent species while ion pairs and free ions are involved for polymerizations with the other initiators.

The data in Table 6-11 show the copolymer composition to be insensitive to the initiator for solvents of high polarity (1,2-dichloroethane and nitrobenzene) and also insensitive to solvent polarity for any initiator except the strongest (SbCl$_5$). The styrene content of the copolymer decreases with increasing solvent polarity when SbCl$_5$ is the initiator. The styrene content also decreases with decreasing initiator strength for the low-polarity solvent (toluene). These results can be interpreted in terms of the effect of solvent and counterion on the identity of the propagating centers and on the extent of preferential solvation of propagating centers by one of the monomers. In the styrene–p-methylstyrene system p-methylstyrene is both the more polar and the more reactive of the two monomers. In the poor solvent (toluene) the monomers compete, against the solvent, with each other to solvate the propagating centers (primarily ion pairs for all initiators other than iodine and trichloroacetic acid). The more polar p-

TABLE 6-10   Effect of Solvent and Initiator on r Values

r_1 Isobutylene	r_2 p-Chlorostyrene	Solvent	Initiator
1.01	1.02	n-C$_6$H$_{14}$ (ϵ 1.8)	AlBr$_3$
14.7	0.15	$\emptyset$-NO$_2$ (ϵ 36)	AlBr$_3$
8.6	1.2	$\emptyset$-NO$_2$ (ϵ 36)	SnCl$_4$

[a]Data from Overberger and Kamath [1959].
[b]Temperature: 0°C.

TABLE 6-11 Effects of Solvent and Counterion on Copolymer Composition in Styrene–*p*-Methylstyrene Copolymerization[a]

	% Styrene in Copolymer[b]		
Initiator System	Toluene (ϵ 2.4)	1,2-Dichloroethane (ϵ 9.7)	Nitrobenzene (ϵ 36)
SbCl$_5$	46	25	28
AlX$_3$	34	34	28
TiCl$_4$, SnCl$_4$, BF$_3$·OEt$_2$, SbCl$_3$	28	27	27
Cl$_3$CCO$_2$H		27	30
I$_2$		17	

[a]Data from O'Driscoll et al. [1966].
[b]Comonomer feed = 1:1 styrene–*p*-methylstyrene.

methylstyrene preferentially solvates the propagating ion pairs and is preferentially incorporated into the copolymer. The selectivity increases in proceeding from SbCl$_5$ to AlCl$_3$ to the other initiatiors, which corresponds to increases in both the amount of ion pairs relative to free ions and amount of tight ion pairs relative to loose ion pairs. For the better solvents the counterion does not appreciably influence the reaction, since the monomers cannot compete with the solvent. In the SbCl$_5$ initiated copolymerization increasing the solvent power of the reaction medium also decreases the ability of the monomers to compete with the solvent to complex with propagating centers. The copolymer composition is then determined primarily by the chemical reactivities of the monomers.

6-4a-3 Effect of Temperature

Temperature has a greater influence on monomer reactivity ratios in cationic copolymerization than in radical copolymerization because of the greater spread of propagation activation energies for the ionic process. The ratio of any two rate constants is expected to tend toward unity with increasing temperature since the smaller rate constant (larger activation energy) will increase faster with increasing temperature than the larger rate constant (smaller activation energy). However, there is no general trend of *r* values tending toward unity (i.e., less selective reaction) in cationic copolymerization with increasing temperature as there is radical copolymerization. Some *r* values increase with temperature and others decrease. Various combinations of effects have been observed for different comonomer pairs [Kennedy and Marechal, 1983]. There are comonomer systems where both r_1 and r_2 tend toward unity as expected, but there are also many systems where an *r* value decreases below or increases above unity with increasing temperature. This unexpected behavior is probably the result of changes in the identities and relative amounts of different propagating species (free ion, ion pair, covalent) either directly as a result of a change in temperature, or indirectly by the effect of temperature on solvent polarity.

6-4b Anionic Copolymerization

6-4b-1 Reactivity

Monomer reactivities in anionic copolymerization are the opposite of those in cationic copolymerization. Reactivity is enhanced by electron-withdrawing substituents that decrease the electron density on the double bond and resonance stabilize the carbanion formed. Although the available data is rather limited [Bywater, 1976; Morton, 1983; Szwarc, 1968], reactivity is generally increased by substituents in the order

$$-CN > -CO_2R > -\phi, -CH{=}CH_2 > -H$$

The reactivity of monomers with electron-releasing substituents in anionic copolymerization is nil. Correlation of reactivity in copolymerization with structure has been achieved in some studies [Favier et al., 1977; Shima et al., 1962].

The general characteristics of anionic copolymerization are very similar to those of cationic copolymerization. There is a tendency toward ideal behavior in most anionic copolymerizations. Steric effects give rise to an alternating tendency for certain comonomer pairs. Thus the styrene–p-methylstyrene pair shows ideal behavior with $r_1 = 5.3$, $r_2 = 0.18$, $r_1 r_2 = 0.95$, while the styrene–α-methylstyrene pair shows a tendency toward alternation with $r_1 = 35$, $r_2 = 0.003$, $r_1 r_2 = 0.11$ [Bhattacharyya et al., 1963; Shima et al., 1962]. The steric effect of the additional substituent in the α-position hinders the addition of α-methylstyrene to α-methylstyrene anion. The tendency toward alternation is essentially complete in the copolymerizations of the sterically hindered monomers 1,1-diphenylethylene and *trans*-1,2-diphenylethylene with 1,3-butadiene, isoprene, and 2,3-dimethyl-1,3-butadiene [Yuki et al., 1964].

6-4b-2 *Effects of Solvent and Counterion*

Monomer reactivity ratios and copolymer compositions in many anionic copolymerizations are altered by changes in the solvent or counterion. Table 6-12 shows data for styrene–isoprene copolymerization at 25°C by n-butyl lithium [Kelley and Tobolsky, 1959]. As in the case of cationic copolymerization, the effects of solvent and counterion cannot be considered independently of each other. For the tightly bound lithium counterion, there are large effects due to the solvent. In poor solvents the copolymer is rich in the less reactive (based on relative rates of homopolymerization) isoprene because isoprene is preferentially complexed by lithium ion. (The complexing of 1,3-dienes with lithium ion is discussed further in Sec. 8-6b). In good solvents preferential solvation by monomer is much less important and the inherent greater reactivity of styrene exerts itself. The quantitative effect of solvent on copolymer composition is less for the more loosely bound sodium counterion.

TABLE 6-12   Effect of Solvent and Counterion on Copolymer Composition in Styrene–Isoprene Copolymerization[a]

	% Styrene in Copolymer for Counterion	
Solvent	Na$^+$	Li$^+$
None	66	15
Benzene	66	15
Triethylamine	77	59
Ethyl ether	75	68
Tetrahydrofuran	80	80

[a]Data from Kelley and Tobolsky [1959].

Copolymerizations of nonpolar monomers with polar monomers such as methyl methacrylate and acrylonitrile are especially complicated. The effects of solvent and counterion may be unimportant compared to the side reactions characteristic of anionic polymerization of polar monomers (Sec. 5-3b-4). In addition, copolymerization is often hindered by the very low tendency of one of the cross-propagation reactions. For example, polystyryl anions easily add methyl methacrylate but there is little tendency for poly(methyl methacrylate) anions to add styrene. Many reports of styrene–methyl methacrylate (and similar comonomer pairs) copolymerizations are not copolymerizations in the sense discussed in this chapter. The initial product is essentially poly(methyl methacrylate) homopolymer. Little styrene is incorporated into copolymer chains until most or all of the methyl methacrylate is exhausted. Reports of significant amounts of styrene in products from anionic copolymerization of styrene–methyl methacrylate are usually artifacts of the particular reaction system, a consequence of heterogeneity of the propagating centers and/or counterion.

The anionic copolymerization of methyl methacrylate and styrene with lithium emulsion and n-butyllithium initiators is interesting [Overberger and Yamamoto, 1966; Richards, 1978; Tobolsky et al., 1958]. Bulk copolymerization of an equimolar mixture of the two monomers with a lithium emulsion yields a copolymer with a high percentage of styrene, whereas n-butyllithium yields a copolymer with essentially no styrene. Further, the product from the lithium emulsion reaction is essentially a block copolymer. The results with lithium emulsion have been attributed to insolubility of lithium counterion. The lithium ion is part of an insoluble lithium particle and propagation takes place on that particle surface. Styrene is more strongly adsorbed than methyl methacrylate on these surfaces because of its dense π-electron system. Reaction occurs with a very high styrene concentration at the reaction site and initial reaction involves a polystyryl homopropagation. At some point the propagating chains detach from the metal surface and become solubilized in the bulk solution where there is a much higher concentration of methyl methacrylate. Polystyryl anions quickly add methyl methacrylate with very little tendency for reverse crossover back to styrene, and the result is a block copolymer. On the other hand, polymerization initiated by n-butyllithium proceeds in solution from the very beginning. The greater reactivity of methyl methacrylate coupled with the very small tendency for crossover from poly(methyl methacrylate) propagating centers to polystyryl results in the product being essentially poly(methyl methacrylate) homopolymer. Copolymer with significant amounts of styrene is obtained only at higher conversion where the feed composition is low in methyl methacrylate.

6-4b-3 Temperature

There are few studies of the effect of temperature on monomer reactivity ratios [Morton, 1983]. For styrene–1,3-butadiene copolymerization by s-butyllithium in n-hexane, there is negligible change in r values with temperature with $r_1 = 0.03$, $r_2 = 13.3$ at $0°C$ and $r_1 = 0.04$, $r_2 = 11.8$ at $50°C$. There is, however, a significant effect of temperature for copolymerization in tetrahydrofuran with $r_1 = 11.0$, $r_2 = 0.04$ at $-78°C$ and $r_1 = 4.00$, $r_2 = 0.30$ at $25°C$. The difference between copolymerization in polar and nonpolar solvents is attributed to preferential complexing of propagating centers and counterion by 1,3-butadiene as described previously. The change in r values in polar solvent is attributed to the same phenomenon. The extent of solvation decreases with increasing temperature, and this results in 1,3-butadiene participating in the solvation process at the higher reaction temperature.

6-5 DEVIATIONS FROM TERMINAL COPOLYMERIZATION MODEL

The derivation of the terminal (or first-order Markov) copolymer composition equation (Eq. 6-12 or 6-15) rests on two important assumptions—one of a kinetic nature and the other of a thermodynamic nature. The first is that the reactivity of the propagating species is independent of the identity of the monomer unit, which precedes the terminal unit. The second is the irreversibility of the various propagation reactions. Deviations from the quantitative behavior predicted by the copolymer composition equation under certain reaction conditions have been ascribed to the failure of one or the other of these two assumptions or the presence of a comonomer complex which undergoes propagation.

6-5a Kinetic Penultimate Behavior

The behavior of some comonomer systems indicates that the reactivity of the propagating species is affected by the next-to-last or penultimate monomer unit. This behavior, referred to as *second-order Markov* or *penultimate* behavior, manifests itself in a particular copolymerization by giving inconsistent values of the monomer reactivity ratios for different comonomer feed compositions. This has been observed in many radical copolymerizations where the monomers contain highly bulky or polar substituents. Thus in the copolymerization of styrene (M_1) and fumaronitrile (M_2), chains rich in fumaronitrile and having styrene as the last added unit show greatly decreased reactivity with fumaronitrile monomer [Fordyce and Ham, 1951]. The effect is due to steric and polar repulsions between the penultimate fumaronitrile unit in the propagating chain and the incoming fumaronitrile monomer.

The mathematical treatment of the penultimate effect [Barb, 1953; Ham, 1964; Merz et al., 1946] in such a copolymerization involves the use of the eight propagating reactions

$$
\begin{aligned}
\sim\sim\sim M_1M_1\cdot + M_1 &\xrightarrow{\ k_{111}\ } \sim\sim\sim M_1M_1M_1\cdot \\
\sim\sim\sim M_1M_1\cdot + M_2 &\xrightarrow{\ k_{112}\ } \sim\sim\sim M_1M_1M_2\cdot \\
\sim\sim\sim M_2M_2\cdot + M_1 &\xrightarrow{\ k_{221}\ } \sim\sim\sim M_2M_2M_1\cdot \\
\sim\sim\sim M_2M_2\cdot + M_2 &\xrightarrow{\ k_{222}\ } \sim\sim\sim M_2M_2M_2\cdot \\
\sim\sim\sim M_2M_1\cdot + M_1 &\xrightarrow{\ k_{211}\ } \sim\sim\sim M_2M_1M_1\cdot \\
\sim\sim\sim M_2M_1\cdot + M_2 &\xrightarrow{\ k_{212}\ } \sim\sim\sim M_2M_1M_2\cdot \\
\sim\sim\sim M_1M_2\cdot + M_1 &\xrightarrow{\ k_{121}\ } \sim\sim\sim M_1M_2M_1\cdot \\
\sim\sim\sim M_1M_2\cdot + M_2 &\xrightarrow{\ k_{122}\ } \sim\sim\sim M_1M_2M_2\cdot
\end{aligned}
\tag{6-75}
$$

with the four reactivity ratios:

$$
\begin{aligned}
r_1 &= \frac{k_{111}}{k_{112}} \quad r_1' = \frac{k_{211}}{k_{212}} \\[6pt]
r_2 &= \frac{k_{222}}{k_{221}} \quad r_2' = \frac{k_{122}}{k_{121}}
\end{aligned}
\tag{6-76}
$$

Each monomer is thus characterized by two monomer reactivity ratios. One monomer reactivity ratio represents the propagating species in which the penultimate and terminal monomer units are the same. The other represents the propagating species in which the penultimate and terminal units differ. The later monomer reactivity ratios are signified by the prime notations (r_1' and r_2').

In a manner similar to that used in deriving Eq. 6-12, the copolymer composition with a kinetic penultimate effect present is obtained as

$$\frac{d[M_1]}{d[M_2]} = \frac{1 + \dfrac{r_1'X(r_1X + 1)}{(r_1'X + 1)}}{1 + \dfrac{r_2'(r_2 + X)}{X(r_2' + X)}} \tag{6-77}$$

where $X = [M_1]/[M_2]$. For the styrene-fumaronitrile system, fumaronitrile is incapable of self-propagation ($r_2 = r_2' = 0$) and Eq. 6-77 simplifies to

$$\frac{d[M_1]}{d[M_2]} - 1 = \frac{r_1'X(r_1X + 1)}{(r_1'X + 1)} \tag{6-78}$$

This equation gives a good fit of the experimental copolymer composition data (Fig. 6-15) with $r_1 = 0.072$ and $r_1' = 1.0$ [Barb, 1953]. It is clear that the copolymerization deviates markedly from the behavior predicted by the first-order Markov model ($r_1 = 0.23$).

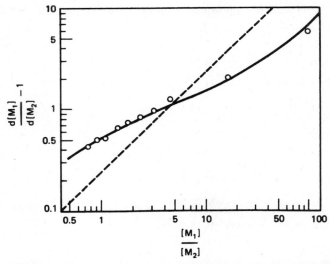

Fig. 6-15 Kinetic penultimate effect in the radical copolymer of styrene (M_1)-fumaronitrile (M_2); plot of copolymer composition as $\{(d[M_1]/d[M_2])-1\}$ versus the comonomer feed composition X. The solid line is the theoretical plot calculated from Eq. 6-78 with $r_1 = 0.072$, $r_1' = 1.0$. The dashed line is calculated assuming first-order Markov behavior (Eq. 6-12) with $r_1 = 0.23$. The circles represent the experimental data. After Barb [1953] (by permission of Wiley, New York).

Penultimate effects have been observed in other systems. Among these are the radical copolymerizations of ethyl methacrylate–styrene, methyl methacrylate–4-vinylpyridine, methyl acrylate–1,3-butadiene, hexafluoroisobutylene–vinyl acetate, 2,4-dicyano-1-butene–isoprene, and other comonomer pairs [Brown and Fujimori, 1987; Cowie et al., 1990; Davis et al., 1990; Fordyce and Ham, 1951; Guyot and Guillot, 1967; Hecht and Ojha, 1969; Hill et al., 1982, 1985; Motoc et al., 1978; Natansohn et al., 1978; Prementine and Tirrell, 1987; Rounsefell and Pittman, 1979; Van der Meer et al., 1979; Wu et al., 1990]. Although ionic copolymerizations have not been as extensively studied, penultimate effects have been found in some cases. Thus in the anionic polymerization of styrene–4-vinylpyridine, 4-vinylpyridine adds faster to chains ending in 4-vinylpyridine if the penultimate unit is styrene [Lee et al., 1963]. When 4-vinylpyridine is the penultimate unit, it apparently acts as an electron sink and hinders bond formation with the approaching monomer. Whether a penultimate effect exists in a particular system cannot always be easily detected. The precision and accuracy of the experimental composition data must be sufficient to allow one to discriminate between the use of Eq. 6-12 (or 6-15) and Eq. 6-77. Penultimate effects are most easily detected in experiments carried out at very low or very high f_1 values.

Data in the styrene (M_1)-fumaronitrile (M_2) system indicates there are effects due to remote monomer units preceding the penultimate unit. The $M_2M_1M_1$ sequence tends to alternate with M_2M_1 as shown by the observation that one cannot obtain copolymers containing more than 40 mole % fumaronitrile. The effect of remote units has been treated by further expansion of the copolymer composition equation by the use of greater numbers of monomer reactivity ratios for each monomer [Ham, 1964]. However, the utility of the resulting expression is limited because of the large number of variables.

6-5b Depropagation during Copolymerization

In contrast to the kinetic approach, Lowry [1960] treated the deviations from the terminal model from a thermodynamic viewpoint. Altered copolymer compositions in certain copolymerizations are accounted for in this treatment in terms of the tendency of one of the monomers (M_2) to depropagate. An essential difference between the kinetic and thermodynamic treatments is that the latter implies that the copolymer composition can vary with the concentrations of the monomers. If the concentration of monomer M_2 falls below its equilibrium value $[M]_c$ at the particular reaction temperature, terminal M_2 units will be prone to depropagate. The result would be a decrease in the amount of this monomer in the copolymer. The kinetic approach does not predict any dependence of the copolymer composition on the monomer concentration. Further, the thermodynamic approach differs from the kinetic approach in that the former emphasizes the temperature dependence of the copolymer composition since the polymerization–depolymerization equilibrium is temperature-dependent. (The *penultimate* model does not, however, predict the copolymer composition to be independent of temperature. The effect of temperature in the penultimate model comes from the variation of r_1, r_1', r_2, and r_2' with temperature.)

The present discussion will be almost completely limited to copolymerizations in which only one of the monomers has a tendency to depropagate. Systems in which both monomers tend to depropagate are difficult to treat mathematically and also involve a large number of unknown parameters. Different types of copolymerization behavior can be considered depending on whether one assumes penultimate effects on the depropagation reaction. Thus Lowry considers two different cases in which monomer M_1 has absolutely no tendency to depropagate irrespective of the preceding units in the chain, while monomer M_2 has no tendency to depropagate if it is attached to an M_1 unit. The two cases differ in the different tendencies of monomer M_2 to depropagate. In Case I, M_2 tends to depropagate if it is attached to another M_2 unit.

$$\sim\sim\sim M_1M_2M_2^* \rightleftharpoons \sim\sim\sim M_1M_2^* + M_2 \tag{6-79}$$

In Case II, M_2 tends to depropagate only when it is attached to a sequence of two or more M_2 units,

$$\sim\sim\sim M_1M_2M_2M_2^* \rightleftharpoons \sim\sim\sim M_1M_2M_2^* + M_2 \tag{6-80}$$

Thus $\sim\sim\sim M_1M_2^*$ does not depropagate in Case I, while neither $\sim\sim M_1M_2^*$ nor $\sim\sim M_1M_2M_2^*$ depropagate in Case II.

The copolymer composition for Case I is given by [Kruger et al., 1987; Lowry, 1960; Szymanski, 1987]

$$\frac{d[M_1]}{d[M_2]} = \frac{(r_1[M_1] + [M_2])(1 - \alpha)}{[M_2]} \tag{6-81}$$

with α defined by

$$\alpha = \frac{1}{2}\left(\left\{1 + K[M_2] + \left(\frac{K[M_1]}{r_2}\right)\right\} - \left[\left\{1 + K[M_2] + \left(\frac{K[M_1]}{r_2}\right)\right\}^2 - 4K[M_2]\right]^{1/2}\right) \tag{6-82}$$

where K is the equilibrium constant for the equilibrium in Eq. 6-79.

The copolymer composition for Case II is given by

$$\frac{d[M_1]}{d[M_2]} = \frac{\left\{\left(\frac{r_1[M_1]}{[M_2]}\right) + 1\right\}\left\{\alpha\gamma + \left[\frac{\alpha}{(1 - \alpha)}\right]\right\}}{\alpha\gamma - 1 + \left\{\frac{1}{(1 - \alpha)}\right\}^2} \tag{6-83}$$

with α defined by Eq. 6-82 and γ by

$$\gamma = \frac{\left\{K[M_2] + \frac{K[M_1]}{r_2} - \alpha\right\}}{K[M_2]} \tag{6-84}$$

where K is now the equilibrium constant for the equilibrium in Eq. 6-80.

The *depropagation* model, described by Eqs. 6-81 and 6-83, has been experimentally tested in several copolymerizations [Cais and Stuk, 1978; Florjanczyk and Krawiec, 1989; Hinton and Spencer, 1976; Ivin and Spensley, 1967; Motoc and Vancea, 1980; O'Driscoll and Gasparro, 1967; Sawada, 1976]. The systems studied included the radical copolymerizations of N-phenylmaleimide–styrene, 1,1-diphenylethylene–methyl acrylate, styrene–α-methylstyrene, styrene–methyl methacrylate, and acrylonitrile–α-methylstyrene and the cationic copolymerization of vinylmesitylene–α-methylstyrene. (Vinylmesitylene is 2,4,6-trimethylstyrene.) Most of these copolymerizations were studied over a range of comonomer feed compositions, reaction temperatures, and monomer concentrations. There is a transition from behavior consistent with the terminal model to behavior described by the depropagation model as the reaction conditions are altered (increased temperature, decreased monomer concentration) to favor depropagation. Thus in the radical copolymerization of styrene–α-methylstyrene, one observes a decrease in the α-methylstyrene content of the copolymer as the reaction temperature is increased from 0 to 100°C [O'Driscoll and Gasparro, 1967]. With increased temperature, there is increased depropagation of α-methylstyrene due to its low ceiling temperature (Sec. 3-9c). The effect is greatest for comonomer feed compositions rich in α-methylstyrene. The data in this system followed the quantitative behavior expected for Case II depropagation.

The anionic copolymerization of vinylmesitylene–α-methylstyrene is an interesting system. The copolymerization was first studied at -78°C at high monomer concentrations to determine the monomer reactivity ratios under conditions where depropagation was negligible. At the higher reaction temperature of 0°C, depropagation was still not important as long as the concentration of vinylmesitylene $[M_2]$ was sufficiently above the value of $[M_2]_c$ at that temperature. $[M_2]_c$ is 0.75 moles/liter at 0°C. When the vinylmesitylene concentration decreased below $[M_2]_c$ at constant $[M_1]/[M_2]$, depropagation became significant and the vinylmesitylene content of the copolymer decreased (see curved line in Fig. 6-16). The theoretical curves for this copolymerization for the Case I and Case II mechanisms were calculated from Eqs. 6-81 and 6-83 using the values of $[M_1]$, $[M_2]$, r_1, r_2, and K. The K value used in such calculations is, by necessity, that obtained from the polymerization–depolymerization equilibrium data for the homopolymerization of monomer M_2. The results are shown as the broken

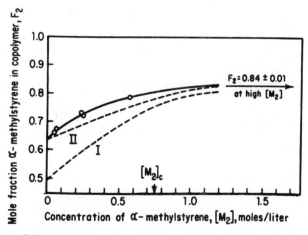

Fig. 6-16 Effect of depropagation on copolymer composition in the anionic copolymerization of vinylmesitylene–(M_1)-α-methylstyrene (M_2) at 0°C for f_2 constant at 0.91. The broken-line plots are the calculated plots for Lowry's Cases I and II (with $r_1 = 0.20$ and $r_2 = 0.72$); the experimental data follow the solid-line plot. After Ivin and Spensley [1967] (by permission of Marcel Dekker, New York).

curves in Fig. 6-16. It is seen that the Case II mechanism fits the experimental data much more closely than the Case I mechanism. This is the general behavior that has been observed for most of the systems studied. It has been noted that the theoretical curve would fit the data even more closely if one assumed that monomer M_1 also tends to propagation. Copolymerization with both M_1 and M_2 undergoing depropagation has been mathematically treated but the result is difficult to apply [Lowry, 1960].

6-5c Copolymerization with Complex Participation

Another model used to describe deviations from the terminal model involves the participation of a comonomer complex (Sec. 6-3b-3) [Seiner and Litt, 1971]. The comonomer complex competes with each of the individual monomers in propagation. The *complex participation* model involves the four propagation steps described by Eqs. 6-2 through 6-5 and the four propagation steps,

$$
\begin{aligned}
M_1^* + \overline{M_2M_1} &\xrightarrow{k_{\overline{121}}} M_1^* \\
M_1^* + \overline{M_1M_2} &\xrightarrow{k_{\overline{112}}} M_2^* \\
M_2^* + \overline{M_2M_1} &\xrightarrow{k_{\overline{221}}} M_1^* \\
M_2^* + \overline{M_1M_2} &\xrightarrow{k_{\overline{212}}} M_2^*
\end{aligned}
\tag{6-85}
$$

with a total of six reactivity ratios,

$$
r_1 = \frac{k_{11}}{k_{12}} \qquad r_2 = \frac{k_{22}}{k_{21}}
$$

$$
r_1' = \frac{k_{\overline{112}}}{k_{\overline{121}}} \qquad r_2' = \frac{k_{\overline{221}}}{k_{\overline{212}}}
\tag{6-86}
$$

$$
r_1'' = \frac{k_{\overline{121}}}{k_{12}} \qquad r_2'' = \frac{k_{\overline{212}}}{k_{21}}
$$

where $\overline{M_2M_1}$ and $\overline{M_1M_2}$ represent the complex adding to a propagating center at the M_2 and M_1 ends, respectively.

The copolymer composition is obtained by the statistical approach (Sec. 6-2b) as

$$
\frac{d[M_1]}{d[M_2]} = \frac{(1 - p_{12})(p_{21} + p_{\overline{221}}) + (1 - p_{22})(p_{12} + p_{\overline{112}})}{(1 - p_{11})(p_{21} + p_{\overline{221}}) + (1 - p_{21})(p_{12} + p_{\overline{112}})}
\tag{6-87}
$$

where the transition probabilities are defined by

$$
p_{11} = \frac{r_1[M_1]}{\Sigma R_1} \qquad\qquad p_{12} = \frac{[M_2]}{\Sigma R_1}
$$

$$
p_{\overline{121}} = \frac{r_1''[M_1M_2]}{\Sigma R_1} \qquad p_{\overline{112}} = \frac{r_1''r_1'[M_1M_2]}{\Sigma R_1}
$$

$$
p_{21} = \frac{[M_1]}{\Sigma R_2} \qquad\qquad p_{22} = \frac{r_2[M_2]}{\Sigma R_2}
\tag{6-88}
$$

$$
p_{\overline{221}} = \frac{r_2''r_2'[M_1M_2]}{\Sigma R_2} \qquad p_{\overline{212}} = \frac{r_2''[M_1M_2]}{\Sigma R_2}
$$

ΣR_1 and ΣR_2 are the sum of the rates of reaction of M_1^* and M_2^*, respectively, defined by

$$\Sigma R_1 = r_1[M_1] + [M_2] + r_1''[\overline{M_1M_2}](1 + r_1')$$
$$\Sigma R_2 = [M_1] + r_2[2] + r_2''[\overline{M_1M_2}](1 + r_2') \tag{6-89}$$

where $[\overline{M_1M_2}]$ is defined from the equilibrium constant of Eq. 6-55 as

$$[\overline{M_1M_2}] = K[M_1][M_2] \tag{6-90}$$

The complex participation model, like the depropagation model, predicts a variation of the copolymer composition with temperature and monomer concentration. The effect of temperature comes from the change in K, resulting in a decrease in the concentration of the comonomer complex with increasing temperature. Increasing monomer concentration at a constant f_1 increases the comonomer complex concentration.

The complex participation model has been tested in the radical copolymerizations of 1,1-diphenylethylene–methyl acrylate, styrene–β-cyanoacrolein, vinyl acetate–hexafluoroacetone, N-vinylcarbazole–diethyl fumarate, N-vinylcarbazole–fumaronitrile, maleic anhydride–vinyl acetate, and styrene–maleic anhydride [Cais et al., 1979; Dodgson and Ebdon, 1977; Fujimori and Craven, 1986; Georgiev and Zubov, 1978; Litt, 1971; Litt and Seiner, 1971; Yoshimura et al., 1978].

A variation of the complex participation model, referred to as the *complex dissociation* model, involves disruption of the complex during reaction with a propagating chain end [Hill et al., 1983; Karad and Schneider, 1978]. Reaction of the propagating center with the complex results in the addition of only one of the monomers with liberation of the unreacted monomer. The overall result is that the complex alters monomer reactivities.

6-5d Discrimination between Models

The ability to determine which copolymerization model best describes the behavior of a particular comonomer pair depends on the quality of the experimental data. There are many reports in the literature where different workers conclude that a different model describes the same comonomer pair. This occurs when the accuracy and precision of the composition data are insufficient to easily discriminate between the different models or composition data are not obtained over a wide range of experimental conditions (feed composition, monomer concentration, temperature). There are comonomer pairs where the behavior is not sufficiently extreme in terms of depropagation or complex participation or penultimate effect such that even with the best composition data it may not be possible to conclude that only one model fits the composition data [Hill et al., 1985; Moad et al., 1989].

The sequence distributions expected for the different models have been adequately described [Hill et al., 1982, 1983; Howell et al., 1970; Tirrell, 1986]. Sequence distributions obtained by ^{13}C NMR are often more useful than composition data for discriminating between different copolymerization models. For example, while composition data for the radical copolymerization of styrene–acrylonitrile is consistent with either the penultimate or complex participation model, sequence distributions show the penultimate model to give the best fit.

6-6   COPOLYMERIZATIONS INVOLVING DIENES

6-6a   Crosslinking

Diene monomers are often used in copolymerizations to obtain a crosslinked structure in the final product. The reaction is generally analogous to step polymerizations involving tri- and tetrafunctional reactants (Sec. 2-10). Crosslinking occurs early or late in the copolymerization depending on the relative reactivities of the two double bonds of the diene. The extent of crosslinking depends on the latter and on the amount of diene relative to the other monomer. There is an extensive literature on the mathematical treatment of the crosslinking process [Dotson et al., 1988; Flory, 1947, 1953; Macosko and Miller, 1976; Scranton and Peppas, 1990; Shultz, 1966; Tobita and Hamielec, 1989; Williams and Vallo, 1988]. Several different cases can be distinguished depending on the type of diene. In most instances it is assumed that the diene is present at low concentrations.

The first case is the copolymerization of monomer A with diene BB where all of the double bonds (i.e., the A double bond and both B double bonds) have the same reactivity. Methyl methacrylate–ethylene glycol dimethacrylate (EGDM), vinyl acetate–divinyl adipate (DVA), and styrene–p- or m-divinylbenzene (DVB) are examples of this type of copolymerization system [Landin and Macosko, 1988; Li et al., 1989; Storey, 1965; Ulbrich et al., 1977]. Since $r_1 = r_2$, $F_1 = f_1$. At the extent of reaction p (defined as the fraction of A and B double bonds reacted), there are $p[A]$ reacted A double bonds, $p[B]$ reacted B double bonds, and $p^2[BB]$ reacted BB monomer units. [A] and [B] are the concentrations of A and B double bonds, [BB] is the concentration of BB, and $[B] = 2[BB]$. The number of crosslinks is simply the number of BB monomer molecules in which both B double bonds are reacted, that is, $p^2[BB]$. The number of polymer chains is the total number of A and B double bonds reacted divided by the degree of polymerization, $([A] + [B])p/\overline{X}_w$ (the weight-average degree of polymerization is employed for reasons previously described in Sec. 2-10). The critical extent of reaction at the gel point p_c occurs when the number of crosslinks per chain is $\frac{1}{2}$ and thus is given by

$$p_c = \frac{[A] + [B]}{[B]\overline{X}_w} \tag{6-91}$$

$\overline{X}_w$ in Eq. 6-91 is essentially the weight-average degree of polymerization that would be observed in the polymerization of monomer A in the absence of diene BB. Equation 6-91 predicts that extensive crosslinking occurs during this type of copolymerization. Thus gelation is observed at 12.5% reaction in the methyl methacrylate–ethylene glycol dimethacrylate system containing 0.05 mole % EGDM [Walling, 1945; Yoshimura et al., 1978]. The use of Eq. 6-91 for the styrene–p-divinylbenzene (DVB) system is shown in Table 6-13. The equation holds best for systems containing low concentrations of the diene monomer; its utility decreases as the concentration of diene increases. With increasing diene concentration, Eq. 6-91 predicts gel points at conversions that are increasingly lower than those found experimentally. This general behavior has been attributed to the wastage of the diene monomer due to intramolecular cyclization (Sec. 6-6b). Also, there is an indication that the reactivity of the second double bond in BB is decreased on reaction of the first double bond as a consequence of its presence in a polymer chain [Hild and Okasha, 1985].

TABLE 6-13   Crosslinking in the Copolymerization of Styrene–Divinylbenzene

| | Gel Point (p_c) | |
Mole Fraction DVB	Calculated from Eq. 6-91	Observed[a]
0.004	0.21	0.16
0.008	0.10	0.14
0.02	0.042	0.076
0.032	0.026	0.074
0.082	0.010	0.052
0.30	0.0042	0.045

[a]Data from Storey [1965].

An alternate expression for p_c is given by

$$p_c = \frac{(1 - q)}{[af(f - 2)(q + \xi/2]} \tag{6-92}$$

where q is the ratio of the propagation rate to the sum of the propagation rate and the rates of all termination and transfer reactions, ξ is given by the ratio of the rate of termination by coupling to the sum of all termination and transfer reactions, f is the functionality of the diene and is equal to 4 (i.e., each double bond is bifunctional), and a is the fraction of all functional groups in the reaction mixture belonging to the diene [Macosko and Miller, 1976; Williams and Vallo, 1988].

A second case is the copolymerization of A and BB in which the reactivities of the two groups are not equal but are, instead, r_1 and r_2, respectively. In this case the critical extent of reaction at gelation is given by

$$p_c = \frac{(r_1[A]^2 + 2[A][B] + r_2[B]^2)^2}{\overline{X}_w[B]([A] + [B])(r_2[B] + [A])^2} \tag{6-93}$$

which reduces to

$$p_c = \frac{[A]r_1^2}{[B]\overline{X}_w} \tag{6-94}$$

for [A] >> [B]. When the double bonds of the diene are more reactive than that of the other monomer ($r_2 > r_1$), crosslinking occurs in the early stages of the copolymerization. Crosslinking is delayed until the later stages if $r_1 > r_2$. For the system where $r_1 > r_2$ and $r_1 > 1$ crosslinking is not as extensive at a given extent of reaction as that taking place in copolymerizations of the type governed by Eq. 6-91, where $r_1 = r_2$.

The third case is the copolymerization of a monomer A with the diene BC where groups A and B have equal reactivity, but group C has a much lower reactivity. An example of such a case would be methyl methacrylate-allyl methacrylate, where A and B are the two methacrylate groups and C is the allyl group. If r is the reactivity ratio between the C and B groups, then the following hold for the radicals derived from the A and B groups:

$$r = \frac{k_{AC}}{k_{AA}} = \frac{k_{AC}}{k_{AB}} = \frac{k_{BC}}{k_{BA}} = \frac{k_{BC}}{k_{BB}} \tag{6-95}$$

For such a system the copolymer will consist of copolymerized A and B groups with pendant, unreacted C groups until the later stages of reaction. Crosslinking does not occur until relatively late in the reaction due to the low reactivity of the C group. The critical extent of reaction at gelation in this case is given by

$$p_c = 1 - \exp \frac{-1}{2q\overline{X}_w r} \tag{6-96}$$

where q is the mole fraction of the diene in the initial comonomer feed.

In selecting the makeup of a crosslinking system one has a number of variables that can be used to control the process. Thus gelation can be delayed with the production of a high-conversion uncrosslinked product, which is amenable to subsequent crosslinking. Gelation can be delayed by reducing the amount of the diene, the degree of polymerization by using chain-transfer agents, or the reactivity of one of the double bonds of the diene by proper choice of the diene reactant. The extent of crosslinking in the final product is also controlled by these variables. Extensive crosslinking, with the formation of tight network structures, is obtained by avoiding chain transfer and using increased amounts of dienes whose double bonds have similar reactivities.

A special situation that is often encountered is that where the reaction of one of the double bonds of the diene results in a decrease in the reactivity of the remaining double bond. If the decrease in reactivity is large, the effect is to markedly delay the crosslinking reaction. This case then becomes very similar to the last one where the C group has a much lower reactivity than the A and B groups. The most notable case in which there is a large drop in reactivity of one group on reaction of the other is in the copolymerization of 1,3-dienes where 1,4-polymerization leads to residual 2,3-double bonds which have lowered reactivity. These are subsequently used to bring about crosslinking by reactions dicussed in Sec. 9-2b.

6-6b   Alternating Intra–Intermolecular Polymerization; Cyclopolymerization

The polymerization of unconjugated dienes [e.g., diallyl phthalate, diethylene glycol bis(allyl carbonate), diallyl maleate] and trienes (e.g., triallyl cyanurate) to form highly crosslinked thermosetting products is a commercially important process. This is also true of the use of such monomers to bring about crosslinking in copolymerization systems (Sec. 6-6a). The crosslinking reaction is almost always found to be somewhat inefficient in that the experimental gel points occur at higher extents of reaction than those predicted by theory. There is a wastage of the diene or triene monomer which has been ascribed to intramolecular cyclization of the diene or triene. This phenomenon is encountered not only in homo- and copolymerizations involving dienes and trienes but also in step polymerizations with trifunctional reactants (Sec. 2-10c).

The extent of cyclization varies considerably depending on the particular reaction system. There are some reactants for which cyclization is very extensive. The competition between the intermolecular and intramolecular reactions in the polymerization of a diene is depicted in Eqs. 6-97 through 6-99, where Z is a structural unit such as the benzene ring in divinylbenzene or the trimethylene group in 1,6-heptadiene. Intermolecular propagation proceeds according to the horizontal set of reactions (Eq. 6-97) to produce a linear polymer with pendant double bonds. The pendant double bonds of a propagating or terminated polymer chain give rise to a crosslinked final product by copolymerizing with either unreacted monomer or the pendant double

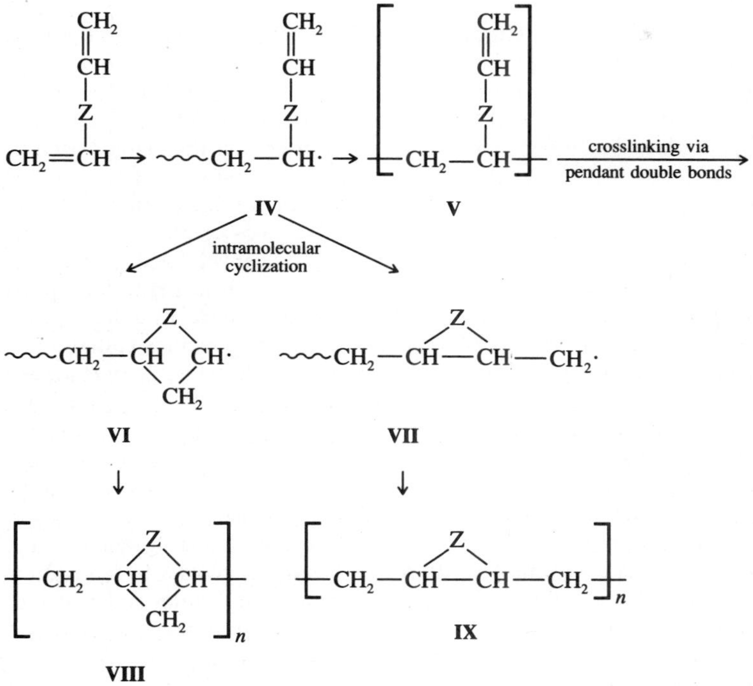

$$
\begin{array}{c}
CH_2 \\
\parallel \\
CH \\
\mid \\
Z \\
\mid \\
CH_2{=}CH
\end{array}
\;\rightarrow\;
\sim\sim\!CH_2{-}CH\cdot
\;\rightarrow\;
\left[\begin{array}{c}
CH_2 \\
\parallel \\
CH \\
\mid \\
Z \\
\mid \\
{-}CH_2{-}CH{-}
\end{array}\right]
\xrightarrow[\text{pendant double bonds}]{\text{crosslinking via}}
\qquad (6\text{-}97)
$$

IV   V

intramolecular
cyclization

$$
\sim\sim\!CH_2{-}\overset{\displaystyle Z}{\underset{\displaystyle CH_2}{CH}}\,CH\cdot
\qquad\qquad
\sim\sim\!CH_2{-}CH{-}CH{-}CH_2\cdot
\qquad (6\text{-}98)
$$

VI   VII

$$
\left[\begin{array}{c}
{-}CH_2{-}\overset{\displaystyle Z}{\underset{\displaystyle CH_2}{CH}}\,CH{-}
\end{array}\right]_n
\qquad
\left[\begin{array}{c}
{-}CH_2{-}CH{-}CH{-}CH_2{-}
\end{array}\right]_n
\qquad (6\text{-}99)
$$

VIII   IX

bonds of other chains. Cyclization occurs when propagating species **IV** reacts intramolecularly with its own pendant double bond in preference to intermolecular propagation. Attachment of the radical center at the methylene and methine carbons of the pendant double bond yields different-sized ring structures as shown in **VIII** and **IX**, respectively. Similar reaction sequences would apply for the competition between inter- and intramolecular propagations in polymerizations involving trienes or copolymerizations in which one of the monomers is a diene or triene. Copolymerizations are more complicated since cyclization can occur before as well as after the addition of the second monomer to a propagating species such as **IV**.

The importance of intramolecular cyclization was emphasized when Butler and coworkers [Butler and Angelo, 1957; Butler and Ingley, 1951] found that the radical polymerizations of diallyl quaternary ammonium salts gave soluble, uncrosslinked polymers with little or no residual unsaturation. There is no tendency for radical **IV** to propagate intermolecularly and undergo crosslinking. The exclusive reaction is intramolecular cyclization and the product is a completely linear polymer with cyclic structures in the backbone. The reaction is referred to as *alternating intra–intermolecular polymerization* or *cyclopolymerization*.

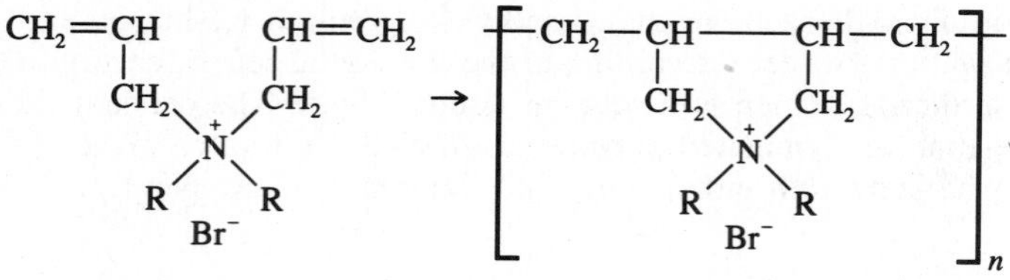

$$
\begin{array}{c}
\text{CH}_2\!\!=\!\!\text{CH} \quad \text{CH}\!\!=\!\!\text{CH}_2 \\
\text{CH}_2 \quad \text{CH}_2 \\
\text{N}^+ \\
\text{R} \quad \text{R} \\
\text{Br}^-
\end{array}
\quad\rightarrow\quad
\left[\!\!
\begin{array}{c}
\text{CH}_2\!-\!\text{CH}\!-\!\!-\!\text{CH}\!-\!\text{CH}_2 \\
\text{CH}_2 \quad \text{CH}_2 \\
\text{N}^+ \\
\text{R} \quad \text{R} \\
\text{Br}^-
\end{array}
\!\!\right]_n
\qquad (6\text{-}100)
$$

The size of the ring structure that can be formed determines whether intermolecular polymerization or intramolecular cyclization is the predominant reaction for a particular monomer [Butler, 1986, 1989; Marvel and Vest, 1957, 1959]. The extent of cyclization generally increases in the following order: 5- and 6-membered rings appreciably greater than 7-membered rings and the latter greater than larger sized rings. 1,6-Dienes (Z in Eqs. 6-97 through 6-99 contains three ring atoms) such as acrylic and methacrylic anhydrides, diallyl quaternary ammonium salts, methyl allyl maleate and fumarate, and allyl methacrylate cyclopolymerize to 5- or 6-membered rings or mixtures of the two. Formation of the 5-membered ring involves kinetic control of the cyclization process proceeding through the less stable radical **VII** to form polymer **IX** instead of thermodynamic control, proceeding through radical **VI** to form the more stable 6-membered ring structure **VIII**. Interestingly, diallyl amines and ammonium salts and divinyl formal form 5-membered rings almost exclusively. Acrylic anhydride forms a mixture of 5- and 6-membered rings while methacrylic anhydride forms only 6-membered rings. Methyl allyl maleate forms mostly 5-membered rings while methyl allyl fumarate forms mostly the 6-membered ring. Diallyldimethylsilane forms mostly 6-membered rings [Saigo et al., 1988]. Dimethacrylamide forms mostly the 6-membered ring, while N-substituted dimethacrylamides form mostly the 5-membered ring. Some systems show a considerable effect of reaction conditions on ring size. Decreasing reaction temperature and solvent polarity result in large decreases in the content of 5-membered rings in the polymerization of acrylic anhydride [Butler and Matsumoto, 1981]. The ring size for methacrylic anhydride is unaffected by reaction temperature and solvent polarity. The variety of behaviors for different diene structures and the sensitivity of some systems to reaction conditions indicate a complex competition between kinetic control and thermodynamic control.

Five-membered rings are usually formed from 1,5-dienes such as 1,5-hexadiene; there are no reports of the formation of the 4-membered ring structure (Z = two ring atoms). However, cyclopolymerization of o-divinylbenzene forms 7-membered rings via a ring closure reaction involving three double bonds per each pair of diene mol-

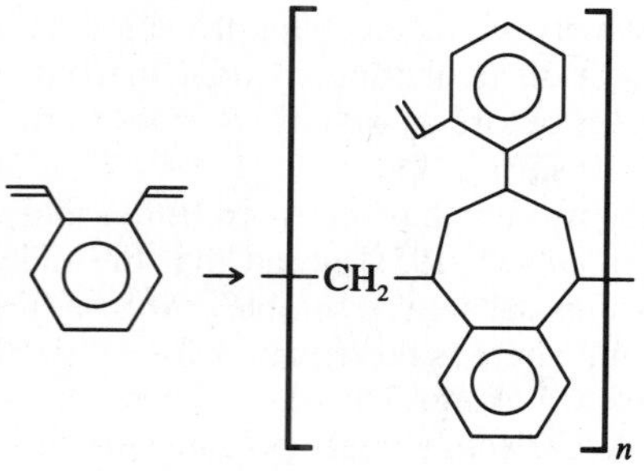

$$ \qquad (6\text{-}101) $$

ecules [Costa et al., 1978]. Cyclopolymerization is observed for ionic polymerizations as well as radical, although the former have been much less studied. Other types of monomers have also been studied, including diynes via coordination catalysts (Chap. 8) [Kim et al., 1988] and dialdehydes and diisocyanates via ionic initiation.

The competition between the rates of intermolecular propagation R_p and intra-molecular cyclization R_c can be expressed in terms of the fraction of cyclized units f_c defined by

$$f_c = \frac{R_c}{R_c + R_p} \tag{6-102}$$

where

$$R_p = 2k_p[M^*][M] \tag{6-103}$$

$$R_c = k_c[M^*] \tag{6-104}$$

The factor of 2 in Eq. 6-103 is due to [M] being defined as the concentration of diene molecules instead of double bonds. Combination of Eqs. 6-101 through 6-103 yields

$$\frac{1}{f_c} = 1 + \frac{2k_p[M]}{k_c} \tag{6-105}$$

Equation 6-105 shows that the extent of cyclization is greater at low monomer concentration. Intermolecular propagation is increasingly favored with increasing [M]. The cyclization ratio k_c/k_p is obtained from the slope of a plot of f_c versus [M]. Values of k_c/k_p are in the range 2–20 moles/liter for most symmetrical 1,6-dienes but some values are even higher; for instance, divinyl formal has a value of 200. 1,5-Dienes generally have lower k_c/k_p values than their 1,6 counterparts. The high tendency toward cyclization is due to a favorable entropy factor. Cyclization has a higher activation energy than does intermolecular propagation. However, cyclization proceeds with a considerably smaller decrease in activation entropy than does intermolecular propagation. For example, for methacrylic anhydride, the activation energy factor favors intermolecular propagation by 10.9 kJ/mole (2.6 kcal/mole), while cyclization is favored by the Arrhenius frequency factor by 256 moles/liter. The overall result is a k_c/k_p value of 2.4 moles/liter. The tendency toward cyclization increases with increasing reaction temperature since cyclization has a higher activation energy than does intermolecular propagation. Cyclization is also increased by using more polar solvents but the mechanism for the solvent effect is not understood [Matsumoto et al., 1987].

The tendency toward cyclization decreases considerably (lower k_c/k_p value) for unsymmetrical 1,6-dienes, such as allyl methacrylate, where the two double bonds have significantly different reactivities. The polymer contains linear repeat units, rings, and pendant double bonds in relative amounts determined by k_c/k_p. The pendant double bonds eventually react to yield a crosslinked structure. Reactants with more than two double bonds per molecule, whether symmetrical such as triallylcyanurate and tetraallyl ammonium bromide or unsymmetrical such as diallyl maleate, behave similarly with the formation of crosslinked network structures.

The extent of cyclization decreases quite sharply as one goes to ring sizes of 7 or more atoms. However, contrary to expectations, the extent of cyclization is still quite significant for many monomers. Thus the extent of cyclization (measured as the percent of monomer units that are cyclized) is 15–20% for diallyl esters giving ring structures containing up to 17 atoms [Holt and Simpson, 1956]. The polymerization of *o*-diallyl phthalate yields a polymer with more than 40% cyclization [Eaton et al., 1989; Matsumoto et al., 1980].

Cyclopolymerizations yielding more complex ring structures have also been reported [Butler, 1986, 1989]. For example, 1-4-dienes such as divinyl ether yield uncrosslinked products with little or no saturation. The product from divinyl benzene has several different bicyclic structures in the polymer chain. The formation of one of the bicyclic structures is shown in the sequence in Eqs. 6-106 through 6-108 [Tsukino and Kunitake, 1979].

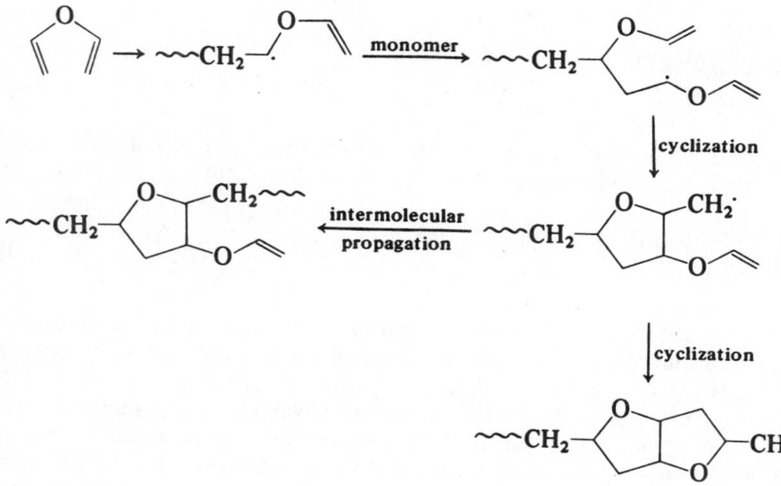

$$(6\text{-}106)$$

$$(6\text{-}107)$$

$$(6\text{-}108)$$

Recall the discussion in Sec. 2-3 concerning the competition between linear polymerization and cyclization in step polymerizations. Cyclization is not competitive with linear polymerization for ring sizes greater than 7 atoms. Further, even for most of the reactants, which would yield rings of 5, 6, or 7 atoms if they cyclized, linear polymerization can be made to predominate because of the interconvertibility of the cyclic and linear structures. The difference in behavior between chain and step polymerizations arises because the cyclic structures in chain polymerization do not depropagate under the reaction condition; that is, the cyclic structure does not interconvert with the linear structure.

6-6c   Interpenetrating Polymer Networks

An interpenetrating polymer network (IPN) (Sec. 2-13c-3) is obtained by carrying out a polymerization with crosslinking in the presence of another already crosslinked polymer [Klempner and Berkowski, 1987]. For example, a crosslinked polyurethane is swollen with a mixture of methyl methacrylate, trimethylolpropane trimethacrylate, and benzoyl peroxide and heated. The methacrylate system polymerizes to a crosslinked network which interpenetrates the polyurethane network. An IPN has the potential for combining the properties of two different crosslinked polymers.

6-7 OTHER COPOLYMERIZATIONS

6-7a Miscellaneous Copolymerizations of Alkenes

A variety of reactants—including sulfur dioxide, carbon monoxide, oxygen, and quinones, which do not homopolymerize because of polar and/or steric considerations—undergo radical copolymerization with alkenes to form polymeric sulfones [Bae et al., 1988; Cais and O'Donnell, 1976; Dainton and Ivin, 1958; Florjanczyk et al., 1987; Tsonis and Ali, 1989], ketones [Starkweather, 1987; Steinberg et al., 1979], peroxides [Cais and Bovey, 1977; Mukundan and Kishore, 1987; Nukui et al., 1982], and ethers [Hauser and Zutty, 1967].

$$CH_2\!=\!CHR + SO_2 \longrightarrow -\!(\!-CH_2CHR\!-\!SO_2\!-\!)_n \qquad\qquad (6\text{-}109)$$

$$CH_2\!=\!CHR + CO \longrightarrow -\!(\!-CH_2CHR\!-\!CO\!-\!)_n \qquad\qquad (6\text{-}110)$$

$$CH_2\!=\!CHR + R'\!-\!NO \longrightarrow \left[-CH_2CHR\!-\!\overset{\overset{\displaystyle R'}{|}}{N}\!-\!O \right]_n \qquad (6\text{-}111)$$

$$CH_2\!=\!CHR + O_2 \longrightarrow -\!(\!-CH_2CHR\!-\!O\!-\!O\!-\!)_n \qquad\qquad (6\text{-}112)$$

The reaction with sulfur dioxide is the most studied of these copolymerizations. Only alkenes such as ethylene, α-olefins, vinyl chloride, and vinyl acetate, without strong electron-withdrawing substituents and that yield highly reactive radicals, undergo facile copolymerization with sulfur dioxide. Many of these monomers yield 1:1 alternating copolymers as a result of the polar effect between the electropositive sulfur dioxide and electronegative alkene. Copolymers with greater than 50% sulfur dioxide are not formed as a result of polar repulsions between adjacent SO_2 units. Monomers without strong electron-withdrawing substituents but that yield relatively stable radicals, such as styrene and 1,3-butadiene, are less reactive toward sulfur dioxide or form copolymers containing less sulfur dioxide than the alternating structure. Monomers with strong electron-withdrawing substituents, such as acrylonitrile and methyl methacrylate, generally do not form copolymers due to polar repulsions with the electrophilic sulfur dioxide. Many copolymerizations with sulfur dioxide show a significant tendency toward depropagation. Depropagation is increasingly important as the substituent on the alkene monomer becomes bulkier or more electropositive.

There is a tendency toward alternation in the copolymerization of ethylene with carbon monoxide. Copolymerizations of carbon monoxide with tetrafluoroethylene, vinyl acetate, vinyl chloride, and acrylonitrile have been reported but with few details [Starkweather, 1987]. The reactions of alkenes with oxygen and quinones are not well defined in terms of the stoichiometry of the products. These reactions are better classified as retardation or inhibition reactions because of the very slow copolymerization rates (Sec. 3-7a). Other copolymerizations include the reactions of alkene monomers with sulfur and nitroso compounds [Green et al., 1967; Miyata and Sawada, 1988].

6-7b Copolymerization of Carbonyl Monomers

Although the homopolymerization of carbonyl monomers has been studied fairly extensively (Sec. 5-5), there are only a few reported studies on the copolymerization of these monomers

$$\text{RCHO} + \text{R'CHO} \rightarrow \text{---(---CHR---O---CHR'---O---)}_{\overline{n}} \qquad (6\text{-}113)$$

Cationic copolymerizations of acetaldehyde with formaldehyde and other higher aldehydes have been carried out [Furukawa and Saegusa, 1963; Vogl, 1967]. Anionic copolymerizations have been reported with formaldehyde, acetaldehyde, and various chloro-substituted acetaldehydes [Furukawa and Saegusa, 1963]. Many of these studies are too fragmentary to give a clear picture of the reaction. In some cases it is unclear whether block or statistical copolymers were obtained. The identity of the initiator appeared to be important in other instances. Depending on the initiator, a comonomer pair might yield homopolymer, block copolymer, or statistical copolymer. The general tendency in the copolymerization of carbonyl monomers is ideal behavior. However, there is a trend toward alternation for comonomer pairs in which one of the monomers has a bulky substituent. Thus the anionic copolymerization of acetaldehyde (M_1)-chloral (M_2) at $-78°C$ shows an alternation tendency with $r_1 = 0.18$ and $r_2 = 0$ [Iwata et al., 1964].

The copolymerization of carbonyl monomers with alkenes has been even less studied than that between different carbonyl monomers. The radiation-initiated copolymerization of styrene with formaldehyde proceeds by a cationic mechanism with a trend toward ideal behavior, $r_1 = 52$ and $r_2 = 0$ at $-78°C$ [Castille and Stannett, 1966]. Hexafluoroacetone undergoes radiation-initiated copolymerization with ethylene, propylene, and other α-olefins [Watanabe et al., 1979]. Anionic copolymerizations of aldehydes with isocyanates have also been reported [Odian and Hiraoka, 1972].

6-8 APPLICATIONS OF COPOLYMERIZATION

6-8a Styrene

Most polystyrene products are not homopolystyrene since the latter is relatively brittle with low impact and solvent resistance (Secs. 3-13b-2, 6-1a). Various combinations of copolymerization and blending are used to improve the properties of polystyrene [Moore, 1989]. Copolymerization of styrene with 1,3-butadiene imparts sufficient flexibility to yield elastomeric products [styrene–1,3-butadiene rubbers (SBR)]. Most SBR rubbers (trade names: *Buna, GR-S, Philprene*) are about 25% styrene–75% 1,3-butadiene copolymer produced by emulsion polymerization; some are produced by anionic polymerization. About 2 billion pounds per year are produced in the United States. SBR is similar to natural rubber in tensile strength, has somewhat better ozone resistance and weatherability but has poorer resilience and greater heat buildup. SBR can be blended with oil (referred to as *oil-extended* SBR) to lower raw material costs without excessive loss of physical properties. SBR is also blended with other polymers to combine properties. The major use for SBR is in tires. Other uses include belting, hose, molded and extruded goods, flooring, shoe soles, coated fabrics, and electrical insulation.

Styrene–1,3-butadiene copolymers with higher styrene contents (50–70%) are used in latex paints. Styrene and 1,3-butadiene terpolymerized with small amounts of an unsaturated carboxylic acid are used to produce latices that can be crosslinked through the carboxyl groups. These *carboxylated* SBR products are used as backing material for carpets. Styrene copolymerized with divinyl benzene yields crosslinked products, which find use in size-exclusion chromatography and as ion-exchange resins (Sec. 9-6).

Radical copolymerization of styrene with 10–40% acrylonitrile yields styrene–acrylonitrile (SAN) polymers (trade name: *Fostacryl*). Acrylonitrile, by increasing the intermolecular forces, imparts solvent resistance, improved tensile strength, and raises the upper use temperature of polystyrene although impact resistance is only slightly improved. SAN finds applications in houseware (refrigerator shelves and drawers, coffee mugs), packaging (bottle closures and sprayers), furniture (chair backs and shells), and electronics (battery cases, cassette parts). About 200 million pounds per year of SAN products are produced in the United States.

Acrylonitrile–butadiene–styrene (ABS) polymers (trade names: *Abson, Abafil, Blendix, Dylel, Kralac, Novodur, Terluran, Tybrene, Uscolite*) combine the properties of SAN with greatly improved resistance to impact. ABS is produced by emulsion, suspension, or bulk copolymerization of styrene–acrylonitrile in the presence of a rubber. The rubber is either poly(1,3-butadiene) or SBR. NBR (also referred to as *nitrile rubber*), a copolymer of 1,3-butadiene and acrylonitrile, is also used. The product of the reaction is a physical mixture of styrene–acrylonitrile copolymer and the graft copolymer of styrene–acrylonitrile onto the rubber. Additionally, SAN is often blended into that mixture. The final product, ABS, consists of a glassy polymer (SAN) dispersed in a rubbery matrix (grafted rubber). About 2 billion pounds per year of ABS are produced in the United States. Applications for ABS include housewares (refrigerator doors, sewing machine and hair-dryer housings, luggage, furniture frames, margarine tubs), housing and construction (pipe, conduit, fittings, bathtubs and shower stalls), transportation (automotive instrument panels, light housings, grilles), business machine housings (telephone, calculator), and recreation (golf clubs, boat hulls, camper top or shell, snowmobile shroud). Mixtures of styrene–methyl methacrylate copolymer and the graft copolymer onto a rubber provide higher heat resistance and improved adhesion to fiberglass compared to ABS.

High-impact polystyrene (HIPS) is produced by polymerizing styrene in the presence of a rubber, usually poly(1,3-butadiene). HIPS has improved impact resistance compared to polystyrene and competes with ABS products at low cost end applications such as fast-food cups, lids, takeout containers, toys, kitchen appliances, and personal-care product containers. HIPS as well as ABS and SMA are used in physical blends with other polymers, such as polycarbonates, polyesters, and polyamides, to improve impact resistance (Sec. 2-13c-3).

6-8b   Ethylene

A significant fraction, perhaps as high as 20–25%, of the low-density polyethylene (LDPE) (Sec. 3-13b-1) produced by radical polymerization consists of various copolymers of ethylene. LDPE has come under increasing economic pressure in recent years because of a combination of factors [Doak, 1986]. High-density polyethylene (HDPE) has displaced LDPE in applications such as blow-molded bottles and thin films where the increased strength of HDPE is preferred over the clarity of LDPE. Linear low-density polyethylene (LLDPE) (Sec. 8-4g-3) competes effectively with LDPE in terms of both cost and properties. New producers of ethylene have entered the LDPE market because of a lack of alternatives for their feedstocks. Many LDPE producers use copolymerization as a strategy to obtain products more resistant to displacement by HDPE and LLDPE.

Ethylene–vinyl acetate (EVA) copolymers (trade names: *Elvax, Escorene, Ultrathene*) represent the largest-volume segment of the ethylene copolymer market, with an annual production of about 1 billion pounds in the United States. As the vinyl acetate content increases, there are decreases in crystallinity, glass and crystalline melting temperatures, and chemical resistance coupled with increases in optical clarity, impact and stress crack resistance, flexibility, and adhesion to a variety of substrates. Copolymers containing 2–18% vinyl acetate are used in meat, poultry, and frozen-food packaging, stretch and shrink films, drum liners, extrusion coating on aluminum foil and polyester film, and heat-sealable coextruded pouches. Copolymers containing up to 20% vinyl acetate are used for molding or extruding squeeze toys, hose, tubing, gaskets, and insulation for electrical wire and cable. Copolymers containing 20–30% vinyl acetate are used as blends with paraffin wax and elastomers (carpet backing, hot-melt adhesives) and bitumen (highway asphalt). Hydrolysis of EVA copolymers yields ethylene–vinyl alcohol copolymers (EVOH). EVOH has exceptional gas barrier properties as well as oil and organic solvent resistance. The poor moisture resistance of EVOH is overcome by coating, coextrusion, and lamination with other substrates. Applications include containers for food (ketchup, jelly, mayonnaise) as well as chemicals and solvents.

Ethylene copolymers with methyl methacrylate and ethyl, butyl, and methyl acrylates are similar to EVA products but have improved thermal stability during extrusion and increased low-temperature flexibility. The commercial products generally contain 15–30% of the acrylate or methacrylate comonomer. Applications include medical packaging, disposable gloves, hose, tubing, gaskets, cable insulation, and squeeze toys.

Terpolymers in which the acrylate monomer is the major component are useful as ethylene–acrylate elastomers (trade name: *Vamac*) [Hagman and Crary, 1985]. A small amount of an alkenoic acid is present to introduce sites ($C{=}C$) for subsequent crosslinking via reaction with primary diamines (Sec. 9-2d). These elastomers have excellent oil resistance and stability over a wide temperature range (-50 to 200°C). They are superior to nitrile and chloroprene rubbers. Although not superior to silicone and fluorocarbon elastomers, they are less costly; uses include automotive (hydraulic system seals, hoses) and wire and cable insulation.

Copolymers of ethylene with up to 15–20% acrylic or methacrylic acid offer improved adhesion, abrasion resistance, toughness, and low-temperature flexibility compared to EVA. Applications include extrusion coatings on aluminum foil for pouches, wire and cable, packaging film, laminations with metal and glass fibers (building and automotive products) and polyurethane (carpet backing).

Neutralization of ethylene copolymers containing up to 5–10% acrylic or methacrylic acid copolymer with a metal salt such as the acetate or oxide of zinc, sodium, magnesium, barium, or aluminum yields products referred to as *ionomers* (trade name: *Surlyn*). Ionomers act like reversibly crosslinked thermoplastics as a result of microphase separation between ionic metal carboxylate and nonpolar hydrocarbon segments. The behavior is similar to the physical crosslinking in thermoplastic elastomers (Secs. 2-13c-2, 5-4a). Processing by extrusion and molding can be accomplished at higher temperatures where aggregation of the ionic segments from different polymer chains is destroyed (or made mobile). Subsequent to processing, the product becomes crosslinked on cooling to ambient temperature since aggregation of ionic segments is reestablished. Ionomers possess outstanding low-temperature flexibility, abrasion and impact resistance, good optical clarity, and adhesion to metals. Applications include the heat-sealable layer in packaging composites for food, skin packaging for hardware and electronics products, golf-ball and bowling ball covers, automotive bumper pads, insulating covers for hot-water tanks, and wrestling mats.

Copolymerization is also important for high-density polyethylene (Sec. 8-4g-2).

6-8c Unsaturated Polyesters

The crosslinking of unsaturated polyesters (Sec. 2-12a) is carried out by copolymerization [Selley, 1988]. Low-molecular-weight unsaturated polyester (prepolymer) and radical initiator are dissolved in a monomer, the mixture poured, sprayed, or otherwise shaped into the form of the desired final product, and then transformed into a thermoset by heating. Styrene is the most commonly used monomer. Vinyltoluene, methyl methacrylate, diallyl phthalate, α-methylstyrene, and triallyl cyanurate are also used, often together with styrene. The crosslinking process involves copolymerization of the added monomer with the double bonds of the unsaturated polyester

$$\sim\sim O_2C-CH{=}CH-CO_2\sim\sim \; + \; CH_2{=}CHA \longrightarrow$$

$$
\begin{array}{ccc}
\qquad\qquad\quad \xi & & \qquad\qquad \xi \\
\sim\sim O_2C-CH-CH-CO_2\sim\sim\sim\sim\sim\sim O_2C-CH-CH-CO_2\sim\sim & & \\
\qquad\qquad\quad\; (CH_2-CHA)_n & & \qquad\;\; (CH_2-CHA)_n \\
\sim\sim O_2C-CH-CH-CO_2\sim\sim\sim\sim\sim\sim O_2C-CH-CH-CO_2\sim\sim & & \\
\qquad\qquad\quad \xi & & \qquad\qquad \xi
\end{array}
\qquad (6\text{-}114)
$$

The mechanical properties of the crosslinked product depend on the average number of crosslinks between polyester chains (crosslink density) and the average length of the crosslinks. The crosslink density depends on the relative amounts of saturated and unsaturated acids used in synthesizing the prepolymer. The average length of the crosslinks depends not only on the relative amounts of prepolymer and monomer but also on the copolymerization behavior of the two double bonds. Thus for a polyester containing fumarate double bonds crosslinking by copolymerization with styrene yields a harder and tougher product than when methyl methacrylate is used. The fumarate–styrene system shows more of an alternating copolymerization behavior than the fumarate–methyl methacrylate system. Methyl methacrylate tends to form a small number of long crosslinks (large value of n in Eq. 6-115), while styrene forms a larger number of short crosslinks (small value of n). Allyl monomers such as diallyl phthalate are useful for producing high densities of short crosslinks due to degradative chain transfer.

6-8d Allyl Resins

Diallyl and triallyl monomers are used in various formulations, together with unsaturated polyesters (Sec. 6-8c) or monomers containing one double bond per molecule

(Sec. 6-6a), to produce a range of thermoset products. Several diallyl monomers are also used alone to form thermosets referred to as *allyl resins* [Schildknecht, 1986]. Diallyl phthalate and isophthalate (the *o*- and *p*-isomers, respectively) (DAP, DAIP) are used in moldings and coatings for connectors and insulators in communication, computer, and aerospace systems where high reliability is needed under adverse environmental conditions. Other applications include impregnated glass cloth for radomes, missile and aircraft parts, and impregnated textiles and papers for decorative stain- and heat-resistant top layers for wall panels and furniture. Diallyl diglycol carbonate (also referred to as *diethylene glycol bis[allyl carbonate]*) (DADC) is used in applications requiring optical transparency, such as plastic lenses for eyewear, safety shields, camera filters, instrument panel covers, and nuclear-track detectors.

Polymers of *N,N,N,N*-dimethyldiallylammonium chloride (DADM) are allyl resins in terms of the monomer used but are very different in properties since they are not crosslinked. Cyclopolymerization is the exclusive mode of reaction (Sec. 6-6b) and the polymers are water-soluble. Applications include paper additives (antistatic, fluorescent, whitener, and reinforcement), waste treatment (coagulation aid in potable and wastewater), zinc, tin, and lead electroplating industries, and detergent additive.

6-8e Other Copolymers

Many other copolymers of commercial importance have been discussed previously; see Secs. 3-13b-3 (vinyl acetate, vinylidene chloride), 3-13b-4 (acrylic and methacrylic acids and esters, acrylonitrile, acrylamide), and 3-13b-5 (fluoropolymers). Other copolymers include fluorocarbon elastomers such as vinylidene fluoride with hexafluoropropylene or 1-hydropentafluoropropylene (with and without tetrafluoroethylene as a third monomer) and perfluoro(methyl vinyl ether) with vinylidene fluoride or tetrafluoroethylene (trade names: *Fluorel, Kalrez, Tecnoflon, Viton*) [Lynn and Worm, 1987]. The fluorocarbon elastomers show superior performance in hostile environments (high and low temperature, chemicals, oils, fuels). Applications include engine oil and drivetrain seals, fuel system hoses and O-rings, and a variety of similar parts in the petroleum and chemical industry as well as in other high-performance machinery. Copolymerization of tetrafluoroethylene with **X** yields perfluorosulfonate ionomers

$$CF_2{=}CF(OCF_2\overset{\displaystyle CF_3}{\underset{\displaystyle |}{CF}})_mOCF_2CF_2SO_2F$$

X

(trade name: *Nafion*) after hydrolysis to the sulfonic acid [Moore and Martin, 1988]. These materials have interesting properties as fuel cell, battery, and electrochemical membranes as well as acid catalysts.

Nitrile rubber (NBR), a copolymer of 1,3-butadiene with 20–40% acrylonitrile, is noted for its oil resistance. More than 150 million pounds are produced annually in the United States. Applications include fuel tanks, gasoline hoses, and creamery equipment. *Nitrile resin* is made by copolymerizing acrylonitrile with about 20–30% styrene or methyl methacrylate in the presence of NBR or SBR rubber to yield a blend of the graft terpolymer and homocopolymer. Applications include extruded and blow-molded containers for household, automotive, and other products as well as some nonbeverage foods (spices, vitamins, candy).

Copolymers of methyl methacrylate and tributyltin methacrylate are used in formulating antifouling and self-polishing coatings for underwater structures [Manders et al., 1987]. Terpolymers of alkyl acrylate, acrylonitrile, and 1–5% of a monomer with groups capable of subsequent crosslinking (e.g., 2-chloroethyl vinyl ether) (Sec. 9-2d) are useful as elastomers for applications requiring good resistance to heat and oils, such as transmission and rear-axle seals. These acrylate elastomers as well as the ethylene–acrylate elastomers (Sec. 6-8b) are higher-performance elastomers than are the nitrile rubbers in terms of resistance to a combination of heat and oil.

REFERENCES

Alfrey, T., Jr. and G. Goldfinger, *J. Chem. Phys.*, **12**, 115, 205, 332 (1944).

Alfrey, T., Jr. and C. C. Price, *J. Polym. Sci.*, **2**, 101 (1947).

Atherton, J. N. and A. M. North, *Trans. Faraday Soc.*, **58**, 2049 (1962).

Athey, R. D., Jr., *Makromol. Chem.*, **179**, 2323 (1978).

Bae, H.-J., T. Miyashita, M. Iino, and M. Matsuda, *Macromolecules*, **21**, 26 (1988).

Bailey, D. B. and P. M. Henrichs, *J. Polym. Sci. Polym. Chem. Ed.*, **16**, 3185 (1978).

Baldwin, M. G., *J. Polym. Sci.*, **A3**, 703 (1965).

Bamford, C. H. and A. D. Jenkins, *J. Polym. Sci.*, **53**, 149 (1965).

Barb, W. G., *J. Polym. Sci.*, **11**, 117 (1953).

Bataille, P. and H. Bourassa, *J. Polym. Sci. Poly. Chem. Ed.*, **27**, 357 (1989).

Bhattacharyya, D. N., C. L. Lee, J. Smid, and M. Szwarc, *J. Am. Chem. Soc.*, **85**, 533 (1963).

Billmeyer, F. W., Jr., "Textbook of Polymer Science," 3rd ed., Wiley-Interscience, New York, 1984.

Boudevska, H. and O. Todorova, *Makromol. Chem.*, **186**, 1711 (1985).

Bovey, F. A., "High Resolution NMR of Macromolecules," Chaps. X and XI, Academic Press, New York, 1972.

Braun, D. and G. Cei, *Makromol. Chem.*, **188**, 171 (1987).

Braun, D. and W. K. Czerwinski, *Makromol. Chem. Macromol. Symp.*, **10/11**, 415 (1987).

Braun, D. and W. K. Czerwinski, "Rates of Copolymerization," pp. 207–218 in "Comprehensive Polymer Science," Vol. 3, G. C. Eastman, A. Ledwith, S. Russo, and P. Sigwalt, Eds., Pergamon Press, Oxford, 1989.

Brown, A. S. and K. Fujimori, *Makromol. Chem.*, **188**, 2177 (1987).

Burkhart, R. D. and N. L. Zutty, *J. Polym. Sci.*, **57**, 593 (1962).

Butler, G. B., "Cyclopolymerization," pp. 543–598 in "Encyclopedia of Polymer Science and Engineering," Vol. 4, 2nd ed., H. F. Mark, N. M. Bikales, C. G. Overberger, and G. Menges, Eds., Wiley-Interscience, New York, 1986.

Butler, G. B., "Cyclopolymerization," pp. 423–451 in "Comprehensive Polymer Science," Vol. 4, G. C. Eastman, A. Ledwith, S. Russo, and P. Sigwalt, Eds., Pergamon Press, Oxford, 1989.

Butler, G. B. and R. J. Angelo, *J. Am. Chem. Soc.*, **79**, 3128 (1957).

Butler, G. B. and C. H. Do, *Makromol. Chem. Suppl.*, **15**, 93 (1989).

Butler, G. B. and F. L. Ingley, *J. Am. Chem. Soc.*, **73**, 894 (1951).

Butler, G. B. and A. Matsumoto, *J. Polym. Sci. Polym. Lett. Ed.*, **19**, 167 (1981).

Bywater, S., "Anionic Polymerization of Olefins," Chap. 1 in "Comprehensive Chemical Kinetics," Vol. 15, C. H. Bamford and C. F. H. Tipper, Eds., Elsevier, New York, 1976.

Cais, R. E. and F. A. Bovey, *Macromolecules*, **10**, 169 (1977).

Cais, R. E., R. G. Farmer, D. J. T. Hill, and J. H. O'Donnell, *Macromolecules*, **12**, 835 (1979).

Cais, R. E. and J. H. O'Donnell, *J. Polym. Sci. Symp.*, **55**, 75 (1976).

Cais, R. E. and G. J. Stuk, *Polymer*, **19**, 179 (1978).

Capek, I., V. Juranicova, and J. Bartoh, *Makromol. Chem.*, **184**, 1597 (1983).

Castille, Y. P. and V. Stannett, *J. Polym. Sci.*, **A-1**(4), 2063 (1966).

Chiang, T. C., Ch. Graillat, J. Guillot, Q. T. Pham, and A. Guyot, *J. Polym. Sci. Polym. Chem. Ed.*, **15**, 2961 (1977).

Chien, J. C. W. and A. L. Finkenaur, *J. Polym. Sci. Polym. Chem. Ed.*, **23**, 2247 (1985).

Colthup, N. B., *J. Polym. Sci. Polym. Chem. Ed.*, **20**, 3167 (1982).

Costa, L., O. Chiantore, and M. Guaita, *Polymer*, **19**, 197, 202 (1978).

Cowie, J. M. G., "Alternating Copolymerization," pp. 377–422 in "Comprehensive Polymer Science," Vol. 4, G. C. Eastman, A. Ledwith, S. Russo, and P. Sigwalt, Eds., Pergamon Press, Oxford, 1989.

Cowie, J. M. G., S. H. Cree, and R. Ferguson, *J. Polym. Sci. Polym. Chem. Ed.*, **28**, 515 (1990).

Dainton, F. S. and K. J. Ivin, *Quart. Rev. (London)*, **12**, 61 (1958).

Davis, T. P., K. F. O'Driscoll, M. C. Piton, and M. A. Winnik, *Macromolecules*, **23**, 2113 (1990).

Dawson, T. L., R. D. Lundberg, and F. J. Welch, *J. Polym. Sci.*, **A-1**(7), 173 (1969).

Dionisio, J. M. and K. F. O'Driscoll, *J. Polym. Sci. Polym. Lett. Ed.*, **17**, 701 (1979).

Disselhoff, G., *Polymer*, **19**, 111 (1978).

Doak, K. W., "Low Density Polyethylene (High Pressure)," pp. 386–429 in "Encyclopedia of Polymer Science and Engineering," Vol. 6, 2nd ed., H. F. Mark, N. M. Bikales, C. G. Overberger, and G. Menges, Eds., Wiley-Interscience, New York, 1986.

Dodgson, K. and J. R. Ebdon, *Eur. Polym. J.*, **13**, 791 (1977).

Doiuchi, T. and Y. Minoura, *Macromolecules*, **10**, 261 (1977).

Dotson, N. A., R. Galvan, and C. W. Macosko, *Macromolecules*, **21**, 2560 (1988).

Eastman, G. C. and E. G. Smith, "Reactivity Ratios in Free Radical Copolymerizations," pp. 333–418 in "Comprehensive Chemical Kinetics," Vol. 14A, C. H. Bamford and C. F. H. Tipper, Eds., Elsevier, New York, 1976.

Eaton, D. R., M. Mlekuz, B. G. Sayer, A. E. Hamielec, and L. K. Kostanski, *Polymer*, **30**, 1989.

Farina, M., *Makromol. Chem.*, **191**, 2795 (1990).

Favier, J. C., P. Sigwalt, and M. Fontanille, *J. Polym. Sci. Polym. Chem. Ed.*, **15**, 2373 (1977).

Fineman, M. and S. D. Ross, *J. Polym. Sci.*, **5**, 259 (1950).

Florjanczyk, Z., T. Florjanczyk, and B. B. Ktopotek, *Makromol. Chem.*, **188**, 2811 (1987).

Florjanczyk, Z. and W. Krawiec, *Makromol. Chem.*, **190**, 2141 (1989).

Flory, P. J., *J. Am. Chem. Soc.*, **69**, 2893 (1947).

Flory, P. J., "Principles of Polymer Chemistry," Chap. IX, Cornell University Press, Ithaca, N.Y., 1953.

Fordyce, R. G. and E. C. Chapin, *J. Am. Chem. Soc.*, **69**, 581 (1947).

Fordyce, R. G. and G. E. Ham, *J. Polym. Sci.*, **3**, 891 (1948); *J. Am. Chem. Soc.*, **73**, 1186 (1951).

Fujimori, K. and I. E. Craven, *J. Polym. Sci. Polym. Chem. Ed.*, **24**, 559 (1986).

Furukawa, J., "Alternating Copolymers," pp. 233–261 in "Encyclopedia of Polymer Science and Engineering," Vol. 4, 2nd ed., H. F. Mark, N. M. Bikales, C. G. Overberger, and G. Menges, Eds., Wiley-Interscience, New York, 1986.

Furukawa, J. and T. Saegusa, "Polymerization of Aldehydes and Oxides," Wiley-Interscience, New York, 1963.

Galvan, R. and M. Tirrell, *J. Poly. Sci. Polym. Chem. Ed.*, **24**, 803 (1986).

Gaylord, N. G., *J. Macromol. Sci. Chem.*, **A6**, 259 (1972).

Georgiev, G. S. and V. P. Zubov, *Eur. Polym. J.*, **14**, 93 (1978).

Goldfinger, G. and T. Kane, *J. Polym. Sci.*, **3**, 462 (1948).

Green, J., N. Mayes, and E. Cottrill, *J. Macromol. Sci. Chem.*, **A1**, 1387 (1967).

Greenley, R. Z. "Free Radical Copolymerization Reactivity Ratios," pp. 153–266 in Chap. II in "Polymer Handbook," 3rd ed., J. Brandrup and E. H. Immergut, Eds., Wiley-Interscience, New York, 1989a.

Greenley, R. Z., "*Q* and *e* Values for Free Radical Copolymerizations of Vinyl Monomers and Telogens," pp. 267–274 in Chap. II in "Polymer Handbook," 3rd ed., J. Brandrup and E. H. Immergut, Eds., Wiley-Interscience, New York, 1989b.

Gumbs, R., S. Penczek, J. Jagur-Grodzinski, and M. Szwarc, *Macromolecules*, **2**, 77 (1969).

Guyot, A. and J. Guillot, *J. Macromol. Sci. Chem.*, **A1**, 793 (1967).

Hagman, J. F. and J. W. Crary, "Ethylene–Acrylic Elastomers," pp. 325–334 in "Encyclopedia of Polymer Science and Engineering," Vol. 1, 2nd ed., H. F. Mark, N. M. Bikales, C. G. Overberger, and G. Menges, Eds., Wiley-Interscience, New York, 1985.

Hall, H. K. Jr., A. B. Padias, A. Pandya, and H. Tanaka, *Macromolecules*, **20**, 247 (1987).

Ham, G. E., "Theory of Copolymerization," Chap. 1 in "Copolymerization," G. E. Ham, Ed., Wiley-Interscience, New York, 1964.

Ham, G. E., *J. Polym. Sci.*, **A2**, 2735, 4191 (1964); *J. Macromol. Chem.*, **1**, 403 (1966).

Ham, G. E., "General Aspects of Free Radical Polymerization," Chap. 1 in "Vinyl Polymerization," Vol. 1, Part I, G. E. Ham, Ed., Marcel Dekker, New York, 1967.

Ham, G. E., *J. Macromol. Sci. Chem.*, **A11**, 227 (1977).

Hamielec, A. E., J. F. Macgregor, and A. Penlidis, "Copolymerization," pp. 17–31 in "Comprehensive Polymer Science," Vol. 3, G. C. Eastmond, A. Ledwith, S. Russo, and P. Sigwalt, Eds., Pergamon Press, London, 1989.

Harwood, H. J., *Makromol. Chem. Macromol. Symp.*, **10/11**, 331 (1987).

Hatada, K., K. Nagata, T. Hasegawa, and H. Yuki, *Makromol. Chem.*, **178**, 2413 (1977).

Hauser, C. F. and N. L. Zutty, *Polym. Prepr.*, **8**(1), 369 (1967).

Hautus, F. L. M., H. N. Linssen, and A. L. German, *J. Polym. Sci. Polym. Chem. Ed.*, **22**, 3487, 3661 (1984).

Hecht, J. K. and N. D. Ojha, *Macromolecules*, **2**, 94 (1969).

Hild, G. and R. Okasha, *Makromol. Chem.*, **186**, 93, 389 (1985).

Hill, D. J. T., J. H. O'Donnell, and P. W. O'Sullivan, *Macromolecules*, **15**, 960 (1982); **16**, 1295 (1983); **18**, 9 (1985).

Hinton, C. V. and H. G. Spencer, *Macromolecules*, **9**, 864 (1976).

Hirai, H., *J. Polym. Sci. Macromol. Rev.*, **11**, 47 (1976).

Holt, T. and W. Simpson, *Proc. Roy. Soc.* (*London*), **A238**, 154 (1956).

Howell, J. A., M. Izu, and K. F. O'Driscoll, *J. Polym. Sci.*, **A-1**(8), 699 (1970).

Ito, K., *J. Polym. Sci. Polym. Chem. Ed.*, **16**, 2725 (1978).

Ivin, K. J. and R. H. Spensley, *J. Macromol. Sci. Chem.*, **A1**, 653 (1967).

Iwata, T., G. Wasai, T. Saegusa, and J. Furukawa, *Makromol. Chem.*, **77**, 229 (1964).

Janovic, Z., K. Saric, and O. Vogl, *J. Polym. Sci. Polym. Chem. Ed.*, **21**, 2713 (1983).

Jenner, G., *J. Polym. Sci. Polym. Chem. Ed.*, **17**, 237 (1979).

Jenner, G. and S. Aieche, *J. Polym. Sci. Polym. Chem. Ed.*, **16**, 1017 (1978).

Johnson, M., T. S. Karmo, and R. R. Smith, *Eur. Polym. J.*, **14**, 409 (1978).

Jones, S. A. and D. A. Tirrell, *Macromolecules*, **19**, 2080 (1986); *J. Polym. Sci. Polym. Chem. Ed.*, **25**, 3177 (1987).

Karad, P. and C. Schneider, *J. Polym. Sci. Polym. Chem. Ed.*, **16**, 1137 (1978).

Kelen, T. and F. Tudos, *React. Kinet. Catal. Lett.*, **1**, 487 (1974); *J. Macromol. Sci.-Chem.*, **A9**, 1 (1975); *Makromol. Chem.*, **191**, 1863 (1990).

Kelley, D. J. and A. V. Tobolsky, *J. Am. Chem. Soc.*, **81**, 1597 (1959).

Kennedy, J. P. and E. Marechal, "Carbocationic Polymerizations," Wiley-Interscience, New York, 1983.

Kim, Y.-H., Y.-S. Gal, U.-Y. Kim, and S.-K. Choi, *Macromolecules*, **21**, 1991 (1988).

Klemner, D. and L. Berkowski, "Interpenetrating Polymer Networks," pp. 279–341 in "Encyclopedia of Polymer Science and Engineering," Vol. 8, 2nd ed., H. F. Mark, N. M. Bikales, C. G. Overberger, and G. Menges, Eds., Wiley-Interscience, New York, 1987.

Kruger, H., J. Bauer, and J. Rubner, *Makromol. Chem.*, **188**, 2163 (1987).

Kruse, R. L., *J. Polym. Sci.*, **B5**, 437 (1967).

Kuntz, I., N. F. Chamberlain, and F. J. Stehling, *J. Polym. Sci. Polym. Chem. Ed.*, **16**, 1747 (1978).

Landin, D. T. and C. W. Macosko, *Macromolecules*, **21**, 846 (1988).

Landler, Y., *Compt. Rend.*, **230**, 539 (1950).

Lebduska, J., J. Snuparek, Jr., K. Kaspar, and V. Cermak, *J. Polym. Sci. Polym. Chem. Ed.*, **24**, 777 (1986).

Ledwith, A., G. Galli, E. Chiellini, and R. Solaro, *Polym. Bull.*, **1**, 491 (1979).

Lee, C., J. Smid, and M. Szwarc, *Trans. Faraday Soc.*, **59**, 1192 (1963).

Leicht, R. and J. Fuhrmann, *J. Polym. Sci. Polym. Chem. Ed.*, **21**, 2215 (1983).

Li, W.-H., A. E. Hamielec, and C. M. Crowe, *Polymer*, **30**, 1513, 1518 (1989).

Litt, M., *Macromolecules*, **4**, 312 (1971).

Litt, M. and J. A. Seiner, *Macromolecules*, **4**, 314, 316 (1971).

Lowry, G. G., *J. Polym. Sci.*, **42**, 463 (1960).

Lowry, T. H. and K. S. Richardson, "Mechanism and Theory in Organic Chemistry," 2nd ed., Harper and Row, New York, 1987, Chap. 2.

Lynn, M. M. and A. T. Worm, "Fluorocarbon Elastomers," pp. 257–269 in "Encyclopedia of Polymer Science and Engineering," Vol. 7, 2nd ed., H. F. Mark, N. M. Bikales, C. G. Overberger, and G. Menges, Eds., Wiley-Interscience, New York, 1987.

Macosko, C. W. and D. R. Miller, *Macromolecules*, **9**, 199 (1976).

Manders, W. F., J. M. Bellama, R. B. Johannesen, E. J. Parks, and F. E. Brinkman, *J. Polym. Sci. Polym. Chem. Ed.*, **25**, 3469 (1987).

Marvel, C. S. and R. D. Vest, *J. Am. Chem. Soc.*, **79**, 5771 (1957); **81**, 984 (1959).

Masuda, S., M. Tanaka, and T. Ota, *Makromol. Chem.*, **188**, 371 (1987a); *Polymer*, **28**, 1945 (1987b).

Matsumoto, A., K. Iwanami, and M. Oiwa, *J. Polym. Sci. Polym. Lett. Ed.*, **18**, 307 (1980).

Matsumoto, A., S. Okuda, and M. Oiwa, *Makromol. Chem. Rapid. Commun.*, **10**, 25 (1989).

Matsumoto, A., T. Terada, and M. Oiwa, *J. Polym. Sci. Poly. Chem. Ed.*, **25**, 775 (1987).

Mayo, F. R. and F. M. Lewis, *J. Am. Chem. Soc.*, **66**, 1594 (1944).

Mayo, F. R. and C. Walling, *Chem. Rev.*, **46**, 191 (1950).

Melville, H. W., B. Noble, and W. F. Watson, *J. Polym. Sci.*, **2**, 229 (1947).

Merz, E., T. Alfrey, Jr., and G. Goldfinger, *J. Polym. Sci.*, **1**, 75 (1946).

Meyer, V. E., *J. Polym. Sci.*, **A-1**(4), 2819 (1966).

Meyer, V. E. and R. K. S. Chan, *J. Polym. Sci.*, **C25**, 11 (1968).

Meyer, V. E. and G. G. Lowry, *J. Polym. Sci.*, **A3**, 2843 (1965).

Mirabella, F. M., Jr., *Polymer*, **18**, 705, 925 (1977).

Miyata, Y. and M. Sawada, *Polymer*, **29**, 1495 (1988).

Moad, G., D. H. Solomon, T. H. Spurling, and R. A. Sone, *Macromolecules*, **22**, 1145 (1989).

Moore, E. R., Ed., "Styrene Polymers," pp. 1–246 in "Encyclopedia of Polymer Science and Engineering," Vol. 16, 2nd ed., H. F. Mark, N. M. Bikales, C. G. Overberger, and G. Menges, Eds., Wiley-Interscience, New York, 1989.

Moore, R. B., III and C. R. Martin, *Macromolecules*, **21**, 1334 (1988).

Morton, M., "Anionic Polymerization: Principles and Practice," Academic Press, New York, 1983.

Motoc, I. and R. Vancea, *J. Polym. Sci. Polym. Chem. Ed.*, **18**, 1559 (1980).

Motoc, I., R. Vancea, and St. Holban, *J. Polym. Sci. Polym. Chem. Ed.*, **16**, 1587, 1595, 1601 (1978).

Mukundan, T. and K. Kishore, *Macromolecules*, **20**, 2382 (1987).

Natansohn A., S. Maxim, and D. Feldman, *Eur. Polym. J.*, **14**, 283 (1978).

North, A. M., *Polymer*, **4**, 134 (1963).

Nukui, M., Y. Ohkatsu, and T. Tsuruta, *Makromol. Chem.*, **183**, 1457 (1982).

Odian, G. and L. S. Hiraoka, *J. Macromol. Sci.-Chem.*, **A6**, 109 (1972).

O'Driscoll, K. F. and F. P. Gasparro, *J. Macromol. Sci. Chem.*, **A1**, 643 (1967).

O'Driscoll, K. F., L. T. Kale, L. H. Garcia-Rubio, and P. M. Reilly, *J. Polym. Sci. Polym. Chem. Ed.*, **22**, 2777 (1984).

O'Driscoll, K. F. and P. M. Reilly, *Makromol. Chem. Macromol. Symp.*, **10/11**, 355 (1987).

O'Driscoll, K. F. and T. Yonezawa, *J. Macromol. Sci. Rev. Macromol. Chem.*, **1**, 1 (1966).

O'Driscoll, K. F., W. Wertz, and A. Husar, *J. Polym. Sci.*, **A-1**(5), 2159 (1967).

O'Driscoll, K F., T. Yonezawa, and T. Higashimura, *J. Macromol. Chem.*, **1**, 17 (1966).

Okasha, R., G. Hild, and P. Rempp, *Eur. Polym. J.*, **15**, 975 (1979).

Olson, K. G. and G. B. Butler, *Macromolecules*, **17**, 2486 (1984).

Otsu, T., *Makromol. Chem. Macromol. Symp.*, **10/11**, 235 (1987).

Overberger, C. G., L. H. Arnold, and J. J. Taylor, *J. Am. Chem. Soc.*, **73**, 5541 (1951).

Overberger, C. G., R. J. Ehrig, and D. Tanner, *J. Am. Chem. Soc.*, **76**, 772 (1954).

Overberger, C. G. and V. G. Kamath, *J. Am. Chem. Soc.*, **81**, 2910 (1959).

Overberger, C. G. and Y. Yamamoto, *J. Polym. Sci.*, **A-1**(4), 3101 (1966).

Overberger, C. G., D. H. Tanner, and E. M. Pearce, *J. Am. Chem. Soc.*, **80**, 4566 (1958).

Pepper, D. C., *Quart. Rev. (London)*, **8**, 88 (1954).

Phillips, G. and W. Carrick, *J. Am. Chem. Soc.*, **84**, 920 (1962a); *J. Polym. Sci.*, **59**, 401 (1962b).

Pichot, C. and Q.-T. Pham, *Makromol. Chem.*, **180**, 2359 (1979).

Pichot, C., A. Guyot, and C. Strazielle, *J. Polym. Sci. Polym. Chem. Ed.*, **17**, 2269 (1979).

Plaumann, H. P. and R. E. Branston, *J. Polym. Sci. Polym. Chem. Ed.*, **27**, 2819 (1989).

Plochocka, K., *J. Macromol. Sci.-Rev. Macromol. Chem.*, **C20**, 67 (1981).

Ponratnam, S. and S. K. Kapur, *Makromol. Chem.*, **178**, 1029 (1977).

Prementine, G. S., S. A. Jones, and D. A. Tirrell, *Macromolecules*, **22**, 770 (1989).

Prementine, G. S. and D. A. Tirrell, *Macromolecules*, **20**, 3034 (1987).

Price, C. C., *J. Polym. Sci.*, **3**, 772 (1948).

Prochazka, O. and P. Kratochvil, *J. Polym. Sci. Polym. Chem. Ed.*, **21**, 3269 (1983).

Quella, F., *Makromol. Chem.*, **190**, 1445 (1989).

Randall, J. C., "Polymer Sequence Determination," Academic Press, New York, 1977.

Randall, J. C., *J. Macromol. Sci.-Rev. Macromol. Chem. Phys.*, **C29**, 153 (1989).

Randall, J. C. and C. J. Ruff, *Macromolecules*, **21**, 3446 (1988).

Rao, S. P., S. Ponratnam, and S. K. Kapur, *J. Polym. Sci. Polym. Lett. Ed.*, **14**, 513 (1976).

Richards, D. H., *Polymer*, **19**, 109 (1978).

Ring, W., I. Mita, A. D. Jenkins, and N. M. Bikales, *Pure Appl. Chem.*, **57**, 1427 (1985).

Rounsefell, T. D. and C. U. Pittman, Jr., *J. Macromol. Sci. Chem*, **A13**, 153 (1979).

Saigo, K., K. Tateishi, and H. Adachi, *J. Polym. Sci. Polym. Chem. Ed.*, **26**, 2085 (1988).

Sato, T., M. Abe, and T. Otsu, *Makromol. Chem.*, **178**, 1061 (1971).

Sawada, H., "Thermodynamics of Polymerization," Chaps. 9 and 10, Marcel Dekker, New York, 1976.

Schildknecht, C. E., "Diallyl and Related Polymers," pp. 779–811 in "Encyclopedia of Polymer Science and Engineering," Vol. 4, 2nd ed., H. F. Mark, N. M. Bikales, C. G. Overberger, and G. Menges, Eds., Wiley-Interscience, New York, 1986.

Scranton, A. B. and N. A. Peppas, *J. Poly. Sci. Polym. Chem. Ed.*, **28**, 39 (1990).

Seiner, J. A. and M. Litt, *Macromolecules*, **4**, 308 (1971).

Selley, J., "Polyesters, Unsaturated," pp. 256–290 in "Encyclopedia of Polymer Science and Engineering," Vol. 12, 2nd ed., H. F. Mark, N. M. Bikales, C. G. Overberger, and G. Menges, Eds., Wiley-Interscience, New York, 1988.

Shima, M., D. N. Bhattacharyya, J. Smid, and M. Szwarc, *J. Am. Chem. Soc.*, **85**, 1306 (1962).

Shultz, A. R., "Crosslinking," pp. 331–441 in "Encyclopedia of Polymer Science and Technology," Vol. 4, H. F. Mark, N. G. Gaylord, and N. M. Bikales, Eds., Wiley-Interscience, New York, 1966.

Skeist, I., *J. Am. Chem. Soc.*, **68**, 1781 (1946).

Smith, W. V., *J. Am. Chem. Soc.*, **70**, 2177 (1948).

Starkweather, H., "Olefin–Carbon Monoxide Copolymers," pp. 369–373 in "Encyclopedia of Polymer Science and Engineering," Vol. 10, 2nd ed., H. F. Mark, N. M. Bikales, C. G. Overberger, and G. Menges, Eds., Wiley-Interscience, New York, 1987.

Staudinger, H. and J. Schneiders, *Ann. Chim.* (*Paris*), **541**, 151 (1939).

Steinberg, M., R. Johnson, W. Cordel, and D. Goodman, *Radiat. Phys. Chem.*, **14**, 613 (1979).

Stejskal, J., P. Kratochvil, D. Strakova, and O. Prochazka, *Macromolecules*, **19**, 1575 (1986).

Stockmayer W. H., *J. Chem. Phys.*, **13**, 199 (1945).

Storey, B. T., *J. Polym. Sci.,* **A3**, 265 (1965).

Suggate, J. R., *Makromol. Chem.*, **179**, 1219 (1978); **180**, 679 (1979).

Szwarc, M., "Carbanions, Living Polymers, and Electron Transfer Processes," Wiley-Interscience, New York, 1968, Chap. IX.

Szymanski, R., *Makromol. Chem.*, **188**, 2605 (1987).

Tacx, J. C. J. F., H. N. Linssen, and A. L. German, *J. Polym. Sci. Poly. Chem. Ed.*, **26**, 61 (1988).

Teramachi, S., A. Hasegawa, F. Sato, and N. Takemoto, *Macromolecules*, **18**, 347 (1985).

Tidwell, P. W. and G. A. Mortimer, *J. Polym. Sci.*, **A3**, 369 (1965); *J. Macromol. Sci. Rev. Macromol. Chem.*, **C4**, 281 (1970).

Tirrell, D. A., "Copolymerization," pp. 192–233 in "Encyclopedia of Polymer Science and Engineering," Vol. 4, 2nd ed., H. F. Mark, N. M. Bikales, C. G. Overberger, and G. Menges, Eds., Wiley-Interscience, New York, 1986.

Tobita, H. and A. E. Hamielec, *Macromolecules*, **22**, 3098 (1989).

Tobolsky, A. V., D. J. Kelley, M. F. O'Driscoll, and C. E. Rogers, *J. Polym. Sci.*, **31**, 425 (1958).

Tomescu, M. and C. Simionescu, *Makromol. Chem.*, **177**, 3221 (1976).

Tosi, C., *Adv. Polym. Sci.*, **5**, 451 (1967–1968).

Truong, N. D., J. C. Galin, J. Francois, and Q. T. Pham, *Polymer*, **27**, 467 (1986).

Tsonis, C. P. and S. A. Ali, *Makromol. Chem. Rapid Commun.*, **10**, 641 (1989).

Tsukino, M. and T. Kunitake, *Macromolecules*, **12**, 387 (1979).

Tudos, F. and T. Kelen, *J. Macromol. Sci.-Chem.*, **A16**, 1283 (1981).

Ulbrich, K., M. Ilvasky, K. Dusek, and J. Kopecek, *Eur. Polym. J.*, **13**, 579 (1977).

Valvassori, A. and G. Sartori, *Adv. Polym. Sci.*, **5**, 28 (1967).

Van der Meer, R., A. L. German, and D. Heikens, *J. Polym. Sci. Polym. Chem. Ed.*, **15**, 1765 (1977a).

Van der Meer, R., H. N. Linssen, and A. L. German, *J. Polym. Sci. Polym. Chem. Ed.*, **16**, 2915 (1978).

Van der Meer, R., E. H. M. Van Gorp, and A. L. German, *J. Polym. Sci. Polym. Chem. Ed.*, **15**, 1489 (1977b).

Van der Meer, R., J. M. Alberti, A. L. German, and H. N. Linssen, *J. Polym. Sci. Polym. Chem. Ed.*, **17**, 3349 (1979).

Vogl, O., *J. Macromol. Sci. Chem.*, **A1**, 243 (1967).

Vollmert, B., "Polymer Chemistry," Springer-Verlag, New York, 1973, pp. 94–147.

Wall, F. T., *J. Am. Chem. Soc.*, **66**, 2050 (1944).

Walling, C., *J. Am. Chem. Soc.*, **67**, 441 (1945); **71**, 1930 (1949).

Walling, C., "Free Radicals in Solution," Wiley, New York, 1957, Chap. 4.

Walling, C. and E. R. Briggs, *J. Am. Chem. Soc.*, **67**, 1774 (1945).

Walling, C., E. R. Briggs, K. B. Wolfstern, and F. R. Mayo, *J. Am. Chem. Soc.*, **70**, 1537, 1544 (1948).

Watanabe, S., O. Matsuda, J. Okamoto, S. Machi, K. Ishigure, and Y. Tabata, *J. Polym. Sci. Polym. Chem. Ed.*, **17**, 551 (1979).

Watts, D. G., H. N. Linsen, and J. Schrijner, *J. Polym. Sci. Polym. Chem. Ed.*, **18**, 1285 (1980).

Williams, R. J. J. and C. I. Vallo, *Macromolecules*, **21**, 2571 (1988).

Wood, K. B., V. T. Stannett, and P. Sigwalt, *Makromol. Chem. Suppl.*, **15**, 71 (1989).

Wu, C., R. Brambilla, and J. T. Yardley, *Macromolecules*, **23**, 997 (1990).

Yoshimura, M., H. Mikawa, and Y. Shirota, *Macromolecules*, **11**, 1085 (1978).

Young, L. J., "Copolymerization Reactivity Ratios," pp. 105–386 in Chap. II in "Polymer Handbook," J. Brandrup and E. H. Immergut, Eds. (with W. McDowell), Wiley-Interscience, New York, 1975.

Yuki, H., K. Kosai, S. Murahashi, and J. Hotta, *J. Polym. Sci.*, **B2**, 1121 (1964).

PROBLEMS

6-1 Discuss the differences in the structures of random, alternating, graft, and block copolymers.

6-2 What is the difference between the ideal and alternating behaviors in copolymerization?

6-3 Consider the following monomer reactivity ratios for the copolymerization of various pairs of monomers:

Case	r_1	r_2
1	0.1	0.2
2	0.1	10
3	0.1	3
4	0	0.3
5	0	0
6	0.8	2
7	1	15

What is the composition of the copolymer that would be formed at low conversion from equimolar mixtures of the two monomers in each case?

6-4   Using the r_1 and r_2 values from Table 6-2, construct plots showing the initial copolymer composition as a function of the comonomer feed composition for the radical copolymerizations of methyl acrylate–methyl methacrylate and styrene–maleic anhydride. Are these examples of ideal or alternating copolymerization?

6-5   Calculate the composition of the initial terpolymer which would be produced from the radical polymerization of a solution containing acrylonitrile, styrene, and 1,3-butadiene in mole fractions of 0.47, 0.47, and 0.06, respectively.

6-6   Ferrocenylmethyl acrylate (FMA) and 2-ferrocenylethyl acrylate (FEA) have been synthesized and copolymerized with styrene, methyl acrylate, and vinyl acetate [C. U. Pittman, Jr., *Macrmolecules*, **4**, 298 (1971)]. The following monomer reactivity ratios were found:

M_1	M_2	r_1	r_2
FEA	Styrene	0.41	1.06
FEA	Methyl acrylate	0.76	0.69
FEA	Vinyl acetate	3.4	0.074
FMA	Styrene	0.020	2.3
FMA	Methyl acrylate	0.14	4.4
FMA	Vinyl acetate	1.4	0.46

a. Predict whether FEA or FMA will have the higher k_p in homopolymerization. Explain the basis of your prediction. How can the difference in k_p values be explained in relation to the structures of FEA and FMA?

b. Which of the above comonomer pairs could lead to azeotropic copolymerization?

c. Is styrene monomer a more or less reactive monomer than FMA monomer toward the FMA propagating center? By what factor?

d. List styrene, methyl acrylate, and vinyl acetate in order of increasing reactivity toward the FEA propagating center? Is the trend in reactivity toward the FMA propagating center the same?

e. List the styrene, methyl acrylate, and vinyl acetate propagating centers in order of increasing reactivity toward FEA monomer.

f. Is the above copolymerization data indicative of radical, cationic, or anionic copolymerization. Explain.

6-7   Consider the radical copolymerization of a benzene solution that is 1.5 M in styrene and 3.0 M in methyl acrylate.

a. What is the initial copolymer composition if the polymerization is carried out at 60°C using benzoyl peroxide at a concentration of 5.0 × 10^{-4} M? How is the copolymer composition affected if 3.0 × 10^{-3} M benzoyl peroxide is used?

b. How will the presence of 5.0 × 10^{-5} M n-butyl mercaptan affect the initial copolymer composition?

c. What would you expect (qualitatively) for the copolymer composition if the reaction were initiated by n-butyllithium? by BF_3 plus water?

6-8   List the following monomers in order of their increasing tendency toward alternation with 1,3-butadiene in radical copolymerization:

a. n-Butyl vinyl ether
b. Methyl methacrylate
c. Methyl acrylate
d. Styrene
e. Maleic anhydride
f. Vinyl acetate
g. Acrylonitrile

Explain the relative alternating tendencies in thse copolymerizations.

6-9   If the copolymerizations in Problem 6-8 were carried out using cationic initiation, what would be expected qualitatively for the copolymer compositions? List the copolymers in order of their increasing butadiene content. Would copolymers be formed from each of the comonomer pairs? Explain. What would be observed if one used anionic initiation?

6-10   Using the Q and e values in Table 6-7, calculate the monomer reactivity ratios for the comonomer pairs styrene–butadiene and styrene–methyl methacrylate. Compare the results with the r_1 and r_2 values in Table 6.2.

6-11   Discuss the general effects of temperature, solvent, and catalyst on the monomer reactivity ratios in ionic copolymerizations. How do these compare with the corresponding effects in radical copolymerizations?

6-12   What are the differences between the two treatments (kinetic penultimate effect and depropagation) used to account for the deviations observed in the copolymer composition equation?

6-13   Discuss qualitatively the course of the radical copolymerization for each of the following comonomer pairs in terms of the degree of reaction at which gelation would be expected to occur:

a. Styrene–divinylbenzene
b. Methyl methacrylate–allyl methacrylate
c. Vinyl acetate–ethylene glycol dimethacrylate
d. Methyl methacrylate–divinyl adipate
e. Styrene–butadiene

6-14   The product obtained in the polymerization of 4-methyl-1,6-heptadiene contains no residual unsaturation. What is its chemical structure?

CHAPTER 7

RING-OPENING POLYMERIZATION

In addition to step and chain polymerizations, another mode of polymerization is of importance. This is the *ring-opening polymerization* of cyclic monomers such as cyclic ethers, acetals, amides (lactams), esters (lactones), and siloxanes. Ring-opening polymerization is of commercial interest in a number of systems, including the polymerizations of ethylene oxide,

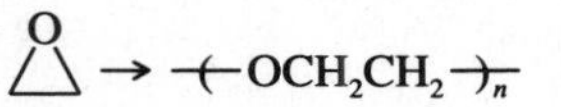

$$\overset{O}{\underset{\triangle}{}} \rightarrow \,-\!\!\left(OCH_2CH_2\right)_n\!\!- \tag{7-1}$$

trioxane,

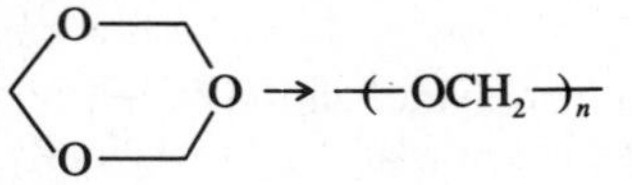

$$\rightarrow \,-\!\!\left(OCH_2\right)_n\!\!- \tag{7-2}$$

ϵ-caprolactam,

$$\rightarrow \,-\!\!\left(NHCOCH_2CH_2CH_2CH_2CH_2\right)_n\!\!- \tag{7-3}$$

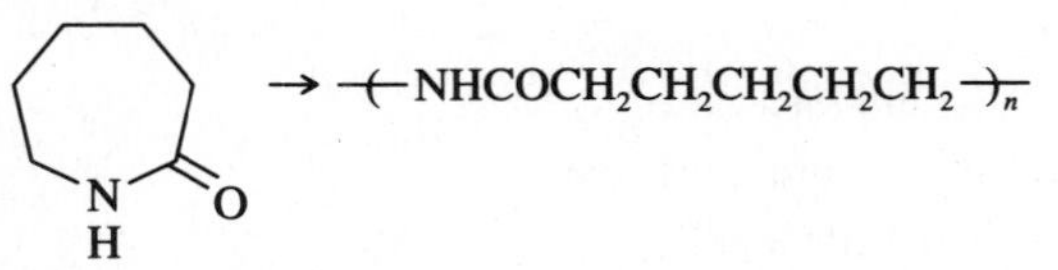

and octamethylcyclotetrasiloxane

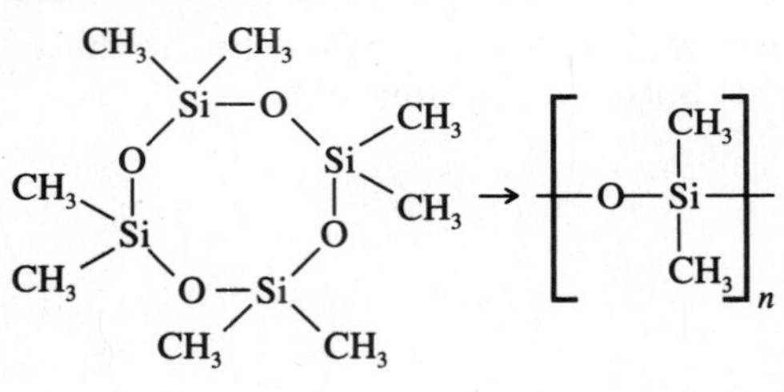

$$\rightarrow \left[O\!-\!\underset{\underset{CH_3}{|}}{\overset{\overset{CH_3}{|}}{Si}}\right]_n \tag{7-4}$$

7-1 GENERAL CHARACTERISTICS

7-1a Scope; Polymerizability

A wide variety of cyclic monomers have been successfully polymerized by the ring-opening process [Frisch and Reegan, 1969; Ivin and Saegusa, 1984; Saegusa and Goethals, 1977]. This includes cyclic amines, sulfides, olefins, cyclotriphosphazenes, and N-carboxy-α-amino acid anhydrides, in addition to those classes of monomers mentioned above. The ease of polymerization of a cyclic monomer depends on both thermodynamic and kinetic factors as previously discussed in Sec. 2-5.

The single most important factor that determines whether a cyclic monomer can be converted to linear polymer is the thermodynamic factor, that is, the relative stabilities of the cyclic monomer and linear polymer structure [Allcock, 1970; Sawada, 1976]. Table 7-1 shows the semiempirical enthalpy, entropy, and free-energy changes for the conversion of cycloalkanes to the corresponding linear polymer (polymethylene in all cases) [Dainton and Ivin, 1958; Finke et al., 1956]. The lc (denoting liquid–crystalline) subscripts of ΔH, ΔS, and ΔG indicate that the values are those for the polymerization of liquid monomer to crystalline polymer.

Polymerization is favored thremodynamically for all except the 6-membered ring. Ring-opening polymerization of 6-membered rings is generally not observed. The order of thermodynamic feasibility is $3,4 > 8 > 5,7$, which follows from the previous discussion (Sec. 2-5c) on bond angle strain in 3- and 4-membered rings, eclipsed conformational strain in the 5-membered ring, and transannular strain in 7- and 8-membered rings. One notes that ΔH_{lc} is the major factor in determining ΔG_{lc} for the 3- and 4-membered rings, while ΔS_{lc} is very important for the 5- and 6-membered rings. The enthalpy and entropy factors ΔH_{lc} and ΔS_{lc} contribute about equally for larger-sized rings. Since both ΔH_{lc} and ΔS_{lc} are negative, ΔG_{lc} becomes less negative with increasing temperature. Above some temperature (the ceiling temperature) ΔG_{lc} becomes positive, and polymerization is no longer favorable. For all-sized rings the presence of substituents decreases thermodynamic feasibility for polymerization. Interactions between substituents are more severe in the linear polymer than in the cyclic monomer [Cubbon, 1964; Sawada, 1976]; ΔH_{lc} is less negative, while ΔS_{lc} is more negative. Exceptions to this generalization occur when the substituents are linked to each other to form a second ring in such a manner that there is increased strain in the ring containing the polymerizable functional group. Structures **I** and **II**, *cis*- and *trans*-8-oxabicyclo[4.3.0]nonane, illustrate this point. The 5-membered ring in the cis isomer (**I**), is almost completely free of strain and does not polymerize. However, the trans isomer undergoes polymerization since the tetrahydrofuran ring is twisted and highly strained [Kops and Spangaard, 1975].

TABLE 7-1   Thermodynamics of Polymerization of Cycloalkanes at 25°C

$(CH_2)_n$ n	ΔH_{lc} (kJ/mole)	ΔS_{lc} (J/mole-°C)	ΔG_{lc} (kJ/mole)
3	−113.0	−69.1	−92.5
4	−105.1	−55.3	−90.0
5	−21.2	−42.7	−9.2
6	+2.9	−10.5	+5.9
7	−21.8	−15.9[a]	−16.3
8	−34.8	−3.3[a]	−34.3

[a]Data from Cubbon [1964]; all other data are from Dainton and Ivin [1958] and Finke et al. [1956].

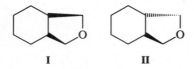

I   **II**

Although ring-opening polymerization is thermodynamically favored for all except the 6-membered cycloalkane, polymerization of cycloalkanes has been achieved in very few cases, almost exclusively with cyclopropane derivatives and only oligomers are obtained [Pinazzi et al., 1971; Sogo et al., 1978]. This points out that thermodynamic feasibility does not guarantee the actual polymerization of a cyclic monomer. Polymerization requires that there be a kinetic pathway for the ring to open and undergo reaction. The cycloalkanes do not have a bond in the ring structure that is easily prone to attack by an initiator species. The lactams, lactones, cyclic ethers and acetals, and other cyclic monomers stand in marked contrast to the cycloalkanes. The presence of a heteroatom in the ring provides a site for nucleophilic or electrophilic attack by an initiator species, resulting in initiation and subsequent propagation by ring opening. Such monomers polymerize, since both thermodynamic and kinetic factors are favorable. Overall, one observes that polymerizability (a combination of thermodynamic and kinetic feasibility) is higher for rings of 3, 4, and 7 to 11 members, lower for rings of 5 members, and much lower for rings of 6 members. In actual practice ring-opening polymerizations are usually limited to monomers of less than 9 members due to the general unavailability of larger-sized cyclic monomers. Also, 3-membered rings are not available for a number of classes of compounds (lactams, lactones, cycloalkenes, cyclic siloxanes, and acetals).

7-1b Polymerization Mechanism and Kinetics

Ring-opening polymerizations are generally initiated by the same types of ionic initiators previously described for the cationic and anionic polymerizations of monomers with carbon–carbon and carbon–oxygen double bonds (Chap. 5). Most cationic ring-opening polymerizations involve the formation and propagation of oxonium ion centers. Reaction involves the nucleophilic attack of monomer on the oxonium ion

$$\sim\!\!\sim\!\!\overset{+}{Z}\!\!\bigcirc + Z\!\!\bigcirc \rightarrow \sim\!\!\sim\!\!\overset{+}{Z}\!\!\bigcirc \qquad (7\text{-}5)$$

The typical anionic ring-opening polymerization involves the formation and propagation of anionic centers. Reaction proceeds by nucleophilic attack of the propagating anion on monomer

$$\sim\!\!\sim\!\!Z^- + Z\!\!\bigcirc \rightarrow \sim\!\!\sim\!\!Z^- \qquad (7\text{-}6)$$

In Eq. 7-5 Z represents a reactive functional group, such as O—C, N—C, Si—O, CO—O, CO—NH in ethers, amines, siloxanes, esters, amides, respectively. In Eq. 7-6 Z^- represents an anionic center, such as alkoxide or carboxylate, derived from the cyclic monomer.

Ionic ring-opening polymerizations show most of the characteristics described in Chap. 5. There is a minimal discussion in this chapter of those characteristics that are the same for ring-opening polymerization as for C=C and C=O polymerizations. There are few specifics mentioned for the initiation process unless polymerization of a particular cyclic monomer is achieved with initiators not discussed previously. Ionic ring-opening polymerizations show effects of solvent and counterion, propagation by different species (covalent, ion pairs, free ions), and association phenomena analogous to those discussed in Chap. 5.

The nature of the growth process in ring-opening polymerization bears a superficial resemblance to that in chain polymerization. Only monomer adds to the growing chains in a propagation step. Species larger than monomer do not react with each other. However, ring-opening polymerizations are not necessarily—and often are not—chain polymerizations. The classification of a ring-opening polymerization as a chain or step polymerization can be made on the basis of two criteria: (1) the experimentally observed kinetic laws that describe polymerization and (2) the relationship between polymer molecular weight and conversion. The second criterion is the prime characteristic that distinguishes chain and step polymerizations. High polymer is formed throughout the course of a chain polymerization in contrast to the slow buildup of polymer molecular weight in step polymerization. Most, but not all, ring-opening polymerizations behave as step polymerizations in that the polymer molecular weight increases relatively slowly with conversion. This is a consequence of the types of reactions involved in ring-opening polymerizations. The rate constants for ring-opening reactions of cyclic monomers, such as ethers, amines, siloxanes, amides, and esters, have values much closer to those for the reactions of step polymerization (e.g., esterification, amidation) than for chain polymerization (addition of radical, carbocation, or carbanion to C=C).

Irrespective of whether a particular ring-opening polymerization is a step or chain reaction in its buildup of polymer molecular weight, its kinetics usually are described by expressions resembling those of chain polymerizations since growing chains add only monomer. Many ring-opening polymerizations are complicated by the occurrence of polymerization–depolymerization equilibria. Various situations will be illustrated in this chapter.

7-2   CYCLIC ETHERS

Cyclic ethers can be named simply as oxacycloalkanes, such as oxacyclopropane, oxacyclobutane, oxacyclopentane, oxacyclohexane, where the prefix *oxa* indicates the replacement of CH_2 by O in corresponding cycloalkanes. Most cyclic ethers, however, are known by other names. The 3-, 4-, 5-, and 6-membered rings are oxirane, oxetane, oxolane, and oxane, respectively, or ethylene oxide (or epoxide), trimethylene oxide, tetrahydrofuran, and tetrahydropyran.

The carbon–oxygen bond in ethers is a strong bond and the ether oxygen is basic in the Lewis sense. The result is that ring-opening polymerization of cyclic ethers is initiated only by cationic species except for epoxides. Epoxides are polymerized by both anionic and cationic initiators because of the high degree of strain in the 3-membered ring.

The polymerization of simple cyclic ethers (i.e., those with a single ether linkage) has been generally limited to those of 3, 4, and 5 members, although some work has been done with the 7-membered (oxepane) ring. The study of larger-sized rings has been carried out mostly with cyclic acetals (Sec. 7-2b-4). The reactivity of different-sized cyclic ethers follows the generally expected order. Cyclic ethers of less than 5 members or more than 6 members are relatively easily polymerized. The 5-membered cyclic ethers polymerize with more difficulty. Substituted 5-membered cyclic ethers are usually unreactive, although some cyclic acetals undergo polymerization. The 6-membered cyclic ethers such as tetrahydropyran (**III**) and 1,4-dioxane (**IV**) are unreactive under a wide range of reaction conditions, but the 6-membered cyclic acetal, trioxane (**V**), undergoes polymerization.

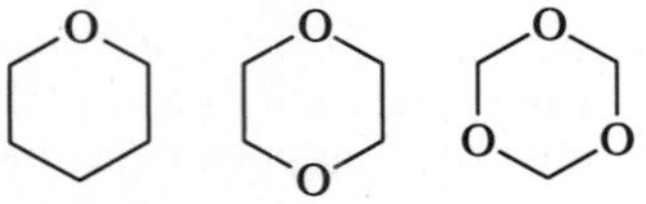

III   **IV**   **V**

7-2a   Anionic Polymerization of Epoxides

7-2a-1   Reaction Characteristics

The anionic polymerization of epoxides such as ethylene and propylene oxides can be initiated by hydroxides, alkoxides, oxides, and metal alkyls and aryls, including radical–anion species such as sodium naphthalene [Boileau, 1989; Dreyfuss and Dreyfuss, 1976; Inoue and Aida, 1984; Ishii and Sakai, 1969]. Thus the polymerization of ethylene oxide by M^+A^- involves initiation,

$$H_2C\overset{O}{-\!\!-\!\!-}CH_2 + M^+A^- \rightarrow A-CH_2CH_2O^-M^+ \tag{7-7}$$

followed by propagation,

$$A-CH_2CH_2O^-M^+ + H_2C\overset{O}{-\!\!-\!\!-}CH_2 \rightarrow A-CH_2CH_2OCH_2CH_2O^-M^+ \tag{7-8a}$$

which may be generalized as

$$A\text{---}(CH_2CH_2O)_{n}^{-}CH_2CH_2O^-M^+ + H_2C\overset{O}{-\!\!-\!\!-}CH_2 \rightarrow \tag{7-8b}$$
$$A\text{---}(CH_2CH_2O)_{(n+1)}^{-}CH_2CH_2O^-M^+$$

A number of initiators are responsible for epoxide polymerization occurring through an *anionic coordination* mechanism. These initiators include a ferric chloride–propylene oxide adduct $ClFe[OCH(CH_3)CH_2Cl]_2$ (referred to as the *Pruitt–Baggett initiator*), adducts such as $Zn(OCH_3)_2$ and $([Zn(OCH_3)_2]_2 \cdot [C_2H_5ZnOCH_3]_6)$ derived by reaction of dialkylzinc with alcohol, bimetallic μ-oxoalkoxides such as $[(RO)_2AlO]_2Zn$, and metalloporphyrin derivatives of zinc, aluminum, and maganese (e.g., **VI**, where Z is Cl, OR, R, OOCR) [Colak and Alyuruk, 1989; Hagiwara et al., 1981; Hasebe and Tsurata, 1988; Inoue, 1988; Kasperczyk and Jedlinski, 1986; Kuroki et al., 1988a, 1988b]. There are also other less well-defined initiators, such as $Zn(C_2H_5)_2/H_2O$ and $Al(C_2H_5)_3/H_2O/acetylacetone$. Propagation in these systems involves a concerted or

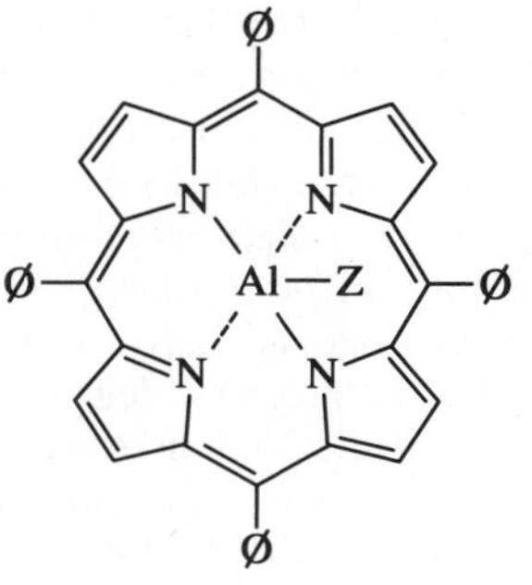

VI

coordinated process in which the epoxide monomer is inserted into a metal–oxygen bond

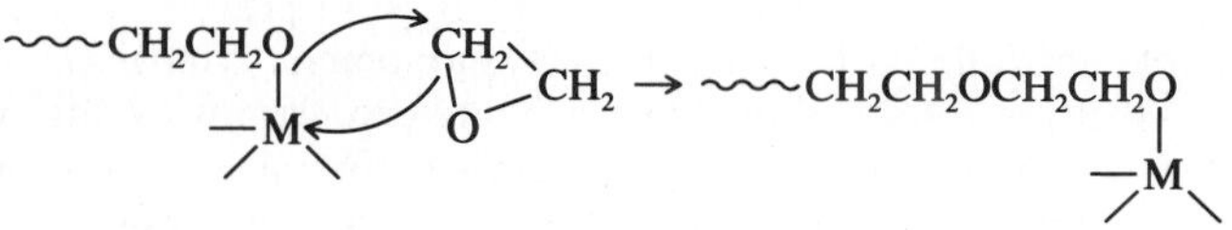

$$\text{(7-9)}$$

The propagation is categorized as an anionic coordination process as one can visualize the formation of an incipient alkoxide anion on cleavage of the oxygen–metal bond in the propagating chain. (There is some controversy that polymerizations with a few coordination initiators, e.g., AlR_3/H_2O, proceed as cationic, not anionic, processes [Bansleban et al., 1984].) Many anionic coordination polymerizations proceed with stereochemical consequences, and these will be discussed further in Chap. 8.

The coordination process described above is similar to several propagations described in Chap. 5 (pseudocationic, living cationic, group transfer polymerizations) in which there is propagation through a covalent species.

The polymerization of an unsymmetrical epoxide such as propylene oxide involves the possibility of two different sites (at carbons 1 and 2 or α and β) on the epoxide ring for the nucleophilic ring-opening reaction. Two different propagating species are then possible

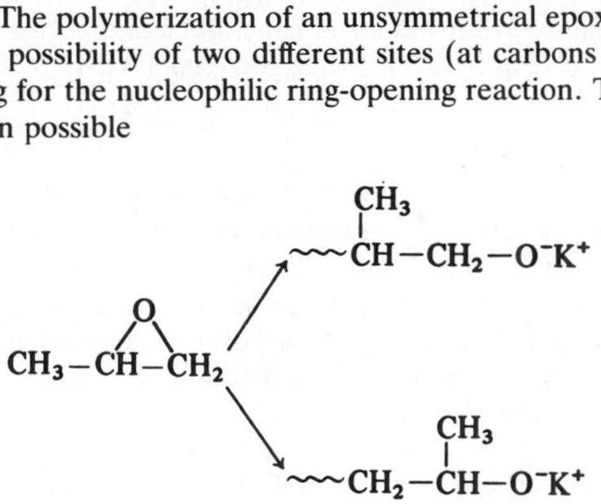

$$\text{(7-10)}$$

$$\text{(7-11)}$$

The polymer has a predominately head-to-tail structure with propagation occurring almost exclusively by attack at the β-carbon—the less sterically hindered site (Eq. 7-11), that is, an S_N2 attack [Kasperczyk and Jedlinski, 1986; Oguni et al., 1973; Price and Osgan, 1956]. (Ring opening is somewhat less regioselective for cationic and coordination polymerizations [Inoue and Aida, 1984; Jedlinski et al., 1979; Price, 1974]).

Anionic polymerizations of epoxides, including coordination processes, typically proceed as step polymerizations with the polymer molecular weight increasing relatively slowly with conversion. However, most epoxide polymerizations have the characteristics of living polymerizations—the ability to polymerize successive monomer charges and form block copolymers. The expressions for the rate and degree of polymerization are essentially those used in living chain polymerizations (Sec. 5-3e). Thus, in the sodium methoxide initiated polymerization of ethylene oxide [Gee et al., 1959], the polymerization rate is given by Eq. 5-92, where [M$^-$] is the total concen-

$$R_p = k_p^{app}[M^-][M] \tag{5-92}$$

tration of all living anionic propagating centers (free ions and ion pairs). Effects of reaction media on the polymerization rate are similar to those previously described in Chap. 5 for ionic polymerizations of C=C and C=O monomers. Changes in solvent and counterion affect reaction rates and the observed rate expressions by altering the relative amounts of free ion and also altering the ion pair propagating species and association of initiator. Using the approach described in Sec. 5-3e-2-a, one can obtain the individual rate constants for propagation of free ions and ion pairs, although extensive studies have not been carried out [Sigwalt and Boileau, 1978]. Reported values of k_p^- are in the range of 10^{-2}–10 liters/mole-sec with $k_p^{\pm}$ values usually being lower by 1–2 orders of magnitude [Boileau, 1989].

The degree of polymerization at time t in the reaction is given by the concentration of monomer that has reacted divided by the initial initiator concentration

$$\overline{X}_n = \frac{[M]_0 - [M]_t}{[I]} \tag{7-12}$$

where [M]$_0$ and [M]$_t$ are the monomer concentrations at times zero and t. Equation 7-12 becomes exactly the equivalent of Eq. 5-107 at 100% conversion ([M]$_t$ = 0).

7-2a-2 *Exchange Reactions*

Epoxide polymerizations taking place in the presence of protonic substances such as water or alcohol involve the presence of exchange reactions. Examples of such polymerizations are those initiated by metal alkoxides and hydroxides that require the presence of water or alcohol to produce a homogeneous system by solubilizing the initiator. Such substances increase the polymerization rate not only by solubilizing the initiator but probably also by increasing the concentration of free ions and loose ion pairs. In the presence of alcohol the exchange reaction

$$R\text{---}(OCH_2CH_2\text{---})_n\text{---}O^-Na^+ + ROH \rightleftharpoons R\text{---}(OCH_2CH_2\text{---})_n\text{---}OH + RO^-Na^+ \tag{7-13}$$

between a propagation chain and the alcohol is possible. Similar exchange reactions are possible between the newly formed polymeric alcohol in Eq. 7-13 and other propagating chains,

$$R\text{---}(\text{---}OCH_2CH_2\text{---})_n\text{---}OH + R\text{---}(\text{---}OCH_2CH_2\text{---})_m\text{---}O^-Na^+ \rightleftharpoons$$

$$R\text{---}(\text{---}OCH_2CH_2\text{---})_n\text{---}O^-Na^+ + R\text{---}(\text{---}OCH_2CH_2\text{---})_m\text{---}OH \tag{7-14}$$

These exchange reactions lower the polymer molecular weight. The number-average degree of polymerization is given by

$$\overline{X}_n = \frac{[M]_0 - [M]_t}{[I] + [ROH]} \tag{7-15}$$

since each alcohol molecule contributes equally with an initiator species to determining the number of progagating chains. The exchange reactions appear equivalent to chain-transfer reactions, but they are not. Any polymeric alcohol formed via exchange is not dead but simply dormant. All alcohol and alkoxide molecules in the reaction system are in dynamic equilibrium. Each polymer chain alternates between the active propagating alkoxide and dormant alcohol forms.

The exchange reaction places an upper limit on the polymer molecular weight that is possible for polymerizations performed in the presence of alcohols or other protonic substances. There are very few reports of polymers with molecular weights above 10,000 for ethylene oxide polymerizations initiated by alkoxides or hydroxides in alcohol, although molecular weights as high as 50,000 appear (very rarely) in the literature [Boileau, 1989; Clinton and Matlock, 1986]. (Another reason for the upper limit on molecular weight is dehydration from the hydroxy end groups of the dormant polymeric alcohol [Clinton, and Matlock, 1986]. This would adversely affect the living nature of the polymerization by converting dormant alcohol species to dead polymer.) Polymerizations initiated by alkoxides and hydroxides in aprotic polar solvents do not have this limitation. This limitation also does not apply to polymerizatons initiated by the other initiators, including metal alkyls and aryls and the various coordination initiators, since those initiators are soluble in solvents such as benzene or tetrahydrofuran. Homogeneous reactions can be obtained without the need for a solvent such as alcohol. Molecular weights as high as 10^5–10^6 have been achieved in a number of systems. However, the addition of alcohol or other protonic substance is useful for control of polymer molecular weight. Equation 7-15 allows one to calculate the amount of added alcohol or other substance required to achieve some desired value of the number-average molecular weight.

A number of different situations are possible depending on the relative rates of initiation, propagation, and exchange. When exchange is absent and initiation is much faster than propagation, initiation is essentially complete before propagation begins. All polymer chains start growing at the same time and grow for the same period with the result that the molecular-weight distribution is very narrow as in living chain polymerizations; that is, the distribution is Poisson as defined by Eq. 5-59. When initiation is slow some chains are growing while others have not yet been initiated. The polymerization proceeds with an initial period in which the reaction rate shows an acceleration as initiator is converted to propagating species; thereafter, the rate is constant. The molecular-weight distribution broadens since chains grow for different periods of time. This effect is observed in most anionic coordination polymerizations because the initiators are generally aggregated and possess different initiator sites with different reactivities. An exception to this generalization are the metalloporphyrin initiators, which yield living polymers with narrow molecular weights.

The effect of exchange depends on the relative acidities of the alcohol (or other protonic substance) and the polymeric alcohol. The exchange reaction occurs throughout the course of the polymerization if the acidities of the two alcohols are approximately the same. The polymerization rate is unaffected while the molecular weight decreases (Eq. 7-15), but the molecular-weight distribution (MWD) is Poisson.

If the added alcohol ROH is much more acidic than the polymeric alcohol, most of it will undergo reaction with the first-formed propagating species,

$$ROCH_2CH_2O^-Na^+ + ROH \rightarrow ROCH_2CH_2OH + RO^-Na^+ \tag{7-16}$$

before polymerization begins. Reinitiation by RO^-Na^+ would usually be slower, since ROH is relatively acidic. This results in a decreased polymerization rate and a broadening of the polymer molecular weight. The rate of polymerization will be relatively unaffected during most of the polymerization for the case in which ROH is less acidic than the polymeric alcohol. Exchange will occur in the later stages of reaction with a broadening of the MWD.

The use of protonic compounds such as HCl or RCOOH in place of ROH or H_2O yields a different result in most systems. When such substances take part in the exchange reaction the result is not exchange as described above but inhibition or retardation since an anion, such as Cl^- or $RCOO^-$, possesses little or no nucleophilicity. Reinitiation does not occur or is very slow. The polymeric alcohols are no longer dormant but are dead. Both the polymerization rate and polymer molecular weight decrease along with a broadening of the polymer molecular weight. Hydrogen chloride and RCOOH act as the equivalent of inhibitors or retarders and the polymerization rate can be treated by the expressions developed in Sec. 3-7a. The number-average degree of polymerization is analyzed by the expression used to handle chain transfer (Eq. 3-129). Polymerizations initiated by the metalloporphyrins do not follow this pattern. Reinitiation occurs even when HCl or RCOOH is present, that is, **VI** with Z = Cl, and RCOO is an effective initiator, although the reasons are not known. Polymerizations initiated by the metalloporphyrins are difficult to "kill" and have been termed *immortal polymerizations* [Aida et al., 1988].

7-2a-3   Chain Transfer to Monomer

Excluding polymerizations with anionic coordination initiators, the polymer molecular weights are low for anionic polymerizations of propylene oxide (<5000) [Clinton and Matlock, 1986; Boileau, 1989; Gagnon, 1986; Ishii and Sakai, 1969; Sepulchre et al., 1979]. Polymerization is severely limited by chain transfer to monomer. This involves proton abstraction from the methyl group attached to the epoxide ring followed by rapid ring cleavage to form the allyl alkoxide anion **VII**, which isomerizes partially to the enolate anion **VIII**. Species **VII** and **VIII** reinitiate polymerization of propylene

$$\tag{7-17}$$

$$CH_2\overset{O}{\overbrace{\qquad}}CH-CH_2^-Na^+ \rightarrow CH_2=CH-CH_2O^-Na^+ \qquad (7\text{-}18)$$

VII

$$\Updownarrow$$

$$CH_3-CH=CHO^-Na^+ \qquad (7\text{-}19)$$

VIII

oxide as evidenced by the presence of infrared absorptions for allyl ether and 1-propenyl ether groups in the polymer. The monomer chain-transfer constant has values of 0.013 and 0.027 at 70 and 93°C, respectively, in the sodium methoxide polymerization of propylene oxide [Gee et al., 1961]. These values are larger by factors of 10^2–10^4 than the usual monomer transfer constants (Tables 3-5 and 5-4). The extent of this transfer reaction increases with increasing number of methyl groups on the epoxide ring.

Chain transfer to monomer is much less prevalent for polymerizations with most of the anionic coordination initiators. Much higher polymer molecular weights are possible in these polymerizations. For example, molecular weights in the 10^5 range are reported for polymerization of propylene oxide by an initiator derived from diphenyltin sulfide and bis(3-dimethylaminopropyl)zinc [Bots et al., 1987].

7-2b Cationic Polymerization

7-2b-1 Propagation

Propagation in the cationic polymerization of cyclic ethers is generally considered as proceeding via a tertiary *oxonium ion*, for example, for the polymerization of 3,3-bis(chloromethyl)oxetane (R = CH_2Cl)

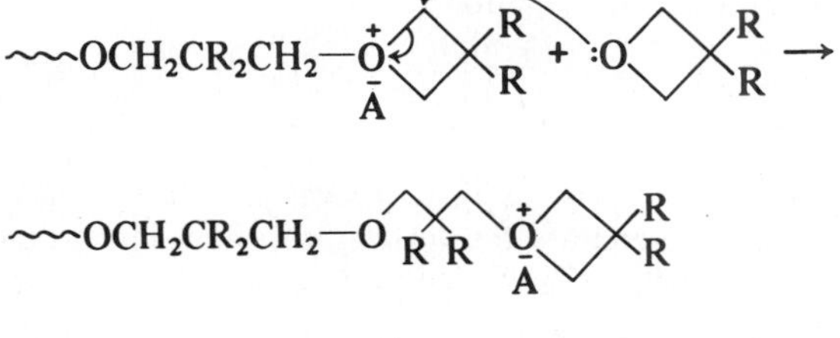

$$\sim\sim\sim OCH_2CR_2CH_2-\overset{+}{\underset{A^-}{O}}\overset{R}{\diagup\diagdown}\overset{R}{\diagup}_R + \overset{}{:}O\overset{R}{\diagup\diagdown}_R \rightarrow$$

$$\sim\sim\sim OCH_2CR_2CH_2-O\overset{}{\underset{R\ R}{\diagup\diagdown}}\overset{+}{\underset{A^-}{O}}\overset{R}{\diagup\diagdown}_R \qquad (7\text{-}20)$$

where A^- is the counterion [Frisch and Reegan, 1969; Inoue and Aida, 1984; Penczek and Kubisa, 1989a, 1989b].

The α-carbon of the oxonium ion is electron-deficient because of the adjacent positively charged oxygen. Propagation is a nucleophilic attack of the oxygen of a monomer molecule on the α-carbon of the oxonium. The nucleophilic reaction is an S_N2 reaction for most cyclic ethers as shown by studying the stereochemistry of appropriate reactants. For example, the polymer from *endo*-2-*t*-butyl-7-oxabicyclo-[2.2.1]heptane has the ether linkages in the 1,4-positions trans to each other, indicating inversion at either carbon 1 or carbon 4 [Saegusa et al., 1976]. The polymerization is pictured as

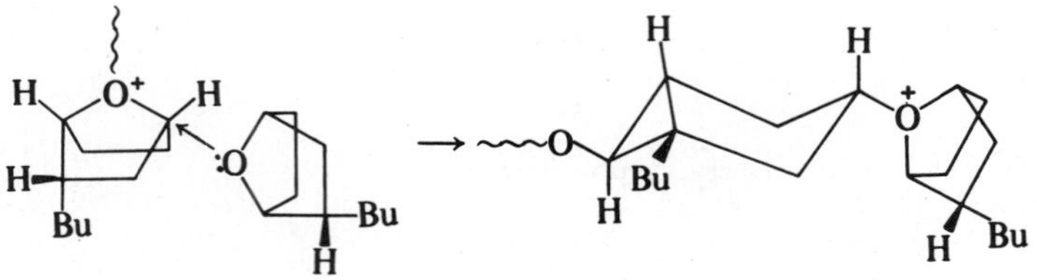

$$(7\text{-}21)$$

with propagation involving S_N2 attack and inversion at carbon 4 of the ring. Attack at carbon 1 is sterically hindered compared to attack at carbon 4. An S_N1 mechanism has been suggested for a monomer containing two alkyl substituents at an α-position (e.g., 2,2-dimethyloxetane) [Dreyfuss and Dreyfuss, 1987; Kops and Spanggaard, 1982]. S_N1 ring opening at the substituted α-carbon would result in a relatively stable tertiary carbocation.

Most cationic ring-opening polymerizations are regioselective with the formation of head-to-tail structures, although varying amounts of head-to-head and tail-to-tail structures are found in some systems [Dreyfuss and Dreyfuss, 1987].

The propagating species in cationic ring-opening polymerization may involve varying amounts of free ion (**IX**) and ion pair (**X**) as well as covalent ester (**XI**) depending

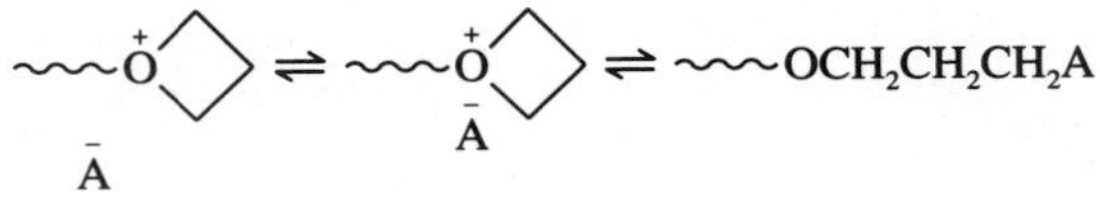

$$(7\text{-}22)$$

on reaction conditions and the monomer, similar to the situation described previously for ionic polymerizations of C=C and C=O monomers (Secs. 5-1, 5-2e-3, 5-2f-1). There is usually no difference in reactivity between free ion and ion pair, but the covalent ester has a significantly lower propagation rate constant.

7-2b-2 Initiation

A variety of initiator systems, of the types used in the cationic polymerization of alkenes (Sec. 5-2a), can be used to generate the tertiary oxonium ion propagating species [Dreyfuss and Dreyfuss, 1969, 1976; Inoue and Aida, 1984; Penczek and Kubisa, 1989a, 1989b].

Strong protonic acids such as sulfuric, trifluoroacetic, fluorosulfonic, and trifluoromethanesulfonic (triflic) acids initiate polymerization via the initial formation of a secondary oxonium ion,

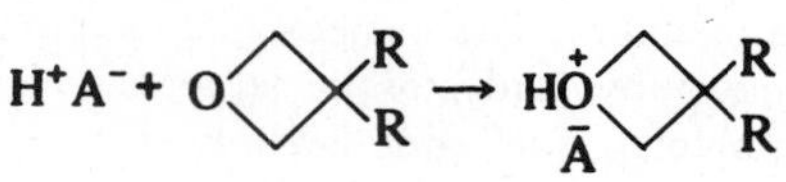

$$(7\text{-}23)$$

which reacts with a second monomer molecule to form the tertiary oxonium ion,

$$
\text{HO}\overset{+}{\underset{\bar{\text{A}}}{\diagdown}}\diagdown\!\!\diagup\!\!{<}^{\text{R}}_{\text{R}} + \text{O}\diagdown\!\!\diagup\!\!{<}^{\text{R}}_{\text{R}} \rightarrow \text{HOCH}_2\text{CR}_2\text{CH}_2-\overset{+}{\underset{\bar{\text{A}}}{\text{O}}}\diagdown\!\!\diagup\!\!{<}^{\text{R}}_{\text{R}} \tag{7-24}
$$

This type of initiation is limited by the nucleophilicity of the anion A^- derived from the acid. For acids other than the very strong acids such as fluorosulfonic and triflic acids, the anion is sufficiently nucleophilic to compete with monomer for either the proton or secondary and tertiary oxonium ions. Only very-low-molecular-weight products are possible. The presence of water can also directly disrupt the polymerization since its nucleophilicity allows it to compete with monomer for the oxonium ions.

Lewis acids such as BF_3 and $SbCl_5$, almost always in conjuction with water or some other protogen, initiate polymerization of cyclic ethers. The initiator and coinitiator form an initiator–coinitiator complex [e.g., $BF_3 \cdot H_2O$, $H^+(SbCl_6)^-$], which acts as a proton donor in an initiation sequence similar to Eqs. 7-23 and 7-24. Similar proton donors are formed by photolysis of diaryliodonium and triarylsulfonium salts (Sec. 5-2a-3-b).

The use of a cationogen such as an alkyl or acyl halide with a Lewis acid generates carbocations and acylium ions, either *in situ* or as isolable salts, which can initiate polymerization, for example,

$$
\emptyset_3\text{CCl} + \text{AgSbF}_6 \rightarrow \text{AgCl} + \emptyset_3\text{C}^+(\text{SbF}_6)^- \tag{7-25}
$$

$$
\overset{\text{O}}{\underset{\parallel}{\emptyset_3\text{CCl}}} + \text{SbCl}_5 \rightarrow \overset{\text{O}}{\underset{\parallel}{\emptyset\text{C}^+}}(\text{SbCl}_6)^- \tag{7-26}
$$

Carbocations are also obtained by ionization of the esters of very strong acids

$$
\text{F}_3\text{CSO}_3\text{CH}_3 \rightarrow \text{CH}_3^+(\text{F}_3\text{CSO}_3)^- \tag{7-27}
$$

The initiation with some carbocations, especially trityl, does not involve direct addition to monomer. The carbocation abstracts a hydride ion from the α-carbon of monomer and the newly formed carbocation initiates polymerization [Afsar-Taromi et al., 1978; Kuntz, 1967]. This hydride ion abstraction is so facile with 1,3-dioxolane that it is used to preform stable 1,3-dioxolan-2-ylium salts (**XII**) that can be used

$$
\emptyset_3\text{C}^+\text{A}^- + \text{O}\diagup\!\diagdown\text{O} \rightarrow \emptyset_3\text{CH} + \overset{\text{H} \quad \text{A}^-}{\underset{}{\text{O}\diagup\!\overset{+}{\diagdown}\text{O}}} \tag{7-28}
$$

XII

subsequently as initiators [Jedlinski et al., 1985].

Since the tertiary oxonium ion is the propagating species, preformed tertiary oxonium ions such as triethyloxonium tetrafluoroborate are useful for initiating polymerization (Eq. 7-29) [Meerwein et al., 1960].

Many combinations of a Lewis acid and a reactive cyclic ether (e.g., an epoxide

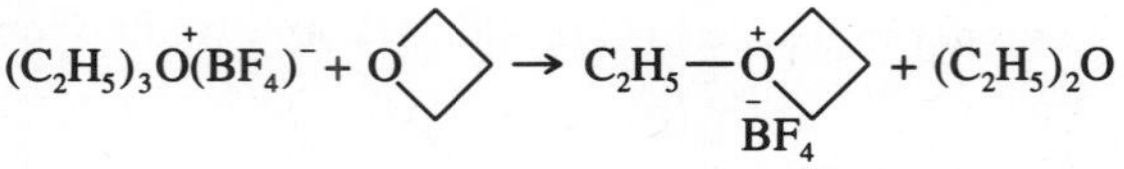

$$(C_2H_5)_3\overset{+}{O}(BF_4)^- + O\!\!\!\bigtriangledown \rightarrow C_2H_5-\overset{+}{\underset{BF_4}{O}}\!\!\!\bigtriangledown + (C_2H_5)_2O \tag{7-29}$$

or oxetane), usually together with a protogen or cationogen, have been used to initiate the polymerization of less reactive cyclic ethers such as tetrahydrofuran [Saegusa and Matsumoto, 1968]. Initiation occurs by formation of the secondary and tertiary oxonium ions of the more reactive cyclic ether, which then act at the initiating species for polymerization of the less reactive cyclic ether. e.g., for the use of an epoxide in the polymerization of tetrahydrofuran (THF)

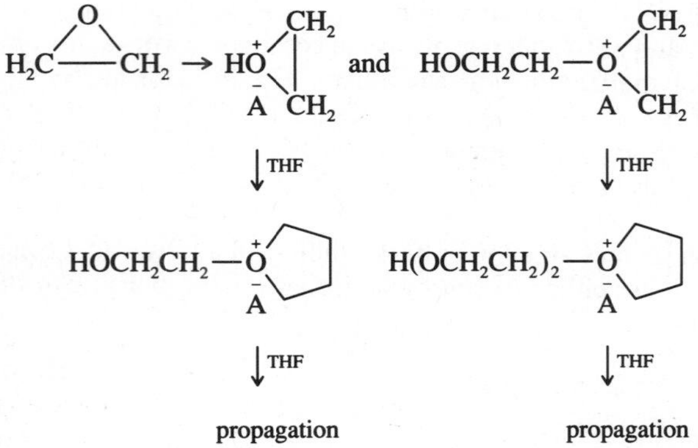

$$H_2C\overset{O}{\underset{}{-\!\!\!-\!\!\!-}}CH_2 \rightarrow HO\overset{+}{\underset{A}{\diagup}}\overset{CH_2}{\underset{CH_2}{\diagdown}} \quad and \quad HOCH_2CH_2-\overset{+}{\underset{A}{O}}\overset{CH_2}{\underset{CH_2}{\diagdown}}$$

$$\downarrow THF \qquad\qquad\qquad\qquad \downarrow THF$$

$$HOCH_2CH_2-\overset{+}{\underset{A}{O}}\!\!\!\bigcirc \qquad H(OCH_2CH_2)_2-\overset{+}{\underset{A}{O}}\!\!\!\bigcirc \tag{7-30}$$

$$\downarrow THF \qquad\qquad\qquad\qquad \downarrow THF$$

$$propagation \qquad\qquad\qquad propagation$$

The reactive cyclic ether is often referred to as a *promoter*. The promoter, used in small amounts relative to the cyclic ether being polymerized, increases the ability of the latter to form the tertiary oxonium ion.

7-2b-3 Termination and Transfer Processes.

Under certain conditions, cationic cyclic ether polymerizations have the characteristics of living polymerizations in that the propagating species are long-lived and narrow MWDs are obtained. Living polymerizations occur when initiation is fast relative to propagation and there is an absence of termination processes. These conditions are found for polymerizations initiated with acylium and 1,3-dioxolan-2-ylium salts containing very stable counterions such as AsF_6^-, PF_6^-, and $SbCl_6^-$ or with very strong acids (fluorosulfonic and trifluoromethanesulfonic) or their esters whose anions may form covalent ester bonds but in a reversible manner [Jedlinski et al., 1985; Penczek and Kubisa, 1989a, 1989b; Penczek and Matyjaszewski, 1976]. The rate and degree of polymerization are given by the expressions previously described (Eqs. 5-92, 7-12).

7-2b-3-a Transfer Reactions. Transfer reactions occur in many instances by a variety of reactions, some of which are analogous to those in the cationic polymerization of alkenes. *Chain transfer to polymer* is a common mode by which a propagating chain is terminated, although the kinetic chain is unaffected. The reaction involves nucleophilic attack by the ether oxygen in a polymer chain on the oxonium ion propagating center, the same type of reaction involved in propagation, to form the tertiary oxonium ion **XIII**. Subsequent nucleophilic attack on **XIII** by monomer yields **XIV** and regen-

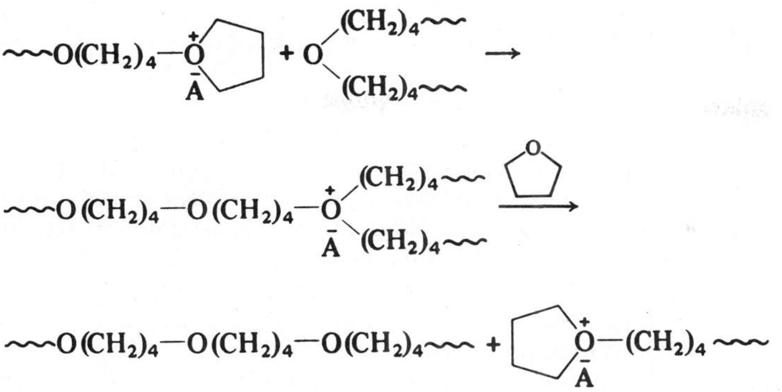

$$\sim\sim O(CH_2)_4 - \overset{+}{\underset{\bar{A}}{O}} \bigotimes + O \overset{(CH_2)_4\sim}{\underset{(CH_2)_4\sim}{\diagup}} \longrightarrow$$

$$\sim\sim O(CH_2)_4 - O(CH_2)_4 - \overset{+}{\underset{\bar{A}}{O}} \overset{(CH_2)_4\sim}{\underset{(CH_2)_4\sim}{\diagup}} \xrightarrow{\quad\quad} \qquad (7\text{-}31)$$

$$\sim\sim O(CH_2)_4 - O(CH_2)_4 - O(CH_2)_4 \sim\sim + \left[\underset{\bar{A}}{\overset{+}{O}} - (CH_2)_4 \sim\sim\right.$$

erates the propagating species. The overall effect is an exchange of polymer chain segments with a broadening of the MWD from the narrow distribution for a living polymerization [Goethals, 1977]. The MWD in some reaction systems is close to the distribution expected for a step polymerization ($\overline{X}_w/\overline{X}_n \sim 2$). Chain transfer to polymer occurs as an intramolecular reaction as well as intermolecular reaction. The intramolecular reaction results in the formation of cyclic oligomers instead of linear polymer; in other words, the chain ends in **XIV** are connected to each other. The competition between propagation and chain transfer to polymer depends on several factors. Propagation is favored on steric grounds since attack by monomer is less hindered than attack by the ether oxygen in a polymer chain. The relative nucleophilic activities of the two different ether oxygens is important and varies considerably depending on monomer ring size. Finally, intramolecular chain transfer to polymer (but not intermolecular transfer) becomes progressively more important at lower monomer concentrations.

The distribution of different-sized cyclic oligomers formed in cyclic ether polymerizations generally follows what is expected from the relative stabilities of different-sized rings (Secs. 2-5, 7-1a). The presence of a polymer–cyclic oligomer equilibrium (often in addition to a polymer–monomer equilibrium) is well documented in many different cyclic ether polymerizations [Bucquoye and Goethals, 1978; Dreyfuss and Dreyfuss, 1969, 1987; Dreyfuss et al., 1989; McKenna et al., 1977; Penczek and Slomkowski, 1989a, 1989b]. Polymerization of an oxirane yields more cyclic oligomer than obtained from any other-sized cyclic ether. The relative nucelophilicity of a cyclic ether compared to a polymer chain ether is lowest for the 3-membered ring. Propagation for an oxirane involves an increase in bond angle strain on conversion of the sp^3-hydridized oxygen in monomer to the sp^2-hybridized oxygen in the oxonium ion. The cyclic dimer 1,4-dioxane is the major cyclic oligomer formed in the polymerization of ethylene oxide. In fact, 1,4-dioxane is the major product in the typical polymeri-

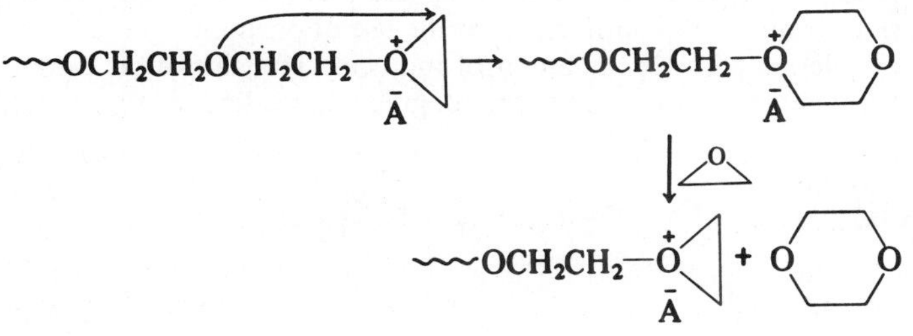

$$\sim\sim OCH_2CH_2\overset{\frown}{O}CH_2CH_2 - \overset{+}{\underset{\bar{A}}{O}} \triangleleft \longrightarrow \sim\sim OCH_2CH_2 - \overset{+}{\underset{\bar{A}}{O}} \bigotimes O$$

$$\Bigg\downarrow \overset{O}{\triangleleft}$$

$$\sim\sim OCH_2CH_2 - \overset{+}{\underset{\bar{A}}{O}} \triangleleft \; + \; O \bigotimes O \qquad (7\text{-}32)$$

zation of ethylene oxide, with yields as high as or higher than 80% [Libiszowski et al., 1989]. Propylene oxide gives less cyclic dimer than ethylene oxide for steric reasons; cyclic tetramer predominates.

The nucleophilicity of a cyclic ether relative to the ether oxygen in a polymer chain increases with ring size so that cyclic oligomer formation is somewhat less for oxetane and much less for tetrahydrofuran compared to oxirane. The cyclic tetramer is the most abundant cyclic oligomer in oxetane polymerization with lesser amounts of trimer and sizes from pentamer up to nonamer. No cyclic dimer is found. Cyclic dimers through octamers are found in tetrahydrofuran polymerization. The tetramer is the most abundant species followed by the pentamer and hexamer; only very small amounts of the other-sized cyclics are formed. The total amount of cyclic oligomers is less than 3% in the typical tetrahydrofuran polymerization.

Recent work indicates that some epoxide polymerizations can be carried out under reaction conditions whereby cyclic oligomer formation can be greatly suppressed. Reaction is carried out with an acid initiator such as BF_3 or triflic acid in the presence of alcohol and at very low monomer concentrations. Monomer is added to the initiator and alcohol in a continuous manner at a rate equal to its rate of consumption. Polymerization proceeds by an *activated monomer* (AM) mechanism. Under the AM reaction conditions most of the monomer is present in its protonated form **XV**, while the propagating centers are hydroxyl groups instead of oxonium ions. The "normal" cationic process of nucleophilic attack by monomer on oxonium ion centers is minimized and the major reaction involves nucleophilic attack by hydroxyl end groups of polymer on protonated monomer. This approach allows one to minimize the extent

$$\overset{+}{\triangleright}\text{OH} + \text{HO}\sim\sim \rightarrow \text{HOCH}_2\text{CH}_2\overset{+}{\underset{\text{H}}{\text{O}}}\sim\sim \qquad (7\text{-}33)$$

$$\textbf{XV} \qquad \textbf{XVI} \qquad\qquad \textbf{XVII}$$

of cyclic oligomer formation [Bednarek et al., 1989; Biedron et al., 1990; Penczek et al., 1986; Wojtania et al., 1986]. To date the activated monomer approach has been successfully demonstrated for the synthesis of telechelic polymers of propylene oxide and epichlorohydrin (number-average molecular weights of 1000–2500). Apparently, the large amount of alcohol required to suppress the normal propagation mode and accentuate the activated monomer pathway precludes synthesis of higher-molecular-weight polymers from oxiranes. Application of the AM pathway to other monomers has not been reported.

7-2b-3-b Termination Reactions. Termination occurs to varying degrees by combination of the propagating oxonium ion with either the counterion or an anion derived from the counterion. It has been mentioned that the use of protonic acids as initiators is limited by the nucleophilicity of the anion of the acid. Transfer of an anion from the counterion, for example, occurs to varying degrees depending on the stability of

$$\sim\sim\text{OCH}_2\text{CH}_2-\overset{+}{\underset{\bar{\text{B}}\text{F}_3\text{OH}}{\text{O}}}\triangle \rightarrow \sim\sim\text{OCH}_2\text{CH}_2\text{OCH}_2\text{CH}_2\text{OH} + BF_3 \qquad (7\text{-}34)$$

the counterion. Thus, counterions such as $(PF_6)^-$ and $(SbCl_6)^-$ have little tendency to bring about termination by transfer of a halide ion, while counterions of aluminum and tin have appreciable transfer tendencies; others such as $(BF_4)^-$ and $(FeCl_4)^-$ are intermediate in behavior.

Termination may also occur by chain transfer with the initiator (e.g., water or alcohol) or a deliberately added chain-transfer agent. Deliberate termination of growth is carried out to produce polymers with specific molecular weights or, more often, telechelic polymers with specific end groups. Hydroxyl and amine end groups are obtained by using water and ammonia as chain-transfer agents. Carboxyl-ended telechelics can be obtained by termination with ketene silyl acetal followed by hydrolysis with base [Kobayashi et al., 1989].

The polymerization of 3,3-bis(chloromethyl)oxetane is complicated by the insolubility of the polymer in the reaction mixture. Termination may involve burial of the propagating oxonium ion centers in the solid polymer, resulting in their inaccessibility to further propagation.

7-2b-4 Cyclic Acetals

Cyclic acetals contain at least one 1,1-dialkoxy grouping [i.e., $(RO)_2CH_2$ or $(RO)_2CHR$] as part of a cyclic structure. A variety of cyclic acetals undergo facile cationic polymerization [Penczek and Kubisa, 1977, 1989b; Schulz et al., 1984]. This includes various 1,3-dioxacycloalkanes such as 1,3-dioxolane ($m = 2$), 1,3-dioxepane ($m = 4$), and

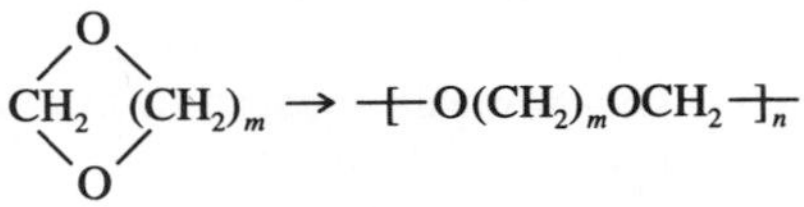

$$\begin{array}{c} O \\ / \quad \backslash \\ CH_2 \quad (CH_2)_m \\ \backslash \quad / \\ O \end{array} \rightarrow \quad -\!\!\left[O(CH_2)_mOCH_2\right]_n\!\!- \qquad (7\text{-}35)$$

1,3-dioxocane ($m = 5$) [Kawakami and Yamashita, 1977a, 1977b; Matyjaszewski et al., 1980; Okada et al., 1978; Plesch, 1976]. The polymers can be considered as copolymers of the $O(CH_2)_m$ and OCH_2 units. 1,3-Dioxane ($m = 3$) does not polymerize as a result of the stability of the 6-membered ring.

Other cyclic acetals that have been studied are 1,3,5-trioxane, 1,3,5-trioxepane, 1,3,6,9-tetraoxacycloundecane, and 1,3,5,7-tetroxocane (also referred to as 1,3,5,7-tetroxane) [Kawakami and Yamashita, 1979; Munoz-Escalona, 1978; Schulz et al., 1984; Szwarc and Perrin, 1979]. Polymerization of bicyclic acetals has been of interest

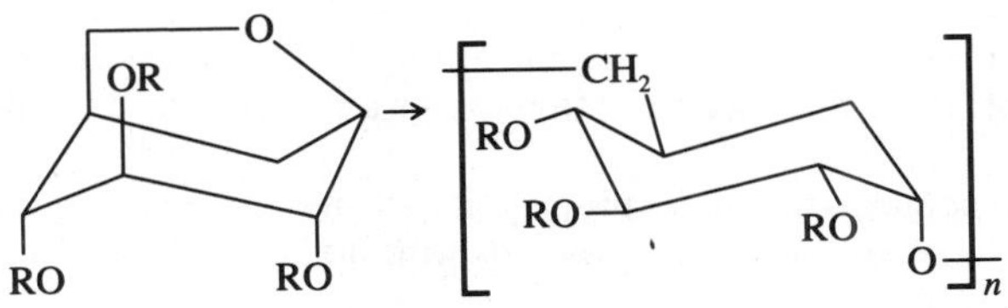

$$(7\text{-}36)$$

for synthesizing polysaccharides [Good and Schuerch, 1985; Hirasawa et al., 1988; Okada et al., 1989; Sumitomo and Okada, 1984].

There has been considerable debate on the details of cyclic acetal polymerizations [Plesch, 1976; Szymanski et al., 1983]. First, it was thought that polymerization proceeded exclusively to yield cyclics, but this is not the case. As with all ring-opening polymerizations, reaction conditions can be manipulated to maximize the formation of either polymer or cyclics. (Study of the relative extents of cyclic and polymer has been complicated by the experimental difficulties involved in identifying the presence of cyclics. One needs to bring a variety of experimental tools and skills to bear on the problem, including a combination of fractional precipitation or solubilization, size-exclusion chromatography, and high-pressure liquid chromatography to separate the different components of a reaction mixture followed by high-resolution NMR and mass spectroscopy for chemical identification.) Second, there has been controversy over whether the propagating species in cyclic acetal polymerization are oxonium ions (**XVIII**) as in cyclic ether polymerization or oxycarbocations (also referred to as *alkoxycarbocations*) (**XIX**). One assumes the oxycarbocation **XIX** is more stable than

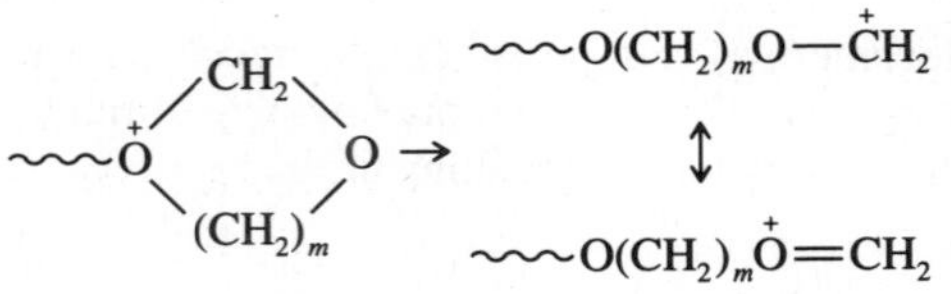

$$\sim\sim O \underset{(CH_2)_m}{\overset{CH_2}{<}} O \rightarrow \quad \begin{matrix} \sim\sim O(CH_2)_mO\overset{+}{-}\overset{+}{C}H_2 \\ \updownarrow \\ \sim\sim O(CH_2)_m\overset{+}{O}\!=\!CH_2 \end{matrix} \tag{7-37}$$

$$\text{XVIII} \qquad\qquad \text{XIX}$$

its carbocation counterpart in cyclic ether polymerization since **XIX** is stabilized via charge dispersal by the adjacent oxygen. However, this stabilization is apparently insufficient to have the oxycarbocation as the propagating species in preference to the oxonium ion. Both experimental and computational data indicate that more than 99.9% of the propagating species are oxonium ions for unsubstituted 1,3-dioxacycloalkanes [Lahti et al., 1990; Szymanski et al., 1983; Xu et al., 1987]. Since oxonium ions are apparently only 100-fold lower in reactivity toward propagation than oxycarbocations, propagation is predominately carried by oxonium ions [Penczek and Kubisa, 1989a, 1989b]. However, propagation by oxycarbocations is significant for 2-alkyl-1,3-dioxacycloalkanes because of additional stabilization of the positive charge center by the 2-alkyl substituent coupled with steric hindrance toward propagation through oxonium ions.

1,3,5-Trioxane, the cyclic trimer of formaldehyde, yields the same polymer, polyoxymethylene (Eq. 7-2), as obtained by the ionic polymerization of formaldehyde (Sec. 5-6). This polymerization is carried out on an industrial scale using boron trifluoride etherate [Dolce and Grates, 1985]. The presence of water is required for polymerization as noted by the lack of polymer formation in the complete absence of water [Collins et al., 1979]. That 1,3,5-trioxane polymerizes in spite of containing the stable 6-membered ring is a consequence of the polymerization occurring with simultaneous precipitation of the polymer in crystalline form. The heat of crystallization makes the total process of converting monomer to polymer a significantly exothermic process.

In addition to the usual polymer–monomer propagation–depropagation equilibrium that may be present, trioxane polymerization proceeds with the occurrence of a polymer–formaldehyde equilibrium

$$\sim\sim\sim OCH_2OCH_2O\overset{+}{C}H_2 \rightleftharpoons \sim\sim\sim OCH_2O\overset{+}{C}H_2 + CH_2O \tag{7-38}$$

The formation of formaldehyde probably occurs through the small concentration of oxycarbocations present, as shown above, and not through the more abundant oxonium ions.

1,3,5-Trioxane polymerizations proceed with induction periods, which correspond to the buildup of the equilibrium concentration of formaldehyde [Lu et al., 1990]. This also corresponds to a buildup in 1,3,5,7-tetroxocane, apparently by insertion of formaldehyde into 1,3,5-trioxane. Polymer is not formed until after both formaldehyde and 1,3,5,7-tetroxocane appear. Formaldehyde and, to a lesser extent, 1,3,5,7-te-troxocane decrease the induction period by increasing the rate of formation of propagating species of sufficient size to precipitate from solution. Subsequent polymerization by addition of 1,3,5-trioxane is accelerated as the reaction proceeds with conversion of monomer to crystalline polymer. The relative importance of participation by formaldehyde and 1,3,5,7-tetroxocane in propagation relative to 1,3,5-trioxane after the induction period is unclear.

Polymerization of 1,3,5-trioxepane involves more complicated propagation–depropagation equilibria. The initially formed oxycarbocation **XX** loses formaldehyde to yield oxycarbocation **XXI**, which in turn loses 1,3-dioxolane to regenerate **XX**

$$\sim\!\!\sim\!\!OCH_2CH_2OCH_2O\overset{+}{C}H_2 \;\rightleftharpoons\; CH_2O + \sim\!\!\sim\!\!OCH_2CH_2O\overset{+}{C}H_2 \qquad (7\text{-}39)$$
$$\quad\quad\quad \textbf{XX} \quad\quad\quad\quad\quad\quad\quad\quad\quad\quad \textbf{XXI}$$

$$\sim\!\!\sim\!\!OCH_2CH_2O\overset{+}{C}H_2 \;\rightleftharpoons\; O\!\!\diagdown\!\!\diagup\!\!O + \sim\!\!\sim\!\!OCH_2CH_2OCH_2O\overset{+}{C}H_2 \qquad (7\text{-}40)$$
$$\quad \textbf{XXI} \quad\quad\quad\quad\quad\quad\quad\quad \textbf{XX}$$

Both **XX** and **XXI** undergo propagation (presumably through the corresponding oxonium ions). The two equilibria (Eqs. 7-39 and 7-40) do not proceed to the same extent, with the result that the copolymer structure deviates from that of the monomer [Szwarc and Perrin, 1979].

Transfer and termination occur by the modes described previously for cyclic ether polymerizations. Chain transfer to polymer (both inter- and intramolecular) is facilitated in cyclic acetal polymerization compared to cyclic ethers since acetal oxygens in the polymer chain are more basic than the corresponding ether oxygens [Penczek and Kubisa, 1989a, 1989b]. An additional termination in the polymerization of trioxane is chain transfer to monomer by hydride ion transfer, which results in terminating the

$$\sim\!\!\sim\!\!OCH_2OCH_2O\overset{+}{C}H_2 + \underset{}{CH_2}\!\!\diagup\!\!\overset{O-CH_2}{\diagdown}\!\!\overset{}{O} \;\rightarrow$$

$$\sim\!\!\sim\!\!OCH_2OCH_2OCH_3 + \overset{+}{C}H\!\!\diagup\!\!\overset{O-CH_2}{\diagdown}\!\!\overset{}{O} \qquad (7\text{-}41)$$
$$\quad\quad\quad\quad\quad\quad\quad\quad\quad\quad\quad\quad \textbf{XXI}$$

propagating chain with a methoxyl group while carbocation **XXI** reinitiates polymerization [Kern et al., 1966; Weissermel et al., 1967].

7-2b-5 Kinetics

7-2b-5-a Rate of Polymerization. The rate laws that describe the cationic ring-opening polymerizations take several forms. Some polymerizations can be described by kinetic expressions similar to those used in alkene polymerizations (Sec. 5-2d). In some polymerizations where there is little or no termination one can employ kinetic expressions similar to those of living polymerizations (Sec. 5-3e), for example,

$$R_p = k_p[M^*][M] \tag{7-42}$$

where $[M^*]$ is the concentration of propagating oxonium ions.

Ring-opening polymerizations that take place without termination and with a propagation–depropagation equilibrium are described in a different manner [Afshar-Taromi et al., 1978; Beste and Hall, 1964; Kobayashi et al., 1974; Szwarc, 1979]. (The following treatment for reversible ring-opening polymerization is also applicable to other reversible polymerizations such as those of alkenes or carbonyl monomers.) The propagation–depropagation equilibrium can be expressed by

$$M_n^* + M \underset{k_{dp}}{\overset{k_p}{\rightleftharpoons}} M_{n+1}^* \tag{7-43}$$

which is analogous to Eq. 3-192. The polymerization rate is given by the difference between the rates of the propagation and depropagation reactions

$$R_p = \frac{-d[M]}{dt} = k_p[M^*][M] - k_{dp}[M^*] \tag{7-44}$$

At equilibrium, the polymerization rate is zero and Eq. 7-44 becomes

$$k_p[M]_c = k_{dp} \tag{7-45}$$

where $[M]_c$ is the equilibrium monomer concentration (as in Eq. 3-196). (The derivations in Sec. 3-9c for $[M]_c$ and ceiling temperature T_c as a function of ΔS° and ΔH° are applicable to the present system.) Combination of Eqs. 7-44 and 7-45 gives the polymerization rate as

$$\frac{-d[M]}{dt} = k_p[M^*]([M] - [M]_c) \tag{7-46}$$

which can be integrated to yield

$$\ln\left(\frac{[M]_0 - [M]_c}{[M] - [M]_c}\right) = k_p[M^*]t \tag{7-47}$$

where $[M]_0$ is the initial monomer concentration. (A corresponding consideration of the degree of polymerization in an equilibrium polymerization is given in Sec. 7-2b-4-c.)

Equations 7-44 and 7-45 can be used to determine the propagation rate constant. The equilibrium monomer concentration is obtained by direct analysis or as the intercept of a plot of polymerization rate versus initial monomer concentration. Figure 7-1 shows such a plot for the polymerization of tetrahydrofuran in dichloroethane solution at 0°C using triethyloxonium tetrafluoroborate as the initiator [Vofsi and Tobolsky, 1965]. The polymerization data are then plotted in accordance with Eq.

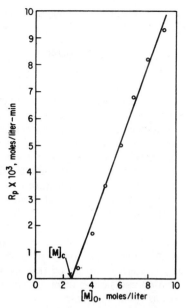

Fig. 7-1 Determination of the equilibrium monomer concentration $[M]_c$ for the $(C_2H_5)_3O^+$ $(BF_4)^-$ initiated polymerization of tetrahydrofuran in dichloroethane at 0°C. After Vofsi and Tobolsky [1965] (by permission of Wiley-Interscience, New York).

7-47) as the left side of that equation versus time to yield a straight line (Fig. 7-2) whose slope is $k_p[M^*]$. Since $[M^*]$ for a living polymer can be obtained from measurements of the number-average molecular weight, one can evaluate the propagation rate constant. The concentration of the propagating species and the polymerization rate in a polymerization will be determined by the concentrations of initiator and coinitiator.

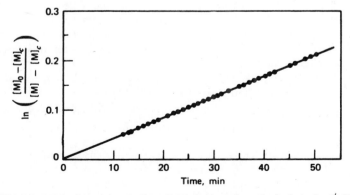

Fig. 7-2 Disappearance of monomer in the polymerization of tetrahydrofuran by $\phi_2CH^+(SbCl_6)^-$ at 25°C; a plot of Eq. 7-47. After Afshar-Taromi et al. [1978] (by permission of Huthig and Wepf Verlag, Basel).

For the case where the concentration of propagating centers changes with time, intergration of Eq. 7-46 yields

$$\ln \left(\frac{[M]_1 - [M]_c}{[M]_2 - [M]_c} \right) = k_p \int_{t_1}^{t_2} [M^*]t \tag{7-48}$$

where $[M]_1$ and $[M]_2$ are the monomer concentrations at times t_1 and t_2, respectively.

7-2b-5-b Values of Kinetic Parameters.

The availability of reliable kinetic and thermodynamic data is far less for ring-opening polymerizations than for step and chain polymerizations. However, the general similarity of ring-opening and step polymerizations is clearly apparent when one compares the available data on propagation rate constants. For various oxirane, oxetane, tetrahydrofuran, 1,3-dioxepane, and 1,3,6-trioxocane polymerizations, k_p is in the range 10^{-1}–10^{-3} liters/mole-sec [Chien et al., 1988; Dreyfuss et al., 1989; Matyjasewski et al., 1984; Mijangos and Leon, 1983; Penczek, 1974; Penczek and Kubisa, 1989; Saegusa, 1972; Xu et al., 1987]. Oxepane has a k_p of 1.3×10^{-4} [Mayyjasewski et al., 1984]. These values can be compared to the corresponding values for step and chain polymerizations (Tables 2-8, 3-11, 5-3, and 5-11). The k_p values for cyclic ether and acetal polymerizations are close to the rate constants for polyesterification and similar reactions and much smaller than those for various chain polymerizations. (The concentrations of propagating centers in the typical ring-opening polymerization is 10^{-2}–10^{-3} M, which is comparable to cationic chain polymerizations of alkenes.) There is one glaring exception to this generalization—the k_p value for 1,3-dioxolane is reported to be 10^2–10^4 liters/mole-sec [Penczek and Kubisa, 1989a, 1989b; Szymanski et al., 1983]. The reason for the much higher reactivity of this cyclic acetal is not understood.

A number of interesting observations have been made for tetrahydrofuran polymerizations. The observed rate constants k_p^+ and $k_p^\pm$ for propagation by free ions and ion pairs are the same [Baran et al., 1983; Matyjaszewski et al., 1979; Penczek and Kubisa, 1989a, 1989b]. (The equal reactivity of free ions and ion pairs has been observed in various solvents (CCl_4, CH_2Cl_2, CH_3NO_2) where the relative concentration of free ions varied from a few percent to over 95%.) This is only slightly different from the situation in cationic polymerizations of alkenes where the ion pair is about an order of magnitude less reactive than the free ion (Sec. 5-2e-3). The ion pair in THF polymerization is apparently even looser than in alkene polymerizations. The equal reactivity of free ions and ion pairs is generally assumed in all cyclic ether (as well as most ring-opening) polymerizations, but there are data for very few systems other than tetrahydrofuran [Gandini and Cheradame, 1985].

The value of k_p^+ ($= k_p^\pm$) decreases with increasing solvent polarity for mixtures of THF with CCl_4 and CH_3NO_2 [Dreyfuss et al., 1989]. For free ions or their equivalent, loose and highly solvated ion pairs, the transition state **XXIII** involves greater charge

$$\tag{7-49}$$

XXIII

delocalization than the initial state. Polar solvents stabilize the initial state more than the transition state. This is the typical behavior observed for reaction between a charged substrate and neutral nucleophile or between a neutral substrate and charged nucleophile in S_N2 reactions.

Propagation by covalent ester species has been observed in polymerizations of THF by very strong acids and their esters (CF_3SO_3H, $C_2H_5OSO_2CF_3$) in solvents of low polarity such as CCl_4 and CH_2Cl_2. The evidence that covalent macroester propagating species exist in addition to ion pair and free ion species comes from NMR spectroscopy. Propagation by covalent macroesters is analogous to the previously described pseudocationic and living cationic polymerizations of alkenes (Secs. 5-2f-1, 5-2g). The rate constant for propagation by macroesters is about 10^2 lower than the rate constant for the corresponding ion pairs [Kobayashi et al., 1975; Penczek and Kubisa, 1989a, 1989b]. The lowered reactivity of covalent esters relative to ion pairs and free ions is generally assumed, but data are available for very few systems other than THF. Interestingly, the only other cyclic ether for which there is data, oxepane, shows the propagation rate constant for the ion pair–free ion to be only twice that for the covalent species [Baran et al., 1983].

7-2b-5-c *Degree of Polymerization.* The quantitative dependence of the degree of polymerization on various reaction parameters has been described [Hirota and Fukuda, 1987; Tobolsky, 1957, 1958; Tobolsky and Eisenberg, 1959, 1960] for an equilibrium polymerization involving initiation

$$I + M \xrightleftharpoons{K_i} M^* \tag{7-50}$$

followed by propagation (Eq. 7-43), where I is the initiating species. The equilibrium constants for initiation K_i and propagation K_p ($= k_p/k_{dp}$) are each assumed to be independent of the size of the propagating species. This is another instance of the concept of functional group reactivity independent of size. The concentration $[M_n^*]$ of propagating chains of size n is given by

$$[M_n^*] = K_i[I]_c[M]_c(K_p[M]_c)^{n-1} \tag{7-51}$$

where the c subscripts refer to equilibrium concentrations. The total concentration of polymer molecules of all sizes [N] is then the summation of Eq. 7-51 over chains of all sizes:

$$[N] = \sum_{n=1}^{\infty} [M_n^*] = \frac{K_i[I]_c[M]_c}{1 - K_p[M]_c} \tag{7-52}$$

while the total concentration [W] of monomer segments incorporated into the polymer is given by

$$[W] = \sum_{n=1}^{\infty} n[M_n^*] = \frac{K_i[I]_c[M]_c}{(1 - K_p[M]_c)^2} \tag{7-53}$$

The degree of polymerization is obtained as

$$\overline{X}_n = \frac{[\mathrm{W}]}{[\mathrm{N}]} = \frac{1}{1 - K_p[\mathrm{M}]_c} \tag{7-54}$$

Material balances on the monomer and initiator are represented by

$$[\mathrm{M}]_0 = [\mathrm{M}]_c + [\mathrm{W}] \tag{7-55}$$

$$[\mathrm{I}]_0 = [\mathrm{I}]_c + [\mathrm{N}] \tag{7-56}$$

where the 0 subscripts refer to initial concentrations. Combinations of Eqs. 7-53 and 7-55 and Eqs. 7-52 and 7-56 yield

$$[\mathrm{M}]_0 = [\mathrm{M}]_c(1 + K_i\overline{X}_n^2[\mathrm{I}]_c) \tag{7-57}$$

$$[\mathrm{I}]_0 = [\mathrm{I}]_c(1 + K_i\overline{X}_n[\mathrm{M}]_c) \tag{7-58}$$

Combination of Eqs. 7-54 through 7-56 also yields the obvious relationship

$$\overline{X}_n = \frac{[\mathrm{M}]_0 - [\mathrm{M}]_c}{[\mathrm{I}]_0 - [\mathrm{I}]_c} \tag{7-59}$$

The various relationships show the dependence of the degree of polymerization on the initial and equilibrium concentrations of monomer and initiator and on the equilibrium constants K_i and K_p. The polymer molecular weight increases with decreasing K_i and $[\mathrm{I}]_0$ and increasing K_p and $[\mathrm{M}]_0$.

7-2b-6 Energetic Characteristics

7-2b-6-a Effect of Temperature on Rate and Degree of Polymerization.
The effects of temperature on the rate and degree of polymerization of cyclic ethers and acetals vary considerably depending on the specific reaction system. Variations due to different monomers, solvents, initiators, and coinitiators are generally analogous to those observed in ionic polymerizations of alkenes (Secs. 5-2f, 5-3e-2). Increased temperature almost always results in an increase in the rate of polymerization; that is, E_{R_p} is positive. Few E_{R_p} or individual activation energies for initiation, propagation, and termination–transfer are available. Typical values of E_{R_p} or E_p are in the range 20–80 kJ/mole with most values in the upper part of that range. The E_p value is 25, 47, 61, 49, 75, and 86 kJ/mole for epichlorohydrin, oxetane, THF, 1,3-dioxolane, oxepane, and 1,3-dioxepane, respectively [Chien et al., 1988; Dreyfuss and Dreyfuss, 1976; Sims, 1966]. The E_{R_p} value is 72 kJ/mole for 3,3-bis(chloromethyl)oxetane [Chapiro and Penczek, 1962; Penczek and Penczek, 1963]. As with ionic polymerizations of alkenes, the activation energies can vary considerably with reaction conditions. Thus E_{R_p} for 3,3-bis(chloromethyl)oxetane is only 12–16 kJ/mole for radiation-initiated polymerization in the crystalline state. The lower activation energy may be a consequence of a favorable orientation of the monomer for propagation in the crystalline state.

The effect of temperature on the degree of polymerization is more complex. For most polymerizations, increasing the temperature decreases the polymer molecular weight as a result of increases in the rates of transfer and termination relative to propagation. Table 7-2 shows this effect for the polymerization of oxetane by boron

TABLE 7-2   Effect of Temperature on Polymerization of Oxetane[a]

Temperature (°C)	Intrinsic Viscosity of Polymer (dl/g)	Ultimate Conversion of Monomer (%)	Proportion of Tetramer (%)
−80	2.9	95	4
0	2.1	94	10
50	1.3	64	66
100	1.1	62	62

[a]Data from Rose [1956].

trifluoride [Rose, 1956]. (The intrinsic viscosity, in deciliters per gram, is a measure of polymer molecular weight.) The decrease in polymer molecular weight in this polymerization is due primarily to an increase in intramolecular chain transfer to polymer as indicated by the corresponding increase in the yield of cyclic tetramer with increasing temperature. In other polymerizations the termination and transfer reactions may not be appreciably affected by increasing temperature while the propagation reaction rate increases. The result is an increase in polymer molecular weight with increasing temperature. Figure 7-3 shows a THF polymerization in which the polymer molecular weight increases initially with increasing temperature up to about −5°C and then subsequently decreases at higher temperatures. Termination is relatively unaffected at the lower temperatures but increases with temperature at the higher temperatures.

Table 7-2 shows a decrease in the ultimate conversion of monomer to polymer for the oxetane polymerization as the reaction temperature is increased. In this polymerization the effect appears to be due to thermal destruction of the initiator [Rose, 1956]. The same trend is observed for most cyclic ether and acetal polymerizations but for a different reason. Increased temperature usually decreases the conversion of monomer to polymer by a ceiling temperature effect. The propagation–depropagation equilibrium (Eq. 7-43) is shifted to the left with increasing temperature. The equilibrium monomer concentration increases with increasing temperature in accord with Eq.

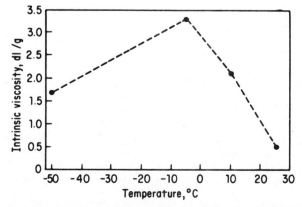

Fig. 7-3 Effect of temperature on polymer molecular weight in the BF_3 polymerization of tetrahydrofuran. After Rose [1956] (by permission of Wiley-Interscience, New York).

3-196 (keep in mind that $\Delta H°$ is negative). Figure 7-4 shows this effect for THF polymerization in bulk using benzene diazonium hexafluorophosphate as the initiator.

7-2b-6-b   Thermodynamics of Polymerization.

The enthalpies and entropies of polymerization for various cyclic ethers and acetals are shown in Table 7-3. Data for formaldehyde and chloral are also included. A comparison of the ΔH and ΔS values for alkenes, carbonyl monomers, and the cyclic monomers can be made from Tables 7-3 and 3-14. The ΔH for the 3- and 4-membered cyclics are comparable to those for alkenes, and both are appreciably larger than the values for the two carbonyl monomers. The conversion of a carbonyl π-bond to a σ-bond is not as exothermic as the corresponding conversion of an alkene π-bond. The ΔH values for the larger-sized cyclic monomers are much lower than those for alkenes. The ΔS value is considerably smaller for most cyclic monomers compared to the alkenes and carbonyls. The cyclic monomers, having less degrees of freedom to begin with, have ΔS values indicating that the loss in disorder on polymerization is less than for noncyclic monomers. Ethylene oxide is an exception because of the highly strained nature of the 3-membered ring. The ΔS values for a few of the cyclic monomers, THF and 3,3-bis(chloromethyl)oxetane, are similar to the lowest values observed for alkene monomers.

The ΔH values for the various cyclic ethers and acetals follow closely the order expected based on considerations of relative stability as a function of ring size (Secs. 7-1a, 2-5). The 3- and 4-membered cyclic monomers undergo the most exothermic polymerizations. There is a fast decrease in the exothermicity of polymerization for the larger-sized rings. Trioxane, the only 6-membered ring monomer to undergo polymerization, has a ΔH very close to zero. Substituents on a ring structure decrease the tendency to polymerize, as noted by the less exothermic polymerizations for the substituted 1,3-dioxolane, 1,3-dioxepane, and 1,3,6-trioxocane monomers relative to the corresponding unsubstituted monomers. Substituted oxetanes are an exception to this generalization.

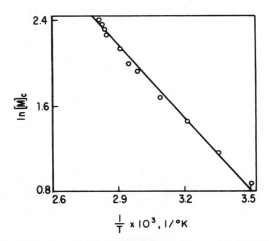

Fig. 7-4   Temperature dependence of the equilibrium monomer concentration in THF polymerization by $\emptyset N_2^+ PF_6^-$. After Dreyfuss and Dreyfuss [1966] (by permission of Wiley-Interscience, New York).

TABLE 7-3 ΔH and ΔS of Polymerization of Cyclic Ethers and Acetals[a]

Monomer	Ring Size	$-\Delta H$ (kJ/mole)	$-\Delta S$ (J/K-mole)
Ethylene oxide	3	94.5	174
Oxetane	4	81	
3,3-Bis(chloromethyl)oxetane	4	84.5	83
Tetrahydrofuran[b]	5	23.4	82.4
1,3-Dioxolane[c]	5	17.6	47.7
4-Methyl-1,3-dioxolane[c]	5	13.4	53.1
Trioxane	6	4.5	18
1,3-Dioxepane[d]	7	15.1	48.1
2-Methyl-1,3-dioxepane[d]	7	8.8	37.2
4-Methyl-1,3-dioxepane[d]	7	9.2	38.9
1,3-Dioxocane	8	18.3	
1,3,6-Trioxocane[e]	8	13.0	21.3
2-n-Butyl-1,3,6-trioxocane[e]	8	7.9	16.3
1,3,6,9-Tetraoxacycloundecane[f]	11	8.0	6.2
Formaldehyde		31.1	79.2
Chloral		20	95

[a]Data from Busfield [1989] unless otherwise noted.
[b]Dreyfuss and Dreyfuss [1989].
[c]Okada et al. [1976].
[d]Okada et al. [1978].
[e]Xu et al. [1987].
[f]Rentsch and Schulz [1978].

7-2b-7   Commercial Applications

There are a number of commercial applications of ring-opening polymerization of cyclic ethers. Polymerizations and copolymerizations of ethylene and propylene oxides as well as polymerization of tetrahydrofuran are used to produce polyether macrodiols, specifically, telechelic polyethers having hydroxyl end groups [Clinton and Matlock, 1986; Dreyfuss et al., 1989; Gagnon, 1986]. The commercial materials are typically in the molecular-weight range 500–6000 and are used to produce polyurethane and polyester block copolymers, including the thermoplastic polyurethane and polyester elastomers (Secs. 2-12e, 2-13b-3, 2-13c-2). The polyether macrodiols are also referred to as *polyether glycols* or *polyols* (trade names: *Carbowax, Jeffox, Plurocol, Polyglycol, Polymeg, Terathane, Vibrathane*). Ethylene oxide and propylene oxide polymers and copolymers are also available in molecular weights of up to about 20,000 and are used as hydraulic fluids and lubricants, additives in cosmetics, and binders in ceramic and powder metallurgy. Ethylene oxide polymers with molecular weights of 10^5–10^6, referred to as *poly(ethylene oxide)*, find applications that take advantage of the high viscosity of their aqueous solutions. This includes flocculation, denture adhesives, packaging films (pesticides, herbicides, seed tapes), thickening of acid cleaners and water-based paints, and friction reduction. There is considerable effort aimed at using ethylene oxide polymers and copolymers complexed with ionic salts as the electrolyte in all-solid rechargeable batteries.

Polyepichlorohydrin and copolymers and terpolymers of epichlorohydrin with ethylene oxide and allyl glycidyl ether are useful elastomers [Body and Kyllinstad, 1986]. These materials (trade names: *Epichlomer, Herclor, Hydrin*) have good resistance to

$$CH_2=CHCH_2OCH_2\overset{\displaystyle O}{\overbrace{CH-CH_2}}$$

Allyl glycidyl ether

fuel, oil, ozone, and heat combined with low-temperature flexibility and resistance to vapor permeation by air, hydrocarbons, and fluorocarbons.

Polyoxymethylene, also referred to as *acetal resin* or POM, is obtained either by anionic polymerization of formaldehyde or cationic ring-opening copolymerization of trioxane with a small amount of a cyclic ether or acetal (e.g., ethylene oxide or 1,3-dioxolane) [Cherdron et al., 1988; Dolce and Grates, 1985]. The properties and uses of POM have been discussed previously (Sec. 5-6e).

Both anionic and cationic initiation are used to cure epoxy resins (Sec. 2-12d) by ring-opening polymerization of the oxirane end groups. The cationic process is less frequently used as its more rapid rate is more difficult to control. Photochemically initiated cationic curing is becoming of importance for coating applications.

7-3 LACTAMS

The polymerization of lactams (cyclic amides) can be initiated by bases, acids, and

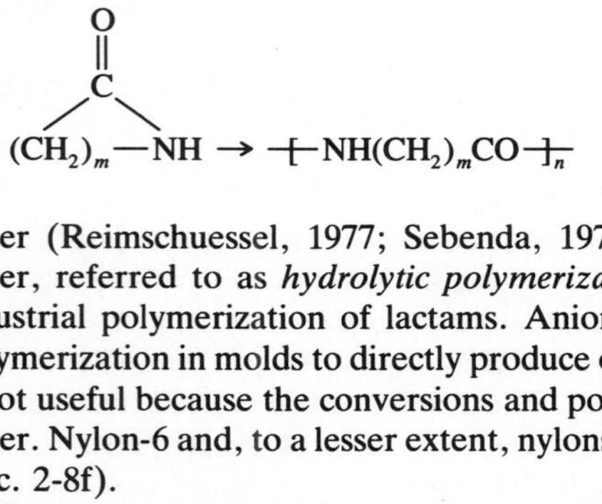

$$\underset{\displaystyle (CH_2)_m-NH}{\overset{\displaystyle \overset{O}{\underset{\displaystyle \|}{C}}}{\diagup\diagdown}} \rightarrow -\!\!\!\left[NH(CH_2)_mCO\right]\!\!\!_n \tag{7-60}$$

water (Reimschuessel, 1977; Sebenda, 1976, 1978; Sekiguchi, 1984]. Initiation by water, referred to as *hydrolytic polymerization*, is the most often used method for industrial polymerization of lactams. Anionic initiation is also practiced, especially polymerization in molds to directly produce objects from monomer. Cationic initiation is not useful because the conversions and polymer molecular weights are considerably lower. Nylon-6 and, to a lesser extent, nylons-11 and 12, are of commercial importance (Sec. 2-8f).

Lactams are named in several ways. They are named as alkanolactams by the IUPAC substitutive system, such as 3-propanolactam, 4-butanolactam, 5-pentanolactam, and 6-hexanolactam, respectively, for the 4-, 5-, 6-, and 7-membered rings, respectively. An alternate IUPAC method, the specialist heterocyclic nomenclature system, names these lactams as 2-azetidinone, 2-pyrrolidinone, 2-piperidinone, and hexahydro-2H-azepin-2-one, respectively. These lactams are also known by the trivial names β-propiolactam, α-pyrrolidone (γ-butyrolactam), α-piperidone (δ-valerolactam), and ε-caprolactam, respectively.

7-3a Cationic Polymerization

A variety of protonic and Lewis acids initiate the cationic polymerization of lactams [Bertalan et al., 1988a, 1988b; Puffr and Sebenda, 1986; Sebenda, 1988]. The reaction generally follows the mechanism of acid-catalyzed nucleophilic substitution reactions of amides. Initiation involves nucleophilic attack of monomer on protonated monomer (**XXIV**) to form an ammonium salt (**XXV**) that undergoes proton exchange with monomer to yield **XXVI** and protonated monomer. The conversion of **XXIV** to **XXV**

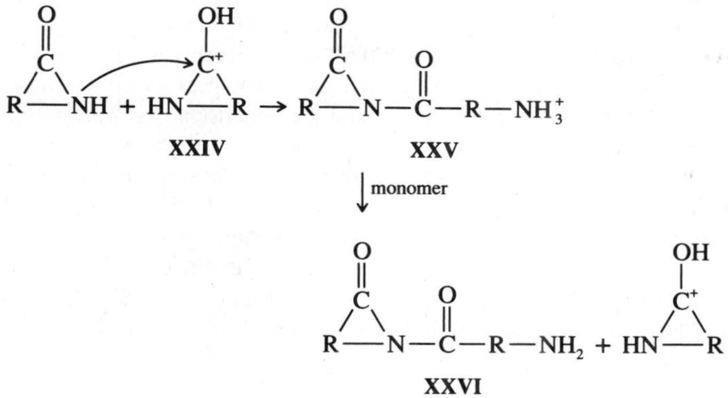

$$\underset{\textbf{XXIV}}{R-NH} + \underset{\textbf{}}{HN-R} \rightarrow \underset{\textbf{XXV}}{R-N-\overset{O}{\overset{\|}{C}}-R-NH_3^+}$$

$$\Big\downarrow \text{monomer}$$

$$\underset{\textbf{XXVI}}{R-N-\overset{O}{\overset{\|}{C}}-R-NH_2} + HN-R \qquad (7\text{-}61)$$

involves several steps—attachment of nitrogen to C^+, proton transfer from one N to the other N, ring opening of the HO-containing ring, and transfer from $HO-C^+$ to N with formation of **XXV**. Some literature references present variations of the preceding equation whereby protonation of lactam is shown as occurring at nitrogen. This does not correspond to the generally accepted mechanism for acid-catalyzed substitution reactions of amides, such as hydrolysis of acetamide. The available data on such reactions are consistent with protonation occurring almost exclusively at the carbonyl oxygen [Lowry and Richardson, 1987].

Propagation follows in a similar manner as a nucleophilic attack by the primary amine end group of a growing polymer chain (**XXVII**) on protonated monomer to yield **XXVIII**, which then undergoes

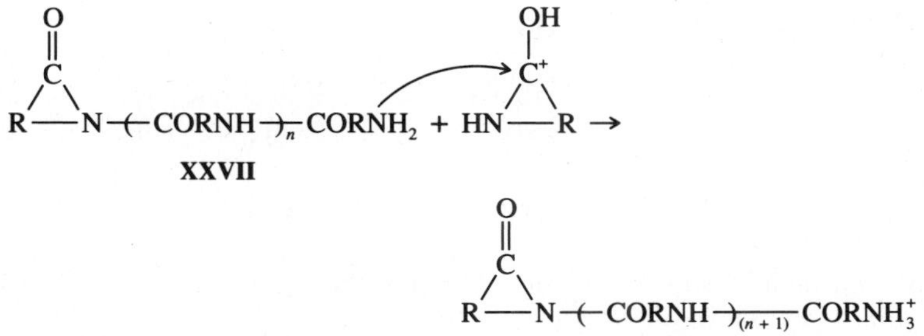

$$\underset{\textbf{XXVII}}{R-N+CORNH)_{\overline{n}}CORNH_2} + HN-R \rightarrow$$

$$\underset{\textbf{XXVIII}}{R-N+CORNH)_{\overline{(n+1)}}CORNH_3^+} \qquad (7\text{-}62)$$

proton exchange with monomer

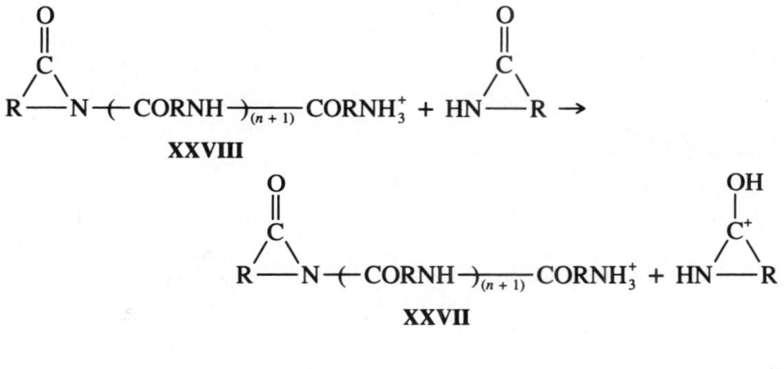

$$\underset{\textbf{XXVIII}}{R-N+CORNH)_{\overline{(n+1)}}CORNH_3^+} + HN-R \rightarrow$$

$$\underset{\textbf{XXVII}}{R-N+CORNH)_{\overline{(n+1)}}CORNH_3^+} + HN-R \qquad (7\text{-}63)$$

Simultaneous with propagation by Eqs. 7-62 and 7-63, there is propagation by self-reaction of **XXVII**. **XXVII** is an AB type of monomer, and step polymerization occurs by reaction between amine and lactam end groups. The reaction is the same as above—nucleophilic attack of amine on protonated lactam. Although this reaction is only a minor contribution to the overall conversion of lactam to polymer, it determines the final degree of polymerization. There are also exchange reactions that involve inter-molecular nucleophilic attack of amine on amide linkages within a polymer chain. Intramolecular versions of the various reactions result in cyclic oligomer formation. Overall, one generally observes the MWD of the polymer (excluding cyclic oligomer and monomer) to be close to the Flory most probable distribution.

For cationic polymerization with an acid whose anion Z^- is nucleophilic, initiation involves the sequence described by Eq. 7-61 plus the formation of **XXIX**. Species **XXIX** propagates by a sequence similar to that described by Eqs. 7-62 and 7-63 except

$$
\begin{array}{c}
\text{OH} \\
| \\
\text{C}^+ \\
/ \quad \backslash \\
\text{Z}^- + \text{HN}\!\!-\!\!\text{R} \rightarrow \text{ZCORNH}_2 \\
\textbf{XXIX}
\end{array}
\qquad (7\text{-}64)
$$

that a growing polymer chain possesses a $Z\!-\!CO\!-$ end group instead of a lactam end group.

Various side reactions greatly limit the conversions and polymer molecular weights that can be achieved in cationic polymerization of lactams. The highest molecular weights obtained in these polymerizations are 10,000–20,000. The most significant side reaction is amidine (**XXXI**) formation [Bertalan et al., 1984]. Propagation of the

$$
\sim\!\!\sim\!\!\text{NH}_2 + \sim\!\!\sim\!\!\underset{+}{\overset{\overset{\displaystyle\text{OH}}{|}}{\text{C}}}\!\!-\!\!\overset{a}{\text{N}}\!\!< \rightarrow \sim\!\!\sim\!\!\underset{\underset{+}{\overset{\displaystyle|}{\sim\!\!\sim\!\!\text{NH}_2}}}{\overset{\overset{\displaystyle\text{OH}}{|}}{\text{C}}}\!\!-\!\!\overset{a}{\text{N}}\!\!< \xrightarrow[-\text{H}^+]{-\text{H}_2\text{O}} \sim\!\!\sim\!\!\underset{\underset{\sim\!\!\sim\!\!\text{N}}{\|}}{\text{C}}\!\!-\!\!\overset{a}{\text{N}}\!\!< \qquad (7\text{-}65)
$$

$$
\qquad\qquad\qquad\qquad\quad \textbf{XXX} \qquad\qquad\qquad \textbf{XXXI}
$$

polymer chain and amidine formation proceed through the common intermediate **XXX** formed by attachment of an amine group to the protonated carbonyl of any amide group (that in lactam monomer or the lactam end group in polymer or the amide groups in the repeat units of the polymer). Propagation involves cleavage of the $C\!-\!\overset{a}{N}$ bond (overall result is substitution at carbonyl), while amidine formation involves loss of water from **XXX** (overall result is addition to carbonyl followed by dehydration). Amidine formation decreases the concentration of amine groups in the reaction system, and this leads to a slower rate of reaction. The water released during amidine formation can initiate polymerization (Sec. 7-3b), but at a slower rate. Fur-thermore, amidine groups react rapidly with acidic initiators to give relatively un-reactive salts and this limits the rate and degree of polymerization.

7-3b   Hydrolytic Polymerization

Hydrolytic polymerization of ε-caprolactam to form nylon-6 (Sec. 2-8f) is carried out commercially in both batch and continuous processes by heating the monomer in the presence of 5–10% water to temperatures of 250–270°C for periods of 12 hr to more than 24 hr [Zimmerman, 1988]. Several equilibria are involved in the polymerization [Bertalan et al., 1984; Sekiguchi, 1984]. These are hydrolysis of the lactam to ε-aminocaproic acid

$$(CH_2)_5 \!-\! NH + H_2O \;\rightarrow\; HO_2C(CH_2)_5NH_2 \tag{7-66}$$

step polymerization of the amino acid with itself

$$\text{\textasciitilde\textasciitilde COOH} + H_2N\text{\textasciitilde\textasciitilde} \;\rightleftharpoons\; \text{\textasciitilde\textasciitilde CO}\!-\!\text{NH}\text{\textasciitilde\textasciitilde} + H_2O \tag{7-67}$$

and initiation of ring-opening polymerization of lactam by the amino acid. The aminoacid is the equivalent of **XXIX** in Eq. 7-64 (Z = OH). The COOH group of the aminoacid protonates the lactam to form **XXVI** followed by nucleophilic attack of amine on the protonated lactam. The propagation process follows in the same manner with nucleophilic attack of **XXXII** on protonated lactam to form **XXXIII** [R = (CH$_2$)$_5$].

$$^-\text{OOCRNH}\text{\textbf{-(-}}\text{CORNH}\text{\textbf{-)}}_{\!n}\text{CORNH}_2 + \text{HN}\!-\!\text{R} \;\rightarrow$$

$$\textbf{XXXII} \qquad\qquad\qquad \textbf{XXVI}$$

$$\text{HOOCRNH}\text{\textbf{-(-}}\text{CORNH}\text{\textbf{-)}}_{\overline{(n+1)}}\text{CORNH}_2 \tag{7-68}$$

$$\textbf{XXXIII}$$

Species **XXXIII** subsequently protonates lactam and propagation continues. The sequence is the same as in Eqs. 7-62 and 7-63 except for the propagating chain possessing a carboxyl end group instead of lactam end group. Hydrolytic polymerization is simply a special case of cationic polymerization. [The commercial processes often add ε-aminocaproic acid or the 1:1 ammonium carboxylate salt from a diamine and diacid (Eq. 2-133) along with water. This is a faster route to initiation as it supplies amine and carboxyl groups at the very beginning instead of waiting for hydrolysis of the lactam to supply those groups.]

The overall rate of conversion of ε-caprolactam to polymer is higher than the polymerization rate of ε-aminocaproic acid by more than an order of magnitude [Hermans et al., 1958, 1960]. Step polymerization of ε-aminocaproic acid with itself (Eq. 7-67) accounts for only a few percent of the total polymerization of ε-caprolactam. Ring-opening polymerization (Eq. 7-68) is the overwhelming route for polymer formation. Polymerization is acid-catalyzed as indicated by the observations that amines and sodium ε-aminocaproate are poor initiators in the absence of water and the polymerization rate in the presence of water is first-order in lactam and second-order in COOH end groups [Majury, 1958].

Although step polymerization of ε-aminocaproic acid with itself is only a minor contribution to the overall conversion of lactam to polymer, it does determine the final degree of polymerization at equilibrium since the polymer undergoes self-condensation. The final degree of polymerization is dependent in large part on the equilibrium water concentration (Sec. 2-4). Most of the water used to initiate polymerization is removed after about 80–90% conversion in order to drive the system to high molecular weight. Molecular-weight control at a desired level also requires control of the initial water and monomer concentrations (7-2b-4-c) and the addition of small but specific amounts of a monofunctional acid (Sec. 2-6). The MWD is essentially the Flory most probable distribution except for the presence of monomer and cyclic oligomers.

The final product in the industrial polymerization of ε-caprolactam contains about 8% monomer and 2% cyclic oligomer [DiSilvestro et al., 1987; Reimschuessel, 1977; Zimmerman, 1988]. About one-half of the cyclic fraction is the cyclic dimer with the remaining one-half consisting mostly of the trimer and tetramer, although small amounts of cyclics up to the nonamer have been identified [Rothe, 1958; Zahn, 1957]. Cyclic oligomers are formed when any of the propagations discussed above occur intramolecularly (both for cationic and hydrolytic polymerizations). The monomer and cyclic oligomer are removed in the industrial production of poly(ε-caprolactam) by extraction with hot water (or vacuum). The water content of the final product is lowered to less than about 0.1% by vacuum drying at 100–200°C and about 1 torr pressure.

Amidines are formed in hydrolytic polymerizations of lactams but do not limit the polymer molecular weight. Molecular-weight buildup is not impeded since the carboxy end groups of growing polymer are quite reactive toward amidine groups [Bertalan et al., 1984].

7-3c Anionic Polymerization

7-3c-1 Use of Strong Base Alone

Strong bases such as alkali metals, metal hydrides, metal amides, metal alkoxides, and organometallic compounds initiate the polymerization of a lactam by forming the lactam anion **XXXIV** [Sebenda, 1989; Sekiguchi, 1984], for example, for ε-caprolactam with a metal

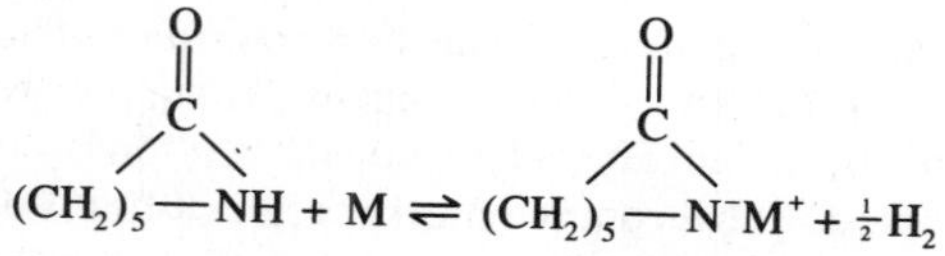

$$(CH_2)_5{-}NH + M \rightleftharpoons (CH_2)_5{-}N^-M^+ + \tfrac{1}{2}H_2 \tag{7-69a}$$
$$\textbf{XXXIV}$$

or with a metal derivative,

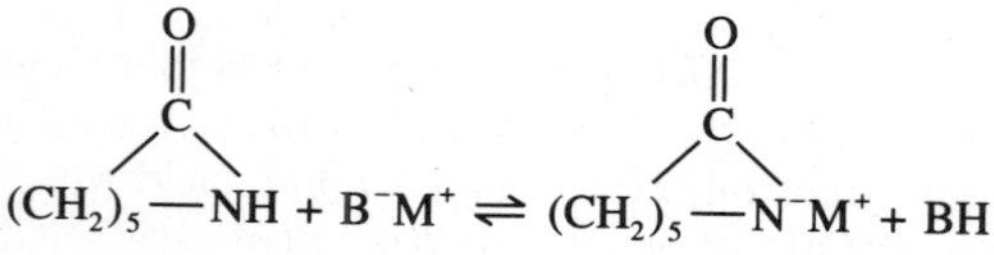

$$(CH_2)_5{-}NH + B^-M^+ \rightleftharpoons (CH_2)_5{-}N^-M^+ + BH \tag{7-69b}$$

There are advantages and disadvantages with almost any choice of base. The use of weaker bases produces a lower concentration of lactam anion unless the equilibrium is pushed to the right. The use of an alkali metal or metal hydride gives a high concentration of lactam anion, but side reactions contaminate the product with amines and water that destroy reactive initiating and propagating species. The use of most bases contaminates the system with the conjugate acid (BH) of the base, and this also destroys reactive species. The preferred method of initiation consists of preforming and purifying the lactam anion followed by addition to the reaction system.

The lactam anion reacts with monomer in the second step of the initiation process by a ring-opening transamidation to form the primary amine anion **XXXV**. Species

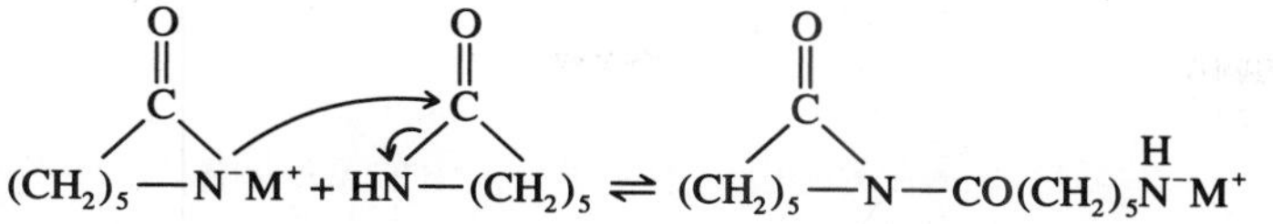

$$(CH_2)_5-N^-M^+ + HN-(CH_2)_5 \rightleftharpoons (CH_2)_5-N-CO(CH_2)_5N^-M^+ \quad (7\text{-}70)$$

$$\textbf{XXXV}$$

XXXV, unlike the lactam anion, is not stabilized by conjugation with a carbonyl group. The primary amine anion is highly reactive and rapidly abstracts a proton from monomer to form the imide dimer **XXXVI**, N-(ε-aminocaproyl)caprolactam, and regenerate the lactam anion.

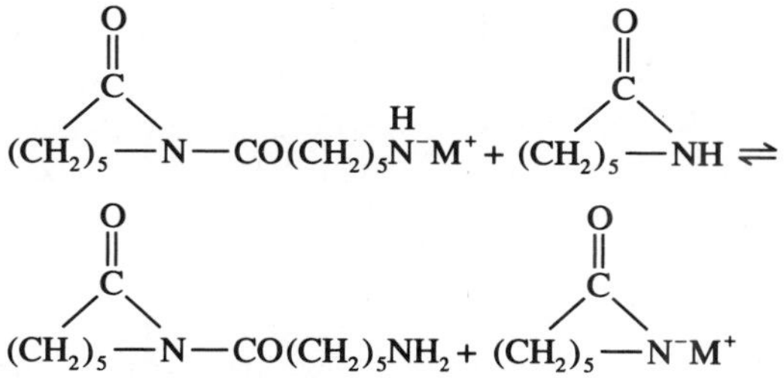

$$(CH_2)_5-N-CO(CH_2)_5N^-M^+ + (CH_2)_5-NH \rightleftharpoons$$

$$(CH_2)_5-N-CO(CH_2)_5NH_2 + (CH_2)_5-N^-M^+ \quad (7\text{-}71)$$

$$\textbf{XXXVI}$$

The imide dimer has been isolated and is the actual initiating species necessary for the onset of polymerization [Hall, 1958; Rothe et al., 1962]. Lactam polymerization is characterized by an initial induction period of low reaction rate as the concentration of imide dimer builds up slowly. The imide dimer is necessary for polymerization because the amide linkage in the lactam is not sufficiently reactive (i.e., not sufficiently electron deficient) toward transamidation by lactam anion. The presence of the *exo*-carbonyl group attached to the nitrogen in the N-acyllactam increases the electron deficiency of the amide linkage. This increases the reactivity of the amide ring structure toward nucleophilic attack by the lactam anion. Propagation follows in the same manner as the reaction of a propagating N-acyllactam species (**XXXVII**) and the lactam anion

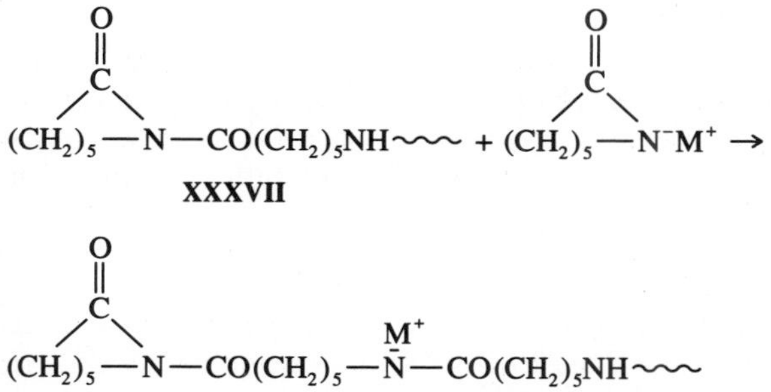

$$(CH_2)_5-N-CO(CH_2)_5NH\sim\sim + (CH_2)_5-N^-M^+ \rightarrow$$

$$\textbf{XXXVII}$$

$$(CH_2)_5-N-CO(CH_2)_5-N-CO(CH_2)_5NH\sim\sim \quad (7\text{-}72)$$

followed by fast proton exchange with monomer

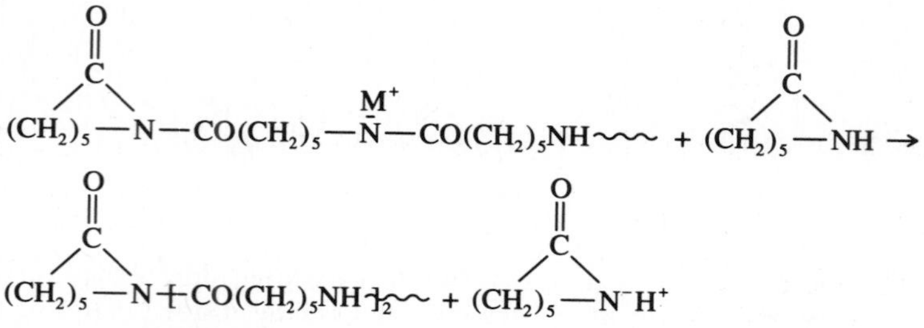

$$
\underset{\textbf{XXXVIII}}{(CH_2)_5 - N \overset{\overset{\displaystyle O}{\parallel}}{\underset{}{C}} - CO(CH_2)_5 - \overset{M^+}{\underset{}{N}} - CO(CH_2)_5NH \rightsquigarrow + (CH_2)_5 - \overset{\overset{\displaystyle O}{\parallel}}{\underset{}{C}} NH \rightarrow \\
(CH_2)_5 - N \overset{\overset{\displaystyle O}{\parallel}}{\underset{}{C}} \{CO(CH_2)_5NH\}_2 \rightsquigarrow + (CH_2)_5 - \overset{\overset{\displaystyle O}{\parallel}}{\underset{}{C}} N^- H^+ \qquad (7\text{-}73)
$$

to regenerate the lactam anion and the propagating N-acyllactam **XXXVIII**.

The anionic polymerization of lactams proceeds by a mechanism analogous to the activated monomer mechanism for anionic polymerization of acrylamide (Sec. 5-7b) and some cationic polymerizations of epoxides (Sec. 7-2b-3-a). The propagating center is the cyclic amide linkage of the N-acyllactam. Monomer does not add to the propagating chain; it is the monomer anion (lactam anion), often referred to as *activated monomer*, which adds to the propagating chain [Szwarc, 1965, 1966]. The propagation rate depends on the concentrations of lactam anion and N-acyllactam, both of which are determined by the concentrations of lactam and base.

7-3c-2 *Addition of N-Acyllactam*

The use of a strong base alone for lactam polymerization is limiting. There are the induction periods previously noted and, more importantly, only the more reactive lactams, such as ε-caprolactam and 7-heptanolactam (ζ-enantholactam), readily undergo polymerization. The less reactive lactams, 2-pyrrolidinone and 2-piperidinone, are much more sluggish toward polymerization by strong base alone. Formation of the imide dimer is difficult from these relatively unreactive lactams. Both limitations are overcome by forming an imide by reaction of lactam with an acylating agent such as acid chloride or anhydride, isocyanate, and others. Thus ε-caprolactam can be rapidly converted to an N-acylcaprolactam **XXXIX** by reaction with an acid chloride. The

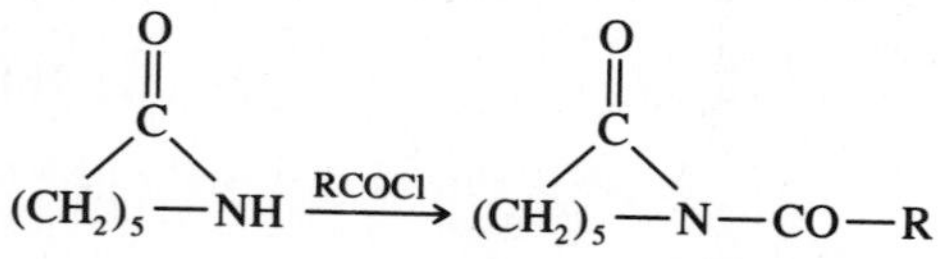

$$
(CH_2)_5 - \overset{\overset{\displaystyle O}{\parallel}}{\underset{}{C}} NH \xrightarrow{RCOCl} \underset{\textbf{XXXIX}}{(CH_2)_5 - N \overset{\overset{\displaystyle O}{\parallel}}{\underset{}{C}} - CO - R} \qquad (7\text{-}74)
$$

N-acyllactam can be synthesized *in situ* or preformed and then added to the reaction system.

Initiation consists of the reaction of the N-acyllactam with activated monomer followed by fast proton exchange with monomer

$$
(CH_2)_5 - N \overset{\overset{\displaystyle O}{\parallel}}{\underset{}{C}} - CO - R + (CH_2)_5 - \overset{\overset{\displaystyle O}{\parallel}}{\underset{}{C}} N^- M^+ \rightarrow
$$

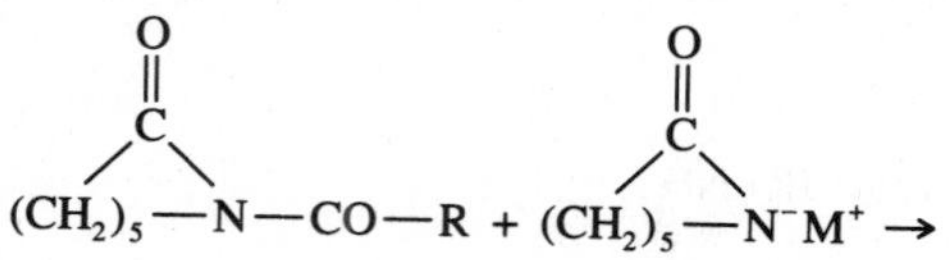

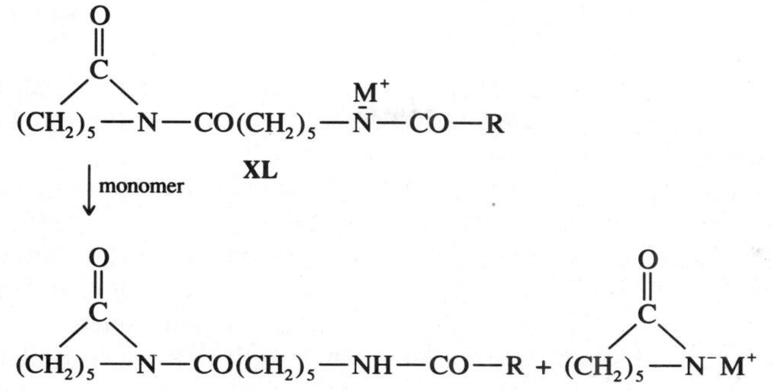

$$
\underset{\textbf{XL}}{\text{(CH}_2)_5-\text{N}-\text{CO(CH}_2)_5-\overset{\text{M}^+}{\overset{|}{\text{N}}}-\text{CO}-\text{R}} \tag{7-75}
$$

↓ monomer

$$
\underset{\textbf{XLI}}{\text{(CH}_2)_5-\text{N}-\text{CO(CH}_2)_5-\text{NH}-\text{CO}-\text{R}} + \text{(CH}_2)_5-\text{N}^-\text{M}^+ \tag{7-76}
$$

Species **XL** and **XLI** correspond to species **XXXV** and **XXXVI** for polymerization in the absence of an acylating agent. The acylating agent achieves facile polymerization of many lactams by substituting the fast initiation sequence in Eqs. 7-75 and 7-76 for the slower sequence in Eqs. 7-70 and 7-71. The use of an acylating agent is advantageous even for the more reactive lactams as induction periods are absent, polymerization rates are higher, and lower reaction temperatures can be used.

Polymerizations in the absence and presence of an acylating agent are often referred to as *nonassisted* and *assisted* polymerizations, respectively. The terms *nonactivated* and *activated* polymerization are also used, but this is confusing since both reactions involve the lactam anion, which is usually referred to as *activated monomer*. The literature is also confusing with regard to terms used to describe the acylating agent. *Promoter*, *catalyst*, and *initiator* have been used. Terminology for the strong base is also a problem since its role is not the same for activated and nonactivated polymerizations. The base is needed in both polymerizations for forming the lactam anion (activated monomer). For nonassisted polymerization, the lactam anion is not only the "real" monomer but is also the species required to form the initiating species (imide dimer). This text will avoid the use of any of these terms for the strong base and acylating agent.

Propagation follows in the same manner as for propagation of species **XXXVII**

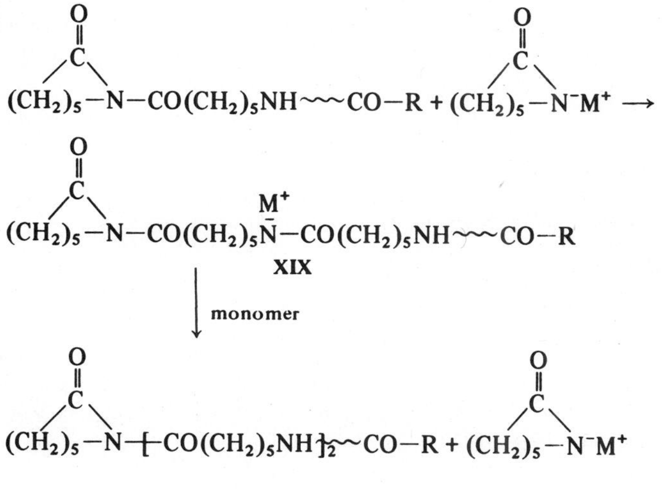

$$
\text{(CH}_2)_5-\text{N}^-\text{CO(CH}_2)_5\text{NH}\sim\sim\text{CO}-\text{R} + \text{(CH}_2)_5-\text{N}^-\text{M}^+ \rightarrow
$$

$$
\underset{\textbf{XIX}}{\text{(CH}_2)_5-\text{N}-\text{CO(CH}_2)_5\overset{\text{M}^+}{\overset{|}{\text{N}}}-\text{CO(CH}_2)_5\text{NH}\sim\sim\text{CO}-\text{R}} \tag{7-77}
$$

↓ monomer

$$
\text{(CH}_2)_5-\text{N}-\{\text{CO(CH}_2)_5\text{NH}\}_2\sim\text{CO}-\text{R} + \text{(CH}_2)_5-\text{N}^-\text{M}^+ \tag{7-78}
$$

except that the propagating chain has an acylated end group instead of the amine end group.

Initiation in activated lactam polymerization may involve contributions from both the imide dimer (**XXXVI**) and *N*-acyllactam (**XXXIX**). However, in the usual case the latter is by far the most important and the former may be ignored. Few detailed kinetic studies of these lactam polymerizations have been achieved, but some of the main reaction characteristics are indicated [Kuskova et al., 1978; Sebenda, 1989; Tani and Konomi, 1966]. The polymerization rate is dependent on the concentrations of base and *N*-acyllactam, which determine the concentrations of activated monomer and propagating chains, respectively. The degree of polymerization increases with conversion and with increasing concentration of monomer or decreasing *N*-acyllactam concentration. These characteristics are qualitatively similar to those of living polymerizations, but lactam polymerizations seldom are living. There is a fairly rapid decay of both the activated monomer and propagating chain concentrations due to a complex set of side reactions [Roda et al., 1976; Sebenda and Kouril, 1971, 1972; Sekiguchi, 1984]. These side reactions include hydrogen abstraction from C—H bonds α- to amide carbonyls (which result in the formation of various β-ketoimides and β-ketoamides via nucleophilic substitution by the carbanions on lactam end groups of propagating chains) and subsequent Claisen-type condensation reactions of the β-ketoimides and β-ketoamides. The latter reaction yields water as a by-product, and water reacts with and decreases the concentrations of both activated monomer and *N*-acyllactam.

Lactam polymerizations (nonassisted as well as assisted) are usually complicated by heterogeneity, usually when polymerization is carried out below the melting point of the polymer [Fries et al., 1987; Karger-Kocsis and Kiss, 1979; Malkin et al., 1982; Roda et al., 1979]. (This is probably the main reason why there are so few reliable kinetic studies of lactam polymerizations.) An initially homogeneous reaction system quickly becomes heterogeneous at low conversion, for example, 10–20% conversion (attained at a reaction time of no more than 1 min) for 2-pyrrolidinone polymerization initiated by potassium *t*-butoxide and *N*-benzoyl-2-pyrrolidinone. The (partially) crystalline polymer starts precipitating from solution (which may be molten monomer) and subsequent polymerization occurs at a lower rate as a result of decreased mobility of *N*-acyllactam propagating species.

The MWD is usually broader than the most probable distribution as a result of branching, which occurs in the later stages of reaction. As the monomer and lactam anion concentrations decrease, there is an increasing tendency for the polymeric amide anion **XLII** to attack the lactam end group of another polymer chain. Further, branch-

$$\sim\sim\sim\bar{N}-CO\sim\sim + (CH_2)_5-N-CO\sim\sim \rightarrow \sim\sim N-CO(CH_2)_5\bar{N}-CO\sim\sim$$

XLII

$$(7\text{-}79)$$

ing, and to some extent crosslinking, result from the Claisen-type condensations mentioned above. The MWD is considerably broader when an *N*-acyllactam is not used, since initiation is slow and polymer chains initiated early grow for longer times than those initiated later [Costa et al., 1979].

7-3d Reactivity

Lactam reactivity toward polymerization depends on ring size and the type of initiation. The slower cationic and hydrolytic polymerizations result in a wider spread of reactivities than does anionic polymerization. The 6-membered lactam undergoes polymerization in anionic polymerization, whereas most 6-membered cyclic monomers are unreactive toward ring-opening polymerization. Substituted lactams show the expected lower reactivity relative to unsubstituted lactams, a consequence of steric hindrance at the reaction site [Cubbon, 1964]. However, the effects of substitution can be complex. In the polymerization of α,α-dialkyl-β-propiolactams (**XLIII**), the propagation rate constant increased about twofold as R increased from methyl to *n*-butyl. The

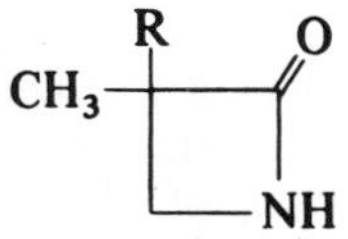

$$CH_3 \begin{array}{c} R \quad\quad O \\ \diagdown\quad\diagup \\ \\ \diagdown\;NH \end{array}$$

XLIII

enhanced reactivity may result from the larger hydrophobic R group preventing strong interaction of the counterion and propagating center, resulting in a looser ion pair.

7-4 *N*-CARBOXY-α-AMINO ACID ANHYDRIDES

N-Carboxy-α-amino acid anhydrides, also referred to as *4-substituted oxazolidine-2,5-diones, Leuchs's anhydrides*, or *N-carboxyanhydrides* (NCA), are polymerized by bases such as metal alkoxides, hydroxides, and hydrides and amines [Bamford and Block, 1976; Imanishi, 1984; Kricheldorf, 1989; Shalitin, 1969]. Polymerization proceeds with simultaneous decarboxylation to produce a polyamide

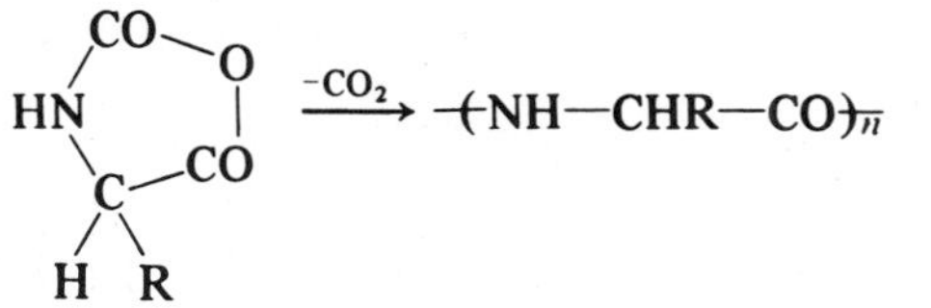

$$HN \begin{array}{c} CO \diagdown \\ \diagup \quad\quad O \\ \diagdown\quad | \\ C - CO \\ \diagup \diagdown \\ H \quad R \end{array} \xrightarrow{-CO_2} +NH-CHR-CO +_n \qquad (7\text{-}80)$$

The polyamide is a substituted nylon-2, that is, derived from an α-amino acid. NCA polymerizations have been studied for the synthesis of polypeptides, both homopolymers and copolymers.

Two different polymerization mechanisms are operative depending on the relative nucleophilicity and basicity of the initiator [Giannakidis and Harwood, 1978; Goodman et al., 1977; Imanishi et al., 1977; Kricheldorf, 1986]. Primary aliphatic amines, which have a high nucleophilicity relative to basicity, polymerize *N*-carboxyanhydrides by the *amine* or *normal* mechanism involving nucleophilic attack by nitrogen at C-5 of NCA with evolution of carbon dioxide. Studies with ^{14}C-labeled NCA show that the carbon dioxide is derived exclusively from the C-2 carbonyl. The polymerization rate is directly dependent on the concentrations of monomer and amine. Most NCA polymerizations have similar but not exactly identical characteristics as living polymerizations. The degree of polymerization is proportional but not always equal to the

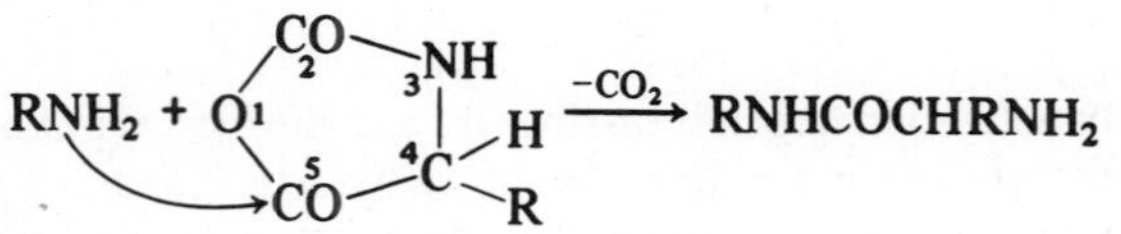

$$RNH_2 + \underset{\underset{\overset{\displaystyle 5}{CO}}{\overset{\displaystyle CO_2}{|}}}{O_1} \underset{\overset{\displaystyle 4}{C}}{\overset{\displaystyle 3}{NH}} \xrightarrow{-CO_2} RNHCOCHRNH_2 \qquad (7\text{-}81)$$

$$RNH \backsim\backsim COCHRNH_2 + O \underset{CO}{\overset{CO}{\diagup}} \underset{\overset{\displaystyle C}{|}}{\overset{NH}{\diagdown}} \xrightarrow{-CO_2}$$

$$RNH \backsim\backsim COCHRNHCOCHRNH_2 \qquad (7\text{-}82)$$

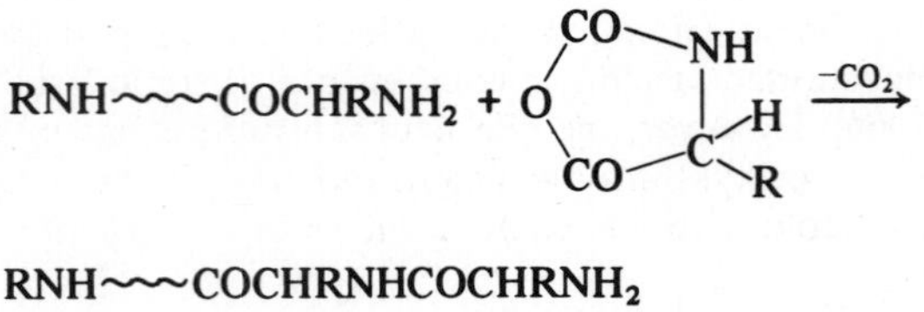

monomer/amine ratio, and the MWD may be broadened or bimodal. There are two sources for these deviations from the usual pattern of living polymerization. Most NCA polymerizations proceed with precipitation and/or the formation of polypeptide secondary structure, and this may result in physical termination of propagating chains. Some chemical termination reactions are also indicated, such as intramolecular reaction of the amine group of a propagating chain with a functional group in the R side chain and formation of hydantoic acid end groups [Kricheldorf, 1989].

Polymerization by aromatic primary amines proceeds with molecular-weight broadening due to a slower rate of initiation relative to propagation. Water has also been used as an initiator, but the molecular weights are considerably lower, no higher than 10,000–20,000, as a result of slow initiation.

Polymerizations initiated by strong bases (R^-, HO^-, RO^-) and tertiary amines (which are poor nucleophiles) proceed at much faster rates than do polymerizations initiated by primary amines. Also, unlike the latter, where each polymer chain contains one initiator fragment (i.e., RNH—), these polymerizations do not result in incorporation of the initiator into the polymer chain. Polymerization proceeds by an activated monomer mechanism similar to that in the anionic polymerization of lactams. The reacting monomer is the NCA anion **XLIV** formed by proton transfer from monomer to base

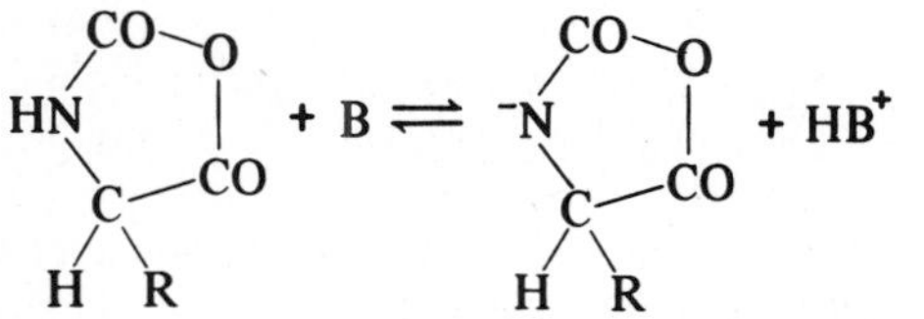

$$HN \underset{\underset{H \quad R}{C}}{\overset{CO}{\diagup}} \underset{CO}{\overset{O}{\diagdown}} + B \rightleftharpoons \ ^-N \underset{\underset{H \quad R}{C}}{\overset{CO}{\diagup}} \underset{CO}{\overset{O}{\diagdown}} + HB^+ \qquad (7\text{-}83)$$

$$\textbf{XLIV}$$

Initiation involves nucleophilic attack by NCA anion on NCA followed by proton transfer with another monomer molecule to regenerate an anion. Propagation proceeds similarly:

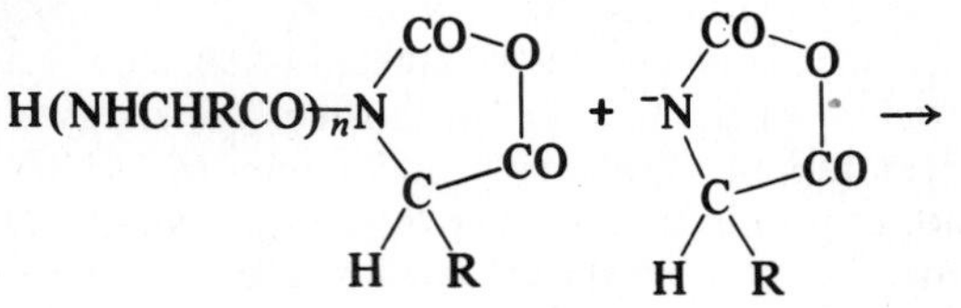

$$H(NHCHRCO)_{\overline{n}}N \underset{\underset{H \quad R}{C}}{\overset{CO}{\diagup}} \underset{CO}{\overset{O}{\diagdown}} + \ ^-N \underset{\underset{H \quad R}{C}}{\overset{CO}{\diagup}} \underset{CO}{\overset{O}{\diagdown}} \rightarrow$$

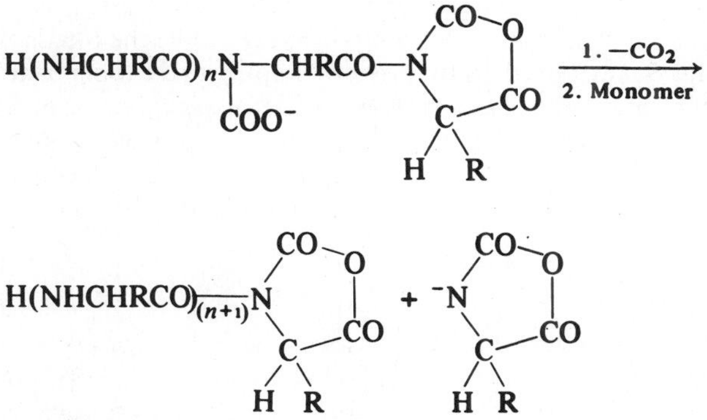

$$H(NHCHRCO)_n N-CHRCO-N\underset{\underset{H\quad R}{C}}{\overset{CO-O}{\diagdown}}CO \xrightarrow[\text{2. Monomer}]{1.-CO_2}$$

$$H(NHCHRCO)_{\overline{(n+1)}}N\underset{\underset{H\quad R}{C}}{\overset{CO-O}{\diagdown}}CO \quad + \quad {}^-N\underset{\underset{H\quad R}{C}}{\overset{CO-O}{\diagdown}}CO \qquad (7\text{-}84)$$

Polymerization by trialkylamines is useful for synthesizing polypeptides of molecular weights of up to about 0.5 million, and the polymerization has many characteristics similar to those of living polymerizations. Polymerizations by the most powerful bases, especially organometallic compounds, are not as useful for polymerizations to such high molecular weights because of side reactions [Imanishi, 1984; Kricheldorf, 1989].

Interestingly, an apparently neutral salt such as lithium chloride is an effective initiator for polymerization by the activated monomer mechanism [Ballard et al., 1960]. Lithium ion is known to coordinate strongly with amide groups, and this probably increases the acidity of the NH proton sufficiently for abstraction by chloride ion. Polymerization of NCA polymerization by secondary amines may involve the amine or activated monomer mechanism or a combination thereof. Unhindered secondary amines such as dimethylamine and piperidine react like primary amines, and polymerization occurs by the amine mechanism. Polymerization by slightly hindered amines such as diethylamine, N-methylbenzylamine, and di-n-propylamine involves a combination of the amine and activated monomer mechanisms. More hindered secondary amines, such as di-n-isopropylamine and dicyclohexylamine, react almost exclusively via the activated monomer mechanism.

NCA polymerizations in aprotic solvents such as 1,4-dioxane and THF proceed with acceleration in rate at about 20–30% conversion where the propagating species have grown to the hexamer to dodecamer size [Kim et al., 1979]. Since this is approximately the size at which α-helical conformations become established for many poly(α-aminoacids), the autoacceleration is generally attributed to a conformational enhancement of reactivity; the reacting species are bound to the α-helix in a steric arrangement that enhances reactivity. Molecular-weight broadening occurs under these conditions.

7-5 LACTONES

Lactones (cyclic esters) undergo anionic and cationic polymerization to form polyesters [Johns et al., 1984; Lundberg and Cox 1969; Young et al., 1977]. The reactivity of

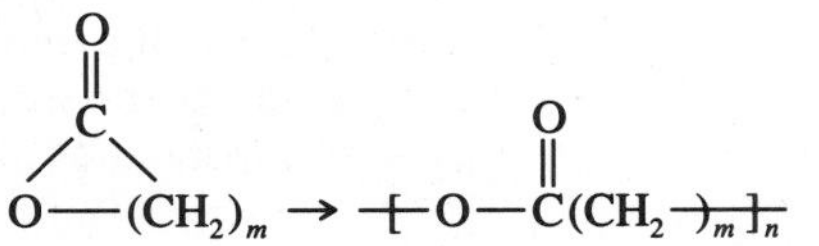

$$\underset{O-(CH_2)_m}{\overset{\overset{O}{\|}}{\overset{C}{\diagup\diagdown}}} \rightarrow \quad \begin{array}{c} O \\ \| \\ \text{---}O\text{---}C(CH_2\text{---})_m\text{---}]_n \end{array} \qquad (7\text{-}85)$$

different lactones generally follows the pattern for other cyclic monomers, except that the 5-membered lactone (γ-butyrolactone) does not polymerize while the 6-membered δ-valerolactone does. The 3-membered lactone is not known or is too reactive to isolate. Polymerization of ε-caprolactone (also referred to as *2-oxepanone*) is practiced commercially. Telechelic poly(ε-caprolactone), with hydroxyl end groups, is used in synthesizing polyurethane block copolymers. Poly(ε-caprolactone) is also blended with other polymers to improve stress crack resistance, dyeability, and adhesion.

7-5a Anionic Polymerization

A variety of anionic initiators, similar to those previously described for cyclic ethers (Sec. 7-2a), have been used to polymerize lactones [Jerome and Teyssie, 1989]. Much of the recent activity involves the use of anionic coordination initiators such as metalloporphyrins, metal alkoxides such as R_2AlOCH_3 and $Al(Oi\text{-}Pr)_3$, and aluminoxanes such as oligomeric $[Al(CH_3)\text{-}O]_n$ [Duda et al., 1990; Endo et al., 1987a, 1987b; Gross et al., 1988; Kricheldorf et al., 1990; Shimasaki et al., 1987].

Anionic polymerization proceeds by acyl–oxygen cleavage for almost all lactones, consistent with the mechanism for alkaline saponification of esters [Hofman et al., 1984; Kricheldorf et al., 1988]. For example, initiation by methoxide ion proceeds as

$$
CH_3O^- + \underset{O\text{---}R}{\overset{\displaystyle\overset{O}{\underset{\|}{C}}}{\diagdown}} \rightarrow CH_3O\text{---}CO\text{---}R\text{---}O^- \tag{7-86}
$$

with propagation as

$$
CH_3O\text{---}(CO\text{---}R\text{---}O)_n\text{---}CO\text{---}R\text{---}O^- + \underset{O\text{---}R}{\overset{\displaystyle\overset{O}{\underset{\|}{C}}}{\diagdown}} \rightarrow
$$
$$
CH_3O\text{---}(CO\text{---}R\text{---}O)_{(n+1)}CO\text{---}R\text{---}O^- \tag{7-87}
$$

For coordination initiators, the metal coordinates with the carbonyl oxygen followed by insertion of an alkoxy (or other anionic fragment depending on the initiator) into the acyl–oxygen bond.

The experimental evidence for the acyl–oxygen cleavage mechanism comes from end-group analysis of the polymer. For example, the polymer produced by the sequence in Eqs. 7-86 to 7-87 possesses methyl ester and hydroxyl end groups after termination with water or other acidic terminating agent. If polymerization proceeded by alkyl–oxygen cleavage, the end groups would be different—methoxyl and COOH. NMR analysis of the polymer clearly indicates that acyl–oxygen cleavage occurs. Additional support for the acyl–oxygen cleavage comes from trapping experiments in which specific terminating agents are used to identify the propagating centers [Hofman et al., 1987a, 1987b]. Polymerization of optically active lactones where the chiral carbon is attached to the alkyl oxygen are especially useful for detecting the mode of ring opening. Thus the polymerization of β-butyrolactone (**XLV**) proceeds with retention of configuration, indicating acyl–oxygen cleavage (bond b) instead of alkyl–

CH₃ structure (XLV):

$$O=\langle\overset{b\ a}{\underset{O\ \ H}{\ \ }}\rangle\overset{CH_3}{|}$$

XLV

oxygen cleavage (bond a) [Le Borgne and Spassky, 1989]. Cleavage at the alkyl–oxygen bond would result in racemization instead of retention.

Many of these polymerizations, especially with the less active coordination initiators, proceed with the characteristics of living polymerizations. The polymerization rate is first-order in both monomer and initiator, $\overline{X}_n$ is given by the ratio of monomer to initiator, and narrow MWDs are obtained. However, significant transesterification occurs and is responsible for molecular-weight broadening and cyclization when the more active (more ionic) initiators are used, for example, alkoxides of Mg, Zn, Ti, Zr instead of aluminum isopropoxide or $(CH_3)_3SiONa$ instead of $CH_3OAl(C_2H_5)_2$ [Hofman et al., 1987a, 1987b; Kricheldorf et al., 1988]. Termination by abstraction of α-protons also occurs but is not well documented.

Polymerization of β-propiolactone (also referred to as *2-oxetanone*) shows a number of interesting features due to the high strain in the 4-membered ring [Jerome and Teyssie, 1989]. β-Propiolactone, unlike other lactones, undergoes polymerization with weakly nucleophilic initiators such as metal carboxylates, tertiary amines, and phosphines. Ring opening occurs at the alkyl–oxygen bond instead of the acyl–oxygen bond. Polymerization with tertiary amines and phosphines involves zwitterion propagating species as in the corresponding anionic polymerizations of alkenes (Sec. 5-3a-1). Initiation by stronger nucleophiles occurs by acyl–oxygen as well as alkyl–oxygen cleavage to give a mixture of carboxyl and alkoxide propagating centers. However, the carboxylate centers are unable to propagate by cleavage of acyl–oxygen bonds while alkoxide propagating centers cleave both alkyl–oxygen and acyl–oxygen bonds. Alkoxide propagating centers convert to carboxylates with time, and carboxylate centers constitute more than 95% of all propagating centers after only several propagations.

Polymerization of β-propiolactone in a highly polar solvent (DMF) in the presence of a crown ether shows the unexpected feature that ion-pair reactivity decreases more slowly with decreasing temperature compared to free-ion reactivity [Slomkowski, 1986]. Solvation of free ions becomes so extensive at 20°C that they are less reactive than ion pairs.

7-5b Cationic Polymerization

Cationic polymerization of lactones is achieved with the range of initiators used for cyclic ethers (Sec. 7-2b) [Hofman et al., 1987; Kricheldorf and Sumbel, 1988a, 1988b; Kricheldorf et al., 1986, 1987a, 1987b; Penczek and Slomkowski, 1989a, 1989b]. Initiation was formerly thought to involve attack of a positive species on the endocyclic oxygen to form an oxonium ion followed by propagation through acyl–oxygen cleavage. However, polymer end-group analysis combined with trapping of propagating centers by reaction with triphenylphosphine indicates that this is not the mechanism. Initiation involves attack of a positive center on the exocyclic oxygen (the more basic oxygen) to form a dioxocarbocation (**XLVI**). For example, for initiation by methyl carbocation derived from $CH_3OSO_2CF_3$ or $(CH_3)_2I^+SbF_6^-$

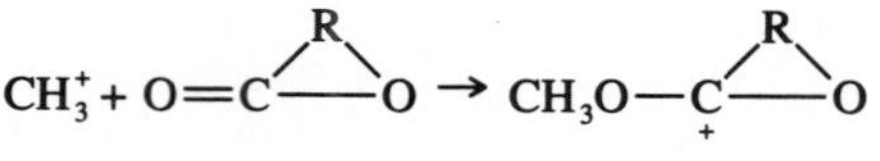

$$CH_3^+ + O=C \overset{R}{\underset{}{\diagup\!\!\!\diagdown}} O \rightarrow CH_3O - \underset{+}{C} \overset{R}{\underset{}{\diagup\!\!\!\diagdown}} O \qquad (7-88)$$

XLVI

Propagation follows in a similar manner with alkyl–oxygen cleavage

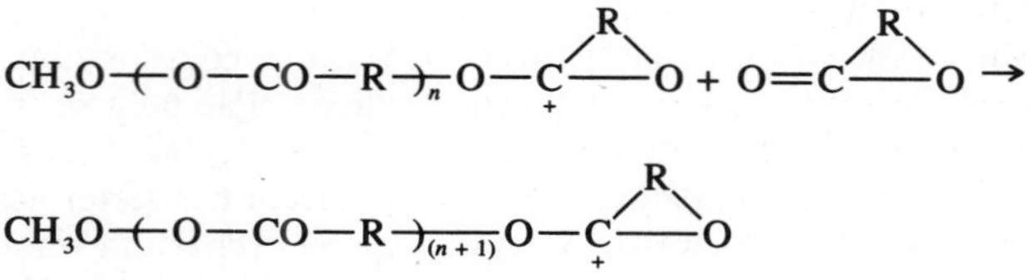

$$CH_3O +\!\!\!-O-CO-R\,)_{\overline{n}}O-\underset{+}{C}\overset{R}{\underset{}{\diagup\!\!\!\diagdown}}O + O=C\overset{R}{\underset{}{\diagup\!\!\!\diagdown}}O \rightarrow$$

$$CH_3O +\!\!\!-O-CO-R\,)_{\overline{(n+1)}}O-\underset{+}{C}\overset{R}{\underset{}{\diagup\!\!\!\diagdown}}O \qquad (7-89)$$

Cationic polymerization is not nearly as useful as anionic polymerization for synthesizing high-molecular-weight polyesters. The cationic route appears to be limited by intramolecular transesterification (cyclization) as well as other chain transfer to polymer reactions (including proton or hydride transfer) although there are few details in the literature. However, molecular weights as high as 100,000 have been reported in some polymerizations of the highly reactive monomer β-propiolactone. These include polymerizations initiated by acetyl chloride + SbCl$_5$, AlCl$_3$ + trifluoroacetic anhydride, and acetyl perchlorate + triflic acid. The highly strained β-propiolactone undergoes reaction with a mixture of alkyl–oxygen and acyl–oxygen cleavages.

7-5c   Other Cyclic Esters

Polymerization and copolymerization of the two 1,4-dioxane-2,5-diones (dilactones), glycolide and lactide (**XLVII** with R = H and CH$_3$, respectively), proceed using anionic

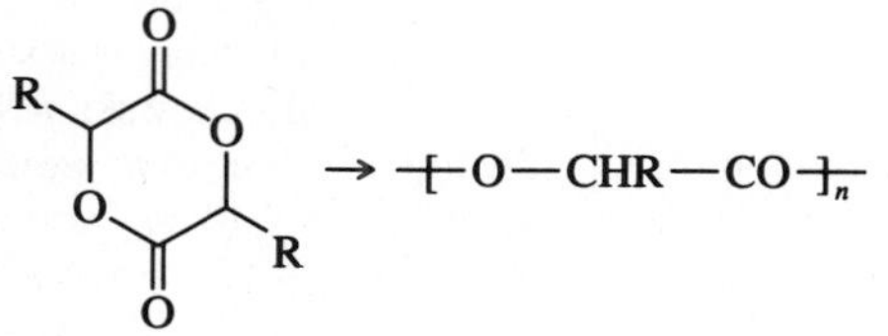

$$\rightarrow +\!\!\!-O-CHR-CO-)_{\overline{n}} \qquad (7-90)$$

XLVII

or cationic initiators [Bero et al., 1990; Kricheldorf and Kreiser-Saunders, 1990; Kricheldorf et al., 1987a, 1987b; Leenslag and Pennings, 1987]. The polymerization rates are generally lower than those for lactones. The polymerizations are of considerable interest because the polymers are both biocompatible and biodegradable [Vert, 1986]. Biodegradation occurs because the polymers are biologically hydrolyzed to the corresponding α-hydroxy acids, which are then eliminated through existing metabolic processes. Potential uses of these polymers include drug encapsulation for slow drug release and absorbable sutures.

Various bicyclic oxalactones have been polymerized to linear polyesters containing cyclic ether rings in the polymer chain, for example, the polymerization of 2,5-dioxabicyclo[2.2.2]octan-3-one [Okada et al., 1987].

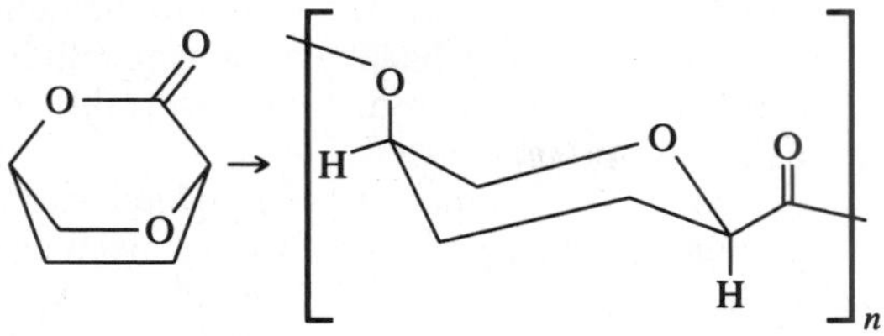

$$(7\text{-}91)$$

Polymerization of a cyclic carbonate ester yields a linear polycarbonate [Kuhling et al., 1989]. For example, the cyclic oligomer ($m = 2\text{-}20$ in Eq. 7-92) of the carbonate

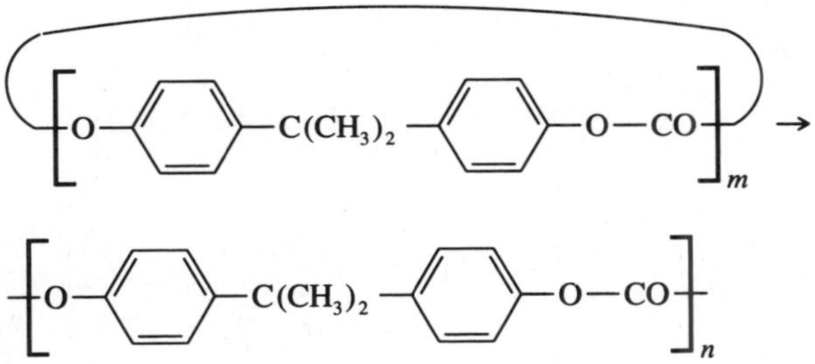

$$(7\text{-}92)$$

derived from bisphenol A, 2,2'-bis(4-hydroxyphenyl)propane, offers an alternate route to polycarbonates other than step polymerization of bisphenol A with phosgene or ester interchange with diphenyl carbonate (Sec. 2-8e) [Brunelle et al., 1989; Stewart, 1989].

The ring-opening route to polycarbonates offers potential advantages relative to the step-polymerization process. One advantage is that higher polymer molecular weights can be more easily achieved. Ring-opening polymerization by anionic initiators approximates a living system and the polymer molecular weight is determined by the ratio of cyclic monomer to initiator. Molecular-weight control in step polymerization is determined by conversion. Molecular weights as high as 100,000–300,000 are reported for the ring-opening polymerization, while the highest molecular weights achieved by step polymerization are 40,000–60,000. It is more difficult to achieve the very high conversions needed in step polymerization to reach the 10^5 range of molecular weights than to control the monomer/initiator ratio (and impurity levels) in ring-opening polymerization. Another advantage of the ring-opening route to polycarbonates is the absence of any by-products in the polymerization process. This allows the use of reactive processing techniques in which cyclic monomer is directly polymerized into final objects by extrusion or molding. The ring-opening route is also being studied for other high-performance polymers presently synthesized by step polymerization—polyarylates, polyetherimides, polyetherketones, and aromatic polyamides. The ring-opening route is viable, however, only when the cyclic oligomer can be synthesized and polymerized in high yield.

7-6 NITROGEN HETEROCYCLICS

7-6a Cyclic Amines

Cyclic amines (referred to as *imines*) are polymerized by acids and other cationic initiators [Goethals, 1984, 1989a, 1989b; Hauser, 1969; Tomalia and Killat, 1985]. The

3-membered imines (IUPAC: aziridines) are the most studied of the cyclic amines. Polyethyleneimine [IUPAC name: poly(iminoethylene)] had been commercially available and used in the treatment of paper and textiles. It is no longer available in the United States because of the high toxicity of the monomer.

The high degree of ring strain results in an extremely rapid polymerization for ethyleneimine. Initiation involves protonation or cationation of ethyleneimine followed by nucleophilic attack by monomer on the iminium C—N^+ bond. Propagation follows in the same manner. The propagating species is an iminium ion and the reaction

$$\underset{H}{N}\!\!\triangleleft \overset{H^+}{\longrightarrow} H\overset{+}{\underset{H}{N}}\!\!\triangleleft \overset{\overset{N\triangleleft}{H}}{\longrightarrow} H_2NCH_2CH_2\!-\!\overset{+}{\underset{H}{N}}\!\!\triangleleft \qquad (7\text{-}93)$$

$$H(NHCH_2CH_2\!\!\xrightarrow{}_{n}\overset{+}{\underset{H}{N}}\!\!\triangleleft + \underset{H}{N}\!\!\triangleleft \longrightarrow H\!\!-\!\!(NHCH_2CH_2\!\!\xrightarrow{}_{\overline{(n+1)}}\overset{+}{\underset{H}{N}}\!\!\triangleleft \qquad (7\text{-}94)$$

is analogous to the cationic polymerization of cyclic ethers. Extensive branching occurs during polymerization as evidenced by the presence of primary, secondary, and tertiary amine groups in the approximate ratio 1:2:1. Tertiary amine groups result from intermolecular nucleophilic attack of secondary amine nitrogens in polymer repeat units on iminium propagating centers. This reaction simultaneously increases the primary amine group content of a polymer chain. The detailed mechanism is quite complicated since there are many equilibria present involving proton transfers among the different types of amine groups present.

Polyethyleneimine is also extensively cyclized as a result of intramolecular nucleophilic attack of primary and secondary amines on the iminium group. This results in cyclic oligomer as well as polymer molecules containing large-sized rings as part of their structure.

Substitution on the aziridine ring hinders polymerization [Baklouti et al., 1989; Van de Velde, 1986]. The 1,2- and 2,3-disubstituted aziridines do not polymerize; 1- and 2-substituted aziridines undergo polymerization but the yield of polymer relative to low-molecular-weight linear and cyclic oligomers and the molecular weight of the polymer depend on the substituent (both electronic and steric effects are important).

Cationic polymerization of 4-membered imines (IUPAC: azetidines) generally follows the same patterns as the aziridines [Matyjaszewski, 1984a, 1984b; Muhlbach and Schulz, 1988]. Imines are generally unreactive toward anionic polymerization presumably because of the instability of an amine anion (which would constitute the propagating species). The exception occurs with N-acylaziridines as a result of the electron deficiency of the nitrogen coupled with the highly strained 3-membered ring.

7-6b Other Nitrogen Heterocyclics

Various *endo*-imino cyclic ethers (**XLVIII**) undergo cationic polymerization to yield a poly(N-acylalkyleneimine) (**XLIX**). The most studied monomers are the 2-substi-

$$nR\!\!-\!\!\underset{O}{\overset{N}{C}}\!\!\diagup(CH_2)_m \longrightarrow \left[\underset{\underset{}{\overset{\overset{COR}{|}}{N}}\!\!-\!\!(CH_2)_m}\right]_n \xrightarrow{\text{hydrolysis}} \left[NH\!\!-\!\!(CH_2)_m\right]_n \qquad (7\text{-}95)$$

$$\quad\quad\quad\textbf{XLVIII}\quad\quad\quad\quad\quad\quad\textbf{XLIX}\quad\quad\quad\quad\quad\quad\quad\quad\quad\quad\textbf{L}$$

tuted-2-oxazolines ($m = 2$) (also referred to as *2-substituted-1,3-oxazolin-2-enes*) [Goethals, 1989a, 1989b; Kobayashi and Saegusa, 1984, 1986; Kobayashi et al., 1990a, 1990b; Tomalia and Killat, 1985]. Propagation proceeds via nucleophilic attack of monomer on the C—O bond of an oxazolinium ion

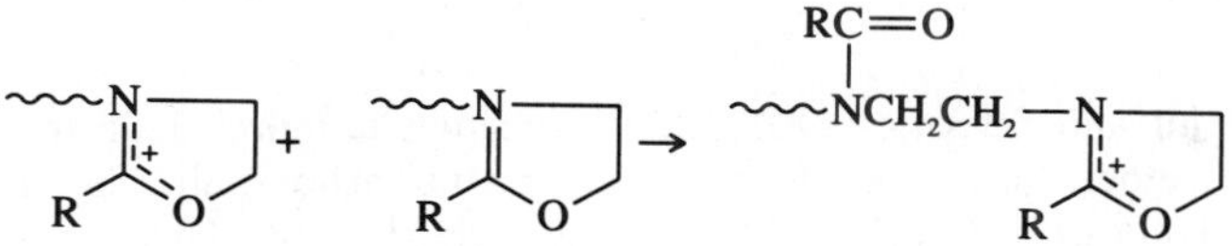

$$(7\text{-}96)$$

2-Substituted-5,6-dihydro-4H-1,3-oxazines ($m = 3$) have also been polymerized. Polymer **XLIX** can be hydrolyzed to the corresponding polyamine **L**. This is the only route to linear polyethyleneimine ($m = 2$) [Tanaka et al., 1983]. The polymerization of ethyleneimine yields a highly branched and cyclic product.

exo-Imino cyclic compounds (**LI**) such as iminocarbonates (Y = O), 2-imino-1,3-oxazolidines (Y = NR), and 2-iminotetrahydrofurans (Y = CH$_2$) have also been polymerized.

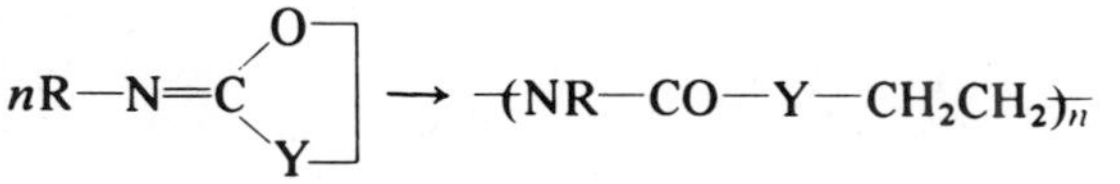

$$(7\text{-}97)$$

An alternate approach to forming polymers by ring-opening polymerization of 2-oxazoline structures involves the reactant **LII**; **LII** undergoes polymerization to **LV**

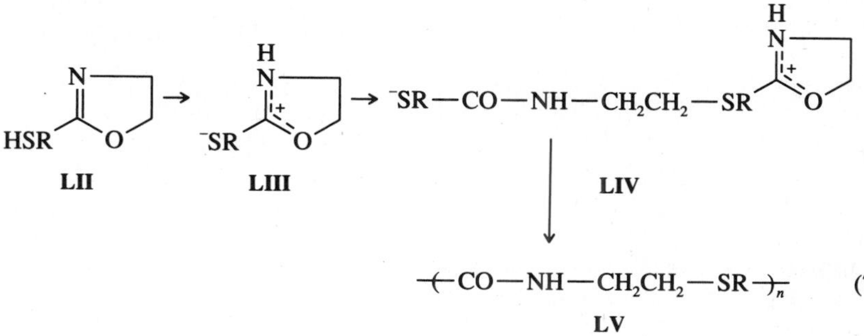

$$-\!(CO\!-\!NH\!-\!CH_2CH_2\!-\!SR\,)_{\overline{n}} \qquad (7\text{-}98)$$

LV

on heating [Gunatillake et al., 1988]. The reaction is postulated to occur by zwitterion intermediate **LIII** formed by proton transfer from sulfur to nitrogen. The zwitterion reacts with itself (nucleophilic attack of S⁻ on oxazolinium ring to break the C—O bond) to form a larger zwitterion (**LIV**) and propagation continues in a like manner to form the final product **LV**.

7-7 SULFUR HETEROCYCLICS

The 3- and 4-membered cyclic sulfides, referred to as *thiiranes* (also episulfides) and *thietanes*, respectively, are easily polymerized by both cationic and anionic initiators, such as polymerization of ethylene sulfide to poly(ethylene sulfide) (also referred to as *polythiirane*; IUPAC: polythioethylene) [Aida et al., 1990; Boileau, 1989; Goethals,

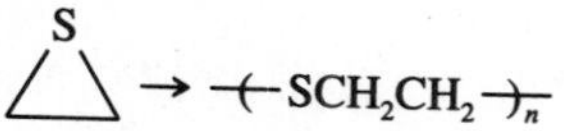

$$\underset{\triangle}{\overset{S}{\triangle}} \rightarrow \text{--}(\text{SCH}_2\text{CH}_2)_n\text{--} \tag{7-99}$$

1989a, 1989b; Sigwalt and Spassky, 1984]. Polymerization is more facile than the corresponding cyclic ethers because of the more polarizable carbon–sulfur bond. This also explains the polymerization of the thietanes by anionic initiators. However, cyclic sulfides are less strained than their oxygen analogs because of the larger size of the sulfur atom, and the 5-membered ring tholane (tetrahydrothiophene), unlike tetrahydrofuran, does not undergo polymerization. There are no reported polymerizations of larger-sized cyclic sulfides. By analogy to the cyclic ethers, the propagating species in cyclic sulfide polymerizations are considered to be the cyclic sulfonium (**LVI**) and

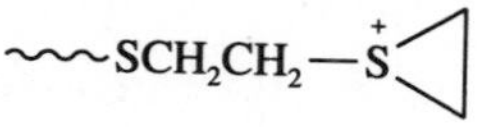

$$\text{\textasciitilde\textasciitilde\textasciitilde SCH}_2\text{CH}_2\text{--}\overset{+}{\text{S}}\overset{\triangleleft}{} \qquad \text{\textasciitilde\textasciitilde\textasciitilde CH}_2\text{CH}_2\text{SCH}_2\text{CH}_2\text{S}^-$$

LVI    **LVII**

sulfide anion (**LVII**) for cationic and anionic polymerization, respectively.

Although there are no reported polymerizations of simple cyclic sulfides of ring size 5 or higher, polymerization of 1,3,5-trithane (the cyclic trimer of thioformaldehyde) (**LVIII**), disulfides such as 1-oxa-4,5-dithiacycloheptane (**LIX**), and trisulfides such as norbornene trisulfide (**LX**) have been achieved [Andrzejewski et al., 1988; Baran et al., 1984; Moore et al., 1977; Zuk and Jeczalik, 1979]. The polymerization

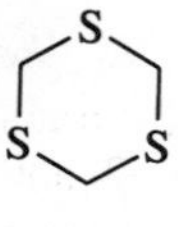

LVIII 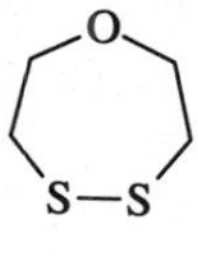   **LIX** 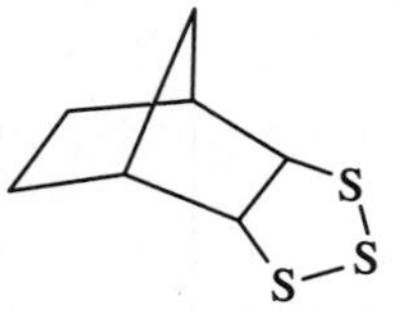   **LX**

of **LXI**, a stable isolable zwitterion, on heating involves nucleophilic ring-opening

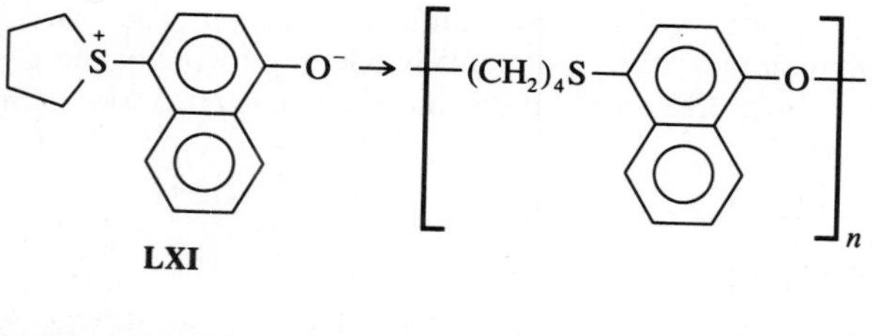

$$\underset{\text{LXI}}{\left[\text{C}\right]\overset{+}{\text{S}}\text{--}\langle O \rangle\text{--}O^-} \rightarrow \left[\text{--}(\text{CH}_2)_4\text{S}\text{--}\langle O \rangle\text{--}O\text{--}\right]_n \tag{7-100}$$

attack by phenoxide on the cyclic sulfonium ring [Odian et al., 1990].

7-8 CYCLOALKENES

Cycloalkenes, in the presence of coordination catalysts based on transition metals, undergo ring-opening polymerization to yield polymers containing a double bond, such as polymerization of cyclopentene to polypentenamer (IUPAC: poly[1-penten-

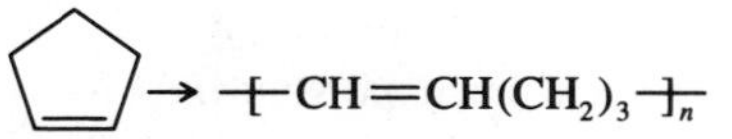

$$\text{(cyclopentene)} \rightarrow -\!\!\left[\text{CH}\!=\!\text{CH(CH}_2)_3\right]\!\!_n- \tag{7-101}$$

ylene]) [Amass, 1989; Amass et al., 1987; Doherty et al., 1986; Ivin, 1984, 1987; Leymet and Siove, 1989; Ofstead, 1988; Patton and McCarthy, 1987]. The polymerization is often referred to as *metathesis polymerization* by analogy to the olefin metathesis reaction. The latter reaction results in a transalkylidenation between two alkenes

$$\text{RCH}\!=\!\text{CHR} + \text{R}'\text{CH}\!=\!\text{CHR}' \rightarrow 2\text{RCH}\!=\!\text{CHR}' \tag{7-102}$$

Both olefin metathesis and metathesis polymerization require the same catalysts and proceed by the same reaction mechanism. The initiating and propagating species are metal-carbene complexes present in or generated from catalysts based on transition-metal compounds, usually molybdenum or tungsten. Rhenium and ruthenium catalysts are also used. The catalyst systems include stable metal–carbene complexes such as $\emptyset_2\text{C}\!=\!\text{W(CO)}_5$ and two-component systems, for example, WCl_6 with $\text{Sn(CH}_3)_4$ and MoO_3 with Al_2O_3 or $\text{C}_2\text{H}_5\text{AlCl}_2$. For the two-component systems that contain alkyl groups, carbene ligands are generated from those alkyl groups. For the other two-component systems, carbenes are generated by interaction between monomer and the transition-metal centers in the catalyst system. The catalyst systems often include a third component such as water, alcohol, oxygen, or other oxygen-containing compound. There is a further consideration of this type of catalyst system, referred to as a *coordination catalyst*, as well as metathesis polymerization in Chap. 8.

The propagating center is a metal–carbene bond. Propagation involves coordination of the double bond of monomer with the transition metal (via a vacant coordination site), cleavage of the π-bond with formation of a 4-membered intermediate, followed by rearrangement. The overall result is insertion of the halves of the double bond

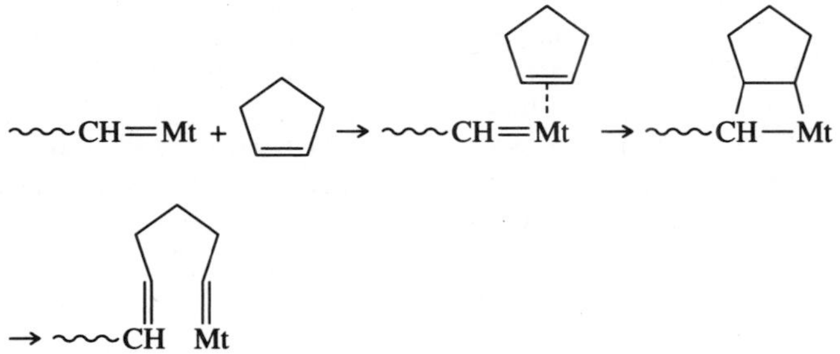

$$\sim\!\!\sim\!\!\text{CH}\!=\!\text{Mt} + \text{(cyclopentene)} \rightarrow \sim\!\!\sim\!\!\text{CH}\!=\!\text{Mt} \rightarrow \sim\!\!\sim\!\!\text{CH}\!-\!\text{Mt}$$

$$\rightarrow \sim\!\!\sim\!\!\text{CH} \quad \text{Mt} \tag{7-103}$$

into the metal–carbene bond. Proof for cleavage of the double bond instead of an adjacent single bond comes from analysis of the ozonolysis products from a copolymer of cyclooctene and 1-^{14}C-cyclopentene. The description of the metal center is over-simplified in Eq. 7-103. In addition to the carbene bond and empty site for coordination to the cycloalkene, the transition-metal center typically possesses three ligands whose identity depends on the catalyst components used for the particular polymerization. Metathesis polymerization proceeds as a living polymerization for most catalyst systems.

Aside from cyclohexene, which yields only very-low-molecular-weight oligomers, a wide range of cycloalkenes and bicycloalkenes have been polymerized to high-molecular-weight products. There are some commercial applications of metathesis polymerization [Feast, 1989; Ofstead, 1988; Streck, 1989]. Poly(1-octenylene) (trade name: *Vestenamer*) is used as the minor component (10–30%) in elastomer blends with SBR and natural rubbers for gaskets, sealing profiles, brake hoses, and printing rollers. Polynorbornene (**LXII**) (IUPAC: poly[1,3-cyclopentylenevinylene]) (trade name: *Norsorex*) is a specialty rubber that can take up to several times its own weight of

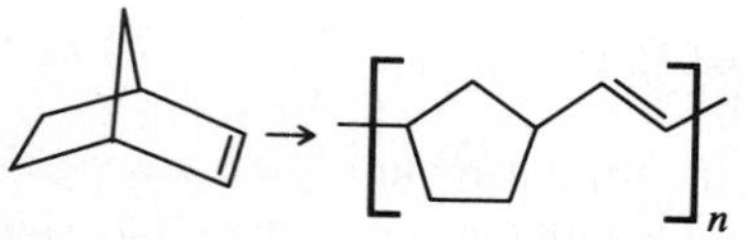

$$(7\text{-}104)$$

LXII

plasticizer oil and still retain useful properties. It has high tear strength and high dynamic damping properties and finds applications in noise control (e.g., under the hoods of diesel-powered autos) and vibration damping (auto-body mounts and instrument isolation pads). Polymerization of *endo*-dicyclopentadiene yields a commercially useful plastic (trade name: *Metton*). Polymerization initially yields **LXIII**,

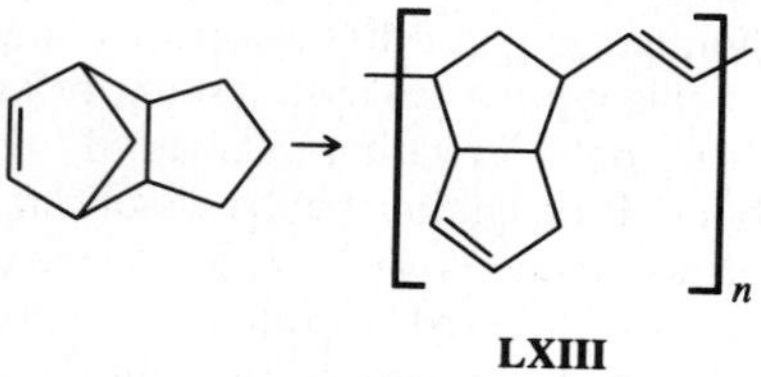

$$(7\text{-}105)$$

LXIII

but ring-opening polymerization through the pendant cyclopentene ring results in crosslinking. Polymerization and crosslinking occur in the same process, and this is useful for producing products such as satellite antenna dishes, snowmobile bodies, and auto-body parts by reaction injection molding.

Two different metathesis polymerizations offer alternate routes to polyacetylene. One is the polymerization of 1,3,5,7-cyclooctatetraene and the other is the polym-

$$\rightarrow \ \ {-}\!\!\left(\text{CH}\!=\!\text{CH}\right)_{\!n}\!{-} \qquad\qquad (7\text{-}106a)$$

erization of benzvalene followed by isomerization of **LXIV** to polyacetylene [Grubbs et al., 1988; Klavetter and Grubbs, 1988].

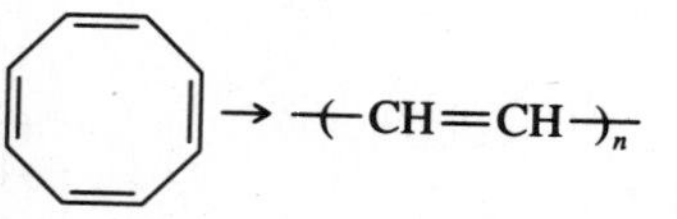

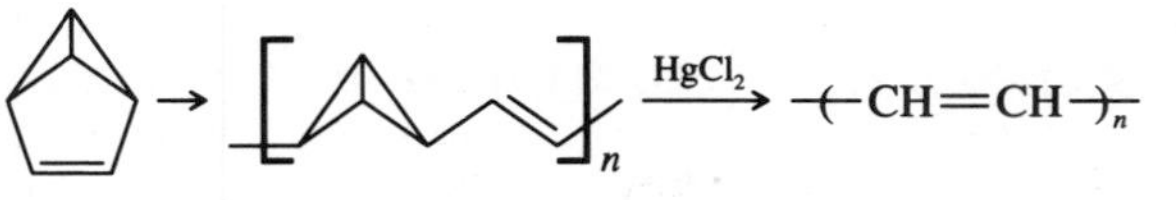

$$\text{LXIV} \quad \overset{\text{HgCl}_2}{\longrightarrow} \quad \left(\text{CH}=\text{CH}\right)_n \qquad \text{(7-106b)}$$

The corresponding ring-opening polymerization of cyclic alkynes also occurs [Krouse and Schrock, 1989], for example, the polymerization of cyclooctyne to polyoctynamer or poly(1-octynylene):

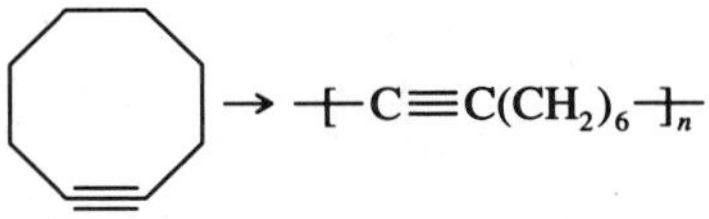

$$\longrightarrow \quad \left(\text{C}\equiv\text{C}(\text{CH}_2)_6\right)_n \qquad \text{(7-107)}$$

7-9 MISCELLANEOUS OXYGEN HETEROCYCLICS

Polymerizations of various bicyclic and spiro orthoesters and orthocarbonates proceed with either no volume shrinkage or an expansion in volume due to the simultaneous opening of two or more rings. All other ring-opening polymerizations as well as chain and step polymerizations proceed with volume shrinkage. Polymerizations without volume shrinkage, and especially those with volume expansion, are of interest for applications such as high-strength adhesives, dental fillings, and prestressed castings.

Cationic polymerization of a spiro orthocarbonate (**LXV**) proceeds by initial formation of carbocation **LXVI**. The carbocation center is stabilized by interaction with the three adjacent oxygens. Depending on ring size and the identity of substitutents on the ring, **LXXI** propagates by cleavage at bond **a**, **b**, or **c** to yield polymers **LXVII**, **LXVIII**, and **LXIX**, respectively [Bailey and Endo, 1978; Takata and Endo, 1988].

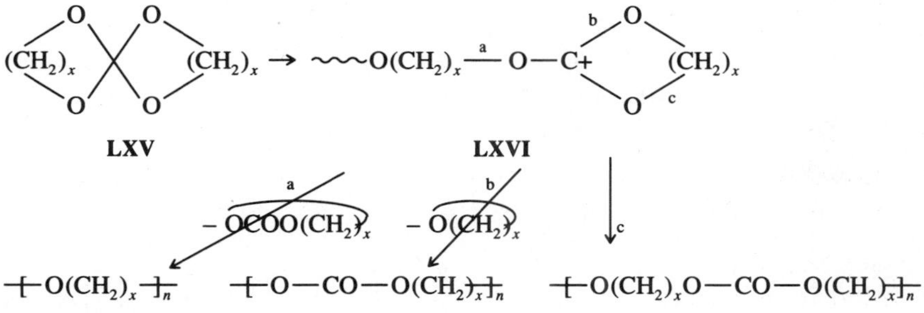

LXV   **LXVI**

$$\left(\text{O}(\text{CH}_2)_x\right)_n \qquad \left(\text{O}-\text{CO}-\text{O}(\text{CH}_2)_x\right)_n \qquad \left(\text{O}(\text{CH}_2)_x\text{O}-\text{CO}-\text{O}(\text{CH}_2)_x\right)_n$$

LXVII   **LXVIII**   **LXIX**

$$\text{(7-108)}$$

A cyclic carbonate and cyclic ether are eliminated as by-products in reaction pathways **a** and **b**, respectively. Polymer **LXIX** contains both ether and carbonate functional groups in the polymer chain.

Spiro orthoesters such as 1,4,6-trioxaspiro[4.4]nonane (**LXX**) undergo cationic polymerization to polymers containing both ether and ester groups in the polymer

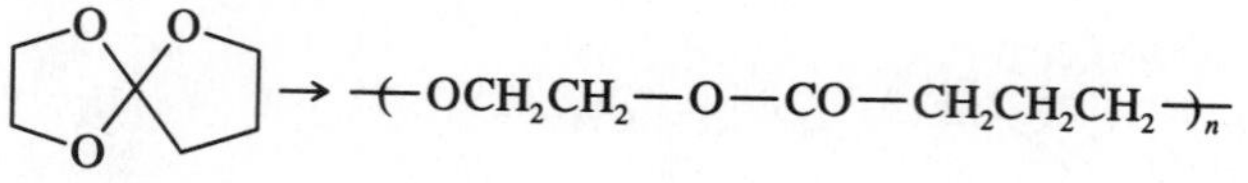

$$\text{LXX} \quad \rightarrow \; -\!\!\left(\!OCH_2CH_2\!-\!O\!-\!CO\!-\!CH_2CH_2CH_2\!\right)_{\overline{n}} \qquad (7\text{-}109)$$

chain [Bailey, 1975; Matyjaszewski, 1984a, 1984b].

The bicyclo orthoester, 1,4-dialkyl-2,6,7-trioxabicyclo[2.2.2]octane (**LXXI**), undergoes cationic polymerization to a polyether with pendant ester groups [Saigo et al., 1983].

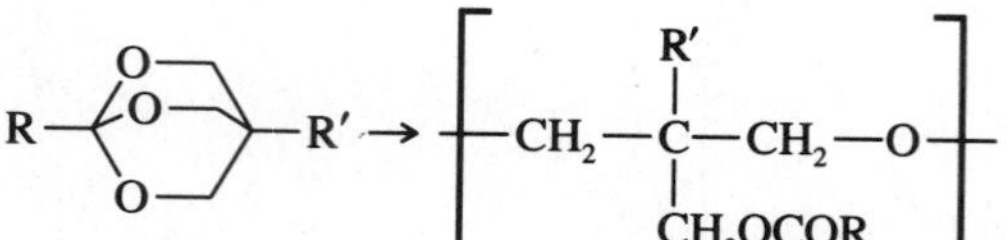

$$\text{LXXI} \quad R\!-\!\left\langle \begin{smallmatrix} O \\ O \\ O \end{smallmatrix} \right\rangle\!-\!R' \rightarrow \left[-CH_2-\underset{\underset{CH_2OCOR}{|}}{\overset{\overset{R'}{|}}{C}}-CH_2-O- \right]_n \qquad (7\text{-}110)$$

The polymerization of various oxygen heterocyclics containing more than one oxygen and a pendant carbon–carbon double bond have been studied. The behavior of such monomers is essentially the sum of the behaviors of the two polymerizable moieties (ring and C=C) when neither carbon of the double bond is part of the ring. For example, compounds such as **LXV** and **LXXI** containing pendant allyl or acryloxy groups undergo radical polymerization through the carbon–carbon double bond and cationic polymerization by ring opening [Endo et al., 1987a, 1987b; Herweh, 1989]. Further, crosslinked polymers are obtained by sequential radical and cationic polymerizations. An unexpected result is obtained for monomers such as 4-phenyl-2-methylene-1,3-dioxepane (**LXXII**), a cyclic ketene acetal [Bailey, 1985; Bailey et al., 1988; Endo et al., 1985]. Radical polymerization results in a concerted reaction involving both the C=C and ring opening. The initially formed radical **LXXIII** undergoes ring opening to form the benzyl radical, which subsequently propagates in the same manner

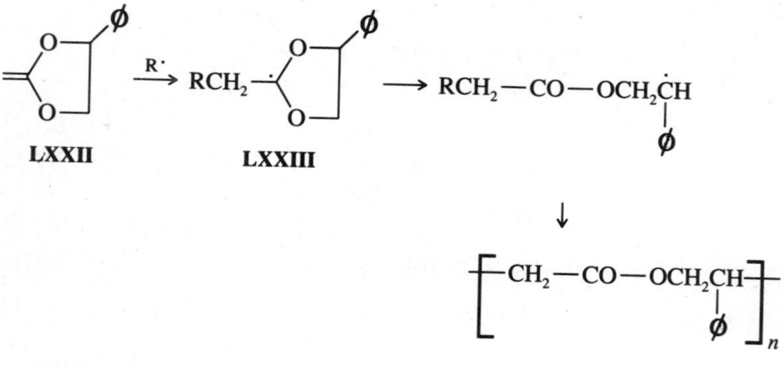

$$\overset{O-}{\underset{O-}{\diagdown}}\!\!\diagup^{\phi} \;\;\xrightarrow{R^{\cdot}}\;\; RCH_2-\!\!\overset{O-}{\underset{O-}{\diagdown}}\!\!\diagup^{\phi} \;\;\longrightarrow\;\; RCH_2-CO-OCH_2\overset{\cdot}{C}H$$

$$\text{LXXII} \qquad\qquad \text{LXXIII} \qquad\qquad\qquad\qquad \phi$$

$$\downarrow$$

$$\left[-CH_2-CO-OCH_2\underset{\phi}{CH}- \right]_n \qquad (7\text{-}111)$$

to form a polyester. This is one of the very few ring-opening polymerizations proceeding via a radical process. The extent of ring opening is dependent on the presence of a ring substituent such as phenyl to stabilize the radical formed on ring opening and on the extent of strain in the ring structure. Varying degrees of ring-opening polymerization with radical initiators have been observed with methylene spiro orthocarbonates and orthoesters [Endo and Bailey, 1975, 1980; Pan et al., 1988; Tagoshi and Endo, 1989].

7-10 OTHER RING-OPENING POLYMERIZATIONS

There is considerable literature on efforts to polymerize cycloalkanes [Hall and Snow, 1984]. With very few exceptions, ring-opening polymerization by cleavage of σ–σ bonds does not occur or produces very-low-molecular-weight products. The few exceptions to this generalization involve monomers that possess some structural feature conducive to ring-opening reaction. Many vinylcyclopropane derivatives (**LXXIV**) undergo a radical 1,5-polymerization. Radical addition occurs initially at the double

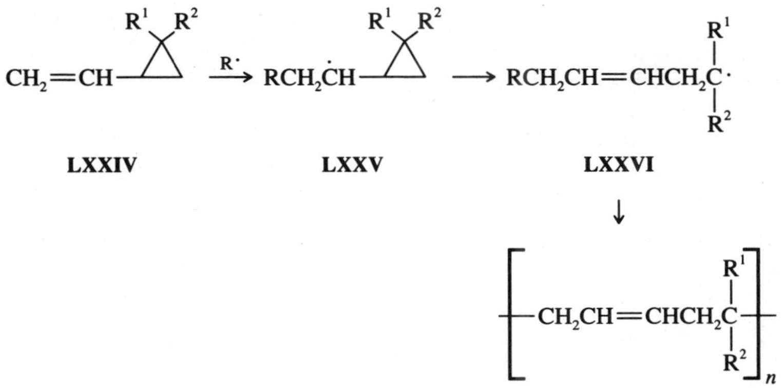

$$CH_2{=}CH\!-\!\!\overset{R^1\ R^2}{\triangle}\ \xrightarrow{R\cdot}\ RCH_2\overset{\cdot}{C}H\!-\!\!\overset{R^1\ R^2}{\triangle}\ \longrightarrow\ RCH_2CH{=}CHCH_2\overset{\overset{R^1}{|}}{\underset{\underset{R^2}{|}}{C}}\cdot$$

$$\textbf{LXXIV}\qquad\qquad\textbf{LXXV}\qquad\qquad\textbf{LXXVI}$$

$$\downarrow$$

$$\left[\!\!-CH_2CH{=}CHCH_2\overset{\overset{R^1}{|}}{\underset{\underset{R^2}{|}}{C}}\!\!-\!\!\right]_n \qquad (7\text{-}112)$$

bond to form radical **LXXV** followed by ring opening to radical **LXXVI**. High-molecular-weight polymers are obtained only when R^1 and/or R^2 are substituents such as COOR or CN, which are capable of radical stabilization [Endo and Suga, 1989; Endo et al., 1989]. Vinylcyclopropanes undergo polymerization exclusively through the double bond by cationic and coordination initiators. Anionic polymerization of cyclopropanes with electron-withdrawing substituents proceeds exclusively through opening of the ring.

Bicyclobutanes undergo radical and anionic polymerization

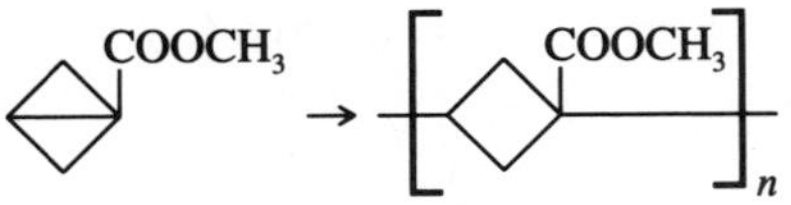

$$\overset{COOCH_3}{\bowtie} \rightarrow \left[\overset{COOCH_3}{\diamondsuit}\right]_n \qquad\qquad (7\text{-}113)$$

Anionic polymerization of 1,1-dimethylsilacyclopent-3-enes proceeds by anionic addition to Si to form a pentacoordinate siliconate propagating center that undergoes ring opening on addition to monomer [Zhang et al., 1988; Zhou and Weber, 1990].

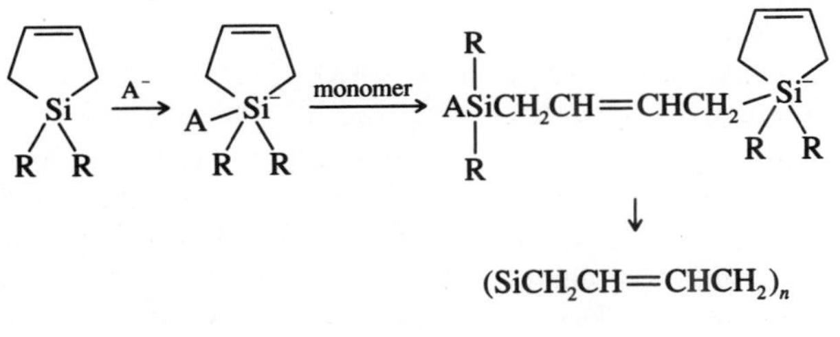

$$\underset{R\ \ R}{Si}\ \xrightarrow{A^-}\ \underset{R\ \ R}{A{-}Si^-}\ \xrightarrow{monomer}\ A\underset{\underset{R}{|}}{\overset{\overset{R}{|}}{Si}}CH_2CH{=}CHCH_2\underset{R\ \ R}{Si^-}$$

$$\downarrow$$

$$(SiCH_2CH{=}CHCH_2)_n \qquad\qquad (7\text{-}114)$$

7-11 INORGANIC AND PARTIALLY INORGANIC POLYMERS

As noted in Sec. 2-15, there has been considerable interest in the synthesis of inorganic and partially inorganic polymers. Some of the polymers studied have been obtained by ring-opening polymerization.

7-11a Cyclosiloxanes

The commercial importance of polysiloxanes was previously discussed in Sec. 2-12f. The higher-molecular-weight polysiloxanes are synthesized by anionic or cationic polymerization of cyclic siloxanes [Bostick, 1969; Noll, 1968; Kendrick et al., 1989; Saam, 1989; Wright, 1984]. Both cationic and anionic polymerizations often have the characteristics of living polymerizations in that one can achieve sequential polymerizations and form block copolymers. The most commonly encountered polymerization is that of the cyclic tetramer octamethylcyclotetrasiloxane

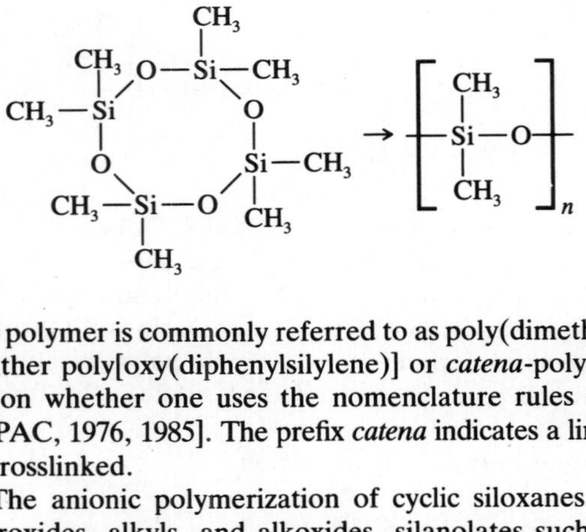

$$\tag{7-115}$$

The polymer is commonly referred to as poly(dimethylsiloxane), but the IUPAC name is either poly[oxy(diphenylsilylene)] or *catena*-poly[(diphenylsilicon)-μ-oxo] depending on whether one uses the nomenclature rules of organic or inorganic polymers [IUPAC, 1976, 1985]. The prefix *catena* indicates a linear polymer, that is, not branched or crosslinked.

The anionic polymerization of cyclic siloxanes can be initiated by alkali metal hydroxides, alkyls, and alkoxides, silanolates such as potassium trimethylsilanoate, $(CH_3)_3SiOK$, and other bases. Both initiation

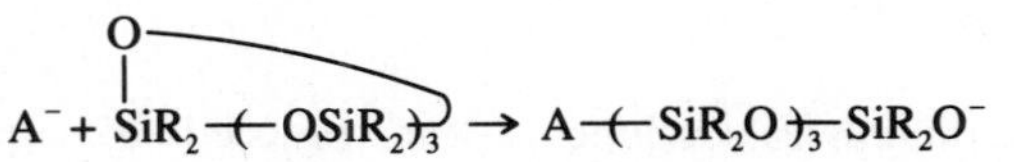

$$A^- + SiR_2 \overbrace{(OSiR_2)_3} \rightarrow A(SiR_2O)_3 SiR_2O^- \tag{7-116}$$

and propagation

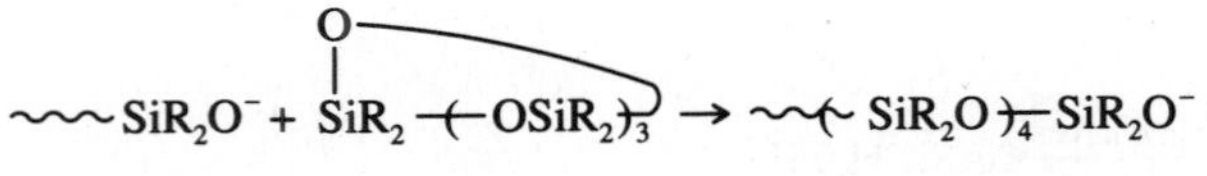

$$\sim\sim\sim SiR_2O^- + SiR_2 \overbrace{(OSiR_2)_3} \rightarrow \sim\sim (SiR_2O)_4 SiR_2O^- \tag{7-117}$$

involve a nucleophilic attack on monomer in a manner analogous to the anionic polymerization of epoxides [Mazurek and Chojnowski, 1977; Mazurek et al., 1980].

An interesting aspect of this polymerization is that ΔH is nearly zero and, amazingly, ΔS is positive by 6.7 J/mole K [Lee and Johannson, 1966, 1976]. The driving force in this polymerization is the increase in entropy (disorder) on polymerization.

A positive value of ΔS is very rare for a polymerization process. The only other reported instances of positive ΔS values are those for the polymerizations of the cyclic octamers of sulfur and selenium and cyclic carbonate oligomers (Sec. 7-5c) [Brunelle et al., 1989]. All other polymerizations involve a decrease in entropy because of the decreased disorder for a polymer relative to its monomer. The positive ΔS values for the cyclic siloxane, S, and Se probably result from the high degree of flexibility of the linear polymer chains due to the large-sized atoms comprising them. This flexibility leads to greater degrees of freedom in the linear polymer compared to the cyclic monomer.

Cationic polymerization has been initiated by a variety of protonic and Lewis acids. The cationic process is more complicated and less understood than the anionic. Polymerization under most reaction conditions involves the presence of a step polymerization simultaneously with the ring-opening chain polymerization. This appears to be the only way to reconcile the observed (complicated) kinetics for the overall process [Chojnowski and Wilczek, 1979; Sigwalt, 1987; Wilczek et al., 1986].

Initiation consists of protonation of monomer followed by subsequent reaction with monomer to form the tertiary oxonium ion **LXXVII**. Propagation for the ring-opening

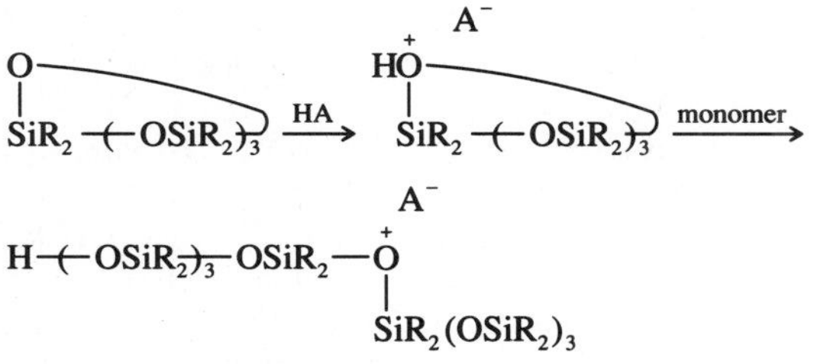

$$(7\text{-}118)$$

LXXVII

chain polymerization process follows in the same manner with nucleophilic attack of monomer on the tertiary oxonium ion

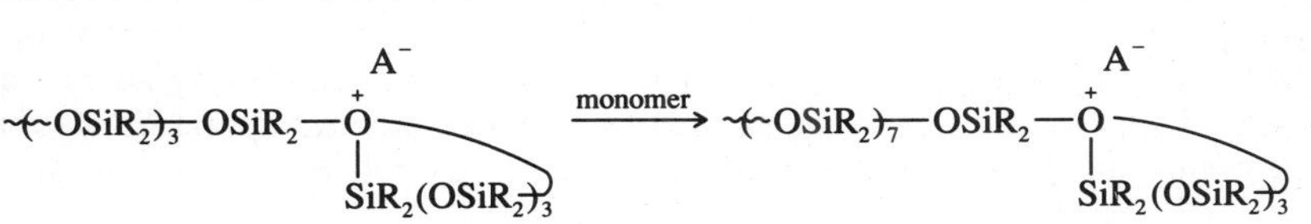

$$(7\text{-}119)$$

(This is analogous to the cationic polymerization of cyclic ethers; see Eq. 7-20.)

Some workers have proposed the covalent ester (**LXXVIII**) and/or silicenium ion (**LXXIX**) as alternatives to the oxonium ion as the propagating center [Sigwalt and

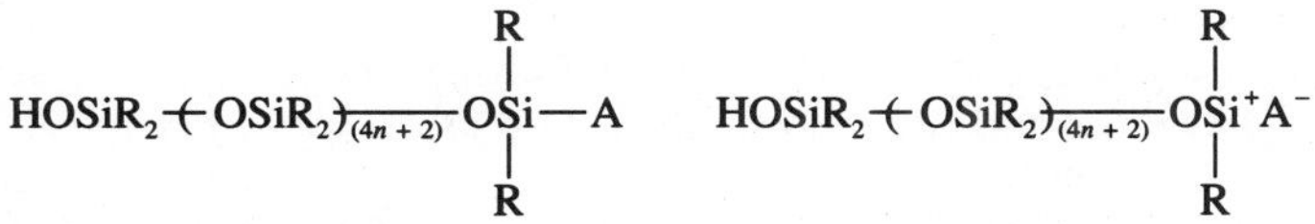

LXXVIII   **LXXIX**

Stannett, 1990]. Participation of such species would proceed by step polymerization involving self-reaction of the covalent ester or silicenium ion with elimination of HA.

7-11b   Poly(organophosphazenes)

Thermal polymerization of hexachlorocyclotriphosphazene (**LXXX**) (also referred to as *phosphonitrilic chloride*) yields poly(dichlorophosphene) (**LXXXI**) (IUPAC:

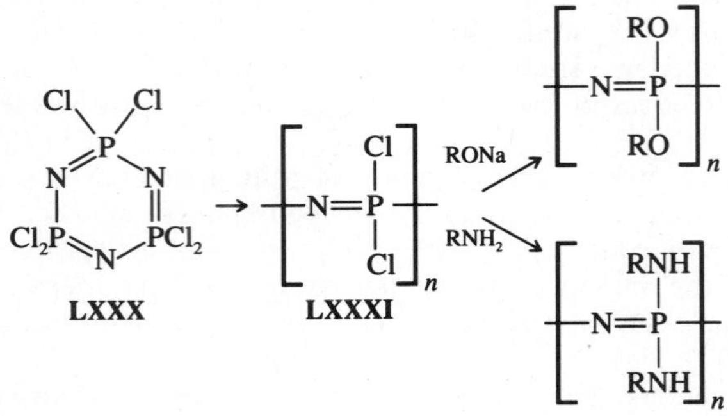

(7-120a)

(7-120b)

poly[nitrilo(dichlorophosphoranylidyne)] or *catena*-poly[(dichlorophosphorus)-μ-nitrido]) [Allcock, 1976, 1986; Allcock and Connolly, 1985; Liu and Stannett, 1990; Majumdar et al., 1989, 1990; Potts et al., 1989; Scopelianos and Allcock, 1987; Sennett et al., 1986].

Although poly(dichlorophosphazene) is a hydrolytically unstable elastomer, it can be converted to stable alkoxy and amino derivatives by nucleophilic substitution reactions with RONa (R can aliphatic or aromatic) and amines (1° or 2°) (Eqs. 7-120a,b). The polyphosphazene backbone has a high inherent flexibility (T_g is -60 to $-100°C$) combined with photolytic, oxidative, and solvent resistance. A wide range of properties can be imparted to the polymer by varying the organic moiety. Most poly-(organophosphazenes) are elastomers, and, indeed, the first commercial application involves the use of fluorinated alkoxy derivatives as high-performance elastomers in O-rings, pipelines, and seals used in oil and fuel delivery and storage systems. Some of the derivatives are microcrystalline (due to the R groups) and form flexible films and fibers. The incorporation of CH_3NH or glucose units as side groups imparts water solubility. Many of the derivatives are biocompatible, some are biodegradable, and others are bioactive and are being studied for possible medical applications. Some poly(organophosphazenes) have potential as solid electrolytes for lightweight rechargeable batteries.

Polymerization of hexachlorocyclotriphosphazene is typically carried out as a melt polymerization at 220–250°C. Conversions must be limited to about 70% to prevent branching and crosslinking since the crosslinked product is insoluble and cannot be derivatized. High yields of soluble polymer require purified monomer, but exceptionally pure and dry monomer shows very low reactivity except in polymerizations initiated by ionizing radiation. Various Lewis acids (BCl_3 is the most studied) show a significant catalytic effect and allow higher conversions, without crosslinking, to be achieved at lower reaction temperatures (150–210°C). Polymerization in inert solvents such as chlorinated aromatics and carbon disulfide is also useful for this purpose.

A main feature of the melt polymerization is the sharp increase in electrical conductivity at temperatures (>220°C) where polymerization occurs. This observation together with those described above support a cationic polymerization mechanism based on ionization of a P—Cl bond (Eq. 7-121) followed by electrophilic attack of P^+ on monomer (Eq. 7-122). Boron trichloride and trace amounts of water accelerate

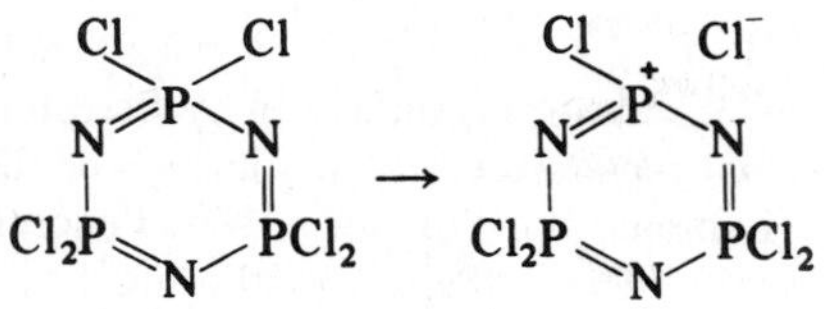

$$(7\text{-}121)$$

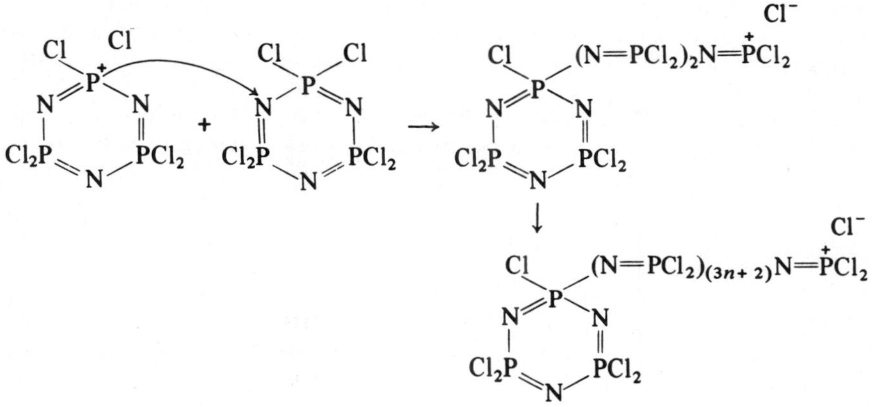

$$(7\text{-}122)$$

the polymerization by facilitating ionization of the P—Cl bond. Other reaction characteristics further support the mechanism. The inability of cyclotriphosphazenes without any halogens (e.g., hexamethylcyclotriphosphazene or hexamethylaminocyclotriphosphazene) to polymerize is ascribed to the absence of an easily ionizable bond to phosphorus. However, polymerization is possible for partially substituted chlorocyclotriphosphazenes, although the reaction is less efficient than that for the fully halogenated monomer.

Many mechanistic features of $(NPCl_2)_3$ polymerization are unclear. Termination may involve reaction of propagating centers with PCl_5 (which is used in monomer preparation and may be present as a trace impurity) or other Lewis acid (including water) present as a catalyst. The branching and crosslinking that can occur during polymerization are increased by the presence of water concentrations above about 0.1%, although the water also acts as a retarder. Below the 0.1% level water accelerates the polymerization rate. This is reminiscent of the effect of water in many cationic polymerizations of alkenes (Sec. 5-2a-2).

Alkyl and aryl derivatives of poly(dichlorophosphazene) are not efficiently synthesized by nucleophilic reaction of **LXXXI** with metal alkyls or aryls. The halogen substitution reaction occurs but is accompanied by polymer chain cleavage. Use of poly(difluorophosphazene) or introduction of aryl and alkyl groups at the monomer stage offer some improvement, but neither method is fully satisfactory. The best route to alkyl and aryl derivatives is thermal polymerization of N-(trimethylsilyl)-P,P-di-alkyl-P-(trifluoroethoxy) diaphosphoranimines [Neilson and Wisian-Neilson, 1988].

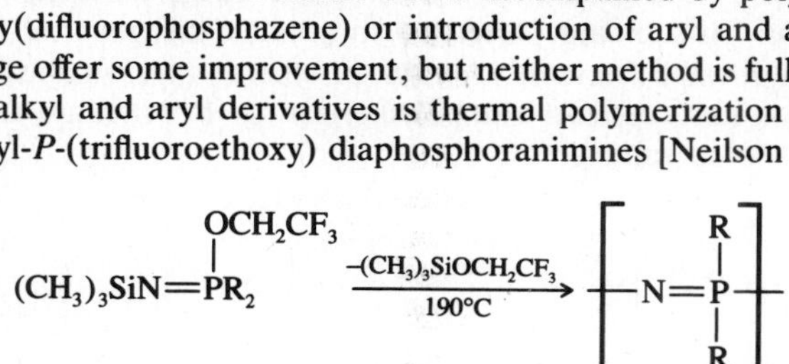

$$(7\text{-}123)$$

7-11c Phosphorus-Containing Cyclic Esters

There has been interest in polymerizing various phosphorus-containing cyclic esters with a view toward obtaining polymer chains that have structural similarities to the naturally occurring nucleic and teichoic acids [Lapienis and Penczek, 1984; Penczek and Libiszowski, 1988; Penczek and Slomkowski, 1989a, 1989b; Wodzki and Kaluzynski, 1989]. The biopolymers are complex materials in which the polymer chain consists of alternating diol (from sugars) and phosphoric acid units; that is, they are poly(phosphate esters). Some of these materials have been studied for use in drug-delivery systems.

Cyclic phosphates (**LXXXII**) and phosphonates (**LXXXIII**) are among the most studied of these monomers. Anionic polymerization yields polymers of high molecular

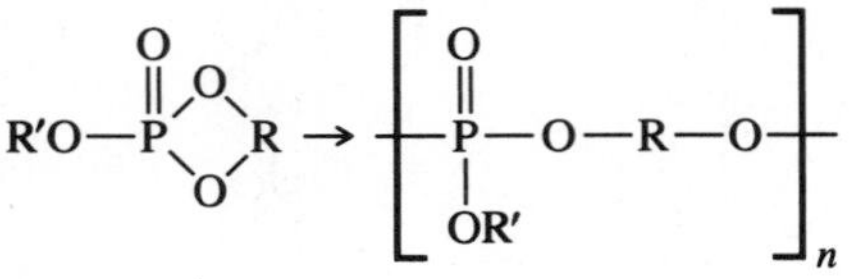

$$R'O-P\diagdown_O^{O}\diagup R \rightarrow \left[\begin{matrix} O \\ \| \\ -P-O-R-O- \\ | \\ OR' \end{matrix}\right]_n \tag{7-124}$$

LXXXII

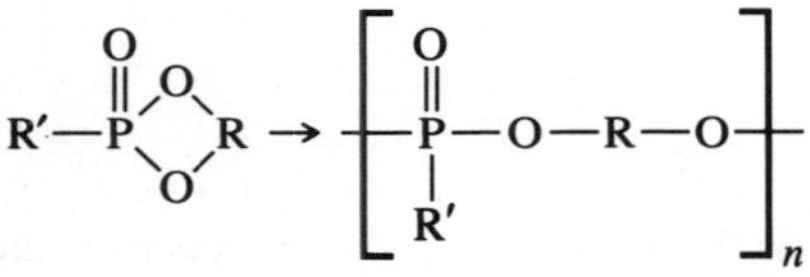

$$R'-P\diagdown_O^{O}\diagup R \rightarrow \left[\begin{matrix} O \\ \| \\ -P-O-R-O- \\ | \\ R' \end{matrix}\right]_n \tag{7-125}$$

LXXXIII

weight while cationic polymerization yields molecular weights no higher than a few thousand. The cationic polymerizations are limited because of extensive chain-transfer reactions.

Other monomers which have been studied include cyclic phosphonites and phosphites as well as some sulfur analogues and nitrogen derivatives [Baran et al., 1989; Kobayashi et al., 1986].

7-11d Sulfur

The stable form of sulfur under normal conditions is crystalline cyclooctasulfur, referred to as *rhombic sulfur*. Heating rhombic sulfur causes it to change into *monoclinic sulfur* (which is still cyclooctasulfur but with a different crystalline structure) at about 95°C. Further heating causes melting at about 119°C and ring-opening polymerization to *catena*-polysulfur at 150–180°C via a biradical propagating species [Laitinen et al., 1987; Meyer, 1976; Ray, 1978]. *Catena* polysulfur is elastomeric but is unstable—it slowly reverts to rhombic sulfur on standing. Attempts to prevent this reversal by various means such as addition of small amounts of P, As, or alkenes (presumably to end-cap the polymer) have been unsuccessful. The polymerizations of selenium and tellurium proceed in a similar way as sulfur.

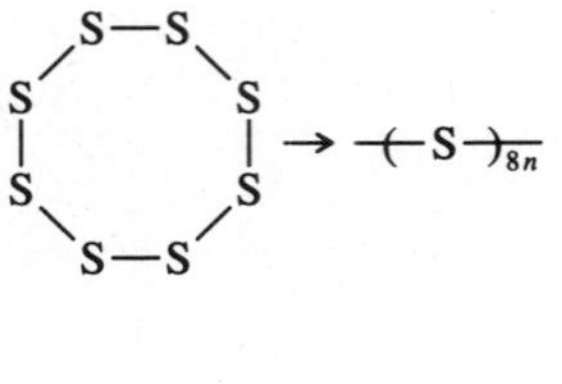

$$\begin{matrix} S-S \\ S \qquad S \\ | \qquad | \\ S \qquad S \\ S-S \end{matrix} \rightarrow +S+_{8n} \tag{7-126}$$

7-11e Polymeric Sulfur Nitride

Solid-state (topochemical) polymerization of disulfur dinitride to poly(sulfur nitride) (or polythiazyl) occurs on standing at 0°C or higher [Labes et al., 1979; Ray, 1978].

$$\begin{array}{c} \text{S—N} \\ |\quad| \\ \text{N—S} \end{array} \rightarrow \; -\!\!+\!\text{SN}\!\rightarrow_{\overline{2n}} \qquad (7\text{-}127)$$

Disulfur dinitride is obtained by sublimation of tetrasulfur tetranitride. Polythiazyl is a potentially useful material, since it behaves like a metal. It has an electrical conductivity at room temperature about the same order of magnitude as an ordinary metal like mercury and is a superconductor at 0.3°C. Polythiazyl also has high light reflectivity and good thermal conductivity. However, it is insoluble and infusible, which prevents its practical utilization.

7-12 COPOLYMERIZATION

Considerable work has been published on the copolymerizations of various cyclic monomers reacting through the same type of functional group (e.g., two cyclic ethers or two cyclic siloxanes) or through different types of functional groups (e.g., copolymerization of a cyclic ether with a lactone), and also the copolymerizations of cyclic monomers with alkenes and other compounds. Some copolymerizations of cyclic monomers are of commercial importance. These include the various cyclic ether and acetal copolymerizations described previously (Sec. 7-2b-6). Copolymerization is used extensively for silicone polymers. Varying amounts of substituents such as phenyl or vinyl are introduced into the basic poly(dimethylsiloxane) structure to modify properties of the final product. Phenyl groups improve the low-temperature flexibility of silicone polymers by decreasing their low-temperature crystallinity. High-temperature performance and miscibility with organic solvents is also improved. Vinyl groups allow more efficient crosslinking by either hydrosilation or peroxides.

Although the literature contains much on copolymerizations of cyclic monomers, including data on monomer reactivity ratios, the reader is cautioned that most of the data are less reliable than corresponding data for radical copolymerizations of alkenes (Chap. 6). The situation is much more like that which occurs in ionic copolymerizations. Ring-opening copolymerizations are complicated in several ways. For copolymerizations proceeding by the activated monomer mechanism (anionic polymerizations of lactams and N-carboxy-α-amino acid anhydrides, possibly some cationic polymerizations of cyclic ethers), the "real" monomers are the activated monomers. The relative concentrations of the two activated monomers (e.g., the lactam anions from two lactams undergoing copolymerization) may be different from the relative concentrations of the monomers in the feed. Calculation of monomer reactivity ratios using the feed composition of comonomers will be incorrect.

Most ring-opening copolymerizations involve propagation–depropagation equilibria, which require that experimental data be handled in the appropriate manner (Sec. 6-5b), but this is rarely done. The situation is much more complicated in many systems because there are additional equilibria—between polymer and cyclic oligomer and between different polymer chains—which result in reshuffling of monomer units. The reshuffling equilibria can be avoided by using mild initiators, low conversions, and highly reactive (more strained) monomers. However, the situation in industrial practice usually involves conditions that result in near complete reshuffling of monomer units. The effect of reshuffling is evident in the anionic copolymerization of octamethylcyclotetrasiloxane (M_1) and 1,3,5,7-tetramethyl-1,3,5,7-tetravinylcyclotetrasiloxane (M_2) (feed ratio 9:1) by potassium silanolate at 130°C [Ziemelis and Saam, 1989]. The initial copolymer is very rich in M_2, and this monomer is exhausted early. Further reaction results not only in conversion of M_1 but also redistribution of the M_2 units. Silicon-29 NMR shows the M_2 units to be in blocks at short reaction times but statistically distributed at long reaction times.

Counterion effects similar to those in ionic chain copolymerizations of alkenes (Sec. 6-4) are observed in ring-opening polymerizations. Thus, copolymerizations of cyclopentene and norbornene with rhenium and ruthenium-based initiators yield copolymers very rich in norbornene, while a more reactive (less discriminating) tungsten-based initiator yields a copolymer with comparable amounts of the two comonomers [Irvin, 1987]. Monomer reactivity ratios are also sensitive to solvent and temperature.

Other complicating features of ring-opening polymerizations include the simultaneous operation of two different polymerization mechanisms; for example, NCA copolymerizations initiated by some secondary amines proceed with both the amine and activated monomer mechanisms. The monomer reactivity ratios for any pair of monomers are unlikely to be the same for the two different propagations. Any experimentally determined monomer reactivity ratios are each composites of two different r values. Polymer conformational effects on reactivity have been observed in NCA copolymerizations where the particular polymer chain conformation, which is usually solvent-dependent, results in different interactions with each monomer [Imanishi, 1984].

The initiator employed is especially important for copolymerizations between monomers containing different polymerizing functional groups. Basic differences in the propagating centers (carbocation, oxonium ion, imide anion, carbanion, etc.) for different types of monomers preclude some copolymerizations.

7-12a   Monomers with Same Functional Group

Copolymerizations between pairs of cyclic esters, acetals, sulfides, siloxanes, alkenes, lactams, lactones, N-carboxy-α-amino acid anhydrides, imines, and other cyclic monomers have been studied [Frisch and Reegan, 1969; Ivin, 1987; Ivin and Saegusa, 1984; Kendrick et al., 1989; Sebenda, 1989; Tomalia and Killat, 1985]. Copolymerizations have also been achieved between closely related types of monomers such as cyclic ethers with cyclic acetals (Sec. 7-2b-6) or lactones with lactides and cyclic carbonates [Keul et al., 1988; Kricheldorf et al., 1985].

The interpretation of experimental monomer reactivity ratios is subject to the cautions described above. When copolymerization proceeds with minimal complications, the r values can often be analyzed by considering the effects of ring size on the general tendency toward ring opening (see considerations in Sec. 2-5b) and the reactivity toward attack by the particular propagating species. For those polymerizations involving a cyclic propagating center (e.g., oxonium ions in cyclic ether polymerization), there is also the need to consider the effect of ring size on formation of the propagating center.

Some monomers with no tendency toward homopolymerization are found to have some (not high) activity in copolymerization. This behavior is found in cationic copolymerizations of tetrahydropyran, 1,3-dioxane, and 1,4-dioxane with 3,3-bis-

(chloromethyl)oxetane [Dreyfuss and Dreyfuss, 1969]. These monomers are formally similar in their unusual copolymerization behavior to the radical copolymerization behavior of sterically hindered monomers such as maleic anhydride, stilbene, and diethyl fumarate (Sec. 6-3b-3), but not for the same reason. The copolymerizability of these otherwise unreactive monomers is probably a consequence of the unstable nature of their propagating centers. Consider the copolymerization in which M_2 is the

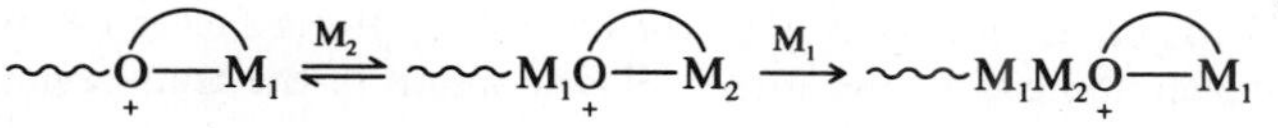

$$\sim\sim\underset{+}{O}-M_1 \overset{M_2}{\rightleftharpoons} \sim\sim M_1\underset{+}{O}-M_2 \overset{M_1}{\longrightarrow} \sim\sim M_1M_2\underset{+}{O}-M_1 \tag{7-128}$$

LXXXIV

cyclic monomer with no tendency to homopolymerize. In homopolymerization, the propagation–depropagation equilibrium for M_2 is completely toward monomer. However, with the second monomer M_1 present, the species **LXXXIV** adds M_1 before depropagation can occur and copolymerization of M_1 and M_2 results. Similar behavior has been observed for the comonomer pairs β-propiolactone–γ-butyrolactone and cyclopentene–cyclohexene [Ivin et al., 1979; Tada et al., 1964]. The second monomer in each pair does not homopolymerize, but copolymerization occurs.

The anionic copolymerization (using an acetylated initiator) of ε-caprolactam (M_1) and α-pyrrolidinone (M_2) illustrates the problem of interpreting r_1 and r_2 values obtained by employing the standard copolymerization equation (Eq. 6-12). The monomer reactivity ratios have been calculated as $r_1 = 0.75$ and $r_2 = 5.0$ with the conclusion that α-caprolactam is about eight times more reactive than ε-pyrrolidinone [Kobayashi and Matsuya, 1963]. However, a detailed study of the system indicates that the copolymer composition is not determined by the rate constants k_{11}, k_{12}, k_{22}, and k_{21} for the various homo- and cross-propagation reactions [Bar-Zakay et al., 1967]. Transacylation involving an exchange of the monomer segments in the initiating and prop-

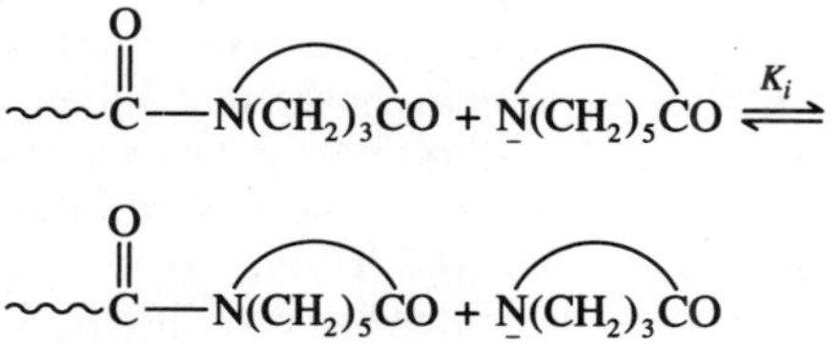

$$\sim\sim\overset{\overset{O}{\parallel}}{C}-N(CH_2)_3CO + \overset{\frown}{N}(CH_2)_5CO \overset{K_i}{\rightleftharpoons}$$

$$\sim\sim\overset{\overset{O}{\parallel}}{C}-N(CH_2)_5CO + \overset{\frown}{N}(CH_2)_3CO \tag{7-129}$$

agating imides, occurs at a much faster rate than any of the various propagation reactions. The copolymer composition is thus dependent on the transacylation equilibrium according to

$$\frac{d[M_1]}{d[M_2]} = \frac{K_i[M_1^-]}{[M_2^-]} \tag{7-130}$$

where $[M_1^-]$ and $[M_2^-]$ represent the concentrations of the lactam anions. The ratio $[M_1^-]/[M_2^-]$ is determined by the relative acidities of the two monomers by the equilibrium

$$M_2^- + M_1 \overset{K_a}{\rightleftharpoons} M_2 + M_1^- \tag{7-131}$$

Combining the equilibrium expression for Eq. 7-131 with Eq. 7-130 yields

$$\frac{d[M_1]}{d[M_2]} = \frac{K_a K_t [M_1]}{[M_2]} \tag{7-132}$$

where K_a and K_t have been independently determined as 0.4 and 0.3, respectively. The greater reactivity of α-pyrrolidone in the copolymerization with ϵ-caprolactam can then be attributed to the greater acidity of α-pyrrolidone (by a factor of about 2.5) and the greater nucleophilicity of the α-pyrrolidone anion in the transacylation equilibrium (by a factor of about 3.3).

7-12b Monomers with Different Functional Groups

Copolymerization between monomers containing different functional groups is selective. One of the cross-propagation steps in a contemplated copolymerization may be highly unfavorable because of the wide variations in the types of propagating centers involved and, often, in the energies of the bonds being broken during propagation. Copolymerization between certain types of monomers does not occur and that between others is difficult to achieve, while some copolymerizations occur with relative ease. Thus, the combination of a lactam with any other type of monomer, such as a lactone, epoxide, or alkene, is incompatible because of differences in their propagation mechanisms. For example, anionic copolymerization of a lactam with an epoxide does not occur since the anion derived from the epoxide terminates by proton abstraction from lactam monomer.

Epoxides readily undergo anionic copolymerization with lactones and cyclic anhydrides because the propagating centers are similar—alkoxide and carboxylate [Aida et al., 1985; Cherdron and Ohse, 1966; Inoue and Aida, 1989; Luston and Vass, 1984]. Most of the polymerizations show alternating behavior, with the formation of polyester, but the mechanism for alternation is unclear. There are few reports of cationic copolymerizations of lactones and cyclic ethers other than the copolymerizations of β-propiolactone with tetrahydrofuran and 3,3-bis(chloromethyl)oxetane, probably indicative of the difference in propagating centers (dioxocarbocation and oxonium ion, respectively) [Yamashita et al., 1966]. The copolymerizations tend toward ideal behavior with the product containing large amounts of the cyclic ether.

Anionic copolymerization of an episulfide and epoxide shows extreme ideal behavior with only small amounts of epoxide incorporated into the copolymer, since episulfides are much more reactive than epoxides. Cationic copolymerization does not occur to any extent because sulfonium ion centers do not add epoxide, although oxonium ion centers do add episulfide.

There are very few reported copolymerizations between cyclic monomers and carbon–carbon double-bond monomers. Such copolymerizations would require a careful selection of the monomers and reaction conditions to closely match the reactivities of the different monomers and propagating centers. The almost complete absence of successes indicates that the required balancing of reactivities is nearly impossible to achieve. There are a few reports of copolymerizations between carbon–carbon double-bond monomers and cyclic ethers or acetals [Higashimura et al., 1967; Inoue and Aida, 1984; Yamashita et al., 1966].

Some copolymerizations have been studied where one of the reactants is a compound not usually considered as a monomer. These include copolymerizations of epoxides and higher cyclic ethers with carbon dioxide, episulfides with carbon dioxide and carbon disulfide, and epoxides with sulfur dioxide [Aida et al., 1986; Baran et al., 1984; Inoue and Aida, 1989; Soga et al., 1977]. The copolymers are reported to be either 1:1 alternating copolymers or contain 1:1 alternating sequences together with blocks of the cyclic monomer.

7-12c Block Copolymers

A wide range of different block copolymers have been synthesized by using the sequential addition method applicable to living polymer systems. These include various copolymers in which the different blocks are derived from two different cyclic monomers containing the same functional group, such as two cycloalkenes, cyclic ethers, lactones, N-carboxy-α-amino acid anhydrides, and episulfides [Cannizzo and Grubbs, 1988; Dreyfuss and Dreyfuss, 1989; Greene et al., 1988; Imanishi, 1984; Inoue and Aida, 1989; Jedlinski et al., 1987; Sigwalt and Spassky, 1984; Yasuda et al., 1984]. There are very few reports of block copolymers of lactams or cyclic siloxanes produced by sequential addition, probably indicative of the exchange reactions that occur in polymerizations of these monomers.

The sequential addition method also allows the synthesis of many different block copolymers in which the two monomers have different functional groups, such as epoxide with lactone, lactide or cyclic anhydride, cyclic ether with 2-methyl-2-oxazoline, imine or episulfide, lactone with lactide or cyclic carbonate, and cycloalkene with acetylene [Aida et al., 1985; Dreyfuss and Dreyfuss, 1989; Farren et al., 1989; Inoue and Aida, 1989; Keul et al., 1988; Kobayashi et al., 1990a, 1990b, 1990c; Yasuda et al., 1984].

Further, there have been a number of successes in synthesizing block copolymers by sequential polymerization of a cyclic monomer and an alkene monomer. Examples include the combination of methyl methacrylate with epoxide, episulfide or lactone, polystyrene with hexamethylcyclotrisiloxane, lactone or cyclic carbonate, and 2-isopropenylnaphthalene with hexamethylcyclotrisiloxane [Keul and Hocker, 1986; Krause et al., 1982; Kuroki et al., 1988a, 1988b; Rhein and Schulz, 1985; Sigwalt and Spassky, 1984]. It turns out to be much easier to obtain block copolymers from these combinations than to obtain the corresponding statistical copolymers. Synthesis of a statistical copolymer requires that there not be a large imbalance between the various homo- and cross-propagation rates. Synthesis of a block copolymer by sequential addition requires that the system be living with one or the other cross-propagation proceeding at a reasonable rate. Thus, anionic copolymerization of styrene with ethylene oxide does not occur because the addition of alkoxide ion to the carbon–carbon double bond and homopropagation of polystyryl anion are much less favored relative to addition of polystyryl anion to ethylene oxide and homopropagation of alkoxide anion. However, block copolymerization is easily achieved by adding ethylene oxide to living polystyrene. Triblock polymers can be obtained in some systems by using appropriate difunctional initiators.

Other routes to block copolymers include the use of a telechelic homopolymer with appropriate end group to initiate the polymerization of the second monomer. For example, polystyrene containing terminal amino groups, synthesized by radical polymerization using a bis-amino azo initiator, is used to initiate the polymerization of an N-carboxy-α-amino acid anhydride by the amine mechanism [Janssen et al., 1988]. Polystyrene and polysiloxane with p-toluenesulfonyl ester end groups initiate the homopolymerizations of 2-methyl-2-oxazoline and N-t-butylaziridine, respectively [Ishizu et al., 1985; Kazama et al., 1988]. Coupling reactions have also been studied for synthesizing block copolymer.

7-12d Zwitterion Polymerization

Certain combinations of nucleophiles and electrophiles undergo polymerization without the need for initiator [Kobayashi and Saegusa, 1985; Saegusa, 1977, 1979, 1981]. The polymerization, referred to as *zwitterion polymerization*, proceeds via zwitterion intermediates. For example, polymerization between 2-oxazoline and β-propiolactone involves nucleophilic attack of 2-oxazoline on β-propiolactone to form the dimer zwitterion **LXXXV**, which reacts with itself to form the tetramer zwitterion **LXXXVI**. The latter reacts with itself and **LXXXV** to form octamer and hexamer zwitterions,

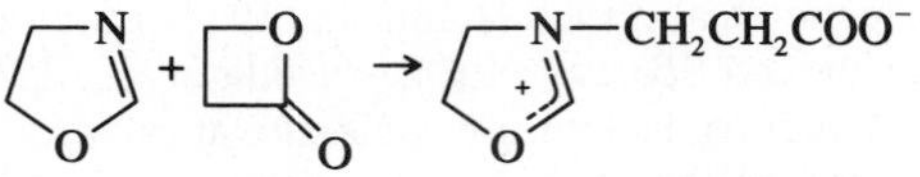

LXXXV

↓

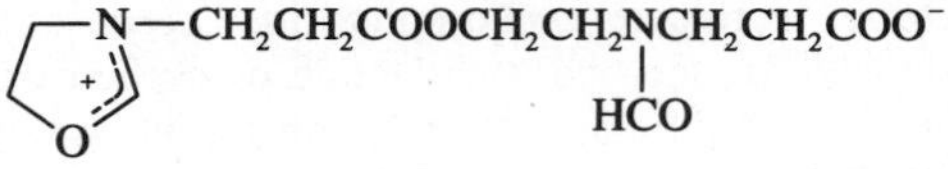

LXXXVI

↓

$$\left[\begin{array}{c} -CH_2CH_2NCH_2COO- \\ | \\ HCO \end{array}\right]_n \qquad (7\text{-}133)$$

respectively. Propagation continues in this manner as various-sized zwitterion react with each other to form larger zwitterions.

Among the nucleophilic monomers studied are 5,6-dihydro-4H-1,3-oxazines, cyclic phosphites and phosphonites, iminodioxolanes, and imines; electrophilic monomers include cyclic anhydrides, lactones and sultones as well as acyclic compounds such as acrylic acid and acrylamide. With only a very few exceptions, these polymerizations do not yield polymer molecular weights higher than 1000–3000. Low molecular weights are a consequence of the low concentration of the reacting species, the zwitterions, coupled with facile termination of the zwitterion + and − centers by reaction with the nucleophilic and electrophilic monomers [Odian and Gunatillake, 1984; Odian et al., 1990]. High molecular weights are obtained in zwitterion polymerizations only when the zwitterion concentrations are high relative to other materials present that can act as terminating agents. This has been successfully achieved in several systems. One of the systems (Eq. 7-98) has a low concentration of zwitterion intermediate, but there are no materials present (such as the electrophilic and nucleophilic monomers) to act as terminating agents. The other system (Eq. 7-100) contains only zwitterion since the zwitterion is isolated and is stable. Polymerization occurs when the zwitterion is heated.

REFERENCES

Afshar-Taromi, F., M. Scheer, P. Rempp, and E. Franta, *Makromol. Chem.*, **179**, 849 (1978).

Aida, T., M. Ishikawa, and S. Inoue, *Macromolecules*, **19**, 8 (1986).

Aida, T., K. Kawaguchi, and S. Inoue, *Macromolecules*, **23**, 3887 (1990).

Aida, T., Y. Maekawa, S. Asano, and S. Inoue, *Macromolecules*, **21**, 1195 (1988).

Aida, T., K. Sanuki, and S. Inoue, *Macromolecules*, **18**, 1049 (1985).

Allcock, H. R., *J. Macromol. Sci. Revs. Macromol. Chem.*, **C4**, 149 (1970).

Allcock, H. R., *Science*, **193**, 1214 (1976); *Makromol. Chem. Macromol. Symp.*, **6**, 101 (1986).

Allcock, H. R., "Polyphosphazenes," pp. 31–42 in "Encyclopedia of Polymer Science and Engineering," Vol. 13, 2nd ed., H. F. Mark, N. M. Bikales, C. G. Overberger, and G. Menges, Eds., Wiley-Interscience, New York, 1988.

Allcock, H. R. and M. S. Connolly, *Macromolecules*, **18**, 1330 (1985).

Amass, A. J., "Metathesis Polymerization: Chemistry," Chap. 6 in "Comprehensive Polymer Science," Vol. 4, G. C. Eastmond, A. Ledwith, S. Russo, and P. Sigwalt, Eds., Pergamon Press, London, 1989.

Amass, A. J., M. Lotfipour, J. A. Zurimendi, B. J. Tighe, and C. Thompson, *Makromol. Chem.*, **188**, 2121 (1987).

Andrzejewska, E., A. Zuk, and J. Garbarczyk, *J. Polym. Sci. Polym. Chem. Ed.*, **26**, 3151 (1988).

Bailey, W. J., *J. Macromol. Sci.-Chem.*, **9**, 849 (1975).

Bailey, W. J., *Makromol. Chem. Suppl.*, **13**, 171 (1985).

Bailey, W. J., J. L. Chou, P.-Z. Feng, B. Issari, V. Kuruganti, and L.-L. Zhou, *J. Macromol. Sci.-Chem.*, **A25**, 781 (1988).

Bailey, W. J. and T. Endo, *J. Polym. Sci. Polym. Symp.*, **64**, 17 (1978).

Baklouti, M., R. Chaabouni, J. Sledz, and F. Schue, *Polym. Bull.*, **21**, 243 (1989).

Ballard, D. G. H., C. H. Bamford, and A. Elliot, *Makromol. Chem. Suppl. 2*, **35**, 222 (1960).

Bamford, C. H. and H. Block, "The Polymerization of *N*-Carboxy-α-Amino Acid Anhydrides," Chap. 8 in "Comprehensive Chemical Kinetics," Vol. 15, C. H. Bamford and C. F. H. Tipper, Eds., Elsevier, New York, 1976.

Bansleban, D. A., M. J. Hersman, and O. Vogl, *J. Polym. Sci. Polym. Chem. Ed.*, **22**, 2489 (1984).

Baran, T., K. Brzezinska, K. Matyjaszewski, and S. Penczek, *Makromol. Chem.*, **184**, 2497 (1983).

Baran, T., A. Duda, and S. Penczek, *J. Polym. Sci. Polym. Chem. Ed.*, **22**, 1085 (1984).

Baran, T., P. Klosinski, and S. Penczek, *Makromol. Chem.*, **190**, 1903 (1989).

Bar-Zakay, S., M. Levy and D. Vofsi, *J. Polym. Sci.*, **A-1**(5), 965 (1967).

Bednarek, M., P. Kubisa, and S. Penczek, *Makromol. Chem. Suppl.*, **15**, 49 (1989).

Bero, M., J. Kasperczyk, and Z. J. Jedlinski, *Makromol. Chem.*, **191**, 2287 (1990).

Bertalan, G., I. Rusznak, and P. Anna, *Makromol. Chem.*, **185**, 1285 (1984).

Bertalan, G., I. Rusznak, P. Anna, M. Boros-Ivicz, and G. Marosi, *Polym. Bull.*, **19**, 539 (1988a).

Bertalan, G., T. T. Nagy, P. Valko, A. Boros, M. Boros-Ivicz, and P. Anna, *Polym. Bull.*, **19**, 547 (1988b).

Beste, L. F. and H. K. Hall, Jr., *J. Phys. Chem.*, **68**, 269 (1964).

Biedron, T., R. Szymanski, P. Kubisa, and S. Penczek, *Makromol. Chem. Macromol. Symp.*, **32**, 155 (1990).

Body, R. W. and V. L. Kyllinstad, "Polyether Elastomers," pp. 307–322 in "Encyclopedia of

Polymer Science and Engineering," Vol. 6, 2nd ed., H. F. Mark, N. M. Bikales, C. G. Overberger, and G. Menges, Eds., Wiley-Interscience, New York, 1986.

Boileau, S., "Anionic Ring-Opening Polymerizations; Epoxides and Episulfides," pp. 467–487 in "Comprehensive Polymer Science," Vol. 3, G. C. Eastmond, A. Ledwith, S. Russo, and P. Sigwalt, Eds., Pergamon Press, London, 1989.

Bostick, E. E., "Cyclic Siloxanes and Silazanes," Chap. 8 in "Ring-Opening Polymerization," K. C. Frisch and S. L. Reegen, Eds., Marcel Dekker, New York, 1969.

Bots, J. G., L. van der Does, A. Bantjes, and J. Boersma, *Makromol. Chem.*, **188**, 1665 (1987).

Brunelle, D. J., T. L. Evans, T. G. Shannon, E. P. Boden, K. R. Stewart, L. P. Fontana, and D. K. Bonauto, *Polym. Prep.*, **30**(2), 569 (1989).

Bucquoye, M. and E. J. Goethals, *Makromol. Chem.*, **179**, 168 (1978).

Busfield, W. K., "Heats and Entropies of Polymerization, Ceiling Temperatures, Equilibrium Monomer Concentrations; and Polymerizability of Heterocyclic Compounds," Chap. II, pp. 295–334 in "Polymer Handbook," 2nd ed., J. Brandrup and E. H. Immergut, Eds., Wiley-Interscience, New York, 1989.

Cannizzo, L. F. and R. H. Grubbs, *Macromolecules*, **21**, 1961 (1988).

Chapiro, A. and S. Penczek, *J. Chim. Phys.*, **59**, 696 (1962).

Cherdron, V. H. and H. Ohse, *Makromol. Chem.*, **92**, 213 (1966).

Cherdron, H., K. Burg, and F. Kloos, *Makromol. Chem. Macromol. Symp.*, **13/14**, 289 (1988).

Chien, J. C. W., Y.-G. Cheun, and C. P. Lillya, *Macromolecules*, **21**, 870 (1988).

Chojnowski, J. and L. Wilczek, *Makromol. Chem.*, **180**, 117 (1979).

Clinton, N. and P. Matlock, "1,2-Epoxide Polymers," pp. 225–273 in "Encyclopedia of Polymer Science and Engineering," Vol. 6, 2nd ed., H. F. Mark, N. M. Bikales, C. G. Overberger, and G. Menges, Eds., Wiley-Interscience, New York, 1986.

Colak, N. and K. Alyuruk, *Polymer*, **30**, 1709, (1989).

Collins, G. L., R. K. Greene, F. M. Berardinelli, and W. V. Garruto, *J. Polym. Sci. Polym. Lett. Ed.*, **17**, 667 (1979).

Costa, G., E. Pedemonte, S. Russo, and E. Sava, *Polymer*, **20**, 713 (1979).

Cubbon, R. C. P., *Makromol. Chem.*, **80**, 44 (1964).

Dainton, F. S. and K. J. Ivin, *Quart. Rev. (London)*, **12**, 61 (1958).

DiSilvestro, G., P. Sozzani, S. Bruckner, L. Malpezzi, and C. Guaita, *Makromol. Chem.*, **188**, 2745 (1987).

Doherty, M., A. Siove, A. Parlier, H. Rudler, and M. Fontanille, *Makromol. Chem. Macromol. Symp.*, **6**, 33 (1986).

Dolce, T. J. and J. A. Grates, "Acetal Resins," pp. 42–61 in "Encyclopedia of Polymer Science and Engineering," Vol. 1, 2nd ed., H. F. Mark, N. M. Bikales, C. G. Overberger, and G. Menges, Eds., Wiley-Interscience, New York, 1985.

Dreyfuss, M. P. and P. Dreyfuss, *J. Polym. Sci.*, **A-1**(4), 2179 (1966).

Dreyfuss, P. and M. P. Dreyfuss, "1,3 Epoxides and Higher Epoxides," Chap. 2 in "Ring-Opening Polymerization," K. C. Frisch and S. L. Reegan, Eds., Marcel Dekker, New York, 1969.

Dreyfuss, P. and M. P. Dreyfuss, "Polymerization of Cyclic Ethers and Sulphides," Chap. 4 in "Comprehensive Chemical Kinetics," Vol. 15, C. H. Bamford and C. F. H. Tipper, Eds., Elsevier, New York, 1976.

Dreyfuss, M. P. and P. Dreyfuss, "Oxetane Polymers," pp. 653–670 in "Encyclopedia of Polymer Science and Engineering," Vol. 10, 2nd ed., H. F. Mark, N. M. Bikales, C. G. Overberger, and G. Menges, Eds., Wiley-Interscience, New York, 1987.

Dreyfuss, P. and M. P. Dreyfuss, "Cationic Ring-Opening Polymerization: Copolymerization," Chap. 53 in "Comprehensive Polymer Science," Vol. 3, G. C. Eastmond, A. Ledwith, S. Russo, and P. Sigwalt, Eds., Pergamon Press, London, 1989.

Dreyfuss, P., M. P. Dreyfuss, and G. Pruckmayr, "Tetrahydrofuran Polymers," pp. 649–681 in "Encyclopedia of Polymer Science and Engineering," Vol. 16, 2nd ed., H. F. Mark, N. M. Bikales, C. G. Overberger, and G. Menges, Eds., Wiley-Interscience, New York, 1989.

Duda, A., Z. Florjanczyk, A. Hofman, S. Stomkowski, and S. Penczek, *Macromolecules*, **23**, 1640 (1990).

Endo, T., T. Aida, and S. Inoue, *Macromolecules*, **20**, 2982 (1987a).

Endo, T. and W. J. Bailey, *J. Polym. Sci. Polym. Lett. Ed.*, **13**, 193 (1975); **18**, 25 (1980).

Endo, T., S. Maruoka, T. Yokozawa, *Marcomolecules*, **20**, 2690 (1987b).

Endo, T. and K. Suga, *J. Polym. Sci. Polym. Chem. Ed.*, **27**, 1831 (1989).

Endo, T., M. Watanabe, K. Suga, and T. Yokozawa, *Makromol. Chem.*, **190**, 691 (1989).

Endo, T., N. Yako, K. Azuma, and K. Nate, *Makromol. Chem.*, **186**, 1543 (1985).

Farren, T. R., A. J. Amass, M. S. Beevers, and J. A. Stowell, *Polymer*, **30**, 1008 (1989).

Feast, W. J., "Metathesis Polymerization: Applications," Chap. 7 in "Comprehensive Polymer Science," Vol. 4, G. C. Eastmond, A. Ledwith, S. Russo, and P. Sigwalt, Eds., Pergamon Press, London, 1989.

Finke, H. L., D. W. Scott, M. E. Gross, and G. Waddington, *J. Am. Chem. Soc.*, **78**, 5469 (1956).

Fries, T., J. Belohlavkova, O. Novakova, J. Roda, and J. Kralicek, *Makromol. Chem.*, **188**, 239 (1987).

Frisch, K. C. and S. L. Reegan, Eds., "Ring-Opening Polymerization," Marcel Dekker, New York, 1969.

Gagnon, S. D., "Propylene Oxide and Higher 1,2-Epoxide Polymers," pp. 273–307 in "Encyclopedia of Polymer Science and Engineering," Vol. 6, 2nd ed., H. F. Mark, N. M. Bikales, C. G. Overberger, and G. Menges, Eds., Wiley-Interscience, New York, 1986.

Gandini, A. and H. Cheradame, "Cationic Polymerization," pp. 729–814 in "Encyclopedia of Polymer Science and Engineering," Vol. 2, 2nd ed., H. F. Mark, N. M. Bikales, C. G. Overberger, and G. Menges, Eds., Wiley-Interscience, New York, 1985.

Gee, G., W. C. E. Higginson, and G. T. Merrall, *J. Chem. Soc.*, 1345 (1959).

Gee, G., W. C. E. Higginson, K. J. Taylor, and M. W. Trenholme, *J. Chem. Soc.*, 4298 (1961).

Giannakidis, D. and H. J. Harwood, *J. Polym. Sci. Polym. Lett. Ed.*, **16**, 491 (1978).

Goethals, E. J., *Adv. Polym. Sci.*, **23**, 103 (1977).

Goethals, E. J., "Cyclic Amines," Chap. 10 in "Ring-Opening Polymerization," Vol. 2, K. J. Ivin and T. Saegusa, Eds., Elsevier, London, 1984.

Goethals, E. J., "Cationic Ring-Opening Polymerization: Amines and N-Containing Heterocycles," Chap. 52 in "Comprehensive Polymer Science," Vol. 3, G. C. Eastmond, A. Ledwith, S. Russo, and P. Sigwalt, Eds., Pergamon Press, London, 1989a.

Goethals, E. J., "Cationic Ring-Opening Polymerization: Sulfides," Chap. 51 in "Comprehensive Polymer Science," Vol. 3, G. C. Eastmond, A. Ledwith, S. Russo, and P. Sigwalt, Eds., Pergamon Press, London, 1989b.

Good, F. J. Jr. and C. Schuerch, *Macromolecules*, **18**, 595 (1985).

Goodman, M., E. Peggion, M. Szwarc, and C. H. Bamford, *Macromolecules*, **10**, 1299 (1977).

Greene, R. M. E., K. J. Ivin, J. J. Rooney, J. Kress, and J. A. Osborn, *Makromol. Chem.*, **189**, 2797 (1988).

Gross, R. A., Y. Zhang, G. Konrad, and R. W. Lenz, *Macromolecules*, **21**, 2657 (1988).

Grubbs, R. H., D. A. Dougherty, and T. M. Swager, *J. Am. Chem. Soc.*, **110**, 2973 (1988).

Gunatillake, P. A., G. Odian, and D. A. Tomalia, *Macromolecules*, **21**, 1556 (1988).

Hagiwara, T., M. Ishimori, and T. Tsuruta, *Makromol. Chem.*, **182**, 501 (1981).

Hall, H. K., Jr., *J. Amer. Chem. Soc.*, **80**, 6404 (1958).

Hall, H. K., Jr. and L. G. Snow, "Ring-Opening Polymerization via Carbon–Carbon σ-Bond Cleavage," Chap. 2 in Vol. 1, K. J. Ivin and T. Saegusa, Eds., Elsevier, London, 1984.

Hasebe, Y. and T. Tsuruta, *Makromol. Chem.*, **189**, 1915 (1988).

Hauser, M., "Alkylenimines," Chap. 5 in "Ring-Opening Polymerization," K. C. Frisch and S. L. Reegen, Eds., Marcel Dekker, New York, 1969.

Hermans, P. H., D. Heikens, and P. F. van Velden, *J. Polym. Sci.*, **30**, 81 (1958); **44**, 437 (1960).

Herweh, J. E., *J. Polym. Sci. Polym. Chem. Ed.*, **27**, 333 (1989).

Higashimura, T., A. Tanaka, T. Miki, and S. Okamura, *J. Polym. Sci.*, **A-1**(5), 1927, 1937 (1967).

Hirasawa, T., M. Okada, and H. Sumitomo, *Macromolecules*, **21**, 1566 (1988).

Hirota, M. and H. Fukuda, *Makromol. Chem.*, **188**, 2259 (1987).

Hofman, A., S. Slomkowski, and S. Penczek, *Makromol. Chem.*, **188**, 2027 (1987a); *Makromol. Chem. Rapid Commun.*, **8**, 387 (1987b).

Hofman, A., R. Szymanski, S. Slomkowski, and S. Penczek, *Makromol. Chem.*, **185**, 655 (1985).

Imanishi, Y., "*N*-Carboxyanhydrides," Chap. 8 in Ring-Opening Polymerization," Vol. 1, K. J. Ivin and T. Saegusa, Eds., Elsevier, London, 1984.

Imanishi, Y., A. Aoyama, Y. Hashimoto, and T. Higashimura, *Biopolymers*, **16**, 187 (1977).

Inoue, S., *J. Macromol. Sci.-Chem.*, **A25**, 571 (1988).

Inoue, S. and T. Aida, "Cyclic Ethers," Chap. 4 in "Ring-Opening Polymerization," Vol. 1, K. J. Ivin and T. Saegusa, Eds.; Elsevier, London, 1984.

Inoue, S. and T. Aida, "Anionic Ring-Opening Polymerization: Copolymerization," Chap. 37 in "Comprehensive Polymer Science," Vol. 3, G. C. Eastmond, A. Ledwith, S. Russo, and P. Sigwalt, Eds., Pergamon Press, London, 1989.

Ishii, Y. and S. Sakai, "1,2 Epoxides," Chap. 2 in "Ring-Opening Polymerization," K. C. Frisch and S. L. Reegen, Eds., Marcel Dekker, New York, 1969.

Ishizu, K., S. Ishikawa, and T. Fukutomi, *J. Polym. Sci. Polym. Chem. Ed.*, **23**, 445 (1985).

IUPAC, *Pure Appl. Chem.*, **48**, 375 (1976); **57**, 149 (1985).

Ivin, K. J., "Cycloalkenes and Bicycloalkenes," Chap. 3 in "Ring-Opening Polymerization," Vol. 1, K. J. Ivin and T. Saegusa, Eds., Elsevier, London, 1984.

Ivin, K. J., "Metathesis Polymerization," pp. 634–668 in "Encyclopedia of Polymer Science and Engineering," Vol. 9, 2nd ed., H. F. Mark, N. M. Bikales, C. G. Overberger, and G. Menges, Eds., Wiley-Interscience, New York, 1987.

Ivin, K. J., G. Lapienis, J. J. Rooney, and C. D. Stewart, *Polymer*, **20**, 1308 (1979).

Ivin, K. J. and T. Saegusa, Eds., "Ring-Opening Polymerization," Vols. 1, 2, Elsevier, London, 1984.

Janssen, K., M. Van Beylen, and C. Samyn, *Polymer*, **29**, 1513 (1988).

Jedlinski, Z., A. Dworak, and M. Bero, *Makromol. Chem.*, **180**, 949 (1979).

Jedlinski, Z., M. Kowalczuk, P. Kurcok, L. Brzoskowski, and J. Franek, *Makromol. Chem.*, **188**, 1575 (1987).

Jedlinski, Z., A. Wolinska, and J. Lukaszcsyk, *Macromolecules*, **18**, 1648 (1985).

Jerome, R. and P. Teyssie, "Anionic Ring-Opening Polymerization: Lactones," Chap. 34 in "Comprehensive Polymer Science," Vol. 3, G. C. Eastmond, A. Ledwith, S. Russo, and P. Sigwalt, Eds., Pergamon Press, London, 1989.

Johns, D. B., R. W. Lenz, and A. Luecke, "Lactones," Chap. 7 in "Ring-Opening Polymerization," Vol. 1, K. J. Ivin and T. Saegusa, Eds., Elsevier, London, 1984.

Karger-Kocsis, J. and L. Kiss, *Makromol. Chem.,* **180**, 1593 (1979).

Kasperczyk, J. and Z. J. Jedlinski, *Macromolecules*, **187**, 2215 (1986).

Kawakami, Y. and Y. Yamashita, *Macromolecules*, 10, 837 (1977a); *J. Polym. Sci. Polym. Lett. Eds.*, **15**, 214 (1977b); *Macromolecules*, **12**, 399 (1979).

Kazama, H., Y. Tezuka, K. Imai, and E. J. Goethals, *Makromol. Chem.*, **189**, 985 (1988).

Kendrick, T. C., B. M. Parbhoo, and J. W. White, "Polymerization of Cyclosiloxanes," Chap. 25 in "Comprehensive Polymer Science," Vol. 4, G. C. Eastmond, A. Ledwith, S. Russo, and P. Sigwalt, Eds., Pergamon Press, London, 1989.

Kern, W., H. Deibig, A. Geifer, and V. Jaacks, *Pure Appl. Chem.*, **12**, 371 (1966).

Keul, H. and H. Hocker, *Makromol. Chem.*, **187**, 2833 (1986).

Keul, H., H. Hocker, E. Leitz, K.-H. Ott, and L. Morbitzer, *Makromol. Chem.*, **189**, 2303 (1988).

Kim, K. Y., T. Komoto, and T. Kawai, *Makromol. Chem.*, **180**, 465 (1979).

Klavetter, F. L. and R. H. Grubbs, *J. Am. Chem. Soc.*, **110**, 7807 (1988).

Kobayashi, F. and K. Matsuya, *J. Polym. Sci.*, **A1**, 111 (1963).

Kobayashi, S., H. Danda, and T. Saegusa, *Macromolecules*, **7**, 415 (1974).

Kobayashi, S., K. Morikawa, and T. Saegusa, *Macromolecules*, **8**, 386 (1975).

Kobayashi, S. and T. Saegusa, "Cyclic 1,3-Oxaza Compounds," Chap. 11 in "Ring-Opening Polymerization," Vol. 2, Ivin, K. J. and T. Saegusa, Eds., Elsevier, London, 1984.

Kobayashi, S. and T. Saegusa, "Alternating Copolymerization Involving Zwitterions," Chap. 22 in "Alternating Copolymers," J. M. G. Cowie, Ed., Plenum Press, New York, 1985.

Kobayashi, S. and T. Saegusa, "Cyclic Imino Ethers," Polymerization," pp. 525–537 in "Encyclopedia of Polymer Science and Engineering," Vol. 4, 2nd ed., H. F. Mark, N. M. Bikales, C. G. Overberger, and G. Menges, Eds., Wiley-Interscience, New York, 1986.

Kobayashi, S., M. Suzuki, and T. Saegusa, *Macromolecules*, **19**, 462 (1986).

Kobayashi, S., Y. Tsukamoto, and T. Saegusa, *Macromolecules*, **23**, 2609 (1990a).

Kobayashi, S., H. Uyama, and T. Saegusa, *Macromolecules*, **23**, 352 (1990b).

Kobayashi, S., H. Uyama, E. Ihara, and T. Saegusa, *Macromolecules*, **23**, 1586 (1990c).

Kobayashi, S., H. Uyama, M. Ogaki, T. Yoshida, and T. Saegusa, *Macromolecules*, **22**, 4412 (1989).

Kops, J. and H. Spangaard, *Makromol. Chem.*, **176**, 299 (1975).

Kops, J. and H. Spangaard, *Macromolecules*, **15**, 1225 (1982).

Krause, S., M. Iskandar, and M. Iqbal, *Macromolecules*, **15**, 105 (1982).

Kricheldorf, H. R., *Makromol. Chem. Macromol. Symp.*, **6**, 165 (1986).

Kricheldorf, H. R., "Anionic Ring-Opening Polymerization of *N*-Carboxyanhydrides," Chap. 36 in "Comprehensive Polymer Science," Vol. 3, G. C. Eastmond, A. Ledwith, S. Russo, and P. Sigwalt, Eds., Pergamon Press, London, 1989.

Kricheldorf, H. R., M. Berl, and N. Scharnagl, *Macromolecules*, **21**, 286 (1988).

Kricheldorf, H. R., R. Dunsing, and A. S. Albet, *Makromol. Chem.*, **188**, 2453 (1987a).

Kricheldorf, H. R., R. Dunsing, and A. Serra, *Macromolecules*, **20**, 2050 (1987b).

Kricheldorf, H. R., J. M. Jonte, and R. Dunsing, *Makromol. Chem.*, **187**, 771 (1986).

Kricheldorf, H. R. and I. Kreiser-Saunders, *Makromol. Chem.*, **191**, 1057 (1990).

Kricheldorf, H. R., I. Kreiser-Saunders, and N. Scharnagl, *Makromol. Chem. Macromol. Symp.*, **32**, 285 (1990).

Kricheldorf, H. R., T. Mang, and J. M. Jonte, *Makromol. Chem.*, **186**, 955 (1985).

Krouse, S. A. and R. R. Schrock, *Macromolecules*, **22**, 2569 (1989).

Kricheldorf, H. R. and M.-V. Sumbel, *Makromol. Chem. Macromol. Symp.*, **13/14**, 81 (1988a); *Makromol. Chem.*, **189**, 317 (1988b).

Kuhling, S., H. Keul, and H. Hocker, *Makromol. Chem.*, *Suppl.*, **15**, 9 (1989).

Kuntz, I., *J. Polym. Sci.*, **A-1**(5), 193 (1967).

Kuroki, M., T. Aida, and S. Inoue, *Makromol. Chem.*, **189**, 1305 (1988a).

Kuroki, M., S. Nashimoto, T. Aida, and S. Inoue, *Macromolecules*, **21**, 3114 (1988b).

Kuskova, M., J. Roda, and J. Kralicek, *Makromol. Chem.*, **179**, 337 (1978).

Labes, M. M., P. Love, and L. F. Nichols, *Chem. Rev.*, **79**, 1 (1979).

Lahti, P. M., C. P. Lillya, and J. C. W. Chien, *Macromolecules*, **23**, 1214 (1990).

Laitinen, R. S., T. A. Padkanen, and R. S. Steudel, *J. Am. Chem. Soc.*, **109**, 710 (1987).

Lapienis, G. and S. Penczek, "Cyclic Compounds Containing Phosphorus Atoms in the Ring," Chap. 13 in "Ring-Opening Polymerization," Vol. 2, K. J. Ivin and T. Saegusa, Eds., Elsevier, London, 1984.

Le Borgne, A. and N. Spassky, *Polymer*, **30**, 2312 (1989).

Lee, C. L. and O. K. Johannson, *J. Polym. Sci.*, **A-1**(4), 3013 (1966); *J. Polym. Sci. Polym. Chem. Ed.*, **14**, 729 (1976).

Leenslag, J. W. and A. J. Pennings, *Makromol. Chem.*, **188**, 1809 (1987).

Leymet, I. and A. Siove, *Makromol. Chem.*, **190**, 2397 (1989).

Libiszowski, J., R. Szymanski, and S. Penczek, *Makromol. Chem.*, **190**, 1225 (1989).

Liu, H. Q. and V. T. Stannett, *Macromolecules*, **23**, 140 (1990).

Lowry, T. H. and K. S. Richardson, "Mechanism and Theory in Organic Chemistry," 3rd ed., Harper and Row, New York, 1987, p. 715.

Lu, N., G. L. Collins, and N.-L. Yang, *Makromol. Chem. Macromol. Symp.*, **42/43**, 425 (1991).

Lundberg, R. D. and E. F. Cox, "Lactones," Chap. 6 in "Ring-Opening Polymerization," K. C. Frisch and S. L. Reegen, Eds., Marcel Dekker, New York, 1969.

Luston, J. and F. Vass, *Adv. Polym. Sci.*, **56**, 91 (1984).

Majumdar, A. N., S. G. Scott, R. L. Merker, and J. H. Magill, *Makromol. Chem.*, **190**, 2293 (1989); *Macromolecules*, **23**, 14 (1990).

Majury, T. G., *J. Polym. Sci.*, **31**, 383 (1958).

Malkin, A. Ya., S. L. Ivanova, V. G. Frolov, A. N. Ivanova, and Z. S. Andrianova, *Polymer*, **23**, 1791 (1982).

Matyjaszewski, K., *J. Polym. Sci. Polym. Chem. Ed.*, **22**, 29 (1984a).

Matyjaszewski, K., *Makromol. Chem.*, **185**, 37, 51 (1984b).

Matyjaszewski, K., S. Slomkowski, and S. Penczek, *J. Polym. Sci. Polym. Chem. Ed.*, **17**, 69, 2413 (1979).

Matyjaszewski, K., R. Szymanski., P. Kubisa, and S. Penczek, *Acta Polym.*, **35**, 14 (1984).

Matyjaszewski, K., M. Zielinski, P. Kubisa, S. Stomkowski, J. Chojnowski, and S. Penczek, *Makromol. Chem.*, **181**, 1469 (1980).

Mazurek, M. and J. Chojnowski, *Makromol. Chem.*, **10**, 1005 (1977).

Mazurek, M., M. Scibiorek, J. Chojnowski, B. G. Zavin, and A. A. Zdanov, *Eur. Polym. J.*, **16**, 57 (1980).

McKenna, J. M., T. K. Wu, and G. Pruckmayr, *Macromolecules*, **10**, 877 (1977).

Meerwein, H., D. Delfs, and H. Morshel, *Angew. Chem.*, **72**, 927 (1960).

Meyer, B., *Chem. Rev.*, **76**, 367 (1976).

Mijangos, F. and L. M. Leon, *J. Polym. Sci. Polym. Lett. Ed.*, **21**, 885 (1983).

Moore, J. A., J. E. Kelly, D. N. Harpp, and T. G. Back, *Macromolecules*, **10**, 718 (1977).

Muhlbach and R. C. Schulz, *Makromol. Chem.*, **189**, 1267 (1988).

Muñoz-Escalona, A., *Makromol. Chem.*, **179**, 219, 1083 (1978).

Neilson, R. H. and P. Wisian-Neilson, *Chem. Rev.*, **88**, 541 (1988).

Noll, W., "Chemistry and Technology of Silicones," Academic Press, New York, 1968.

Odian, G. and P. A. Gunatillake, *Macromolecules*, **17**, 1297 (1984).

Odian, G., M. P. O'Callaghan, C.-K. Chien, P. Gunatillake, M. Periyasamy, and D. L. Schmidt, *Macromolecules*, **23**, 918 (1990).

Ofstead, E. A., "Polyalkenamers," pp. 287–314 in "Encyclopedia of Polymer Science and Engineering," Vol. 11, 2nd ed., H. F. Mark, N. M. Bikales, C. G. Overberger, and G. Menges, Eds., Wiley-Interscience, New York, 1988.

Oguni, N., S. Watanabe, M. Maki, and H. Tani, *Macromolecules*, **6**, 195 (1973).

Okada, M., T. Hirasawa, and H. Sumitomo, *Makromol. Chem.*, **190**, 1289 (1989).

Okada, M., T. Hisada, and H. Sumitomo, *Makromol. Chem.*, **179**, 959 (1978).

Okada, M., K. Mita, and H. Sumitomo, *Makromol. Chem.*, **177**, 2055 (1976).

Okada, M., H. Sumitomo, M. Atsumi, and H. K. Hall, Jr., *Macromolecules*, **20**, 1199 (1987).

Pan, C.-Y., Y. Wang, and W. J. Bailey, *J. Poilym. Sci. Polym. Chem. Ed.*, **26**, 2737 (1988).

Patton, P. A. and T. J. McCarthy, *Macromolecules*, **20**, 778 (1987).

Penczek, S. and P. Kubisa, "Progress in Polymerization of Cyclic Acetals," Chap. 5 in "Ring-Opening Polymerization," T. Saegusa and E. Goethals, Eds., American Chemical Society, Washington, D.C., 1977.

Penczek, S. and P. Kubisa, "Cationic Ring-Opening Polymerization: Ethers," Chap. 48 in "Comprehensive Polymer Science," Vol. 3, G. C. Eastmond, A. Ledwith, S. Russo, and P. Sigwalt, Eds., Pergamon Press, London, 1989a.

Penczek, S. and P. Kubisa, "Cationic Ring-Opening Polymerization: Acetals," Chap. 49 in "Comprehensive Polymer Science," Vol. 3, G. C. Eastmond, A. Ledwith, S. Russo, and P. Sigwalt, Eds., Pergamon Press, London, 1989b.

Penczek, S., P. Kubisa, and R. Szymanski, *Makromol. Chem. Macromol. Symp.*, **3**, 203 (1986).

Penczek, S. and J. Libiszowski, *Makromol. Chem.*, **189**, 1765 (1988).

Penczek, S. and K. Matyjaszewski, *J. Polym. Sci. Symp.*, **56**, 255 (1976).

Penczek, I. and S. Penczek, *Makromol. Chem.*, **67**, 203 (1963).

Penczek, S. and S. Slomkowski, "Cationic Ring-opening Polymerization: Cyclic Esters," Chap. 50 in "Comprehensive Polymer Science," Vol. 3, G. C. Eastmond, A. Ledwith, S. Russo, and P. Sigwalt, Eds., Pergamon Press, London, 1989a.

Penczek, S. and S. Slomkowski, "Cationic Ring-opening Polymerization: Formation of Cyclic Oligomers," Chap. 47 in "Comprehensive Polymer Science," Vol. 3, G. C. Eastmond, A. Ledwith, S. Russo, and P. Sigwalt, Eds., Pergamon Press, London, 1989b.

Pinazzi, C. P., J. Brossas, J. C. Brosse, and A. Pleurdeau, *Makromol. Chem.*, **144**, 155 (1971); *Polym. Prep.*, **13**(1), 445 (1972).

Plesch, P. H., *Pure Appl. Chem.*, **48**, 287 (1976).

Potts, M. K., G. L. Hagnauer, M. S. Sennett, and G. Davies, *Macromolecules*, **22**, 4235 (1989).

Price, C. C., *Acct. Chem. Res.*, **7**, 294 (1974).

Price, C. C. and M. Osgan, *J. Am. Chem. Soc.*, **78**, 4787 (1956).

Puffr, R. and J. Sebenda, *Makromol. Chem. Macromol. Symp.*, **13/14**, 249 (1986).

Ray, N. H., "Inorganic Polymers," Academic Press, New York, 1978.

Reimschuessel, H. K., *J. Polym. Sci. Macromol. Rev.*, **12**, 65 (1977).

Rentsch, C. and R. C. Schulz, *Makromol. Chem.*, **179**, 1131, 1403 (1978).

Rhein, T. and R. C. Schulz, *Makromol. Chem.*, **186**, 2301 (1985).

Roda, J., J. Kralicek, and K. Sanda, *Eur. Polym. J.*, **12**, 729 (1976).

Roda, J., O. Kucera, and J. Kralicek, *Makromol. Chem.*, **180**, 89 (1979).

Rose, J. B., *J. Chem. Soc.*, 542, 547 (1956).

Rothe, M., *J. Polym. Sci.*, **30**, 227 (1958).

Rothe, V. M., G. Reinisch, W. Jaeger, and I. Schopov, *Makromol. Chem.*, **54**, 183 (1962).

Saam, J. C. "Formation of Linear Siloxane Polymers" Chap. 3 in "Silicon Based Polymer Science," Am. Chem. Soc. Adv. Chem. Ser., 224, Washington, D.C., 1989.

Saegusa, T., *Angew. Chem. Int. Ed. Engl.*, **16**, 826 (1977); *Makromol. Chem. Suppl.*, **3**, 157 (1979); *Pure Appl. Chem.*, **53**, 691 (1981).

Saegusa, T. and E. Goethals, Eds., "Ring-Opening Polymerization," American Chemical Society, Washington, D.C., 1977.

Saegusa, T. and S. Matsumoto, *Macromolecules*, **1**, 442 (1968).

Saegusa, T., M. Motoi, and H. Suda, *Macromolecules*, **9**, 231 (1976).

Saigo, K., W. J. Bailey, T. Endo, and M. Okawara, *J. Polym. Sci. Polym. Chem. Ed.*, **21**, 1435 (1983).

Sawada, H., "Thermodynamics of Polymerization," Marcel Dekker, New York, 1976, Chap. 7.

Schulz, R. C., W. Hellermann, and J. Nienburg, "Cyclic Compounds Containing Two or More Oxygen Atoms in the Ring," Chap. 6 in "Ring-Opening Polymerization," Vol. 1, K. J. Ivin and T. Saegusa, Eds., Elsevier, London, 1984.

Scopelianos, A. G. and H. R. Allcock, *Macromolecules*, **20**, 432 (1987).

Sebenda, J., "Lactams," Chap. 6 in "Comprehensive Chemical Kinetics," Vol. 15, C. H. Bamford and C. F. H. Tipper, Eds., Elsevier, New York, 1976.

Sebenda, J., *Prog. Polym. Sci.*, **6**, 123 (1978).

Sebenda, J., *Makromol. Chem. Macromol. Symp.*, **13/14**, 97 (1988).

Sebenda, J., "Anionic Ring-Opening Polymerization: Lactams," Chap. 35 in "Comprehensive Polymer Science," Vol. 3, G. C. Eastmond, A. Ledwith, S. Russo, and P. Sigwalt, Eds., Pergamon Press, London, 1989.

Sebenda, J. and J. Kouril, *Eur. Polym. J.*, **7**, 1637 (1971); **8**, 437 (1972).

Sekiguchi, H., "Lactams and Cyclic Imides," Chap. 12 in "Ring-Opening Polymerization," Vol. 2, K. J. Ivin and T. Saegusa, Eds., Elsevier, London, 1984.

Sennett, M. S., G. L. Hagnauer, R. E. Singler, and G. Davies, *Macromolecules*, **19**, 959 (1986).

Sepulchre, M., A. Khalil, and N. Spassky, *Makromol. Chem.*, **180**, 131 (1979).

Shalitin, Y., "*N*-Carboxy-α-Amino Acid Anhydrides," Chap. 10 in "Ring-Opening Polymerization," K. C. Frisch and S. L. Reegen, Eds., Marcel Dekker, New York, 1969.

Shimasaki, K., T. Aida, and S. Inoue, *Macromolecules*, **20**, 3076 (1987).

Sigwalt, P., *Polym. J.*, **19**, 567 (1987).

Sigwalt, P. and S. Boileau, *J. Polym. Sci. Polym. Symp.*, **62**, 51 (1978).

Sigwalt, P. and N. Spassky, "Cyclic Compounds Containing Sulfur in the Rings," Chap. 9 in "Ring-Opening Polymerization," Vol. 2, K. J. Ivin and T. Saegusa, Eds., Elsevier, London, 1984.

Sigwalt, P. and V. Stannett, *Makromol. Chem. Macromol. Symp.*, **32**, 217 (1990).

Sims, D., *Makromol. Chem.*, **98**, 235 (1966).

Slomkowski, S., *Polymer*, **27**, 71 (1986).

Sogo, K., I. Hattori, S. Ikeda, and S. Kambara, *Makromol. Chem.*, **179**, 2559 (1978).

Sogo, K., I. Hattori, J. Kinoshita, and S. Ikeda, *J. Polym. Sci. Polym. Chem. Ed.*, **15**, 745 (1977).

Stewart, K. R., *Polym. Prep.*, **30**(2), 575 (1989).

Stolarcyk, A., *Makromol. Chem.*, **187**, 745 (1986).

Streck, R., *Chemtech*, 498 (1989).

Sumitomo, H. and M. Okada, "Sugar Anhydrides and Related Bicyclic Acetals," Chap. 5 in "Ring-Opening Polymerization," Vol. 1, K. J. Ivin and T. Saegusa, Eds., Elsevier, London, 1984.

Szwarc, M., *Adv. Polym. Sci.*, **4**, 1 (1965); *Pure Appl. Chem.*, **12**, 127 (1966).

Szwarc, M., *Makromol. Chem. Suppl.*, **3**, 327 (1979).

Szwarc, M. and C. L. Perrin, *Macromolecules*, **12**, 699 (1979).

Szymanski, R., P. Kubisa, and S. Penczek, *Macromolecules*, **16**, 1000 (1983).

Tada, K., Y. Numata, T. Saegusa, and J. Furukawa, *Makromol. Chem.*, **77**, 220 (1964).

Tagoshi, H. and T. Endo, *J. Polym. Sci. Polym. Chem. Ed.*, **27**, 1415 (1989).

Takata, T. and T. Endo, *Macromolecules*, **21**, 900 (1988).

Tanaka, R., I. Ueoka, Y. Takari, K. Kataoka, and S. Saito, *Macromolecules*, **16**, 849 (1983).

Tani, H. and T. Konomi, *J. Polym. Sci.*, **A-1**(4), 301 (1966).

Tobolsky, A. V., *J. Polym. Sci.*, **25**, 220 (1957); **31**, 126 (1958).

Tobolsky, A. V. and A. Eisenberg, *J. Am. Chem. Soc.*, **81**, 2302 (1959); **82**, 289 (1960).

Tomalia, D. A. and G. R. Killat, "Alkyleneimine Polymers," pp. 680–739 in "Encyclopedia of Polymer Science and Engineering," Vol. 1, 2nd ed., H. F. Mark, N. M. Bikales, C. G. Overberger, and G. Menges, Eds., Wiley-Interscience, New York, 1985.

Van de Velde, M. and E. J. Goethals, *Makromol. Chem. Macromol. Symp.*, **6**, 271 (1986).

Vert, M., *Makromol. Chem. Macromol. Symp.*, **6**, 109 (1986).

Vofsi, D. and A. V. Tobolsky, *J. Polym. Sci.*, **A3**, 3261 (1965).

Weissermel, K., E. Fischer, K. Gutweiler, H. D. Hermann, and H. Cherdron, *Angew. Chem. Int. Ed. Engl.*, **6**, 526 (1967).

Wilczek, L., S. Rubinsztajn, and J. Chojnowski, *Makromol. Chem.*, **187**, 39 (1986).

Wodzki, R. and K. Kaluzynski, *Makromol. Chem.*, **190**, 107 (1989).

Wojtania, M., P. Kubisa, and S. Penczek, *Makromol. Chem. Macromol. Symp.*, **6**, 201 (1986).

Wright, P. V., "Cyclic Siloxanes," Chap. 14 in "Ring-Opening Polymerization," Vol. 2, K. J. Ivin and T. Saegusa, Eds., Elsevier, London, 1984.

Xu, B., C. P. Lillya, and J. C. W. Chien, *Macromolecules*, **20**, 1445 (1987).

Yamashita, Y., T. Tsuda, M. Okada, and S. Iwatsuki, *J. Polym. Sci.*, **A-1**(4), 2121 (1966).

Yasuda, T., T. Aida, and S. Inoue, *Macromolecules*, **17**, 2217 (1984).

Young, R. H., M. Matzner, and L. A. Pilata, "Ring-Opening Polymerizations: Mechanism of Polymerization of ε-Caprolactone," Chap. 11 in "Ring-Opening Polymerization," T. Saegusa and E. Goethals, Eds., American Chemical Society, Washington, D.C., 1977.

Zahn, H., *Agnew. Chem.*, **69**, 270 (1957).

Zhang, X., Q. Zhou, W. P. Weber, R. F. Horvath, T. H. Chan, and G. Manuel, *Macromolecules*, **21**, 1563 (1988).

Zhou, S. Q. and W. P. Weber, *Macromolecules*, **23**, 1915 (1990).

Ziemelis, M. J. and J. C. Saam, *Macromolecules*, **22**, 2111 (1989).

Zimmerman, J., "Polyamides," pp. 315–381 in "Encyclopedia of Polymer Science and Engineering," Vol. 11, 2nd ed., H. F. Mark, N. M. Bikales, C. G. Overberger, and G. Menges, Eds., Wiley-Interscience, New York, 1988.

Zuk, A. and J. Jeczalik, *J. Polym. Sci. Polym. Chem. Ed.*, **17**, 2233 (1979).

PROBLEMS

7-1   Consider the following monomers and initiating systems:

Initiating System	Monomer
n-C_4H_9Li	Propylene oxide
$BF_3 + H_2O$	α-Pyrrolidone
H_2SO_4	δ-Valerolactam
$NaOC_2H_5$	Ethylenimine
H_2O	Octamethylcyclotetrasiloxane
	Propylene sulfide
	Trioxane
	Oxacyclobutane

Which initiating system(s) can be used to polymerize each of the various monomers? Show the mechanism of each polymerization by chemical equations.

7-2   Give the cyclic monomer(s), initiator, and reaction conditions necessary to synthesize each of the following polymers:

a. $\leftmoon NHCO(CH_2)_4 \rightmoon_n$

b. $\leftmoon NH-CH-CO \rightmoon_n$

$\qquad\qquad\qquad |$

$\qquad\qquad\quad C_2H_5$

c. $\leftmoon N-CH_2CH_2CH_2 \rightmoon_n$

$\qquad\quad |$

$\qquad\;\; CHO$

d. $\leftmoon O(CH_2)_2OCH_2 \rightmoon_n$

e. $\leftmoon CH=CH(CH_2)_2 \rightmoon_n$

f. $\leftmoon Si(CH_3)_2O \rightmoon_n$

7-3   Discuss the effect of ring size on the tendency of a cyclic monomer toward ring-opening polymerization.

7-4   Anionic polymerization of propylene oxide is usually limited to producing a relatively low-molecular-weight polymer. Discuss the reasons for this occurrence.

7-5   T. Saegusa, H. Fujii, S. Kobayashi, H. Ando, and R. Kawase, *Macromolecules*, **6**, 26 (1973), have found that the polymerization rate for the BF_3 polymerization of oxetane follows Eq. 7-42:

$$\frac{-d[M]}{dt} = k_p[P^*][M]$$

a. Describe by an equation the mechanism for the propagation step in this polymerization.
b. The values of k_p at $-20°C$ are 0.18 and 0.019 liter/mole-sec, respectively, for polymerizations carried out in methylcyclohexane and methylene chloride. Explain why propagation is faster in the less polar solvent.
c. The following rate constants and activation parameters for propagation were observed for oxetane and 3-methyloxetane:

	Oxetane	3-Methyloxetane
k_p at $-20°C$ (liters/mole-sec)	0.019	0.11
$\Delta G_p^{\ddagger}$ (kJ/mole)	71.2	65.7
$\Delta H_p^{\ddagger}$ (kJ/mole)	57.4	65.3
$\Delta S_p^{\ddagger}$ (J/K-mole)	-50.7	-5.4

Discuss why 3-methyloxetane is more reactive than oxetane. Is a negative entropy of activation for propagation consistent with the proposed propagation mechanism?

7-6   The polymerization of an epoxide by hydroxide or alkoxide ion is often carried out in the presence of an alcohol. Why? How is the degree of polymerization affected by alcohol? Discuss how the presence of alcohol affects both the polymerization rate and molecular weight distribution.

7-7   Explain the following observations:

    a. A small amount of epichlorohydrin greatly increases the rate of the polymerization of tetrahydrofuran by BF_3 even though epichlorohydrin is much less basic than tetrahydrofuran.

    b. The addition of small amounts of water to the polymerization of oxetane by BF_3 increases the polymerization rate but decreases the degree of polymerization.

7-8   An equilibrium polymerization is carried out with an initial concentration of 12.1 M tetrahydrofuran, $[M^*] = 2.0 \times 10^{-3}\ M$, and $k_p = 1.3 \times 10^{-2}$ liter/ mole-sec. Calculate the initial polymerization rate if $[M]_c = 1.5\ M$. What is the polymerization rate at 20% conversion?

7-9   What are the roles of an acylating agent and activated monomer in the anionic polymerization of lactams?

7-10   Consider the equilibrium polymerization of ε-caprolactam initiated by water at 220°C. For the case where $[I]_0 = 0.352$, $[M]_0 = 8.79$ and $[M]_c = 0.484$, the degree of polymerization at equilibrium is 152, calculate the values of K_i and K_p at equilibrium.

7-11   Discuss by means of equations the occurrence of backbiting, ring-expansion reactions in the polymerizations of cyclic ethers, acetals, and amines.

CHAPTER 8

STEREOCHEMISTRY OF POLYMERIZATION

Constitutional (formerly: *structural*) *isomerism* is encountered when polymers have the same overall chemical composition but atoms or groups of atoms are connected in different orders. Polyacetaldehyde, poly(ethylene oxide), and poly(vinyl alcohol) are constitutional isomers. The first two polymers are obtained from isomeric mon-

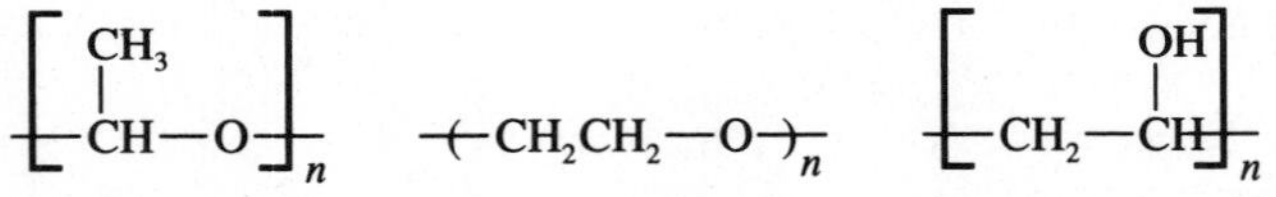

$$\left[\begin{array}{c} CH_3 \\ | \\ CH-O \end{array}\right]_n \qquad \left(CH_2CH_2-O\right)_n \qquad \left[\begin{array}{c} OH \\ | \\ CH_2-CH \end{array}\right]_n$$

Polyacetaldehyde   Poly(ethylene oxide)   Poly(vinyl alcohol)

omers, acetaldehyde and ethylene oxide. The third is obtained by hydrolysis of poly(vinyl acetate) since vinyl alcohol does not exist (it is the enol form of acetaldehyde). Poly(methyl methacrylate) and poly(ethyl acrylate) are isomeric polymers formed from isomeric

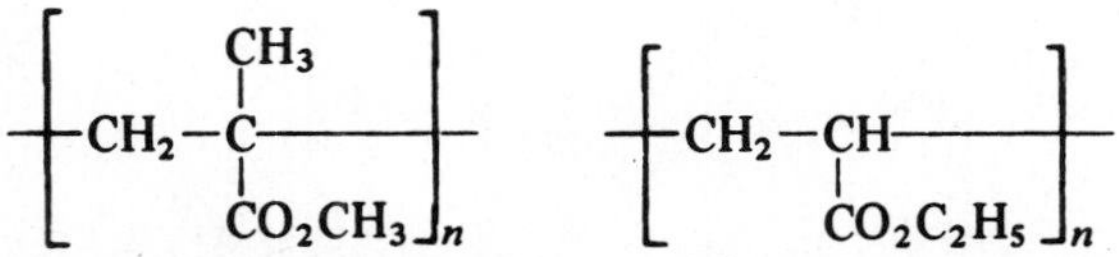

$$\left[\begin{array}{c} CH_3 \\ | \\ CH_2-C \\ | \\ CO_2CH_3 \end{array}\right]_n \qquad \left[\begin{array}{c} \\ CH_2-CH \\ | \\ CO_2C_2H_5 \end{array}\right]_n$$

Poly(methyl methacrylate)   Poly(ethyl acrylate)

monomers. Poly(ε-caprolactam) and poly(hexamethylene adipamide) are isomeric even though the monomers are not directly isomeric. In this case, ε-caprolactam has the

$$\left[CO(CH_2)_5NH\right]_n \qquad \left[NH(CH_2)_6NHCO(CH_2)_4CO\right]_n$$

Poly(ε-caprolactam)   Poly(hexamethylene adipamide)

same overall composition as an equimolar mixture of hexamethylene diamine and adipic acid. Similarly, the 1:1 copolymer of ethylene and 1-butene is isomeric with polypropylene.

Isomeric polymers can also be obtained from a single monomer if there is more than one polymerization route. The head-to-head placement that can occur in the polymerization of a vinyl monomer is isomeric with the normal head-to-tail placement (see structures **III** and **IV** in Sec. 3-2a). Isomerization during carbocation polymerization is another instance whereby isomeric structures can be formed (Sec. 5-2b). Monomers with two polymerizable groups can yield isomeric polymers if one or other of the two alternate polymerization routes is favored. Examples of this type of isomerism are the 1,2- and 1,4-polymers from 1,3-dienes (Secs. 3-13b-6 and 8-6), the separate polymerizations of the alkene and carbonyl double bonds in ketene and acrolein (Sec. 5-7a), and the synthesis of linear or cyclized polymers from nonconjugated dienes (Sec. 6-6b). The different examples of constitutional isomerism are important to note from the practical viewpoint, since the isomeric polymers usually differ considerably in their properties.

This chapter is concerned primarily with a different type of isomerism in polymers—that which arises as a result of stereoisomerism in the structure of polymers as a consequence of the polymerization reaction. This is an important topic because of the significant effect that stereoisomerism has on many polymer properties. Considerations of stereoisomerism in chain polymerizations of vinyl monomers were recognized early [Staudinger et al., 1929]. However, the possibility that each of the propagation steps in the growth of a polymer chain could give rise to stereoisomerism was not fully appreciated until more than two decades later. Further, the synthesis of polymers with ordered configurations in which the stereoisomerism of repeating units in the chain would show a regular or ordered arrangement instead of a random one was essentially not considered. One of the major developments in the polymer field has been the elucidation of the occurrence of stereoisomerism in polymers. More importantly, the pioneering works of Ziegler and Natta and their co-workers led to the convenient synthesis of polymers with highly stereoregular structures [Boor, 1979; Ketley, 1967; Natta, 1955a, 1955b, 1965; Natta et al., 1955; Ziegler et al., 1955]. Large-scale commercial applications include high-density and linear low-density polyethylenes (HDPE, LLDPE), polypropylene, ethylene–propylene co- and terpolymers, and polymers from 1,3-dienes (Sec. 8-6). It is interesting to note that we are now doing what nature has been doing for eons. Stereoregular polymers of different kinds are commonly found in nature; these include natural rubber, cellulose, starch, polypeptides, and nucleic acids.

8-1 TYPES OF STEREOISOMERISM IN POLYMERS

Stereoisomerism in polymers arises from different spatial arrangements (*configurations*) of the atoms or substituents in a molecule [Farina, 1987; Farina and Bressan, 1968; Goodman, 1967]. Cis–trans (*geometric*) isomers arise from different configurations of substituents on a double bond or on a cyclic structure. *Enantiomers* arise from different configurations of substituents on a saturated carbon or other tetravalent atom. The term *configuration* should not be confused with *conformation*. Conformation refers to the different arrangements of atoms and substituents in a molecule that result from rotations around single bonds. Examples of different polymer conformations are the fully extended planar zigzag, randomly coiled, helical, and folded-chain arrangements. Conformational isomers can be interconverted one into the other by bond rotations. Configurational isomers differ in the spatial arrangements of their atoms and substituents in a manner such that they can be interconverted only by breaking and reforming primary chemical bonds.

8-1a Monosubstituted Ethylenes

8-1a-1 Site of Steric Isomerism

Isomerism is observed in the polymerization of alkenes whenever one of the carbon atoms of the double bond is at least monosubstituted. The polymerization of a mono-substituted ethylene, $CH_2{=}CHR$ (where R is any substituent other than H), leads to polymers in which every tertiary carbon atom in the polymer chain is a *stereocenter* (or *stereogenic center*). The stereocenter in each repeating unit is denoted as C^* in **Ia** and **Ib**

$$
\begin{array}{cc}
\text{H} & \text{R} \\
| & | \\
\text{\textasciitilde\textasciitilde C*\textasciitilde\textasciitilde} & \text{\textasciitilde\textasciitilde C*\textasciitilde\textasciitilde} \\
| & | \\
\text{R} & \text{H} \\
\textbf{Ia} & \textbf{Ib}
\end{array}
$$

A stereocenter is defined as an atom bearing several groups whose identities are such that an interchange of two of the groups produces a stereoisomer. Thus, an interchange of the R and H groups converts **Ia** into **Ib** and vice versa.

Each stereocenter C^* is a site of steric isomerism in the polymerization of $CH_2{=}CHR$. Each such site in a polymer chain can exhibit either of two different configurations (**Ia** or **Ib**). Considering the main carbon—carbon chain of the polymer $-(-CH_2CHR-)_n-$ to be stretched out in its fully extended planar zigzag conformation, two different configurations are possible for each stereocenter since the R group may be situated on either side of the plane of the carbon–carbon polymer chain. If the plane of the carbon–carbon chain is considered as being in the plane of this page, the R groups are located on one side or the other of this plane. The two configurations are usually referred to as *R* and *S*, although the Cahn–Ingold–Prelog priority rules cannot be applied strictly since that requires knowledge of the relative lengths of the two polymer chain segments. Without knowledge of the lengths of the polymer chain segments, designation of *R* and *S* configurations can be done only in an arbitrary manner. One of the configurations (**Ia** or **Ib**) is designated as *R* and the other as *S*. The lowercase designations, *r* and *s*, are often used in place of the uppercase letters to emphasize this aspect. (Other designations of the two configurations appear in the literature: D and L, *d* and *l*, + and −.)

There are two constitutional repeat units (Sec. 1-2c) from a stereochemical view-point, one with *R* configuration for the stereocenter and the other with *S* configuration for the stereocenter (corresponding to **Ia** and **Ib**). These are referred to as the two *configurational base units* and have an enantiomeric relationship [IUPAC, 1981].

The classification of a C^* stereocenter in the polymer planar zigzag conformation as *chirotopic* or *achirotopic* presents a difficulty [Mislow and Siegel, 1984]. A stereocenter is chirotopic if it lies in a chiral environment. A chiral environment is one that is not superposable on its mirror image. The literal definition of the term *chirotopic* would classify each C^* as chirotopic since the two polymer chain segments, ∿∿ and ∿∿∿, are not equivalent. The nonequivalence of the two chain segments is due to differences in length. (There may also be differences in end groups depending on the mode of termination.) However, classification of C^* as chirotopic implies optical activity and this is clearly not true from a practical viewpoint. Optical activity is a short-range phenomenon with its magnitude determined by the differences in the groups attached to the stereocenter. The first few atoms of the two polymer chain segments attached to C^* are the same and such stereocenters do not contribute measurably to the optical activity of the polymer. The only stereocenters with significant optical activity are those at the ends of the polymer chain, but their population is negligible for a high-molecular-weight polymer. The sum of contributions from all the stereocenters is such that optical activity is below the limits of detection. (This discussion of optical activity excludes optical activity arising from the existence of a polymer molecule exclusively in a chiral conformation, a very rare occurrence except for biological macromolecules; see Sec. 8-8b. A typical polymer in solution exists in the random coil conformation. Many of the conformations are chiral, but on a time-averaged basis there are equal numbers of conformations of opposite rotation and the net rotation is zero.) Thus, it is more useful to classify stereocenters such as those in **Ia** and **Ib** as *achirotopic*.

The terms *stereocenter*, *chirotopic*, and *achirotopic* will be used in this text in line with the most recent terminology used by organic chemists [Mislow and Siegel, 1984]. However, these terms are not used extensively in the polymer literature at the present time [IUPAC, 1966, 1981]. Most polymer papers and texts refer to C^* as a *chiral* or *pseudochiral center*. The term *pseudochiral center* is based on the same convention used above to classify C^* as achirotopic instead of chirotopic. (The terms *asymmetric* or *pseudoasymmetric center* in place of chiral or pseudochiral are found in older literature.)

8-1a-2 Tacticity

The regularity in the configurations of successive stereocenters determines the overall order (*tacticity*) of the polymer chain. If the R groups on successive stereocenters are randomly distributed on the two sides of the planar zigzag polymer chain, the polymer does not have order and is termed *atactic*. Two types of ordered or *tactic* structures or placements can occur: *isotactic* and *syndiotactic*. An isotactic structure occurs when the stereocenter in each repeating unit in the polymer chain has the same configuration. All the R groups will be located on one side of the plane of the carbon–carbon polymer chain—either all above or all below the plane of the chain. A syndiotactic polymer structure occurs when the configuration of the stereocenter alternates from one repeating unit to the next with the R groups located alternately on the opposite sides of the plane of the polymer chain. These different polymer structures are shown in Fig. 8-1. The various polymer structures are described by three different pictorial representations. The ones on the far left side of Fig. 8-1 show the polymer chain in the plane of this page with the H and R groups above (triangular bonds) and below (dotted bonds) that plane. The representations in the middle are the corresponding Fischer projections. Vertical lines in the Fischer projections correspond to bonds projecting behind the plane of this page; horizontal lines represent bonds projecting in front of the plane. This is the usual convention for Fischer projections [Bovey, 1982]. The configuration at each carbon atom in the polymer chain is drawn in the

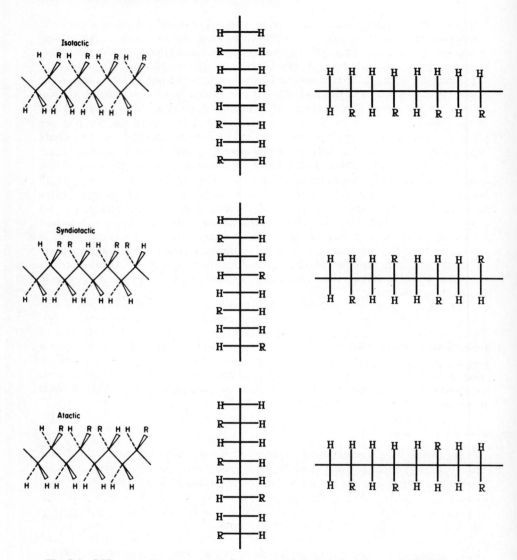

Fig. 8-1 Different polymer structures from a monosubstituted ethylene, $-(\text{CH}_2\text{CHR})_n$.

Fischer projection by imagining the rotation of each carbon–carbon bond in the polymer chain into an eclipsed conformation as opposed to the staggered conformation that actually exists. The representations on the far right side of Fig. 8-1, recommended by IUPAC [1981], are obtained by rotating the previous Fischer projections 90° counterclockwise. The resulting Fischer projections do have the advantage of showing the polymer chain going in the same direction as the far left representations. Their disadvantage is that one must always keep in mind that the usual Fischer convention for horizontal (forward) and vertical (back) bonds is reversed. These rotated Fischer projections will be used in the remainder of this chapter.

The Fischer projections show that isotactic placement corresponds to *meso-* or *m*-placement for a pair of consecutive stereocenters. Syndiotactic placement corresponds to *racemo-* (for racemic) or *r*-placement for a pair of consecutive stereocenters. The *configurational repeating unit* is defined as the smallest set of configurational base units that describe the configurational repetition in the polymer. For the isotactic polymer from a monosubstituted ethylene, the configurational repeating unit and configurational base unit are the same. For the syndiotactic polymer, the configurational repeating unit is a sequence of two configurational base units, an *R* unit followed by an *S* unit or vice versa.

Polymerizations that yield tactic structures (either isotactic or syndiotactic) are termed *stereospecific polymerizations*. Polymerizations yielding isotactic and syndiotactic are *isospecific* and *syndiospecific polymerizations*, respectively. The polymer structures are termed *stereoregular polymers*. The terms *isotactic* and *syndiotactic* are placed before the name of a polymer to indicate the respective tactic structures, such as isotactic polypropylene and syndiotactic polypropylene. The absence of these terms denotes the atactic structure; thus, "polypropylene" means atactic polypropylene. The prefixes "it-" and "st-" together with the formula of the polymer, have been suggested for the same purpose: it-$[CH_2CH(CH_3)]_n$ and st-$[CH_2CH(CH_3)]_n$ [IUPAC, 1966].

Both isotactic and syndiotactic polypropylenes are achiral as a result of a series of mirror planes (i.e., planes of symmetry) perpendicular to the polymer chain axis. Neither exhibits optical activity.

8-1b Disubstituted Ethylenes

8-1b-1 1,1-Disubstituted Ethylenes

For disubstituted ethylenes, the presence and type of tacticity depends on the positions of substitution and the identity of the substituents. In the polymerization of a 1,1-disubstituted ethylene, $CH_2=CRR'$, stereoisomerism does not exist if the R and R' groups are the same (e.g., isobutylene and vinylidene chloride). When R and R' are different (e.g., —CH_3 and —$COOCH_3$ in methyl methacrylate), stereoisomerism occurs exactly as in the case of a monosubstituted ethylene. The methyl groups can be located all above or all below the plane of the polymer chain (isotactic), alternately above and below (syndiotactic), or randomly (atactic). The presence of the second substituent has no effect on the situation since steric placement of the first substituent automatically fixes that of the second. The second substituent is isotactic if the first is isotactic, syndiotactic if the first substituent is syndiotactic, and atactic if the first is atactic.

8-1b-2 1,2-Disubstituted Ethylenes

The polymerization of 1,2-disubstituted ethylenes, RCH=CHR', such as 2-pentene (R = —CH_3, R' = —C_2H_5), presents a different situation. Polymerization yields a

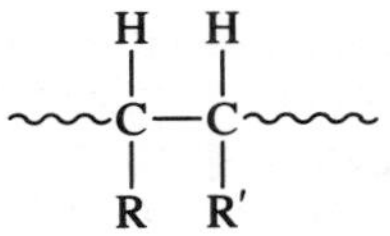

II

polymer structure **II** in which there are two different stereocenters in each repeating unit. Several possibilities of *ditacticity* exist that involve different combinations of tacticity for the two stereocenters. Various stereoregular structures can be defined as shown in Fig. 8-2. *Diisotactic* structures occur when placement at each of the two stereocenters is isotactic. A *disyndiotactic* structure occurs when placement at each of the two different stereocenters is syndiotactic. Two diisotactic structures are possible. These are differentiated by the prefixes *threo* and *erythro*. The meaning of these prefixes corresponds to their use in carbohydrate chemistry. Considering the planar zigzag polymer chain, the erythro structure is the one in which like groups on adjacent

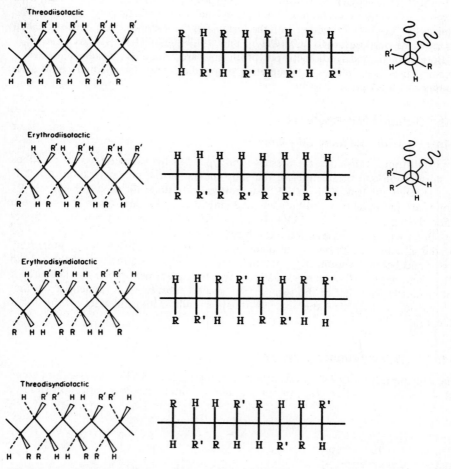

Fig. 8-2 Stereoregular polymers from a 1,2-disubstituted ethylene, $-(CHR-CHR')_n$.

carbons are anti to each other (i.e., R and R' are anti to each other and H and H are anti to each other). The threo structure has an anti arrangement of unlike groups on adjacent carbons (i.e., R and H are anti and R' and H are also anti). The difference between the erythro and threodiisotactic structures can also be shown by Newman representations of the eclipsed conformation of two consecutive carbon atoms in the polymer chain. These are shown at the far right side of Fig. 8-2. Like groups are eclipsed on like groups (H on H, R on R', polymer chain segment on polymer chain segment) in the erythro structure; unlike groups are eclipsed in the threo structure. For the *threodiisotactic* polymer, the two stereocenters have opposite configurations. In the zigzag pictorial representation, both R and R' are on the same side of the plane of the polymer chain. In the Fischer projection, R and R' are on opposite sides of the line representing the polymer chain. For the *erthyrodisisotactic* polymer, the configurations at the two stereocenters are the same.

The corresponding *threodisyndiotactic* and *erythrodisyndiotactic* polymers are drawn in Fig. 8-2. However, a close examination of the two disyndiotactic structures shows that they are identical except for a difference in the end groups. Ignoring end groups, the two structures are superposable. Thus, from the practical viewpoint, there is only one disyndiotactic polymer. The nomenclature for ditactic polymers follows in the same manner as for monotactic polymers. Thus the various stereoregular poly(2-pentene) polymers would be threodiisotactic poly(2-pentene), erythrodiisotactic poly-(2-pentene), and disyndiotactic poly(2-pentene) with the prefixes "tit-," "eit-," and "st-," respectively, used before the formula $[CH(CH_3)CH(C_2H_5)]_n$.

The stereocenters in all three stereoregular polymers are achirotopic. The polymers are achiral and do not possess optical activity. The diisotactic polymers contain mirror planes perpendicular to the polymer chain axis. The disyndiotactic polymer has a mirror glide plane of symmetry. The latter refers to superposition of the disyndiotactic structure with its mirror image after one performs a glide operation. A glide operation involves movement of one structure relative to the other by sliding one polymer chain axis parallel to the other chain axis.

It should be noted that other polymer structures can be postulated—those where one substituent is atactic while the other is either isotactic or syndiotactic or those where one substituent is isotactic while the other is syndiotactic. However, these possibilities appear to be trivial at this time since their syntheses have not been achieved. The factors present during polymerization that lead to ordering or disordering of one substituent generally have the same effect on the other substituent. The term *stereoregular* polymer is not used for those structures where one or the other substituent is atactic. *Stereoregular* is reserved for structures where all stereocenters are ordered.

8-1c Carbonyl and Ring-Opening Polymerizations

Several other types of monomers are capable of yielding stereoisomeric polymer structures. Ordered structures are possible in the polymerization of carbonyl monomers (RCHO and RCOR') and the ring-opening polymerizations of certain monomers. Thus, for example, the polymers from acetaldehyde and propylene oxide can have isotactic and syndiotactic structures as shown in Figs. 8-3 and 8-4.

Polyacetaldehyde (or poly[oxyethylidene]), like the systems described previously,

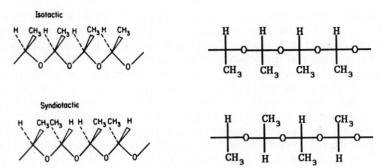

Fig. 8-3 Stereoregular polymers of polyacetaldehyde, $\text{-}\!\!+\!CH(CH_3)O\!\!\text{+}_n$.

contains stereocenters that are achirotopic. Both the isotactic and syndiotactic poly-mers are achiral and do not possess optical activity.

The repeating unit (**III**) in poly(propylene oxide) (or poly[oxypropylene]), on the other hand, possesses stereocenters that are chirotopic since the two chain segments

$$\sim\!\!\sim\!\!O-\underset{\underset{CH_3}{|}}{\overset{\overset{H}{|}}{C^*}}-CH_2\!\!\sim\!\!\sim$$

III

attached to C* are very different—one is attached via an oxygen and the other via a methylene. Isotactic poly(propylene oxide) contains no symmetry element, is not superposable on its mirror image, and possesses optical activity. The syndiotactic polymer is incapable of optical activity since it possesses a mirror glide plane of symmetry.

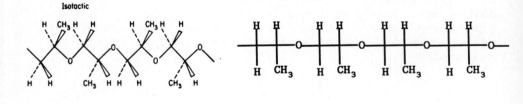

Fig. 8-4 Stereoregular polymers of poly(propylene oxide), $\text{-}\!\!+\!CH_2CH(CH_3)O\!\!\text{+}_n$.

One should note that stereoregularity in the products of ring-opening polymerization do not typically arise from the formation of a stereocenter as a consequence of bond formation in the polymerization reaction as is the case in the polymerization of vinyl and carbonyl monomers. The stereocenter is present in the cyclic monomer and remains intact on polymerization. This is the usual case as ring-opening polymerization of a monomer such as propylene oxide does not involve bond breaking at the stereocenter. The term *stereospecific polymerization* is not employed for such reactions (but see Sec. 8-8c).

The stereochemistry of step polymerization is considered at this point. Bond formation during step polymerization almost never results in the formation of a stereocenter. Thus, neither the ester nor amide groups in polyesters and polyamides, respectively, possess stereocenters. Stereoregular polymers are possible from step polymerizations in the same manner as from ring-opening polymerization, when there is a chirotopic stereocenter in the monomer(s). An example would be the polymerization of (R) or (S)-$H_2NCHRCOOH$. Naturally occurring polypeptides are stereoregular polymers formed from optically active α-amino acids.

8-1d 1,3-Butadiene and 2-Substituted 1,3-Butadienes

8-1d-1 1,2- and 3,4-Polymerizations

The polymerization of a 1,3-diene such as butadiene, isoprene, and chloroprene,

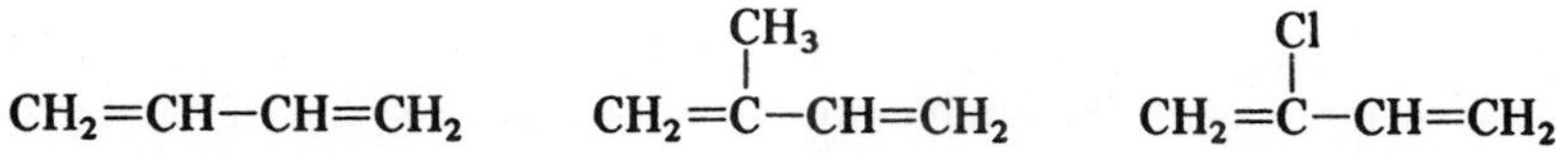

$$CH_2{=}CH{-}CH{=}CH_2 \qquad CH_2{=}\overset{\overset{\displaystyle CH_3}{|}}{C}{-}CH{=}CH_2 \qquad CH_2{=}\overset{\overset{\displaystyle Cl}{|}}{C}{-}CH{=}CH_2$$

1,3-Butadiene   Isoprene   Chloroprene

can lead to cis and trans isomers when polymerization occurs by a 1,4-polymerization. On the other hand, 1,2-polymerization through one or the other double bond can yield isotactic, syndiotactic, or atactic polymers, for example, for 1,3-butadiene,

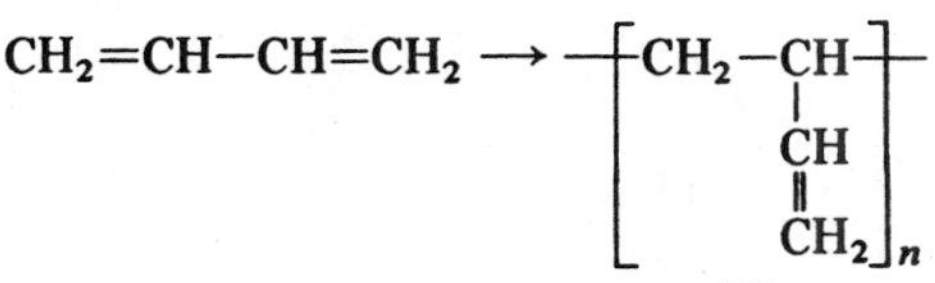

$$CH_2{=}CH{-}CH{=}CH_2 \longrightarrow \left[CH_2{-}\underset{\displaystyle \underset{\displaystyle CH_2}{\overset{\displaystyle \|}{CH}}}{CH} \right]_n \tag{8-1}$$

IV

The situation is exactly analogous to the polymerization of mono-substituted alkenes; the various polymer structures would be those in Fig. 8-1 with R = $-CH{=}CH_2$. With chloroprene and isoprene, the possibilities are enlarged since the two double bonds are substituted differently. Polymerizations through the 1,2- and 3,4-double bonds do not yield the same product as they would in butadiene polymerization. There are, therefore, a total of six structures possible—corresponding to isotactic, syndiotactic, and atactic structures for both 1,2- and 3,4-polymerizations; for instance, for isoprene, one has the situation shown in Eq. 8-2 with polymers **V** and **VI** capable of exhibiting the structures in Fig. 8-1 with R = $-CH{=}CH_2$ and $-C(CH_3){=}CH_2$, respectively.

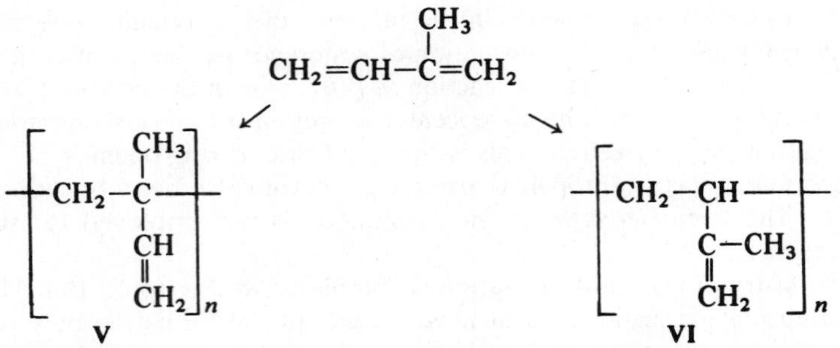

$$CH_2=CH-\underset{\underset{CH_3}{|}}{C}=CH_2$$

$\begin{bmatrix} & CH_3 \\ & | \\ -CH_2-C- \\ & | \\ & CH \\ & \| \\ & CH_2 \end{bmatrix}_n$	$\begin{bmatrix} -CH_2-CH- \\ \quad\quad\quad | \\ \quad\quad\quad C-CH_3 \\ \quad\quad\quad \| \\ \quad\quad\quad CH_2 \end{bmatrix}_n$
V	**VI**
1,2-Polymerization	3,4-Polymerization

(8-2)

8-1d-2 1,4-Polymerization

From the practical viewpoint, the more important products from 1,3-dienes are those that occur by 1,4-polymerization of the conjugated diene system, for example, for isoprene,

$$CH_2=\underset{\underset{CH_3}{|}}{C}-CH=CH_2 \rightarrow \left[CH_2-\underset{\underset{CH_3}{|}}{C}=CH-CH_2 \right]_n$$

VII

(8-3)

1,4-Polymerization leads to a polymer structure (**VII**) with a repeating alkene double bond in the polymer chain. The double bond in each repeating unit of the polymer chain is a site of steric isomerism since it can have either a cis or a trans configuration. The polymer chain segments on each carbon atom of the double bond are located on the same side of the double bond in the cis configuration (**VIII**) and on opposite sides in the trans configuration (**IX**), for example, for isoprene.

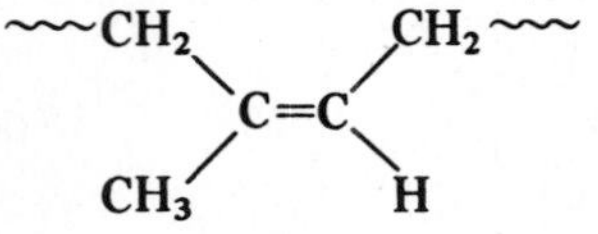

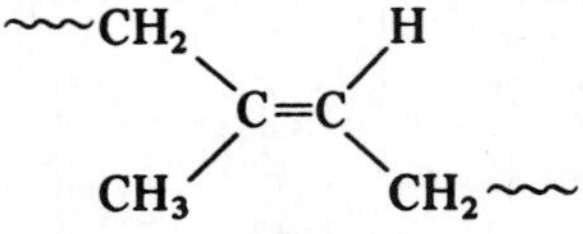

Cis	Trans
VIII	**IX**

When all of the double bonds in the polymer molecule have the same configuration, the result is two different ordered polymer structures—*transtactic* and *cistactic*. Figure 8-5 shows the structures of the completely cis and completely trans polymers of isoprene. The stereochemistry of these polymers is indicated in their names. For example, the trans polymer (**IX**) is named as trans-1,4-polyisoprene or transtactic poly(1-methyl-1-butenylene) or poly(*trans*-1-methyl-1-butenylene). The first name is the IUPAC-recommended trivial name; the second two names are IUPAC structure-based names (Sec. 1-2c) [IUPAC, 1966, 1981; Jenkins, 1979]. There is the convention that stereochemical terms such as trans are italicized in names only when they appear after the prefix "poly."

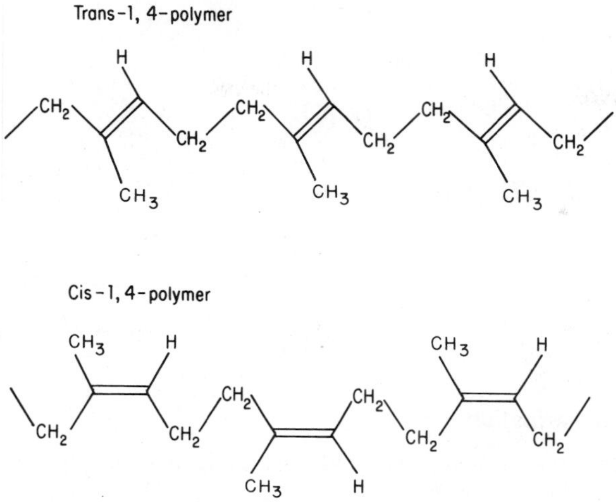

Trans-1, 4-polymer

Cis-1, 4-polymer

Fig. 8-5 Cis- and trans-1,4-polymers from isoprene, $+CH_2CH(CH_3)=CHCH_2+_n$.

It should be noted that both carbon atoms of the double bond in **VIII** and **IX** are achirotopic stereocenters. Interchange of the groups attached to one or the other carbon of the double bond converts **VIII** into **IX** and vice versa.

8-1e 4-Substituted and 1,4-Disubstituted 1,3-Butadienes

8-1e-1 1,2- and 3,4-Polymerizations

The polymerizations of 4-substituted (**X**) and 1,4-disubstituted (**XI**) 1,3-butadienes,

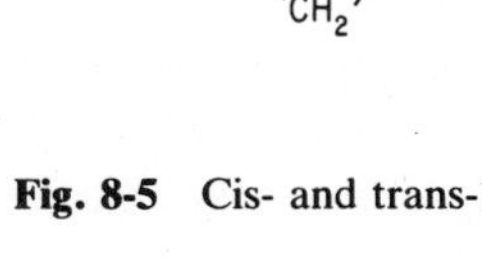

$$\overset{R}{\underset{|}{CH}}=CH-CH=CH_2 \qquad \overset{R}{\underset{|}{CH}}=CH-CH=\overset{R'}{\underset{|}{CH}}$$

$$\text{X} \qquad\qquad\qquad \text{XI}$$

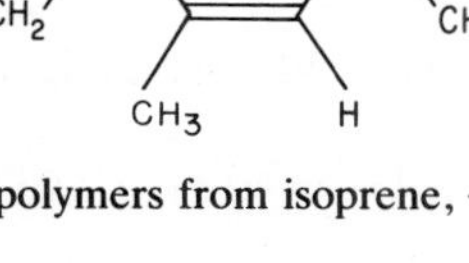

involve several possibilities. 1,2-Polymerization of a 4-substituted 1,3-butadiene proceeds with the same possibilities as the polymerization of a monosubstituted ethylene such as propylene. 1,2-Polymerization of a 4-substituted 1,3-butadiene (Eq. 8-4) as well as 1,2- and 3,4-polymerizations of 1,4-disubstituted 1,3-dienes (Eqs. 8-5 and 8-6) proceed with the same possibilities as the polymerization of a 1,2-disubstituted ethylene.

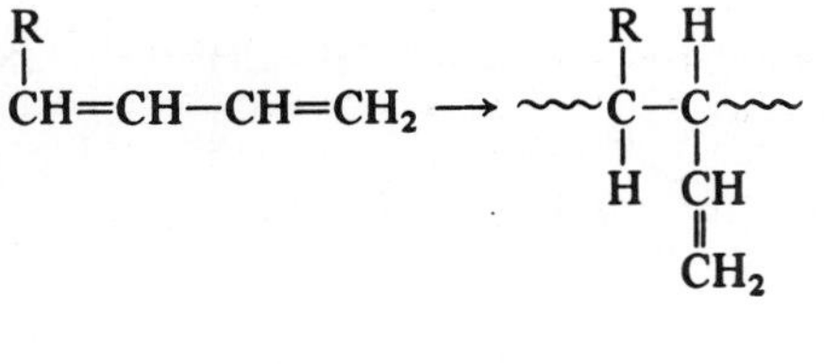

$$\overset{R}{\underset{|}{CH}}=CH-CH=CH_2 \rightarrow \text{~~~}\overset{R}{\underset{|}{\underset{H}{C}}}-\overset{H}{\underset{|}{\underset{\underset{\underset{CH_2}{\|}}{CH}}{C}}}\text{~~~} \qquad\qquad (8\text{-}4)$$

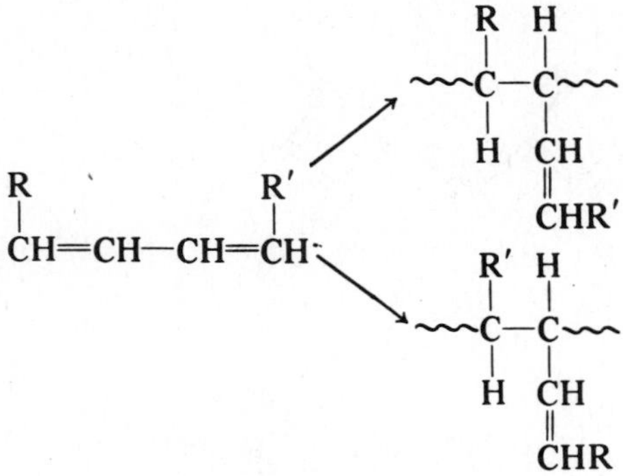

$$\begin{array}{cc}
& \overset{R\ H}{\underset{H\ \underset{\|}{CH}}{\sim\sim C-C\sim\sim}} \\
& \overset{}{\underset{CHR'}{}}
\end{array} \tag{8-5}$$

$$R \qquad\qquad R'$$
$$\overset{|}{CH}=CH-CH=\overset{|}{CH}\text{.}$$

$$\begin{array}{cc}
& \overset{R'\ H}{\underset{H\ \underset{\|}{CH}}{\sim\sim C-C\sim\sim}} \\
& \overset{}{\underset{CHR}{}}
\end{array} \tag{8-6}$$

8-1e-2 1,4-Polymerization

The 1,4-polymerization of a 4-substituted 1,3-butadiene can yield four stereoregular

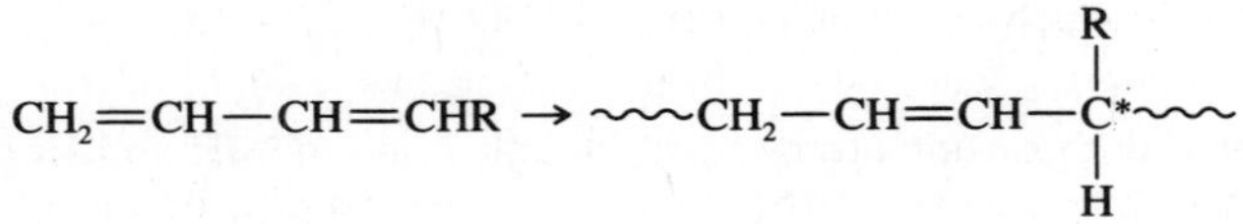

$$CH_2=CH-CH=CHR \rightarrow \sim\sim\sim CH_2-CH=CH-\overset{R}{\underset{H}{\overset{|}{C^*}}}\sim\sim\sim \tag{8-7}$$

XII

polymer structures. The double bond can have cis or trans configurations, and each of these can be combined with isotactic or syndiotactic placement of R groups. (Other polymer structures, not referred to as stereoregular polymers, are possible with either random placement of double bonds or atactic placement of R groups.)

The all-trans–all-isotactic and all-trans–all-syndiotactic structures for the 1,4-polymerization of 1,3-pentadiene are shown in Fig. 8-6. In naming polymers with both types of stereoisomerism, that due to cis–trans isomerism is named first unless it is indicated after the prefix "poly." Thus, the all-trans–all-isotactic polymer is named as transisotactic 1,4-poly(1,3-pentadiene) or transisotactic poly(3-methyl-1-butenylene) or isotactic poly(3-methyl-*trans*-1-butenylene).

The sp^3 stereocenter (i.e., C*) in **XII** is chirotopic, like the case of poly(propylene oxide), since the first couple of atoms of the two chain segments are considerably different. The isotactic structures are optically active while the syndiotactic structures are not optically active.

The 1,4-polymerization of 1,4-disubstituted 1,3-butadienes leads to structure (**XIII**),

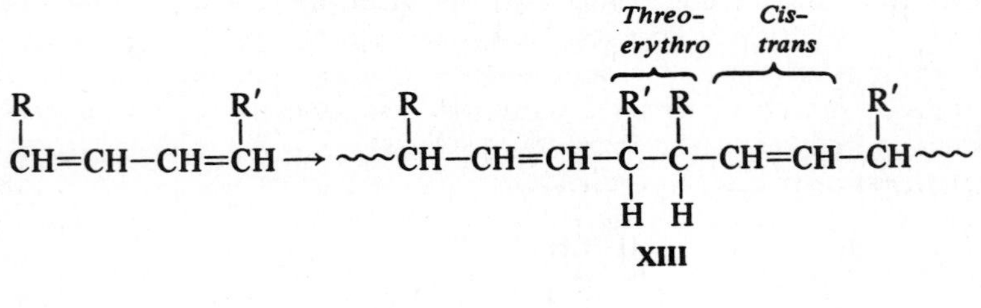

$$R \qquad\qquad R' \qquad\qquad R \qquad \overset{Threo-}{\overset{erythro}{\overbrace{}}}\ \overset{Cis-}{\overset{trans}{\overbrace{}}}$$
$$\overset{|}{CH}=CH-CH=\overset{|}{CH} \rightarrow \sim\sim\sim\overset{|}{CH}-CH=CH-\overset{R'}{\underset{H}{\overset{|}{C}}}-\overset{R}{\underset{H}{\overset{|}{C}}}-CH=CH-\overset{R'}{\underset{}{\overset{|}{CH}}}\sim\sim\sim \tag{8-8}$$

XIII

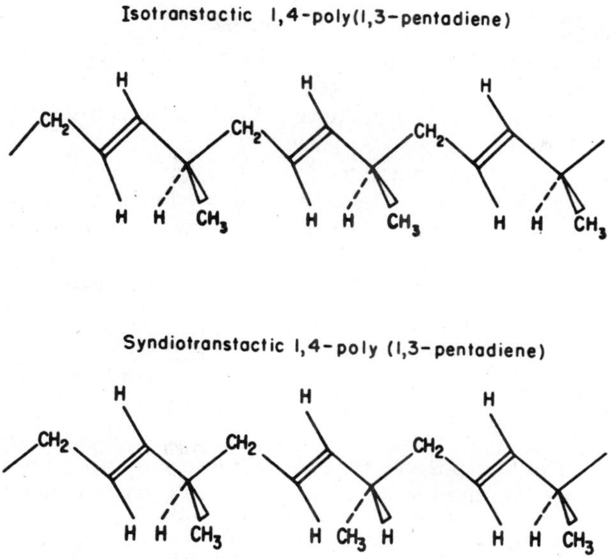

Isotranstactic 1,4-poly(1,3-pentadiene)

Syndiotranstactic 1,4-poly (1,3-pentadiene)

Fig. 8-6 Two of the stereoregular forms of 1,4-poly(1,3-pentadiene), $-\!\!+\!CH_2CH\!\!=\!\!CHCH (CH_3)\!\!+_n$.

which can exhibit tritacticity since the repeating unit contains three sites of steric isomerism—a double bond and the carbons holding R and R′ substituents. Eight different stereoregular polymers are possible with various combinations of ordered arrangements at the three sites. For example, polymer **XIV** possesses an erythrodi-

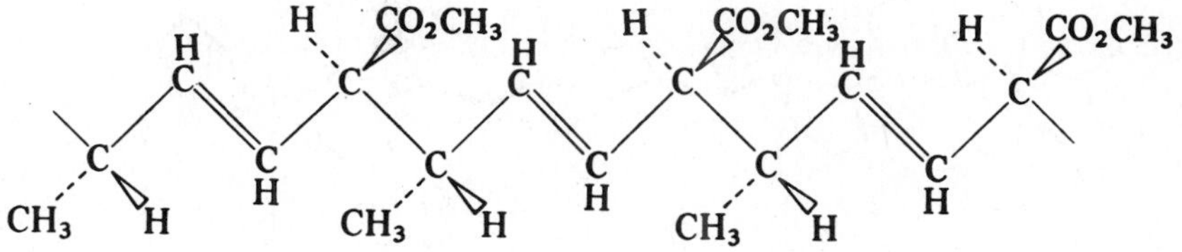

XIV

isotactic arrangement of R and R′ groups and an all-trans arrangement of the double bonds. The polymer is named transerythrodiisotactic 1,4-poly(methyl sorbate) or trans-erythrodiisotactic poly[3-(methoxycarbonyl)-4-methyl-1-butenylene] or diisotactic poly[*erythro*-3-(methoxycarbonyl)-4-methyl-*trans*-1-butenylene]. All four diisotactic polymers (cis and trans erythro and threo) are chiral and possess optical activity. Each of the four disyndiotactic polymers possesses a mirror glide plane and is achiral. For symmetrical 1,4-disubstituted 1,3-butadienes (R = R′), only the cis and transthreo-diisotactic structures are chiral. Each of the erythrodiisotactic and threodisyndiotactic polymers has a mirror plane. Each of the erythrodisyndiotactic polymers has a mirror glide plane.

8-1f   Other Polymers

The polymerization of the alkyne triple bond (Secs. 5-7d and 8-4d-2) and ring-opening methathesis polymerization of a cycloalkene (Secs. 7-8 and 8-4d-1) yield polymers containing double bonds in the polymer chain. Cis–trans isomerism is possible analogous to the 1,4-polymerization of 1,3-dienes.

Polymers containing rings incorporated into the main chain (e.g., by double-bond polymerization of a cycloalkene) are also capable of exhibiting stereoisomerism. Such polymers possess two stereocenters—the two atoms at which the polymer chain enters and leaves each ring. Thus the polymerization of cyclopentene to polycyclopentene (poly[1,2-cyclopentylene]) is considered in the same manner as that of a 1,2-disub-

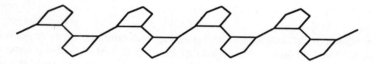

$$(8-9)$$

stituted ethylene. The four possible stereoregular structures are shown in Fig. 8-7. The erythro polymers are those in which there is a cis configuration of the polymer

Erythrodiisotactic polycyclopentene

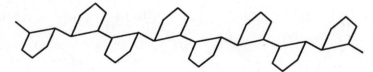

Thredoiisotactic polycyclopentene

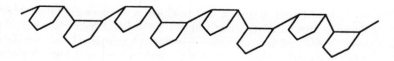

Erythrodisyndiotactic polycyclopentene

Threodisyndiotactic polycyclopentene

Fig. 8-7   Stereoregular forms of polycyclopentene.

chain bonds entering and leaving each ring; the threo polymers have a trans configuration of the polymer chain bonds entering and leaving each ring. The threodiisotactic
and threodisyndiotactic polymers are chiral while the two erythro polymers are achiral.
The erythrodiisotactic polymer has a mirror glide plane while the erythrodisyndiotactic
polymer has mirror planes perpendicular to the polymer chain axis. The situation is
different for an asymmetric cycloalkene such as 2-methylcyclopentene where the diisotactic structures are chiral while the disyndiotactic structures are achiral.

8-2 PROPERTIES OF STEREOREGULAR POLYMERS

8-2a Significance of Stereoregularity

The occurrence of stereoisomerism in polymers plays a major role in their practical
utilization. There are very significant differences in the properties of unordered and
ordered polymers as well as in the properties of ordered polymers of different types
(cis versus trans, isotactic versus syndiotactic). The ordered polymers are dramatically
different from the corresponding unordered structures in morphological and physical
properties.

8-2a-1   Isotactic, Syndiotactic, and Atactic Polypropylene

The regularity or lack of regularity in polymers affects their properties by way of large
differences in their abilities to crystallize. Atactic polymers are amorphous (noncrystalline), soft ("tacky") materials with little or no physical strength. The corresponding
isotactic and syndiotactic polymers are usually obtained as highly crystalline materials.
The ordered structures are capable of packing into a crystal lattice, while the unordered
structures are not. Crystallinity leads to high physical strength and increased solvent
and chemical resistance as well as differences in other properties that depend on
crystallinity. The prime example of the industrial utility of stereoregular polymers is
polypropylene. Isotactic polypropylene is a high-melting, strong, crystalline polymer,
which finds large-scale use as both a plastic and fiber (Sec. 8-4g-4) [Juran, 1989;
Lieberman and Barbe, 1988]. The annual United States production of isotactic polypropylene exceeds 7 billion pounds. Atactic polypropylene is an amorphous material
that finds some use in asphalt blends and formulations for sealants and adhesives but
the volumes are very much lower than for isotactic polypropylene.

While the properties and utility of isotactic polymers have and are being extensively
studied, those of syndiotactic polymers have received little attention. The reason is
the relative ease of forming isotactic polymers; stereospacific polymerizations that
yield syndiotactic structures are much less frequently encountered compared to those
that yield isotactic structures. However, in the case of polypropylene, the properties
of the syndiotactic polymer have been studied to a considerable extent [Youngman
and Boor, 1967]. Syndiotactic polypropylene, like the isotactic structure, is easily
crystallized. However, it has a lower T_m by about 20°C and is more soluble in ether
and hydrocarbons than isotactic polypropylene.

8-2a-2   cis- and trans-1,4-Poly-1,3-Dienes

Stereoisomerism in 1,4-poly-1,3-dienes results in significant differences in the properties of the cis and trans polymers. The trans isomer crystallizes to a greater extent

TABLE 8-1 Crystalline Melting and Glass Transition Temperatures[a]

Polymer	Isomer	T_g (°C)	T_m (°C)
1,4-Polybutadiene	Cis	-95	6
	Trans	-83	145
1,4-Polyisoprene	Cis	-73	28
	Trans	-58	74

[a]Data from Brandrup and Immergut [1989].

as a result of higher molecular symmetry and has higher T_m and T_g values (Table 8-1). These trends are the same as for small molecules such as *cis*- and *trans*-2-butenes. The differences in the cis and trans polymers leads to major differences in their properties and utilization. Thus, cis-1,4-polyisoprene has very low crystallinity, low T_g and T_m values, and is an excellent elastomer over a considerable temperature range, which includes ambient temperatures. About 2 billion pounds per year of cis-1,4-polyisoprene are used in the United States for tires, footwear, gloves, molded objects, adhesives, rubber bands, and other typical elastomer applications. trans-1,4-Polyisoprene is a much harder, much less rubbery polymer since it has relatively high T_g and T_m values and significant crystallinity. It is less an elastomer and more like a thermoplastic. trans-1,4-Polyisoprene has very good resistance to abrasion and is used in golf-ball covers and orthopedic devices. 1,4-Polyisoprene is found in nature in both the cis and trans forms. Hevea rubber, obtained from the *Hevea brasiliensis* tree (in Brazil, Indonesia, Malaysia, Sri Lanka), is the major naturally occurring 1,4-polyisoprene. It contains more than 98% of the double bonds in the cis configuration. (Over 90% of all cis-1,4-polyisoprene used commercially is the natural Hevea rubber.) cis-1,4-Polyisoprene is also found in the guayule bush. Other trees, mostly in Central America and Malaysia, yield gutta percha or balata rubber, which is predominantly the trans isomer.

8-2a-3 Cellulose and Amylose

The isomeric polysaccharides amylose (a component of starch) and cellulose also show the significance of stereoisomerism on polymer properties. Cellulose and amylose have the structures shown in Fig. 8-8. (Amylopectin, the other component of starch, has the same structure as amylose except that it is branched at carbon 6.) Both are polymers of glucose in which the glucose units are joined together by glucoside linkages at carbons 1 and 4. The two polysaccharides differ only in the configuration at carbon 1.

In terms of the nomenclature used for stereoregular polymers, cellulose has a threodisyndiotactic structure while amylose is erythrodiisotactic. In the nomenclature of carbohydrate chemistry, cellulose consists of β-1,4-linked D-glucopyranose chains and amylose of α-1,4-linked D-glucopyranose chains. The 1,4-linkage is trans (diequatorial) in cellulose and cis (equatorial–axial) in amylose. The difference in structure results in amylose existing in a random-coil conformation while cellulose exists in an extended-chain conformation. This leads to closer packing of polymer molecules with much greater intermolecular attractive forces and crystallinity in cellulose relative to amylose (and also amylopectin). Cellulose, compared to amylose, has good physical strength and mechanical properties, decreased solubility, and increased stability to hydrolysis. The result is that cellulose is used as a structural material both in nature

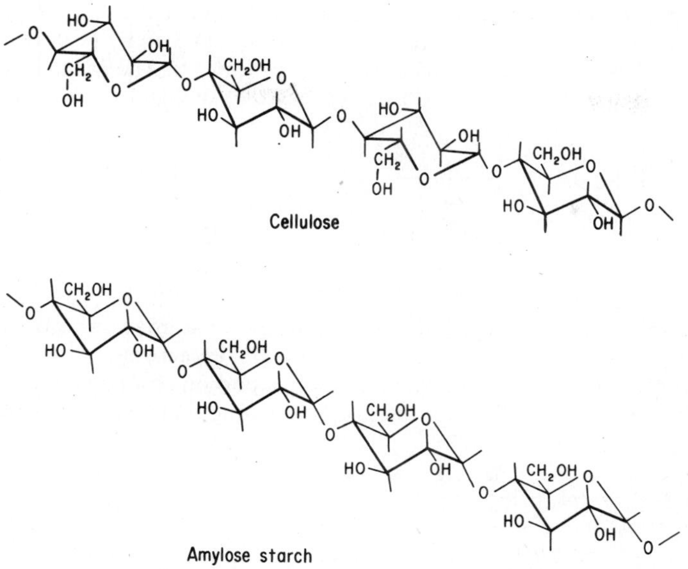

Fig. 8-8   Structures of cellulose and amylose starch.

(plants) and by people. Although starch is not useful as a structural material, it does serve important functions as a food source and storage form of energy for both plants and animals. The structure of each polysaccharide is uniquely fitted to its biological function in nature.

8-2b   Analysis of Stereoregularity

Infrared spectroscopy has been used for quantitatively measuring the amounts of 1,2-, 3,4-, cis-1,4-, and trans-1,4-polymers in the polymerization of 1,3-dienes; its use for analysis of isotactic and syndiotactic polymer structures is very limited [Coleman et al., 1978; Tosi and Ciampelli, 1973]. Nuclear magnetic resonance spectroscopy is the most powerful tool for detecting both types of stereoisomerism in polymers. High-resolution proton NMR and especially ^{13}C NMR allow one to obtain considerable detail about the sequence distribution of stereoisomeric units within the polymer chain [Bovey, 1972, 1982; Tonelli, 1989; Zambelli and Gatti, 1978].

Consider the description of the sequence distribution of isotactic and syndiotactic placements in the polymerization of a monosubstituted ethylene. The approach is general and can be applied with appropriate modification to the 1,4-polymerization of a 1,3-diene. *Dyad tacticity* is defined as the fractions of pairs of adjacent repeating units that are isotactic or syndiotactic to one another. The isotactic and syndiotactic dyads, more frequently referred to as *meso* and *racemic dyads*, can be depicted as

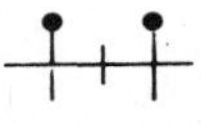

Isotactic   Syndiotactic
XVa   **XVb**

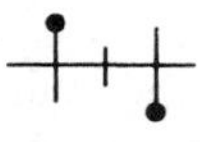

where the horizontal line represents a segment of the polymer chain, ↑ represents the configuration at the stereocenter in each repeating unit, and | represents the two hydrogens at the carbon in between adjacent stereocenters. The fractions of isotactic and syndiotactic dyads are referred to as (m) and (r) or i and s, respectively. *Triad*

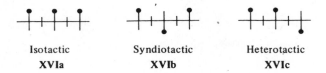

Isotactic	Syndiotactic	Heterotactic
XVIa	**XVIb**	**XVIc**

tacticity describes isotactic, syndiotactic, and heterotactic triads whose fractions are designated as (mm), (rr), and (mr), or I, S, and H, respectively.

These definitions are clarified by considering a portion of a polymer chain such as **XVII**. Chain segment **XVII** has a total of 9 repeating units but only 8 dyads and 7

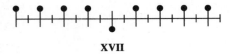

XVII

triads. There are 6 meso dyads and 2 racemic dyads; $(m) = \frac{3}{4}$, $(r) = \frac{1}{4}$. There are 4 isotactic, 2 heterotactic, and 1 syndiotactic triads; $(mm) = \frac{4}{7}$, $(rr) = \frac{2}{7}$, $(mr) = \frac{1}{7}$.

The dyad and triad fractions each total unity by definition, that is,

$$(m) + (r) = 1 \tag{8-10}$$

$$(mm) + (rr) + (mr) = 1 \tag{8-11}$$

and the two are related by

$$(m) = (mm) + 0.5(mr) \tag{8-12}$$

$$(r) = (rr) + 0.5(mr) \tag{8-13}$$

It is clear that a determination of any two triad fractions allows a complete definition of both the triad and dyad structures of a polymer via Eqs. 8-10 through 8-13. An atactic polymer is one in which $(r) = (m) = 0.5$ and $(mm) = (rr) = 0.25$, $(mr) = 0.5$ with a random distribution of dyads and triads. The (all-) isotactic polymer has $(m) = (mm) = 1$. The (all-) syndiotactic polymer is defined by $(r) = (rr) = 1$. For random distributions with $(m) \neq (r) \neq 0.5$ or $(mm) \neq (rr) \neq 0.25$, one has different degrees of syndiotacticity or isotacticity. Isotacticity predominates when $(m) > 0.5$ and $(mm) > 0.25$ and syndiotacticity predominates when $(r) > 0.5$ and $(rr) > 0.25$. These polymers are *random tactic polymers*, containing random placement of isotactic and syndiotactic dyads and triads. When the distribution of dyads and triads is less than completely random, the polymer is a *stereoblock polymer* in which there are blocks (which may be short or long) of isotactic and syndiotactic dyads and triads.

The advent of high-resolution NMR allows the determination of tetrad, pentad, and even higher sequence distributions in many polymers [Bovey, 1972; Farina, 1987]. The tetrad distribution consists of the isotactic sequence *mmm*, the syndiotactic sequence *rrr*, and the heterotactic sequences *mmr*, *rmr*, *mrm*, *rrm*. The sum of the tetrad fractions is unity, and the following relationships exist:

$$(mm) = (mmm) + 0.5(mmr) \tag{8-14}$$

$$(rr) = (rrr) + 0.5(mrr) \tag{8-15}$$

$$(mr) = (mmr) + 2(rmr) = (mrr) + 2(mrm) \tag{8-16}$$

The pentad distribution consists of the isotactic sequence *mmmm*, the syndiotactic sequence *rrrr*, and the heterotactic sequences *rmmr*, *mmmr*, *mmrr*, *rmrm*, *rmrr*, *mrrm*, *rrmm*. The sum of the pentad sequences is unity, and the following relationships exist:

$$(mmmr) + 2(rmmr) = (mmrm) + (mmrr) \tag{8-17}$$

$$(mrrr) + 2(mrrm) = (rrmr) = (rrmm) \tag{8-18}$$

$$(mmm) = (mmmm) + 0.5(mmmr) \tag{8-19}$$

$$(mmr) = (mmmr) + 2(rmmr) = (mmrm) + (mmrr) \tag{8-20}$$

$$(rmr) = 0.5(mrmr) + 0.5(rmrr) \tag{8-21}$$

$$(mrm) = 0.5(mrmr) + 0.5(mmrm) \tag{8-22}$$

$$(rrm) = 2(mrrm) + (mrrr) = (mmrr) + (rmrr) \tag{8-23}$$

$$(rrr) = (rrrr) + 0.5(mrrr) \tag{8-24}$$

8-3 FORCES OF STEREOREGULATION IN ALKENE POLYMERIZATIONS

Having discussed the types of stereoisomerism that are possible in polymers, one can consider the forces that are actually operating toward or against stereospecificity in a polymerization. The stereochemistry of alkene polymerizations will be considered first. The extent of stereospecificity in a polymerization is determined by the relative rate at which an incoming monomer molecule is added with the same configuration as the preceding monomer unit compared to its rate of addition with the opposite configuration. Different forces determine the relative rates of these two modes of addition depending on whether the active end of a propagation polymer chain is a free species or one that is coordinated (associated) with the initiator. For freely propagating species that are not coordinated in any manner both modes of addition are possible. The stereoregularity of the final polymer product is dependent primarily on the polymerization temperature, which determines the relative rates of the two modes of addition. The situation is quite different when the propagating species is coordinated with the initiator. One or the other of the modes of addition may be prevented from taking place by the configuration of the coordinated complex (usually consisting of the propagating chain end, initiator, and monomer). Under these circumstances coordination directs the mode of monomer addition in a stereospecific manner.

8-3a Radical Polymerization

Radical polymerizations are generally considered to involve freely propagating species. The planar or nearly planar trigonal carbon atom does not have a specified configuration since there is free rotation about the terminal carbon–carbon bond (as indicated in Eq. 8-25 by the circular arrows). The configuration of a monomer unit in the polymer chain is not determined during its addition to the radical center but only after the next monomer unit adds to it. The situation can be depicted as

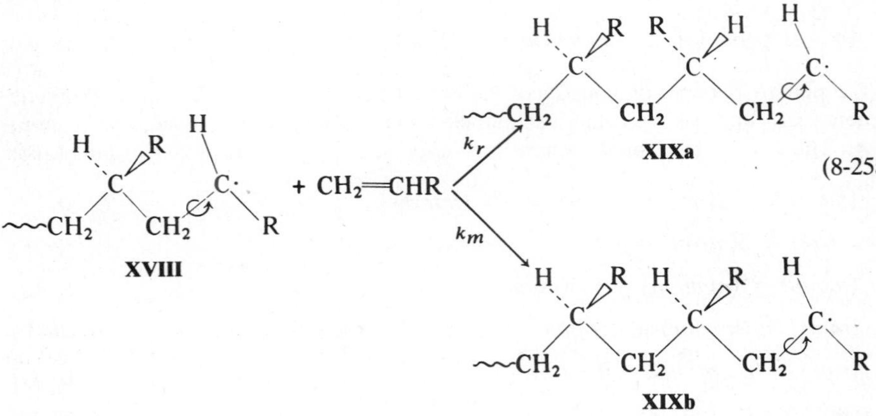

(8-25a)

(8-25b)

where the two placements—syndiotactic (Eq. 8-25a) and isotactic (Eq. 8-25b)—take place with the rate constants k_r and k_m, respectively. Whether the same placement is propagated through successive additions of monomer units determines the stereoregularity of the final polymer molecule. The amount and type of stereoregularity are determined by the value of k_r/k_m. Isotactic polymer is produced if this ratio is zero, syndiotactic polymer if it is infinity, and atactic polymer if it is unity. For k_r/k_m values between unity and infinity, the polymer is partially atactic and partially syndiotactic. For k_r/k_m values between zero and unity, the polymer is partially atactic and partially isotactic.

The value of k_r/k_m is determined by the difference $\Delta\Delta G^{\ddagger}$ in the free energies of activation between the syndiotactic $\Delta G_r^{\ddagger}$ and isotactic $\Delta G_m^{\ddagger}$ placements

$$\frac{k_r}{k_m} = \exp\left(\frac{-\Delta\Delta G^{\ddagger}}{RT}\right) = \exp\left\{\left(\frac{\Delta\Delta S^{\ddagger}}{R}\right) - \left(\frac{\Delta\Delta H^{\ddagger}}{RT}\right)\right\} \qquad (8\text{-}26)$$

with

$$\Delta\Delta G^{\ddagger} = \Delta G_r^{\ddagger} - \Delta G_m^{\ddagger} \qquad (8\text{-}27a)$$

$$\Delta\Delta S^{\ddagger} = \Delta S_r^{\ddagger} - \Delta S_m^{\ddagger} \qquad (8\text{-}27b)$$

$$\Delta\Delta H^{\ddagger} = \Delta H_r^{\ddagger} - \Delta H_m^{\ddagger} \qquad (8\text{-}27c)$$

where $\Delta G_r^{\ddagger}$, $\Delta H_r^{\ddagger}$, $\Delta S_r^{\ddagger}$ are the activation free energy, enthalpy, and entropy for syndiotactic placement and $\Delta G_m^{\ddagger}$, $\Delta H_m^{\ddagger}$, $\Delta S_m^{\ddagger}$ are the corresponding quantities for isotactic placement. Stereoregularity should be temperature-dependent with ordered structures favored at low temperatures [Huggins, 1944]. Calculations based on small molecules indicated that the differences in activation enthalpy $\Delta\Delta H^{\ddagger}$ and entropy

$\Delta\Delta S^{\ddagger}$ between syndiotactic and isotactic placements are both expected to be small; $\Delta\Delta H^{\ddagger}$ is about -4 to -8 kJ/mole and $\Delta\Delta S^{\ddagger}$ about 0 to -4 J/mole-K [Fordham, 1959]. Syndiotactic placement is favored over isotactic placement primarily because of the enthalpy difference. The small energy differences between syndiotactic and isotactic placements have been confirmed for several monomers by studying the temperature dependence of tacticity [Bovey, 1972; Pino and Suter, 1976, 1977]. Thus, $\Delta\Delta H^{\ddagger}$ and $\Delta\Delta S^{\ddagger}$ are -4.5 kJ/mole and -4.2 J/mole-K, respectively, for methyl methacrylate and -1.3 kJ/mole and -2.5 J/mole-K, respectively, for vinyl chloride [Bovey et al., 1967; Fox and Schnecko, 1962].

The slight preference for syndiotactic placement over isotactic placement is a consequence of steric and/or electrostatic repulsions between substituents in the polymer chain. Repulsions between R groups, more specifically, between R groups on the terminal and penultimate units of the propagating chair, are minimized in the transition state of the propagation step (and also in the final polymer molecule) when they are located in the alternating arrangement of syndiotactic placement. The steric and electrostatic repulsions between R groups are maximum for isotactic placement.

The difference in activation free energies for syndiotactic and isotactic placements leads to an increasing tendency toward syndiotacticity with decreasing polymerization temperature. Figure 8-9 shows the increase in the fraction of syndiotactic dyads, (r), of poly(vinyl chloride from 0.51 to 0.67 as the reaction temperature decreases from 120 to $-78°C$ [Talamini and Vidotto, 1967]. Corresponding data for methyl methacrylate shows an increase in (r) from 0.64 at 250°C to 0.87 at $-78°C$ [Fox and Schnecko, 1962; Otsu et al., 1966]. With decreasing temperature, the energy difference between syndiotactic and isotactic placements exerts a progressively increasing influence on the stereospecificity of the polymerization; at high temperatures, their effects are progressively diminished.

Since radical polymerizations are generally carried out at moderately high temperatures, most of the resulting polymers are highly atactic. This does not mean that there is a complete absence of syndiotacticity. There is a considerable difference in the extent of syndiotacticity from one polymer to another. Thus, methyl methacrylate has a much greater tendency toward syndiotactic placement than vinyl chloride. Whereas the poly(vinyl chloride) produced at the usual commercial polymerization temperatures (ca. 60°C) is essentially completely atactic, that is, $(r) \simeq (m) \simeq 0.5$, this is not

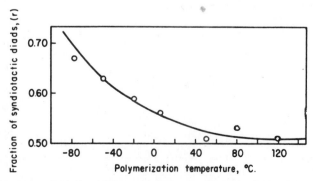

Fig. 8-9 Dependence of syndiotacticity on temperature for the radical polymerization of vinyl chloride. After Talamini and Vidotto [1967] (by permission of Huthig and Wepf Verlag, Basel).

the case for poly(methyl methacrylate). The polymerization of methyl methacrylate, usually carried out at temperatures up to 100°C, yields polymers with appreciable syndiotacticity—(r) is 0.73 at 100°C. The difference is a consequence of methyl methacrylate being a 1,1-disubstituted ethylene, leading to greater repulsions between substituents in adjacent monomer units.

8-3b   Ionic and Coordination Polymerization

8-3b-1   Effect of Coordination

For ionic chain polymerizations in solvents with high solvating power where solvent-separated ion-pair or free-ion propagating species are involved, the factors governing the stereochemistry of the reaction are similar to those for radical polymerization. Syndiotactic polymerization is increasingly favored as the polymerization temperature is lowered. When, however, polymerizations are carried out in solvents with poor solvating power, there is extensive coordination among the initiator, propagating chain end, and monomer, and the stereochemical result may be quite different. Heterogeneous reaction conditions are often necessary to achieve the needed level of coordination. In the usual case, propagation is prevented from taking place by one or the other of the two placements ($R-$ or $S-$) as coordination becomes the dominant driving force for stereospecificity in the polymerization. The k_r/k_m ratio tends towards a value of zero and stereospecific polymerization occurs with isotactic polymerization. Coordination can also yield syndiotactic polymers in certain systems, although the usual result is isotactic placement. Table 8-2 shows a few examples of highly stereospecific polymerizations. The stereoregularity of these systems is greater than 90–95% in terms of syndiotactic or isotactic dyads. The methyl methacrylate and i-butyl vinyl ether polymerizations involve initiators and reaction conditions discussed previously (Chap. 5) without consideration of stereochemistry.

The first reported instance of coordination-directed stereospecific polymerization was probably the cationic polymerization of i-butyl vinyl ether in 1947 [Schildknecht et al., 1947]. A semicrystalline polymer was obtained when the reaction was carried out at -80 to $-60°C$ using boron trifluoride etherate as the initiator with propane as the solvent. The full significance of the polymerization was not realized at the time

TABLE 8-2   Stereospecific Polymerizations

Monomer	Reaction Conditions	Polymer Structure
i-Butyl vinyl ether[a]	BF$_3$ etherate in propane at -80 to $-60°C$	Isotactic
Methyl methacrylate[b]	ØMgBr (30°C) or n-C$_4$H$_9$Li (-78 to 0°C) in ØCH$_3$	Isotactic
Methyl methacrylate[b]	n-C$_4$H$_9$Li in THF at $-78°C$	Syndiotactic
Propene[c]	TiCl$_4$, (C$_2$H$_5$)$_3$Al in n-heptane at 50°C	Isotactic
Propene[d]	VCl$_4$, (C$_2$H$_5$)$_3$Al, ØOCH$_3$ in ØCH$_3$ at $-78°C$	Syndiotactic

[a]Data from Schildknecht et al. [1947].
[b]Data from Yuki et al. [1975].
[c]Data from Natta et al. [1958].
[d]Data from Zambelli et al. [1963].

as the crystallinity was attributed to a syndiotactic structure. X-Ray diffraction studies in 1956 indicated that the polymer was isotactic [Natta et al., 1956a, 1956b]. The field of stereospecific polymerization came into existence in the mid-1950s with the work of Ziegler in Germany and Natta in Italy. Ziegler was studying the reactions of ethylene catalyzed by trialkylaluminium at high temperatures and pressures. Both oligomerization to higher 1-alkenes and polymerization occurred. The highest polymer molecular weights achieved were 5000. The presence of a transition-metal compound in combination with the trialkylaluminum had a dramatic effect on the polymerization. High-molecular-weight polyethylene was formed at low temperatures (50–100°C) and low pressures [Ziegler, 1964; Ziegler et al., 1955]. (The polyethylene produced by this route is much less branched and has property enhancements compared to polyethylene produced by radical polymerization of ethylene; see Sec. 8-4g.) Natta employed Ziegler's initiator system to achieve stereospecific polymerizations (both isotactic and syndiotactic) of various 1-alkenes (α-olefins) [Natta, 1965]. The scientific and practical significance of their work earned Natta and Ziegler the joint award of the Nobel Price in Chemistry in 1963.

The initial work of Natta and Ziegler has led to the development of a large number of initiator systems obtained by the interaction of an organometallic compound or hydride of a Group I–III metal with a halide, hydroxide, alkoxide or other derivative of a Group IV–VIII transition metal [Pasquon, et al. 1989]. The polymerizations are carried out in hydrocarbon solvents such as n-heptane. Some of the compounds used as the Group I–III metal component are triethylaluminum, diethylaluminum chloride, and diethylzinc; titanium trichloride, vanadium trichloride, dicyclopentadienyltitanium dichloride, and chromium triacetylacetonate are examples of the transition metal component [Choi and Ray, 1985; Corradini et al., 1989; Pino et al., 1987; Tait, 1989]. The most commonly used system is that based on titanium and aluminum components. These initiator systems are referred to as *Ziegler–Natta* or *coordination initiators*. The last term is the most appropriate since it also includes other initiators that are not of the same identity as the Ziegler–Natta initiators but that bring about stereospecific polymerization by a common mechanism. Examples of coordination initiators that are not of the Ziegler–Natta type are n-butyllithium, phenylmagnesium bromide, and boron trifluoride. As with other initiators, including ionic and radical, the term *catalyst* is usually used instead of *initiator*. Polymerizations brought about by coordination initiators are referred to as *coordination polymerizations*. The terms *isospecific* and *syndiospecific* are often used to describe initiators and polymerizations producing isotactic and syndiotactic polymers, respectively.

8-3b-2 Mechanism of Stereospecific Placement

Coordination initiators perform two functions. First, they supply the species that initiates the polymerization. Second, the fragment of the initiator aside from the initiating portion has unique coordinating powers. Coordination of this fragment (which may be considered as the counterion of the propagating center) with both the propagating chain end and the incoming monomer occurs so as to orient the monomer with respect to the growing chain end and bring about stereospecific addition. Many different mechanisms have been advanced to explain the usual isotactic placement obtained with coordination initiators [Corradini et al., 1989; Tait and Watkins, 1989]. Stereospecific polymerization is best considered as a concerted, multicentered reaction. Figure 8-10 depicts the general situation for an anionic coordination polymerization proceeding with isotactic placement. The polymer chain end has a partial negative

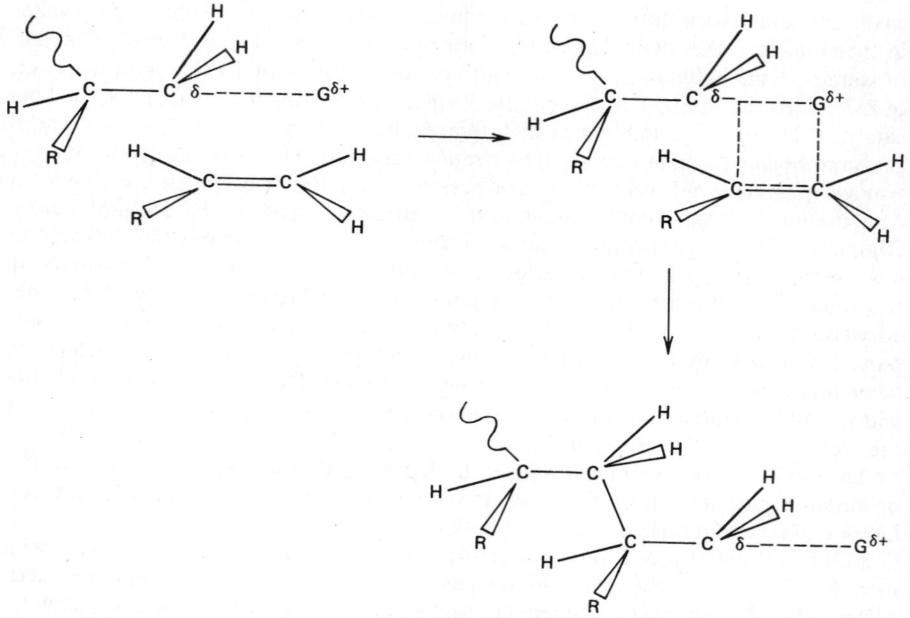

Fig. 8-10   Mechanism of stereospecific polymerization with isotactic placement.

charge with the initiator fragment G (the counterion) having a partial positive charge. (Cationic coordination polymerization involves a similar mechanism except for reversal of the signs of the partial charges.) The initiator fragment G is coordinated with both the propagating chain end and the incoming monomer molecule.

Monomer is oriented and "held in place" by coordination during addition to the polymer chain. Coordination between the initiator fragment and the propagating center is broken simultaneously with the formation of bonds between the propagating center and the new monomer unit and between the initiator fragment and the new monomer unit. Propagation proceeds in the four-center cyclic transition state by the insertion of monomer between the initiator fragment and the propagating center. The initiator fragment essentially acts as a template or mold for the successive orientations and isotactic placements of the incoming monomer units. Isotactic polymerization occurs because the initiator fragment forces each monomer unit to approach the propagating center with the same face. This mechanism is referred to as *catalyst site control* or *enantiomorphic site control*. For the polymerization described in Fig. 8-10, monomer approaches the propagating center with its *re face* facing the propagating center. The face of a monomer molecule is labeled as the *re face* if viewing the molecule from that face shows the groups attached to the CHR carbon decreasing in priority in a clockwise manner ($=C > R > H$). The opposite face is the *si face*, in which priorities decrease in a counterclockwise manner when the face is viewed from the propagating center. The re and si faces of a monomer are *enantiotopic*; that is, the products formed by additions of a propagating center to the two faces have opposite configurations. (The term *prochiral* has been used for a carbon atom such as the CHR carbon of a monomer molecule to indicate additions to the si and re faces yield, respectively, one enantiomer or the other of a pair of enantiomers.) The description in Fig. 8-10 for isotactic polymerization could also have been shown with the R and H groups of the last monomer unit in the polymer chain reversed (R in back and H in front), in which case isotactic propagation consists of monomer always approaching the propagating center with its si face.

The property of the initiator fragment that forces successive placements of monomer to occur with the same face is chirality. There is a stereochemical "fit" between initiator and monomer, which overrides the usual tendency toward moderate syndiotacticity. The initiator in isotactic polymerization is usually a racemic mixture of two enantiomers. One of the enantiomers yields isotactic polymer by forcing all propagations via the re face of monomer; the other enantiomer, via the si face. The isotactic polymer structures formed from the enantiomeric initiators are the same polymer; that is, they are superposable (ignoring the effects of end groups) for most monomers. (Exceptions are some ring-opening polymerizations, e.g., propylene oxide, and 1,4-polymerizations of some substituted 1,3-dienes, e.g., 1,3-pentadiene, where the two isotactic polymers are enantiomeric.) Most isospecific initiators are heterogeneous, with chirality attributable to the chiral crystal lattice in the vicinity of the initiator active center. The chirality of homogeneous isospecific initiators resides in their molecular structures; that is, the initiator molecule is chiral (Sec. 8-4f).

When the initiator is achiral, it can coordinate equally well (more or less) with either face of the incoming monomer. This results in either atactic or syndiotactic polymerizations depending on the specific interaction of the initiator fragment G, polymer chain end, and incoming monomer as well as the reaction temperature. Syndiotactic polymerization occurs in the relatively few systems where coordination accentuates the repulsive interactions between substituents on the polymer chain end and incoming monomer. The driving force for syndiotactic placement is the same as previously described for low-temperature polymerizations but the initiator exaggerates the situation. Instead of obtaining a polymer with moderate syndiotacticity, say, 60–70% syndiotactic dyads, one obtains 90% or higher syndiotactic dyads. This mechanism for syndiospecific polymerization is referred to as *chain end control*.

The exact nature of the initiator required to bring about isospecific polymerization differs considerably depending on the monomer. The ease of imposing catalyst site control on the entry of a monomer unit into the polymer chain increases with the ability of monomer to coordinate with initiator. The coordinating ability of a monomer depends on its polarity. This is greatest for those monomers (e.g., acrylates, methacrylates, vinyl ethers) with polar functional groups capable of taking an active and strong role in the coordination process. Ethylene, α-olefins (e.g., propene, 1-butene), and other alkenes without polar substituents have poor coordinating power. The result is that nonpolar monomers require the use of initiators with very strong coordinating and stereoregulating powers to produce isospecific polymerization. This almost always has necessitated the use of heterogeneous Ziegler–Natta initiators to impose the most severe hindrance on the approach of a monomer molecule to the propagating chain end. With few exceptions, the use of homogeneous (soluble) initiators with nonpolar monomers results in atactic polymers. There are two exceptions to this generalization. A few soluble Ziegler–Natta initiators (e.g., the vanadium system in Table 8-2) are syndiospecific for propene. Certain rigid chiral metallocene derivatives were recently found to be isospecific for α-olefin (1-alkene) monomers (Sec. 8-4f).

For polar monomers, heterogeneity is seldom a requirement for isotactic polymerization; syndiotactic polymers are obtained only with soluble initiators. Styrene and

1,3-dienes are intermediate in behavior between the polar and nonpolar monomers since the phenyl and alkenyl substituents are not strongly polar. These monomers are polymerized to isotactic structures with both homogeneous and heterogeneous initiators.

8-4   ZIEGLER–NATTA POLYMERIZATION OF NONPOLAR ALKENE MONOMERS

The Ziegler–Natta initiators (also referred to as *catalysts*) are a remarkable group of initiators. They are the only initiators that polymerize α-olefins such as propene and 1-butene. Recall that α-olefins are not polymerized by either radical or ionic initiators (Chaps. 3, 5). Ziegler–Natta initiators are important not only for obtaining stereoregular polymers but also for the polymerization of ethylene. Polyethylene obtained by these initiators is different in structure and properties from that obtained by radical polymerization (Sec. 8-4g).

Since the original discoveries of Ziegler and Natta, there have been literally thousands of different combinations of transition and Group I–III metal components, often together with other compounds such as electron donors, studied for use in alkene polymerizations. The resulting initiators exhibit a range of behaviors in terms of activity and stereospecificity. The term *activity* as used in this text and in most literature references applies exclusively to the rate of polymerization. Activity is often expressed in terms of kilograms of polymer formed per gram of initiator component. Modification of an initiator system to increase activity has often come at the expense of stereospecificity and vice versa. The great utility of the Ziegler–Natta initiator system is the ability to make changes in one or another of the components to obtain very high stereospecificity with high activity. The choice of the initiator components has evolved in an empirical manner due to a lack of understanding of the detailed structure of almost all Ziegler–Natta initiators and the mechanism of their stereoregulating action.

8-4a   Mechanism of Ziegler–Natta Polymerization

Many mechanisms have been proposed to explain the stereospecificity of Ziegler–Natta initiators [Boor, 1979; Carrick, 1973; Corradini et al., 1989; Cossee, 1967; Ketley, 1967a, 1967b; Tait and Watkins, 1989; Zambelli and Tosi, 1974]. Most mechanisms contain considerable details that distinguish them from each other but cannot be verified. All too often, the detailed mechanisms are not applicable to initiator systems other than the specific system for which they are proposed. In this section the mechanistic features of Ziegler–Natta initiators and polymerizations are considered with emphasis on those characteristics that hold for the large bulk of initiator systems. The major interest will be on the titanium–aluminum systems, more specifically, $TiCl_3$ with $Al(C_2H_5)_2Cl$ and $TiCl_4$ with $Al(C_2H_5)_3$—probably the most studied systems.

It is useful to review the evolution of the titanium–aluminum initiator system before proceeding with the mechanistic consideration. The original initiator used by Ziegler for ethylene polymerization was obtained in situ as a precipitate on mixing the soluble components $TiCl_4$ with $Al(C_2H_5)_3$ in a hydrocarbon solvent. This mixture was then used directly for initiating polymerization. Natta, recognizing that the major product of the reaction was β-$TiCl_3$ (brown in color), explored various methods of preforming it outside the polymerization system, for example, by reduction of $TiCl_4$ with hydrogen, aluminum, and various alkylaluminium compounds, including $Al(C_2H_5)_2Cl$. The stereospecificity of these early initiator systems was low with *isotactic indices* of 20–40%. (The isotactic index, a measure of the isotactic content of a polymer, is the percentage of the sample insoluble in a hydrocarbon solvent such as boiling *n*-heptane. This is not as informative a technique as high-resolution NMR since insoluble molecules may contain some syndiotactic and atactic sequences and soluble molecules may contain some isotactic sequences. It does, however, give a simple measure of isotacticity that is usually within about 10% of the value obtained from NMR.)

There was a dramatic improvement in stereospecificity, isotactic index = 80–95%, when the α-, δ-, or γ-crystalline form of $TiCl_3$ (all violet in color) was used directly. Also, the initiator activity was enhanced by various ball milling and heat treatments of the initiator components before and after mixing. (Ball milling not only increases surface area but also facilitates reactions between the components of the initiator system.) However, the activity was low in comparison to the present initiator systems, with considerably less than 1% of the Ti being active in polymerization. Subsequent generations of initiators involved large increases in activity without sacrificing stereospecificity. The effective surface area of the active component of the initiator system was increased by more than two orders of magnitude by using a magnesium chloride (solid) support. Stereospecificity was kept high by the presence of electron-donor additives such as ethyl benzoate. Thus, a typical recipe for a present-day *superactive* or *high-mileage* initiator system involves initial ball milling (mechanical grinding or mixing) of magnesium chloride (or the alkoxide) and $TiCl_4$ followed by the addition of $Al(C_2H_5)_3$ with an organic Lewis base usually added in each of the steps of initiator preparation [Chien et al, 1982; Hu and Chien, 1988]. These initiator systems have very high activity (50–2000 kg polypropylene/g Ti) coupled with high isospecificity (>90–98% isotactic dyads). High activity is important not so much for the savings in amount of initiator required but for minimizing the expensive task of initiator removal from the polymer product.

A number of general questions arise concerning the mechanism of Ziegler–Natta polymerizations. What is the chemical nature of the propagating species? Does propagation take place at a carbon–to-transition-metal bond or at a carbon–to-Group I–III-metal bond? What is the chemical and physical identity of the actual or active initiator species that initiates polymerization?

8-4a-1 *Chemical Nature of Propagating Species*

There was some confusion in the early literature concerning the nature of the propagating species. Some polymerizations reported as Ziegler–Natta polymerizations were conventional free-radical, cationic, or anionic polymerizations proceeding without high stereospecificity. Some Ziegler–Natta initiators contain components that are capable of initiating conventional ionic polymerizations of certain monomers. Thus the use of alkyllithium as the Group I–III metal component can result in anionic polymerization of monomers such as acrylates and methacrylates. Monomers with electron-rich double bonds, such as vinyl ethers, can undergo cationic polymerization when one uses an initiator component such as $TiCl_4$ or R_2AlCl.

Most Ziegler–Natta components participate in a complex set of reactions involving alkylation and reduction of the transition-metal component by the Group I–III component as shown below for $TiCl_4$ + AlR_3:

$$TiCl_4 + AlR_3 \rightarrow TiCl_3R + AlR_2Cl \tag{8-28}$$

$$TiCl_4 + AlR_2Cl \rightarrow TiCl_3R + AlRCl_2 \tag{8-29}$$

$$TiCl_3R + AlR_3 \rightarrow TiCl_2R_2 + AlRCl_2 \tag{8-30}$$

$$TiCl_3R \rightarrow TiCl_3 + R\cdot \tag{8-31}$$

$$TiCl_3 + AlR_3 \rightarrow TiCl_2R + AlR_2Cl \tag{8-32}$$

$$R\cdot \rightarrow \text{combination} + \text{disproportionation} \tag{8-33}$$

Radicals produced in Eq. 8-31 are capable of initiating radical polymerizations with some monomers, for example, vinyl chloride.

Ionic as well as radical coordination mechanisms were originally proposed for the Ziegler–Natta polymerizations of ethylene, propylene, 1,3-dienes, and styrene. The experimental evidence favors an ionic mechanism [Boor, 1979]. Radical chain-transfer agents have no effect on polymer molecular weight. Changes in the initiator components often result in different monomer reactivity ratios in copolymerization.

The generally accepted mechanism for the stereospecific polymerization of α-olefins and other nonpolar alkenes is a π-complexation of monomer and transition metal (utilizing the latter's d-orbitals) followed by a four-center anionic coordination insertion process in which monomer is inserted into a metal–carbon bond as described in Fig. 8-10. Support for the initial π-complexation has come from ESR, NMR, and IR studies [Burfield, 1984]. The insertion reaction has both cationic and anionic features. There is a concerted nucleophilic attack by the incipient carbanion polymer chain end on the α-carbon of the double bond together with an electrophilic attack by the cationic counterion on the alkene π-electrons.

The anionic character of the polymerization is consistent with the polymerization rate decreasing in the order ethylene > propylene > 1-butene [Bier, 1961; Boor, 1967]. The reverse order is expected for a polymerization involving the conversion of a monomer into the corresponding carbocation. For addition of a carbanion to the monomers, attack occurs at the α-carbon to form the less substituted (and more stable) carbanion. Further, α-substituents sterically hinder the approach of a carbanion and/ or counterion with the result that reactivity decreases with increasing substituent size. Evidence for the anionic nature of propagation also comes from studies in which labeled methanol is used to terminate chain growth. The terminated polymer is radioactive when CH_3O^3H is used, while termination by $^{14}CH_3OH$ yields a nonradioactive polymer [Burfield and Savariar, 1979; Zakharov et al., 1977]. Additional verification comes from experiments with ^{14}CO and $^{14}CO_2$ [Mejzlik and Lesna, 1977].

8-4a-2 *Primary vs Secondary Insertion; Regiospecificity*

The insertion reaction shown in Fig. 8-10 is *primary insertion* (or *1,2-addition*)—the unsubstituted end of the double bond carries the partial negative charge and is attached to the counterion G. The other possibility is *secondary insertion* (or *2,1-addition*) where the substituted end of the double bond becomes attached to G. The two modes of insertion are described by

$$\sim\sim\sim G + RCH{=}CH_2 \rightarrow \sim\sim\sim CHR{-}CH_2{-}G \qquad (8\text{-}34)$$

$$\sim\sim\sim G + CH_2{=}CHR \rightarrow \sim\sim\sim CH_2{-}CHR{-}G \qquad (8\text{-}35)$$

Analysis for the mode of insertion involves the simultaneous determination of the degree of regiospecificity. Carbon-13 NMR analysis shows isospecific polymerizations of propylene to be completely regiospecific [Doi, 1979a, 1979b; Doi et al., 1979; Natta et al., 1956a, 1956b; Zambelli et al., 1979, 1982a, 1982b]. There are no detectable regioirregular placements in the isotactic fractions (the fractions insoluble in boiling *n*-heptane). The soluble fractions, containing varying degrees of syndiotactic and atactic sequences, show 0.1–5% regioirregular placements. The insertion mode is determined by analysis of the monomer units adjacent to the polymer end group derived from the initiator. The NMR analysis, greatly enhanced by the use ^{13}C-enriched $Al(^{13}CH_2CH_3)_3$, shows that primary insertion occurs exclusively in isospecific polymerization. This is what is expected for a propagating carbanion—the less substituted carbanion is more stable.

Syndiospecific polymerizations of propylene are somewhat less regiospecific than the isospecific reactions, with the typical syndiotactic polymer showing a few percent of the monomer units in head-to-head placement [Doi, 1979a, 1979b; Doi et al., 1984a, 1984b, 1985; Zambelli et al., 1974, 1987]. The mode of insertion is secondary, contrary to what is expected for a carbanion propagating center. Apparently, steric requirements imposed by the counterion derived from the initiator force propagation to proceed by secondary insertion.

8-4a-3 Propagation at Carbon–Transition-Metal Bond

Both the transition metal–carbon bond and the Group I–III metal–carbon bond have been proposed as the site at which propagation occurs. However, the available evidence clearly points to propagation at the transition-metal–carbon bond. The most significant evidence is that the Group I–III metal component alone does not initiate polymerization, but the transition-metal component alone does [Ballard, 1973; Giannini et al., 1970; Karol et al., 1978; Kohara et al., 1979; Matlack and Breslow, 1965; Soga and Yanagihara, 1988a, 1988b; Soga et al., 1977]. Among the heterogeneous initiators that achieve polymerization of ethylene and α-olefins are those obtained by ball milling of a transition metal alone or together with an alkyl halide, I_2, or H_2. Others include transition-metal hydrides or halides, the latter with a Lewis base such as an amine or phosphine. Homogeneous initiators include benzyl, η^3-allyl, η^5-cyclopentadienyl, and η^5-indenyl derivatives of transition metals. Polymerizations with these initiators are generally lower in activity and/or stereospecificity than when a Group I–III metal component is present.

The obvious question arises as to the function of the Group I–III metal component, since polymerization does occur in its absence. The Group I–III metal component increases the activity and/or stereospecificity of the initiator. The mechanism usually proposed for this action is the alkylation (often accompanied by reduction) of the transition-metal component to form more active and stereospecific reaction sites. However, the Group I–III metal component may also be involved in stabilizing the active transition metal sites by complexation.

8-4a-4 *Bimetallic versus Monometallic Mechanisms*

A number of structures have been proposed for the active species (sites) in Ziegler–Natta initiator systems. The diversity of proposed species arises from the multitude of products that have been observed or can be postulated in the interaction of the two components of a Ziegler–Natta initiator. The proposed active species fall into either of two general categories: monometallic and bimetallic [Allegra, 1971; Arlman and Cossee, 1964; Corradini et al., 1989; Natta, 1960a, 1960b; Patat and Sinn, 1958; Tait and Watkins, 1989]. The two types are illustrated by structures **XX** and **XXI** for

the active species from titanium chloride and alkylaluminum components, such as TiCl$_4$ or TiCl$_3$ with AlR$_3$ or AlR$_2$Cl. The □ in the structures represents an unoccupied (vacant) octahedral orbital of titanium.

Structure **XX** represents an active titanium site at the surface of a TiCl$_3$ crystal. The titanium atom shares four chloride ligands with its neighboring titanium atoms and has an alkyl ligand (incorporated through exchange of alkyl from the alkylaluminum for chloride) and a vacant orbital. The high mileage supported catalysts previously mentioned involve species such as **XX** and **XXI** that are part of a magnesium lattice structure, usually involving Mg—Cl—Ti bonds; thus, Ti and Mg are interchangeable within the metal halide lattice. Some workers postulate Mg—O—Ti bonds formed by reaction of Mg—OH bonds (present or formed in the magnesium chloride lattice during catalyst preparation) with Ti—Cl bonds.

A typical propagation mechanism based on such active sites is shown in Fig. 8-11. Monomer is coordinated at the vacant orbital of titanium and then inserted into the polymer chain at the transition metal–carbon bond. This results in regeneration of the vacant orbital with a configuration opposite from the original. If propagation continued with this species, the result would be syndiotactic propagation. Isotactic propagation requires migration of the polymer chain to its original position with regeneration of the original configuration of the vacant orbital.

Monometallic mechanisms stress the importance of the transition metal. However, careful reading of papers depicting monometallic mechanisms usually reveal the authors' assumptions that some of the ligands of the transition metal are shared with the Group I–III metal. The active sites are actually bimetallic in nature as shown in **XXI** [Rodriguez and van Looy, 1966]. A bimetallic mechanism for **XXI** is exactly like that in Fig. 8-11 (monomer complexation at Ti, insertion at Ti—C bond, ligand migration), except both Al and Ti are involved. The carbon of the Ti—C propagation center is coordinated to Al as well as to Ti, although insertion occurs at the Ti—C bond.

The literature contains many different variations on the monometallic and bimetallic active sites and propagation mechanisms. These often appear different from what has

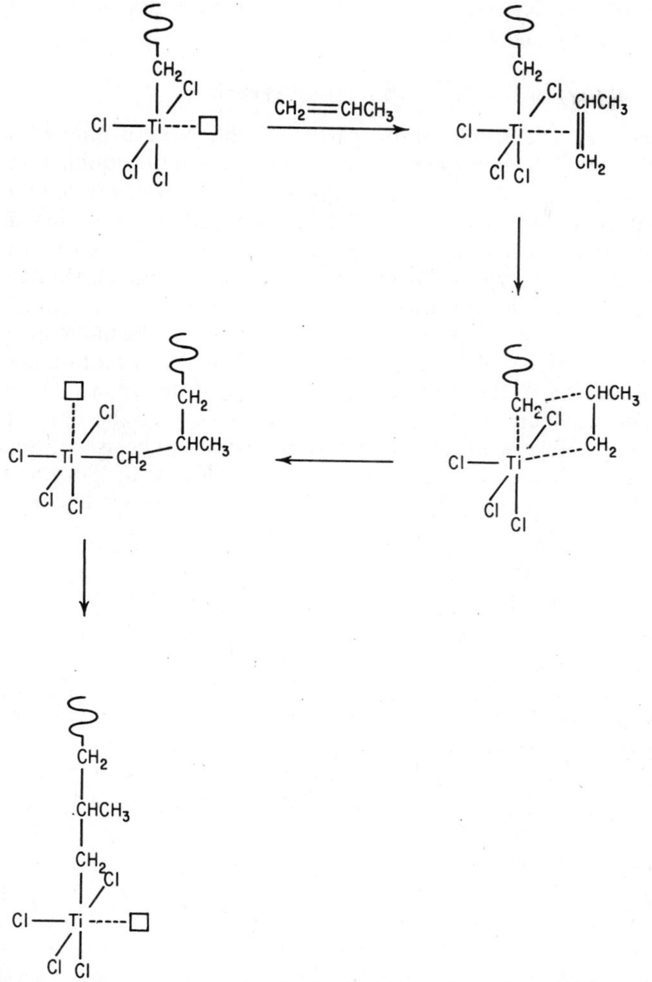

Fig. 8-11 Monometallic mechanism for catalyst site control model of isotactic polymerization. After Cossee [1964] (by permission of Academic Press, New York).

been described above, but a careful reading shows them to be very similar. For example, a bimetallic species such as **XXII** is often used (as it was in the previous edition of this text) to describe the bimetallic mechanism. Structure **XXII** differs from **XXI** only in that it shows neither the fourth chloride ligand nor the vacant orbital.

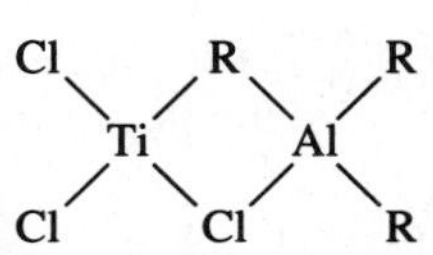

XXII

This omission is probably an attempt to streamline the drawings required in writing out the propagation mechanism.

8-4a-5   Direction of Double-Bond Opening

A syn or cis addition to the carbon–carbon double bond is implied in the Ziegler–Natta mechanism. The effect of the mode of addition to the double bond has no effect on the stereochemical structure of the polymers from 1-substituted ethylenes. However, the ditacticity of polymers from 1,2-disubstituted ethylenes depends not only on the mode of addition but also on whether the monomer is a cis or trans monomer. For diisotactic polymerization, syn addition to a trans monomer would give the threo-diisotactic structure. This is shown in Fig. 8-12, where it is proposed that the carbon–carbon bond in the monomer unit rotates after addition of the monomer to the polymer chain so as to avoid the 1,2-interactions in the fully eclipsed conformation [Goodman, 1967]. Syn addition to the cis monomer would yield the erythrodiisotactic polymer. Trans or anti addition to the two monomers would give the opposite results. Syn addition to the double bond in isotactic polymerization has been confirmed from studies with deuterium-labeled propylenes [Miyazawa and Ideguchi, 1964; Natta et al., 1960; Zambelli et al., 1968]. The cis- and trans-1-d-propylenes gave the erythro and threo-diisotactic polymers, respectively.

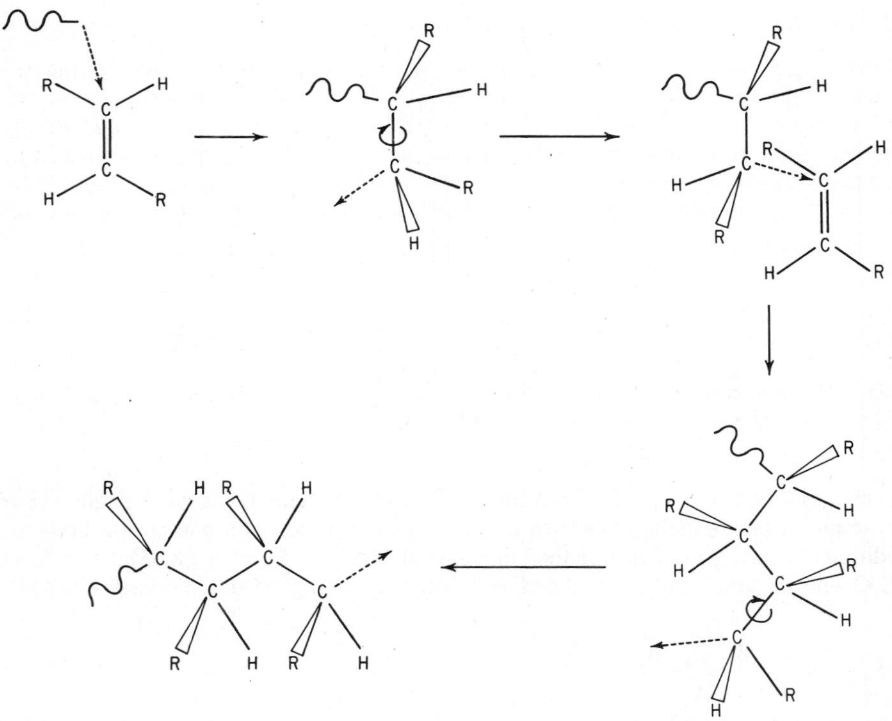

Fig. 8-12   Syn addition to a *trans*-1,2-disubstituted ethylene to yield the threodiisotactic polymer. After Goodman [1967] (by permission of Wiley-Interscience, New York).

The syn addition mechanism described in Fig. 8-12 for isotactic polymerization is based on monomer always approaching the propagating center with the same face (either always re or always si)—a reasonable assumption for a stereospecific process leading to isotactic polymerization. It is also assumed that incoming monomer is aligned with the propagating chain end so as to minimize steric repulsions between the R group of the chain end and the R groups on the carbon of the double bond that bonds to the propagating chain end; in other words, the R groups are opposed to each other. Syn addition leads to syndiotactic polymerization when successive monomer molecules alternate their two faces in approaching the propagating center. Syn addition in syndiotactic polymerization has been established, since copolymerizations of *cis*- and *trans*-1-*d*-propylenes with perdeuteropropylene yield the *trans*- and *gauche*-syndiotactic structures **XXIII** and **XXIV**, respectively [Zambelli et al., 1968]. Similar results have been observed in the syndiotactic polymerization of styrene [Longo et al., 1988].

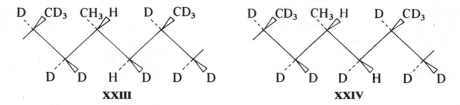

XXIII   **XXIV**

8-4a-6 *Mechanism of Isotactic Control*

Isotactic polymerization depends intimately on the crystal structure of the initiator surface. For a coordination lattice, as opposed to a molecular crystal lattice, the crystal contains a number of ligand vacancies in order to achieve overall electrical neutrality of the crystal. α-TiCl$_3$ crystals are made up of elementary crystal sheets of alternating titanium and chlorine layers aligned along the principal crystal axis [Natta et al., 1961a, 1961b, 1961c, 1961d]. Figure 8-13 is a representation of a portion of the crystal [Cor-

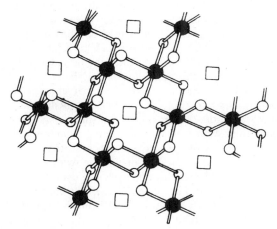

Fig. 8-13 Crystal structure of α-TiCl$_3$. Ti and Cl atoms are represented by large black circles and small empty circles, respectively; octahedral vacancies, by open squares. After Corradini et al. [1989] (by permission of Pergamon Press, London).

radini et al., 1989]. The titanium and chlorine atoms are represented by solid black and open spheres, respectively. The titaniums are at the octahedral interstices of the chlorine lattice while chlorines are hexagonally close-packed. Every third titanium in the lattice is missing; that is, there is a vacancy in between pairs of titanium atoms. Vacancies are represented by □ in Fig. 8-13.

Polymerization occurs at active sites found on the edges (surfaces) of elementary sheets of the crystal and not in the basal planes. This is supported by microscopic observations of polymer growth at the crystal edges [Cossee, 1967; Kollar et al., 1968a, 1968b; Rodriguez and van Looy, 1966]. A titanium atom at the surface is bonded to only five chlorines instead of six because of the imposed requirement of electroneutrality. Of the five chlorines, four are more strongly bonded since they are bridged to other titanium atoms (or other metal atoms for bimetallic species). The fifth, nonbridged chlorine may be replaced by an alkyl group when the titanium component interacts with the Group I–III component. The octahedral vacancy remains as a vacancy.

The polymerization-active titanium sites have structures such as **XX** or **XXI**. Figure 8-13 is drawn with the active sites of the monometallic type (**XX**) but could just as easily be drawn with sites of the bimetallic type (**XXI**). Neighboring metal atoms (bridged by two chlorines) have opposite chirality [Allegra, 1971]. The two enantiotopic titaniums can be represented as in **XXV** and **XXVI** where one of the chlorine

radini et al., 1989]. The titanium and chlorine atoms are represented by solid black and open spheres, respectively. The titaniums are at the octahedral interstices of the chlorine lattice while chlorines are hexagonally close-packed. Every third titanium in the lattice is missing; that is, there is a vacancy in between pairs of titanium atoms. Vacancies are represented by □ in Fig. 8-13.

XXV   XXVI

ligands has been replaced by R via alkylation by the Group I–III metal component. Propagation occurs by coordination of monomer at the vacancy, insertion into the Ti—R bond, and migration of the newly formed bond to regenerate the vacancy in the original configuration. Considering either **XXV** or **XXVI**, coordination of monomer can occur through either of the two monomer faces (si or re), which results in two diasteromeric intermediates or transition states. The two diasteromeric situations give rise to the two different placements of monomer units in the polymer chain— *meso* versus *racemo* [Corradini et al., 1984, 1985; Zambelli et al., 1978, 1980].

The driving force for isotactic propagation results from steric and electrostatic interactions between the substituent of the incoming monomer and the ligands of the transition metal. The chirality of the active site dictates that monomer coordinate to the transition-metal vacancy through only one of the two faces. Structures **XXV** and **XXVI** each yield isotactic polymer through exclusive coordination with the re and si monomer face, respectively, or vice versa. That is, we cannot state which face will coordinate with **XXV** and which face with **XXVI**, but we can state that only one of the faces will coordinate with **XXV** while the opposite face will coordinate with **XXVI**. This is the *catalyst site control* or *enantiomorphic site control model* for isospecific polymerization.

The enantiomorphic model attributes stereocontrol in isotactic polymerization to the initiator active site with no influence of the structure of the propagating polymer chain end. The mechanism is supported by several observations:

1. Carbon-13 NMR analysis of isotactic polypropylene shows the main type of error to be pairs of racemic dyads (**XXVII**) instead of isolated dyads (**XXVIII**) [Heatley

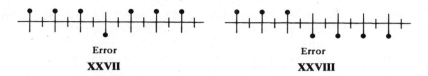

Error   Error

XXVII   **XXVIII**

et al., 1969; Wolfsgruber et al., 1975]. An error in addition of a monomer molecule is immediately corrected when stereocontrol is by the chiral active site. If stereocontrol was due to the propagating chain end, an error would propagate in an isotactic manner to yield a polymer, referred to as *isotactic stereoblock* (**XXVIII**), containing long isotactic all-*R* and all-*S* blocks on each side of the error.

2. Carbon-13 NMR analysis of ethylene–propylene copolymers of low ethylene content produced by initiators that yield isotactic polypropylene shows that the isotactic propylene units on each side of an ethylene unit have predominantly structure **XXIX**, consistent with catalyst site control [Zambelli et al., 1971, 1978, 1979]. Stereocontrol by the propagating chain end would yield about equal amounts of **XXIX** and **XXX**.

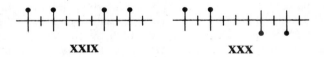

XXIX   **XXX**

3. Stereoselective and stereoelective polymerizations (Sec. 8-8c) and statistical analysis of the stereochemical sequence distributions (Sec. 8-10) also support the enantiomorphic site control model.

Further insights into isospecific polymerization come from ^{13}C NMR examination of the configurations of the first monomer unit inserted into the Ti—R active site [Ammendola et al., 1986; Tritto et al., 1986; Zambelli et al., 1982a, 1982b]. The NMR analysis is enhanced by using ^{13}C-enriched alkyl groups in the Group I–III metal component. There is some minimum amount of steric bulk required at the active site in order to achieve high enantioselectivity between the two monomer faces. The monomer and various transition-metal ligands contribute to the steric bulk at the active site. For propylene polymerization, the enantioselectivity in the first mono-mer unit added is low for R = methyl, appreciable for R = ethyl, and very high for R = *i*-butyl or phenyl. Enantioselectivity of the first addition is enhanced for larger-sized monomers such as styrene and vinylcyclohexane compared to propylene and for larger-sized halogen ligands. The various steric effects act in a cooperative manner. Thus, addition of the first monomer unit for propylene polymerization is as isospecific as all subsequent monomer units when TiI_3 and $Al(C_2H_5)_3$ are the initiator components.

8-4a-7 *Mechanism of Syndiotactic Polymerization*

The synthesis of highly syndiotactic polymers by Ziegler–Natta initiators has been successful only with propylene, sytrene, and some 1,3-dienes [Pasquon et al., 1989; Youngman and Boor, 1967]. Syndiotactic polymerizations have been most studied for propylene. Only soluble initiators yield highly syndiospecific polymerizations. Soluble Ziegler–Natta initiators for producing syndiotactic polypropylene have been limited almost exclusively to those based on vanadium compounds.

The initiator formed from VCl_4 and $Al(C_2H_5)_2Cl$ is one of the most efficient for syndiotactic polymerization of propylene, especially in the presence of a Lewis base such as anisole (methoxybenzene) [Doi, 1979a, 1979b; Natta et al., 1962; Zambelli et al., 1978, 1980]. Other vanadium compounds such as vanadium acetylacetonate and various vanadates [$VO(OR)_xCl_{(3-x)}$ where $x = 1,2,3$] can be used in place of VCl_4 but are more limited in their stereospecificity [Doi et al., 1979a, 1979b]. The tendency toward syndiotacticity increases with decreasing temperature; most syndiospecific polymerizations are carried out below $-40°C$ and usually at $-78°C$. The initiators usually must be prepared at the low temperatures since most of them become heterogeneous (and no longer produce syndiotactic polymer) when prepared at or warmed up to temperatures above about 40°C.

The driving force for syndiotactic placement with Ziegler–Natta initiators is similar to that (Sec. 8-3) for low-temperature radical and ionic, noncoordinated polymerizations—the repulsive interaction between substituents of the terminal unit of the propagating chain and incoming monomer. This is the *polymer chain end control mechanism* for stereocontrol. Figure 8-14 shows this model for syndiotactic placement with secondary insertion. (The original reference [Boor and Youngman, 1966] for this figure incorrectly showed propagation proceeding by primary insertion; see Sec. 8-4a-2.) The polymer chain end control model for syndiotactic placement is similar to the catalyst site control model for isotactic placement in that both are based on an octahedral transition-metal complex that has an alkyl ligand as the propagating site and a coordination vacancy for complexing monomer. (A mechanism based on pentacoordinated vanadium instead of hexacoordinated vanadium has also been proposed [Zambelli and Allegra, 1980]).

There are important differences between the catalyst site and polymer chain end control models. The homogeneous syndiospecific initiator allows coordination of the monomer (and insertion into the polymer chain) via either of the monomer faces. The initiator achieves its stereospecificity in essentially the same manner as lowered temperature in a noncoordination polymerization. Hindrance between the methyl group of the last unit of the propagating chain and the ligand(s) attached to vanadium prevent rotation about the transition metal–carbon bond. This brings into play the repulsive interaction between methyl groups of the terminal monomer unit and incoming monomer. Syndiotactic placement is energetically favored as methyl–methyl interactions force the monomer to be coordinated at its opposite face at each successive propagation step. On the other hand, monomer–monomer interactions are minimized with isospecific initiators by rotation about the transition metal–carbon bond. Isotactic placement occurs since only one configuration is allowed for coordination and addition of monomer to the propagating chain. Isotactic placement proceeds with migration of the polymer chain to its original ligand position prior to the next propagation step. Syndiotactic propagation occurs alternately at the two ligand positions.

In summary, syndiospecific initiators exaggerate the inherent tendency toward syndiotactic placement by accentuating the methyl–methyl repulsive interactions between

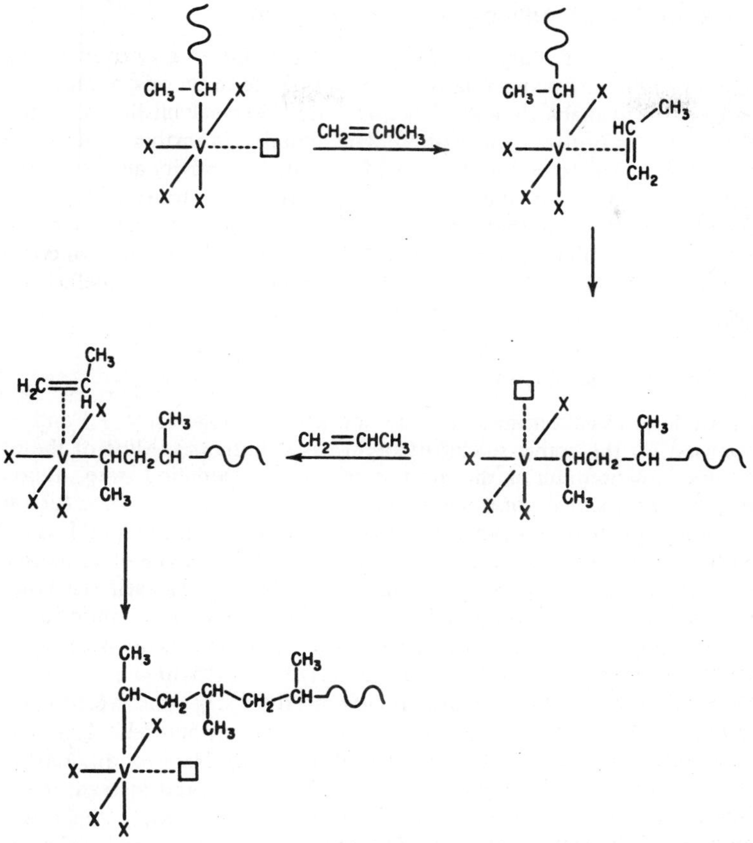

Fig. 8-14 Polymer chain end control model for syndiotactic placement. After Boor and Youngman [1966] (by permission of Wiley-Interscience, New York).

the propagating chain end and incoming monomer. Isotactic placement occurs against this inherent tendency when chiral active sites force monomer to coordinate with the same face at each propagation step.

The polymer chain end control model is supported by the observation that highly syndiotactic polypropylene is obtained only at low temperatures (about $-78°C$). Syndiotacticity is significantly decreased even by raising the temperature to $-40°C$ [Boor, 1979]. Carbon-13 NMR analysis of propylene–ethylene copolymers of low ethylene content produced by vanadium initiators indicates that a syndiotactic block formed after an ethylene unit enters the polymer chain is just as likely to start with an S- as with an R-placement of the first propylene unit in that block [Bovey et al., 1974; Zambelli et al., 1971, 1978, 1979]. Stereocontrol is not exerted by chiral sites as in isotactic placement, which allows only one type of placement (either S- or R-, depending on the chirality of the active site). Stereocontrol is exerted by the chain end. An ethylene terminal unit has no preference for either placement, since there are no differences in repulsive interactions.

8-4b   Effect of Components of Ziegler–Natta Initiator

The stereospecificity and activity of Ziegler–Natta initiators vary over a wide range depending on the identity and relative amounts of the initiator components [Boor, 1979]. Much of the available data are difficult to interpret mechanistically as the identity of the active sites is not well established. The situation is further complicated since changes in the initiator components often affect initiator activity and stereospecificity in opposite directions. Also, the trends observed from one transition- (or Group I–III-) metal component to another may be different. Some generalizations are presented below within these restrictions with emphasis on the effects of the initiator components on stereospecificity. These generalizations are also restricted to α-olefins; they may not apply to other monomers such as 1,3-dienes.

8-4b-1   Transition-Metal Component

The most studied transition metal is titanium. Oxidation states of $+3$ and $+2$ have been proposed for the active site of titanium-based initiators. Most of the evidence points to trivalent titanium as the most stereospecific oxidation state, although not necessarily the most active nor the only one [Chien et al., 1982]. (Data for vanadium systems indicate that trivalent vanadium sites are the syndiospecific sites [Lehr, 1968].) Initiators based on the α-, γ-, and δ-titanium trihalides are much more stereospecific (isospecific) than those based on the tetrahalide or dihalide. By itself, $TiCl_2$ is inactive as an initiator but is activated by ball milling due to disproportionation to $TiCl_3$ and Ti [Werber et al., 1968]. The overall order of stereospecificity is usually α-, γ-, δ-$TiCl_3$ > $TiCl_2$ > $TiCl_4$ ≃ β-$TiCl_3$ [Natta et al., 1957a, 1957b].

The low stereospecificity of β-$TiCl_3$ relative to the α-, γ-, and δ-forms is a consequence of the different crystalline structure. While α-, γ-, and δ-$TiCl_3$ are similar in that all contain a layered structure [Natta, 1960a, 1960b; Natta et al., 1961a, 1961b, 1961c, 1961d], in γ-$TiCl_3$ the chlorines are cubic close-packed instead of hexagonal close-packed as in α-$TiCl_3$, δ-$TiCl_3$ has a mixed hexagonal and cubic close-packed layered structure, and β-$TiCl_3$ consists of bundles of linear $TiCl_3$ chains. The structure of β-$TiCl_3$ results in a surface in which half the titanium atoms have two vacancies and the other half have one vacancy. The titanium sites containing two vacancies each are responsible for the low stereospecificity of β-$TiCl_3$. The two vacancies yield a site with a loose configuration of ligands since one of the chlorines and/or the growing polymer chain will be loosely bound. This is exactly opposite to the situation with the other titanium trichlorides (one vacancy per site) in which all chlorides and the polymer chain have fixed positions in a rigid configuration on the titanium site. That some isotactic polymer is formed with β-$TiCl_3$ is a result of the presence of titanium sites that have only one vacancy each.

The oxidation state of the transition-metal active sites is dependent not only on the transition-metal component but also on the Group I–III-metal component. For example, $TiCl_4$ is usually used instead of $TiCl_3$ in initiator recipes because it is a liquid and more conveniently handled but is usually employed under conditions (in combination with AlR_3 or AlR_2Cl), which reduces a large fraction of titanium to the trivalent state. Further, the extent of reduction depends on the amount of the Group I–III metal component relative to the transition metal [Kollar et al., 1968a, 1968b; Schindler, 1968]. At Al/Ti ratios greater than 3 there is extensive reduction to divalent titanium and a significant decrease in stereospecificity. The optimum ratio is usually near or less than 1. However, the situation is very different for the $MgCl_2$-supported initiators. The typical catalyst contains about 0.5–2 wt % Ti and the optimum Al/Ti ratio is often as high as 10–100 [Chien et al., 1982; Dumas and Hsu, 1984].

Changes in the ligands of the transition-metal component and the transition metal can greatly affect stereospecificity [Natta et al., 1957a, 1957b; Rishina et al., 1976]. For propylene polymerization by various titanium compounds in combination with triethylaluminum, the extent of isotacticity increases in the orders

$$\alpha\text{-}TiCl_3 > TiBr_3 > TiI_3$$

$$TiCl_4 \simeq TiBr_4 \simeq TiI_4 \tag{8-36}$$

$$TiCl_4 > TiCl_2(OC_4H_9)_2 >> Ti(OC_4H_9)_4 \simeq Ti(OH)_4$$

Changes in the transition metal itself lead to the following differences in stereospecificity

$$\alpha\text{-}TiCl_3 > CrCl_3 > VCl_3 > FeCl_3$$

$$TiCl_4 \simeq VCl_4 \simeq ZrCl_4 \tag{8-37}$$

Differences in initiator stereospecificity and activity are greatest for the more stereospecific or active transition-metal components (trivalent oxidation state). Initiators based on the less stereospecific or active transition metals (tetravalent oxidation state) are much less affected by structural changes. The crystal structure, steric size, and electronegativity of the ligands and transition metal are all involved in determining initiator stereospecificity and activity. Decreasing the electronegativity of the transition metal (leading to a more easily polarized metal growth bond) by changes in the transition metal, or its valence state, or its ligands, increases the stereospecificity of the Ziegler-Natta initiator. Increasing the size of the ligands decreases stereospecificity due probably to a decreased coordinating ability. The quantitative interrelationship of these and other factors with the crystal structure of the transition metal and the active sites in determining stereospecificity and activity is not clear at present.

8-4b-2 *Group I–III Metal Component*

Although the Group I–III-metal component is not an absolute necessity for obtaining a Ziegler–Natta initiator, its presence has a very significant effect on both initiator activity and stereospecificity [Boor, 1967, 1979; Coover et al., 1967; Diedrich, 1975; Natta, 1960a, 1960b]. The Group I–III-metal component is almost always required in combination with the transition-metal component to obtain high stereospecificity and high activity. Active initiators for ethylene or α-olefin polymerization have been found using Li, Na, and K from Group I, Be, Mg, Zn, and Cd from Group II and Al and Ga from Group III. Ziegler–Natta initiators based on other Group I–III metals have far lower activity and stereospecificity. Aluminum compounds are by far the most often used, a consequence of their ready availability and ease of handling (once in solution). Gallium is also active but expensive. Zinc and magnesium alkyls are the most thoroughly studied Group II metal component [Greco et al., 1979]; beryllium compounds have received little attention as a result of their toxicity. Lithium, sodium, and potassium alkyls are the most studied Group I metal component. They are generally not as attractive as the Group II and III metals because of their lower solubility in the hydrocarbon solvents normally used. (Lithium alkyls, although soluble, are highly associated in hydrocarbon solvents.)

For initiators in which the Group I–III metal is unchanged, isospecificity generally decreases as the size of the organic group increases, although the opposite effect has also been reported. When titanium is the transition metal, the replacement of an alkyl group in AlR_3 by any halogen other than fluorine results in increased stereospecificity and decreased activity with the effect of halogen being I > Br > Cl [Danusso, 1964; Doi and Keii, 1978]. The replacement of a second alkyl group in the aluminum component by halogen leads to a further decrease in activity as well as a moderate decrease in stereospecificity. The behavior with vanadium compounds is very different. Trialkylaluminum yields isotactic polypropylene [Chien et al., 1989], while AlR_2Cl yields syndiotactic polypropylene. In fact, syndiotactic polypropylene is obtained in high yield only with AlR_2Cl in combination with a soluble vanadium salt.

Attempts to interpret the effect of the Group I–III component in terms of the monometallic mechanism are unsuccessful. The order of stereospecificity or activity does not correlate with the exact order expected for reduction or alkylation of the transition metal component. The trends are more readily correlated in terms of the bimetallic species **XXI**. One expects the stability of that species to be maximum when the Group I–III metal is similar to the transition metal in size and electronegativity as in the case of Be and Al, which are similar to Ti. Increasing the size of the bridging alkyl group decreases the stability. A single halogen in the aluminum alkyl component apparently alters the electronegativity of aluminum to more closely match that of titanium; a second halogen makes aluminum too electropositive.

8-4b-3 Third Component; Lewis Base

A variety of different compounds have been added to recipes for Ziegler–Natta initiators [Boor, 1979; Coover et al., 1967]. These include oxygen, water, inorganic halides (KCl, NaF), organic halides, phenols, ethers, esters, amines, phosphines, aromatics, carbon disulfide, and hexamethylphosphoramide. The effects of these third components on initiator activity and stereospecificity vary considerably depending on the additive and the other two components of the initiator system. Some additives increase stereospecificity and/or activity, while others have the opposite effect. Some increase stereospecificity while decreasing initiator activity or vice versa. Some additives such as oxygen and water have deleterious effects on both activity and stereospecificity. Still other third components affect polymer molecular weight either exclusively or along with changes in initiator activity and/or stereospecificity.

The present generation of high mileage or superactive initiators achieve high activity by supporting the transition-metal component on magnesium chloride or other magnesium compound. This increases the surface area of transition-metal sites. High stereospecificity is maintained by the presence of appropriate third components [Barbe et al., 1987; Chien and Hu, 1987; Chien and Bres, 1986; Galli et al., 1981; Pino and Mulhaupt, 1980; Soga et al., 1990]. These are Lewis bases (electron donors) with esters, such as ethyl benzoate, di-n-butyl phthalate, and methyl-p-toluate, which are used most often. Phenols such as p-cresol and amines such as 2,2,6,6-tetramethylpiperidine are also used. Various mechanisms have been proposed for the action of Lewis bases in improving stereospecificity. The mechanisms, all involving electron donation by the third component, include decreasing the reactivity of less stereospecific sites, increasing the number of stereospecific sites by assisting in their stabilization and/or dispersal in the support, and improving the stereospecificity of active sites by altering the identities of attached ligands. The latter effect may involve increasing the steric bulk at the reaction site.

The superactive initiators are produced using both *internal* and *external Lewis bases*. A Lewis base (referred to as the *internal base*) is typically ball milled with the support material and the transition-metal component subsequently added with further mixing. This is followed by the addition of a mixture of the Group I–III-metal component with a Lewis base (referred to as the *external base*) [Chien et al., 1982]. Many initiator recipes use the same Lewis base, usually an ester, as both internal and external base. Many recent patents show an ester as the internal base and an organosilicon compound such as phenyltriethoxysilane as the external base [Soga et al., 1988]. The individual roles of the internal and external bases are not clear. Both bases affect the stereo-specificity of an initiator as shown from ^{13}C NMR analysis of the stereochemistry of the first monomer units added at active sites [Sacchi et al., 1990]. The enhancement of stereospecificity by the external base may result from its displacing the internal base and/or augmenting its effect.

8-4c Kinetics

8-4c-1 Observed Rate Behavior

The kinetics of Ziegler–Natta polymerization, like other aspects of the reaction, are complex. The relatively few polymerizations that are homogeneous behave in a manner generally similar to noncoordination ionic polymerizations (Chap. 5). The heterogeneous systems usually exhibit complicated behavior as shown in Fig. 8-15 [Boor, 1979; Burfield, 1984; Burfield et al., 1976; Cooper, 1976; Keii, 1972; Tait and Watkins, 1989]. The behavior described by plot 1 is usually observed when the particle size of the transition-metal component is relatively large. The particles of the transition-metal component consist of aggregates of smaller crystals. The mechanical pressure of the growing polymer chains cleaves these aggregates with the result that the initiator surface area, number of active sites, and polymerization rate increase with time. After this initial period, referred to as a *buildup* or *settling period*, a steady-state rate is reached, which corresponds to cleavage of the initial particles to the smallest-sized particles. When the initial particle size is decreased (by ball milling), the time required to reach the steady-state polymerization rate is decreased (plot 2). Other factors responsible for the buildup period include the time needed for the formation of the

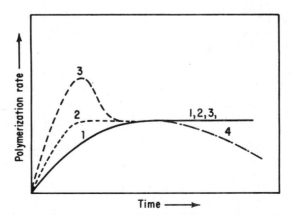

Fig. 8-15 Types of rate versus time behavior in Ziegler–Natta polymerizations.

active sites by reaction of the two metal components, slow initiation, and the presence of impurities. Most of these factors are moderated for the typical situation where the initiator system is preformed and allowed to age prior to use in initiating polymerization. The superactive supported initiators often show very little or no buildup period [Busico et al., 1986; Chien and Hu, 1987].

Many polymerizations show a settling period with a relatively rapid rise in rate to a maximum value followed by a decay to the steady-state rate (plot 3). This behavior indicates the presence of active sites of differing activities with some of the active sites decaying with time. This behavior is avoided in some instances by aging of the initiator. Ziegler–Natta polymerizations may exhibit a continuous rate decrease (plot 4) after the settling period (which may be of either type 1, 2, or 3) due to active site destruction. This can be due to thermal deactivation or further reduction of the transition metal by the Group I–III metal component. Diffusion control of the propagation reaction has also been postulated. Diffusion of the monomer through the formed polymer to the propagation centers may become rate-determining at higher conversions. This has been substantiated in some systems where the polymerization rate increases with increased rate of stirring, but data in other systems indicate diffusion control to be absent [Chien et al., 1985].

8-4c-2   Termination

Ziegler–Natta polymerizations have the characteristics of living polymerizations with regard to active sites but not individual propagating chains. The lifetime of propagating chains is of the order of seconds or minutes at most, while active sites have lifetimes of hours or days. Each active site produces many polymer molecules. Propagating chains terminate by various reactions:

1. Spontaneous intramolecular β-hydride transfer:

$$\text{Ti—CH}_2\text{—CH}\!\!\sim\!\!\sim \xrightarrow{k_s} \text{Ti—H} + \text{CH}_2\!\!=\!\!\text{C}\!\!\sim\!\!\sim \tag{8-38}$$
$$\qquad\quad \overset{|}{\text{CH}_3} \qquad\qquad\qquad\qquad \overset{|}{\text{CH}_3}$$

where Ti represents the transition-metal active site at which propagation occurs.

2. Chain transfer to monomer:

$$\text{Ti—CH}_2\text{—CH}\!\!\sim\!\!\sim + \text{CH}_3\text{CH}\!\!=\!\!\text{CH}_2 \xrightarrow{k_{tr,M}} \text{Ti—CH}_2\text{CH}_2\text{CH}_3$$
$$\qquad\quad \overset{|}{\text{CH}_3}$$
$$+ \text{CH}_2\!\!=\!\!\text{C}\!\!\sim\!\!\sim \tag{8-39a}$$
$$\qquad\;\; \overset{|}{\text{CH}_3}$$

and

$$\text{Ti—CH}_2\text{—CH}\!\!\sim\!\!\sim + \text{CH}_3\text{CH}\!\!=\!\!\text{CH}_2 \xrightarrow{k_{tr,M}} \text{Ti—CH}\!\!=\!\!\text{CH—CH}_3$$
$$\qquad\quad \overset{|}{\text{CH}_3}$$
$$+ \text{CH}_3\text{—CH}\!\!\sim\!\!\sim \tag{8-39b}$$
$$\qquad\;\;\;\; \overset{|}{\text{CH}_3}$$

3. Chain transfer to the Group I–III metal alkyl:

$$Ti-CH_2-CH\!\!\sim\!\!\sim + Al(C_2H_5)_3 \xrightarrow{\ k_{tr,A}\ } Ti-CH_2CH_3$$
$$\hspace{1.0cm}\Big|$$
$$\hspace{1.0cm}CH_3$$

$$+ (C_2H_5)_2Al-CH_2-CH\!\!\sim\!\!\sim \hspace{2cm}(8\text{-}40)$$
$$\hspace{2.5cm}\Big|$$
$$\hspace{2.5cm}CH_3$$

4. Chain transfer to an active hydrogen compound such as molecular hydrogen:

$$Ti-CH_2-CH\!\!\sim\!\!\sim + H_2 \xrightarrow{\ k_{tr,H_2}\ } Ti-H + CH_3-CH\!\!\sim\!\!\sim \hspace{1cm}(8\text{-}41)$$
$$\hspace{1.0cm}\Big| \hspace{5.5cm}\Big|$$
$$\hspace{1.0cm}CH_3 \hspace{5.3cm}CH_3$$

None of the above reactions terminate the kinetic chain. All are chain-transfer reactions in that there is reinitiation of new propagating chains. The extents to which the various termination reactions occur depend on the monomer, identity and concentrations of the initiator components, temperature, and other reaction conditions [Boor, 1979; Cooper, 1976; Longi et al., 1963]. There are considerable differences in the efficiencies of transfer for different Group I–III metal components; for instance, diethylzinc is much more effective in chain transfer compared to triethylaluminum [Natta et al., 1961a, 1961b, 1961c, 1961d]. Molecular hydrogen is a highly effective chain-transfer agent and is used for molecular-weight control in the typical industrial polymerization of propylene.

Termination is more complicated than described above. Polyethylene produced by Ziegler–Natta polymerization contains vinyl ($CH_2{=}CHR$), vinylidene ($CH_2{=}CRR'$), and *trans*-vinylene ($RCH{=}CHR'$) end groups in amounts of approximately 60, 20, and 20%, respectively [Beach and Kissin, 1986]. Vinyl groups are formed by β-hydride transfer and chain transfer to monomer, the equivalent of Eqs. 8-38 and 8-39a,b for ethylene polymerization. The formation of vinylidene and *trans*-vinylene end groups is less clear. One possibility is participation of vinyl end groups in chain transfer analogous to Eq. 8-39b. Transfer to the methylene and methine protons of vinyl end groups would generate *trans*-vinylene and vinylidene end groups, respectively. This mechanism also predicts the formation of small amounts of $CHR{=}CR'R''$ end groups in propylene polymerization. Transfer to vinylidene and vinylene end groups (formed by Eqs. 8-38 and 8-39) would both yield trisubstituted double-bond end groups. This has not been reported for propylene itself, but trisubstituted double bonds have been observed in the polymerizations of 1-butene and 1-octene [Rossi and Odian, 1990].

8-4c-3 Rate Expression

Homogeneous kinetics are applicable to some Ziegler–Natta polymerizations, when adsorption of initiator components or monomer is not important. The polymerization rate is expressed as

$$R_p = k_p[C^*][M] \hspace{3cm}(8\text{-}42)$$

where $[C^*]$ is the concentration of active sites expressed in moles per liter.

Adsorption phenomena are important in most Ziegler–Natta polymerizations, and this requires treatment by heterogeneous kinetics [Bohm, 1978; Boor, 1979; Chien and Ang, 1987; Chien and Hu, 1987; Cooper, 1976; Keii, 1972; Tait and Watkins, 1989]. The exact form of the resulting kinetic expressions differs depending on the specific adsorption phenomena, that are important in the particular reaction system. Consider a Langmuir-Hinschelwood model where reaction occurs only after monomer is adsorbed from solution onto the transition metal active sites. Further, we assume that the Group I–III-metal component is present in solution and competes with monomer for the same sites; that is, there is excess Group I–III-metal component over and above the amount needed to activate (reduce/alkylate) the transition-metal sites. The fractions Θ_A and Θ_M of the transition-metal sites covered with the Group I–III-metal component and monomer, respectively, are given by

$$\Theta_A = \frac{K_A[A]}{1 + K_A[A] + K_M[M]} \tag{8-43}$$

$$\Theta_M = \frac{K_M[M]}{1 + K_A[A] + K_M[M]} \tag{8-44}$$

where [A] and [M] are the concentrations of the Group I–III-metal component and monomer in solution, respectively, and K_A and K_M are the respective equilibrium constants for their adsorption [Adamson, 1986].

Propagation occurs by reaction of adsorbed monomer at the active sites at a rate given by

$$R_p = k_p[C^*]\Theta_M \tag{8-45}$$

Combination of Eqs. 8-43 to 8-45 yields the polymerization rate as

$$R_p = \frac{k_p K_M[M][C^*]}{1 + K_M[M] + K_A[A]} \tag{8-46}$$

If the Group I–III-metal component does not compete with monomer for the active sites, then $K_A[A] = 0$, and Eq. 8-46 reduces to

$$R_p = \frac{k_p K_M[M][C^*]}{1 + K_M[M]} \tag{8-47}$$

The degree of polymerization is obtained as

$$\frac{1}{\overline{X}_n} = \frac{k_{tr,M}}{k_p} + \frac{k_s}{k_p K_M[M]} + \frac{k_{tr,A} K_A[A]}{k_p K_M[M]} + \frac{k_{tr,H_2}[H_2]}{k_p K_M[M]} \tag{8-48}$$

by dividing the propagation rate by the sum of the rates of all termination (transfer) reactions. The derivations of Eqs. 8-46 through 8-48 assume that transfer to hydrogen does not involve adsorption of hydrogen at the active sites prior to transfer. If hydrogen competes with monomer and the Group I–III-metal component for adsorption at the active sites, the above treatment requires modification of Θ_A and Θ_M and the introduction of Θ_{H_2}.

8-4c-4 Values of Kinetic Parameters

Evaluation of the various kinetic parameters requires a determination of the active-site concentration. [C*] is usually determined from experiments in which the active sites are quenched (made inactive) with CH_3O^3H, ^{14}CO, or $^{14}CO_2$ [Jaber and Fink, 1989; Mejzlik et al., 1986, 1988; Tait and Watkins, 1989; Vozka and Mejzlik, 1990]. Other methods include the use of number-average molecular weight (combined with polymer yield) and ^{14}C-labeled Group I–III-metal component. Each of the techniques has limitations that require care if reliable results are to be obtained. For example, quenching with CH_3O^3H gives high values for [C*] since the polymer chains terminated by chain transfer to the Group I–III-metal component are reactive toward methanol. Values of [C*] obtained at different conversions need to be extrapolated to zero conversion.

Literature values of the active-site concentrations range from tenths or hundredths of a percent to tens of percents of the transition-metal concentration [Lieberman and Barbe, 1988; Tait and Watkins, 1989]. Much of this is indicative of the range of activity of different initiators, especially when comparing older initiators to the recent high-mileage initiators. However, some of the variation is due to the problems inherent in measurement of [C*]. Literature values of k_p and other rate constants also show a considerable range.

Table 8-3 shows various kinetic parameters in propylene polymerization (0.65 M in n-heptane) using a $TiCl_4/Al(C_2H_5)_3$ initiator ([Ti] = 0.0001 M) supported on $MgCl_2$ with ethyl benzoate and methyl-p-toluate as internal and external Lewis bases. The concentrations of isospecific and aspecific active sites were obtained by a combination of quenching experiments for the total concentration of active sites and ESR (as well as isotactic index) measurements for the isospecific active sites. The amounts of isospecific and aspecific active sites as percentages of total titanium are expressed by $(C^*)_i$ and $(C^*)_a$, respectively; k_{pi} and k_{pa} represent the propagation rate constants for isospecific and aspecific active sites, respectively. The external base is critical to achieving a highly isospecific polymerization. With an internal base but no external base,

TABLE 8-3 Kinetic Parameters in Polymerization of Propylene by $MgCl_2/TiCl_4/Al(C_2H_5)_3$ at 50°C[a]

Kinetic Parameter[b]	Internal Base Only	Internal and External Bases
$(C^*)_i$	6.0	2.3
$[C^*]_i$	6.0×10^{-6}	2.3×10^{-6}
$(C^*)_a$	24	2.4
k_{pi}	138	133
k_{pa}	16.6	5.1
Isotactic index	68.2	96.0
$k_{tr,A}{}^c$	4.0×10^{-4}	1.2×10^{-4}
$k_{tr,M}{}^c$	9.1×10^{-3}	7.2×10^{-3}
$k_s{}^c$	8.2×10^{-3}	9.7×10^{-3}

[a]Data from Chien and Hu [1987].
[b]Units: (C*) = % Ti; [C*]$_i$ = moles/liter; k_s = sec^{-1}; all other rate constants = liters/mole-sec; isotactic index = % sample insoluble in refluxing n-heptane.
[c]Values are for the isospecific sites.

isospecific polymerization is favored over aspecific by a factor of 2.15 as $(C^*)_i/(C^*)_a$ favors aspecific polymerization by a factor of 4 while k_{pi}/k_{pa} favors isospecific polymerization by a factor of 8.3. With both internal and external bases present, the concentration of aspecific active sites decreases, as does their reactivity, while there is negligible effect on the isospecific sites. Isospecific polymerization becomes favored by a factor of 25. Thus, the isotactic index increases from 68.2 to 96.0% for polymerizations with and without external base. The $k_{tr,A}$, $k_{tr,M}$, and k_s values in Table 8-3 are those for the isospecific active sites. These are lower than k_{pi} by a factor of 10^4–10^6, which results in high polymer molecular weights ($\overline{M}_n = 2.0$ and 1.5×10^5, respectively, with and without external base). Lower molecular weights are achieved only in the presence of H_2, whose rate constant for transfer is much higher. The aspecific active sites yield lower molecular weights ($\overline{M}_n = 1.1$ and 0.9×10^5, respectively, with and without external base) than the isospecific active sites.

The molecular-weight distributions obtained with heterogeneous Ziegler–Natta initiators vary considerably depending on the specific initiator. Whereas $\overline{X}_w/\overline{X}_n$ is slightly above 3 for the polymerization in Table 8-3, most literature reports show values in the 5–30 range. The broader distributions are found for the less isospecific polymerizations. Homogeneous Ziegler–Natta polymerizations typically have narrower distributions than do the heterogeneous systems since there is a narrower distribution of active sites of different reactivity.

The overall activation energies for the rates of most Ziegler–Natta polymerizations fall in the range 20–70 kJ/mole. The term E_R is a composite of the activation energy for propagation and the heat of adsorption of monomer. Although polymerization rates increase with temperature, reaction temperatures above 70–100°C seldom are employed. High temperatures result in loss of stereospecificity as well as lowered polymerization rates as a result of the decreased stability of the initiator.

8-4d Scope of Ziegler–Natta Initiator

The Ziegler–Natta initiator polymerizes a variety of monomers, including ethylene and α-olefins such as propylene, 1-butene, 4-methyl-1-pentene, vinylcyclohexane, and styrene. 1,1-Disubstituted ethylenes such as isobutylene and α-methylstyrene are polymerized, but the reaction proceeds by a noncoordination cationic mechanism. Except for the 1-deuteropropylenes (Sec. 8-4a-5) and cycloalkenes (Sec. 8-4d-1), 1,2-disubstituted ethylenes have not been polymerized because of steric hindrance. Polymers are obtained from some 1,2-disubstituted ethylenes, but the reactions involve isomerization of the monomer to a 1-substituted ethylene prior to polymerization; for example, 2-butene yields poly(1-butene) [Endo et al., 1979]. Alkyne polymerization is discussed below. The polymerization of conjugated dienes is discussed in Sec. 8-6.

Many polar monomers such as vinyl acetate, vinyl chloride, acrylates, and methacrylates have been polymerized by Ziegler–Natta initiators but not by the anionic coordination mechanism. The reactions proceed by noncoordinated radical or ionic mechanisms. Many polar monomers, especially those containing electron-donor atoms such as nitrogen and oxygen, cannot be polymerized by Ziegler–Natta initiators. These monomers inactivate the initiator either by strongly complexing or reacting with one or both metal components. The stereospecific polymerization of some polar monomers by Ziegler–Natta initiators has been achieved in a few instances. For those monomers that inactivate the initiator by coordination, stereospecific polymerization is sometimes achieved by using a solvent (such as DMF or THF) that complexes with the initiator but that is displaced by monomer. Another possibility is to use a less active form of the initiator. Polar monomers can also be polymerized if the polar atom or group is shielded by sterically hindered substituents. Thus monomers with hydroxyl or amino groups are polymerized by converting the groups to $OSiR_3$ and NR_2, respectively [Giannini et al., 1967].

8-4d-1 Cycloalkenes

Cycloalkenes, unlike acyclic 1,2-disubstituted ethylenes, undergo Ziegler–Natta polymerization because of the presence of ring strain [Boor, 1979; Dall'Asta et al., 1962; Natta et al., 1966; Pasquon et al., 1989]. Two polymerization routes are possible: polymerization through the double bond or ring-opening olefin metathesis (Sec. 7-8), for example, for cyclobutene

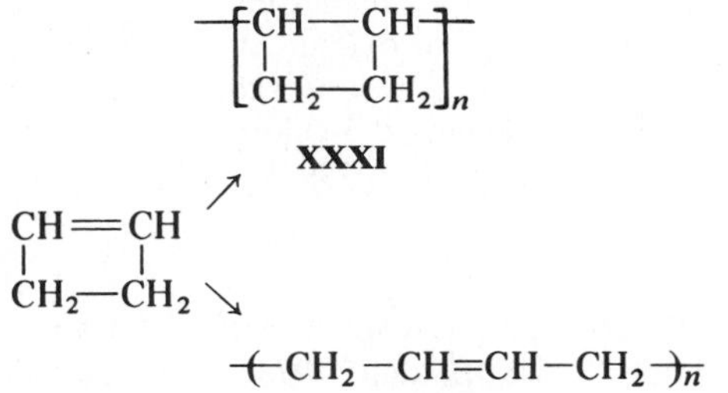

$$\left[\begin{array}{c} CH-CH \\ | \qquad | \\ CH_2-CH_2 \end{array}\right]_n \qquad (8\text{-}49)$$

XXXI

$$CH=CH$$
$$| \qquad |$$
$$CH_2-CH_2$$

$$+CH_2-CH=CH-CH_2 +_{\overline{n}} \qquad (8\text{-}50)$$

XXXII

Each polymerization route can yield different stereoisomers. Polymer **XXXI**, poly(1,2-cyclobutylene), has four different stereoisomers as described in Sec. 8-1f. Cis and trans isomers are possible for polymer **XXXII**, poly(1-butenylene) (**XXXII** is the same polymer as obtained by the 1,4-polymerization of 1,3-butadiene). Initiators based on vanadium yield polymerization almost exclusively through the double bond. Some vanadium initiators yield the erythrodiisotactic polymer, while others yield the erythrodisyndiotactic polymer. Initiators based on tungsten, titanium, and ruthenium result in polymerization predominantly by ring-opening metathesis reaction with varying amounts of cis and trans isomers.

Exclusive polymerization through the double bond does not occur with any other cycloalkene. Cyclopentene yields mixtures of double bond and ring-opening polymerizations with some titanium and vanadium initiators. Ring-opening polymerization occurs exclusively with molybdenum and tungsten initiators, as well as some Re, Nb, and Ta initiators. The relative amounts of cis and trans structures vary with the composition of the initiator and temperature. Thus WCl_6 and $MoCl_5$ with $Al(C_2H_5)_3$ at $-30°C$ yield the trans and cis structures, respectively [Dall'Asta et al., 1962]. The combination of WCl_6 with $Sn(C_2H_5)_4$ yields the cis and trans polymers, respectively, at -30 and $0°C$ [Pampus and Lehnert, 1974].

Cyclohexene does not polymerize by either route except when the cyclohexene ring is part of a bicyclic structure as in norbornene. Stereochemistry in polynorbornene (structure **LXII** in Sec. 7-8) is more complicated since there is the possibility of isomerism at the ring as well as the double bond. Most polymerizations yield cis stereochemistry at the cyclopentane ring and varying amounts of cis and trans isomers at the double bond depending on the initiator and reaction conditions [Ivin, 1987].

Little is known about the R/S isomerism (i.e., erythro and threo ditactic structures are possible) at the stereocenters on the cyclopentane ring. Cycloheptene and higher cycloalkenes undergo only ring-opening polymerization—double-bond polymerization does not occur as the larger rings can accommodate the double bond without being highly strained.

8-4d-2 Alkynes

Acetylene is polymerized to polyacetylene or poly(vinylene)

$$CH\equiv CH \rightarrow -\!\!(\!-CH\!=\!CH\!-\!)_n- \tag{8-51}$$

by Ziegler–Natta initiators such as $Ti(O\text{-}i\text{-}C_4H_9)_4$ with $Al(C_2H_5)_3$ [Ito et al., 1974; Shelburne and Baker, 1987; Theophilou and Naarman, 1989]. Low reaction temperature ($-40°C$) favors formation of the cis structure, while higher temperature ($40°C$) favors the trans structure. Polyacetylene, doped with an oxidant such as I_2 or a reductant such as AsF_5, shows promise as a polymeric semiconductor [Chien, 1984]. Various substituted acetylenes such as phenylacetylene and 3-chloro-1-propyne have also been studied [Furlani et al., 1989; Kunzler and Percec, 1990].

8-4d-3 Copolymerization

Statistical copolymerization occurs among ethylene and α-olefins [Baldwin and Ver Strate, 1972; Cooper, 1976; Pasquon et al., 1967; Randall, 1978]. The reactivities of monomers in copolymerization parallel their homopolymerization behavior: ethylene > propylene > 1-butene > 1-hexene [Soga et al., 1989]. As one might expect, the monomer reactivity ratios are somewhat sensitive to the identity of the initiator. Table 8-4 shows r_1 and r_2 for several copolymerizations.

There have been some reports of block copolymers formed by Ziegler–Natta initiators [Coover et al., 1966; Prabhu et al., 1980]. However, the block structures have not been substantiated by fractionation or NMR analysis. It appears unlikely that block copolymers are possible in view of the short lifetimes of individual propagating chains.

TABLE 8-4 Monomer Reactivity Ratios in Ziegler–Natta Copolymerizations

M_1	M_2	Initiator	r_1	r_2
Ethylene	Propylene	$TiCl_3/Cp_2Ti(CH_3)_2$[a]	9.9	0.22
		$TiCl_3/Al(n\text{-}C_6H_{13})_3$[b]	15.7	0.032
		$VOCl_3/Al(C_2H_5)_2Cl$[c]	12.1	0.018
Ethylene	1-Butene	$TiCl_3/Cp_2Ti(CH_3)_2$[a]	72	0.11
		$MgH_2/TiCl_4/Al(C_2H_5)_3$[d]	55	0.02
Ethylene	1-Hexene	$TiCl_3/Cp_2Ti(CH_3)_2$[a]	68	0.024
Propylene	1-Butene	$VCl_4/Al(n\text{-}C_6H_{13})_3$[e]	4.4	0.23
Propylene	Styrene	$TiCl_3/Al(C_2H_5)_3$[a]	130	0.18

[a]Data from Soga and Yanagihara [1989]; Soga et al. [1989].
[b]Data from Natta et al. [1961d].
[c]Data from Cozewith and Ver Strate [1971].
[d]Data from Ojala and Fink [1988].
[e]Data from Mazzanti et al. [1960].

8-4e Transition-Metal Oxide Initiators

Various supported transition-metal oxides, such as CrO_3 and MoO_3, initiate the polymerization of ethylene. The most active initiator is chromium oxide [Beach and Kissin, 1986; Pino et al., 1987; Tait, 1989; Tait and Watkins, 1989; Witt, 1974]. Silica (SiO_2) or aluminosilicates (mixed SiO_2/Al_2O_3) are used as the support material. The support is sometimes modified with titania (TiO_2). The CrO_3 catalyst, referred to as a *Phillips catalyst* or *initiator*, is prepared by impregnating the finely divided support with an aqueous solution of CrO_3. The chromium loading is in the range 0.5–5 wt %, with 1% being typical. The initiator is fixed on the support surface by heating at 500–800°C and higher. This probably results in the reaction of surface hydroxy groups in the support material with CrO_3 to form chromate (**XXXIII**) and dichromate (**XXXIV**)

$$(8\text{-}52)$$

species. The details of initiation with the Phillips initiator are not understood. Both Cr(II) and Cr(III) have been proposed as the active oxidation state of chromium. Initiation involves the formation of chromium–carbon bonds by reaction of ethylene with the active sites. Initiation is accelerated by carrying out the heat treatment of the catalyst in a reducing atmosphere of CO, H_2, or metal hydride or treatment with AlR_3 or $Al(OR)_3$. Although highly active for ethylene polymerization, the Phillips catalyst is not useful for propylene and other α-olefins since it does not bring about a stereospecific polymerization.

Active initiators for ethylene polymerization have also been obtained by depositing various organometallic derivatives of chromium and other transition metals, such as di-η^5-cyclopentadienylchromium (chromocene), (η^3-allyl)$_4$Zr, and bis(triphenylsilyl)chromate on silica and other supports [Choi and Ray, 1985].

8-4f Homogeneous, Isospecific Polymerization by Rigid Chiral Metallocenes

There has been considerable effort to obtain soluble initiators that bring about isospecific polymerization. A variety of soluble initiators, such as di-η^5-cyclopentadienyldiphenyltitanium and tetrabenzylzirconium, are active but aspecific initiators. However, the combination of some of these soluble initiators with methylaluminoxane (oligomeric $[Al(CH_3)\text{-}O]_n$) are partially effective for initiating isospecific polymerization [Cam et al., 1990; Ewen, 1984; Oliva et al., 1988; Zambelli et al., 1986]. Isotactic indices as high as 85% were obtained in the polymerization of propylene at -45°C with di-η^5-cyclopentadiendiphenyltitanium and methylaluminoxane. The polymer has an isotactic stereoblock structure (**XXVIII** in Sec. 8-4a-6). Methylaluminoxane apparently interacts with the transition-metal compound to form an initiator (which may be chiral?) with a considerable tendency to isospecific polymerization. However, polymerization at higher temperature (25°C) yields the atactic structure.

Attention turned to chiral analogs of the various initiators to obtain very highly isospecific initiators. Chiral initiators such as η^5-cyclopentadienyl-η^5-indenylmethylzirconium chloride are active but are still not exceptionally high in isospecificity [Couturier et al., 1980; Martin et al., 1975]. A major breakthrough occurred with the preparation of rigid chiral metallocene initiators, such as *racemic* 1,1'-ethylenedi-η^5-indenylzirconium dichloride (**XXXV** and **XXXVI**), the titanium and hafnium analogs,

XXXV   XXXVI

and the corresponding *racemic* 1,1'-ethylenedi-η^5-4,5,6,7-tetrahydroindenyl compounds [Cheng and Ewen, 1989; Grassi et al.; 1988; Kaminsky, 1983; Kaminsky et al., 1988; Rieger et al., 1990; Soga et al., 1987] These metallocenes are rigid as a result of the ethylene bridge between the two 5-membered rings. Similar metallocenes have been synthesized with dimethylsilyl, $Si(CH_3)_2$, bridges between the two rings.

The rigid chiral metallocenes in the presence of methylaluminoxane are very highly isospecific. In the typical polymerization recipe, the ratio of aluminum to transition metal is 10^3–10^4. Without methylaluminoxane, the rigid chiral metallocenes are not highly isospecific. This indicates that the rigidity of the rigid metallocenes is insufficient to impart high isospecificity. The methylaluminoxane may function by increasing the rigidity of the rigid chiral metallocenes by complexation (in addition to alkylation of the transition metal). Isotactic indices of greater than 90% are easily achieved in the polymerization of propylene and other α-olefins by the rigid chiral metallocenes together with methyl aluminoxane. Isotactic indices of 98–99% have been obtained in a number of polymerizations. These initiators are about as highly isospecific as the best of the heterogeneous Ziegler–Natta initiators. Further, the rigid chiral metallocenes are very highly active, often surpassing the best of the heterogeneous initiators (the superactive, high-mileage initiators; see Sec. 8-4a) by an order of magnitude. Activities as high as 43,000 kg of polypropylene per gram of Zr have been achieved. This is somewhat more than an order of magnitude higher than the most active heterogeneous initiators.

Structures **XXXV** and **XXXVI** are enantiomers. The typical synthesis of these initiators produces the racemic mixture. Each enantiomer produces isotactic polymer just as isotactic polymer is produced by each adjacent enantiotopic active site in heterogeneous initiators. The high activity and isospecificity of the rigid chiral metallocenes is achieved only in the presence of methylaluminoxane, similar to the situation with the heterogeneous Ziegler–Natta initiators that require a Group I–III metal compound.

There is a general resemblance of **XXXV** and **XXXVI** to the chiral species of heterogeneous initiators (**XXV** and **XXVI**). Chirality in both systems is not due to the presence of chiral atoms (atoms with four different substituents). The chirality of the metallocene initiator results from the chirality of the molecule as a whole. The chirality of the heterogeneous initiator results from the chirality of the crystal structure. For the soluble isospecific initiators, the requirements for very high isospecificity include both chirality and structural rigidity. Neither chirality nor structural rigidity alone is sufficient. Thus, rigid achiral initiators such as *meso* 1,1'-ethylenedi-η^5-indenyl zirconium (IV) dichloride (**XXXVII**) are not highly isospecific, nor are nonrigid

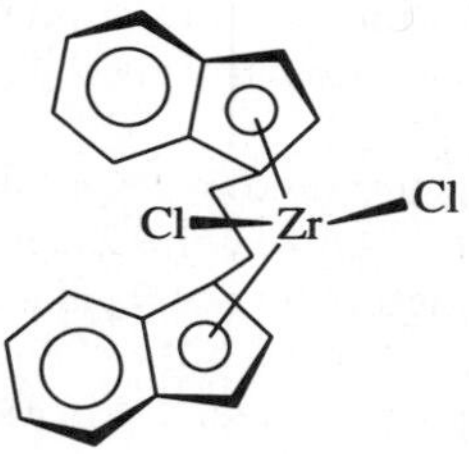

XXXVII

chiral initiators such as those previously mentioned or the nonbridged analogs of **XXV** and **XXVI**.

Very highly syndiotactic polypropylene (86% racemic pentads) was obtained by using *i*-propyl(η^5-cyclopentadienyl-η^3-fluorenyl)zirconium dichloride (**XXXVIII**) with methylaluminoxane to initiate polymerization [Asanuma et al., 1991; Ewen et al.,

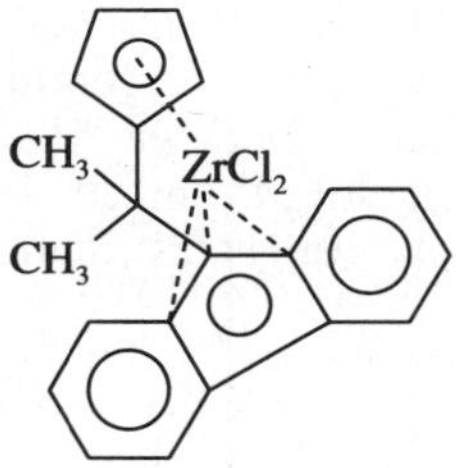

XXXVIII

1988]. Structure **XXXVIII** is the only initiator besides the vanadium-based Ziegler–Natta initiators (Sec. 8-4a-7) to give such highly syndiospecific polymerizations.

8-4g Commercial Utilization

The importance of coordination polymerization of alkenes is evident when it is noted that more than 19 billion pounds of polymers are produced annually in the United States by this route. This corresponds to 40–45% of the total industrial production of polymers from monomers containing the carbon–carbon double bond.

8-4g-1 Process Conditions

Commercial polymerizations of ethylene, propylene, and other α-olefins are carried out as slurry (suspension) and gas-phase processes [Beach and Kissin, 1986; Diedrich, 1975; Lieberman and Barbe, 1988; Magovern, 1979; Vandenberg and Repka, 1977; Weissermel et al., 1975]. Solution polymerization has been used in the past for ethylene polymerization at 140–150°C, pressures of up to about 8 MPa (1 MPa = 145 psi = 9.869 atm), using a solvent such as cyclohexane. The solution process with its higher temperatures was employed for polymerization with the relatively low efficiency early

Phillips initiators. (Polyethylene, but not the initiator, is soluble in the reaction medium under the process conditions.) The development of a variety of high-efficiency initiators has allowed their use in lower-temperature suspension and gas-phase processes, which are more advantageous from many viewpoints (energy consumption, product workup). Solution polymerization is limited at present to producing low-molecular-weight polyethylene.

Both Ziegler–Natta and metal oxide Phillips-type initiators are used in suspension polymerization. Both types of initiators are used for ethylene, but only the Ziegler–Natta initiators are used for propylene since Phillips-type initiators do not yield stereospecific polymerizations.

The use of gas-phase processes has increased greatly since their initial use in 1968 for ethylene polymerization. The process has been extended to ethylene copolymers and, very recently, to polypropylene. The absence of solvent accomplishes major economies for the gas-phase process relative to suspension polymerization. Pressures of about 2–3 MPa and temperatures in the range 70–105°C are employed in both fluidized-bed and stirred-bed reactors [Brockmeier, 1987]. The reaction medium is a well-stirred mixture of initiator and polymer powders together with gaseous monomer. After emerging from the reactor, polymer is separated from unreacted monomer, and the latter is recycled. Temperature control and temperature homogeneity throughout the reactor are critically important. The temperature must be maintained below the softening temperature of the polymer to prevent agglomeration of the polymer product into large lumps. Agglomeration leads to an inability to control reactor temperature, and this results in deterioration of the product as well as reactor shutdown. Highly active initiators based on both chromium and titanium are employed. For ethylene polymerization titanium-based initiators yield narrower-molecular-weight distributions than do chromium-based initiators [Karol, 1989].

8-4g-2 *High-Density (Linear) Polyethylene*

Polyethylene produced by Ziegler–Natta and Phillips initiators differs structurally from that obtained by radical polymerization (Sec. 3-13b-1) in having a much lower degree of branching (0.5–3 versus 15–30 methyl groups per 500 monomer units). The two polyethylenes are often referred to as *linear* and *branched polyethylenes*, respectively. The lower degree of branching results in higher crystallinity (70–90% versus 40–60%), higher density (0.94–0.96 versus 0.91–0.93 g/cm^3), and higher crystalline melting temperature (133–138 vs. 105–115°C) for linear polyethylene. The polyethylenes produced by coordination and radical polymerizations are also referred to as *high-density* (HDPE) and *low-density* (LDPE) *polyethylenes*, respectively. Compared to LDPE, HDPE has increased tensile strength, stiffness, chemical resistance, and upper use temperature combined with decreased low-temperature impact strength, elongation, permeability, and resistance to stress cracking. The property enhancements result in the annual production of over 8 billion pounds of HDPE in the United States. The two polyethylenes complement each other as over 9 billion pounds of LDPE are also produced annually in the United States.

Most HDPEs have number-average molecular weights of 50,000–250,000. These materials are used in a wide range of applications [Beach and Kissin, 1986; Juran, 1989]. The largest market (40%) consists of blow-molded products such as bottles (milk, food, detergent), housewares, toys, and pails. Injection-molded objects similar to those produced by blow molding constitute about 30% of the total market. Extruded products comprise most of the remainder of the market for HDPE. This includes film for producing grocery and merchandise bags and food packaging, sheet for truck bed liners and luggage, pipe, tubing, and wire and cable. Very-low-molecular-weight (several thousand) polyethylenes are used as wax substitutes in paper and spray coatings, crayons, and polishes.

Various specialty HDPEs are produced by polymerization to higher molecular weights. Increased molecular weight results in increased tensile strength, elongation, low-temperature impact resistance, and stress crack resistance, although processing is more expensive because of increased melt viscosity. *High-molecular-weight-high-density polyethylene* (0.25–1.5 million molecular weight) is used for pressure piping in mining, industrial, sewer, gas, oil, and water applications. Other uses include large blow-molded parts such as shipping containers and bulk storage tanks and blown film for grocery and merchandise bags and can liners. *Ultra-high-molecular-weight high-density polyethylene* (>1.5 million molecular weight) has very high abrasion resistance and impact strength, the highest of any thermoplastic material. It can be processed without additives and stabilizers because the longer chains are resistant to mechanical scission, and even when chain scission occurs the molecular weight is sufficiently high to retain mechanical strength. Applications include low-speed bearings, gears for snowmobile drives, impellers for snow blowers, and the sliding surfaces in chutes and hoppers in the mining and freight industries as well as in agricultural and earth moving machinery.

8-4g-3 Linear Low-Density Polyethylene

Copolymerization of ethylene with small amounts of an α-olefin such as 1-butene, 1-hexene, or 1-octene results in the equivalent of the branched, low-density polyethylene produced by radical polymerization. The polyethylene has controlled amounts of ethyl, *n*-butyl, and *n*-hexyl, respectively. Copolymerization with propylene and 4-methyl-1-pentene is also practiced. There was little effort to commercialize linear low-density polyethylene (LLDPE) until 1978, when gas-phase technology made the economics of the process very competitive with the high-pressure radical polymerization process [James, 1986]. The expansion of this technology has been rapid. Almost no new high-pressure plants have been built in recent years. New capacity for LDPE has involved new plants for the low-pressure gas-phase process, which allows the production of HDPE and LDPE as well as polypropylene. There is even some indication of replacements of high-pressure plants by low-pressure plants. Thirty percent of the annual U.S. production of LDPE is LLDPE. This means that 60–65% of the total of all polyethylenes of all densities is produced by anionic coordination processes.

8-4g-4 Polypropylene

Isotactic polypropylene has the lowest density (0.90–0.91 g/cm^3) of the major plastics and possesses a very high strength:weight ratio. It has a crystalline melting point of 165–175°C and is usable to 120°C; both temperatures are higher than the corresponding values for HDPE. More than 7 billion pounds of polypropylene are produced annually in the United States. About 20% of this volume consists of copolymers, mostly copolymers containing 2–5% ethylene, which imparts increased clarity, toughness, and flexibility. Injection-molded products account for about 40% of the total polypropylene volume [Juran, 1989; Lieberman and Barbe, 1988]. This includes durable goods (housings and parts for small and large appliances, furniture and office equipment, battery cases, automobile, interior trim, and air ducts) and semirigid packaging (yogurt and margarine tubs, caps and closures for medicine).

Packaging films (extruded, blown, and cast) are used in pressure-sensitive tapes and electrical applications and as replacements for cellophane and glassine films in liners for cereal boxes, wraps for snack foods, cigarettes, bread, and cheese. Blow-molded containers are used in applications where the higher use temperature of polypropylene is needed (compared to HDPE), such as in packaging of syrups that are hot-filled.

Fiber products account for about 15% of polypropylene consumption. The products range from continuous filaments for carpeting and rope to melt-blown fibers for non-woven goods. Specific applications include outdoor carpets, yarns for upholstery and automobile seats, and replacements for canvas in luggage and shoes, disposable goods (diapers, surgical gowns), ropes, and cords.

8-4g-5 Ethylene–Propylene Elastomers

Copolymers and terpolymers of ethylene and propylene, referred to as *EPM* and *EPDM*, respectively, form useful elastomers [Ver Strate, 1986]. (EPM and EPDM are acronyms for ethylene–propylene monomers and ethylene–propylene–diene monomers, respectively.) The terpolymers contain up to about 4 mole % of a diene such as 5-ethylidene-2-norbornene (**XXXIXa**), dicyclopentadiene (**XXXIXb**), or 1,4-hex-

XXXIXa   **XXXIXb**

adiene. A wide range of products are available, containing 40–90 mole % ethylene. The diene, reacting through one of its double bonds, imparts a pendant double bond to the terpolymer for purposes of subsequent crosslinking.

More than 500 million pounds of EPM and EPDM are produced annually in the United States. Their volume ranks EPM and EPDM third behind styrene–1,3-butadiene copolymers and poly(1,4-butadiene) as synthetic rubbers. EPM and EPDM have good chemical resistance, especially toward ozone. They are very cost-effective products since physical properties are retained when blended with large amounts of fillers and oil. Applications include automobile radiator hoses, weather stripping, and roofing membrane.

8-4g-6 Other Polymers

Isotactic poly(1-butene) and poly(4-methyl-1-pentene) are useful in applications that take advantage of their higher melting and use temperatures compared to polypropylene and HDPE. Poly(1-butene) is used in hot-water plumbing pipe (both commercial and residential) and large diameter pipe for transporting abrasive materials at high temperatures in the mining, chemical, and power-generation industries. Poly(4-methyl-1-pentene) is used to produce various laboratory and medical ware, cook-in containers for both hot-air (conventional) and microwave ovens, and various other items for the lighting, automobile, appliance, and electronics industries.

8-5 STEREOSPECIFIC POLYMERIZATION OF POLAR VINYL MONOMERS

Polar monomers such as methacrylates and vinyl ethers undergo stereospecific polymerization when there is an appropriate balance between monomer, initiator, solvent, and temperature. Under conditions where the propagating species is free (uncoordinated), syndiotacticity is increasingly favored with decreasing reaction temperature (Sec. 8-3). This is the case for all radical polymerizations and for those ionic polymerizations that take place in highly solvating media. Isospecific polymerization can occur in poorly solvating media where the propagating species is coordinated to the counterion. Many of the homogeneous polymerizations described in Chap. 5 result in isospecific polymerization, although the degree of stereospecificity rarely matches that of the Ziegler–Natta polymerizations.

8-5a Methyl Methacrylate

The stereospecific polymerization of various acrylates and methacrylates has been studied using initiators such as alkyllithium [Bywater, 1989; Pasquon et al., 1989]. Table 8-5 illustrates the effects of counterion, solvent, and temperature on the stereochemistry of the anionic polymerization of methyl methacrylate. In polar solvents (pyridine and THF versus toluene), the counterion is removed from the vicinity of the propagating center and does not exert a stereoregulating influence on entry of the next monomer unit. The tendency is toward syndiotactic placement. The extent of syndiotacticity decreases in the order Li > Na > K corresponding to the relative

Table 8-5 Effect of Counterion, Solvent, and Temperature on Polymerization of Methyl Methacrylate

Solvent	Counterion	Temperature (°C)	Triad Tacticity		
			(mm)	(mr)	(rr)
Toluene[a]	Li	0	0.72	0.17	0.11
Pyridine[a]	Li	0	0.08	0.32	0.60
Toluene[b]	Li	−78	0.87	0.10	0.03
Toluene[b]	Mg	−78	0.97	0.03	0
Toluene[b]	Mg	−78	0.23	0.16	0.61
THF[b]	Li	−85	0.01	0.15	0.84
THF[c]	Li	−78	0.05	0.33	0.61
Toluene[c]	Li	−78	0.78	0.16	0.06
Toluene[c]	Li[d]	−78	0	0.10	0.90
Toluene[a]	Na	0	0.57	0.31	0.12
Pyridine[a]	Na	0	0.12	0.46	0.42
Toluene[a]	K	0	0.35	0.42	0.23
Pyridine[a]	K	0	0.14	0.53	0.33
THF[e]	Cs	20	0.10	0.56	0.34
THF[e]	Cs	−100	0	0.40	0.60

[a]Data from Braun et al. [1962].
[b]Data from Bywater [1989].
[c]Data from Kitayama et al. [1989].
[d]Initiator = t-C_4H_9Li + $Al(C_2H_5)_3$.
[e]Data from Muller et al. [1977], Kraft et al. [1980].

extents of solvation of different ions. The smallest ion Li^+ is the most highly solvated and furthest removed from the propagating center. (This is analogous to the effect of polar solvent on propagation rate constants for anionic polymerization of styrene with different counterions. The ion pair propagation rate constant for styrene is largest for Li and smallest for Cs, which is the reverse of the order in less polar solvent; see Sec. 5-3e-2-b.) The extent of syndiotacticity increases with decreasing temperature, as is evident in the data for polymerization with cesium as the counterion.

When nonpolar solvents are employed, polymerization proceeds by an anionic coordination mechanism. The counterion directs isotactic placement of entering monomer units into the polymer chain. The extent of isotactic placement increases with the coordinating ability of the counterion (Li > Na > K). The smaller lithium ion has the greater coordinating and stereoregulating power. Increased reaction temperature decreases the extent of isospecificity.

Mechanisms proposed to explain the isospecific polymerization of methyl methacrylate involve rigidization of the propagating center by coordination of counterion with both the terminal and penultimate monomer units [Braun et al., 1962; Wiles and Bywater, 1965]. The rigid propagating chain end imposes stereospecificity on the propagation in a manner generally analogous to that present for a Ziegler–Natta propagating center. One mechanism suggests a 6-membered cyclic chain end and monomer coordinated to the counterion as in **XL** with propagation occurring by the sequence indicated with the curved arrows [Leitereg and Cram, 1968].

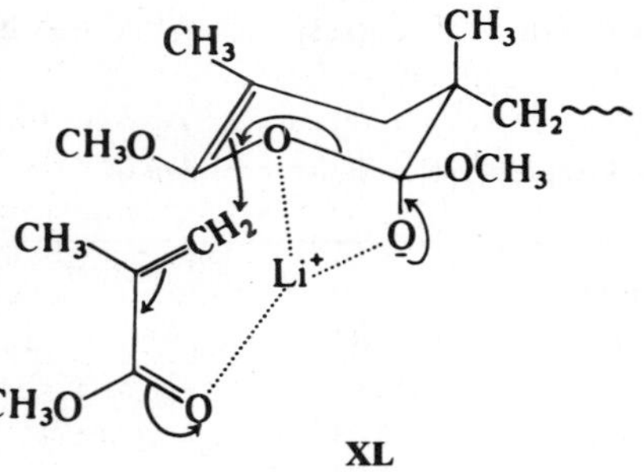

XL

That methacrylate polymerization is not well understood is clear from experiments with ethyl-(Z)-β-d-methacrylate (**XLI**). Highly isotactic polymer of the diisotactic type

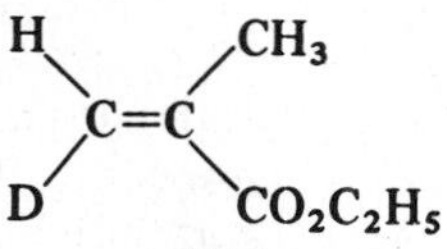

XLI

is formed using fluorenyllithium or $\emptyset$MgBr at $-78°C$ [Fowells et al., 1967]. When polymerization is carried out in toluene, $\emptyset$MgBr yields the erythrodiisotactic polymer, while fluorenyllithium yields the threodiisotactic polymer. Polymerization in toluene containing ether leads to some decrease in isotacticity. However, the polymer (which is still high in isotacticity) becomes threo for $\emptyset$MgBr and erythro for fluorenyllithium. The erythro product in this polymerization corresponds to the syn addition mechanism described in Fig. 8-12. The threo product arises either from a corresponding anti addition in which the monomer approach to the propagating center is always the same and involves the deuterium of the incoming monomer being on the opposite side to the ester group of the propagating chain end or a syn addition in which the monomer approach involves the deuterium of the incoming monomer being aligned with the ester group of the propagating chain end. Irrespective of which mechanism is responsible for the threo product, it is unclear why fluorenyllithium and ϕMgBr are so different and why the stereochemical result reverses with solvent in each case. Further, it is generally found that polymerizations initiated by magnesium Grignard reagents are poorly understood. Polymerization carried out under apparently similar conditions often yield very different results [Bywater, 1989]. Consider the two entries in Table 8-5 for Mg counterion. These involve polymerizations under the same nominal conditions (t-C_4H_9MgCl, toluene, $-78°C$), but one yields a very highly isotactic polymer, $(mm) = 0.97$, while the other yields a moderately syndiotactic polymer, $(rr) = 0.61$. There must be subtle but important differences in the reaction conditions that are responsible for the different stereochemical results.

8-5b   Vinyl Ethers

The isotactic polymerization of a vinyl ether requires a *cationic coordination* process. The cationic process is analogous to the anionic coordination process except that the propagating center is a carbocation instead of a carbanion and the counterion is an anion instead of a cation. Various initiators, of both the homogeneous and heterogeneous types, yield varying degrees of isotactic placement [Ketley, 1967a, 1967b; Pasquon et al., 1989]. This includes boron trifluoride and other Lewis acids, including components (sometimes only one, sometimes both of the two different metal components) used in Ziegler–Natta formulations. Some of the polymerizations proceed with very high isospecificity; for instance, ethylaluminum dichloride and diethylaluminum chloride yield 96–97% isotactic polymer for polymerization of i-butyl vinyl ether at $-78°C$ in toluene, whereas aluminium tribromide yields mostly atactic polymer [Natta et al., 1959c]. Not all vinyl ethers give the same result with the same initiator. Thus the polymerization of t-butyl vinyl ether is only mildly isospecific under the same conditions described above for the highly isospecific polymerization of i-butyl vinyl ether. In general, the effects of solvent, temperature, and other reaction conditions on the extent of isospecificity are similar to those previously described for other types of monomer.

Highly syndiotactic polymers have been obtained in only a few instances—with some monomers containing bulky substituents, such as α-methylvinyl methyl ether, trimethylvinyloxysilane, and menthyl vinyl ether, in polar solvents under homogeneous conditions [Goodman and Fan, 1968; Ledwith et al., 1979; Murahashi et al., 1966]. That less hindered monomers in polar solvents do not yield highly syndiotactic polymers may be indicative of the involvement of ether monomers in intramolecular solvation of propagating centers. The polar solvents such as THF may not be sufficiently polar to displace monomer as a solvating species to yield the highly solvated, relatively free propagating centers that lead to syndiotactic placement. One of the mechanisms proposed to explain isotactic placement of vinyl ethers is consistent with this consideration [Cram and Kopecky, 1959]. Propagation involves a 6-membered cyclic propagating chain end (**XLII**) formed by the *antepenultimate* (the second repeat unit behind the last unit) ether group of the propagating chain solvating the carbocation center.

$$\text{\small\textasciitilde\textasciitilde\textasciitilde} CH_2-\underset{\overset{|}{OR}}{CH}-CH_2-\underset{\overset{|}{OR}}{CH}-CH_2-\underset{\overset{|}{OR}}{\overset{+}{CH}} \rightarrow \text{\small\textasciitilde\textasciitilde\textasciitilde} CH_2 \quad (8\text{-}53)$$

XLII

8-5c Styrene

Styrene is slightly polar compared to ethylene and α-olefins. The lack of a strongly polar functional group allows styrene to undergo highly (>95–98%) isospecific polymerization with many of the heterogeneous Ziegler–Natta initiators effective for α-olefins [Longo et al., 1990; Pasquon et al., 1989; Soga et al., 1988]. Highly syndiotactic polystyrene is obtained using soluble Ziegler–Natta initiators, such as tetrabenzyltitanium, tetrabenzylzirconium, and tetraethoxytitanium or cyclopentadienyltitanium trichloride with methylalumoxane [Ishihara et al., 1988; Pellecchia et al., 1987; Zambelli et al., 1989]. Further, there are recent reports of highly syndiospecific polymerizations of styrene with heterogeneous initiators based on tetra-*n*-butoxytitanium and methylaluminoxane supported on silica or magnesium hydroxide [Soga and Monoi, 1990; Soga and Nakatani, 1990]. This is the only report of the syndiospecific polymerization of any monomer with heterogeneous initiators.

Partially isotactic polystyrenes are obtained with *n*-butyllithium in toluene at −40°C and the heterogeneous *alfin initiator* (allylsodium + sodium *i*-propoxide + NaCl) in hexane at −20°C [Braun et al., 1960; Kern, 1960]. Partially syndiotactic polystyrenes are obtained with many different initiators, including *n*-butyllithium in toluene at −25°C and higher temperatures and cesium naphthalene in toluene at 0°C or THF at −78°C [Kawamura et al., 1982]. There is little tendency toward stereospecificity with cationic initiators, although polymerizations of α-methylstyrene by BF₃, SnCl₄, and other Lewis acids give moderate syndiospecificity [Wicke and Elgert, 1977].

8-6 STEREOSPECIFIC POLYMERIZATION OF 1,3-DIENES

8-6a Radical Polymerization

The polymerization of 1,3-dienes is more complicated than that of alkenes because of the greater number of stereoisomers (Secs. 8-1d and 8-1e). Table 8-6 shows the polymers obtained in radical polymerization of 1,3-butadiene, isoprene, and chloroprene (2-chloro-1,3-butadiene). 1,4-Polymerization occurs in preference to 1,2- or 3,4-polymerization and trans-1,4-polymerization in preference to cis-1,4-polymerization. (Both 1,2- and 3,4-polymerizations yield the same polymer for the symmetrical 1,3-butadiene.) The temperature dependence of the structure of polybutadiene shows trans-1,4-addition to have a more favorable enthalpy of activation by about 12 kJ/mole relative to the cis-1,4-addition, although the latter has a more favorable entropy of activation by about 29 J/mole-K. Trans-1,4-addition is favored over 1,2-addition by an enthalpy of activation by about 4 kJ/mole with no difference in the entropies of activation.

The polymerization of a 1,3-diene involves delocalization of the radical over carbons 2 and 4 of the terminal monomer unit (**XLIII**). The predominance of 1,4-propagation

$$\text{\small\textasciitilde\textasciitilde\textasciitilde} CH_2-CH{=}CH-CH_2{\cdot} \leftrightarrow \text{\small\textasciitilde\textasciitilde\textasciitilde} CH_2-\overset{\cdot}{CH}-CH{=}CH_2$$

XLIIIa   **XLIIIb**

Table 8-6 Stereochemistry of Radical Polymerization of 1,3-Dienes

Monomer	Temperature (°C)	Polymer Structure (%)			
		Cis-1,4	Trans-1,4	1,2	3,4
1,3-Butadiene[a]	−20	6	77	17	
	20	22	58	20	
	100	28	51	21	
	233	43	39	18	
Isoprene[a]	−20	1	90	5	4
	10	11	79	5	5
	100	23	66	5	6
	203	19	69	3	9
Chloroprene[b]	−40	1	97	1	1
	20	3	93	2	2
	90	8	85	3	4

[a]Data from Condon [1953], Richardson [1954], Pollock et al. [1955].
[b]Data from Coleman and Brame [1978], Coleman et al. [1977], Ebdon [1978].

over 1,2-propagation is primarily a consequence of the lower steric hindrance at carbon 4 relative to carbon 2 for bonding to an incoming monomer molecule. This difference increases as the 2-position becomes substituted with the result that 1,4-polymerization is even more favored in the polymerizations of isoprene and chloroprene (Table 8-6). The stability of the polymer is probably also a factor, since the 1,4-polymer (a 1,2-disubstituted alkene) is generally more stable than the 1,2-polymer (a monosubstituted alkene).

Several factors favor the preference of trans-1,4-polymerization over cis-1,4-polymerization. 1,3-Butadiene exists predominately in the *transoid* or *s-trans* conformation (**XLIVa**) as opposed to the *cisoid* or *s-cis* conformation (**XLIVb**). This is

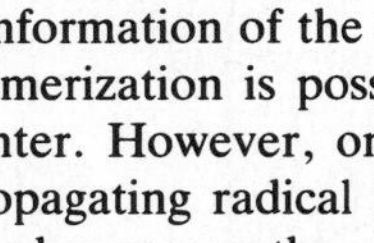

XLIVa 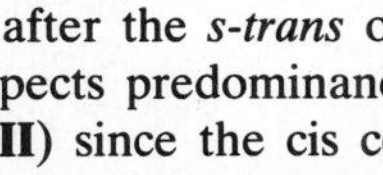   **XLIVb**

probably also the situation for isoprene [Craig et al., 1961; Hsu et al., 1969]. The conformation of the monomer need not translate into polymer stereochemistry since isomerization is possible after the *s-trans* or *s-cis* conformer adds to a propagating center. However, one expects predominance of the trans configuration of the allyl propagating radical (**XLIII**) since the cis configuration involves steric hindrance to overlap among the *p*-orbitals of the carbon atoms of the allyl radical. Another factor is the greater stability of the trans-1,4-polymer over the cis-1,4-polymer. The configuration of the last monomer unit in the propagating chain is determined (with trans favored over cis) when the next monomer unit adds.

The stereochemistry of the 1,3-diene units in copolymerization is expected to be the same as in homopolymerization of 1,3-dienes. This expectation has been verified in styrene-1,3-butadiene copolymerizations at polymerization temperatures of −33 to 100°C [Binder, 1954].

8-6b   Anionic and Anionic Coordination Polymerizations

The anionic polymerization of 1,3-dienes yields different polymer structures depending on whether the propagating center is free or coordinated with a counterion [Morton, 1983; Senyek, 1987; Tate and Bethea, 1985; Van Beylen et al., 1988; Young et al., 1984]. Table 8-7 shows typical data for 1,3-butadiene and isoprene polymerizations. Polymerization of 1,3-butadiene in polar solvents, proceeding via the free anion and/ or solvent-separated ion pair, favors 1,2-polymerization over 1,4-polymerization. The anionic center at carbon 2 is not extensively delocalized onto carbon 4 since the double bond is not a strong electron-acceptor. The same trend is seen for isoprene except that 3,4-polymerization occurs instead of 1,2-polymerization. The 3,4-double bond is sterically more accessible and has a lower electron density relative to the 1,2-double bond. Polymerization in nonpolar solvents takes place with an increased tendency toward 1,4-polymerization. The effect is most pronounced with lithium ion, which has the greatest coordinating power of the counterions. Further, there is a strong tendency toward cis-1,4-polymerization over trans-1,4-polymerization with lithium even though the trans-1,4-polymer is more stable than the cis-1,4-polymer. This effect is much stronger for isoprene than 1,3-butadiene.

There is no mechanism that adequately describes all features of the anionic polymerization of 1,3-dienes. NMR data indicate the presence of π- and σ-bonded propagating chain ends (**XLV** and **XLVI**). When reaction occurs in polar solvent, the

$$\overset{\delta-}{\sim\!\!\sim\!\!\sim} CH_2 \overset{Li^+}{-} CH \overset{\delta-}{-} CH - CH_2 \qquad \sim\!\!\sim\!\!\sim CH_2 - CH = CH - CH_2 Li$$

$$\textbf{XLV} \qquad\qquad\qquad\qquad \textbf{XLVI}$$

Table 8-7   Effect of Solvent and Counterion on Stereochemistry in Anionic Polymerization of 1,3-Dienes[a]

Counterion	Solvent	Cis-1,4	Trans-1,4	3,4	1,2
		(Structure of Polymer (%))			

Counterion	Solvent	Cis-1,4	Trans-1,4	3,4	1,2
(1,3-Butadiene (at 0°C))					
Li	n-Pentane	35	52		13
Na	n-Pentane	10	25		65
K	n-Pentane	15	40		45
Rb	n-Pentane	7	31		62
Cs	n-Pentane	6	35		59
Li	THF	0	4		96
Na	THF	0	9		91
K	THF	0	18		82
Rb	THF	0	25		75
(Isoprene (at 25°C))					
Li	n-Hexane	93	0	7	0
Na	n-Hexane	0	47	45	8
Cs	None	4	51	37	8
Li	THF	0	30	54	16
Na	THF	0	38	49	13

[a]Data from Tobolsky and Rogers [1959], Rembaum et al. [1962].

carbanion center is delocalized as both anionic and cationic centers are solvated. Delocalization also occurs in nonpolar solvent when the counterion does not form a strong covalent bond to carbon. The situation is different for lithium counterion in nonpolar solvent. The allyl carbanion becomes localized in the absence of solvation as long as the counterion (lithium) forms a strong covalent bond to carbon. Localization yields **XLVI** instead of **XLVII**, presumably because of its greater stability since **XLVI**

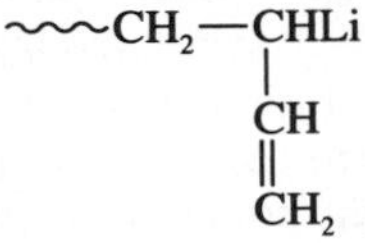

$$\sim\sim CH_2 - CHLi$$
$$|$$
$$CH$$
$$||$$
$$CH_2$$

XLVII

is the more substituted alkene. The formation of **XLVI** results in 1,4-polymerization in nonpolar solvent when lithium is the counterion. The predominance of cis-1,4-polymerization over trans-1,4-polymerization has been ascribed to the greater reactivity of cis-**XLVI**; that is, **XLVI** is formed as a mixture of cis and trans isomers (which are in equilibrium) but the cis isomer is more reactive [Van Beylen et al., 1988]. The preference for 1,2-polymerization of 1,3-butadiene indicates that delocalized species **XLV** is preferentially attacked by monomer at carbon 2 instead of carbon 4. The reason for this is not evident.

The relative extents of the different structural units in 1,3-diene polymerization are not strongly dependent on polymerization temperature in the range -20 to $50°C$ [Morton, 1983]. Table 8-8 shows other features of 1,3-diene polymerization in nonpolar solvent with lithium counterion. The extent of cis-1,4-placement is greatest in the absence of solvent and at low initiator concentration [Worsfold and Bywater, 1978]. Isomerization of cis-**XLVI** to the more stable trans-**XLVI** increases faster than propagation of cis-**XLVI** with an increase in initiator concentration since propagation proceeds only through unassociated species while isomerization proceeds through both associated and unassociated species. Increasing the concentration of initiators such as alkyllithium increases the concentration of associated species much more than unassociated species (Sec. 5-3e-5). The decrease in cis-1,4-placement observed for polymerization in a solvent relative to the neat polymerization is a consequence of the effect of monomer concentration. Increased monomer concentration increases the polym-

Table 8-8   Effect of Solvent and Initiator Concentration on Stereochemistry in Anionic Polymerization of Isoprene at 20°C[a]

Solvent	[RLi] (moles/liter)	Structure of Polymer (%)			
		Cis-1,4	Trans-1,4	3,4	1,2
None	8×10^{-6}	96	0	4	0
None	3×10^{-3}	77	18	5	0
n-Hexane	1×10^{-5}	86	11	3	0
n-Hexane	1×10^{-2}	70	25	5	0
Benzene	4×10^{-5}	70	24	6	0
Benzene	9×10^{-3}	69	25	6	0

[a]Data from Morton [1983].

erization rate without any effect on isomerization. The extent of isomerization of cis-**XLVI** to trans-**XLVI** relative to propagation of cis-**XLVI** decreases with increasing monomer concentrations.

A variety of other initiators have been studied for the stereospecific polymerizations of 1,3-dienes [Cooper, 1979; Pasquon et al., 1989; Porri and Giarrusso, 1989; Tate and Bethea, 1985; Senyek, 1987]. One of the earliest systems was the heterogeneous alfin initiator, consisting of allylsodium, sodium isopropoxide, and sodium chloride. The alfin initiator is strongly but not overwhelmingly stereospecific. Polymerization of 1,3-butadiene proceeds with 68% trans-1,4-, 17% cis-1,4-, and 15% 1,2-placements.

Ziegler–Natta and related initiators yield truly remarkable results, surpassing the stereospecificity exhibited by lithium initiators. Table 8-9 shows the polymer structures obtained from 1,3-butadiene and isoprene using different initiator systems. For the data in Table 8-9, the results shown are those for the crude polymer product, not a purified fraction of the total product. The reader is cautioned that not all similar tables of data in books and articles have the same meaning. One might see a notation that a 91% trans-1,4-polymer is formed in a reaction, but careful reading of the original data shows that this is the structure of the crystalline fraction of the crude product.

The exceptional stereospecificity of the Ziegler–Natta initiators is even more evident here than in the polymerization of alkenes. Four different stereoregular polymers (cis-1,4-, trans-1,4-, st-1,2-, it-1,2-) are possible for 1,3-butadiene. Table 8-9 shows

Table 8-9 Stereospecificity in 1,3-Diene Polymerizations

Initiator	Polymer Structure
1,3-Butadiene	
$TiCl_4/AlR_3$:[a] Al/Ti = 1.2	50% cis-1,4, 45% trans-1,4
Al/Ti = 0.5	91% trans-1,4
TiI_4/AlR_3^b	95% cis-1,4
VCl_3 or VCl_4/AlR_3 or AlR_2Cl^a	94–98% trans-1,4
$(\eta^3\text{-allylNiI})_2^c$	95% trans-1,4
$(\eta^3\text{-allyNiCl})_2^c$	95% cis-1,4
$Rh(NO_3)_3^d$	99% trans-1,4
$Ni(octanoate)_2/AlR_3/HF^e$	97% cis-1,4
$CoCl_2/AlR_2Cl/pyridine^e$	98% cis-1,4
$Co(acetylacetonate)_3/AlR_3/CS_2^e$	99% st-1,2
$(\eta^3\text{-cyclooctadienyl})Co(C_4H_6)/CS_2^e$	99–100% st-1,2
Isoprene	
$TiCl_4/AlR_3$ or AlH_3^e	96% cis-1,4
VCl_3/AlR_3^e	97% trans-1,4
$Ti(OR)_4/AlR_3^e$	60–70% 3,4

[a]Data from Natta et al. [1959a, 1959b, 1959c].
[b]Data from Moyer [1965].
[c]Data from Taube et al. [1987].
[d]Data from Tate and Bethea [1985].
[e]Data from Porri and Giarrusso [1989].

that except for the it-1,2-structure, each of the other three polymers has been obtained in exceptionally high stereospecificity by appropriate choice of initiator. For isoprene, the cis-1,4- and trans-1,4-products are obtained in very high stereospecificity, but not the various stereoregular 3,4- or 1,2-products.

There is also considerable work reported on the polymerizations of various 4-substituted and 1,4- and 2,4-disubstituted 1,3-butadienes. Many more different stereoregular structures are possible in each of these cases (Sec. 8-1e). The result is that very few, if any, of the completely stereoregular polymers have been obtained in very high yield coupled with very high stereoregularity. For example, (E)-2-methyl-1,3-pentadiene (**XLVIII**) was polymerized by $Nd(OOCC_7H_{13})_3/Al(C_2H_5)_2Cl/Al(i\text{-}C_4H_9)_3$

$$CH_2{=}\underset{\underset{CH_3}{|}}{C}{-}CH{=}CH{-}CH_3 \;\rightarrow\; {\underset{\underset{\underset{CH_3}{|}}{}}{-}{+}CH_2{-}\underset{\underset{CH_3}{|}}{C}{=}CH{-}CH(CH_3){)}{+}_n \qquad (8\text{-}54)$$

XLVIII

to a polymer consisting of 98–99% cis-1,4-structure but different fractions (obtained by solvent extraction) differed in the degree of isotacticity at carbon 4 in the repeat unit [Cabassi et al., 1988]. One of the isolated fractions was found to have the cis-1,4-isotactic structure. Similarly, samples of highly stereoregular trans-1,4-syndiotactic and 1,2-syndiotactic polymers of 1,3-pentadiene have been isolated by fractionation of the crude products of various polymerizations [Pasquon et al., 1989; Porri and Giarrusso, 1989].

Although detailed mechanisms have been proposed to explain the stereospecificity in particular diene polymerizations, these processes are less understood than the polymerizations of alkenes. The ability to obtain cis-1,4-, trans-1,4-, and st-1,2-polymers from 1,3-butadiene, each in very high stereoregularity, using different initiators has great practical utility for polymer synthesis. However, why a particular initiator gives a particular stereoregular polymer is not understood. The high stereospecificity observed with various initiators indicates that propagation is a highly restrictive process, involving stereochemical restraints imposed on both the chain end and incoming monomer by the initiator. It is difficult, otherwise, to expect different initiators to be so different in their stereospecificity. Steric and electronic interactions among the counterion, monomer, and propagating chain end are the dominant factors that result in stereospecificity. Various characteristics of the transition metal are important in these interactions: size, electron density, and identity of ligands. Most mechanisms involve considerable discussion of whether coordination of monomer to the transition metal occurs through one or both double bonds, whether this complexation is cis or trans, and whether bond rotations are allowed during or subsequent to monomer addition to the propagating center.

8-6c Cationic Polymerization

The cationic polymerization of 1,3-dienes is not of practical interest as the products are usually low-molecular-weight with cyclized structures [Cooper, 1963; Kennedy, 1975]. Polymerizations have been carried out with conventional cationic initiators as well as Ziegler–Natta initiators under conditions (usually high ratio of transition metal to Group I–III metal) where the latter act as cationic initiators. There is a general tendency toward 1,4-polymerization with trans-1,4 favored over cis-1,4. Cyclization occurs during propagation with the extent of unsaturation in the product varying from about 80% to almost zero depending on initiator, monomer, and reaction conditions [Gaylord and Svestka, 1969; Gaylord et al., 1968; Hasegawa and Asami, 1978; Hasegawa et al., 1977].

Various mechanisms have been proposed to account for cyclization of 1,3-dienes. Cyclization probably occurs by attack of the propagating carbocation on trans-1,4-double bonds, for example

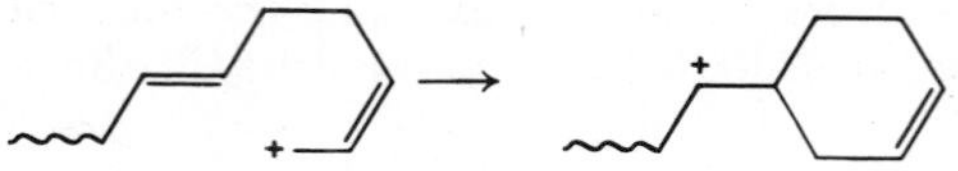

$$(8\text{-}55a)$$

Extensive cyclization could occur by a corresponding sequential process,

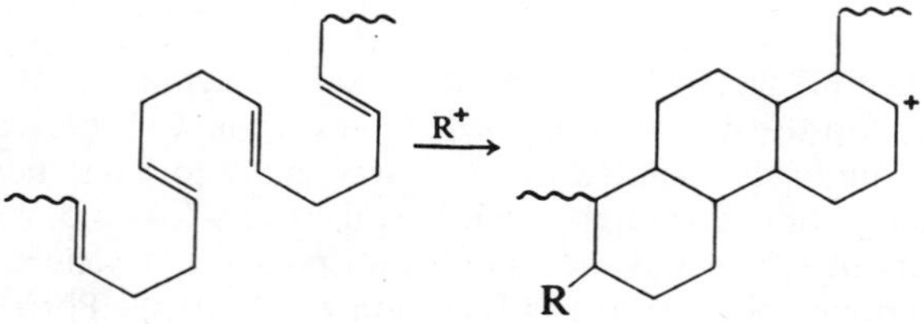

$$(8\text{-}55b)$$

IL

where R^+ is either an initiator species or the propagating carbocation **IL**.

8-6d Other Polymerizations

There have been various attempts to achieve stereospecific polymerization of 1,3-dienes and alkenes by imposing physical restraints on the monomer by means other than the initiator, including polymerizations of crystalline monomer, monomer monolayers, canal complexes, and liquid crystals [Allcock and Levin, 1985; Audisio et al., 1984; Bowden et al., 1978; DiSilvestro et al., 1987; Finkelmann et al., 1978; Miyata et al., 1977; Naegele and Ringsdorf, 1977]. However, these polymerizations have been considerably less stereospecific than those achieved using Ziegler–Natta initiators.

8-6e Commercial Polymers

Several polymers based on 1,3-dienes are used as elastomers. These include styrene–1,3-butadiene (SBR), styrene–1,3-butadiene terpolymer with an unsaturated carboxylic acid (carboxylated SBR), acrylonitrile-1,3-butadiene (NBR or nitrile rubber) (Secs. 6-8a, 6-8e), isobutylene–isoprene (butyl rubber) (Sec. 5-2i-1), and block copolymers of isoprene or 1,3-butadiene with styrene (Sec. 5-4a).

cis-1,4-Polyisoprene is produced using alkyllithium and Ti/Al Ziegler–Natta initiators. About 100 million pounds are produced annually in the United States to supplement the 2 billion pounds of natural rubber that are used. trans-1,4-Polyisoprene is a specialty material produced in small amounts by means of V/Al Ziegler–Natta initiators. The uses for these materials were discussed in Sec. 8-2a-2.

About one billion pounds of cis-1,4-poly(1,3-butadiene) is produced annually in the United States using Ziegler–Natta initiators such as $TiCl_4/I_2/Al(i-C_4H_9)_3$ and $Ni(RCOO)_2/BF_3/Al(C_2H_5)_3$. cis-1,4-Polybutadiene has a lower T_g and, therefore, higher resilience but poorer tear resistance and tensile strength than natural rubber. For this reason, cis-1,4-polybutadiene is not used alone but is blended with either natural rubber or SBR to produce tires for trucks and passenger automobiles. Considerable amounts of cis-1,4-polybutadiene are also used in producing ABS materials (Sec. 6-8a).

About 200 million pounds of polychloroprene (trade name: *Neoprene*) are produced annually in the United States by the radical (emulsion) polymerization of chloroprene. Polychloroprene, highly trans-1,4 in structure, is surpassed in oil and fuel resistance only by nitrile rubber, while its strength is superior to all except cis-1,4-polyisoprene. The high cost of polychloroprene limits its use to applications requiring its unique combination of properties. About 200 million pounds are produced annually in the United States and used for wire and cable coating and jackets, industrial belts and hoses, seals for buildings and highway joints, roof coatings, adhesives, gloves, and coated fabrics.

8-7   ALDEHYDES

The isospecific polymerization of acetaldehyde has been achieved using initiators such as zinc and aluminum alkyls, Grignard reagents, and lithium alkoxides [Kubisa et al., 1980; Pasquon et al., 1989; Pregaglia and Binaghi, 1967; Tani, 1973]. The isospecificity is high in some systems with isotactic indices of 80–90%. Cationic initiators such as BF_3 etherate are less isospecific; isotactic dyads of 70% are the maximum achieved in any polymerization. There are few reports of other aldehydes undergoing stereospecific polymerization; the one exception is chloral (Sec. 8-8b).

8-8   OPTICAL ACTIVITY IN POLYMERS

Optically active polymers are rarely encountered. Most, but not all, syndiotactic polymers are optically inactive since they are achiral. Some isotactic polymers, such as polypropylene, poly(methyl methacrylate), are also achiral. The various situations in which optical activity in polymers is possible have been described (Sec. 8-1). Optically active polymers have been obtained in some situations, and these are discussed below.

8-8a   Optically Active Monomers

Isospecific polymerization of one enantiomer or the other of a pair of enantiomers results in an optically active polymer [Ciardelli, 1987; Delfini et al., 1985; Pino et al., 1963]. For example, polymerization of (*S*)-3-methyl-1-pentene yields the all-*S* poly-

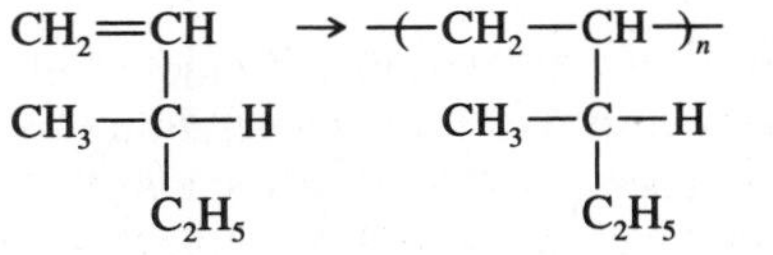

$$
\begin{array}{ccc}
CH_2{=}CH & \rightarrow & {+}CH_2{-}CH{\rightarrow_n} \\
| & & | \\
CH_3{-}C{-}H & & CH_3{-}C{-}H \\
| & & | \\
C_2H_5 & & C_2H_5
\end{array}
\qquad (8\text{-}56)
$$

S-Monomer   *S*-Polymer

mer. The optical activity of the polymer would be maximum for the 100% isotactic polymer. Each racemic placement of the S-monomer decreases the observed optical activity in the polymer.

8-8b   Chiral Conformation

There are few instances of polymer optical activity arising from a chiral conformation. Many isotactic polymers, including polypropylene, are composed of equimolar amounts of right- and left-handed helices, since the initiator is composed of equal numbers of the two enantiomorphic sites. This is also the case for polymerization of chloral (trichloroacetaldehyde) in hexane solution at 58°C with an achiral initiator such as lithium t-butoxide. The bulky CCl_3 group forces monomer units placed in the meso arrangement in the first few propagation steps to form a helical conformation; the helical conformation is subsequently propagated through the isospecific polymerization. The product is optically inactive since there is an equal probability of propagating via right- and left-handed helices. However, polymerization yields an optically active product when one uses a chiral initiator, such as lithium salts of methyl ($+$)- or ($-$)-mandelate or (R)- or (S)-octanoate [Corley et al., 1988; Jaycox and Vogl, 1990]. The chiral initiator forces propagation to proceed to form an excess of one of the two enantiomeric helices. The same driving force has been observed in the polymerization of triphenylmethyl methacrylate at -78°C in toluene using a chiral complex formed from n-butyllithium and an optical active amine, such as 2,2'-diamino-1,1'-binaphthyl, as the initiator [Kanoh et al., 1987a, 1987b].

8-8c   Stereoselection and Stereoelection

Consider the isospecific polymerization of a racemic mixture of monomers. Isospecific polymerization can proceed in two ways depending on initiator, monomer, and reaction conditions. *Stereoselective* (or *enantiosymmetric*) *polymerization* involves both the R and S monomers polymerizing at the same rate but without any cross-propagation. A racemic monomer mixture polymerizes to a racemic mixture of all-R and all-S polymer molecules [Pino, 1965; Sigwalt, 1976, 1979; Tsuruta, 1972]. This is

$$S\text{-Monomer} \rightarrow \sim\!\!\sim\!\!\sim S—S—S—S—S—S—S—S\sim\!\!\sim\!\!\sim \tag{8-57}$$

$$R\text{-Monomer} \rightarrow \sim\!\!\sim\!\!\sim R—R—R—R—R—R—R—R\sim\!\!\sim\!\!\sim \tag{8-58}$$

consistent with the mechanism for isospecific polymerization that attributes steric control to the initiator. The initiator contains R and S enantiomeric polymerization sites in equal numbers such that R sites polymerize only R monomer and S sites polymerize only S monomer. If the isospecificity of R and S sites is less than complete, a modified stereoselective behavior is seen with some cross-propagation of the two enantiomeric monomers and a decrease in the overall isotacticity of the reaction product.

As discussed in Sec. 8-8a, an optically active polymer sample, composed of all-R or all-S polymer molecules, can be synthesized by isospecific polymerization of a pure enantiomer, the pure R or pure S monomer, respectively. The direction of optical rotation of the polymer is usually the same as the corresponding monomer.

Stereoelective (or *asymmetric stereoselective* or *enantioasymmetric*) *polymerization* occurs when one of the enantiomers polymerizes faster than the other. In the extreme case one enantiomer (e.g., the R monomer) is unreactive and polymerization of racemic monomer yields the optically active all-S polymer with the optically active R monomer left unreacted. Stereoelective polymerization does not occur unless the initiator is an (optically active) enantiomer itself. An optically active Ziegler–Natta initiator is obtained by using an optically active Group I–III metal in combination with the transition-metal compound, such as $Zn[(S)\text{-2-methyl-1-butyl}]_2$ plus $TiCl_4$. The use of a chiral electron donor in a Ziegler–Natta initiator has also been used to achieve stereoelective polymerization [Carlini et al., 1977]. Stereoelective polymerizations have also been studied with vinyl ethers, acrylates, and methacrylates [Okamoto et al., 1979; Villiers et al., 1978]. For example, the S monomer is preferentially consumed in the polymerization of (R, S)-1,2-diphenylethyl methacrylate at $-75°C$ in toluene using $C_2H_5MgBr\text{-}(-)$-sparteine as the initiator [Okamoto et al., 1984].

Stereoelection results from the presence of only one (either R or S) of two enantiomeric polymerization sites in the initiator—a consequence of using a chiral initiator component. Only one of the enantiomeric monomers is able to propagate. The reactive monomer is usually the enantiomer having the same absolute configuration as the initiator in terms of the ordering of similar groups. Similar groups are those having similar steric and polar effects. This may not always correspond to both monomer and initiator having the same Cahn–Ingold–Prelog designation. Such stereoelection is referred to as *homosteric stereoelection*. *Antisteric stereoelection* occurs when the reactive enantiomer has a configuration opposite to that of the initiator. Stereoelection is highly sensitive to the positions of chiral centers in both monomer and initiator [Carlini et al., 1974; Chiellini and Marchetti, 1973; Pino et al., 1967]. Maximum stereoelection occurs when the chiral centers are as close as possible to the bonds (in both monomer and initiator) through which reaction takes place. Stereoelection is maximum for a monomer (e.g., 3-methyl-1-pentene) when the chiral center is α- to the double bond, decreases sharply when the chiral carbon is in the β-position (e.g., 4-methyl-1-pentene), and essentially disappears when the chiral carbon is γ- (e.g., 5-methyl-1-heptene). Stereoelection generally does not occur if the chiral carbon in the initiator is γ- to the metal instead of in the β-position.

As with stereoselection, the polymerization sites may not be perfectly isospecific. This results in a modified stereoelective behavior with some cross-propagation of the two enantiomeric monomers and a decrease in the overall isotacticity of the reaction product.

The extent of stereoelection in an isospecific polymerization is experimentally evaluated by measuring the optical activity of unreacted monomer α as a function of the extent of reaction p [Sepulchre et al., 1972]. The rates of reaction of R and S enantiomers are given by

$$\frac{-d[R]}{dt} = k_r[C^*][R] \tag{8-59}$$

$$\frac{-d[S]}{dt} = k_s[C^*][S] \tag{8-60}$$

where k_r and k_s are the rate constants for reaction of R and S enantiomers and $[C^*]$ is the concentration of polymerization sites. This treatment assumes that stereoelection is not dependent on whether the last monomer added to the propagating chain is R or S. Dividing Eq. 8-59 by Eq. 8-60 followed by integration yields

$$\frac{[R]}{[R]_0} = \left(\frac{[S]}{[S]_0}\right)^r \tag{8-61}$$

where $r = k_r/k_s$ and the 0 subscripts indicate initial concentrations.

The variables α and p are related to various concentration terms by

$$\alpha = \alpha_0 \frac{[R] - [S]}{[R] + [S]} \tag{8-62}$$

$$p = 1 - \frac{[R] + [S]}{[R]_0 + [S]_0} \tag{8-63}$$

where α_0 is the absolute value (without sign) of α for the pure enantiomer.

Combination of Eqs. 8-61 through 8-63 yields

$$(1 - p)^{(r-1)} = \frac{1 + (\alpha/\alpha_0)}{1 - (\alpha/\alpha_0)^r} \frac{2^{(r-1)}[S]_0^r}{[R]_0([R]_0 + [S]_0)^{(r-1)}} \tag{8-64}$$

which simplifies to

$$(1 - p)^{(r-1)} = \frac{1 + (\alpha/\alpha_0)}{1 - (\alpha/\alpha_0)^r} \tag{8-65}$$

for the case where the starting monomer mixture is racemic, i.e., $[R]_0 = [S]_0$. Using this treatment, r was obtained as 10.6 for the isospecific polymerization of (R,S)-α-methylbenzyl methacrylate in toluene at $-30°C$ by cyclohexylmagnesium bromide with an optically active diamine [Kanoh et al., 1987a, 1987b].

8-8d Asymmetric Induction

A special case of stereoelection is the isospecific copolymerization of optically active 3-methyl-1-pentene with racemic 3,7-dimethyl-1-octene by $TiCl_4$–$Zn(i$-$C_4H_9)_2$ [Ciardelli et al., 1969]. The copolymer is optically active with respect to both comonomer units as the incorporated optically active 3-methyl-1-pentene directs the preferential entry of only one enantiomer of the racemic monomer. The directing effect of a chiral center in one monomer unit on the second monomer, referred to as *asymmetric induction*, is also observed in radical and ionic copolymerizations. The radical copolymerization of optically active α-methylbenzyl methacrylate with maleic anhydride yields a copolymer that is optically active even after hydrolytic cleavage of the optically active α-methylbenzyl group from the polymer [Kurokawa and Minoura, 1979]. Similar results were obtained in the copolymerizations of mono- and di-l-menthyl fumarate and $(-)$-3-(β-styryloxy)menthane with styrene [Kurokawa et al., 1982].

8-9 RING-OPENING POLYMERIZATION

The stereochemistry of ring-opening polymerizations has been studied for epoxides, episulfides, lactones, cycloalkenes (Sec. 8-4d-1), and other cyclic monomers [Pasquon et al., 1989; Tsuruta and Kawakami, 1989]. Epoxides have been studied more than any other type of monomer. Only chiral cyclic monomers such as propylene oxide are capable of yielding stereoregular polymers. Polymerization of either of the two pure enantiomers yields an isotactic polymer when the reaction proceeds in a regiospecific manner with bond cleavage at bond 1. This is generally the case for the typical

$$\underset{1}{\overset{O}{\diagup}}\underset{2}{\diagdown}CH_3$$
$$\overset{|}{H}$$

L

polymerization of epoxides as well as other types of cyclic monomers, although the regiospecificity is not 100%. The degree of isospecificity is generally decreased in proportion to the extent of ring opening at bond 2. There are no initiator systems that are even moderately isospecific subsequent to bond cleavage at bond 2. The extent of regiospecificity at bond 1 is greater than 90% for many, but not all, anionic and anionic coordination polymerizations. It is very difficult to give a generalization that holds up well in this respect. There are a number of polymerizations where significant amounts of cleavage at bond 2 are reported, such as propylene oxide polymerizations by $Zn(C_2H_5)_2$ or $Al(C_2H_5)_3$ with H_2O or $Zn(C_2H_5)_2$ with CH_3OH, and styrene oxide polymerization by $Al(O\text{-}i\text{-}C_3H_7)_3$ [Tsuruta and Kawakami, 1989]. The extent of regiospecificity depends markedly on reaction conditions in some systems. Thus polymerization by $Zn(C_2H_5)_2/H_2O$ proceeds exclusively by cleavage at bond 1 in many polymerizations, but significant extents of cleavage occur at bond 2 in other polymerizations with the differences depending on the relative amounts of $Zn(C_2H_5)_2$ and H_2O. Cationic polymerizations of epoxides are significantly less regiospecific and isospecific compared to anionic polymerizations.

Polymerization of racemic propylene oxide can proceed in three basic modes: without stereoselection to yield atactic polymer, with stereoselection or with stereoelection to yield isotactic polymer. Polymerization by potassium hydroxide or alkoxide proceeds with better than 95% regioselectivity of cleavage at bond 1 but the product is atactic [Tsuruta and Kawakami, 1989]. Both (R)- and (S)-propylene oxide react at the same rate, as shown by the invariance of the optical rotation of unreacted monomer with conversion in a polymerization where the initial ratio of the two enantiomers was unequal. The atacticity of the product indicates that the initiator is unable to distinguish between the R and S enantiomers of propylene oxide and both enantiomers react at the same rate. This is a case of stereoselection with complete cross-propagation. Polymerizations by $Zn(OCH_3)_2$ and $\alpha,\beta,\gamma,\delta$-tetraphenylporphyrin-aluminum chloride (structure **VI** in Sec. 7-2a-1) proceed in a completely regiospecific manner (cleavage at bond 1) and with modest isospecificity (67 and 68% isotactic dyads, respectively) [Le Borgne et al., 1988; Tsuruta, 1981]. For polymerization by $\alpha,\beta,\gamma,\delta$-tetraphenylporphyrin-aluminum chloride, the degree of isotacticity varied with the ratio of (R)- and (S)-propylene oxide in the feed. Polymerization proceeds in a stereoelective manner but with significant cross-propagation.

There has been considerable work over the years with initiators formed from dialkylzinc and alcohol. The identity of the initiator and its activity–stereospecificity depends on the method of preparation. Many of the reported polymerizations have involved $Zn(OR)_2$, although again polymerization behavior is dependent on the details of initiator preparation. More recently, well-defined organozinc complexes of structures **LI** and **LII** have been prepared by control of reaction conditions [Hasebe and Tsuruta, 1988; Tsurata, 1986; Tsuruta and Kawakami, 1989; Yoshino et al., 1988].

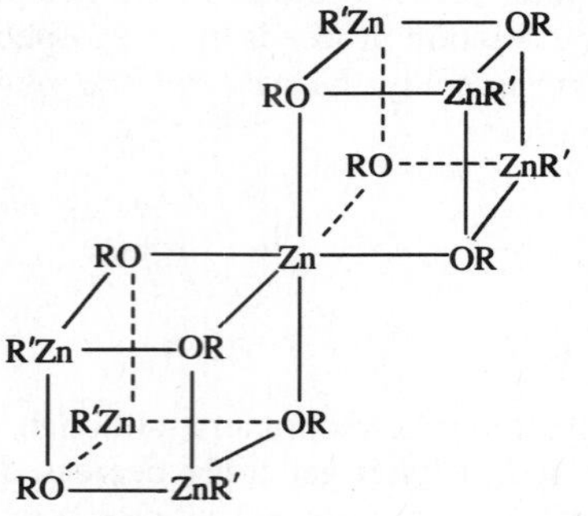

$$(RO)_2Zn \cdot (R'ZnOR)_6 \qquad\qquad (RO)_2Zn \cdot (R'ZnOR)_2$$

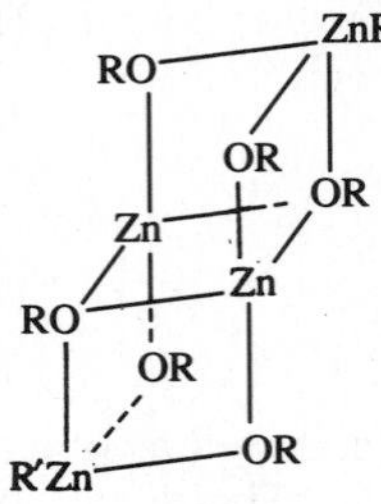

LI   **LII**

Nominally, these are $1:6$ and $1:2$ complexes, respectively, of zinc dialkoxide and alkylzinc monoalkoxide. Structures **LI** and **LII** are formed when methanol and (R,S)-1-methoxy-2-propanol, respectively, are used as the alcohol in the reaction with diethylzinc ($R' = C_2H_5$). Structures **LI** and **LII** have been isolated as crystalline substances whose structures have been obtained by X-ray diffraction. Structure **LI** is a centrosymmetric complex of two enantiomorphic distorted cubes that share a corner (Zn). The two cubes would be equivalent (i.e., superimposable), not enantiomorphic, if they were not distorted. Structure **LII**, also centrosymmetric, consists of two enantiomorphic distorted "chairs without legs" that share a common "seat." The structures of **LI** and **LII** are retained in solution (benzene) as determined by cryoscopic and NMR measurements. The structures of these complexes show remarkable similarities to those of the Ziegler–Natta initiators (structures **XXV** and **XXVI** in Sec. 8-4a-6), including the rigid, chiral metallocenes (structures **XXXV** and **XXXVI** in Sec. 8-4f).

Both *chair* and *cube* complexes are active for initiating polymerization of propylene oxide. Each complex has two enantiomorphic sites for polymerization, at the bonds indicated with arrows. Propylene oxide complexes at Zn and becomes inserted at the Zn—O bond with polymerization proceeding in the same manner. The proposed

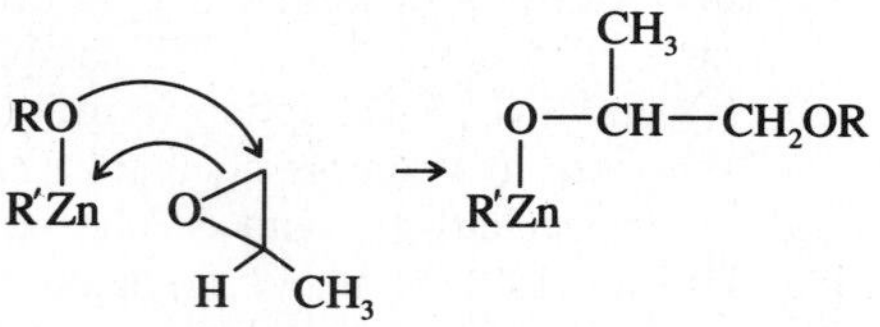

$$(8\text{-}66)$$

reaction mechanism for polymerization by these complexes indicates that each of the two types of polymerization sites for each complex is isospecific for one of the enantiomers of propylene oxide. One site polymerizes the R monomer, and the other site polymerizes the S monomer. The overall stoichiometry showed that only one polymer chain is initiated by each complex. One then expects the formation of equal amounts of R and S polymers since there is equal probability of initiation by each of the active sites. The experimental results show 63 and 81% isotactic dyads for the crude product of the polymerization of (R,S)-propylene oxide in benzene at 80°C using the cube and chair complexes, respectively. The result with the chair complex represents the highest reported isospecificity of any propylene oxide polymerization by any initiator. (There have been reports of more highly isotactic samples of polymer but only for an isolated fraction of the total product, not for the total unfractionated product.) The lower stereospecificity of the cube complex may indicate that its structure in solution is not sufficiently rigid to impose maximum restraints on cross-propagation. One should recall that the chiral, nonrigid metallocene initiators have poor stereospecificity relative to the rigid analogs (Sec. 8-4f). The lack of rigidity in the initiator is probably also responsible for the lack of any exceptionally strong stereoelective and isospecific tendencies in polymerizations of epoxides by chiral initiators such as diethylzinc with (R)-3,3-dimethyl-1,2-butanediol or $(+)$-borneol [Kassamaly et al., 1988; Tsuruta and Kawakami, 1989].

There are studies of the stereochemistry of ring-opening polymerizations of episulfides, β-lactones, and N-carboxy-α-amino acid anhydrides, but these are far less numerous than those of epoxides [Chatani et al., 1979; Elias et al., 1975; Guerin et al., 1980; Imanishi and Hasimoto, 1979; Inoue, 1976; Spassky et al., 1978; Zhang et al., 1990]. There are no instances of exceptionally high stereospecificity in any of the polymerizations.

8-10 STATISTICAL MODELS OF PROPAGATION

The polymer stereosequence distributions obtained by NMR analysis are often analyzed by statistical propagation models to gain insight into the propagation mechanism [Bovey, 1972, 1982; Doi, 1979a, 1979b, 1982; Farina, 1987; Inoue et al., 1984; LeBorgne et al., 1988; Randall, 1977; Shelden et al., 1965, 1969]. Propagation models have been developed for both initiator (enantiomorphic) site control and polymer chain and control. The Bernouilli and Markov models describe polymerizations where stereochemistry is determined by polymer chain end control. The enantiomorphic site model describes polymerizations where stereochemistry is determined by the initiator.

8-10a Polymer Chain End Control

8-10a-1 Bernoulli Model

The Bernoulli model (also referred to as the *zero-order Markov model*) assumes that only the last monomer unit in the propagating chain end is important in determining polymer stereochemistry. Polymer stereochemistry is not affected by the penultimate unit or units further back. Two different propagation events are possible—meso or racemic:

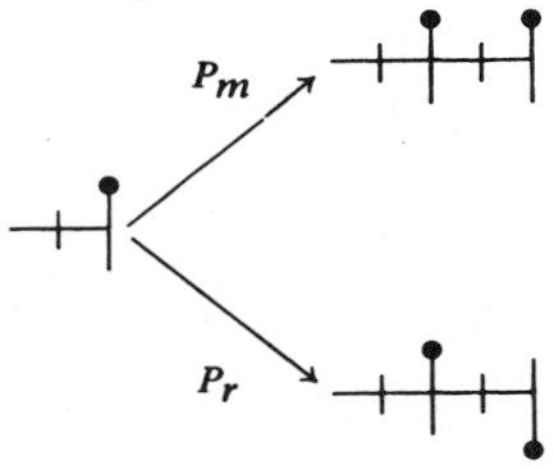

$$P_m \qquad (8\text{-}67)$$

$$P_r \qquad (8\text{-}68)$$

Equations 8-67 and 8-68 are general in representing both the case where the stereo-chemistry of the polymer chain end unit is determined relative to the penultimate unit when the next monomer unit adds and the case where the stereochemistry of the polymer chain end unit is determined relative to the incoming monomer unit when the latter adds.

Probabilities P_m and P_r, the *transition* or *conditional probabilities* of forming meso and racemic dyads, respectively, are defined by

$$P_m = \frac{R_m}{R_m + R_r} \qquad P_r = \frac{R_r}{R_m + R_r} \qquad P_m + P_r = 1 \tag{8-69}$$

where R_m and R_r are the rates of meso and racemic dyad placements, respectively.

Probabilities P_m and P_r are synonymous with the dyad tactic fractions (m) and (r) defined in Sec. 8-2b. Triad probabilities, synonymous with the triad fractions, follow as

$$(mm) = P_m^2 \qquad (mr) = 2P_m(1 - P_m) \qquad (rr) = (1 - P_m)^2 \tag{8-70}$$

The probability of forming a particular triad is the product of the probabilities of forming the two dyads comprising the triad. The coefficient of 2 for the heterotactic triad is required since the heterotactic triad is produced in two ways; thus, *mr* is produced as *mr* and *rm*.

Tetrad probabilities are given by

$$
\begin{aligned}
(mmm) &= P_m^3 & (mrm) &= P_m^2(1 - P_m) \\
(mmr) &= 2P_m^2(1 - P_m) & (rrm) &= 2P_m(1 - P_m)^2 \\
(rmr) &= P_m(1 - P_m)^2 & (rrr) &= (1 - P_m)^3
\end{aligned}
\tag{8-71}
$$

and pentad probabilities by

$$
\begin{aligned}
(mmmm) &= P_m^4 & (rmrm) &= 2P_m^2(1 - P_m)^2 \\
(mmmr) &= 2P_m^3(1 - P_m) & (rmrr) &= 2P_m(1 - P_m)^3 \\
(rmmr) &= P_m^2(1 - P_m)^2 & (mrrm) &= P_m^2(1 - P_m)^2 \\
(mmrm) &= 2P_m^3(1 - P_m) & (rrrm) &= 2P_m(1 - P_m)^3 \\
(mmrm) &= 2P_m^2(1 - P_m)^2 & (rrrr) &= (1 - P_m)^4
\end{aligned}
\tag{8-72}
$$

8-10a-2   First-Order Markov Model

The first-order Markov model describes a polymerization where the penultimate unit is important in determining subsequent stereochemistry. Meso and racemic dyads can each react in two ways:

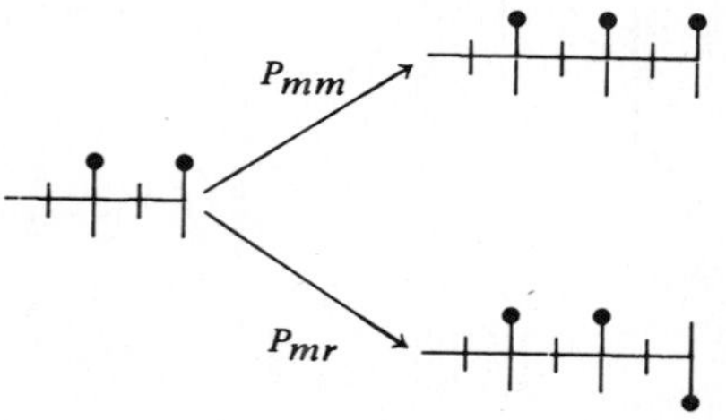

$$P_{mm} \tag{8-73}$$

$$P_{mr} \tag{8-74}$$

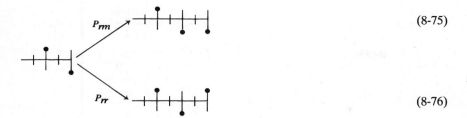

$$(8\text{-}75)$$

$$(8\text{-}76)$$

Meso (isotactic) and racemic (syndiotactic) triads result from reactions 8-73 and 8-76, respectively. Reactions 8-74 and 8-75 yield the heterotactic triad.

Four probabilities, P_{mm}, P_{mr}, P_{rm}, and P_{rr}, characterize this model for propagation with the conservation relationships

$$P_{mr} + P_{mm} = 1 \qquad P_{rm} + P_{rr} = 1 \tag{8-77}$$

The dyad fractions are given by

$$(m) = \frac{P_{rm}}{P_{mr} + P_{rm}} \qquad (r) = \frac{P_{mr}}{P_{mr} + P_{rm}} \tag{8-78}$$

The triad fractions are given by

$$(mm) = \frac{(1 - P_{mr})P_{rm}}{(P_{mr} + P_{rm})}$$

$$(mr) = \frac{2P_{mr}P_{rm}}{(P_{mr} + P_{rm})} \tag{8-79}$$

$$(rr) = \frac{(1 - P_{rm})P_{mr}}{(P_{mr} + P_{rm})}$$

the tetrad fractions by

$$(mmm) = \frac{P_{rm}(1 - P_{mr})^2}{(P_{mr} + P_{rm})} \qquad (mrr) = \frac{2P_{mr}P_{rm}(1 - P_{rm})}{(P_{mr} + P_{rm})}$$

$$(mmr) = \frac{2P_{mr}P_{rm}(1 - P_{mr})}{(P_{mr} + P_{rm})} \qquad (rmr) = \frac{P_{mr}^2 P_{rm}}{(P_{mr} + P_{rm})} \tag{8-80}$$

$$(mrm) = \frac{P_{mr}P_{rm}^2}{(P_{mr} + P_{rm})} \qquad (rrr) = \frac{P_{mr}(1 - P_{rm})^2}{(P_{mr} + P_{rm})}$$

and the pentad fractions by

$$(mmmm) = \frac{P_{rm}(1 - P_{mr})^3}{(P_{mr} + P_{rm})} \qquad (rmrm) = \frac{2P_{mr}^2 P_{rm}^2}{(P_{mr} + P_{rm})}$$

$$(mmmr) = \frac{2P_{mr}P_{rm}(1 - P_{mr})^2}{(P_{mr} + P_{rm})} \qquad (rmrr) = \frac{2P_{mr}^2 P_{rm}(1 - P_{rm})}{(P_{mr} + P_{rm})}$$

$$(rmmr) = \frac{P_{mr}^2 P_{rm}(1 - P_{mr})}{(P_{mr} + P_{rm})} \qquad\qquad (mrrm) = \frac{P_{mr} P_{rm}^2(1 - P_{rm})}{(P_{mr} + P_{rm})}$$

$$(mmrm) = \frac{2P_{mr} P_{rm}^2(1 - P_{mr})}{(P_{mr} + P_{rm})} \qquad\qquad (mrrr) = \frac{2P_{mr} P_{rm}(1 - P_{rm})^2}{(P_{mr} + P_{rm})}$$

$$(mmrr) = \frac{2P_{mr} P_{rm}(1 - P_{mr})(1 - P_{rm})}{(P_{mr} + P_{rm})} \qquad (rrrr) = \frac{P_{mr}(1 - P_{rm})^3}{(P_{mr} + P_{rm})} \qquad (8\text{-}81)$$

A second-order Markov model has also been described to show the effect on stereochemistry of the monomer unit behind the penultimate unit [Bovey, 1972].

8-10b Initiator (Enantiomorphic Site) Control

Control of polymer stereochemistry by the catalyst (initiator) to yield isotactic polymer is described by the enantiomorphic (or catalyst) site control model in terms of our previous described mechanism of polymerizations with Ziegler–Natta initiators, including the rigid, chiral metallocenes, and the organozinc initiators used for propylene oxide polymerization. The initiator has enantiomeric propagation sites, R and S sites, at which propagation occurs through the re and si faces of monomer (R and S monomer, respectively, if the monomer is optically active). The model is described in terms of the single parameter σ [Doi, 1979a, 1979b, 1982; Farina, 1987; Inoue et al., 1984; Le Borgne et al., 1988; Shelden et al., 1965, 1969]. Parameter σ is the probability of an R or re monomer unit adding at the R site; σ is also the probability of an S or si monomer unit adding at the S site.

The dyad fractions are given by

$$(m) = \sigma^2 + (1 - \sigma)^2 \qquad (s) = 2\sigma(1 - \sigma) \qquad\qquad (8\text{-}82)$$

the triad fractions by

$$(mm) = 1 - 3\sigma(1 - \sigma)$$
$$(mr) = 2\sigma(1 - \sigma) \qquad\qquad\qquad (8\text{-}83)$$
$$(rr) = \sigma(1 - \sigma)$$

the tetrad fractions by

$$(mmm) = 2\sigma^4 - 4\sigma^3 + 6\sigma^2 - 4\sigma + 1$$
$$(mmr) = (mrr) = -4\sigma^4 + 8\sigma^3 - 6\sigma^2 + 2\sigma \qquad\qquad (8\text{-}84)$$
$$(mrm) = (rmr) = (rrr) = 2\sigma^4 - 4\sigma^3 + 2\sigma^2$$

and the pentad fractions by

$$(mmmm) = 5\sigma^4 - 10\sigma^3 + 10\sigma^2 - 5\sigma + 1$$
$$(mrrm) = -3\sigma^4 + 6\sigma^3 - 4\sigma^2 + \sigma$$
$$(mmmr) = (mmrr) = -6\sigma^4 + 12\sigma^3 - 8\sigma^2 + 2\sigma \qquad\qquad (8\text{-}85)$$
$$(rmmr) = (rrrr) = \sigma^4 - 2\sigma^3 + \sigma^2$$
$$(mmrm) = (rmrr) = (rmrm) = (mrrr) = 2\sigma^4 - 4\sigma^3 + 2\sigma^2$$

8-10c Application of Propgation Statistics

Experimentally obtained sequence distributions in a particular polymerization are used to determine whether a particular polymerization follows the Bernoulli, Markov, or enantiomorphic site model. The general approach is to fit data on the dyad, triad, and higher-sequence fractions to the appropriate equations for the different models. One needs to recognize the difference between *fitting* and *testing* of data. Fitting of data involves calculating the value(s) of the appropriate probability term(s) from the sequence distributions. Testing involves determining that the sequence data are consistent or inconsistent with a particular mode. All the models can be fitted to dyad data; none of the models can be tested by dyad data. The Bernoulli and enantiomorphic site models require triad data as a minimum for testing; the first-order Markov model requires tetrad data. The appropriate level of sequence data tests a model by showing the consistency or inconsistency of the P_m values (Bernoulli model) or P_{mr} and P_{rm} values (first-order Markov), or σ values (enantiomorphic site model).

There are alternate criteria for testing the different models [Bovey, 1972; Chujo, 1967; Inoue et al., 1971, 1984]. The Bernoulli model requires

$$\frac{4(mm)(rr)}{(mr)^2} = 1 \tag{8-86}$$

The term on the left is extremely sensitive, and this criterion should be used only with sufficiently accurate triad data. This is especially important if the polymer is very highly isotactic or syndiotactic, that is, with very small value of either (rr) or (mm). The term $4(mm)(rr)/(mr)^2$ is considerably larger than one for the Markov and enantiomorphic site models.

The first-order Markov model requires

$$\frac{4(mmm)(rmr)}{(mmr)^2} = 1 \qquad \frac{4(mrm)(rrr)}{(mrr)^2} = 1 \tag{8-87}$$

while the enantiomorphic site model has the criteria

$$\frac{2(rr)}{(mr} = 1 \qquad 1 - \frac{4}{(mr) + 2(rr)} + \frac{1}{(rr)} = 1 \tag{8-88}$$

Having established that a particular polymerization follows Bernoullian or first-order Markov or enantiomorphic site behavior, we can better understand the mechanism by which polymer stereochemistry is determined. The Bernoulli model describes those polymerizations in which the chain end determines stereochemistry, due to interactions either between the last two units in the chain or between the last unit in the chain and the entering monomer. This corresponds to the generally accepted mechanism for polymerizations proceeding in a noncoordinated manner to give mostly atactic polymer—ionic polymerizations in polar solvents and free-radical polymerizations. Highly syndiospecific polymerizations, including those proceeding with Ziegler–Natta initiators, usually follow Markov behavior, which is indicative of a more complex form of chain end control. Highly isospecific polymerizations follow the enantiomorphic site model as expected. (Some isospecific polymerizations are reported to follow Markov behavior, which may be indicative of the difficulty of obtaining sufficiently accurate data of the higher sequences needed to differentiate between

Markov and enantiomorphic site models.) More sophisticated experimental and theoretical analysis of isospecific polymerizations have been performed by using a *two-site model* for propagation [Inoue et al., 1984; Wu et al., 1990]. The polymer product is fractionated into the highly isotactic, insoluble and atactic, soluble fractions, and the respective stereosequence determinations are usually found to fit the enantiomorphic and Bernoulli models.

Although the above treatment has been limited to polymerizations proceeding by meso and racemic placements, the various models are also applicable for describing the stereochemistry of 1,4-propagation in 1,3-dienes.

REFERENCES

Allcock, H. R. and M. L. Levin, *Macromolecules*, **18**, 1324 (1985).

Allegra, G., *Makromol. Chem.*, **145**, 235 (1971).

Ammendola, P., T. Tancredi, and A. Zambelli, *Macromolecules*, **19**, 307 (1986).

Arlman, E. J. and P. Cossee, *J. Catalysis*, **3**, 99 (1964).

Asanuma, T., Y. Nishimori, M. Ito, N. Uchikawa, and T. Shiomura, *Polym. Bull*, **25**, 567 (1991).

Audisio, G., A. Silvani, and L. Zetta, *Macromolecules*, **17**, 29 (1984).

Baldwin, F. P. and G. Ver Strate, *Rubber Chem. Technol.*, **45**, 709 (1972).

Ballard, D. G. H., *Adv. Catal.*, **23**, 263 (1973).

Barbe, P. C., G. Cecchin, and L. Noristi, *Adv. Polym. Sci.*, **81**, 1 (1987).

Beach, D. L. and Y. V. Kissin, "High Density Polyethylene," pp. 454–490 in "Encyclopedia of Polymer Science and Engineering," Vol. 6, 2nd ed., H. F. Mark, N. M. Bikales, C. G. Overberger, and G. Menges, Eds., Wiley-Interscience, New York, 1986.

Bier, G., *Angew. Chem.*, **73**, 186 (1961).

Binder, J. L., *Ind. Eng. Chem.*, **46**, 1727 (1954).

Bohm, L. L., *Polymer*, **18**, 545, 553, 562 (1978).

Boor, J., Jr., *J. Polym. Sci. Macromol. Rev.*, **2**, 115 (1967).

Boor, J., Jr., "Ziegler–Natta Catalysts and Polymerizations," Academic Press, New York, 1979.

Boor, J., Jr. and E. A. Youngman, *J. Polym. Sci.*, **A-1**(4), 1861 (1966).

Bovey, F. A., "High Resolution NMR of Macromolecules," Academic Press, New York, 1972.

Bovey, F. A., "Chain Structure and Conformation of Macromolecules," Academic Press, New York, 1982.

Bovey, F. A., F. P. Hood, E. W. Anderson, and R. L. Kornegay, *J. Phys. Chem.*, **71**, 312 (1967).

Bovey, F. A., M. C. Sacchi, and A. Zambelli, *Macromolecules*, **7**, 752 (1974).

Bowden, M. J., C. H. L. Kennard, J. H. O'Donnell, R. D. Sothman, N. W. Isaacs, and G. Smith, *J. Macromol. Sci. Chem.*, **A12**, 63 (1978).

Brandrup, J. and E. H. Immergut, Eds., "Polymer Handbook," Wiley-Interscience, New York, 1989.

Braun, D., M. Herner, and W. Kern, *Makromol. Chem.*, **36**, 232 (1960).

Braun, D., M. Herner, U. Johnson, and W. Kern, *Makromol. Chem.*, **51**, 15 (1962).

Brockmeier, N. F., "Gas-Phase Polymerization," pp. 480–488 in "Encyclopedia of Polymer Science and Engineering," Vol. 7, 2nd ed., H. F. Mark, N. M. Bikales, C. G. Overberger, and G. Menges, Eds., Wiley-Interscience, New York, 1987.

Burfield, D. R., *Polymer*, **25**, 1647 (1984).

Burfield, D. R., I. D. McKenzie, and P. J. T. Tait, *Polymer*, **17**, 130 (1976).

Burfield, D. R. and C. M. Savariar, *Macromolecules*, **12**, 243 (1979).

Busico, V., P. Corradini, A. Ferraro, and A. Proto, *Makromol. Chem.*, **187**, 1125 (1986).

Bywater, S., "Carbanionic Polymerization: Polymer Configuration and the Stereoregulation Process," Chap. 28 in "Comprehensive Polymer Science," Vol. 3, G. C. Eastman, A. Ledwith, S. Russo, and P. Sigwalt, Eds., Pergamon Press, Oxford, 1989.

Cabassi, F., W. Porzio, G. Ricci, S. Bruckner, S. V. Meille, and L. Porri, *Makromol. Chem.*, **189**, 2135 (1988).

Cam, D., E. Albizzati, and P. Cinquina, *Makromol. Chem.*, **191**, 1641 (1990).

Carlini, C., F. Ciardelli, and D. Pini, *Makromol. Chem.*, **174**, 15 (1974).

Carlini, C., R. Nocci, and F. Ciardelli, *J. Polym. Sci. Polym. Chem. Ed.*, **15**, 767 (1977).

Carrick, W. L., *Adv. Polym. Sci.*, **12**, 65 (1973).

Chatani, Y., M. Yokouchi, and H. Tadokoro, *Macromolecules*, **12**, 823 (1979).

Cheng, H. N. and J. A. Ewen, *Makromol. Chem.*, **190**, 1931 (1989).

Chiellini, E. and M. Marchetti, *Makromol. Chem.*, **169**, 59 (1973).

Chien, J. C. W., "Polyacetylene: Chemistry, Physics, and Materials Science," Academic Press, New York, 1984.

Chien, J. C. W. and T. Ang, *J. Polym. Sci. Polym. Chem. Ed.*, **25**, 1011 (1987).

Chien, J. C. W. and P. Bres, *J. Polym. Sci. Polym. Chem. Ed.*, **24**, 1967 (1986).

Chien, J. C. W. and Y. Hu, *J. Polym. Sci. Polym. Chem. Ed.*, **25**, 897, 2847, 2881 (1987).

Chien, J. C. W., C.-I. Kuo, and T. Ang., *J. Polym. Sci. Polym. Chem. Ed.*, **23**, 723 (1985).

Chien, J. C. W., J.-C. Wu, and C.-I. Kuo, *J. Polym. Sci. Polym. Chem. Ed.*, **20**, 2019 (1982).

Chien, J. C. W., X. Zhou, and S. Lin, *Macromolecules*, **22**, 4136 (1989).

Choi, K.-Y. and W. H. Ray, *J. Macromol. Sci.-Rev. Macromol. Chem. Phys.*, **C25**, 1, 57 (1985).

Chujo, R., *Makromol. Chem.*, **107**, 142 (1967).

Ciardelli, F., "Optically Active Polymers," pp. 463–493 in "Encyclopedia of Polymer Science and Engineering," Vol. 10, 2nd ed., H. F. Mark, N. M. Bikales, C. G. Overberger, and G. Menges, Eds., Wiley-Interscience, New York, 1987.

Ciardelli, F., C. Carlini, and G. Montagnoli, *Macromolecules*, **2**, 296 (1969).

Coleman, M. M. and E. G. Brame, *Rubber Chem. Technol.*, **51**, 668 (1978).

Coleman, M. M., R. J. Petcavich, and P. C. Painter, *Polymer*, **19**, 1243 (1978).

Coleman, M. M., D. L. Tabb, and E. G. Brame, Jr., *Rubber Chem. Technol.*, **50**, 49 (1977).

Condon, F. E., *J. Polym. Sci.*, **11**, 139 (1953).

Cooper, W., "Polyenes," Chap. 8 in "The Chemistry of Cationic Polymerization," P. H. Plesch, Ed., Pergamon Press, Oxford, 1963.

Cooper, W., "Kinetics of Polymerization Initiated by Ziegler-Natta and Related Catalysts," Chap. 3 in Comprehensive Chemical Kinetics," Vol. 15, C. H. Bamford and C. F. H. Tipper, Eds., Elsevier, New York, 1976.

Cooper, W., "Recent Advances in the Polymerization of Conjugated Dienes," Chap. 3 in "Developments in Polymerization—1," R. N. Howard, Ed., Applied Science Publishers, London, 1979.

Coover, H. W., Jr., R. L. McConnell, and F. B. Joyner, "Relationship of Catalysts Composition to Catalyst Activity for the Polymerization of α-Olefins," in "Macromolecular Reviews," Vol. 1, pp. 91–118, A. Peterlin, M. Goodman, S. Okamura, B. H. Zimm, and H. F. Mark, Eds., Wiley-Interscience, New York, 1967.

Coover, H. W., Jr., R. L. McConnell, F. B. Joyner, D. F. Slonaker, and R. L. Combs, *J. Polym. Sci.*, **A-1**(4), 2563 (1966).

Corley, L. S., G. D. Jaycox, and O. Vogl, *J. Macromol. Sci.-Chem.*, **A25**, 519 (1988).

Corradini, P., V. Busico, and G. Guerra, "Monoalkene Polymerization: Stereospecificity," Chap. 3 in "Comprehensive Polymer Science," Vol. 4, G. C. Eastman, A. Ledwith, S. Russo, and P. Sigwalt, Eds., Pergamon Press, Oxford, 1989.

Corradini, P., G. Guerra, and V. Barone, *Eur. Polym. J.*, **20**, 1177 (1984).

Corradini, P., G. Guerra, and V. Villiani, *Macromolecules*, **18**, 1401 (1985).

Cossee, P., *J. Catalysis*, **3**, 80 (1964).

Cossee, P., "The Mechanism of Ziegler-Natta Polymerization. II. Quantum-Chemical and Crystal–Chemical Aspects," Chap. 3 in "The Stereochemistry of Macromolecules," Vol. 1, A. D. Ketley, Ed., Marcel Dekker, New York, 1967.

Couturier, S., G. Tainturier, and B. Gautheron, *J. Organomet. Chem.*, **195**, 291 (1980).

Cozewith, C. and G. Ver Strate, *Macromolecules*, **4**, 482 (1971).

Craig, D., J. J. Shipman, and R. B. Fowler, *J. Am. Chem. Soc.*, **83**, 2885 (1961).

Cram, D. J. and K. R. Kopecky, *J. Am. Chem. Soc.*, **81**, 2748 (1959).

Dall'Asta, G., G. Mazzanti, G. Natta, and L. Porri, *Makromol. Chem.*, **56**, 224 (1962).

Danusso, F., *J. Polym. Sci.*, **C4**, 1497 (1964).

Delfini, M., M. E. Dicocco, M. Paci, M. Aglietto, C. Carlini, L. Crisci, and G. Ruggeri, *Polymer*, **26**, 1459 (1985).

Diedrich, P., *J. Polym. Sci. Appl. Polym. Sym.*, **26**, 1 (1975).

DiSilvestro, G., P. Sozzani, and M. Farina, *Macromolecules*, **20**, 999 (1987).

Doi, Y., *Macromolecules*, **12**, 248, 1012 (1979a); *Makromol. Chem.*, **180**, 2447 (1979b); *Makromol. Chem. Rapid. Commun.*, **3**, 635 (1982).

Doi, Y. and T. Keii, *Makromol. Chem.*, **179**, 2117 (1978).

Doi, Y., T. Koyama, K. Soga, and T. Asakura, *Makromol. Chem.*, **185**, 1827 (1984a).

Doi, Y., F. Nozawa, M. Murata, S. Suzuki, and K. Soga, *Makromol. Chem.*, **186**, 1825 (1984b).

Doi, Y., F. Nozawa, and K. Soga, *Makromol. Chem.*, **186**, 2529 (1985).

Doi, Y., S. Ueki, and T. Keii, *Macromolecules*, **12**, 814 (1979).

Dreyfuss, P. and M. P. Dreyfuss, "Polymerization of Cyclic Ethers and Sulphides," Chap. 4 in "Comprehensive Chemical Kinetics," Vol. 15, C. H. Bamford and C. F. H. Tipper, Eds., Elsevier, New York, 1976.

Dumas, C. and C. C. Hsu, *J. Macromol. Sci.-Rev. Macromol. Chem. Phys.*, **C24**, 355 (1984).

Ebdon, J. R., *Polymer*, **19**, 1232 (1978).

Elias, H.-G. H. G. Buehrer, and J. Semen, *J. Polym. Sci. Appl. Polym. Symp.*, **26**, 269 (1975).

Endo, K., H. Nagahama, and T. Otsu, *J. Polym. Sci. Polym. Chem. Ed.*, **17**, 3647 (1979).

Ewen, J. A., *J. Am. Chem. Soc.*, **106**, 6355 (1984).

Ewen, J. A., R. L. Jones, A. Razavi, and J. D. Ferrara, *J. Am. Chem. Soc.*, **110**, 6255 (1988).

Farina, M., "The Stereochemistry of Linear Macromolecules," pp. 1–112 in "Topics in Stereochemistry," Vol. 17, E. L. Eliel and S. H. Wilen, Eds., Wiley, New York, 1987.

Farina, M. and G. Bressan, "Optically Active Stereoregular Polymers, Chap. 4 in "The Stereochemistry of Macromolecules," Vol. 3, A. D. Ketley, Ed., Dekker, New York, 1968.

Finkelmann, H., M. Happ, M. Portugal, and H. Ringsdorf, *Makromol. Chem.*, **179**, 2541 (1978).

Fordham, J. W. L., *J. Polym. Sci.*, **39**, 321 (1959).

Fowells, W., C. Schuerch, F. A. Bovey, and F. P. Hood, *J. Am. Chem. Soc.*, **89**, 1396 (1967).

Fox, T. G. and H. W. Schnecko, *Polymer*, **3** 575 (1962).

Furlani, A., C. Napoletano, M. V. Russo, A. Camus, and N. Marsich, *J. Polym. Sci. Polym. Chem. Ed.*, **27**, 75 (1989).

Galli, P., L. Luciani, and G. Cecchin, *Angew. Makromol. Chem.*, **94**, 63 (1981).

Gaylord, N. G., I. Kossler, and M. Stolka, *J. Macromol. Sci. Chem.*, **A2**, 421 (1968).

Gaylord, N. G. and M. Svestka, *J. Polym. Sci.*, **B7**, 55 (1969).

Giannini, U., U. Zucchini, and E. Albizzati, *J. Polym. Sci. Polym. Lett. Ed.*, **8**, 405 (1970).

Giannini, U., G. Brucker, E. Pellino, and A. Cassata, *J. Polym. Sci. Polym. Lett. Ed.*, **5**, 527 (1967).

Goodman, M., "Concepts of Polymer Stereochemistry," pp. 73–156 in "Topics in Stereochemistry," Vol. 2, N. L. Allinger and E. L. Eliel, Ed., Wiley-Interscience, New York, 1967.

Goodman, M. and Y. L. Fan, *Macromolecules*, **1**, 163 (1968).

Grassi, A., A. Zambelli, L. Resconi, E. Albizzati, and R. Mazzocchi, *Macromolecules*, **21**, 617 (1988).

Greco, A., G. Bertolini, M. Bruzzone, and S. Cesca, *J. Appl. Polym. Sci.*, **23**, 1333 (1979).

Guerin, P., S. Boileau, and P. Sigwalt, *Eur. Polym. Sci.*, **16**, 121, 129 (1980).

Hasebe, Y. and T. Tsuruta, *Makromol. Chem.*, **189**, 1915 (1988).

Hasegawa, K. and R. Asami, *J. Polym. Sci. Polym. Chem. Ed.*, **16**, 1449 (1978).

Hasegawa, K., R. Asami, and T. Higashimura, *Macromolecules*, **10**, 585, 592 (1977).

Heatley, F., R. Salovey, and F. A. Bovey, *Macromolecules*, **2**, 619 (1969).

Hsu, S. L., M. K. Kemp, J. M. Pochan, R. C. Benson, and W. H. Flygare, *J. Chem. Phys.*, **50**, 1482 (1969).

Hu, Y. and J. C. W. Chien, *J. Polym. Sci. Polym. Chem. Ed.*, **26**, 2003 (1988).

Huggins, M. L., *J. Am. Chem. Soc.*, **66**, 1991 (1944).

Imanishi, Y. and Y. Hashimoto, *J. Macromol. Sci. Chem.*, **A13**, 673 (1979).

Inoue, S., *Adv. Polym. Sci.*, **21**, 77 (1976).

Inoue, Y., Y. Itabashi, R. Chujo, and Y. Doi, *Polymer*, **25**, 1640 (1984).

Inoue, Y., A. Nishioka, and R. Chujo, *Polym. J.*, **2**, 535 (1971).

Ishihara, N., M. Kuramoto, and M. Uoi, *Macromolecules*, **21**, 3356 (1988).

Ito, T., H. Shirakawa, and S. Ikeda, *J. Polym. Sci. Polym. Chem. Ed.*, **12**, 11 (1974).

IUPAC, *Pure Appl. Chem.*, **12**, 643 (1966); **53**, 733 (1981).

Ivin, K. J., "Metathesis Polymerization," pp. 634–668 in "Encyclopedia of Polymer Science and Engineering," Vol. 9, 2nd ed., H. F. Mark, N. M. Bikales, C. G. Overberger, and G. Menges, Eds., Wiley-Interscience, New York, 1987.

Jaber, I. A. and G. Fink, *Makromol. Chem.*, **190**, 2427 (1989).

James, D. E., "Linear Low Density Polyethylene," pp. 429–454 in "Encyclopedia of Polymer Science and Engineering," Vol. 6, 2nd ed., H. F. Mark, N. M. Bikales, C. G. Overberger, and G. Menges, Eds., Wiley-Interscience, New York, 1986.

Jaycox, G. D. and O. Vogl, *Makromol. Chem. Rapid Commun.*, **11**, 61 (1990).

Jenkins, A. D., *Pure Appl. Chem.*, **51**, 1101 (1979).

Juran, R., Ed., Modern Plastics Encyclopedia, 1989.

Kaminsky, "Polymerization and Copolymerization with a Highly Active, Soluble Ziegler-Natta Catalyst," pp. 225–244 in "Transition Metal Catalyzed Polymerization," R. P. Quirk, Ed., Harwood Academic, New York, 1983.

Kaminsky, W., A. Bark, R. Spiehl, N. Moller-Lindenhof, and S. Niedoba, "Isotactic Polymerization of Olefins with Homogeneous Zirconium Catalyst," pp. 292–301 in "Transition Metals and Organometallics as Catalysts for Olefin Polymerization," W. Kaminsky and H. Sinn, Eds., Springer-Verlag, Berlin, 1988.

Kanoh, S., N. Kawaguchi, and H. Suda, *Makromol. Chem.*, **188**, 463 (1987a).

Kanoh, S., N. Kawaguchi, T. Sumino, Y. Hongoh, and H. Suda, *J. Polym. Sci. Polym. Chem. Ed.*, **25**, 1603 (1987b).

Karol, F. J., *Chemist*, 12 (March 1989).

Karol, F. J., W. L. Munn, G. L. Goeke, B. E. Wagner, and N. J. Maraschin, *J. Polym. Sci. Polym. Sci. Ed.*, **16**, 771 (1978).

Kassamaly, A., M. Sepulchre, and N. Spassky, *Polym. Bull.*, **19**, 119 (1988).

Kawamura, T., T. Uryu, and K. Matsuzaki, *Makromol. Chem.*, **183**, 153 (1982).

Keii, T., "Kinetics of Ziegler–Natta Polymerization," Kodansha Ltd., Japan, 1972.

Kennedy, J. P., "Cationic Polymerization of Olefins: A Critical Inventory," Wiley, New York, 1975.

Kern, R. J., *Nature*, **187**, 410 (1960).

Ketley, A. D., "Stereospecific Polymerization of Vinyl Ethers," Chap. 2 in "The Stereochemistry of Macromolecules," Vol. 2, A. D. Ketley, Ed., Marcel Dekker, New York, 1967a.

Ketley, A. D., Ed., "The Stereochemistry of Macromolecules," Vols. 1 and 2, Marcel Dekker, New York, 1967b.

Kitayama, T., T. Shinozaki, T. Sakamoto, M. Yamamoto, and K. Hatada, *Makromol. Chem. Suppl.*, **15**, 167 (1989).

Kohara, T., M. Shinoyama, Y. Doi, and T. Keii, *Makromol. Chem.*, **180**, 2139 (1979).

Kollar, L., A. Simon, and A. Kallo, *J. Polym. Sci.*, **A-1**(6), 937 (1968a).

Kollar, L., A. Simon, and J. Osvath, *J. Polym. Sci.*, **A-1**(6), 919 (1968b).

Kraft, R., A. H. E. Muller, H. Hocker, and G. V. Schulz, *Makromol. Chem. Rapid Commun.*, **1**, 363 (1980).

Kubisa, P., K. Neeld, J. Starr, and O. Vogl, *Polymer*, **21**, 1433 (1980).

Kunzler, J. and V. Percec, *J. Polym. Sci. Polym. Chem. Ed.*, **28**, 1043 (1990).

Kurokawa, M. and Y. Minoura, *J. Polym. Sci. Polym. Chem. Ed.*, **17**, 3297 (1979).

Kurokawa, M., H. Yamguchi and Y. Minoura, *Makromol. Chem.*, **183**, 115 (1982).

Le Borgne, A., N. Spassky, C. L. Jun, and A. Momtaz, *Makromol. Chem.*, **189**, 637 (1988).

Ledwith, A., E. Chiellini, and R. Solaro, *Macromolecules*, **12**, 241 (1979).

Lehr, M. H., *Macromolecules*, **1**, 178 (1968).

Leitereg, T. J. and D. J. Cram, *J. Am. Chem. Soc.*, **90**, 4019 (1968).

Lieberman, R. B. and P. C. Barbe, "Propylene Polymers," pp. 464–531 in "Encyclopedia of Polymer Science and Engineering," Vol. 13, 2nd ed., H. F. Mark, N. M. Bikales, C. G. Overberger, and G. Menges, Eds., Wiley-Interscience, New York, 1988.

Longi, P., G. Mazzanti, A. Roggero, and A. M. Lachi, *Makromol. Chem.*, **61**, 63 (1963).

Longo, P., A. Grassi, L. Oliva, and P. Ammendola, *Makromol. Chem.*, **191**, 237 (1990).

Longo, P., A. Grassi, C. Pellecchia, and A. Zambelli, *Macromolecules*, **20**, 1015 (1987).

Longo, P., A. Grassi, A. Proto, and P. Ammendola, *Macromolecules*, **21**, 24 (1988).

Magovern, R. L., *Polym. Plast. Technol. Eng.*, **13**, 1 (1979).

Martin, M. L., J. Tirouflet, and B. Gautheron, *J. Organomet. Chem.*, **97**, 261 (1975).

Matlack, A. S. and D. S. Breslow, *J. Polym. Sci.*, **A3**, 2853 (1965).

Mazzanti, G., A. Valvassori, G. Sartori, and G. Pajaro, *Chim. Ind. (Milan)*, **42**, 468 (1960).

Mejzlik, J. and M. Lesna, *Makromol. Chem.*, **178**, 261 (1977).

Mejzlik, J., M. Lesna, and J. Kratochvila, *Adv. Polym. Sci.*, **81**, 83 (1986).

Mejzlik, J., P. Vozka, J. Kratochvila, and M. Lesna, "Recent Developments in the Determination of Active Centers in Olefin Polymerization," pp. 79–90 in "Transition Metals and Organometallics as Catalysts for Olefin Polymerization," W. Kaminsky and H. Sinn, Eds., Springer-Verlag, Berlin, 1988.

Mislow, K. and J. Siegel, *J. Am. Chem. Soc.*, **106**, 3319 (1984).

Miyata, M., K. Morioka, and K. Takemoto, *J. Polym. Sci. Polym. Chem. Ed.*, **15**, 2987 (1977).

Miyazawi, T. and T. Ideguchi, *Makromol. Chem.*, **79**, 89 (1964).

Morton, M., "Anionic Polymerization: Principles and Practice," Academic Press, New York, 1983, Chap. 7.

Moyer, P. H., *J. Polym. Sci.*, **A3**, 209 (1965).

Muller, A. H. E., H. Hocker, and G. V. Schulz, *Macromolecules*, **10**, 1087 (1977).

Murahashi, S., S. Nozakura, M. Sumi, H. Yuki, and K. Hatada, *J. Polym. Sci.*, **B4**, 59, 65 (1966).

Naegele, D. and H. Ringsdorf, *J. Polym. Sci. Polym. Chem. Ed.*, **15**, 2821 (1977).

Natta, G., *Makromol. Chem.*, **16**, 213 (1955a); *J. Polym. Sci.*, **16**, 143 (1955b); *J. Polym. Sci.*, **48**, 219 (1960a); *Chim. Ind. (Milan)*, **42**, 1207 (1960b); *Science*, **147**, 261 (1965).

Natta, G., I. Bassi, and P. Corradini, *Makromol. Chem.*, **18–19**, 455 (1956a).

Natta, G., P. Corradini, and G. Allegra, *J. Polym. Sci.*, **51**, 399 (1961a).

Natta, G., M. Farina, and M. Peraldo, *Chim. Ind., (Milan)*, **42**, 255 (1960).

Natta, G., I. Pasquon, and L. Giuffre, *Chim. Ind. (Milan)*, **43**, 871 (1961b).

Natta, G., I. Pasquon, and A. Zambelli, *J. Am. Chem. Soc.*, **84**, 1488 (1962).

Natta, G., P. Pino, and G. Mazzanti, *Gazz. Chim. Ital.*, **87**, 528 (1957a).

Natta, G., L. Porri, and A. Mazzei, *Chim. Ind. (Milan)*, **41** 116 (1959a).

Natta, G., G. Dall'Asta, I. Bassi, and G. Carella, *Makromol. Chem.*, **91**, 87 (1966).

Natta, G., I. Pasquon, A. Zambelli, and G. Gatti, *J. Polym. Sci.*, **51**, 387 (1961c).

Natta, G., P. Pino, G. Mazzanti, and P. Longi, *Gazz. Chim. Ital.*, **87**, 549, 570 (1957b); 88, 220 (1958).

Natta, G., L. Porri, A. Mazzei, and D. Morero, *Chim. Ind. (Milan).*, **41**, 398 (1959b).

Natta, G., G. Dall'Asta, G. Mazzanti, U. Giannini, and S. Cesca, *Angew. Chem.*, **71**, 205 (1959c).

Natta, G., G. Mazzanti, A. Valvassori, G. Sartori, and A. Barbagallo, *J. Polym. Sci.*, **51**, 429 (1961d).

Natta, G., P. Pino, E. Mantica, F. Danusso, G. Mazzanti, and M. Peraldo, *Chim. Ind. (Milan)*, **38**, 124 (1956b).

Natta, G., P. Pino, P. Corradini, F. Danusso, E. Mantica, G. Mazzanti, and G. Moraglio, *J. Am. Chem. Soc.*, **77**, 1708 (1955).

Ojala, T. A. and G. Fink, *Makromol. Chem. Rapid Commun.*, **9**, 85 (1988).

Okamoto, Y., K. Suzuki, K. Ohta, and H. Yuki, *J. Polym. Sci. Polym. Chem., Ed.*, **17**, 293 (1979).

Okamoto, Y., E. Yashima, K. Hatada, H. Yuki, H. Kageyama, K. Miki, and N. Kasai, *J. Polym. Sci. Polym. Chem. Ed.* **22**, 1831 (1984).

Oliva, L., P. Longo, and C. Pellecchia, *Makromol. Chem. Rapid Commun.*, **9**, 51 (1988).

Otsu, T., B. Yamada, and M. Imoto, *J. Macromol. Chem.*, **1**, 61 (1966).

Pampus, G. and G. Lehnert, *Makromol. Chem.*, **175**, 2605 (1974).

Pasquon, I., L. Porri, and U. Giannini, "Stereoregular Polymers," pp. 632–763 in "Encyclopedia of Polymer Science and Engineering," Vol. 15, 2nd ed., H. F. Mark, N. M. Bikales, C. G. Overberger, and G. Menges, Eds., Wiley-Interscience, New York, 1989.

Pasquon, I., A. Valvassori, and G. Sartori, "The Copolymerization of Olefins by Ziegler–Natta Catalysts," Chap. 4 in "Stereochemistry of Macromolecules," Vol. 1, A. D. Ketley, Ed., Marcel Dekker, New York, 1967.

Patat, F. and H. Sinn, *Angew. Chem.*, **70**, 496 (1958).

Pellecchia, C., P. Longo, A. Grassi, P. Ammendola, and A. Zambelli, *Makromol. Chem. Rapid Commun.*, **8**, 277 (1987).

Pino, P., *Adv. Polym. Sci.*, **4**, 393 (1965).

Pino, P., F. Ciarelli, G. P. Lorenzi, and G. Montagnoli, *Makromol. Chem.*, **61**, 209 (1963).

Pino, P., F. Ciardelli, and G. Montagnoli, *J. Polym. Sci.*, **C16**, 3265 (1967).

Pino, P., U. Giannini, and L. Porri, "Insertion Polymerization," pp. 147–220 in "Encyclopedia of Polymer Science and Engineering," Vol. 8, 2nd ed., H. F. Mark, N. M. Bikales, C. G. Overberger, and G. Menges, Eds., Wiley-Interscience, New York, 1987.

Pino, P. and R. Mulhaupt, *Angew. Chem. Int. Ed. Engl.*, **19**, 857 (1980).

Pino, P. and U. W. Suter, *Polymer*, **17**, 977 (1976); **18**, 412 (1977).

Pollock, D. J., L. J. Elyash, and T. W. DeWitt, *J. Polym. Sci.*, **15**, 86 (1955).

Porri, L. and A. Giarruso, "Conjugated Diene Polymerization," Chap. 5 in "Comprehensive Polymer Science," Vol. 4, G. C. Eastman, A. Ledwith, S. Russo, and P. Sigwalt, Eds., Pergamon Press, Oxford, 1989.

Prabhu, P., A. Shindler, M. H. Theil, and R. D. Gilbert, *J. Polym. Sci. Polym. Lett. Ed.*, **18**, 389 (1980).

Pregaglia, G. F. and M. Binaghi, "Ionic Polymerization of Aldehydes, Ketones, and Ketenes," Chap. 3 in "The Stereochemistry of Macromolecules," Vol. 2, Marcel Dekker, New York, 1967.

Randall, J. C., "Polymer Sequence Determination Carbon-13 NMR Method," Academic Press, New York, 1977.

Randall, J. C., *Macromolecules*, **11**, 592 (1978).

Rembaum, A., F. R. Ells, R. C. Morrow, and A. V. Tobolsky, *J. Polym. Sci.*, **61**, 155 (1962).

Richardson, W. S., *J. Polym. Sci.*, **13**, 229 (1954).

Rieger, B., X. Mu, D. T. Mallin, M. D. Rausch, and J. C. W. Chien, *Macromolecules*, **23**, 3559 (1990).

Rishina, L. A., Yu. V. Kissin, and F. S. Dyachkovsky, *Eur. Polymer J.*, **12**, 727 (1976).

Rodriguez, L. A. M. and H. M. van Looy, *J. Polym. Sci.*, **A-1**(4), 1951, 1971 (1966).

Rossi, A. and G. Odian, unpublished results (1990).

Sacchi, M. C., C. Shan, P. Locatelli, and I. Tritto, *Macromolecules*, **23**, 383 (1990).

Schildknecht, C. E., A. O. Zoss, and C. McKinley, *Ind. Eng. Chem.*, **39**, 180 (1947).

Schindler, A., *Makromol. Chem.*, **118**, 1 (1968).

Senyek, M. L., "Isoprene Polymers," pp. 487–564 in "Encyclopedia of Polymer Science and Engineering," Vol. 8, 2nd ed., H. F. Mark, N. M. Bikales, C. G. Overberger, and G. Menges, Eds., Wiley-Interscience, New York, 1987.

Sepulchre, M., N. Spassky, and P. Sigwalt, *Macromolecules*, **5**, 92 (1972).

Shelburne, J. A., III, and G. L. Baker, *Macromolecules*, **20**, 1212 (1987).

Shelden, R. A., T. Fueno, and J. Furukawa, *J. Polym. Sci.*, **A-2**(7), 763 (1969).

Shelden, R. A., T. Fueno, T. Tsunetsugu, and J. Furukawa, *J. Polym. Sci. Polym. Lett.*, **3**, 23 (1965).

Sigwalt, P., *Pure Appl. Chem.*, **48**, 257 (1976); *Makromol. Chem.*, *Suppl.*, 3, 69 (1979).

Soga, K., K. Izumi, S. Ikeda, and T. Keii, *Makromol. Chem.*, **178**, 337 (1977).

Soga, K. and T. Monoi, *Macromolecules*, **23**, 1558 (1990).

Soga, K. and H. Nakatani, *Macromolecules*, **23**, 957 (1990).

Soga, K., J. R. Park, R. Shiono, and N. Kashiwa, *Makromol. Chem. Rapid Commun.*, **11**, 117 (1990).

Soga, K., T. Shiono, and Y. Doi, *Makromol. Chem.*, **189**, 1531 (1988).

Soga, K., T. Shiono, S. Takemura, and W. Kaminsky, *Makromol. Chem. Rapid Commun.*, **8**, 305 (1987).

Soga, K. and H. Yanagihara, *Makromol. Chem.*, **189**, 2839 (1988a); *Makromol. Chem. Rapid Commun.*, **9**, 23 (1988b); *Macromolecules*, **22**, 2875 (1989).

Soga, K., H. Yanagihara, and D. Lee, *Makromol. Chem.*, **190**, 37 (1989).

Spassky, N., A. Leborgne, M. Reix, R. E. Prud'homme, E. Bigdeli, and R. W. Lenz, *Macromolecules*, **11**, 716 (1978).

Staudinger, H., A. A. Ashdown, M. Brunner, H. A. Bruson, and S. Wehrli, *Helv. Chim. Acta.*, **12**, 934 (1929).

Tait, P. J. T., "Monoalkene Polymerization: Ziegler–Natta and Transition Metal Catalysts," Chap. 1 in "Comprehensive" Polymer Science," Vol. 4, G. C. Eastman, A. Ledwith, S. Russo, and P. Sigwalt, Eds., Pergamon Press, Oxford, 1989.

Tait, P. J. T. and N. D. Watkins, "Monoalkene Polymerization: Mechanisms," Chap. 2 in "Comprehensive Polymer Science," Vol. 4, G. C. Eastman, A. Ledwith, S. Russo, and P. Sigwalt, Eds., Pergamon Press, Oxford, 1989.

Talamini, G. and G. Vidotto, *Makromol. Chem.*, **100**, 48 (1967).

Tani, H., *Adv. Polym. Sci.*, **11**, 57 (1973).

Tate, D. P. and T. W. Bethea, "Butadiene Polymers," pp. 537–590 in "Encyclopedia of Polymer Science and Engineering," Vol. 2, 2nd ed., H. F. Mark, N. M. Bikales, C. G. Overberger, and G. Menges, Eds., Wiley-Interscience, New York, 1985.

Taube, R., J.-P. Gehrke, and P. Bohme, *Wiss. Zeitschr. THLM*, **29**, 310 (1987).

Theophilou, N. and H. Naarmann, *Makromol. Chem. Macromol. Symp.*, **24**, 115 (1989).

Tobolsky, A. V. and C. E. Rogers, *J. Polym. Sci.*, **40**, 73 (1959).

Tonelli, A. E., "NMR Spectroscopy and Polymer Microstructure," VCH, New York, 1989.

Tosi, C. and F. Ciampelli, *Adv. Polym. Sci.*, **12**, 87 (1973).

Tritto, I., M. C. Sacchi, and P. Locatelli, *Makromol. Chem.*, **187**, 2145 (1986).

Tsuruta, T., *J. Polym. Sci.*, **D6**, 180 (1972); *J. Polym. Sci. Polym. Symp.*, **67**, 73 (1980); *Pure Appl. Chem.*, **53**, 1745 (1981); *Makromol. Chem. Macromol. Symp.*, **6**, 23 (1986).

Tsuruta, T. and Y. Kawakami, "Anionic Ring-Opening Polymerization: Stereospecificity for Epoxides, Episulfides and Lactones," Chap. 33 in "Comprehensive Polymer Science," Vol. 3, G. C. Eastman, A. Ledwith, S. Russo, and P. Sigwalt, Eds., Pergamon Press, Oxford, 1989.

Van Beylen, M., S. Bywater, G. Smets, M. Szwarc, and D. J. Worsfold, *Adv. Polym. Sci.*, **86**, 87 (1988).

Vandenberg, E. J. and B. C. Repka, "Ziegler-Type Polymerizations," Chap. 11 in "Polymerization Processes," C. E. Schildknecht, Ed. (with I. Skeist), Wiley-Interscience, New York, 1977.

Ver Strate, G., "Ethylene–Propylene Elastomers," pp. 522–564 in "Encyclopedia of Polymer Science and Engineering," Vol. 6, 2nd ed., H. F. Mark, N. M. Bikales, C. G. Overberger, and G. Menges, Eds., Wiley-Interscience, New York, 1986.

Villiers, C., C. Braud, M. Vert, E. Chiellini, and M. Marchetti, *Eur. Polym. J.*, **14**, 211 (1978).

Vozka, P. and J. Mejzlik, *Makromol. Chem..*, **191**, 589 (1990).

Weissermel, K., H. Cherdron, J. Berthold, B. Diedrich, K. D. Keil, K. Rust, H. Strametz, and T. Toth, *J. Polym. Sci. Symp.*, **51**, 187 (1975).

Werber, F. X., C. J. Benning, W. R. Wszolek, and G. E. Ashby, *J. Polym. Sci.*, **A-1**(6), 743 (1968).

Wicke, R. and K. F. Elgert, *Makromol. Chem.*, **178**, 3085 (1977).

Wiles, D. M. and S. Bywater, *Trans. Faraday Soc.*, **61**, 150 (1965).

Witt, D. R., "Reactivity and Mechanism with Chromium Oxide Polymerization Catalysts," Chap. 13 in "Reactivity, Mechanism, and Structure in Polymer Chemistry," A. D. Jenkins and A. Ledwith, Eds., Wiley, New York, 1974.

Wolfsgruber, C., G. Zannoni, E. Rigamonti, and A. Zambelli, *Makromol. Chem.*, **176**, 2765 (1975).

Wu, Q., N.-L. Yang, and S. Lin, *Makromol. Chem.*, **191**, 89 (1990).

Yoshino, N., C. Suzuki, H. Kobayashi, and T. Tsuruta, *Makromol. Chem.*, **189**, 1903 (1988).

Young, R. N., R. P. Quirk, and L. J. Fetters, *Adv. Polym. Sci.*, **56**, 1 (1984).

Youngman, E. A. and J. Boor, Jr., "Syndiotactic Polypropylene," pp. 33–69 in "Macromolecular Reviews," Vol. 2, A. Peterlin, M. Goodman, S. Okamura, B. H. Zimm, and H. F. Mark, Eds., Wiley-Interscience, New York, 1967.

Yuki, H., K. Hatada, K. Ohta, and Y. Okamoto, *J. Macromol. Sci.*, **A9**, 983 (1975).

Zakharov, V. A., G. D. Bukatov, N. B. Chumaevskii, and Y. I. Yermakov, *Makromol. Chem.*, **178**, 967 (1977).

Zambelli, A. and G. Allegra, *Macromolecules*, **13**, 42 (1980).

Zambelli, A., P. Ammendola, A. Grassi, P. Longo, and A. Proto, *Macromolecules*, **19**, 2703 (1986).

Zambelli, A. and G. Gatti, *Macromolecules*, **11**, 485 (1978).

Zambelli, A., G. Gatti, C. Sacchi, W. O. Crain, and J. D. Roberts, *Macromolecules*, **4**, 475 (1971).

Zambelli, A., M. G. Giongo, and G. Natta, *Makromol. Chem.*, **112**, 183 (1968).

Zambelli, A., P. Locatelli, and E. Rigamonte, *Macromolecules*, **12**, 157 (1979).

Zambelli, A., P. Locatelli, M. C. Sacchi, and E. Rigamonti, *Macromolecules*, **13**, 798 (1980).

Zambelli, A., P. Locatelli, M. C. Sacchi, and I. Tritto, *Macromolecules*, **15**, 831 (1982a).

Zambelli, A., P. Locatelli, G. Zannoni, and F. A. Bovey, *Macromolecules*, **11**, 923 (1978).

Zambelli, A., P. Longo, C. Pellecchia, and A. Grassi, *Macromolecules*, **20**, 2035 (1987).

Zambelli, A., G. Natta, and I. Pasquon, *J. Polym. Sci.*, **C4**, 411 (1963).

Zambelli, A., L. Oliva, and C. Pellecchia, *Macromolecules*, **22**, 2129 (1989).

Zambelli, A., M. C. Sacchi, P. Locatelli, and G. Zannoni, *Macromolecules*, **15**, 211 (1982b).

Zambelli, A. and C. Tosi, *Adv. Polym. Sci.*, **15**, 31 (1974).

Zambelli, A., C. Wolfsgruber, G. Zannoni, and F. A. Bovey, *Macromolecules*, **7**, 750 (1974).

Zhang, Y., R. A. Gross, and R. W. Lenz, *Macromolecules*, **23**, 3206 (1990).

Ziegler, K., *Angew. Chem.*, **76**, 545 (1964).

Ziegler, K., E. Holzkamp, H. Breil, and H. Martin, *Angew. Chem.*, **67**, 426, 541 (1955).

PROBLEMS

8-1 Show by structural drawings the various (if any) stereoregular polymers that might possibly be obtained from each of the monomers.

    a. $CH_2=CH-CH_3$

    b. $CH_2=C(CH_3)_2$

    c. $CH_3-CH=CH-CH_3$

    d. $CH_2=C(CH_3)(C_2H_5)$

    e. $CH_3-CH=CH-C_2H_5$

    f. $CH_2=CH-CH=CH_2$

    g. $CH_3-CH=CH-CH=CH_2$

    h. $CH_2=C(CH_3)-CH=CH_2$

    i. $CH_3-CH=C(Cl)-CH=CH_2$

    j. $CH_3-CH=CH-C(CH_3)=CH_2$

    k. $CH_3-CH=CH-CH=CH-CH_3$

    l. $CH_2=C(CH_3)-C(CH_3)=CH_2$

    m. $H_2N-(CH_2)_6-NH_2 + HO_2C-(CH_2)_4-CO_2H$

n. 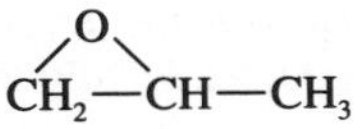 $CH_2\overset{\displaystyle O}{\overbrace{}}CH\!-\!CH_3$

o. $CH_3\!-\!CH\overset{\displaystyle O}{\overbrace{}}CH\!-\!CH_3$

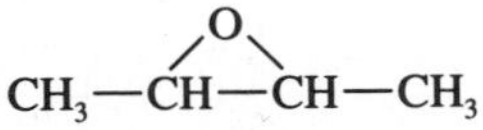

p. 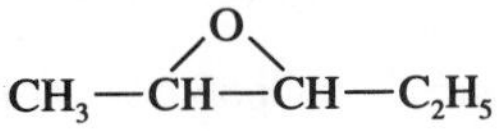 $CH_3\!-\!CH\overset{\displaystyle O}{\overbrace{}}CH\!-\!C_2H_5$

q. $CH_3\!-\!CHO$

r. $(CH_3)_2CO$

s. $CH_3\!-\!CO\!-\!CCl_3$

t.

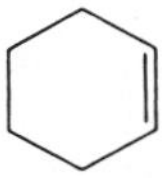

u.

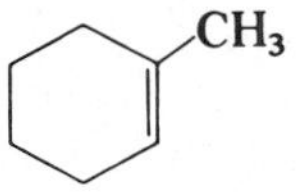

v.

Name the various polymer structures using the nomenclature described in this chapter.

8-2 What are the mechanisms for syndiotactic and isotactic placements in the polymerization of propylene? Describe the reaction conditions that favor each type of stereospecific placement.

8-3 Discuss the use of homogeneous versus heterogeneous reaction conditions for the coordination and Ziegler-Natta polymerizations of propylene, isoprene, styrene, methyl methacrylate, and *n*-butyl vinyl ether.

8-4 What reaction conditions determine the relative amounts of 1,2-, cis-1,4-, and trans-1,4-polymerization in the radical and anionic polymerizations of 1,3-butadiene? Clearly indicate the effect of solvent and counterion in these polymerizations. What polymer structures are obtained using Ziegler-Natta initiators?

8-5 Which of the stereoregular polymer structures in the answers to Question 8-1 are capable of exhibiting optical activity? How would you synthesize optically active polymers of those structures? Explain the inability of certain of the stereoregular polymers to exhibit optical activity.

8-6 The polymerization of *cis*-1-*d*-propylene by Ziegler-Natta initiators in hydrocarbon solvents yields the erythrodiisotactic structure, while under similar solvent conditions anionic polymerization of *cis*-β-*d*-methyl acrylate yields the threodiisotactic polymer. Explain the factor(s) responsible for this difference.

8-7 The polymerization of optically pure propylene oxide by FeCl$_3$ derived initiators yields an optically active polymer. E. J. Vandenberg [*J. Polym. Sci.*, **A-1**(7), 525 (1969)] attempted to gain insight into the mechanism by polymerizing the 2,3-epoxybutanes. The optically active *trans*-2,3-epoxybutane was polymerized to an optically inactive crystalline polymer. Evaluate this result and discuss its implications on the propagation mechanism for propylene oxide polymerization.

8-8 Briefly explain each of the following:

a. There is only one disyndiotactic structure for $\mathrm{-(\!-CHRCHR'\!-)_{\overline{n}}}$, whereas there are erythro- and threodiisotactic structures.

b. Addition of aluminum trialkyl to a Ziegler-Natta polymerization system (e.g., β-$\mathrm{TiCl_3}$, propylene) increases the polymerization rate up to a maximum after which R_p either remains constant or decreases.

c. Hydrogen lowers the molecular weight of polyethylene or polypropylene synthesized using Ziegler-Natta initiators.

d. Low temperatures often lead to syndiotactic polymers.

e. Metal-polymer bonds are sometimes found in polymers synthesized using Ziegler-Natta initiators.

8-9 Distinguish between stereoselective and stereoelective polymerizations. Give examples of each.

8-10 NMR analysis of polypropylene sample A showed $(m) = 0.78$, $(r) = 0.22$, (mm) 0.62, $(mr) = 0.33$, $(rr) = 0.05$, $(mmm) = 0.44$, $(mmr) = 0.27$, $(rmr) = 0.06$, $(rrm) = 0.04$, $(mrm) = 0.13$, $(rrr) = 0.01$. Polypropylene sample B showed $(m) = 0.80$, $(r) = 0.20$, $(mm) = 0.71$, $(mr) = 0.19$, $(rr) = 0.10$. What type of propagation mechanism (Bernoulli, first-order Markov, enantiomorphic site) is indicated for the syntheses of samples A and B?

CHAPTER 9

REACTIONS OF POLYMERS

The preceding chapters dealt with the synthesis of polymers by polymerization reactions. This chapter describes the synthesis of new polymers by modification of existing polymers using a variety of chemical reactions [Fettes, 1964; Moore, 1973]. These include the esterification of cellulose, crosslinking of polyisoprene, hydrolysis of poly(vinyl acetate), and chlorination of polyethylene. A second aspect of polymer reactions is the use of a polymer as a *carrier* or *support* for some component of a reaction system [Benham and Kinstle, 1988; Manecke and Storck, 1978; Mathur et al., 1980]. Examples include polymeric brominating and Wittig reagents, the Merrifield solid-phase synthesis of polypeptides, and polymeric catalysts.

9-1 PRINCIPLES OF POLYMER REACTIVITY

Polymers undergo the same reactions as their low-molecular-weight homologs. Acetylation of the hydroxyl groups of cellulose is basically the same reaction as acetylation of ethanol, chlorination of polyethylene follows the same mechanism as chlorination of n-hexane. It is usually assumed that the reactivity of a functional group in a polymer and a small organic molecule are the same. This is the familiar concept of functional group reactivity independent of molecular size (Sec. 2-1), which is the basis for analyzing polymerization kinetics. However, in many instances, the reaction rates and maximum conversions observed in the reactions of polymer functional groups differ significantly from those for the corresponding low-molecular-weight homolog. Polymer reaction rates and conversions are usually lower, although higher rates are also found in some reactions [Morawetz, 1975; Plate, 1976; Rempp, 1976]. The reaction conditions under which polymer reactivity differs from that of a low-molecular-weight homolog are discussed below.

9-1a Yield

Yield or conversion in reactions of polymers means something quite different than in small molecule reactions when the conversion is less than 100%. For example, 80% yield in the hydrolysis of methyl propanoate has no effect on the purity of the propanoic acid that can be obtained assuming the appropriate techniques are available to separate starting material and product. The 80% yield simply limits the maximum amount of pure propanoic acid that can be obtained to 80% of the theoretical yield. However, 80% yield in the corresponding hydrolysis of poly(methyl acrylate) does not result in 80% yield of poly(acrylic acid) with 20% unreacted poly(methyl acrylate). The product contains copolymer molecules, each of which, on the average, contains 80% acrylic acid repeating units and 20% methyl acrylate units randomly placed along the polymer

$$\xleftarrow{} CH_2 - \underset{\underset{COOCH_3}{|}}{CH} \xrightarrow{}_n \quad \rightarrow \quad \xleftarrow{} CH_2 - \underset{\underset{COOH}{|}}{CH} \xrightarrow{}_{0.8n} \xleftarrow{} CH_2 - \underset{\underset{COOCH_3}{|}}{CH} \xrightarrow{}_{0.2n} \qquad (9\text{-}1)$$

chain. Unlike the corresponding small molecule reaction, the unreacted ester groups cannot be separated from the product since both are part of the same molecule.

9-1b Isolation of Functional Groups

When a polymer reaction involves the random reaction of a pair of neighboring functional groups (either with each other or simultaneously with a common small molecule reagent), the maximum conversion is limited due to the isolation of single functional groups between pairs of reacted functional groups. This effect was studied for the dechlorination of poly(vinyl chloride) when heated in the presence of zinc. The maximum conversion for irreversible reactions has been calculated as 86%, which is in excellent agreement with that observed in most systems [Alfrey et al., 1951; Flory, 1939]. This limitation applies equally to polymers in which the functional groups react with each other, such as anhydride formation in poly(acrylic acid) or with a small reactant, for example, poly(vinyl chloride) with zinc or acetal formation in poly(vinyl alcohol) by reaction with an aldehyde. The conversions are higher for reversible reactions but the reaction times required for complete conversion may be quite long.

9-1c Concentration

The reaction of a polymer in solution involves a considerably higher *local concentration* of functional groups than that indicated by the overall polymer concentration. Polymer molecules are generally present in solution as random coil conformations. The concentration of functional groups is high within the polymer coils and zero outside. The difference between the overall and local concentrations can be illustrated by considering a 1% solution of poly(vinyl acetate) of molecular weight 10^6. The overall concentration of acetate groups is about $0.11\ M$, while the local concentration is estimated as being fivefold higher [Elias, 1984]. The high local concentration of polymer functional groups gives a high local rate for the reaction of a polymer with a small molecule reactant. The reaction rate outside the polymer coils is zero since the concentration of polymer functional groups is zero. The observed overall reaction rate is an average of the rates inside and outside the polymer coils [Sherrington, 1988]. The overall rate may be the same, higher or lower than the corresponding reaction with a low-molecular-weight homolog of the polymer depending on the concentration of the small molecule reactant inside the polymer coils relative to its concentration outside.

The concentrations of the small molecule reactant inside and outside are the same for soluble polymers unless there is some special effect responsible for attracting or repulsing the reactant from the polymer coils. Such situations are described in the remainder of Sec. 9-1. The concentration of a small molecule reactant inside the polymer coils can be lower than outside when one uses a poor solvent for the polymer. This results in lower local and overall reaction rates. In the extreme, a poor solvent results in reaction occurring only on the surfaces of a polymer. Surface reactions are advantageous for applications requiring modification of surface properties without affecting the bulk physical properties of a polymer, such as modification of surface dyeability, biocompatibility, adhesive and frictional behavior, and coatability [Ward and McCarthy, 1989].

9-1d   Crystallinity

The reactions of semicrystalline (and crystalline) polymers proceed differently depending on the reaction temperature and solvent. When reaction proceeds under conditions where portions of the polymer remain crystalline, the product is heterogeneous since only functional groups in the amorphous regions are available for reaction. Such behavior has been observed in many systems, including the chlorination of polyethylene, acetylation of cellulose, and aminolysis of poly(ethylene terephthalate). The functional groups in the crystalline regions are generally inaccessible to chemical reagents, and reaction is limited to the amorphous regions. Thus the rate of chlorosulfonation of different polyethylenes increases directly with the amorphous content of the polymer [Bikson et al., 1979]. The reaction of a polymer is exactly like its low-molecular-weight homolog only when carried out under homogeneous conditions by the appropriate choice of temperature and/or solvent.

The situation is not so clear-cut in some polymer reactions. Reaction often occurs at crystal surfaces, and this leads to subsequent penetration of a small reactant into the crystalline regions. Complete penetration of the crystalline regions occurs if the reaction times are sufficiently long, although the extent of reaction may be lower in the crystalline regions compared to the amorphous regions.

Although homogeneous polymer reactions are generally more desirable than heterogeneous reactions, this is not necessarily always the case. The properties of the products of heterogeneous and homogeneous reactions will clearly be different. For example, homogeneous chlorination of polyethylene yields a less crystalline product with lower T_g and stiffness than heterogeneous chlorination [Fettes, 1964]. Whether one product or the other is more desirable depends on the specific application. Heterogeneous reaction is more suited for applications requiring modification of surface properties without alteration of bulk properties.

9-1e   Changes in Solubility

Abnormal behavior can occur in the reaction of an initially homogeneous system if there is a change in the physical nature of the system on reaction of the polymer. Partial reaction may yield a polymer that is no longer soluble in the reaction medium or that forms a highly viscous system. The solubility changes can be quite complex, as shown in the chlorination of polyethylene when carried out in solution using aliphatic or aromatic hydrocarbon solvents at about 80°C [Fettes, 1964]. The polymer solubility increases with extent of chlorination until about 30% by weight of chlorine has been added and then decreases with further chlorination. After 50–60% chlorination, solubility again increases with further chlorination. Such solubility behavior presents problems in carrying out a polymer reaction. As a minimum, one must take into account the rate changes accompanying the solubility changes when planning the synthesis. In the extreme case, insolubility of a polymer may limit the maximum conversion if the small molecule reagent cannot diffuse into and through the polymer.

Enhanced reaction rates have also been observed in some instances of decreased polymer solubility with conversion when the precipitating polymer absorbs the small molecule reactant. Adsorption results in an increased concentration of the reactant at the actual polymerization site (the polymer coils). Similar effects are observed if a catalyst is adsorbed onto the precipitating polymer [Beresniewicz, 1959].

9-1f Crosslinking

A consideration of the effect of crosslinking on polymer reactivity is important since some polymer reactions are performed to achieve crosslinking for specific applications (Sec. 9-2), while other polymer reactions are carried out on previously crosslinked polymers (Secs. 9-10 through 9-13). For crosslinked polymers, the concentration of small molecule reactant inside the polymer domains can be lower than outside because of a low degree of swelling. This results from a high degree of crosslinking and/or the use of a poor solvent for the polymer. The diffusion coefficients for solvent and small molecule reactant often also decrease under these conditions. The effects of different solvents on polymer reactivity can be quite dramatic. For the S_N2 reaction of pyridine with various alkyl halides, the ratios of the reaction rates in 2-pentanone, toluene, and n-heptane are approximately $7:2:1$. This is the expected order for an S_N2 reaction between neutral substrate and reactant molecules, where increased solvent polarity assists in the development of charge separation in the transition state. For the corresponding reaction of crosslinked poly(4-vinylpyridine), the 2-pentanone:toluene:n-heptane reaction rate ratio is approximately $10:10:1$ [Grieg and Sherrington, 1978]. The high reactivity in toluene is a consequence of this solvent being a good solvent (as is 2-pentanone) for poly(4-vinylpyridine), while n-heptane is a poor solvent. The extent of swelling of the crosslinked polymer is high in 2-pentanone and toluene, which results in higher local concentrations of the small molecule reactant in the polymer domains. Presumably, the diffusion coefficients are also higher in these solvents relative to n-heptane.

The rate of epoxidation of cyclohexene with perbenzoic acid decreases with increasing solvent polarity. The epoxidation by poly(peracrylic acid) shows the opposite trend. A polar solvent causes the polar polymer to swell to a greater extent and the reaction rate is increased due to a higher local concentration of cyclohexene [Takagi, 1975].

High crosslink densities may severely depress polymer reactivity as a result of large decreases in swelling and diffusion rate within the polymer. Diffusion control in a polymer reaction can be detected by the inverse dependence of rate on polymer particle size (radius for spherical particle, thickness for film or sheet) [Imre et al., 1976; Sherrington, 1988].

9-1g Steric Effects

Polymer reactivity can be sterically hindered under certain conditions—when the functional group is close to the polymer chain or in a sterically hindered environment or reaction involves a small molecule reactant that is bulky. Several examples illustrate this effect:

1. The α-chymotrypsin-catalyzed hydrolysis of the p-nitroanilide group in acrylamide copolymers with monomer **I** proceeds with a rate comparable to that of the

$$CH_2=CH-CONH(CH_2)_n CONHCHCONH-\langle O \rangle-NO_2$$
$$\underset{\displaystyle CH_2\phi}{|}$$

I

monomer when $n = 5$. The rate decreases sharply with decreasing value of n as the reactive site becomes progressively closer to the polymer backbone and less accessible to the bulky α-chymotrypsin catalyst [Fu and Morawetz, 1976]. The reaction rate is not, however, affected by the molecular weight of the copolymer [Drobnik et al., 1976].

2. Hydrogenation of cyclododecene using the polymeric rhodium catalyst **II** occurs at a rate five times slower than does cyclohexene [Mathur and Williams, 1976; Mathur

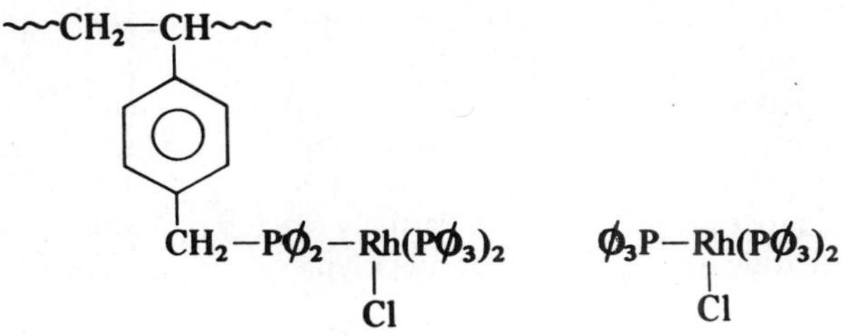

$$\sim\!CH_2-CH\!\sim$$

$$CH_2-P\phi_2-Rh(P\phi_3)_2 \qquad \phi_3P-Rh(P\phi_3)_2$$
$$\underset{\displaystyle Cl}{|} \qquad\qquad\qquad \underset{\displaystyle Cl}{|}$$

$$\textbf{II} \qquad\qquad\qquad\qquad \textbf{III}$$

et al., 1980]. The low-molecular-weight homolog **III** shows no difference in catalytic activity toward the two cycloalkenes.

3. Compounds n-C_4H_9I and n-$C_{18}H_{37}I$ react at the same rate with pyridine but n-C_4H_9I is almost fourfold more reactive than n-$C_{18}H_{37}I$ toward poly(4-vinylpyridine) [Grieg and Sherrington, 1978].

9-1h Electrostatic Effects

Polymer reactions involving the conversion of uncharged functional groups to charged groups often exhibit a decrease in reactivity with conversion. Thus the quaternization of poly(4-vinylpyridine) becomes progressively slower with increasing conversion, since reaction proceeds with the buildup and concentration of charge on the polymer molecule [Noah et al., 1974; Sawage and Loucheux, 1975]. An unreactive pyridine group is less reactive by a factor of 3 when neighboring pyridine groups have reacted. This autoretardation behavior is not observed for the small molecule homolog, such

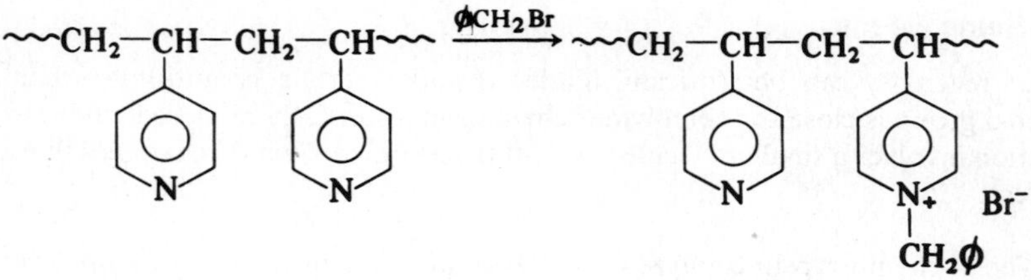

$$\text{~~CH}_2\text{—CH—CH}_2\text{—CH~~} \xrightarrow{\phi CH_2 Br} \text{~~CH}_2\text{—CH—CH}_2\text{—CH~~} \qquad (9\text{-}2)$$

as 4-ethylpyridine, since unreacted molecules are separated from reacted 4-ethylpyridine by solvent.

The same phenomenon is responsible for the decrease in the ionization constant of a polymeric acid such as poly(acrylic acid) with increasing degree of ionization

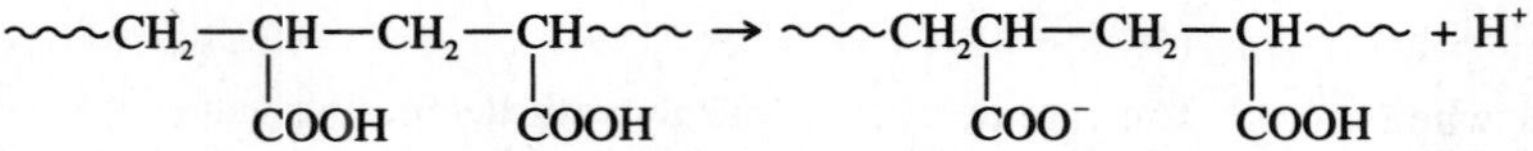

$$\text{~~CH}_2\text{—CH—CH}_2\text{—CH~~} \rightarrow \text{~~CH}_2\text{CH—CH}_2\text{—CH~~} + H^+ \qquad (9\text{-}3)$$

[Kawaguchi et al., 1990; Morawetz and Wang, 1987]. This effect is somewhat moderated for acrylic acid–ethylene copolymers, since charge density on the ionized polymer is decreased. The basicity of a polymeric base such as poly(4-vinylpyridine) or poly(vinyl amine) decreases with increasing protonation of nitrogen [Lewis et al., 1984; Satoh et al., 1989]. These autoretardation effects are partially alleviated by the addition of salts that shield the charge on the polymer.

Charge on a polymer molecule can also affect reactivity by altering the concentration of the small molecule reactant within the polymer domains. The reaction of a charged polymer with a charged reactant results in acceleration for oppositely charged species and retardation when the charges are the same. For example, the rate constant for the KOH saponification of poly(methyl methacrylate) decreases by about an order of magnitude as the reaction proceeds [Plate, 1976]. Partially reacted poly(methyl methacrylate) (**IV**) repells hydroxide ion, while unreacted polymer does not. The concen-

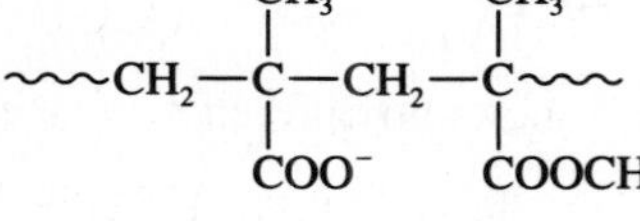

$$\text{~~CH}_2\text{—}\underset{\underset{COO^-}{|}}{\overset{\overset{CH_3}{|}}{C}}\text{—CH}_2\text{—}\underset{\underset{COOCH_3}{|}}{\overset{\overset{CH_3}{|}}{C}}\text{~~}$$

IV

tration of OH$^-$ within the polymer domains decreases progressively with increasing conversion. Acceleration of reactivity is observed for the S$_N$2 reaction of poly(4-

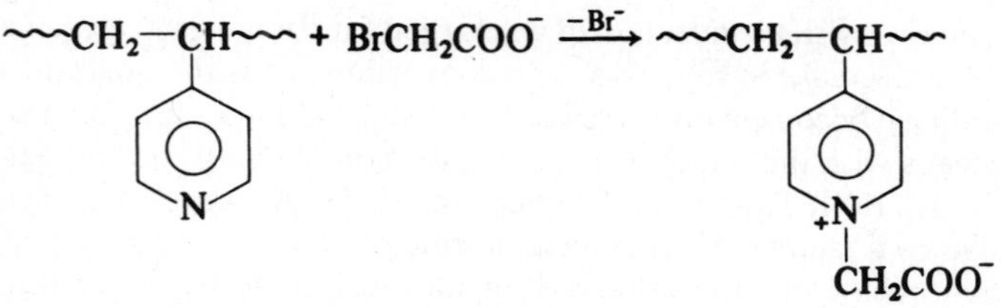

$$\text{~~CH}_2\text{—CH~~} + BrCH_2COO^- \xrightarrow{-Br^-} \text{~~CH}_2\text{—CH~~} \qquad (9\text{-}4)$$

vinylpyridine) and α-bromoacetate ion when the polymer is partially protonated [Ladenheim et al., 1959]. The positive charge on the polymer attracts and concentrates α-bromoacetate ion within the polymer domains.

Polymer charge can also affect polymer reactivity toward neutral reagents. The nucleophilic reactivity of the carboxylate anions of a partially ionized poly(methacrylic acid) toward α-bromoacetamide decreases with increasing charge on the polymer when the reaction is carried out in water [Ladenheim and Morawetz, 1959]. Water is preferentially concentrated within the polymer coils with the partial exclusion of the neutral reagent, resulting in a decreased reaction rate. Similar retardation is observed when partially protonated poly(4-vinylpyridine) acts as the catalyst in hydrolysis of 2,4-dinitrophenyl acetate [Letsinger and Savereide, 1962]. These retardation effects are decreased or eliminated when reaction is carried out in an organic solvent where there is less tendency for preferential concentration of the solvent within the polymer domains.

Related phenomena include the effect of a polymer on the reaction between two small reactants. Acceleration occurs when the polymer attracts both reactants (both reactants have the same charge and both are opposite in sign to the charge on the polymer backbone). An example is the redox reaction

$$Co(NH_3)_5Cl^{2+} + Hg^{2+} + H_2O \rightarrow Co(NH_3)_5H_2O^{3+} + HgCl^+ \tag{9-5}$$

performed in the presence of poly(sodium vinyl sulfonate). The rate increases with polyanion concentration, reaches a maximum, and then decreases with further increase in polyanion concentration [Morawetz and Vogel, 1969]. Retardation occurs for reactions between ions of opposite charge because the polymer attracts one ion but repels the other.

9-1i Neighboring-Group Effects

The reactivity of a functional group on a polymer is sometimes affected directly by an adjacent neighboring group. (The effects described in the beginning of Sec. 9-1h are indirect effects of neighboring groups.) Thus, the saponification of poly(methyl methacrylate) and related polymers proceeds with autoacceleration when carried out with weak bases such as pyridine or low concentrations of strong bases [Gaetjens and Morawetz, 1961; Kheradmand et al., 1988]. (Saponifications at high OH$^-$ concentrations involve retardation due to hydroxide–carboxyl anion repulsions as described in Sec. 9-1h). After the initial formation of some carboxylate anions, subsequent hydrolysis of ester groups occurs not directly by hydroxide ion but by neighboring carboxylate anions. The reaction proceeds by the sequence in Eq. 9-6 with the intermediate formation of a cyclic acid anhydride. Direct rate enhancement by a neighboring group, referred to as *anchimeric assistance*, occurs primarily when the cyclic anhydride intermediate is the favored 5- or 6-membered ring. Such effects are also encountered in low-molecular-weight bifunctional compounds, as in succinic esters.

Neighboring-group effects are dependent not only on the functional group and reaction involved but also on the stereochemistry of the neighboring groups. Thus, whereas there is a rate enhancement of 1–2 orders of magnitude in the saponification of isotactic poly(methyl methacrylate) by pyridine–water at 145°C, no effect is observed for the syndiotactic polymer [Barth and Klesper, 1976; Plate, 1976]. The isotactic structure has neighboring functional groups in the optimum orientation for interaction with each other to form the cyclic anhydride intermediate. The difference between

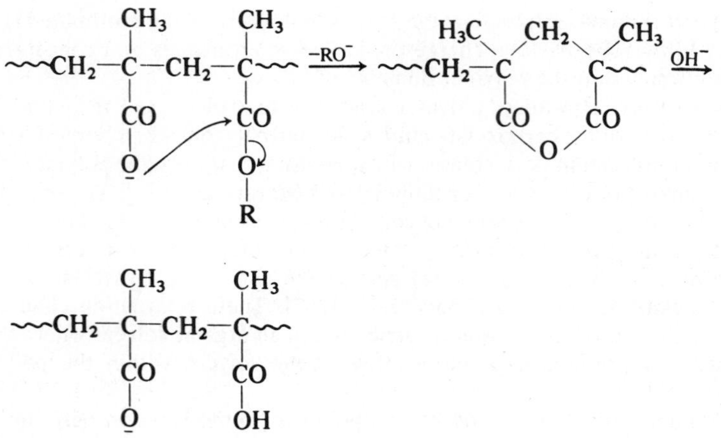

$$\text{(9-6)}$$

the two stereoisomers is evident even for saponification with strong base as the reaction rate constant for syndiotactic PMMA falls much more rapidly with conversion compared to isotactic PMMA.

9-1j Hydrophobic Interactions

Hydrophobic attractive interactions between a polymer and a small molecule reactant often occur in reactions carried out in aqueous media [Imanishi, 1979; Morawetz, 1975; Overberger and Guterl, 1978, 1979]. Acid catalysis of ester hydrolysis by polysulfonic acids is most effective when either the ester or polysulfonic acid contains hydrophobic groups. The reaction rate increases because the small reagent becomes concentrated within the polymer domains where there is a high local hydrogen ion concentration. Similar hydrophobic effects have been observed in the hydrolysis of p-nitrophenyl acetate by partially quaternized poly(4-vinylpyridine) and the aminolysis of 4-nitrophenyl laurate by polythyleneimine containing lauroyl groups [Kirsh et al., 1968; Royer and Klotz, 1969; Spetnagel and Klotz, 1977].

Overberger and co-workers carried out an interesting study on the hydrolysis of various 3-nitro-4-acyloxybenzoic acid substrates (**V**) catalyzed by imidazole (**VI**) and poly[4(5)-vinylimidazole] (**VII**) in ethanol–water mixtures [Overberger et al., 1973].

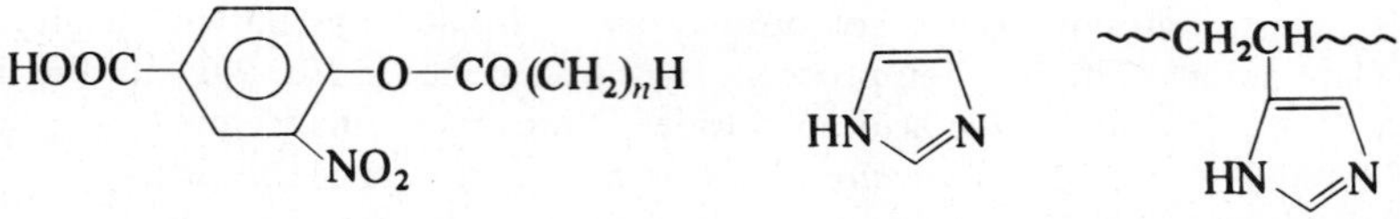

The reaction rate constant is independent of the size of the alkyl group (i.e., value of n) in the substrate for the imidazole-catalyzed reaction. Catalysis by poly[4(5)-vinylimidazole] (PVIm) is more effective than by imidazole and its effectiveness increased sharply with the hydrophobicity of the substrate. Thus, the PVIm-catalyzed reaction rate for $n = 11$ is 30-fold larger than for $n = 1$ and almost 400-fold larger than for the imidazole-catalyzed reaction. Hydrophobic attraction between the imidazole group and the alkyl group of the substrate increases the substrate concentration within the polymer domains. Such hydrophobic attractions decrease considerably and even disappear if the water content of the reaction medium is sufficiently decreased. When the solvent and substrate have similar polarities, there is no preferential attraction of the polymer for substrate. Hydrophobic interactions have been the subject of considerable study with polymers containing the imidazole group [Overberger and Meenakshi, 1984; Overberger and Smith, 1975; Septnagel and Klotz, 1977]. PVIm has received attention as a model for a biological enzyme.

9-1k Other Considerations

Some other considerations relative to polymer reactivity are:

1. Morawetz and co-workers have discussed the reaction between functional groups on different polymer molecules [Cho and Morawetz, 1973; Morawetz, 1975]. Reaction occurs only if the two polymers are sufficiently similar so that there is interpenetration of the two polymer coils. However, most polymers do not fall in this category, polymer mixing is quite endothermic, and reactions of functional groups do not occur. Reaction occurs with dissimilar polymers only when polymer mixing is highly exothermic, for example, an acidic polymer with a basic polymer [Letsinger and Wagner, 1966].

2. Polymer reactivity can also be affected by the conformation of polymer chains [Imanishi, 1979; Overberger et al., 1973; Overberger and Morimoto, 1971; Pshezetsky et al., 1968]. Whether the polymer chain exists in a tight or expanded coil can influence the accessibility of polymer functional groups and the local concentration of a small molecule reactant.

3. The functional groups in a number of polymers are not of the same reactivity. For example, the hydroxyl groups in cellulose (**VIII**) are of different reactivities. The

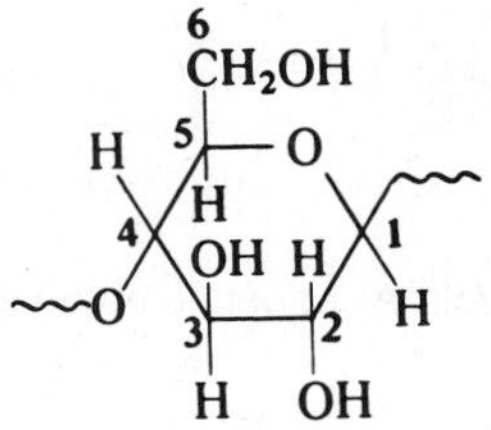

VIII

hydroxyl at C-2 is reported to be slightly more reactive than that at C-6, which is about four times more reactive than the hydroxyl at C-3 [Lenz, 1960]. Reaction of cellulose leads to a nonuniform distribution of reacted groups. Another report shows no reactivity difference between the secondary hydroxyls at C-2 and C-3 [Jain et al., 1985]. The situation is more complicated. For xanthation of hydroxyl groups (Sec. 9-3a), the hydroxyls at C-2 and C-3 are reported to be favored kinetically but the hydroxyl at C-6 is favored thermodynamically [Turbak, 1988].

9-2 CROSSLINKING

A number of crosslinking reactions have been discussed—crosslinking of monomers with functionality greater than 2 (Sec. 2-10), including the crosslinking of prepolymers such as epoxy prepolymers (Sec. 2-12d) and unsaturated polyesters (Sec. 6-8c) and crosslinking via copolymerization of vinyl-divinyl systems (Sec. 6-6). The crosslinking of high-molecular-weight polymers such as 1,4-polyisoprene and polyethylene will be discussed in this section. The reaction by which a polymer is converted into a cross-linked structure is usually difficult to analyze kinetically because of the insolubility of the network system. Decreased reaction rates are encountered at high crosslink densities.

9-2a Alkyds

Unsaturated polyesters in which the alkene double bond resides in a fatty acid component such as oleic (**IX**) and linoleic (**X**) acids are referred to as *alkyds* or *alkyd re-*

$$CH_3(CH_2)_7CH\!\!=\!\!CH(CH_2)_7COOH$$

IX

$$CH_3(CH_2)_7CH\!\!=\!\!CHCH\!\!=\!\!CH(CH_2)_5COOH$$

X

sins (Sec. 2-12a). Alkyds are crosslinked via oxidation by atmospheric oxygen [Marshall et al., 1987; Solomon, 1967]. The process is usually referred to as *drying* or *air-drying*. Varnishes and other surface coatings based on these polymers are crosslinked by standing in air.

The crosslinking process is different depending on whether the unsaturation is an unconjugated double bond as in oleic acid or a conjugated double bond as in linoleic acid. Unconjugated double bonds undergo crosslinking by the initial formation of an allylic hydroperoxide

$$\sim\!\!\sim\!\!CH_2\!\!-\!\!CH\!\!=\!\!CH\!\!\sim\!\!\sim\;\xrightarrow{\;O_2\;}\;\sim\!\!\sim\!\!\underset{\underset{OOH}{|}}{CH}\!\!-\!\!CH\!\!=\!\!CH\!\!\sim\!\!\sim \qquad (9\text{-}7)$$

followed by decomposition of the hydroperoxide. The reaction sequence involves

$$\sim\!\!\sim\!\!OOH \rightarrow \sim\!\!\sim\!\!O\!\cdot + HO\!\cdot \qquad (9\text{-}8)$$

$$2\sim\!\!\sim\!OOH \rightarrow \sim\!\!\sim\!\!O\!\cdot + \sim\!\!\sim\!OO\!\cdot + H_2O \qquad (9\text{-}9)$$

$$\sim\!\!\sim\!\!O\!\cdot + \sim\!\!\sim\!\!H \rightarrow \sim\!\!\sim\!\cdot + \sim\!\!\sim\!OH \qquad (9\text{-}10)$$

$$HO\!\cdot + \sim\!\!\sim\!\!H \rightarrow \sim\!\!\sim\!\cdot + H_2O \qquad (9\text{-}11)$$

$$2\sim\!\!\sim\!\cdot \rightarrow \sim\!\!\sim\!\!\sim\!\!\sim \qquad (9\text{-}12)$$

$$\sim\!\!\sim\!\!O\!\cdot + \sim\!\!\sim\!\cdot \rightarrow \sim\!\!\sim\!\!O\!\sim\!\!\sim \qquad (9\text{-}13)$$

$$2\sim\!\!\sim\!\!O\!\cdot \rightarrow \sim\!\!\sim\!\!O\!\!-\!\!O\!\sim\!\!\sim \qquad (9\text{-}14)$$

where $\sim\!\!\sim\!\!OOH$ and $\sim\!\!\sim\!\!H$ represent the oxidized and unoxidized polymer molecules. In order to achieve practical crosslinking rates, the reaction is catalyzed by hydrocarbon-soluble salts (usually octanoates, naphthenates, and linoleates) of metals (cobalt, lead, and manganese). The metal ions accelerate hydroperoxide decomposition in a manner analogous to the redox initiator systems discussed in Sec. 3-4b.

The relative amounts of carbon–carbon (Eq. 9-12), ether (Eq. 9-13), and peroxide (Eq. 9-14) crosslinks depend on the reaction conditions.

Crosslinking of alkyds containing conjugated double bonds results in more carbon–carbon bonds in the crosslinks than in the alkyds containing unconjugated double bonds. The crosslinking mechanism involves the formation and decomposition of cyclic

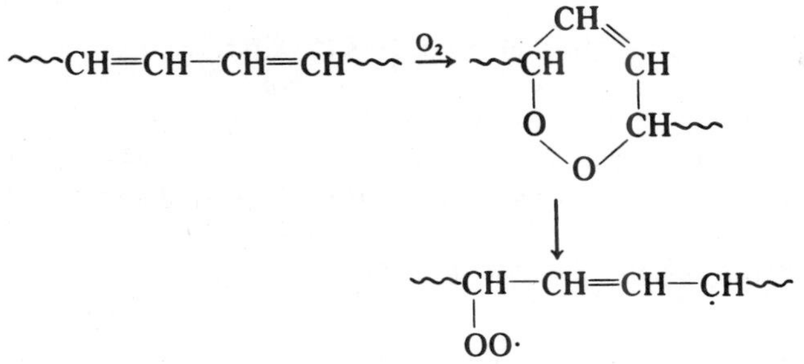

$$\sim\sim CH=CH-CH=CH\sim\sim \xrightarrow{O_2} \quad \text{(cyclic structure)}$$

$$\downarrow$$

$$\sim\sim CH-CH=CH-CH\sim\sim \qquad (9\text{-}15)$$
$$\mid$$
$$OO\cdot$$

peroxides, to yield radicals which initiate crosslinking by 1,4-polymerization of the polymer molecules.

9-2b Elastomers Based on 1,3-Dienes

Various homo- and copolymers of 1,3-dienes make up the large bulk of polymers used as elastomers. Crosslinking is an absolute requirement if elastomers are to have their essential property of rapidly and completely recovering from deformations. The term *vulcanization* is used synonymously with crosslinking in elastomer technology. Crosslinking or vulcanization can be achieved by using sulfur, peroxides, other reagents, or ionizing radiation [Alliger and Sjothun, 1964; Bateman, 1963; Coran, 1978; Morrison and Porter, 1984; Roberts, 1988]. Heating with sulfur is used almost exclusively for the commercial crosslinking of most elastomers based on 1,3-dienes. This includes 1,4-polyisoprene (synthetic and natural), 1,4-poly(1,3-butadiene), ethylene–propylene–diene terpolymer (EPDM), 1,3-butadiene copolymers with styrene (SBR) and acrylonitrile (nitrile rubber or NBR), and isobutylene-isoprene copolymer (butyl rubber or BR). Poly(1-octenylene) and polynorbornene are also vulcanized by sulfur.

9-2b-1 Sulfur Alone

Although sulfur vulcanization has been studied since its discovery in 1839 by Goodyear, its mechanism is not well understood. Free radical mechanisms were originally assumed but most evidence points to an ionic reaction [Bateman, 1963]. Neither radical initiators nor inhibitors affect sulfur vulcanization and radicals have not been detected by ESR spectroscopy. On the other hand, sulfur vulcanization is accelerated by organic acids and bases as well as by solvents of high dielectric constant. The ionic process can be depicted as a chain reaction involving the initial formation of a sulfonium ion (**XI**) by

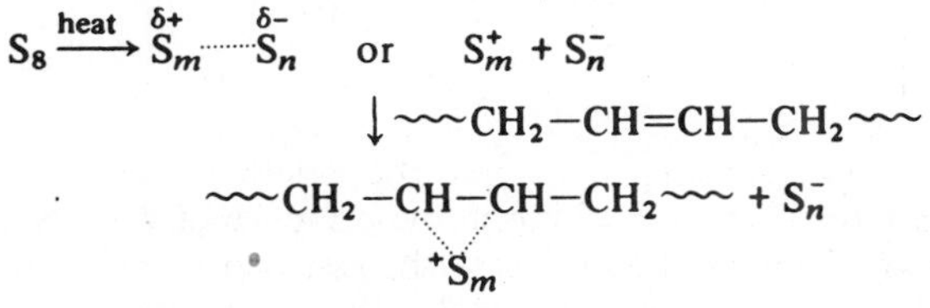

$$S_8 \xrightarrow{\text{heat}} \overset{\delta+}{S_m}\cdots\overset{\delta-}{S_n} \quad \text{or} \quad S_m^+ + S_n^-$$

$$\downarrow \sim\sim CH_2-CH=CH-CH_2\sim\sim$$

$$\sim\sim CH_2-CH-CH-CH_2\sim\sim + S_n^- \qquad (9\text{-}16)$$
$$\mid$$
$$^+S_m$$

XI

reaction of the polymer with polarized sulfur or a sulfur ion pair. The sulfonium ion reacts with a polymer molecule by hydride abstraction to yield the polymeric (allylic) carbocation **XII**, which undergoes crosslinking by reacting with sulfur followed by

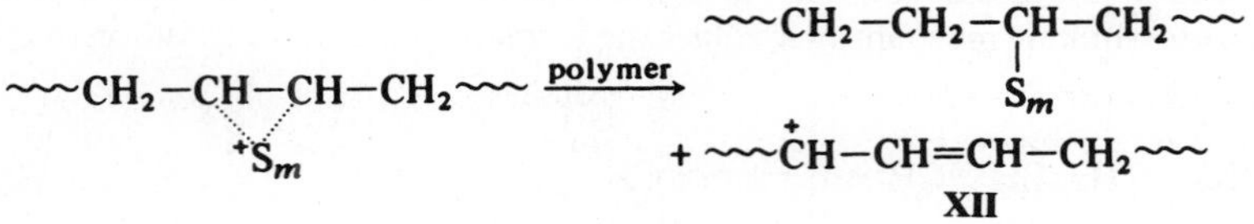

$$
\sim\sim CH_2-CH-CH-CH_2\sim\sim \xrightarrow{\text{polymer}}
\begin{array}{c}
\sim\sim CH_2-CH_2-CH-CH_2\sim\sim \\
| \\
S_m \\
+ \sim\sim \overset{+}{C}H-CH=CH-CH_2\sim\sim \\
\mathbf{XII}
\end{array}
\tag{9-17}
$$

addition to a polymer double bond. A subsequent reaction with polymer by hydride abstraction regenerates the polymer carbocation as shown in the sequence

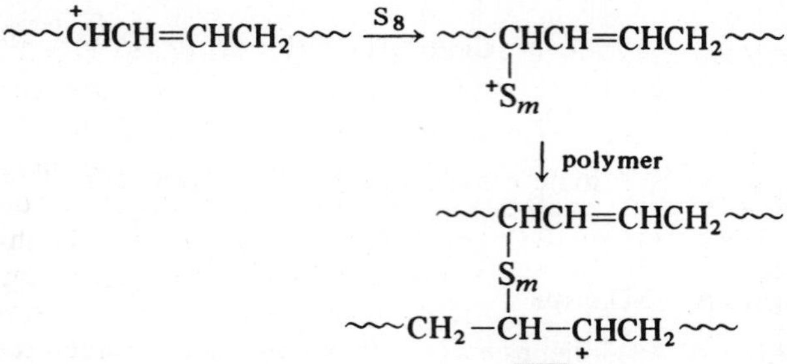

$$
\sim\sim\overset{+}{C}HCH=CHCH_2\sim\sim \xrightarrow{S_8} \sim\sim CHCH=CHCH_2\sim\sim
$$
$$
\underset{\overset{|}{+S_m}}{}
$$

$$\downarrow \text{polymer}$$

$$
\begin{array}{c}
\sim\sim CHCH=CHCH_2\sim\sim \\
| \\
S_m \\
| \\
\sim\sim CH_2-\overset{}{C}H-CHCH_2\sim\sim \\
\mathbf{XIII} \quad +
\end{array}
\tag{9-18a}
$$

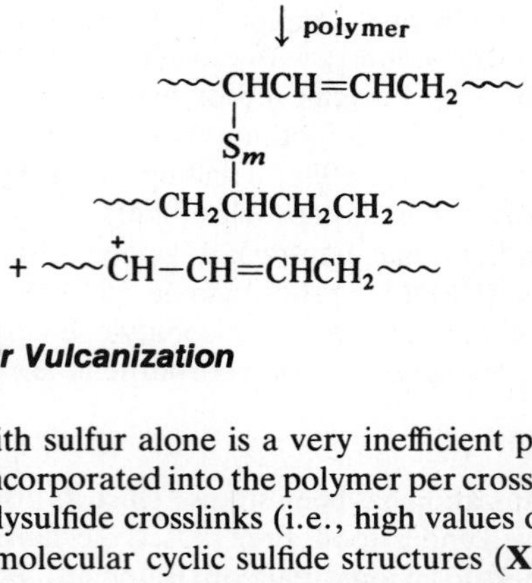

$$\downarrow \text{polymer}$$

$$
\begin{array}{c}
\sim\sim CHCH=CHCH_2\sim\sim \\
| \\
S_m \\
| \\
\sim\sim CH_2CHCH_2CH_2\sim\sim \\
+ \sim\sim \overset{+}{C}H-CH=CHCH_2\sim\sim
\end{array}
\tag{9-18b}
$$

9-2b-2 *Accelerated Sulfur Vulcanization*

Vulcanization by heating with sulfur alone is a very inefficient process with approximately 40–50 sulfur atoms incorporated into the polymer per crosslink. Sulfur is wasted by the formation of long polysulfide crosslinks (i.e., high values of m in **XIII**), vicinal crosslinks (**XIV**), and intramolecular cyclic sulfide structures (**XV**). (Structures **XIV** and **XV** do not contribute significantly to the physical properties of the polymer.)

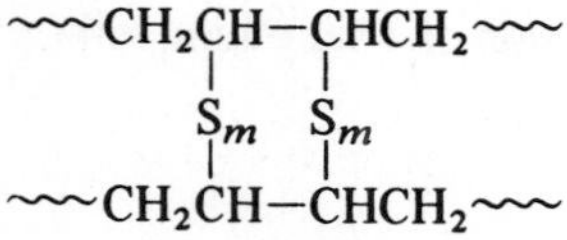

$$
\begin{array}{c}
\sim\sim CH_2CH-CHCH_2\sim\sim \\
| \quad\quad | \\
S_m \quad S_m \\
| \quad\quad | \\
\sim\sim CH_2CH-CHCH_2\sim\sim \\
\mathbf{XIV}
\end{array}
\qquad
\begin{array}{c}
\quad CH \\
\quad / \ \backslash \\
\sim\sim CH_2CH \quad CH \\
| \quad\quad\quad | \\
S-\!-\!-\!-CH-CH_2\sim\sim \\
\mathbf{XV}
\end{array}
$$

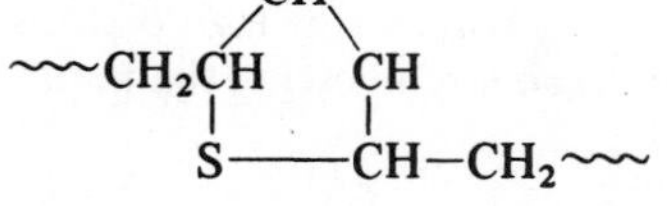

Commercial sulfur vulcanizations are carried out in the presence of various additives, referred to as *accelerators*, which greatly increase the rate and efficiency of the process [Kuczkowski, 1988; Morrison and Porter, 1984]. The most used accelerators

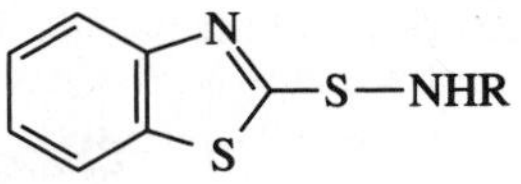

XVI

are sulfenamide derivatives of 2-mercaptobenzothiazole (**XVI**). Actually, the use of accelerators alone usually gives only small increases in crosslinking efficiency. Maximum efficiency is achieved by using accelerators together with a metal oxide and fatty acid. The latter substances are referred to as the *activator*. Zinc oxide and stearic acid are the most commonly used. The fatty acid solubilizes the zinc oxide by forming the zinc carboxylate salt. Vulcanization is achieved in minutes using the accelerator–activator combination compared to hours for sulfur alone. Analysis of the crosslinked product shows a large decrease in the extent of the wastage reactions. The crosslinking efficiency in some systems is increased to slightly less than two sulfur atoms per crosslink. Most of the crosslinks are monosulfide or disulfide with very little vicinal or cyclic sulfide units.

Studies with model alkenes indicate that the action of accelerators is to increase the extent of sulfur substitution (crosslinking) at the allylic positions of the diene polymer [Skinner, 1972]. The mechanism of vulcanization by **XVI** involves the initial formation of 2,2'-dithiobisbenzothiazole (**XVII**) via cleavage of **XVI** to 2-mercapto-benzothiazole followed by oxidative coupling. 2,2'-Dithiobisbenzothiazole reacts with sulfur to form the accelerator polysulfide (**XVIII**). The accelerator polysulfide reacts

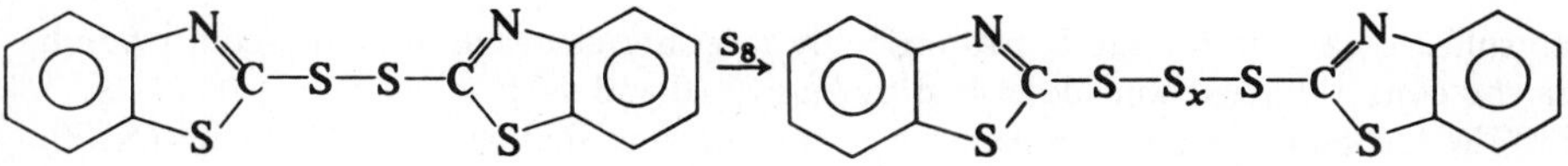

XVII   **XVIII**

$$(9\text{-}19)$$

with the polydiene at an allylic hydrogen in a concerted process to form a rubber polysulfide (**XIX**):

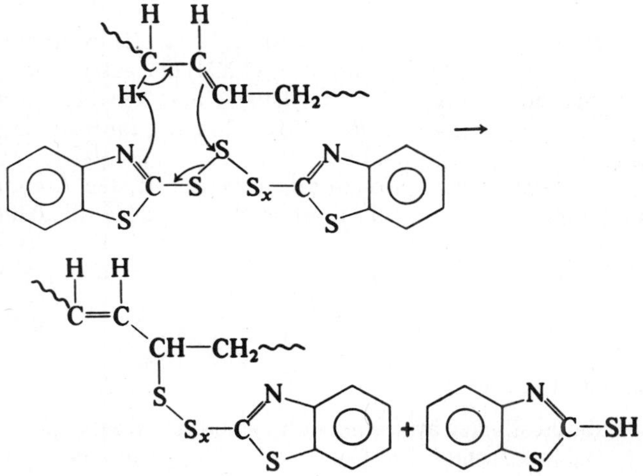

$$(9\text{-}20)$$

XIX

Crosslinking occurs by the corresponding reaction between **XIX** and the polydiene

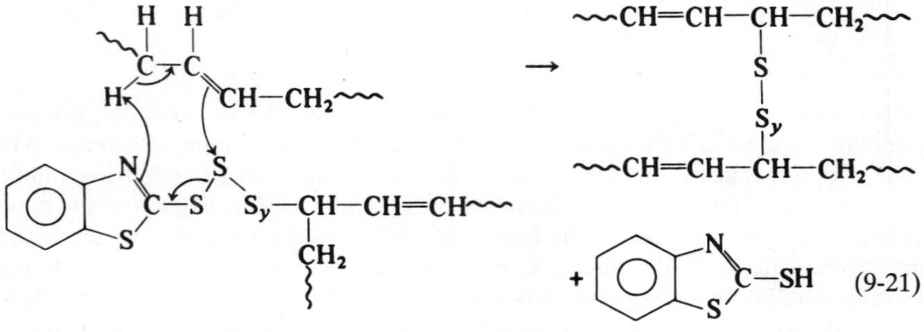

$$(9\text{-}21)$$

The action of zinc in increasing the efficiency and rate of crosslinking is thought to involve chelation of zinc with the accelerator as well as species **XVIII** and **XIX**. Zinc polysulfide compounds such as **XX** are also likely intermediates. Zinc chelated

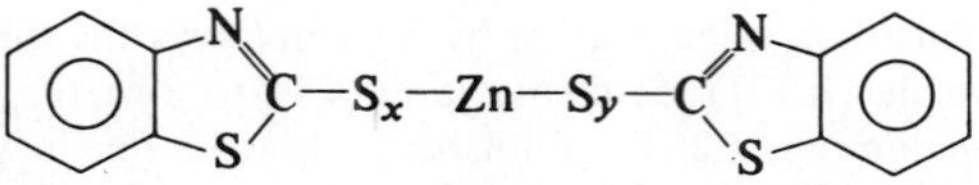

XX

to sulfur or as zinc sulfide bonds probably facilitate cleavage of sulfur–sulfur bonds in the concerted reactions described by Eqs. 9-20 and 9-21.

The second important class of accelerators are zinc dialkyldithiocarbamates (**XXI**). These are more active with faster vulcanization rates than the 2-mercaptobenzothiazole

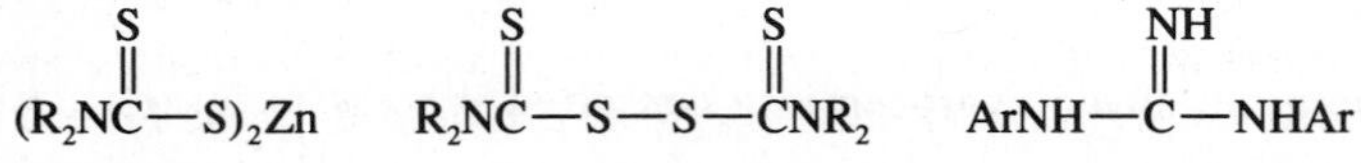

sulfenamides. Also, activators are not usually needed since the zinc is incorporated into the accelerator molecule. Tetralkylthiuram disulfides (**XXII**) in combination with activators give equivalent vulcanization rates. 1,3-Diarylguanidines (**XXIII**) are sometimes used as secondary accelerators in combination with the 2-mercaptobenzothiazole sulfenamides to obtain increased vulcanization rates.

It should be noted that all elastomers, prior to the vulcanization step, are compounded (mixed) with a reinforcing agent such as carbon black or silica (in addition to the crosslinking agents) in industrial applications. The reinforcing agents increase the strength and elastomeric properties of the final crosslinked product (often referred to as a *vulcanizate*).

9-2b-3   Other Vulcanizations

Polydiene rubbers can also be crosslinked by heating with *p*-dinitrosobenzene, phenolic resins, or maleimides [Coran, 1978; Gan and Chew, 1979; Gan et al., 1977, 1978;

Sullivan, 1966]. The crosslinking mechanism is similar to that for accelerated sulfur vulcanization, for example, for vulcanization by *p*-dinitrosobenzene

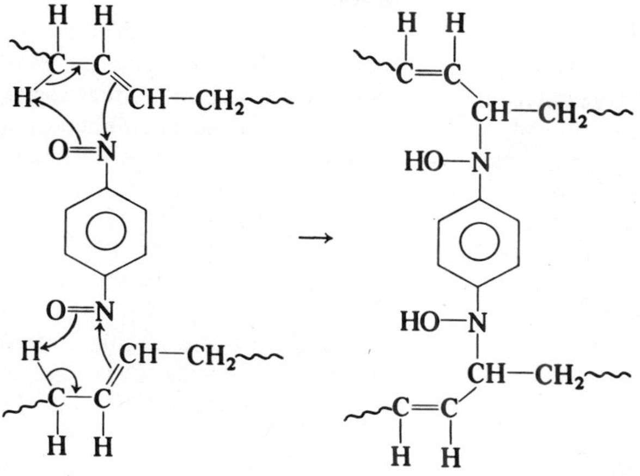

(9-22)

These products find use in specialty applications requiring better thermal stability than available in the sulfur vulcanized elastomers. Other processes are also used to crosslink polydiene rubbers (Secs. 9-2c and 9-2d).

9-2c Peroxide and Radiation Crosslinking

Many polymers are crosslinked by compounding with a peroxide such as dicumyl peroxide or di-*t*-butyl peroxide and then heating the mixture [Alliger and Sjothun, 1964; Coran, 1978; Keller, 1988; Labana, 1986; Peacock, 1987]. Peroxide crosslinking of diene polymers works with all except butyl rubber, which undergoes chain scission. The crosslinks formed via peroxides are more thermally stable than those formed via sulfur vulcanization. However, peroxide crosslinking is not economically competitive with sulfur crosslinking because of the high cost of peroxides. Peroxide crosslinking is primarily used for those polymers that cannot be easily crosslinked by sulfur, such as polyethylene and other polyolefins, ethylene–propylene (no diene) rubbers (EPM), and polysiloxanes. Crosslinking of polyethylene increases its strength properties and extends the upper temperature limit at which this plastic can be used. Uses include electrical wire and cable insulation, pipe, and hose. For EPM and polysiloxanes, crosslinking is essential to their use as elastomers.

Peroxide crosslinking involves the formation of polymer radicals via hydrogen abstraction by the peroxy radicals formed from the thermal decomposition of the peroxide. Crosslinking occurs by coupling of the polymer radicals

$$ROOR \rightarrow 2RO\cdot \qquad (9\text{-}23)$$

$$RO\cdot + \sim\sim\sim CH_2CH_2\sim\sim\sim \rightarrow ROH + \sim\sim\sim CH_2\dot{C}H\sim\sim\sim \qquad (9\text{-}24)$$

$$2\sim\sim\sim CH_2\dot{C}H\sim\sim\sim \rightarrow \begin{array}{c} \sim\sim\sim CH_2CH\sim\sim\sim \\ | \\ \sim\sim\sim CH_2CH\sim\sim\sim \end{array} \qquad (9\text{-}25)$$

The maximum crosslinking efficiency in this process is one crosslink per molecule of peroxide decomposed—far less than that of a chain polymerization where one radical converts large numbers of monomer molecules to product. Further, the actual crosslinking efficiency is often considerably less than one because of various side reactions of the initiator (Sec. 3-4g-2) and the polymer radical. Thus, if a polymer is not formed in the close vicinity of another polymer radical, crosslinking may be impaired since the process takes place in a highly viscous system. Side reactions such as chain scission, hydrogen atom abstraction, or expulsion, and combination with initiator radicals become possible. These considerations, together with the high cost of peroxides, limit the practical utility of the peroxide crosslinking process.

The crosslinking efficiency of the peroxide process can be increased for some systems by incorporating small amounts of a comonomer containing vinyl groups into the polymer. This approach is used for polysiloxanes by copolymerization with small amounts of vinyltrimethylsilanol

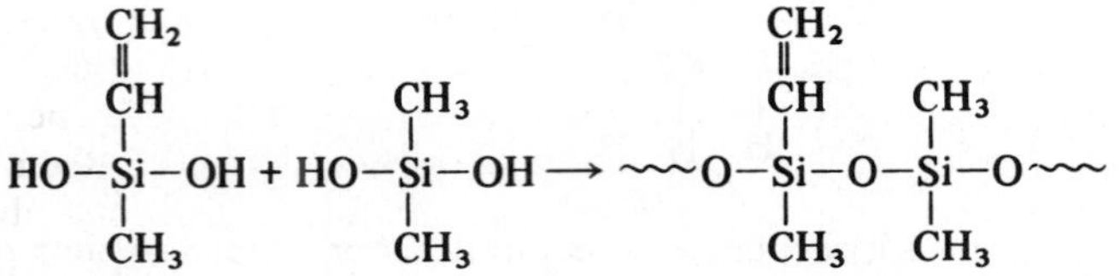

$$
\begin{array}{cc}
CH_2 & CH_2 \\
\parallel & \parallel \\
CH \quad CH_3 & CH \quad CH_3 \\
| \quad | & | \quad | \\
HO-Si-OH + HO-Si-OH \rightarrow \; \sim\!\!\sim\!\!\sim O-Si-O-Si-O\sim\!\!\sim\!\!\sim \\
| \quad | & | \quad | \\
CH_3 \quad CH_3 & CH_3 \quad CH_3
\end{array}
\tag{9-26}
$$

Peroxide crosslinking of the copolymer is more efficient than that of the homopolymer (Table 9-1). The process becomes a chain reaction (but with short kinetic chain length) involving polymerization of the pendant vinyl groups on the polysiloxane chains in combination with coupling of polymeric radicals. The crosslinking of EPDM rubbers is similarly more efficient when compared to EPM rubbers since the former contain double bonds in the polymer chain.

Radiation crosslinking of various polymers has been studied [Chapiro, 1962; McGinniss, 1986; Pappas, 1989; Wilson, 1974]. The reaction is essentially the same as in peroxide crosslinking except that polymer radicals are formed by the interaction of ionizing radiation (Sec. 3-4d) with polymer. Ultraviolet radiation can also be used but is more limited in that the depth of penetration of a sample by UV is considerably less than by ionizing radiation. Radiation crosslinking has found commercial utilization in the crosslinking of polyethylene, other polyolefins, and poly(vinyl chloride) for electrical wire and cable insulation and heat-shrinkable products (tubing, packaging

TABLE 9-1 Efficiency of Crosslinking of Polydimethylsiloxane by Bis(2,4-dichlorobenzoyl) Peroxide[a]

Vinyl Comonomer Content (mole %)	Number of Crosslinks per Peroxide Molecule Decomposed for Peroxide Concentration of	
	0.74%	1.47%
0.0	0.31	0.19
0.1	0.80	0.42
0.2	1.0	0.63

[a]Data from Bobear [1967].

film, and bags). Curing of coatings and adhesives are other applications of radiation crosslinking.

Ultraviolet and ionizing radiation play a major role in integrated-circuit technology. For *negative resists*, a pattern is formed in a polymer placed on a substrate such as silicon by a controlled electron beam or using a masking technique with UV radiation. The portion of the polymer exposed to radiation undergoes crosslinking, and the corresponding image can be developed by dissolving away the unexposed, uncrosslinked polymer with an appropriate solvent. The pattern is etched into the silicon substrate by treatment with acid. The portions of the silicon covered with polymer are protected during the etching process. [Positive resists use a polymer, such as poly(methyl methacrylate), which undergoes degradation on irradiation.] The same principle is used in photoimaging applications.

9-2d   Other Crosslinking Processes

Many other crosslinking reactions are used in commercial applications. A variety of halogen-containing elastomers are crosslinked by heating with a basic oxide (e.g., MgO or ZnO) and a primary diamine [Labana, 1986; Schmiegel, 1979]. This includes poly(epichlorohydrin) (Sec. 7-2b-6); various co- and terpolymers of fluorinated monomers such as vinylidene fluoride, hexafluoropropylene, perfluoro(methyl vinyl ether), and tetrafluoroethylene (Sec. 6-8e); and terpolymers of alkyl acrylate, acrylonitrile, and 2-chloroethyl vinyl ether (Sec. 6-8e).

Crosslinking involves dehydrohalogenation followed by addition of the diamine to the double bond (Eq. 9-27) with the metal oxide acting as an acid acceptor. Some

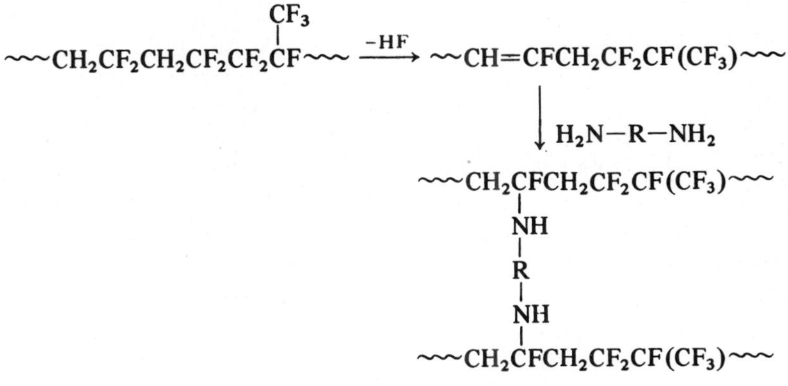

$$\begin{array}{c} \overset{\displaystyle CF_3}{\underset{\displaystyle |}{}} \\ \sim\!\sim\!CH_2CF_2CH_2CF_2CF_2CF\!\sim\!\sim \xrightarrow{-HF} \sim\!\sim\!CH\!=\!CFCH_2CF_2CF(CF_3)\sim\!\sim \\ \\ \downarrow H_2N\!-\!R\!-\!NH_2 \\ \\ \sim\!\sim\!CH_2CFCH_2CF_2CF(CF_3)\sim\!\sim \\ \underset{\displaystyle |}{NH} \\ \underset{\displaystyle |}{R} \\ \underset{\displaystyle |}{NH} \\ \sim\!\sim\!CH_2CFCH_2CF_2CF(CF_3)\sim\!\sim \end{array} \qquad (9\text{-}27)$$

vulcanizations employ a dithiol instead of a diamine. Elastomeric terpolymers of alkyl acrylate, ethylene, and alkenoic acid (Sec. 6-8b) are vulcanized by addition of diamine.

The vulcanization of polychloroprene (Neoprene) is carried out in different ways. Vulcanization by sulfur, even with an accelerator, is not practiced to a large extent. Vulcanizations by metal oxides (without diamine), either alone or in combination with sulfur (sometimes together with an accelerator), give the best physical properties for the crosslinked product. Halogenated butyl rubber is crosslinked in a similar manner. The mechanism for crosslinking by metal oxide alone is not established [Stewart et al., 1985; Vukov, 1984].

Cellulosic fibers (cotton, rayon) are crosslinked by reaction of the hydroxyl groups of cellulose with formaldehyde, diepoxides, diisocyanates, and various methylol compounds such as urea–formaldehyde prepolymers, N,N'-dimethylol-N,N'-dimethylene urea, and trimethylolmelamine [Marsh, 1966]. Crosslinking imparts crease and wrinkle resistance and results in iron-free fabrics.

9-3 REACTIONS OF CELLULOSE

Cellulose is an abundant naturally occurring polymer, representing about one-third of all plants. Although found widely in nature, commercial cellulose is derived almost entirely from cotton and wood [Arthur, 1989; Bikales and Segal, 1971; Billmeyer, 1984; Brydson, 1982; Ott et al., 1954, 1955]. Cotton, the hairs on the seed of the cotton plant, is composed of about 89% cellulose and 7% water, with the remaining portion consisting mostly of waxes, pectic and other organic acids, and protein. The seed hairs contain both long fibers (cotton or lint) and short fibers (linters). The long fibers are used directly to manufacture cotton textiles with more than 2 billion pounds produced annually in the United States. Cotton fabrics hold up well to laundering and have an excellent feel on the human body. The latter is a consequence of cotton possessing high moisture absorption and heat conduction. The short fibers, after treatment with 2–5% aqueous sodium hydroxide to remove most impurities, are used as a source of cellulose (99% pure) for producing a variety of polymeric materials as described below.

Wood contains about 40–50% cellulose with the remaining being lignin and lower-molecular-weight polysaccharides. Treatment of wood pulp with acid and steam followed by basic sodium sulfide yields a product that is 92–98% cellulose.

The properties of the cellulose obtained from cotton short fibers and wood are such that it cannot be directly formed into useful products. It is highly crystalline and insoluble and decomposes at high temperatures without flowing or melting—all a consequence of the extremely strong hydrogen-bonding present. However, cellulose is used in very large quantities (>0.75 billion pounds annually in the United States) for both fiber and plastics applications by employing various chemical reactions to impart processability.

9-3a Dissolution of Cellulose

Cellulose (**VIII**) is spun into fiber or cast into film by using a chemical reaction to convert it into a soluble xanthate derivative (Turbak, 1988). This is achieved by treating cellulose with 18–20% aqueous sodium hydroxide solution at 25–30°C for about 0.5–1 hr. Much of the sodium hydroxide is physically absorbed into the swollen polymer; some of it may be in the form of cellulose alkoxides. The excess alkali is pressed out of the cellulose pulp and the mass aged to allow oxidative degradation of the polymer chains to the desired molecular weight. The alkali cellulose is then treated with carbon disulfide at about 30°C and the resulting mass dissolved in dilute sodium hydroxide to form the sodium xanthate derivative of cellulose (**XXIV**).

The degree of xanthation for the commercial process is about 0.5 xanthate group per repeat unit. This is sufficient to solubilize the cellulose. (Although **XXIV** shows the xanthate group only at C-6, it should be understood that there is a distribution of xanthate groups among C-2, C-3, and C-6 as per the discussion in Sec. 9-1k.) Fibers or films are produced by spinning or casting the viscous cellulose sodium xanthate

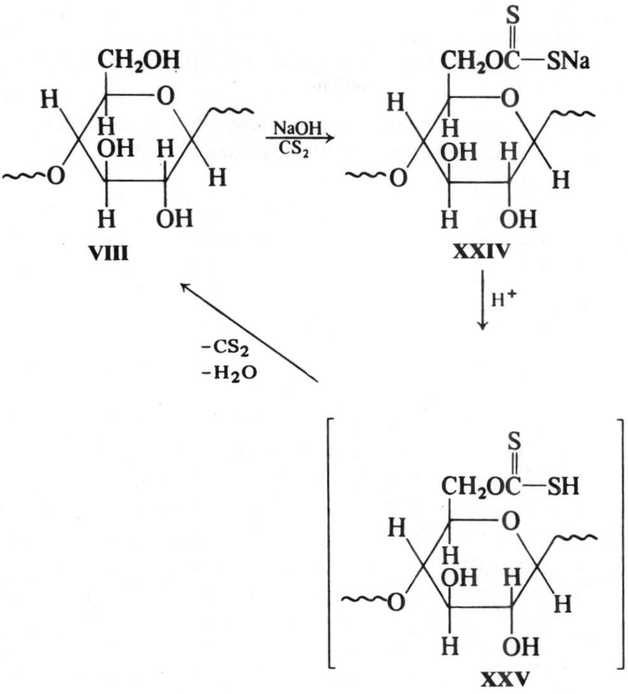

$$(9\text{-}28)$$

solution into a coagulation bath containing about 10% sulfuric acid at 35–40°C. The acid hydrolyzes the sodium xanthate derivative to the xanthic acid (**XXV**), which is unstable and decomposes (without isolation). This regenerates cellulose that is insoluble in the aqueous medium, and the result is a solid cellulose fiber or film product, referred to as *viscose rayon* and *cellophane*, respectively. Both products are also referred to as *regenerated cellulose*. Viscose rayon is not as good a fiber as cotton (sometimes referred to as *native cotton*). The mechanical strength is lower as a result of a somewhat lower crystallinity and lower molecular weight (chain scission occurs during the various processing steps). However, it is an important fiber with more than 400 million pounds produced annually in the United States. Cellophane is widely used in packaging applications (wrap for food and tobacco products).

An alternate procedure used in a few specialty applications is the *cuprammonium* process. This involves stabilization of cellulose in an ammonia solution of cupric oxide. Solubilization occurs by complex formation of cupric ion with ammonia and the hydroxyl groups of cellulose. Regeneration of cellulose, after formation of the desired products, is accomplished by treatment with acid. The main application of the cuprammonium process is for the synthesis of films and hollow fibers for use in artificial kidney dialysis machines. The cuprammonium process yields products with superior permeability and biocompatibility properties compared to the xanthation process.

9-3b Esterification of Cellulose

Acetate, mixed acetate–propionate and acetate–butyrate, and nitrate esters of cellulose are produced commercially [Brewer and Boagan, 1985]. Most products fall into either of two categories—about 2.4 or close to 3 ester groups per repeat unit. As a result of decreased hydrogen-bonding and crystallinity relative to cellulose, these esters are thermoplastic (i.e., they flow when heated) and can be manufactured into products by extrusion, molding, and similar processes. All cellulose esters except the triacetate generally require plasticization to achieve the required flow behavior. Plasticization is insufficient for the triacetate and processing is usually carried out in solutions of methylene chloride–methanol.

Cellulose acetate is the most important ester derivative of cellulose. It is produced by acetylation of cellulose using acetic anhydride in acetic acid in the presence of a strong acid catalyst (usually sulfuric acid). In Eq. 9-29 the symbol Ⓟ is a general

$$
\text{Ⓟ—OH} + \text{HO}\overset{\overset{\displaystyle O}{\|}}{C}\text{CH}_3 \rightleftharpoons \text{Ⓟ—O}\overset{\overset{\displaystyle O}{\|}}{C}\text{CH}_3 + \text{H}_2\text{O}
\tag{9-29}
$$

means of representing a polymer molecule minus the functional group of interest and Ⓟ—OH specifically represents the cellulose molecule, although it could also represent poly(vinyl alcohol) or any other polymer containing a hydroxyl group.

Partially acetylated cellulose (i.e., cellulose with less than three ester groups per repeat unit) is produced by an indirect route. Direct synthesis yields an inhomogeneous product due to insolubility of cellulose in the reaction mixture. Some chains are completely acetylated while others may be completely unreacted. A partially acetylated product is usually produced by controlled hydrolysis of the triacetate. The triacetate is soluble in the reaction mixture and complete solubility ensures that the final product will be more homogeneous. Hydrolysis of the triacetate is carried out by controlled reversal of the esterification reaction by the addition of water or dilute acetic acid.

Cellulose acetate and other esters are tough, strong materials with good optical clarity. The acetate has the best strength and hardness while the acetate–butyrate is better in weatherability, dimensional stability, and low-temperature impact strength. The acetate–propionate has properties in between the other two materials. The largest volume application of cellulose acetate is as a fiber for textiles (clothes, draperies, and bedspreads) and cigarette filters. Plastic applications for cellulose acetate and the mixed ester derivatives include eyeglass frames, photographic film, combs, brush handles, film and sheet for decorative signs, and lacquers and protective coatings for automobiles and wood furniture. Cellulose nitrate is used for lacquer finishes (automobiles, wood furniture), explosives, and propellants.

9-3c   Etherification of Cellulose

Various ether derivatives of cellulose are synthesized by reaction of cellulose with sodium hydroxide followed by an alkyl halide. The methyl and carboxymethyl

$$
\text{Ⓟ—OH} \xrightarrow[\text{2. RCl}]{\text{1. NaOH}} \text{Ⓟ—OH}
\tag{9-30}
$$

(R = CH_3, CH_2COOH, respectively) ethers are the most important commercial products [Just and Majewicz, 1985]. Applications include lacquers, additives for detergents, food, drilling muds, and paints.

9-4 REACTIONS OF POLY(VINYL ACETATE)

In addition to its use as a plastic, poly(vinyl acetate) is used to produce two polymers that cannot be synthesized directly since their monomers do not exist. Poly(vinyl alcohol) is obtained by alcoholysis of poly(vinyl acetate) with methanol

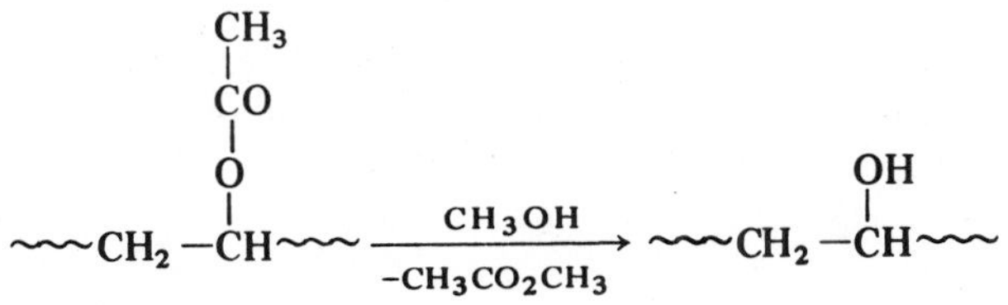

$$
\begin{array}{c}
CH_3 \\
| \\
CO \\
| \\
O \\
| \\
\sim\!\!\sim\!CH_2-CH\!\sim\!\!\sim
\end{array}
\xrightarrow[-CH_3CO_2CH_3]{CH_3OH}
\begin{array}{c}
OH \\
| \\
\sim\!\!\sim\!CH_2-CH\!\sim\!\!\sim
\end{array}
\qquad (9\text{-}31)
$$

Both acids and bases catalyze the reaction, but base is usually employed because of the more rapid rates and freedom from side reactions.

Reaction of poly(vinyl alcohol) with an aldehyde yields the corresponding poly(vinyl acetal)

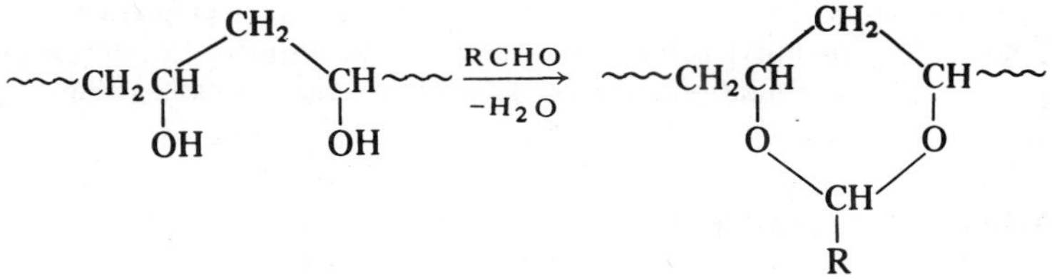

$$
\begin{array}{c}
\quad\quad CH_2 \\
\sim\!\!\sim\!CH_2\,CH\quad\quad\quad CH\!\sim\!\!\sim \\
|\quad\quad\quad\quad\quad | \\
OH\quad\quad\quad\quad OH
\end{array}
\xrightarrow[-H_2O]{RCHO}
\begin{array}{c}
\quad\quad CH_2 \\
\sim\!\!\sim\!CH_2\,CH\quad\quad\quad CH\!\sim\!\!\sim \\
|\quad\quad\quad\quad\quad | \\
O\quad\quad\quad\quad\quad O \\
\quad\quad\!\!\backslash\;CH\;/ \\
\quad\quad\quad | \\
\quad\quad\quad R
\end{array}
\qquad (9\text{-}32)
$$

The reaction is usually carried out with an acid catalyst. Acetal formation does not proceed to completion because of isolation of single hydroxyl groups between pairs of acetal structures. The two most important acetals are the formal and butyral (R = H and C_3H_7, respectively). The applications of poly(vinyl alcohol) and its acetals are described in Sec. 3-13b-3-b.

9-5 HALOGENATION

9-5a Natural Rubber

Chlorination and hydrochlorination of natural rubber are industrial processes carried out on solutions of uncrosslinked rubber in chlorinated solvents [Allen, 1972; Ceresa, 1978; Subramaniam, 1988]. Hydrochlorination, usually carried out at about 10°C, proceeds by electrophilic addition to give the Markownikoff product with chlorine on the tertiary carbon (Eq. 9-33) [Golub and Heller, 1964; Tran and Prud'homme, 1977]. Some cyclization of the intermediate carbocation (**XXVI**) also takes place (Sec. 9-7). The product, referred to as *rubber hydrochloride*, has low permeability to water vapor and is resistant to many aqueous solutions (but not bases or oxidizing acids). Applications include packaging film laminates with metal foils, paper, and cellulose films, although it has been largely replaced by cheaper packaging materials such as polyethylene.

Chlorination of natural rubber (NR) is carried out with chlorine in carbon tetra-

$$\sim\!\!\sim\!\!CH_2\overset{\overset{\displaystyle CH_3}{\displaystyle |}}{C}\!\!=\!\!CH\!-\!CH_2\sim\!\!\sim \xrightarrow{H^+} \sim\!\!\sim\!\!CH_2-\overset{\overset{\displaystyle CH_3}{\displaystyle |}}{\underset{\displaystyle +}{C}}\!-\!CH_2-CH_2\sim\!\!\sim$$

$$\mathbf{XXVI}$$

$$\Big\downarrow Cl^-$$

$$\sim\!\!\sim\!\!CH_2-\overset{\overset{\displaystyle CH_3}{\displaystyle |}}{\underset{\underset{\displaystyle Cl}{\displaystyle |}}{C}}\!-\!CH_2-CH_2\sim\!\!\sim \qquad (9\text{-}33)$$

chloride solution at 60–90°C to yield a chlorinated rubber containing about 65% chlorine, which corresponds to 3.5 chlorine atoms per repeat unit. The process is complex and includes chlorine addition to the double bond, substitution at allylic positions, and cyclization. Chlorinated rubber has high moisture resistance and is resistant to most aqueous reagents (including mineral acids and bases). It is used in chemical- and corrosion-resistant paints, printing inks, and textile coatings.

Butyl rubber, containing only 0.5–2.5% isoprene units, is not efficiently crosslinked by sulfur. Chlorination of butyl rubber is carried out to improve its vulcanization efficiency by allowing a combination of sulfur and metal oxide vulcanizations.

9-5b Saturated Hydrocarbon Polymers

The chlorination of polyethylene, poly(vinyl chloride), and other saturated polymers has been studied [Favre et al., 1978; Lukas et al., 1978; McGuchan and McNeil, 1968]. The reaction is a free-radical chain process catalyzed by radical initiators. Chlorinated

$$\sim\!\!\sim\!\!CH_2\sim\!\!\sim + Cl\cdot \rightarrow \sim\!\!\sim\!\!\overset{\displaystyle \cdot}{C}H\sim\!\!\sim + HCl \qquad (9\text{-}34)$$

$$\sim\!\!\sim\!\!\overset{\displaystyle \cdot}{C}H\sim\!\!\sim + Cl_2 \rightarrow \sim\!\!\sim\!\!CHCl\sim\!\!\sim + Cl\cdot \qquad (9\text{-}35)$$

poly(vinyl chloride) (CPVC) has increased T_g compared to PVC, and this increases its upper use temperature. Applications include hot- and cold-water pipe as well as pipe for the handling of industrial chemical liquids. Chlorinated polyethylene (CPE) finds use as roofing and other vapor-barrier membranes, pond liners, and as an additive to improve the impact strength of PVC.

The reaction of polyethylene with chlorine in the presence of sulfur dioxide yields an elastomer containing both chloro and chlorosulfonyl groups

$$\sim\!\!\sim\!\!CH_2\sim\!\!\sim\!\!CH_2\sim\!\!\sim \xrightarrow[-HCl]{Cl_2,\ SO_2} \sim\!\!\sim\!\!\overset{\overset{\displaystyle Cl}{\displaystyle |}}{C}H\sim\!\!\sim\!\!\overset{\overset{\displaystyle SO_2Cl}{\displaystyle |}}{C}H\sim\!\!\sim \qquad (9\text{-}36)$$

The commercial products contain one chlorine atom per 2–3 repeat units and about one chlorosulfonyl group per 70 repeat units. The chlorosulfonyl groups allow the elastomer to be vulcanized with metal oxides such as lead or magnesium oxide by the formation of metal sulfonate linkages

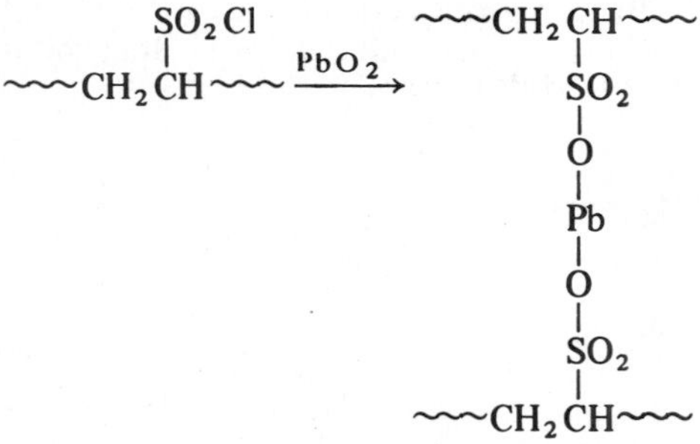

$$\underset{\underset{\displaystyle SO_2Cl}{\mid}}{\sim\!\sim\!\sim CH_2CH\sim\!\sim} \xrightarrow{\ PbO_2\ } \underset{\underset{\underset{\underset{\underset{\underset{\underset{\sim\!\sim\!\sim CH_2CH\sim\!\sim}{\mid}}{SO_2}}{\mid}}{O}}{\mid}}{\underset{\displaystyle Pb}{\mid}}}{\overset{\overset{\overset{\overset{\sim\!\sim\!\sim CH_2CH\sim\!\sim}{\mid}}{SO_2}}{\mid}}{O}}} \tag{9-37}$$

Additional curing is often achieved with sulfur, peroxide, or maleimide formulations. Chlorosulfonated polyethylene has improved resistance to oil, ozone, and heat compared to other elastomers. Applications include barrier membranes and liners, surface coatings on fabrics, automobile air-conditioner hose, electrical cable insulation, and spark-plug boots [Andrews and Dawson, 1986].

9-6 AROMATIC SUBSTITUTION

Aromatic electrophilic substitution is used commercially to produce styrene polymers with ion-exchange properties by the incorporation of sulfonic acid or quaternary ammonium groups [Brydson, 1982; Luca et al., 1980; Miller et al., 1963]. Crosslinked styrene–divinylbenzene copolymers are used as the starting polymer to obtain insoluble final products, usually in the form of beads but also membranes. The use of polystyrene itself would yield soluble ion-exchange products. An anion-exchange product is obtained by chloromethylation followed by reaction with a tertiary amine (Eq. 9-38) while sulfonation yields a cation-exchange product (Eq. 9-39).

$$\tag{9-38}$$

$$\tag{9-39}$$

9-7 CYCLIZATION

Natural rubber and other 1,4-poly-1,3-dienes are cyclized by treatment with strong protonic acids or Lewis acids [Golub, 1969; Subramaniam, 1988]. The reaction involves protonation of the double bond (Eq. 9-40) followed by cyclization via attack of the carbocation on the double bond of an adjacent monomer unit (Eq. 9-41). Some bicyclic

$$\text{~~CH}_2\overset{\overset{\displaystyle CH_3}{|}}{C}=\text{CHCH}_2\,\text{CH}_2\overset{\overset{\displaystyle CH_3}{|}}{C}=\text{CHCH}_2\text{ ~~} \xrightarrow{H^+}$$

$$\text{~~CH}_2\overset{\overset{\displaystyle CH_3}{|}}{\underset{+}{C}}-\text{CH}_2\,\text{CH}_2\,\text{CH}_2\overset{\overset{\displaystyle CH_3}{|}}{C}=\text{CHCH}_2\text{ ~~} \qquad (9\text{-}40)$$

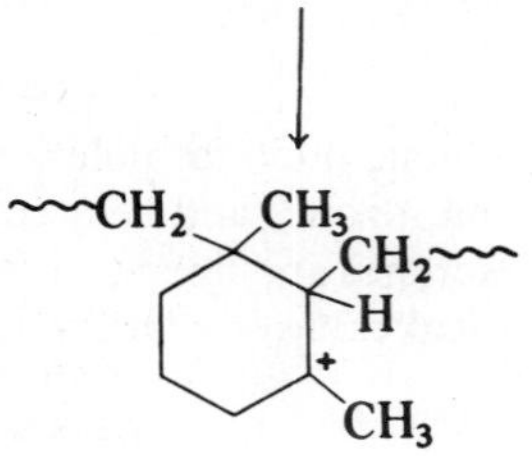

$$(9\text{-}41)$$

and polycyclic fused ring structures are formed by propagation of the cyclization reaction through more than two successive monomer units. However, the average number of fused rings in a sequence is only 2–4 because of the steric restrictions involved in the polymer reaction as well as competing transfer reactions and the isolation of unreacted monomer units between pairs of cyclized units. Cyclized rubbers have been used for shoe soles and heels, adhesives, hard moldings, and coatings.

Cyclization is a key reaction in the production of *carbon fibers* from polyacrylonitrile (PAN) (acrylic fiber; see Sec. 3-13b-4-b). The acrylic fiber used for this purpose usually contains no more than 0.5–5% comonomer (usually methyl acrylate or methacrylate or methacrylic acid). Highly drawn (oriented) fibers are subjected to successive thermal treatments—initially 200–300°C in air followed by 1200–2000°C in nitrogen [Riggs, 1985]. PAN undergoes cyclization via polymerization through the nitrile groups to form a ladder structure (**XXVII**). Further reaction results in aromatization to the

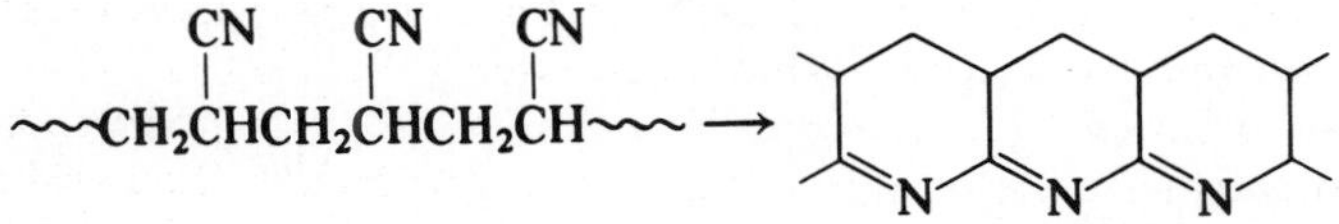

$$\text{~~CH}_2\overset{\overset{\displaystyle CN}{|}}{C}\text{HCH}_2\overset{\overset{\displaystyle CN}{|}}{C}\text{HCH}_2\overset{\overset{\displaystyle CN}{|}}{C}\text{H~~} \longrightarrow \qquad (9\text{-}42)$$

XXVIII

polyquinizarine structure (**XXVIII**) while some of the rings remain as hydrogenated polyquinizarine structures (**XXIX**). Further heating at temperatures above 2500°C in nitrogen or argon for brief periods eliminates all elements except carbon to yield

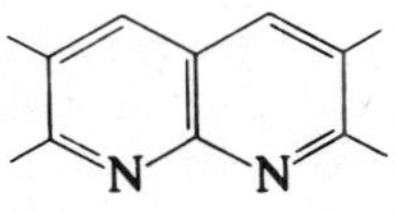

XXVIII 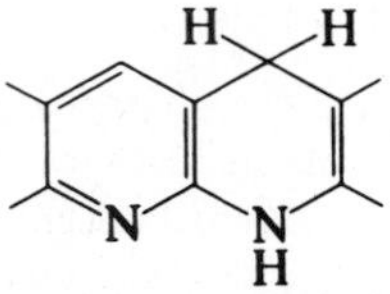   **XXIX**

carbon fibers with graphite-like morphology. The carbon fibers are subjected to a final oxidizing atmosphere such as a mixture of gaseous nitrogen and oxygen at 350–1100°C to oxidize the surface for purposes of imparting adhesive properties. The process for producing carbon fibers is expensive because of the large energy expenditures required as well as the need to treat the volatile by-products (which include toxic HCN).

Carbon fibers combine ultrahigh strength and low density when used in composite formulations with epoxy and other resins. More than 50% of carbon fiber use is in air and spacecraft applications where weight and fuel savings are achieved without sacrificing performance. A variety of components of air and spacecraft are fabricated from carbon fiber–epoxy composites. Carbon fiber–epoxy composites have been used to produce the entire aircraft structure of a business jet (Lear 2100) and the payload bay doors of the Space Shuttle. Carbon fibers combined with higher performance matrix materials such as polyimides or polybenzimidazole offer the potential for further improvements. Civilian applications include sporting goods—golf-club shafts, tennis racquets, fishing rods, and boat masts. The lighter weight of the carbon fiber composite (compared to the metal component it replaces) allows a redesign of the item for improved performance. The lighter golf-club shaft allows a heavier head design. Other applications include high-quality loudspeakers and various medical applications, including artificial limbs and surgical implants.

Other polymers undergo cyclization, but there are no commercial applications. Poly(methacrylic acid) cyclizes by anhydride formation and poly(methyl vinyl ketone) by condensation (with dehydration) between methyl and carbonyl groups.

9-8 GRAFT COPOLYMERS

Graft and block copolymers contain long sequences of each of two different monomers in the copolymer chain. A block copolymer contains the sequences of the two monomers in a continuous arrangement along the copolymer chain while a graft copolymer contains a long sequence of one monomer (often referred to as the *backbone polymer*) with one or more branches (grafts) of long sequences of a second monomer (Secs. 2-13a, 6-1b). A variety of techniques for synthesizing such structures have been studied. Some of the more useful or promising techniques for producing graft copolymers will be reviewed in this section; block copolymers are discussed in Sec. 9-9.

Most methods of synthesizing graft copolymers involve radical polymerization, although ionic graft polymerizations are receiving increasing attention [Battaerd and Tregear, 1967; Ceresa, 1973, 1976; Dreyfuss and Quirk, 1986; Rempp and Lutz, 1989]. Graft polymerization systems are carried out in heterogeneous as well as homogeneous systems.

9-8a Radical Graft Polymerization

9-8a-1 Chain Transfer and Copolymerization

The radical polymerization of a monomer, in the presence of a dissolved polymer, by thermal decomposition of an initiator, results in a mixture of homopolymerization and graft polymerization. Consider the polymerization of styrene in the presence of 1,4-poly-1,3-butadiene [Brydon et al., 1973, 1974; Ludwico and Rosen, 1975, 1976]. Polymer radicals (**XXX**), formed by chain transfer between the propagating radical and polymer, initiate graft polymerization of styrene. The product (**XXXI**) consists of polystyrene grafts on the 1,4-poly-1,3-butadiene backbone. Polymer radicals are

$$\sim\!\!\sim\!CH_2CH\!\!=\!\!CHCH_2 \sim\!\!\sim + \sim\!\!\sim CH_2\dot{C}H\phi \longrightarrow$$

$$\sim\!\!\sim\!\dot{C}HCH\!\!=\!\!CHCH_2 \sim\!\!\sim + \sim\!\!\sim CH_2CH_2\phi \qquad (9\text{-}43)$$

$$\text{XXX}$$

$$\sim\!\!\sim\!\dot{C}HCH\!\!=\!\!CHCH_2 \sim\!\!\sim + n CH_2\!\!=\!\!CH\phi \longrightarrow \sim\!\!\sim\!CHCH\!\!=\!\!CHCH_2 \sim\!\!\sim$$
$$\overset{|}{(CH_2CH\phi)_n}\!\!\sim\!\!\sim \qquad (9\text{-}44)$$

$$\text{XXXI}$$

also formed by attack on polymer by primary radicals from the initiator. Although chain transfer usually involves hydrogen abstraction, some polymers contain more labile atoms; for instance, chlorine atom abstraction occurs with chlorinated rubber [Kaleem et al., 1979].

For polymers containing double bonds as in 1,4-poly-1,3-butadiene, graft polymerization also involves copolymerization between the polymerizing monomer and the double bonds of the polymer in addition to grafting initiated by chain transfer. The

$$\sim\!\!\sim\!CH_2CH\!\!=\!\!CHCH_2\!\!\sim\!\!\sim + \sim\!\!\sim CH_2\dot{C}H\phi \longrightarrow \sim\!\!\sim\!CH_2\dot{C}H\!-\!CHCH_2\!\!\sim\!\!\sim$$
$$\overset{|}{CH\phi}$$
$$\overset{|}{CH_2}\!\!\sim\!\!\sim$$

$$(9\text{-}45)$$

relative amounts of the two processes depend on the identity of the double bond. Grafting via chain transfer predominates for 1,4-poly-1,3-dienes containing relatively unreactive 1,2-disubstituted double bonds. Grafting via copolymerization of the double bond predominates when the double bond is more reactive, such as polybutadienes with high contents of vinyl groups (produced by 1,2-polymerization).

Although this method yields a mixture of homopolymer and graft copolymer, and probably also ungrafted backbone polymer, some of the systems have commercial utility. These are high-impact polystyrene (HIPS) [styrene polymerized in the presence of poly(1,3-butadiene)], ABS and MBS [styrene–acrylonitrile and methyl methacrylate–styrene, respectively, copolymerized in the presence of either poly(1,3-butadiene) or SBR] (Sec. 6-8a).

9-8a-2 Ionizing Radiation

Polymer radicals can also be produced by the irradiation of a polymer–monomer mixture with ionizing radiation. Thus, the interaction of ionizing radiation with polyethylene–styrene produces radical centers on polyethylene, and these initiate graft polymerization of styrene to produce poly(ethylene–*graft*-styrene) [Rabie and Odian, 1977].

Most radiation graft polymerizations are carried out as heterogeneous reactions. The polymer is swollen by monomer but does not dissolve in the monomer. (For

$$\sim\sim CH_2CH_2\sim\sim \xrightarrow{\text{radiation}} \sim\sim CH_2\dot{C}H\sim\sim + H\cdot \tag{9-46}$$

$$\sim\sim CH_2\dot{C}H\sim\sim + nCH_2{=}CH\phi \rightarrow \sim\sim CH_2CH\sim\sim \tag{9-47}$$
$$| $$
$$(CH_2CH\phi)_n\sim\sim$$

semicrystalline polymers, swelling and grafting take place only in the amorphous regions.) The typical reaction system involves equilibration of polymer with monomer followed by irradiation of the monomer-swollen polymer while immersed in excess monomer. Whether graft polymerization occurs uniformly throughout the volume of the polymer or mainly at its surfaces depends on the rate of monomer diffusion into the polymer relative to the grafting rate [Imre and Odian, 1979; Odian et al., 1980]. For reaction systems with slow diffusion rates and/or fast grafting rates, the overall process becomes diffusion-controlled, with the observed grafting rate limited by the rate of diffusion of monomer into polymer. Anomolous kinetic behavior is observed under these reaction conditions; for example, the reaction rate becomes greater than first-order in monomer, less than half-order in initiation rate, and inversely dependent on the thickness of the polymer sample.

Most graft polymerizations, irrespective of the initiation process, yield mixtures of the graft copolymer, ungrafted backbone polymer, and homopolymer of the monomer. The relative amounts of the three species depend on the monomer–polymer combination and the initiation process. For initiation by chain transfer, the efficiency of the grafting reaction relative to homopolymerization is dependent on the tendency of a propagating radical to transfer to polymer compared to propagation. A consideration of C_P values (Sec. 3-6d) indicates that the extent of homopolymerization will be very significant during graft polymerization. The grafting efficiency of the irradiation process is often discussed in terms of the relative extents of radical formation in the polymer and monomer. Combinations of a polymer such as polyethylene (relatively susceptible to radiolytic bond cleavage) and a monomer such as styrene (relatively resistant to radiolytic bond cleavage) are generally considered to involve minimal homopolymerization relative to graft polymerization. However, this conclusion is somewhat naive as it assumes that hydrogen atoms formed in the radiolytic initiation step (Eq. 9-46) will mostly form molecular hydrogen. This is unlikely in the presence of monomer, which should scavenge at least some of the hydrogen radicals. The extent of homopolymerization equals that of graft polymerization if all hydrogen radicals initiate homopolymerization [Machi and Silverman, 1968].

The range of radiation grafting systems studied is enormous; almost all of the more common monomers such as styrene, methyl methacrylate, vinyl chloride, and acrylonitrile have been grafted to most of the commercially available addition and condensation polymers [Chapiro, 1962; Wilson, 1974]. One commercial application of radiation processing is the curing of wood impregnated with monomers for high-performance commercial flooring. The product involves a combination of homopolymerization and graft polymerization of a monomer such as methyl methacrylate to wood.

Graft polymerization can also be achieved by irradiation with ultraviolet radiation, often in the presence of a photosensitizer such as benzophenone or benzoin [Guthrie et al., 1979; Tazuke and Kimura, 1978]. Photolytic grafting is similar to radiation grafting, except that the depth of penetration by UV is far less than by ionizing radiation.

9-8a-3 Redox Initiation

Redox initiation is often an efficient method for graft polymerization. Hydroxyl-containing polymers such as cellulose and poly(vinyl alcohol) undergo redox reaction with ceric ion or other oxidizing agents to form polymer radicals capable of initiating

$$\sim\sim CH_2CH\sim\sim + Ce^{4+} \rightarrow \sim\sim CH_2\dot{C}\sim\sim + H^+ + Ce^{3+} \qquad (9\text{-}48)$$
$$\quad\quad |\qquad\qquad\qquad\qquad\qquad\quad |$$
$$\quad\quad OH\qquad\qquad\qquad\qquad\qquad OH$$

polymerization [Graczyk and Hornof, 1988; Leza et al., 1990; Odian and Kho, 1970; Storey and Goff, 1989]. Redox initiation usually results in grafting with a minimum of homopolymerization since only the polymer radical is formed. It is, however, limited to polymers containing the necessary functional group. There are some reports of polymers with amide, urethane, and nitrile groups undergoing redox initiation [Feng et al., 1985; Lee et al., 1976; Nayak et al., 1979a, 1979b; Sengupta and Palit, 1978].

9-8b Anionic Graft Polymerization

Metallation of a polymer by treatment with strong base; for instance, the reaction of 1,4-poly(1,3-butadiene) with t-butyllithium or a polyamide (or polyurethane) with

$$\sim\sim CH_2CH{=}CHCH_2\sim\sim \xrightarrow{\text{BuLi}} \sim\sim\overset{=}{C}HCH{=}CHCH_2\sim\sim \qquad (9\text{-}49)$$

sodium yields polymeric anions which initiate the graft polymerization of monomers

$$\sim\sim NH{-}CO\sim\sim \xrightarrow{\text{Na}} \sim\sim\overset{=}{N}{-}CO\sim\sim \qquad (9\text{-}50)$$

such as styrene, acrylonitrile, and ethylene oxide [Adibi et al., 1979, 1981; Hadjichristidis and Roovers, 1978; Kashani et al., 1978; Takayangi and Katayose, 1983].

Polymers containing carboxylate anion groups (COOH and COOR groups are precursors) have been used to initiate graft polymerization [Sundet, 1978]. Carboxylate

$$\sim\!\!\!\sim\!\!\!\sim \atop COO^- \quad + \quad \underset{O}{\overset{}{\square}}\!-\!\!\!\underset{O}{\overset{}{\diagdown}} \quad \rightarrow \quad \sim\!\!\!\sim\!\!\!\sim \atop COO(CH_2CH_2COO)_n\!\!\sim\!\!\sim \qquad (9\text{-}51)$$

anions can be introduced into polymers for the specific purpose of introducing initiation sites for graft polymerization. Copolymerization of i-butylene with a small amount of p-methylstyrene is an example. The copolymer is metallated at the benzyl hydrogen, followed by reaction with carbon dioxide and base to form the carboxylate anion, which initiates anionic polymerization of monomers such as lactones [Harris and Sharkey, 1986].

Coupling of polymers to form graft copolymers is accomplished by nucleophilic reaction between living polystyryl carbanion and various chlorine-containing polymers

$$\sim\sim CH_2CH\sim\sim + \sim\sim CH_2\overset{=}{C}H\phi \rightarrow \sim\sim CH_2CH\sim\sim \qquad (9\text{-}52)$$
$$\quad\quad |\qquad\qquad\qquad\qquad\qquad\qquad\qquad\qquad |$$
$$\quad\quad Cl\qquad\qquad\qquad\qquad\qquad\qquad\qquad CH\phi CH_2\sim\sim$$

such as poly(vinyl chloride) [Kucera, 1983, 1985; Majid et al., 1982]. Elimination is a side reaction in this process.

9-8c Cationic Graft Polymerization

Polymeric carbocations have been formed from a chlorine-containing polymer such as chlorinated SBR and polystyrene, poly(vinyl chloride), and polychloroprene, by reaction with $(C_2H_5)_2AlCl$ or $AgClO_4$, for example

$$\sim\!\sim\!CH_2CH\!\sim\!\sim + (C_2H_5)_2AlCl \rightarrow \sim\!\sim\!CH_2\overset{+}{C}H\!\sim\!\sim \tag{9-53}$$
$$\underset{Cl}{|} \qquad\qquad\qquad\qquad (C_2H_5)_2AlCl_2^-$$

The polymeric carbocations can initiate polymerization of i-butylene, tetrahydrofuran, and other monomers [Cai and Yan, 1987; Cameron and Sarmouk, 1990; Kennedy and Marechal, 1982].

A hydroxy-containing polymer can be coupled with living polytetrahydrofuran carbocation [Cameron and Duncan, 1983].

$$\boxed{P}\!-\!OH + \left[\bigcirc\!\overset{+}{O}\!-\!\!\{(CH_2)_4O\}_n\!\sim\!\sim \longrightarrow \boxed{P}\!-\!O\!-\!\{(CH_2)_4O\}_{(n+1)}\!\sim\!\sim \tag{9-54}$$

9-8d Other Approaches to Graft Copolymers

Nonionic coupling reactions between appropriate polymers is another approach to graft copolymers. An example is the use of a telechelic polymer such as polystyrene containing one carboxyl end group per molecule. Graft copolymers are obtained by coupling of carboxyl groups with the pendant epoxy groups of a copolymer of methyl methacrylate and glycidyl methacrylate [Miyauchi et al., 1988]. The carboxyl-terminated polystyrene can be obtained by radical polymerization of styrene in the presence of a chain-transfer agent such as 3-mercaptopropionic acid.

A variation of this approach is the synthesis of telechelic polymers containing two carboxyl groups at one end of a polymer (**XXXII**) by using a chain-transfer agent such as 2-mercaptosuccinic acid

$$(CH_2CH)_n^{\cdot} + HSCHCOOH \rightarrow (CH_2CH)_n SCHCOOH \tag{9-55}$$
$$\underset{\phi}{|} \quad \underset{CH_2COOH}{|} \qquad \underset{\phi}{|} \quad \underset{CH_2COOH}{|}$$

$$\textbf{XXXII}$$

Telechelic polymers such as **XXXII**, which contain more than one functional group at one end of the molecule, are referred to as *macromonomers* since they are capable of participating in polymerization reactions. Polymer **XXXII** is used as a reactant in a step polymerization with a diol (or diamine) and diacid to yield a polyester (or polyamide) containing branches of polystyrene. The typical situation involves macromonomers that are in the 1000–2000-molecular-weight range so that one obtains a graft copolymer with relative short branches. The graft copolymer is referred to as a *comb polymer*.

A macromonomer containing a carbon–carbon double bond at one chain end can be used to synthesize comb polymers by copolymerization with monomers such as styrene or methyl methacrylate. Such macromonomers are obtained by reaction of a hydroxy-containing polymer or a living anionic polymer with acroyl chloride or reaction of a living cationic polymer with sodium p-vinylphenolate [Asami et al., 1983; Gnanou and Rempp, 1987; Schulz and Milkovich, 1982]. Alternately, one can initiate polymerization by using an initiator containing the double bond [Sierra-Vargas et al., 1980], for example,

$$CH_2{=}CH{-}CO{-}Cl + AgSbF_6 \rightarrow CH_2{=}CH{-}CO^+(SbF_6)^-$$

$$\downarrow THF$$

$$CH_2{=}CH{-}CO{-}(THF)_n \qquad (9\text{-}56)$$

9-9 BLOCK COPOLYMERS

Most of the methods of synthesizing block copolymers have been described previously. Block copolymers can be obtained by step copolymerization of polymers with functional end groups capable of reacting with each other (Sec. 2-13c-2). Sequential addition methods by anionic, group transfer, and cationic propagation have been described in Secs. 5-4a and 7-12c. The use of telechelic polymers, coupling reactions, and transformation reactions were described in Secs. 5-4b, 5-4c, and 5-4d. The few methods not previously discussed are considered here.

Block copolymers have been obtained from polymeric radicals, which—unlike those employed for graft polymerization—have the radical centers at the ends of the polymer chains. This can be achieved by breaking chemical bonds in the polymer backbone by mastication (mixing) of the polymer [Ceresa, 1973, 1976, 1978; Sakaguchi and Sohma, 1978]. If the shearing forces are sufficiently high during mastication, their concentration at individual bonds in the polymer chain results in bond rupture. Block copolymers are obtained by the mastication of either a mixture of two homopolymers

$$\left.\begin{array}{c} \sim\!\!\sim\!M_1M_1\!\sim\!\!\sim \\ \\ \sim\!\!\sim\!M_2M_2\!\sim\!\!\sim \end{array}\right\} \xrightarrow{\text{milling}} \begin{array}{c} \sim\!\!\sim\!M_1{\cdot} \\ \\ \sim\!\!\sim\!M_2{\cdot} \end{array} \left.\begin{array}{c} \\ \\ \end{array}\right\} \longrightarrow \begin{array}{c} \sim\!\!\sim\!M_1M_2\!\sim\!\!\sim \\ \sim\!\!\sim\!M_1M_1\!\sim\!\!\sim \\ \sim\!\!\sim\!M_2M_2\!\sim\!\!\sim \end{array} \qquad (9\text{-}57)$$

or a mixture of a polymer and a monomer

$$\sim\!\!\sim\!M_1M_1\!\sim\!\!\sim \xrightarrow{\text{milling}} \sim\!\!\sim\!M_1{\cdot} \xrightarrow{M_2} \sim\!\!\sim\!M_1M_2\!\sim\!\!\sim \qquad (9\text{-}58)$$

The former gives a mixture of the block copolymer with the two homopolymers since the polymer radicals combine randomly. The latter yields the block copolymer along with the homopolymer of M_1.

The synthesis of special initiators offers possibilities for block copolymer synthesis. For example, reaction of **XXXIII** with a hydroxyl-terminated

$$\overset{\displaystyle CH_3 \qquad\quad CH_3}{\underset{\displaystyle CN \qquad\quad CN}{Cl{-}CO(CH_2)_2\overset{|}{\underset{|}{C}}{-}N{=}N{-}\overset{|}{\underset{|}{C}}(CH_2)_2CO{-}Cl}}$$

XXXIII

polyester or polyether yields a polymeric azonitrile, which initiates polymerization of an alkene [Laverty and Gardlund, 1977; Walz and Heitz, 1978].

9-10 POLYMERS AS CARRIERS OR SUPPORTS

The use of a polymer as a carrier or support for some component of a reaction system is discussed in the remainder of this chapter. Three classes of polymer supports are encountered: *polymer reagents*, *polymer catalysts*, and *polymer substrates*. Polymers that serve as reagents, catalysts, or substrates are *functional polymers*; that is, they contain a functional group that performs a particular chemical function. The term *reactive polymer* is also used. A polymer reagent is a polymer containing a functional group that acts as the reagent to bring about a chemical transformation on some small (i.e., low-molecular-weight) molecule. A polymer substrate has an attached molecule on which some transformation is carried out using a small molecule reagent. A polymer catalyst contains a group (or groups) that performs a catalytic function in some reaction—usually a reaction between small molecules. This area of polymer science is one of considerable activity.

9-10a Synthesis

There are two general approaches to the synthesis of a polymer reagent, catalyst, or substrate [Hodge, 1978; Mathur et al., 1980; Neckers, 1978; Sherrington, 1980, 1988]. In one approach the required chemical group, which is to perform the reagent, catalyst, or substrate function, is attached to a readily available polymer by an appropriate reaction. The attachment is usually through a covalent bond, but ionic bonding has also been employed. The second approach to obtaining a functional polymer is to synthesize an appropriate monomer with the desired group and then polymerize or copolymerize that monomer.

9-10a-1 *Attachment of Group to Polymer*

This approach can be illustrated by describing the preparation of the polymer rhodium catalyst **II** (Sec. 9-1g). The synthesis is based on a nucleophilic substitution reaction of chloromethylated polystyrene [Grubbs and Kroll, 1971]

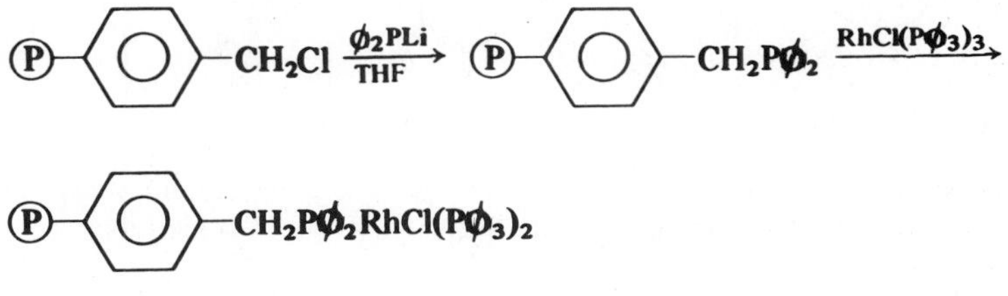

$$\tag{9-59}$$

II

The chloromethylated polystyrene is usually obtained by chloromethylation of polystyrene (Eq. 9-38), although polymerization of p-vinyl benzyl chloride has also been used [Arshady et al., 1976]. A less desirable variation of this approach is the physical entrapment of a catalyst, substrate, or reagent within a polymer.

There are a number of considerations in the choice of the polymer to be used as a support—availability and cost, mechanical, thermal, and chemical stability, porosity, and compatibility with reagents and solvents. The use of a commercially available, inexpensive polymer is clearly preferable to a more expensive polymer or one that must be specifically synthesized for the purpose. The support polymer must be chemically and thermally inert under the conditions necessary for its functionalization and subsequent use. Heterogeneous polymer catalysts, reagents, and substrates are generally more advantageous than homogeneous ones (Sec. 9-10a-4), and this requires the support to possess a reasonable degree of mechanical strength. Some degree of crosslinking is usually employed for this purpose. The physical form of the functionalized polymer is also a consideration in terms of its application. The polymer support as well as the functionalized polymer should be porous to allow access of the reagent(s) and solvent(s), if any, to achieve a homogeneously active polymer. A polymer reagent, catalyst, or substrate with only surface activity would result if the polymer were nonporous. Compatibility of the functionalized polymer (more specifically, the regions in the vicinity of the functional groups) with reagents and solvents is also needed to achieve high reaction rates and high yields.

Polystyrene is presently the overwhelming choice as the polymer support [Frechet and Farrall, 1977; Frechet et al., 1988; Messing, 1974]. The polystyrene used is a crosslinked polymer prepared by copolymerization of styrene with divinylbenzene (DVB). (The divinylbenzene is usually the commercially available mixture containing the *m*- and *p*-isomers in a ratio of about 2:1.) Crosslinked polystyrene fulfills many of the requirements of a support—it is an inexpensive, readily available polymer with good mechanical, chemical, and thermal stability. Very importantly, polystyrene can be functionalized by many routes, such as chloromethylation, lithiation, carboxylation, acylation, and sulfonation. The chloromethyl and lithio derivatives are the most useful. The two complement each other by reacting with nucleophilic and electrophilic reagents, respectively, to yield a wide range of functionalized polystyrenes. For example, ammonium, carboxylate, aldehyde, thiol, cyanide, and diphenyl phosphine groups can be introduced by reacting chloromethylated polystyrene (Sec. 9-6) with an amine, RCOOK, NaHCO$_3$, RSK, KCN, and LiPϕ_2, respectively. Lithiated polystyrene can be obtained by direct lithiation of polystyrene with *n*-butyllithium in the presence of tetramethylethylenediamine or indirectly by reaction of brominated polystyrene with *n*-butyllithium

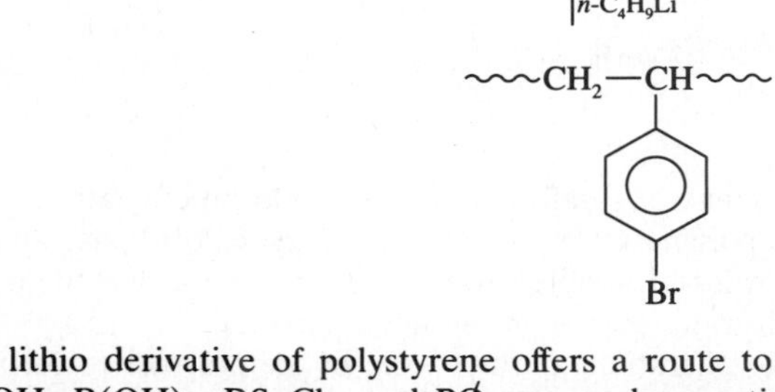

$$(9\text{-}60)$$

The lithio derivative of polystyrene offers a route to polystyrenes containing OH, COOH, B(OH)$_2$, RSnCl$_2$, and Pϕ_2 groups by reaction with ethylene oxide, CO$_2$, B(OR)$_3$, MgBr$_2$ followed by RSnCl$_3$, and ϕ_2PCl, respectively.

The usual polystyrene support contains 1–2% DVB, as this yields a mechanically strong support that is *microporous*; that is, it is highly swollen by solvents for polystyrene. Swelling is accompanied with a significant expansion in volume. Much higher degrees of crosslinking, up to and sometimes exceeding 20% DVB, yield rigid polymers that are *macroporous* or *macroreticular*. They are produced by carrying out the copolymerization of styrene–DVB in the presence of significant amounts of diluent. This keeps the polymer in an expanded form during preparation and results in the macroporous structure. The diluents used are always solvents for the monomers but may or may not be solvents for the polymer. The macroporous structure allows a macroreticular resin to take up solvent with little or no change in volume. Microporous supports offer certain advantages relative to macroreticular supports. Microporous supports are less fragile, require less care in handling, react faster in the functionalization and applications reactions, and possess higher loading capacities. Macroreticular supports are less frequently employed but do have certain advantages, including ease of removal from a reaction system (due to greater rigidity) and the lack of diffusional limitations on reaction rates (since the pore sizes are usually larger and there is no microporosity). Macroreticular supports can be used with almost any solvent irrespective of whether it is a good solvent for the uncrosslinked polymer. The behavior of microporous supports varies considerably depending on the solvent used. Solvents for polystyrene are optimal; examples are benzene, toluene, dioxane, tetrahydrofuran, chloroform, and methylene chloride. Although the overwhelming majority of supports used are based on polystyrene, polystyrene has the significant limitation of not being optimally suitable for use with systems involving polar and hydrophilic reactants or solvents. Other polymers studied as supports include poly(acrylic acid), polyamides, bisphenol A-epichlorohydrin copolymer, polyacrylamide, cellulose, dextran, poly(glycidyl methacrylate), and polymaleimide copolymers. Inorganic polymers such as silica have also been used; organic groups are attached via hydroxyl groups on the inorganic surface.

9-10a-2 *Polymerization of a Functional Monomer*

The approach of synthesizing a monomer containing the desired functional group followed by polymerization or copolymerization can be illustrated for the poly[4(5)-vinylimidazole] catalyst (**VII**) described in Sec. 9-1j. Synthesis involves the sequence of reactions starting from histidine (**XXXIV**) to yield 4(5)-vinylimidazole (**XXXV**),

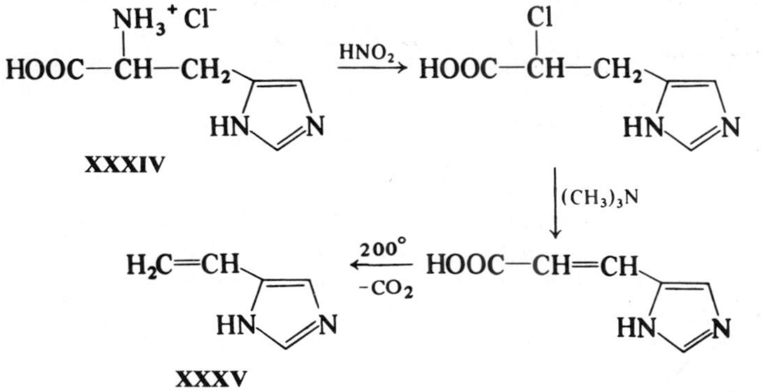

$$(9\text{-}61)$$

which is subsequently polymerized by radical initiation [Overberger and Vorchheimer, 1963].

The key to the utility of this approach is whether the required functional group can be incorporated into a monomer that undergoes polymerization. Two methods have been used. One is the synthesis of a unique monomer such as 4(5)-vinylimidazole that contains the required functional group with an attached polymerizable linkage. The second method involves the incorporation of the required group as part of one of the more common types of monomers. The most successfully employed have been functional derivatives of acrylic and methacrylic esters and styrene where the functional group is part of the ester group and aromatic ring, respectively. Monomers suitable for step polymerizations have been much less frequently employed.

9-10a-3 Comparison of the Two Approaches

One cannot generalize that either approach is more useful than the other even though the functionalization of existing polymers (especially polystyrene) is the most often used approach. This may only reflect the present status of a relatively immature area of polymer chemistry. The two approaches should be considered as complementary. The functionalization of a polymer may be more advantageous for a particular system due to the availability of the appropriate polymer and the ease of accomplishing the required functionalization reaction in high yield with a minimum of side reactions. It may be completely unsuitable for a system where the appropriate polymer is not readily available and/or the required functionalization reaction does not proceed cleanly to high yield. The polymerization of a functional monomer will be advantageous for a system if synthesis of the required monomer can be accomplished in high yield and purity and polymerization or copolymerization proceeds to yield a high polymer of the required mechanical strength with good thermal and chemical resistance. The approach may be impractical if either monomer synthesis or polymer formation does not proceed satisfactorily. For any specific polymer reagent, catalyst, or substrate, one approach may be more suitable than the other.

9-10b Advantages of Polymer Reagents, Catalysts, and Substrates

Polymer reagents, catalysts, and substrates have practical advantages compared to their small molecule analogs. An insoluble polymer reagent, catalyst, or substrate can be easily separated from the other (i.e., the small molecule) components of a reaction system by filtration. This ease of separation allows the synthesis of high-purity products and the recovery of the polymeric species, with resulting economies. Reactions with functional polymers are sometimes advantageously carried out in a manner similar to column chromatography. The polymeric species is packed in a column and the small molecule species poured through the column. Some of the more specific details are described in Secs. 9-11 through 9-13. A soluble polymer reagent, catalyst, or substrate also offers the advantage of ease of separation since it can be selectively precipitated by an appropriate nonsolvent and then filtered off.

When the polymer reagent, catalyst, or substrate is easily recoverable, it can be economically used in large excess to achieve high yields in a reaction. Small molecule reagents, catalysts, or substrates that are highly reactive, toxic, or malodorous can be handled much more safely and easily in the form of the corresponding polymers. Other advantages of functional polymers may arise from the enhancement of rate or equilibrium due to factors discussed in Sec. 9-1. These factors may also lead to negative effects in some systems.

There are disadvantages that limit the commercial utilization of polymer reagents, catalysts, and substrates at this time. Functional polymers are more expensive than their small molecule analogs. Many applications that work well in the laboratory fail the economics of industrial use. One needs large-scale uses where one or another of the advantages of functional polymers cannot be achieved by other means. Filtration of a polymer reagent, catalyst, or substrate is often not easy. Fine particles slough off from larger beads or ground resin because of the stresses resulting from mechanical stirring or shaking. These "fines" interfere with filtration by clogging the filters. For a few functional polymers, the functional group(s) of importance may not be permanently anchored; in other words, the functionalization reaction is reversible. This is the case with some polymer catalysts containing transition-metal groups. The transition-metal components leach out with time. Organic supports are of lesser stability than inorganic supports.

9-11   POLYMER REAGENTS

A wide range of polymer reagents have been studied [Akelah and Sherrington, 1983; Blossey and Ford, 1989; Ford, 1986a, 1986b]. The epoxidation of an alkene by a polymer peracid illustrates the use of a polymer reagent [Frechet and Hague, 1975]. Chloromethylated polystyrene is treated with potassium bicarbonate in dimethylsulfoxide to yield the formylated derivative (**XXXVI**), which is oxidized to the peracid (**XXXVII**)

$$\boxed{P}-\phi-CH_2Cl \xrightarrow{KHCO_3} \boxed{P}-\phi-CHO \xrightarrow{H_2O_2,H^+} \boxed{P}-\phi-CO_3H \qquad (9\text{-}62)$$
$$\qquad\qquad\qquad\qquad\quad \mathbf{XXXVI} \qquad\qquad\qquad \mathbf{XXXVII}$$

The polymer peracid is used to epoxidize an alkene by the scheme shown in Fig. 9-1. A mixture of the polymer reagent (peracid) and low-molecular-weight substrate (alkene), often with a solvent for the latter, are allowed to react and then filtered. Solvent is evaporated from the filtrate to yield the low-molecular-weight product (epoxide); purification steps such as distillation may also be involved. From the economic viewpoint, it is desirable if the polymer reagent can be regenerated from the polymer by-product (polymer acid). This is easily accomplished in the particular example as the polymer acid yields the peracid on treatment with hydrogen peroxide. It is not possible to obtain the economy of polymer regeneration in all polymer reagent systems. Polymer peracids have also been used for the oxidation of sulfides to sulfoxides and sulfones [Grieg et al., 1980; Harrison and Hodge, 1976].

Oxidation of alcohols to aldehydes or ketones is accomplished with N-haloamides, such as poly(N-bromoacrylamide) and N-chloronylon-66 (**XXXVIII**), which act as

$$\sim\sim\sim CO(CH_2)_4CON(CH_2)_6N\sim\sim\sim$$
$$\qquad\qquad\quad | \qquad\qquad |$$
$$\qquad\qquad\quad Cl \qquad\quad Cl$$
$$\qquad\qquad\qquad \mathbf{XXXVIII}$$

$$\sim\sim CH_2-CH\sim\sim$$
$$\qquad\qquad |$$

$$CH_2\overset{+}{N}R_3(A^-)$$
$$\qquad\mathbf{XXXIX}$$

sources of hypohalite in aqueous systems [George and Pillai, 1988; Sato et al., 1977; Schuttenberg et al., 1973]. Oxidation of alcohols can also be accomplished with chromate bound to an anion exchange resin (**XXXIX** with $A^- = HCrO_4^-$) [Frechet et al., 1981; Taylor, 1986].

Polymer reducing agents for reducing aldehydes and ketones to alcohols are ob-

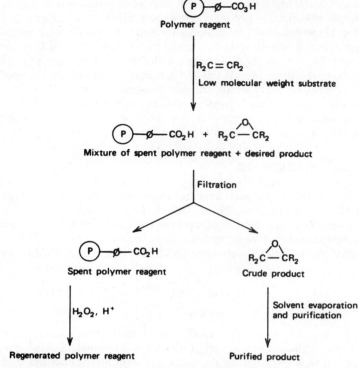

Fig. 9-1   Scheme for utilization of a polymer reagent. The epoxidation of an alkene by a polymer peracid.

tained by complexing AlH_3 or BH_3 with a nitrogen-containing polymer such as poly(4-vinylpyridine), such as **XXXX** [Boga et al., 1989; Menger et al., 1980]. Another useful

reducing reagent is **XXXIX** with $A^- = BH_4^-$, which involves ionic bonding of boron hydride anion to a quaternary ammonium polymer [Sande et al., 1984]. The use of a chiral polymer instead of the achiral polymers in **XXXIX** and **XXXX** allows the stereoselective reduction of an aldehyde or ketone, such as the reduction of aceto-

(9-63)

phenone to (R)-1-phenylethanol in 76–97% enantiomeric excess [Itsuno et al., 1985]. The symbol Ⓟ in Eq. 9-63 represents polystyrene.

Pyridinium bromide, useful for bromine addition to alkenes and α-substitution on aldehydes and ketones, is dangerous to handle. The polymer analog (**XXXXI**) is easy

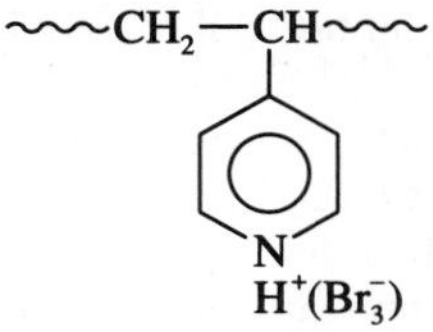

XXXXI

to handle and accomplishes the same reaction as the small molecule reagent [Frechet et al., 1977; Johar et al., 1982]. Similar halogenating reagents are obtained from **XXXIX** (e.g., $A^- = Br_3^-$, $BrCl_2^-$, ICl_2^-) [Sket and Zupan, 1984]. An ethylene-N-bromomaleimide copolymer has been used for bromine substitution at allylic and benzyl positiòns [Yaroslavsky et al., 1970].

Polymer Wittig reagents are useful for the conversion of an aldehyde or ketone to an alkene according to the sequence in Eq. 9-64 [Ford, 1986a, 1986b].

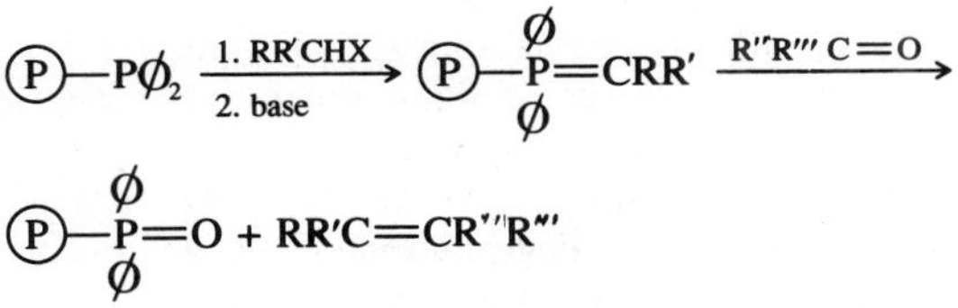

$$(9\text{-}64)$$

Several applications of functional polymers for achieving chemical separations have been described. Ion-complexing polymers allow the removal of ions from solution for analytical or preparative purposes [Akelah and Sherrington, 1983; Braun and Farag, 1974; Lundgren and Schilt, 1977]. For example, polymers containing the 8-hydroxyquinoline unit (**XXXXII**) are useful for chelating nickel, cobalt, and copper ions. Che-

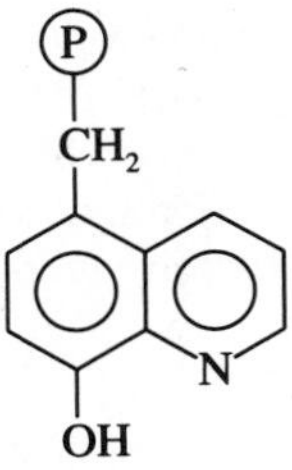

XXXXII

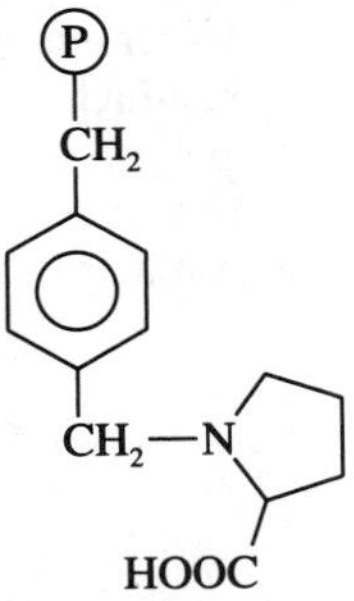

XXXXIII

lation occurs through the hydroxyl and ring nitrogen. The metal is separated by filtration of the polymer and can be recovered from the polymer by changing the pH. Polymers containing crown ether or cryptand (see structures **XXVIII** and **XXIX** in Sec. 5-3e-5) macrocyclic ligands bind and separate various cations [Jaycox and Smid,

1981; Mathias and Al-Jumar, 1980]. A variety of polymers containing chiral groups are useful for resolving racemic mixtures into the individual enantiomers [Kiniwa et al., 1987; Mathur et al., 1980; Wulff et al., 1980]. For example, the copper(II) complex of **XXXXIII** resolves racemates of amino acids [Sugden et al., 1980]. The separation is based on the formation of a pair of diastereomeric complexes from the reaction of the polymer reagent with the two enantiomers. One of the enantiomers is complexed more strongly than the other and this achieves separation of the enantiomers.

A *polymer drug* can be synthesized by the covalent bonding of a drug to a polymer or the synthesis and (co)polymerization of a monomer containing the drug moiety [Callant and Schacht, 1990; Donaruma and Vogl, 1978; Hsieh, 1988; Levenfeld et al., 1990; Ottenbrite and Sunamoto, 1986]. (Ionic bonding or complexing of a drug to the polymer are other approaches.) Polymer drugs are of interest since they offer a number of potential advantages relative to the corresponding low-molecular-weight drugs. The action of a polymer drug usually depends on *in vivo* hydrolytic or enzymatic cleavage of the drug moiety from the polymer. This gives the advantage of a delayed and sustained release of the drug over a longer time period with a corresponding decrease in side effects such as irritation, intolerance, and toxicity. There is the potential to tailor-make a polymer drug with a specific required solubility, rate of diffusion, and increased or decreased activity by appropriate choice of the polymer and drug. Other possibilities include the synthesis of polymer drugs with negligible absorptivity for situations where localized drug action is desired (e.g., skin treatment), the ability to design drugs capable of getting to a specific organ (via appropriate affinity groups) in the desired amounts at the required time, and the coupling of two or more drugs onto the same polymer to achieve enhanced or synergistic activity. The term *polymer drug* also includes the physical encapsulation of a low-molecular-weight drug by a polymer. Drug action requires the hydrolytic or enzymatic degradation of the polymer. Such materials can show many of the advantages of polymer drugs in which the drug is chemically bound to a polymer. Also of interest are polymer drugs that do not depend on release of a low-molecular-weight drug moiety for activity but possess activity as polymers.

Polymeric agricultural chemicals such as polymer herbicides, fertilizers, and insecticides have also been studied [Lohmann and d'Hondt, 1987; Scher, 1977; Shambhu et al., 1976]. Controlled and slow release of the herbicide, fertilizer, or insecticide moiety has the advantages of prolonged action and decreased indiscriminant pollution of the environment. Other polymer reagents of interest include polymer antioxidants, flame retardants, and UV stabilizers as additives to polymers [Scott, 1987, 1989; Vogl et al., 1985]. The advantage of the polymeric additive is decreased migration or leaching from the polymer compared to the corresponding low-molecular-weight additive.

9-12 POLYMER CATALYSTS

Some examples of polymer catalysts have been described previously—hydrogenation by the polymer rhodium catalyst (**II**) (Sec. 9-1g) and ester hydrolysis by poly[4(5)-vinylimidazole] (**VII**) (Sec. 9-1j). Many other types of polymer catalysts have been described [Akelah and Sherrington, 1983; Blossey and Ford, 1989; Hodge and Sherrington, 1980; Mathur et al., 1980]. The physical method of carrying out a reaction using a polymer catalyst is the same as the scheme described in Fig. 9-1, with the exception that two low-molecular-weight substances, instead of only one, are usually involved. Many of the general advantages of polymer catalysts have been described (Sec. 9-10b). Polymer catalysts are often easier to handle, less toxic, and more stable toward moisture and atmospheric contaminants. The products of reactions catalyzed by low-molecular-weight catalysts, especially metal catalysts, are often contaminated by the catalysts. Catalyst removal can be difficult in such systems but necessary if the properties of the products are adversely affected by the presence of catalyst. Catalyst removal is much simpler with the use of a polymer catalyst. Further, the recovery and regeneration of an expensive catalyst is facilitated since the polymer catalyst can be separated from the reaction mixture by filtration.

Various polymer acids are used as polymer catalysts. Sulfonated polystyrene (Eq. 9-39) has been used to catalyze a variety of acid-catalyzed reactions, including acetal and ketal formation, esterification, hydrolysis of amides and esters, aliphatic and aromatic alkylations, hydration of alkynes and alcohols, and the synthesis of bisphenol A from phenol and acetone [Delmas and Gaset, 1980; Hasegawa and Higashimura, 1981; Jerabek, 1980; Moxley and Gates, 1981]. A large-scale commercial application is the acid-catalyzed addition of methanol to i-butylene to form methyl t-butyl ether, which is used as a gasoline additive. Poly(acrylic acid) and the hydrochloride of poly(4-vinylpyridine) have also been used as polymer acid catalysts [Yosida et al., 1981]. Various *polymer superacids* have been described. Sulfonated polystyrene complexed with $AlCl_3$ catalyzes the cracking and isomerization of n-hexane. Perfluorosulfonic acid copolymers (trade name: *Nafion*) (Sec. 6-8e) catalyze the addition of alcohols to ethylene oxide and the alkylation of aromatics [Olah et al., 1980, 1981; Waller, 1986]. Base-catalyzed reactions such as dehydrohalogenation, ester hydrolysis, and condensation of active methylene compounds with aldehydes and ketones have been carried out using polymer bases such as poly(4-vinylpyridine), poly(2-dimethylaminoethyl acrylate), and poly(p-styrylmethyltrimethylammonium hydroxide) (**XXXIX** with $A^- = OH^-$) [Bao and Pitt, 1990; Chiellini et al., 1981].

Various transition-metal catalysts, including those based on Rh, Pt, Pd, Co, and Ti, have been bound to polymer supports—mainly through the phosphenation reaction described by Eq. 9-59 for polystyrene but also including other polymers, such as silica and cellulose, and also through other reactions (e.g., alkylation of titanocene by chloromethylated polystyrene). Transition-metal polymer catalysts have been studied in hydrogenation, hydroformylation, and hydrosilation reactions [Chauvin et al., 1977; Mathur et al., 1980].

The covalent bonding of photosensitizer moieties such as benzophenone, rose bengal, and eosin to polystyrene yields polymeric photosensitizers that can be used to bring about excitation and subsequent chemical reaction in low-molecular-weight substrates [Neckers, 1986; Nishikubo et al., 1989].

The esterolytic activity of poly[4(5)-vinylimidazole] and its copolymers has been studied in an attempt to understand and mimic the activity of the naturally occurring enzyme α-chymotrypsin [Imanishi, 1979; Kunitake, 1980; Overberger and Guterl, 1978, 1979]. The activity of α-chymotrypsin is generally ascribed to the imidazole and OH groups of the histidine and serine residues, respectively. Copolymers of 4(5)-vinylimidazole containing OH groups (from phenol moieties) were found to have increased esterolytic activity relative to poly[4(5)-vinylimidazole] or polymers containing only hydroxyl functions. The catalysis by the copolymers is ascribed to cooperative involvement of hydroxyl and imidazole groups. Among the proposed mechanisms are an acylation of the ester by the hydroxyl group followed by the imidazole moiety assisting the attack of water on the acylated intermediate (**XXXXV**)

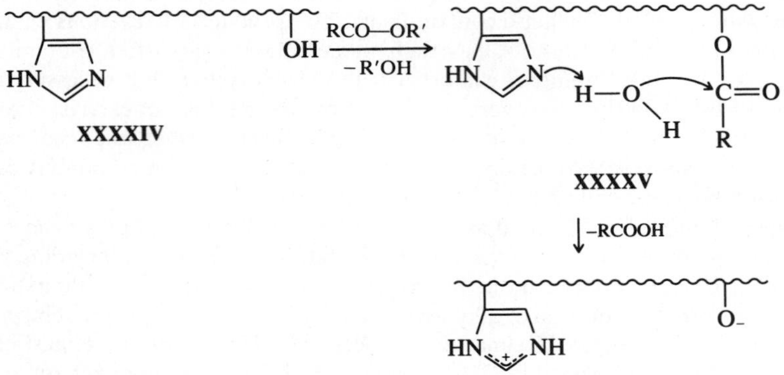

$$(9\text{-}65)$$

A catalyst such as **XXXXIV** is referred to as a *bifunctional polymer catalyst*—two different catalyst moieties function simultaneously in the same reaction.

Polymer phase-transfer catalysts (also referred to as *triphase catalysts*) are useful in bringing about reaction between a water-soluble reactant and a water-insoluble reactant [Akelah and Sherrington, 1983; Ford and Tomoi, 1984; Regen, 1979; Tomoi and Ford, 1988]. Polymer phase transfer catalysts (usually insoluble) act as the meeting place for two immiscible reactants. For example, the reaction between sodium cyanide (aqueous phase) and 1-bromooctane (organic phase) proceeds at an accelerated rate in the presence of polymeric quaternary ammonium salts such as **XXXIX** [Regen, 1975, 1976]. Besides the ammonium salts, polymeric phosphonium salts, crown ethers and cryptates, poly(ethylene oxide), and quaternized polyethylenimine have been studied as phase-transfer catalysts [Hirao et al., 1978; Ishiwatari et al., 1980; Molinari et al., 1977; Tundo, 1978].

An *immobilized enzyme* is obtained by insolubilizing an enzyme on a polymer support [Chibata, 1978; Manecke and Vogt, 1978; Wingard et al., 1979]. (Enzymes are naturally occurring polypeptides that catalyze biochemical reactions.) The naturally occurring polysaccharide agarose, an alternating copolymer of D-galactose and 3,6-anhydro-L-galactose, is the most commonly used support. Cellulose, crosslinked dextran, polyacrylamide, and microporous glass are also used. For *in vitro* use, immobilized enzymes have the advantage of increased stability to temperature, pH, and other environmental conditions compared to the native enzymes. Immobilized enzymes are of interest for carrying out industrial biochemical and organic synthesis because of their generally high efficiency and specificity. The largest industrial application is the production of fructose corn syrup from corn by an immobilized form of the enzyme glucose isomerase. The annual production exceeded 8 billion pounds in 1984 [Keyes and Albert, 1986]. Immobilization can be achieved in many instances by covalent bonding of the enzyme to the polymer support. Enzymes containing amine groups can be immobilized by reaction with polymer supports containing hydroxyl groups (e.g., polysaccharide supports) through a coupling reaction with cyanogen bromide, carbonyldiimidazole, or *p*-nitrophenyl chloroformate. Immobilization by physical adsorption on the polymer support is also possible. Another approach is the polymerization of a mixture of a monomer and the enzyme. Polymerization of the monomer results in entrapment of the enzyme within the polymer matrix. The monomer used is often one that undergoes crosslinking during the process, such as by using a mixture of acrylamide and methylene bisacrylamide [Conlon and Walt, 1986]. This approach yields the equivalent of a semiinterpenetrating network (Secs. 2-13c-3 and 6-6c). Immobilization of whole cells has also received attention since it has the advantage of not requiring the separation and purification of the active enzyme from the rest of the cell in order to produce the immobilized enzyme. Also, in some instances the active enzyme is active only in the presence of other species (coenzymes, cofactors). The use of whole cells avoids the need to establish the identity of the other required species followed by their separation and incorporation in the immobilized enzyme.

Affinity chromatography, a chromatographic method of separation and purification, is an application of immobilized enzymes [Chaiken et al., 1984; Cuatrecasas and Anfinsen, 1971; Jakoby and Wilchek, 1974]. An immobilized enzyme constitutes the adsorbent in a chromatography column. When a mixture of biochemicals, such as a mixture of proteins, is passed through the column, a specific interaction (often through covalent bonding) between the immobilized enzyme and one of the components of the mixture separates that component from the mixture. Subsequent passage of the appropriate reagent through the column reverses the interaction and frees the desired component from the column. Immobilized enzymes in the form of affinity chromatography, electrodes, and other sensing devices are of interest for analytical purposes (including immunoassays).

9-13 POLYMER SUBSTRATES

The general scheme for utilizing a polymer substrate for synthetic purposes is shown in Fig. 9-2. A substrate S is covalently bonded to an appropriate polymer support to yield the polymer substrate Ⓟ—S (which is usually of the insoluble type). The polymer substrate is treated with a low-molecular-weight reagent (and/or catalyst) R to yield the polymer-bound product Ⓟ—S' and spent reagent R'. Filtration separates the polymer-bound product. An appropriate reaction is carried out to cleave the desired product S' from the polymer. The spent polymer support is filtered off, leaving the crude product S' in the filtrate. The purified product S' is obtained from the filtrate after solvent evaporation and appropriate purification.

9-13a Solid-Phase Synthesis of Peptides

One of the most important applications of the polymer substrate technique is the *solid-phase synthesis of polypeptides* developed by Merrifield [Barany et al., 1988; Erickson and Merrifield, 1976; Kent, 1988; Merrifield, 1978, 1985, 1988; Stewart, 1980]. Merrifield received the 1984 Nobel Prize in Chemistry for this achievement. Polypeptide synthesis is of interest for two reasons. First, it is useful in confirming the structure of naturally occurring biological macromolecules. Second, it is a source of these substances as well as analogs that may show more desirable biological activity. Polypeptide synthesis involves the formation of an amide linkage between the amino and carboxyl groups of successive α-amino acid monomers. In reacting two different amino acids with each other, it is necessary to protect the amino acid of one amino

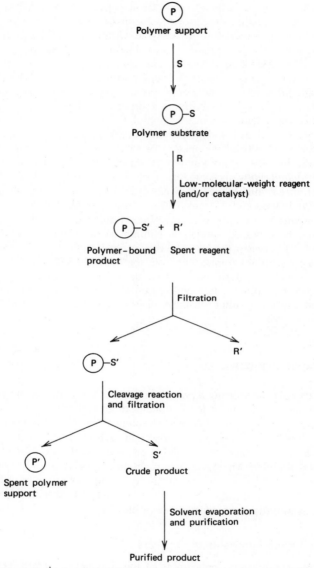

Fig. 9-2 Scheme for utilization of a polymer substrate. After Mathur and Williams [1976] (by permission of Marcel Dekker, Inc., New York).

acid and protect the carboxyl of the other amino acid. This prevents self-reaction of each amino acid with itself and directs the synthesis in the desired cross-reaction route. There are two copolymerization routes, and only one of these is the desired route. For example, alanylglycine (Eq. 9-66) and glycylalanine (Eq. 9-67) are possible from the cross-reaction of alanine and glycine. Synthesis of alanylglycine requires that the

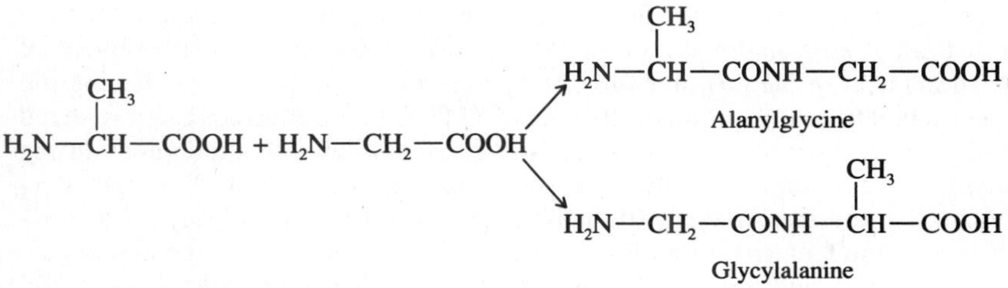

$$\underset{\text{Alanylglycine}}{H_2N-\overset{\overset{\displaystyle CH_3}{|}}{C}H-CONH-CH_2-COOH} \quad (9\text{-}66)$$

$$\underset{\substack{CH_3\\ \\ \\ \\H_2N-CH-COOH}}{} + H_2N-CH_2-COOH$$

$$\underset{\text{Glycylalanine}}{H_2N-CH_2-CONH-\overset{\overset{\displaystyle CH_3}{|}}{C}H-COOH} \quad (9\text{-}67)$$

amine end of alanine and the carboxyl end of glycine be protected prior to reaction. Synthesis of glycylalanine requires that the amine and carboxyl ends of glycine and alanine, respectively, be protected.

The solid-phase synthesis of a polypeptide is described in Fig. 9-3. Chloromethylated polystyrene (usually based on 1% DVB) is commonly used as the polymer support. Reactions are carried out in nonaqueous media, usually methylene chloride, although DMF has also been used. The first amino acid monomer (often referred to as a *unit* or *residue*) is attached to the polymer support by nucleophilic displacement of the benzylic chloride by carboxylate anion. Reasonable reaction rates are achieved

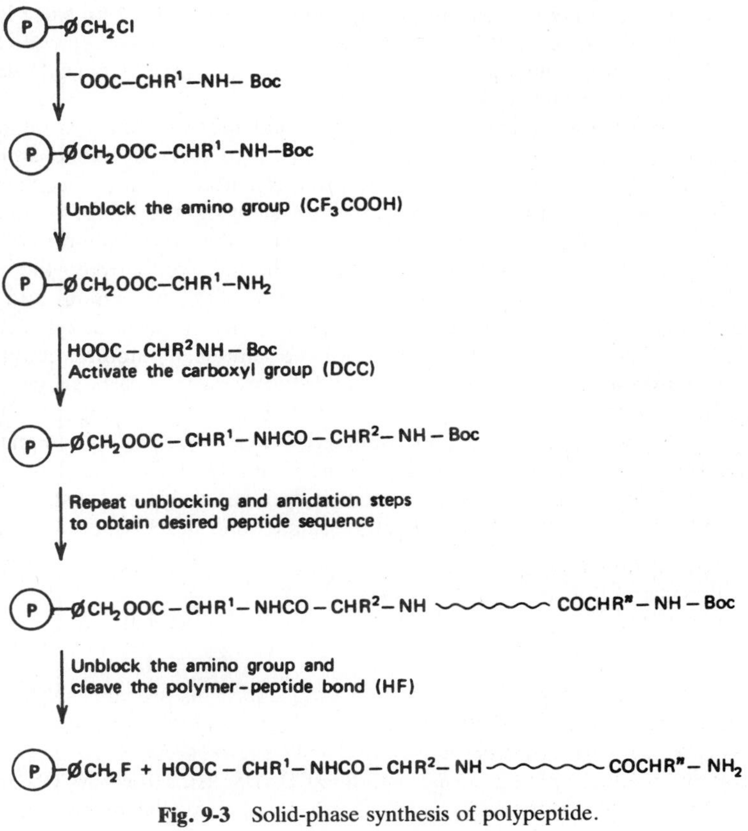

Fig. 9-3 Solid-phase synthesis of polypeptide.

by having high concentrations of the carboxylate anion present by using either the cesium or trialkyl ammonium salt instead of the carboxylic acid. The attachment of the first amino acid to the polymer support also serves the purpose of protecting the carboxyl group. The amino group of the amino acid must be protected (often referred to as *N-blocking* or *N-protecting*) to prevent self-condensation of that monomer during attachment to the support. A variety of protecting groups were studied, including carbobenzoxy, *p*-toluene sulfonyl, triphenylmethyl, and *t*-butoxycarbonyl. The *t*-butoxycarbonyl group (referred to as *Boc*) is usually formed by reacting the amino acid with 2-(*t*-butoxycarbonyloxyimino)-2-phenylacetonitrile. Di-*t*-butyl dicarbonate and

$$\text{(CH}_3)_3\text{CO}-\overset{\overset{\textstyle O}{\|}}{\text{C}}-\text{O}-\text{N}{=}\underset{\phi}{\text{C}}-\text{CN} + \text{H}_2\text{N}-\text{CHR}-\text{COOH} \rightarrow$$

$$\text{(CH}_3)_3\text{CO}-\overset{\overset{\textstyle O}{\|}}{\text{C}}-\text{NH}-\text{CHR}-\text{COOH} + \text{HO}-\text{N}{=}\underset{\phi}{\text{C}}-\text{CN} \qquad (9\text{-}68)$$

t-butoxycarbonyl azide have also been used for synthesizing Boc-protected amino acids. The Boc group is the most frequently used *N*-protecting group since it can be hydrolyzed, usually with 25–50% trifluoroacetic acid (TFA) in methylene chloride, without affecting the polymer support–polypeptide bond (i.e., the benzyl ester bond formed in the first step in Fig. 9-3).

After attachment of the first amino acid to the polymer support, the amino group must be deprotected (i.e., the Boc group removed) to allow a second amino acid residue to be added. The resulting ammonium salt is neutralized with a tertiary amine such as *N,N*-diisopropylethylamine, and the free amino group of the bound amino acid is ready for coupling with a second amino acid. The second amino acid is also used in the form of the *N*-protected derivative. The reaction between the carboxyl group of the second amino acid and the amino group of the first amino acid residue is carried out under conditions that increase the reactivity of the carboxyl group since this reactivity is rather low at the ambient or moderate temperatures employed in solid-phase synthesis. A variety of more active derivatives of the carboxyl group (e.g., acid chloride, anhydride, ester) are possible. Dicyclohexylcarbodiimide (DCC) is commonly used as an activator for the condensation reaction since it results in high reaction rates and yields with a general absence of side reactions. DCC is added to the *N*-protected amino acid and the mixture reacted with the polymer substrate without isolation of the intermediate. The active intermediate is the *O*-acyl isourea (**XXXXVI**),

$$\text{RCOOH} + \text{C}_6\text{H}_{13}-\text{N}{=}\text{C}{=}\text{N}-\text{C}_6\text{H}_{13} \rightarrow \text{C}_6\text{H}_{13}-\text{NH}-\underset{\text{OCOR}}{\overset{\textstyle |}{\text{C}}}{=}\text{N}-\text{C}_6\text{H}_{13} \qquad (9\text{-}69)$$

XXXXVI

$$\downarrow \text{R'NH}_2$$

$$\text{R}-\text{CONHR}' + \text{C}_6\text{H}_{13}-\text{NHCONH}-\text{C}_6\text{H}_{13}$$

$$(9\text{-}70)$$

formed by reaction with the carboxyl group, which undergoes nucleophilic attack by the amino group of an amino acid

The synthesis is continued in a similar manner until the desired sequence of amino acid residues is added. The addition of each amino acid involves amidation with an N-protected amino acid in the presence of DCC followed by deprotection by reaction with TFA. The final step in the overall process is to cleave the peptide from the polymer support. Treatment with HF accomplishes this cleavage simultaneously with deprotection of the last Boc protecting group, without cleavage of any of the amide linkages. The solid-phase method (also referred to as the *Merrifield synthesis*) has been successfully applied to various polypeptides such as bradykinin, oxytocin, insulin, and ribonuclease. The synthesis of ribonuclease comprises a formidible task since there are 124 amino acid residues in the ribonuclease molecule, which corresponds to a total of 369 chemical reactions (protection, coupling, deprotection). Such a synthesis involves an unreasonable length of time to accomplish in view of the much larger number (11,931) of procedural steps involved (the total of chemical reactions and separation steps) if carried out in the usual manner. The use of the polymer substrate approach brings this synthesis into the realm of possibility by allowing automation of the various chemical and separation steps.

The present commercially available automated peptide synthesizers allow synthesis of a polypeptide in a time of approximately one hour per amino acid unit. The ribonuclease synthesis takes about 1–2 weeks.

The Merrifield synthesis is not without limitations since anything less than quantitative conversion in each chemical step with the complete absence of side reactions and the complete removal of reagents prior to the next chemical step would yield an impure polypeptide compared to the naturally occurring biological macromolecule. Thus, for the synthesis of a polypeptide containing 100 amino acid residues, an error of only 1% in adding each amino acid unit would lower the overall yield of the desired product to 36.6%. Even more significantly, the small errors lead to impurities (i.e., polypeptides that are missing one or more of the amino acid residues in the desired final product) that are carried along throughout the complete synthetic process and result in contamination of the final product. The ribonuclease synthesized by Merrifield was a relatively impure product with far less biological activity than the naturally occurring ribonuclease. At the present time, the Merrifield synthesis is most efficient for synthesizing polypeptides containing no more than 30–50 amino acid residues.

The major limitation in the solid-phase synthesis of polypeptides involves the deprotecting reactions. The polymer support–peptide bond (the bond between the polypeptide and the polymer support) and all amide linkages in the polypeptide chain must be completely stable during the successive deprotecting reactions. Also, side-chain protecting groups for those amino acid residues with a third functional group (e.g., lysine contains a second amino group and aspartic acid contains a second carboxyl group) must be stable to the normal deprotecting conditions but must be capable of being deprotected at the end of the synthesis without adversely affecting any of the amide bonds.

Considerable success has been achieved in overcoming the problems associated with the protecting and deprotecting steps by improvements in several areas—the blocking groups used, the type of chemical bond used for attaching the first amino acid residue to the support and careful control of the deprotecting and final cleavage conditions. The bond between the peptide and the polymer support is less stable toward trifluoroacetic acid than desired. There is 1–2% loss of polypeptide from the support per cycle (i.e., per amino acid residue added) [Kent, 1988]. A much more stable means of anchoring the first amino acid residue to the support is accomplished by using a phenylacetamidomethyl (PAM) group

$$\textcircled{P}-\phi-CH_2NH_2 + HOOCCH_2\phi CH_2OOC-CHR-NH-Boc \rightarrow$$

$$\textcircled{P}-\phi-\underbrace{CH_2NH-OCCH_2\phi CH_2OOC}-CHR-NH-Boc \qquad (9\text{-}71)$$
$$\text{PAM linker}$$

The PAM linker increases the acid stability of the anchored first amino acid residue by a factor of more than 100 compared to the simple benzyl ester. This approach uses an aminomethylated polystyrene support (usually referred to as *PAM resin*) instead of the chloromethylated polystyrene. Most solid-phase syntheses of polypeptides are now performed using the PAM resin.

Protecting groups for the side chain COOH or NH_2 groups are chosen for high stability under conditions used for hydrolysis of Boc end groups. For example, *p*-toluenesulphonyl (arginine, histidine), 2-chlorobenzyloxycarbonyl (lysine), 2,4-dinitrophenyl (histidine), cyclohexyl ester (aspartic and glutamic acids), 2-bromobenzyloxycarbonyl (tyrosine), *N*-formyl (tryptophan), and 4-methylbenzyl (cysteine) are used. These protecting groups are quite stable to the conditions (25–50% anhydrous TFA) used for hydrolysis of Boc groups. Side reactions encountered during deprotection of the side groups and cleavage of the polypeptide from the polymer support are minimized by using a two-step hydrolytic sequence, referred to as the *low–high HF procedure*. The initial step involves 25% HF, 65% dimethylsulfide, and 5% each of *p*-cresol and *p*-thiocresol, conditions that foster S_N2 reaction over S_N1 and thus avoids undesirable acylation and alkylation side reactions. The subsequent hydrolytic step employs 90% HF and 5% each of *p*-cresol and *p*-thiocresol to remove the more resistant protecting groups (*p*-toluenesulphonyl, 4-methylbenzyl, and cyclohexyl ester).

An alternate strategy has developed to deprotect the amino group of each amino acid residue without affecting the side group protecting groups and the polymer support–polypeptide bond. This strategy involves the use of the 9-fluorenylmethoxycarbonyl (Fmoc) protecting group instead of the Boc group. The key to this strategy is that the Fmoc group can be cleaved with base instead of acid (piperidine in DMF or methylene chloride). The protecting groups used for amino acid side groups are mostly ether, ester, and urethane derivatives based on *t*-butyl alcohol. The side group protecting groups and the polymer support–polypeptide bond, which are stable toward base, are subsequently cleaved by TFA instead of HF. The strategy based on the Fmoc protecting group is second only to that based on the Boc group for solid-phase synthesis of polypeptides.

Another limitation to the solid-phase of polypeptides is that the maximum yield of coupling and deprotection reactions is 99.5–99.8% instead of 100%. This has been ascribed to the less than complete compatibility of the polymer support with the reagents and/or growing polypeptide chain, although not all workers accept this explanation [Kent, 1988]. Efforts to overcome this problem have included the introduction of spacer groups on the benzene ring, for example, $CH_2CH_2CH_2$ instead of CH_2. This results in greater compatibility and flexibility of the reaction site for attachment of the first amino acid residue and of the growing polypeptide chain. Another solution, the use of polymers more polar than polystyrene (e.g., polyacrylamide), has not been successfully executed. Polystyrene remains the polymer support of choice.

9-13b Other Applications

The success of solid-phase synthesis of polypeptides has stimulated efforts to use polymer substrates in other biochemical and organic syntheses:

1. The solid-phase approach has been successfully extended to the synthesis of nucleic acids (polynucleotides) and oligosaccharides [Frechet, 1980a, 1980b; Gait, 1980; Itakura et al., 1984; Narang, 1983]. The interest in synthesizing nucleic acids is usually coupled with recombinant DNA technology to synthesize polypeptides (both naturally occurring polypeptides and their analogs). The very large effort in this area at present makes this one of the most important applications of the polymer substrate method.

2. Conversion of one functional group in a molecule containing two (or more) functional groups can be achieved by covalently bonding the molecule to a polymer support through one (or more) of the functional groups. The latter group(s) is(are) protected while reactions are carried out on the unprotected group [Blossey and Ford, 1989; Hodge, 1988]. An example is the selective acylation of **XXXXVII** at the hydroxyl group on C-3 [Frechet, 1980a, 1980b]. Treatment of **XXXXVI** with poly(*p*-styryl-

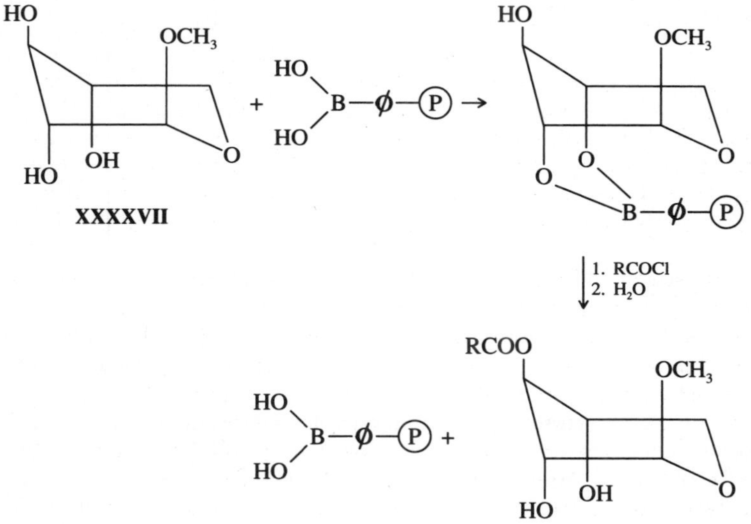

$$(9\text{-}72)$$

boronic acid) protects the hydroxyls at C-2 and C-4 and allows acylation at C-3. The polymer support is hydrolytically cleaved from the product with regeneration of the original protecting group. The same approach has been described for the monoderivitization of diacids, dialdehydes, diamines, and other difunctional compounds [Leznoff, 1978].

3. Undesirable intermolecular reactions can be avoided during certain synthetic conversions. Thus it is often useful to carry out *C*-alkylation and *C*-acylation of compounds which form enolate anions, for example, esters with α-hydrogens. Such reactions are often complicated by self-condensation since the enolate anion can attack the carbonyl group of a second ester molecule. Attachment of the enolizable ester to a polymer support at low loading levels allows the alkylation and acylation reactions

(Eq. 9-73) to be performed under the equivalent of high dilution conditions, where ester molecules are not sufficiently close to undergo intermolecular condensation

$$\text{(P)}-\phi CH_2-OCOCH_2R \xrightarrow{\text{base}} \text{(P)}-\phi CH_2-OCO\overset{..}{C}HR \xrightarrow[\substack{2.\ HBr \\ 3.\ Heat\,(-CO_2)}]{1.\ R'COX} RCH_2COR' \qquad (9\text{-}73a)$$

$$\downarrow \substack{1.\ R'X \\ 2.\ HBr}$$

$$HOOCCHRR' \qquad (9\text{-}73b)$$

[Kraus and Patchornik, 1971, 1974]. In a similar manner, one can favor various intramolecular cyclization reactions of supported diesters, cyanoesters, and dinitriles, as well as the cyclization reaction of a supported malonic ester with an α,ω-dihalide [Crowley et al., 1973; Mohanraj and Ford, 1985; Patchornik and Kraus, 1970].

4. Other applications of the polymer substrate technique include the synthesis of threaded macrocyclic systems (hooplanes, catenanes, knots), the retrieval of a minor component from a reaction system, and the trapping of reaction intermediates [Frechet, 1980a, 1980b; Hodge, 1988; Hodge and Sherrington, 1980; Mathur et al., 1980].

REFERENCES

Adibi, K., M. H. George, and J. A. Barrie, *Polymer*, **20**, 483 (1979); *J. Polym. Sci. Polym. Chem. Ed.*, **19**, 57 (1981).

Akelah, A. and D. C. Sherrington, *Polymer*, **24**, 1369 (1983).

Alfrey, T., Jr., H. C. Haas, and C. W. Lewis, *J. Am. Chem. Soc.*, **73**, 2851 (1951).

Allen, P. W., "Natural Rubber and the Synthetics," Crosby Lockwood, London, 1972.

Alliger, G. and I. J. Sjothun, Eds., "Vulcanization of Elastomers," Van Nostrand Reinhold, New York, 1964.

Andrews, G. D. and R. L. Dawson, "Chlorosulfonated Polyethylene," pp. 513–522 in "Encyclopedia of Polymer Science and Engineering," Vol. 6, 2nd ed., H. F. Mark, N. M. Bikales, C. G. Overberger, and G. Menges, Eds., Wiley-Interscience, New York, 1986.

Arshady, R., G. W. Kenner, and A. Ledwith, *Makromol. Chem.*, **177**, 2911 (1976).

Arthur, J. C., Jr., "Chemical Modification of Cellulose and Its Derivatives," Chap. 2 in "Comprehensive Polymer Science," Vol. 6, G. C. Eastmond, A. Ledwith, S. Russo, and P. Sigwalt, Eds., Pergamon Press, Oxford, 1989.

Asami, R., M. Takaki, K. Kyuda, and E. Asakura, *Polymer J.*, **15**, 139 (1983).

Bao, Y. T. and C. G. Pitt, *J. Polym. Sci. Polym. Chem. Ed.*, **28**, 741 (1990).

Barany, G., N. Kneib-Cordonier, and N. G. Mullen, "Polypeptide Synthesis," pp. 811–858 in "Encyclopedia of Polymer Science and Engineering," Vol. 12, 2nd ed., H. F. Mark, N. M. Bikales, C. G. Overberger, and G. Menges, Eds., Wiley-Interscience, New York, 1988.

Barth, V. and E. Klesper, *Polymer*, **17**, 777, 787, 893 (1976).

Bateman, L., "The Chemistry and Physics of Rubber-Like Substances," Wiley-Interscience, New York, 1963.

Battaerd, H. A. and G. W. Tregear, "Graft Copolymers," Wiley-Interscience, New York, 1967.

Benham, J. L. and J. F. Kinstle, "Chemical Reactions on Polymers," *Am. Chem. Soc. Symp. Ser.*, **364**, Washington, D.C., 1988.

Beresniewicz, A., *J. Polym. Sci.*, **39**, 63 (1959).

Bikales, N. M. and L. Segal, Eds., "Cellulose and Cellulose Derivatives," Wiley-Interscience, New York, Parts IV and V, 1971.

Bikson, B., J. Jagur-Grodzinski, and D. Vofsi, *Polymer*, **20**, 215 (1979).

Billmeyer, F. W., Jr., "Textbook of Polymer Science," 3rd ed., Wiley-Interscience, New York, 1984.

Blossey, E. C. and W. T. Ford, "Polymeric Reagents," Chap. 3 in "Comprehensive Polymer Science," Vol. 6, G. C. Eastmond, A. Ledwith, S. Russo, and P. Sigwalt, Eds., Pergamon Press, Oxford, 1989.

Bobear, W. J., *Rubber Chem. Technol.*, **40**, 1560 (1967).

Boga, C., M. Contento, and F. Manescachi, *Makromol. Chem. Rapid Commun.*, **10**, 303 (1989).

Braun, T. and A. B. Farag, *Anal. Chim. Acta*, **72**, 133 (1974).

Brewer, R. J. and R. T. Bogan "Cellulose Esters, Inorganic," pp. 139–157 and "Cellulose Esters, Organic," pp. 158–181 in "Encyclopedia of Polymer Science and Engineering," Vol. 3, 2nd ed., H. F. Mark, N. M. Bikales, C. G. Overberger, and G. Menges, Eds., Wiley-Interscience, New York, 1985.

Brydon, A., G. M. Burnett, and G. G. Cameron, *J. Polym. Sci. Polym. Chem. Ed.*, **11**, 3255 (1973); **12**, 1011 (1974).

Brydson, J. A., "Plastics Materials," 4th ed., Butterworth, London, 1982.

Cai, G.-F. and D.-Y. Yan, *Makromol. Chem.*, **188**, 1005 (1987).

Callant, D. and E. Schacht, *Makromol. Chem.*, **191**, 529 (1990).

Cameron, G. G. and A. W. S. Duncan, *Makromol. Chem.*, **184**, 1153 (1983).

Cameron, G. G. and K. Sarmouk, *Makromol. Chem.*, **191**, 17 (1990).

Ceresa, R. J., "The Chemical Modification of Polymers," Chap. 11 in "Science and Technology of Rubber," F. R. Eirich, Ed., Academic Press, New York, 1978.

Ceresa, R. J., Ed., "Block and Graft Copolymerization," Vols. 1 and 2, Wiley-Interscience, New York, 1973, 1976.

Chaiken, I., M. Wilchek, and I. Parikh, Eds., "Affinity Chromatography and Biological Recognition," Academic Press, New York, 1984.

Chang, C., F. Fish, D. D. Muccio, and T. St. Pierre, *Macromolecules*, **20**, 621 (1987).

Chapiro, A., "Radiation Chemistry of Polymeric Systems," Chaps. IX and X, Wiley-Interscience, New York, 1962.

Chauvin, Y., D. Commereuc, and F. Dawans, *Prog. Polym. Sci.*, **5**, 95 (1977).

Chibata, I., "Immobilized Enzymes," Halstead Press (Wiley), New York, 1978.

Chiellini, E., R. Solaro, and S. D'Antone, *Makromol. Chem. Suppl.*, **5**, 82 (1981).

Cho, J. R. and H. Morawetz, *Macromolecules*, **6**, 628 (1973).

Conlon, H. D. and D. R. Walt, *J. Chem. Ed.*, **63**, 369 (1986).

Coran, A. Y., "Vulcanization," Chap. 7 in "Science and Technology of Rubber," F. R. Eirich, Ed., Academic Press, New York, 1978.

Crowley, J. I., T. B. Harvey, and H. Rapoport, *J. Macromol. Sci. Chem.*, **7**, 1117 (1973).

Cuatrecasas, P. and C. B. Anfinsen, *Ann. Rev. Biochem.*, **40**, 259 (1971).

Delmas, M. and A. Gaset, *Synthesis*, **871** (1980).

Donaruma, L. G. and O. Vogl, "Polymeric Drugs," Academic Press, New York, 1978.

Dreyfuss, P. and R. P. Quirk, "Graft Copolymers," pp. 551–579 in "Encyclopedia of Polymer Science and Engineering," Vol. 7, 2nd ed., H. F. Mark, N. M. Bikales, C. G. Overberger, and G. Menges, Eds., Wiley-Interscience, New York, 1986.

Drobnik, J., J. Kopecek, J. Labsky, P. Rejmanova, J. Exner, V. Saudek, and J. Kalal, *Makromol. Chem.*, **177**, 2833 (1976).

Elias, H.-G., "Macromolecules," Vol. 2, 2nd ed., Plenum Press, New York, 1984, Chap. 23.

Erickson, B. W. and R. B. Merrifield, "Solid-Phase Peptide Synthesis," Chap. 3 in "The Proteins," Vol. 2, 3rd ed., H. Neurath and R. L. Hill, Eds., Academic Press, New York, 1976.

Favre, R., P. Berticat, and P. Q. Tho, *Eur. Polym. J.*, **14**, 51, 157 (1978).

Feng, X. D., Y. H. Sun, and K. Y. Qui, *Macromolecules*, **18**, 2105 (1985).

Fettes, E. M., Ed., "Chemical Reactions of Polymers," Wiley-Interscience, New York, 1964.

Fettes, E. M., "Chemical Modification," Chap. 6 in "Crystalline Olefin Polymers," Part II, R. A. V. Raff and K. W. Doak, Eds., Wiley-Interscience, New York, 1964.

Flory, P. J., *J. Am. Chem. Soc.*, **61**, 1518 (1939).

Ford, W. T., Ed., "Polymeric Reagents and Catalysts," Am. Chem. Soc. Symp. Ser., **308**, Washington, D.C., 1986a.

Ford, W. T., "Wittig Reactions on Polymer Supports," Chap. 8 in "Polymeric Reagents and Catalysts," Ed., W. T. Ford, *Am. Chem. Soc. Symp. Ser.*, **308**, Washington, D.C., 1986b.

Ford, W. T. and M. Tomoi, *Adv. Polym. Sci.*, **55**, 49 (1984).

Frechet, J. M. J., "Synthesis Using Polymer-Supported Protecting Groups," Chap. 6 in "Polymer-Supported Reactions in Organic Synthesis," P. Hodge and D. C. Sherrington, Eds., Wiley, New York, 1980a.

Frechet, J. M. J., "Polymer-Supported Synthesis of Oligosaccharides," Chap. 8 in "Polymer-Supported Reactions in Organic Synthesis," P. Hodge and D. C. Sherrington, Eds., Wiley, New York, 1980b.

Frechet, J. M. J., G. D. Darling, and M. J. Farrall, *J. Org. Chem.*, **46**, 1728 (1981).

Frechet, J. M. J., G. D. Darling, S. Itsuno, P.-Z. Lu, M. V. de Meftahi, and W. A. Rolls, Jr., *Pure Appl. Chem.*, **60**, 352 (1988).

Frechet, J. M. J. and M. J. Farrall, "Functionalization of Crosslinked Polystyrene Resins by Chemical Modification: A Review," pp. 59–83 in "Chemistry and Properties of Crosslinked Polymers," S. S. Labana, Ed., Academic Press, New York, 1977.

Frechet, J. M. J., M. J. Farrall, and L. J. Nuyens, *J. Macromol. Sci. Chem.*, **A11**, 507 (1977).

Frechet, J. M. J. and K. E. Haque, *Macromolecules*, **8**, 130 (1975).

Fu, T.-Y. and H. Morawetz, *J. Biol. Chem.*, **251**, 2087 (1976).

Gaetjens, E. and H. Morawetz, *J. Am. Chem. Soc.*, **83**, 1738 (1961).

Gait, M. J., "Polymer-Supported Synthesis of Oligonucleotides," Chap. 9 in "Polymer-Supported Reactions in Organic Synthesis," P. Hodge and D. C. Sherrington, Eds., Wiley, New York, 1980.

Gan, L. M. and C. H. Chew, *J. Appl. Polym. Sci.*, **24**, 371 (1979).

Gan, L. M., G. B. Soh, and K. L. Ong, *J. Appl. Polym. Sci.*, **21**, 1771 (1977); *Rubber Chem. Technol.*, **51**, 267 (1978).

George, B. K. and V. N. R. Pillai, *Macromolecules*, **21**, 1867 (1988).

Gnanou, Y. and P. Rempp, *Makromol. Chem.*, **188**, 2111 (1987).

Golub, M. A., "Cyclized and Isomerized Rubber," Chap. 10A in "Polymer Chemistry of Synthetic Elastomers," Part II, J. P. Kennedy and E. G. Tornqvist, Eds., Wiley-Interscience, New York, 1969.

Golub, M. A. and J. Heller, *J. Polym. Sci.*, **B2**, 723 (1964).

Graczyk, T. and V. Hornof, *J. Polym. Sci. Polym. Chem. Ed.*, **26**, 2019 (1988).

Grieg, J. A., R. D. Hancock, and D. C. Sherrington, *Eur. Polym. J.*, **16**, 293 (1980).

Grieg, J. A., and D. C. Sherrington, *Polymer*, **19**, 1963 (1978).

Grubbs, R. H. and L. C. Kroll, *J. Am. Chem. Soc.*, **93**, 3062 (1971).

Guthrie, J. T., M. Ryder, and F. I. Abdel-Hay, *Polym. Bull.*, **1**, 501 (1979).

Hadjichristidis, N. and J. E. L. Roovers, *J. Polym. Sci. Polym. Phys. Ed.*, **16**, 851 (1978).

Harris, J. F., Jr. and W. H. Sharkey, *Macromolecules*, **19**, 2903 (1986).

Harrison, C. R. and P. Hodge, *J. Chem. Soc., Perkin I*, **605**, 2252 (1976).

Hasegawa, H. and T. Higashimura, *Polym. J.*, **13**, 915 (1981).

Hirao, A., S. Nakahama, M. Takahashi, H. Mochizuki, and N. Yamazaki, *Makromol. Chem.*, **179**, 2343 (1978).

Hodge, P., *Chem. Br.*, **14**, 237 (1978).

Hodge, P., "Polymers as Chemical Reagents," pp. 618–658 in "Encyclopedia of Polymer Science and Engineering," Vol. 12, 2nd ed., H. F. Mark, N. M. Bikales, C. G. Overberger, and G. Menges, Eds., Wiley-Interscience, New York, 1988.

Hodge, P. and D. C. Sherrington, Eds., "Polymer-Supported Reactions in Organic Synthesis," Wiley, New York, 1980.

Hsieh, D. S. T., Ed., "Controlled Release Systems: Fabrication Technology," CRC Press, Boca Raton, Fla., 1988.

Imanishi, Y., *J. Polym. Sci. Macromol. Rev.*, **14**, 1 (1979).

Imre, K. and G. Odian, *J. Polym. Sci. Polym. Chem. Ed.*, **17**, 2601 (1979).

Imre, K., G. Odian, and A. Rabie, *J. Polym. Sci. Polym. Chem. Ed.*, **14**, 3045 (1976).

Ishiwatari, T., T. Okubo, and N. Ise, *Macromolecules*, **13**, 53 (1980).

Itakura, K., J. J. Rossi, and R. B. Wallace, *Ann. Rev. Biochem.*, **53**, 323 (1984).

Itsuno, S., M. Nakano, K. Ito, A. Hirao, M. Owa, N. Kanda, and S. Nakahama, *J. Chem. Soc. Perkin Trans. I*, 2615 (1985).

Jain, R. K., S. L. Agnish, K. Lal, and H. L. Bhatnagar, *Makromol. Chem.*, **186**, 2501 (1985).

Jakoby, W. B. and M. Wilchek, Eds., "Affinity Techniques," Vol. XXXIV in "Methods in Enzymology," Academic Press, New York, 1974.

Jaycox, G. D. and J. Smid, *Makromol. Chem. Rapid Commun.*, **2**, 299 (1981).

Jerabek, K., *J. Polym. Sci. Polym. Chem. Ed.*, **18**, 65 (1980).

Johar, Y., M. Zupan, and B. Sket, *J. Chem. Soc. Perkin Trans. I*, 2059 (1982).

Just, E. K. and T. G. Majewicz, "Cellulose Ethers," pp. 226–269 in "Encyclopedia of Polymer Science and Engineering," Vol. 3, 2nd ed., H. F. Mark, N. M. Bikales, C. G. Overberger, and G. Menges, Eds., Wiley-Interscience, New York, 1985.

Kaleem, K., C. R. Reddy, S. Rajadurai, and M. Santappa, *Makromol. Chem.*, **180**, 851 (1979).

Kashani, H. A., J. A. Barrie, and M. H. George, *J. Polym. Sci. Polym. Chem. Ed.*, **16**, 533 (1978).

Kawaguchi, S., Y. Nishikawa, T. Kitano, K. Ito, and A. Minakata, *Macromolecules*, **23**, 2710 (1990).

Keller, R. C., *Rubber Chem. Technol.*, **61**, 238 (1988).

Kennedy, J. P. and E. Marechal, "Carbocationic Polymerization," Wiley-Interscience, New York, 1982, Chap. 8.

Kent, S. B. H., *Ann. Rev. Biochem.*, **57**, 957 (1988).

Keyes, M. H. and D. Albert, "Enzymes, Immobilized," pp. 189–209 in "Encyclopedia of Polymer Science and Engineering," Vol. 6, 2nd ed., H. F. Mark, N. M. Bikales, C. G. Overberger, and G. Menges, Eds., Wiley-Interscience, New York, 1986.

Kheradmand, H., J. Francois, and V. Plazanet, *Polymer*, **29**, 860 (1988).

Kiniwa, H., Y. Doi, and T. Nishikaji, *Makromol. Chem.* **188**, 1841, 1851 (1987).

Kirsh, T. E., V. A. Kabanov, and V. A. Kargin, *Vysokomol. Soedin.*, **A10**, 349 (1968).

Kraus, M. A. and A. Patchornik, *J. Am. Chem. Soc.*, **93**, 7325 (1971); *J. Polym. Sci. Polym. Symp.*, **47**, 11 (1974).

Kucera, M., Z. Salajka, K. Majerova, and M. Navratil, *Makromol. Chem.*, **184**, 527 (1983); *Polymer*, **26**, 1575 (1985).

Kuczkowski, J. A., "Rubber Chemicals," pp. 716–762 in "Encyclopedia of Polymer Science and Engineering," Voll. 14, 2nd ed., H. F. Mark, N. M. Bikales, C. G. Overberger, and G. Menges, Eds., Wiley-Interscience, New York, 1988.

Kunitake, T., "Enzyme-like Catalysis by Synthetic Linear Polymers," Chap. 4 in "Polymer-Supported Reactions in Organic Synthesis," P. Hodge and D. C. Sherrington, Eds., Wiley, New York, 1980.

Labana, S. S., "Cross-Linking," pp. 350–375 in "Encyclopedia of Polymer Science and Engineering," Vol. 4, 2nd ed., H. F. Mark, N. M. Bikales, C. G. Overberger, and G. Menges, Eds., Wiley-Interscience, New York, 1986.

Ladenheim, H., E. M. Loebl, and H. Morawetz, *J. Am. Chem. Soc.*, **81**, 20 (1959).

Ladenheim, H. and H. Morawetz, *J. Am. Chem. Soc.*, **81**, 4860 (1959).

Laverty, J. J. and Z. G. Gardlund, *J. Polym. Sci. Polym. Chem. Ed.*, **15**, 2001 (1977).

Lee, M., H. Nakamura, and Y. Minoura, *J. Polym. Sci. Polym. Chem. Ed.*, **14**, 961 (1976).

Lenz, R. W., *J. Am. Chem. Soc.*, **82**, 182 (1960).

Letsinger, R. L. and T. J. Savereide, *J. Am. Chem. Soc.*, **84**, 3122 (1962).

Letsinger, R. L. and T. E. Wagner, *J. Am. Chem. Soc.*, **88**, 2062 (1966).

Levenfeld, B., J. San Roman, and E. L. Madruga, *Polymer*, **31**, 160 (1990).

Lewis, E. A., T. J. Barkley, R. R. Reams, L. D. Hansen, and T. St. Pierre, *Macromolecules*, **17**, 2874 (1984).

Leza, M. L., I. Casinos, and G. M. Guzman, *Angew. Makromol. Chem.*, **178**, 119 (1990).

Leznoff, C. C., *Acct. Chem. Res.*, **11**, 327 (1978).

Lohmann, D. and C. d'Hondt, *Makromol. Chem.*, **188**, 295 (1987).

Luca, C., S. Dragan, V. Barboiu, and M. Dima, *J. Polym. Sci. Polym. Chem. Ed.*, **18**, 449 (1980).

Ludwico, W. A. and S. L. Rosen, *J. Appl. Polym. Sci.*, **19**, 757 (1975); *J. Polym. Sci. Polym. Chem. Ed.*, **14**, 2121 (1976).

Lukas, R., M. Kolinsky, and D. Doskocilova, *J. Polym. Sci. Polym. Chem. Ed.*, **16**, 889 (1978).

Lundgren, J. L. and A. A. Schilt, *Anal. Chem.*, **49**, 974 (1977).

Machi, S. and J. Silverman, *J. Polym. Sci.*, **A-1**(7), 273 (1968).

Majid, M. A., M. H. George, and J. A. Barrie, *Polymer*, **23**, 57 (1982).

Manecke, G. and W. Storck, *Angew. Chem. Int. Ed. Engl.*, **17**, 657 (1978).

Manecke, G. and H.-G. Vogt, *Pure Appl. Chem.*, **50**, 655 (1978).

Marsh, J. T., "An Introduction to Textile Finishing," Chapman and Hall, London, 1966.

Marshall, G. L., M. E. A. Cudby, K. Smith, T. H. Stevenson, K. J. Packer, and R. K. Harris, *Polymer*, **28**, 1093 (1987).

Mathias, L. J. and K. Al-Jumar, *J. Polym. Sci. Polym. Chem. Ed.*, **18**, 2911 (1980).

Mathur, N. K., C. K. Narang, and R. E. Williams, "Polymers as Aids in Organic Chemistry," Academic Press, New York, 1980.

Mathur, N. K. and R. E. Williams, *J. Macromol. Sci. Rev. Macromol. Chem.*, **C15**, 117 (1976).

McGinniss, V. D., "Cross-Linking with Radiation," pp. 418–449 in "Encyclopedia of Polymer Science and Engineering," Vol. 4, 2nd ed., H. F. Mark, N. M. Bikales, C. G. Overberger, and G. Menges, Eds., Wiley-Interscience, New York, 1986.

McGuchan, R. and I. C. McNeil, *J. Polym. Sci.*, **A-1**(6), 205 (1968).

Menger, F. M., H. Sinozaki, and L. C. Lee, *J. Org. Chem.*, **45**, 272 (1980).

Merrifield, R. B., *Pure Appl. Chem.*, **50**, 643 (1978). *Angew. Chem. Int. Ed. Engl.*, **24**, 799 (1985); *Makromol. Chem. Macromol. Symp.*, **19**, 31 (1988).

Messing, R. A., *Biotechnol. Bioeng.*, **16**, 897 (1974).

Miller, J. R., D. G. Smith, W. E. Marr, and T. R. E. Kressman, *J. Chem. Soc.*, 218 (1963).

Miyauchi, N., I. Kirikihira, X. Li, and M. Akashi, *J. Polym. Sci. Polym. Chem. Ed.*, **26**, 1561 (1988).

Mohanraj, S. and W. T. Ford, *J. Org. Chem.*, **50**, 1616 (1985).

Molinari, H., F. Montanari, and P. Tundo, *J. Chem. Soc. Chem. Commun.*, 639 (1977).

Moore, J. A., Ed., "Reactions on Polymers," D. Reidel, Boston, 1973.

Morawetz, H., "Macromolecules in Solution," 2nd ed., Chaps. VIII and IX, Wiley-Interscience, New York, 1975.

Morawetz, H. and B. Vogel, *J. Am. Chem. Soc.*, **91**, 563 (1969).

Morawetz, H. and Y. Wang, *Macromolecules*, **20**, 194 (1987).

Morrison, N. J. and M. Porter, "Crosslinking of Rubbers," Chap. 6 in "Comprehensive Polymer Science," Vol. 6, G. C. Eastmond, A. Ledwith, S. Russo, and P. Sigwalt, Eds., Pergamon Press, Oxford, 1989.

Moxley, T. T. and B. C. Gates, *J. Mol. Catal.*, **12**, 389 (1981).

Narang, S. A., *Tetrahedron*, **39**, 3 (1983).

Nayak, P. L., S. Lenka, and N. C. Pati, *Angew. Makromol. Chem.*, **75**, 29 (1979a); *J. Polym. Sci. Polym. Chem. Ed.*, **17**, 3425 (1979b).

Neckers, D. C., *Chem. Tech.*, 108 (Feb. 1978).

Neckers, D. C., "Polymeric Photosensitizers," Chap. 6 in "Polymeric Reagents and Catalysts," W. T. Ford, Ed., *Am. Chem. Soc. Symp. Ser.*, **308**, Washington, D.C., 1986.

Nishikubo, T., T. Kondo, and K. Inomata, *Macromolecules*, **22**, 3827 (1989).

Noah, O. V., A. D. Litmanovich, and N. A. Plate, *J. Polym. Sci.*, **12**, 1711 (1974).

Odian, G., A. Derman and K. Imre, *J. Polym. Sci. Polym. Chem. Ed.*, **18**, 737 (1980).

Odian, G. and J. H. T. Kho, *J. Macromol. Sci. Chem.*, **A4**, 317 (1970).

Olah, G. A., A. P. Fung, and D. Meidar, *Synthesis*, 280, (1981).

Olah, G. A., D. Meidar, R. Malhotra, J. A. Olah, and S. C. Narang, *J. Catal.*, **61**, 96 (1980).

Ott, E., H. M. Spurlin, and M. W. Grafflin, Eds., "Cellulose and Cellulose Derivatives," Wiley-Interscience, New York, Parts I and II, 1954; Part III, 1955.

Ottenbrite, R. M. and J. Sunamoto, *Polym. Sci. Technol.*, **34**, 333 (1986).

Overberger, C. G., R. C. Glowaky, and P. H. Vandewyer, *J. Am. Chem. Soc.*, **95**, 6008 (1973).

Overberger, C. G. and A. C. Guterl, Jr., *J. Polym. Sci. Polym. Symp.*, **62**, 13 (1978); *J. Polym. Sci. Polym. Chem. Ed.*, **17**, 1887 (1979).

Overberger, C. G. and A. Meenakshi, *J. Polym. Sci. Polym. Chem. Ed.*, **22**, 1923 (1984).

Overberger, C. G. and M. Morimoto, *J. Am. Chem. Soc.*, **93**, 3222 (1971).

Overberger, C. G. and T. W. Smith, *Macromolecules*, **8**, 401, 407, 416 (1975).

Overberger, C. G. and N. Vorchheimer, *J. Am. Chem. Soc.*, **85**, 951 (1963).

Pappas, S. P., "Photocrosslinking," Chap. 5 in "Comprehensive Polymer Science," Vol. 6, G. C. Eastmond, A. Ledwith, S. Russo, and P. Sigwalt, Eds., Pergamon Press, Oxford, 1989.

Patchornik, A. and M. A. Kraus, *J. Am. Chem. Soc.*, **92**, 7587 (1970).

Peacock, A. J., *Polymer Commun.*, **28**, 259 (1987).

Plate, N. A., *Pure Appl. Chem.*, **46**, 49 (1976).

Pshezetsky, V. S., I. Massouh, and V. A. Kabanov, *J. Polym. Sci.*, **C22**, 309 (1968).

Rabie, A. and G. Odian, *J. Polym. Sci. Polym. Chem. Ed.*, **15**, 469 (1977).

Regen, S. L., *J. Am. Chem. Soc.*, **97**, 5956 (1975); **98**, 6270 (1976); *Angew. Chem. Int. Ed. Engl.*, **18**, 421 (1979).

Rempp, P., *Pure Appl. Chem.*, **46**, 9 (1976).

Rempp, P. F. and P. J. Lutz, "Synthesis of Graft Copolymers," Chap. 12 in "Comprehensive Polymer Science," Vol. 6, G. C. Eastmond, A. Ledwith, S. Russo, and P. Sigwalt, Eds., Pergamon Press, Oxford, 1989.

Riggs, J. P., "Carbon Fibers," pp. 640–685 in "Encyclopedia of Polymer Science and Engineering," Vol. 2, 2nd ed., H. F. Mark, N. M. Bikales, C. G. Overberger, and G. Menges, Eds., Wiley-Interscience, New York, 1985.

Roberts, A. D., Ed., "Natural Rubber Science and Technology," Oxford University Press, Oxford, 1988.

Royer, G. P. and I. M. Klotz, *J. Am. Chem. Soc.*, **91**, 5885 (1969).

Sakaguchi, M. and J. Sohma, *J. Appl. Polym. Sci.*, **22**, 2915 (1978).

Sande, A. R., M. H. Jagdale, R. R. Mare, and M. M. Salunke, *Tetrahedron Lett.*, **25**, 3501 (1984).

Sato, Y., N. Kunieda, and M. Kinoshita, *Makromol. Chem.*, **178**, 683 (1977).

Satoh, M., E. Yoda, T. Hayashi, and J. Komiyama, *Macromolecules*, **22**, 1808 (1989).

Sawage, J. M. and C. Loucheux, *Makromol. Chem.*, **176**, 315 (1975).

Scher, H. B., Ed., "Controlled Release Pesticides," *Am. Chem. Soc. Symp. Ser.*, **53**, Washington, D.C., 1977.

Schmiegel, W. W., *Angew. Makromol. Chem.*, **76/77**, 39 (1979).

Schulz, G. O. and R. Milkovich, *J. Appl. Polym. Sci.*, **27**, 4773 (1982).

Schuttenberg, H., G. Klump, V. Kaczmar, S. R. Turner, and R. C. Schulz, *J. Macromol. Sci. Chem.*, **7**, 1085 (1973).

Scott, G., Ed., "Developments in Polymer Stabilization-8," Elsevier, London, 1987.

Scott, G., *Makromol. Chem. Macromol. Symp.*, **28**, 59 (1989).

Sengupta, T. K. and S. R. Palit, *J. Polym. Sci. Polym. Chem. Ed.*, **16**, 713 (1978).

Septnagel, W. J. and I. M. Klotz, *J. Polym. Sci. Polym. Chem. Ed.*, **15**, 621 (1977).

Shambhu, M. B., G. A. Digenis, D. K. Gulati, K. Bowman, and P. S. Sabharwal, *J. Agric. Food Chem.*, **24**, 666 (1976).

Sherrington, D. C., "Preparation, Functionalization, and Characteristics of Polymer Supports," Chap. 1 in "Polymer-Supported Reactions in Organic Synthesis," P. Hodge and D. C. Sherrington, Eds., Wiley-Interscience, Chichester, U.K., 1980.

Sherrington, D. C., "Reactions of Polymers," pp. 101–169 in "Encyclopedia of Polymer Science and Engineering," Vol. 14, 2nd ed., H. F. Mark, N. M. Bikales, C. G. Overberger, and G. Menges, Eds., Wiley-Interscience, New York, 1988.

Sierra-Vargas, J., J. G. Zilliox, P. F. Rempp, and E. Franta, *Polym. Bull.*, **3**, 83 (1980).

Sket, B., and M. Zupan, *Tetrahedron*, **40**, 2865 (1984).

Skinner, T. D., *Rubber Chem. Technol.*, **45**, 182 (1972).

Solomon, D. H., "The Chemistry of Organic Film Formers," Chaps. 2, 3, and 5, Wiley-Interscience, New York, 1967.

Spetnagel, W. J. and I. M. Klotz, *J. Polym. Sci. Polym. Chem. Ed.*, **15**, 621 (1977).

Stewart, J. M., "Polymer-Supported Synthesis and Degradation of Peptides," Chap. 7 in "Polymer-Supported Reactions in Organic Synthesis," P. Hodge and D. C. Sherrington, Eds., Wiley, New York, 1980.

Stewart, C. A. Jr., T. Takeshita, and M. L. Coleman, "Chloroprene Polymers," pp. 441–462 in "Encyclopedia of Polymer Science and Engineering," 2nd ed., Vol. 3, H. F. Mark, N. M. Bikales, C. G. Overberger, and G. Menges, Eds., Wiley-Interscience, New York, 1985.

Storey, R. F. and L. J. Goff, *Macromolecules*, **22**, 1058 (1989).

Subramaniam, A., "Rubber Derivatives," pp. 762–786 in "Encyclopedia of Polymer Science and Engineering," Vol. 14, 2nd ed., H. F. Mark, N. M. Bikales, C. G. Overberger, and G. Menges, Eds., Wiley-Interscience, New York, 1988.

Sugden, K. C. Hunter, and G. L. Jones, *J. Chromatography*, **192**, 228 (1980).

Sullivan, A. B., *J. Org. Chem.*, **31**, 2811 (1966).

Sundet, S. A., *Macromolecules*, **11**, 146 (1978).

Takagi, T., *J. Appl. Polym. Sci.*, **19**, 1649 (1975).

Takayanagi, M. and T. Katayose, *J. Polym. Sci. Polym. Chem. Ed.*, **21**, 31 (1983).

Taylor, R. T., "Polymer-Bound Oxidizing Agents," Chap. 7 in "Polymeric Reagents and Catalysts," W. T. Ford, Ed., *Am. Chem. Soc. Symp. Ser.*, **308**, Washington, D.C., 1986.

Tazuke, S. and H. Kimura, *J. Polym. Sci. Polym. Lett. Ed.*, **15**, 93 (1977); *Makromol. Chem.*, **179**, 2603 (1978).

Tomoi, M. and W. T. Ford, "Polymeric Phase Transfer Catalysts," Chap. 5 in "Synthesis and Separations Using Functional Polymers," D. C. Sherrington and P. Hodge, Eds., Wiley, New York, 1988.

Tran, A. and J. Prud'homme, *Macromolecules*, **10**, 149 (1977).

Tundo, P., *Synthesis*, 315 (1978).

Turbak, A., "Rayon," pp. 45–72 in "Encyclopedia of Polymer Science and Engineering," Vol. 14, 2nd ed., H. F. Mark, N. M. Bikales, C. G. Overberger, and G. Menges, Eds., Wiley-Interscience, New York, 1988.

Vogl, O., A. C. Albertsson, and Z. Janovic, *Polymer*, **26**, 1288 (1985).

Vukov, R., *Rubber Chem. Technol.*, **57**, 275 (1984).

Waller, F. J., "Catalysis with a Perfluorinated Ion-Exchange Polymer," Chap. 3 in "Polymeric Reagents and Catalysts," W. T. Ford, Ed., *Am. Chem. Soc. Symp. Ser.*, **308**, Washington, D.C., 1986.

Walz, R. and W. Heitz, *J. Polym. Sci. Polym. Chem. Ed.*, **16**, 1807 (1978).

Ward, W. J. and T. J. McCarthy, "Surface Modification," pp. 674–689 in "Encyclopedia of Polymer Science and Engineering," 2nd ed., Supplement, H. F. Mark, N. M. Bikales, C. G. Overberger, and G. Menges, Eds., Wiley-Interscience, New York, 1989.

Wilson, J. E., "Radiation Chemistry of Monomers, Polymers, and Plastics," Marcel Dekker, New York, 1974, Chap. 7.

Wingard, L. B., Jr., E. Katchalski-Katzir, and L. Goldstein, Eds., "Enzyme Technology," Vol. 2 in "Applied Biochemistry and Bioengineering," Academic Press, New York, 1979.

Wulff, G., I. Schulze, and K. Zabrocki, *Makromol. Chem.*, **181**, 531 (1980).

Yaroslavsky, C., A. Patchornik, and E. Katchalski, *Tetrahedron Lett.*, 3629 (1979).

Yoshida, J. I. J. Hashimoto, and N. Kawabata, *Bull. Chem. Soc. Jpn.*, **54**, 309 (1981).

PROBLEMS

9-1   What reactions can be used to crosslink the following polymers:

  a. Polyester from ethylene glycol and maleic anhydride.
  b. Polyester from ethylene glycol, phthalic acid, and oleic acid.
  c. *cis*-1,4-Polyisoprene.
  d. Polydimethylsiloxane.
  e. Polyethylene.
  f. Cellulose.
  g. Chlorosulfonated polyethylene.

  Describe the crosslinking reactions by equations.

9-2   Show by equations the synthesis of each of the following:

  a. Cellulose acetate.
  b. Cellulose nitrate.

    c. Methyl cellulose.

    d. Poly(vinyl formal).

9-3 Show by equations each of the following:

    a. Chlorination of polyethylene.

    b. Chlorination of 1,4-polyisoprene.

    c. Chlorosulfonation of polyethylene.

    d. Synthesis of an anionic ion-exchange polymer from polystyrene.

    e. Cyclization of 1,4-polyisoprene by HBr.

9-4 Show the structure of each of the following:

    a. Poly(propylene-*b*-styrene).

    b. Poly(styrene-*g*-methyl methacrylate).

    c. Poly(methyl methacrylate-*g*-styrene).

    d. Poly(methyl methacrylate-*alt*-styrene).

Describe by equations the methods of synthesizing each of these copolymers.

9-5 Explain each of the following observations.

    a. The sulfonation of polystyrene can result in an insoluble (and brittle) product instead of the cation-exchange derivative shown in Eq. 9-39.

    b. Poly(vinyl sulfonic acid) increases the rate of the reaction between $Co(NH_3)_5Cl^{2+}$ and Hg^{2+} (Eq. 9-5) to a greater extent than does poly-(methacryloxy ethyl sulfonic acid).

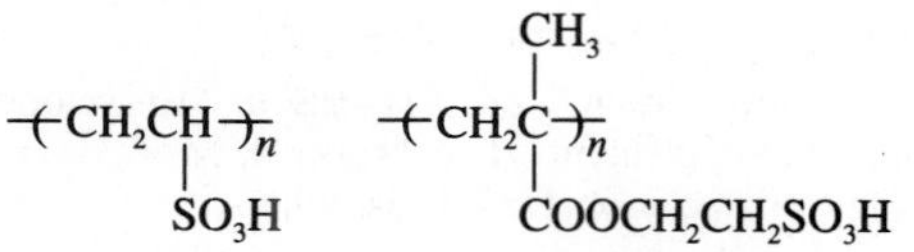

$$\text{+CH}_2\text{CH+}_n \quad\quad \text{+CH}_2\overset{\overset{\displaystyle CH_3}{|}}{\underset{\underset{\displaystyle COOCH_2CH_2SO_3H}{|}}{C}}\text{+}_n$$
$$\quad\quad\underset{\displaystyle SO_3H}{|}$$

Poly(vinyl   Poly(methacryloxy
sulfonic acid)   ethyl sulfonic acid)

    c. It is difficult to synthesize homogeneous cellulose diacetate by the direct acetylation of cellulose.

9-6 α-Chymotrypsin, which contains several NH_2 substituents, can be immobilized on an ethylene–maleic anhydride copolymer. Show by equations this synthesis of immobilized α-chymotrypsin. The enzymatic activity of the immobilized enzyme shows a maximum at pH = 9.5 compared to pH = 8 for α-chymotrypsin. Addition of sodium chloride to the immobilized enzyme causes the maximum activity to shift back to pH = 8. Explain.

9-7 Show by equations how the Merrifield method can be used to synthesize the tripeptide glycylvalylphenylalanine (Gly·Val·Phe)

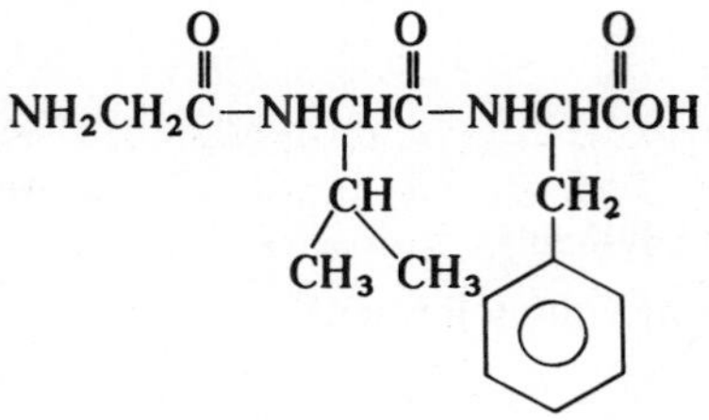

9-8   Synthesize $HOOC\!-\!CH(CH_3)CH_2CH_3$ using the polymer substrate technique. Your synthesis should include the preparation of the appropriate polymer support.

9-9   Show by equations how to synthesize each of the following polymer reagents:
  a. Polyamide support with $-CO\!-\!N(COCF_3)-$ functional group.
  b. Polystyrene support with $-\varnothing SnH_2(n\text{-}C_4H_9)$ functional group.
  c. Polystyrene support with $-\varnothing CH_2N^+(CH_3)_3(CN^-)$ functional group.

9-10   Show by equations how to synthesize each of the following polymer catalysts:
  a. Polystyrene support with $-P\varnothing_2PtCl_2$ catalyst group.

  b. Poly(4-vinyl pyridine) support with $-\!\!\left\langle\!\!\bigcirc\!\!\right\rangle\!\!NCu(OH)Cl$ catalyst group.

9-11   A suggested route to the administration of small doses of a drug over relatively long periods of time is to place the desired drug as a side chain on a high-molecular-weight polymer. The polymer is expected to be only slowly cleared from the patient and if the drug were attached by a hydrolyzable bond it might be slowly released over an extended time. Consider the following candidates for such a polymeric drug

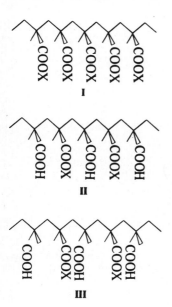

where X is the drug moiety. Which polymeric drug (**I**, **II**, or **III**) would be the best candidate for the stated objective. Explain and discuss your answer in terms of the relative rates of release of the drug X from **I**, **II**, and **III**.

INDEX